T0331233

Uncovering Quantum Field Theory and the Standard Model

This textbook provides an accessible introduction to quantum field theory and the Standard Model of particle physics. It adopts a distinctive pedagogical approach with clear, intuitive explanations to complement the mathematical exposition.

The book begins with basic principles of quantum field theory, relating them to quantum mechanics, classical field theory, and statistical mechanics, before building toward a detailed description of the Standard Model. Its concepts and components are introduced step by step, and their dynamical roles and interactions are gradually established. Advanced topics of current research are woven into the discussion, and key chapters address physics beyond the Standard Model, covering subjects such as axions, technicolor, and Grand Unified Theories.

This book is ideal for graduate courses and as a reference and inspiration for experienced researchers. Additional material is provided in appendices, while numerous end-of-chapter problems and quick questions reinforce the understanding and prepare students for their own research.

Wolfgang Bietenholz is Professor of Physics at the National Autonomous University of Mexico (UNAM). His research focuses on lattice field theory with applications to elementary particles and condensed matter. He also writes articles on physics, mathematics, and the history of science for a broader audience. He studied at ETH Zürich and the University of Bern. He previously held positions at Centro Brasileiro de Pesquisas Físicas in Rio de Janeiro, the Massachusetts Institute of Technology (MIT), Deutsches Elektronen-Synchrotron (DESY), Nordic Institute for Theoretical Physics, and the Humboldt University in Berlin.

Uwe-Jens Wiese is Professor at the University of Bern, and his research is focused on strongly interacting quantum systems in particle and condensed matter physics. He studied at the University of Hannover, and he previously held positions at DESY and MIT, where his work has been recognized by an Alfred P. Sloan Fellowship. He was a recipient of an ERC Advanced Grant for the duration from 2014 to 2019 and of a Humboldt Research Award in 2022.

Uncovering Quantum Field Theory and the Standard Model

From Fundamental Concepts to Dynamical Mechanisms

WOLFGANG BIETENHOLZ

Universidad Nacional Autónoma de México

UWE-JENS WIESE

Universität Bern

CAMBRIDGE
UNIVERSITY PRESS

Shaftesbury Road, Cambridge CB2 8EA, United Kingdom

One Liberty Plaza, 20th Floor, New York, NY 10006, USA

477 Williamstown Road, Port Melbourne, VIC 3207, Australia

314–321, 3rd Floor, Plot 3, Splendor Forum, Jasola District Centre, New Delhi – 110025, India

103 Penang Road, #05–06/07, Visioncrest Commercial, Singapore 238467

Cambridge University Press is part of Cambridge University Press & Assessment, a department of the University of Cambridge.

We share the University's mission to contribute to society through the pursuit of education, learning and research at the highest international levels of excellence.

www.cambridge.org
Information on this title: www.cambridge.org/9781108472333

DOI: 10.1017/9781108657037

First published 2024

A catalogue record for this publication is available from the British Library

Library of Congress Cataloging-in-Publication Data
Names: Bietenholz, Wolfgang, 1962– author. | Wiese, Uwe-Jens, 1958– author.
Title: Uncovering quantum field theory and the standard model : from fundamental concepts to dynamical mechanisms / Wolfgang Bietenholz, Universidad Nacional Autónoma de México, Uwe-Jens Wiese, Universität Bern, Switzerland.
Description: Cambridge, United Kingdom ; New York, NY : Cambridge University Press, 2024. | Includes bibliographical references and indexes.
Identifiers: LCCN 2023028733 (print) | LCCN 2023028734 (ebook) | ISBN 9781108472333 (hardback) | ISBN 9781108657037 (ebook)
Subjects: LCSH: Quantum field theory. | Particle Physics | Standard Model of Particle Physics
Classification: LCC QC174.45 .B53 2024 (print) | LCC QC174.45 (ebook) | DDC 530.14/3–dc23/eng/20230928
LC record available at https://lccn.loc.gov/2023028733
LC ebook record available at https://lccn.loc.gov/2023028734

ISBN 978-1-108-47233-3 Hardback

Additional resources for this publication at www.cambridge.org/bietenholz-wiese.

Contents

Preface

Intention of this Book

This book is an introduction to *quantum field theory* and the *Standard Model* of particle physics, which is a relativistic quantum field theory that incorporates the basic principles of quantum physics and special relativity. Quantum field theory provides a systematic, universal framework that allows us to understand the local interactions of field degrees of freedom attached to the different points of space at all times. Quantum field theories are of central importance in many areas of physics, ranging from condensed matter physics, atomic, molecular, and optical physics, to nuclear and particle physics. At the most fundamental level, quantum field theory describes the interactions between elementary particles, which are nothing but quantized field excitations, in the framework of the Standard Model.

The Standard Model of particle physics is one of the greatest achievements of science in the second half of the twentieth century, and – in light of its high-precision predictions – of all history. It summarizes all we know today about the fundamental structure of matter, forces, and symmetries, by describing the *electromagnetic, weak*, and *strong* interactions of *Higgs particles, leptons*, and *quarks,* mediated by *photons, W- and Z-bosons*, and *gluons*. It describes further interactions by Yukawa couplings between the Higgs field and the fermion fields, and by the self-interaction of the Higgs field. It is our goal to explain these topics sufficiently well, such that a deep understanding becomes possible. Achieving profound insight into a complex subject such as quantum field theory takes time, but leaves us with a sense of empowerment and an urge to progress to more advanced topics. Empowering the curious reader and encouraging him or her to think about Nature's biggest puzzles at a deep level is a major goal of this book.

This book provides a detailed description of those features of the Standard Model and some subjects beyond it that should be of general interest to any physicist of the twenty-first century, irrespective of his or her specialization. It concentrates on the model's symmetries, on its hierarchies of energy scales and the related puzzles, on its predictive power and its limitations, as well as on its possible extensions to even higher energies. Our goal is to expose the structure of the Standard Model in a language that is accessible to physicists with just a basic background in special relativity, quantum mechanics, statistical mechanics, and classical – but not necessarily quantum – field theory. This book introduces quantum field theory at a level that is of interest for applications not only in particle but also in nuclear and condensed matter physics.

The enormous importance and robustness of the Standard Model, which has maintained its validity in the era of the Large Hadron Collider (LHC) at CERN (near Geneva, Switzerland), has intrigued many physicists. Irrespective of whether or not an extended Standard

Model (perhaps with additional right-handed neutrino fields) may be valid all the way up to the Planck scale, the Standard Model will stay with us as the most fundamental description of Nature up to the TeV energy scale. Familiarizing the interested reader, either with a particle physics or with a more general physics orientation, with its fascinating dynamical structures is a main intention of this book. The broad relevance of the Standard Model is also reflected in the structure of M.Sc. and Ph.D. programs. While all physicists should become familiar with the basic features of the Standard Model, many physicists will want to gain a deeper understanding of its fundamental structures and the underlying dynamical mechanisms. The present book fulfills this purpose.

Victor Weisskopf's teaching at MIT was characterized by his motto:

It is better to uncover a little than to cover a lot.

As the title of this book suggests, it is our intention to uncover some fundamental structures of quantum field theory and the Standard Model, which are usually not strongly emphasized in most of the textbook literature. Exposing some deeper layers of the subject, whose understanding is sometimes obstructed by the involved mathematical structure of quantum field theory or by the rich particle phenomenology of the Standard Model, is a central intention of the book. Lectures based on this book should be able to follow Weisskopf's motto. In particular, we aim at facilitating a deep understanding of a subject by going through each individual step of an extensive explanation. While the variety of subjects discussed in the book is not small, each one is presented in sufficient detail, in order to clarify it at a deep level. In this way, somewhat like in archeology, we aim at uncovering some hidden layers, which are obstructed from the more common perspective of large parts of the textbook literature.

Subjects of this Book

The book begins with an overview of the fundamental concepts underlying quantum field theory. A basic understanding of quantum field theory is facilitated by relating it to quantum mechanics and to classical field theory, as well as (in its Euclidean functional integral formulation) to classical statistical mechanics. Canonical quantization as well as the perturbative quantization applying *dimensional regularization*, and the non-perturbative *lattice regularization* are then introduced, using scalar field theory as a simple framework. In this context, we address the very nature of *"particles"* in quantum field theory, which arise as quantized wave excitations of the corresponding quantum fields.

Other fields are introduced one by one, starting with *Abelian gauge fields* and then moving on to fermion fields. We put an emphasis on *Weyl fermions*, which are basic building blocks of the Standard Model, and then relate them to *Dirac and Majorana fermions*. We discuss the chiral symmetries of fermions, both in continuum and in lattice quantum field theory. We then move on to *non-Abelian gauge fields*.

In order to elucidate the interplay of its various dynamical ingredients, the Standard Model is then constructed step by step, starting from the Higgs sector, proceeding further to the electroweak and strong gauge fields, and finally adding fermionic lepton and quark matter fields. Important dynamical mechanisms, such as the spontaneous break-down of a global symmetry, the *Higgs mechanism* describing the "spontaneous break-ing" of a gauge symmetry, as well as the quantum-induced anomalous breakdown of symmetries, and the requirement of *gauge anomaly cancellation* are each addressed at the appropriate stage of the construction. For example, the consistency of the Standard Model as a quantum field theory, which is tied to the cancellation of perturbative and non-perturbative anomalies, is discussed in detail and is related to the issue of charge quantization.

The book concentrates on the fundamental structures of the Standard Model, its symme-tries and their various realizations, its basic dynamical mechanisms, and some of its less obvious beautiful structures, more than on particle phenomenology or advanced pertur-bative techniques. The book also emphasizes non-perturbative physics and uses the lattice regularization when appropriate, but it is by no means focused on lattice field theory (when it is possible, we use continuum notation).

The lattice regularization is more physical than dimensional regularization (by analytic continuation of the space–time dimension). In particular, it also arises naturally in con-densed matter physics (in the form of spatial crystal lattices); thus it provides a bridge between the different disciplines. Most important, the lattice regularization leads to *non-perturbative* insights into essential dynamical mechanisms including the *confinement of quarks and gluons* and the spontaneous breakdown of the quarks' *chiral symmetry*.

Along with its impressive success, the Standard Model also gives rise to puzzles and open questions, which may hint to new physics. The discussion of theories that go beyond the Standard Model, such as *technicolor*, *axion models*, and *Grand Unified Theories*, is embedded in the construction of the Standard Model itself. Such ideas are addressed as soon as it is suitable; certain aspects of these theories are discussed even before addressing the theory of the strong interaction – Quantum Chromodynamics (QCD).

Special attention is given to the *number of quark colors N_c* (which is 3 in the real world) as a discrete parameter of the Standard Model. Via anomaly cancellation, it affects the electric quark charges, with consequences, *e.g.*, for the electromagnetic decay of the neutral pion, which is not addressed correctly in large parts of the textbook literature.

Again in contrast to most standard textbooks, this book puts an emphasis on *topolog-ical aspects* of the Standard Model and its low-energy effective theories. Topological ef-fects are related to anomalies and play an important role in the dynamics of pions and the other Nambu–Goldstone bosons of the strong interaction, as well as for the $U(1)_A$-*problem*, the *strong CP-problem,* and for the generation of the *baryon asymmetry.* The topology of quantum fields is again of general interest, with numerous links to condensed matter physics.

Finally, the book supplements nine appendices, which address, for instance, the histor-ical development of experimental and theoretical high-energy physics, as well as units, energy scales, and fundamental parameters in particle physics. Other appendices describe Minkowski space–time, the Lorentz-covariant formulation of classical electrodynamics, as well as the Monte Carlo method and second-order phase transitions, in order to facilitate a smooth transition beyond B.Sc. knowledge. More advanced mathematical topics includ-ing group theory and homotopy theory, which are again of general interest beyond particle physics, are addressed in two more appendices.

Structure of this Book

This book has emerged from numerous lectures at the Master and Ph.D. student level, held over three decades at the Rheinisch-Westfälische Technische Hochschule (RWTH) Aachen, the Humboldt University in Berlin, the University of Bern, the Massachusetts Institute of Technology (MIT), Potsdam University, the Universidad Nacional Autónoma de México (UNAM), and at Wuppertal University.

Part I of this book can be covered in a 1-semester introduction to quantum field theory. It begins with an Ouverture that provides an overview of the fundamental concepts underlying quantum field theory. It continues with a concise presentation of the theoretical framework of quantum field theory in the functional integral approach, provided in Chapter 1. Chapters 2–4 discuss scalar field theory, starting at the classical level, and progressing from canonical quantization to the perturbative quantization using the Euclidean functional integral in dimensional regularization. In this context, the nature of "particles" in a quantum field theory is clarified: they are quantized waves, which we temporarily address as "wavicles". Chapter 5 provides an introduction to the renormalization group from a Wilsonian perspective. The quantization of the electromagnetic field, and the characterization of electrically charged particles as "infraparticles" are the subjects of Chapters 6 and 7. Chapters 8–10 discuss Weyl, Dirac, and Majorana fermions, as well as their various symmetries, both in canonical quantization and using a functional integral approach. Part I ends with Chapter 11, in which non-Abelian gauge fields and their quantization are introduced. Part I serves as a prerequisite for studying the other parts of the book.

Part II can be the subject of a 1-semester introduction to the Standard Model. It begins with an Intermezzo that summarizes the fundamental concepts and dynamical mechanisms that underlie the Standard Model. Chapters 12–14 introduce the bosonic Higgs and gauge fields with their dynamics, including the Higgs mechanism, as well as confinement and Coulomb phases. The gauge interactions of lepton and quark fields of the first fermion generation are introduced in Chapter 15. The fermions are endowed with mass by coupling them to the Higgs field in Chapter 16, while Chapter 17 discusses the physics related to the other fermion generations.

Part III corresponds to a course on the strong interaction, from a fundamental Standard Model point of view (Chapters 18–20), from the perspective of effective models (Chapters 21 and 22), and in the framework of systematic low-energy effective theories (Chapters 23 and 24). Part IV, which includes Chapter 25 on the strong CP-problem and Chapter 26 on Grand Unified Theories, addresses selected topics beyond the Standard Model, and can be taught as part of a special topics course. Both Part III and Part IV can be presented in another 1-semester course, relying on the material of Part I and II as a prerequisite. The appendices provide some background for the various chapters of the main text.

All chapters end with a set of exercises. In addition, "Quick Questions", which allow to test the reader's immediate understanding, are embedded in the text. It cannot be stressed enough that theoretical physics can be learned in depth only by working through a large number of problems. This means a lot of work for the student which, however, is well worth investing. The skills that one learns by solving the exercises are very useful when working on actual research problems.

As an additional motivation for the curious reader, the book connects the Standard Model with some deep fundamental questions (inserted in boxes) that we hope to be answered in the course of the twenty-first century. For example, one such question is known as the *hierarchy problem*, *i.e.* the question: *Why is the electroweak scale so much lower than the Planck scale?* The authors encourage the reader to face such fundamental questions and think about them at a deep level. This should serve as a strong motivation to penetrate the subject of the Standard Model in a profound manner. Although quantum field theory is almost a century old, it remains one of the most promising tools that will allow us to push the boundaries of current knowledge further into the unknown.

Readers who find mistakes of any kind are encouraged to kindly report them to us. Outstanding lists of mistakes will be awarded. Relevant corrections will be listed – if necessary – on the book website.

Acknowledgments

First of all, we like to thank the numerous students who have helped to improve the lecture notes and exercises on which this book is based by their constructive feedback. Their enthusiasm provided the motivation for turning the lecture notes into a regular textbook.

It is impossible to thank all the colleagues and numerous collaborators who have influenced our thinking about quantum field theory and the Standard Model. Very special thanks go to Oliver Bär, Debasish Banerjee, Detlev Buchholz, Wilfried Buchmüller, Klaus Fredenhagen, Urs Gerber, Carlo Giunti, Kieran Holland, Gurtej Kanwar, Martin Lüscher, Alessandro Mariani, Colin Morningstar, Mike Peardon, Michele Pepe, João Pinto Barros, Lilian Prado, Simona Procacci, Christopher Smith, Rainer Sommer, Youssef Tammam, Christiane Tretter, Christof Wetterich, and Edward Witten, who have read parts of this book or have substantially improved it through their invaluable comments.

We like to thank Aline Guevara Villegas for her help with the figures in Chapters 4, 17, and 19, as well as Lucian Harland-Lang for preparing Figure 22.8. We are indebted to Nadiia Vlasii for realizing the picture on the cover page with chalk on the blackboard. We also thank Sarah Armstrong, Jane Chan, Nicholas Gibbons, Shanthy Jaganathan, Suresh Kumar, Sunantha Ramamoorthy, Stephanie Windows, and the other staff at Cambridge University Press for their competent support of this project.

Wolfgang Bietenholz would like to thank Klaus Hepp and Heinrich Leutwyler, who supervised his diploma thesis at ETH Zürich and his Ph.D. thesis at the University of Bern, respectively. He further thanks Hans-Jürg Gerber, Mikhail Terentyev, José Abdalla Helayël-Neto, John Negele, Klaus Schilling, Michael Müller-Preußker, and Gerrit Schierholz for their support in the course of his career (in chronological order).

It was his great pleasure to publish research articles with numerous authors from 24 countries, which underscores the cosmopolitan spirit of science, which builds bridges between peoples and cultures.

He would like to particularly highlight the pleasant collaboration with David Adams, Kostas Anagnostopoulos, Alejandro Ayala, Richard Brower, Shailesh Chandrasekharan, Hermann Dilger, Philippe de Forcrand, José Antonio García-Hernández, Urs Gerber, Juan

José Giambiagi, Ivan Hip, Frank Hofheinz, Roger Horsely, Edgar López-Contreras, Julieta Medina, Héctor Mejía-Díaz, Jaime Fabián Nieto Castellanos, Jun Nishimura, Denjoe O'Connor, Kostas Orginos, Marco Panero, João Pinto Barros, Paul Rakow, Fernando Rejón-Barrera, Gerrit Schierholz, Luigi Scorzato, Stanislav Shcheredin, Jan Volkholz, Marc Wagner, and Urs Wenger.

He acknowledges inspiring communication and helpful advice by Poul Damgaard, Stefan Dürr, Peter Hasenfratz, Christian Hoelbling, Martin Lüscher, Peter Minkowski, Mike Peardon, Christian Schubert, and Rainer Sommer. He also thanks for the support during his sabbatical periods at Bern University – which boosted the work on this book – by the *Programa de Apoyos para la Superación del Personal Académico de la UNAM* and by the *Swiss National Science Foundation*.

Finally, W. B. is grateful to Diego, Marlene, Philippe, and Tatiana for their moral support and patience during various stages of the work on this book.

Uwe-Jens Wiese is deeply indebted to his Ph.D. advisor, Peter Sauer, who created an exciting scientific atmosphere at the Institute for Theoretical Physics at Hannover University, were quarks, gluons, or W- and Z-bosons were discussed at a profound level, before formal courses on particle physics or quantum field theory were offered there. U.-J. W. also thanks Gerrit Schierholz, who ignited his interest in lattice field theory and co-advised his Ph.D. thesis over the distance between DESY and Hannover, for creating a wonderful research group and providing it with exciting and challenging problems. During his years as a postdoc, U.-J. W. has benefited tremendously from guidance, encouragement, as well as constructive criticism from Peter Hasenfratz, Jiři Jersák, Hans Kastrup, Heinrich Leutwyler, and Martin Lüscher. As an Assistant and later Associate Professor at MIT, he received invaluable support from Robert Jaffe, John Negele, and many other colleagues, during one of the most exciting periods in his career.

He had the privilege to discuss quantum field theory and the Standard Model with many colleagues at the University of Bern and at MIT, including Thomas Becher, Matthias Blau, Gilberto Colangelo, Jean-Pierre Derendinger, Edward Farhi, Jürg Gasser, Jeffrey Goldstone, Christoph Greub, Roman Jackiw, Mikko Laine, Patrick Lee, Peter Minkowski, Ferenc Niedermayer, Domenico Orlando, Krishna Rajagopal, Susanne Reffert, Urs Wenger, and Frank Wilczek. Other people who had an impact on his understanding of the subject, either through collaboration, discussion, or through their published work include Richard Brower, Shailesh Chandrasekharan, Jürg Fröhlich, Meinulf Göckeler, David Kaplan, Andreas Kronfeld, Morten Laursen, and Jean Zinn-Justin.

He is very grateful to all these people for making theoretical physics a very exciting and satisfying experience over several decades.

Finally, U.-J. W. thanks Nadiia Vlasii for her love and support which sustain him and were vital for finishing this book.

Notations and Conventions

In the following, we list various notations and conventions to be used throughout this book.

- Minkowski and Euclidean space–time:
  In Minkowski space–time, we use the metric

  $$g_{\mu\nu} = g^{\mu\nu} = \text{diag}(1, -1, -1, -1).$$

  Co- and contra-variant vectors describing a space–time point are given by

  $$x^{\mu} = \left(x^0, \vec{x}\right), \quad x_{\mu} = g_{\mu\nu}x^{\nu} = (x_0, -\vec{x}), \quad x^0 = x_0 = ct,$$

  and the corresponding space–time derivatives take the form

  $$\partial_{\mu} = \left(\frac{\partial}{\partial x^0}, \frac{\partial}{\partial x^1}, \frac{\partial}{\partial x^2}, \frac{\partial}{\partial x^3}\right) = \left(\frac{1}{c}\partial_t, \vec{\nabla}\right), \quad \partial^{\mu} = \left(\frac{1}{c}\partial_t, -\vec{\nabla}\right).$$

  Minkowski and Euclidean space–time are related by the Wick rotation $x_4 = ix_0$. We distinguish the imaginary unit i from a generic index i. In Euclidean space–time, we exclusively use lower indices, with $x_{\mu} = (x_1, x_2, x_3, x_4)$ and the standard Euclidean metric $g_{\mu\nu} = \delta_{\mu\nu}$. The totally anti-symmetric Levi-Civita symbol $\epsilon_{\mu\nu\rho\sigma}$ obeys $\epsilon_{1234} = 1$.

  Both in Minkowski and in Euclidean space–time, we denote a Lagrange density or Lagrangian by $\mathcal{L}$, while a Lagrange function is denoted by L. A Hamiltonian density is denoted by $\mathcal{H}$, and a Hamilton operator by $\hat{H}$. We dress the operators that act in a Hilbert or Fock space with a hat.

- Internal symmetries:
  The generators T^a of a Lie algebra obey

  $$[T^a, T^b] = if_{abc}T^c, \quad \text{Tr}[T^a T^b] = \frac{1}{2}\delta_{ab}.$$

  Examples of internal symmetries are $\text{SU}(2)_I$ isospin, which is generated by $T^a = \tau^a/2$ ($a \in \{1, 2, 3\}$), with the Pauli matrices

  $$\vec{\tau} = (\tau^1, \tau^2, \tau^3) = \left(\begin{pmatrix} 0 & 1 \\ 1 & 0 \end{pmatrix}, \begin{pmatrix} 0 & -i \\ i & 0 \end{pmatrix}, \begin{pmatrix} 1 & 0 \\ 0 & -1 \end{pmatrix}\right),$$

  $\text{SU}(3)_c$ color, which is generated by $T^a = \lambda^a/2$, with the Gell-Mann matrices λ^a ($a \in \{1, 2, \ldots, 8\}$), which are written down in Appendix F.7, or the flavor symmetry $\text{SU}(N_f)$, which is generated by $T^a = \eta^a/2$ ($a \in \{1, 2, \ldots, N_f^2 - 1\}$).

- Gauge fields:
  Electrodynamics is formulated with the 4-vector potential $A^{\mu}(x) = \left(\phi(\vec{x}, t), \vec{A}(\vec{x}, t)\right)$. The electromagnetic interaction is governed by the Abelian gauge group $\text{U}(1)_{\text{em}}$. We use Lorentz–Heaviside units in which the potential of a static point charge carrying the elementary electric charge e is given by

  $$\phi(\vec{x}) = \frac{e}{4\pi|\vec{r}|}.$$

The 4-vector potential as well as a complex field $\Phi(x) \in \mathbb{C}$, which carries $Q \in \mathbb{Z}$ units of the elementary charge e, transform under gauge transformations as

$$A'_\mu(x) = A_\mu(x) - \partial_\mu \alpha(x), \quad \Phi'(x) = \exp(\mathrm{i}Qe\alpha(x)) \, \Phi(x).$$

The gauge fields of the Standard Model transform under the gauge group $\mathrm{SU}(3)_{\mathrm{c}} \times \mathrm{SU}(2)_L \times \mathrm{U}(1)_Y$. The Abelian hypercharge gauge field $B_\mu(x) \in \mathbb{R}$ is associated with the symmetry $\mathrm{U}(1)_Y$. It couples with the strength g' to a scalar field that carries the weak hypercharge Y, such that

$$B'_\mu(x) = B_\mu(x) - \partial_\mu \varphi(x), \quad \Phi'(x) = \exp(\mathrm{i}Yg'\varphi(x)) \, \Phi(x).$$

Non-Abelian vector potentials mediate the weak and the strong gauge interaction with the fundamental coupling constants g and g_{s}, and with the matrix-valued gauge transformations $L(x) \in \mathrm{SU}(2)_L$ and $\Omega(x) \in \mathrm{SU}(3)_{\mathrm{c}}$. They are defined by the anti-Hermitian gauge fields W_μ and G_μ,

$$W_\mu(x) = \mathrm{i}gW_\mu^a(x)\frac{\tau^a}{2}, \quad W_\mu^a(x) \in \mathbb{R}, \quad W'_\mu(x) = L(x)\left(W_\mu(x) + \partial_\mu\right)L(x)^\dagger,$$

$$G_\mu(x) = \mathrm{i}g_{\mathrm{s}}G_\mu^a(x)\frac{\lambda^a}{2}, \quad G_\mu^a(x) \in \mathbb{R}, \quad G'_\mu(x) = \Omega(x)\left(G_\mu(x) + \partial_\mu\right)\Omega(x)^\dagger.$$

Occasionally, we will extend the Abelian gauge symmetry $\mathrm{U}(1)_Y$ to a non-Abelian symmetry with $R(x) \in \mathrm{SU}(2)_R$, or the $\mathrm{SU}(3)_{\mathrm{c}} \times \mathrm{SU}(2)_L \times \mathrm{U}(1)_Y$ symmetry to a grand unified symmetry with $\Upsilon(x) \in \mathrm{SU}(5)$, such that

$$X_\mu(x) = \mathrm{i}g'X_\mu^a(x)\frac{\tau^a}{2}, \quad X_\mu^3(x) = B_\mu(x), \quad X'_\mu(x) = R(x)\left(X_\mu(x) + \partial_\mu\right)R(x)^\dagger,$$

$$V_\mu(x) = \mathrm{i}g_5 V_\mu^a(x)\frac{\eta^a}{2}, \quad V_\mu^a(x) \in \mathbb{R}, \quad V'_\mu(x) = \Upsilon(x)\left(V_\mu(x) + \partial_\mu\right)\Upsilon(x)^\dagger.$$

In the context of the Higgs mechanism, we always write "spontaneous symmetry breaking" in inverted commas, because a gauge symmetry (which reflects a redundancy) can, in fact, not break.

- Fermion fields:
  The spin $\vec{S} = \vec{\sigma}/2$ of elementary fermions is described by the same set of Pauli matrices

$$\vec{\sigma} = (\sigma^1, \sigma^2, \sigma^3) = \left(\begin{pmatrix} 0 & 1 \\ 1 & 0 \end{pmatrix}, \begin{pmatrix} 0 & -\mathrm{i} \\ \mathrm{i} & 0 \end{pmatrix}, \begin{pmatrix} 1 & 0 \\ 0 & -1 \end{pmatrix} \right).$$

In the context of canonical quantization, fermion fields are described by field operators $\hat{\psi}(\vec{x})$ and $\hat{\psi}(\vec{x})^\dagger$. In the framework of the functional integral, fermions are described by independent, anti-commuting Grassmann fields $\psi(x)$ and $\bar{\psi}(x)$.

In Minkowski space–time, we follow the γ-matrix conventions of the books of Peskin and Schroeder (1997), Srednicki (2007), Zee (2010), Schwartz (2014), and others,

$$\{\gamma^\mu, \gamma^\nu\} = 2g^{\mu\nu}, \quad \gamma^5 = \mathrm{i}\gamma^0\gamma^1\gamma^2\gamma^3, \quad \{\gamma^\mu, \gamma^5\} = 0,$$

which in the chiral basis amount to

$$\gamma^0 = \begin{pmatrix} 0 & \mathbb{1} \\ \mathbb{1} & 0 \end{pmatrix}, \quad \gamma^i = \begin{pmatrix} 0 & \sigma^i \\ -\sigma^i & 0 \end{pmatrix}, \quad \gamma^5 = \begin{pmatrix} -\mathbb{1} & 0 \\ 0 & \mathbb{1} \end{pmatrix}.$$

In order to distinguish them from γ-matrices in Euclidean space–time, we arrange things such that γ-matrices in Minkowski space–time occur exclusively with upper indices.

A 4-component Dirac spinor $\psi(x)$ is built from a left- and a right-handed 2-component Weyl spinor, $\psi_L(x)$ and $\psi_R(x)$,

$$\psi(x) = \begin{pmatrix} \psi_L(x) \\ \psi_R(x) \end{pmatrix}, \quad \bar{\psi}(x) = (\bar{\psi}_R(x), \bar{\psi}_L(x)), \quad P_L = \frac{1}{2}(1 - \gamma^5), \quad P_R = \frac{1}{2}(1 + \gamma^5),$$

$$\begin{pmatrix} \psi_L(x) \\ 0 \end{pmatrix} = P_L \psi(x), \quad \begin{pmatrix} 0 \\ \psi_R(x) \end{pmatrix} = P_R \psi(x).$$

Left- and right-handed Weyl spinors are associated with the matrices $\bar{\sigma}^\mu$ and σ^μ, respectively,

$$\gamma^\mu = \begin{pmatrix} 0 & \sigma^\mu \\ \bar{\sigma}^\mu & 0 \end{pmatrix}, \quad \sigma^\mu = (\mathbb{1}, \vec{\sigma}), \quad \bar{\sigma}^\mu = (\mathbb{1}, -\vec{\sigma}).$$

A Majorana spinor results from a Dirac spinor by imposing the constraints

$$\psi_L(x) = -i\sigma^2 \bar{\psi}_R(x)^\mathsf{T}, \quad \bar{\psi}_L(x) = \psi_R(x)^\mathsf{T} i\sigma^2.$$

Here, as well as in other places in the book, T denotes transpose.

The γ-matrices in Euclidean space–time result from a Wick rotation. They are Hermitian, $\gamma_\mu^\dagger = \gamma_\mu$, and obey the relations

$$\{\gamma_\mu, \gamma_\nu\} = 2\delta_{\mu\nu}, \quad \gamma_5 = -\gamma_1\gamma_2\gamma_3\gamma_4, \quad \{\gamma_\mu, \gamma_5\} = 0.$$

In the chiral basis, they take the form

$$\gamma_i = \begin{pmatrix} 0 & -i\sigma^i \\ i\sigma^i & 0 \end{pmatrix}, \quad \gamma_4 = \begin{pmatrix} 0 & 1 \\ 1 & 0 \end{pmatrix}, \quad \gamma_5 = \begin{pmatrix} -1 & 0 \\ 0 & 1 \end{pmatrix}.$$

Euclidean γ-matrices will be used with lower indices only.

The Euclidean variants of the matrices σ^μ and $\bar{\sigma}^\mu$ again carry lower Lorentz indices only,

$$\gamma_\mu = \begin{pmatrix} 0 & \sigma_\mu \\ \bar{\sigma}_\mu & 0 \end{pmatrix}, \quad \sigma_\mu = (-i\vec{\sigma}, \mathbb{1}), \quad \bar{\sigma}_\mu = (i\vec{\sigma}, \mathbb{1}).$$

• Discrete symmetries C, P, and T:

In Euclidean space–time, charge conjugation C, parity P, and time reversal T act on left- and right-handed Weyl fermion fields as

$$^C\psi_R(x) = i\sigma^2 \bar{\psi}_L(x)^\mathsf{T} = {}^c\psi_L(x), \quad {}^C\bar{\psi}_R(x) = -\psi_L(x)^\mathsf{T} i\sigma^2 = {}^c\bar{\psi}_L(x),$$

$$^C\psi_L(x) = -i\sigma^2 \bar{\psi}_R(x)^\mathsf{T} = {}^c\psi_R(x), \quad {}^C\bar{\psi}_L(x) = \psi_R(x)^\mathsf{T} i\sigma^2 = {}^c\bar{\psi}_R(x),$$

$$^P\psi_R(x) = \psi_L(-\vec{x}, x_4), \quad {}^P\bar{\psi}_R(x) = \bar{\psi}_L(-\vec{x}, x_4),$$

$$^P\psi_L(x) = \psi_R(-\vec{x}, x_4), \quad {}^P\bar{\psi}_L(x) = \bar{\psi}_R(-\vec{x}, x_4),$$

$$^T\psi_R(x) = i\sigma^2 \bar{\psi}_R(\vec{x}, -x_4)^\mathsf{T}, \quad {}^T\bar{\psi}_R(x) = \psi_R(\vec{x}, -x_4)^\mathsf{T} i\sigma^2,$$

$$^T\psi_\mathsf{l}(x) = i\sigma^2 \bar{\psi}_\mathsf{l}(\vec{x}, -x_4)^\mathsf{T}, \quad {}^T\bar{\psi}_\mathsf{l}(x) = \psi_\mathsf{l}(\vec{x}, -x_4)^\mathsf{T} i\sigma^2.$$

The resulting, discrete transformations for Euclidean Dirac spinor fields are

$$
{}^C\psi(x) = C\bar{\psi}(x)^\mathsf{T}, \quad {}^C\bar{\psi}(x) = -\psi(x)^\mathsf{T}C^{-1}, \quad C = \begin{pmatrix} -\mathrm{i}\sigma^2 & 0 \\ 0 & \mathrm{i}\sigma^2 \end{pmatrix},
$$

$$
{}^P\psi(x) = P\psi(-\vec{x}, x_4), \quad {}^P\bar{\psi}(x) = \bar{\psi}(-\vec{x}, x_4)P^{-1}, \quad P = \begin{pmatrix} 0 & \mathbb{1} \\ \mathbb{1} & 0 \end{pmatrix},
$$

$$
{}^T\psi(x) = T\bar{\psi}(\vec{x}, -x_4)^\mathsf{T}, \quad {}^T\bar{\psi}(x) = -\psi(\vec{x}, -x_4)^\mathsf{T}T^{-1}, \quad T = \begin{pmatrix} 0 & \mathrm{i}\sigma^2 \\ \mathrm{i}\sigma^2 & 0 \end{pmatrix}.
$$

- Higgs field:

  We use three equivalent parametrizations of the Standard Model Higgs field, as a complex doublet,

  $$
  \Phi(x) = \begin{pmatrix} \Phi^+(x) \\ \Phi^0(x) \end{pmatrix}, \quad \Phi^+(x),\ \Phi^0(x) \in \mathbb{C},
  $$

as a 4-component, real field

$$
\vec{\phi}(x) = (\phi_1(x), \phi_2(x), \phi_3(x), \phi_4(x)) \in \mathbb{R}^4,
$$
$$
\Phi^+(x) = \phi_2(x) + \mathrm{i}\,\phi_1(x), \quad \Phi^0(x) = \phi_4(x) - \mathrm{i}\,\phi_3(x),
$$

and as a 2×2 matrix, proportional to an SU(2) matrix,

$$
\Phi(x) = \begin{pmatrix} \Phi^0(x)^* & \Phi^+(x) \\ -\Phi^+(x)^* & \Phi^0(x) \end{pmatrix}
$$
$$
= \phi_4(x)\,\mathbb{1} + \mathrm{i}\left[\phi_1(x)\tau^1 + \phi_2(x)\tau^2 + \phi_3(x)\tau^3\right].
$$

Glossary

The following acronyms are frequently used throughout the book:

QED	Quantum Electrodynamics
QCD	Quantum Chromodynamics
IR	infrared
UV	ultraviolet

Some important physical scales are:

G	Newton's constant
M_{Planck}	Planck scale
Λ_{c}	cosmological constant
Λ_{QCD}	QCD scale
v	vacuum expectation value of the Higgs field

Relevant integer-valued parameters include:

N_c	number of quark colors
N_f	number of quark flavors
N_g	number of fermion generations

Some important physical parameters are:

λ	self-coupling of the Higgs field
g_s	strong $SU(3)_c$ color gauge coupling
g	weak $SU(2)_L$ gauge coupling
g'	$U(1)_Y$ hypercharge gauge coupling
e	unit of the electric charge
θ_W	Weinberg angle
θ	QCD vacuum-angle
θ_{QED}	QED vacuum-angle
θ_C	Cabibbo angle
$f_u,\ f_d,\ f_c,\ f_s,\ f_t,\ f_b$	quark Yukawa couplings
$m_u,\ m_d,\ m_c,\ m_s,\ m_t,\ m_b$	quark masses
$f_e,\ f_\mu,\ f_\tau$	charged lepton Yukawa couplings
$m_e,\ m_\mu,\ m_\tau$	charged lepton masses
F_π	pion decay constant

The fields of the Standard Model, including only the first fermion generation, are:

$\Phi(x),\ \mathbf{\Phi}(x),\ \vec{\phi}(x)$	three equivalent forms of the Higgs field
$G_\mu(x)$	$SU(3)_c$ gluon field
$W_\mu(x)$	$SU(2)_L$ electroweak gauge field
$B_\mu(x)$	$U(1)_Y$ weak hypercharge gauge field
$A_\mu(x)$	$U(1)_{em}$ photon field
$Z_\mu(x)$	neutral Z-boson field
$W_\mu^\pm(x)$	charged W-boson field
$e_R(x),\ \bar{e}_R(x)$	right-handed $SU(2)_L$-singlet electron fields
$l_L(x) = \begin{pmatrix} \nu_L(x) \\ e_L(x) \end{pmatrix},\quad \bar{l}_L(x) = (\bar{\nu}_L(x), \bar{e}_L(x))$	left-handed $SU(2)_L$-doublet lepton fields
$u_R(x), \bar{u}_R(x), d_R(x), \bar{d}_R(x)$	right-handed $SU(2)_L$-singlet quark fields
$q_L(x) = \begin{pmatrix} u_L(x) \\ d_L(x) \end{pmatrix},\quad \bar{q}_L(x) = (\bar{u}_L(x), \bar{d}_L(x))$	left-handed $SU(2)_L$-doublet quark fields

PART I

QUANTUM FIELD THEORY

Ouverture: Concepts of Quantum Field Theory

In this Ouverture, we anticipate fundamental concepts and basic principles of quantum field theory, in order to pave the way for a systematic exposition of the subject in the continuation of Part I of this book. In particular, we highlight the fundamental roles of *locality* and *symmetries*.

We emphasize, however, that understanding this introduction is by no means necessary for getting started with the book. Chapter 1 begins at a simple level, assuming only basic knowledge of quantum mechanics and special relativity. A reader who may be overwhelmed by the concepts mentioned in this Ouverture is encouraged to simply skip it and to proceed right away to Chapter 1. We recommend consulting this introductory overview repeatedly while working through Part I.

Point Particles versus Fields at the Classical Level

Theoretical physics in the modern sense was initiated by Isaac Newton who published his book Philosophiæ Naturalis Principia Mathematica in 1687. This eruption of genius provided us with the description of *classical point particle mechanics,* in terms of *ordinary* differential equations for the position vectors $\vec{x}_a(t)$ of N individual point particles ($a \in \{1, \ldots, N\}$) as functions of the time t. Classical mechanics is *local in time,* because Newton's equation contains time-derivatives, $d^2\vec{x}_a(t)/dt^2$, but no finite time-differences $t - t'$. On the other hand, Newtonian mechanics is *non-local in space,* because the finite distances $|\vec{x}_a - \vec{x}_b|$ between different particles determine instantaneous forces, in particular in Newtonian gravity. Hence in classical mechanics, there are conceptual differences between space and time. In point particle theories, the fundamental degrees of freedom – namely, the particle positions $\vec{x}_a(t)$ – are *mobile:* They roam around in space. As a consequence, at almost all points space is empty, *i.e.* nothing is there, except when a point particle occupies that position.

In contrast, the fundamental degrees of freedom of a *field theory* – the field values $\phi(\vec{x}, t)$ – are *immobile;* they are attached to a given space point $\vec{x}$ at all times t. In this case, it is the field value ϕ – not the position $\vec{x}$ – which changes as a function of time. In a field theory, space plays a role which is very different from point particle mechanics. In particular, it is *not empty* anywhere because field degrees of freedom exist at all points $\vec{x}$, at any time t. Fluid dynamics is an example for a non-relativistic classical field theory in which the mass density enters as a scalar field $\phi(\vec{x}, t)$. The classical field equations are *partial* differential equations – involving both space- and time-derivatives of $\phi(\vec{x}, t)$ – which determine the evolution of the fields. Hence, in contrast to point particle theories, field theories are *local in both space and time*.

The most fundamental classical field theory is James Clerk Maxwell's electrodynamics of electric and magnetic fields, $\vec{E}(\vec{x}, t)$ and $\vec{B}(\vec{x}, t)$ (Clerk Maxwell, 1865). Although this was not known until the end of the 19th century, Maxwell's electrodynamics is a *relativistic* classical field theory: It is invariant against space–time translations and rotations forming the *Poincaré symmetry group*. On the other hand, Newton's point particle mechanics is invariant under Galilean instead of Lorentz boosts. Thus, it is non-relativistic and therefore inconsistent with the relativistic Minkowski space–time underlying Maxwell's electrodynamics.

Albert Einstein's *special theory of relativity* (Einstein, 1905) modifies Newton's point particle mechanics in such a way that it becomes Poincaré-invariant. Indeed, in the framework of special relativity, charged point particles can interact with classical electromagnetic fields in a Poincaré-invariant manner. On the other hand, relativistic point particles cannot interact *directly* with each other; thus, they remain necessarily free in the absence of a mediating (*e.g.*, electromagnetic) field. This follows from Heinrich Leutwyler's *non-interaction theorem* for relativistic systems of N point particles (Leutwyler, 1965), which extends an earlier result for the 2-particle case (Currie *et al.*, 1963).

Indeed, in a relativistic quantum field theory, such as the Standard Model, the point particle concept is abandoned and all "particles" are just *field excitations,* which Frank Wilczek sometimes denotes as "wavicles" (Wilczek, 2012). This is a very useful distinction, which allows us to avoid confusion that might otherwise arise quite easily. In particular, while a Newtonian point particle has a completely well-defined position $\vec{x}_a$, a wavicle does not.

Particles versus Waves in Quantum Theory

Quantum mechanics, as formulated in the 1920s by Niels Bohr, Max Born, Werner Heisenberg, Erwin Schrödinger, and others (see Hund (1974) for a historic account), applies the basic principles of quantum theory – such as unitarity, which implies the conservation of probability – to Newton's point particles. As a consequence, the particle positions are affected by quantum uncertainty, but they still exist conceptually. This situation is described in terms of a wave function $\Psi(\vec{x}_1, \vec{x}_2, \ldots, \vec{x}_N, t)$, which obeys a non-relativistic Schrödinger equation – a partial differential equation containing derivatives with respect to the time t and to the N particle positions $\vec{x}_a$. It is important to note that (unlike $\phi(\vec{x}, t)$) $\Psi(\vec{x}_1, \vec{x}_2, \ldots, \vec{x}_N, t)$ is *not* a field in space–time, but just a time-dependent complex function over the N-particle configuration space $(\vec{x}_1, \vec{x}_2, \ldots, \vec{x}_N)$. A time-dependent state in a quantum field theory, on the other hand, can be described by a complex-valued wave functional $\Psi(\phi(\vec{x}), t) = \Psi([\phi], t)$, which depends on time and on the field configuration $[\phi]$ (at all space points) and which obeys a *functional* Schrödinger equation (Stückelberg, 1938; Tomonaga, 1946).

When one discusses quantum mechanical experiments of the *double-slit* type, one assigns the observed interference pattern to the wave properties of quantum particles. This does, however, not mean that such a particle is understood as a quantized wave excitation of a field. It is just considered as a point particle (that is indeed endowed with the concept of a position), which is, however, affected by quantum uncertainty. As long as its position is not measured, it can "pass through both slits" (symbolically speaking), until it hits the detection screen that registers its position (which is then unambiguous, within the resolution of the detector). Only after repeating such a single-particle experiment a large number of times, the detected positions of the individual particles give rise to the well-known interference pattern. In the context of quantum mechanics, particle-wave duality just means that the spatial probability distribution of point particles results from a quantum mechanical wave function $\Psi(\vec{x}_1, \vec{x}_2, \ldots, \vec{x}_N, t)$.

When a classical (*e.g.,* electromagnetic) wave is diffracted at a double slit, it shows an interference pattern for different reasons. As a field excitation, the wave exists simultaneously at all points in some region of space; it does not even conceptually have a sharp position. In contrast to the experiment with quantum mechanical point particles, the interference pattern arises immediately, as soon as the classical wave reaches the detection screen.

When one interprets this experiment with light at the quantum level and refers to individual photons, then the interference pattern again emerges only after the experiment has been repeated a large number of times. The complementary "particle" character of the photon is usually emphasized in the context of the Compton effect. However, while we may be used to thinking of an electron as a point particle (with some position, affected by quantum uncertainty), we should definitely not think of a photon in a similar way (and of the electron neither; see below). As a quantized wave excitation of the electromagnetic field, a photon does not even conceptually have a well-defined point-like position in space.

So what does it mean when we refer to the photon as a "particle"? Unfortunately, in our casual language the term particle is associated with the idea of a point-like (or small) object, which is not what the photon is like. Wilczek's term *wavicle* serves its purpose when it prevents us from thinking of a photon as a tiny ball. At the end of the day, only mathematics provides an appropriate and accurate description of "particles", like a photon. In the formalism of quantum field theory, "particle-wave duality" refers to the fact that "particles" actually are wavicles, *i.e.* quantized wave excitations of fields.

When Paul Adrien Maurice Dirac discovered his relativistic equation for the electron (Dirac, 1928), the 4-component *Dirac spinor* was initially interpreted as the wave function of an electron or a positron with spin up or down. However, relativity also requires the possibility of electron–positron pair creation at sufficiently high energy, which is indeed observed experimentally, along with pair annihilation. As a consequence, the Dirac equation does not have a consistent single-particle interpretation. Actually, the Dirac spinor is not a wave function, but a fermion field, whose quantized wave excitations manifest themselves as electrons or positrons (or other fermions). In other words, not only photons but *all elementary "particles"* are, in fact, wavicles. When Dirac's electron–positron field is coupled to the electromagnetic field, one arrives at Quantum Electrodynamics (QED), the quantized version of Maxwell's theory. Its construction was pioneered by Freeman Dyson, Richard Feynman, Julian Schwinger, and Sin-Itiro Tomonaga (Tomonaga, 1946; Schwinger, 1948; Dyson, 1949a,b; Feynman, 1949a,b). QED is an integral part of the Standard Model, in which all elementary "particles", including quarks, leptons, and the Higgs particle, are quantized wave excitations of the corresponding fields. Unlike point particles, Higgs, quark, and lepton fields can interact in a relativistic manner, with or without the mediation by gauge fields.

Although in the Standard Model all "particles" are, in fact, wavicles, one often reads the statement that quarks and leptons are "point-like". What could this possibly mean for a wavicle that does not even have a conceptually well-defined position in space? Again, this is a deficiency of our casual language, which is properly resolved by the unambiguous mathematics of quantum field theory. What the above statement actually means is that even the highest energy experiments have, at least until now, not revealed any substructure of quarks or leptons; *i.e.* they seem to be truly elementary.[1] This does not apply to strongly interacting particles – known as hadrons – like protons, neutrons, or pions, which (essentially) consist of quarks, anti-quarks, and gluons. Interestingly, while being "point-like" in the above sense, an electron is at the same time "infinitely extended", due to the Coulomb cloud that surrounds it. Such "dressed" electrically charged particles are known as *infraparticles*. In the real world, however, their Coulomb field is usually screened by some opposite charge nearby.

This section should have convinced the reader that particle physics does *not* deal with point particles. Perhaps it would be better to call it "wavicle physics". However, as long as we are aware that our casual language is not sufficiently precise in this respect, the terminology is secondary. So, we will generally use the term "particle", keeping in mind that in quantum field theory this is actually a quantized, extended field excitation.

Classical and Quantum Gauge Fields

In the Standard Model, *gauge fields* play a key role (although it also contains non-gauge-field-mediated couplings between quark, lepton, and Higgs fields): Gauge fields mediate the fundamental strong, weak, and electromagnetic interactions. The (classical) Maxwell equations can be expressed entirely in terms of the electromagnetic field strengths $\vec{E}$ and $\vec{B}$, which form the anti-symmetric field strength tensor $F_{\mu\nu}$, with $F_{0i} = E_i$, $\epsilon_{ijk}F_{jk} = 2B_i$. In relativistic quantum field theory, Abelian gauge fields are described by the 4-vector potential $A^\mu = (A^0, \vec{A})$. Together with the ordinary derivative ∂_μ, the 4-vector potential A_μ forms a *covariant derivative* $D_\mu = \partial_\mu + ieA_\mu$ (where e is the electric charge) and $F_{\mu\nu} = \partial_\mu A_\nu - \partial_\nu A_\mu$. In particular, the Aharonov–Bohm effect (Aharonov and Bohm, 1959, 1961; Berry, 1980; Peshkin and Tonomura, 1989) is naturally expressed through line integrals of the 3-vector potential $\vec{A}$. It is part of the 4-vector potential, which can be gauge transformed to $A_\mu{}' = A_\mu - \partial_\mu\alpha$, where $\alpha(\vec{x}, t)$ is an arbitrary (differentiable) gauge transformation function of space and time. In contrast to A_μ, the field strength tensor $F_{\mu\nu}$ is gauge-invariant and thus physical.

When we work with vector potentials, we employ redundant gauge-variant variables to describe gauge-invariant physical observables. While this is a matter of choice in classi-

---

[1] It could also be interpreted such that an elementary particle is always detected in a single pixel or cell of a detector, no matter how fine its resolution is. However, here enters the notorious quantum measurement problem, which we are not going to discuss.

cal theories, it seems inevitable in quantum theories. In particular, already in the quantum mechanics of a charged point particle, the complex phase ambiguity of the wave function turns into a local gauge freedom. Similarly, in quantum field theory the complex phase of a Dirac spinor field is gauge-*variant,* but it can be combined with the gauge-variant vector potential to form the gauge-*invariant* QED Lagrange density or *Lagrangian. Gauge invariance* is a local symmetry which must be maintained *exactly* in order to exclude unphysical effects due to the redundant gauge variables.

Since a gauge symmetry – which we consider synonymous to a *local symmetry* – just reflects a redundancy in our theoretical description of the gauge-invariant physics, it entails physical consequences which differ from those of a global symmetry. Both for global and for local symmetries, the Hamilton operator of the theory is invariant under symmetry transformations. In case of a *global* symmetry (at least in the absence of spontaneous symmetry breaking), this implies that energy eigenstates belong to (generally non-trivial) multiplets of the symmetry group. As a consequence, there are degeneracies in the spectrum whenever an irreducible representation is more than 1-dimensional. In case of a *gauge symmetry,* on the other hand, all physical states are gauge-invariant; *i.e.* they belong to a trivial 1-dimensional representation of the gauge group. Hence, gauge symmetries do not lead to degeneracies in the spectrum of physical states. Indeed, the gauge-variant eigenstates of the gauge-invariant Hamilton operator are exiled from the physical Hilbert space, by imposing the Gauss law as a constraint on physical states.

Ultraviolet Divergences, Regularization, and Renormalization

Field theories have a fixed number of fundamental field degrees of freedom attached to each point in space. In continuous space, the total number of field degrees of freedom is thus uncountably large. While this is not a problem in classical field theory, where the solutions of field equations are smooth functions of space and time, quantum fields perform fluctuations – even at infinitesimally close points – which give rise to ultraviolet (UV) divergences. In order to obtain meaningful finite values for physical quantities, quantum field theories must be *regularized*, for example, by introducing a UV cut-off.

One may speculate that regularization would no longer be necessary if one could work directly with the ultimate degrees of freedom that may exist at ultra-short distances, say of the order of the *Planck length* $l_{\text{Planck}} \simeq 1.62 \cdot 10^{-35}$ m. The corresponding energy scale is the *Planck mass* $M_{\text{Planck}} = \hbar/(c \, l_{\text{Planck}}) \simeq 1.22 \cdot 10^{19}$ GeV, at which gravity (which is extremely weak at low energies) is expected to become strongly coupled. Although string theory provides a framework for its formulation, an established non-perturbative theory of quantum gravity, valid even at the Planck scale, is not known until now. The necessity for regularization apparently suggests that quantum fields in continuous space–time may ultimately not be the correct degrees of freedom that Nature is built of at ultra-short distances.

Fortunately, we do not need to know the *Theory of Everything* before we can address the physics in the TeV energy regime that is accessible to present-day accelerator experiments, which have tested the Standard Model with great scrutiny. The currently accessible TeV energy physics is insensitive to the details of more fundamental, underlying degrees of freedom, be they strings or something else.

In order to mimic the effects of the unknown, ultimate, ultra-short distance degrees of freedom, one can introduce a regularization in different ways. It is crucial that the regularization does not violate important symmetries, in particular, gauge symmetries; otherwise, it is hard to restore them in the final limit, and unphysical redundant variables may contaminate the result. In perturbation theory, the most efficient way to introduce a gauge-invariant regularization is *dimensional regularization*, *i.e.* analytic continuation in the space–time dimension d (Bollini and Giambiagi, 1972a,b; 't Hooft and Veltman, 1972). Dimensional regularization actually leads to finite perturbative results without explicitly introducing a UV cut-off.

Beyond perturbation theory, the *lattice regularization,* which reduces space–time to a 4-dimensional (usually hyper-cubic) grid of discrete lattice points, provides a natural cut-off that allows us to maintain gauge invariance (Wilson, 1974); for lattice textbooks, see Creutz (1983); Rothe (1992); Montvay and Münster (1994); Smit (2002); DeGrand and DeTar (2006); Gattringer and Lang (2009); Knechtli *et al.* (2017). In this case, the lattice spacing a, *i.e.* the distance between nearest-neighbor lattice sites, acts as an inverse UV cut-off π/a. In contrast to continuous space–time, in lattice field theory the number of degrees of freedom becomes *countable,* which removes the divergences in physical observables. Still, in order to obtain meaningful physical results, one must extrapolate to the *continuum limit,* $a \rightarrow 0$. This is achieved by tuning the coupling constants in the Lagrangian in such a way that the long-distance continuum physics becomes insensitive to the lattice spacing. Such systems can be described using the terminology of classical statistical mechanics, where the continuum limit corresponds to a *critical point*. Approaching criticality is related to the process of *renormalization*.

Euclidean Quantum Field Theory versus Classical Statistical Mechanics

The quantization of field theories is a subtle mathematical problem. The functional integral approach (*i.e.* Feynman's path integral applied to field theory) offers an attractive alternative to canonical quantization. When real Minkowski time is analytically continued to purely imaginary Euclidean time, the functional integral becomes mathematically much better behaved. As an extra benefit, Euclidean quantum field theory, in particular when formulated on a space–time lattice, is mathematically analogous to a system of classical statistical mechanics. The Euclidean fields correspond to generalized spin variables, and the classical Hamilton function is analogous to the Euclidean lattice action of quantum field theory. The temperature T, which controls the thermal fluctuations in statistical mechanics, is analogous to $\hbar$, which controls the strength of quantum fluctuations. A spin correlation function is analogous to a Euclidean 2-point function. Usually, it exhibits an exponential decay, which determines a *correlation length* $\xi = \hbar/(Mc)$, where M is a particle mass.

The analogy between classical statistical mechanics and Euclidean quantum field theory has far-reaching consequences, because the theory of *critical phenomena* can be translated to quantum field theory. In particular, a critical point, where the correlation length diverges in units of the lattice spacing, $\xi/a \to \infty$, corresponds to the continuum limit of a Euclidean lattice field theory with $Ma \to 0$. The insensitivity of low-energy physics to the details of the regularization of quantum field theory is a manifestation of *universality*. *Relevant, marginal,* and *irrelevant* couplings are identified based on the *renormalization group* (Wilson and Kogut, 1974; Wilson, 1975). Furthermore, Monte Carlo methods, which were originally developed for classical statistical mechanics, can be applied to lattice field theory where they provide systematically controlled, non-perturbative results.

1 Basics of Quantum Field Theory

This chapter presents an introduction to quantum field theory in the *functional integral formalism*. Classical field theories are introduced as an extension of point particle mechanics to systems with infinitely many degrees of freedom – a fixed number at each space point. Similarly, quantum field theories formally correspond to quantum mechanical systems with infinitely many degrees of freedom. In the same way as point particle mechanics, classical field theories can be quantized by means of the path integral – or *functional integral* – method. A schematic overview is sketched in Figure 1.1.

The transition from the physical real time to Euclidean time, by means of a *Wick rotation,* is highly favorable for the convergence of functional integrals. The resulting quantum field theories in Euclidean space–time can be related to classical statistical mechanics. In this context, we also address the *lattice regularization,* which provides a formulation of quantum field theories beyond perturbation theory.

1.1 From Point Particle Mechanics to Classical Field Theory

Point particle mechanics describes the dynamics of classical, non-relativistic point particles. The coordinates of the particles represent a finite number of degrees of freedom. In the simplest case – a single particle moving on a line – this degree of freedom is just given by the particle position[1] x, as a function of the time t. The dynamics of a particle of mass M moving in an external potential $V(x)$ obeys *Newton's equation of motion*

$$M\partial_t^2 x = F(x) = -\frac{dV(x)}{dx}. \qquad (1.1)$$

Once the initial conditions are specified, this ordinary second-order differential equation determines the path of the particle, $x(t)$. To compute it, one proceeds in infinitesimal time steps.

An alternative approach considers a *finite* time interval with a given initial and final particle position and identifies the classical path connecting them. In this setting, Newton's equation is obtained from the *variational principle* by minimizing the *action,*

$$S[x] = \int dt\, L(x, \partial_t x), \qquad (1.2)$$

---

[1] For the considerations here, and in Sections 1.2 and 1.3, the space dimension hardly matters. For simplicity we set it to 1, but a generalization to higher dimensions is straightforward; one just replaces $x(t)$ by $\vec{x}(t)$ everywhere.

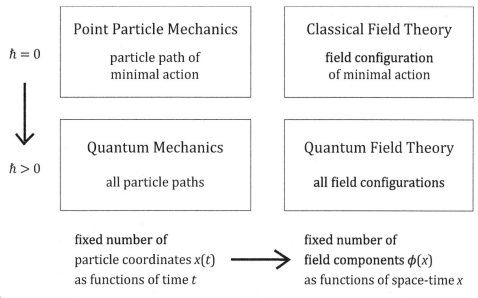

Fig. 1.1 Overview of different branches of physics: We proceed from point particle mechanics to field theory (left to right) and from classical physics to quantum physics (top to bottom).

in the set of all paths $[x]$ with the given end-points.[2] The action is a functional[3] that results from the time-integral of the *Lagrange function*

$$L(x, \partial_t x) = \frac{M}{2}(\partial_t x)^2 - V(x) \tag{1.3}$$

over a path with fixed end-points in space and time. Now the variational condition $\delta S = 0$ implies the *Euler–Lagrange equation*

$$\partial_t \frac{\delta L}{\delta \partial_t x} - \frac{\delta L}{\delta x} = 0, \tag{1.4}$$

which coincides with Newton's equation (1.1) at any instant t.

Classical field theories are a generalization of point particle mechanics to systems with infinitely many degrees of freedom – a fixed number N at each space point $\vec{x}$. Here the degrees of freedom are the field values $\phi_a(\vec{x}, t)$, $a \in \{1, 2, \ldots, N\}$, where ϕ represents a generic field. We anticipate a few examples:

- In the case of a *neutral scalar field*, the value of ϕ is a single real number, $\phi \in \mathbb{R}$, representing $N = 1$ degree of freedom per space point.
- A *charged scalar field* is described by a complex number, $\Phi = \phi_1 + i\phi_2 \in \mathbb{C}$ (with $\phi_i \in \mathbb{R}$). Hence, it represents $N = 2$ real degrees of freedom per space point.
- The *Higgs field* $\phi_a(\vec{x}, t)$ ($a \in \{1, \ldots, 4\}$), which is part of the Standard Model, is a complex doublet; it has $N = 4$ real degrees of freedom per space point.

---

[2] More precisely, one identifies a stationary point in the set of possible paths connecting fixed end-points.

[3] We use the term "functional" for a function whose argument is itself a function, which is integrated (or summed) over. Hence the functional depends on the entire function in its argument, not just on its value at a specific point; the latter is the case for the Lagrange function $L(x, \partial_t x)$.

- An *Abelian gauge field* $A_\mu(\vec{x}, t)$ (with index $\mu \in \{0, 1, 2, 3\}$) – in particular, the photon field in electrodynamics – is a neutral vector field, which seems to have 4 real degrees of freedom per space point. However, two of them are unphysical due to the U(1) gauge symmetry, as we will discuss in Chapter 6. Hence, the Abelian gauge field has only $N = 2$ physical degrees of freedom per space point, which correspond to the two polarization states of the massless photon.

- A *non-Abelian gauge field* $G_\mu^a(\vec{x}, t)$ has an additional index a. For example, the *gluon field* in Quantum Chromodynamics (QCD) with a color index $a \in \{1, 2, \ldots, 8\}$ represents $N = 2 \times 8 = 16$ physical degrees of freedom per space point, again because of some redundancy (in this case due to the so-called color gauge group SU(3)).

  The field $W_\mu^a(\vec{x}, t)$ that represents the *W-bosons* in the Standard Model has an index $a \in \{1, 2, 3\}$ and transforms under another gauge group SU(2). Thus, it seems to represent $2 \times 3 = 6$ physical degrees of freedom. However, in contrast to the photon and the gluons, the *W*-bosons are massive due to the Higgs mechanism, to be discussed in Chapter 12. Therefore they are equipped with three (not just two) polarization states, which leads to $N = 9$. The three extra degrees of freedom are provided by the Higgs field, which is then left with only $N = 1$ physical degree of freedom at each space point. That remaining degree of freedom gives rise to the Higgs boson.

The analogue of Newton's equation in field theory is the *classical field equation*. For instance, for a neutral scalar field, it is given in terms of the d'Alembert operator $\Box = \partial_\mu \partial^\mu = \partial_t^2/c^2 - \sum_{i=1}^3 \partial_i^2$ and the derivative of the potential,

$$\Box \phi = \partial_\mu \partial^\mu \phi = -\frac{dV(\phi)}{d\phi}. \tag{1.5}$$

Again, after specifying appropriate initial conditions, it determines the classical field configuration $[\phi]$, *i.e.* the values of the field ϕ at all space–time points $x = (x_0, \vec{x}) = (ct, \vec{x})$. Hence, the role of *time* in point mechanics is played by *space–time* in field theory, and the role of the point particle *coordinates* is now played by the *field values*. As before, the classical equation of motion emerges by minimizing the action, which now takes the form

$$S[\phi] = \int d^4x \, \mathcal{L}(\phi, \partial_\mu \phi), \qquad d^4x = d(ct) \, d^3x = c \, dt \, d^3x. \tag{1.6}$$

The integral over time in eq. (1.2) is extended to an integral over space–time. The Lagrange function L of point particle mechanics, given in eq. (1.3), is replaced by the Lagrange density, or *Lagrangian*,[4]

$$\mathcal{L}(\phi, \partial_\mu \phi) = \frac{1}{2} \partial_\mu \phi \partial^\mu \phi - V(\phi). \tag{1.7}$$

A prominent, interacting field theory is the $\lambda \phi^4$ model with the potential[5]

$$V(\phi) = \frac{m^2}{2} \phi^2 + \frac{\lambda}{4!} \phi^4. \tag{1.8}$$

Classically, m is the mass of the scalar field ϕ and λ is the coupling strength of its self-interaction. The mass term[6] corresponds to a harmonic oscillator potential in the point

---

[4] Throughout this book, the derivatives ∂_μ and ∂^μ act only on the immediately following field.

[5] Part of the literature restricts "potentials" to the interaction terms, so the mass term is not included, but this is just a matter of terminology.

[6] Here we assume m^2 to be positive. Later, in particular in Chapter 12, we will also consider the case $m^2 < 0$.

Table 1.1 A dictionary that translates 1-d point particle mechanics into classical field theory in 4-d space–time. As we proceed from point particle mechanics to field theory, we promote the number of degrees of freedom from one to infinite. This dictionary refers to field theories with a Lorentz-invariant Lagrangian $\mathcal{L}$.

Point Particle Mechanics	Classical Field Theory
time t	space–time $x = (ct, \vec{x})$
particle coordinate x	field value ϕ
particle path $x(t)$	field configuration $\phi(x)$
Lagrange function $L(x, \partial_t x) = M(\partial_t x)^2/2 - V(x)$	Lagrangian $\mathcal{L}(\phi, \partial_\mu \phi) = \partial_\mu \phi \partial^\mu \phi/2 - V(\phi)$
action $S[x] = \int dt\, L(x, \partial_t x)$	action $S[\phi] = \int d^4 x\, \mathcal{L}(\phi, \partial_\mu \phi)$
equation of motion $\partial_t\, \delta L/\delta(\partial_t \phi) - \delta L/\delta x = 0$	classical field equation $\partial_\mu\, \delta \mathcal{L}/\delta(\partial_\mu \phi) - \delta \mathcal{L}/\delta \phi = 0$
Newton's equation $M\partial_t^2 x = -dV(x)/dx$	scalar field equation $\partial_\mu \partial^\mu \phi = -dV(\phi)/d\phi$
kinetic energy $M(\partial_t x)^2/2$	derivative contribution to $\mathcal{L}$ $\partial_\mu \phi \partial^\mu \phi/2$
harmonic oscillator potential $M\omega^2 x^2/2$	mass term $m^2 \phi^2/2$
anharmonic term $\lambda x^4/4!$	quartic self-interaction term $\lambda \phi^4/4!$

particle mechanics analogue, while the interaction term corresponds to an anharmonic contribution.

Here the condition $\delta S = 0$ leads to the Euler–Lagrange equation

$$\partial_\mu \frac{\delta \mathcal{L}}{\delta(\partial_\mu \phi)} - \frac{\delta \mathcal{L}}{\delta \phi} = 0, \tag{1.9}$$

which is the classical field equation for ϕ. In particular, based on the Lagrangian (1.7), we arrive at the scalar field equation (1.5). The analogies between point particle mechanics and classical field theory are summarized in Table 1.1.

1.2 Quantum Mechanical Path Integral

In many cases, the quantization of field theories can be conveniently performed by using the *path integral* – or *functional integral* – approach (Dirac, 1933; Feynman, 1948). We first discuss the path integral in quantum mechanics = quantized point particle mechanics   in the real-time formalism. A mathematically solid formulation employs an analytic continuation to the so-called Euclidean time. The Euclidean time path integral will be addressed in Section 1.3.

We use the *Dirac notation*, where a *ket* $|\Psi\rangle$ describes some state as a unit-vector in a Hilbert space and a *bra* $\langle\Psi|$ its Hermitian conjugate. Thus the *bracket* $\langle\Psi'|\Psi\rangle$ is a scalar product. The corresponding wave functions in (1-dimensional, cf. footnote 1) coordinate space and in momentum space are obtained as

$$\Psi(x,t) = \langle x|\Psi(t)\rangle, \qquad \Psi(p,t) = \langle p|\Psi(t)\rangle, \tag{1.10}$$

where $|x\rangle$ and $|p\rangle$ are coordinate and momentum eigenstates, $\hat{x}|x\rangle = x|x\rangle$, $\hat{p}|p\rangle = p|p\rangle$. We further denote the energy eigenstates as $|n\rangle$, *i.e.* $\hat{H}|n\rangle = E_n|n\rangle$, where $\hat{H}$ is the *Hamilton operator* and E_n are the energy eigenvalues. The energy eigenfunctions (in coordinate representation) are then given by $\langle x|n\rangle$.

The eigenstates $|x\rangle$, $|p\rangle$, and $|n\rangle$ build three complete orthonormal sets,

$$\int dx\, |x\rangle\langle x| = \frac{1}{2\pi\hbar}\int dp\, |p\rangle\langle p| = \sum_n |n\rangle\langle n| = \hat{\mathbb{1}}. \tag{1.11}$$

(It is understood to take sums or integrals whenever the set of states is discrete or continuous, respectively.) So we can write a scalar product as

$$\langle\Psi'|\Psi\rangle = \int dx\, \langle\Psi'|x\rangle\langle x|\Psi\rangle = \int dx\, \Psi'(x)^*\, \Psi(x). \tag{1.12}$$

The wave functions in coordinate and momentum space can be converted into one another by the Fourier transform and its inverse,

$$\Psi(p,t) = \int dx\, \langle p|x\rangle\, \langle x|\Psi(t)\rangle = \int dx\, \exp(-ipx/\hbar)\, \Psi(x,t),$$

$$\Psi(x,t) = \frac{1}{2\pi\hbar}\int dp\, \langle x|p\rangle\, \langle p|\Psi(t)\rangle = \frac{1}{2\pi\hbar}\int dp\, \exp(ipx/\hbar)\, \Psi(p,t), \tag{1.13}$$

such that $\hat{p}\Psi(x,t) = -i\hbar\,\partial_x\Psi(x,t)$ and $\hat{x}\Psi(p,t) = i\hbar\,\partial_p\Psi(p,t)$.

A Hermitian operator, $\hat{O} = \hat{O}^\dagger$, takes the expectation value

$$\langle\Psi|\hat{O}|\Psi\rangle = \int dx\, \Psi(x)^*\, \hat{O}\, \Psi(x). \tag{1.14}$$

The time evolution of a quantum system – described by a Hamilton operator $\hat{H}$ – is determined by the time-dependent *Schrödinger equation*

$$i\hbar\partial_t\, |\Psi(t)\rangle = \hat{H}\, |\Psi(t)\rangle. \tag{1.15}$$

Like Newton's equation, the Schrödinger equation describes the evolution over an infinitesimal time. As in Section 1.1, we proceed to its integrated form, *i.e.* to *a finite time interval*, for which we write the ansatz

$$|\Psi(t')\rangle = \hat{U}(t',t)\, |\Psi(t)\rangle, \qquad t' \geq t, \tag{1.16}$$

where $\hat{U}(t',t)$ is the unitary *evolution operator*. For a Hamilton operator without explicit time-dependence, it takes the simple form[7]

$$\hat{U}(t',t) = \exp\left(-\frac{i}{\hbar}\hat{H}(t'-t)\right). \qquad (1.17)$$

Let us consider the evolution of a wave function in coordinate space. Using eqs. (1.10) and (1.11), we obtain

$$\Psi(x',t') = \int dx \, \langle x'|\hat{U}(t',t)|x\rangle \, \Psi(x,t), \qquad (1.18)$$

i.e. $\langle x'|\hat{U}(t',t)|x\rangle$ acts as a *propagator* for the wave function, if we assume $t' \geq t$.

The propagator is a quantity of primary physical interest. In particular, it contains information about the *energy spectrum*. Let us consider the propagation from an initial position eigenstate $|x\rangle$ back to itself,

$$\langle x|\hat{U}(t',t)|x\rangle = \langle x|\exp\left(-\frac{i}{\hbar}\hat{H}(t'-t)\right)|x\rangle = \sum_n |\langle x|n\rangle|^2 \exp\left(-\frac{i}{\hbar}E_n(t'-t)\right), \quad (1.19)$$

where we applied eq. (1.11). Hence the propagator reveals the energy spectrum, as well as the probability density of the energy eigenstates.

Inserting now a complete set of position eigenstates at some intermediate time t_1, with $t < t_1 < t'$, we obtain

$$\begin{aligned}
\langle x'|\hat{U}(t',t)|x\rangle &= \langle x'|\exp\left(-\frac{i}{\hbar}\hat{H}(t'-t_1)\right)\exp\left(-\frac{i}{\hbar}\hat{H}(t_1-t)\right)|x\rangle \\
&= \int dx_1 \langle x'|\exp\left(-\frac{i}{\hbar}\hat{H}(t'-t_1)\right)|x_1\rangle\langle x_1|\exp\left(-\frac{i}{\hbar}\hat{H}(t_1-t)\right)|x\rangle \\
&= \int dx_1 \langle x'|\hat{U}(t',t_1)|x_1\rangle\langle x_1|\hat{U}(t_1,t)|x\rangle.
\end{aligned} \qquad (1.20)$$

Obviously we can repeat this process an arbitrary number of times. This is exactly what we do in the formulation of the path integral. Let us divide the time interval $[t,t']$ into N equidistant steps of size ε such that

$$t' - t = N\varepsilon. \qquad (1.21)$$

Inserting a complete set of position eigenstates at all intermediate times $t_n = t + n\varepsilon$, $n = 1, 2, \ldots, N-1$, we arrive at

$$\begin{aligned}
\langle x'|\hat{U}(t',t)|x\rangle &= \int dx_1 \int dx_2 \ldots \int dx_{N-1} \langle x'|\hat{U}(t',t_{N-1})|x_{N-1}\rangle \ldots \\
&\times \ldots \langle x_2|\hat{U}(t_2,t_1)|x_1\rangle\langle x_1|\hat{U}(t_1,t)|x\rangle.
\end{aligned} \qquad (1.22)$$

Now we are integrating over *all* paths (in discrete time), as depicted schematically in Figure 1.2.

In the next step, we focus on one of the factors in eq. (1.22). We consider a single non-relativistic point particle moving in an external potential $V(x)$ such that

$$\hat{H} = \frac{\hat{p}^2}{2M} + \hat{V}(\hat{x}). \qquad (1.23)$$

---

[7] In the general case of a time-dependent Hamilton operator, the evolution operator can be expanded into a Dyson series. In the present case, we could just write $\hat{U}(t'-t)$. We stay with the general notation $\hat{U}(t',t)$, however, since the crucial decomposition in eq. (1.24) and the central result (1.26) hold generally.

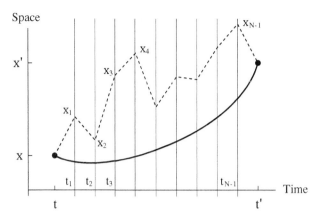

Fig. 1.2 An illustration of the transition of a point particle from (x, t) to (x', t'). The classical particle follows the path of minimal action, represented by the smooth line in this plot. For the quantum mechanical particle, the transition amplitude is obtained by integrating over all possible positions at all intermediate times $t_1, \ldots, t_{N-1}$. As an example, a sequence of such points is symbolically connected by the dashed line.

Using the Baker–Campbell–Hausdorff formula, and neglecting terms of order ε^2, we obtain[8]

$$
\begin{aligned}
\langle x_{n+1}|\hat{U}(t_{n+1}, t_n)|x_n\rangle &= \langle x_{n+1}| \exp\left(-\frac{i\varepsilon}{2\hbar}\hat{V}(\hat{x})\right) \exp\left(-\frac{i\varepsilon}{\hbar}\frac{\hat{p}^2}{2M}\right) \exp\left(-\frac{i\varepsilon}{2\hbar}\hat{V}(\hat{x})\right)|x_n\rangle \\
&= \frac{1}{2\pi\hbar}\int dp\, \langle x_{n+1}| \exp\left(-\frac{i\varepsilon}{2\hbar}\hat{V}(\hat{x})\right) \\
&\quad \times \exp\left(-\frac{i\varepsilon}{\hbar}\frac{\hat{p}^2}{2M}\right)|p\rangle\langle p| \exp\left(-\frac{i\varepsilon}{2\hbar}\hat{V}(\hat{x})\right)|x_n\rangle \\
&= \frac{1}{2\pi\hbar}\int dp\, \exp\left(-\frac{i\varepsilon}{\hbar}\frac{p^2}{2M}\right) \exp\left(-\frac{i}{\hbar}p(x_{n+1}-x_n)\right) \\
&\quad \times \exp\left(-\frac{i\varepsilon}{2\hbar}[V(x_{n+1}) + V(x_n)]\right).
\end{aligned}
\tag{1.24}
$$

We have arranged the terms such that time reversal symmetry holds even at finite ε.

The above integral over the momentum p is subtle, because the integrand is a rapidly oscillating function. To make this expression convergent, we replace the time step ε by $\varepsilon - i\delta$, $\delta > 0$, *i.e.* we step into the complex time plane by a small distance δ. After performing the integral, we take the limit $\delta \to 0$; a cleaner derivation is postponed to

---

[8] This factorization of $\exp(-i\hat{H}\varepsilon/\hbar)$ is also known as *Trotter's formula* (Trotter, 1959). The fact that it holds only up to $\mathcal{O}(\varepsilon^2)$ is the reason why we need the decomposition into infinitesimal time steps. This factorization is obvious for bounded operators, say $\exp\left(\varepsilon(\hat{A} + \hat{B})\right) = \exp(\varepsilon\hat{A})\exp(\varepsilon\hat{B}) + \mathcal{O}(\varepsilon^2)$ with $\hat{A}$, $\hat{B}$ bounded. However, it is non-trivial and highly relevant for the path integral that it also holds if $\hat{A}$ and $\hat{B}$ are only *bounded from below*.

Problem 1.3. We keep in mind that the definition of the path integral requires an analytic continuation in time. We obtain

$$\langle x_{n+1}|\hat{U}(t_{n+1},t_n)|x_i\rangle = \left(\frac{M}{2\pi\,\mathrm{i}\,\hbar\varepsilon}\right)^{1/2}\exp\left(\frac{\mathrm{i}}{\hbar}\varepsilon\left[\frac{M}{2}\left(\frac{x_{n+1}-x_n}{\varepsilon}\right)^2 - \frac{1}{2}\Big(V(x_n)+V(x_{n+1})\Big)\right]\right).$$
(1.25)

Inserting this term back into the expression for the propagator, we arrive at the celebrated path integral formula (Feynman, 1948)

$$\langle x'|\hat{U}(t',t)|x\rangle = \int \mathcal{D}x\,\exp\left(\frac{\mathrm{i}}{\hbar}S[x]\right).$$
(1.26)

The action has been identified in the formal time continuum limit as

$$S[x] = \lim_{\varepsilon\to 0}\varepsilon\sum_n\left[\frac{M}{2}\left(\frac{x_{n+1}-x_n}{\varepsilon}\right)^2 - \frac{1}{2}\Big(V(x_{n+1})+V(x_n)\Big)\right]$$

$$= \int dt\,\left[\frac{M}{2}(\partial_t x)^2 - V(x)\right].$$
(1.27)

The integration measure in eq. (1.26) is given by

$$\int \mathcal{D}x = \lim_{\varepsilon\to 0}\left(\frac{M}{2\pi\,\mathrm{i}\,\hbar\varepsilon}\right)^{N/2}\int dx_1\int dx_2\ldots\int dx_{N-1}.$$
(1.28)

This means that we integrate over all possible particle positions at each intermediate time t_n. In this way, we integrate over *all possible paths* of a particle moving from position eigenstate $|x\rangle$ at time t to $|x'\rangle$ at $t' > t$. Each path is weighted with a phase factor $\exp(\mathrm{i}S[x]/\hbar)$.

Note that the significant contributions to the path integral are due to the (numerous) paths that are continuous but nowhere differentiable (Coleman, 1985). As an intuitive argument, we see from eq. (1.27) that finite contributions only require the magnitude of the time step, $\mathcal{O}(\varepsilon)$, to coincide with $\mathcal{O}(x_{n+1}-x_n)^2$. Such paths are continuous indeed, but the ratio $(x_{n+1}-x_n)/\varepsilon$ naturally diverges. Hence we arrive at a perplexing situation: The path integral formalism has been constructed based on the Lagrange function, but it turns out that this function is not really defined for the paths that dominate the integral. The divergence of the action in the continuum limit suppresses these contributions, but this effect is compensated by the huge multiplicity of this type of paths. Hence, the last line in eq. (1.27) is only symbolic, but – in agreement with the literature – we will keep using this notation here and later in quantum field theory.

As in classical point particle mechanics, the propagation over a finite time interval is handled by referring to the *action*. In quantum mechanics, this formulation eliminates the operators, but it employs a path integral measure $\mathcal{D}x$, which is non-trivial to deal with.

If the path is varied at a macroscopic scale, the phase factor $\exp(\mathrm{i}S[x]/\hbar)$ undergoes extremely rapid oscillations, because $\hbar$ is very small ($\hbar \simeq 6.6 \times 10^{-34}$ kg m^2/s). The vicinity of the classical path of minimal action[9] performs the least oscillations, hence this vicinity

---

[9] For simplicity, we keep on assuming a unique solution to $\delta S = 0$ at the minimum of the action, although this does not hold in all systems.

provides the largest contribution to the path integral. In the limit $\hbar \to 0$ only the contribution of the classical path persists, and we are back at the Euler–Lagrange equation (1.4). At non-zero $\hbar$ (which is tiny on macroscopic scales), the contributions of non-classical paths are still suppressed (or "washed out") by the rapidly oscillating phase; their remaining contributions to the path integral give rise to the quantum effects.

Eq. (1.26) is the key result of the path integral formulation of quantum mechanics. It provides a transparent transition from classical physics to quantum physics as we turn on $\hbar$ to include fluctuations around the path of minimal action. This transition has an analogue in classical optics, if we proceed from Fermat's principle of geometrical optics to the more fundamental Huygens principle of wave optics. Along this line, also the transition behavior of quantum particles through double (or multiple) slits is very intuitive in view of the path integral description. More detailed presentations of the path integral can be found, *e.g.*, in the textbooks by Feynman and Hibbs (1965); Schulman (1981); Glimm and Jaffe (1987); Popov (1983); Kleinert (1990); Swanson (1992); and Roepstorff (1994).

1.3 Path Integral in Euclidean Time

As we have seen, it takes at least a small excursion into the complex time plane to render the path integral mathematically well-defined. Now we perform a radical step into that plane and consider the purely imaginary, so-called *Euclidean time*. Remarkably, this formulation has a direct physical interpretation in the framework of quantum statistical mechanics, as reviewed comprehensively by Roepstorff (1994).

Let us consider the central quantity of statistical mechanics, the canonical *partition function*

$$Z = \text{Tr } \exp(-\beta \hat{H}), \tag{1.29}$$

where $\beta = 1/T$ is the inverse temperature (in units of the Boltzmann constant; *i.e.* we set $k_B = 1$). By analytic continuation, β is related to the time that we discussed in the real-time path integral. In particular, the operator $\exp(-\beta \hat{H})$ turns into the time evolution operator $\hat{U}(t', t)$ of eq. (1.17) if we identify

$$\beta = \frac{i}{\hbar}(t' - t). \tag{1.30}$$

In this sense, the system at finite temperature corresponds to a propagation in purely imaginary time, *i.e.* in *Euclidean time*

$$t_E = i\, t. \tag{1.31}$$

This rotation of the time coordinate by $\pi/2$ in the complex time plane is denoted as the *Wick rotation*, named after the Italian physicist Gian Carlo Wick. It transforms Minkowski's metric into a Euclidean metric (justifying the term "Euclidean time"), as we will describe explicitly in Chapter 4.

By dividing the Euclidean time interval into N equidistant steps, *i.e.* by writing $\beta = Na/\hbar$ – and by inserting again complete sets of position eigenstates – we now arrive at the *Euclidean time path integral*[10]

$$Z = \mathrm{Tr}\exp(-\beta\hat{H}) = \int \mathcal{D}x \exp\left(-\frac{1}{\hbar}S_E[x]\right). \qquad (1.32)$$

Here the action takes the Euclidean form

$$S_E[x] = \lim_{a\to 0}\sum_n a\left[\frac{M}{2}\left(\frac{x_{n+1}-x_n}{a}\right)^2 + V(x_n)\right] = \int_0^\beta dt_E\left[\frac{M}{2}(\partial_{t_E}x)^2 + V(x)\right]. \qquad (1.33)$$

Unlike in the real-time case, the measure now involves N integrals,

$$\int \mathcal{D}x = \lim_{a\to 0}\left(\frac{M}{2\pi\hbar a}\right)^{N/2}\int dx_1 \int dx_2 \ldots \int dx_N. \qquad (1.34)$$

The additional integration over $x_N = x'$ is due to the trace in eq. (1.29). Note that there is no extra integration over $x_0 = x$ because the trace implies periodic boundary conditions in Euclidean time, $x_0 = x_N$.

The Euclidean path integral allows us to evaluate *thermal expectation values*. For example, let us consider an operator $\hat{\mathcal{O}}(\hat{x})$ that is diagonal in the position state basis $\{|x\rangle\}$. By inserting this operator into the path integral, we obtain an expression for its expectation value,

$$\langle\hat{\mathcal{O}}(\hat{x})\rangle = \frac{1}{Z}\mathrm{Tr}\left[\hat{\mathcal{O}}(\hat{x})\exp(-\beta\hat{H})\right] = \frac{1}{Z}\int \mathcal{D}x\,\mathcal{O}(x(0))\exp\left(-\frac{1}{\hbar}S_E[x]\right). \qquad (1.35)$$

Since the theory is translation invariant in Euclidean time, one can place $\mathcal{O}(x(t_E))$ at any time t_E, *e.g.*, at $t_E = 0$, as it has been done here.

When we approach the low-temperature limit, $\beta \to \infty$, the thermal fluctuations are suppressed and only the quantum ground state $|0\rangle$ (with energy E_0) contributes to the partition function, $Z \to \exp(-\beta E_0)$. In this limit, the path integral is formulated in a very long Euclidean time interval, which determines the *ground state expectation value*. For instance, for the 1-point function it reads

$$\langle 0|\hat{\mathcal{O}}(\hat{x})|0\rangle = \lim_{\beta\to\infty}\frac{1}{Z}\int \mathcal{D}x\,\mathcal{O}(x(0))\exp\left(-\frac{1}{\hbar}S_E[x]\right). \qquad (1.36)$$

In order to extract the energy spectrum, it is of interest to consider *2-point functions* of operators at different instances in Euclidean time,

---

[10] Note that here the momentum integral corresponding to eq. (1.24) is well-defined from the beginning.

$$\langle \hat{\mathcal{O}}(\hat{x}(0))\, \hat{\mathcal{O}}(\hat{x}(t_E))\rangle = \frac{1}{Z}\mathrm{Tr}\left[\,\exp(t_E\hat{H}/\hbar)\,\hat{\mathcal{O}}(\hat{x}(0))\,\exp(-t_E\hat{H}/\hbar)\,\hat{\mathcal{O}}(\hat{x}(0))\,\exp(-\beta\hat{H})\right]$$

$$= \frac{1}{Z}\sum_{n,m}\langle n|\exp(t_E\hat{H}/\hbar)\,\hat{\mathcal{O}}(\hat{x}(0))\,\exp(-t_E\hat{H}/\hbar)|m\rangle\,\langle m|\hat{\mathcal{O}}(\hat{x}(0))\,\exp(-\beta\hat{H})|n\rangle$$

$$= \frac{1}{Z}\sum_{n,m}|\langle n|\hat{\mathcal{O}}(\hat{x}(0))|m\rangle|^2\,\exp(-E_n\beta - t_E[E_m - E_n]/\hbar)$$

$$\overset{!}{=} \frac{1}{Z}\int \mathcal{D}x\,\mathcal{O}(x(t_E))\,\mathcal{O}(x(0))\,\exp\left(-\frac{1}{\hbar}S_E[x]\right). \tag{1.37}$$

Again we consider the limit $\beta \to \infty$, but we also assume a large Euclidean time separation between the operators; *i.e.* we also let $t_E \to \infty$ (while maintaining $t_E/\hbar \ll \beta$). Then the leading contribution is $|\langle 0|\mathcal{O}(x)|0\rangle|^2$, the product of two independent 1-point functions. In order to reveal the actual *correlation,* we subtract this part and build the *connected 2-point function*. Asymptotically one obtains

$$\lim_{\beta,\,t_E\to\infty}\langle \hat{\mathcal{O}}(x(t_E))\hat{\mathcal{O}}(x(0))\rangle - |\langle \hat{\mathcal{O}}(x)\rangle|^2\bigg|_{\beta\gg t_E} = |\langle 1|\hat{\mathcal{O}}(x)|0\rangle|^2\,\exp\left(-t_E[E_1 - E_0]/\hbar\right). \tag{1.38}$$

Here $|1\rangle$ is the first excited state of the quantum system, with energy E_1 (we assume that there is no degeneracy). The connected 2-point function decays exponentially at large Euclidean time separations. This decay is governed by the *energy gap* $E_1 - E_0$.

As an appetizer, we anticipate that in a quantum field theory, E_1 corresponds to the energy of the lightest particle in the spectrum. Its mass is determined by the energy gap $E_1 - E_0$ above the vacuum, *i.e.* the difference between the presence or absence of one particle. Hence in Euclidean field theory, particle masses are evaluated from the exponential decay of connected 2-point functions.

1.4 Spin Models in Classical Statistical Mechanics

So far, we have considered quantum systems, both at zero and at finite temperature. We have represented their partition functions by Euclidean path integrals over configurations on a time lattice of extent β. We will now take a new start and consider lattice systems of *classical statistical mechanics at finite temperature*. We will see that their mathematical description is very similar to the path integral formulation of quantum systems. The physical interpretation, however, is different. In the next section, we will set up another dictionary that allows us to translate quantum physics language into the terminology of classical statistical mechanics. Some aspects of phase transitions and critical phenomena are sketched in Appendix I. For further reading, we recommend textbooks devoted to this subject, such as Ma (1976); Pfeuty and Toulouse (1977); Amit (1978); and Herbut (2007).

Let us concentrate on *classical spin models*. Here, the term "spin" does not mean that we deal with quantized angular momenta. All we do is work with classical vectors as

Table 1.2 $O(N)$-symmetric classical spin models and their global symmetries.

	Model	Global symmetry
$N = 1$	Ising model	$\mathbb{Z}(2)$
$N = 2$	XY model	$O(2)$
$N = 3$	Heisenberg model	$O(3)$
$N = \infty$	Spherical model	$O(\infty)$

statistical variables.[11] We denote these vectors as $\vec{s}_x = (s_x^1, \dots, s_x^N)$.[12] In these models, the spins reside on the sites of a d-dimensional spatial grid, so x denotes a corresponding lattice site. If it represents a *crystal lattice*, then the lattice spacing has a physical meaning. This is in contrast to the time step that we have introduced before as a regularization in order to render the path integral mathematically well-defined and that one finally sends to zero in order to reach the temporal continuum limit.

One often normalizes the spin vectors to $|\vec{s}_x| = 1$ at each site x. The simplest spin model of this kind is the *Ising model* with classical spin variables $s_x = \pm 1$. In Table 1.2, we list some spin models, all of them with the constraint $|\vec{s}_x| = 1$ and a global $O(N)$ spin rotation symmetry (which turns into $\mathbb{Z}(2)$ for $N = 1$), cf. Appendix I. The *XY model* captures universal aspects of superfluids. The (classical) *Heisenberg model* can be thought of as a simplified description of ferromagnets, where the electron spins in some subvolume are summed up to act collectively like a classical spin (see, *e.g., Ma (1976)*). The large N limit enables analytical calculations, which are in general not feasible at finite N. Along with a $1/N$ expansion, one hopes to capture some features of the physically more relevant cases at small N.

These $O(N)$ spin models are characterized by a classical *Hamilton function* $\mathcal{H}$ (not a quantum Hamilton operator), which specifies the energy of any spin configuration. The couplings between different spins are often limited to nearest-neighbor sites, which we denote as $\langle xy \rangle$. The standard form of the Hamilton function reads

$$\mathcal{H}[s] = -J \sum_{\langle xy \rangle} \vec{s}_x \cdot \vec{s}_y - \vec{B} \cdot \sum_x \vec{s}_x. \tag{1.39}$$

It is *ferromagnetic* for a coupling constant $J > 0$ (which favors parallel spins) and anti-ferromagnetic for $J < 0$. Moreover, the spins prefer to be aligned with the external "magnetic field" (or "ordering field") $\vec{B} = (B^1, \dots, B^N)$.

In particular, the partition function of the Ising model takes the form

$$Z = \prod_x \sum_{s_x = \pm 1} \exp(-\mathcal{H}[s]/T) = \int \mathcal{D}s \, \exp(-\mathcal{H}[s]/T), \tag{1.40}$$

---

[11] Quantum spin models – such as the quantum XY or Heisenberg model – also exist, but they are more complicated: For instance, in those models it is often impossible to analytically identify the ground state.

[12] The number of spin components N should not be confused with the number of time steps in the path integral, which was also denoted by N.

where we still set the Boltzmann constant $k_B = 1$. The *sum over all spin configurations* corresponds to the summation over all possible orientations of individual spins. For $N \geq 2$, the measure $\mathcal{D}s$ can be written as

$$\mathcal{D}s = \prod_x \int_{S^{N-1}} d^{N-1}s_x = \prod_x \int_{-1}^{1} ds_x^1 \dots \int_{-1}^{1} ds_x^N \, \delta(|\vec{s}_x| - 1). \qquad (1.41)$$

Thermal averages are computed by inserting the observable of interest in the integrand. For example, the *magnetization density* is given by

$$\langle \vec{s}_x \rangle = \frac{1}{Z} \int \mathcal{D}s \, \vec{s}_x \exp(-\mathcal{H}[s]/T). \qquad (1.42)$$

Due to translation invariance,[13] the result does not depend on x. So, we can simply write the magnetization density as $\langle \vec{s} \rangle$, in analogy to the time-independence of the 1-point function of eq. (1.35).

Similarly, the spin–spin *correlation function* is defined as

$$\langle \vec{s}_x \cdot \vec{s}_y \rangle = \frac{1}{Z} \int \mathcal{D}s \, \vec{s}_x \cdot \vec{s}_y \, \exp(-\mathcal{H}[s]/T), \qquad (1.43)$$

which (again due to translation invariance) only depends on the distance vector $x - y$. Subtracting the disconnected contribution – as in eq. (1.38) – we obtain the *connected* spin–spin correlation function. At large distances, it tends to decay exponentially,

$$\langle \vec{s}_x \cdot \vec{s}_y \rangle_c = \langle \vec{s}_x \cdot \vec{s}_y \rangle - \langle \vec{s}_x \rangle \cdot \langle \vec{s}_y \rangle = \langle \vec{s}_x \cdot \vec{s}_y \rangle - \langle \vec{s} \rangle^2 \sim \exp\left(-\frac{|x - y|}{\xi}\right), \qquad (1.44)$$

which defines the *correlation length* ξ. (Here, the relation $\sim$ means that – for large $|x - y|$ – the decay of the connected correlation function is asymptotically dominated by this exponential factor, although there may be a polynomial prefactor depending on $|x-y|$.)

The correlation length is analogous to the inverse energy gap $E_1 - E_0$, *i.e.* the inverse gap between the ground state energy and the minimal excitation energy. A point in the phase diagram where ξ diverges is known as a *critical point*. Next to a critical point, ξ is a natural length scale of the system; all dimensioned quantities can be measured by their ratios with the suitable power of ξ. At high temperature, strong thermal noise tends to destroy long-range correlations, such that the correlation length can well be just a few lattice spacings.

The *Ising model* was originally motivated as a simple description of ferromagnetism. Moreover, it is equivalent to a *lattice gas model*, where each lattice site can be occupied by 0 or 1 "gas molecules" or by one or the other state of a binary liquid. As a model for such complex systems, the Ising model might appear as a drastic over-simplification, for example, because real magnets also involve couplings beyond nearest-neighbor spins. However, the details of the Hamilton function at the scale of the lattice spacing do not always matter. There may be a *critical temperature* T_c at which ξ diverges (in infinite volume), and in its vicinity a universal behavior emerges. At this temperature, a *second-order phase transition* takes place.[14] Then, at $T \approx T_c$ the details of the model at the scale

---

[13] Translation invariance holds in infinite volume (which we assume at this point) or in a finite volume with periodic boundary conditions.

[14] Strictly speaking, the order can be 2 or higher. For increasing order, phase transitions become smoother. Paul Ehrenfest postulated that a phase transition is of order k if k is the lowest derivative of the free energy F (given

of a few lattice spacings are irrelevant for the long-range physics that takes place at the scale ξ.

In fact, the *critical behavior* of the Ising model (its behavior in the vicinity of the critical temperature, with a small or vanishing external field) attracts a lot of interest. It was solved in $d = 1$ by Ernst Ising (1925) and in $d = 2$ by Lars Onsager (1944) (this means that observables like correlation functions have been computed analytically). In higher dimensions, an analytic solution has not been found so far, but there are analytic approximation techniques as well as accurate numerical results.

In $d = 1$, there is no non-zero critical temperature T_c (ξ diverges only at $T = 0$), and Ising concluded that this model is an over-simplification. There are, however, non-zero critical temperatures in $d > 1$, as first argued qualitatively by Rudolf Peierls (1936) for the 2-d and 3-d case (see also Polyakov (1987)). Hence, this model *is* of interest, and it does have a wide variety of applications.

The Ising model is just a very simple member of a large *universality class* of different models, which all share the same critical behavior. This does not mean that they have the same critical temperatures. However, as the temperature T approaches T_c from below, their magnetization,

$$M = \sum_x s_x, \tag{1.45}$$

vanishes with the same power of $T_c - T$. In thermodynamic terminology, one would write

$$\langle M \rangle = -\left.\frac{\partial F}{\partial B}\right|_{T=\text{const.}}, \tag{1.46}$$

where

$$F = -T \log Z \tag{1.47}$$

is the *free energy*.

The universal behavior (at $B = 0$) is characterized by the *critical exponent* β (not to be confused with the inverse temperature β),

$$\lim_{T \nearrow T_c} \langle M \rangle \propto (T_c - T)^\beta. \tag{1.48}$$

Alternatively, if we fix the temperature T_c and include an external magnetic field B, which is then gradually turned off, the magnetization converges to zero as $\langle M \rangle \sim B^{1/\delta}$, which defines another critical exponent $\delta > 0$.

On the other hand, the magnetic *susceptibility* χ diverges for $T \to T_c$, and the same holds for the *specific heat C*. The corresponding definitions and power-laws read

$$\chi = \left.\frac{1}{V}\frac{\partial \langle M \rangle}{\partial B}\right|_{B=0} = -\left.\frac{1}{V}\frac{\partial^2 F}{\partial B^2}\right|_{B=0} = \frac{1}{V}\left(\langle M^2 \rangle - \langle M \rangle^2\right) \propto |T - T_c|^{-\gamma}$$

$$C = \left.\frac{T}{V}\frac{\partial S}{\partial T}\right|_{B=\text{const.}} = -\left.\frac{T}{V}\frac{\partial^2 F}{\partial T^2}\right|_{B=\text{const.}} = \frac{1}{VT^2}\left(\langle \mathcal{H}^2 \rangle - \langle \mathcal{H} \rangle^2\right) \propto |T - T_c|^{-\alpha},$$

$$\tag{1.49}$$

in eq. (1.47)), which is discontinuous. For the modern classification, we refer, *e.g.*, to Blundell and Blundell (2008).

where $S = -\partial F/\partial T|_{B=\text{const.}}$ is the entropy. Here, V symbolizes the volume, although – in a mathematical sense – phase transitions require an infinite volume; we see that χ and C are intensive quantities, whereas $\mathcal{H}$, F, M, and S are all extensive.

The parameters β, δ, γ, and α all are *critical exponents*. They characterize the behavior in the vicinity of a critical point, and they coincide exactly within a universality class. Therefore, these values *characterize a universality class*.

Note that dimensioned quantities – like T_c – will obviously change if, for instance, the lattice spacing of the crystal is altered, as it happens for materials composed of different kinds of molecules. On the other hand, the critical exponents are dimensionless; hence, these are suitable parameters for agreement within a universality class.

1.5 Quantum Mechanics versus Classical Statistical Mechanics

We notice a close analogy between the Euclidean time path integral for a quantum mechanical system and a classical statistical mechanics system. A list of the corresponding terms in these two frameworks is given in Table 1.3.

The path integral for a quantum system is defined on a 1-dimensional Euclidean time lattice, while a spin model can be defined on a d-dimensional spatial lattice. In the path integral, we integrate over all "paths", *i.e.* over all "trajectories" $[x]$ that are defined by the intermediate points $x_n = x(t_n)$. In the spin model, we sum over all spin configurations $[s]$ that are composed of the oriented spins $\vec{s}_x$. Paths are weighted by their Euclidean action $S_E[x]$, while spin configurations are weighted with their Boltzmann factors based on the classical Hamilton function $\mathcal{H}[s]$.

The action has the prefactor $1/\hbar$, and the Hamilton function has the prefactor $1/T$. Thus, $\hbar$ determines the strength of quantum fluctuations, while the temperature T controls the strength of thermal fluctuations. The classical limit $\hbar \to 0$ (in which only the path with the least action contributes) corresponds to the limit $T \to 0$ (where only the configuration of minimal Hamilton function contributes). A difference is, of course, that T is variable in Nature, whereas $\hbar$ is a fixed, fundamental constant.

The kinetic energy $M(x_{n+1} - x_n)^2/(2a^2)$ in the path integral is analogous to the nearest-neighbor spin coupling $-J\vec{s}_x \cdot \vec{s}_{x+\hat{i}}$ (where $\hat{i}$ is a vector in the i-direction with the length of one lattice unit), due to $J(\vec{s}_{x+\hat{i}} - \vec{s}_x)^2/2 = J(1 - \vec{s}_x \cdot \vec{s}_{x+\hat{i}})$, where the additive constant is unimportant. The potential term $V(x_n)$ is similar to the coupling $\vec{B} \cdot \vec{s}_x$ to an external magnetic field.

The magnetization density $\langle \vec{s} \rangle$ corresponds to the vacuum expectation value of an operator $\langle \hat{\mathcal{O}}(\hat{x}) \rangle$, also denoted as a *1-point function* or *condensate*,[15] and the correlation function $\langle \vec{s}_x \cdot \vec{s}_y \rangle$ corresponds to the 2-point function $\langle \hat{\mathcal{O}}(\hat{x}(0)) \, \hat{\mathcal{O}}(\hat{x}(t_E)) \rangle$.

The inverse correlation length $1/\xi$ is analogous to the energy gap $E_1 - E_0$ (and hence to a particle mass in quantum field theory). Finally, the Euclidean time continuum limit $a \to 0$ corresponds to a second-order phase transition, where $\xi/a \to \infty$. The lattice spacing in the path integral is an artifact of the regularized description. We send it to zero at the

---

[15] Of course, the potential and the vacuum expectation value of an arbitrary operator are more general than the corresponding specific quantities that we mention for the spin models.

Table 1.3 A dictionary that translates quantum mechanics into the terminology of classical statistical mechanics. The points x are located in physical space, whereas $\vec{s}_x \in S^{N-1}$ is a unit-vector in an (abstract) spin space (the index x represents a site on some crystal lattice). The continuum limit corresponds to a second-order phase transition, where $a/\xi \longrightarrow 0$.

Quantum Mechanics	Classical Statistical Mechanics
1-d Euclidean time lattice	d-dimensional spatial lattice
elementary time step	crystal lattice spacing
particle position $x_n = x(t_{\mathrm{E},n})$	classical spin variable $\vec{s}_x$
particle path $[x] = \{x_n\}$ $(n = 1, \ldots, N)$	spin configuration $[s] = \{\vec{s}_x\}$ $(x \in \text{lattice})$
path integral $\int \mathcal{D}x$	integral over all spin configurations $\int \mathcal{D}s$
Euclidean action $S_{\mathrm{E}}[x]$	classical Hamilton function $\mathcal{H}[s]$
weight of a path $\exp\left(-S_{\mathrm{E}}[x]/\hbar\right)$	Boltzmann factor $\exp\left(-\mathcal{H}[s]/T\right)$
Planck's constant $\hbar$	temperature T
quantum fluctuations	thermal fluctuations
classical limit	zero-temperature limit
kinetic energy $M\,(x_{n+1} - x_n)^2/(2a^2)$	nearest-neighbor coupling $-J\,\vec{s}_x \cdot \vec{s}_{x+\hat{\imath}}$
potential energy $V(x_n)$	external field energy $\vec{B} \cdot \vec{s}_x$
1-point function $\langle \hat{\mathcal{O}}(\hat{x}(t_{\mathrm{E}}))\rangle$	magnetization density $\langle \vec{s}_x \rangle$
2-point function $\langle \hat{\mathcal{O}}(\hat{x}(0))\,\hat{\mathcal{O}}(\hat{x}(t_{\mathrm{E}}))\rangle$	correlation function $\langle \vec{s}_x \cdot \vec{s}_y \rangle$
energy gap $E_1 - E_0$	inverse correlation length $1/\xi$
continuum limit $a \to 0$	critical point $\xi \to \infty$

end, and the physical quantities emerge asymptotically in this limit. In classical statistical mechanics, on the other hand, the lattice spacing may be physical and hence fixed, while the correlation length ξ/a diverges at a second-order phase transition. Hence, a second-order phase transition is mathematically equivalent to a continuum limit.

1.6 Transfer Matrix

Let us consider the classical statistical mechanics analogue of the quantum mechanical Hamilton operator, which is known as the *transfer matrix*. The Hamilton operator of a quantum system is the generator of infinitesimal time translations. On a Euclidean time lattice, there are no infinitesimal time steps, but only steps of finite extent a. The corresponding evolution operator is the transfer matrix

$$\hat{\mathcal{T}} = \exp\left(-a\hat{H}\right) \quad \Rightarrow \quad Z = \mathrm{Tr}\exp\left(-\beta\hat{H}\right) = \mathrm{Tr}\,\hat{\mathcal{T}}^N, \quad \beta = Na. \tag{1.50}$$

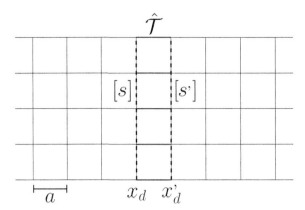

The transfer matrix $\hat{\mathcal{T}}$ describes the transition of the spin configuration $[s]$ in a $(d-1)$-dimensional slice Λ_{x_d} at position x_d to the configuration $[s']$ in the adjacent slice $\Lambda_{x'_d}$ at $x'_d = x_d + a$. The nearest-neighbor spin interactions within a slice are indicated by the bold dashed lines, while the ones between Λ_{x_d} and $\Lambda_{x'_d}$ are indicated by the bold solid lines.

Let us now consider a classical statistical mechanics system of spins on a d-dimensional spatial lattice with periodic boundary conditions in the d-direction of extent Na. As in Section 1.4, we assume the spins to be coupled only to their nearest neighbors. The corresponding partition function can then again be expressed as

$$Z = \prod_x \int_{S^{N-1}} d^{N-1} s_x \exp\left(-\mathcal{H}[s]/T\right) = \mathrm{Tr}\,\hat{\mathcal{T}}^N. \tag{1.51}$$

The transfer matrix $\hat{\mathcal{T}}^N$ now describes the transition of the classical spin configuration $[s]$ in a $(d-1)$-dimensional spatial slice Λ_{x_d} at position x_d to the configuration $[s']$ in the neighboring slice $\Lambda_{x'_d}$ at $x'_d = x_d + a$. Here, the spatial d-direction is analogous to Euclidean time in the quantum system. The elements of the transfer matrix $\langle s|\hat{\mathcal{T}}|s'\rangle$ receive contributions from the spin interactions $\mathcal{H}_{x_d}[s]$ and $\mathcal{H}_{x'_d}[s']$ within slices and from $\mathcal{H}_{x_d,x'_d}[s,s']$ between the two adjacent slices Λ_{x_d} and $\Lambda_{x'_d}$ (cf. Figure 1.3)

$$\langle s|\hat{\mathcal{T}}|s'\rangle = \exp\left(-\frac{1}{2}\mathcal{H}_{x_d}[s]/T\right) \exp\left(-\mathcal{H}_{x_d,x'_d}[s,s']/T\right) \exp\left(-\frac{1}{2}\mathcal{H}_{x'_d}[s']/T\right),$$

$$\mathcal{H}_{x_d}[s] = -J \sum_{\langle xy\rangle,x,y\in\Lambda_{x_d}} \vec{s}_x \cdot \vec{s}_y, \quad \mathcal{H}_{x'_d}[s'] = -J \sum_{\langle xy\rangle,x,y\in\Lambda_{x'_d}} \vec{s}'_x \cdot \vec{s}'_y,$$

$$\mathcal{H}_{x_d,x'_d}[s,s'] = -J \sum_{\langle xy\rangle,x\in\Lambda_{x_d},y\in\Lambda_{x'_d}} \vec{s}_x \cdot \vec{s}'_y. \tag{1.52}$$

The spin interactions within a slice contribute with a factor $1/2$ because they are shared between two adjacent transfer matrix steps. This renders the transfer matrix symmetric against reflections at the hyperplane in the center between the slices Λ_{x_d} and $\Lambda_{x'_d}$. One can show that the eigenvalues of $\hat{\mathcal{T}}$ are non-negative. As a result, the transfer matrix is reflection-positive. In a quantum theory, *reflection positivity* is vital for showing that the

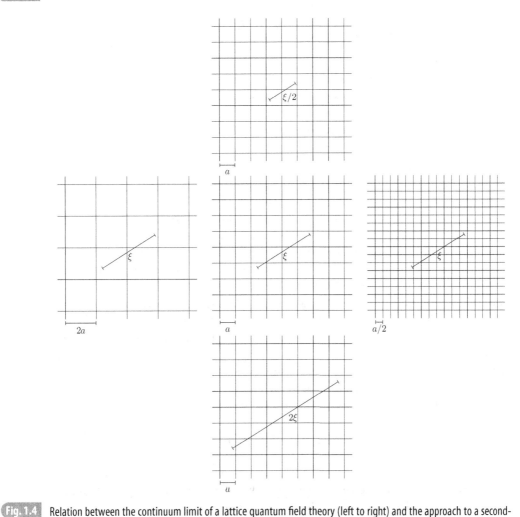

Fig. 1.4 Relation between the continuum limit of a lattice quantum field theory (left to right) and the approach to a second-order phase transition in a classical lattice spin model (top to bottom). The continuum limit is approached by sending the lattice spacing $a \rightarrow 0$ at fixed correlation length ξ, while at a second-order phase transition $\xi \rightarrow \infty$ at fixed a. Both cases are mathematically equivalent, regarding the crucial limit $\xi/a \rightarrow \infty$.

theory in Euclidean time can be analytically continued (*i.e.* Wick-rotated) to the physical real time (Osterwalder and Schrader, 1973, 1975). The ratio of the two largest eigenvalues of the transfer matrix is given by $\exp(-a(E_1 - E_0)) = \exp(-a/\xi)$ and thus determines the correlation length ξ.

Squaring the transfer matrix proceeds by integrating over the spins in the slice $\Lambda_{x'_d}$

$$\langle s|\hat{\mathcal{T}}^2|s''\rangle = \prod_{x \in \Lambda_{x'_d}} \int_{S^{N-1}} d^{N-1}s'_x \, \langle s|\hat{\mathcal{T}}|s'\rangle \langle s'|\hat{\mathcal{T}}|s''\rangle, \qquad (1.53)$$

such that eq. (1.51) follows in a straightforward manner.

1.7 Lattice Field Theory

So far, we have considered quantum mechanics and classical statistical mechanics. The former was defined by a path integral on a 1-dimensional Euclidean time lattice, while the latter involved spin models on a spatial lattice.

Now we quantize field theories on the lattice by formulating a theory on a d-dimensional *space–time lattice, i.e.* usually the lattice is 4-dimensional. Just as we integrated over all paths $x(t_E)$ of a quantum mechanical particle, we now integrate over *all configurations* $\phi(x)$ of a quantum field defined at any Euclidean space–time point $x = (\vec{x}, x_4)$.[16] For a schematic overview, we refer once more to Figure 1.1. The relation between the continuum limit $a \to 0$ in a lattice quantum field theory and the approach to a second-order phase transition in a classical lattice spin model is illustrated in Figure 1.4. In a lattice quantum field theory, the continuum limit $a \to 0$ is taken by tuning the couplings, while keeping the correlation length $\xi = 1/m$, which corresponds to an inverse particle mass m, fixed in physical units. This reflects the fact that the lattice spacing serves as a UV regulator that should ultimately be removed. In a classical lattice spin model, on the other hand, the lattice is considered physical and may, *e.g.*, correspond to a crystal lattice with spacing a. At a second-order phase transition, the correlation length then diverges in physical units, $\xi \to \infty$. The common mathematical property that the correlation length diverges in units of the lattice spacing, $\xi/a \to \infty$, is thus interpreted in two physically different ways.

The weight factor in the functional integral is again given by the Euclidean action $S_E[\phi]$. Let us consider a free neutral scalar field $\phi(x) \in \mathbb{R}$, with

$$S_E[\phi] = \int d^4x \, \mathcal{L}_E(\phi, \partial_\mu \phi), \quad \mathcal{L}_E(\phi, \partial_\mu \phi) = \frac{1}{2} \partial_\mu \phi(x) \, \partial_\mu \phi(x) + \frac{m^2}{2} \phi(x)^2. \quad (1.54)$$

The transition from Minkowski to Euclidean space–time will be discussed in more detail in Chapter 4. Interactions could be included, for example, by adding a $\lambda \phi^4$ term, as we have seen before. The partition function for this system is symbolically written as

$$Z = \int \mathcal{D}\phi \exp(-S_E[\phi]). \quad (1.55)$$

Note that we have set

$$\hbar = c = 1, \quad (1.56)$$

i.e. from now on we use *natural units*, which we discuss in Appendix B. The physical units can be reconstructed unambiguously by inserting the powers of $\hbar$ and c which match the dimensions. In natural units, we only deal with one scale, which can represent, *e.g.*, length or time, or its inverse that corresponds, *e.g.*, to mass, energy, momentum, or temperature. For instance, $\phi(x)$ has the mass dimension $(d-2)/2$, because the action (being the argument of an exponential) must be dimensionless.

The integral $\int \mathcal{D}\phi$ extends over *all field configurations,* which is an undefined expression, unless a *regularization* is imposed. One can make the expression mathematically

---

[16] As in Section 1.3, we denote the Euclidean time coordinate as $x_4 = i x_0 = i \, ct$, along with $p_4 = i p_0 = i E/c$, following the standard convention. In Euclidean space–time, we only write lower indices, cf. Chapter 4. In fact, the Wick rotation erases the distinction between co- and contra-variant vectors.

well-defined by using the *lattice regularization*. Relying on well-defined expressions is essential from a conceptual point of view. Moreover, this formulation extends to the interacting case, including strong interactions. Thus, it is also essential in practice, if the interaction does not happen to be small. That situation occurs, in particular, in QCD (the acronym for Quantum Chromodynamics) at low and moderate energies.

On the lattice, the continuum field $\phi(x)$ is replaced by a lattice field ϕ_x, which is restricted to the sites x of a d-dimensional hyper-cubic space–time lattice with some spacing a, $x_\mu/a \in \mathbb{Z}$. Then, the continuum action (1.54) can be transcribed into discretized derivatives,

$$S_{\mathrm{E}}[\phi] = \frac{1}{2} a^d \sum_x \left[\sum_{\mu=1}^{d} \left(\frac{\phi_{x+\hat{\mu}} - \phi_x}{a} \right)^2 + m^2 \phi_x^2 \right], \tag{1.57}$$

where $\hat{\mu}$ is the vector of length a in the μ-direction. This is the *standard lattice action*, but many different discretized derivatives (with couplings beyond nearest-neighbor sites) are equivalent in the continuum limit. The corresponding lattice actions belong to the same universality class.

The integral over all field configurations now becomes a high-dimensional integral over all field values at all lattice sites,

$$Z = \prod_x \sqrt{\frac{a^{d-2}}{2\pi}} \int_{-\infty}^{\infty} d\phi_x \exp(-S_{\mathrm{E}}[\phi]). \tag{1.58}$$

We actually integrate over the dimensionless variable $a^{(d-2)/2}\phi_x$, such that also Z is dimensionless. For convenience, it also incorporates a factor 2π that arises in Gaussian integrals. This factor is irrelevant because it cancels in all expectation values. For a free field theory, the partition function is just given by Gaussian integrals. We write its lattice action as

$$S_{\mathrm{E}}[\phi] = \frac{1}{2} a^d \sum_{x,y} \phi_x \Delta_{xy} \phi_y, \tag{1.59}$$

with the symmetric matrix

$$\Delta_{xy} = \frac{1}{a^2} \left[\sum_\mu \left(2\delta_{xy} - \delta_{x,y-\hat{\mu}} - \delta_{x,y+\hat{\mu}} \right) + (ma)^2 \delta_{xy} \right], \tag{1.60}$$

which contains the couplings between the field variables at the lattice sites (the sum over μ represents a discrete Laplacian). For instance, in a 1-d lattice (with periodic boundary conditions) it takes the form

$$\Delta = \frac{1}{a^2} \begin{pmatrix} 2+(ma)^2 & -1 & 0 & . & . & 0 & -1 \\ -1 & 2+(ma)^2 & -1 & . & . & 0 & 0 \\ 0 & -1 & 2+(ma)^2 & . & . & 0 & 0 \\ . & . & . & . & . & . & . \\ . & . & . & . & . & . & . \\ 0 & 0 & 0 & . & . & 2+(ma)^2 & -1 \\ -1 & 0 & 0 & . & . & -1 & 2+(ma)^2 \end{pmatrix}.$$

We can diagonalize Δ by an orthogonal transformation matrix Ω,

$$\Delta = \Omega^{\mathsf{T}} D \, \Omega, \quad D = \mathrm{diag}\,(D_1, D_2, \ldots, D_N), \tag{1.61}$$

where N is the number of lattice sites. We choose $\Omega \in SO(N)$, and with the substitution

$$\phi'_x = \Omega_{xy}\phi_y \tag{1.62}$$

(note that we are now using the summation convention for the lattice sites and that the Jacobian is $\det \Omega = 1$), we arrive at

$$Z = \prod_x \sqrt{\frac{a^{d-2}}{2\pi}} \int_{-\infty}^{\infty} d\phi'_x \exp\left(-\frac{a^d}{2}\sum_x \phi'_x D_{xx}\phi'_x\right) = \prod_x \left(a^2 D_{xx}\right)^{-1/2} = \frac{1}{\sqrt{\det(a^2 D)}}$$

$$= \frac{1}{\sqrt{\det(a^2 \Delta)}}. \tag{1.63}$$

To extract the energy eigenvalues of the corresponding (quantum) Hamilton operator, we explore the *2-point function* of the lattice field,

$$\langle \phi_x \phi_y \rangle = \frac{1}{Z} \int \mathcal{D}\phi \; \phi_x \phi_y \exp(-S_{\mathrm{E}}[\phi]). \tag{1.64}$$

This can be achieved elegantly by introducing a real-valued *source field* j_x in the partition function,

$$Z[j] = \int \mathcal{D}\phi \exp(-S_{\mathrm{E}}[\phi] + a^d j^\mathsf{T}\phi), \tag{1.65}$$

where we use the short-hand notation $j^\mathsf{T}\phi = \phi^\mathsf{T}j = \sum_x j_x\phi_x$. Similarly, we are going to write below $\phi^\mathsf{T}\Delta\phi = \sum_{x,y}\phi_x\Delta_{xy}\phi_y$, etc. The field j_x has mass dimension $(d+2)/2$; it can be viewed as a generalization of the external magnetic field in Section 1.4. In quantum field theory, $Z[j]$ is known as the *generating functional*.

The 2-point function and the connected 2-point function are given by

$$\langle \phi_x \phi_y \rangle = \frac{1}{Z}\frac{\delta^2}{\delta j_x \delta j_y} Z[j]\bigg|_{j=0},$$

$$\langle \phi_x \phi_y \rangle_{\mathrm{c}} = \langle \phi_x \phi_y \rangle - \langle \phi_x \rangle \langle \phi_y \rangle = \frac{\delta^2}{\delta j_x \delta j_y} \log Z[j]\bigg|_{j=0}. \tag{1.66}$$

In our case, $\langle \phi_x \rangle$ and $\langle \phi_y \rangle$ vanish, *i.e.* $\langle \phi_x \phi_y \rangle_{\mathrm{c}} = \langle \phi_x \phi_y \rangle$, due to the symmetry under $\phi \to -\phi$. The derivative δ represents

$$\frac{\delta}{\delta j_x} = \frac{1}{a^d}\frac{\partial}{\partial j_x}. \tag{1.67}$$

The factor a^{-d} guarantees that it corresponds to the usual functional derivative in the continuum limit.

We eliminate the term $a^d j^\mathsf{T}\phi$ in the exponent of eq. (1.65) by another substitution

$$\phi' = \phi - \Delta^{-1}j, \tag{1.68}$$

such that the Boltzmann factor for $Z[j]$ is (up to a factor a^d) given by the exponent

$$-\frac{1}{2}\phi^\mathsf{T}\Delta\phi + j^\mathsf{T}\phi = -\frac{1}{2}\phi'^\mathsf{T}\Delta\phi' + \frac{1}{2}j^\mathsf{T}\Delta^{-1}j, \tag{1.69}$$

where $j^\mathsf{T}\Delta^{-1}j = \sum_{x,y}j_x(\Delta^{-1})_{xy}j_y$. Performing now the functional integral over ϕ', we obtain

$$Z[j] = \frac{1}{\sqrt{\det(a^2\Delta)}}\exp\left(\frac{a^d}{2}j^\mathsf{T}\Delta^{-1}j\right), \tag{1.70}$$

and from eq. (1.66) we infer

$$\langle \phi_x \phi_y \rangle_c = \frac{\delta}{\delta j_x} \frac{\delta}{\delta j_y} \frac{a^d}{2} j_u (\Delta^{-1})_{uv} j_v = \frac{1}{2} \frac{\delta}{\delta j_x} \left[\delta_{yu} (\Delta^{-1})_{uv} j_v + j_u (\Delta^{-1})_{uv} \delta_{vy} \right]$$

$$= \frac{a^{-d}}{2} \left[(\Delta^{-1})_{yv} \delta_{vx} + \delta_{xu} (\Delta^{-1})_{uy} \right] = a^{-d} (\Delta^{-1})_{xy}. \tag{1.71}$$

To make this result more explicit, we have to invert the matrix Δ. This can be accomplished in momentum space,

$$\phi(p) = a^d \sum_x \phi_x \exp(-ipx), \quad \phi_x = \frac{1}{(2\pi)^d} \int_B d^d p \, \phi(p) \exp(ipx), \tag{1.72}$$

where $px = \sum_{\mu=1}^d p_\mu x_\mu$ and $\phi(p)$ has mass dimension $-(d+2)/2$. The momentum integration on the lattice is restricted to the (first) *Brillouin zone*[17]

$$B = \left] -\frac{\pi}{a}, \frac{\pi}{a} \right]^d. \tag{1.73}$$

In fact, the structure of a hyper-cubic lattice imposes periodicity of the field values in momentum space; they are periodic over the Brillouin zone B. (Analogously, a finite spatial volume, say L^d, with periodic boundary conditions, corresponds to a lattice structure in momentum space; *i.e.* $\phi(p)$ is only defined for discrete momenta $p = 2\pi n/L$, $n \in \mathbb{Z}^d$.)

The virtue of momentum space is that the free action becomes *diagonal*,

$$S_E[\phi] = \frac{1}{2} \frac{1}{(2\pi)^d} \int_B d^d p \, \phi(-p) \Delta(p) \phi(p),$$

$$\Delta(p) = \sum_{\mu=1}^d \left(\frac{2}{a} \sin\left(\frac{p_\mu a}{2} \right) \right)^2 + m^2. \tag{1.74}$$

(We use the notation $\Delta(p)$ in analogy to Δ_{xy}, although they are not related by a Fourier transform.) The function $\Delta(p)$ must be periodic over B, which is achieved by the "lattice momentum" $(2/a) \sin(p_\mu a/2)$.

A treatment of the term $\langle \phi(k) \phi(p) \rangle_c$ by means of functional derivatives with respect to $j(q)$ – similar to eq. (1.71) – (with k, p, q, s being momenta) leads to

$$\langle \phi(k) \phi(p) \rangle_c = \frac{\delta^2}{\delta j(k) \, \delta j(p)} \frac{1}{2} \frac{1}{(2\pi)^d} \int_{B^2} d^d q \, d^d s \, j(q) \Delta(s)^{-1} j(s) \, \delta(q+s)$$

$$= \frac{\delta}{\delta j(k)} \frac{1}{2} \left[\int_B d^d s \, \Delta(s)^{-1} j(s) \, \delta(p+s) + \int_B d^d q \, j(q) \Delta(p)^{-1} \delta(p+q) \right]$$

$$= \Delta(p)^{-1} (2\pi)^d \delta(k+p) \quad \Rightarrow$$

$$\frac{1}{(2\pi)^d} \int d^d k \, \langle \phi(k) \phi(p) \rangle_c = \left[\sum_\mu \left(\frac{2}{a} \sin(p_\mu a/2) \right)^2 + m^2 \right]^{-1} \xrightarrow{a \to 0} \frac{1}{p^2 + m^2}. \tag{1.75}$$

In the last step, we have taken the continuum limit, which leads to the (field theoretic) propagator

---

[17] The notation $] - \pi/a, \pi/a]$ implies that the momentum components $p_\mu = \pi/a$ belong to B, whereas $p_\mu = -\pi/a$ do not. It could also be opposite; what matters is that the (periodic) boundaries of B are included exactly once.

$$\langle \phi(x)\phi(y)\rangle_c = \frac{1}{(2\pi)^d} \int d^d p \, \frac{\exp(\mathrm{i}p(x-y))}{p^2 + m^2}. \tag{1.76}$$

This term is also known as the *Green's function* of the Klein–Gordon operator $(-\partial^2 + m^2)$, since

$$(-\partial^2 + m^2) \, \langle \phi(0)\phi(x)\rangle_c = \delta(x). \tag{1.77}$$

Based on the propagator Δ^{-1}, we can deduce the energy spectrum of the lattice theory. For this purpose, we construct a lattice field with definite spatial momentum $\vec{p}$ located in a specific Euclidean time-slice,

$$\phi(\vec{p}\,)_{x_d} = a^{d-1} \sum_{\vec{x}} \phi_{\vec{x},x_d} \exp(-\mathrm{i}\vec{p}\cdot\vec{x}) = \frac{1}{2\pi} \int_{-\pi/a}^{\pi/a} dp_d \, \phi(p) \exp(\mathrm{i}p_d x_d). \tag{1.78}$$

Next we consider its 2-point function

$$\langle \phi(-\vec{p}\,)_0 \phi(\vec{p}\,)_{x_d}\rangle_c = \frac{1}{2\pi} \int_{-\pi/a}^{\pi/a} dp_d \, \langle \phi(-p)\phi(p)\rangle \, \exp(\mathrm{i}p_d x_d). \tag{1.79}$$

Inserting the lattice propagator of eq. (1.74), we can compute this integral. With $p_d = \mathrm{i}E$, we encounter poles in the propagator when

$$\left[\frac{2}{a}\sinh\left(\frac{Ea}{2}\right)\right]^2 = \sum_{i=1}^{d-1}\left(\frac{2}{a}\sin\left(\frac{p_i a}{2}\right)\right)^2 + m^2. \tag{1.80}$$

In the complex plane, the integral (1.79) circles x_d/a times around the pole with $E > 0$. Hence, the 2-point function decays exponentially with rate E,

$$\langle \phi(-\vec{p}\,)_0 \phi(\vec{p}\,)_{x_d}\rangle_c \propto \exp(-E x_d), \tag{1.81}$$

This allows us to identify E as the energy of the lattice scalar particle with spatial momentum $\vec{p}$. In the continuum limit, we arrive at the familiar relativistic dispersion relation

$$E = \sqrt{\vec{p}^{\,2} + m^2}. \tag{1.82}$$

At finite a, the lattice dispersion relation (1.80) differs from the continuum result; *i.e.* we are confronted with *lattice artifacts*.

The lattice literature often uses "lattice units", where one sets the lattice spacing $a = 1$. In these units, agreement with continuum physics is found in the limit where E, $\vec{p}$, and m are small. In particular, $E(\vec{p} = \vec{0}) = m$ corresponds to the inverse correlation length $1/\xi$, and we pointed out before that $\xi/a \to \infty$ characterizes the continuum limit, as illustrated in Figure 1.4. To express this in different terms, $|\vec{p}\,|$ indicates the resolution with which we investigate the system: At low $|\vec{p}\,|$ lattice artifacts are modest, while at large $|\vec{p}\,|$ they are sizable.

Eq. (1.82) coincides with the dispersion relation that results from the Klein–Gordon classical field equation. (We are going to refer to the Klein–Gordon equation in a problem to this chapter and again in Section 2.1.)

It is straightforward to extend this consideration to general n-point functions. Due to the symmetry under a sign flip of ϕ, they vanish whenever n is odd. For even $n \geq 4$, they *factorize into 2-point functions,* as the above technique with the source derivatives shows. Let us illustrate this for the case of a 4-point function,

$$\langle \phi_x \phi_y \phi_z \phi_w \rangle = \frac{1}{Z[j]} \frac{\delta^4}{\delta j_x \delta j_y \delta j_z \delta j_w} Z[j] \Big|_{j=0} = \frac{\delta}{\delta j_x} \frac{\delta}{\delta j_y} \frac{\delta}{\delta j_z} \frac{\delta}{\delta j_w} \exp\left(\frac{a^d}{2} j^{\mathsf{T}} \Delta^{-1} j \right) \Big|_{j=0}$$
$$= a^{-2d} \left[(\Delta^{-1})_{xy} (\Delta^{-1})_{zw} + (\Delta^{-1})_{xz} (\Delta^{-1})_{yw} + (\Delta^{-1})_{xw} (\Delta^{-1})_{yz} \right]. \quad (1.83)$$

The corresponding generic property is known as the *Wick contraction rule.* It will be addressed in Problem 1.10.

Hence, in the free theory all n-point functions are assembled from 2-point functions. However, the continuum limit of such a building block, which is given in eq. (1.76), may diverge as it stands in $d \geq 2$ – this happens in particular at $x = y$. Here, we see the necessity for an *ultraviolet (UV) regularization.*

One method is the lattice regularization that we just described, which restricts the momenta to the Brillouin zone. An alternative approach would be to restrict the integration to momenta $|p| < \Lambda$, where Λ acts as a UV cut-off, and in $d > 2$ we obtain a divergence $\propto \Lambda^{d-2}$ for large Λ (for $d = 2$ it is logarithmic). This is sometimes practical for illustrative purposes, and this is how we are also going to use it occasionally, but calculations in this scheme often run into conceptual problems. A superior method is the *Pauli–Villars regularization* (Pauli and Villars, 1949), which substitutes

$$\frac{\exp(\mathrm{i}p(y-x))}{p^2 + m^2} \to \exp(\mathrm{i}p(y-x)) \left[\frac{1}{p^2 + m^2} - \frac{1}{p^2 + M_{\mathrm{PV}}^2} \right], \quad (1.84)$$

where M_{PV} is a very large mass, which implements the UV regularization. At the end of the calculation, M_{PV} is sent to infinity. This method preserves Lorentz invariance, and it was popular until the early 1970s. Then, *dimensional regularization* was formulated, which enables perturbative calculations with non-Abelian gauge fields (where the Pauli–Villars method runs into trouble). Chapter 4 will concentrate on dimensional regularization, which is now dominant in perturbation theory, while non-perturbative results are obtained with the lattice regularization, often by means of Monte Carlo simulations.

Exercises

1.1 Klein–Gordon equation

The Lagrangian of a neutral scalar field $\phi(x)$ is given by

$$\mathcal{L}(\phi, \partial_\mu \phi) = \frac{1}{2} \partial_\mu \phi(x) \partial^\mu \phi(x) - \frac{m^2}{2} \phi^2(x) - \frac{\lambda}{4!} \phi^4(x).$$

Derive the (classical) equation of motion for the field $\phi(x)$.

Assume $\lambda > 0$, but $m^2 \in \mathbb{R}$ may be positive or negative. Identify the solutions to the equation of motion with a constant field, $\phi(x) = \phi_0$, and discuss their stability (which is relevant in the context of spontaneous symmetry breaking, to be discussed in Chapter 12).

1.2 Baker–Campbell–Hausdorff formula

We consider two bounded operators $\hat{A}$ and $\hat{B}$, and an infinitesimal parameter ε. For the product of the exponential functions $\exp(\varepsilon\hat{A}) \exp(\varepsilon\hat{B})$, we make the following ansatz,

$$\exp(\varepsilon\hat{A}) \exp(\varepsilon\hat{B}) = \exp\left(\varepsilon\hat{X} + \varepsilon^2\hat{Y} + \varepsilon^3\hat{Z} + \mathcal{O}(\varepsilon^4)\right).$$

Compute the bounded operators $\hat{X}$, $\hat{Y}$, and $\hat{Z}$, and express them in a neat form, in terms of $\hat{A}$ and $\hat{B}$, by using appropriate commutators.

1.3 Imaginary Gaussian integral

a) Compute the imaginary Gaussian integral

$$\int_{-\infty}^{\infty} \exp(-i\alpha x^2)\, dx, \quad \alpha > 0,$$

by considering an integration contour in the complex plane in the form of a polygon connecting the points: $\{-R, R(-1+i), R(1-i), R\}$, and taking the limit $R \to \infty$.

b) Which integration contour is suitable to calculate the corresponding result for $\alpha < 0$?

1.4 Free propagator

a) A free, non-relativistic, quantum mechanical particle of mass M is moving on a line. Compute the transition amplitude between an initial state $|x\rangle$ at time $t = 0$ and a final state $|x'\rangle$ at time $t = T$, *with the path integral formalism*.

b) Confirm the result by calculating the transition amplitude directly (in one step), and compare it to a simplification, which only considers the classical path.

1.5 Harmonic oscillator

The Lagrange function of the harmonic oscillator is given by $L(x, \partial_t x) = (M(\partial_t x)^2 - M\omega^2 x^2)/2$.

a) Use a suitable textbook (for instance, Schulman (1981) or Kleinert (1990)) to review how the harmonic oscillator can be treated with the path integral formulation of quantum mechanics. Re-derive in this framework the well-known result for the energy spectrum, $E_n = \hbar\omega(n + 1/2)$, $n = 0, 1, 2, \ldots$.

b) Also consider the 2-point function $C(t) = \langle 0|x(0)x(t)|0\rangle$. First, compute $C(t)$ using the quantum mechanical operator formalism with creation and annihilation operators $a^\dagger$ and a. Then, also use the real-time path integral to compute $C(t)$.

1.6 Quantum rotor

A non-relativistic, quantum mechanical particle of mass M moves freely on a circle of circumference L. Compute the transition amplitude for the particle to return to its initial position after time T, using the path integral formalism. Determine the energy spectrum for this particle.

Hints: The result of Problem 1.4 can be used. Moreover, it is helpful to express $\sum_{n\in\mathbb{Z}} f(n)$ as $\sum_{n\in\mathbb{Z}} \int_{-\infty}^{\infty} dx\,\delta(x-n)f(x)$ and then to apply *Poisson's summation formula* for a periodic δ-function (or "Dirac comb")

$$\sum_{n\in\mathbb{Z}} \delta(x-n) = \sum_{n\in\mathbb{Z}} \exp(2\pi\, \mathrm{i} nx).$$

1.7 Transfer matrix of the 1-dimensional Ising model

Consider the $\mathbb{Z}(2)$-symmetric Ising model with spin variables $s_x = \pm 1$ on a 1-dimensional periodic lattice with N points and with the nearest-neighbor Hamilton function

$$\mathcal{H}[s] = -J \sum_{\langle xy\rangle} s_x s_y.$$

Evaluate the partition function

$$Z = \int \mathcal{D}s \, \exp\left(-\mathcal{H}[s]/T\right),$$

and analyze the N-dependence in order to extract the spectrum of the transfer matrix

$$\langle s|\hat{T}|s'\rangle = \exp(Jss'/T).$$

Determine the correlation length as a function of the temperature T. Is there a second-order phase transition, and if so at which temperature?

1.8 Spectrum of the 1-dimensional XY model

Consider the O(2)-symmetric XY model with 2-component unit-vectors $\vec{s}_x$ on a 1-dimensional lattice with the nearest-neighbor Hamilton function

$$\mathcal{H}[s] = -J \sum_{\langle xy\rangle} \vec{s}_x \cdot \vec{s}_y.$$

Construct the transfer matrix

$$\langle s|\hat{T}|s'\rangle = \exp(J\vec{s} \cdot \vec{s}\,'/T)$$

and derive its spectrum. Determine the correlation length as a function of T. Is there a second-order phase transition, and if so at which temperature?

1.9 Connected n-point functions

A connected n-point function is built from its unconnected counterpart by subtracting all connected subsets, *e.g.*,

$$\langle \phi_x\rangle_c = \langle \phi_x\rangle$$
$$\langle \phi_x\phi_y\rangle_c = \langle \phi_x\phi_y\rangle - \langle \phi_x\rangle\,\langle \phi_y\rangle$$
$$\langle \phi_x\phi_y\phi_z\rangle_c = \langle \phi_x\phi_y\phi_z\rangle - \langle \phi_x\phi_y\rangle_c\,\langle \phi_z\rangle - \langle \phi_x\phi_z\rangle_c\,\langle \phi_y\rangle - \langle \phi_y\phi_z\rangle_c\,\langle \phi_x\rangle - \langle \phi_x\rangle\,\langle \phi_y\rangle\,\langle \phi_z\rangle.$$

In the presence of a source field j, the partition function of some quantum field theoretical model on the lattice is given by

$$Z[j] = \int \mathcal{D}\phi \, \exp\left(-S[\phi] + a^d j^\mathsf{T}\phi\right).$$

Show that the connected 3-point function at $j = 0$ is obtained as

$$\langle \phi_x \phi_y \phi_z \rangle_\mathrm{c} = \frac{\delta}{\delta j_x} \frac{\delta}{\delta j_y} \frac{\delta}{\delta j_z} \log Z[j]|_{j=0},$$

without or with assuming translation invariance.

Following this recipe, formulate a simple induction rule to proceed from $\langle \phi_{x_1} \phi_{x_2} \ldots \phi_{x_n} \rangle_\mathrm{c}$ to $\langle \phi_{x_1} \phi_{x_2} \ldots \phi_{x_{n+1}} \rangle_\mathrm{c}$, and apply it to $n = 3$.

Now identify the coefficients in $\langle \phi_{x_1} \phi_{x_2}, \ldots, \phi_{x_n} \rangle_\mathrm{c}$ (for any $n \in \mathbb{N}$) of the terms $\langle \phi_{x_1} \phi_{x_2} \ldots \phi_{x_{n-1}} \rangle \langle \phi_{x_n} \rangle$, $\langle \phi_{x_1} \phi_{x_2} \rangle \langle \phi_{x_3} \rangle \ldots \langle \phi_{x_n} \rangle$, and $\langle \phi_{x_1} \rangle \langle \phi_{x_2} \rangle \ldots \langle \phi_{x_n} \rangle$.

1.10 Wick contractions

In Section 1.7, we used the following short-hand notation for the generating functional of a free lattice scalar field ϕ_x in the presence of a source field j_x,

$$Z[j] = \int D\phi \, \exp\left(-\frac{a^d}{2}\phi^\mathsf{T}\Delta\phi + a^d j^\mathsf{T}\phi\right) = \int D\phi' \, \exp\left(-\frac{a^d}{2}\phi'^\mathsf{T}\Delta\phi' + \frac{a^d}{2} j^\mathsf{T}\Delta^{-1}j\right),$$

$(\phi' = \phi - \Delta^{-1}j)$.

Derive a general expression – in terms of the symmetric matrix Δ – for an n-point function

$$\langle \phi_{x_1} \phi_{x_2} \ldots \phi_{x_n} \rangle = \frac{1}{Z[j]} \frac{\delta^n}{\delta j_{x_1} \delta j_{x_2} \ldots \delta j_{x_n}} Z[j] \Bigg|_{j=0}.$$

The feature observed here is denoted as the *Wick contraction rule*.

1.11 Lattice propagator for a free scalar field

We have considered the case of a neutral scalar field $\phi_x \in \mathbb{R}$ on an infinite lattice of spacing a, and we computed its propagator.

a) At which order of a is the corresponding dispersion relation plagued by lattice artifacts?

b) Plot the dispersion relations for masses $ma = 0, 1$, and 2 for the field on the lattice and in the continuum.

2 Scalar Field Theory and Canonical Quantization

Scalar fields play an important role in various areas of physics. For example, the *Higgs field* of the Standard Model of particle physics is a scalar field that gives rise to the "spontaneous breakdown" of the $SU(2)_L \times U(1)_Y$ gauge symmetry to the $U(1)_{em}$ gauge symmetry of electromagnetism.[1] This is how elementary particles obtain their masses in the Standard Model.

The lightest strongly interacting particles are the *pions*. The pions are so-called pseudo-Nambu–Goldstone bosons associated with the spontaneous breakdown of the approximate global chiral symmetry of QCD. In the low-energy effective field theory for QCD, a pion is described by a scalar field.[2]

In addition, effective scalar fields are used in condensed matter physics to describe *Cooper pairs* of electrons in superconductors. In that case, the scalar field dynamics leads to the "spontaneous breaking" of the $U(1)_{em}$ gauge symmetry of electromagnetism.

Besides being physically relevant, scalar fields are simpler to handle theoretically than fermion fields or gauge fields.

On the other hand, there are reasons to believe (or speculate) that truly elementary scalar fields may not even exist. For example, the scalar field describing Cooper pairs is actually composed of electron fields. Similarly, at a more fundamental level pions, and other light mesons, which are described at low energies by effective scalar fields, are composed of quarks, anti-quarks, and gluons. It is conceivable that the Higgs particle is not truly fundamental either. For example, in technicolor extensions of the Standard Model, the Higgs field is replaced by tightly bound fermions.

This chapter refers to Minkowski space–time with the metric $g_{\mu\nu} = \mathrm{diag}(1, -1, -1, -1)$. Its structure is described in Appendix C.

2.1 Scalar Fields

Scalar fields $\phi(x)$ transform trivially under space–time transformations. Concretely, under Lorentz transformations Λ (cf. Appendix C) the field values remain unaltered,

$$\phi'(x') = \phi(x) = \phi(\Lambda^{-1}x'), \quad \Lambda_\mu^\rho \Lambda_\nu^\sigma g_{\rho\sigma} = g_{\mu\nu}. \tag{2.1}$$

Only the space–time point, to which the field value ϕ' is attached, undergoes a space–time rotation $x' = \Lambda x$. While they transform trivially under the Lorentz group, scalar

---

[1] Strictly speaking, a gauge symmetry *cannot* break (spontaneously), since it represents a mathematical identity. We will be more precise in Chapter 13.

[2] Pion fields are actually pseudo-scalar: They change sign under a parity transformation, cf. Chapter 21.

fields may transform non-trivially under certain internal symmetries. If such symmetries are continuous, they give rise to conserved charges which could be coupled to gauge fields.

The simplest scalar field is *neutral* and has no additional quantum numbers. It is described by one single real number per space–time point x. A *charged* scalar field, such as the one representing a Cooper pair of electrons, is described by a complex number per space–time point. The scalar Higgs field of the Standard Model is a *complex doublet*. It is described by two complex (or alternatively four real) numbers per space–time point. These examples are included in our overview of Section 1.1.

Let us now consider an N-component real scalar field

$$\vec{\phi}(x) = (\phi_1(x), \phi_2(x), \ldots, \phi_N(x)) \in \mathbb{R}^N. \tag{2.2}$$

A neutral scalar field corresponds to $N = 1$, a charged scalar field to $N = 2$, and the Standard Model Higgs field to $N = 4$. The standard Lagrangian of the corresponding scalar field theory takes the form

$$\mathcal{L}(\vec{\phi}, \partial_\mu \vec{\phi}) = \frac{1}{2} \partial_\mu \vec{\phi} \cdot \partial^\mu \vec{\phi} - V(\vec{\phi}), \tag{2.3}$$

where $V(\vec{\phi})$ is a self-interaction potential. In particular,

$$V(\vec{\phi}) = \frac{m^2}{2} \vec{\phi} \cdot \vec{\phi} + \frac{\lambda}{4!} (\vec{\phi} \cdot \vec{\phi})^2 \tag{2.4}$$

is a scalar potential that contains the mass parameter m of the field[3], as well as the coupling constant λ of its self-interaction. This theory is also known as a *linear σ-model*. In order to guarantee that the potential is bounded from below, we must require $\lambda \geq 0$. As long as $m^2 \geq 0$, the classical vacuum configuration, *i.e.* the configuration of lowest energy, is simply given by $\vec{\phi}(x) = \vec{0}$. (So far, we are at the classical level. Both m and λ get renormalized in the quantized theory.)

Due to the uncertainty principle, the vacuum of scalar *quantum* field theory (*i.e.* its ground state) cannot just be given by $\vec{\phi}(x) = \vec{0}$. This is analogous to an anharmonic oscillator with potential $V(x) = m\omega^2 x^2/2 + \lambda x^4/4!$. While the energy of the classical oscillator is minimized for $x = 0$, the quantum ground state contains quantum fluctuations around the classical vacuum. For example, for the harmonic oscillator (with $\lambda = 0$) these fluctuations are described by a Gaussian wave function.

We begin our discussion of the theory at the classical level. The (classical) equation of motion for scalar field theory is given by

$$\partial_\mu \frac{\delta \mathcal{L}}{\delta \partial_\mu \phi_i} - \frac{\delta \mathcal{L}}{\delta \phi_i} = \partial_\mu \partial^\mu \phi_i + \frac{\partial V(\vec{\phi})}{\partial \phi_i} = \left[\partial_\mu \partial^\mu + m^2 + \frac{\lambda}{3!} \phi^2 \right] \phi_i = 0, \quad i \in \{1, 2, \ldots, N\}. \tag{2.5}$$

Let us look for non-zero solutions of the equation of motion (2.5). Due to the non-linearity of this equation, this is a non-trivial task (at least if $\phi(x)$ is not constant). The situation simplifies significantly when we skip the self-interaction of the scalar field by setting $\lambda = 0$. Then, the classical equation of motion of the resulting free field theory,

---

[3] A parameter $m^2 > 0$ represents a bare mass squared; its perturbative renormalization will be discussed (for the 1-component scalar field) in Chapter 4. In Chapter 12, we will also consider the case of $m^2 < 0$, which implies spontaneous symmetry breaking. In that case, m does not play the role of a particle mass.

$$\partial_\mu \partial^\mu \phi_i + m^2 \phi_i = 0, \tag{2.6}$$

is linear and known as the *Klein–Gordon equation*. It admits simple plane-wave solutions

$$\vec{\phi}(x) = \vec{\phi}_0 \cos(\vec{k} \cdot \vec{x} - \omega t + \varphi), \tag{2.7}$$

which obey the relativistic dispersion relation

$$\omega^2 = \vec{k}^2 + m^2. \tag{2.8}$$

At this point, it is instructive to insert the constants $\hbar$ and c. Then, for instance, the d'Alembert operator takes the form $\partial_\mu \partial^\mu = \partial_t^2/c^2 - \Delta$, where Δ is the Laplace operator. Upon quantization, the relations $E = \hbar\omega$ and $\vec{p} = \hbar\vec{k}$ will represent the energy and the momentum of a free, relativistic quantum particle, with the energy–momentum relation

$$E^2 = \vec{p}^2 c^2 + (mc^2)^2. \tag{2.9}$$

This is consistent with eq. (2.8), where we have put $\hbar = c = 1$. In particular, the parameter m in the Lagrangian determines the mass $m\hbar/c$ of the resulting quantized wave, which manifests itself as a free particle.

2.2 Noether Current

According to a theorem derived by Emmy Noether, continuous symmetries of a Lagrangian give rise to conserved quantities (Noether, 1918a,b). This theorem is rigorously reviewed for instance in the textbook by Arnold (1978), and two books are entirely devoted to it (Kosmann-Schwarzbach, 2011; Sardanashvily, 2016).[4]

The Lagrangian of eqs. (2.3) and (2.4) has an O(N) symmetry; *i.e.* it is invariant against rotations

$$\vec{\phi}'(x) = \Omega \vec{\phi}(x), \quad \Omega \in O(N), \tag{2.10}$$

where Ω is an $N \times N$ (real-valued) orthogonal rotation matrix, *i.e.* $\Omega^{\mathsf{T}} = \Omega^{-1}$. The matrices Ω either have determinant 1 or -1. While for $N \geq 2$ the O(N) symmetry is continuous, for $N = 1$ it reduces to the discrete group $\mathbb{Z}(2) = \{1, -1\}$.

The O(N) symmetry of these scalar field theories is *global*; *i.e.* a symmetry transformation Ω does not depend on space or time. This is in contrast to gauge theories in which symmetries are realized locally. Gauge theories have conserved charges; *e.g.,* electric charge is conserved both in classical and in quantum electrodynamics. Charge conservation is encoded in the *continuity equation* $\partial_\mu j^\mu(x) = 0$ for the electromagnetic current $j^\mu(x) = (\rho(x), \vec{j}(x))$. Here, $\rho(x)$ and $\vec{j}(x)$ are the charge and current densities, respectively.

The O(N) symmetry of scalar field theory with a Lagrangian of the form (2.3), (2.4) also gives rise to a conserved current. Just for the purpose of deriving this current, we now consider a *local* O(N) transformation

$$\vec{\phi}'(x) = \Omega(x) \vec{\phi}(x). \tag{2.11}$$

---

[4] They distinguish between Noether's first and second theorem; here, we refer to the first one.

The potential contribution to the Lagrangian, $V(\vec{\phi})$ of eq. (2.4), is invariant even against this local O(N) transformation. This follows simply from

$$|\vec{\phi}'(x)|^2 = \vec{\phi}(x)\,\Omega(x)^{\mathsf{T}} \cdot \Omega(x)\,\vec{\phi}(x) = |\vec{\phi}(x)|^2. \tag{2.12}$$

On the other hand, the derivative term is invariant under global but *not* under local O(N) transformations. The variation of the action under local transformations will allow us to read off the Noether current. Here, we address classical field theory, so the variational principle guarantees that the action remains invariant to first order of an infinitesimal modification of the field configuration. We will see that this implies the conservation of the Noether current.

Therefore, we compute the variation of the derivative contribution to the Lagrangian with respect to *infinitesimal* local O(N) transformations, for $N \geq 2$,

$$\Omega(x) = \exp(\mathrm{i}\omega(x)) \approx \mathbb{1} + \mathrm{i}\omega(x). \tag{2.13}$$

Here, $\omega(x) = \omega_a(x)T^a$ is an infinitesimal $N \times N$ matrix which can be expressed as a linear combination of the Hermitian generators T^a of the O(N) algebra, with real-valued coefficients $\omega_a(x)$.[5] The orthogonality of $\Omega(x)$ implies

$$\mathbb{1} = \Omega(x)^{\mathsf{T}}\Omega(x) \approx \left[\mathbb{1} + \mathrm{i}\omega(x)^{\mathsf{T}}\right]\left[\mathbb{1} + \mathrm{i}\omega(x)\right] \approx \mathbb{1} + \mathrm{i}\left[\omega(x) + \omega(x)^{\mathsf{T}}\right] \quad \Rightarrow$$
$$\omega(x)^{\mathsf{T}} = -\omega(x). \tag{2.14}$$

From relation (2.13), we infer $\det\Omega(x) = 1$ and $\mathrm{Tr}\,\omega(x) = 0$. Just like the matrices $\omega(x)$ themselves, the generators T^a are imaginary, traceless, anti-symmetric $N \times N$ matrices. There are $N(N-1)/2$ linearly independent matrices of this kind; hence, the dimension of the group O(N) is $N(N-1)/2$. The infinitesimally transformed scalar field takes the form

$$\vec{\phi}'(x) \approx \left[\mathbb{1} + \mathrm{i}\omega(x)\right]\vec{\phi}(x), \tag{2.15}$$

and therefore

$$\partial^{\mu}\vec{\phi}'(x) \approx \partial^{\mu}\vec{\phi}(x) + \mathrm{i}\partial^{\mu}\omega(x)\,\vec{\phi}(x) + \mathrm{i}\omega(x)\partial^{\mu}\vec{\phi}(x). \tag{2.16}$$

By using the relation (2.14), we obtain to $\mathcal{O}(\omega)$ (without explicitly writing the space–time dependence)

$$\partial_{\mu}\vec{\phi}' \cdot \partial^{\mu}\vec{\phi}' \approx \left(\partial_{\mu}\vec{\phi} - \mathrm{i}\vec{\phi}\,\partial_{\mu}\omega - \mathrm{i}\partial_{\mu}\vec{\phi}\omega\right) \cdot \left(\partial^{\mu}\vec{\phi} + \mathrm{i}\partial^{\mu}\omega\,\vec{\phi} + \mathrm{i}\omega\partial^{\mu}\vec{\phi}(x)\right)$$
$$= \partial_{\mu}\vec{\phi} \cdot \partial^{\mu}\vec{\phi} + \mathrm{i}\partial_{\mu}\vec{\phi} \cdot \partial^{\mu}\omega\,\vec{\phi} - \mathrm{i}\vec{\phi}\,\partial_{\mu}\omega \cdot \partial^{\mu}\vec{\phi}. \tag{2.17}$$

Assuming that there are no boundary contributions, the variation of the action under local infinitesimal O(N) transformations reads

$$S[\phi'] - S[\phi] = \int d^4x\,\left[\mathcal{L}(\phi',\partial_{\mu}\phi') - \mathcal{L}(\phi,\partial_{\mu}\phi)\right] = -\mathrm{i}\int d^4x\,\partial_{\mu}\omega_{ij}j^{\mu}_{ij}$$
$$= \mathrm{i}\int d^4x\,\omega_{ij}\partial_{\mu}j^{\mu}_{ij}, \tag{2.18}$$

---

[5] In this case, a real, anti-symmetric generating matrix would be a more obvious choice, without dealing with complex numbers. We choose the notation of eq. (2.13), where ω is purely imaginary, since it naturally extends to unitary transformations, as discussed in Appendix F.

with the O(N) *Noether current*

$$j_{ij}^{\mu}(x) = \phi_i(x)\partial^{\mu}\phi_j(x) - \phi_j(x)\partial^{\mu}\phi_i(x). \tag{2.19}$$

This current is conserved, *i.e.* $\partial_{\mu}j_{ij}^{\mu} = 0$. At the classical level, this follows directly from the variational principle $\delta S[\vec{\phi}] = 0$ applied to relation (2.18).[6] Alternatively, it can be derived from the equations of motion (2.5), which follow from the same principle,

$$\begin{aligned}
\partial_{\mu}j_{ij}^{\mu} &= \partial_{\mu}\left[\phi_i\partial^{\mu}\phi_j - \phi_j\partial^{\mu}\phi_i\right] = \phi_i\partial_{\mu}\partial^{\mu}\phi_j - \phi_j\partial_{\mu}\partial^{\mu}\phi_i \\
&= -\phi_i\frac{\partial V(\vec{\phi})}{\partial\phi_j} + \phi_j\frac{\partial V(\vec{\phi})}{\partial\phi_i} \\
&= \phi_i\left[m^2\phi_j + \frac{\lambda}{3!}\phi^2\phi_j\right] - \phi_j\left[m^2\phi_i + \frac{\lambda}{3!}\phi^2\phi_i\right] = 0.
\end{aligned} \tag{2.20}$$

This is the conserved current which is generally predicted by the Noether theorem in the presence of any continuous symmetry. Here, we obtain a conserved current of the form of the second line in eq. (2.20) for the Lagrangian (2.3), with any potential of the form $V(\vec{\phi}^2)$, due to the O(N) symmetry.

2.3 From the Lagrangian to the Hamilton Density

We now consider a 1-component real scalar field $\phi(x)$ with the Lagrangian (or Lagrange density)

$$\mathcal{L}(\phi, \partial_{\mu}\phi) = \frac{1}{2}\partial_{\mu}\phi(x)\partial^{\mu}\phi(x) - \frac{m^2}{2}\phi(x)^2. \tag{2.21}$$

The field, which is canonically conjugate to $\phi(x)$, reads

$$\Pi(x) = \frac{\delta\mathcal{L}}{\delta\partial_0\phi(x)} = \partial^0\phi(x), \tag{2.22}$$

which is just the time-derivative of $\phi(x)$. The classical *Hamilton density* is given by the (field theoretical) *Legendre transform*

$$\mathcal{H}(\phi, \Pi) = \Pi(x)\,\partial^0\phi(x) - \mathcal{L} = \frac{1}{2}\Pi(x)^2 + \frac{1}{2}\partial_i\phi(x)\,\partial_i\phi(x) + \frac{m^2}{2}\phi(x)^2. \tag{2.23}$$

The index i is summed over the spatial directions only. The classical *Hamilton functional* is the spatial integral of the Hamilton density

$$H[\phi, \Pi] = \int d^3x\,\mathcal{H}(\phi, \Pi) = \int d^3x\,\frac{1}{2}\left[\Pi(\vec{x})^2 + \partial_i\phi(\vec{x})\,\partial_i\phi(\vec{x}) + m^2\phi(\vec{x})^2\right]. \tag{2.24}$$

This quantity is a functional of the classical field configuration $[\phi]$ and its canonically conjugate momentum field configuration $[\Pi]$. Upon quantization, the Hamilton functional will turn into the Hamilton operator of the corresponding quantum field theory.

---

[6] If one only considers a global O(N) transformation, then δS vanishes trivially, as we see from eq. (2.17), and we cannot derive the form of the Noether current along these lines.

2.4 Commutation Relations for the Scalar Field Operators

Canonical quantization of field theory is a somewhat tedious approach. Therefore, for most of this book we will concentrate on the functional integral formalism. However, canonical quantization has the advantage that it is rather similar to Schrödinger's (or Heisenberg's) formulation of quantum mechanics, which we are familiar with. Moreover, it provides a more transparent notion of *particles* in quantum field theory, as we are going to see in Chapter 3.

Here, we consider the canonical quantization of a free 1-component scalar field, which is straightforward to carry out. The complications of the quantization set in only when interactions are included.

In the canonical quantization of field theory, the field values and their conjugate momenta become operators acting in a Hilbert space. As we discussed before, the field value $\phi(\vec{x})$ is analogous to the particle coordinate $\vec{x}(t)$ in quantum mechanics. Similarly, $\Pi(\vec{x})$ is analogous to the momentum $\vec{p}(t)$ of a particle. In quantum mechanics, position and momentum operators do not commute,

$$[\hat{x}_i, \hat{p}_j] = \mathrm{i}\delta_{ij}, \quad [\hat{x}_i, \hat{x}_j] = [\hat{p}_i, \hat{p}_j] = 0. \tag{2.25}$$

Similarly, one postulates the following *commutation relations for the field operators* $\hat{\phi}(\vec{x})$ *and* $\hat{\Pi}(\vec{y})$,

$$[\hat{\phi}(\vec{x}), \hat{\Pi}(\vec{y})] = \mathrm{i}\delta(\vec{x} - \vec{y}), \quad [\hat{\phi}(\vec{x}), \hat{\phi}(\vec{y})] = [\hat{\Pi}(\vec{x}), \hat{\Pi}(\vec{y})] = 0. \tag{2.26}$$

It is important that these commutation relations are *local*. In particular, any field variables at different points in space, $\vec{x} \neq \vec{y}$, commute.

In quantum mechanics, the momentum operator is represented as the derivative with respect to the position

$$\hat{p}_i = -\mathrm{i}\frac{\partial}{\partial x_i}. \tag{2.27}$$

Similarly, the conjugate field operator $\hat{\Pi}(\vec{x})$ can be expressed as

$$\hat{\Pi}(\vec{x}) = -\mathrm{i}\frac{\delta}{\delta\phi(\vec{x})}, \tag{2.28}$$

i.e. as a functional derivative with respect to the field.

2.5 Hamilton Operator in Scalar Field Theory

When we replace the classical fields by operators, it is straightforward to turn the classical Hamilton functional $H[\phi, \Pi]$ into the quantum *Hamilton operator*

$$\hat{H} = \int d^3x \, \frac{1}{2}\left[\hat{\Pi}(\vec{x})^2 + \partial_i\hat{\phi}(\vec{x})\,\partial_i\hat{\phi}(\vec{x}) + m^2\hat{\phi}(\vec{x})^2\right]. \tag{2.29}$$

It is now obvious that quantum field theory is formally equivalent to quantum mechanics with infinitely many degrees of freedom (in this case one for each spatial point $\vec{x}$). However, we like to stress again that the concepts of a particle are different in the two theories.

As usual, *solving* this quantum theory amounts to diagonalizing the Hamilton operator. For this purpose, it is convenient to move to momentum space. Hence, we introduce Fourier transformed field operators[7]

$$\hat{\phi}(\vec{p}) = \int d^3x\, \hat{\phi}(\vec{x}) \exp(-\mathrm{i}\,\vec{p}\cdot\vec{x}), \quad \hat{\Pi}(\vec{p}) = \int d^3x\, \hat{\Pi}(\vec{x}) \exp(-\mathrm{i}\,\vec{p}\cdot\vec{x}). \tag{2.30}$$

Note that, unlike $\hat{\phi}(\vec{x})$ and $\hat{\Pi}(\vec{x})$, the operators $\hat{\phi}(\vec{p})$ and $\hat{\Pi}(\vec{p})$ are *not* Hermitian, but they obey

$$\hat{\phi}(\vec{p})^\dagger = \hat{\phi}(-\vec{p}), \quad \hat{\Pi}(\vec{p})^\dagger = \hat{\Pi}(-\vec{p}). \tag{2.31}$$

Using the commutation relations for $\hat{\phi}(\vec{x})$ and $\hat{\Pi}(\vec{y})$, one derives the commutation relations between $\hat{\phi}(\vec{p})$ and $\hat{\Pi}(\vec{q})$,

$$[\hat{\phi}(\vec{p}), \hat{\Pi}(\vec{q})] = \mathrm{i}(2\pi)^3 \delta(\vec{p}+\vec{q}), \quad [\hat{\phi}(\vec{p}), \hat{\phi}(\vec{q})] = [\hat{\Pi}(\vec{p}), \hat{\Pi}(\vec{q})] = 0, \tag{2.32}$$

and we can write the Hamilton operator as

$$\hat{H} = \frac{1}{(2\pi)^3} \int d^3p\, \frac{1}{2} \left[\hat{\Pi}(\vec{p})^\dagger \hat{\Pi}(\vec{p}) + (\vec{p}^2 + m^2)\hat{\phi}(\vec{p})^\dagger \hat{\phi}(\vec{p}) \right]. \tag{2.33}$$

This Hamilton operator is reminiscent of the one for the harmonic oscillator in quantum mechanics, with $\omega = \sqrt{\vec{p}^2 + m^2}$ playing the role of the angular frequency. This suggests the introduction of *annihilation and creation operator fields* $\hat{a}(\vec{p})$ and $\hat{a}(\vec{p})^\dagger$ as

$$\hat{a}(\vec{p}) = \frac{1}{\sqrt{2}} \left[\sqrt{\omega}\,\hat{\phi}(\vec{p}) + \frac{\mathrm{i}}{\sqrt{\omega}}\,\hat{\Pi}(\vec{p}) \right], \quad \hat{a}(\vec{p})^\dagger = \frac{1}{\sqrt{2}} \left[\sqrt{\omega}\,\hat{\phi}(-\vec{p}) - \frac{\mathrm{i}}{\sqrt{\omega}}\,\hat{\Pi}(-\vec{p}) \right], \tag{2.34}$$

which obey the commutation relations

$$\left[\hat{a}(\vec{p}), \hat{a}(\vec{q})^\dagger \right] = \frac{\mathrm{i}}{2} \left[\hat{\Pi}(\vec{p}), \hat{\phi}(-\vec{q}) \right] - \frac{\mathrm{i}}{2} \left[\hat{\phi}(\vec{p}), \hat{\Pi}(-\vec{q}) \right] = (2\pi)^3 \delta(\vec{p}-\vec{q}),$$
$$\left[\hat{a}(\vec{p}), \hat{a}(\vec{q}) \right] = \left[\hat{a}(\vec{p})^\dagger, \hat{a}(\vec{q})^\dagger \right] = 0. \tag{2.35}$$

In terms of $\hat{a}(\vec{p})$ and $\hat{a}(\vec{p})^\dagger$, the Hamilton operator takes the form

$$\hat{H} = \frac{1}{(2\pi)^3} \int d^3p\, \sqrt{\vec{p}^2 + m^2} \left(\hat{a}(\vec{p})^\dagger \hat{a}(\vec{p}) + \frac{V}{2} \right). \tag{2.36}$$

Strictly speaking, the term $\delta(\vec{0})$ which emerges here is undefined. We have assumed it to be regularized by putting the system in a large volume V (although we deal with continuous momenta), such that

$$\delta(\vec{p}) = \frac{1}{(2\pi)^3} \int d^3x\, \exp(-\mathrm{i}\,\vec{p}\cdot\vec{x}) \quad \Rightarrow \quad (2\pi)^3 \delta(\vec{0}) = \int d^3x\, 1 = V. \tag{2.37}$$

---

[7] In this chapter, we distinguish the original operators from their Fourier transforms only by their arguments $\vec{x}$ versus $\vec{p}$.

2.6 Vacuum State and Vacuum Energy

In analogy to a single harmonic oscillator, the vacuum state $|0\rangle$ of the scalar field theory is determined by

$$\hat{a}(\vec{p})|0\rangle = 0, \tag{2.38}$$

for all momenta $\vec{p}$. The vacuum is indeed an eigenstate of the Hamilton operator of eq. (2.36) with the energy

$$E_0 = \frac{1}{(2\pi)^3} \frac{V}{2} \int d^3p \, \sqrt{\vec{p}^2 + m^2}. \tag{2.39}$$

The volume factor represents a harmless infrared divergence. It is natural in a field theory that the energy of the vacuum is proportional to the spatial volume.

However, even the vacuum energy *density*,

$$\rho = \frac{E_0}{V} = \frac{1}{(2\pi)^3} \frac{1}{2} \int d^3p \, \sqrt{\vec{p}^2 + m^2} = \frac{1}{4\pi^2} \int_0^\infty d|\vec{p}| \, \vec{p}^{\,2} \sqrt{\vec{p}^2 + m^2}, \tag{2.40}$$

is divergent in the ultraviolet (UV). This is a typical short-distance (*i.e.* high-momentum) divergence of quantum field theory. The theory must be regularized in order to make the vacuum energy density finite. This can be achieved, for example, by introducing a momentum cut-off $\Lambda \gg m$.[8] Then, the regularized vacuum energy density reads

$$\rho = \frac{1}{4\pi^2} \int_0^\Lambda d|\vec{p}| \, \vec{p}^{\,2} \sqrt{\vec{p}^2 + m^2} \sim \mathcal{O}(\Lambda^4). \tag{2.41}$$

2.7 Cosmological Constant Problem

The vacuum energy density of field theory gives rise to one of the greatest mysteries in physics – the cosmological constant problem. When one couples classical gravity, *i.e.* Albert Einstein's General Relativity, to quantum field theory, the vacuum energy manifests itself as a *cosmological constant*.

As observations of very distant supernova explosions and detailed investigations of the cosmic microwave background radiation have shown, matter (including dark matter) alone can hardly explain the composition of the Universe. Indeed, there is evidence for vacuum energy – or dark energy – which, unlike matter, does not cluster inside galaxies but fills all of space homogeneously. This form of energy counteracts the gravitational pull of the matter and leads to an *accelerated* expansion of the Universe (Riess *et al.*, 1998; Perlmutter *et al.*, 1999), instead of the deceleration expected for a matter-dominated Universe.

The major fraction of the total energy of the Universe is vacuum energy, filling all of space homogeneously; currently, its contribution is estimated as ≈ 72 percent. This brings up a

---

[8] In general, this regularization is problematic: In gauge theories, it breaks gauge invariance, and even without the presence of gauge fields it sometimes leads to conceptual problems in the perturbative expansion. However, in this context it is sufficient for illustrative purposes.

Box 2.1   Deep question

What is the nature of the vacuum energy?

In order to obtain a static Universe, Einstein had included a cosmological constant Λ_c in the equations of General Relativity

$$G_{\mu\nu} = 8\pi G T_{\mu\nu} + \Lambda_c g_{\mu\nu}. \tag{2.42}$$

Here, $G_{\mu\nu}$ is the Einstein tensor, G is Newton's constant, $T_{\mu\nu}$ is the energy momentum tensor of matter and radiation, and $g_{\mu\nu}$ is the metric.

After Edwin Hubble and Milton Humason had observed the cosmic expansion in 1929, Einstein withdrew this term from his theory,[9] and he accepted the cosmological solutions that Alexander Friedmann and George Lemaître had derived. As we have just seen, in quantum field theory the vacuum fluctuations of the fields indeed give rise to vacuum energy. However, that energy is formally divergent. A simple consideration might associate the cut-off Λ with the Planck scale

$$M_{\text{Planck}} = \frac{1}{\sqrt{G}} \approx 1.2 \times 10^{19} \, \text{GeV}, \tag{2.43}$$

at which quantum effects of gravity become important (G is Newton's constant). Hence, a naive field theoretical estimate of the vacuum energy density is of the magnitude

$$\rho_{\text{naive}} \approx M_{\text{Planck}}^4. \tag{2.44}$$

The observed vacuum energy density, on the other hand, is only

$$\rho_{\text{observation}} \approx (2 \times 10^{-3} \, \text{eV})^4, \tag{2.45}$$

and, as a rough estimate, one obtains the exorbitant ratio

$$\frac{\rho_{\text{naive}}}{\rho_{\text{observation}}} \approx \mathcal{O}(10^{120}). \tag{2.46}$$

This is presumably the worst discrepancy between "theory" and observation ever encountered in physics. The "theoretical" estimate ρ_{naive} is far too naive, because it lacks any understanding of how gravity responds to the vacuum fluctuations of quantum fields. Only a consistent, unified description of gravity with the fundamental interactions of the Standard Model could allow us to address this problem in a meaningful way.

The naive estimate from above drastically illustrates that we have presently no idea how to compute the tiny but non-zero vacuum energy density. Indeed, the cosmological constant problem is one of the greatest puzzles in physics today. We still do not have any established theory of quantum gravity, and hence, we do not even know the rules from which one could possibly derive the value of the cosmological constant $\Lambda_c = 8\pi G \rho_{\text{observation}}$. Perhaps, the correct theory of quantum gravity has some peculiar property which entails a naturally small value of $\Lambda_c^{1/2} \approx 10^{-60} M_{\text{Planck}}$. For reviews of the cosmological constant problem, we refer to Carroll (2001), Peebles and Ratra (2003), and Padmanabhan (2003).

---

[9] It is popular to quote Einstein as saying that the introduction of the cosmological constant was his "biggest blunder" (or "größte Eselei" in German). However, this citation is not documented, and its credibility is controversial (Livio, 2013; O'Raifeartaigh and Mitton, 2018).

2.8 Particle States and their Energies and Statistics

From a pure particle physics point of view, *i.e.* ignoring the gravitational effect of the vacuum energy, the divergence of ρ is not dramatic. In particular, the energies of particles, which are excitations above the vacuum, are *differences* between the energy of an excited state and the ground state (the vacuum). In these energy differences, the divergent contribution drops out. In particular, the single-particle states of the theory are obtained as

$$|\vec{p}\rangle = \hat{a}(\vec{p})^\dagger |0\rangle, \tag{2.47}$$

with an energy $E(\vec{p})$ given by

$$E(\vec{p}) - E_0 = \omega = \sqrt{\vec{p}^2 + m^2}. \tag{2.48}$$

This is indeed the energy of a free particle with mass m and momentum $\vec{p}$.

Multi-particle states can be generated by acting with more than one particle creation operator on the vacuum state. For example, the 2-particle states are obtained as

$$|\vec{p}_1;\vec{p}_2\rangle = \hat{a}(\vec{p}_1)^\dagger \hat{a}(\vec{p}_2)^\dagger |0\rangle. \tag{2.49}$$

From $[\hat{a}(\vec{p}_1)^\dagger, \hat{a}(\vec{p}_2)^\dagger] = 0$, one infers

$$|\vec{p}_2;\vec{p}_1\rangle = |\vec{p}_1;\vec{p}_2\rangle, \tag{2.50}$$

i.e. the 2-particle state is symmetric under particle permutation. The same is true for multi-particle states. This shows that the particles of scalar field theory are indeed identical and indistinguishable bosons, which obey Bose–Einstein statistics.

Strictly speaking, in quantum field theory the number of particles is not a well-defined quantity, in contrast to the usual setup in quantum mechanics. In particular, once interactions are included, quantum field theory allows for the creation and annihilation of particle–anti-particle pairs. The space spanned by the particle states with all possible occupation numbers is known as the *Fock space*. Generic states are superpositions of energy eigenstates.

2.9 Momentum Operator

Let us finally consider the momentum carried by a scalar field. At the classical level, energy and momentum are components of the energy–momentum tensor

$$\mathcal{T}_{\mu\nu}(x) = \partial_\mu \phi(x) \partial_\nu \phi(x) - g_{\mu\nu}\mathcal{L}. \tag{2.51}$$

The energy-momentum tensor results from deriving the Noether theorem for the symmetry of space–time translations. The conserved "charge" that results from time-translation invariance is the total energy. Similarly, the "charges" associated with translation invariance in the different spatial directions correspond to the components of the total momentum vector. The local conservation of energy and momentum is described by the continuity equation

$$\partial^\mu \mathcal{T}_{\mu\nu}(x) = 0. \tag{2.52}$$

The Hamilton density (2.23) corresponds to

$$\mathcal{H}(x) = \mathcal{T}_{00}(x) = \partial_0\phi(x)\partial_0\phi(x) - g_{00}\mathcal{L} = \Pi(x)^2 - \mathcal{L}. \tag{2.53}$$

Similarly, the momentum density is given by

$$\mathcal{P}_i(x) = \mathcal{T}_{0i}(x) = \partial_0\phi(x)\partial_i\phi(x) - g_{0i}\mathcal{L} = \Pi(x)\partial_i\phi(x). \tag{2.54}$$

Accordingly, in quantum field theory, a Hermitian momentum operator is constructed as

$$\begin{aligned}
\hat{P}_i &= \int d^3x \, \frac{1}{2} \left[\hat{\Pi}(\vec{x}) \, \partial_i\hat{\phi}(\vec{x}) + \partial_i\hat{\phi}(\vec{x}) \, \hat{\Pi}(\vec{x}) \right] \\
&= \frac{1}{(2\pi)^3} \int d^3p \, \frac{ip_i}{2} \left[\hat{\Pi}(\vec{p})\hat{\phi}(-\vec{p}) + \hat{\phi}(-\vec{p})\hat{\Pi}(\vec{p}) \right] \\
&= \frac{1}{(2\pi)^3} \int d^3p \, p_i \, \hat{a}(\vec{p})^\dagger \hat{a}(\vec{p}).
\end{aligned} \tag{2.55}$$

The vacuum $|0\rangle$ is an eigenstate of the momentum operator with eigenvalue $\vec{0}$. Hence, the vacuum has zero momentum, as expected, and it is therefore translation invariant. The single-particle states are eigenstates as well,

$$\hat{P}_i|\vec{p}\rangle = p_i|\vec{p}\rangle, \tag{2.56}$$

which confirms that $\vec{p}$ is indeed the momentum of the particle in the momentum eigenstate $|\vec{p}\rangle$.

Exercises

2.1 Noether theorem

Consider a complex scalar field $\Phi(x) = \phi_1(x) + i\phi_2(x))$ ($\phi_i \in \mathbb{R}$) with the potential

$$V(\Phi) = \frac{m^2}{2}|\Phi|^2 + \frac{\lambda}{4!}|\Phi|^4.$$

a) What is the global symmetry of this theory?

b) Use the Noether theorem to derive the conserved current that results from the corresponding symmetry.

c) Use the equations of motion to show explicitly that the current is indeed conserved at the classical level.

2.2 Momentum space commutation relations of field operators

Derive the momentum space commutation relations of the field operators (2.32), as well as of the creation and annihilation operators (2.35).

2.3 Hamilton operator of the free scalar field

Derive eq. (2.36) for the Hamilton operator of the free scalar field in terms of creation and annihilation operators, as well as the relation (2.48).

2.4 Momentum operator of the scalar field and its eigenvalues

Derive the expression (2.55) for the momentum operator of the scalar field, as well as eq. (2.56).

2.5 For the beginning graduate student

Solve the problem addressed in Box 2.2.

3   From Particles to Wavicles and Back

From *classical Hamiltonian mechanics,* we are familiar with the concept of point particles which are characterized by their position $x(t)$ and momentum $p(t)$ as functions of time. In *canonical quantum mechanics,* position and momentum are represented by Hermitian operators $\hat{x}$ and $\hat{p}$, which do not commute, $[\hat{x}, \hat{p}] = i\hbar$.[1] Consequently, as reflected in Heisenberg's uncertainty relation, $\Delta x \Delta p \geq \hbar/2$, position and momentum cannot be measured simultaneously with arbitrary precision. The fact that quantum mechanical particles are described by a wave function implies that particles also display wave features, such as interference. Interference patterns of single quantum mechanical particles are indeed observed, for example, in double-slit experiments. This gives rise to the notion of particle-wave duality, which is addressed in the quantum mechanics textbook literature.

When we study *field theory*, we encounter fields and not particle coordinates as the fundamental degrees of freedom, as we discussed in Chapter 1. The field concept is familiar from classical electrodynamics, but also from classical continuum mechanics, which is applied to solids and fluids. For example, just like the electromagnetic fields $\vec{E}(\vec{x}, t)$ and $\vec{B}(\vec{x}, t)$, the mass density $\rho(\vec{x}, t)$ (of a liquid or gas) also represents a field that attaches a local degree of freedom $\rho \in \mathbb{R}$ to each space point $\vec{x}$, at any time t. Just as classical electromagnetic waves (fluctuations of the electromagnetic field) can propagate in vacuum, acoustic waves (density fluctuations) can propagate inside a medium. Unlike classical particles, classical waves are certainly not point-like objects. In fact, they do not even have a well-defined position but exist in entire regions of space simultaneously.

When classical field theories are treated with the principles of canonical quantum physics, *i.e.* when we study *canonical quantum field theory* (as we did in Chapter 2), then the fields themselves are operator-valued. After quantization, the excitations of the fields manifest themselves as "particles". In particular, the quanta of electromagnetic waves are known as *photons*, while the quanta of acoustic waves in solids, namely, the quantized vibrations of a crystal lattice, are known as *phonons*.

Photons and phonons are *not* particles in the sense of classical or quantum point particle mechanics. Unlike for classical point particles, and just like for classical waves, *a priori,* the position of a photon or phonon is not even accurately defined. In order to avoid confusion with the point particle concept, we will temporarily use the term *wavicle* (following Wilczek (2012)) to denote "particles" that result from the quantization of fields. In the Standard Model not only photons, but *all* elementary "particles" (including leptons and quarks, as well as gluons, W- and Z-bosons, and the Higgs boson) result from the quantization of the corresponding fields. In this sense, all these "particles" are actually wavicles. Consequently, following this terminology, one might hence speak of wavicle instead of

---

[1] For the purpose of this chapter, it is instructive to write the constant $\hbar$ explicitly.

particle physics. Since this would sound too weird for the rest of this book, only in this chapter we distinguish between particles and wavicles.

The *localization* of wavicles is a tricky issue, and the position of a photon is a subtle concept. In order to address this question, in this chapter we will investigate a simple quantum mechanical model for the *ions in a crystal*. As we will see, phonons (*i.e.* wavicles) will emerge from the underlying ion (*i.e.* particle) dynamics, as quantized lattice vibrations. In fact, this quantum mechanical model admits both a particle and a wavicle interpretation, and thus displays what one might call *"particle-wavicle complementarity"* (not to be confused with the common notion of particle-wave duality).

Applying just the standard rules of non-relativistic quantum mechanics, we will then be able to derive the properties of wavicles. It turns out that, just like photons, phonons have a *linear* energy–momentum dispersion relation, however, only for small momenta. Hence, phonons are massless and propagate "relativistically", except that the speed of light is replaced by the speed of sound. Using a simple ion-phonon (*i.e.* particle-wavicle) model, we will even be able to understand how an effective particle description (in terms of point particles) arises for massive non-relativistic wavicles.

Once the difference between particles and wavicles has been understood, in the forthcoming chapters of this book we will return to standard high-energy physics terminology and denote wavicles as particles.

3.1 Model for Ions Forming a Crystal

Let us consider a simple model of non-relativistic quantum mechanics for N ions forming a crystal. The ions are treated as point particles with position operators $\hat{x}_n$ aligned along a 1-dimensional chain.[2] Here, the index $n \in \{1, 2, \ldots, N\}$ enumerates the ions according to their order in the crystal. The ions are coupled to their nearest neighbors by harmonic oscillator forces, as illustrated in Figure 3.1. The equilibrium distance between neighboring ions determines the lattice spacing a. The quantized vibrations of the resulting ion crystal manifest themselves as *phonons,* which are wavicles. Using the familiar rules of non-relativistic point particle quantum mechanics, we will then derive the properties of the emergent wavicles.

Fig. 3.1   Linear chain of ions coupled harmonically to their nearest neighbors.

---

[2] Actually, in a 1-dimensional system of ions, crystalline order does not persist up to infinite distances. The same is true in two spatial dimensions, at least at non-zero temperature. Only in three or more spatial dimensions, crystalline order may persist up to infinite distances, at non-zero temperature. This is a consequence of the *Mermin–Wagner–Hohenberg–Coleman theorem*, which will be discussed in Chapter 12. Although strict crystalline order does not exist in one or two spatial dimensions, we will still refer to the 1-dimensional system of ions as a "crystal".

The Hamilton operator of the ion model is given by

$$\hat{H} = \sum_{n=1}^{N} \left[\frac{\hat{p}_n^2}{2M} + \frac{M\omega_0^2}{2} (\hat{x}_{n+1} - \hat{x}_n - a)^2 \right]. \tag{3.1}$$

Here, $\hat{p}_n$ and $\hat{x}_n$ are the momentum and position operators of the ion with label n, M is the ion mass, and ω_0 is the angular frequency of the harmonic oscillators connecting neighboring ions. In order to simplify the Fourier analysis of the model, we assume the periodic boundary condition

$$\hat{x}_{N+1} = \hat{x}_1 + Na = \hat{x}_1 + L, \tag{3.2}$$

i.e. we also couple the first and the last ion in the chain as if they were nearest neighbors (this can be realized if the ion chain forms a ring). Ultimately, we are most interested in the large volume limit, *i.e.* an infinite crystal of length $L = Na \to \infty$, and the effects of the periodic boundary condition disappear in this limit.

Next we denote the *displacement* of each ion as

$$\hat{y}_n = \hat{x}_n - na, \tag{3.3}$$

such that the Hamilton operator now takes the form

$$\hat{H} = \sum_{n=1}^{N} \left[\frac{\hat{p}_n^2}{2M} + \frac{M\omega_0^2}{2} (\hat{y}_{n+1} - \hat{y}_n)^2 \right]. \tag{3.4}$$

Due to the periodic boundary condition, which now implies $\hat{y}_{N+1} = \hat{y}_1$, the Hamilton operator has a discrete cyclic $\mathbb{Z}(N)$ *translation symmetry* which replaces $\{\hat{y}_1, \hat{y}_2, \ldots, \hat{y}_N\}$ by $\{\hat{y}_N, \hat{y}_1, \ldots, \hat{y}_{N-1}\}$. As a consequence of this $\mathbb{Z}(N)$ symmetry, all energy eigenstates can be characterized by a conserved *Bloch wave number*

$$k = \frac{2\pi l}{L} = \frac{2\pi l}{Na} \in B = \left] -\frac{\pi}{a}, \frac{\pi}{a} \right], \tag{3.5}$$

which takes values in the Brillouin zone B. Here, l is an integer, which (for even N) takes values $l \in \{-\frac{N}{2} + 1, -\frac{N}{2} + 2, \ldots, \frac{N}{2}\}$.[3] Motivated by the $\mathbb{Z}(N)$ symmetry, we perform a discrete Fourier transform of the variables $\hat{y}_n$ with respect to their index n, *i.e.*

$$\hat{y}(k) = \sum_{n=1}^{N} \hat{y}_n \exp(-ikna). \tag{3.6}$$

The property $\hat{y}(k + 2\pi/a) = \hat{y}(k)$ reflects the periodicity over the Brillouin zone B, due to the spatial discretization. Since the operators $\hat{y}_n$ are Hermitian, one obtains $\hat{y}(k)^\dagger = \hat{y}(-k)$, in analogy to eq. (2.31). The inverse Fourier transform reads

$$\hat{y}_n = \frac{1}{N} \sum_{k \in B} \hat{y}(k) \exp(ikna). \tag{3.7}$$

In the Hilbert space, the discrete translation by one lattice spacing is implemented by a unitary operator which acts on energy eigenstates with Bloch wave number k by multiplication with $\exp(ika)$; this can be seen from eq. (3.7).

---

[3] At this point, one might prefer to denote the Bloch wave numbers as k_l, but in the following that notation would become impractical.

In addition to the discrete $\mathbb{Z}(N)$ translation invariance, there is a *continuous translation invariance* under a shift $\hat{x}'_n = \hat{x}_n + d$ of all ions, which amounts to an overall translation of the entire crystal by a distance d. The corresponding conserved quantity, in the sense of Noether's (first) theorem (cf. Section 2.2), is the total momentum

$$\hat{P} = \sum_{n=1}^{N} \hat{p}_n \tag{3.8}$$

of the entire crystal.

Let us now rewrite the potential energy of the ion model in terms of the Fourier transformed operators $\hat{y}(k)$,

$$\hat{V} = \frac{M\omega_0^2}{2} \sum_{n=1}^{N} (\hat{y}_{n+1} - \hat{y}_n)^2 = \frac{M\omega_0^2}{2} \frac{1}{N^2} \sum_{k,k' \in B} \hat{y}(k')^{\dagger} \hat{y}(k)$$

$$\times \sum_{n=1}^{N} \Big[\exp(-ik'(n+1)a) - \exp(-ik'na) \Big] \Big[\exp(ik(n+1)a) - \exp(ikna) \Big]$$

$$= \frac{1}{N} \frac{M}{2} \sum_{k \in B} \omega(k)^2 \hat{y}(k)^{\dagger} \hat{y}(k), \tag{3.9}$$

with

$$\omega(k) = 2\omega_0 \left| \sin \frac{ka}{2} \right|. \tag{3.10}$$

As in Problem 1.6, we have used the *Poisson summation formula*

$$\frac{1}{N} \sum_{n=1}^{N} \exp\left(i(k - k')na \right) = \delta_{kk'}, \tag{3.11}$$

as well as the identity

$$2 - \exp(ika) - \exp(-ika) = 2\Big(1 - \cos(ka)\Big) = \left(2\sin \frac{ka}{2} \right)^2, \tag{3.12}$$

and the absolute value in eq. (3.10) will prove useful in forthcoming formulae.

Applying the discrete Fourier transform also to the momenta, we obtain

$$\hat{p}(k) = \sum_{n=1}^{N} \hat{p}_n \exp(-ikna), \quad \hat{p}(k)^{\dagger} = \hat{p}(-k), \tag{3.13}$$

and correspondingly

$$\hat{p}_n = \frac{1}{N} \sum_{k \in B} \hat{p}(k) \exp(ikna). \tag{3.14}$$

It should be noted that two types of momenta are involved. The momenta $\hat{p}_n$ of the ions are associated with the continuous translation invariance of the space $\mathbb{R}$, while the Fourier "momentum" k refers to the discrete $\mathbb{Z}(N)$ translation symmetry. The kinetic energy can now be written as

$$\hat{T} = \sum_{n=1}^{N} \frac{\hat{p}_n^2}{2M} = \frac{1}{N^2} \sum_{k,k' \in B} \frac{\hat{p}(k')\hat{p}(k)}{2M} \sum_{n=1}^{N} \exp(\mathrm{i}(k'+k)na)$$

$$= \frac{1}{N} \sum_{k \in B} \frac{\hat{p}(k)^\dagger \hat{p}(k)}{2M}, \tag{3.15}$$

in agreement with Parseval's theorem. Expressed in terms of Fourier transformed coordinates and momenta, the Hamilton operator thus takes the form

$$\hat{H} = \hat{T} + \hat{V} = \frac{1}{N} \sum_{k \in B} \left[\frac{\hat{p}(k)^\dagger \hat{p}(k)}{2M} + \frac{M\omega(k)^2}{2} \hat{y}(k)^\dagger \hat{y}(k) \right]$$

$$= \frac{\hat{P}^2}{2NM} + \frac{1}{N} \sum_{k \in B, k \neq 0} \left[\frac{\hat{p}(k)^\dagger \hat{p}(k)}{2M} + \frac{M\omega(k)^2}{2} \hat{y}(k)^\dagger \hat{y}(k) \right]. \tag{3.16}$$

The zero-mode (with $k = 0$) is associated with the overall translational motion of the crystal. It is generated by the conserved total momentum

$$\hat{P} = \sum_{n=1}^{N} \hat{p}_n = \hat{p}(k = 0), \tag{3.17}$$

cf. eq. (3.8). The kinetic energy of the center-of-mass depends on the total mass NM of the crystal. The non-zero-modes represent a set of harmonic oscillators – one for each non-zero value of k, given by eq. (3.5) – with the k-dependent angular frequency $\omega(k)$. While two ions ($N = 2$) would just oscillate against each other with the fixed frequency $2\omega_0$, an infinite crystal supports vibrations of arbitrary frequency $\omega(k) \in [0, 2\omega_0]$.

3.2 Phonon Creation and Annihilation Operators

The quantized vibrations of a solid are known as *phonons*. In the context of a single harmonic oscillator, we are familiar with lowering and raising operators for the energy. This guides us to the introduction of the *phonon annihilation and creation operators* $\hat{a}$ and $\hat{a}^\dagger$ (they are not related to the lattice spacing a),

$$\hat{a}(k) = \frac{1}{\sqrt{2N}} \left(\alpha(k)\hat{y}(k) + \frac{\mathrm{i}\hat{p}(k)}{\alpha(k)\hbar} \right),$$

$$\hat{a}(k)^\dagger = \frac{1}{\sqrt{2N}} \left(\alpha(k)\hat{y}(k)^\dagger - \frac{\mathrm{i}\hat{p}(k)^\dagger}{\alpha(k)\hbar} \right), \tag{3.18}$$

with

$$\alpha(k) = \sqrt{\frac{M\omega(k)}{\hbar}}. \tag{3.19}$$

It should be noted that the phonon annihilation and creation operators are well-defined only for $k \neq 0$ because $\omega(0) = 0$ would give rise to a singularity in the expressions for $\hat{a}(0)$ and $\hat{a}(0)^\dagger$. For $k, k' \neq 0$, we obtain

$$[\hat{a}(k), \hat{a}(k')^\dagger] = -\frac{\mathrm{i}\alpha(k)}{2N\alpha(k')\hbar}[\hat{y}(k), \hat{p}(-k')] + \frac{\mathrm{i}\alpha(k')}{2N\alpha(k)\hbar}[\hat{p}(k), \hat{y}(-k')]. \tag{3.20}$$

The commutation relations of the coordinates and momenta are given by

$$[\hat{y}_n, \hat{y}_{n'}] = 0, \quad [\hat{p}_n, \hat{p}_{n'}] = 0, \quad [\hat{y}_n, \hat{p}_{n'}] = i\hbar\delta_{nn'}, \tag{3.21}$$

which implies $[\hat{y}(k), \hat{y}(k')] = 0$, $[\hat{p}(k), \hat{p}(k')] = 0$, and

$$[\hat{y}(k), \hat{p}(-k')] = \sum_{n,n'=1}^{N} [\hat{y}_n, \hat{p}_{n'}] \exp(-ikna + ik'n'a)$$

$$= i\hbar \sum_{n=1}^{N} \exp(i(k' - k)na) = i\hbar N\delta_{kk'}. \tag{3.22}$$

This finally leads to

$$[\hat{a}(k), \hat{a}(k')^{\dagger}] = \delta_{kk'}. \tag{3.23}$$

Similarly, one can derive the commutation relations

$$[\hat{a}(k), \hat{a}(k')] = 0, \quad [\hat{a}(k)^{\dagger}, \hat{a}(k')^{\dagger}] = 0. \tag{3.24}$$

Using

$$\hat{a}(k)^{\dagger}\hat{a}(k) = \frac{1}{2N}\left(\alpha(k)\hat{y}(k)^{\dagger} - \frac{i\hat{p}(k)^{\dagger}}{\alpha(k)\hbar}\right)\left(\alpha(k)\hat{y}(k) + \frac{i\hat{p}(k)}{\alpha(k)\hbar}\right)$$

$$= \frac{1}{2N}\left(\alpha(k)^2\hat{y}(k)^{\dagger}\hat{y}(k) + \frac{\hat{p}(k)^{\dagger}\hat{p}(k)}{\alpha(k)^2\hbar^2}\right) + \frac{i}{2N\hbar}\left[\hat{y}(-k)\hat{p}(k) - \hat{p}(-k)\hat{y}(k)\right], \tag{3.25}$$

the Hamilton operator of eq. (3.16) takes the form

$$\hat{H} = \frac{\hat{P}^2}{2NM} + \sum_{k\in B, k\neq 0} \hbar\omega(k)\left(\hat{a}(k)^{\dagger}\hat{a}(k) - \frac{i}{2N\hbar}[\hat{y}(k), \hat{p}(-k)]\right)$$

$$= \frac{\hat{P}^2}{2NM} + \sum_{k\in B, k\neq 0} \hbar\omega(k)\left(\hat{v}(k) + \frac{1}{2}\right). \tag{3.26}$$

Here, we have introduced the *phonon number operator*

$$\hat{v}(k) = \hat{a}(k)^{\dagger}\hat{a}(k), \tag{3.27}$$

which determines the number of phonons occupying a given mode with Bloch wave number k.

Obviously, these formulae are similar to Section 2.5, where we dealt with continuous coordinates and momenta.

3.3 Quantum States of a Vibrating Crystal

An energy eigenstate of the vibrating solid is characterized by its *total momentum P* and by the *phonon occupation numbers* $v(k) \in \{0, 1, \dots, \infty\}$ of all modes with $k \neq 0$.[4] The

---

[4] Correspondingly, the states of the electromagnetic field are characterized by the number of photons occupying each mode, as we will discuss in Chapter 6.

ground state $|0\rangle$ of the solid has $P = 0$ and contains no phonons, *i.e.* $v(k) = 0$ for all modes, hence

$$\hat{P}|0\rangle = 0, \quad \hat{a}(k)|0\rangle = 0. \tag{3.28}$$

Even in the absence of phonons, *i.e.* when $v(k) = 0$ for all k, the solid has a non-vanishing zero-point energy. The ground state energy of the solid is given by

$$E_0 = \langle 0|\hat{H}|0\rangle = \sum_{k \in B} \frac{1}{2} \hbar \omega (k) . \tag{3.29}$$

In the infinite volume limit, $N \to \infty$, the sum over discrete wave numbers turns into an integral over the continuous Brillouin zone $B =] - \pi/a, \pi/a]$. The integration measure is obtained as $dk = 2\pi/L$, and thus, the ground state energy density takes the form

$$\rho_0 = \frac{E_0}{L} = \frac{1}{2\pi} \int_B dk \, \frac{\hbar \omega (k)}{2} = \frac{\hbar}{\pi} \int_0^{\pi/a} dk \, \omega_0 \sin \frac{ka}{2} = \frac{2\hbar\omega_0}{\pi a}. \tag{3.30}$$

In the following, we will only consider energy *differences* and therefore ignore the ground state energy because it just corresponds to an overall constant energy shift, *i.e.* to a "vacuum energy", cf. Section 2.6, which diverges in the continuum limit $a \to 0$.

In the infinite volume limit, the total mass NM of the solid diverges and the kinetic energy $P^2/(2NM)$ of the center-of-mass vanishes for any finite momentum P. Hence, there is a whole set of *degenerate ground states*. Degenerate ground states are characteristic for *spontaneous symmetry breaking,* which will be discussed in Chapter 12. In this case, the affected symmetry would be the continuous translation invariance. However, there is a subtlety: According to the Mermin–Wagner–Hohenberg–Coleman theorem,[5] infinite-range crystalline order cannot establish itself in one spatial dimension. Still, in the infinite volume limit, $N \to \infty$, our model of a 1-dimensional solid does develop degenerate ground states.

The single-phonon states describing a phonon of wave number k result from

$$|k\rangle = \hat{a}(k)^\dagger |0\rangle. \tag{3.31}$$

These states have an energy difference to the ground state of

$$E(k) = \hbar \omega (k) = 2\hbar\omega_0 \left| \sin \frac{ka}{2} \right|. \tag{3.32}$$

One may associate a "de Broglie momentum" $p_{dB} = \hbar k$ with the phonon of wave number k (it is not an eigenvalue of $\hat{p}(k)$). We recall that this momentum is related to the discrete $\mathbb{Z}(N)$ translation invariance, not to the continuous translation invariance generated by the momentum operator $\hat{P}$. Interestingly, at least for small wave numbers $|k|$, just like photons, phonons also have a *linear* energy–momentum dispersion relation, *i.e.* $E(k) \simeq \hbar |k| c$, as eq. (3.32) shows. At low energy, phonons behave like massless relativistic "particles". However, in this case c is not the speed of light; instead,

$$c = \omega_0 a \tag{3.33}$$

is the *speed of sound.* After all, phonons are just quantized sound waves propagating inside a medium. Unlike for photons, the phonon energy–momentum dispersion relation becomes

---

[5] We mentioned this theorem before in footnote 2, and it will be discussed in Section 12.5.

flat for large momenta at the edge of the Brillouin zone. As a result, the *group velocity* of sound

$$c(k) = \left| \frac{dE}{dp} \right| = \omega_0 a \cos \frac{ka}{2} \tag{3.34}$$

is k-dependent. It is largest (and equal to c) for small $|k|$, but it even vanishes for the wave numbers $k = \pm \pi/a$ at the edge of the Brillouin zone.

States of two phonons, with de Broglie momenta $p_{dB,1} = \hbar k_1$ and $p_{dB,2} = \hbar k_2$, are given as

$$|k_1; k_2\rangle = \hat{a}(k_1)^\dagger \hat{a}(k_2)^\dagger |0\rangle. \tag{3.35}$$

Since $[\hat{a}(k_1)^\dagger, \hat{a}(k_2)^\dagger] = 0$, as established in eq. (3.24), one obtains

$$|k_1; k_2\rangle = |k_2; k_1\rangle, \tag{3.36}$$

i.e. the state is symmetric under the exchange of the two phonons. This implies that phonons are *bosons*. Correspondingly, the phonon occupation number of a given mode is unrestricted; it can take values $\nu(k) \in \{0, 1, 2, \ldots, \infty\}$, as in Section 2.8.

3.4 Phonons as Wavicles

Let us attempt to define the *position* of a wavicle, in this case of a phonon. Since we have defined momentum eigenstates $|k\rangle$ for a phonon, quantum mechanics suggests to define a corresponding position eigenstate as[6]

$$|n\rangle = \frac{a}{2\pi} \int_B dk \, |k\rangle \exp(ikna). \tag{3.37}$$

We recall that n enumerates the ions in the crystal. Defining the creation operators

$$\hat{a}_n^\dagger = \frac{a}{2\pi} \int_B dk \, \hat{a}(k)^\dagger \exp(ikna), \tag{3.38}$$

one obtains

$$|n\rangle = \hat{a}_n^\dagger |0\rangle. \tag{3.39}$$

Also introducing the annihilation operator

$$\hat{a}_n = \frac{a}{2\pi} \int_B dk \, \hat{a}(k) \exp(-ikna), \tag{3.40}$$

one arrives at the commutation relation

$$[\hat{a}_n, \hat{a}_{n'}^\dagger] = \left(\frac{a}{2\pi}\right)^2 \int_B dk \int_B dk' \, [\hat{a}(k), \hat{a}(k')^\dagger] \exp(i(k'n' - kn)a)$$

$$= \frac{a}{2\pi} \int_B dk \, \exp(ik(n' - n)a) = \delta_{nn'}, \tag{3.41}$$

as well as

$$[\hat{a}_n, \hat{a}_{n'}] = 0, \quad [\hat{a}_n^\dagger, \hat{a}_{n'}^\dagger] = 0. \tag{3.42}$$

---

[6] For convenience, we keep referring to infinite volume.

This looks as if we had indeed found a useful definition of an operator that creates a phonon in the position eigenstate $|n\rangle$. However, this is strange: How can just one oscillator be excited, while leaving all the rest in their ground states, although the ions are coupled?

In fact, a closer look at the operator $\hat{a}_n^\dagger$ shows that this is not the case. Inserting the definition of $\hat{a}(k)^\dagger$, we obtain

$$\hat{a}_n^\dagger = \frac{a}{2\pi} \int_B dk \, \frac{1}{\sqrt{2N}} \left(\alpha(k)\hat{y}(k)^\dagger - \frac{i\hat{p}(k)^\dagger}{\alpha(k)\hbar} \right) \exp(ikna). \tag{3.43}$$

Since $\hat{a}_n^\dagger$ is the Fourier transform of the sum of two products, it can be expressed as a convolution in coordinate space,

$$\hat{a}_n^\dagger = \frac{1}{\sqrt{N}} \sum_{n'\in\mathbb{Z}} \left(f_{n-n'}\hat{y}_{n'} - g_{n-n'}\frac{i}{\hbar}\hat{p}_{n'} \right). \tag{3.44}$$

This expression involves a sum over all ion labels n', and the quantities

$$f_n = f_{-n} = \frac{a}{2\pi} \int_B dk \, \frac{\alpha(k)}{\sqrt{2}} \exp(ikna) = \frac{a}{\pi}\sqrt{\frac{M\omega_0}{2\hbar}} \int_0^{\pi/a} dk \, \sqrt{\sin(ka/2)}\cos(kna),$$

$$g_n = g_{-n} = \frac{a}{2\pi} \int_B dk \, \frac{1}{\sqrt{2}\alpha(k)} \exp(ikna) = \frac{a}{\pi}\sqrt{\frac{\hbar}{2M\omega_0}} \int_0^{\pi/a} dk \, \frac{\cos(kna)}{\sqrt{\sin(ka/2)}}, \tag{3.45}$$

are non-zero for any n. Hence, the operator $\hat{a}_n^\dagger$ is by no means localized at a single position n in the crystal lattice.

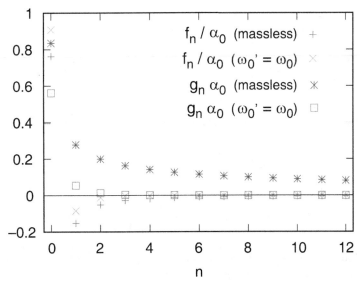

Fig. 3.2 The decay of the quantities f_n and g_n, defined in eq. (3.45), in units of $\alpha_0 = \sqrt{M\omega_0/\hbar}$, for increasing distance $x = na$. For the massless wavicle, the decay is slow, in particular for the momentum-related term g_n. Both quantities follow the power-laws of relation (3.46), which correspond to non-locality. Exponential decays, and therefore locality, are obtained in the massive case, with $\omega_0' > 0$ included, in accordance with relation (3.52). This is also illustrated in this plot, for the example where $\omega_0' = \omega_0$.

For large $|n| \to \infty$, these functions follow a *power-law decay*

$$f_n \sim \frac{1}{\sqrt{|n|}^3}, \quad g_n \sim \frac{1}{\sqrt{|n|}}, \tag{3.46}$$

and they are therefore *non-local* (locality requires at least an exponential decay). The values for f_n and g_n (in units of $\alpha_0 = \sqrt{M\omega_0/\hbar}$), for $n = 0, \ldots, 12$, are plotted in Figure 3.2.

In contrast to particles which are point-like objects (but do respect the position-momentum uncertainty relation), massless wavicles are inherently non-local objects. In particular, if we want to create a phonon at a position n in the crystal, we must excite ion oscillations everywhere in the crystal in a peculiarly coordinated manner. Hence, when we use the standard terminology and refer to phonons as "particles", we should keep in mind that they are not localizable in the usual point particle sense.

3.5 Explicit Breaking of Continuous Translation Symmetry

As we will understand later, phonons are massless as a consequence of the spontaneous symmetry breaking of the continuous translation invariance. In higher dimensions, they are *Nambu–Goldstone bosons*. As we will discuss in Chapter 12, Nambu–Goldstone bosons are massless excitations that arise when a continuous, global symmetry breaks spontaneously. We will now see how phonons pick up a mass when the continuous translation symmetry is *explicitly* broken. For this purpose, we add a term

$$\hat{V}' = \sum_{n=1}^{N} \frac{M\omega_0'^2}{2}(\hat{x}_n - na)^2 = \sum_{n=1}^{N} \frac{M\omega_0'^2}{2}\hat{y}_n^2 = \frac{1}{N}\sum_{k\in B} \frac{M\omega_0'^2}{2}\hat{y}(k)^\dagger \hat{y}(k) \tag{3.47}$$

to the potential energy of the solid. This additional term couples the ions harmonically to their equilibrium positions na with another angular frequency ω_0'. The Hamilton operator then takes the form

$$\hat{H} = \hat{T} + \hat{V} + \hat{V}' = \frac{1}{N}\sum_{k\in B}\left[\frac{\hat{p}(k)^\dagger \hat{p}(k)}{2M} + \frac{M}{2}\left(\left(2\omega_0 \sin \frac{ka}{2}\right)^2 + \omega_0'^2\right)\hat{y}(k)^\dagger \hat{y}(k)\right]$$

$$= \sum_{k\in B}\hbar\omega(k)\left(\hat{v}(k) + \frac{1}{2}\right), \tag{3.48}$$

where we now have

$$\omega(k) = \sqrt{\left(2\omega_0 \sin \frac{ka}{2}\right)^2 + \omega_0'^2}. \tag{3.49}$$

For small de Broglie momenta $p_{dB} = \hbar k$, the energy–momentum dispersion relation then reduces to the one of a massive relativistic particle

$$E(k) = \hbar\omega(k) = \sqrt{(\hbar kc)^2 + (\hbar\omega_0')^2} = \sqrt{(p_{dB}c)^2 + (mc^2)^2}, \tag{3.50}$$

still with $c = \omega_0 a$. Here, we have identified the *phonon mass*

$$m = \frac{\hbar\omega_0'}{c^2} = \frac{\hbar\omega_0'}{\omega_0^2 a^2}. \tag{3.51}$$

Since $\omega(0) = \omega'_0 = mc^2/\hbar > 0$, in contrast to the massless case, the functions f_n and g_n are now *exponentially suppressed* at large $|n|$,[7]

$$f_n, g_n \sim \exp\left(-\frac{mc|n|a}{\hbar}\right) = \exp\left(-\frac{|x|}{\lambda_\phi}\right). \tag{3.52}$$

Here, $x = na$ and

$$\lambda_\phi = \frac{\hbar}{mc} \tag{3.53}$$

is the *Compton wavelength* of the massive phonon, which sets the length scale for the localization of a wavicle. This implies that massive wavicles, although they still do not behave like point particles, are more localized than massless wavicles, which are truly non-local objects (they only decay with power-laws, as in relation (3.46)). As an example, the decays of f_n and g_n are also illustrated in Figure 3.2, for the case $\omega'_0 = \omega_0$.

> **Quick Question 3.5.1   Size of a wavicle**
> Assume a neutrino to have a mass of 1 eV. Compare the size of the neutrino wavicle and of a cell of the human body.

3.6 Debye Field Theory of the Vibrating Solid

Since wavicles arise as quantum fluctuations of fields, it should be possible to describe the vibrating solid by a field theory. This is indeed what Peter Debye did already in the early 20th century (Debye, 1912). We can take two complementary points of view on the dynamics of the vibrating solid:

- On the one hand, we can use a (quantum mechanical) *particle description*, in which we work with a wave function that depends on the coordinates of the individual ions in the crystal.
- On the other hand, we can also use a *wavicle description*, in which we characterize a state of the vibrating solid by specifying the phonon occupation numbers $\nu(k)$ for all modes k.

The two views on the vibrating solid are a manifestation of what one might call *particle-wavicle complementarity*. Remarkably, in this case one can describe the same physical phenomena using either a particle or a wavicle picture. In fact, one can also interpret the ion particle model as a lattice regularized quantum field theory. By taking the continuum limit, one recovers the Debye field theory of the vibrating solid.

In order to re-interpret the ion model as a lattice-regularized quantum field theory, we identify the ion coordinates $\hat{y}_n$ with quantum field operators $\hat{\phi}(x)$, where $x = na$. Correspondingly, we identify the ion momentum operators $\hat{p}_n$ with conjugate field momentum operators $\hat{\Pi}(x)$, such that

---

[7] In eq. (3.45), the substitution $\kappa = kn$ provides a plausibility argument for the power-law decays (3.46). If we use instead eq. (3.49), then $\sqrt{\omega(\kappa)}$ can be Taylor expanded, and an analytic behavior in Fourier space corresponds to locality.

$$\hat{\phi}(x = na) = \sqrt{M\omega_0^2 a}\, \hat{y}_n,$$

$$\hat{\Pi}(x = na) = \frac{\hat{p}_n}{\sqrt{Mc^2 a}} = -\frac{i\hbar}{a\sqrt{M\omega_0^2 a}}\frac{\partial}{\partial y_n} = -\frac{i\hbar}{a}\frac{\delta}{\delta\phi(x)}. \tag{3.54}$$

The constant square root factors are arranged such that the fields $\hat{\phi}$ and $\hat{\Pi}$ have the appropriate dimension, and that the commutation relation matches eq. (2.26),

$$[\hat{\phi}(x), \hat{\Pi}(x')] = \frac{i\hbar}{a}\delta_{nn'} \longrightarrow i\hbar\delta(x - x'). \tag{3.55}$$

We have identified $x = na$, $x' = n'a$, and we have taken the continuum limit $a \to 0$. The Hamilton operator of the ion model now takes the form

$$\hat{H} = \sum_{n=1}^{N}\left[\frac{\hat{p}_n^2}{2M} + \frac{M\omega_0^2}{2}(\hat{y}_{n+1} - \hat{y}_n)^2 + \frac{M\omega_0'^2}{2}\hat{y}_n^2\right]$$

$$= a\sum_x \frac{1}{2}\left[c^2\hat{\Pi}(x)^2 + \left(\frac{\hat{\phi}(x+a) - \hat{\phi}(x)}{a}\right)^2 + \left(\frac{mc}{\hbar}\right)^2\hat{\phi}(x)^2\right]$$

$$\to \int_0^L dx\, \frac{1}{2}\left[c^2\hat{\Pi}(x)^2 + \partial_x\hat{\phi}(x)\partial_x\hat{\phi}(x) + \left(\frac{mc}{\hbar}\right)^2\hat{\phi}(x)^2\right]. \tag{3.56}$$

In the last step, we have again taken the continuum limit $a \to 0$. This requires a "renormalization" of the angular frequency ω_0, such that the speed of sound $c = \omega_0 a$ remains constant, or at least finite. Then, the angular frequency ω_0' does not need to be renormalized since the parameter $mc/\hbar = \omega_0'/c$ automatically remains finite in the continuum limit. The corresponding Lagrangian

$$\mathcal{L}(\phi, \partial_\mu\phi) = \frac{1}{2}\left[\frac{1}{c^2}\partial_t\phi(x)\partial_t\phi(x) - \partial_x\phi(x)\partial_x\phi(x) - \left(\frac{mc}{\hbar}\right)^2\phi(x)^2\right], \tag{3.57}$$

is the one of a free massive scalar field in $(1 + 1)$ dimensions.

3.7 From Wavicles Back to Particles

If elementary "particles" are actually quantized waves, why can we describe the electron surrounding the proton in the hydrogen atom by a non-relativistic point particle Schrödinger equation? Of course, the non-relativistic Schrödinger equation does not capture all details of the hydrogen spectrum, such as the fine-structure which is a relativistic effect or the Lamb shift which requires a quantum field theoretical explanation. Still, the Schrödinger equation correctly describes the leading contribution to the hydrogen spectrum. Hence, the underlying point particle description should not be completely misleading for the electron wavicle inside the hydrogen atom.

The Compton wavelength

$$\lambda_e = \frac{\hbar}{m_e c} \simeq 3.9 \times 10^{-13}\text{ m}, \tag{3.58}$$

where m_e is the electron mass, sets the scale for the localization of the electron wavicle. The size of the hydrogen atom, on the other hand, is determined by the Bohr radius

$$r_B = \frac{4\pi\hbar^2}{e^2 m_e} = \frac{\hbar}{\alpha m_e c} = \frac{\lambda_e}{\alpha} \simeq 5.3 \cdot 10^{-11} \text{ m} = 0.53 \text{ Å} , \quad \alpha = \frac{e^2}{4\pi\hbar c} \simeq \frac{1}{137.036}.$$ (3.59)

Here, $-e$ is the electric charge of the electron and α is the fine-structure constant, which determines the strength of the electromagnetic interaction, and which relates the length scales, $\lambda_e = \alpha r_B$. Since electromagnetism is a rather weak force, the Bohr radius is two orders of magnitude larger than the localization scale λ_e of the electron wavicle. Consequently, at the low energy scales of the hydrogen atom, treating the electron as a point particle is a rather good approximation. In processes at higher energies, on the other hand, the wavicle features become more pronounced and only a field theoretical description reveals the true nature of the electron.

How can we turn from a field theoretical wavicle description to a Schrödinger equation for point particles? In the massive case, the single-phonon states $|k\rangle$ have the energy

$$E(k) = \hbar\omega(k) = \sqrt{\left(\frac{2\hbar c}{a} \sin\frac{ka}{2}\right)^2 + \left(mc^2\right)^2}.$$ (3.60)

One can implicitly define a single-particle Hamilton operator $\hat{\mathbf{H}}$ by demanding the plane wave

$$\Phi_n(k) = A \exp(ikna)$$ (3.61)

to be an energy eigenstate, *i.e.*

$$\hat{\mathbf{H}}\,\Phi_n(k) = E(k)\Phi_n(k).$$ (3.62)

The single-particle Hamilton operator $\hat{\mathbf{H}}$ then correctly describes the time evolution of an initial wave packet (defined at time $t = 0$)

$$\Psi_n(t = 0) = \frac{1}{2\pi} \int_B dk\, A(k) \exp(ikna),$$ (3.63)

by the time-dependent Schrödinger equation

$$i\hbar\, \partial_t \Psi_n(t) = \hat{\mathbf{H}}\,\Psi_n(t),$$ (3.64)

which is solved by

$$\Psi_n(t) = \frac{1}{2\pi} \int_B dk\, A(k) \exp(ikna - i\omega(k)t).$$ (3.65)

Only if the localization length scale λ_ϕ is negligible compared to the de Broglie wavelength

$$\lambda_{dB} = \frac{2\pi}{k} = \frac{2\pi\hbar}{p_{dB}},$$ (3.66)

the wavicle can be treated approximately like a point particle. This condition implies

$$\lambda_{dB} \gg \lambda_\phi \quad \Rightarrow \quad \frac{2\pi\hbar}{p_{dB}} \gg \frac{\hbar}{mc} \quad \Rightarrow \quad p_{dB} \ll mc.$$ (3.67)

It is remarkable that this condition just matches the *non-relativistic limit*. We conclude that, at least at sufficiently low energies, *massive wavicles behave approximately like point particles*.

The physics discussed in this book is mostly relativistic, and wavicle features are thus important. Still, for the rest of this book we return to standard terminology and refer to wavicles as *particles*. Of course, one should keep in mind that such "particles" are different from the point particles one deals with in classical or quantum mechanics, because they are not localizable in the usual sense. (In quantum mechanics, the uncertainty Δx can be made arbitrarily small if Δp is sufficiently large to respect the Heisenberg uncertainty relation.)

3.8   What is Space?

The title (but not the content) of this section is borrowed from an Einstein Lecture that Frank Wilczek gave upon invitation of the Albert Einstein Center for Fundamental Physics at the University of Bern in 2009.

It is worth pointing out that there are different concepts of space underlying non-relativistic classical or quantum point particle mechanics versus classical or quantum field theory, and versus General Relativity. Unraveling the true nature of space remains one of the biggest challenges in physics, whose solution is likely to require a deeper understanding of quantum gravity:

Box 3.1	Deep question

What is the microscopic structure of space?

In classical point particle mechanics, space is essentially empty, except at those points $\vec{x}_a(t)$ where a particle resides at time t. Newton's concept of space thus corresponds to that of an essentially empty rigid vessel in which point particles roam around.

In field theory, on the other hand, space is not empty anywhere. Instead, it provides a substrate on which the field values $\phi_a(\vec{x}, t)$ reside. The points $\vec{x}$ essentially provide the address where a particular field value can be found. Locality, which is one of the cornerstones of field theory, demands that field values interact only with their infinitesimal neighbors in space. In contrast, non-relativistic point particles can interact instantaneously over finite distances (as, *e.g.*, in Newtonian gravity), thus violating spatial locality. In relativistic theories, this is incompatible with causality. In field theories without gravity, while space is nowhere empty, it is still rigid in the sense that it does not have its own dynamics.

In General Relativity, *i.e.* in the relativistic theory of gravity, the space–time metric itself becomes a field, thus endowing space and time with their own dynamics. Even a hypothetical "empty" space (which is not filled with any other fields) thus becomes a dynamical medium, whose energy density and pressure are determined by the cosmological constant that manifests itself as vacuum energy (also known as dark energy).

The simple ion-phonon model discussed in this chapter may shed some light on the nature of space in field theory. As we have seen, the coordinates of the ion point particles $\hat{y}_n$ (relative to their reference position na) manifest themselves as field operators $\hat{\phi}_n = \sqrt{M\omega_0^2 a}\,\hat{y}_n$ of a lattice field theory, whose continuum limit assigns the operator $\hat{\phi}(x) = \hat{\phi}_n$ to the point $x = na$. Hence, in the ion-phonon model the space underlying the resulting effective Debye field theory of phonons has its origin in the crystal lattice formed by the ion point particles.

Is it conceivable that the space underlying the quantum field theory of the Standard Model has a similar origin? In other words, is it possible that space itself consists of "space atoms" (analogous to the ions in our simple model) and that space has emerged from some crystallization process? While this is clearly a wild speculation, we like to entertain it a bit further, if only to motivate the reader to think deeply about the question addressed in this section.

We treated the ions in our model as distinguishable particles, labeled by their position na along the ion chain. This is justified only after the lattice has been formed in a crystallization process, in which every ion has settled down at its equilibrium position. It would be straightforward to extend the model by introducing more realistic Van der Waals interactions between the ions and by treating the ions as indistinguishable particles subject to the Pauli principle or its bosonic analogue. The formation of the crystal itself could then be addressed in the extended model. Crystallization now amounts to the spontaneous breakdown of translation invariance, and the discrete $\mathbb{Z}(N)$ shift symmetry would be replaced by an S_N permutation symmetry of N indistinguishable ions. The Pauli principle then demands that the wave function of the entire crystal is totally symmetric or anti-symmetric, depending on whether the ions are bosons or fermions.

Is it conceivable that the space that we live in has a similar structure at ultra-short distances, consisting of bosonic or fermionic "space atoms"? These would have found their equilibrium positions only after space, as we know it, has emerged from a crystallization process immediately after the Big Bang. Could the phonons of such a space lattice manifest themselves as gravitons?

While it is tempting to entertain such speculations, one should not forget that, due to relativity, understanding the deep nature of space may require to also understand the true nature of *time*. In the dynamical model of space sketched above, time is absolute, just as in Newtonian mechanics. Remarkably, as we have seen before, the dynamics of a non-relativistic ion chain can then still give rise to an emergent low-energy effective Debye field theory of massless phonons with a relativistic energy–momentum dispersion relation. Then, the speed of sound plays the role of the speed of light, as it was the case in this chapter. When more than one field is involved, which is certainly the case for the Standard Model, a simple dynamical model of space, such as the one that we sketched above, would have a hard time to explain naturally why the velocities of sound would be the same for the different types of fields.

We may conclude that understanding the true nature of space–time (and not just of space alone) remains one of the great open questions that may or may not be accessible in the foreseeable future. We hope that we have motivated the reader to think deeply about such questions.

Exercises

3.1 A 1-dimensional "solid" consisting of only two ions

Consider the following Hamilton operator for two ions moving in one dimension

$$\hat{H} = \frac{\hat{p}_1^2}{2M} + \frac{\hat{p}_2^2}{2M} + \frac{1}{2}M\omega_0^2(\hat{x}_2 - \hat{x}_1 - a)^2.$$

a) Introduce the variables $\hat{y}_1 = \hat{x}_1 - a$ and $\hat{y}_2 = \hat{x}_2 - 2a$, as well as $\hat{Y} = (\hat{y}_1 + \hat{y}_2)/2$ and $\hat{y} = \hat{y}_1 - \hat{y}_2$. Rewrite $\hat{H}$ in terms of the coordinates $\hat{Y}$ and $\hat{y}$.

b) Determine the partition function $Z(\beta)$ of the corresponding canonical ensemble.

c) Determine the average energy $E = \langle \hat{H} \rangle$.

3.2 Ion Hamilton operator in momentum space

Derive the ion Hamilton operators of eqs. (3.16) and (3.48) in momentum space from the corresponding coordinate space expressions eqs. (3.1) and (3.47).

3.3 Phonons in a 2-dimensional solid

Consider a model for a 2-dimensional solid with a quadratic lattice (spacing a) of ions (mass M) coupled harmonically to their nearest neighbors.

a) Derive the energy–momentum dispersion relation $E(\vec{k})$ for phonons with momenta in the Brillouin zone $\vec{k} \in B =] - \pi/a, \pi/a]^2$.

b) Construct the canonical partition function for the lattice vibrations, and determine the energy density ρ as a function of temperature T both in the low- and in the high-temperature limit.

4 Perturbative Scalar Field Functional Integral in Dimensional Regularization

The material presented in this chapter is partly inspired by unpublished lecture notes of Jürg Gasser and Heinrich Leutwyler. We thank for their permission to consult their notes as a source of inspiration.

In this chapter (as in most of the continuation of this book), we will refer to Euclidean space–time by default, so we suppress the subscript E. *On the other hand, quantities in Minkowski space–time will be indicated by a subscript* M.

In this chapter, we return to the use of natural units; i.e. we set $\hbar = c = 1$.

As we have seen in Chapter 1, it is mathematically favorable to define the quantum mechanical path integral in Euclidean time. In addition, the Euclidean path integral is intimately related to quantum statistical mechanics. Hence, we now consider quantum field theory in the framework of the Euclidean functional (or path) integral.

Just like the paths of a particle contributing to the quantum mechanical path integral, field configurations contributing to the functional integral are not directly physical objects. Instead, the fields serve as integration variables which allow us to derive physical quantities, such as particle masses and coupling constants. There is a rigorous connection between the Euclidean-time functional integral and the functional integral formulated in Minkowski space–time; the physical results coincide.

More explicitly, the use of the Euclidean signature is justified because the expectation values of n-point functions – which contain the physical information – can be analytically continued to Minkowski space–time, if four conditions are satisfied. These conditions are known as the *Osterwalder–Schrader axioms* (Osterwalder and Schrader, 1973, 1975), which are reviewed for instance by Roepstorff (1994). Two of them are denoted as *analyticity* and *regularity*; from a physical perspective, they are related to locality and unitarity. Regarding *O(4) invariance* and *reflection positivity*, the physical interpretation is even more direct: They ensure Lorentz invariance and a probabilistic interpretation of the observables.

One should, however, not confuse the Euclidean field configurations with time evolutions of physical fields in real time.

4.1 From Minkowski to Euclidean Space–Time

Let us start from the Lagrangian of a scalar field in Minkowski space–time,

$$\mathcal{L}_{\mathrm{M}}(\phi, \partial_\mu\phi) = \frac{1}{2}\partial_\mu\phi\partial^\mu\phi - V(\phi) = \frac{1}{2}(\partial_t\phi\partial_t\phi - \partial_i\phi\partial_i\phi) - V(\phi), \qquad (4.1)$$

which gives rise to the action

$$S_{\mathrm{M}}[\phi] = \int dt\, d^3x\, \mathcal{L}_{\mathrm{M}}(\phi, \partial_\mu\phi). \qquad (4.2)$$

The functional integral in Minkowski space–time is given by the formal expression

$$Z_M = \int \mathcal{D}\phi \, \exp(\mathrm{i}\, S_M[\phi]). \tag{4.3}$$

As in Section 1.3, we analytically continue the time coordinate to purely imaginary values; *i.e.* we perform the Wick rotation

$$x_4 = \mathrm{i}\, t. \tag{4.4}$$

The Lagrangian (4.1) transforms to

$$-\left[\frac{1}{2}(\partial_4\phi\partial_4\phi + \partial_i\phi\partial_i\phi) + V(\phi)\right] = -\left[\frac{1}{2}\partial_\mu\phi\partial_\mu\phi + V(\phi)\right] = -\mathcal{L}(\phi, \partial_\mu\phi), \tag{4.5}$$

where $\mathcal{L}$ is the Euclidean Lagrangian. In Euclidean space–time, the distinction between co- and contra-variant indices ceases to exist because the metric is simply given by $g_{\mu\nu} = \delta_{\mu\nu}$. Therefore, the Einstein summation convention now applies to repeated lower indices. Due to the time-integration, the action picks up a factor $-\mathrm{i}$,

$$dt\, d^3x = -\mathrm{i}\, d^3x\, dx_4 = -\mathrm{i}\, d^4x. \tag{4.6}$$

The functional integral in Euclidean space–time thus takes the form

$$Z = \int \mathcal{D}\phi \, \exp(-S[\phi]), \tag{4.7}$$

with the Euclidean action

$$S[\phi] = \int d^4x\, \mathcal{L}(\phi, \partial_\mu\phi) = \int d^3x \int_0^\beta dx_4 \left[\frac{1}{2}\partial_\mu\phi\partial_\mu\phi + V(\phi)\right]. \tag{4.8}$$

We have introduced a finite extent $\beta = 1/T$ of a periodic Euclidean time dimension, $\phi(\vec{x}, x_4 + \beta) = \phi(\vec{x}, x_4)$, which exposes the field theory to the finite temperature T. The functional integral Z is just the corresponding partition function of quantum statistical mechanics. In the following, we will consider the limit of zero temperature $T \to 0$, *i.e.* of infinite Euclidean time extent $\beta \to \infty$.

As it stands, the functional integral is still a highly divergent, formal expression which needs to be regularized and properly renormalized. We have already sketched in Section 1.7 how this can be achieved by using the lattice regularization. While that scheme has the great virtue of capturing non-perturbative effects, for perturbative calculations the lattice regularization is not the most convenient option. It is much simpler to use *dimensional regularization*, *i.e.* an analytic continuation in the dimension of space–time (Bollini and Giambiagi (1972a,b); 't Hooft and Veltman (1972); for a historical account, see Bietenholz and Prado (2014)).

This is what we will concentrate on in this chapter (except for Section 4.13).[1] In the following, we will keep the space–time dimension d as a continuous, complex parameter. Then, the formal expression for the Euclidean action reads

$$S[\phi] = \int d^dx \left[\frac{1}{2}\partial_\mu\phi\partial_\mu\phi + V(\phi)\right]. \tag{4.9}$$

---

[1] We repeat that dimensional regularization is *limited to perturbation theory;* it does not define quantum field theory at the non-perturbative level.

For a non-integer dimension d, it is not clear what the integration measure $d^d x$ or the summation over repeated indices μ actually means. Indeed, the formal expressions will obtain a definite meaning only later, when we evaluate individual Feynman diagram contributions to the perturbative expansion. The corresponding partition function – including an external source field $j(x)$ – takes the form

$$Z[j] = \int \mathcal{D}\phi \, \exp\left(-S[\phi] + \int d^d x \, j\phi\right), \tag{4.10}$$

with $Z[0] = Z$. The generalization $Z[j]$ is known as a *generating functional*.

4.2 Euclidean Propagator and Contraction Rule

In Euclidean field theory, physical information is extracted from n-point correlation functions, also known as Schwinger functions,

$$\langle \phi(x_1)\phi(x_2)\ldots\phi(x_n)\rangle = \frac{1}{Z}\int \mathcal{D}\phi \, \phi(x_1)\phi(x_2)\ldots\phi(x_n)\exp(-S[\phi]). \tag{4.11}$$

They are analytic continuations of Wightman functions, which correspond to vacuum expectation values of time-ordered products of field operators in Minkowski space–time $\langle 0|T\hat{\phi}(x_1)\hat{\phi}(x_2)\ldots\hat{\phi}(x_n)|0\rangle$. Here, the state $|0\rangle$ denotes the vacuum, or ground state, which is attained in the limit $\beta \to \infty$, and T is the *time-ordering operator* (not to be confused with the temperature).

For example, the 2-point function can be obtained as

$$\frac{1}{Z}\int \mathcal{D}\phi \, \phi(x_1)\phi(x_2)\exp(-S[\phi]) = \frac{1}{Z}\frac{\delta^2 Z[j]}{\delta j(x_1)\delta j(x_2)}\bigg|_{j=0}. \tag{4.12}$$

This formulation illustrates the role of $Z[j]$ as a generating functional, from which the n-point functions can be extracted by means of functional differentiation.

In complete analogy to the lattice calculation discussed in Section 1.7, for a free scalar field, with $V(\phi) = m^2\phi^2/2$, one arrives at

$$Z[j] = Z \exp\left[\frac{1}{2}\int d^d x \, d^d y \, j(x)G(x - y)j(y)\right], \tag{4.13}$$

where $G(x)$ is the Euclidean propagator. In momentum space, it takes the form[2]

$$\widetilde{G}(p) = \int d^d x \, G(x)\exp(-i\,px) = \frac{1}{p^2 + m^2}. \tag{4.14}$$

Alternatively, we can write

$$G(x) = \frac{1}{(2\pi)^d}\int d^d p \, \frac{\exp(i\,px)}{p^2 + m^2}, \tag{4.15}$$

which implies

$$(-\partial_\mu\partial_\mu + m^2)G(x) = \delta^d(x). \tag{4.16}$$

---

[2] Here, we indicate the functions in momentum space by a tilde. In Chapter 3, we abstained from this notation because it is not elegantly compatible with the "hats" as the operator symbol.

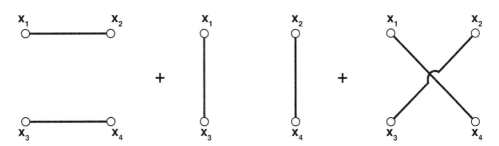

Fig. 4.1   The Feynman diagrams corresponding to the sum of terms in eq. (4.18).

This identifies $G(x)$ as the *Green's function* of the Euclidean Klein-Gordon operator $-\partial_\mu\partial_\mu + m^2$.

For a free field, the n-point functions are related to the 2-point function in a simple manner. For example, the 4-point function takes the form

$$\langle\phi(x_1)\phi(x_2)\phi(x_3)\phi(x_4)\rangle = \frac{1}{Z}\int \mathcal{D}\phi\ \phi(x_1)\phi(x_2)\phi(x_3)\phi(x_4)\exp(-S[\phi])$$

$$= \frac{1}{Z}\frac{\delta^4 Z[j]}{\delta j(x_1)\delta j(x_2)\delta j(x_3)\delta j(x_4)}\bigg|_{j=0}. \tag{4.17}$$

Using eq. (4.13), one obtains

$$\langle\phi(x_1)\phi(x_2)\phi(x_3)\phi(x_4)\rangle =$$
$$G(x_1 - x_2)G(x_3 - x_4) + G(x_1 - x_3)G(x_2 - x_4) + G(x_1 - x_4)G(x_2 - x_3). \tag{4.18}$$

In a *Feynman diagram*, the propagators are represented as *lines* connecting the points x_i. The right-hand side of eq. (4.18) corresponds to the three diagrams in Figure 4.1.

This is an example of the general *Wick contraction rule* (Wick, 1950), for n-point functions with even n (the n-point functions for odd n simply vanish); see also Peskin and Schroeder (1997),

$$\langle\phi(x_1)\phi(x_2)\dots\phi(x_n)\rangle = \sum_{\text{contractions}} G(x_{i_1} - x_{i_2})G(x_{i_3} - x_{i_4})\dots G(x_{i_{n-1}} - x_{i_n}). \tag{4.19}$$

The sum extends over all partitions of the indices $1, 2, \dots, n$ into pairs (i_1, i_2), (i_3, i_4), $\dots$, (i_{n-1}, i_n).

Quick Question 4.2.1   Combinatorics of Wick contractions
    How many contractions does the sum in eq. (4.19) involve?

4.3 Perturbative Expansion of the Functional Integral

Perturbation theory splits the action into a free part – to be left in the exponent – and an interacting part – to be expanded as a power series.

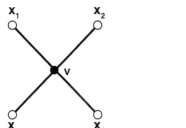 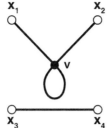 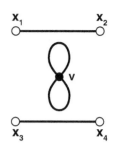

Fig. 4.2 Examples for diagrams with external points $(x_1, \ldots, x_4)$, connected by external lines, as well as one internal point (v). The second and third diagrams include closed internal lines, which give rise to the UV singularities $G(0)$ and $G(0)^2$, respectively. Only the first diagram in this set is connected (cf. Section 4.7).

Let us divide the Euclidean action of the $\lambda\phi^4$ model accordingly,

$$S[\phi] = S_f[\phi] + S_i[\phi],$$
$$S_f[\phi] = \int d^d x \, \frac{1}{2} \left(\partial_\mu \phi \partial_\mu \phi + m^2 \phi^2 \right), \quad S_i[\phi] = \int d^d x \, \frac{\lambda}{4!} \phi^4. \tag{4.20}$$

The interaction term makes it impossible to compute the full functional integral analytically, because it is no longer Gaussian. In perturbation theory, one assumes the coupling constant to be small, $\lambda \gtrsim 0$, and one expands

$$\exp(-S[\phi]) = \exp(-S_f[\phi]) \left(1 - S_i[\phi] + \frac{1}{2} S_i[\phi]^2 + \ldots \right). \tag{4.21}$$

Inserting this expansion into the functional integral, the expression for the n-point function takes the form

$$\langle \phi(x_1)\phi(x_2) \ldots \phi(x_n) \rangle = \frac{1}{Z} \int \mathcal{D}\phi \, \phi(x_1)\phi(x_2) \ldots \phi(x_n)$$
$$\times \left[\sum_{k=0}^{\infty} \frac{(-\lambda)^k}{(4!)^k k!} \int d^d v_1 d^d v_2 \ldots d^d v_k \, \phi(v_1)^4 \phi(v_2)^4 \ldots \phi(v_k)^4 \right] \exp(-S_f[\phi]). \tag{4.22}$$

This expression is similar to the one for the n-point function in the free theory. However, in addition to the *external points* $x_1, x_2, \ldots, x_n$, we now also have k *internal points* $v_1, v_2, \ldots, v_k$, also known as *vertices*. Their positions are independently integrated over space–time. These vertices are the points at which the field ϕ experiences its self-interaction.

As before, we can apply the Wick contraction rule. However, we now have four fields $\phi(v_i)$ at each vertex. Correspondingly, there are now contractions that connect vertices and even a vertex back to itself via a propagator $G(v_i - v_i) = G(0)$. In a Feynman diagram, such propagators appear as *internal lines,* while the propagators connected to external points x_i are denoted as *external lines.* Examples are shown in Figure 4.2.

In two or more dimensions, the propagator at zero distance, $G(0)$, is ultraviolet (UV) divergent, as we mentioned in Section 1.7. This divergence can be regularized by analytically continuing the dimension of space–time.

4.4 Dimensional Regularization

Let us consider the free propagator in d-dimensional space–time

$$G(x) = \frac{1}{(2\pi)^d} \int d^d p \, \frac{\exp(ipx)}{p^2 + m^2}. \tag{4.23}$$

Using the identity

$$\frac{1}{p^2 + m^2} = \int_0^\infty dt \, \exp\left(-t(p^2 + m^2)\right), \tag{4.24}$$

we can write

$$
\begin{aligned}
G(x) &= \frac{1}{(2\pi)^d} \int_0^\infty dt \, \exp\left(-tm^2\right) \int d^d p \, \exp\left(ipx - tp^2\right) \\
&= \frac{1}{(2\pi)^d} \int_0^\infty dt \, \exp\left(-tm^2 - \frac{x^2}{4t}\right) \int d^d q \, \exp\left(-tq^2\right) \\
&= \frac{1}{(4\pi)^{d/2}} \int_0^\infty dt \, t^{-d/2} \exp\left(-tm^2 - \frac{x^2}{4t}\right) \\
&= \frac{1}{(2\pi)^{d/2}} m^{d-2} (m|x|)^{1-d/2} K_{1-d/2}(m|x|).
\end{aligned}
\tag{4.25}
$$

In the last step, we have applied an identity for the modified Bessel function $K_\nu(z)$. For $|x| \neq 0$, the propagator is finite for any value of d.[3]

However, at $|x| = 0$ we have

$$G(0) = \frac{1}{(4\pi)^{d/2}} \int_0^\infty dt \, t^{-d/2} \exp\left(-tm^2\right) = \frac{1}{(4\pi)^{d/2}} m^{d-2} \Gamma\left(1 - \frac{d}{2}\right), \tag{4.26}$$

which diverges for $d = 2, 4, 6, \ldots$ For all other space–time dimensions d, however, we can assign a finite value to the Γ-function, although the above integral over t converges only for $d < 2$. The result of the integral can be analytically continued to general $d \in \mathbb{C}$, except to even integer dimensions.[4] The domain of definition of the (analytically continued) Γ-function is shown in Figure 4.3. Near $d = 4$, the Γ-function takes the form

$$\Gamma\left(1 - \frac{d}{2}\right) = \frac{2}{d-4} + \gamma - 1 + \mathcal{O}(d-4), \tag{4.27}$$

which reveals the UV singularity as a pole in the space–time dimension at $d = 4$. In the finite part, Euler's constant $\gamma = -\Gamma'(1) = 0.577\ldots$ appears, cf. Problem 4.1.

Of course, the analytic continuation in the space–time dimension is just a mathematical trick that renders the propagator well-defined in the UV limit. Unlike the lattice regularization, dimensional regularization can hardly be interpreted physically. Still, it is an elegant way to make the formal expressions of continuum field theory mathematically well-defined – at least in perturbation theory. Furthermore, it yields the same results in the

---

[3] The singularity at $m \to 0$ is removable, since the power series of $K_\nu(z)$ starts with a term $\propto z^\nu$, as for other types of Bessel functions.

[4] This can be argued by starting from $\Gamma(z)$ with $\mathrm{Re}\, z > 0$ (where the integral $\int_0^\infty dt \, t^{z-1} \exp(-t)$ converges) and iteratively applying the rule $\Gamma(z-1) = \Gamma(z)/(z-1)$.

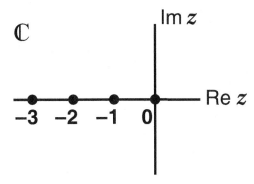

The domain of definition of the function $\Gamma(z)$, which is analytically extended over the complex plane, up to the points $z = 0, -1, -2, -3 \ldots$. Therefore, the loop propagator $G(0)$ is defined in all dimensions $d \in \mathbb{C}$, except for $d = 2, 4, 6, 8 \ldots$

perturbative continuum limit as the lattice regularization (Reisz, 1988a) or other regularization schemes, but it is easier to handle, and other schemes are not always applicable. In particular, perturbation theory is more difficult to handle in the lattice regularization, *e.g.*, because the momentum integrations are restricted to the Brillouin zone, cf. Section 1.7, which is a manifestation of Lorentz invariance breaking at the regularized level (for a review of lattice perturbation theory, see Capitani (2003)). It is remarkable and reassuring that the physics is ultimately regularization-independent.

4.5 2-Point Function to 1 Loop

We will now use dimensional regularization to evaluate the 2-point function to $\mathcal{O}(\lambda)$. To this end, up to corrections of $\mathcal{O}(\lambda^2)$, we have to evaluate the expression

$$\langle \phi(x_1)\phi(x_2)\rangle = \frac{1}{Z} \int \mathcal{D}\phi \, \phi(x_1)\phi(x_2) \left[1 - \frac{\lambda}{4!} \int d^d v \, \phi(v)^4 \right] \exp(-S_f[\phi]). \quad (4.28)$$

The partition function itself must be expanded as well,

$$Z = \int \mathcal{D}\phi \left[1 - \frac{\lambda}{4!} \int d^d v \, \phi(v)^4 \right] \exp(-S_f[\phi]), \quad (4.29)$$

which implies

$$\begin{aligned}
\langle \phi(x_1)\phi(x_2)\rangle &= \frac{1}{Z_f} \int \mathcal{D}\phi \, \phi(x_1)\phi(x_2) \exp(-S_f[\phi]) \\
&\quad - \frac{1}{Z_f} \int \mathcal{D}\phi \, \phi(x_1)\phi(x_2) \frac{\lambda}{4!} \int d^d v \, \phi(v)^4 \exp(-S_f[\phi]) \\
&\quad + \frac{1}{Z_f} \int \mathcal{D}\phi \, \frac{\lambda}{4!} \int d^d v \, \phi(v)^4 \exp(-S_f[\phi]) \\
&\quad \times \frac{1}{Z_f} \int \mathcal{D}\phi \, \phi(x_1)\phi(x_2) \exp(-S_f[\phi]), \quad (4.30)
\end{aligned}$$

where

$$Z_f = \int \mathcal{D}\phi \, \exp(-S_f[\phi]) \tag{4.31}$$

is the partition function of the free theory. The first term on the right-hand side of eq. (4.30) is the free propagator, and the other terms are corrections of $\mathcal{O}(\lambda)$. We use the Wick contraction rule to evaluate these terms. In the second term, $\phi(x_1)$ can be contracted either with $\phi(x_2)$ or with one of the factors in $\phi(v)^4$, for which there are four possibilities. Hence, we obtain

$$\int \mathcal{D}\phi \, \phi(x_1)\phi(x_2) \int d^d v \, \phi(v)^4 \exp(-S_f[\phi])$$

$$= G(x_1 - x_2) \int \mathcal{D}\phi \int d^d v \, \phi(v)^4 \exp(-S_f[\phi])$$

$$+ 4 \int d^d v \, G(x_1 - v) \int \mathcal{D}\phi \, \phi(x_2)\phi(v)^3 \exp(-S_f[\phi]). \tag{4.32}$$

In the last term, there are three ways to contract $\phi(x_2)$ with one of the factors in $\phi(v)^3$ such that

$$\frac{1}{Z_f} \int \mathcal{D}\phi \, \phi(x_2)\phi(v)^3 \exp(-S_f[\phi]) = 3G(v - x_2)G(0). \tag{4.33}$$

By substituting $x_2 = v$ in this expression, we obtain

$$\frac{1}{Z_f} \int \mathcal{D}\phi \, \phi(v)^4 \exp(-S_f[\phi]) = 3G(0)^2. \tag{4.34}$$

Inserting these relations into eq. (4.32), we arrive at

$$\frac{1}{Z_f} \int \mathcal{D}\phi \, \phi(x_1)\phi(x_2) \int d^d v \, \phi(v)^4 \exp(-S_f[\phi]) =$$

$$3G(x_1 - x_2)G(0)^2 \int d^d v \, 1 + 12G(0) \int d^d v \, G(x_1 - v)G(v - x_2). \tag{4.35}$$

In an analogous (but simpler) manner, the last term in eq. (4.30) leads to

$$\frac{1}{Z_f} \int \mathcal{D}\phi \int d^d v \, \phi(v)^4 \exp(-S_f[\phi]) \frac{1}{Z_f} \int \mathcal{D}\phi \, \phi(x_1)\phi(x_2) \exp(-S_f[\phi])$$

$$= 3G(x_1 - x_2)G(0)^2 \int d^d v \, 1. \tag{4.36}$$

Thus, the terms to $\mathcal{O}(\lambda)$ are complete; their diagrams are illustrated in Figure 4.4.

It should be noted that the expression (4.36), as well as the first contribution to eq. (4.35), is infrared (IR) divergent; *i.e.* they go to infinity with the space–time volume $\int d^d v \, 1$. However, the two IR divergent terms *cancel*.

These two terms arise from so-called *vacuum bubble diagrams*. In those diagrams, there are pieces completely disconnected from any external point x_i. In that case, the integral over the corresponding vertex v_k is unsuppressed and diverges with the space–time volume. One can prove that the contributions of vacuum bubbles to general n-point functions *always* cancel. We will not present a proof,[5] but we will use this result and always drop vacuum bubble diagrams.

---

[5] Intuitively, it is plausible that all vacuum bubbles that occur in the expansion of the denominator Z are canceled by the additional expansion $1/Z = 1/Z_f + \dots$; see, *e.g.*, Kleinert and Schulte-Frohlinde (2001).

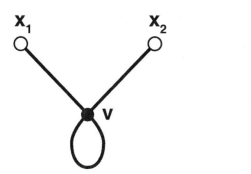 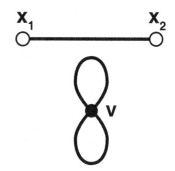

Fig. 4.4 The Feynman diagrams which contribute to the 2-point function of the $\lambda\phi^4$ model in $\mathcal{O}(\lambda)$. The diagram on the left is connected.

Putting everything together, one obtains

$$\langle\phi(x_1)\phi(x_2)\rangle = G(x_1 - x_2) - \frac{\lambda}{2}G(0)\int d^d\mathrm{v}\, G(x_1 - \mathrm{v})G(\mathrm{v} - x_2). \qquad (4.37)$$

The integral over v corresponds to a convolution in coordinate space, which (after a Fourier transform) turns into a simple product in momentum space. Hence, in momentum space eq. (4.37) takes the form

$$\widetilde{G}_\mathrm{r}(p) = \frac{1}{p^2 + m^2} - \frac{\lambda}{2}G(0)\frac{1}{(p^2 + m^2)^2} + \mathcal{O}(\lambda^2) = \frac{1}{p^2 + m_\mathrm{r}^2} + \mathcal{O}(\lambda^2). \qquad (4.38)$$

Here, we have introduced the *renormalized propagator* $\widetilde{G}_\mathrm{r}(p)$ and the *renormalized mass* m_r, which have now both been corrected to $\mathcal{O}(\lambda)$. One reads off

$$m_\mathrm{r}^2 = m^2 + \frac{\lambda}{2}G(0) + \mathcal{O}(\lambda^2) = m^2 + \frac{\lambda}{2}\frac{1}{(4\pi)^{d/2}}m^{d-2}\Gamma\left(1 - \frac{d}{2}\right) + \mathcal{O}(\lambda^2). \qquad (4.39)$$

4.6 Mass Renormalization

We have seen that the mass m of the free scalar particle changes to m_r when the ϕ^4 interaction is taken into account to leading-order in λ. Although eq. (4.39) yields a finite value for m_r for almost all values of d, m_r still diverges when one approaches the physical space–time dimension $d = 4$. This is a manifestation of the UV divergences of quantum field theory which, unlike IR divergences, do not simply cancel. In particular, as it stands, the expression for m_r still diverges in the physical limit. To cure this problem, in addition to regularizing the UV divergence (in this case, by an analytic continuation in the dimension of space–time), we must now *renormalize* the mass.

The physical mass is indeed m_r and not m. Hence, m_r should be finite, while m is the so-called *bare* (*i.e.* unrenormalized) parameter that appears in the Lagrangian, but has no direct physical meaning. Renormalization of the mass means that we let the unphysical bare parameter m depend on the regulator (in this case on $d - 4$), such that the renormalized mass m_r remains fixed,

$$m^2 = m_r^2 - \frac{\lambda}{2}G(0) = m_r^2 - \frac{\lambda}{2}\frac{1}{(4\pi)^{d/2}}m_r^{d-2}\Gamma\left(1 - \frac{d}{2}\right) + \mathcal{O}(\lambda^2)$$

$$= m_r^2 - \frac{\lambda}{2}\frac{1}{(4\pi)^{d/2}}m_r^{d-2}\left[\frac{2}{d-4} + \gamma - 1 + \mathcal{O}(d-4)\right] + \mathcal{O}(\lambda^2), \qquad (4.40)$$

which means that m^2 depends on d. In particular, now m^2 diverges (it has a pole at $d = 4$), while the physical mass m_r remains finite.

We have traded an unphysical bare parameter m for a physical mass m_r. However, the theory itself does not predict the physical value of the mass. Just as we could choose any value of m in the free theory, we can now choose any value of m_r for the interacting theory.

4.7 Connected, Disconnected, and 1-Particle Irreducible Diagrams

It is useful to classify Feynman diagrams according to their structure, which is denoted as their *topology*. An n-point function $\langle\phi(x_1)\phi(x_2)\ldots\phi(x_n)\rangle$ is represented by diagrams with n external points x_i, plus internal vertices at $v_1, \ldots, v_k$, and a number of propagators represented by lines connecting these points. There is one line emanating from each external point, and there are four lines running into each vertex. Vertices that are not connected directly or indirectly to any external points belong to the vacuum bubbles. As we mentioned in Section 4.5, Feynman diagrams that contain vacuum bubbles cancel. The remaining diagrams decompose into various connected pieces.

A connected piece is characterized by the set of external points that it contains (their number must be even). Those are connected to each other by propagators, either directly or indirectly via some vertices. All vertices belong to some connected piece; otherwise, there would be vacuum bubbles. A Feynman diagram that contains more than one connected piece is called *disconnected*. The contribution of such a diagram to the n-point function factorizes into contributions from each individual connected piece. Hence, the issue of computing a general n-point function reduces to the evaluation of *connected diagrams*. For example, the 2-point function $\langle\phi(x_1)\phi(x_2)\rangle$ automatically receives contributions from connected diagrams only. The 4-point function $\langle\phi(x_1)\phi(x_2)\phi(x_3)\phi(x_4)\rangle$, on the other hand, receives contributions from various connected pieces corresponding to the partitions $[(x_1, x_2), (x_3, x_4)]$, $[(x_1, x_3), (x_2, x_4)]$, $[(x_1, x_4), (x_2, x_3)]$, as well as from contributions in which all four external points are connected with each other (x_1, x_2, x_3, x_4). In Figures 4.2 and 4.4, only the diagram on the left is connected.

An important subset of the connected diagrams contains the *1-particle irreducible* ones. Those remain connected when any single internal propagator line in the diagram is cut. Diagrams that fall apart into two disconnected pieces under the cutting operation are 1-particle reducible. Examples are shown in Figure 4.5.

The full 2-point function obtains contributions from chains of 1-particle irreducible diagrams connected by single propagators. These terms form a geometric series. The first term of the series is the free propagator $(p^2 + m^2)^{-1}$. The second term contains 1-particle irreducible diagrams of the form $-(p^2 + m^2)^{-1}\Sigma(p^2)(p^2 + m^2)^{-1}$, where $\Sigma(p^2)$ is the so-called *self-energy*. The third term is given by $(p^2 + m^2)^{-1}\Sigma(p^2)(p^2 + m^2)^{-1}\Sigma(p^2)(p^2 + m^2)^{-1}$, etc. The fifth term is sketched in Figure 4.6. The composed propagator takes the form

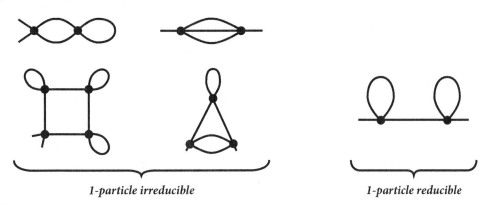

1-particle irreducible   1-particle reducible

Fig. 4.5 Examples for 1-particle irreducible parts of diagrams ("sub-diagrams"), along with a reducible counter-example.

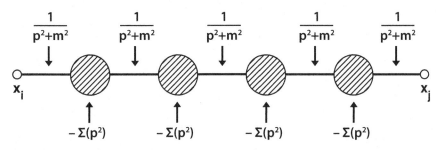

Fig. 4.6 A schematic illustration of the fifth term in the series (4.41). A shaded area represents a 1-particle irreducible sub-diagram with self-energy $\Sigma(p^2)$.

$$\widetilde{G}_r(p) = \frac{1}{p^2 + m^2} - \frac{1}{p^2 + m^2}\Sigma(p^2)\frac{1}{p^2 + m^2}$$

$$+ \frac{1}{p^2 + m^2}\Sigma(p^2)\frac{1}{p^2 + m^2}\Sigma(p^2)\frac{1}{p^2 + m^2} - \dots$$

$$= \frac{1}{p^2 + m^2}\frac{1}{1 + \Sigma(p^2)/(p^2 + m^2)} = \frac{1}{p^2 + m^2 + \Sigma(p^2)}. \tag{4.41}$$

One can now determine the physical mass m_r of the particle from the value $p^2 = -m_r^2$ for which the full propagator has a pole, *i.e.*

$$m_r^2 = m^2 + \Sigma(-m_r^2). \tag{4.42}$$

The perturbative calculation from Section 4.5 implies

$$\Sigma(p^2) = \frac{\lambda}{2}\frac{1}{(4\pi)^{d/2}}m^{d-2}\Gamma\left(1 - \frac{d}{2}\right) + \mathcal{O}(\lambda^2). \tag{4.43}$$

If we expand the full propagator around the pole at $p^2 = -m_r^2$, we obtain

$$\widetilde{G}_r(p) = \frac{Z_\phi}{p^2 + m_r^2}, \tag{4.44}$$

where Z_ϕ is the residue of this pole. With a first-order Taylor expansion

$$\Sigma(p^2) = \Sigma(-m_r^2) + (p^2 + m_r^2)\Sigma'(-m_r^2) + \dots \tag{4.45}$$

one identifies

$$Z_\phi = \frac{1}{1 + \Sigma'(-m_r^2)}. \tag{4.46}$$

The factor $\sqrt{Z_\phi}$ is often denoted as a "wave function renormalization constant", although it refers to a field rather than a wave function, so we prefer to call it a "field renormalization constant".

4.8   Feynman Rules for the $\lambda\phi^4$ Model

It is convenient to consider the Fourier transform of the n-point function

$$\int d^d x_1 d^d x_2 \dots d^d x_n \, \langle \phi(x_1)\phi(x_2)\dots\phi(x_n)\rangle \, \exp[-\mathrm{i}(p_1 x_1 + p_2 x_2 + \dots + p_n x_n)]$$
$$= (2\pi)^d \delta(p_1 + p_2 + \dots + p_n) \, \Gamma(p_1, p_2, \dots, p_n). \tag{4.47}$$

The δ-function results from translation invariance. Dropping Feynman diagrams containing vacuum bubbles, perturbation theory leads to

$$(2\pi)^d \delta(p_1 + p_2 + \dots + p_n) \, \Gamma(p_1, p_2, \dots, p_n) =$$
$$\frac{1}{Z_f} \int \mathcal{D}\phi \, \tilde{\phi}(p_1)\tilde{\phi}(p_2)\dots\tilde{\phi}(p_n) \exp(-S_f[\phi]) \times$$
$$\sum_{k=0}^{\infty} \frac{(-\lambda)^k}{(4!)^k k!} \int d^d v_1 d^d v_2 \dots d^d v_k \, \phi(v_1)^4 \phi(v_2)^4 \dots \phi(v_k)^4, \tag{4.48}$$

where $\tilde{\phi}(p)$ is the Fourier transform of the field $\phi(x)$. Since we are interested in connected diagrams only, we must contract any field $\tilde{\phi}(p_i)$ (carrying an external momentum p_i) with a field $\phi(v_j)$ at an internal vertex v_j. The resulting contraction takes the form

$$\frac{1}{Z_f} \int \mathcal{D}\phi \, \tilde{\phi}(p_i)\phi(v_j) \exp(-S_f[\phi]) = \frac{\exp(-\mathrm{i}p_i v_j)}{p_i^2 + m^2}. \tag{4.49}$$

In addition, the remaining fields $\phi(v_i)$ must be contracted among themselves, which yields

$$\frac{1}{Z_f} \int \mathcal{D}\phi \, \phi(v_i)\phi(v_j) \exp(-S_f[\phi]) = \frac{1}{(2\pi)^d} \int d^d q \, \frac{\exp(-\mathrm{i}q(v_i - v_j))}{q^2 + m^2}. \tag{4.50}$$

The momentum q associated with the internal line connecting the vertices v_i and v_j must be integrated over. After performing the Wick contractions, the positions of the vertices v_i appear only in exponential factors. Hence, when one integrates over v_i one generates a δ-function for the four momenta flowing into that vertex.

These observations lead to the following *Feynman rules* for the evaluation of connected n-point functions in the $\lambda\phi^4$ model:

1. Consider all pairwise Wick contractions in the product of internal and external fields

$$\widetilde{\phi}(p_1)\widetilde{\phi}(p_2)\ldots\widetilde{\phi}(p_n)\phi(v_1)^4\phi(v_2)^4\ldots\phi(v_k)^4$$

   that generate a connected Feynman diagram. Each diagram has a multiplicity factor that counts the number of pairings leading to the same topology of the diagram. One must also keep track of the factor $1/k!$ from the expansion of the exponential function.

2. Write down a propagator $(p_i^2 + m^2)^{-1}$ for each external line.

3. Associate a momentum q_i with each internal line, and write a factor $(q_i^2 + m^2)^{-1}$.

4. Write down a momentum conserving δ-function $(2\pi)^d\delta(k_1 + k_2 + k_3 + k_4)$ for each vertex v_i taking into account the orientation of the internal or external lines. Here, the momenta k_i can be internal or external. Moreover, a factor $-\lambda/4!$ is attached to each vertex.

5. Finally, integrate the expression over all internal momenta,

$$\int d^d q_1 \ldots d^d q_I / (2\pi)^{Id}.$$

The number I of internal lines can be determined as follows. There are $4k$ lines emanating from the k vertices, n of which are external. Each internal line connects two vertices and thus

$$I = \frac{1}{2}(4k - n). \tag{4.51}$$

There are I integrations and k δ-functions. However, one of the δ-functions reflects translation invariance and will remain in the final result, cf. eq. (4.47). Hence, only $(k - 1)$ δ-functions can be used to perform some integrations trivially. The remaining

$$l = I - (k - 1) = k + 1 - \frac{n}{2} \tag{4.52}$$

integrations determine the number of *loops* l in the diagram. Simple examples with $l = 1$ (one loop), $l = 0$ (tree level), and $l = -1$ (disconnected) are shown in Figure 4.7.

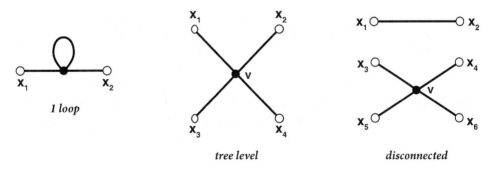

Examples for the loop numbers $l = 1$ (1-loop diagram), $l = 0$ (tree diagram), and $l = -1$ (disconnected diagram).

The number of loops increases with the order k of the expansion, but it decreases with the number of external momenta n of the n-point function. For example, to $\mathcal{O}(\lambda)$, *i.e.* for $k = 1$, a 2-point function (with $n = 2$) receives a 1-loop contribution, while at the same order a 4-point function has no contributions from loop diagrams. Diagrams without loops (*i.e.* with $l = 0$) are denoted as *tree diagrams*.[6] The main difficulty in evaluating Feynman diagrams is to perform the loop integrations over internal momenta. Tree diagrams are easy to evaluate, whereas multi-loop diagrams are hard to deal with.

4.9 4-Point Function to 1 Loop

Let us begin with something simple: the tree diagram for the 4-point function. There is just one vertex and all four external momenta flow into that vertex. There are four possible contractions for the external line with momentum p_1, three remaining possible contractions for the external line with momentum p_2, two remaining possible contractions for the external line with momentum p_3, and finally only one possible contraction for the external line with momentum p_4. Hence, the multiplicity factor is $4 \cdot 3 \cdot 2 = 4!$. Following the Feynman rules, we obtain

$$(2\pi)^d \delta(p_1 + p_2 + p_3 + p_4)\Gamma(p_1, p_2, p_3, p_4) =$$
$$4! \frac{1}{(p_1^2 + m^2)(p_2^2 + m^2)(p_3^2 + m^2)(p_4^2 + m^2)} \frac{(-\lambda)}{4!}(2\pi)^d \delta(p_1 + p_2 + p_3 + p_4), \quad (4.53)$$

such that

$$\Gamma(p_1, p_2, p_3, p_4) = -\frac{\lambda}{(p_1^2 + m^2)(p_2^2 + m^2)(p_3^2 + m^2)(p_4^2 + m^2)}. \quad (4.54)$$

In the next order, $k = 2$, we are confronted with 1-loop diagrams. Now there are two vertices. If all external lines would flow into the same vertex, the diagram would contain a vacuum bubble. Hence, either one, two, or three external lines may run into the same vertex.

First, we consider the case where the external line with the momentum p_1 runs into one vertex and the other external lines (with momenta p_2, p_3, and p_4) run into the other vertex. This diagram is shown is Figure 4.8. There are 8 possible contractions for the first external line with momentum p_1. Then, there are $4 \cdot 3 \cdot 2$ possible contractions of the three other external lines at the other vertex. Finally, there are 3 contractions for an internal line with momentum q_1 connecting the two vertices and one remaining internal line with momentum q_2 leading from the first vertex back to itself. Hence, the total multiplicity factor amounts to $8 \cdot 4 \cdot 3 \cdot 2 \cdot 3 = (4!)^2$. Following the Feynman rules, one obtains the contribution

---

[6] Diagrams with $l < 0$ are disconnected; an example is shown in Figure 4.7.

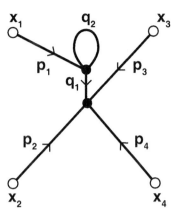

Fig. 4.8 1-loop diagram for the term (4.55), which contributes to the 4-point function in $\mathcal{O}(\lambda^2)$.

$$\frac{1}{(2\pi)^{2d}} \int d^d q_1 d^d q_2 \, (4!)^2 \frac{1}{2!} \frac{1}{(p_1^2 + m^2)(p_2^2 + m^2)(p_3^2 + m^2)(p_4^2 + m^2)}$$

$$\times \frac{1}{(q_1^2 + m^2)(q_2^2 + m^2)} \left(\frac{-\lambda}{4!}\right)^2 (2\pi)^{2d} \delta(p_1 - q_1)\delta(q_1 + p_2 + p_3 + p_4)$$

$$= \frac{\lambda^2 G(0)/2}{(p_1^2 + m^2)^2 (p_2^2 + m^2)(p_3^2 + m^2)(p_4^2 + m^2)} (2\pi)^d \delta(p_1 + p_2 + p_3 + p_4). \quad (4.55)$$

The integration over q_1 was performed trivially using a δ-function, while the integration over the loop-momentum q_2 yields the divergent (but dimensionally regularized) factor $G(0)$. Similar expressions exist for diagrams where the loop is attached to the external lines with momenta p_2, p_3, and p_4. These four 1-loop diagrams just give rise to the renormalization of the mass of the incoming particles. Together with the tree diagram, they lead to

$$\Gamma(p_1, p_2, p_3, p_4) = -\frac{\lambda}{(p_1^2 + m_r^2)(p_2^2 + m_r^2)(p_3^2 + m_r^2)(p_4^2 + m_r^2)}, \quad (4.56)$$

with $m_r^2 = m^2 + \frac{1}{2}\lambda G(0) + \mathcal{O}(\lambda^2)$, according to eq. (4.40).

However, this is not the full answer to order λ^2. We also need to take into account the 1-loop diagrams where two external lines run into the same vertex, as shown in Figure 4.9.

There are again 8 possible contractions for the external line with momentum p_1. If the external line with momentum p_2 runs into the same vertex, there are 3 remaining possible contractions. The external line with momentum p_3 has 4 possible contractions at the other vertex, and the external line with momentum p_4 has 3 remaining contractions also at that other vertex. Finally, there are 2 possible contractions for internal lines with momenta q_1 and q_2 connecting the two vertices.[7] Hence, the total multiplicity factor for this diagram is again $8 \cdot 3 \cdot 4 \cdot 3 \cdot 2 = (4!)^2$.

---

[7] Alternatively, we could connect x_2 to the same vertex as x_1; following the same scheme, we again end up with the multiplicity factor $8 \cdot 3 \cdot 4 \cdot 3 \cdot 2 = (4!)^2$.

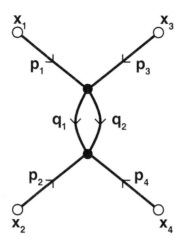

1-loop diagram for the term (4.58), which contributes to the 4-point function at $\mathcal{O}(\lambda^2)$.

At this point, we introduce the *Mandelstam variables*

$$s = (p_1 + p_2)^2, \quad t = (p_1 + p_3)^2, \quad u = (p_1 + p_4)^2. \tag{4.57}$$

For the diagram under consideration, the Feynman rules lead to

$$\frac{1}{(2\pi)^{2d}} \int d^d q_1 d^d q_2 \, (4!)^2 \frac{1}{2!} \frac{1}{(p_1^2 + m^2)(p_2^2 + m^2)(p_3^2 + m^2)(p_4^2 + m^2)}$$

$$\times \frac{1}{(q_1^2 + m^2)(q_2^2 + m^2)} \left(\frac{-\lambda}{4!} \right)^2 (2\pi)^{2d} \delta(p_1 + p_2 - q_1 - q_2) \delta(q_1 + q_2 + p_3 + p_4)$$

$$= \frac{\lambda^2 J(s)/2}{(p_1^2 + m^2)(p_2^2 + m^2)(p_3^2 + m^2)(p_4^2 + m^2)} (2\pi)^d \delta(p_1 + p_2 + p_3 + p_4), \tag{4.58}$$

where

$$J(p^2) = \frac{1}{(2\pi)^d} \int d^d q \, \frac{1}{(q^2 + m^2)((p - q)^2 + m^2)}. \tag{4.59}$$

There are two more diagrams with similar topology which depend on the Mandelstam variables t and u. The total contribution to the 4-point function to $\mathcal{O}(\lambda^2)$ amounts to

$$\Gamma(p_1, p_2, p_3, p_4) = -\frac{\lambda - \lambda^2 [J(s) + J(t) + J(u)]/2}{(p_1^2 + m_r^2)(p_2^2 + m_r^2)(p_3^2 + m_r^2)(p_4^2 + m_r^2)}. \tag{4.60}$$

4.10 Dimensional Regularization of $J(p^2)$

Just like $G(0)$, $J(p^2)$ is divergent in four space–time dimensions and must be regularized. As before, we choose dimensional regularization. In order to evaluate the corresponding integral, we use a trick due to Feynman,

$$\frac{1}{AB} = \int_0^1 d\tau \, \frac{1}{[(1 - \tau)A + \tau B]^2}. \tag{4.61}$$

Based on eq. (4.59), we identify

$$
J(p^2) = \frac{1}{(2\pi)^d} \int_0^1 d\tau \int d^d q \, \frac{1}{[(1-\tau)(q^2+m^2) + \tau((p-q)^2+m^2)]^2},
$$
$$
= \frac{1}{(2\pi)^d} \int_0^1 d\tau \int d^d \overline{q} \, \frac{1}{(\overline{q}^2 + \overline{m}^2)^2}. \tag{4.62}
$$

In the last step, we have shifted the integration variable from q to $\overline{q} = q - \tau p$, and we have introduced $\overline{m}^2 = m^2 + \tau(1-\tau)p^2$. In addition, we use the Schwinger parameterization

$$
\frac{1}{(\overline{q}^2 + \overline{m}^2)^2} = \int_0^\infty dt \, t \exp\left(-t(\overline{q}^2 + \overline{m}^2)\right), \tag{4.63}
$$

and perform the d-dimensional Gaussian integral

$$
\frac{1}{(2\pi)^d} \int d^d \overline{q} \, \exp\left(-t\overline{q}^2\right) = (4\pi t)^{-d/2}, \tag{4.64}
$$

to obtain

$$
J(p^2) = \int_0^1 d\tau \int_0^\infty dt \, (4\pi t)^{-d/2} t \exp\left(-t\overline{m}^2\right)
$$
$$
= \frac{\Gamma(2-\frac{d}{2})}{(4\pi)^{d/2}} \int_0^1 d\tau \, \overline{m}^{d-4} = -\frac{1}{8\pi^2(d-4)} + \ldots \tag{4.65}
$$

The last expression reveals a pole at $d = 4$, and the dots include the remaining finite piece. It is remarkable that the divergent piece is *independent* of p^2, and that the order of the pole is the same as for the power divergent term $G(0)$. Generally, in dimensional regularization all divergences are given by simple poles of the Γ-function. This is in contrast to other regularizations, which lead to a hierarchy of divergences with different powers of the cut-off, as we are going to discuss in Section 4.13.

4.11 Renormalization of the Coupling

Since the divergence of the 4-point function is momentum-independent, it can be absorbed into a redefinition – a renormalization – of the coupling constant λ. The coupling depends on the momenta of the interacting particles. Hence, in order to define the physical coupling in a unique way, one must specify a so-called *renormalization condition*.

The specific form of the renormalization condition is a matter of convention. It is usual to define the physical coupling at the center of the "Mandelstam triangle",

$$
s = t = u = \frac{4}{3}m_{\rm r}^2, \tag{4.66}
$$

where $m_{\rm r}$ is the renormalized mass. We define the finite part of the function $J(p^2)$ as

$$
\overline{J}(p^2) = J(p^2) - J(4m_{\rm r}^2/3), \tag{4.67}
$$

and we obtain

$$\Gamma(p_1, p_2, p_3, p_4) = -\frac{\lambda_r - \lambda_r^2[\overline{J}(s) + \overline{J}(t) + \overline{J}(u)]/2}{(p_1^2 + m_r^2)(p_2^2 + m_r^2)(p_3^2 + m_r^2)(p_4^2 + m_r^2)}, \tag{4.68}$$

where the physical renormalized coupling is given by

$$\lambda_r = \lambda - \frac{3}{2}J(4m_r^2/3)\lambda^2 + \mathcal{O}(\lambda^3). \tag{4.69}$$

The cut-off is now removed by demanding that the coupling constant λ_r remains at its physical value, while the bare coupling λ diverges. Actually, eq. (4.69) suggests $|\lambda| \to \infty$ in the final limit $d \to 4$, which might seem strange in view of the initial assumption of the perturbative expansion. It is important to transfer this assumption to the *renormalized* parameters; *i.e.* λ_r is positive and small.

It is a notable property that the renormalizations of m and λ are sufficient to render *all* higher n-point functions finite without any further adjustments. In other words, after fixing the renormalized mass m_r and coupling λ_r, all other results of the theory are completely fixed. This implies a large predictive power of the theory. Just like the classical $\lambda\phi^4$ model, the corresponding quantum field theory has only two free parameters.

Box 4.1	Definition of renormalizability

Quantum field theories with a finite number of parameters, which are capable of absorbing all the divergences, are called *renormalizable*.

4.12 Renormalizability of Scalar Field Theories

Now we consider a scalar field theory in d space–time dimensions with a general polynomial self-interaction potential

$$V(\phi) = \sum_\nu g_\nu \phi^\nu. \tag{4.70}$$

The corresponding Euclidean action takes the form of eq. (4.9),

$$S[\phi] = \int d^d x \left[\frac{1}{2} \partial_\mu \phi \partial_\mu \phi + V(\phi) \right]. \tag{4.71}$$

Since the action enters the Boltzmann factor $\exp(-S[\phi])$ in the functional integral as an exponent, it must be dimensionless. Consequently, the field ϕ has the mass dimension

$$d_\phi = \frac{d - 2}{2}. \tag{4.72}$$

For example, in $d = 2$, a scalar field is dimensionless ($d_\phi = 0$), while in $d = 4$ it has the dimension of a mass ($d_\phi = 1$). Similarly, the coupling constant g_ν has the dimension

$$d_{g_\nu} = d - \nu d_\phi = d\left(1 - \frac{\nu}{2}\right) + \nu. \tag{4.73}$$

In particular, the coefficient in front of the ϕ^2-term has dimension $d_{g_2} = 2$, irrespective of the dimension d. Of course, the prefactor $g_2 = m^2/2$ of this term is indeed given by the bare mass squared of the scalar particle.

As we will point out in the next section, theories containing couplings with negative mass dimension (*i.e.* $d_{g_\nu} < 0$) are not renormalizable. Hence, renormalizability requires

$$d\left(1 - \frac{\nu}{2}\right) + \nu \geq 0 \quad \Rightarrow \quad \nu \leq \frac{2d}{d-2}. \tag{4.74}$$

As a result, in $d = 2$ space–time dimensions *any* polynomial potential, with ν up to ∞, leads to a renormalizable scalar field theory. In $d = 3$, on the other hand, renormalizability requires $\nu \leq 6$, and in $d = 4$ dimensions $\nu \leq 4$. In $d = 6$, we have $\nu \leq 3$, *i.e.* a ϕ^3 interaction is formally still renormalizable. However, in the absence of a stabilizing term, like ϕ^4 or ϕ^6, such a model does not have a stable vacuum (or ground state) and is thus not well-defined beyond perturbation theory.

Generally, in dimensions $d > 4$ renormalizable scalar field theories are necessarily *trivial*; *i.e.* they are free field theories because the potential is limited to at most quadratic terms, which correspond to simple Gaussian integrals. While there are only a few renormalizable field theories in $d = 4$ space–time dimensions, for $d = 2$ field theory is as varied as, let us say, botany in a rainforest, because there is an unlimited variety of species of renormalizable terms in the action. In this regard, $d \geq 4$ is rather desert-like.

4.13 Condition for Renormalizability

Let us now derive eq. (4.74) – *i.e.* the renormalizability condition $d_{g_\nu} \geq 0$ – by analyzing the degree of divergence of Feynman diagrams. For this purpose, we choose a more transparent regularization than dimensional regularization, which is nothing more (and nothing less) than a very neat mathematical device. An appropriate regularization would be the one using a space–time lattice. Here, we choose a method which is simpler to handle in this context, namely, a regularization using a momentum cut-off Λ. This means that the integrals over internal momenta q_i are restricted to $|q_i| \leq \Lambda$ (we commented on that regularization in footnote 8 of Chapter 2).

Let us consider a connected Feynman diagram for an n-point function in a d-dimensional scalar field theory with general polynomial self-interactions. There are n external lines, k_ν vertices of type $\nu > 2$ (where interactions take place, with ν lines emanating), and I internal lines. The total number of vertices is $k = \sum_\nu k_\nu$, and the total number of lines emanating from these vertices is $\sum_\nu k_\nu \nu$. Hence, the number of internal lines is given by

$$I = \frac{1}{2}\left(\sum_\nu k_\nu \nu - n\right), \tag{4.75}$$

which generalizes formula (4.51). As before, each vertex is associated with a momentum-conserving δ-function, and – as a manifestation of translation invariance – for an n-point function there is one remaining overall δ-function, $\delta(p_1 + p_2 + \cdots + p_n)$. Again, there are

$$l = I - k + 1 \tag{4.76}$$

non-trivial loop integrations.

Each integration over a momentum q_i of an internal line contains a measure factor $\int d|q_i| \, |q_i|^{d-1}$, and each internal line propagator takes the form $(q_i^2 + m^2)^{-1}$. We are interested in an upper limit on the degree of UV divergence of the diagram. Hence, we neglect m^2 in the propagator (whose UV contributions are dominated by $q_i^2 \approx \Lambda^2$), so we replace the propagator by q_i^{-2}. The so-called *superficial degree of divergence* (without considering possible cancellations) is then given by

$$\delta = dl - 2I = \frac{d-2}{2}\left(\sum_\nu k_\nu \nu - n\right) - d\sum_\nu k_\nu + d. \qquad (4.77)$$

The integral in the corresponding Feynman diagram is convergent if $\delta < 0$, *i.e.* if

$$\sum_\nu \left[k_\nu\left(\frac{d-2}{2}\nu - d\right)\right] < \frac{d-2}{2}n - d. \qquad (4.78)$$

In order to prevent the superficial degree of divergence to become arbitrarily large for a large number k_ν of vertices, *i.e.* in order for a model to be renormalizable, one requires

$$\frac{d-2}{2}\nu - d \le 0 \quad \Rightarrow \quad \nu \le \frac{2d}{d-2}. \qquad (4.79)$$

This is just the condition of eq. (4.74).

Theories with $\nu < 2d/(d-2)$ for all vertices are called *super-renormalizable*. For example, the $\lambda\phi^4$ model (with $\nu = 4$) is super-renormalizable in $d = 3$ space–time dimensions. In four dimensions, the $\lambda\phi^4$ model is still renormalizable because then $\nu = 2d/(d-2)$. In such a renormalizable (but not super-renormalizable) model, Feynman diagrams are superficially divergent if

$$n \le \frac{2d}{d-2}, \qquad (4.80)$$

as we infer from the right-hand side of eq. (4.78). For example, in $d = 4$, 2- and 4-point functions diverge, as we have seen, but all higher n-point functions converge, while in 3 dimensions 6-point functions are still divergent.

A full proof of the renormalizability of a quantum field theory, even only at the level of perturbation theory, is a subtle mathematical problem, which requires the understanding of overlapping divergences in nested Feynman diagrams. In the framework of the cut-off regularization in momentum space, this was pioneered by Nikolai Bogoliubov and Ostap Parasiuk (Bogoliubov and Parasiuk, 1957). A proof to all orders of perturbation theory was provided by Klaus Hepp (1966) and later simplified by Wolfhart Zimmermann (1969), leading to the *BPHZ theorem*.

Exercises

4.1 Euler's constant

We have used the following Laurent expansion of the Γ-function

$$\Gamma(\varepsilon) = \frac{1}{\varepsilon} - \gamma + \mathcal{O}(\varepsilon),$$

where $\gamma = \lim_{N \to \infty} \left(\sum_{k=1}^{N} \frac{1}{k} - \ln N \right) \approx 0.577216\ldots$ is the Euler (or Euler-Mascheroni) constant.

  a) Reproduce the derivation of this relation, or check it numerically.

  b) Now compute also $\Gamma(-N + \varepsilon)$ for any $N \in \mathbb{N}$ up to $\mathcal{O}(\varepsilon)$.

4.2 Mixed propagator

Derive the relation for the "mixed" free propagator

$$\langle \tilde{\phi}(p_i)\phi(v_j) \rangle = \frac{\exp(-ip_iv_j)}{p_i^2 + m^2},$$

which we have used in eq. (4.49) for the external lines in the Feynman diagram of scalar fields.

4.3 Negative loop number

Justify with some plausibility argument the statement in footnote 6: A diagram with loop number $l < 0$ cannot be connected.

4.4 Product of two propagators

Show that the Fourier transform of the product of two propagators is given by

$$J(q) = \int d^d x \, \exp(iqx) \, G(x)^2 = \int \frac{d^d p}{(2\pi)^d} \frac{1}{p^2 + m^2} \frac{1}{(p-q)^2 + m^2}.$$

Show that $J(q)$ has a pole at $d = 4$, *i.e.*

$$J(q) = -\frac{1}{8\pi^2(d-4)} + \text{finite terms.}$$

It is helpful to use the Feynman parametrization eq. (4.61).

4.5 Feynman parameterization

The reader who is interested in advanced perturbative expansions (beyond the considerations in this book) is encouraged to study the generalization of the identity (4.61) to

$$\frac{1}{A_1 A_2 \ldots A_n} = (n-1)! \int_0^1 d\tau_1 \ldots \int_0^1 d\tau_n \, \delta\left(1 - \sum_{k=1}^{n} \tau_k\right) \left[\sum_{k=1}^{n} \tau_k A_k\right]^{-n}.$$

4.6 Disconnected diagrams

In Section 4.9, we discussed the connected diagrams that contribute to the 4-point function of the $\lambda\phi^4$ model to $\mathcal{O}(\lambda^2)$. Which disconnected diagrams contribute to this order?

  a) Draw these diagrams and write down the corresponding terms.

  b) Which are the (non-contributing) vacuum bubble diagrams at $\mathcal{O}(\lambda^2)$?

4.7 4-point function in the ϕ^3 model

The Lagrangian of a scalar ϕ^3 model is given by

$$\mathcal{L}(\phi, \partial_\mu \phi) = \frac{1}{2} \partial_\mu \phi \partial_\mu \phi + \frac{m^2}{2} \phi^2 + g \phi^3.$$

Ignore the fact that the vacuum of this theory is unstable (because the potential $V(\phi) = \frac{1}{2} m^2 \phi^2 + g \phi^3$ is unbounded from below), and compute the 4-point function up to $\mathcal{O}(g^2)$.

4.8 $\lambda \phi^6$ model

We consider the $\lambda \phi^6$ model, with

$$\mathcal{L}(\phi, \partial_\mu \phi) = \frac{1}{2} \partial_\mu \phi \partial_\mu \phi + \frac{m^2}{2} \phi^2 + \frac{\lambda}{6!} \phi^6.$$

a) Compute the 2-point function $\Gamma(p_1, p_2)$ to $\mathcal{O}(\lambda)$.

b) Draw all the connected diagrams that contribute to $\mathcal{O}(\lambda^2)$. For each of these diagrams, count the number of loops, determine the multiplicity factor, and write down the corresponding term.

5   Renormalization Group

Why is physics so successful in describing natural phenomena over a vast span of distance scales, ranging from elementary particles, atomic nuclei, atoms and molecules, quantum and classical fluids and solids, planets and stars, all the way up to the entire cosmos? Obviously, Isaac Newton was extremely successful in understanding classical mechanics, without any knowledge about the fundamental building blocks of matter, let alone quantum physics. Fortunately, one need not know the hypothetical "Theory of Everything", before one can address the physics at large distances and low energy scales.

In the 1960s, Kenneth Wilson – partly based on previous work of André Petermann and Ernst Stückelberg, Murray Gell-Mann and Francis Low, Nikolai Bogoliubov and Dmitry Shirkov, and Leo Kadanoff – developed the ideas of the *renormalization group* (which is not a group in the mathematical sense): It naturally explains why systematic low-energy effective theories – although limited in their range of applicability – are extremely powerful. The basic concepts underlying the renormalization group are *space–time locality* and the existence of *hierarchies of energy scales*.

5.1   Locality and Hierarchies of Energy Scales

Although we do not understand the origin of space and time, physics benefits tremendously from the fact that both are local; *i.e.* they are endowed with infinitesimal neighborhood relations. While vastly different energy scales exist in Nature, we often do not understand their origin. Despite this fact, in physics we do not hesitate to take advantage of this situation. What are the relevant local degrees of freedom that dominate the physics at a given energy scale? How do these degrees of freedom interact with each other? Are some effects small at this energy scale and can be neglected at the leading orders of a systematic expansion? Does this give rise to some approximate global symmetry that may simplify our theoretical description?

These are questions to be addressed in the construction of a low-energy effective field theory. Based on the concept of *universality*, the renormalization group explains why the effective theory method, which underlies a lot of physics, is so successful. Universality states that the fine details of the short-distance, high-energy dynamics have only an indirect influence on the long-distance, low-energy physics. In particular, the low-energy physics can be accurately expressed in terms of effective low-energy degrees of freedom. Only the values of the parameters in their Lagrangian are determined by the short-distance physics.

This means that, at least to a large extent, the different effective theories can stand on their own and do not require prior knowledge of any underlying theory. Still, if we gain a deeper understanding of shorter distance scales, an effective theory can benefit, because its intrinsic parameters can now be predicted based on the newly understood short-distance physics. The history of physics is full of success stories that went exactly like this. The macroscopic properties of matter were deduced from atoms, whose inner structure was understood in terms of electrons surrounding a nucleus. The atomic nucleus was identified as a system of protons and neutrons, which in turn consist of quarks and gluons. This is were we stand today. Further steps deeper into the ultraviolet would not invalidate our current knowledge, but would put it on a presently unknown, new foundation.

Since the unknown short-distance physics has only indirect effects at low energies, as the flip side of the coin of universality, it is very difficult to infer new high-energy physics from low-energy data. As we have seen in Chapter 4, some quantum field theories are *renormalizable*. As we will discuss later, these include the Standard Model. This might suggest that it could, at least in principle, be considered at arbitrarily high energy scales. Despite its renormalizability, since the Standard Model does not include quantum gravity, it is expected to break down at the Planck scale, if not earlier. The Planck scale hence plays the role of a physical ultraviolet cut-off, for which there is no mathematical necessity to be sent to infinity. From the Wilsonian point of view, the UV cut-off is indeed something that does not need to be removed. We just need to understand the physics that emerges well below that energy scale (Wilson and Kogut, 1974).

The divergences in quantum field theory may indicate that we are not using the truly adequate degrees of freedom at ultra-short-distance scales. Be they superstrings, qubits, or some tiny wheels turning around at the Planck scale, irrespective of the precise nature of the ultimate fundamental degrees of freedom, universality makes the low-energy physics insensitive to short-distance details and facilitates the success of physics at low energy scales, without any prior knowledge of a "Theory of Everything".

5.2 Renormalization Group Blocking and Fixed Points

Let us discuss the renormalization group in the non-perturbative context of N-component scalar field theory on a d-dimensional Euclidean space–time lattice. For the purpose of the present discussion, we imagine the inverse lattice spacing as a physical high-momentum cut-off – let us say at the Planck scale – rather than as a purely mathematical UV regulator. Therefore, we do not send the cut-off to infinity. Instead, we are taking the continuum limit by considering the long-distance, low-energy physics at length scales much longer than the lattice spacing a.

This point of view is common in classical statistical mechanics, which is mathematically equivalent to quantum field theory in the Euclidean functional integral formulation. In that case, a spatial lattice mimics short atomic length scales that are beyond the classical description and represent a UV cut-off. For example, when one uses the Ising model to describe the universal features of binary fluids, the lattice has no specific physical significance, but is still necessary to make the description mathematically well-defined. Similarly, when we mimic the Planck scale by a Euclidean lattice spacing, we do not make any statement about the nature of quantum gravity, except that some qualitatively new degrees of freedom are expected to appear at this length scale.

We consider a d-dimensional, hypercubic, Euclidean space–time lattice $\Lambda = (a\mathbb{Z})^d$, with a real-valued N-component scalar field $\Phi_x^a \in \mathbb{R}$ residing at the lattice sites x. The functional integral, or partition function, then takes the form[1]

$$Z = \int \mathcal{D}\Phi \, \exp(-S[\Phi]) = \prod_{x \in \Lambda} \prod_{a=1}^{N} \sqrt{\frac{a^{d-2}}{2\pi}} \int_{-\infty}^{\infty} d\Phi_x^a \, \exp(-S[\Phi]). \tag{5.1}$$

As in Section 1.7, the square root prefactor compensates the dimension of the field Φ in the integration measure. We will assume the Euclidean action to be a polynomial functional of the lattice field

$$S[\Phi] = \sum_{x \in \Lambda} c_x^a \Phi_x^a + \sum_{x,y \in \Lambda} c_{xy}^{ab} \Phi_x^a \Phi_y^b + \sum_{x,y,z \in \Lambda} c_{xyz}^{abc} \Phi_x^a \Phi_y^b \Phi_z^c + \sum_{x,y,z,w \in \Lambda} c_{xyzw}^{abcd} \Phi_x^a \Phi_y^b \Phi_z^c \Phi_w^d + \cdots \tag{5.2}$$

For instance, a mass term corresponds to $c_{xx}^{aa} = m^2/2$ and the standard self-interaction term is given by $c_{xxxx}^{aabb} = \lambda/4!$. We assume that the action is invariant against discrete space–time translations and rotations, and perhaps also against internal O(N) rotations. Most important, we assume the action to be *local*; *i.e.* the couplings $c_{xyz...}^{abc...} \in \mathbb{R}$ decay at least exponentially when the points $x, y, z, \ldots$ are separated from each other. If the couplings are strictly zero beyond a finite lattice distance, a lattice action is called *ultra-local*. On the other hand, if the couplings decay only power-like, we characterize the action as *non-local*. A sufficient degree of locality is an important prerequisite for universality.

The action functionals of eq. (5.2) exist in an abstract, infinite-dimensional space spanned by the couplings $c_{xyz...}^{abc...}$, with each point in this space representing a different microscopic model. In each of these theories, one can define the 2-point correlation function

$$\langle \Phi_x^a \Phi_y^b \rangle = \frac{1}{Z} \int \mathcal{D}\Phi \, \Phi_x^a \Phi_y^b \exp(-S[\Phi]) \,, \tag{5.3}$$

and investigate the decay properties of the connected 2-point function $\langle \Phi_x^a \Phi_y^b \rangle - \langle \Phi_x^a \rangle \langle \Phi_y^b \rangle$ at large distances, cf. Chapter 1. If the decay is exponential with a finite correlation length ξ, the model has a mass gap $m = 1/\xi$ at a finite fraction of the cut-off, *i.e.* $ma = a/\xi > 0$. On the other hand, if the decay is power-like, the correlation length is infinite, $\xi/a = \infty$ (in units of the lattice spacing) and there is non-trivial physics at arbitrarily low energies. In statistical mechanics, this corresponds to a *second-order phase transition*, and in quantum field theory on the lattice one can then define a *continuum limit*. In both cases, universality makes the low-energy continuum physics insensitive to the details of the model at the cut-off scale. Models with infinite correlation length reside on a so-called *critical surface* in the infinite-dimensional space of action functionals.

The dimension of the critical surface is typically almost as large as the dimension of the entire space of couplings; they differ only by a small co-dimension. This co-dimension decides how many coupling parameters must be tuned in order to reach the critical surface. In classical statistical mechanics, usually the temperature is one such parameter, whose fine-tuning is required to accurately adjust to a second-order phase transition. In quantum field theory applied to elementary particle physics, similar fine-tuning often seems less natural, because we may not understand why a fundamental theory of Nature is located near such a point in the space of all possible models. For example, in the Standard Model the bare mass

---

[1] The index a should not be confused with the lattice spacing.

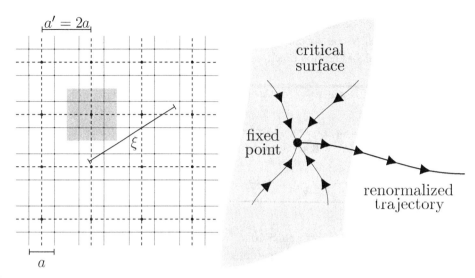

Fig. 5.1 Left: Factor 2 renormalization group blocking transformation. The average of the field values over the 2^d sites within a block (shaded square) on the fine lattice (with spacing a) determines the field value at the block center on the coarse lattice (with spacing $a' = 2a$). After the blocking, the correlation length ξ, which remains unchanged in physical units, is a factor of 2 smaller in units of the new lattice spacing a', *i.e.* $\xi/a' = \frac{1}{2}\xi/a$. Right: Sketch of a critical surface in the infinite-dimensional space of all couplings. Iterated renormalization group blocking transformations within the critical surface (on which $\xi/a = \infty$) converge to a fixed point. A renormalized trajectory leaves the critical surface in a relevant direction. Theories on the renormalized trajectory are completely free of lattice artifacts — they have a perfect lattice action — even though for them ξ/a is finite.

of the Higgs field must be fine-tuned in order to strongly separate the renormalized mass of the Higgs particle from the Planck scale. This fine-tuning issue, which generates the hierarchy between the electroweak scale and the Planck scale, is known as the *hierarchy problem*.

In order to characterize critical behavior, Wilson introduced the concept of *renormalization group transformations* which provide a systematic way to integrate out short-distance ultraviolet degrees of freedom. On a lattice, this is most naturally achieved in a local manner by *blocking transformations*, which were first introduced by Kadanoff (1966). For this purpose, we collect the 2^d lattice sites x at the corners of an elementary hypercube into a block $c_{x'}$, with x' being the block center (which itself does not belong to Λ). As illustrated in the left panel of Figure 5.1, in this way the lattice is decomposed into disjoint blocks whose centers x' form a lattice Λ' of twice the lattice spacing $a' = 2a$. We then introduce new field variables $\Phi_{x'}^{\prime a}$ that are associated with the block centers x', and we construct a Gaussian blocking kernel

$$\exp(-T[\Phi', \Phi]) = \prod_{x' \in \Lambda'} \frac{a'}{\sqrt{\alpha_2}} \exp\left(-\frac{a'^d}{2\alpha_2}\left(\Phi_{x'}^{\prime a} - \frac{\beta_2}{2^d}\sum_{x \in c_{x'}} \Phi_x^a\right)^2\right), \qquad (5.4)$$

that couples the blocked field Φ' to the original field Φ on the fine lattice. Here, α_2 and β_2 (with the indices referring to the blocking factor 2) are adjustable parameters that characterize the details of the blocking transformation, which is then used to transform the functional integral into

$$\exp(-S'[\Phi']) = \int \mathcal{D}\Phi \ \exp(-T[\Phi', \Phi]) \exp(-S[\Phi]). \tag{5.5}$$

By integrating out the original field Φ, one obtains an *effective action* $S'[\Phi']$ for the blocked field Φ'. The invariance of the partition function is guaranteed by

$$\int \mathcal{D}\Phi' \ \exp(-T[\Phi', \Phi]) =$$

$$\prod_{x' \in \Lambda'} \prod_{a=1}^{N} \sqrt{\frac{a'^{d-2}}{2\pi}} \int_{-\infty}^{\infty} d\Phi'^a_{x'} \frac{a'}{\sqrt{\alpha_2}} \exp\left(-\frac{a'^d}{2\alpha_2}\left(\Phi'^a_{x'} - \frac{\beta_2}{2^d}\sum_{x \in c_{x'}} \Phi^a_x\right)^2\right) = 1, \tag{5.6}$$

which implies

$$Z' = \int \mathcal{D}\Phi' \ \exp(-S'[\Phi']) = \int \mathcal{D}\Phi \int \mathcal{D}\Phi' \ \exp(-T[\Phi', \Phi]) \exp(-S[\Phi])$$

$$= \int \mathcal{D}\Phi \ \exp(-S[\Phi]) = Z. \tag{5.7}$$

When the original field Φ is integrated out exactly, by construction the physics remains unchanged, except that it now manifests itself on a lattice that is two times coarser, with $a' = 2a$. Consequently, the correlation length of the blocked system in units of the coarse lattice spacing $\xi/a' = \xi/(2a)$ is half of the correlation length in units of the fine lattice spacing a.

When we apply the blocking transformation to a theory on the critical surface (with $\xi/a = \infty$), the blocked theory *remains on the critical surface*. In that case, we can iterate the blocking transformation an infinite number of times and generate a sequence of blocked actions that, for an appropriate value of β_2, converges to a renormalization group fixed point

$$S[\Phi] \ \rightarrow \ S'[\Phi'] \ \rightarrow \ \cdots \rightarrow S^{(\nu)}[\Phi^{(\nu)}] \ \rightarrow \ \cdots \rightarrow S^*[\Phi^*]. \tag{5.8}$$

The *fixed point action* $S^*[\Phi^*]$, which results when the number ν of blocking steps goes to infinity, depends on the parameter α_2 of the renormalization group blocking transformation. In general, the fixed point action is rather complicated and usually impossible to calculate exactly. However, it is often still local (in the sense of exponentially decaying couplings). The fixed point action is remarkable because, by construction, it represents the same continuum physics as any other theory on the critical surface that is in the basin of attraction of this fixed point. Actually, this basin of attraction contains all members of a given *universality class*, which hence share the same continuum limit, despite the fact that their actions may differ significantly at the lattice spacing scale. The theories that belong to the same universality class share the same symmetries and, of course, exist in the same space–time dimension. Remarkably, the theory at the fixed point is completely *free of any cut-off artifacts*.

The renormalization group flow in the vicinity of a fixed point is characteristic of the continuum field theory that is defined there. Let us consider infinitesimal deviations from the fixed point that reproduce themselves under renormalization group blocking, up to a multiplicative factor, *i.e.*

$$S[\Phi] = S^*[\Phi] + \delta S[\Phi] \quad \Rightarrow \quad S'[\Phi'] = S^*[\Phi'] + 2^\Delta \delta S[\Phi']. \tag{5.9}$$

When one moves away from the fixed point infinitesimally in some direction in the infinite-dimensional coupling space, most directions remain within the critical surface. Upon blocking, they are pulled back to the fixed point. Such directions, which are associated with $\Delta < 0$, are known as *irrelevant*. *Relevant directions*, on the other hand, have $\Delta > 0$ and are characterized by the fact that they are pushed away from the fixed point under blocking transformations. In particular, directions that lead out of the critical surface to theories with finite correlation lengths, $\xi/a < \infty$, are relevant. Directions in which an infinitesimal shift remains unchanged under blocking have $\Delta = 0$ and are known as *marginal*. In that case, a higher-order analysis of the renormalization group flow can clarify whether such a direction is actually marginally relevant or irrelevant. The scaling dimensions Δ of the relevant operators $\delta S[\Phi]$ provide an important characteristic of a given fixed point.

For example, the relevant direction that is associated with the mass leads out of the critical surface. Iterated renormalization group transformations then follow a so-called *renormalized trajectory* (see the right panel of Figure 5.1). Theories on the renormalized trajectory have *perfect actions* that are completely free of lattice artifacts, despite the fact that for them ξ/a is finite.

5.3 Gaussian Fixed Points of Lattice Scalar Field Theory

In order to gain analytic insight into the origin of universality and into the mechanisms of the renormalization group, we will now consider a set of fixed points of a free 1-component scalar lattice field theory. Since the corresponding functional integrals are Gaussian and can hence be performed analytically, the corresponding fixed points are known as *Gaussian fixed points*.

We begin with the standard nearest-neighbor lattice action of a massless scalar field

$$S[\Phi] = a^d \sum_{x,\mu} \frac{1}{2a^2} \left(\Phi_{x+\hat\mu} - \Phi_x \right)^2. \tag{5.10}$$

Here, $\mu \in \{1, 2, \ldots, d\}$ denotes a Euclidean space–time direction and $\hat\mu$ is a vector pointing in the μ-direction, with a length that corresponds to the lattice spacing a. The standard action is a finite-difference discretization of the continuum action $S[\phi] = \int d^d x\, \frac{1}{2} \partial_\mu \phi \partial_\mu \phi$. It represents a simple (but otherwise not very special) point on the critical surface. Let us parametrize a general, lattice translation invariant, quadratic action as

$$S[\Phi] = \frac{a^d}{2} \sum_{x,y} \Phi_x \Delta(x - y) \Phi_y. \tag{5.11}$$

Parity symmetry implies $\Delta(-z) = \Delta(z)$ with $z = x - y \in (a\mathbb{Z})^d$. The standard action is characterized by

$$\Delta(z) = \frac{1}{a^2} \sum_{\mu=1}^{d} (\delta_{z-\hat\mu,0} - 2\delta_{z,0} + \delta_{z+\hat\mu,0}), \tag{5.12}$$

which corresponds to the nearest-neighbor lattice Laplacian. Going to momentum space (with a momentum p in the Brillouin zone $B =]-\pi/a, \pi/a]^d$), we obtain

$$\Phi(p) = a^d \sum_x \Phi_x \exp(-ipx) , \quad \Phi_x = \frac{1}{(2\pi)^d} \int_B d^d p\, \Phi(p) \exp(ipx) ,$$

$$S[\Phi] = \frac{1}{(2\pi)^d} \int_B d^d p\, \Phi(-p)\Delta(p)\Phi(p) , \quad \Delta(p) = \sum_{\mu=1}^d \left(\frac{2}{a} \sin \frac{p_\mu a}{2}\right)^2 , \quad (5.13)$$

cf. Section 1.7. The Fourier transform of the block average $\sum_{x \in c_{x'}} \Phi_x/2^d$ takes the form $\Phi(p)\Pi_2(pa)$, with the factor 2 blocking function in momentum space given by

$$\Pi_2(pa) = \prod_{\mu=1}^d \frac{\sin(p_\mu a)}{2\sin(p_\mu a/2)} = \prod_{\mu=1}^d \cos(p_\mu a/2). \quad (5.14)$$

Quick Question 5.3.1   Blocking function in momentum space
Verify that the block average $\sum_{x \in c_{x'}} \Phi_x/2^d$ leads to the blocking function of eq. (5.14).

Next, we perform renormalization group blocking transformations using eqs. (5.4) and (5.5)

$$
\begin{aligned}
\exp(-S'[\Phi']) =& \int \mathcal{D}\Phi \mathcal{D}\Omega' \exp\left\{ -\frac{1}{(2\pi)^d} \int_B d^d p\, \frac{1}{2}\Phi(-p)\Delta(p)\Phi(p) \right. \\
& - \frac{1}{(2\pi)^d} \int_{B'} d^d p' \left[i\Omega'(-p') \left(\Phi'(p') - \beta_2 \sum_l \Phi\left(p' + \frac{2\pi l}{a'}\right) \Pi_2(p'a + \pi l) \right) \right. \\
& \left. \left. + \frac{\alpha_2}{2}\Omega'(-p')\Omega'(p') \right] \right\} \\
=& \int \mathcal{D}\Phi \mathcal{D}\Omega' \exp\left\{ -\frac{1}{(2\pi)^d}\int_B d^d p \left[\frac{1}{2}\Phi(-p)\Delta(p)\Phi(p) - i\Omega'(-p)\beta_2\Phi(p)\Pi_2(pa) \right] \right. \\
& \left. - \frac{1}{(2\pi)^d} \int_{B'} d^d p' \left[\frac{\alpha_2}{2}\Omega'(-p')\Omega'(p') + i\Omega'(-p')\Phi'(p') \right] \right\} . \quad (5.15)
\end{aligned}
$$

Here, we have introduced an auxiliary scalar field $\Omega'_{x'}$ on the coarse lattice ($x' \in \Lambda'$) and its Fourier transform $\Omega'(p')$ over the corresponding Brillouin zone $B' =]-\pi/a', \pi/a']^d$. The sum over the vector l extends over its components $l_\mu \in \{1, 2\}$. In the last step, this sum has been combined with the integral over B' to an integral over the larger Brillouin zone B. We now integrate out the field Φ on the fine lattice. Since the integral is Gaussian for $\alpha_2 > 0$, it reduces to minimizing the action. The corresponding classical solution,

$$\Phi_c(p) = i\beta_2 \Delta(p)^{-1}\Pi_2(pa)\Omega'(p) , \quad (5.16)$$

is then plugged back into the exponent, and we obtain

$$
\begin{aligned}
\exp(-S'[\Phi']) =& \int \mathcal{D}\Omega' \exp\left\{ -\frac{1}{(2\pi)^d} \int_{B'} d^d p' \left[i\Omega'(-p')\Phi'(p') \right. \right. \\
& \left. \left. + \frac{1}{2}\Omega'(-p') \left(\beta_2^2 \sum_l \Delta\left(p' + \frac{2\pi l}{a'}\right)^{-1} \Pi_2(p'a + \pi l)^2 + \alpha_2 \right) \Omega'(p') \right] \right\} .
\end{aligned}
$$

$$(5.17)$$

We now integrate out the auxiliary field, again by solving its classical equation of motion,

$$\Omega'_c(p') = -i\Delta'(p')\Phi'(p'), \quad \Delta'(p')^{-1} = \beta_2^2 \sum_l \Delta\left(p' + \frac{2\pi l}{a'}\right)^{-1} \Pi_2(p'a + \pi l)^2 + \alpha_2.$$

(5.18)

Inserting this result back into the exponent, we obtain

$$S'[\Phi'] = \frac{1}{(2\pi)^d} \int_{B'} d^d p' \frac{1}{2} \Phi'(-p')\Delta'(p')\Phi'(p').$$

(5.19)

In order to complete the first renormalization group step, we identify $\Delta^{(1)}(p) = 2^d \Delta'(p/2)$ and we rescale $a' \to a$.

Iterating this procedure ν times results in the action $S^{(\nu)}[\Phi^{(\nu)}]$ that is given in terms of

$$\Delta^{(\nu)}(p)^{-1} = \left(\frac{\beta_2^2}{2^d}\right)^\nu \sum_l \Delta\left(\frac{p + 2\pi l/a}{2^\nu}\right)^{-1} \prod_{\mu=1}^d \left(\frac{\sin(p_\mu a/2)}{2^\nu \sin((p_\mu a + 2\pi l_\mu)/2^{\nu+1})}\right)^2$$
$$+ \alpha_2 \frac{1 - (\beta_2^2/2^d)^\nu}{1 - \beta_2^2/2^d}.$$

(5.20)

The summation now extends over vectors l with integer components $l_\mu \in \{1, 2, 3, \ldots, 2^\nu\}$. In the limit $\nu \to \infty$, one may expect to approach a fixed point of the renormalization group with the action $S^*[\Phi^*]$ that is characterized by

$$\Delta^*(p)^{-1} = \lim_{\nu \to \infty} \left(\frac{\beta_2^2}{2^d}\right)^\nu \sum_l \frac{2^{2\nu}}{(p + 2\pi l/a)^2} \prod_{\mu=1}^d \left(\frac{\sin(p_\mu a/2)}{p_\mu a/2 + \pi l_\mu}\right)^2 + \alpha_2 \frac{1 - (\beta_2^2/2^d)^\nu}{1 - \beta_2^2/2^d}.$$

(5.21)

The renormalization group transformations converge to a fixed point only if

$$\left(\frac{\beta_2^2}{2^d}\right)^\nu 2^{2\nu} = 1 \quad \Rightarrow \quad \beta_2 = 2^{(d-2)/2}.$$

(5.22)

Here, $(d-2)/2$ is the canonical dimension of a scalar field in d space–time dimensions, and β_2 ensures the appropriate rescaling of the blocked field. The resulting fixed point action is given by the propagator (Bell and Wilson, 1975)

$$\Delta^*(p)^{-1} = \sum_{l \in \mathbb{Z}^d} \frac{1}{(p + 2\pi l/a)^2} \prod_{\mu=1}^d \left(\frac{\sin(p_\mu a/2)}{p_\mu a/2 + \pi l_\mu}\right)^2 + \frac{4\alpha_2}{3}.$$

(5.23)

The parameter α_2 of the blocking transformation leads to a family of fixed points, one for each value of $\alpha_2 \geq 0$ (the limit $\alpha_2 \to 0$ corresponds to δ-function blocking; see below). These fixed points are all physically equivalent and belong to the same universality class.

5.4 Blocking from the Continuum to the Lattice

Next, we follow Wilson (1976) and derive the fixed point lattice action as well as the entire renormalized trajectory that emerges from the fixed point by blocking directly from the continuum. For this purpose, we average the continuum field $\phi(y)$ over hypercubes c_x of size a^d, centered at the lattice sites,

$$\Phi_x = \frac{1}{a^d} \int_{c_x} d^d y \, \phi(y). \tag{5.24}$$

In momentum space, this corresponds to

$$\Phi(p) = \sum_{l \in \mathbb{Z}^d} \phi(p + 2\pi l/a) \Pi(pa + 2\pi l), \quad \Pi(pa) = \prod_{\mu=1}^{d} \frac{2 \sin(p_\mu a/2)}{p_\mu a}. \tag{5.25}$$

Thanks to the summation over $l \in \mathbb{Z}^d$, the lattice field $\Phi(p)$ is periodic over the Brillouin zone B, as it must be. The lattice propagator is then given in terms of the continuum propagator as

$$\langle \Phi(-p)\Phi(p) \rangle = \sum_{l \in \mathbb{Z}^d} \left\langle \phi(-p - 2\pi l/a)\phi(p + 2\pi l/a) \right\rangle \Pi(pa + 2\pi l)^2$$

$$= \sum_{l \in \mathbb{Z}^d} \frac{\Pi(pa + 2\pi l)^2}{(p + 2\pi l/a)^2} = \Delta^*(p)^{-1}, \tag{5.26}$$

which corresponds to the fixed point action of eq. (5.23) with $a = a^*$ and $\alpha_2 = 0$. The corresponding factor 2 renormalization group transformation enforces a δ-function constraint between the field Φ averaged over a block on the fine lattice and the blocked field Φ' on the coarse lattice.

As a next step, we introduce a mass $m \neq 0$, thus deviating from the fixed point, and we perform a renormalization group blocking transformation with a general parameter α directly out of the continuum. The resulting action, which is situated on the renormalized trajectory, is *perfect* in the sense that it is completely free of lattice artifacts

$$\exp(-S_{\text{perfect}}[\Phi]) = \int \mathcal{D}\phi \mathcal{D}\Omega \exp \left\{ -\frac{1}{(2\pi)^d} \int d^d p \, \frac{1}{2} \phi(-p) \left(p^2 + m^2\right) \phi(p) \right\}$$

$$\times \exp \left\{ -\frac{1}{(2\pi)^d} \int_B d^d p \left[i\Omega(-p) \left(\Phi(p) - \sum_{l \in \mathbb{Z}^d} \phi(p + 2\pi l/a)\Pi(pa + 2\pi l) \right) \right. \right.$$

$$\left. \left. + \frac{\alpha}{2} \Omega(-p)\Omega(p) \right] \right\}$$

$$= \int \mathcal{D}\phi \mathcal{D}\Omega \exp \left\{ -\frac{1}{(2\pi)^d} \int d^d p \left[\frac{1}{2} \phi(-p) \left(p^2 + m^2\right) \phi(p) \right. \right.$$

$$\left. \left. - i\Omega(-p)\phi(p)\Pi(pa) \right] \right\}$$

$$\times \exp \left\{ -\frac{1}{(2\pi)^d} \int_B d^d p \left[\frac{\alpha}{2} \Omega(-p)\Omega(p) + i\Omega(-p)\Phi(p) \right] \right\}. \tag{5.27}$$

We have introduced an auxiliary lattice field Ω that enforces eq. (5.25) as a δ-function constraint for $\alpha = 0$ and smears the δ-function to a Gaussian for $\alpha > 0$. In the last step, we have combined the integration over the Brillouin zone B and the summation over $l \in \mathbb{Z}^d$ to an integration over the continuum momentum space $\mathbb{R}^d$.

Performing the Gaussian integral over the continuum field ϕ can be achieved by solving a classical equation of motion, *i.e.* by minimizing the action, which yields

$$\phi_c(p) = i(p^2 + m^2)^{-1} \Pi(pa)\Omega(p), \tag{5.28}$$

and by plugging the result back into the exponent such that

$$
\exp(-S_{\text{perfect}}[\Phi]) = \int \mathcal{D}\Omega \exp \left\{ -\frac{1}{(2\pi)^d} \int d^d p \left[i\Omega(-p)\Phi(p) \right. \right.
$$
$$
\left. \left. + \frac{1}{2}\Omega(-p) \left(\sum_{l \in \mathbb{Z}^d} \frac{\Pi(pa + 2\pi l)^2}{(p + 2\pi l/a)^2 + m^2} + \alpha \right) \Omega(p) \right] \right\}. \tag{5.29}
$$

Finally, we integrate out the auxiliary field, which again amounts to solving its classical equation of motion,

$$
\Omega_{\text{c}}(p) = -i\,\Delta_{\text{perfect}}(p)\Phi(p). \tag{5.30}
$$

Plugging this result back into the exponent, one obtains

$$
S_{\text{perfect}}[\Phi] = \frac{1}{(2\pi)^d} \int d^d p \, \frac{1}{2} \Phi(-p) \Delta_{\text{perfect}}(p) \Phi(p), \tag{5.31}
$$

with the *perfect lattice propagator*

$$
\Delta_{\text{perfect}}(p)^{-1} = \langle \Phi(-p)\Phi(p) \rangle = \sum_{l \in \mathbb{Z}^d} \frac{\Pi(pa + 2\pi l)^2}{(p + 2\pi l/a)^2 + m^2} + \alpha. \tag{5.32}
$$

For $m = 0$, this corresponds to the fixed point action of eq. (5.23) with $\alpha_2 = 3\alpha/4$.

5.5 Perfect Lattice Actions on the Renormalized Trajectory

For $m \neq 0$, the resulting lattice action is located on the renormalized trajectory that emanates from the fixed point in the relevant direction of the mass parameter. If one takes an infinitesimal step in this direction, thus leaving the critical surface, and iterates the factor 2 blocking transformation, one traces out the renormalized trajectory. Since they are connected to the fixed point by exact renormalization group blocking transformations, the lattice actions on the entire renormalized trajectory are completely free of lattice artifacts. In particular, the corresponding lattice theory has exactly the same energy spectrum as the continuum theory.

Let us demonstrate this amazing property by investigating the energy spectrum of the theory on the renormalized trajectory. For this purpose, we consider the 2-point correlation function of a scalar field with definite spatial momentum $\vec{p} \in]-\pi/a, \pi/a]^{d-1}$ at a position x_d in Euclidean time, similar to Section 1.7,

$$
\Phi(\vec{p})_{x_d} = \frac{1}{2\pi} \int_{-\pi/a}^{\pi/a} dp_d \, \Phi(p) \exp(ip_d x_d), \tag{5.33}
$$

and the field $\Phi_{\vec{x},0}$ at the spatial point $\vec{x}$ at Euclidean time 0. The corresponding correlation function is given by

$$
\begin{aligned}
\langle \Phi_{\vec{x},0} \Phi(\vec{p})_{x_d} \rangle &= \frac{1}{2\pi} \int_{-\pi/a}^{\pi/a} dp_d \, \Delta_{\text{perfect}}(p)^{-1} \exp(\mathrm{i} p_d x_d) \\
&= \frac{1}{2\pi} \int_{-\pi/a}^{\pi/a} dp_d \left[\sum_{l \in \mathbb{Z}^d} \frac{\Pi(pa + 2\pi l)^2}{(p + 2\pi l/a)^2 + m^2} + \alpha \right] \exp(\mathrm{i} p_d x_d) \\
&= \frac{1}{2\pi} \int_{-\infty}^{\infty} dp_d \sum_{\vec{l} \in \mathbb{Z}^{d-1}} \frac{1}{\left(\vec{p} + 2\pi \vec{l}/a \right)^2 + p_d^2 + m^2} \\
&\quad \times \prod_{j=1}^{d-1} \left(\frac{2\sin(p_j a/2)}{p_j a + 2\pi l_j} \right)^2 \left(\frac{2\sin(p_d a/2)}{p_d a} \right)^2 \exp(\mathrm{i} p_d x_d) + \alpha \frac{\delta_{x_d,0}}{a} \\
&= \sum_{\vec{l} \in \mathbb{Z}^{d-1}} C(\vec{p} + 2\pi \vec{l}/a) \exp\left(-E(\vec{p} + 2\pi \vec{l}/a) x_d \right) + \alpha \frac{\delta_{x_d,0}}{a}. \quad (5.34)
\end{aligned}
$$

Here, $C(\vec{p} + 2\pi \vec{l}/a)$ is a momentum-dependent, but x_d-independent amplitude. The sum over l_d has been combined with the integral of p_d over the interval $] - \pi/a, \pi/a]$ to the integral over $\mathbb{R}$ of the corresponding continuum theory. The sum over the spatial vectors $\vec{l} \in \mathbb{Z}^{d-1}$ gives rise to infinitely many poles of the integrand, each contributing an exponential decay to the 2-point function. The energies of the corresponding states are determined by the locations of the poles, $E(\vec{p} + 2\pi \vec{l}/a) = \mathrm{i} p_d$, with

$$
E(\vec{p} + 2\pi \vec{l}/a)^2 = -p_d^2 = \left(\vec{p} + 2\pi \vec{l}/a \right)^2 + m^2. \quad (5.35)
$$

This is indeed exactly the correct energy of a particle in the continuum theory with momentum $\vec{p} + 2\pi \vec{l}/a$. The integer vectors $\vec{l}$ extend the Brillouin zone $] - \pi/a, \pi/a]^{d-1}$ of the lattice theory to the entire momentum space $\mathbb{R}^{d-1}$ of the continuum theory. The energy spectrum of the lattice theories on the renormalized trajectory is hence *identical* to that of the continuum theory. Cut-off lattice artifacts are completely eliminated, so these actions are indeed *perfect*. In particular, despite the fact that continuous Poincaré invariance is not manifest in these actions, the resulting physics is fully covariant.

These remarkable features persist for interacting theories. However, other than for a free theory, an analytic computation of the perfect action will then in general not be possible, and even a numerical determination tends to be very tedious.

One can adjust the free parameter α in order to optimize the locality of the perfect action. This can even be done analytically for $d = 1$. The perfect propagator then takes the form

$$
\begin{aligned}
\Delta_{\text{perfect}}(p)^{-1} &= \sum_{l \in \mathbb{Z}} \frac{\Pi(pa + 2\pi l)^2}{(p + 2\pi l/a)^2 + m^2} + \alpha \\
&= \frac{1}{m^2} - \frac{2}{m^3 a} \frac{\coth(ma/2)}{\cot^2(pa/2) + \coth^2(ma/2)} + \alpha. \quad (5.36)
\end{aligned}
$$

Setting

$$\alpha = \frac{\widetilde{m} - m}{m^3}\,, \quad \widetilde{m} = \frac{1}{a}\sinh(ma)\,, \quad \tilde{m} = \frac{2}{a}\sinh(ma/2), \tag{5.37}$$

onc obtains

$$\Delta_{\text{perfect}}(p) = \frac{m^3}{\widetilde{m}\tilde{m}^2}\left[\left(\frac{2}{a}\sin(pa/2)\right)^2 + \tilde{m}^2\right], \tag{5.38}$$

which corresponds to the ultra-local standard nearest-neighbor action. In the massless case, the optimal choice is $\alpha = a^2/6$. For the above choice of α, in higher dimensions the perfect action still remains very local (in the sense of fast exponential decays of the couplings between distant neighbors).

5.6 Wilson–Fisher Fixed Points

As we have seen in Chapter 1, the functional integral approach to quantum field theory in Euclidean space–time is mathematically equivalent to classical statistical mechanics on a spatial lattice. The renormalization group therefore also applies to classical physics. In classical statistical mechanics on a 3-dimensional spatial lattice, N-component $\lambda\phi^4$ theory is related to the Ising model for $N = 1$, the XY model for $N = 2$, the Heisenberg model for $N = 3$, all the way up to the so-called spherical model at $N = \infty$, cf. Table 1.2. All these models (which for $N \geq 2$ are known as σ-models) possess second-order phase transitions at some critical temperature, which are governed by the so-called *Wilson–Fisher fixed points*, whose detailed features depend on N.

In 3-d $\lambda\phi^4$ theory, the Gaussian fixed point (of the free theory) has two relevant directions, one corresponding to the mass and one corresponding to the quartic self-coupling λ. Staying within the critical surface of massless theories, there is a trajectory that leaves the Gaussian fixed point in the relevant λ direction. This implies that in 3 dimensions the Gaussian fixed point is unstable. Iterated blocking leads away from it into the stable Wilson–Fisher fixed point, which has only one relevant direction associated with the mass. The corresponding renormalization group flow is illustrated in Figure 5.2.

The characteristics of the Wilson–Fisher fixed point manifest themselves as the critical exponents of the statistical mechanics models mentioned above. These exponents are *universal* and hence insensitive to the microscopic details of a specific model. They are the same for the entire universality class associated with the corresponding fixed point. Kenneth Wilson and Michael Fisher have used dimensional regularization and developed an analytic perturbative ε-expansion around the free theory in 4 dimensions (Wilson and Fisher, 1972). The $d = 4 - \varepsilon = 3$ theory is then reached by putting the expansion parameter to $\varepsilon = 1$.

Expansions around $d = 2$ have been worked out to very high precision by Jean Zinn-Justin and Edouard Brézin (Brézin and Zinn-Justin, 1976). Very precise values of the critical exponents have also been obtained from numerical Monte Carlo simulations of specific statistical mechanics models (Campostrini *et al.*, 2002) or via the so-called bootstrap approach (El-Showk *et al.*, 2012).

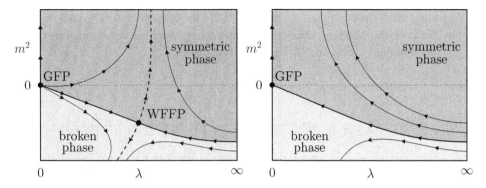

Fig. 5.2 Renormalization group flow in scalar field theory, projected into the λ-m^2 plane. The free massless theory resides at the Gaussian fixed point (GFP). Left: In 3 dimensions, it has two relevant directions. The one associated with the mass corresponds to $\lambda = 0$. It leads out of the critical surface, which is the solid line that separates the symmetric from the broken phase. The one associated with the quartic coupling lies within the critical surface and leads from the trivial (*i.e.* non-interacting) Gaussian fixed point to the non-trivial Wilson–Fisher fixed point (WFFP). The Wilson–Fisher fixed point has only one relevant direction associated with the mass (the dashed line), which leads out of the critical surface. Right: In 4 dimensions, there is only the trivial Gaussian fixed point because the quartic coupling turns out to be marginally irrelevant. Then, the mass defines the only relevant direction, which leads out of the critical surface at $\lambda = 0$.

5.7 Renormalization of Scalar Field Theory in a Cut-off Regularization

The rest of this chapter is partly inspired by unpublished lecture notes of Peter Hasenfratz.

Let us now consider a 1-component real scalar field $\phi(x) \in \mathbb{R}$ in the continuum with the Euclidean Lagrangian

$$\mathcal{L}(\psi, \partial_\mu \psi) = \frac{1}{2} \partial_\mu \phi \partial_\mu \phi + \frac{m^2}{2} \phi^2 + \frac{\lambda}{4!} \phi^4. \tag{5.39}$$

Here, m and λ are the bare mass and the bare coupling. For the following considerations, we impose a momentum cut-off Λ; *i.e.* all momentum integrals will be restricted to $|q| \leq \Lambda$. We aim at a critical theory with vanishing renormalized mass $m_r = 0$, which requires fine-tuning of the bare mass m. We will work in perturbation theory to one loop and use the Feynman rules presented in Chapter 4. We recall that the free propagator is given by

$$\widetilde{G}(p) = \frac{1}{p^2 + m^2}, \tag{5.40}$$

while the renormalized propagator takes the form

$$\widetilde{G}_r(p) = \frac{1}{p^2 + m^2} - \frac{\lambda}{2} G(0) \frac{1}{(p^2 + m^2)^2} + \mathcal{O}(\lambda^2) = \frac{1}{p^2 + m_r^2} + \mathcal{O}(\lambda^2),$$

$$m_r^2 = m^2 + \frac{\lambda}{2} G(0) + \mathcal{O}(\lambda^2). \tag{5.41}$$

We have derived these expressions in eq. (4.40) in the context of dimensional regularization. They remain valid here since they are not yet regularized in the present form.

So we now need to evaluate $G(0)$ in cut-off regularization

$$
\begin{aligned}
G(0) &= \frac{1}{(2\pi)^4} \int d^4p \frac{1}{p^2 + m^2} \\
&= \frac{1}{8\pi^2} \int_0^\Lambda dp \frac{p^3}{p^2 + m^2} = \frac{1}{16\pi^2} \left(\Lambda^2 - m^2 \log \left(1 + \frac{\Lambda^2}{m^2} \right) \right).
\end{aligned}
\tag{5.42}
$$

We adjust the bare mass as a function of the bare coupling such that the renormalized mass vanishes

$$
m_r^2 = m^2 + \frac{\lambda}{2} \frac{1}{16\pi^2} \left(\Lambda^2 - m^2 \log \left(1 + \frac{\Lambda^2}{m^2} \right) \right) = 0 \quad \Rightarrow \quad m^2 = -\frac{\lambda}{32\pi^2} \Lambda^2 + \mathcal{O}(\lambda^2).
\tag{5.43}
$$

It is no problem that the bare mass squared m^2 is negative; only the renormalized mass $m_r = 0$ is physically relevant.

At tree-level, the 4-point function takes the form

$$
\Gamma(p_1, p_2, p_3, p_4) = -\frac{\lambda}{(p_1^2 + m^2)(p_2^2 + m^2)(p_3^2 + m^2)(p_4^2 + m^2)}.
\tag{5.44}
$$

The 1-loop diagram shown in Figure 4.8 just renormalizes the mass and gives rise to

$$
\Gamma'(p_1, p_2, p_3, p_4) = -\frac{\lambda}{(p_1^2 + m_r^2)(p_2^2 + m_r^2)(p_3^2 + m_r^2)(p_4^2 + m_r^2)} = -\frac{\lambda}{p_1^2 p_2^2 p_3^2 p_4^2}.
\tag{5.45}
$$

Finally, also taking into account the diagram shown in Figure 4.9, the 4-point function to 1 loop can be expressed in terms of the Mandelstam variables,

$$
\begin{aligned}
\Gamma(p_1, p_2, p_3, p_4) &= -\frac{\lambda - \lambda^2 (J(s) + J(t) + J(u))/2}{p_1^2 p_2^2 p_3^2 p_4^2}, \\
s &= (p_1 + p_2)^2, \quad t = (p_1 + p_3)^2, \quad u = (p_1 + p_4)^2.
\end{aligned}
\tag{5.46}
$$

As a next step, we renormalize the field as well as the coupling constant,

$$
\phi_r(x) = \sqrt{Z_\phi} \phi(x), \quad \lambda = Z_\lambda \lambda_r.
\tag{5.47}
$$

In the massless 4-d theory at the classical level, there is no dimensioned parameter. Hence, the classical theory is scale-invariant. However, scale invariance is *anomalous*; *i.e.* it is explicitly broken by quantum effects. In particular, the momentum cut-off Λ that enters at the quantum level is a dimensioned quantity. In the process of renormalization, the cut-off Λ is traded for a renormalization scale μ.

In order to fix Z_ϕ, we demand for the 2-point function

$$
\tilde{G}_r(p) \big|_{p^2 = \mu^2} = \frac{1}{\mu^2}.
\tag{5.48}
$$

Furthermore, we fix reference momenta k_1, k_2, k_3, k_4 such that

$$
k_i k_j = \mu^2 \left(\delta_{ij} - \frac{1}{4} \right) \quad \Rightarrow \quad k_i^2 = \frac{3}{4}\mu^2, \quad k_i k_j = -\frac{1}{4}\mu^2, \ i \neq j.
\tag{5.49}
$$

This definition is consistent with momentum conservation since $(k_1 + k_2 + k_3 + k_4)k_i = 0$. For the renormalized coupling, we now demand

$$\Gamma(k_1, k_2, k_3, k_4) = -\frac{\lambda_r}{k_1^2 k_2^2 k_3^2 k_4^2}. \tag{5.50}$$

This is a condition on λ or alternatively on Z_λ.

5.8 Callan–Symanzik Equation

The dimensionless constants Z_λ and Z_ϕ can only depend on λ and on Λ/μ. As a consequence of renormalizability, in the limit $\Lambda \to \infty$, the cut-off dependence of all n-point Green's functions can be absorbed in λ_r such that

$$\Gamma_r^{(n)}(p_1, p_2, \ldots, p_n; \lambda_r, \mu) = Z_\phi^{n/2}(\lambda, \Lambda/\mu)\, \Gamma^{(n)}(p_1, p_2, \ldots, p_n; \lambda, \Lambda). \tag{5.51}$$

While the bare n-point functions are (by definition) independent of the renormalization scale μ, again as a consequence of renormalizability, the renormalized n-point functions are independent of the cut-off Λ. In the limit $\Lambda \to \infty$, we obtain

$$\mu \frac{d}{d\mu} \Gamma^{(n)}(p_1, p_2, \ldots, p_n; \lambda, \Lambda) = 0 \quad \Rightarrow$$

$$\mu \frac{d}{d\mu} \left[Z_\phi^{-n/2}(\lambda, \Lambda/\mu)\, \Gamma_r^{(n)}(p_1, p_2, \ldots, p_n; \lambda_r, \mu) \right] = 0. \tag{5.52}$$

Using the chain rule, this implies the *Callan–Symanzik equation*

$$\left[\mu \frac{\partial}{\partial \mu} + \beta(\lambda_r) \frac{\partial}{\partial \lambda_r} - n\gamma(\lambda_r) \right] \Gamma_r^{(n)}(p_1, p_2, \ldots, p_n; \lambda_r, \mu) = 0,$$

$$\beta(\lambda_r) = \mu \frac{\partial}{\partial \mu} \lambda_r(\mu),$$

$$\gamma(\lambda_r) = \frac{1}{2} \mu \frac{\partial}{\partial \mu} \log Z_\phi(\lambda, \Lambda/\mu), \tag{5.53}$$

which was derived independently by Curtis Callan (1970) and Kurt Symanzik (1970). The expressions β and γ are denoted as the *β-function* and the *anomalous dimension*, respectively. We have used

$$\mu \frac{d}{d\mu} Z_\phi^{-n/2}(\lambda, \Lambda/\mu) = -\mu \frac{n}{2} Z_\phi^{-n/2-1}(\lambda, \Lambda/\mu) \frac{\partial}{\partial \mu} Z_\phi(\lambda, \Lambda/\mu)$$

$$= -n Z_\phi^{-n/2}(\lambda, \Lambda/\mu) \frac{1}{2} \mu \frac{\partial}{\partial \mu} \log Z_\phi(\lambda, \Lambda/\mu). \tag{5.54}$$

It is important that, once the cut-off Λ is removed to infinity, the β-function $\beta(\lambda_r)$ and the anomalous dimension $\gamma(\lambda_r)$ do not depend on μ explicitly, because they can only depend on the dimensionless ratio Λ/μ. The Callan–Symanzik eq. (5.53) is valid to all orders of perturbation theory and is expected to remain valid even beyond perturbation theory.

5.9 β-Function and Anomalous Dimension to 1 Loop

At the 1-loop level, the renormalized 2-point function takes the form $\widetilde{G}_r(p) = 1/p^2$, which implies $Z_\phi = 1$, thus leading to the anomalous dimension $\gamma(\lambda_r) = 0$.

According to Section 4.11, the renormalized coupling is again given in terms of the Mandelstam variables s, t, and u,

$$\lambda_r = \lambda - \frac{\lambda^2}{2}(J(s) + J(t) + J(u)). \tag{5.55}$$

At the renormalization scale μ, we obtain

$$s = (k_1+k_2)^2 = t = (k_1+k_3)^2 = u = (k_1+k_4)^2 = \mu^2 \quad \Rightarrow \quad \lambda_r = \lambda - \frac{3\lambda^2}{2}J(\mu^2). \tag{5.56}$$

Using eq. (4.62), but now applying cut-off regularization, one finds

$$J(\mu^2) = \frac{1}{(2\pi)^4} \int_0^1 d\tau \int d^4\bar{q} \, \frac{1}{(\bar{q}^2 + \overline{\mu}^2)^2}, \quad \overline{\mu}^2 = \tau(1-\tau)\mu^2, \quad \bar{q} = q - \tau\mu. \tag{5.57}$$

First performing the $\bar{q}$-integration in the limit of large Λ results in[2]

$$\frac{1}{(2\pi)^4} \int d^4\bar{q} \, \frac{1}{(\bar{q}^2 + \overline{\mu}^2)^2} = \frac{2\pi^2}{(2\pi)^4} \int_0^\Lambda d\bar{q} \, \frac{\bar{q}^3}{(\bar{q}^2 + \overline{\mu}^2)^2} \sim \frac{1}{16\pi^2} \log\left(\frac{\Lambda^2}{\overline{\mu}^2}\right). \tag{5.58}$$

As in Section 4.10, we only keep the leading terms for large Λ. Thus, the τ-integration yields

$$J(\mu^2) = \int_0^1 d\tau \, \frac{1}{16\pi^2} \log\left(\frac{\Lambda^2}{\tau(1-\tau)\mu^2}\right) \sim \frac{1}{8\pi^2} \log\left(\frac{\Lambda}{\mu}\right), \tag{5.59}$$

where $\mu > 0$. Hence, the final result for the renormalized coupling is

$$\lambda_r = \lambda + \frac{3\lambda^2}{16\pi^2} \log\left(\frac{\mu}{\Lambda}\right). \tag{5.60}$$

For the β-function, this implies

$$\beta(\lambda_r) = \mu \frac{\partial}{\partial\mu} \lambda_r = \frac{3}{16\pi^2}\lambda^2 = \frac{3}{16\pi^2}\lambda_r^2 + \mathcal{O}(\lambda_r^3). \tag{5.61}$$

The *sign* of the β-function is crucial.

- A positive sign indicates that the coupling increases with the energy scale μ.
- A negative sign, on the other hand, implies *asymptotic freedom*; *i.e.* the interaction becomes weaker in the UV.

Since the β-function of eq. (5.61) is positive, 4-d $\lambda\phi^4$ theory is not asymptotically free.

The β-function can be calculated systematically to higher loop orders in perturbation theory. For example, at the 2-loop level one obtains

---

[2] If one starts from $|q| \leq \Lambda$, then the integration region for $\bar{q}$ is shifted, but this does not alter the (approximate) final term in eq. (5.58).

$$\beta(\lambda_r) = \beta_0 \lambda_r^2 + \beta_1 \lambda_r^3 + \mathcal{O}(\lambda_r^4), \quad \beta_0 = \frac{3}{(4\pi)^2}, \quad \beta_1 = -\frac{17}{3(4\pi)^4}. \tag{5.62}$$

The constants β_0 and β_1 are known as the 1- and 2-loop coefficients of the β-function. They are *universal* in the sense that they do not depend on the UV regularization or the renormalization scheme or scale.

5.10 Running Coupling

Let us use the renormalization group to determine the evolution of the coupling λ_r with the energy scale μ,

$$\mu \frac{\partial}{\partial \mu} \lambda_r = \beta(\lambda_r) = \beta_0 \lambda_r^2 + \mathcal{O}(\lambda_r^3) \quad \Rightarrow \quad \int_{\lambda_r(\mu_0)}^{\lambda_r(\mu)} \frac{d\lambda_r}{\beta_0 \lambda_r^2} = \int_{\mu_0}^{\mu} \frac{d\mu}{\mu} + \mathcal{O}(\lambda_r^3) \quad \Rightarrow$$

$$\lambda_r(\mu) = \frac{\lambda_r(\mu_0)}{1 - \lambda_r(\mu_0)\beta_0 \log(\mu/\mu_0)} + \mathcal{O}(\lambda_r^3). \tag{5.63}$$

We see that the running coupling $\lambda_r(\mu)$ *increases* with increasing energy scale μ. In fact, eq. (5.63) seems to suggest that $\lambda_r(\mu)$ even diverges when $\log(\mu/\mu_0) = 1/(\lambda_r(\mu_0)\beta_0)$. However, this so-called *Landau pole* should not be taken too seriously, because 1-loop perturbation theory cannot be trusted for large values of $\lambda_r(\mu)$. In fact, one might argue that eq. (5.63) should be expanded to $\mathcal{O}(\lambda_r^2)$ such that

$$\lambda_r(\mu) = \lambda_r(\mu_0) + \lambda_r(\mu_0)^2 \beta_0 \log(\mu/\mu_0) + \mathcal{O}(\lambda_r^3). \tag{5.64}$$

Remarkably, this point of view is too pessimistic. Let us consider the 2-loop running of the coupling

$$\int_{\lambda_r(\mu_0)}^{\lambda_r(\mu)} \frac{d\lambda_r}{\beta_0 \lambda_r^2 + \beta_1 \lambda_r^3} + \mathcal{O}(\lambda_r^4) = \int_{\mu_0}^{\mu} \frac{d\mu}{\mu} \quad \Rightarrow$$

$$\lambda_r(\mu) = \frac{\lambda_r(\mu_0)}{1 - \lambda_r(\mu_0)\beta_0 \log(\mu/\mu_0) + \lambda_r(\mu_0)(\beta_1/\beta_0) \log[1 - \lambda_r(\mu_0)\beta_0 \log(\mu/\mu_0)]}$$
$$+ \mathcal{O}(\lambda_r^4). \tag{5.65}$$

When we expand this result to $\mathcal{O}(\lambda_r^2)$, we recover eq. (5.64), just as one would expect. However, let us assume that there is a large negative logarithm, *i.e.* despite the fact that $\lambda_r(\mu_0)$ is small, $-\lambda_r(\mu_0)\beta_0 \log(\mu/\mu_0) \gg 1$. In that case,

$$\lambda_r(\mu) \rightarrow \frac{1}{-\beta_0 \log(\mu/\mu_0) + (\beta_1/\beta_0) \log[-\lambda_r(\mu_0)\beta_0 \log(\mu/\mu_0)]}. \tag{5.66}$$

The 1-loop contribution in the denominator (proportional to β_0) is indeed much larger than the 2-loop contribution (proportional to β_1), despite the fact that, at first glance, the convergence of the series seemed doubtful for large values of the logarithm. This indicates that the renormalization group allows us to *sum large logarithms* to all orders of perturbation theory, as it has actually happened in eq. (5.63).

5.11 Infrared and Ultraviolet Fixed Points

The zeros of the β-function correspond to fixed points of the renormalization group. In particular, when $\beta(\lambda_r^*) = 0$, the physics does not change under a variation of the scale μ and thus becomes *scale-invariant* at $\lambda_r = \lambda_r^*$. In 4-d $\lambda\phi^4$ theory, there is a trivial (*i.e.* non-interacting) Gaussian fixed point at $\lambda_r^* = 0$. Since the β-function is positive at $\lambda_r > 0$, the coupling decreases with decreasing energy scale μ. Hence, the fixed point is reached in the infrared limit and is thus known as an *IR fixed point*.

The 2-loop β-function of eq. (5.62) might suggest that 4-d $\lambda\phi^4$ theory has a non-trivial fixed point at $\lambda_r^* = -\beta_0/\beta_1 = 9(4\pi)^2/17$. If that were the case, the β-function would be positive for $\lambda_r < \lambda_r^*$, and the coupling would increase toward λ_r^* with increasing μ. Similarly, for $\lambda_r > \lambda_r^*$, the β-function would be negative and the coupling would decrease toward λ_r^* with increasing μ. Such a fixed point would be UV stable.

Actually, 4-d $\lambda\phi^4$ theory does *not* possess such a non-trivial fixed point. The corresponding β-function is illustrated in the left panel of Figure 5.3. In fact, the perturbative 2-loop β-function cannot be trusted either for large values of the coupling, and the existence of a non-trivial fixed point can only be decided beyond perturbation theory. In the absence of a non-trivial fixed point, 4-d $\lambda\phi^4$ theory is *trivial*; *i.e.* the cut-off Λ can be completely removed only in the free theory. Such a theory cannot be truly fundamental, but it can still be very useful as an effective theory that will eventually break down at some high energy scale that plays the role of the cut-off Λ. As we will discuss in Chapter 12, this is indeed the situation that we encounter in the Higgs sector of the Standard Model. While there was strong evidence for triviality from early investigations of 4-d $\lambda\phi^4$ theory (Aizenman, 1981; Fröhlich, 1982; Lüscher and Weisz, 1987, 1988), it took a long time until triviality was rigorously established by Aizenman and Duminil-Copin (2021).

Remarkably, 3-d $\lambda\phi^4$ theory indeed has a non-trivial, UV-stable fixed point, namely, the one that Wilson and Fisher discovered in the ε-expansion around $d = 4 - \varepsilon$ by bravely inserting $\varepsilon = 1$. The corresponding β-function is illustrated in the right panel of Figure 5.3.

Interestingly, the fixed points that are important for the Standard Model of particle physics are Gaussian fixed points whose vicinity can be explored using perturbation the-

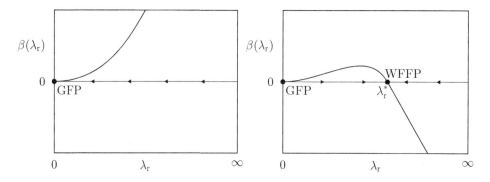

Fig. 5.3 The β-functions and fixed points of $\lambda\phi^4$ theory. The arrows indicate the direction of the renormalization group flow in which the scale μ decreases. Left: 4-d $\lambda\phi^4$ theory is trivial. It only has an IR Gaussian fixed point (GFP). Right: 3-d $\lambda\phi^4$ theory has both a UV Gaussian fixed point and the non-trivial IR Wilson–Fisher fixed point (WFFP).

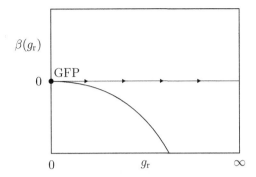

Fig. 5.4 The β-function of 4-d Yang–Mills theory with the running non-Abelian gauge coupling g_r is negative, and the theory is thus asymptotically free. The Gaussian fixed point then has a marginally relevant direction and is a UV fixed point. The arrows indicate the direction of the renormalization group flow for a decreasing scale μ.

ory. The Higgs field in the Standard Model is a 4-component scalar field. The triviality of 4-d $\lambda \phi^4$ theory implies that *one cannot completely remove the cut-off in the Standard Model without switching off its $\lambda \phi^4$ interaction*. Hence, the cut-off should be kept finite, which implies that the Standard Model should be considered as an *effective theory* that will eventually break down. In view of the fact that, in any case, the Planck scale provides a physical cut-off for the Standard Model physics, this is not problematical. There are other interactions mediated by the Higgs field or by an Abelian gauge field, which are also trivial because they are defined near an IR Gaussian fixed point.

Let us also consider an alternative scenario that is realized in non-Abelian gauge theories including Quantum Chromodynamics – the theory that describes the strong interaction between quarks and gluons. The 4-d Yang–Mills theory that describes the self-interaction of non-Abelian gauge fields with a local SU(N) symmetry will be discussed in Chapter 11. As a function of the renormalized gauge coupling g_r, it has the β-function

$$\beta(g_r) = \beta_0 g_r^3 + \beta_1 g_r^5 + \mathcal{O}(g_r^6), \quad \beta_0 = -\frac{11N}{3(4\pi)^2}, \quad \beta_1 = -\frac{34N^2}{3(4\pi)^4}. \tag{5.67}$$

The β-function was computed to 1 loop by Vanyashin and Terentev (1965); Khriplovich (1969); Gross and Wilczek (1973); and Politzer (1973), and to 2 loops by Caswell (1974) and by Jones (1974). In contrast to 4-d $\lambda \phi^4$ theory, as illustrated in Figure 5.4, 4-d Yang–Mills theory has a negative β-function and is thus *asymptotically free*. The correct interpretation of this groundbreaking discovery was first given by David Gross and Frank Wilczek, and independently by David Politzer. It implies that g_r *decreases* with increasing energy scale μ. In this case, the Gaussian fixed point at $g_r^* = 0$ is a UV fixed point. The high-momentum physics is then accessible to perturbation theory, while the low-energy physics is non-perturbative.

The low-energy counterpart of asymptotic freedom (which implies that quarks and gluons behave like free particles at asymptotically high energies) is *confinement*, which permanently binds quarks and gluons inside protons, neutrons, and other hadrons at low energies. The low-energy physics of quarks and gluons is inaccessible to perturbation theory and requires non-perturbative techniques, in particular, the lattice regularization. In contrast to trivial theories, in asymptotically free theories the cut-off can be completely removed to infinity while maintaining strong interactions at low energies. Such theories could potentially be truly fundamental and not "just" effective.

Exercises

5.1   Renormalization group blocking step
Derive eq. (5.19), with $\Delta'(p')$ given in eq. (5.18), by computing the Gaussian integrals, without inserting solutions to the classical equations of motion.

5.2   Locality of the perfect action for $d = 1$
Verify eq. (5.36) for the $d = 1$ perfect propagator. Convince yourself that the choice of α in eq. (5.37) optimizes the locality of the perfect action.

6 Quantization of the Free Electromagnetic Field

In order to understand photons, in this chapter we first consider the canonical quantization of the free, electromagnetic field. In particular, we investigate the energy, momentum, and angular momentum of photons. We also derive Planck's formula for black-body radiation that is essential for the cosmic background radiation of photons, which is a relic of the Big Bang. Finally, we discuss the functional integral quantization of electrodynamics in Euclidean space–time. A summary of the relativistic formulation of classical electrodynamics is provided in Appendix D.

The concept of photons as quanta of light was first used by Max Planck in 1900 in his derivation of the energy distribution of black-body radiation. It was taken even more seriously by Albert Einstein in 1905 when he explained the photoelectric effect. Arthur Compton's investigations of the collisions of photons with electrons may be viewed as the experimental discovery of the photon as a "particle".

Indeed, the photon is the only "particle" that was always associated with the quantization of fields. For example, electrons were first described as point particles in the framework of quantum mechanics before they were identified (together with their positron anti-particles) as the quanta of Paul Dirac's fermion field. In contrast, photons always require a relativistic description, because they are massless and thus move with the velocity of light. Since relativistic point particle quantum mechanics is limited to free particles, it is not surprising that photons cannot be treated as point particles. Like any other "particle" in quantum field theory, they are actually *wavicles* (in the terminology of Chapter 3) with the corresponding localization properties.

6.1 Vector Potential and Gauge Symmetry

We are familiar with the *Maxwell equations* that describe the classical evolution of the electric and magnetic fields, $\vec{E}(\vec{x}, t)$ and $\vec{B}(\vec{x}, t)$, in space and time. The homogeneous Maxwell equations,[1]

$$\vec{\nabla} \times \vec{E}(\vec{x}, t) + \frac{1}{c} \partial_t \vec{B}(\vec{x}, t) = \vec{0}, \quad \vec{\nabla} \cdot \vec{B}(\vec{x}) = 0, \tag{6.1}$$

are automatically satisfied when we introduce the scalar potential $\phi(\vec{x}, t)$ and the vector potential $\vec{A}(\vec{x}, t)$, and write

---

[1] In this section (but not later), we do not set $c = 1$, just to establish direct contact with typical textbook notation as well as with Appendix D.

$$\vec{E}(\vec{x},t) = -\vec{\nabla}\phi(\vec{x},t) - \frac{1}{c}\partial_t\vec{A}(\vec{x},t), \quad \vec{B}(\vec{x},t) = \vec{\nabla} \times \vec{A}(\vec{x},t). \tag{6.2}$$

It is important that the electromagnetic field strengths $\vec{E}(\vec{x},t)$ and $\vec{B}(\vec{x},t)$ are uniquely defined by the scalar and vector potentials, but not vice versa. Instead, there is a redundancy in the potentials, which manifests itself as a *gauge ambiguity*. When we perform a gauge transformation

$$\phi'(\vec{x},t) = \phi(\vec{x},t) - \frac{1}{c}\partial_t\alpha(\vec{x},t), \quad \vec{A}'(\vec{x},t) = \vec{A}(\vec{x},t) + \vec{\nabla}\alpha(\vec{x},t), \tag{6.3}$$

the field strengths remain invariant

$$\begin{aligned}
\vec{E}'(\vec{x},t) &= -\vec{\nabla}\phi'(\vec{x},t) - \frac{1}{c}\partial_t\vec{A}'(\vec{x},t) \\
&= -\vec{\nabla}\phi(\vec{x},t) - \frac{1}{c}\vec{\nabla}\partial_t\alpha(\vec{x},t) - \frac{1}{c}\partial_t\vec{A}(\vec{x},t) + \frac{1}{c}\partial_t\vec{\nabla}\alpha(\vec{x},t) \\
&= \vec{E}(\vec{x},t), \\
\vec{B}'(\vec{x},t) &= \vec{\nabla} \times \vec{A}'(\vec{x},t) = \vec{\nabla} \times \vec{A}(\vec{x},t) + \vec{\nabla} \times \vec{\nabla}\alpha(\vec{x},t) \\
&= \vec{B}(\vec{x},t).
\end{aligned} \tag{6.4}$$

The gauge transformation function $\alpha(\vec{x},t) \in \mathbb{R}$ is an arbitrary, differentiable function of space–time. Hence, unlike global symmetries, gauge symmetries are *local* and amount to a very large set of symmetry transformations. However, it is important to point out that, in contrast to global symmetries, gauge symmetries are not actual symmetries which manifest themselves in the physical spectrum of a theory. They just reflect redundancies in the theoretical description of the physics. The field strengths $\vec{E}(\vec{x},t)$ and $\vec{B}(\vec{x},t)$, but not the scalar and vector potentials $\phi(\vec{x},t)$ and $\vec{A}(\vec{x},t)$, are gauge-invariant and physically meaningful.[2] In particular, the gauge copies $\phi'(\vec{x},t)$ and $\vec{A}'(\vec{x},t)$ give rise to the same physical results. In order to guarantee that the gauge redundancy has no effect on the physics, gauge invariance must be maintained exactly.

The concept of gauge symmetry – in the sense of scale transformations – was introduced by Hermann Weyl (Weyl, 1918). Our modern understanding (to be described in this chapter) deals with space–time-dependent phase factors, which were introduced by Vladimir Fock (he denoted such a factor as a "fifth coordinate", Fock (1926)).

We now turn to the relativistic formulation of electrodynamics by using

$$x^0 = ct, \quad x^\mu = (x^0, \vec{x}), \quad \partial_\mu = \left(\frac{1}{c}\partial_t, \vec{\nabla}\right), \quad A^\mu(x) = (\phi(\vec{x},t), \vec{A}(\vec{x},t)). \tag{6.5}$$

The electric and magnetic fields then define the field strength tensor

---

[2] There are gauge-invariant expressions, in particular Wegner-Wilson loops (to be addressed in Chapter 14), which are also physically relevant, for instance, in the Aharonov–Bohm effect (Aharonov and Bohm, 1959, 1961; Berry, 1980; Peshkin and Tonomura, 1989).

$$F^{\mu\nu}(x) = \partial^\mu A^\nu(x) - \partial^\nu A^\mu(x) = \begin{pmatrix} 0 & -E_x(\vec{x},t) & -E_y(\vec{x},t) & -E_z(\vec{x},t) \\ E_x(\vec{x},t) & 0 & -B_z(\vec{x},t) & B_y(\vec{x},t) \\ E_y(\vec{x},t) & B_z(\vec{x},t) & 0 & -B_x(\vec{x},t) \\ E_z(\vec{x},t) & -B_y(\vec{x},t) & B_x(\vec{x},t) & 0 \end{pmatrix}.$$

$$(6.6)$$

which is invariant against gauge transformations,

$$A'^{\mu}(x) = A^\mu(x) - \partial^\mu \alpha(x), \quad F'^{\mu\nu}(x) = F^{\mu\nu}(x). \tag{6.7}$$

6.2 From the Lagrangian to the Hamilton Density

Let us consider the Lagrange density, or Lagrangian, of the free electromagnetic field,

$$\mathcal{L}(\partial^\mu A^\nu) = -\frac{1}{4}F_{\mu\nu}(x)F^{\mu\nu}(x) = \frac{1}{2}\left(\vec{E}(x)^2 - \vec{B}(x)^2\right). \tag{6.8}$$

In order to isolate the physical, gauge-invariant degrees of freedom of the gauge field, we now fix to the so-called *temporal gauge* by choosing

$$A^0(x) = \phi(\vec{x},t) = 0. \tag{6.9}$$

This leaves a remnant invariance against time-independent – but still space-dependent – gauge transformations. In particular, after a time-independent gauge transformation with $\alpha(x) - \alpha(\vec{x})$, we still have

$$A'^0(x) = A^0(x) - \partial^0 \alpha(x) = A^0(x) - \partial^0 \alpha(\vec{x}) = A^0(x) = 0. \tag{6.10}$$

Then, the electric field reduces to $\vec{E}(x) = -\partial^0 \vec{A}(x)$, and the momentum, which is canonically conjugate to the vector potential $A^i(x)$, is just the electric field (up to a minus sign),

$$\Pi_i(x) = \frac{\delta\mathcal{L}}{\delta\partial^0 A^i(x)} = \partial^0 A^i(x) = -E_i(x). \tag{6.11}$$

It should be pointed out that $A_i(x) = -A^i(x)$, where the ordinary 3-vector potential is given by

$$\vec{A}(\vec{x},t) = (A_x(\vec{x},t), A_y(\vec{x},t), A_z(\vec{x},t)) = (A^1(x), A^2(x), A^3(x)) = -(A_1(x), A_2(x), A_3(x)). \tag{6.12}$$

In order to avoid confusion about minus signs, we are going to use upper indices $A^i(x)$ for the 3-vector potential. Similarly, we write the nabla operator as

$$\vec{\nabla} = (\partial_x, \partial_y, \partial_z) = (\partial_1, \partial_2, \partial_3) = -(\partial^1, \partial^2, \partial^3), \tag{6.13}$$

which is the reason why we will use lower indices ∂_i for spatial derivatives. The objects $\Pi_i(x) = -E_i(x)$ and $B_i(x)$ are not components of 4-vectors and will always be written with lower indices. In the following, repeated spatial indices will always be summed (following Einstein's summation convention), but they are not necessarily pairs of one upper and one lower index. In particular, in what follows we just encounter the standard Euclidean spatial metric.

Canonical quantization singles out the temporal direction and is thus not manifestly covariant. The classical Hamilton density is given by

$$\mathcal{H}(A^i, \Pi_i) = \Pi_i(x)\partial^0 A^i(x) - \mathcal{L} = \frac{1}{2}\Big(\Pi_i(x)\Pi_i(x) + B_i(x)B_i(x)\Big)$$
$$= \frac{1}{2}\Big(E_i(x)E_i(x) + B_i(x)B_i(x)\Big). \tag{6.14}$$

The classical Hamilton function is the spatial integral of the Hamilton density,

$$H[A^i, \Pi_i] = \int d^3x\, \mathcal{H} = \int d^3x\, \frac{1}{2}\Big[\Pi_i(\vec{x})\Pi_i(\vec{x}) + \epsilon_{ijk}\partial_j A^k(\vec{x})\epsilon_{ilm}\partial_l A^m(\vec{x})\Big]. \tag{6.15}$$

The Hamilton function is a functional of the classical field configuration $A^i(\vec{x})$ and its canonically conjugate momentum field $\Pi_i(\vec{x}) = -E_i(\vec{x})$. Upon quantization, the Hamilton function will turn into the Hamilton operator of the corresponding quantum field theory.

In analogy to the canonical quantization of scalar fields that we discussed in Chapter 2, we now postulate commutation relations between the gauge field operators $\hat{A}^i(x)$ and their canonically conjugate momentum operators $\hat{\Pi}_i(x) = -\hat{E}_i(x)$,

$$[\hat{A}^i(\vec{x}), \hat{\Pi}_j(\vec{y})] = i\delta_{ij}\,\delta(\vec{x} - \vec{y}),$$
$$[\hat{A}^i(\vec{x}), \hat{A}^j(\vec{y})] = [\hat{\Pi}_i(\vec{x}), \hat{\Pi}_j(\vec{y})] = 0. \tag{6.16}$$

Again, these commutation relations are *local*; *i.e.* field operators at different points in space commute with each other. The field operator $\hat{\Pi}_i(\vec{x})$ takes the form

$$\hat{\Pi}_i(\vec{x}) = -i\frac{\delta}{\delta A^i(\vec{x})}. \tag{6.17}$$

6.3 Hamilton Operator for the Photon Field

We now turn the classical Hamilton function H of eq. (6.15) into the *Hamilton operator*

$$\hat{H} = \int d^3x\, \frac{1}{2}\Big[\hat{\Pi}_i(\vec{x})\hat{\Pi}_i(\vec{x}) + \epsilon_{ijk}\partial_j \hat{A}^k(\vec{x})\epsilon_{ilm}\partial_l \hat{A}^m(\vec{x})\Big]. \tag{6.18}$$

As usual, in order to diagonalize the Hamilton operator it is convenient to move to momentum space. Hence, we introduce Fourier transforms of the operator-valued fields

$$\hat{A}^i(\vec{p}) = \int d^3x\, \hat{A}^i(\vec{x}) \exp(-i\vec{p} \cdot \vec{x}), \quad \hat{\Pi}_i(\vec{p}) = \int d^3x\, \hat{\Pi}_i(\vec{x}) \exp(-i\vec{p} \cdot \vec{x}), \quad (6.19)$$

which obey

$$\hat{A}^i(\vec{p})^\dagger = \hat{A}^i(-\vec{p}), \quad \hat{\Pi}_i(\vec{p})^\dagger = \hat{\Pi}_i(-\vec{p}). \quad (6.20)$$

The corresponding commutation relations take the form

$$[\hat{A}^i(\vec{p}), \hat{\Pi}_j(\vec{q})] = i(2\pi)^3 \delta_{ij}\, \delta(\vec{p} + \vec{q}), \quad [\hat{A}^i(\vec{p}), \hat{A}^j(\vec{q})] = [\hat{\Pi}_i(\vec{p}), \hat{\Pi}_j(\vec{q})] = 0, \quad (6.21)$$

and the Hamilton operator can be written as

$$\hat{H} = \frac{1}{(2\pi)^3} \int d^3p\, \frac{1}{2} \left[\hat{\Pi}_i(\vec{p})^\dagger \hat{\Pi}_i(\vec{p}) + \epsilon_{ijk} p_j \hat{A}^k(\vec{p})^\dagger \epsilon_{ilm} p_l \hat{A}^m(\vec{p}) \right]. \quad (6.22)$$

Thus, the forms (6.18) and (6.22) of the Hamilton operator are consistently related according to Parseval's theorem.

6.4 Gauss Law

At the classical level, the *Gauss law* for the electric field in the absence of electric charges, $\vec{\nabla} \cdot \vec{E}(\vec{x}, t) = 0$, is one of the Maxwell equations. At the quantum level, on the other hand, it is not satisfied as an operator identity. Instead, it provides a constraint on the physically admissible quantum states $|\Psi\rangle$ of the free electromagnetic field,

$$\vec{\nabla} \cdot \hat{\vec{E}}(\vec{x}) |\Psi\rangle = 0 \quad \Rightarrow \quad p_i \hat{\Pi}_i(\vec{p}) |\Psi\rangle = 0. \quad (6.23)$$

It is sufficient to impose the Gauss law at some initial instant of time, because, as we shall see, this constraint commutes with the Hamilton operator. Let us rewrite the magnetic term in the Hamilton operator as

$$\epsilon_{ijk} p_j \hat{A}^k(\vec{p})^\dagger \epsilon_{ilm} p_l \hat{A}^m(\vec{p}) = \hat{A}^i(\vec{p})^\dagger \mathcal{M}(\vec{p})_{ij} \hat{A}^j(\vec{p}) =$$

$$(\hat{A}^1(\vec{p})^\dagger, \hat{A}^2(\vec{p})^\dagger, \hat{A}^3(\vec{p})^\dagger) \begin{pmatrix} \vec{p}^2 - p_1^2 & -p_1 p_2 & -p_1 p_3 \\ -p_2 p_1 & \vec{p}^2 - p_2^2 & -p_2 p_3 \\ -p_3 p_1 & -p_3 p_2 & \vec{p}^2 - p_3^2 \end{pmatrix} \begin{pmatrix} \hat{A}^1(\vec{p}) \\ \hat{A}^2(\vec{p}) \\ \hat{A}^3(\vec{p}) \end{pmatrix}. \quad (6.24)$$

The symmetric matrix $\mathcal{M}(\vec{p})$ has the form

$$\mathcal{M}(\vec{p})_{ij} = \vec{p}^2 \left(\delta_{ij} - e_{p_i} e_{p_j} \right), \quad \vec{e}_p = \frac{\vec{p}}{|\vec{p}|}. \quad (6.25)$$

The unit-vector $\vec{e}_p$ points in the direction of the momentum and is an eigenvector of $\mathcal{M}(\vec{p})$ with zero eigenvalue. The other two unit-eigenvectors, $\vec{e}_1, \vec{e}_2 \in \mathbb{R}^3$, have the same eigenvalue $\vec{p}^2$ and obey the orthogonality relations

$$\vec{e}_1 \cdot \vec{e}_p = \vec{e}_2 \cdot \vec{e}_p = 0, \quad \vec{e}_1 \cdot \vec{e}_2 = 0, \quad \vec{e}_1 \times \vec{e}_2 = \vec{e}_p. \quad (6.26)$$

Quick Question 6.4.1   Spectrum of the matrix $\mathcal{M}(\vec{p})$
Solve the eigenvalue problem of the matrix $\mathcal{M}(\vec{p})$.

From classical electrodynamics, we know that electromagnetic waves are transversely polarized; *i.e.* the electric and magnetic fields are perpendicular to the propagation direction. We will see that the directions $\vec{e}_1$ and $\vec{e}_2$, which are perpendicular to $\vec{p}$, correspond to two transverse polarization directions of a photon with momentum $\vec{p}$.

The projection of the angular momentum vector onto the direction of the momentum is simultaneously measurable with the momentum itself; see Appendix E. For a massless particle, like a photon, which always moves with the velocity of light, this projection is the same for all observers. As we will see, this gives rise to two spin projections, $+$ and $-$, which correspond to two circular polarizations of the corresponding classical electromagnetic waves. The corresponding unit-vectors obey

$$\vec{e}_\pm = \frac{1}{\sqrt{2}}(\vec{e}_1 \pm i\vec{e}_2), \quad \vec{e}_\pm^* \cdot \vec{e}_\pm = 1, \quad \vec{e}_\pm \cdot \vec{e}_p = 0, \quad \vec{e}_-^* \cdot \vec{e}_+ = 0, \quad \vec{e}_- \times \vec{e}_+ = i\vec{e}_p. \quad (6.27)$$

We now introduce the unitary transformation

$$U(\vec{p}) = \begin{pmatrix} e_{+1} & e_{+2} & e_{+3} \\ e_{p1} & e_{p2} & e_{p3} \\ e_{-1} & e_{-2} & e_{-3} \end{pmatrix}, \quad U(\vec{p})\mathcal{M}(\vec{p})U(\vec{p})^\dagger = \vec{p}^{\,2} \begin{pmatrix} 1 & 0 & 0 \\ 0 & 0 & 0 \\ 0 & 0 & 1 \end{pmatrix}, \quad (6.28)$$

which diagonalizes the matrix $\mathcal{M}(\vec{p})$ and is associated with the basis change

$$\begin{pmatrix} \hat{A}_+(\vec{p}) \\ \hat{A}_p(\vec{p}) \\ \hat{A}_-(\vec{p}) \end{pmatrix} = U(\vec{p}) \begin{pmatrix} \hat{A}^1(\vec{p}) \\ \hat{A}^2(\vec{p}) \\ \hat{A}^3(\vec{p}) \end{pmatrix}, \quad \begin{pmatrix} \hat{\Pi}_+(\vec{p}) \\ \hat{\Pi}_p(\vec{p}) \\ \hat{\Pi}_-(\vec{p}) \end{pmatrix} = U(\vec{p}) \begin{pmatrix} \hat{\Pi}_1(\vec{p}) \\ \hat{\Pi}_2(\vec{p}) \\ \hat{\Pi}_3(\vec{p}) \end{pmatrix}. \quad (6.29)$$

It should be noted that $\hat{A}_\pm(\vec{p})$ describes *transverse photons,* while $\hat{A}_p(\vec{p})$ is associated with unphysical, longitudinally polarized states. After the unitary basis change, the operators still obey canonical commutation relations, and the Hamilton operator takes the form

$$\hat{H} = \frac{1}{(2\pi)^3} \int d^3p \, \frac{1}{2} \Big[\hat{\Pi}_+(\vec{p})^\dagger \hat{\Pi}_+(\vec{p}) + \vec{p}^{\,2}\hat{A}_+(\vec{p})^\dagger \hat{A}_+(\vec{p})$$
$$+ \hat{\Pi}_-(\vec{p})^\dagger \hat{\Pi}_-(\vec{p}) + \vec{p}^{\,2}\hat{A}_-(\vec{p})^\dagger \hat{A}_-(\vec{p}) + \hat{\Pi}_p(\vec{p})^\dagger \hat{\Pi}_p(\vec{p}) \Big]. \quad (6.30)$$

The operator $\hat{\Pi}_p(\vec{p})$ is the one that imposes the Gauss law, $\vec{p} \cdot \hat{\vec{E}}(\vec{p})|\Psi\rangle = 0$, because

$$\hat{\Pi}_p(\vec{p}) = U(\vec{p})_{pi}\hat{\Pi}_i(\vec{p}) = -e_{pi}\hat{E}_i(\vec{p}) = -\frac{\vec{p}}{|\vec{p}\,|} \cdot \hat{\vec{E}}(\vec{p}). \quad (6.31)$$

Since $\hat{H}$ does not depend on $\hat{A}_p(\vec{p})$, the Gauss law constraint, $\hat{\Pi}_p(\vec{p})|\Psi\rangle = 0$, indeed commutes with the Hamilton operator. This constraint eliminates longitudinally polarized states from the physical Hilbert space. While they are mathematical eigenstates of $\hat{H}$, they do not belong to the physical spectrum because they are not gauge-invariant.

In complete analogy to the canonical quantization of a scalar field, using $\omega = |\vec{p}\,|$, we now introduce *photon creation and annihilation operators*, $\hat{a}_{\pm}(\vec{p})^{\dagger}$ and $\hat{a}_{\pm}(\vec{p})$,

$$\hat{a}_{\pm}(\vec{p}) = \frac{1}{\sqrt{2}}\left[\sqrt{\omega}\,\hat{A}_{\pm}(\vec{p}) + \frac{i}{\sqrt{\omega}}\,\hat{\Pi}_{\pm}(\vec{p})\right],$$

$$\hat{a}_{\pm}(\vec{p})^{\dagger} = \frac{1}{\sqrt{2}}\left[\sqrt{\omega}\,\hat{A}_{\pm}(-\vec{p}) - \frac{i}{\sqrt{\omega}}\,\hat{\Pi}_{\pm}(-\vec{p})\right], \tag{6.32}$$

which obey the commutation relations

$$[\hat{a}_{\pm}(\vec{p}), \hat{a}_{\pm}(\vec{q})^{\dagger}] = \frac{i}{2}[\hat{\Pi}_{\pm}(\vec{p}), \hat{A}_{\pm}(-\vec{q})] - \frac{i}{2}[\hat{A}_{\pm}(\vec{p}), \hat{\Pi}_{\pm}(-\vec{q})] = (2\pi)^{3}\delta(\vec{p} - \vec{q}),$$

$$[\hat{a}_{+}(\vec{p}), \hat{a}_{-}(\vec{q})^{\dagger}] = [\hat{a}_{-}(\vec{p}), \hat{a}_{+}(\vec{q})^{\dagger}] = 0,$$

$$[\hat{a}_{\pm}(\vec{p}), \hat{a}_{\pm}(\vec{q})] = [\hat{a}_{\pm}(\vec{p}), \hat{a}_{\mp}(\vec{q})] = [\hat{a}_{\pm}(\vec{p})^{\dagger}, \hat{a}_{\pm}(\vec{q})^{\dagger}] = [\hat{a}_{\pm}(\vec{p})^{\dagger}, \hat{a}_{\mp}(\vec{q})^{\dagger}] = 0. \tag{6.33}$$

In terms of these operators, in the physical Hilbert space the Hamilton operator takes the form

$$\hat{H} = \frac{1}{(2\pi)^{3}} \int d^{3}p\,|\vec{p}\,|\left(\hat{a}_{+}(\vec{p})^{\dagger}\hat{a}_{+}(\vec{p}) + \hat{a}_{-}(\vec{p})^{\dagger}\hat{a}_{-}(\vec{p}) + V\right), \tag{6.34}$$

where V represents the (large) spatial volume. (As in Chapter 2, we treat V as a finite, additive constant in $\hat{H}$, although we are dealing with continuous momenta, rather than discrete momenta that arise in a finite periodic volume.)

6.5 Vacuum and Photon States

Again in complete analogy to the case of a scalar field, the vacuum state $|0\rangle$ of the electromagnetic field is determined by

$$\hat{a}_{\pm}(\vec{p})|0\rangle = 0, \tag{6.35}$$

for all momenta $\vec{p}$ and for both polarizations $\pm$. The vacuum is the eigenstate of the Hamilton operator with the lowest energy. From the form (6.34), we read off its ground state energy

$$E_0 = \frac{V}{(2\pi)^{3}} \int d^{3}p\,|\vec{p}\,|. \tag{6.36}$$

The corresponding vacuum energy density takes the form

$$\rho = \frac{E_0}{V} = \frac{1}{(2\pi)^{3}} \int d^{3}p\,|\vec{p}\,| = \frac{1}{2\pi^{2}} \int_{0}^{\infty} dp\,p^{3}, \tag{6.37}$$

which is ultraviolet-divergent. When we apply a momentum cut-off $p \leq \Lambda$, the energy density diverges as $\rho \sim \Lambda^{4}$, as for the scalar field.

The photon energies are *differences* between the energy of an excited state and the vacuum. In these energy differences, the divergent constant E_0 drops out. The single-photon states are given by

$$|\vec{p}, \pm\rangle = \hat{a}_{+}(\vec{p})^{\dagger}|0\rangle, \tag{6.38}$$

with an energy

$$E(\vec{p}) - E_0 = \omega = |\vec{p}|. \tag{6.39}$$

This is the energy of a free particle with vanishing rest mass and momentum $\vec{p}$. Multi-photon states – and thus the entire Fock space – are obtained by acting with more than one creation operator on the vacuum state, along the lines of Section 2.8. For instance, the 2-photon states are obtained as

$$|\vec{p}_1, \pm; \vec{p}_2, \pm\rangle = \hat{a}_\pm(\vec{p}_1)^\dagger \hat{a}_\pm(\vec{p}_2)^\dagger |0\rangle. \tag{6.40}$$

The commutation relation $[\hat{a}_\pm(\vec{p}_1)^\dagger, \hat{a}_\pm(\vec{p}_2)^\dagger] = 0$ implies

$$|\vec{p}_2, \pm; \vec{p}_1, \pm\rangle = |\vec{p}_1, \pm; \vec{p}_2, \pm\rangle, \tag{6.41}$$

which confirms that photons are bosons.

6.6 Momentum Operator of the Electromagnetic Field

Let us also construct the *momentum* operator of the photon field. As discussed in Appendix D, the energy–momentum tensor of the classical electromagnetic field takes the form

$$T_{\mu\nu}(x) = -F_\mu{}^\rho(x)F_{\nu\rho}(x) - g_{\mu\nu}\mathcal{L}. \tag{6.42}$$

This is consistent with the Hamilton density of eq. (6.14),

$$\begin{aligned}
\mathcal{H} = T_{00}(x) &= -F_0{}^i(x)F_{0i}(x) - g_{00}\mathcal{L} = E_i(x)E_i(x) - \mathcal{L} \\
&= \frac{1}{2}\Big(E_i(x)E_i(x) + B_i(x)B_i(x)\Big).
\end{aligned} \tag{6.43}$$

The classical momentum density is given by

$$\mathcal{P}_i(x) = T_{0i}(x) = -F_0{}^\rho(x)F_{i\rho}(x) - g_{0i}\mathcal{L} = -F_0{}^j(x)F_{ij}(x) = \epsilon_{ijk}E_j(x)B_k(x), \tag{6.44}$$

which is the well-known *Poynting vector.* Accordingly, the Hermitian momentum operator of the photon field theory is given by

$$\begin{aligned}
\hat{P}_i &= \int d^3x\, \epsilon_{ijk}\frac{1}{2}\Big(\hat{E}_j(\vec{x})\hat{B}_k(\vec{x}) + \hat{B}_k(\vec{x})\hat{E}_j(\vec{x})\Big) \\
&= \frac{1}{(2\pi)^3}\int d^3p\, p_i \Big(\hat{a}_+(\vec{p})^\dagger \hat{a}_+(\vec{p}) + \hat{a}_-(\vec{p})^\dagger \hat{a}_-(\vec{p})\Big).
\end{aligned} \tag{6.45}$$

This identity shall be derived in Problem 6.4.

Quick Question 6.6.1   Commutator of photon field operators
Using identity (6.45), show that

$$[\hat{P}_i, \hat{a}_\pm(\vec{p})^\dagger] = p_i\, \hat{a}_\pm(\vec{p})^\dagger. \tag{6.46}$$

As expected, the vacuum $|0\rangle$ is an eigenstate of the momentum operator with eigenvalue $\vec{0}$. The single-photon states are eigenstates as well,

$$\hat{P}_i |\vec{p}, \pm\rangle = p_i |\vec{p}, \pm\rangle, \tag{6.47}$$

which confirms that $\vec{p}$ is indeed the momentum of the photon.

6.7 Angular Momentum Operator and Helicity of Photons

The *angular momentum* of the classical electromagnetic field is given by

$$\vec{J} = \int d^3x \, \vec{x} \times \left(\vec{E}(\vec{x}) \times \vec{B}(\vec{x}) \right). \tag{6.48}$$

Correspondingly, the Hermitian angular momentum operator of the quantum field theory takes the form

$$\hat{\vec{J}} = \int d^3x \, \vec{x} \times \frac{1}{2} \left(\hat{\vec{E}}(\vec{x}) \times \hat{\vec{B}}(\vec{x}) - \hat{\vec{B}}(\vec{x}) \times \hat{\vec{E}}(\vec{x}) \right). \tag{6.49}$$

It is a straightforward but somewhat tedious exercise to show that the components of $\hat{\vec{P}}$ and $\hat{\vec{J}}$ obey the standard commutation relations

$$[\hat{P}_i, \hat{P}_j] = 0, \quad [\hat{P}_i, \hat{J}_j] = \mathrm{i}\epsilon_{ijk}\hat{P}_k, \quad [\hat{J}_i, \hat{J}_j] = \mathrm{i}\epsilon_{ijk}\hat{J}_k. \tag{6.50}$$

These relations belong to the Poincaré algebra that is discussed in Appendix E. While all components of the momentum are simultaneously measurable, the components of the angular momentum are not. The projection of the angular momentum onto the momentum, $\hat{\vec{J}} \cdot \hat{\vec{P}}$, plays a special role, because it is the only component of the angular momentum that can be simultaneously measured with the momentum

$$[\hat{P}_i, \hat{P}_j\hat{J}_j] = \mathrm{i}\epsilon_{ijk}\hat{P}_j\hat{P}_k = 0. \tag{6.51}$$

It is well known that the angular momentum depends on the choice of the origin of the coordinate system. In particular, when one shifts the coordinates from $\vec{x}$ to $\vec{x}' = \vec{x} + \vec{d}$, the angular momentum changes from $\hat{\vec{J}}$ to $\hat{\vec{J}}' = \hat{\vec{J}} + \vec{d} \times \hat{\vec{P}}$. This is true for the electromagnetic field as well as for a quantum mechanical point particle, but the *projection* $\hat{\vec{J}}' \cdot \hat{\vec{P}} = \hat{\vec{J}} \cdot \hat{\vec{P}}$ is independent of such a shift.

The projection of the angular momentum onto the direction $\vec{e}_p$ of the momentum of a photon, $\hat{\vec{J}} \cdot \vec{e}_p$, is known as the photon's *helicity*. Since photons are massless and hence move with the velocity of light, their helicity is the same in all reference frames. The helicity of massive particles, on the other hand, depends on the motion of the observer. In particular, when moving in the same direction but faster than a massive particle, an observer perceives the particle with opposite sign of the momentum and hence with helicity opposite to an observer at rest.

It is another straightforward but somewhat tedious exercise to show that

$$[\hat{\vec{J}} \cdot \vec{e}_p, \hat{a}_\pm(\vec{p})^\dagger] = \pm\hat{a}_\pm(\vec{p})^\dagger. \tag{6.52}$$

Using $\hat{\vec{J}} \cdot \vec{e}_p \, |0\rangle = 0$ (which trivially follows from $\hat{\vec{J}} \, |0\rangle = \vec{0}$),[3] this implies

$$\hat{\vec{J}} \cdot \vec{e}_p |\vec{p}, \pm\rangle = \hat{\vec{J}} \cdot \vec{e}_p \, \hat{a}_\pm(\vec{p})^\dagger |0\rangle = [\hat{\vec{J}} \cdot \vec{e}_p, \hat{a}_\pm(\vec{p})^\dagger]|0\rangle = \pm \hat{a}_\pm(\vec{p})^\dagger |0\rangle = \pm |\vec{p}, \pm\rangle. \quad (6.53)$$

Hence, photons have two possible helicities, $\pm$, associated with the complex unit-vectors $\vec{e}_\pm = (\vec{e}_1 \pm i\vec{e}_2)/\sqrt{2}$, which were introduced in eq. (6.27). This is the quantum analogue of left and right circular polarization of classical electromagnetic waves, which are formed by a superposition of two linearly polarized electromagnetic waves with orthogonal polarization directions $\vec{e}_1$ and $\vec{e}_2$ and a 90° phase shift represented by $\pm i$.

Just like classical electromagnetic waves, *photons are transversely polarized;* usually, there are no longitudinal photons.[4] Indeed, the corresponding states are exiled from the physical Hilbert space by the Gauss law constraint. Photons are the quanta of the 4-vector field $A^\mu(x)$. Hence, one might naively expect four resulting degrees of freedom (per space point). However, due to gauge redundancies, the number of physical degrees of freedom is reduced to two. First, $A^0(x)$ was removed by fixing to the temporal gauge. The remaining redundancy related to time-independent gauge transformations is taken care of by imposing the Gauss law constraint $\partial_i \hat{E}_i(\vec{x})|\Psi\rangle = 0$ on the physical states $|\Psi\rangle$.

Photons are vector particles with spin 1. However, as we just discussed, their helicity is limited to ± 1, while the state with helicity 0 is unphysical. In the Standard Model, photons indeed arise as massless particles with just two transverse polarizations. As we will discuss in Chapter 13, other gauge fields are affected by the Higgs mechanism in which a scalar Higgs field endows the gauge bosons W and Z with mass. This implies that they acquire a third longitudinal polarization degree of freedom which is provided by the scalar Higgs field.

6.8 Planck's Formula and the Cosmic Background Radiation

As an application of the theoretical description of the photon field, we will now derive Max Planck's formula for black-body radiation. This provides the basis for our understanding of the *cosmic background radiation of photons* that was created by the pair annihilation of matter and anti-matter particles about 1 second after the Big Bang. The Big Bang itself is usually associated with a singularity of the cosmic evolution that results from Einstein's classical theory of General Relativity, which predicts an expansion of space whose speed is determined by the energy content of the Universe, including the vacuum energy. From a variety of observational data, we conclude that the Big Bang happened about 1.3×10^{10} years ago. It should not be confused with an "explosion" that originated from a point-like center in space. Instead, it happened "everywhere" in the Universe at the same time (for an observer in the cosmic rest frame, in which the cosmic background radiation is isotropic). The standard picture associates it with a singularity in time with an infinite energy density.

---

[3] We use the property $\hat{\vec{J}} \, |0\rangle = \vec{0}$ as an input, since it is physically natural. However, a mathematical demonstration of this vacuum property is quite non-trivial.

[4] As we will discuss in Section 13.1, inside a superconductor photons pick up a mass via the Higgs mechanism. The "Higgs" field, which in this case describes Cooper pairs of electrons, then provides a longitudinal polarization degree of freedom for the photon.

The evolution of the Universe immediately after the Big Bang is a matter of speculation, because we have no well-established understanding of physics at arbitrarily high energy scales.

Remarkably, the Standard Model provides a sound understanding at energies at least up to a few TeV (1000 GeV), which corresponds to times as early as about 10^{-14} seconds after the Big Bang. At that time, the Universe was filled with an ultra-relativistic plasma of electrically charged as well as neutral particles. Until about 1 second after the Big Bang, almost all of the charged particles, for example, protons and electrons, annihilated with their oppositely charged anti-particles, in that case anti-protons and positrons, thus generating an enormous number of photons. At around the same time, the weakly interacting, electrically neutral neutrinos decoupled. They form a *cosmic neutrino background radiation* whose detection is beyond our current technical capabilities (it is extremely difficult to detect neutrinos with such low kinetic energies, of the order of 10^{-4} eV).[5]

The cosmic background radiation of photons was first detected (accidentally) in 1964 by Arno Penzias and Robert Wilson, and was soon identified as solid observational evidence for the hot beginning of the Universe. After their creation about 1 second after the Big Bang, the cosmic photons were still interacting with the tiny fraction of protons and electrons that survived annihilation. Together, they constitute practically all the charged matter that is still present in the Universe today. About 380 000 years after the Big Bang, when the Universe had expanded and cooled down to temperatures in the few eV range, protons and electrons formed neutral hydrogen atoms and the Universe became *transparent*. Since then, the cosmic photons stream freely through the Universe. They reach us from all directions and carry very valuable information about the beginning of the Universe.

At the time when the photons were generated, no atoms and thus no preferred spectral lines existed. Instead, the photons were created in thermal equilibrium with charged matter and anti-matter; thus, they follow a *black-body spectrum*. While traveling through the expanding Universe, the wavelengths of the photons were red-shifted, thus leading to the microwave background radiation that we observe today, with a mean wavelength of 1.9 mm. The present temperature of the cosmic photon background radiation is 2.725 K (while the neutrino background is even colder, around 1.94 K, if one assumes massless neutrinos). It is amazing that the temperature of the photons is to very high precision the same, no matter what corner of the Universe they come from. This was first explained by Alan Guth employing his concept of the *inflationary Universe* (Guth, 1981).

Let us now derive Planck's formula for black-body radiation, which describes the intensity as a function of frequency. For simplicity, we replace the Universe by a large box of spatial size $L \times L \times L$ with periodic boundary conditions. This is only a technical trick that allows us to simplify the calculation; at the end, we let $L \to \infty$. We will proceed in three steps. First, we classify all possible modes of the electromagnetic field in the box. Then, we populate these modes with photons. Finally, we use quantum statistical mechanics by summing over all quantum states using a Boltzmann distribution.

The modes of the electromagnetic field in an L^3 periodic box are classified by their momenta $\vec{p}$, which are restricted to discrete values

$$\vec{p} = \frac{2\pi}{L}\vec{m}, \quad m_i \in \mathbb{Z}. \tag{6.54}$$

---

[5] The neutrino background radiation will be discussed in Section 8.4.

The corresponding angular frequency is given by $\omega = |\vec{p}\,|$. Each of the modes exists in two polarization states $\pm$.

We now populate the modes with photons. We completely classify a quantum state of the electromagnetic field by specifying the number of photons $n_\pm(\vec{p}) \in \{0, 1, 2, \ldots\}$ occupying each mode (characterized by the momentum $\vec{p}$ and the polarization $\pm$). Note that it makes no sense to ask "which photon" occupies which mode. Individual photons with the same momentum and polarization are indistinguishable from each other. Hence, specifying their number per mode determines their state completely.

Now that we have classified all quantum states of the electromagnetic field by specifying the photon occupation numbers for all modes, we turn to quantum statistical mechanics. We evaluate the partition function by summing over all states. Since the modes are completely independent of one another, the partition function factorizes into partition functions for each individual mode. Hence, we consider a single-mode partition function

$$Z_\pm(\vec{p}) = \sum_{n_\pm(\vec{p})=0}^{\infty} \exp(-\beta n_\pm(\vec{p})\,|\vec{p}\,|) = \frac{1}{1 - \exp(-\beta\,|\vec{p}\,|)}. \tag{6.55}$$

Each state is weighted by its Boltzmann factor $\exp(-\beta n_\pm(\vec{p})\,|\vec{p}\,|)$ which is determined by the total energy of photons occupying that mode, $n_\pm(\vec{p})\,|\vec{p}\,|$, and by the inverse temperature β. We are interested in the thermal average of the energy in a particular mode

$$\langle n_\pm(\vec{p})\,|\vec{p}\,|\rangle = \frac{1}{Z_\pm(\vec{p})} \sum_{n_\pm(\vec{p})=0}^{\infty} n_\pm(\vec{p})\,|\vec{p}\,|\exp(-\beta n_\pm(\vec{p})\,|\vec{p}\,|)$$

$$= -\frac{1}{Z_\pm(\vec{p})}\frac{\partial Z_\pm(\vec{p})}{\partial\beta} = -\frac{\partial \log Z_\pm(\vec{p})}{\partial\beta} = \frac{|\vec{p}\,|}{\exp(\beta|\vec{p}\,|) - 1}. \tag{6.56}$$

Finally, we are interested in the average total energy as a sum over all modes

$$\langle \hat{H} \rangle = \sum_{\vec{p},\pm} \langle n_\pm(\vec{p})\,|\vec{p}\,|\rangle \to 2\left(\frac{L}{2\pi}\right)^3 \int d^3p\,\langle n_\pm(\vec{p})\,|\vec{p}\,|\rangle. \tag{6.57}$$

Here, the factor 2 accounts for the two polarization states. In the last step, we have approximated the sum over discrete modes by an integral, which is justified in a large volume. It is not surprising that the result is extensive; *i.e.* the total energy grows in proportion to the volume L^3. So we consider the energy density $u = \langle \hat{H} \rangle / L^3$ in the limit $L \to \infty$. By performing the angular integration over all momentum directions $\vec{e}_p$, and by replacing $|\vec{p}\,| = \omega$, one obtains

$$u = \frac{1}{\pi^2} \int_0^{\infty} d\omega\,\frac{\omega^3}{\exp(\beta\omega) - 1}. \tag{6.58}$$

Before performing the integral, one can read off the energy density per frequency unit for modes of a given frequency ω

$$\frac{du(\omega)}{d\omega} = \frac{1}{\pi^2}\frac{\omega^3}{\exp(\beta\omega) - 1}. \tag{6.59}$$

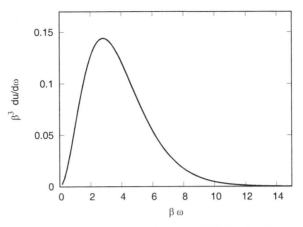

Fig. 6.1 The Planck distribution of the energy density $du/d\omega$, according to eq. (6.59), in units of the temperature $1/\beta$.

This is *Planck's formula* that was at the historical origin of quantum physics; it is illustrated in Figure 6.1. It is remarkable that quantum physics started with such a rather advanced topic that involves the quantum statistical treatment of a quantum field theory.

In the classical limit (which corresponds to small $\beta\omega$), one obtains the *Rayleigh-Jeans law*

$$\frac{du(\omega)}{d\omega} = \frac{\omega^2}{\pi^2 \beta}. \tag{6.60}$$

Integrating over all frequencies leads to a result which diverges due to the high frequency end of the spectrum. This is the so-called ultraviolet Rayleigh-Jeans catastrophe. Thus, the thermodynamics of the classical electromagnetic field gives an unphysical result. Quantum field theory comes to the rescue. Returning to Planck's quantum result (6.59) and performing the integral over all frequencies ω, one obtains the *Stefan-Boltzmann law*

$$u = \frac{\pi^2}{15\beta^4}, \tag{6.61}$$

which is not only finite, but also agrees with experiment. It is interesting that for high temperatures and for low frequencies; *i.e.* for $\beta\omega \ll 1$, Planck's formula reduces to the classical result. Quantum effects become import only in the low-temperature or high-frequency regimes; *i.e.* this happens when the photon energy is high in units of the temperature.

The spectrum of the cosmic photon background was measured to high precision by the COBE satellite. It accurately agrees with Planck's formula (6.59) and therefore with Figure 6.1. In fact, the cosmic photon background represents the most precise black-body radiation that has ever been measured. The average wavelength amounts to 1.9 mm (which is in the microwave regime), and the density is 411 photons per cm^3. (For comparison, one estimates about 112 neutrinos (including anti-neutrinos) for each flavor per cm^3.)

We now understand how satellites like COBE, WMAP, and Planck measure the temperature of the cosmic background radiation. The energy density is measured for various frequencies and is then compared with Planck's formula (6.59) which leads to a high-precision determination of the inverse temperature β. The satellite data tell us a lot about the early Universe. In fact, the early history of the Universe is imprinted on the background photons which are a cosmic relic of the Big Bang. Sometimes, one must understand the very small before one can understand the very big.

6.9   Gauge Fixing and Photon Propagator

The Euclidean action of the free electromagnetic field is given by

$$S[A] = \int d^d x \, \frac{1}{4} F_{\mu\nu} F_{\mu\nu}. \tag{6.62}$$

The field strength tensor $F_{\mu\nu} = \partial_\mu A_\nu - \partial_\nu A_\mu$, and therefore also the action, is invariant under gauge transformations

$$^\alpha A_\mu(x) = A_\mu(x) - \partial_\mu \alpha(x), \tag{6.63}$$

which is the Euclidean version of the transformation (6.3). The functional integral in Euclidean space–time,

$$Z = \int \mathcal{D}A \, \exp\left(-S[A]\right), \tag{6.64}$$

is a formal expression, which must be defined by imposing an appropriate regularization. In addition, in a perturbatively quantized gauge theory, one must also *fix the gauge*. Otherwise, there would be a divergence resulting from functional integration over infinitely many gauge copies with the same Euclidean action.

Without gauge fixing, the free photon propagator $G_{\mu\nu}(p)$ would be obtained from the action (6.62). To this end, we rewrite the Euclidean Lagrangian by means of partial integration,

$$\mathcal{L}(A) = \frac{1}{2} A_\mu(x) \left[-\partial_\rho \partial_\rho \delta_{\mu\nu} + \partial_\mu \partial_\nu \right] A_\nu(x),$$

$$G_{\mu\nu}(p) = \frac{1}{p^2 \delta_{\mu\nu} - p_\mu p_\nu}. \tag{6.65}$$

This expression is a singular matrix (with an infinite eigenvalue) in the space of Lorentz indices because (cf. Quick Question 6.4.1)

$$\left(p^2 \delta_{\mu\nu} - p_\mu p_\nu \right) p_\mu = 0. \tag{6.66}$$

Fortunately, in Abelian gauge theories gauge fixing is relatively easy to implement. Unlike in canonical quantization, in the functional integral quantization it is most convenient to work with a Lorentz-covariant gauge such as the *Lorenz gauge*, or *Landau gauge*,

$$\partial_\mu {}^\alpha A_\mu(x) = \partial_\mu \left(A_\mu(x) - \partial_\mu \alpha(x) \right) = 0 \quad \Rightarrow \quad \alpha(x) = \Box^{-1} \partial_\mu A_\mu(x), \quad \Box = \partial_\mu \partial_\mu. \tag{6.67}$$

Instead of imposing this condition as a δ-function constraint, we proceed in a slightly more general manner by defining

$$C(x) = \partial_\mu {}^\alpha A_\mu(x), \tag{6.68}$$

and by multiplying the functional integral by the irrelevant, ξ-dependent constant

$$\int \mathcal{D}C \, \exp\left(-S_{\text{gf}}[C]\right) = \int \mathcal{D}C \, \exp\left(-\int d^d x \, \frac{1}{2\xi} C^2\right), \quad S_{\text{gf}}[C] = \int d^d x \, \frac{1}{2\xi} \left(\partial_\mu {}^\alpha A_\mu\right)^2, \tag{6.69}$$

which drops out in expectation values. Here, $\xi \geq 0$ is a dimensionless gauge fixing parameter. Only in the limit $\xi \to 0$, the gauge fixing condition is imposed as a δ-function constraint.

For a given gauge field $A_\mu(x)$, we now view $C(x)$ as a function of the gauge transformation $\alpha(x)$, and we replace the functional integration $\mathcal{D}C$ by an integration $\mathcal{D}\alpha$,

$$\mathcal{D}C = \mathcal{D}\alpha \, \det\left(\frac{\delta C}{\delta \alpha}\right) = \mathcal{D}\alpha \, \det\left(\frac{\delta \partial_\mu{}^\alpha A_\mu}{\delta \alpha}\right). \tag{6.70}$$

Here, we have inserted the appropriate functional Jacobi determinant. The gauge variation of $C(x)$ takes the form

$$\delta C(x) = \delta \, \partial_\mu{}^\alpha A_\mu(x) = \partial_\mu{}^\alpha A_\mu(x) - \partial_\mu A_\mu(x) = -\partial_\mu \partial_\mu \alpha(x). \tag{6.71}$$

This leads to the functional Jacobi determinant

$$\det\left(\frac{\delta C}{\delta \alpha}\right) = \det\left(\frac{\delta \partial_\mu{}^\alpha A_\mu}{\delta \alpha}\right) = \det(-\partial_\mu \partial_\mu), \tag{6.72}$$

which is independent of the gauge field. Hence, it is another irrelevant constant that cancels in expectation values.

Altogether, the functional integral now takes the form

$$
\begin{aligned}
Z &= \int \mathcal{D}C \int \mathcal{D}A \, \exp\left(-S[A] - S_{\mathrm{gf}}[{}^\alpha A]\right) \\
&= \int \mathcal{D}\alpha \, \det(-\partial_\mu \partial_\mu) \int \mathcal{D}A \, \exp\left(-S[A] - S_{\mathrm{gf}}[{}^\alpha A]\right) \\
&= \int \mathcal{D}\alpha \, \det(-\partial_\mu \partial_\mu) \int \mathcal{D}^\alpha A \, \exp\left(-S[{}^\alpha A] - S_{\mathrm{gf}}[{}^\alpha A]\right) \\
&= \int \mathcal{D}\alpha \, \det(-\partial_\mu \partial_\mu) \int \mathcal{D}A \, \exp\left(-S[A] - S_{\mathrm{gf}}[A]\right) \\
&\propto \int \mathcal{D}A \, \exp\left(-S[A] - S_{\mathrm{gf}}[A]\right).
\end{aligned}
\tag{6.73}
$$

Here, we have used the fact that the action and the measure are gauge-invariant, *i.e.* $S[A] = S[{}^\alpha A]$ and $\mathcal{D}A = \mathcal{D}^\alpha A$, and we have then renamed the functional integration variable ${}^\alpha A$ by A. Finally, we have dropped the irrelevant A-independent constant $\det(-\partial_\mu \partial_\mu)$ as well as the functional integral $\int \mathcal{D}\alpha$ over all gauge transformations. The latter amounts exactly to the infinite factor that caused problems before gauge fixing.

As a result, the gauge fixing is incorporated as an additional term in the Euclidean action which then becomes

$$S[A] + S_{\mathrm{gf}}[A] = \int d^d x \left[\frac{1}{4} F_{\mu\nu} F_{\mu\nu} + \frac{1}{2\xi}(\partial_\mu A_\mu)^2\right]. \tag{6.74}$$

The gauge fixing term suppresses configurations which deviate from the Lorenz-Landau gauge fixing condition, in particular in the limit $\xi \to 0$. We keep the gauge fixing parameter ξ as a free constant; the physical results do not depend on it.

In momentum space, the Euclidean action takes the form

$$S[A] + S_{\text{gf}}[A] = \frac{1}{(2\pi)^d} \int d^d p \, \frac{1}{2} A_\mu(-p) \, G_{\mu\nu}(p)^{-1} A_\nu(p),$$

$$G_{\mu\nu}(p) = \frac{1}{p^2 \delta_{\mu\nu} - p_\mu p_\nu (1 - 1/\xi)}. \tag{6.75}$$

Unlike the form (6.65), this expression for the photon propagator $G_{\mu\nu}(p)$ is no longer singular, unless $\xi \to \infty$. A particularly simple situation emerges in the so-called *Feynman gauge*, $\xi = 1$, for which the photon propagator reduces to $\delta_{\mu\nu}/p^2$.

We anticipate that in the perturbative functional integral quantization of *non-Abelian* gauge theories gauge fixing is a more involved issue, which will be addressed in Chapter 11. Fortunately, when gauge theories are quantized non-perturbatively, using the *space–time lattice* regularization, gauge fixing is not even necessary. Only when one uses non-compact Abelian lattice gauge fields, which belong to the algebra of the gauge group (as in the continuum), then gauge fixing is still required even on the lattice. However, the good news is that this is *not* required in the lattice formulation with *compact* link variables, which belong to the gauge group, as we will discuss in Chapter 11.

Exercises

6.1 Canonical commutation relations and Hamilton operator

a) Show that the canonical commutation relations of eq. (6.21) follow from those of eq. (6.16).

b) Show that the Hamilton operator of eq. (6.22) follows from the one of eq. (6.18).

6.2 Gauss law

Show that the Hamilton operator $\hat{H}$ of the free electromagnetic field commutes with the Gauss law generator of gauge transformations $\vec{\nabla} \cdot \hat{\vec{E}}$.

6.3 Diagonalization of the gauge theory Hamilton operator

a) Show eqs. (6.24), (6.25), (6.27), and (6.28).

b) Derive the Hamilton operator of eq. (6.30) from eq. (6.22).

c) Derive the commutation relations of eq. (6.33) as well as the Hamilton operator of eq. (6.34).

6.4 Momentum operator of the photon field

Derive identity (6.45) for the momentum operator of the photon field.

6.5 Angular momentum operator of the photon field

Derive the commutation relation (6.52) involving the angular momentum operator of the photon field.

6.6 Commutation relations of momentum and angular momentum of the electromagnetic field

We have seen that the momentum and angular momentum operators of the electromagnetic field take the form

$$\hat{\vec{P}} = \int d^3x \, \frac{1}{2} \left(\hat{\vec{E}}(\vec{x}) \times \hat{\vec{B}}(\vec{x}) - \hat{\vec{B}}(\vec{x}) \times \hat{\vec{E}}(\vec{x}) \right),$$

$$\hat{\vec{J}} = \int d^3x \, \vec{x} \times \frac{1}{2} \left(\hat{\vec{E}}(\vec{x}) \times \hat{\vec{B}}(\vec{x}) - \hat{\vec{B}}(\vec{x}) \times \hat{\vec{E}}(\vec{x}) \right).$$

Show that these operators are consistent with the Poincaré algebra relations

$$[\hat{P}_i, \hat{P}_j] = 0, \quad [\hat{P}_i, \hat{J}_j] = i\epsilon_{ijk}\hat{P}_k, \quad [\hat{J}_i, \hat{J}_j] = i\epsilon_{ijk}\hat{J}_k.$$

6.7 Momentum and helicity of photons

Show that the photon state $|\vec{p}, \pm\rangle = \hat{a}_\pm(\vec{p})^\dagger |0\rangle$ is an eigenstate of the momentum operator $\hat{\vec{P}}$ and of the helicity operator $\hat{\vec{J}} \cdot \vec{e}_p$.

6.8 Density of the cosmic background radiation

a) Compute the average number $\langle n(\vec{p}) \rangle$ and average energy $\langle n(\vec{p})E(\vec{p}) \rangle$ of photons in a mode of the electromagnetic field with momentum $\vec{p}$ as a function of the temperature T.

b) Compute the photon number density $n = \langle N \rangle / L^3$ and the energy density $u = \langle E \rangle / L^3$ by integrating over all modes.

c) What is the average number of photons per cm^3 for the cosmic background radiation, which has a temperature of $T = 2.725$K? What is the average energy per photon, and hence what is its typical wavelength?

6.9 Wien's law in an expanding Universe

a) The cosmic background radiation originated from an annihilation of electrons and positrons immediately after the Big Bang. In such a process, the rest mass $2m_e$ of the electron–positron pair as well as their kinetic energy is converted into photon energy. Ignore the kinetic electron–positron energy, and assume that the average energy of a photon generated in an annihilation event is $2m_ec^2$. Use this together with the results for the photon energy density u and the photon number density n to estimate the initial temperature T_i of the cosmic background radiation at the time it was generated.

b) After the creation of the cosmic background radiation, the Universe has expanded by a factor x and, as a consequence, the photon gas has cooled down. Write down Planck's formula at the initial temperature T_i, and then change the frequency ω according to the red shift. Also account for the fact that the volume of the Universe has increased. According to *Wien's law*, the most probable wave length behaves as $\lambda_{\text{peak}} \propto 1/T$. Convince yourself that this corresponds to Planck's formula, and that this matches the changed energy distribution, now with decreased temperature. How is the final temperature T_f, after the Universe has expanded by a factor x, related to the initial temperature T_i? Then, how much has the Universe expanded since the cosmic background radiation has been generated?

7

Charged States in Scalar Quantum Electrodynamics

We are familiar with charged point particles from classical electrodynamics. They are sources of the Coulomb field that surrounds them. In quantum field theory, charged particles are the quanta of charged fields, which couple to the electromagnetic field. As such, they are wavicles and hence definitely not point-like. In fact, both at the classical and at the quantum level, charged particles are *non-local* objects because the Coulomb field that surrounds them decays only slowly (not exponentially but power-like) at large distances.

In this chapter, we start out with a complex scalar field with a global U(1) symmetry, which we then turn into a gauge symmetry by coupling the scalar field to the electromagnetic field. We finally construct the composite field operator that creates a charged particle together with its surrounding Coulomb field. The resulting non-local object is known as an *infraparticle*. Its structure is sensitive to the long-range conditions: We discuss the cases of an infinite spatial volume and of finite volumes with various boundary conditions.

As in most of this book, in this chapter we operate in Euclidean space–time. However, it should be stressed that the particle interpretation of quantum field theory refers to the physical Minkowski space–time.

7.1 Complex Scalar Field with Global U(1) Symmetry

A charged, spinless particle is described by a complex scalar field $\Phi(x) \in \mathbb{C}$. In fact, it takes two real degrees of freedom to describe both a scalar and an anti-scalar. As we discussed in Chapter 1, a quantum field theory can be defined by a Euclidean functional integral over all field configurations

$$Z = \int \mathcal{D}\Phi \, \exp\left(-S[\Phi]\right), \quad S[\Phi] = \int d^4x \, \mathcal{L}(\Phi, \partial_\mu \Phi). \tag{7.1}$$

Here, Z is the partition function and $S[\Phi]$ is the Euclidean action, which depends on the configurations of the field $\Phi(x) = \phi_1(x) + i\phi_2(x)$ $(\phi_i(x) \in \mathbb{R})$, with the Lagrangian

$$\mathcal{L}(\Phi, \partial_\mu \Phi) = \frac{1}{2}\partial_\mu \Phi^* \partial_\mu \Phi + V(\Phi) = \frac{1}{2}\partial_\mu \phi_1 \partial_\mu \phi_1 + \frac{1}{2}\partial_\mu \phi_2 \partial_\mu \phi_2 + V(\phi_1, \phi_2). \tag{7.2}$$

A simple, renormalizable form of the potential is that of the $\lambda|\Phi|^4$ model,

$$V(\Phi) = \frac{m^2}{2}|\Phi|^2 + \frac{\lambda}{4!}|\Phi|^4, \quad |\Phi|^2 = \Phi^*\Phi = \phi_1^2 + \phi_2^2. \tag{7.3}$$

In the free case, $\lambda = 0$, the classical Euclidean field equation takes the form

$$\partial_\mu \frac{\delta \mathcal{L}}{\delta \partial_\mu \Phi} - \frac{\delta \mathcal{L}}{\delta \Phi} = (\partial_\mu \partial_\mu - m^2)\Phi = 0. \tag{7.4}$$

This is the 2-component Euclidean Klein-Gordon equation for a free, charged scalar field (in Section 2.1, we encountered this equation in Minkowski space–time).

The Lagrangian of eqs. (7.2) and (7.3) has a global symmetry: It is invariant under U(1) transformations

$$\Phi'(x) = \exp(\mathrm{i}e\alpha)\,\Phi(x) \quad \Leftrightarrow \quad \Phi'(x)^* = \exp(-\mathrm{i}e\alpha)\,\Phi(x)^*, \tag{7.5}$$

where $\alpha \in \mathbb{R}$ is a space–time-independent (global) phase.[1] It is also invariant against complex conjugation of the field, which corresponds to the symmetry against charge conjugation C,

$$^C\Phi(x) = \Phi(x)^*, \quad S[^C\Phi] = S[\Phi]. \tag{7.6}$$

For $m^2 \geq 0$, the potential has a single minimum at $\Phi = 0$. Then, the configuration that solves the classical Euclidean field equation of least action, the *classical vacuum*, is simply the constant field $\Phi(x) = 0$. This vacuum configuration is invariant against the U(1) transformations of eq. (7.5). In Chapter 12, we will investigate the case $m^2 < 0$, in which the U(1) symmetry is spontaneously broken, but here we stay with $m^2 \geq 0$.

The states of the theory can be classified according to various conserved quantum numbers, including momentum or angular momentum. The single-particle states with momentum $\vec{p}$ can be generated from the vacuum by the field

$$\Phi(\vec{p}, x_4) = \int d^3x\,\Phi(\vec{x}, x_4)\exp(-\mathrm{i}\vec{p}\cdot\vec{x}),$$

$$\langle\Phi(\vec{p},0)\Phi(\vec{p},x_4)^*\rangle \sim \exp(-E(\vec{p})x_4), \tag{7.7}$$

whose Euclidean-time correlation function decays exponentially with the particle's energy $E(\vec{p}) = \sqrt{\vec{p}^2 + m_r^2}$. Here, m_r is the renormalized particle mass. The U(1) symmetry gives rise to a global "charge" $Q \in \mathbb{Z}$ in units of e, where $-e$ is the charge of the electron. A state with charge Q transforms with a complex phase $\exp(\mathrm{i}Qe\alpha)$. Since under U(1) transformations the scalar field transforms as $\Phi(\vec{p}, x_4)' = \exp(\mathrm{i}e\alpha)\,\Phi(\vec{p}, x_4)$, the single-particle states carry the charge $Q = 1$. Similarly, the single-anti-particle states, which are generated by $^C\Phi(\vec{p}, x_4)$, carry the charge $Q = -1$.

7.2 Scalar Quantum Electrodynamics

We now want to promote the *global* U(1) symmetry to a *local* one. This may seem to be an enormous enlargement of symmetry, since we proceed from one single symmetry parameter to one parameter at each space–time point. However, the resulting gauge symmetry is not actually a symmetry in the usual sense, but rather a huge redundancy in the mathematical description of the physics. Therefore, it does not manifest itself in degenerate multiplets of states.

What we demand is $U(1)_{em}$ invariance under transformations of the form

$$\Phi'(x) = \exp(\mathrm{i}e\alpha(x))\,\Phi(x), \tag{7.8}$$

where $\alpha(x)$ is now a (differentiable) space–time-dependent transformation.

---

[1] At this point, it is not yet obvious why the phase comes with a prefactor $e \in \mathbb{R}$. However, once the U(1) symmetry is gauged (in Section 7.2), the parameter e will be identified as the electric charge of the field Φ

Again we refer to the Lagrangian of the $\lambda |\Phi|^4$ model, with $\Phi(x) \in \mathbb{C}$, given in eqs. (7.2) and (7.3). The potential is invariant already, $V(\Phi') = V(\Phi)$. The derivative term, on the other hand, is not invariant as it stands, because

$$\partial_\mu \Phi'(x) = \exp(ie\alpha(x)) \left[\partial_\mu \Phi(x) + ie\partial_\mu \alpha(x)\, \Phi(x) \right]. \tag{7.9}$$

In order to render it locally invariant, we must modify the derivative. To this end, we introduce the electromagnetic vector potential $A_\mu(x)$ and build a *covariant derivative*

$$D_\mu \Phi(x) = \left[\partial_\mu + ieA_\mu(x) \right] \Phi(x). \tag{7.10}$$

The gauge field transforms such that the term $\partial_\mu \alpha$ in the covariant derivative is canceled,

$$A'_\mu(x) = A_\mu(x) - \partial_\mu \alpha(x) \quad \Rightarrow$$
$$(D_\mu \Phi)'(x) = \left[\partial_\mu + ieA'_\mu(x) \right] \Phi'(x) = \exp(ie\alpha(x)) D_\mu \Phi(x). \tag{7.11}$$

In this sense, the differential operator D_μ is *gauge covariant*. It can therefore be used to formulate a gauge-invariant Lagrangian

$$\mathcal{L}(\Phi, \partial_\mu \Phi, A_\mu) = \frac{1}{2}(D_\mu \Phi)^* D_\mu \Phi + V(\Phi). \tag{7.12}$$

The parameter e represents the electric charge of the scalar field, *i.e.* the strength of its coupling to A_μ. The anti-scalar, represented by the field Φ^*, has the opposite charge $-e$, and its covariant derivative transforms as

$$(D_\mu \Phi)^{*\,\prime}(x) = \exp(-ie\alpha(x)) D_\mu \Phi^*(x). \tag{7.13}$$

Up to now, A_μ has appeared only as an external field; we have not yet introduced a derivative term for it. From the free electromagnetic field – and from Chapter 6 – we know such a term: Its building block is the field strength tensor

$$F_{\mu\nu}(x) = \partial_\mu A_\nu(x) - \partial_\nu A_\mu(x), \tag{7.14}$$

which is the obvious gauge-invariant quantity assembled from first derivatives of $A_\mu(x)$,

$$F'_{\mu\nu}(x) = \partial_\mu A'_\nu(x) - \partial_\nu A'_\mu(x) = \partial_\mu A_\nu(x) - \partial_\mu \partial_\nu \alpha(x) - \partial_\nu A_\mu(x) + \partial_\nu \partial_\mu \alpha(x)$$
$$= F_{\mu\nu}(x). \tag{7.15}$$

The complete *Lagrangian of scalar quantum electrodynamics* (QED), in Euclidean space–time, takes the form

$$\mathcal{L}(\Phi, \partial_\mu \Phi, A_\mu, \partial_\mu A_\nu) = \frac{1}{2}(D_\mu \Phi)^* D_\mu \Phi + V(\Phi) + \frac{1}{4} F_{\mu\nu} F_{\mu\nu}. \tag{7.16}$$

It is *not* allowed to add an explicit mass term $M_\gamma^2 A_\mu A_\mu / 2$, because such a term would violate gauge invariance; *i.e.* it is not invariant under the transformation $A_\mu(x) \to A'_\mu(x)$ according to eq. (7.11). The term $(\partial_\mu A_\mu)^2$ is not gauge-invariant either. As we sketched in Section 6.9, it is usually added to the Lagrangian – with some constant coefficient – if one wants to fix the gauge.

The Lagrangian of scalar QED is still invariant against charge conjugation

$$^C\Phi(x) = \Phi(x)^*, \quad ^CA_\mu(x) = -A_\mu(x) \quad \Rightarrow$$
$$^C(D_\mu\Phi(x)) = \left[\partial_\mu + ie\,^CA_\mu(x)\right]{}^C\Phi(x) = \left[\partial_\mu - ieA_\mu(x)\right]\Phi(x)^* = (D_\mu\Phi(x))^*. \ (7.17)$$

Photons are neutral particles. Furthermore, they are their own anti-particles.[2] Still, the photon field changes sign under charge conjugation, and so does the photon creation operator $\hat{a}_\pm(\vec{p})^\dagger$. As a result, a photon state changes sign under charge conjugation, but it still describes the same particle.

The classical Euclidean equations of motion of the Lagrangian correspond to the inhomogeneous Maxwell equations

$$\partial_\mu F_{\mu\nu}(x) = j_\nu(x) = \frac{ie}{2}\left(D_\nu\Phi^*(x)\,\Phi(x) - \Phi^*(x)D_\nu\Phi(x)\right), \qquad (7.18)$$

while the homogeneous Maxwell equations,

$$\frac{1}{2}\epsilon_{\mu\nu\rho\sigma}\,\partial_\nu F_{\rho\sigma}(x) = \epsilon_{\mu\nu\rho\sigma}\,\partial_\nu\partial_\rho A_\sigma(x) = 0, \qquad (7.19)$$

are automatically implemented by the use of the 4-vector potential A_μ (and the anti-symmetry of the Levi-Civita symbol $\epsilon_{\mu\nu\rho\sigma}$).

For $m^2 \geq 0$, the U(1)$_{\text{em}}$ symmetry is unbroken, and the system is in a *Coulomb phase* with massive scalar particles of charge e and massless photons. In this phase, the electric charge is a conserved quantity, and opposite (equal) charges $\pm e$ at a distance r attract (repel) each other with a Coulomb potential $V(r) = \mp e^2/(4\pi r)$. Indeed, also in the full Standard Model the U(1)$_{\text{em}}$ symmetry of electromagnetism is realized in the Coulomb phase. In this context, Chapter 13 is going to address the case $m^2 < 0$; then, the Higgs mechanism endows the photon with mass. This actually happens inside superconductors, but not in the vacuum.

7.3 Charged Particles as Infraparticles

As we have seen, in the complex scalar field theory with a global U(1) symmetry, the field $\Phi(x)$ generates a particle of charge Qe, with $Q = 1$, while $^C\Phi(x) = \Phi(x)^*$ generates the charge-conjugate anti-particle with $Q = -1$. Like any object in a quantum field theory, the resulting particles and anti-particles are not point-like, but are extended wavicles. In fact, in an interacting quantum field theory, in a heuristic perturbative picture, any particle is surrounded by a cloud of virtual particles and anti-particles and is hence, at least in this sense, a composite object. It should, however, be emphasized that the appearance of particles depends on the specific operations or measurements that one performs on the system.

The particles in the $\lambda|\Phi|^4$ model with unbroken global U(1) symmetry are local objects. This is because the cloud that surrounds them consists of virtual, massive particles and anti-particles. Thus, it exhibits an exponential decay with a range that is determined by the

---

[2] This is a specific property of the photon. In general, if a particle is electrically neutral, this does not mean that it coincides with its anti-particle. Counter-examples are neutrons, Dirac neutrinos, or (as a bosonic case) the neutral kaons. All these particles will be discussed in forthcoming chapters.

inverse particle mass. This changes qualitatively in the Coulomb phase of scalar electrodynamics; because the Coulomb field that surrounds charged particles extends to infinity, its
decay is only power-like. Consequently, charged particles are *non-local* objects, which are
known as *infraparticles* (Schroer, 1963; Fröhlich, 1973; Buchholz, 1982). When disturbed
by external influences, an infraparticle can shed off soft massless photons of arbitrarily low
energy.

Also, electrons fall in the category of infraparticles and hence have non-local features.
However, this does not imply that an electron cannot be localized in a finite pixel of a
detector. When this happens in an experiment, it is inevitable that photons are created.
Therefore, such a measurement turns an original single-particle electron state (which includes the surrounding Coulomb field) into a multi-particle electron-photon state. As a
result, all particles – including infraparticles – can be localized in a finite region, however,
at the expense of generating additional particles.

Yet, electrons are sometimes said to be "point-like", even in the context of quantum field
theory. In that case, what is really meant is that, in contrast to particles like protons and
neutrons (or generally hadrons) – which consist of quarks and gluons – electrons are not
composite. As far as we know, they do not consist of more fundamental building blocks.
Still, as a field excitation in the Standard Model, the physical electron receives contributions from all fundamental quantum fields, not only from Dirac's electron–positron field
and Maxwell's electromagnetic photon field, but also – to a much smaller extent – *e.g.*,
from quark and gluon fields. This is even experimentally verified by the very accurately
measured anomalous magnetic moments of the electron and its heavier cousin, the muon
(Jegerlehner, 2008).

However, what we want to stress here is not so much that particles in interacting quantum
field theories have a complex inner structure, but that electrically charged particles, which
are surrounded by an infinite-range Coulomb field, are characterized by non-local features.
As we mentioned before, this is already the case in classical physics, where electrons are
described as Newtonian point particles that are surrounded by their long-range Coulomb
field. In practice, the Coulomb field of an electron is usually screened by positive charges
in its vicinity, for example, by an atomic nucleus with which the electron forms a neutral
atom. An individual electron in an otherwise empty space indeed behaves qualitatively
differently, because its Coulomb field decays more slowly at large distances.

An energy eigenstate of such a particle has definite momentum and is thus *not normalizable*. Due to Lorentz contraction, the asymptotic Coulomb field of a moving electron depends on the momentum. Since electron states with different momenta have different Coulomb fields at infinity, they cannot be superimposed by any local physical process. As a result, a normalizable wave packet of an individual electron, which is localized
in space, can simply not be formed. These are some of the characteristic features of an
"infraparticle".

In scalar electrodynamics with its local $U(1)_{em}$ gauge symmetry, the field $\Phi(x)$ alone is
inappropriate to generate a charged state because it is not gauge-invariant. The gauge-
invariant field that generates a charged particle must also account for the surrounding
Coulomb cloud. This can be achieved by constructing the gauge transformation $\alpha_C(\vec{x})$ that
turns a general vector potential $A_i(\vec{x})$ into the Coulomb gauge

$$\partial_i A_i'(\vec{x}) = \partial_i \left(A_i(\vec{x}) - \partial_i \alpha_C(\vec{x}) \right) = 0 \quad \Rightarrow \quad \alpha_C(\vec{x}) = \Delta^{-1} \partial_i A_i(\vec{x}). \qquad (7.20)$$

Here, $\Delta = \partial_i \partial_i$ is the spatial Laplace operator. In order for its inverse Δ^{-1} to be well-defined, a constant zero-mode contribution to $\partial_i A_i(\vec{x})$ must be absent. From a momentum-space perspective, we see that the term $p_i A_i(\vec{p})/\vec{p}^{\,2}$ can only be finite at $\vec{p} = \vec{0}$ if $A_i(\vec{p} = \vec{0}) = 0$ holds for $i = 1, 2, 3$.

Under a general gauge transformation $\alpha(\vec{x})$, with $A_i'(\vec{x}) = A_i(\vec{x}) - \partial_i \alpha(\vec{x})$, the transformation $\alpha_C(\vec{x})$, which turns the gauge field into the Coulomb gauge, transforms as

$$\alpha_C'(\vec{x}) = \Delta^{-1} \partial_i A_i'(\vec{x}) = \Delta^{-1} \partial_i \left(A_i(\vec{x}) - \partial_i \alpha(\vec{x}) \right) = \alpha_C(\vec{x}) - \alpha(\vec{x}). \tag{7.21}$$

Following Dirac (1964), this allows us to construct the field

$$\Phi_C(\vec{x}) = \exp\left(i e \alpha_C(\vec{x})\right) \Phi(\vec{x}) = \exp\left(i e \Delta^{-1} \partial_i A_i(\vec{x})\right) \Phi(\vec{x}), \tag{7.22}$$

which can be used to generate a charged particle including its surrounding Coulomb field. The inverse Laplacian Δ^{-1} renders this expression non-local.

Quick Question 7.3.1 Gauge invariance
Show that the field $\Phi_C(\vec{x})$ in eq. (7.22) is gauge-invariant.

This refers to specific, local transformations which do not alter the property $A_i(\vec{p} = \vec{0}) = 0$, which was requested above. In contrast, under a global transformation, the field $\Phi_C(\vec{x})$ transforms non-trivially,

$$\Phi'(\vec{x}) = \exp(i e \alpha) \Phi(\vec{x}), \quad A_i'(\vec{x}) = A_i(\vec{x}) \quad \Rightarrow \quad \Phi_C'(\vec{x}) = \exp(i e \alpha) \Phi_C(\vec{x}). \tag{7.23}$$

The energy $E(\vec{p})$ of a charged particle that moves with momentum $\vec{p}$ can be obtained from the exponential decay of the Euclidean-time correlation function of the spatially non-local (but temporally local) field, following the usual scheme,

$$\Phi_C(\vec{p}, x_4) = \int d^3 x \, \Phi_C(\vec{x}, x_4) \exp(-i\vec{p} \cdot \vec{x}),$$
$$\langle \Phi_C(\vec{p}, 0) \, \Phi_C(\vec{p}, x_4)^* \rangle \sim \exp(-E(\vec{p}) x_4). \tag{7.24}$$

The energy of a moving infraparticle is a subtle quantity. This is due to the presence of soft photons of arbitrarily low energy, which can mix into the charged particle state. As a result, in the framework of axiomatic quantum field theory, the following property has been rigorously proved by Detlev Buchholz (1982):

Box 7.1   **Fundamental property of electrically charged particles**

A state of a charged particle cannot simultaneously have a precise mass and electric charge.

7.4 Superselection Sectors

Charged states reside in so-called *superselection sectors* (Wick *et al.*, 1952). States from different superselection sectors are characterized by the property that they *cannot be entangled* with each other by any local physical process. As a result, they cannot occur in a quantum mechanical superposition, unless they already happen to be entangled by some (perhaps not understood) initial condition. In practice, decoherence, *i.e.* the tendency of a quantum system to interact with the rest of the world, may be viewed as a possible origin of superselection sectors. Schrödinger's cat serves as a famous example. We never encounter a cat that is in an entangled state of "dead" and "alive" (we obviously assume dying to be an irreversible classical process). In particular, decoherence prevents us from actually performing Schrödinger's thought experiment and "dead" and "alive" are, at least in practice, very well-defined "superselection sectors" for the state of a cat.

This is not exactly how superselection sectors are defined in algebraic quantum field theory. One then takes the vacuum, *i.e.* the state of lowest energy, which is characterized by zero momentum, zero angular momentum, and – at least in a Coulomb phase – also by zero electric charge, as a reference state (due to $Q = 0$ it does not contradict Buchholz' rule 7.1).

Then, all states that can be reached by the application of local physical observables belong to the vacuum superselection sector. There are different types of superselection rules. For example, the degenerate vacuum states in a theory with a spontaneously broken, exact, global symmetry belong to different superselection sectors. Indeed, no local physical process can turn one vacuum state into another one, because this would require a non-local change of state everywhere in space. Then, there are global "charge" superselection sectors, which are associated with particle states that carry a conserved quantum number associated with an exact global symmetry. Again, no physical process can change the value of such a quantum number. The creation operators that lead from the vacuum into such a superselection sector do not represent physical observables. They are just mathematical devices that allow us to construct the corresponding particle states.

Finally, there are the more subtle *charge superselection sectors,* associated with the gauge charge of an Abelian gauge theory in the Coulomb phase. Since such charged states can be created from the vacuum only by non-local operators, they define their own superselection sectors. First of all, states with different electric charges reside in different superselection sectors, because the asymptotic forms of their Coulomb fields at spatial infinity differ so much that only a non-local operator can lead from one sector to the other.

However, the situation is even more complicated. As was shown by Jürg Fröhlich, Giovanni Morchio, and Franco Strocci (Fröhlich *et al.*, 1979), as well as by Detlev Buchholz (1982) in the framework of algebraic quantum field theory, even states with the *same* charge belong to different superselection sectors, if the angular distributions of their asymptotic Coulomb fields at spatial infinity are different. As a consequence, even Lorentz invariance is said to be spontaneously broken inside a subspace of the Hilbert space with a fixed charge. These properties were revisited by Kapec *et al.* (2017a,b).

The spontaneous breakdown of Lorentz symmetry is familiar in condensed matter physics. Indeed, a macroscopic piece of condensed matter defines a preferred inertial frame (a condensed matter "ether") in which its center-of-mass is at rest. An interpretation is that the object has "spontaneously" chosen a preferred inertial frame, thus spontaneously breaking boost invariance. Of course, due to particular initial conditions, there is usually a good reason why a macroscopic piece of matter is in a specific state of motion (or at rest). In particular, it hardly ever happens that a macroscopic object is in a quantum mechanical superposition of states with different positions or velocities. In other words, we usually do not encounter classical objects in Schrödinger-cat-like states.

It is a great experimental challenge to realize quantum superposition states of a macroscopic object. True spontaneous Lorentz symmetry breaking in the mathematical sense, with corresponding distinct superselection sectors, would forbid such experiments and would, strictly speaking, require an infinitely large piece of matter.

In a similar sense, the issue of spontaneous Lorentz symmetry breaking inside a charge superselection sector is a rather academic issue. Of course, being academics ourselves, we should not be unwilling to engage in such discussions. For this purpose, we like to point to interesting discussions by Buchholz (2013) and by Buchholz and Roberts (2014). In any case, the problem of distinct superselection sectors, within say the charge 1 sector, is rather subtle. First of all, if we aim at forming a charge 1 state by "removing" a compensating charge -1 "behind the moon", this physical process will not take us out of the total charge-0 superselection sector, because it will not alter the electric field at spatial infinity. Even the angular dependence of a Coulomb field of a moving charge 1, which has existed at all times and decays only slowly at infinity, cannot be altered by any local physical process. This gives rise to a continuous variety of distinct superselection sectors with equal electric charge, which are distinguished by their corresponding momenta.

We conclude that electrically charged particles are rather delicate objects, which are very sensitive in the infrared. An individual electron at rest that had enough time to settle down in its ground state is a non-local infraparticle whose Coulomb field extends (at least in principle) all the way to spatial infinity. Unlike the scattering theory of ordinary local particles, which was developed for quantum field theory by Harry Lehmann, Kurt Symanzik, and Wolfhart Zimmermann (LSZ formalism, Lehmann *et al.* (1955)), as well as in the algebraic approach by Rudolf Haag and David Ruelle (Haag, 1958; Ruelle, 1962), the more complicated *scattering theory of infraparticles* (Fröhlich *et al.*, 1979) is developed to a lesser degree, *e.g.*, because non-interacting asymptotic initial and final states are not a natural concept for non-local objects. It should be noted that inclusive cross sections (ignoring low-energy photons) can still be defined. In the analysis of experimental data, one indeed employs this fact.

What does this imply, for example, for the deep inelastic high-energy electron-proton scattering experiments, which revealed the internal quark and gluon structure of the proton (cf. Chapter 22), or for the low-energy electron-proton scattering experiments that are used to analyze the charge radius of the proton[3] (see, *e.g.*, Gao and Vanderhaeghen (2022))? Does it matter when and how the incoming electron was initially extracted from a neutral atom and how much time it had to build up its long-range Coulomb field before it collided with the proton? Does it matter at what distance the proton's Coulomb field is screened by the walls of the laboratory? At the energy scales of typical particle physics experiments,

---

[3] The proton charge radius is determined by the motion of the electrically charged quarks inside the proton. It is finite, despite the fact that the Coulomb field of an individual proton extends to infinity.

such subtleties are usually irrelevant. However, the fact that soft photons of arbitrarily low energies may escape detection needs to be taken into account when one analyzes such experiments (Bloch and Nordsieck, 1937; Yennie *et al.*, 1961).

The problem of superselection sectors is also related to the so-called black hole *"no-hair theorem"* (Misner *et al.*, 1973) according to which a classical black hole has no other features than its mass, electric (or magnetic) charge, and angular momentum. In fact, these conserved quantities are protected by superselection sectors, and not even a black hole can swallow them without leaving a trace. According to our present understanding, there may be no exact global symmetry at all (in the framework of General Relativity, Lorentz invariance is a local symmetry, and CPT invariance is attached to it). This includes the baryon or lepton number.[4] If this is correct, a black hole cannot possess baryonic or leptonic "hair". In other words, a baryon or lepton that has been swallowed by a black hole does not endow the hole with a conserved quantum number.

7.5 Charged Particles in a Periodic Volume

In order to gain a better understanding of the subtleties that are associated with non-local charged infraparticles, which are very sensitive to the behavior at spatial infinity, it is useful to regularize the theory in the infrared. Adding a photon mass by hand would break the gauge symmetry and is therefore not a good idea. Instead, putting the particle in a finite volume is much better. Fixed boundary conditions explicitly break translation invariance and thus do in general not allow for the construction of a zero-momentum state in order to determine the rest mass of a charged particle.

Imposing *periodic* boundary conditions, *i.e.* replacing infinite space $\mathbb{R}^3$ by a torus T^3 of finite size L^3, maintains translation invariance. However, as a consequence of the Gauss law, a torus is always electrically neutral. This is because no net flux can escape the finite periodic volume

$$Q = \int_{T^3} d^3x \; \partial_i E_i = \int_{\partial T^3} d^2\sigma_i \; E_i = 0, \tag{7.25}$$

simply because the torus has no boundary, $\partial T^3 = \emptyset$.

The periodicity of the torus gives rise to additional symmetries. The gauge transformation

$$\alpha(\vec{x}) = \frac{2\pi}{eL} \, \vec{n} \cdot \vec{x}, \quad n_i \in \mathbb{Z}, \tag{7.26}$$

which is periodic up to integer multiples of $2\pi/e$, changes the vector potential by a constant background field

$$A_i'(\vec{x}) = A_i(\vec{x}) - \partial_i\alpha(\vec{x}) = A_i(\vec{x}) - \frac{2\pi}{eL} \, n_i. \tag{7.27}$$

---

[4] Protons and neutrons fall in the category of baryons, while electrons are leptons. The electromagnetic and strong interactions, but – beyond perturbation theory – not the weak interaction, conserve baryon and lepton number. The Standard Model conserves the difference between these two quantum numbers, but in light of the observation of neutrino masses, that symmetry is broken as well; see Chapter 16.

This leaves the field strength – and hence the action – unchanged. For the charged field, the phase factor $\exp(ie\alpha(\vec{x}))$ is simply a periodic gauge transformation which also leaves the action invariant. The transformations of eq. (7.26) are *"large"*, *i.e. topologically non-trivial*, gauge transformations. The complex phase $\exp(ie\alpha(\vec{x}))$ is a map from the torus into the group U(1). Such maps are topologically characterized by the elements of the homotopy group $\Pi_1[U(1)] = \mathbb{Z}$ (cf. Appendix G). The three integers n_i are winding numbers associated with the three spatial directions. On the physical Hilbert space, the transformations of eq. (7.26) are represented by unitary operators $\hat{U}(n_i)$ which commute with the Hamilton operator. The energy eigenstates can therefore be labeled by the eigenvalues of $\hat{U}(n_i)$ which take the form $\exp(i\vec{n}\cdot\vec{\theta})$. The three angles $\theta_i \in]-\pi,\pi]$ are conserved quantum numbers which *characterize superselection sectors* of QED on the torus. They are an Abelian analogue of topologically non-trivial electric and magnetic flux sectors of non-Abelian SU(N) gauge theories on a torus, which were first described by Gerard 't Hooft (1979).

Since a torus is always neutral, it is not possible to generate a single charged particle from the vacuum by using the charged field $\Phi_C(\vec{x})$ of eq. (7.22). Instead, one can use the field product

$$\prod_{a=1}^{N} \Phi_C(\vec{x}_a)\Phi_C(\vec{y}_a)^* = \prod_{a=1}^{N} \exp\left(ie\Delta^{-1}\partial_i A_i(\vec{x}_a)\right)\Phi(\vec{x}_a)\exp\left(-ie\Delta^{-1}\partial_i A_i(\vec{y}_a)\right)\Phi(\vec{y}_a)^*,$$
(7.28)

which describes an overall neutral state with N charges e located at the points $\vec{x}_a$ and N anti-charges $-e$ located at the points $\vec{y}_a$. It should be noted that Δ^{-1} is now well-defined because, due to charge neutrality, the Laplacian Δ does not have zero-modes. Under the transformations of eq. (7.26), the N-charge–N-anti-charge field transforms as

$$\prod_{a=1}^{N} \Phi'_C(\vec{x}_a)\Phi'_C(\vec{y}_a)^* = \prod_{a=1}^{N} \exp\left(i\frac{2\pi}{L}\vec{n}\cdot(\vec{x}_a-\vec{y}_a)\right)\Phi_C(\vec{x}_a)\Phi_C(\vec{y}_a)^*.$$
(7.29)

Hence, it generates a state that is characterized by

$$\vec{\theta} = \frac{2\pi}{L}\sum_{a=1}^{N}(\vec{x}_a-\vec{y}_a),$$
(7.30)

which may be interpreted as a "center-of-charge" (somewhat analogous to the center-of-mass).

Before we interpret the conservation law related to $\vec{\theta}$ physically, let us introduce the so-called *Wegner-Wilson loop*

$$W_\mathcal{C} = \exp\left(i\int_\mathcal{C} dl_i\, A_i(\vec{x})\right),$$
(7.31)

which contains the line integral of the vector potential along some closed curve $\mathcal{C}$. It is invariant under ordinary periodic gauge transformations. If the curve $\mathcal{C}$ winds $m_i \in \mathbb{Z}$ times around the torus in the i-direction, the Wegner-Wilson loop transforms under the topologically non-trivial gauge transformations of eq. (7.26) as

$$W'_\mathcal{C} = \exp\left(i\frac{2\pi}{e}\vec{m}\cdot\vec{n}\right)W_\mathcal{C}.$$
(7.32)

Hence, when applied to the vacuum, the Wegner-Wilson loop generates a state in the superselection sector $\vec{\theta} = 2\pi\vec{m}/e$. A Wegner-Wilson loop that is closed via the periodic boundary conditions generates a state with non-zero electric flux $\vec{\theta}$.

Altogether, the conserved quantity $\vec{\theta}$ receives contributions from wrapping fluxes as well as from the center-of-charge of particles and anti-particles. When a particle–anti-particle pair that was created at the same point annihilates after the anti-particle moved around the periodic volume, a wrapping flux is left behind such that $\vec{\theta}$ remains unchanged. This peculiar behavior of QED on a torus is due to the presence of $\vec{\theta}$-superselection sectors and the requirement of charge neutrality.

This confirms that, due to the massless photons, the theory is very sensitive in the infrared regime, such that the boundary conditions have drastic effects on the QED dynamics.

7.6 C-periodic Boundary Conditions

An interesting alternative for studying charged particles in a finite volume are so-called *C-periodic boundary conditions* (Polley and Wiese, 1991; Kronfeld and Wiese, 1991), for which the torus is endowed with a charge conjugation twist. C-periodic boundary conditions maintain gauge invariance as well as translation invariance, and (in contrast to periodic boundary conditions) they enable the existence of charged states. In particular, they can help us to better understand why a charged infraparticle is not simultaneously a mass and a charge eigenstate, according to the rule 7.1. They are also of practical importance for numerical lattice simulations of QCD coupled to the electromagnetic field (Patella, 2017).

A C-periodic box is a torus, say of size L^3, with the spatial boundary conditions

$$\Phi(\vec{x} + L\vec{e}_i) = {}^C\Phi(\vec{x}) = \Phi(\vec{x})^*,$$
$$A_\mu(\vec{x} + L\vec{e}_i) = {}^C A_\mu(\vec{x}) = -A_\mu(\vec{x}), \quad \alpha(\vec{x} + L\vec{e}_i) = -\alpha(\vec{x}). \tag{7.33}$$

Here, $\vec{e}_i$ is a unit-vector that points in the spatial i-direction. The theory with C-periodic boundary conditions is still gauge-invariant, but the gauge transformations $\alpha(x)$ must be C-periodic as well. The most general form of C-periodic boundary conditions involves additional gauge twists which ensure that only gauge-invariant quantities (and not the gauge-variant fields $\Phi(x)$ and $A_\mu(x)$) are C-periodic. For our purposes, the boundary conditions of eq. (7.33) are sufficient. At finite temperature, the bosonic fields obey ordinary periodic boundary conditions in the Euclidean time direction.

In a C-periodic spatial volume, the creation operator of a charged state has the form of eq. (7.22), *i.e.* the same form as in the infinite volume. Unlike for periodic boundary conditions, with C-periodic boundary conditions the Laplacian Δ has no zero-modes and eq. (7.22) is automatically well-defined. Hence, unlike a periodic torus, a finite C-periodic volume can contain charged particles. With C-periodic boundary conditions, general global (*i.e.* space–time independent) U(1) transformations $\exp(ie\alpha)$ are reduced to the $\mathbb{Z}(2)$ transformations ± 1. Still, a sign change of the charged field remains a symmetry of the theory. A charged particle that traverses the C-periodic boundary returns as an anti-particle on the other side. Consequently, a charged state in a C-periodic box decomposes into a C-even and a C-odd linear combination of particle and anti-particle.

We can decompose the charged scalar field into its C-even (real) and C-odd (imaginary) components, $\Phi(\vec{x}) = \phi_1(\vec{x}) + i\phi_2(\vec{x})$. In a C-periodic volume, C-even fields are periodic, while C-odd fields experience anti-periodic boundary conditions. In particular, the electromagnetic field $A_\mu(x)$, which is C-odd, perceives anti-periodic boundary conditions. In a C-periodic volume, the C-even states have momenta $p_i = 2\pi n_i/L$, which are quantized in integer units $n_i \in \mathbb{Z}$, while C-odd states have momenta $p_i = 2\pi(n_i + 1/2)/L$ that are quantized in half-odd-integer units. In a C-periodic box, the C-even and C-odd components of a charged particle necessarily have different momenta. This is characteristic of an infraparticle: Due to the mixture of different soft momenta, even the infinite volume charge eigenstate is not a sharp energy eigenstate. For the same reason, a true charged energy eigenstate cannot exist in a finite C-periodic volume either. Because its C-even and C-odd components have different momenta, they also have different energies.

All these subtle infrared effects disappear in the absence of massless photons. In particular, the complex scalar field without a coupling to electromagnetism, and hence with just a global $U(1)$ symmetry, has local particle excitations. These can happily exist as individual particles or anti-particles in an ordinary periodic volume. As a consequence of the Gauss law, this is not possible for the charged infraparticles in QED.

Exercises

7.1 Charge–anti-charge states on a torus

Consider a state with N charges e and N anti-charges $-e$ in scalar QED on a 3-d spatial torus.

a) Show that the field product of eq. (7.28) changes under the "large" gauge transformations of eq. (7.26) according to eq. (7.29).

b) Show that the Wegner-Wilson loop of eq. (7.31) transforms under the large gauge transformations according to eq. (7.32).

8 Canonical Quantization of Free Weyl, Dirac, and Majorana Fermions

In this chapter, we introduce *left- and right-handed Weyl fermions* as well as *Dirac and Majorana fermions*. The basic fermionic building blocks of the Standard Model are indeed Weyl fermions. Here, we investigate how fermions are described in a Hamiltonian formulation using anti-commuting fermion creation and annihilation operators. We also discuss Lorentz and Poincaré invariance as well as the discrete symmetries of parity P and of charge conjugation C. In addition, we address the chiral $U(1)_L \times U(1)_R$ symmetry of massless left- and right-handed Weyl fermions, as well as the $U(1)_F$ and $\mathbb{Z}(2)_F$ fermion number symmetries of massive Dirac and Majorana fermions, respectively. In the next chapter, we are going to relate the Hamiltonian formulation to the Euclidean fermionic functional integral in which fermions are described by anti-commuting Grassmann fields.

8.1 Massless Weyl Fermions

The fermions of the Standard Model are described by left- and right-handed Weyl spinor fields. In the absence of the Higgs field, these fermions would be massless. Hence, it is natural to begin our discussion with *massless Weyl fermions,* a formulation originally derived by Hermann Weyl (Weyl, 1929).

Massless free particles are characterized by their conserved 3-momentum $\vec{p}$, which determines their energy $E(\vec{p}) = |\vec{p}|$. The fermions in the Standard Model are spin-1/2 particles whose spin $\vec{S} = \vec{\sigma}/2$ is described in terms of the Pauli matrices

$$\vec{\sigma} = (\sigma^1, \sigma^2, \sigma^3) = \left(\begin{pmatrix} 0 & 1 \\ 1 & 0 \end{pmatrix}, \begin{pmatrix} 0 & -\mathrm{i} \\ \mathrm{i} & 0 \end{pmatrix}, \begin{pmatrix} 1 & 0 \\ 0 & -1 \end{pmatrix} \right). \tag{8.1}$$

Massless Weyl fermions are energy eigenstates with two different *helicities*. This means that their spin vector $\vec{S}$ is either parallel or anti-parallel to their momentum vector $\vec{p}$; we will see in Section 8.2 that momentum and helicity can be measured simultaneously. Since massless fermions travel with the speed of light, their helicity is independent of the reference frame. (The helicity of massive fermions, on the other hand, depends on the observer.)

Weyl fermions exist with two *chiralities*: left- or right-handed (just as for gloves), which are distinguished by their behavior under Lorentz transformations. In contrast to helicity, which is a property of a quantum state, chirality is a characteristic of a Weyl fermion field that does not depend on the reference frame, even if the field is endowed with mass.

The Hamilton operator of a free, massless, *right-handed* Weyl fermion field reads

$$\hat{H}_R = \int d^3x \, \hat{\psi}_R^\dagger(\vec{x}) \left(-\mathrm{i}\vec{\sigma} \cdot \vec{\nabla} \right) \hat{\psi}_R(\vec{x}). \tag{8.2}$$

The field operators

$$\hat{\psi}_{\mathrm{R}}(\vec{x}) = \begin{pmatrix} \hat{\psi}_{\mathrm{R}}^{1}(\vec{x}) \\ \hat{\psi}_{\mathrm{R}}^{2}(\vec{x}) \end{pmatrix}, \quad \hat{\psi}_{\mathrm{R}}^{\dagger}(\vec{x}) = \left(\hat{\psi}_{\mathrm{R}}^{1\dagger}(\vec{x}), \hat{\psi}_{\mathrm{R}}^{2\dagger}(\vec{x}) \right), \tag{8.3}$$

obey the canonical *anti-commutation relations*

$$\left\{ \hat{\psi}_{\mathrm{R}}^{a}(\vec{x}), \hat{\psi}_{\mathrm{R}}^{b\dagger}(\vec{y}) \right\} = \delta_{ab} \delta(\vec{x} - \vec{y}),$$

$$\left\{ \hat{\psi}_{\mathrm{R}}^{a}(\vec{x}), \hat{\psi}_{\mathrm{R}}^{b}(\vec{y}) \right\} = \left\{ \hat{\psi}_{\mathrm{R}}^{a\dagger}(\vec{x}), \hat{\psi}_{\mathrm{R}}^{b\dagger}(\vec{y}) \right\} = 0, \tag{8.4}$$

with the anti-commutator being defined as $\{\hat{A}, \hat{B}\} = \hat{A}\hat{B} + \hat{B}\hat{A}$. Therefore, the Fourier transformed field operators

$$\hat{\psi}_{\mathrm{R}}(\vec{p}) = \int d^3x \, \hat{\psi}_{\mathrm{R}}(\vec{x}) \exp(-\mathrm{i}\vec{p} \cdot \vec{x}), \quad \hat{\psi}_{\mathrm{R}}^{\dagger}(\vec{p}) = \int d^3x \, \hat{\psi}_{\mathrm{R}}^{\dagger}(\vec{x}) \exp(\mathrm{i}\vec{p} \cdot \vec{x}), \tag{8.5}$$

obey the anti-commutation relations

$$\left\{ \hat{\psi}_{\mathrm{R}}^{a}(\vec{p}), \hat{\psi}_{\mathrm{R}}^{b\dagger}(\vec{q}) \right\} = (2\pi)^3 \delta_{ab} \delta(\vec{p} - \vec{q}),$$

$$\left\{ \hat{\psi}_{\mathrm{R}}^{a}(\vec{p}), \hat{\psi}_{\mathrm{R}}^{b}(\vec{q}) \right\} = \left\{ \hat{\psi}_{\mathrm{R}}^{a\dagger}(\vec{p}), \hat{\psi}_{\mathrm{R}}^{b\dagger}(\vec{q}) \right\} = 0, \tag{8.6}$$

and the *Weyl Hamilton operator* takes the form

$$\hat{H}_{\mathrm{R}} = \frac{1}{(2\pi)^3} \int d^3p \, \hat{\psi}_{\mathrm{R}}^{\dagger}(\vec{p}) \, \vec{\sigma} \cdot \vec{p} \, \hat{\psi}_{\mathrm{R}}(\vec{p}). \tag{8.7}$$

We diagonalize the Hamilton operator by the unitary transformation

$$U(\vec{p}) \, (\vec{\sigma} \cdot \vec{p}) \, U(\vec{p})^{\dagger} = |\vec{p}| \, \sigma^3. \tag{8.8}$$

For the 3-momentum

$$\vec{p} = |\vec{p}| \, \vec{e}_p, \quad \vec{e}_p = (\sin\theta\cos\varphi, \sin\theta\sin\varphi, \cos\theta), \tag{8.9}$$

the diagonalizing matrix is given by[1]

$$U(\vec{p}) = \begin{pmatrix} \cos(\theta/2) & \sin(\theta/2)\exp(-\mathrm{i}\varphi) \\ -\sin(\theta/2)\exp(\mathrm{i}\varphi) & \cos(\theta/2) \end{pmatrix}. \tag{8.10}$$

Quick Question 8.1.1   Diagonalization
Verify that the matrix $U(\vec{p})$ diagonalizes the Hamilton operator, according to eq. (8.8).

---

[1] Strictly speaking, this matrix U only depends on $\vec{e}_p$, but we write $U(\vec{p})$ here and in the following, for the sake of simplicity.

For the transformed fermion field components, we introduce the *annihilation* and *creation operators* $\hat{c}_R$ and $\hat{d}_R^\dagger$,

$$\hat{\psi}_R(\vec{p}) = \begin{pmatrix} \hat{\psi}_R^1(\vec{p}) \\ \hat{\psi}_R^2(\vec{p}) \end{pmatrix} = U(\vec{p})^\dagger \begin{pmatrix} \hat{c}_R(\vec{p}) \\ \hat{d}_R^\dagger(\ \vec{p}) \end{pmatrix}. \tag{8.11}$$

The positive energy eigenstates are associated with *fermions*, while the negative energy states are associated with *anti-fermions*. As we will see, the operator $\hat{c}_R(\vec{p})$ annihilates a fermion with momentum $\vec{p}$, while $\hat{d}_R^\dagger(-\vec{p})$ creates an anti-fermion with momentum $-\vec{p}$. These operators obey the anti-commutation relations

$$\left\{\hat{c}_R(\vec{p}), \hat{c}_R^\dagger(\vec{q})\right\} = (2\pi)^3 \delta(\vec{p} - \vec{q}), \quad \left\{\hat{d}_R(\vec{p}), \hat{d}_R^\dagger(\vec{q})\right\} = (2\pi)^3 \delta(\vec{p} - \vec{q}),$$

$$\left\{\hat{c}_R(\vec{p}), \hat{c}_R(\vec{q})\right\} = \left\{\hat{c}_R^\dagger(\vec{p}), \hat{c}_R^\dagger(\vec{q})\right\} = 0, \quad \left\{\hat{d}_R(\vec{p}), \hat{d}_R(\vec{q})\right\} = \left\{\hat{d}_R^\dagger(\vec{p}), \hat{d}_R^\dagger(\vec{q})\right\} = 0,$$

$$\left\{\hat{c}_R(\vec{p}), \hat{d}_R(\vec{q})\right\} = \left\{\hat{c}_R(\vec{p}), \hat{d}_R^\dagger(\vec{q})\right\} = \left\{\hat{c}_R^\dagger(\vec{p}), \hat{d}_R(\vec{q})\right\} = \left\{\hat{c}_R^\dagger(\vec{p}), \hat{d}_R^\dagger(\vec{q})\right\} = 0. \tag{8.12}$$

Inserting eq. (8.11) into the Hamilton operator of eq. (8.7), we obtain

$$\hat{H}_R = \frac{1}{(2\pi)^3} \int d^3p \, |\vec{p}\,| \left[\hat{c}_R^\dagger(\vec{p})\hat{c}_R(\vec{p}) - \hat{d}_R(\vec{p})\hat{d}_R^\dagger(\vec{p})\right]$$

$$= \frac{1}{(2\pi)^3} \int d^3p \, |\vec{p}\,| \left[\hat{c}_R^\dagger(\vec{p})\hat{c}_R(\vec{p}) + \hat{d}_R^\dagger(\vec{p})\hat{d}_R(\vec{p}) - V\right]. \tag{8.13}$$

As in Section 2.7, the spatial volume arises by identifying $(2\pi)^3 \delta(\vec{0}) \hat{=} V$.

By definition, the vacuum state $|0\rangle_R$ is the state of lowest energy. It is annihilated by all particle or anti-particle annihilation operators

$$\hat{c}_R(\vec{p}) |0\rangle_R = \hat{d}_R(\vec{p}) |0\rangle_R = 0. \tag{8.14}$$

Massless particle or anti-particle excitations with momentum $\vec{p}$ above the vacuum cost the positive energy $|\vec{p}\,|$.

As we will see, right-handed (or right-chirality) Weyl fermions with momentum $\vec{p}$ and energy $E(\vec{p}) - E_0 = |\vec{p}\,|$ have *positive helicity* $\vec{\sigma} \cdot \vec{e}_p = 1$. They are created from the vacuum by $\hat{c}_R^\dagger(\vec{p})$, while their anti-particles are created by $\hat{d}_R^\dagger(\vec{p})$, with *negative helicity* $\vec{\sigma} \cdot \vec{e}_p = -1$,

$$\hat{c}_R^\dagger(\vec{p}) |0\rangle_R = |\vec{p}, \vec{\sigma} \cdot \vec{e}_p = 1\rangle_R, \quad \hat{d}_R^\dagger(\vec{p}) |0\rangle_R = |\vec{p}, \vec{\sigma} \cdot \vec{e}_p = -1\rangle_R. \tag{8.15}$$

Since the creation operators anti-commute, at most one fermion can occupy a given quantum state; this property is known as the *Pauli principle*. For example, a state with two right-handed Weyl fermions is given by

$$|\vec{p}_1, \vec{\sigma} \cdot \vec{e}_{p_1} = 1; \ \vec{p}_2, \vec{\sigma} \cdot \vec{e}_{p_2} = 1\rangle_R = \hat{c}_R^\dagger(\vec{p}_1)\hat{c}_R^\dagger(\vec{p}_2) |0\rangle_R = -\hat{c}_R^\dagger(\vec{p}_2)\hat{c}_R^\dagger(\vec{p}_1) |0\rangle_R$$

$$= -|\vec{p}_2, \vec{\sigma} \cdot \vec{e}_{p_2} = 1; \ \vec{p}_1, \vec{\sigma} \cdot \vec{e}_{p_1} = 1\rangle_R. \tag{8.16}$$

The anti-symmetry implies that a state with two fermions of the same momentum $\vec{p}_1 = \vec{p}_2$ and the same helicity does not exist.

Now let us consider the vacuum energy density

$$\rho = \frac{E_0}{V} = -\frac{1}{(2\pi)^3} \int d^3p \, |\vec{p}\,|. \tag{8.17}$$

As in the case of a free scalar field that we discussed in Section 2.7, it is ultraviolet-divergent, but in the fermionic case ρ is *negative*, as we conclude from eq. (8.13). The vacuum state is denoted as the filled "Dirac sea" (in this case, perhaps better the filled "Weyl sea", although this picture was suggested by Paul Dirac (1930)), in which all negative energy states are occupied and all positive energy states are empty. A missing particle of negative energy then manifests itself as an anti-particle with positive energy, relative to the filled vacuum sea.

Massless, left-handed (or left-chirality) Weyl fermions are very similar to right-handed ones, except that they have negative helicity, $\vec{\sigma} \cdot \vec{e}_p = -1$, while their anti-particles have positive helicity, $\vec{\sigma} \cdot \vec{e}_p = 1$. The corresponding Hamilton operator takes the form

$$\hat{H}_L = \int d^3x\, \hat{\psi}_L^\dagger(\vec{x}) \left(i\vec{\sigma} \cdot \vec{\nabla} \right) \hat{\psi}_L(\vec{x}) = \frac{1}{(2\pi)^3} \int d^3p\, |\vec{p}| \left[\hat{c}_L^\dagger(\vec{p})\hat{c}_L(\vec{p}) + \hat{d}_L^\dagger(\vec{p})\hat{d}_L(\vec{p}) - V \right],$$
(8.18)

in this case with

$$\hat{\psi}_L(\vec{p}) = \begin{pmatrix} \hat{\psi}_L^1(\vec{p}) \\ \hat{\psi}_L^2(\vec{p}) \end{pmatrix} = U(-\vec{p})^\dagger \begin{pmatrix} \hat{c}_L(\vec{p}) \\ \hat{d}_L^\dagger(-\vec{p}) \end{pmatrix}, \quad U(-\vec{p})(-\vec{\sigma} \cdot \vec{p})U(-\vec{p})^\dagger = |\vec{p}|\,\sigma^3,$$
(8.19)

with the unitary matrix $U(-\vec{p})$ obtained from eq. (8.10).

> **Quick Question 8.1.2 Inverse momentum**
> We write the unit-vector in the direction of $-\vec{p}$ as $\vec{e}_{-p}$. What form does it take in terms of the angles θ and φ of eq. (8.9)?

The various creation and annihilation operators again obey canonical anti-commutation relations. The corresponding vacuum state is denoted as $|0\rangle_L$,

$$\hat{c}_L(\vec{p})\,|0\rangle_L = \hat{d}_L(\vec{p})\,|0\rangle_L = 0,$$
(8.20)

and the left-handed single fermion and anti-fermion states are given by

$$\hat{c}_L^\dagger(\vec{p})\,|0\rangle_L = |\vec{p}, \vec{\sigma} \cdot \vec{e}_p = -1\rangle_L, \quad \hat{d}_L^\dagger(\vec{p})\,|0\rangle_L = |\vec{p}, \vec{\sigma} \cdot \vec{e}_p = 1\rangle_L.$$
(8.21)

8.2 Momentum, Angular Momentum, and Helicity of Weyl Fermions

In order to convince ourselves that the theory of free, right-handed Weyl fermions indeed provides a representation of the Poincaré algebra, we need to show that the Hamilton operator $\hat{H}_R$ of eq. (8.2) is accompanied by *momentum, angular momentum*, and *boost operators*, $\hat{\vec{P}}_R$, $\hat{\vec{J}}_R$, and $\hat{\vec{K}}_R$, such that the commutation relations of eq. (E.3) are satisfied. This is indeed the case with the following operators

$$\hat{\vec{P}}_{\text{R}} = \int d^3x \, \hat{\psi}_{\text{R}}^{\dagger}(\vec{x}) \left(-i\vec{\nabla}\right) \hat{\psi}_{\text{R}}(\vec{x}),$$

$$\hat{\vec{J}}_{\text{R}} = \int d^3x \, \hat{\psi}_{\text{R}}^{\dagger}(\vec{x}) \left(\vec{x} \times \left(-i\vec{\nabla}\right) + \frac{1}{2}\vec{\sigma}\right) \hat{\psi}_{\text{R}}(\vec{x}),$$

$$\hat{\vec{K}}_{\text{R}} = \int d^3x \, \hat{\psi}_{\text{R}}^{\dagger}(\vec{x}) \frac{1}{2} \left(\vec{x}\left(-i\vec{\nabla}\cdot\vec{\sigma}\right) + \left(-i\vec{\nabla}\cdot\vec{\sigma}\right)\vec{x}\right) \hat{\psi}_{\text{R}}(\vec{x}). \qquad (8.22)$$

Using eq. (8.11), we rewrite the momentum operator as

$$\hat{\vec{P}}_{\text{R}} = \frac{1}{(2\pi)^3} \int d^3p \, \vec{p} \left[\hat{c}_{\text{R}}^{\dagger}(\vec{p})\hat{c}_{\text{R}}(\vec{p}) + \hat{d}_{\text{R}}^{\dagger}(\vec{p})\hat{d}_{\text{R}}(\vec{p})\right]. \qquad (8.23)$$

As one would expect, this implies that the vacuum has zero momentum, *i.e.* $\hat{\vec{P}}_{\text{R}} |0\rangle_{\text{R}} = \vec{0}$.

> ### Quick Question 8.2.1   Commutation of momentum and creation operators
> Verify the commutation relations
>
> $$\left[\hat{\vec{P}}_{\text{R}}, \hat{c}_{\text{R}}^{\dagger}(\vec{p})\right] = \vec{p}\,\hat{c}_{\text{R}}^{\dagger}(\vec{p}), \quad \left[\hat{\vec{P}}_{\text{R}}, \hat{d}_{\text{R}}^{\dagger}(\vec{p})\right] = \vec{p}\,\hat{d}_{\text{R}}^{\dagger}(\vec{p}). \qquad (8.24)$$

In this way, one readily confirms that $\hat{c}_{\text{R}}^{\dagger}(\vec{p})$ or $\hat{d}_{\text{R}}^{\dagger}(\vec{p})$ indeed creates single-particle or anti-particle states with momentum $\vec{p}$, *e.g.*,

$$\hat{\vec{P}}_{\text{R}} |\vec{p}, \vec{\sigma} \cdot \vec{e}_p = 1\rangle_{\text{R}} = \hat{\vec{P}}_{\text{R}} \, \hat{c}_{\text{R}}^{\dagger}(\vec{p}) |0\rangle_{\text{R}} = \left(\left[\hat{\vec{P}}_{\text{R}}, \hat{c}_{\text{R}}^{\dagger}(\vec{p})\right] + \hat{c}_{\text{R}}^{\dagger}(\vec{p})\hat{\vec{P}}_{\text{R}}\right) |0\rangle_{\text{R}}$$

$$= \vec{p}\,\hat{c}_{\text{R}}^{\dagger}(\vec{p}) |0\rangle_{\text{R}} = \vec{p} \, |\vec{p}, \vec{\sigma} \cdot \vec{e}_p = 1\rangle_{\text{R}}. \qquad (8.25)$$

Let us verify the helicity of the single-particle states in a similar manner. First of all, $\hat{\vec{J}}_{\text{R}}$ and $\hat{\vec{P}}_{\text{R}}$ do not commute and are thus not simultaneously measurable. However, the component of the angular momentum vector in the direction of a particle's momentum, $\hat{\vec{J}}_{\text{R}} \cdot \vec{e}_p$, *is* simultaneously measurable with the momentum. In Problem 8.3, we are going to show that

$$\left[\hat{\vec{J}}_{\text{R}} \cdot \vec{e}_p, \hat{c}_{\text{R}}^{\dagger}(\vec{p})\right] = \frac{1}{2}\hat{c}_{\text{R}}^{\dagger}(\vec{p}), \quad \left[\hat{\vec{J}}_{\text{R}} \cdot \vec{e}_p, \hat{d}_{\text{R}}^{\dagger}(\vec{p})\right] = -\frac{1}{2}\hat{d}_{\text{R}}^{\dagger}(\vec{p}). \qquad (8.26)$$

At this point, we assume the property $\hat{\vec{J}}_{\text{R}} |0\rangle_{\text{R}} = \vec{0}$. This is physically expected, and it is in fact true, but – in contrast to the case of $\hat{\vec{P}}_{\text{R}}$ – it is not so easy to demonstrate. It leads to

$$\hat{\vec{J}}_{\text{R}} \cdot \vec{e}_p |\vec{p}, \vec{\sigma} \cdot \vec{e}_p = 1\rangle_{\text{R}} = \left[\hat{\vec{J}}_{\text{R}} \cdot \vec{e}_p, \hat{c}_{\text{R}}^{\dagger}(\vec{p})\right] |0\rangle_{\text{R}} = \frac{1}{2}\hat{c}_{\text{R}}^{\dagger}(\vec{p}) |0\rangle_{\text{R}} = \frac{1}{2}|\vec{p}, \vec{\sigma} \cdot \vec{e}_p = 1\rangle_{\text{R}},$$

$$\hat{\vec{J}}_{\text{R}} \cdot \vec{e}_p |\vec{p}, \vec{\sigma} \cdot \vec{e}_p = -1\rangle_{\text{R}} = \left[\hat{\vec{J}}_{\text{R}} \cdot \vec{e}_p, \hat{d}_{\text{R}}^{\dagger}(\vec{p})\right] |0\rangle_{\text{R}} = -\frac{1}{2}\hat{d}_{\text{R}}^{\dagger}(\vec{p}) |0\rangle_{\text{R}} = -\frac{1}{2}|\vec{p}, \vec{\sigma} \cdot \vec{e}_p = -1\rangle_{\text{R}}.$$

$$(8.27)$$

Similarly, for left-handed Weyl fermions one obtains

$$\left[\hat{\vec{J}}_L \cdot \vec{e}_p, \hat{c}_L^\dagger(\vec{p})\right] = -\frac{1}{2}\hat{c}_L^\dagger(\vec{p}), \quad \left[\hat{\vec{J}}_L \cdot \vec{e}_p, \hat{d}_L^\dagger(\vec{p})\right] = \frac{1}{2}\hat{d}_L^\dagger(\vec{p}),$$

$$\hat{\vec{J}}_L \cdot \vec{e}_p |\vec{p}, \vec{\sigma} \cdot \vec{e}_p = -1\rangle_L = \left[\hat{\vec{J}}_L \cdot \vec{e}_p, \hat{c}_L^\dagger(\vec{p})\right]|0\rangle_L = -\frac{1}{2}\hat{c}_L^\dagger(\vec{p})|0\rangle_L = -\frac{1}{2}|\vec{p}, \vec{\sigma} \cdot \vec{e}_p = -1\rangle_L,$$

$$\hat{\vec{J}}_L \cdot \vec{e}_p |\vec{p}, \vec{\sigma} \cdot \vec{e}_p = 1\rangle_L = \left[\hat{\vec{J}}_L \cdot \vec{e}_p, \hat{d}_L^\dagger(\vec{p})\right]|0\rangle_L = \frac{1}{2}\hat{d}_L^\dagger(\vec{p})|0\rangle_L = \frac{1}{2}|\vec{p}, \vec{\sigma} \cdot \vec{e}_p = 1\rangle_L. \quad (8.28)$$

8.3 Fermion Number, Parity, and Charge Conjugation

In a free field theory, the total numbers of particles and of anti-particles are constant in time, simply because the individual particles do not interact. However, in this respect free field theory is exceptional. In general, in quantum field theory particles can be created and annihilated. In particular, a particle and its anti-particle can annihilate each other, or they can be pair-created. Hence, unlike in non-relativistic quantum mechanics, in relativistic quantum field theories the particle number is usually not a conserved – or even a meaningful – physical quantity. Particle–anti-particle annihilation and pair creation proceed exclusively via interactions and are hence absent in a free field theory. While the total numbers of fermions or anti-fermions are not separately conserved in a generic quantum field theory, often (though not always) the number of fermions minus the number of anti-fermions, known as the *fermion number,* is a conserved quantity.

For a right-handed Weyl fermion, the corresponding *fermion number operator* is defined as

$$\hat{F}_R = \frac{1}{(2\pi)^3} \int d^3p \left[\hat{c}_R^\dagger(\vec{p})\hat{c}_R(\vec{p}) - \hat{d}_R^\dagger(\vec{p})\hat{d}_R(\vec{p})\right]. \quad (8.29)$$

We see that it counts the number of fermions minus anti-fermions. The conservation of $\hat{F}_R$ is associated with a $U(1)_R$ symmetry. In the multi-particle–multi-anti-particle Hilbert space, *i.e.* in *Fock space*, this symmetry is represented by unitary transformations

$$\hat{U}_R(\chi_R) = \exp(i\chi_R\hat{F}_R), \quad \chi_R \in \mathbb{R}, \quad (8.30)$$

which are constructed by exponentiating the infinitesimal symmetry generator $\hat{F}_R$. When applied to the field operators, this symmetry transformation acts as

$$\hat{U}_R(\chi_R)\hat{\psi}_R(\vec{x})\hat{U}_R(\chi_R)^\dagger = \exp(i\chi_R)\hat{\psi}_R(\vec{x}),$$
$$\hat{U}_R(\chi_R)\hat{\psi}_R^\dagger(\vec{x})\hat{U}_R(\chi_R)^\dagger = \hat{\psi}_R^\dagger(\vec{x})\exp(-i\chi_R). \quad (8.31)$$

This transformation leaves the Hamilton operator of eq. (8.2) invariant, which implies that $\hat{F}_R$ is indeed a conserved quantity.

For left-handed Weyl fermions, there is an analogous conserved fermion number

$$\hat{F}_L = \frac{1}{(2\pi)^3} \int d^3p \left[\hat{c}_L^\dagger(\vec{p})\hat{c}_L(\vec{p}) - \hat{d}_L^\dagger(\vec{p})\hat{d}_L(\vec{p})\right], \quad (8.32)$$

which results from a $U(1)_L$ symmetry.

Table 8.1 Fermion numbers F_R, F_L, and helicity $\vec{\sigma} \cdot \vec{e}_p$ of Weyl fermions				
	fermion (right)	fermion (left)	anti-fermion (right)	anti-fermion (left)
F_R	1	0	-1	0
F_L	0	1	0	-1
$\vec{\sigma} \cdot \vec{e}_p$	1	-1	-1	1

Consequently, we can assign a fermion number $F_{R,L} = \pm 1$ to each of the single-particle and anti-particle states that we constructed before

$$\hat{F}_R|\vec{p}, \vec{\sigma} \cdot \vec{e}_p = 1\rangle_R = |\vec{p}, \vec{\sigma} \cdot \vec{e}_p = 1\rangle_R, \quad \hat{F}_R|\vec{p}, \vec{\sigma} \cdot \vec{e}_p = -1\rangle_R = -|\vec{p}, \vec{\sigma} \cdot \vec{e}_p = -1\rangle_R,$$
$$\hat{F}_L|\vec{p}, \vec{\sigma} \cdot \vec{e}_p = -1\rangle_L = |\vec{p}, \vec{\sigma} \cdot \vec{e}_p = -1\rangle_L, \quad \hat{F}_L|\vec{p}, \vec{\sigma} \cdot \vec{e}_p = 1\rangle_L = -|\vec{p}, \vec{\sigma} \cdot \vec{e}_p = 1\rangle_L.$$
$$(8.33)$$

We hence confirm that right-handed Weyl fermions with $F_R = 1$ have positive helicity ($\vec{\sigma} \cdot \vec{e}_p = 1$), while their anti-particles with $F_R = -1$ have negative helicity. Left-handed Weyl fermions with $F_L = 1$, on the other hand, have negative helicity, and their anti-particles (with $F_L = -1$) have positive helicity. This is summarized in Table 8.1.

Let us now discuss two discrete symmetry transformations:

- A *parity transformation* performs a spatial inversion at the origin, $\vec{x} \to -\vec{x}$. This implies that a state with momentum $\vec{p}$ turns into a state with momentum $-\vec{p}$. Angular momenta (such as $\vec{x} \times \vec{p}$ or spin), however, are pseudo-vectors and thus do not change under a parity transformation. As a result, the helicity of a state (the projection of the spin on the momentum) does change sign under parity. This implies that the parity partner of a left-handed Weyl fermion is right-handed. As we will see in Chapter 15, in the Standard Model there are only left-handed but no right-handed neutrino fields. Consequently, the parity P is not a symmetry of the Standard Model.

- *Charge conjugation* turns particles into anti-particles and *vice versa*, but leaves their spin and momenta and hence their helicity unchanged. The charge conjugation partner of a left-handed Weyl fermion is a right-handed anti-fermion. Again, since the Standard Model includes only left-handed neutrino fields, charge conjugation C is not a symmetry either.

- The *combined transformation CP* turns a left-handed Weyl fermion (with negative helicity) into a left-handed anti-fermion (with positive helicity) and *vice versa*; hence, it does not require the presence of a right-handed Weyl fermion field. Indeed, despite the absence of right-handed neutrino fields, CP would be a symmetry of the Standard Model, if it had fewer than three fermion generations, cf. Chapter 17.

The effects of P, C, and CP on the helicity of left- and right-handed massless Weyl fermion and anti-fermion states are illustrated in Figure 8.1.

Let us now assume a theory in which both left- and right-handed Weyl fermion fields are present. For example, the Standard Model has both left- and right-handed electron fields. Then, the parity transformation is implemented by a unitary transformation $\hat{U}_P$ in

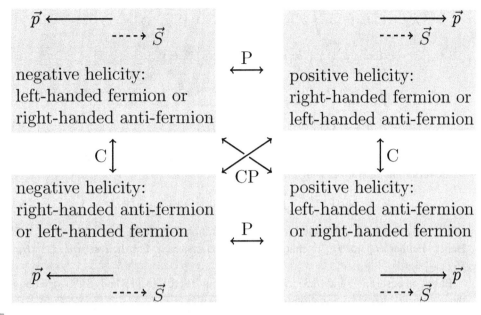

Fig. 8.1 Effect of the discrete symmetries P, C, and CP on the helicity of left- and right-handed massless Weyl fermion and anti-fermion states.

the combined Hilbert space of left- and right-handed fields,

$$
\begin{aligned}
{}^{P}\hat{\psi}_{R}(\vec{x}) &= \hat{U}_{P}\hat{\psi}_{R}(\vec{x})\hat{U}_{P}^{\dagger} = \hat{\psi}_{L}(-\vec{x}),\\
{}^{P}\hat{\psi}_{L}(\vec{x}) &= \hat{U}_{P}\hat{\psi}_{L}(\vec{x})\hat{U}_{P}^{\dagger} = \hat{\psi}_{R}(-\vec{x}).
\end{aligned}
\tag{8.34}
$$

The parity-transformed Hamilton operator of a right-handed Weyl fermion results as

$$
\begin{aligned}
{}^{P}\hat{H}_{R} = \hat{U}_{P}\hat{H}_{R}\hat{U}_{P}^{\dagger} &= \int d^{3}x \, {}^{P}\hat{\psi}_{R}^{\dagger}(\vec{x})\left(-i\vec{\sigma}\cdot\vec{\nabla}\right){}^{P}\hat{\psi}_{R}(\vec{x})\\
&= \int d^{3}x \, \hat{\psi}_{L}^{\dagger}(-\vec{x})\left(-i\vec{\sigma}\cdot\vec{\nabla}\right)\hat{\psi}_{L}(-\vec{x}) = \int d^{3}x \, \hat{\psi}_{L}^{\dagger}(\vec{x})\left(i\vec{\sigma}\cdot\vec{\nabla}\right)\hat{\psi}_{L}(\vec{x}) = \hat{H}_{L}. \tag{8.35}
\end{aligned}
$$

In the last step, we have substituted the integration variable $\vec{x} \to -\vec{x}$. Similarly, one obtains ${}^{P}\hat{H}_{L} = \hat{H}_{R}$. While neither $\hat{H}_{R}$ nor $\hat{H}_{L}$ is parity-invariant, their sum is

$$
[\hat{H}_{R} + \hat{H}_{L}, \hat{U}_{P}] = 0. \tag{8.36}
$$

Charge conjugation is implemented by another unitary transformation $\hat{U}_{C}$ which acts as

$$
\begin{aligned}
{}^{C}\hat{\psi}_{R}(\vec{x}) &= \hat{U}_{C}\hat{\psi}_{R}(\vec{x})\hat{U}_{C}^{\dagger} = i\sigma^{2}\hat{\psi}_{L}^{\dagger}(\vec{x})^{\mathsf{T}},\\
{}^{C}\hat{\psi}_{L}(\vec{x}) &= \hat{U}_{C}\hat{\psi}_{L}(\vec{x})\hat{U}_{C}^{\dagger} = -i\sigma^{2}\hat{\psi}_{R}^{\dagger}(\vec{x})^{\mathsf{T}}.
\end{aligned}
\tag{8.37}
$$

We repeat that here – and throughout this book – T denotes "transpose". The charge-conjugated Hamilton operator then takes the form

$$
\begin{aligned}
{}^{\mathrm{C}}\hat{H}_{\mathrm{R}} = \hat{U}_{\mathrm{C}}\hat{H}_{\mathrm{R}}\hat{U}_{\mathrm{C}}^{\dagger} &= \int d^3x \, {}^{\mathrm{C}}\hat{\psi}_{\mathrm{R}}^{\dagger}(\vec{x}) \left(-\mathrm{i}\vec{\sigma} \cdot \vec{\nabla} \right) {}^{\mathrm{C}}\hat{\psi}_{\mathrm{R}}(\vec{x}) \\
&= \int d^3x \, \hat{\psi}_{\mathrm{L}}(\vec{x})^{\mathsf{T}} \left(\mathrm{i}\sigma^2 \right)^{\dagger} \left(-\mathrm{i}\vec{\sigma} \cdot \vec{\nabla} \right) \mathrm{i}\sigma^2 \hat{\psi}_{\mathrm{L}}^{\dagger}(\vec{x})^{\mathsf{T}} \\
&= \int d^3x \, \hat{\psi}_{\mathrm{L}}(\vec{x})^{\mathsf{T}} \left(\mathrm{i}\vec{\sigma}^{\mathsf{T}} \cdot \vec{\nabla} \right) \hat{\psi}_{\mathrm{L}}^{\dagger}(\vec{x})^{\mathsf{T}} = \int d^3x \left(-\vec{\nabla}\hat{\psi}_{\mathrm{L}}^{\dagger}(\vec{x}) \cdot \mathrm{i}\vec{\sigma} \right) \hat{\psi}_{\mathrm{L}}(\vec{x}) \\
&= \int d^3x \, \hat{\psi}_{\mathrm{L}}^{\dagger}(\vec{x}) \left(\mathrm{i}\vec{\sigma} \cdot \vec{\nabla} \right) \hat{\psi}_{\mathrm{L}}(\vec{x}) = \hat{H}_{\mathrm{L}}.
\end{aligned}
\tag{8.38}
$$

Here, we have used the identity $(\mathrm{i}\sigma^2)^{\dagger}\vec{\sigma}(\mathrm{i}\sigma^2) = -\vec{\sigma}^{\mathsf{T}}$, as well as the anti-commutativity of the fermionic operators, which implies

$$
\hat{\psi}_{\mathrm{L}}(\vec{x})^{\mathsf{T}} \left(\mathrm{i}\vec{\sigma}^{\mathsf{T}} \cdot \vec{\nabla} \right) \hat{\psi}_{\mathrm{L}}^{\dagger}(\vec{x})^{\mathsf{T}} = - \left(\vec{\nabla}\hat{\psi}_{\mathrm{L}}^{\dagger}(\vec{x}) \cdot \mathrm{i}\vec{\sigma} \right) \hat{\psi}_{\mathrm{L}}(\vec{x}),
\tag{8.39}
$$

and, finally, we have performed a partial integration. Similarly, one finds ${}^{\mathrm{C}}\hat{H}_{\mathrm{L}} = \hat{H}_{\mathrm{R}}$. Hence, neither $\hat{H}_{\mathrm{R}}$ nor $\hat{H}_{\mathrm{L}}$ is charge conjugation-invariant, but their sum is, $i.e.$ $[\hat{H}_{\mathrm{R}} + \hat{H}_{\mathrm{L}}, \hat{U}_{\mathrm{C}}] = 0$.

Based on eqs. (8.34) and (8.37), as well as on eqs. (8.10), (8.11), and (8.19), one obtains the parity and charge conjugation transformation rules

$$
\begin{aligned}
\left(\begin{array}{c} {}^{\mathrm{P}}\hat{c}_{\mathrm{R}}(\vec{p}) \\ {}^{\mathrm{P}}\hat{c}_{\mathrm{L}}(\vec{p}) \end{array} \right) = \sigma^1 \left(\begin{array}{c} \hat{c}_{\mathrm{R}}(-\vec{p}) \\ \hat{c}_{\mathrm{L}}(-\vec{p}) \end{array} \right), \qquad &\left(\begin{array}{c} {}^{\mathrm{P}}\hat{d}_{\mathrm{R}}(\vec{p}) \\ {}^{\mathrm{P}}\hat{d}_{\mathrm{L}}(\vec{p}) \end{array} \right) = \sigma^1 \left(\begin{array}{c} \hat{d}_{\mathrm{R}}(-\vec{p}) \\ \hat{d}_{\mathrm{L}}(-\vec{p}) \end{array} \right), \\[2mm]
\left(\begin{array}{c} {}^{\mathrm{C}}\hat{c}_{\mathrm{R}}(\vec{p}) \\ {}^{\mathrm{C}}\hat{c}_{\mathrm{L}}(\vec{p}) \end{array} \right) = \mathrm{i}\sigma^2 \left(\begin{array}{c} \hat{d}_{\mathrm{R}}(\vec{p}) \\ \hat{d}_{\mathrm{L}}(\vec{p}) \end{array} \right), \qquad &\left(\begin{array}{c} {}^{\mathrm{C}}\hat{d}_{\mathrm{R}}(\vec{p}) \\ {}^{\mathrm{C}}\hat{d}_{\mathrm{L}}(\vec{p}) \end{array} \right) = -\mathrm{i}\sigma^2 \left(\begin{array}{c} \hat{c}_{\mathrm{R}}(\vec{p}) \\ \hat{c}_{\mathrm{L}}(\vec{p}) \end{array} \right), \\[2mm]
\left(\begin{array}{c} {}^{\mathrm{CP}}\hat{c}_{\mathrm{R}}(\vec{p}) \\ {}^{\mathrm{CP}}\hat{c}_{\mathrm{L}}(\vec{p}) \end{array} \right) = -\sigma^3 \left(\begin{array}{c} \hat{d}_{\mathrm{R}}(-\vec{p}) \\ \hat{d}_{\mathrm{L}}(-\vec{p}) \end{array} \right), \qquad &\left(\begin{array}{c} {}^{\mathrm{CP}}\hat{d}_{\mathrm{R}}(\vec{p}) \\ {}^{\mathrm{CP}}\hat{d}_{\mathrm{L}}(\vec{p}) \end{array} \right) = \sigma^3 \left(\begin{array}{c} \hat{c}_{\mathrm{R}}(-\vec{p}) \\ \hat{c}_{\mathrm{L}}(-\vec{p}) \end{array} \right).
\end{aligned}
\tag{8.40}
$$

Quick Question 8.3.1   Combining the transformations P and C
We define $\hat{U}_{\mathrm{CP}} = \hat{U}_{\mathrm{C}}\hat{U}_{\mathrm{P}}$ and $\hat{U}_{\mathrm{PC}} = \hat{U}_{\mathrm{P}}\hat{U}_{\mathrm{C}}$. Verify the relation $\hat{U}_{\mathrm{PC}}\hat{\psi}_{\mathrm{L,R}}(\vec{x})\hat{U}_{\mathrm{PC}}^{\dagger} = -\hat{U}_{\mathrm{CP}}\hat{\psi}_{\mathrm{L,R}}(\vec{x})\hat{U}_{\mathrm{CP}}^{\dagger}$.

While they are neither P- nor C-invariant, both $\hat{H}_{\mathrm{R}}$ and $\hat{H}_{\mathrm{L}}$ are individually invariant against the combined operation CP, $i.e.$ ${}^{\mathrm{CP}}\hat{H}_{\mathrm{R}} = {}^{\mathrm{C}}\hat{H}_{\mathrm{L}} = \hat{H}_{\mathrm{R}}$ and ${}^{\mathrm{CP}}\hat{H}_{\mathrm{L}} = {}^{\mathrm{C}}\hat{H}_{\mathrm{R}} = \hat{H}_{\mathrm{L}}$. Under a CP transformation, a right-handed fermion state transforms as

$$
\begin{aligned}
\hat{U}_{\mathrm{CP}} |\vec{p}, \vec{\sigma} \cdot \vec{e}_p = 1\rangle_{\mathrm{R}} = \hat{U}_{\mathrm{CP}}\hat{c}_{\mathrm{R}}^{\dagger}(\vec{p}) |0\rangle_{\mathrm{R}} &= {}^{\mathrm{CP}}\hat{c}_{\mathrm{R}}^{\dagger}(\vec{p})\hat{U}_{\mathrm{CP}} |0\rangle_{\mathrm{R}} = -\hat{d}_{\mathrm{R}}^{\dagger}(-\vec{p}) |0\rangle_{\mathrm{R}} \\
&= -|-\vec{p}, -\vec{\sigma} \cdot \vec{e}_p = -1\rangle_{\mathrm{R}}.
\end{aligned}
\tag{8.41}
$$

Here, we have used the fact that the vacuum is CP-invariant, $i.e.$ $\hat{U}_{\mathrm{CP}} |0\rangle_{\mathrm{R}} = |0\rangle_{\mathrm{R}}$. We conclude that the CP partner of a massless right-handed fermion (which hence has helicity $\vec{\sigma} \cdot \vec{e}_p = 1$) is a right-handed anti-fermion with opposite momentum $-\vec{p}$ and helicity $-\vec{\sigma} \cdot \vec{e}_p = -1$. (Note that the helicity of a state with momentum $-\vec{p}$ is $-\vec{\sigma} \cdot \vec{e}_p$ because the unit-vector pointing in the direction of the momentum $-\vec{p} = -|\vec{p}| \, \vec{e}_p$ is $-\vec{e}_p$.) In the same manner as above, we obtain

$$
\begin{aligned}
\hat{U}_{\mathrm{CP}} |\vec{p}, \vec{\sigma} \cdot \vec{e}_p = \pm 1\rangle_{\mathrm{R}} &= \mp |-\vec{p}, -\vec{\sigma} \cdot \vec{e}_p = \mp 1\rangle_{\mathrm{R}}, \\
\hat{U}_{\mathrm{CP}} |\vec{p}, \vec{\sigma} \cdot \vec{e}_p = \pm 1\rangle_{\mathrm{L}} &= \mp |-\vec{p}, -\vec{\sigma} \cdot \vec{e}_p = \mp 1\rangle_{\mathrm{L}}.
\end{aligned}
\tag{8.42}
$$

8.4 Cosmic Background Radiation of Neutrinos

In the Standard Model, the neutrinos are described by massless left-handed Weyl spinors. Just like the cosmic background radiation of photons, there must also be a cosmic background radiation of neutrinos. Since neutrinos are very weakly interacting, they have decoupled from the rest of the matter as early as 1 second after the Big Bang. After the neutrinos decoupled, electrons and positrons still annihilated into cosmic photons, thus increasing their entropy. As a result, the cosmic background of photons has a higher temperature than the neutrinos. While today the temperature of the cosmic photons is $T_\gamma = 2.725$ K, assuming massless neutrinos, the temperature of the neutrino background is expected to be $T_\nu \approx 1.9$ K. Neutrinos of such low energies are not detectable with current technology.

As a simple application of the quantum field theory of Weyl fermions, let us derive the fermionic analogue of Planck's formula for black-body radiation at some temperature $T = 1/\beta$. Since massless left-handed Weyl fermions have a conserved fermion number $F_L \in \mathbb{Z}$, it is natural to consider them using the *grand canonical ensemble* in which the expectation value of the fermion number is controlled by a *chemical potential* μ. The grand canonical partition function is given by

$$Z = \mathrm{Tr} \exp\left(-\beta \left(\hat{H}_L - \mu \hat{F}_L\right)\right). \tag{8.43}$$

The trace extends over the Fock space of all multi-fermion states. Just as for the derivation of the photon black-body radiation in Section 6.8, we replace the Universe by a large box of spatial size $L \times L \times L$ with periodic boundary conditions. The modes of the Weyl fermion field in the L^3 periodic box are characterized by their momenta $\vec{p} = 2\pi \vec{m}/L$, with $m_i \in \mathbb{Z}$, and by their helicity ± 1. The partition function factorizes into contributions associated with the individual modes, $Z = \prod_{\vec{p}} Z_+(\vec{p}) Z_-(\vec{p})$, with the single-mode partition functions given by

$$Z_\pm(\vec{p}) = \sum_{n=0}^{1} \exp\left(-\beta n \left(|\vec{p}| \pm \mu\right)\right) = 1 + \exp\left(-\beta \left(|\vec{p}| \pm \mu\right)\right). \tag{8.44}$$

Here, we have used the fact that each mode can be occupied with at most one fermion (Pauli principle), that the energy of a particle or anti-particle is $|\vec{p}|$, and that the fermion number associated with the helicity ± 1 is $F_L = \mp 1$.

The expectation value of the energy in a particular mode is given by

$$\langle \hat{H}_L \rangle_\pm(\vec{p}) = -\frac{\partial \log Z_\pm(\vec{p})}{\partial \beta} = \frac{|\vec{p}|}{\exp\left(\beta \left(|\vec{p}| \pm \mu\right)\right) + 1}. \tag{8.45}$$

It should be noted that here we treat β and $\beta\mu$ as independent variables, such that $\beta\mu$ is not affected by the derivative with respect to β. This is necessary in order to generate the appropriate observable by taking the derivative of the partition function. Finally, we are interested in the expectation value of the total energy density as a sum over all modes

$$\rho = \frac{1}{L^3} \langle \hat{H}_L \rangle = \frac{1}{L^3} \sum_{\vec{p}} \left(\langle \hat{H}_L \rangle_+(\vec{p}) + \langle \hat{H}_L \rangle_-(\vec{p}) \right)$$

$$\rightarrow \frac{1}{(2\pi)^3} \int d^3p \left(\frac{|\vec{p}|}{\exp\left(\beta\left(|\vec{p}| + \mu\right)\right) + 1} + \frac{|\vec{p}|}{\exp\left(\beta\left(|\vec{p}| - \mu\right)\right) + 1} \right). \tag{8.46}$$

In the last step, we took the infinite volume limit $L \rightarrow \infty$. Then, the sum over discrete momentum modes turns into an integral. First, we perform the angular integration which just provides a factor 4π. Before we perform the radial integration, we read off the energy density for modes of fixed $\omega = |\vec{p}|$

$$\frac{d\rho(\omega)}{d\omega} = \frac{\omega^3}{2\pi^2} \left(\frac{1}{\exp(\beta(\omega + \mu)) + 1} + \frac{1}{\exp(\beta(\omega - \mu)) + 1} \right). \tag{8.47}$$

This is the fermionic analogue of Planck's formula for bosonic black-body radiation. Also performing the radial integration, one obtains the total energy density. For $\mu = 0$, one finds

$$\rho = \int_0^\infty d\omega \, \frac{d\rho(\omega)}{d\omega} = \frac{1}{\pi^2} \int_0^\infty d\omega \, \frac{\omega^3}{\exp(\beta\omega) + 1} = \frac{7\pi^2}{120\beta^4} = \frac{7\pi^2 T^4}{120}. \tag{8.48}$$

The mode expectation value of the fermion number is

$$\langle \hat{F}_\text{L} \rangle_\pm (\vec{p}) = \frac{\partial \log Z_\pm(\vec{p})}{\partial(\beta\mu)} = \mp \frac{1}{\exp(\beta(|\vec{p}| \pm \mu)) + 1}, \tag{8.49}$$

which leads to the fermion number density

$$f_\text{L} = \frac{1}{L^3} \langle \hat{F}_\text{L} \rangle \rightarrow \frac{1}{2\pi^2} \int_0^\infty d\omega \left(\frac{\omega^2}{\exp(\beta(|\vec{p}| - \mu)) + 1} - \frac{\omega^2}{\exp(\beta(|\vec{p}| + \mu)) + 1} \right). \tag{8.50}$$

The expectation value of the fermion number vanishes for vanishing chemical potential, $\mu = 0$. This is not surprising if one keeps in mind that $\hat{F}_\text{L}$ counts the difference between the number of fermions and anti-fermions.

8.5 Massive Dirac Fermions

Dirac fermions have both a left- and a right-handed Weyl component, which are combined to a *4-component Dirac spinor*,

$$\hat{\psi}(\vec{x}) = \begin{pmatrix} \hat{\psi}_\text{L}(\vec{x}) \\ \hat{\psi}_\text{R}(\vec{x}) \end{pmatrix} = \begin{pmatrix} \hat{\psi}_\text{L}^1(\vec{x}) \\ \hat{\psi}_\text{L}^2(\vec{x}) \\ \hat{\psi}_\text{R}^1(\vec{x}) \\ \hat{\psi}_\text{R}^2(\vec{x}) \end{pmatrix},$$

$$\hat{\psi}^\dagger(\vec{x}) = \left(\hat{\psi}_\text{L}^\dagger(\vec{x}), \hat{\psi}_\text{R}^\dagger(\vec{x}) \right) = \left(\hat{\psi}_\text{L}^{1\dagger}(\vec{x}), \hat{\psi}_\text{L}^{2\dagger}(\vec{x}), \hat{\psi}_\text{R}^{1\dagger}(\vec{x}), \hat{\psi}_\text{R}^{2\dagger}(\vec{x}) \right). \tag{8.51}$$

Let us introduce 4×4 matrices in the so-called *chiral basis* or *Weyl basis* that project out the left- and right-handed components

$$P_\text{L} = \frac{1}{2}(1 - \gamma^5) = \begin{pmatrix} \mathbb{1} & 0 \\ 0 & 0 \end{pmatrix}, \quad P_\text{R} = \frac{1}{2}(1 + \gamma^5) = \begin{pmatrix} 0 & 0 \\ 0 & \mathbb{1} \end{pmatrix}, \quad \gamma^5 = \begin{pmatrix} -\mathbb{1} & 0 \\ 0 & \mathbb{1} \end{pmatrix}. \tag{8.52}$$

Here, 0 and $\mathbb{1}$ are 2×2 zero- and unit-matrices, respectively (for lack of symbols, the 4×4 unit-matrix is written simply as 1).

Based on the transformation properties (8.34) of its Weyl fermion components, under parity a Dirac spinor transforms as

$$ {}^{P}\hat{\psi}(\vec{x}) = \begin{pmatrix} {}^{P}\hat{\psi}_{L}(\vec{x}) \\ {}^{P}\hat{\psi}_{R}(\vec{x}) \end{pmatrix} = \begin{pmatrix} \hat{\psi}_{R}(-\vec{x}) \\ \hat{\psi}_{L}(-\vec{x}) \end{pmatrix} = \gamma^{0}\hat{\psi}(-\vec{x}). \tag{8.53} $$

Here, we have introduced the 4×4 *Dirac matrix* γ^{0}, which – together with the 4×4 matrices $\vec{\gamma}$ – forms the 4-vector γ^{μ},

$$ \gamma^{0} = \begin{pmatrix} 0 & \mathbb{1} \\ \mathbb{1} & 0 \end{pmatrix}, \quad \vec{\gamma} = (\gamma^{1}, \gamma^{2}, \gamma^{3}) = \begin{pmatrix} 0 & \vec{\sigma} \\ -\vec{\sigma} & 0 \end{pmatrix}, \tag{8.54} $$

still in the chiral basis. We follow the γ-matrix conventions of the books of Peskin and Schroeder (1997), Srednicki (2007), Zee (2010), Schwartz (2014), and many others. In any basis, *i.e.* for any valid choice, different γ-matrices, including γ^{5}, anti-commute and form a *Clifford algebra*

$$ \{\gamma^{\mu}, \gamma^{\nu}\} = 2g^{\mu\nu}, \quad \gamma^{5} = i\gamma^{0}\gamma^{1}\gamma^{2}\gamma^{3}, \quad \{\gamma^{\mu}, \gamma^{5}\} = 0, \tag{8.55} $$

where $g^{\mu\nu} = \text{diag}(1, -1, -1, -1)$ is the metric of Minkowski space–time. The first relation is the defining property of the Dirac matrices γ^{μ}.

Quick Question 8.5.1   Dirac matrices
Show that the first two relations of eq. (8.55) imply the third one.

In Chapter 9, we will also introduce γ-matrices in Euclidean space–time. In that case, we will use lower indices only, *i.e.* γ_{μ}. In Minkowski space–time, on the other hand, we will exclusively use γ-matrices with upper indices, *i.e.* γ^{μ}.

According to eq. (8.37), a Dirac spinor transforms under charge conjugation as

$$ {}^{C}\hat{\psi}(\vec{x}) = \begin{pmatrix} {}^{C}\hat{\psi}_{L}(\vec{x}) \\ {}^{C}\hat{\psi}_{R}(\vec{x}) \end{pmatrix} = \begin{pmatrix} -i\sigma^{2}\hat{\psi}_{R}^{\dagger}(\vec{x})^{\mathsf{T}} \\ i\sigma^{2}\hat{\psi}_{L}^{\dagger}(\vec{x})^{\mathsf{T}} \end{pmatrix} = C\gamma^{0}\hat{\psi}^{\dagger}(\vec{x})^{\mathsf{T}}, \quad C = \begin{pmatrix} -i\sigma^{2} & 0 \\ 0 & i\sigma^{2} \end{pmatrix}. \tag{8.56} $$

The Hamilton operator of a free, massive Dirac fermion is the sum of $\hat{H}_{R}$ and $\hat{H}_{L}$ plus a mass term that couples the left- and right-handed components,

$$ \begin{aligned} \hat{H}_{D} &= \int d^{3}x \left[\hat{\psi}_{R}^{\dagger}(\vec{x}) \left(-i\vec{\sigma} \cdot \vec{\nabla} \right) \hat{\psi}_{R}(\vec{x}) + \hat{\psi}_{L}^{\dagger}(\vec{x}) \left(i\vec{\sigma} \cdot \vec{\nabla} \right) \hat{\psi}_{L}(\vec{x}) \right. \\ &\quad \left. + m \left(\hat{\psi}_{R}^{\dagger}(\vec{x})\hat{\psi}_{L}(\vec{x}) + \hat{\psi}_{L}^{\dagger}(\vec{x})\hat{\psi}_{R}(\vec{x}) \right) \right] \\ &= \int d^{3}x \, \hat{\psi}^{\dagger}(\vec{x}) \left(-i\vec{\alpha} \cdot \vec{\nabla} + \beta m \right) \hat{\psi}(\vec{x}), \end{aligned} \tag{8.57} $$

where we have introduced the 4×4 matrices

$$ \vec{\alpha} = \gamma^{0}\vec{\gamma} = \begin{pmatrix} -\vec{\sigma} & 0 \\ 0 & \vec{\sigma} \end{pmatrix}, \quad \beta = \gamma^{0} = \begin{pmatrix} 0 & \mathbb{1} \\ \mathbb{1} & 0 \end{pmatrix}, \tag{8.58} $$

again in the chiral basis. The mass term is parity, charge conjugation, and Lorentz-invariant, but it explicitly breaks the chiral $U(1)_{L} \times U(1)_{R}$ symmetry to its diagonal subgroup $U(1)_{F}$

that is generated by the total fermion number $\hat{F} = \hat{F}_R + \hat{F}_L$ ($\hat{F}_R$ and $\hat{F}_L$ are defined in Section 8.3).

Using eqs. (8.11) and (8.19), in momentum space the *Dirac Hamilton operator* takes the form

$$
\hat{H}_D = \frac{1}{(2\pi)^3} \int d^3 p \left[\left(\hat{c}_R^\dagger(\vec{p}), \hat{d}_L(-\vec{p}) \right) \begin{pmatrix} |\vec{p}| & m\exp(-i\varphi) \\ m\exp(i\varphi) & -|\vec{p}| \end{pmatrix} \begin{pmatrix} \hat{c}_R(\vec{p}) \\ \hat{d}_L^\dagger(-\vec{p}) \end{pmatrix} \right.
$$
$$
\left. + \left(\hat{c}_L^\dagger(\vec{p}), \hat{d}_R(-\vec{p}) \right) \begin{pmatrix} |\vec{p}| & -m\exp(-i\varphi) \\ -m\exp(i\varphi) & -|\vec{p}| \end{pmatrix} \begin{pmatrix} \hat{c}_L(\vec{p}) \\ \hat{d}_R^\dagger(-\vec{p}) \end{pmatrix} \right]
$$
$$
= \frac{1}{(2\pi)^3} \int d^3 p \sqrt{\vec{p}^{\,2} + m^2} \left[\hat{c}_+^\dagger(\vec{p}) \hat{c}_+(\vec{p}) + \hat{d}_+^\dagger(\vec{p}) \hat{d}_+(\vec{p}) \right.
$$
$$
\left. + \hat{c}_-^\dagger(\vec{p}) \hat{c}_-(\vec{p}) + \hat{d}_-^\dagger(\vec{p}) \hat{d}_-(\vec{p}) - 2V \right]. \tag{8.59}
$$

We have diagonalized the Hamilton operator by means of the unitary transformations

$$
\begin{pmatrix} \hat{c}_R(\vec{p}) \\ \hat{d}_L^\dagger(-\vec{p}) \end{pmatrix} = V(\vec{p})^\dagger \begin{pmatrix} \hat{c}_+(\vec{p}) \\ \hat{d}_+^\dagger(-\vec{p}) \end{pmatrix}, \quad \begin{pmatrix} \hat{c}_L(\vec{p}) \\ \hat{d}_R^\dagger(-\vec{p}) \end{pmatrix} = V(-\vec{p})^\dagger \begin{pmatrix} \hat{c}_-(\vec{p}) \\ \hat{d}_-^\dagger(-\vec{p}) \end{pmatrix},
$$
$$
V(\vec{p}) \begin{pmatrix} |\vec{p}| & m\exp(-i\varphi) \\ m\exp(i\varphi) & -|\vec{p}| \end{pmatrix} V(\vec{p})^\dagger = \begin{pmatrix} \sqrt{\vec{p}^{\,2} + m^2} & 0 \\ 0 & -\sqrt{\vec{p}^{\,2} + m^2} \end{pmatrix},
$$
$$
V(\vec{p}) = \begin{pmatrix} \cos(\chi/2) & \sin(\chi/2)\exp(-i\varphi) \\ -\sin(\chi/2)\exp(i\varphi) & \cos(\chi/2) \end{pmatrix}, \quad \cos\chi = \frac{|\vec{p}|}{\sqrt{\vec{p}^{\,2} + m^2}},
$$
$$
V(-\vec{p}) \begin{pmatrix} |\vec{p}| & -m\exp(-i\varphi) \\ -m\exp(i\varphi) & -|\vec{p}| \end{pmatrix} V(-\vec{p})^\dagger = \begin{pmatrix} \sqrt{\vec{p}^{\,2} + m^2} & 0 \\ 0 & -\sqrt{\vec{p}^{\,2} + m^2} \end{pmatrix},
$$
$$
V(-\vec{p}) = \begin{pmatrix} \cos(\chi/2) & -\sin(\chi/2)\exp(-i\varphi) \\ \sin(\chi/2)\exp(i\varphi) & \cos(\chi/2) \end{pmatrix}, \tag{8.60}
$$

where φ is the azimuthal angle of the momentum vector defined in eq. (9.9). The operators $\hat{c}_\pm(\vec{p})$, $\hat{c}_\pm^\dagger(\vec{p})$, $\hat{d}_\pm(\vec{p})$, $\hat{d}_\pm^\dagger(\vec{p})$ again obey canonical anti-commutation relations, as in eq. (8.12).

By definition, the vacuum of Dirac fermions (the filled Dirac sea) is the state of lowest energy, in which all negative energy states are occupied while all positive energy states are empty. Hence, the Dirac vacuum $|0\rangle_D$ fulfills

$$
\hat{c}_\pm(\vec{p}) |0\rangle_D = \hat{d}_\pm(\vec{p}) |0\rangle_D = 0. \tag{8.61}
$$

Single-fermion states (with $F = 1$) and single–anti-fermion states (with $F = -1$) are created from the vacuum by the creation operators $\hat{c}_\pm^\dagger(\vec{p})$ and $\hat{d}_\pm^\dagger(\vec{p})$, respectively,

$$
\hat{c}_\pm^\dagger(\vec{p}) |0\rangle_D = |F = 1, \vec{p}, \vec{\sigma} \cdot \vec{e}_p = \pm 1\rangle_D,
$$
$$
\hat{d}_\pm^\dagger(\vec{p}) |0\rangle_D = |F = -1, \vec{p}, \vec{\sigma} \cdot \vec{e}_p = \pm 1\rangle_D. \tag{8.62}
$$

The momentum operator of Dirac fermions takes the form

$$
\hat{\vec{P}}_D = \hat{\vec{P}}_R + \hat{\vec{P}}_L = \frac{1}{(2\pi)^3} \int d^3 p \, \vec{p} \left[\hat{c}_R^\dagger(\vec{p}) \hat{c}_R(\vec{p}) + \hat{d}_R^\dagger(\vec{p}) \hat{d}_R(\vec{p}) + \hat{c}_L^\dagger(\vec{p}) \hat{c}_L(\vec{p}) + \hat{d}_L^\dagger(\vec{p}) \hat{d}_L(\vec{p}) \right]
$$
$$
= \frac{1}{(2\pi)^3} \int d^3 p \, \vec{p} \left[\hat{c}_+^\dagger(\vec{p}) \hat{c}_+(\vec{p}) + \hat{d}_+^\dagger(\vec{p}) \hat{d}_+(\vec{p}) + \hat{c}_-^\dagger(\vec{p}) \hat{c}_-(\vec{p}) + \hat{d}_-^\dagger(\vec{p}) \hat{d}_-(\vec{p}) \right]. \tag{8.63}
$$

Quick Question 8.5.2   Dirac momentum operator
Show that

$$\left[\hat{\vec{P}}_{\mathrm{D}}, \hat{c}^\dagger_\pm(\vec{p})\right] = \vec{p}\,\hat{c}^\dagger_\pm(\vec{p}), \quad \left[\hat{\vec{P}}_{\mathrm{D}}, \hat{d}^\dagger_\pm(\vec{p})\right] = \vec{p}\,\hat{d}^\dagger_\pm(\vec{p}), \qquad (8.64)$$

and that this implies

$$\hat{\vec{P}}_{\mathrm{D}}\, |F, \vec{p}, \vec{\sigma} \cdot \vec{e}_p = \pm 1\rangle_{\mathrm{D}} = \vec{p}\,|F, \vec{p}, \vec{\sigma} \cdot \vec{e}_p = \pm 1\rangle_{\mathrm{D}}. \qquad (8.65)$$

Similarly, the Dirac angular momentum operator is given by $\hat{\vec{J}}_{\mathrm{D}} = \hat{\vec{J}}_{\mathrm{R}} + \hat{\vec{J}}_{\mathrm{L}}$. In Problem 8.7, we will show that eqs. (8.22) and (8.28) lead to

$$\left[\hat{\vec{J}}_{\mathrm{D}} \cdot \vec{e}_p, \hat{c}^\dagger_\pm(\vec{p})\right] = \pm\frac{1}{2}\hat{c}^\dagger_\pm(\vec{p}), \quad \left[\hat{\vec{J}}_{\mathrm{D}} \cdot \vec{e}_p, \hat{d}^\dagger_\pm(\vec{p})\right] = \pm\frac{1}{2}\hat{d}^\dagger_\pm(\vec{p}), \qquad (8.66)$$

which implies

$$\hat{\vec{J}}_{\mathrm{D}} \cdot \vec{e}_p\, |F, \vec{p}, \vec{\sigma} \cdot \vec{e}_p = \pm 1\rangle_{\mathrm{D}} = \pm\frac{1}{2}|F, \vec{p}, \vec{\sigma} \cdot \vec{e}_p = \pm 1\rangle_{\mathrm{D}}. \qquad (8.67)$$

We conclude that the indices $\pm$ of the Dirac fermion creation and annihilation operators refer to positive and negative helicity. While the helicity of a massless Weyl fermion or anti-fermion is uniquely determined by the chirality (left or right) of the corresponding Weyl fermion field, massive Dirac fermions or anti-fermions exist with both helicities. This follows, because the helicity of a massive particle is not Lorentz-invariant. An observer who moves faster than a massive fermion perceives its momentum – and hence its helicity – with a changed sign relative to an observer at rest. A massless fermion, on the other hand, moves with the velocity of light and has a Lorentz-invariant helicity.

Based on eqs. (8.40) and (8.60), one obtains the parity and charge conjugation transformation rules

$$\begin{aligned}{}^{\mathrm{P}}\hat{c}_\pm(\vec{p}) &= \hat{c}_\mp(-\vec{p}), \quad {}^{\mathrm{P}}\hat{d}_\pm(\vec{p}) = \hat{d}_\mp(-\vec{p}), \\ {}^{\mathrm{C}}\hat{c}_\pm(\vec{p}) &= \pm\hat{d}_\pm(\vec{p}), \quad {}^{\mathrm{C}}\hat{d}_\pm(\vec{p}) = \pm\hat{c}_\pm(\vec{p}).\end{aligned} \qquad (8.68)$$

Applying the unitary transformation $\hat{U}_{\mathrm{P}}$ (which implements parity in Hilbert space) on a single-particle state, one then obtains

$$\hat{U}_{\mathrm{P}}\, |F = 1, \vec{p}, \vec{\sigma} \cdot \vec{e}_p = \pm 1\rangle_{\mathrm{D}} = \hat{U}_{\mathrm{P}}\hat{c}^\dagger_\pm(\vec{p}) |0\rangle_{\mathrm{D}} = {}^{\mathrm{P}}\hat{c}^\dagger_\pm(\vec{p})\hat{U}_{\mathrm{P}} |0\rangle_{\mathrm{D}} = \hat{c}^\dagger_\mp(-\vec{p}) |0\rangle_{\mathrm{D}}$$
$$= |F = 1, -\vec{p}, -\vec{\sigma} \cdot \vec{e}_p = \mp 1\rangle_{\mathrm{D}}. \qquad (8.69)$$

Here, we have used the fact that the vacuum is parity-invariant, $\hat{U}_{\mathrm{P}} |0\rangle_{\mathrm{D}} = |0\rangle_{\mathrm{D}}$. A corresponding relation applies to anti-particle states, such that in general

$$\hat{U}_{\mathrm{P}}\, |F, \vec{p}, \vec{\sigma} \cdot \vec{e}_p = \pm 1\rangle_{\mathrm{D}} = |F, -\vec{p}, -\vec{\sigma} \cdot \vec{e}_p = \mp 1\rangle_{\mathrm{D}}. \qquad (8.70)$$

We conclude that the parity partner of a Dirac fermion or anti-fermion indeed has the opposite momentum and helicity. Similarly, by using the charge conjugation invariance of the vacuum, $\hat{U}_{\mathrm{C}} |0\rangle_{\mathrm{D}} = |0\rangle_{\mathrm{D}}$, one obtains

$$\hat{U}_{\mathrm{C}}\, |F, \vec{p}, \vec{\sigma} \cdot \vec{e}_p = \pm 1\rangle_{\mathrm{D}} = \pm| - F, \vec{p}, \vec{\sigma} \cdot \vec{e}_p = \pm 1\rangle_{\mathrm{D}}. \qquad (8.71)$$

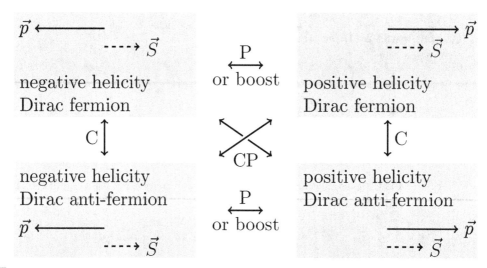

Fig. 8.2 Effect of the discrete transformations P, C, and CP on the helicity of massive Dirac fermion and anti-fermion states. For massive particles, an appropriate boost can also change the helicity.

Hence, as expected, charge conjugation exchanges fermions and anti-fermions, leaving their momentum and helicity unchanged. Based on the previous relations, one arrives at

$$\hat{U}_{\mathrm{CP}} \, |F, \vec{p}, \vec{\sigma} \cdot \vec{e}_p = \pm 1\rangle_{\mathrm{D}} = \mp | -F, -\vec{p}, -\vec{\sigma} \cdot \vec{e}_p = \mp 1\rangle_{\mathrm{D}}. \tag{8.72}$$

It is again easy to convince oneself that $\hat{U}_{\mathrm{PC}} \hat{\psi}(\vec{x}) \hat{U}^{\dagger}_{\mathrm{PC}} = -\hat{U}_{\mathrm{CP}} \hat{\psi}(\vec{x}) \hat{U}^{\dagger}_{\mathrm{CP}}$, as in Quick Question 8.3.1.

The effects of P (or an appropriate boost), C, and CP on the helicity of massive Dirac fermion and anti-fermion states are illustrated in Figure 8.2.

8.6 Massive Majorana Fermions

As Ettore Majorana realized, some fermions – now known as *Majorana fermions* – are indistinguishable from their anti-particles (Majorana, 1937). The corresponding Majorana spinor results from a Dirac spinor by imposing a constraint, which is known as the *Majorana condition*,

$$^{\mathrm{C}}\hat{\psi}(\vec{x}) = \hat{\psi}(\vec{x}) \quad \Rightarrow \quad ^{\mathrm{C}}\hat{\psi}_{\mathrm{R}}(\vec{x}) = \mathrm{i}\sigma^2 \hat{\psi}^{\dagger}_{\mathrm{L}}(\vec{x})^{\mathsf{T}} = \hat{\psi}_{\mathrm{R}}(\vec{x}) \quad \Rightarrow \quad \hat{\psi}^a_{\mathrm{R}}(\vec{x}) = \epsilon_{ab} \hat{\psi}^{b\dagger}_{\mathrm{L}}(\vec{x}). \tag{8.73}$$

Quick Question 8.6.1 Properties of the Majorana condition

Check that condition (8.73) is consistent with the anti-commutation relations (8.4) of the right-handed spinors, and that it implies $\hat{H}_{\mathrm{R}} = \hat{H}_{\mathrm{L}}$.

While the Majorana condition is Lorentz-invariant, it is important that it is not invariant against the chiral $U(1)_L \times U(1)_R$ symmetry transformation,

$$\hat{\psi}_L(\vec{x})' = \exp(i\chi_L)\hat{\psi}_L(\vec{x}), \quad \hat{\psi}_R(\vec{x})' = \exp(i\chi_R)\hat{\psi}_R(\vec{x}). \tag{8.74}$$

This follows because

$$i\sigma^2 \hat{\psi}_L^\dagger(\vec{x})'^\mathsf{T} = \exp(-i\chi_L)i\sigma^2\hat{\psi}_L^\dagger(\vec{x})^\mathsf{T} = \exp(-i\chi_L)\hat{\psi}_R(\vec{x}) = \exp(-i\chi_L - i\chi_R)\hat{\psi}_R(\vec{x})'. \tag{8.75}$$

The Majorana condition remains invariant only if $\chi_L = -\chi_R$, which implies that it breaks the $U(1)_L \times U(1)_R$ symmetry explicitly down to $U(1)_{L=R^*}$. The fermion number symmetry $U(1)_F = U(1)_{L=R}$ is then reduced to $\exp(i\chi_L) = \exp(i\chi_R) = \exp(-i\chi_R) \in \{\pm 1\} = \mathbb{Z}(2)_F$. As a consequence, the number of Majorana fermions (which are indistinguishable from their anti-particles) is conserved only modulo 2.

Interestingly, the Majorana condition is not consistent with parity invariance either, because

$$i\sigma^{2\,P}\hat{\psi}_L^\dagger(\vec{x})^\mathsf{T} = i\sigma^2\hat{\psi}_R^\dagger(-\vec{x})^\mathsf{T} = i\sigma^2(-i\sigma^2)^\mathsf{T}\,\hat{\psi}_L(-\vec{x}) = -{}^P\hat{\psi}_R(\vec{x}). \tag{8.76}$$

On the other hand, when we combine a chiral transformation with parity, we obtain

$$\begin{aligned}
{}^P\hat{\psi}_L(\vec{x})' &= \exp(i\chi_L)^P\hat{\psi}_L(\vec{x}) = \exp(i\chi_L)\hat{\psi}_R(-\vec{x}), \\
{}^P\hat{\psi}_R(\vec{x})' &= \exp(i\chi_R)^P\hat{\psi}_R(\vec{x}) = \exp(i\chi_R)\hat{\psi}_L(-\vec{x}), \\
i\sigma^{2\,P}\hat{\psi}_L^\dagger(\vec{x})'^\mathsf{T} &= \exp(-i\chi_L)i\sigma^2\hat{\psi}_R^\dagger(-\vec{x})^\mathsf{T} = -\exp(-i\chi_L)\hat{\psi}_L(-\vec{x}) \\
&= -\exp(-i\chi_L - i\chi_R)^P\hat{\psi}_R(\vec{x})'.
\end{aligned} \tag{8.77}$$

Hence, the Majorana condition is invariant against P combined with $U(1)_{L=-R^*}$. The corresponding fermion number transformations $U(1)_{L=-R^*=R}$ are characterized by $\exp(i\chi_L) = \exp(i\chi_R) = \pm i$.

We denote the parity transformation P, combined with the phase factor i, as the *modified parity transformation* P',

$$P'\hat{\psi}_L(\vec{x}) = i\hat{\psi}_R(-\vec{x}), \quad P'\hat{\psi}_R(\vec{x}) = i\hat{\psi}_L(-\vec{x}). \tag{8.78}$$

Applying the Majorana condition, the Dirac Hamilton operator reduces to $\hat{H}_D = 2\hat{H}_M$ with the *Majorana Hamilton operator* given by

$$\begin{aligned}
\hat{H}_M &= \int d^3x \left[\hat{\psi}_R^\dagger(\vec{x})\left(-i\vec{\sigma}\cdot\vec{\nabla}\right)\hat{\psi}_R(\vec{x}) + \frac{m}{2}\left(\hat{\psi}_R(\vec{x})^\mathsf{T}i\sigma^2\hat{\psi}_R(\vec{x}) - \hat{\psi}_R^\dagger(\vec{x})i\sigma^2\hat{\psi}_R^\dagger(\vec{x})^\mathsf{T}\right)\right] \\
&= \int d^3x \left[\hat{\psi}_L^\dagger(\vec{x})\left(i\vec{\sigma}\cdot\vec{\nabla}\right)\hat{\psi}_L(\vec{x}) + \frac{m}{2}\left(-\hat{\psi}_L(\vec{x})^\mathsf{T}i\sigma^2\hat{\psi}_L(\vec{x}) + \hat{\psi}_L^\dagger(\vec{x})i\sigma^2\hat{\psi}_L^\dagger(\vec{x})^\mathsf{T}\right)\right].
\end{aligned} \tag{8.79}$$

While the Dirac mass term is invariant against the continuous $U(1)_F$ fermion number symmetry, the *Majorana mass term* is invariant only under the discrete $\mathbb{Z}(2)_F$ symmetry. Just like the Majorana condition itself, the Majorana mass term is invariant, not against P, but against the modified parity transformation P'.

In momentum space, the first contribution to the Majorana mass term is given by

$$\hat{\psi}_{\mathrm{R}}(\vec{p})^{\mathsf{T}} \mathrm{i}\sigma^2 \hat{\psi}_{\mathrm{R}}(-\vec{p}) = \left(\hat{c}_{\mathrm{R}}(\vec{p}), \hat{d}_{\mathrm{R}}^{\dagger}(-\vec{p})\right) U(\vec{p})^* \mathrm{i}\sigma^2 U(-\vec{p})^{\dagger} \left(\begin{array}{c} \hat{c}_{\mathrm{R}}(-\vec{p}) \\ \hat{d}_{\mathrm{R}}^{\dagger}(\vec{p}) \end{array}\right)$$

$$= -\exp(\mathrm{i}\varphi)\hat{c}_{\mathrm{R}}(\vec{p})\hat{c}_{\mathrm{R}}(-\vec{p}) - \exp(-\mathrm{i}\varphi)\hat{d}_{\mathrm{R}}^{\dagger}(-\vec{p})\hat{d}_{\mathrm{R}}^{\dagger}(\vec{p}), \quad (8.80)$$

where we refer again to the unitary matrix $U(\vec{p})$ of eq. (8.10). The Majorana Hamilton operator then takes the form

$$\hat{H}_{\mathrm{M}} = \frac{1}{(2\pi)^3} \int_{\mathbb{R}^3/2} d^3p \left[\left(\hat{c}_{\mathrm{R}}^{\dagger}(\vec{p}), \hat{c}_{\mathrm{R}}(-\vec{p})\right) \left(\begin{array}{cc} |\vec{p}\,| & m\exp(-\mathrm{i}\varphi) \\ m\exp(\mathrm{i}\varphi) & -|\vec{p}\,| \end{array}\right) \left(\begin{array}{c} \hat{c}_{\mathrm{R}}(\vec{p}) \\ \hat{c}_{\mathrm{R}}^{\dagger}(-\vec{p}) \end{array}\right) \right.$$

$$\left. + \left(-\hat{d}_{\mathrm{R}}^{\dagger}(\vec{p}), \hat{d}_{\mathrm{R}}(-\vec{p})\right) \left(\begin{array}{cc} |\vec{p}\,| & -m\exp(-\mathrm{i}\varphi) \\ -m\exp(\mathrm{i}\varphi) & -|\vec{p}\,| \end{array}\right) \left(\begin{array}{c} -\hat{d}_{\mathrm{R}}(\vec{p}) \\ \hat{d}_{\mathrm{R}}^{\dagger}(-\vec{p}) \end{array}\right) \right]. \quad (8.81)$$

Here, the integration is limited to one half of the momentum space. This prevents double-counting, since both $\vec{p}$ and $-\vec{p}$ appear explicitly in the integrand. The Majorana condition implies $^{\mathrm{C}}\hat{c}_{\mathrm{R}}(\vec{p}) = \hat{d}_{\mathrm{L}}(\vec{p}) = \hat{c}_{\mathrm{R}}(\vec{p})$, $^{\mathrm{C}}\hat{c}_{\mathrm{L}}(\vec{p}) = -\hat{d}_{\mathrm{R}}(\vec{p}) = \hat{c}_{\mathrm{L}}(\vec{p})$, as well as $^{\mathrm{C}}\hat{c}_{\pm}(\vec{p}) = \pm\hat{d}_{\pm}(\vec{p}) = \hat{c}_{\pm}(\vec{p})$. We can hence rewrite eq. (8.60) and diagonalize the Majorana Hamilton operator by the unitary transformation

$$\left(\begin{array}{c} \hat{c}_{\mathrm{R}}(\vec{p}) \\ \hat{c}_{\mathrm{R}}^{\dagger}(-\vec{p}) \end{array}\right) = V(\vec{p})^{\dagger} \left(\begin{array}{c} \hat{c}_{+}(\vec{p}) \\ \hat{c}_{+}^{\dagger}(-\vec{p}) \end{array}\right), \quad \left(\begin{array}{c} -\hat{d}_{\mathrm{R}}(\vec{p}) \\ \hat{d}_{\mathrm{R}}^{\dagger}(-\vec{p}) \end{array}\right) = V(-\vec{p})^{\dagger} \left(\begin{array}{c} \hat{c}_{-}(\vec{p}) \\ -\hat{c}_{-}^{\dagger}(-\vec{p}) \end{array}\right).$$

$$(8.82)$$

The two components of the first equation take the form

$$\hat{c}_{\mathrm{R}}(\vec{p}) = \cos(\chi/2)\, \hat{c}_{+}(\vec{p}) - \sin(\chi/2)\, \exp(-\mathrm{i}\varphi)\hat{c}_{+}^{\dagger}(-\vec{p}),$$
$$\hat{c}_{\mathrm{R}}^{\dagger}(-\vec{p}) = \sin(\chi/2)\, \exp(\mathrm{i}\varphi)\hat{c}_{+}(\vec{p}) + \cos(\chi/2)\, \hat{c}_{+}^{\dagger}(-\vec{p}). \quad (8.83)$$

Taking the Hermitian conjugate of the equation for the first component, and replacing $\vec{p}$ by $-\vec{p}$, which implies replacing $\exp(\mathrm{i}\varphi)$ by $-\exp(\mathrm{i}\varphi)$, one indeed obtains the equation for the second component. This shows the consistency of the diagonalizing unitary transformation with the Majorana constraint, which leads to

$$\hat{H}_{\mathrm{M}} = \frac{1}{(2\pi)^3} \int_{\mathbb{R}^3/2} d^3p \sqrt{\vec{p}^{\,2} + m^2} \left[\hat{c}_{+}^{\dagger}(\vec{p})\hat{c}_{+}(\vec{p}) - \hat{c}_{+}(-\vec{p})\hat{c}_{+}^{\dagger}(-\vec{p}) \right.$$

$$\left. + \hat{c}_{-}^{\dagger}(\vec{p})\hat{c}_{-}(\vec{p}) - \hat{c}_{-}(-\vec{p})\hat{c}_{-}^{\dagger}(-\vec{p}) \right]$$

$$= \frac{1}{(2\pi)^3} \int d^3p \sqrt{\vec{p}^{\,2} + m^2} \left[\hat{c}_{+}^{\dagger}(\vec{p})\hat{c}_{+}(\vec{p}) + \hat{c}_{-}^{\dagger}(\vec{p})\hat{c}_{-}(\vec{p}) - V\right]. \quad (8.84)$$

In the last step, we have extended the integration to the entire momentum space $\mathbb{R}^3$.

The vacuum of Majorana fermions is characterized by $\hat{c}_{\pm}(\vec{p}) |0\rangle_{\mathrm{M}} = 0$, and the single-particle states are given by

$$\hat{c}_{\pm}^{\dagger}(\vec{p}) |0\rangle_{\mathrm{M}} = |(-1)^F = -1, \vec{p}, \vec{\sigma} \cdot \vec{e}_p = \pm 1\rangle_{\mathrm{M}}. \quad (8.85)$$

As usual, under a parity transformation this state changes both its momentum and its helicity,

$$\hat{U}_{P'} \, |(-1)^F = -1, \vec{p}, \vec{\sigma} \cdot \vec{e}_p = \pm 1\rangle_M = {}^{P'}\hat{c}_{\pm}^{\dagger}(\vec{p}) \, |0\rangle_M = \hat{c}_{\mp}^{\dagger}(-\vec{p}) \, |0\rangle_M$$
$$= |(-1)^F = -1, -\vec{p}, -\vec{\sigma} \cdot \vec{e}_p = \mp 1\rangle_M. \quad (8.86)$$

Similarly, since the Majorana condition implies ${}^C\hat{c}_{\pm}^{\dagger}(\vec{p}) = \hat{c}_{\pm}^{\dagger}(\vec{p})$, one obtains

$$\hat{U}_C \, |(-1)^F = -1, \vec{p}, \vec{\sigma} \cdot \vec{e}_p = \pm 1\rangle_M = |(-1)^F = -1, \vec{p}, \vec{\sigma} \cdot \vec{e}_p = \pm 1\rangle_M. \quad (8.87)$$

Hence, Majorana fermion states are indeed invariant under charge conjugation; *i.e.* particles and anti-particles are indistinguishable. Since C acts trivially on Majorana fermions, P' and CP' then have the same effect.

8.7 Massive Weyl Fermions

As we have already seen in eq. (8.79), we can reinterpret the Majorana Hamilton operator as a *massive Weyl Hamilton operator*,

$$\hat{H}_W = \int d^3x \left[\hat{\psi}_R^{\dagger}(\vec{x}) \left(-i\vec{\sigma} \cdot \vec{\nabla} \right) \hat{\psi}_R(\vec{x}) + \frac{m}{2} \left(\hat{\psi}_R(\vec{x})^{\mathsf{T}} i\sigma^2 \hat{\psi}_R(\vec{x}) - \hat{\psi}_R^{\dagger}(\vec{x}) i\sigma^2 \hat{\psi}_R^{\dagger}(\vec{x})^{\mathsf{T}} \right) \right].$$
$$(8.88)$$

In this case, one would not assume that there is a Dirac fermion field (consisting of both left- and right-handed components) with a Majorana constraint (that expresses the left-handed field as a function of the right-handed one). Instead, one simply works with a right-handed Weyl spinor *without ever introducing a left-handed component*. As a result, P and C are not symmetries of this Hamilton operator. Not even CP is, but only CP', which is the combination of C with the modified parity transformation P' of eq. (8.78). In addition, the U(1)$_R$ symmetry of a massless Weyl fermion is explicitly broken down to $\mathbb{Z}(2)_R$.

The diagonalization of the massive Weyl Hamilton operator is equivalent to that of the Majorana fermion, except that one now substitutes $\hat{c}_-(\vec{p})$ by $-\hat{d}_-(\vec{p})$ (which are equivalent for Majorana fermions), *i.e.*

$$\begin{pmatrix} \hat{c}_R(\vec{p}) \\ \hat{c}_R^{\dagger}(-\vec{p}) \end{pmatrix} = V(\vec{p})^{\dagger} \begin{pmatrix} \hat{c}_+(\vec{p}) \\ \hat{c}_+^{\dagger}(-\vec{p}) \end{pmatrix}, \quad \begin{pmatrix} -\hat{d}_R(\vec{p}) \\ \hat{d}_R^{\dagger}(-\vec{p}) \end{pmatrix} = V(-\vec{p})^{\dagger} \begin{pmatrix} -\hat{d}_-(\vec{p}) \\ \hat{d}_-^{\dagger}(-\vec{p}) \end{pmatrix}. \quad (8.89)$$

One then obtains

$$\hat{H}_W = \frac{1}{(2\pi)^3} \int d^3p \, \sqrt{\vec{p}^{\,2} + m^2} \left[\hat{c}_+^{\dagger}(\vec{p})\hat{c}_+(\vec{p}) + \hat{d}_-^{\dagger}(\vec{p})\hat{d}_-(\vec{p}) - V \right]. \quad (8.90)$$

The vacuum of massive Weyl fermions is characterized by $\hat{c}_+(\vec{p}) \, |0\rangle_R = \hat{d}_-(\vec{p}) \, |0\rangle_R = 0$, and the corresponding vacuum energy density is given by

$$\rho = \frac{E_0}{V} = -\frac{1}{(2\pi)^3} \int d^3p \, \sqrt{\vec{p}^{\,2} + m^2}. \quad (8.91)$$

This is exactly opposite to two times the positive vacuum energy that we encountered for the free real-valued scalar field theory in eq. (2.40). We conclude that a theory containing a complex-valued scalar field and a Weyl fermion field, with the same mass m, has a zero cosmological constant, because the bosonic and the fermionic contributions to the vacuum energy cancel each other.

In fact, *supersymmetry* relates the bosonic to the fermionic sector, with the complex scalar field and the Weyl fermion field forming the physical components of a so-called chiral super-multiplet. Hence, in an exactly supersymmetric world, the vacuum energy density would indeed vanish. However, we also know that in the real world, *i.e.* at low energy, supersymmetry has to be broken – if it exists at all – since for instance the "selectron" (the bosonic partner of the electron) must be much heavier than the electron; otherwise, it would have been observed already. Imposing the supersymmetry breaking, which is minimally required by phenomenology (even before the LHC results), leads to a naive estimate of the vacuum energy density that is still about 10^{60} times larger than the observed value $\approx (2 \cdot 10^{-3} \text{ eV})^4$ (Carroll, 2001), which accelerates the expansion of the Universe, so supersymmetry does not solve the cosmological constant problem.

The single-particle states of massive Weyl fermions with right-chirality are given by

$$\hat{c}_+^\dagger |0\rangle_R = |(-1)^{F_R} = -1, \vec{p}, \vec{\sigma} \cdot \vec{e}_p = 1\rangle_R,$$
$$\hat{d}_-^\dagger |0\rangle_R = |(-1)^{F_R} = -1, \vec{p}, \vec{\sigma} \cdot \vec{e}_p = -1\rangle_R. \quad (8.92)$$

Under CP′, these states transform as

$$\hat{U}_{CP'} |(-1)^{F_R} = -1, \vec{p}, \vec{\sigma} \cdot \vec{e}_p = \pm 1\rangle_R = \mp|(-1)^{F_R} = -1, -\vec{p}, -\vec{\sigma} \cdot \vec{e}_p = \mp 1\rangle_R \ ; \quad (8.93)$$

i.e. they change both their momentum and their helicity. Since U(1)$_R$ is reduced to $\mathbb{Z}(2)_R$, just as for Majorana fermions, there is no longer any conserved fermion number that distinguishes particles from anti-particles.

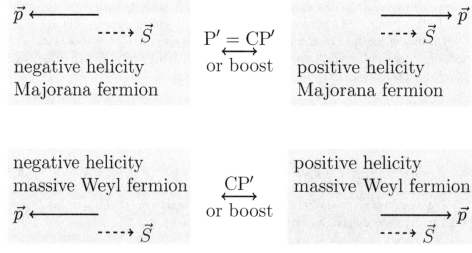

Fig. 8.3   Effect of the discrete symmetries P′ and CP′ on the helicity of massive Majorana and Weyl fermion states. In this case, particles are not distinguished from anti-particles. Since Majorana fermion fields are, by definition, invariant against C, for them CP′ is indistinguishable from P′. Again, an appropriate boost can also change the helicity.

Free massive Weyl and Majorana fermions are mathematically equivalent, but seem to have different physical interpretations. Majorana fermion fields are, by definition, unaffected by charge conjugation. Hence, for them C is a trivial symmetry operation. The Majorana condition is inconsistent with the standard parity transformation P, but consistent with the modified transformation P′. The Hamilton operator of massive Weyl fermions is not invariant under the individual symmetries C, P, or P′, but only under the combination CP′. At the end, both free, massive Weyl and Majorana fermions respond to the CP′ symmetry (which is indistinguishable from the modified parity P′ for Majorana fermions) in the same way. As a result, the corresponding Hamilton operators enjoy the same CP′ symmetry and are indeed physically equivalent.

The effects of P′ (or an appropriate boost) and of CP′ on the helicity of massive Majorana or Weyl fermions are illustrated in Figure 8.3.

8.8 Redundant Particle Labels and the Pauli Principle as a "Gauss Law"

We have seen in eq. (8.16) that – as a result of the anti-commutativity of the fermion field operators – the multi-fermion states in a quantum field theory are automatically antisymmetric against particle (or actually wavicle) exchange. We have also seen in Section 3.7 that, in the non-relativistic limit, massive wavicles behave like indistinguishable quantum mechanical particles. Unlike in quantum field theory, in quantum mechanics the antisymmetry of a multi-fermion wave function must be imposed by hand, by postulating the Pauli principle. In contrast, the field-theoretic anti-commutation relation (8.4) is required for conceptual reasons, as we are going to see in Section 9.9. In this section, we compare how *Fermi-Dirac statistics* manifests itself in quantum mechanics and in quantum field theory.

Let us consider a quantum mechanical non-relativistic multi-particle system that consists of N indistinguishable point particles with spin $1/2$. The position, momentum, and spin operators of the particles are denoted by $\hat{\vec{x}}_a$, $\hat{\vec{p}}_a$, and $\hat{\vec{S}}_a$ with $a \in \{1, 2, \ldots, N\}$. The multi-fermion wave function then takes the form $\Psi(\vec{x}_1, \vec{x}_2, \ldots, \vec{x}_N, S_1^3, S_2^3, \ldots, S_N^3)$, where S_a^3 is the spin projection along the 3-direction for the particle with the label a. In general, the multi-fermion Hamilton operator $\hat{H}(\hat{\vec{x}}_a, \hat{\vec{p}}_a, \hat{\vec{S}}_a)$ is Hermitian and depends on the position, momentum, and spin operators of the various particles. A simple spin-independent example is $\hat{H} = \sum_{a=1}^{N} \hat{\vec{p}}_a^{\,2} / 2m + \sum_{a<b} \hat{V}(|\hat{\vec{x}}_a - \hat{\vec{x}}_b|)$.

As an unavoidable consequence of the quantum mechanical description, we have to assign labels to the particles: We called them a. Since the particles are actually indistinguishable, the label is just a redundant unphysical bookkeeping device which should not have any effect on the physics. If the particles are truly indistinguishable, the Hamilton operator must be invariant against arbitrary permutations P of the particle labels, *i.e.* $[\hat{H}(\hat{\vec{x}}_a, \hat{\vec{p}}_a, \hat{\vec{S}}_a), \hat{P}] = 0$. Here, $\hat{P}$ is the unitary operator that implements a permutation P in the N-fermion Hilbert space. For example, the pair permutation $\hat{P}_{12}$ of the particle labels $a = 1$ and 2 acts as

$$\hat{P}_{12}\Psi(\vec{x}_1, \vec{x}_2, \ldots, \vec{x}_N, S_1^3, S_2^3, \ldots, S_N^3) = \Psi(\vec{x}_2, \vec{x}_1, \ldots, \vec{x}_N, S_2^3, S_1^3, \ldots, S_N^3). \qquad (8.94)$$

The permutations of the N particle labels form the *permutation group* S_N with $N!$ elements. The group theory of the permutation group S_N is discussed in Appendix F.8. Since the Hamilton operator is invariant against permutations of the particle labels, its spectrum decomposes into *degenerate multiplets* forming irreducible representations of S_N. Remarkably, almost all of these states are unphysical, because they are not totally anti-symmetric under particle exchange. Those states would be physical only if the particles were distinguishable. In order to select the physical states, we impose the Pauli principle, which demands for all permutations $\hat{P}$

$$\hat{P} \, \Psi(\vec{x}_1, \vec{x}_2, \ldots, \vec{x}_N, S_1^3, S_2^3, \ldots, S_N^3) = \text{sign}(P) \, \Psi(\vec{x}_1, \vec{x}_2, \ldots, \vec{x}_N, S_1^3, S_2^3, \ldots, S_N^3). \qquad (8.95)$$

Here, $\text{sign}(P) = \pm 1$ is the sign of the permutation. It is 1 if the permutation is composed of an even number of pair permutations, and -1 otherwise. Hence, the Pauli principle identifies the totally anti-symmetric 1-dimensional representation of the permutation group with the physical states. The only other 1-dimensional representation of S_N is totally symmetric. It is realized for indistinguishable bosons. All other representations of S_N are higher-dimensional. Those cannot be realized with indistinguishable particles.

We notice some striking similarities of the implementation of the Pauli principle in quantum mechanics and of gauge symmetry in quantum field theories. Abelian gauge theory was discussed in Chapters 6 and 7, and non-Abelian gauge theory will be introduced in Chapter 11. In both cases, one uses *redundant variables:* the artificial particle labels that are used to "distinguish" indistinguishable particles in quantum mechanics, and the gauge-variant vector potentials that are used to describe gauge-invariant physical quantities in gauge theories. In order to be able to eliminate unphysical effects of the redundant variables, one demands invariance of the Hamilton operator under a gauge symmetry. In gauge field theories, this symmetry is local in space and time and transforms vector potentials into gauge-equivalent ones.

In the quantum mechanics of N indistinguishable fermions, the permutation group S_N plays the role of a "gauge" symmetry. It transforms one arbitrary labeling of particles into another one. In this case, the gauge symmetry is not associated with points in space and is thus not local. Still, it serves the usual purpose of a gauge symmetry, namely, to guarantee that unphysical effects of redundant variables, in this case of the artificial particle labels, can be eliminated.

The elimination itself is realized via the implementation of the *Gauss law.* In gauge field theories, Gauss' law demands that physical states $|\Psi\rangle$ are invariant against infinitesimal gauge transformations, *i.e.* $\hat{G}^a(\vec{x})|\Psi\rangle = 0$. Here, $\hat{G}^a(\vec{x})$ is an infinitesimal generator of a gauge transformation at the point $\vec{x}$. This implies that all physical states transform as singlets under the gauge group. In the quantum mechanics of N indistinguishable fermions, the role of the Gauss law is played by the Pauli principle which demands $\hat{P}|\Psi\rangle = \text{sign}(P)$ $|\Psi\rangle$. Here, $\hat{P}$ is the unitary operator that implements the permutation $P \in S_N$ of the particle labels in the quantum mechanical N-fermion Hilbert space. Again, only singlets under the "gauge" group S_N qualify as physical states, because only for them the artificial particle labels have no effect on the dynamics of indistinguishable fermions.

8.9 Can We Supersede Gauge Symmetry?

This section speculates about the origin of gauge symmetry and can be skipped while still following the context.

As we have just seen in Section 8.8, quantum field theory naturally explains why elementary fermions are indistinguishable. They simply result as excitations of the same quantum field, based on the same creation operator. In addition, the anti-symmetry of multi-fermion states follows from the anti-commutation relations of the fermion field operators, without the necessity to impose the Pauli principle as a constraint by hand.

In quantum mechanics, on the other hand, the indistinguishability of elementary fermions is not naturally encoded in the point particle concept. As a consequence, one introduces artificial particle labels and demands particle permutation symmetry with the "gauge" group S_N. Finally, one imposes the Pauli principle as a "Gauss law" for S_N in order to eliminate the unphysical states that correspond to distinguishable particles. We conclude that the fundamental quantum field theoretical description is more economical than the quantum mechanical one, which plays the role of a non-relativistic effective theory. In particular, the fundamental field theory description does not use redundant variables and hence does not need a gauge symmetry to eliminate them.

As we will see in detail in the rest of the book, the dynamics of the Standard Model are dominated by gauge symmetries, *i.e.* the local transformations of Abelian and non-Abelian vector potentials. At present, the Standard Model is our most fundamental description of Nature and hence we do not understand the deep origin of its gauge symmetry.

Could it be that, just like the S_N "gauge" symmetry of N-fermion quantum mechanics, the gauge symmetry of the Standard Model will no longer appear in a more fundamental underlying theory? To some extent, this is indeed the case in string theory which is a candidate for such a theory. Can the analogy with S_N "gauge" symmetry of N-fermion quantum mechanics provide a hint about the origin of gauge symmetry in the Standard Model? When fermion field operators replace point particle coordinates, one no longer needs S_N "gauge" symmetry. What should replace the redundant vector potentials of the Standard Model in order to supersede its $SU(3)_c \times SU(2)_L \times U(1)_Y$ gauge symmetry?

These are wide-open questions whose answers are unknown, but which are very well worth thinking about at a deep level.

Box 8.1   **Big question**

What is the origin of gauge symmetry?

Exercises

8.1 Canonical anti-commutation relations and Hamilton operator of Weyl fermions

a) Show that the canonical anti-commutation relations of eq. (8.12) follow from those of eq. (8.4).

b) Show that the Hamilton operator of eq. (8.13) follows from the one of eq. (8.2).

8.2 Momentum operator of Weyl fermions

Derive eq. (8.23) from the first line of eq. (8.22).

8.3 Weyl fermion helicity

Derive eqs. (8.26).

8.4 γ-gymnastics

a) Show that solving the relation $\{\gamma^\mu, \gamma^\nu\} = 2g^{\mu\nu}$ of eq. (8.54) requires (at least) 4×4 matrices.

Consider now γ-matrices without specifying any basis; *i.e.* only refer to the above anti-commutation relations and to the definition $\gamma^5 = i\gamma^0\gamma^1\gamma^2\gamma^3$.

b) Show the following properties: $(\gamma^5)^2 = 1$, $\mathrm{Tr}\,\gamma^5 = 0$, $\{\gamma^\mu, \gamma^5\} = 0$, $\gamma^{5\,\dagger} = \gamma^5$. Which are the eigenvalues of γ^5 ?

c) Show that $P_{\mathrm{R,L}} = \frac{1}{2}(1 \pm \gamma^5)$ are projection operators (idempotent), *i.e.* $P_{\mathrm{R,L}}^2 = P_{\mathrm{R,L}}$. Compute $P_+ + P_-$ and P_+P_-.

d) Simplify the expressions $\mathrm{Tr}(\gamma^\mu a_\mu \gamma^\nu b_\nu)$ and $\mathrm{Tr}(\gamma_5 \gamma^\mu a_\mu \gamma^\nu b_\nu)$ for arbitrary 4-vectors a and b.

8.5 Dirac and Majorana basis of the γ-matrices

a) In the Dirac basis, the γ-matrices take the form

$$\gamma^0 = \begin{pmatrix} 1 & 0 \\ 0 & -1 \end{pmatrix}, \quad \vec{\gamma} = (\gamma^1, \gamma^2, \gamma^3) = \begin{pmatrix} 0 & \vec{\sigma} \\ -\vec{\sigma} & 0 \end{pmatrix},$$

and in the Majorana basis $i\vec{\gamma}$ is real,

$$\gamma^0 = \begin{pmatrix} 0 & \sigma_2 \\ \sigma_2 & 0 \end{pmatrix}, \quad \gamma^1 = \begin{pmatrix} i\sigma^3 & 0 \\ 0 & i\sigma^3 \end{pmatrix},$$

$$\gamma^2 = \begin{pmatrix} 0 & -\sigma^2 \\ \sigma^2 & 0 \end{pmatrix}, \quad \gamma^3 = \begin{pmatrix} -i\sigma^1 & 0 \\ 0 & -i\sigma^1 \end{pmatrix}.$$

Which is the form of γ^5 ? Show that the relation (8.55) holds.

b) We stay with the relation $^C\hat{\psi}(\vec{x}) = C\gamma^0\hat{\psi}^\dagger(\vec{x})^\mathsf{T}$, as in eq. (8.56). How do we have to choose the matrix C in the Dirac and in the Majorana basis?

8.6 Diagonalization of the Dirac Hamilton operator, parity, and charge conjugation

a) Derive the Dirac Hamilton operator of eq. (8.59) from eq. (8.57).

b) Use the transformations of eqs. (8.40) and (8.60) to derive eq. (8.68).

c) Show that the parity partner of a Dirac fermion has the opposite momentum, and that the charge conjugation partner of a fermion is an anti-fermion with the same momentum.

8.7 Dirac fermion helicity

Derive eqs. (8.66) and (8.67).

8.8 Clifford algebra

The following set forms a so-called Clifford algebra,

$$\Gamma^c \in \left\{ \mathbb{1}, \ \gamma^\mu, \ \gamma^\mu \gamma^5, \ \sigma^{\mu\nu} = [\gamma^\mu, \gamma^\nu]/(2\mathrm{i}), \ \mathrm{i}\gamma^5 \right\}.$$

The index $c = 1, \ldots, N$ runs over all independent elements in this set.

a) What is N ?

b) The terms $\bar{\psi} \Gamma^c \psi$ can represent fermionic observables.

Show that they are all Hermitian.

c) What is the inverse of Γ^c for each $c = 1, \ldots, N$?

(Use lower Lorentz indices to express $(\Gamma^c)^{-1}$ in a pleasant form.)

Fermionic Functional Integrals

In this chapter, we formulate the *functional integral for fermion fields* using anti-commuting Grassmann variables. We then derive the *Dirac, Weyl, and Majorana equation* from the corresponding Lagrangians and show that the Hamiltonian formulation of the previous chapter arises upon canonical quantization, by replacing Grassmann fields by anti-commuting fermion creation and annihilation operator fields. We then perform a Wick rotation to *Euclidean time* and investigate the Euclidean version of Lorentz invariance, and of the discrete symmetries C and P, as well as time reversal T. This will lead to a discussion of the *CPT theorem* and the *spin–statistics theorem*. Finally, we show in the simple case of massless Weyl fermions that the Euclidean functional integral is equivalent to the Hamiltonian formulation. For this purpose, we construct the *transfer matrix* that emerges from the functional integral on a Euclidean time lattice, and we show that it reduces to the correct Hamilton operator in the continuous time limit.

9.1 Grassmann Algebra, Pfaffian, and Fermion Determinant

In contrast to bosons, fermions cannot be piled up in large numbers in the same quantum state and hence do not manifest themselves directly at the classical level – for instance, we cannot build a "fermionic laser". While for bosonic quantum field theories one uses (real- or complex-valued) classical fields as integration variables in the functional integral, for fermionic theories no corresponding classical fields exist. Instead, the fermionic functional integral uses anti-commuting *Grassmann numbers* as the integration variables.

In the following, we summarize some of the basic rules for operating with Grassmann numbers. This kind of numbers was introduced by Hermann Grassmann (1809–1877)[1] and a century later adopted in quantum field theory, in particular by Felix Berezin (Berezin, 1966).

The generators of a *Grassmann algebra* are anti-commuting variables η_i, $i \in \{1, 2, \ldots, N\}$, that obey

$$\eta_i \eta_j = -\eta_j \eta_i. \tag{9.1}$$

This implies in particular $\eta_i^2 = 0$. An element of the Grassmann algebra is a polynomial in the variables η_i,

$$f(\eta) = f + \sum_i f_i \eta_i + \sum_{ij} f_{ij} \eta_i \eta_j + \sum_{ijk} f_{ijk} \eta_i \eta_j \eta_k + \ldots. \tag{9.2}$$

---

[1] The authentic spelling seems to be "Graßmann", but we adopt the spelling which is internationally used.

The coefficients $f_{ij...l}$ are ordinary commuting numbers (real or complex); they are anti-symmetric in the indices $i, j, \ldots, l$. There can be at most $l = N$ indices, since each variable η_i occurs in each term either with power 0 or 1. Thus, the expansion (9.2) is already the most general "Taylor series" of a function $f(\eta)$; there are only 2^N independent terms.

One defines a formal differentiation, which follows the familiar pattern (regarding the left-most factor). The corresponding rules are

$$\frac{\delta}{\delta \eta_i} \eta_j = \delta_{ij}, \quad \frac{\delta}{\delta \eta_i} \eta_i \eta_j = \eta_j \quad \Rightarrow \quad \frac{\delta}{\delta \eta_i} \eta_j \eta_i = -\eta_j \qquad (i \neq j). \tag{9.3}$$

Since we aim at a functional integral for Grassmann fields (Berezin, 1966), we also need rules for the *integration* of Grassmann variables. As usual, one defines integration as a linear operation

$$\int d\eta_i \, (a + b\eta_i) = a \int d\eta_i + b \int d\eta_i \, \eta_i, \qquad a, b \in \mathbb{C}, \tag{9.4}$$

where the integrand represents the most general function $f(\eta_i)$.

Unlike ordinary numbers, Grassmann numbers cannot take particular values. Hence, it is not meaningful to extend a Grassmann integral over some interval as its integration range. Still, integrals over Grassmann numbers share some features with ordinary integrals $\int_{-\infty}^{\infty} dx$ that extend over all of $\mathbb{R}$. This is, in fact, the range that we extensively used in the bosonic functional integrals for a real-valued scalar field. We often made use of its *invariance under translations* $x \rightarrow x + c$ (where c is a constant), for instance, when we evaluated the partition function and the 2-point function of the free (lattice) scalar field in Section 1.7. Grassmann integrals are formulated such that invariance under translation holds as well,

$$\int d\eta_i \, (a + b\eta_i) = \int d\eta_i \, (a + b(\eta_i + c + d \, \eta_j)) = \int d\eta_i \, (a + b\eta_i + bc + bd \, \eta_j) \quad (i \neq j). \tag{9.5}$$

This provides an argument for the rule that integrals $\int d\eta_i$ over terms that do *not* involve η_i vanish. On the other hand, the integral $\int d\eta_i \, \eta_i$ is defined to be non-zero, and – since we have no scale at hand – we set it to 1.[2] Thus, we have motivated the following integration rules,

$$\int d\eta_i = 0, \quad \int d\eta_i \, \eta_j = \delta_{ij}, \quad \int d\eta_i d\eta_j \, \eta_i \eta_j = -1 \qquad (i \neq j). \tag{9.6}$$

The last rule corresponds to the prescription to carry out the innermost integral first. Again, these integrals are formal expressions and one should not ask over what range of η_i we integrate. Interestingly, Grassmann integration acts just like differentiation, cf. eq. (9.3). After all, we have constructed the rules (9.6) based on translation invariance, which holds for differentiation as well.

Up to permutations, the only non-vanishing integral over the entire set of Grassmann variables is

$$\int d\eta_1 d\eta_2 \ldots d\eta_N \, \eta_N \ldots \eta_2 \eta_1 = 1. \tag{9.7}$$

Let us consider the "Gaussian" integral for N Grassmann variables.

---

[2] Choosing a different constant ($\neq 0$) would not affect any fermionic expectation values.

- $N = 2$

$$\int d\eta_1 d\eta_2 \, \exp(-\eta_1 A_{12}\eta_2) = \int d\eta_1 d\eta_2 \, (1 - \eta_1 A_{12}\eta_2) = A_{12}. \tag{9.8}$$

Note that this holds for any $A_{12} \in \mathbb{C}$. The expansion of the exponential terminates because $\eta_1^2 = \eta_2^2 = 0$.

- $N = 3$

$$\int d\eta_1 d\eta_2 d\eta_3 \, \exp(-\eta_1 A_{12}\eta_2 - \eta_1 A_{13}\eta_3 - \eta_2 A_{23}\eta_3) =$$

$$\int d\eta_1 d\eta_2 d\eta_3 \, (1 - \eta_1 A_{12}\eta_2 - \eta_1 A_{13}\eta_3 - \eta_2 A_{23}\eta_3) = 0. \tag{9.9}$$

The quadratic term in the expansion of the exponential vanishes because at least one of the three Grassmann variables gets squared. Indeed, the corresponding integral

$$\int \mathcal{D}\eta \, \exp\left(-\frac{1}{2}\eta A\eta\right) = \int d\eta_1 d\eta_2 \dots d\eta_N \, \exp\left(-\frac{1}{2}\eta_i A_{ij}\eta_j\right), \tag{9.10}$$

with A being any anti-symmetric matrix (*i.e.* $A_{ij} = -A_{ji}$), vanishes for *all* odd N. This follows because each term in the expansion of the exponential contains an even number of Grassmann variables, while the non-vanishing integral of eq. (9.7) requires N of them.

- $N = 4$

$$\int \mathcal{D}\eta \, \exp\left(-\frac{1}{2}\eta A\eta\right) = \int d\eta_1 d\eta_2 d\eta_3 d\eta_4 \, \frac{1}{2}\left(\frac{1}{2}\eta_i A_{ij}\eta_j\right)^2 = A_{12}A_{34} - A_{13}A_{24} + A_{23}A_{14}. \tag{9.11}$$

Explicit calculation shows $(A_{12}A_{34} - A_{13}A_{24} + A_{23}A_{14})^2 = \det(A)$.

This extends to the result for general N

$$\int \mathcal{D}\eta \, \exp\left(-\frac{1}{2}\eta A\eta\right) = \mathrm{Pf}(A), \tag{9.12}$$

where $\mathrm{Pf}(A)$ is known as the *Pfaffian* of the anti-symmetric matrix A, named after Johann Pfaff, 1765–1825. Generally, its square is the determinant of A,

$$\mathrm{Pf}(A)^2 = \det(A). \tag{9.13}$$

While the square root of the determinant has a sign ambiguity, the Pfaffian is uniquely defined. It is given by a sum over the elements of the permutation group, $P \in S_N$, whose signature depends on whether P is composed of an even ($\mathrm{sign}(P) = 1$) or odd ($\mathrm{sign}(P) = -1$) number of pair permutations,

$$\mathrm{Pf}(A) = \frac{1}{2^{N/2}(N/2)!} \sum_{P \in S_N} \mathrm{sign}(P) \prod_{i=1}^{N/2} A_{P(2i-1),P(2i)}. \tag{9.14}$$

As we will see later, Pfaffians arise in the functional integrals for Majorana or massive Weyl fermions.

Quick Question 9.1.1   From the Pfaffian to the determinant

Consider the case $N = 4$: check the Grassmann integral (9.11), and show that it coincides with Pf(A) as defined in formula (9.14) and that its square is equal to det(A).

In fermionic quantum field theories, two distinct sets of Grassmann numbers η_i and $\bar{\eta}_i$ are associated with fermion creation and annihilation operators. It is important that η_i and $\bar{\eta}_i$ are *independent* Grassmann numbers. In particular, unlike creation and annihilation operators, $\bar{\eta}_i$ is not in any sense an adjoint of η_i. From the point of view of the Grassmann algebra, the bar is just a bookkeeping device that distinguishes two subsets of generators. Hence, if we again introduce η_i with $i \in \{1, 2, \ldots, N\}$, and, in addition, $\bar{\eta}_i$, the total number of generators is now $2N$.

It is instructive to evaluate the Gaussian integral in the $N = 2$ case, which yields

$$\int d\bar{\eta}_1 d\eta_1 d\bar{\eta}_2 d\eta_2 \, \exp\left(-(\bar{\eta}_1, \bar{\eta}_2) \begin{pmatrix} M_{11} & M_{12} \\ M_{21} & M_{22} \end{pmatrix} \begin{pmatrix} \eta_1 \\ \eta_2 \end{pmatrix} \right) = M_{11}M_{22} - M_{12}M_{21}.$$

$$(9.15)$$

Quick Question 9.1.2   Fermion determinant of a 2×2 matrix

Check eq. (9.15).

This result generalizes when we enlarge the Grassmann algebra to an arbitrary, even number $2N$ of generators, and we obtain a *fermion determinant*

$$\int \mathcal{D}\bar{\eta}\mathcal{D}\eta \, \exp(-\bar{\eta}M\eta) = \int d\bar{\eta}_1 d\eta_1 d\bar{\eta}_2 d\eta_2 \ldots d\bar{\eta}_N d\eta_N \, \exp(-\bar{\eta}_i M_{ij} \eta_j)$$

$$= \det(M).$$

$$(9.16)$$

An easy way to make this result plausible proceeds by performing the substitution $\eta' = M\eta$, for which det(M) arises as the Jacobian,

$$\int \mathcal{D}\bar{\eta}\mathcal{D}\eta \, \exp(-\bar{\eta}M\eta) = \int \mathcal{D}\bar{\eta}\mathcal{D}\eta' \det(M) \exp(-\bar{\eta}\eta') =$$

$$\det(M) \prod_i \int d\bar{\eta}_i d\eta'_i \, \exp(-\bar{\eta}_i \eta'_i) = \det(M) \prod_i \int d\bar{\eta}_i d\eta'_i \, (-\bar{\eta}_i \eta'_i) = \det(M). \quad (9.17)$$

One might question, however, whether the usual Jacobian factor really applies to the Grassmann integral. An explicit calculation will be asked for in Problem 9.3.

Just as the Pfaffian, the determinant can also be expressed as a sum over permutations

$$\det(M) = \sum_{P \in S_N} \text{sign}(P) \prod_{i=1}^{N} A_{i, P(i)}. \tag{9.18}$$

The result of eq. (9.16) is consistent with eq. (9.12) because we can write

$$
\int \mathcal{D}\bar{\eta}\mathcal{D}\eta \, \exp(-\bar{\eta}M\eta) = \int \mathcal{D}\bar{\eta}\mathcal{D}\eta \, \exp\left(-\frac{1}{2}\bar{\eta}_i M_{ij}\eta_j + \frac{1}{2}\eta_j M_{ji}^\mathsf{T}\bar{\eta}_i\right)
$$

$$
= \int \mathcal{D}\bar{\eta}\mathcal{D}\eta \, \exp\left(-\frac{1}{2}(\eta,\bar{\eta})\begin{pmatrix} 0 & -M^\mathsf{T} \\ M & 0 \end{pmatrix}\begin{pmatrix} \eta \\ \bar{\eta} \end{pmatrix}\right) = \mathrm{Pf}(A),
$$

where we have introduced an anti-symmetric matrix A, whose Pfaffian squared indeed coincides with $\det(M)^2$,

$$
A = \begin{pmatrix} 0 & -M^\mathsf{T} \\ M & 0 \end{pmatrix}, \quad \mathrm{Pf}(A)^2 = \det(A) = \det(M)\det(M^\mathsf{T}) = \det(M)^2. \tag{9.19}
$$

It is instructive to compare the previous results for fermionic Gaussian integrals with ordinary, *i.e.* bosonic ones:

- First, we consider a real-valued lattice scalar field $\phi_i \in \mathbb{R}$ with $i \in \{1, 2, \dots, N\}$, such that

$$
\int \mathcal{D}\phi \, \exp\left(-\frac{1}{2}\phi A\phi\right) = \int_\mathbb{R} d\phi_1 \int_\mathbb{R} d\phi_2 \dots \int_\mathbb{R} d\phi_N \frac{1}{(2\pi)^{N/2}} \exp\left(-\frac{1}{2}\phi_i A_{ij}\phi_j\right). \tag{9.20}
$$

For convenience, we have absorbed N factors $1/\sqrt{2\pi}$ in the definition of the measure $\mathcal{D}\phi$. We assume A to be real and – without loss of generality – symmetric, *i.e.* $A_{ij} \in \mathbb{R}$ and $A^\mathsf{T} = A$, so it can be diagonalized by an orthogonal transformation $\Omega \in \mathrm{SO}(N)$,

$$
\Omega A \Omega^\mathsf{T} = D = \mathrm{diag}(a_1, a_2, \dots, a_N), \tag{9.21}
$$

where $a_i \in \mathbb{R}$ are the eigenvalues of A. The Gaussian integral exists only if all a_i are positive. Performing the orthogonal rotation $\phi' = \Omega\phi$, we obtain

$$
\int \mathcal{D}\phi \, \exp\left(-\frac{1}{2}\phi A\phi\right) = \int \mathcal{D}\phi' \, \exp\left(-\frac{1}{2}\phi' D\phi'\right) =
$$

$$
\int_\mathbb{R} d\phi'_1 \int_\mathbb{R} d\phi'_2 \dots \int_\mathbb{R} d\phi'_N \prod_{i=1}^N \frac{1}{\sqrt{2\pi}} \exp\left(-\frac{1}{2}a_i \phi_i'^2\right) = \prod_{i=1}^N \frac{1}{\sqrt{a_i}} = \frac{1}{\sqrt{\det(A)}}, \tag{9.22}
$$

as we saw before in Section 1.7. This is similar to the result of eq. (9.12), except that A is now symmetric and the square root of its determinant now appears in the denominator.

- Next, we consider a complex lattice scalar field $\Phi_i \in \mathbb{C}$ with $i \in \{1, 2, \dots, N\}$, such that

$$
\int \mathcal{D}\Phi \, \exp\left(-\frac{1}{2}\Phi^\dagger M\Phi\right) = \int_\mathbb{C} d\Phi_1 \int_\mathbb{C} d\Phi_2 \dots \int_\mathbb{C} d\Phi_N \frac{1}{(2\pi)^N} \exp\left(-\frac{1}{2}\Phi_i^* M_{ij}\Phi_j\right). \tag{9.23}
$$

Here, we assume M to be Hermitian and thus be diagonalizable by a unitary transformation $U \in \mathrm{SU}(N)$

$$
UMU^\dagger = D = \mathrm{diag}(m_1, m_2, \dots, m_N), \quad m_i \in \mathbb{R}. \tag{9.24}
$$

The Gaussian integral again exists only if all eigenvalues m_i are positive. Performing the unitary transformation $\Phi' = U\Phi$, we obtain

$$\int \mathcal{D}\Phi \, \exp\left(-\frac{1}{2}\Phi^\dagger M\Phi\right) = \int \mathcal{D}\Phi' \, \exp\left(-\frac{1}{2}\Phi'^\dagger D\Phi'\right) =$$

$$\int_{\mathbb{C}} d\Phi_1' \int_{\mathbb{C}} d\Phi_2' \cdots \int_{\mathbb{C}} d\Phi_N' \prod_{i=1}^{N} \frac{1}{2\pi} \exp\left(-\frac{1}{2}m_i|\Phi_i'|^2\right) = \prod_{i=1}^{N} \frac{1}{m_i} = \frac{1}{\det(M)}.$$

$$(9.25)$$

This is similar to the result of eq. (9.16), except that M is now Hermitian (rather than being unrestricted as in the fermionic case), and the determinant again appears in the *denominator*. Essentially, it is only the power of the determinant that depends on the type of integration variables.

A quantity of physical interest is the *fermionic 2-point function*. In the two-variable ($N = 2$) case, we readily obtain

$$\int d\eta_1 d\eta_2 \, \eta_1 \eta_2 \exp(-\eta_1 A_{12}\eta_2) = -1,$$

$$(9.26)$$

such that

$$\langle \eta_1 \eta_2 \rangle = \frac{\int d\eta_1 d\eta_2 \, \eta_1 \eta_2 \exp(-\eta_1 A_{12}\eta_2)}{\int d\eta_1 d\eta_2 \, \exp(-\eta_1 A_{12}\eta_2)} = -\frac{1}{A_{12}} = (A^{-1})_{12},$$

$$A^{-1} = \begin{pmatrix} 0 & A_{12} \\ -A_{12} & 0 \end{pmatrix}^{-1} = \begin{pmatrix} 0 & -1/A_{12} \\ 1/A_{12} & 0 \end{pmatrix}.$$

$$(9.27)$$

That result generalizes to

$$\langle \eta_i \eta_j \rangle = \frac{\int \mathcal{D}\eta \, \eta_i \eta_j \exp\left(-\frac{1}{2}\eta A\eta\right)}{\int \mathcal{D}\eta \, \exp\left(-\frac{1}{2}\eta A\eta\right)} = (A^{-1})_{ij}.$$

$$(9.28)$$

This will be demonstrated in Problem 9.4.

Next we consider the case with two distinct sets of Grassmann generators, first for $N = 2$,

$$\int d\bar{\eta}_1 d\eta_1 d\bar{\eta}_2 d\eta_2 \, \eta_1 \bar{\eta}_2 \exp\left(-(\bar{\eta}_1, \bar{\eta}_2) \begin{pmatrix} M_{11} & M_{12} \\ M_{21} & M_{22} \end{pmatrix} \begin{pmatrix} \eta_1 \\ \eta_2 \end{pmatrix}\right) =$$

$$\int d\bar{\eta}_1 d\eta_1 d\bar{\eta}_2 d\eta_2 \, \eta_1 \bar{\eta}_2 (-\bar{\eta}_1 M_{12}\eta_2) = -M_{12},$$

$$(9.29)$$

which implies

$$\langle \eta_1 \bar{\eta}_2 \rangle = (M^{-1})_{12}, \quad M^{-1} = \frac{1}{M_{11}M_{22} - M_{12}M_{21}} \begin{pmatrix} M_{22} & -M_{12} \\ -M_{21} & M_{11} \end{pmatrix}.$$

$$(9.30)$$

For arbitrary N, this result generalizes to

$$\langle \eta_i \bar{\eta}_j \rangle = \frac{\int \mathcal{D}\bar{\eta}\mathcal{D}\eta \, \eta_i \bar{\eta}_j \exp\left(-\bar{\eta}M\eta\right)}{\int \mathcal{D}\bar{\eta}\mathcal{D}\eta \, \exp\left(-\bar{\eta}M\eta\right)} = (M^{-1})_{ij}.$$

$$(9.31)$$

This is consistent with eq. (9.28) because

$$A^{-1} = \begin{pmatrix} 0 & -M^{\mathsf{T}} \\ M & 0 \end{pmatrix}^{-1} = \begin{pmatrix} 0 & M^{-1} \\ -(M^{-1})^{\mathsf{T}} & 0 \end{pmatrix} \quad \Rightarrow$$

$$(M^{-1})_{ij} = (A^{-1})_{i,j+N} \quad \Rightarrow \quad \langle \eta_i \bar\eta_j \rangle = \langle \eta_i \eta_{j+N} \rangle, \tag{9.32}$$

which indeed results when we relabel $\bar\eta_j$ as η_{j+N}. Again there is a remarkable similarity to the scalar field result, in this case to the 2-point function in eq. (1.71).

Let us prove eq. (9.31) by means of the *generating functional*

$$Z[\bar\xi, \xi] = \int \mathcal{D}\bar\eta \mathcal{D}\eta \, \exp\left(-\bar\eta M \eta + \bar\xi \eta_i + \bar\eta_j \xi\right). \tag{9.33}$$

Here, we have introduced Grassmann sources ξ and $\bar\xi$ only at the positions i and j.[3] Differentiating the generating functional with respect to the two Grassmann source variables $\bar\xi$ and ξ, we obtain

$$\frac{\delta}{\delta\bar\xi} \exp\left(\bar\xi \eta_i\right) = \frac{\delta}{\delta\bar\xi}\left(1 + \bar\xi \eta_i\right) = \eta_i, \quad \frac{\delta}{\delta\xi} \exp\left(\bar\eta_j \xi\right) = \frac{\delta}{\delta\xi}\left(1 + \bar\eta_j \xi\right) = -\bar\eta_j \quad \Rightarrow$$

$$\frac{\delta}{\delta\xi}\frac{\delta}{\delta\bar\xi} Z[\bar\xi, \xi] = \int \mathcal{D}\bar\eta\mathcal{D}\eta \, \eta_i \bar\eta_j \exp\left(-\bar\eta M \eta\right). \tag{9.34}$$

Performing the shifts $\bar\eta'_k = \bar\eta_k - \bar\xi M^{-1}_{ik}$ and $\eta'_l = \eta_l - M^{-1}_{lj}\xi$ yields

$$-\bar\eta_k M_{kl}\eta_l + \bar\xi\eta_i + \bar\eta_j\xi = -\left(\bar\eta'_k + \bar\xi M^{-1}_{ik}\right)M_{kl}\left(\eta'_l + M^{-1}_{lj}\xi\right) + \bar\xi\left(\eta'_i + M^{-1}_{ij}\xi\right) + \left(\bar\eta'_j + \bar\xi M^{-1}_{ij}\right)\xi$$

$$= -\bar\eta'_k M_{kl}\eta'_l + \bar\xi M^{-1}_{ij}\xi \quad \Rightarrow$$

$$Z[\bar\xi, \xi] = \det(M) \exp\left(\bar\xi M^{-1}_{ij}\xi\right),$$

$$\frac{\delta}{\delta\xi}\frac{\delta}{\delta\bar\xi} \exp\left(\bar\xi M^{-1}_{ij}\xi\right) = \frac{\delta}{\delta\xi}\frac{\delta}{\delta\bar\xi}\left(1 + \bar\xi M^{-1}_{ij}\xi\right) = M^{-1}_{ij} \quad \Rightarrow$$

$$\frac{\delta}{\delta\xi}\frac{\delta}{\delta\bar\xi} Z[\bar\xi, \xi] = \det(M) M^{-1}_{ij}, \tag{9.35}$$

which – together with eq. (9.16) – proves eq. (9.31).

It is interesting to note that the generating functional can also be evaluated by interpreting the exponential as a Euclidean action $S[\bar\eta, \eta]$,

$$Z[\bar\xi, \xi] = \int \mathcal{D}\bar\eta\mathcal{D}\eta \, \exp(-S[\bar\eta, \eta]), \quad S[\bar\eta, \eta] = \bar\eta M \eta - \bar\xi\eta_i - \bar\eta_j\xi, \tag{9.36}$$

and by varying it with respect to the Grassmann field in order to obtain "classical equations of motion"

$$\frac{\delta S[\bar\eta, \eta]}{\delta\bar\eta_k} = M_{kl}\eta_l - \delta_{kj}\xi = 0 \quad \Rightarrow \quad \eta_l = M^{-1}_{lj}\xi,$$

$$\frac{\delta S[\bar\eta, \eta]}{\delta\eta_l} = -\bar\eta_k M_{kl} + \bar\xi\delta_{il} = 0 \quad \Rightarrow \quad \bar\eta_k = \bar\xi M^{-1}_{ik}. \tag{9.37}$$

---

[3] If one would introduce source fields elsewhere, one would have to put them to zero at the end of the calculation. It seems unnatural to assign zero to a Grassmann number, which does not actually take any value. If one does it anyways, one obtains the same result as the one we will derive now.

Inserting the result back into the action, we obtain its "value" at the "stationary point"

$$S_0[\bar{\eta}, \eta] = \bar{\xi} M_{ik}^{-1} M_{kl} M_{lj}^{-1} \xi - \bar{\xi} M_{ij}^{-1} \xi - \bar{\xi} M_{ij}^{-1} \xi = -\bar{\xi} M_{ij}^{-1} \xi. \qquad (9.38)$$

The result of the Gaussian integral then resembles the outcome of the steepest decent method,

$$Z[\bar{\xi}, \xi] = \int \mathcal{D}\bar{\eta}\mathcal{D}\eta \, \exp(-S[\bar{\eta}, \eta]) = \det(M) \exp(-S_0[\bar{\eta}, \eta])$$

$$= \det(M) \exp(\bar{\xi} M_{ij}^{-1} \xi). \qquad (9.39)$$

This method, which uses the "classical equations of motion", is applicable to all Gaussian Grassmann integrals.

9.2 Dirac Equation

Paul Adrien Maurice Dirac constructed a relativistic equation for the electron (Dirac, 1928)

$$\left(i\gamma^\mu \partial_\mu - m\right) \psi(x) = 0, \quad \psi(x) = \begin{pmatrix} \psi_L(x) \\ \psi_R(x) \end{pmatrix}, \qquad (9.40)$$

which led him to predict the existence of anti-matter. It was a great triumph of Dirac's theory that the positron – the anti-particle of the electron – was discovered with the predicted properties a few years later (Anderson, 1933).

Again we deal with the γ-matrices in the chiral basis (which Dirac was originally not using)

$$\gamma^0 = \begin{pmatrix} 0 & \mathbb{1} \\ \mathbb{1} & 0 \end{pmatrix}, \quad \gamma^i = \begin{pmatrix} 0 & \sigma^i \\ -\sigma^i & 0 \end{pmatrix}, \quad \gamma^5 = i\gamma^0\gamma^1\gamma^2\gamma^3 = \begin{pmatrix} -\mathbb{1} & 0 \\ 0 & \mathbb{1} \end{pmatrix}. \quad (9.41)$$

It took a long time to decipher the meaning of the Dirac equation, in particular, to figure out what the 4-component spinor $\psi(x)$ means. Originally, Dirac interpreted his equation as a relativistic generalization of the Schrödinger equation, with $\psi(x)$ playing the role of a quantum mechanical wave function that describes a single particle or anti-particle. The equation can then be written as

$$i\partial_0 \psi(x) = -i\gamma^0\gamma^i \partial_i \psi(x) + \gamma^0 m \psi(x) = \left(-i\vec{\alpha} \cdot \vec{\nabla} + \beta m\right) \psi(x), \qquad (9.42)$$

with the matrices $\alpha^i = \gamma^0\gamma^i$ and $\beta = \gamma^0$ matching eq. (8.58). The right-hand side of this equation indeed resembles the Dirac Hamilton operator of eq. (8.57),

$$\hat{H}_D = \int d^3x \, \hat{\psi}^\dagger(\vec{x}) \left(-i\vec{\alpha} \cdot \vec{\nabla} + \beta m\right) \hat{\psi}(\vec{x}), \qquad (9.43)$$

where it is sandwiched between fermion creation and annihilation operators $\hat{\psi}^\dagger(\vec{x})$ and $\hat{\psi}(\vec{x})$. It eventually became clear that an interpretation of the Dirac equation with a fixed particle number is inconsistent, and that the equation does not belong to quantum mechanics but to quantum field theory. In other words, $\psi(x)$ is not a wave function but a quantum field. In fact, we can promote the time-independent field operator $\hat{\psi}(\vec{x})$ in the Schrödinger picture to a time-dependent operator in the Heisenberg picture

$$\hat{\psi}(x) = \hat{\psi}(x^0, \vec{x}) = \exp\left(i\hat{H}_D x^0\right) \hat{\psi}(\vec{x}) \exp\left(-i\hat{H}_D x^0\right), \tag{9.44}$$

which obeys the Heisenberg equation of motion

$$i\partial_0 \hat{\psi}(x) = \left[\hat{\psi}(x), \hat{H}_D\right]. \tag{9.45}$$

Using the equal-time anti-commutation relations of eq. (8.4), for both chiralities, we obtain

$$\begin{aligned} \left[\hat{\psi}(x), \hat{H}_D\right] &= \exp\left(i\hat{H}_D x^0\right) \left[\hat{\psi}(\vec{x}), \hat{H}_D\right] \exp\left(-i\hat{H}_D x^0\right) \\ &= \exp\left(i\hat{H}_D x^0\right) \left(-i\vec{\alpha} \cdot \vec{\nabla} + \beta m\right) \hat{\psi}(\vec{x}) \exp\left(-i\hat{H}_D x^0\right) \\ &= \left(-i\vec{\alpha} \cdot \vec{\nabla} + \beta m\right) \hat{\psi}(x). \end{aligned} \tag{9.46}$$

We see that the Dirac equation

$$\left(i\gamma^\mu \partial_\mu - m\right) \hat{\psi}(x) = 0 \tag{9.47}$$

can be identified as the Heisenberg equation of motion for the fermion field operator $\hat{\psi}(x)$ in the Heisenberg picture. Introducing $\hat{\bar{\psi}}(x) = \hat{\psi}^\dagger(x)\gamma^0$, one can similarly derive the "adjoint Dirac equation"

$$-i\partial_\mu \hat{\bar{\psi}}(x)\gamma^\mu - m\hat{\bar{\psi}}(x) = 0. \tag{9.48}$$

Quick Question 9.2.1 Dirac and Heisenberg equations
Verify eqs. (9.46) and (9.48).

Interestingly, there is another interpretation of the Dirac equation as the "classical" equation of motion for an anti-commuting Grassmann field $\psi(x)$. To appreciate this, we introduce the Dirac Lagrangian for a free fermion field,

$$\mathcal{L}_D(\bar{\psi}, \psi) = \bar{\psi}(x) \left(i\gamma^\mu \partial_\mu - m\right) \psi(x), \tag{9.49}$$

$$\psi(x) = \begin{pmatrix} \psi_L(x) \\ \psi_R(x) \end{pmatrix}, \quad \bar{\psi}(x) = \left(\bar{\psi}_R(x), \bar{\psi}_L(x)\right). \tag{9.50}$$

It is important that $\bar{\psi}(x)$ and $\psi(x)$ are *independent* Grassmann fields which – unlike $\hat{\bar{\psi}}(x)$ and $\hat{\psi}(x)$ in the Hamiltonian formulation – are not related via Hermitian conjugation. By using the Grassmann algebra rules for differentiation of eq. (9.3), we obtain the Euler–Lagrange equation

$$\frac{\delta\mathcal{L}_D(\bar{\psi}, \psi)}{\delta\bar{\psi}} - \partial_\mu \frac{\delta\mathcal{L}_D(\bar{\psi}, \psi)}{\delta\partial_\mu\bar{\psi}} = \left(i\gamma^\mu \partial_\mu - m\right) \psi(x) = 0, \tag{9.51}$$

which is nothing else than the Dirac equation. While eq. (9.48) for $\hat{\bar{\psi}}$ is just the Hermitian conjugate of eq. (9.47) for $\hat{\psi}$, in the Grassmann field formulation its "adjoint" is an independent equation of motion for $\bar{\psi}$

$$\frac{\delta\mathcal{L}_D(\bar{\psi}, \psi)}{\delta\psi} - \partial_\mu \frac{\delta\mathcal{L}_D(\bar{\psi}, \psi)}{\delta\partial_\mu\psi} = i\partial_\mu \bar{\psi}(x)\gamma^\mu + m\bar{\psi}(x) = 0. \tag{9.52}$$

The "classical" Euler–Lagrange equations of motion for ψ and $\bar{\psi}$ are just the Dirac equations for the Grassmann fields. It should be clear, however, that these equations make no sense in the context of classical field theory. Even in quantum field theory, they are rather formal relations. What does it mean that a Grassmann field, which does not even take particular values, satisfies a partial differential equation? In the next section, we will use Grassmann fields as integration variables in fermionic functional integrals.

As we have seen in eqs. (9.36) – (9.39), "minimizing" a fermionic action, and thus solving the corresponding Euler–Lagrange equation, is at least useful for performing Gaussian Grassmann integrals. The "classical" equations of motion, eqs. (9.51) and (9.52), can also be used to show that the *fermion current* $j^\mu(x)$ is conserved

$$j^\mu(x) = \bar{\psi}(x)\gamma^\mu\psi(x), \quad \partial_\mu j^\mu(x) = 0. \tag{9.53}$$

Next, let us construct the canonically conjugate momenta of the Grassmann fields

$$\Pi_\psi(x) = \frac{\delta\mathcal{L}_D(\bar{\psi},\psi)}{\delta\partial_0\psi} = -i\bar{\psi}(x)\gamma^0, \quad \Pi_{\bar{\psi}}(x) = \frac{\delta\mathcal{L}_D(\bar{\psi},\psi)}{\delta\partial_0\bar{\psi}} = 0, \tag{9.54}$$

from which we obtain a Dirac Hamilton density function,

$$\begin{aligned}
\mathcal{H}_D(\bar{\psi},\psi) &= \partial_0\bar{\psi}(x)\Pi_{\bar{\psi}}(x) - \Pi_\psi(x)\partial_0\psi(x) - \mathcal{L}_D(\bar{\psi},\psi) \\
&= i\bar{\psi}(x)\gamma^0\partial_0\psi(x) - \bar{\psi}(x)\left(i\gamma^\mu\partial_\mu - m\right)\psi(x) \\
&= \bar{\psi}(x)\left(-i\gamma^i\partial_i + m\right)\psi(x)
\end{aligned} \tag{9.55}$$

(where $\gamma^i\partial_i = -\vec{\gamma}\cdot\vec{\nabla}$). Upon canonical quantization, the Grassmann fields $\psi(x)$ and $\bar{\psi}(x)$ are replaced by fermion field operators $\hat{\psi}(x)$ and $\hat{\psi}^\dagger(x)\gamma^0$, for which one postulates the usual canonical anti-commutation relations,

$$\begin{pmatrix} \psi_L(x) \\ \psi_R(x) \end{pmatrix} \leftrightarrow \begin{pmatrix} \hat{\psi}_L(\vec{x}) \\ \hat{\psi}_R(\vec{x}) \end{pmatrix}, \quad \left(\bar{\psi}_R(x), \bar{\psi}_L(x)\right)\gamma^0 = \left(\bar{\psi}_L(x), \bar{\psi}_R(x)\right) \leftrightarrow \left(\hat{\psi}_L^\dagger(\vec{x}), \hat{\psi}_R^\dagger(\vec{x})\right). \tag{9.56}$$

The Hamilton density $\mathcal{H}_D(\bar{\psi},\psi)$ then turns into the Dirac Hamilton operator $\hat{H}_D$ of eq. (9.43).

9.3 Weyl and Majorana Equations

For the sake of an efficient handling of the γ-matrices in the chiral basis, we introduce the notation

$$\gamma^\mu = \begin{pmatrix} 0 & \sigma^\mu \\ \bar{\sigma}^\mu & 0 \end{pmatrix}, \quad \sigma^\mu = (\mathbb{1}, \vec{\sigma}), \quad \bar{\sigma}^\mu = (\mathbb{1}, -\vec{\sigma}). \tag{9.57}$$

With the matrix

$$\gamma^5 = i\gamma^0\gamma^1\gamma^2\gamma^3 = \begin{pmatrix} -\mathbb{1} & 0 \\ 0 & \mathbb{1} \end{pmatrix} \tag{9.58}$$

we obtain the projection operators

$$P_{\mathrm{L}} = \frac{1}{2}(1 - \gamma^5) = P_{\mathrm{L}}^2, \quad P_{\mathrm{R}} = \frac{1}{2}(1 + \gamma^5) = P_{\mathrm{R}}^2. \tag{9.59}$$

Quick Question 9.3.1   **Chiral projection operators**
Show that $P_{\mathrm{L,R}}$ are idempotent, $P_{\mathrm{L,R}}^2 = P_{\mathrm{L,R}}$, which defines a projection operator.

In the notation of eq. (9.50), $P_{\mathrm{L,R}}$ decompose the Grassmann field into left- and right-handed components,

$$P_{\mathrm{L}}\psi = \begin{pmatrix} \psi_{\mathrm{L}} \\ 0 \end{pmatrix}, \quad P_{\mathrm{R}}\psi = \begin{pmatrix} 0 \\ \psi_{\mathrm{R}} \end{pmatrix}; \quad \bar{\psi}P_{\mathrm{L}} = (\bar{\psi}_{\mathrm{R}}, 0), \quad \bar{\psi}P_{\mathrm{R}} = (0, \bar{\psi}_{\mathrm{L}}). \tag{9.60}$$

In terms of these chiral components, the Dirac equation takes the form

$$\begin{aligned}
i\sigma^\mu \partial_\mu \psi_{\mathrm{R}}(x) - m\psi_{\mathrm{L}}(x) &= 0, \quad i\partial_\mu \bar{\psi}_{\mathrm{R}}(x)\sigma^\mu + m\bar{\psi}_{\mathrm{L}}(x) = 0, \\
i\bar{\sigma}^\mu \partial_\mu \psi_{\mathrm{L}}(x) - m\psi_{\mathrm{R}}(x) &= 0, \quad i\partial_\mu \bar{\psi}_{\mathrm{L}}(x)\bar{\sigma}^\mu + m\bar{\psi}_{\mathrm{R}}(x) = 0.
\end{aligned} \tag{9.61}$$

In the massless case, these equations decouple into independent *Weyl equations*

$$\begin{aligned}
i\sigma^\mu \partial_\mu \psi_{\mathrm{R}}(x) &= 0, \quad i\partial_\mu \bar{\psi}_{\mathrm{R}}(x)\sigma^\mu = 0, \\
i\bar{\sigma}^\mu \partial_\mu \psi_{\mathrm{L}}(x) &= 0, \quad i\partial_\mu \bar{\psi}_{\mathrm{L}}(x)\bar{\sigma}^\mu = 0.
\end{aligned} \tag{9.62}$$

In the Hamiltonian formulation, the *Majorana condition,* eq. (8.73), is given by

$$\hat{\psi}_{\mathrm{L}}(\vec{x}) = -i\sigma^2 \hat{\psi}_{\mathrm{R}}^\dagger(\vec{x})^\mathsf{T} \quad \Rightarrow \quad \hat{\psi}_{\mathrm{L}}^\dagger(\vec{x}) = \hat{\psi}_{\mathrm{R}}(\vec{x})^\mathsf{T} i\sigma^2. \tag{9.63}$$

Replacing the fermion creation and annihilation operators by Grassmann fields according to relation (9.56), one obtains the *Grassmann Majorana constraint*

$$\psi_{\mathrm{L}}(x) = -i\sigma^2 \bar{\psi}_{\mathrm{R}}(x)^\mathsf{T}, \quad \bar{\psi}_{\mathrm{L}}(x) = \psi_{\mathrm{R}}(x)^\mathsf{T} i\sigma^2. \tag{9.64}$$

When we use these constraints to eliminate the left-handed fields from the Dirac Lagrangian, we obtain $\mathcal{L}_{\mathrm{D}}(\bar{\psi}, \psi) = 2\mathcal{L}_{\mathrm{M}}(\bar{\psi}_{\mathrm{R}}, \psi_{\mathrm{R}})$ with the Majorana Lagrangian

$$\begin{aligned}
\mathcal{L}_{\mathrm{M}}(\bar{\psi}_{\mathrm{R}}, \psi_{\mathrm{R}}) = {}& \frac{1}{2}\bar{\psi}_{\mathrm{R}}(x)i\sigma^\mu \partial_\mu \psi_{\mathrm{R}}(x) - \frac{1}{2}\partial_\mu \bar{\psi}_{\mathrm{R}}(x)i\sigma^\mu \psi_{\mathrm{R}}(x) \\
& - \frac{m}{2}\left(\psi_{\mathrm{R}}(x)^\mathsf{T} i\sigma^2 \psi_{\mathrm{R}}(x) - \bar{\psi}_{\mathrm{R}}(x)i\sigma^2 \bar{\psi}_{\mathrm{R}}(x)^\mathsf{T}\right).
\end{aligned} \tag{9.65}$$

The corresponding Euler–Lagrange equations are the *Majorana equations*

$$\begin{aligned}
i\sigma^\mu \partial_\mu \psi_{\mathrm{R}}(x) + i\sigma^2 m\bar{\psi}_{\mathrm{R}}(x)^\mathsf{T} &= 0, \\
i\partial_\mu \bar{\psi}_{\mathrm{R}}(x)\sigma^\mu + m\psi_{\mathrm{R}}(x)^\mathsf{T} i\sigma^2 &= 0.
\end{aligned} \tag{9.66}$$

Quick Question 9.3.2   Majorana equations
Derive eqs. (9.65) and (9.66).

Let us again consider the conjugate momentum fields,

$$\Pi_{\psi_R}(x) = \frac{\delta \mathcal{L}_M(\bar{\psi}_R, \psi_R)}{\delta \partial_0 \psi_R} = -\frac{i}{2}\bar{\psi}_R(x), \quad \Pi_{\bar{\psi}_R}(x) = \frac{\delta \mathcal{L}_M(\bar{\psi}_R, \psi_R)}{\delta \partial_0 \bar{\psi}_R} = -\frac{i}{2}\psi_R(x), \quad (9.67)$$

from which we obtain the *Majorana Hamilton density*

$$\begin{aligned}
\mathcal{H}_M(\bar{\psi}_R, \psi_R) &= \partial_0 \bar{\psi}_R(x) \Pi_{\bar{\psi}_R}(x) - \Pi_{\psi_R}(x) \partial_0 \psi_R(x) - \mathcal{L}_M(\bar{\psi}_R, \psi_R) \\
&= \frac{1}{2}\partial_i \bar{\psi}_R(x) i\sigma^i \psi_R(x) - \frac{1}{2}\bar{\psi}_R(x) i\sigma^i \partial_i \psi_R(x) \\
&\quad + \frac{m}{2}\left(\psi_R(x)^{\mathsf{T}} i\sigma^2 \psi_R(x) - \bar{\psi}_R(x) i\sigma^2 \bar{\psi}_R(x)^{\mathsf{T}}\right).
\end{aligned} \quad (9.68)$$

When we replace Grassmann fields by fermion field operators according to eq. (9.56), the Hamilton density $\mathcal{H}_M(\bar{\psi}_R, \psi_R)$ indeed turns into the Majorana Hamilton operator of eq. (8.79) (or equivalently the massive Weyl Hamilton operator of eq. (8.88)). This also involves partial integration applied to the first term on the right-hand side.

9.4 Euclidean Fermionic Functional Integral

Until now, we have treated fermion fields in Minkowski space–time. The functional integral for free Dirac fermion fields in Minkowski space–time takes the form of a Grassmann integral

$$\int \mathcal{D}\bar{\psi} \mathcal{D}\psi \, \exp\left(iS[\bar{\psi}, \psi]\right) = \int \mathcal{D}\bar{\psi} \mathcal{D}\psi \, \exp\left(i\int dt \, d^3x \, \bar{\psi}\left(i\gamma^\mu \partial_\mu - m\right)\psi\right). \quad (9.69)$$

In the continuation, we will work in Euclidean space–time. For this purpose, we perform once more the Wick rotation $x_4 = it = ix^0$. The Dirac Lagrangian then turns into

$$\begin{aligned}
\bar{\psi}(x)\left(i\gamma^\mu \partial_\mu - m\right)\psi(x) &= -\bar{\psi}(x)\left(\gamma_4 \partial_4 - i\gamma^i \partial_i + m\right)\psi(x) \\
&= -\bar{\psi}(x)\left(\gamma_\mu \partial_\mu + m\right)\psi(x) = -\mathcal{L}_D(\bar{\psi}, \psi).
\end{aligned} \quad (9.70)$$

Here, $\mathcal{L}_D(\bar{\psi}, \psi)$ is the Euclidean Lagrangian (which we denote by the same symbol as previously the Lagrangian in Minkowski space–time). As usual, in Euclidean space–time we no longer distinguish between co- and contra-variant vectors, so we only write lower space–time indices. As we mentioned earlier, we arrange things such that *all* γ-matrices in Minkowski space–time occur with upper indices only. In the following, we introduce Euclidean γ-matrices which will all appear with lower indices only. In the chiral basis, the Euclidean γ-matrices take the form

$$
\gamma_i = -i\gamma^i = \begin{pmatrix} 0 & -i\sigma^i \\ i\sigma^i & 0 \end{pmatrix}, \quad \gamma_4 = \gamma^0 = \begin{pmatrix} 0 & \mathbb{1} \\ \mathbb{1} & 0 \end{pmatrix},
$$

$$
\gamma_5 = -\gamma_1\gamma_2\gamma_3\gamma_4 = \begin{pmatrix} -\mathbb{1} & 0 \\ 0 & \mathbb{1} \end{pmatrix} = \gamma^5. \tag{9.71}
$$

Just like in Minkowski space–time, different γ-matrices anti-commute, but the Euclidean γ-matrices are all Hermitian

$$
\{\gamma_\mu, \gamma_\nu\} = 2\delta_{\mu\nu}, \quad \gamma_\mu^\dagger = \gamma_\mu. \tag{9.72}
$$

Here, the Kronecker symbol $\delta_{\mu\nu}$ is the metric of Euclidean space–time. Using $dt \, d^3x = -i d^3x \, dx_4 = -i d^4x$, the Euclidean functional integral for free Dirac fermions takes the form

$$
Z = \int \mathcal{D}\bar{\psi}\mathcal{D}\psi \, \exp\left(-S_D[\bar{\psi}, \psi]\right) = \int \mathcal{D}\bar{\psi}\mathcal{D}\psi \, \exp\left(-\int d^4x \, \mathcal{L}_D(\bar{\psi}, \psi)\right)
$$

$$
= \int \mathcal{D}\bar{\psi}\mathcal{D}\psi \, \exp\left(-\int d^4x \, \bar{\psi} \left(\gamma_\mu\partial_\mu + m\right)\psi\right). \tag{9.73}
$$

As it stands, the Grassmann measure $\mathcal{D}\bar{\psi}\mathcal{D}\psi$ of the functional integral is a formal expression that needs to be properly regularized. This is a non-trivial issue, in particular when fermions are coupled to gauge fields. In Chapter 18, we will use the lattice regularization to address some of the related subtleties. On a Euclidean space–time lattice Λ, the fermions are described by Grassmann fields ψ_x^a and $\bar{\psi}_x^a$, which are associated with the lattice sites x and have internal (*e.g.*, Lorentz) indices a. The lattice regularized free fermion measure is

$$
\mathcal{D}\bar{\psi}\mathcal{D}\psi = \prod_{x\in\Lambda} \prod_a d\bar{\psi}_x^a d\psi_x^a. \tag{9.74}
$$

As we will discuss in the next section, and as we have already seen for bosonic theories in Chapter 1, the Euclidean functional integral represents the canonical partition function of quantum statistical mechanics,

$$
Z = \mathrm{Tr} \, \exp\left(-\beta\hat{H}_D\right). \tag{9.75}
$$

Here, $\hat{H}_D$ is the Dirac Hamilton operator from eq. (8.57). As we have seen in Section 1.5, the inverse temperature $\beta = 1/T$ manifests itself as the extent of the Euclidean time dimension. Bosonic fields obey periodic boundary conditions in Euclidean time. As we will see later in this chapter, due to the peculiar features of Grassmann integration, in order to obtain eq. (9.75), fermion fields must obey *anti-periodic* Euclidean-time boundary conditions, *i.e.*

$$
\psi(\vec{x}, x_4 + \beta) = -\psi(\vec{x}, x_4), \quad \bar{\psi}(\vec{x}, x_4 + \beta) = -\bar{\psi}(\vec{x}, x_4). \tag{9.76}
$$

Applying the rules of Grassmann integration, one obtains

$$
Z = \int \mathcal{D}\bar{\psi}\mathcal{D}\psi \, \exp\left(-\int d^4x \, \bar{\psi}(x)\left(\gamma_\mu\partial_\mu + m\right)\psi(x)\right) = \int \mathcal{D}\bar{\psi}\mathcal{D}\psi \, \exp\left(-\bar{\psi}D\psi\right)
$$

$$
= \det(D),
$$

where

$$
\bar{\psi}D\psi = \int d^4x \, d^4y \, \bar{\psi}(x)D(x,y)\psi(y), \quad D(x,y) = \delta(x-y)\left(\gamma_\mu\partial_\mu + m\right). \tag{9.77}
$$

Here, D is the *Dirac operator*. It plays the role of the matrix M in eq. (9.16) with the matrix times vector multiplications being realized as integrations.

Let us also reconsider free Weyl and Majorana fermions. First, we introduce Euclidean variants of the matrices σ^μ and $\bar\sigma^\mu$, which again carry lower Lorentz indices only,

$$\gamma_\mu = \begin{pmatrix} 0 & \sigma_\mu \\ \bar\sigma_\mu & 0 \end{pmatrix}, \quad \sigma_\mu = (-i\vec\sigma, \mathbb{1}), \quad \bar\sigma_\mu = (i\vec\sigma, \mathbb{1}). \tag{9.78}$$

The Euclidean Lagrangians of massless Weyl fermions then take the form

$$\mathcal{L}_R(\bar\psi_R, \psi_R) = \bar\psi_R(x)\sigma_\mu\partial_\mu\psi_R(x), \quad \mathcal{L}_L(\bar\psi_L, \psi_L) = \bar\psi_L(x)\bar\sigma_\mu\partial_\mu\psi_L(x). \tag{9.79}$$

Introducing the *Weyl operators*

$$W_R(x, y) = \delta(x - y)\sigma_\mu\partial_\mu, \quad W_L(x, y) = \delta(x - y)\bar\sigma_\mu\partial_\mu, \tag{9.80}$$

the Dirac operator becomes

$$D = \begin{pmatrix} m\delta(x - y) & W_R \\ W_L & m\delta(x - y) \end{pmatrix}. \tag{9.81}$$

Similarly, the Euclidean Lagrangian for Majorana (or equivalently massive Weyl) fermions is given by

$$\mathcal{L}_M(\bar\psi_R, \psi_R) = \frac{1}{2}\bar\psi_R(x)\sigma_\mu\partial_\mu\psi_R(x) - \frac{1}{2}\partial_\mu\bar\psi_R(x)\sigma_\mu\psi_R(x)$$
$$+ \frac{m}{2}\left(\psi_R(x)^\top i\sigma^2\psi_R(x) - \bar\psi_R(x)i\sigma^2\bar\psi_R(x)^\top\right). \tag{9.82}$$

In this case, the functional integral takes the form

$$Z = \int \mathcal{D}\bar\psi_R\mathcal{D}\psi_R \exp\left(-\int d^4x\, \mathcal{L}_M(\bar\psi_R, \psi_R)\right)$$
$$= \int \mathcal{D}\bar\psi_R\mathcal{D}\psi_R \exp\left(-\frac{1}{2}\left(\psi_R^\top, \bar\psi_R\right)A_R\begin{pmatrix} \psi_R \\ \bar\psi_R^\top \end{pmatrix}\right) = \mathrm{Pf}(A_R). \tag{9.83}$$

We identify the anti-symmetric Majorana operator as

$$A_R = \begin{pmatrix} i\sigma^2 m\delta(x - y) & -W_R^\top \\ W_R & -i\sigma^2 m\delta(x - y) \end{pmatrix}, \quad A_R^\top = -A_R. \tag{9.84}$$

9.5 Euclidean Lorentz Group

The 4-dimensional Euclidean space–time is invariant against translations by 4-vectors as well as against SO(4) space–time rotations. Together, this constitutes the Euclidean version of Poincaré invariance. In the presence of spinor fields, it is important to consider the universal covering group

$$\mathrm{Spin}(4) = \mathrm{SU}(2)_L \times \mathrm{SU}(2)_R \tag{9.85}$$

of the Euclidean version of the Lorentz group. In a relativistic quantum field theory, the fields must transform appropriately under space–time rotations. Their transformation behavior can be characterized by specifying the representation of Spin(4), or equivalently

Table 9.1 Transformation of different types of relativistic quantum fields

type of field	representation of Spin(4) = SU(2)$_L$ × SU(2)$_R$
scalar	$(0,0)$
4-vector	$(1/2, 1/2)$
left-handed Weyl fermion	$(1/2, 0)$
right-handed Weyl fermion	$(0, 1/2)$
Dirac fermion	$(1/2, 0) + (0, 1/2)$

of SU(2)$_L$ and SU(2)$_R$. Since SU(2) representations are characterized by a "spin" $S = 0, 1/2, 1, \ldots$, the transformation behavior of relativistic quantum fields under space–time rotations is characterized by a pair (S_L, S_R). The transformation of various types of fields, regarding the representation of SU(2)$_L$ × SU(2)$_R$, is summarized in Table 9.1.

The six generators of the SO(4) algebra[4] can be expressed as commutators of Euclidean γ-matrices

$$\sigma_{\mu\nu} = \frac{1}{2i} [\gamma_\mu, \gamma_\nu] \quad \Rightarrow \quad \sigma_{4i} = \begin{pmatrix} \sigma^i & 0 \\ 0 & -\sigma^i \end{pmatrix}, \quad \sigma_{ij} = \epsilon_{ijk} \begin{pmatrix} \sigma^k & 0 \\ 0 & \sigma^k \end{pmatrix}. \tag{9.86}$$

The generators of SU(2)$_L$ × SU(2)$_R$ take the form

$$L_i = \frac{1}{2} \begin{pmatrix} \sigma^i & 0 \\ 0 & 0 \end{pmatrix}, \quad R_i = \frac{1}{2} \begin{pmatrix} 0 & 0 \\ 0 & \sigma^i \end{pmatrix}. \tag{9.87}$$

From them, we can construct the generators

$$J_i = R_i + L_i = \frac{1}{2} \epsilon_{ijk} \sigma_{jk}, \quad K_i = R_i - L_i = \frac{1}{2} \sigma_{i4}, \tag{9.88}$$

where the J_i generate the vector subgroup SU(2)$_{L=R}$ of spatial rotations (the universal covering group of SO(3)) and the K_i generate the Euclidean boosts.

Under a Euclidean Lorentz transformation $\Lambda \in$ SO(4), a space–time point x rotates into $x' = \Lambda x$, such that $x = \Lambda^{-1} x'$, i.e.

$$x_\nu = \Lambda_{\nu\rho}^{-1} x_\rho' = \Lambda_{\nu\rho}^\mathsf{T} x_\rho' = \Lambda_{\rho\nu} x_\rho' \quad \Rightarrow \quad \partial_\mu' x_\nu = \frac{\partial x_\nu}{\partial x_\mu'} = \Lambda_{\rho\nu} \delta_{\mu\rho} = \Lambda_{\mu\nu}. \tag{9.89}$$

Here, we have used the fact that Λ is an orthogonal rotation matrix, i.e. $\Lambda^{-1} = \Lambda^\mathsf{T}$.

Left- and right-handed Weyl spinor fields transform as

$$\psi_R'(x') = \Lambda_R \psi_R(\Lambda^{-1} x'), \quad \bar{\psi}_R'(x') = \bar{\psi}_R(\Lambda^{-1} x') \Lambda_L^\dagger, \quad \Lambda_R \in \text{SU(2)}_R,$$
$$\psi_L'(x') = \Lambda_L \psi_L(\Lambda^{-1} x'), \quad \bar{\psi}_L'(x') = \bar{\psi}_L(\Lambda^{-1} x') \Lambda_R^\dagger, \quad \Lambda_L \in \text{SU(2)}_L. \tag{9.90}$$

The transformation Λ in the 4-dimensional vector representation of Spin(4) is related to the transformations Λ_R and Λ_L in the two 2-dimensional spinor representations by

---

[4] A rigorous notation distinguishes the groups SO(N) and SU(N) from the corresponding algebras so(N) and su(N). For simplicity, we suppress this distinction here, but we use it in Appendix F, which is specifically devoted to Lie groups and their algebras.

$$\Lambda_{\mu\nu} = \frac{1}{2}\mathrm{ReTr}\left(\Lambda_{\mathrm{L}}^{\dagger}\sigma_{\mu}\Lambda_{\mathrm{R}}\bar{\sigma}_{\nu}\right) \quad \Rightarrow$$

$$\Lambda_{\mathrm{L}}^{\dagger}\sigma_{\mu}\Lambda_{\mathrm{R}} = \Lambda_{\mu\nu}\sigma_{\nu} = \sigma_{\nu}\Lambda_{\nu\mu}^{-1}, \quad \Lambda_{\mathrm{R}}^{\dagger}\bar{\sigma}_{\mu}\Lambda_{\mathrm{L}} = \Lambda_{\mu\nu}\bar{\sigma}_{\nu} = \bar{\sigma}_{\nu}\Lambda_{\nu\mu}^{-1}. \tag{9.91}$$

For spatial rotations in the vector subgroup SU(2)$_{\mathrm{L=R}}$, one has $\Lambda_{\mathrm{R}} = \Lambda_{\mathrm{L}} = \Lambda_{\mathrm{V}}$ and

$$\Lambda_{ij} = \frac{1}{2}\mathrm{ReTr}\left(\Lambda_{\mathrm{V}}^{\dagger}\sigma_{i}\Lambda_{\mathrm{V}}\sigma_{j}\right) = O_{ij}, \quad O \in \mathrm{SO}(3),$$

$$\Lambda_{i4} = 0, \quad \Lambda_{44} = 1, \quad \Lambda_{\mathrm{V}}^{\dagger}\sigma_{i}\Lambda_{\mathrm{V}} = O_{ij}\sigma_{j}. \tag{9.92}$$

We will come back to eqs. (9.91) and (9.92) in Problem 9.7.

Applying the chain rule and using eq. (9.89), we obtain

$$\partial_{\mu}'\psi_{\mathrm{R}}'(x') = \Lambda_{\mathrm{R}}\partial_{\mu}'\psi_{\mathrm{R}}(\Lambda^{-1}x') = \Lambda_{\mathrm{R}}\frac{\partial x_{\nu}}{\partial x_{\mu}'}\partial_{\nu}\psi_{\mathrm{R}}(x) = \Lambda_{\mathrm{R}}\Lambda_{\mu\nu}\partial_{\nu}\psi_{\mathrm{R}}(x),$$

$$\partial_{\mu}'\psi_{\mathrm{L}}'(x') = \Lambda_{\mathrm{L}}\partial_{\mu}'\psi_{\mathrm{L}}(\Lambda^{-1}x') = \Lambda_{\mathrm{L}}\frac{\partial x_{\nu}}{\partial x_{\mu}'}\partial_{\nu}\psi_{\mathrm{L}}(x) = \Lambda_{\mathrm{L}}\Lambda_{\mu\nu}\partial_{\nu}\psi_{\mathrm{L}}(x). \tag{9.93}$$

Hence, the Lagrangians of massless Weyl fermions transform as scalars

$$\bar{\psi}_{\mathrm{R}}'(x')\sigma_{\mu}\partial_{\mu}'\psi_{\mathrm{R}}'(x') = \bar{\psi}_{\mathrm{R}}(x)\Lambda_{\mathrm{L}}^{\dagger}\sigma_{\mu}\Lambda_{\mathrm{R}}\Lambda_{\mu\nu}\partial_{\nu}\psi_{\mathrm{R}}(x) = \bar{\psi}_{\mathrm{R}}(x)\sigma_{\rho}\Lambda_{\rho\mu}^{-1}\Lambda_{\mu\nu}\partial_{\nu}\psi_{\mathrm{R}}(x)$$
$$= \bar{\psi}_{\mathrm{R}}(x)\sigma_{\rho}\delta_{\rho\nu}\partial_{\nu}\psi_{\mathrm{R}}(x) = \bar{\psi}_{\mathrm{R}}(x)\sigma_{\nu}\partial_{\nu}\psi_{\mathrm{R}}(x),$$
$$\bar{\psi}_{\mathrm{L}}'(x')\bar{\sigma}_{\mu}\partial_{\mu}'\psi_{\mathrm{L}}'(x') = \bar{\psi}_{\mathrm{L}}(x)\Lambda_{\mathrm{R}}^{\dagger}\bar{\sigma}_{\mu}\Lambda_{\mathrm{L}}\Lambda_{\mu\nu}\partial_{\nu}\psi_{\mathrm{L}}(x) = \bar{\psi}_{\mathrm{L}}(x)\bar{\sigma}_{\rho}\Lambda_{\rho\mu}^{-1}\Lambda_{\mu\nu}\partial_{\nu}\psi_{\mathrm{L}}(x)$$
$$= \bar{\psi}_{\mathrm{L}}(x)\bar{\sigma}_{\rho}\delta_{\rho\nu}\partial_{\nu}\psi_{\mathrm{L}}(x) = \bar{\psi}_{\mathrm{L}}(x)\bar{\sigma}_{\nu}\partial_{\nu}\psi_{\mathrm{L}}(x). \tag{9.94}$$

Under Euclidean Lorentz transformations, a Dirac spinor transforms as

$$\psi'(x') = \Lambda_{\mathrm{D}}\psi(\Lambda^{-1}x'), \quad \bar{\psi}'(x') = \bar{\psi}(\Lambda^{-1}x')\Lambda_{\mathrm{D}}^{\dagger}, \quad \Lambda_{\mathrm{D}} = \begin{pmatrix} \Lambda_{\mathrm{L}} & 0 \\ 0 & \Lambda_{\mathrm{R}} \end{pmatrix}. \tag{9.95}$$

Eq. (9.91) yields

$$\begin{pmatrix} \Lambda_{\mathrm{L}}^{\dagger} & 0 \\ 0 & \Lambda_{\mathrm{R}}^{\dagger} \end{pmatrix}\begin{pmatrix} 0 & \sigma_{\mu} \\ \bar{\sigma}_{\mu} & 0 \end{pmatrix}\begin{pmatrix} \Lambda_{\mathrm{L}} & 0 \\ 0 & \Lambda_{\mathrm{R}} \end{pmatrix} = \Lambda_{\mu\nu}\begin{pmatrix} 0 & \sigma_{\nu} \\ \bar{\sigma}_{\nu} & 0 \end{pmatrix} \Rightarrow \Lambda_{\mathrm{D}}^{\dagger}\gamma_{\mu}\Lambda_{\mathrm{D}} = \Lambda_{\mu\nu}\gamma_{\nu}, \tag{9.96}$$

which implies that $\bar{\psi}\gamma_{\mu}\partial_{\mu}\psi$ is a *scalar* under Euclidean space–time rotations. The Dirac mass term is a scalar as well, because

$$\bar{\psi}'(x')\psi'(x') = \bar{\psi}(\Lambda^{-1}x')\Lambda_{\mathrm{D}}^{\dagger}\Lambda_{\mathrm{D}}\psi(\Lambda^{-1}x') = \bar{\psi}(x)\psi(x). \tag{9.97}$$

The fermion current $j_{\mu}(x) = \bar{\psi}(x)\gamma_{\mu}\psi(x)$, on the other hand, transforms as a 4-vector field,

$$\bar{\psi}'(x')\gamma_{\mu}\psi'(x') = \bar{\psi}(\Lambda^{-1}x')\Lambda_{\mathrm{D}}^{\dagger}\gamma_{\mu}\Lambda_{\mathrm{D}}\psi(\Lambda^{-1}x') = \Lambda_{\mu\nu}\bar{\psi}(x)\gamma_{\nu}\psi(x), \tag{9.98}$$

while the anti-symmetric tensor field $\bar{\psi}(x)\sigma_{\mu\nu}\psi(x)$ transforms as

$$\bar{\psi}'(x')\sigma_{\mu\nu}\psi'(x') = \bar{\psi}(\Lambda^{-1}x')\Lambda_{\mathrm{D}}^{\dagger}\frac{1}{2\mathrm{i}}[\gamma_{\mu}, \gamma_{\nu}]\Lambda_{\mathrm{D}}\psi(\Lambda^{-1}x')$$

$$= \bar{\psi}(x)\frac{1}{2\mathrm{i}}\left[\Lambda_{\mathrm{D}}^{\dagger}\gamma_{\mu}\Lambda_{\mathrm{D}}, \Lambda_{\mathrm{D}}^{\dagger}\gamma_{\nu}\Lambda_{\mathrm{D}}\right]\psi(x)$$

$$= \Lambda_{\mu\rho}\Lambda_{\nu\sigma}\bar{\psi}(x)\frac{1}{2\mathrm{i}}[\gamma_{\rho}, \gamma_{\sigma}]\psi(x) = \Lambda_{\mu\rho}\Lambda_{\nu\sigma}\bar{\psi}(x)\sigma_{\rho\sigma}\psi(x). \tag{9.99}$$

It is important to note that the Grassmann Majorana constraint eq. (9.64) is Lorentz covariant, *i.e.*

$$\psi'_L(x') = \Lambda_L \psi_L(\Lambda^{-1}x') = -\Lambda_L i\sigma^2 \bar{\psi}_R(\Lambda^{-1}x')^\mathsf{T} = -\Lambda_L i\sigma^2 \Lambda_L^\mathsf{T} \bar{\psi}'_R(x')^\mathsf{T} = -i\sigma^2 \bar{\psi}'_R(x')^\mathsf{T},$$

$$\bar{\psi}'_L(x') = \bar{\psi}_L(\Lambda^{-1}x')\Lambda_R^\dagger = \psi_R(\Lambda^{-1}x')^\mathsf{T} i\sigma^2 \Lambda_R^\dagger = \psi'_R(x')^\mathsf{T}\Lambda_R^* i\sigma^2 \Lambda_R^\dagger = \psi'_R(x')^\mathsf{T} i\sigma^2.$$

$$(9.100)$$

Here, we have used the property that any matrix $U \in \mathrm{SU}(2)$ obeys $U i\sigma^2 U^\mathsf{T} = U U^\dagger i\sigma^2 = i\sigma^2$. This relation is also used to show that the Majorana mass terms transform as scalars under Euclidean space–time rotations,

$$\psi'_R(x')^\mathsf{T} i\sigma^2 \psi'_R(x') = \psi_R(\Lambda^{-1}x')^\mathsf{T}\Lambda_R^\mathsf{T} i\sigma^2 \Lambda_R \psi_R(\Lambda^{-1}x') = \psi_R(x)^\mathsf{T} i\sigma^2 \psi_R(x),$$

$$\bar{\psi}'_R(x')i\sigma^2 \bar{\psi}'_R(x')^\mathsf{T} = \bar{\psi}_R(\Lambda^{-1}x')\Lambda_L^\dagger i\sigma^2 \Lambda_L^* \bar{\psi}_R(\Lambda^{-1}x')^\mathsf{T} = \bar{\psi}_R(x)i\sigma^2 \bar{\psi}_R(x)^\mathsf{T},$$

$$\psi'_L(x')^\mathsf{T} i\sigma^2 \psi'_L(x') = \psi_L(\Lambda^{-1}x')^\mathsf{T}\Lambda_L^\mathsf{T} i\sigma^2 \Lambda_L \psi_L(\Lambda^{-1}x') = \psi_L(x)^\mathsf{T} i\sigma^2 \psi_L(x),$$

$$\bar{\psi}'_L(x')i\sigma^2 \bar{\psi}'_L(x')^\mathsf{T} = \bar{\psi}_L(\Lambda^{-1}x')\Lambda_R^\dagger i\sigma^2 \Lambda_R^* \bar{\psi}_L(\Lambda^{-1}x')^\mathsf{T} = \bar{\psi}_L(x)i\sigma^2 \bar{\psi}_L(x)^\mathsf{T}. \quad (9.101)$$

9.6   Charge Conjugation, Parity, and Time Reversal for Weyl Fermions

As we have seen in Section 8.3, charge conjugation and parity are important discrete symmetries that exchange left- and right-handed Weyl fermions.

Let us first consider charge conjugation, which exchanges particles and anti-particles. Translating the transformation rules of eq. (8.37) from the Hamiltonian formulation into the Euclidean functional integral, charge conjugation acts on the Grassmann fields as

$$^C\psi_R(x) = i\sigma^2 \bar{\psi}_L(x)^\mathsf{T}, \qquad ^C\bar{\psi}_R(x) = -\psi_L(x)^\mathsf{T} i\sigma^2,$$

$$^C\psi_L(x) = -i\sigma^2 \bar{\psi}_R(x)^\mathsf{T}, \qquad ^C\bar{\psi}_L(x) = \psi_R(x)^\mathsf{T} i\sigma^2. \quad (9.102)$$

We see once more that charge conjugation exchanges left- and right-handed fermion fields. We now apply charge conjugation to the action of a right-handed Weyl fermion field,

$$S_R[^C\bar{\psi}_R, {}^C\psi_R] = \int d^4x \, {}^C\bar{\psi}_R(x)\,\sigma_\mu \partial_\mu \, {}^C\psi_R(x) = \int d^4x \, \psi_L(x)^\mathsf{T}(-i\sigma^2)\sigma_\mu \partial_\mu i\sigma^2 \bar{\psi}_L(x)^\mathsf{T}$$

$$= \int d^4x \, \psi_L(x)^\mathsf{T}\bar{\sigma}_\mu^\mathsf{T} \partial_\mu \bar{\psi}_L(x)^\mathsf{T} = \int d^4x \, \bar{\psi}_L(x)\bar{\sigma}_\mu \partial_\mu \psi_L(x)$$

$$= S_L[\bar{\psi}_L, \psi_L]. \quad (9.103)$$

Here, we have used the anti-commutation rules of the Grassmann variables, and we have performed an integration by parts. We see that *charge conjugation exchanges the actions of left- and right-handed Weyl fermions*. The individual Weyl fermion actions are not invariant against C, but their sum (which enters the Dirac Lagrangian (9.70)) is.

In Euclidean space–time, parity acts as a spatial inversion, which replaces $x = (\vec{x}, x_4)$ with $(-\vec{x}, x_4)$. Translating the parity transformation eq. (8.34) from the Hamiltonian formulation to the Euclidean functional integral, for the Grassmann fields one obtains

$$^P\psi_R(x) = \psi_L(-\vec{x}, x_4), \qquad ^P\bar{\psi}_R(x) = \bar{\psi}_L(-\vec{x}, x_4),$$

$$^P\psi_L(x) = \psi_R(-\vec{x}, x_4), \qquad ^P\bar{\psi}_L(x) = \bar{\psi}_R(-\vec{x}, x_4). \quad (9.104)$$

The Lagrangian depends on fields which are functions of x. Since under parity $x = (\vec{x}, x_4)$ turns into $(-\vec{x}, x_4)$, the Lagrangian itself cannot be P-invariant. What may be invariant, however, is the action. Let us now apply the parity transformation to the action of a massless right-handed Weyl fermion field

$$
\begin{aligned}
S_{\mathrm{R}}[{}^{\mathrm{P}}\bar{\psi}_{\mathrm{R}}, {}^{\mathrm{P}}\psi_{\mathrm{R}}] &= \int d^4 x \, {}^{\mathrm{P}}\bar{\psi}_{\mathrm{R}}(x) \, \sigma_\mu \partial_\mu \, {}^{\mathrm{P}}\psi_{\mathrm{R}}(x) \\
&= \int d^4 x \, \bar{\psi}_{\mathrm{L}}(-\vec{x}, x_4)(-\mathrm{i}\sigma_i \partial_i + \partial_4)\psi_{\mathrm{L}}(-\vec{x}, x_4) \\
&= \int d^4 x \, \bar{\psi}_{\mathrm{L}}(\vec{x}, x_4)(\mathrm{i}\sigma_i \partial_i + \partial_4)\psi_{\mathrm{L}}(\vec{x}, x_4) \\
&= \int d^4 x \, \bar{\psi}_{\mathrm{L}}(x)\bar{\sigma}_\mu \partial_\mu \psi_{\mathrm{L}}(x) = S_{\mathrm{L}}[\bar{\psi}_{\mathrm{L}}, \psi_{\mathrm{L}}]
\end{aligned}
\tag{9.105}
$$

(σ_μ and $\bar{\sigma}_\mu$ in Euclidean space are defined in eq. (9.78)). Here, we have changed the coordinates from $-\vec{x}$ to $\vec{x}$. As we see, under parity the action of a right-handed Weyl fermion again turns into the one of a left-handed Weyl fermion and *vice versa*. In particular, each individual Weyl fermion action is not invariant against P, but their sum is.

For the *combination of charge conjugation and parity*, CP, we obtain

$$
\begin{aligned}
{}^{\mathrm{CP}}\psi_{\mathrm{R}}(x) &= {}^{\mathrm{C}}\psi_{\mathrm{L}}(-\vec{x}, x_4) = -\mathrm{i}\sigma^2 \bar{\psi}_{\mathrm{R}}(-\vec{x}, x_4)^{\mathsf{T}}, \\
{}^{\mathrm{CP}}\bar{\psi}_{\mathrm{R}}(x) &= {}^{\mathrm{C}}\bar{\psi}_{\mathrm{L}}(-\vec{x}, x_4) = \psi_{\mathrm{R}}(-\vec{x}, x_4)^{\mathsf{T}} \mathrm{i}\sigma^2, \\
{}^{\mathrm{CP}}\psi_{\mathrm{L}}(x) &= {}^{\mathrm{C}}\psi_{\mathrm{R}}(-\vec{x}, x_4) = \mathrm{i}\sigma^2 \bar{\psi}_{\mathrm{L}}(-\vec{x}, x_4)^{\mathsf{T}}, \\
{}^{\mathrm{CP}}\bar{\psi}_{\mathrm{L}}(x) &= {}^{\mathrm{C}}\bar{\psi}_{\mathrm{R}}(-\vec{x}, x_4) = -\psi_{\mathrm{L}}(-\vec{x}, x_4)^{\mathsf{T}} \mathrm{i}\sigma^2.
\end{aligned}
\tag{9.106}
$$

Since both C and P exchange the actions of left- and right-handed fermions, CP leaves the individual actions invariant.

Let us consider the symmetries of the Grassmann Majorana constraint of eq. (9.64). First, we perform the chiral $\mathrm{U}(1)_{\mathrm{L}} \times \mathrm{U}(1)_{\mathrm{R}}$ transformation

$$
\begin{aligned}
\psi_{\mathrm{L}}'(x) &= \exp(\mathrm{i}\chi_{\mathrm{L}})\psi_{\mathrm{L}}(x), \quad \bar{\psi}_{\mathrm{L}}'(x) = \bar{\psi}_{\mathrm{L}}(x)\exp(-\mathrm{i}\chi_{\mathrm{L}}), \\
\psi_{\mathrm{R}}'(x) &= \exp(\mathrm{i}\chi_{\mathrm{R}})\psi_{\mathrm{R}}(x), \quad \bar{\psi}_{\mathrm{R}}'(x) = \bar{\psi}_{\mathrm{R}}(x)\exp(-\mathrm{i}\chi_{\mathrm{R}}),
\end{aligned}
\tag{9.107}
$$

which implies

$$
\begin{aligned}
-\mathrm{i}\sigma^2 \bar{\psi}_{\mathrm{R}}'(x)^{\mathsf{T}} &= -\exp(-\mathrm{i}\chi_{\mathrm{R}})\mathrm{i}\sigma^2 \bar{\psi}_{\mathrm{R}}(x)^{\mathsf{T}} = \exp(-\mathrm{i}\chi_{\mathrm{R}})\psi_{\mathrm{L}}(x) = \exp(-\mathrm{i}\chi_{\mathrm{R}} - \mathrm{i}\chi_{\mathrm{L}})\psi_{\mathrm{L}}'(x), \\
\psi_{\mathrm{R}}'(x)^{\mathsf{T}} \mathrm{i}\sigma^2 &= \psi_{\mathrm{R}}(x)^{\mathsf{T}} \mathrm{i}\sigma^2 \exp(\mathrm{i}\chi_{\mathrm{R}}) = \bar{\psi}_{\mathrm{L}}(x)\exp(\mathrm{i}\chi_{\mathrm{R}}) = \bar{\psi}_{\mathrm{L}}'(x)\exp(\mathrm{i}\chi_{\mathrm{R}} + \mathrm{i}\chi_{\mathrm{L}}).
\end{aligned}
\tag{9.108}
$$

As a result, the Grassmann Majorana constraint is invariant only against the subgroup $\mathrm{U}(1)_{\mathrm{L=R^*}}$ of $\mathrm{U}(1)_{\mathrm{L}} \times \mathrm{U}(1)_{\mathrm{R}}$ (for which $\mathrm{L} = \exp(\mathrm{i}\chi_{\mathrm{L}}) = \exp(-\mathrm{i}\chi_{\mathrm{R}}) = \mathrm{R^*}$), and hence only against the $\mathbb{Z}(2)_F$ subgroup of $\mathrm{U}(1)_F = \mathrm{U}(1)_{\mathrm{L=R}}$ that is characterized by $\exp(\mathrm{i}\chi_{\mathrm{L}}) = \exp(-\mathrm{i}\chi_{\mathrm{L}}) = \exp(\mathrm{i}\chi_{\mathrm{R}}) = \exp(-\mathrm{i}\chi_{\mathrm{R}}) = \pm 1$. The Grassmann Majorana constraint is not invariant against the parity P. However, it is again invariant against the combination P$'$, which is composed of P and the phase factor i,

$$
\begin{aligned}
{}^{\mathrm{P}'}\psi_{\mathrm{R}}(x) &= \mathrm{i}\psi_{\mathrm{L}}(-\vec{x}, x_4), \quad {}^{\mathrm{P}'}\bar{\psi}_{\mathrm{R}}(x) = -\mathrm{i}\bar{\psi}_{\mathrm{L}}(-\vec{x}, x_4), \\
{}^{\mathrm{P}'}\psi_{\mathrm{L}}(x) &= \mathrm{i}\psi_{\mathrm{R}}(-\vec{x}, x_4), \quad {}^{\mathrm{P}'}\bar{\psi}_{\mathrm{L}}(x) = -\mathrm{i}\bar{\psi}_{\mathrm{R}}(-\vec{x}, x_4),
\end{aligned}
\tag{9.109}
$$

because then

$$-i\sigma^2 \, {}^{P'}\bar{\psi}_R(x)^\mathsf{T} = -\sigma^2 \bar{\psi}_L(-\vec{x}, x_4)^\mathsf{T} = -\sigma^2 (i\sigma^2)^\mathsf{T} \psi_R(-\vec{x}, x_4) = i\psi_R(-\vec{x}, x_4) = {}^{P'}\psi_L(x),$$

$$\, {}^{P'}\psi_R(x)^\mathsf{T} i\sigma^2 = -\psi_L(-\vec{x}, x_4)^\mathsf{T} \sigma^2 = -\bar{\psi}_R(-\vec{x}, x_4)^\mathsf{T}(-i\sigma^2)^\mathsf{T} \sigma^2 = -i\bar{\psi}_R(-\vec{x}, x_4)^\mathsf{T} = {}^{P'}\bar{\psi}_L(x).$$

$$\tag{9.110}$$

Let us also consider the CP transformation of the Weyl fermion mass term

$$\, {}^{CP}\psi_R(x)^\mathsf{T} i\sigma^2 \, {}^{CP}\psi_R(x) - {}^{CP}\bar{\psi}_R(x) i\sigma^2 \, {}^{CP}\bar{\psi}_R(x)^\mathsf{T} =$$

$$\bar{\psi}_R(-\vec{x}, x_4) i\sigma^2 \bar{\psi}_R(-\vec{x}, x_4)^\mathsf{T} - \psi_R(-\vec{x}, x_4)^\mathsf{T} i\sigma^2 \psi_R(-\vec{x}, x_4). \tag{9.111}$$

Interestingly, its contribution to the action is *odd* under CP, but *even* under CP', and the same holds for the left-handed Weyl fermion.

Next we consider *Euclidean time reversal*, which acts as

$$\, {}^{T}\psi_R(x) = i\sigma^2 \bar{\psi}_R(\vec{x}, -x_4)^\mathsf{T}, \quad {}^{T}\bar{\psi}_R(x) = \psi_R(\vec{x}, -x_4)^\mathsf{T} i\sigma^2,$$

$$\, {}^{T}\psi_L(x) = i\sigma^2 \bar{\psi}_L(\vec{x}, -x_4)^\mathsf{T}, \quad {}^{T}\bar{\psi}_L(x) = \psi_L(\vec{x}, -x_4)^\mathsf{T} i\sigma^2. \tag{9.112}$$

Here, the superscript T on the left refers to time reversal and (as usual) the superscript T on the right denotes transpose.

The action of a right-handed Weyl fermion turns out to be T-invariant

$$S_R[{}^{T}\bar{\psi}_R, {}^{T}\psi_R] = \int d^4x \, {}^{T}\bar{\psi}_R(x) \, \sigma_\mu \partial_\mu \, {}^{T}\psi_R(x)$$

$$= \int d^4x \, \psi_R(\vec{x}, -x_4)^\mathsf{T} i\sigma^2 (-i\sigma_i \partial_i + \partial_4) i\sigma^2 \bar{\psi}_R(\vec{x}, -x_4)^\mathsf{T}$$

$$= \int d^4x \, \psi_R(x)^\mathsf{T} (-i\sigma_i^\mathsf{T} \partial_i + \partial_4) \bar{\psi}_R(x)^\mathsf{T}$$

$$= \int d^4x \, \bar{\psi}_R(x)(-i\sigma_i \partial_i + \partial_4) \psi_R(x)$$

$$= \int d^4x \, \bar{\psi}_R(x) \sigma_\mu \partial_\mu \psi_R(x)$$

$$= S_R[\bar{\psi}_R, \psi_R]. \tag{9.113}$$

Quick Question 9.6.1   Mass term of a Weyl fermion under time reversal

Show that the Weyl fermion mass term is odd under the transformation T. Since it is odd under CP as well, it is CPT-invariant. Also show that it is invariant under CP'.

As one would expect, parity, charge conjugation, and time reversal square to the identity, *i.e.*

$$P^2 = C^2 = T^2 = 1, \tag{9.114}$$

while they do not all commute with one another. In particular, one obtains

$$P\,C = -C\,P, \quad C\,T = -T\,C, \quad T\,P = P\,T. \tag{9.115}$$

Quick Question 9.6.2   Commutations of P, C, and T
Derive the commutation relations (9.115).

9.7   C, P, and T Transformations of Dirac Fermions

The properties of Dirac fermions under the discrete symmetries C, P, and T follow from the corresponding properties of the underlying left- and right-handed Weyl fermions,

$$
\begin{aligned}
{}^{\mathrm{C}}\psi(x) &= \begin{pmatrix} {}^{\mathrm{C}}\psi_{\mathrm{L}}(x) \\ {}^{\mathrm{C}}\psi_{\mathrm{R}}(x) \end{pmatrix} = \begin{pmatrix} -\mathrm{i}\sigma^2\bar{\psi}_{\mathrm{R}}(x)^{\mathsf{T}} \\ \mathrm{i}\sigma^2\bar{\psi}_{\mathrm{L}}(x)^{\mathsf{T}} \end{pmatrix} = C\bar{\psi}(x)^{\mathsf{T}}, \\
{}^{\mathrm{C}}\bar{\psi}(x) &= \left({}^{\mathrm{C}}\bar{\psi}_{\mathrm{R}}(x), {}^{\mathrm{C}}\bar{\psi}_{\mathrm{L}}(x) \right) = \left(-\psi_{\mathrm{L}}(x)^{\mathsf{T}}\mathrm{i}\sigma^2, \psi_{\mathrm{R}}(x)^{\mathsf{T}}\mathrm{i}\sigma^2 \right) = -\psi(x)^{\mathsf{T}}C^{-1}, \\
{}^{\mathrm{P}}\psi(x) &= \begin{pmatrix} {}^{\mathrm{P}}\psi_{\mathrm{L}}(x) \\ {}^{\mathrm{P}}\psi_{\mathrm{R}}(x) \end{pmatrix} = \begin{pmatrix} \psi_{\mathrm{R}}(-\vec{x}, x_4) \\ \psi_{\mathrm{L}}(-\vec{x}, x_4) \end{pmatrix} = P\psi(-\vec{x}, x_4), \\
{}^{\mathrm{P}}\bar{\psi}(x) &= \left({}^{\mathrm{P}}\bar{\psi}_{\mathrm{R}}(x), {}^{\mathrm{P}}\bar{\psi}_{\mathrm{L}}(x) \right) = \left(\bar{\psi}_{\mathrm{L}}(-\vec{x}, x_4), \bar{\psi}_{\mathrm{R}}(-\vec{x}, x_4) \right) = \bar{\psi}(-\vec{x}, x_4)P^{-1}, \\
{}^{\mathrm{T}}\psi(x) &= \begin{pmatrix} {}^{\mathrm{T}}\psi_{\mathrm{L}}(x) \\ {}^{\mathrm{T}}\psi_{\mathrm{R}}(x) \end{pmatrix} = \begin{pmatrix} \mathrm{i}\sigma^2\bar{\psi}_{\mathrm{L}}(\vec{x}, -x_4)^{\mathsf{T}} \\ \mathrm{i}\sigma^2\bar{\psi}_{\mathrm{R}}(\vec{x}, -x_4)^{\mathsf{T}} \end{pmatrix} = T\bar{\psi}(\vec{x}, -x_4)^{\mathsf{T}}, \\
{}^{\mathrm{T}}\bar{\psi}(x) &= \left({}^{\mathrm{T}}\bar{\psi}_{\mathrm{R}}(x), {}^{\mathrm{T}}\bar{\psi}_{\mathrm{L}}(x) \right) = \left(\psi_{\mathrm{R}}(\vec{x}, -x_4)^{\mathsf{T}}\mathrm{i}\sigma^2, \psi_{\mathrm{L}}(\vec{x}, -x_4)^{\mathsf{T}}\mathrm{i}\sigma^2 \right) \\
&= -\psi(\vec{x}, -x_4)^{\mathsf{T}}T^{-1}.
\end{aligned}
\tag{9.116}
$$

To be explicit, in the chiral basis, the matrices C, P, and T take the form

$$
C = -C^{-1} = \begin{pmatrix} -\mathrm{i}\sigma^2 & 0 \\ 0 & \mathrm{i}\sigma^2 \end{pmatrix} = -\sigma^3 \otimes \mathrm{i}\sigma^2 = \gamma_2\gamma_4, \quad C^{-1}\gamma_\mu C = -\gamma_\mu^{\mathsf{T}},
$$

$$
P = P^{-1} = \begin{pmatrix} 0 & \mathbb{1} \\ \mathbb{1} & 0 \end{pmatrix} = \sigma_1 \otimes \mathbb{1} = \gamma_4, \qquad P^{-1}\gamma_i P = -\gamma_i, \ P^{-1}\gamma_4 P = \gamma_4,
$$

$$
T = -T^{-1} = \begin{pmatrix} 0 & \mathrm{i}\sigma^2 \\ \mathrm{i}\sigma^2 & 0 \end{pmatrix} = \sigma^1 \otimes \mathrm{i}\sigma^2 = \gamma_5\gamma_2, \quad T^{-1}\gamma_i T = -\gamma_i^{\mathsf{T}}, \ T^{-1}\gamma_4 T = \gamma_4^{\mathsf{T}}.
\tag{9.117}
$$

However, the transformation behavior (9.116) holds generally, *i.e.* for any choice of the Dirac matrices, with the corresponding matrices C, P, and T.

9.8   CPT Invariance in Relativistic Quantum Field Theory

As was first shown by Gerhart Lüders, John Bell, Wolfgang Pauli, and Res Jost, the combination CPT is a symmetry of any relativistic, local quantum field theory; this is the *CPT theorem* (Lüders, 1954; Bell, 1955; Pauli, 1957; Lüders, 1957; Jost, 1957).[5] The combined transformation CPT takes the form

---

[5] While the CPT theorem applies to all relativistic, local quantum field theories, it does not always apply beyond this framework, *e.g.*, in string theory which violates strict locality.

$$\begin{aligned}
^{\text{CPT}}\psi_{\text{R}}(x) &= i\sigma^2 \, {}^{\text{CP}}\bar{\psi}_{\text{R}}(\vec{x},-x_4)^{\top} = i\sigma^2(i\sigma^2)^{\top}\psi_{\text{R}}(-\vec{x},-x_4) = \psi_{\text{R}}(-x), \\
^{\text{CPT}}\bar{\psi}_{\text{R}}(x) &= {}^{\text{CP}}\psi_{\text{R}}(\vec{x},-x_4)^{\top}i\sigma^2 = \bar{\psi}_{\text{R}}(-\vec{x},-x_4)(-i\sigma^2)^{\top}i\sigma^2 = -\bar{\psi}_{\text{R}}(-x), \\
^{\text{CPT}}\psi_{\text{L}}(x) &= i\sigma^2 \, {}^{\text{CP}}\bar{\psi}_{\text{L}}(\vec{x},-x_4)^{\top} = -i\sigma^2(i\sigma^2)^{\top}\psi_{\text{L}}(-\vec{x},-x_4) = -\psi_{\text{L}}(-x), \\
^{\text{CPT}}\bar{\psi}_{\text{L}}(x) &= {}^{\text{CP}}\psi_{\text{L}}(\vec{x},-x_4)^{\top}i\sigma^2 = \bar{\psi}_{\text{L}}(-\vec{x},-x_4)(i\sigma^2)^{\top}i\sigma^2 = \bar{\psi}_{\text{L}}(-x).
\end{aligned} \quad (9.118)$$

For a Dirac fermion field, this implies

$$^{\text{CPT}}\psi(x) = \begin{pmatrix} {}^{\text{CPT}}\psi_{\text{L}}(x) \\ {}^{\text{CPT}}\psi_{\text{R}}(x) \end{pmatrix} = \begin{pmatrix} -\psi_{\text{L}}(-x) \\ \psi_{\text{R}}(-x) \end{pmatrix} = \gamma_5\psi(-x),$$

$$^{\text{CPT}}\bar{\psi}(x) = \left({}^{\text{CPT}}\bar{\psi}_{\text{R}}(x), {}^{\text{CPT}}\bar{\psi}_{\text{L}}(x) \right) = \left(-\bar{\psi}_{\text{R}}(-x), \bar{\psi}_{\text{L}}(-x) \right) = \bar{\psi}(-x)\gamma_5. \quad (9.119)$$

While we will not review the proof of the CPT theorem,[6] we would like to explain why the CPT symmetry is closely related to Lorentz invariance. In four Euclidean space–time dimensions, the SO(4) rotation $\Lambda_{\mu\nu} = -\delta_{\mu\nu}$ turns x into $-x$. If we choose the $\text{SU(2)}_{\text{L}} \times \text{SU(2)}_{\text{R}}$ transformation $\Lambda_{\text{R}} = \mathbb{1}$ and $\Lambda_{\text{L}} = -\mathbb{1}$, this induces the Euclidean Lorentz transformation

$$\Lambda_{\mu\nu} = \frac{1}{2}\text{ReTr}\left(\Lambda_{\text{L}}^{\dagger}\sigma_{\mu}\Lambda_{\text{R}}\bar{\sigma}_{\nu}\right) = -\frac{1}{2}\text{ReTr}\left(\sigma_{\mu}\bar{\sigma}_{\nu}\right) = -\delta_{\mu\nu}, \qquad \Lambda_{\text{D}} = \begin{pmatrix} \Lambda_{\text{L}} & 0 \\ 0 & \Lambda_{\text{R}} \end{pmatrix} = \gamma_5.$$
$$(9.120)$$

This implies that for a relativistic fermion field (be it of Weyl, Dirac, or Majorana type), the combined transformation CPT is indistinguishable from a specific Euclidean Lorentz transformation. This already shows that any relativistic quantum field theory with fermion fields only is automatically CPT-invariant.

As we will discuss in Section 15.2, for complex scalar fields $\Phi(x)$, Abelian gauge fields $A_{\mu}(x)$, and non-Abelian gauge fields $G_{\mu}(x)$, CPT acts as

$$^{\text{CPT}}\Phi(x) = \Phi(-x), \qquad ^{\text{CPT}}A_{\mu}(x) = -A_{\mu}(-x), \qquad ^{\text{CPT}}G_{\mu}(x) = -G_{\mu}(-x). \quad (9.121)$$

Here, the matrix-valued non-Abelian gauge field $G_{\mu}(x) = ig_sG_{\mu}^a(x)T^a$, which will be investigated in detail in Chapter 11, is constructed from the Hermitian generators T^a of a Lie algebra and the fields $G_{\mu}^a(x) \in \mathbb{R}$, along with a coupling constant g_s. For the bosonic fields, CPT is again equivalent to the Euclidean Lorentz transformation $\Lambda_{\mu\nu} = -\delta_{\mu\nu}$. This proves that, for theories of Weyl fermions, scalars, and gauge fields (which includes the Standard Model), CPT invariance indeed follows from the Euclidean variant of Lorentz invariance.

We restrict our discussion to four space–time dimensions. For an extension to Euclidean spaces of other dimensions, we refer to Wetterich (2011).

9.9 Connections between Spin and Statistics

In quantum mechanics, the Pauli principle is imposed by hand. In quantum field theory, on the other hand, Fermi-Dirac statistics is naturally incorporated by the anti-commutativity

---

[6] For the interested reader, we particularly recommend the proof by Jost (1957), who considers an arbitrary local and covariant action, and demonstrates by means of analytic continuation – in the framework of a complex Lorentz group – that any n-point function is CPT-invariant. A caveat is that this paper was written in German; a review in English is contained in Bietenholz (2011).

of Grassmann fields. The *spin–statistics theorem*, which was first proved by Markus Fierz, states that fields with half-odd-integer spin obey Fermi-Dirac statistics, while fields with integer spin obey Bose–Einstein statistics. The spin–statistics theorem follows from relativistic quantum field theory, where the Lagrangian transforms as a Lorentz scalar. It also requires the existence of a stable vacuum state (Fierz, 1939; Pauli, 1940).

Let us investigate the statistics of a generic field $\phi_R(x)$ that transforms as $(0, 1/2)$ under the Euclidean Lorentz group $SU(2)_L \times SU(2)_R = \text{Spin}(4)$, *i.e.*

$$\phi_R'(x') = \Lambda_R \phi_R(\Lambda^{-1}x'). \tag{9.122}$$

At this point, it is undecided whether $\phi_R(x)$ is bosonic or fermionic. We now want to construct a Lagrangian systematically, by considering terms in the order of their dimension. Obviously, a term that is linear in $\phi_R(x)$ cannot be Lorentz-invariant. What about quadratic terms without derivatives, *i.e.* mass terms? Two factors of $\phi_R(x)$ transform as

$$(0, 1/2) \times (0, 1/2) = (0, 0) + (0, 1). \tag{9.123}$$

As we know, two spins $1/2$ couple to a singlet in an anti-symmetric manner; *i.e.* the singlet combination of two right-handed doublets is

$$\epsilon_{ab}\phi_R^a(x)\phi_R^b(x) = \phi_R(x)^\top i\sigma^2\phi_R(x) = \left(\phi_R^1(x), \phi_R^2(x)\right)\begin{pmatrix} 0 & 1 \\ -1 & 0 \end{pmatrix}\begin{pmatrix} \phi_R^1(x) \\ \phi_R^2(x) \end{pmatrix}. \tag{9.124}$$

This is one of the two contributions to a Majorana mass term. If $\phi_R(x)$ would be a commuting bosonic field, the Majorana mass term would simply vanish due to the anti-symmetry of ϵ_{ab}. We conclude that massive Weyl fields must necessarily be fermionic.

A derivative ∂_μ is a 4-vector that transforms as $(1/2, 1/2)$. Hence, in order to incorporate terms with a single derivative in the Lagrangian, we must also introduce a field that transforms as $(1/2, 0)$. We thus introduce a generic field $\bar{\phi}_R(x)$ that transforms as

$$\bar{\phi}_R'(x') = \bar{\phi}_R(\Lambda^{-1}x')\Lambda_L^\dagger. \tag{9.125}$$

First of all, $\bar{\phi}_R(x)$ has its own Majorana mass term that can be arranged to be the Hermitian conjugate of the other mass term in eq. (9.124). Hence, the existence of $\bar{\phi}_R(x)$ already follows from the Hermiticity of the Hamilton operator. Bilinears that contain one factor of $\bar{\phi}_R(x)$ and one factor of $\phi_R(x)$ transform as $(1/2, 0) \times (0, 1/2) = (1/2, 1/2)$. They can thus be made Lorentz-invariant when they are combined with a derivative. We know that $\Lambda_L\sigma_\mu\Lambda_R^\dagger = \sigma_\nu\Lambda_{\nu\mu}^{-1}$ and that $\partial_\mu' = \Lambda_{\mu\nu}\partial_\nu$. We can hence construct a Euclidean Lorentz scalar as $\bar{\phi}_R(x)\sigma_\mu\partial_\mu\phi_R(x)$ (this follows from eq. (9.91)). Such a Lagrangian describes massless right-handed Weyl fields, and we have already seen explicitly that such a theory is consistent with fermionic fields.

What happens if $\phi_R(x)$ and $\bar{\phi}_R(x)$ were bosonic fields? Returning briefly to Minkowski space–time, the Lagrangian would take the form $\mathcal{L}(\bar{\phi}_R, \phi_R) = \bar{\phi}_R(x)i\sigma^\mu\partial_\mu\phi_R(x)$. The corresponding (classical) Hamilton density would then result from the canonically conjugate momenta

$$\Pi_{\phi_R}(\vec{x}) = \frac{\delta\mathcal{L}(\bar{\phi}_R, \phi_R)}{\delta\partial_0\phi_R(\vec{x})} = \bar{\phi}_R(\vec{x}), \quad \Pi_{\bar{\phi}_R}(\vec{x}) = \frac{\delta\mathcal{L}(\bar{\phi}_R, \phi_R)}{\delta\partial_0\bar{\phi}_R(\vec{x})} = 0 \quad \Rightarrow$$

$$\mathcal{H}(\bar{\phi}_R, \phi_R) = \partial_0\phi_R(\vec{x})\Pi_{\phi_R}(\vec{x}) + \partial_0\bar{\phi}_R(\vec{x})\Pi_{\bar{\phi}_R}(\vec{x}) - \mathcal{L}(\bar{\phi}_R, \phi_R)$$

$$= \bar{\phi}_R(\vec{x})\left(-i\vec{\sigma}\cdot\vec{\nabla}\right)\phi_R(\vec{x}) \quad \Rightarrow$$

$$\hat{H} = \int d^3x \, \hat{\bar{\phi}}_R(\vec{x})\left(-i\vec{\sigma}\cdot\vec{\nabla}\right)\hat{\phi}_R(\vec{x}). \tag{9.126}$$

Upon canonical quantization, *i.e.* by promoting the classical fields to field operators, and by postulating bosonic commutation relations between the fields and their canonically conjugate momenta, one arrives at the Hamilton operator $\hat{H}$ in the last line of eq. (9.126). While at first glance there may seem nothing wrong with it, it is easy to see that, in contrast to the fermionic Weyl Hamilton operator, eq. (8.7), it does not have a stable vacuum state. This is because an infinite number of bosons can occupy each negative energy state.[7] Hence, invoking vacuum stability, we conclude that massless Weyl fields also require fermionic statistics.

While the above considerations do not constitute a proof of the spin statistics theorem, they show that Lorentz invariance combined with vacuum stability establishes intimate connections between the spin and the statistics of quantum fields.

9.10 Euclidean Time Transfer Matrix

We have already shown that, upon canonical quantization and by imposing anti-commutation relations, the fermionic functional integral in Minkowski space–time leads back to the Hamiltonian formulation. In this section, we will demonstrate explicitly that the partition function that results from the Euclidean fermionic functional integral is the same as the one that results from the Hamiltonian formulation.

As they stand, functional integrals are formal expressions that require regularization. As we will discuss in Chapter 19, the regularization of Weyl fermions is a subtle issue because they may be afflicted by anomalies. These amount to explicit symmetry breaking, *e.g.*, of the chiral $U(1)_L \times U(1)_R$ symmetry, due to quantum effects, in particular in the presence of gauge fields. The subtleties that arise in the regularization of Weyl fermions are most apparent in the lattice regularization, where they manifest themselves in the so-called fermion doubling problem (see Chapter 10), but they also arise in dimensional regularization.

We do not yet address these subtleties and concentrate entirely on working out the Euclidean functional integral for a single momentum mode of a left-handed Weyl neutrino. The corresponding calculation in the Hamiltonian formulation, along the lines of Chapter 8, leads to a product of single-mode partition functions

$$Z = \text{Tr}\exp\left(-\beta(\hat{H}_L - \mu\hat{F}_L)\right) = \prod_{\vec{p}} Z(\vec{p}). \tag{9.127}$$

The Euclidean functional integral of a left-handed Weyl fermion field at inverse temperature β, coupled to the chemical potential μ (not to be confused with a space–time index), takes the form

---

[7] For the Dirac field, this property is discussed in Section 3.5 of Peskin and Schroeder (1997).

$$Z = \int \mathcal{D}\bar{\psi}_{\mathrm{L}} \mathcal{D}\psi_{\mathrm{L}} \, \exp\left(-S_{\mathrm{L}}[\bar{\psi}_{\mathrm{L}}, \psi_{\mathrm{L}}]\right),$$

$$S_{\mathrm{L}}[\bar{\psi}_{\mathrm{L}}, \psi_{\mathrm{L}}] = \int_0^\beta dx_4 \int d^3x \left(\bar{\psi}_{\mathrm{L}}(x)\bar{\sigma}_\mu \partial_\mu \psi_{\mathrm{L}}(x) - \mu\bar{\psi}_{\mathrm{L}}(x)\psi_{\mathrm{L}}(x)\right). \quad (9.128)$$

The chemical potential μ couples to the conserved fermion number $\hat{F}$, which is associated with the time-component of a conserved current. The $U(1)_F$ fermion number is a global symmetry which could, at least in principle, be gauged, by turning the derivative ∂_μ in the Euclidean action into a covariant derivative $D_\mu = \partial_\mu + ieA_\mu(x)$. As a general rule, the chemical potential μ enters the Euclidean action in the form of a constant, purely imaginary time-component of a vector potential $eA_4(x) = i\mu$, such that $D_4 = \partial_4 - \mu$. Therefore, using $\bar{\sigma}_4 = \mathbb{1}$, we obtain

$$\bar{\psi}_{\mathrm{L}}(x)\bar{\sigma}_4 D_4 \psi_{\mathrm{L}}(x) = \bar{\psi}_{\mathrm{L}}(x)\partial_4 \psi_{\mathrm{L}}(x) - \mu\bar{\psi}_{\mathrm{L}}(x)\psi_{\mathrm{L}}(x). \quad (9.129)$$

As we anticipated in eq. (9.76), and as we will soon understand, the Grassmann fields must obey *anti-periodic boundary conditions in Euclidean time*,

$$\psi_{\mathrm{L}}(\vec{x}, x_4 + \beta) = -\psi_{\mathrm{L}}(\vec{x}, x_4), \quad \bar{\psi}_{\mathrm{L}}(\vec{x}, x_4 + \beta) = -\bar{\psi}_{\mathrm{L}}(\vec{x}, x_4). \quad (9.130)$$

As usual, we also impose periodic spatial boundary conditions over a box of size $L \times L \times L$. Performing the spatial Fourier transform

$$\psi_{\mathrm{L}}(\vec{p}, x_4) = \int d^3x \, \psi_{\mathrm{L}}(\vec{x}, x_4) \exp(-i\vec{p} \cdot \vec{x}),$$

$$\bar{\psi}_{\mathrm{L}}(\vec{p}, x_4) = \int d^3x \, \bar{\psi}_{\mathrm{L}}(\vec{x}, x_4) \exp(i\vec{p} \cdot \vec{x}), \quad (9.131)$$

and introducing the short-hand notation $\psi_{\mathrm{L}}(\vec{p}, x_4) = \psi(x_4)$, $\bar{\psi}_{\mathrm{L}}(\vec{p}, x_4) = \bar{\psi}(x_4)$, the partition function for a single 3-momentum mode then takes the form of a quantum mechanical fermionic path integral (the operators $\hat{H}$ and $\hat{F}$ will be defined below)

$$Z(\vec{p}) = \mathrm{Tr}\exp\left(-\beta\left(\hat{H} - \mu\hat{F}\right)\right) = \int \mathcal{D}\bar{\psi}\mathcal{D}\psi \, \exp\left(-S[\bar{\psi}, \psi]\right),$$

$$S[\bar{\psi}, \psi] = \int_0^\beta dx_4 \, \bar{\psi}(x_4)\left(\partial_4 + \vec{\sigma} \cdot \vec{p} - \mu\right)\psi(x_4). \quad (9.132)$$

Upon canonical quantization and by imposing the anti-commutation relations

$$\{\hat{\psi}^a, \hat{\psi}^{b\dagger}\} = \delta_{ab}, \quad \{\hat{\psi}^a, \hat{\psi}^b\} = \{\hat{\psi}^{a\dagger}, \hat{\psi}^{b\dagger}\} = 0, \quad a, b \in \{1, 2\}, \quad (9.133)$$

this yields the single-mode Hamilton operator $\hat{H} = -\hat{\psi}^\dagger \, \vec{\sigma} \cdot \vec{p} \, \hat{\psi}$ as well as the fermion number operator $\hat{F} = \hat{\psi}^\dagger \hat{\psi}$. The corresponding Fock space is spanned by the four states $|0\rangle$, $|1\rangle$, $|2\rangle$, and $|12\rangle$, such that

$$\hat{\psi}^1|0\rangle = \hat{\psi}^2|0\rangle = 0, \quad \hat{\psi}^{1\dagger}|0\rangle = |1\rangle, \quad \hat{\psi}^{2\dagger}|0\rangle = |2\rangle, \quad \hat{\psi}^{2\dagger}\hat{\psi}^{1\dagger}|0\rangle = |12\rangle. \quad (9.134)$$

It should be noted that $|0\rangle$ is *not* the physical vacuum state of lowest energy, but just the empty Fock state that serves as a reference state. As a consequence, the physical vacuum energy is not subtracted from $\hat{H}$. In addition, the fermion number F is also measured with respect to the empty Fock state $|0\rangle$.

Next we introduce *coherent Grassmann states* $|\psi\rangle$ which, just like the familiar bosonic coherent states, are eigenstates of the annihilation operators $\hat{\psi}^a$ with Grassmann number eigenvalues ψ^a

$$\hat{\psi}^u|\psi\rangle = \psi^u|\psi\rangle, \quad |\psi\rangle = |0\rangle - \psi^1|1\rangle - \psi^2|2\rangle + \psi^1\psi^2|12\rangle. \tag{9.135}$$

The Grassmann numbers are treated as not only anti-commuting among each other, but also with the fermion creation and annihilation operators, such that indeed

$$\hat{\psi}^1|\psi\rangle = \hat{\psi}^1|0\rangle + \psi^1\hat{\psi}^1|1\rangle + \psi^2\hat{\psi}^1|2\rangle + \psi^1\psi^2\hat{\psi}^1|12\rangle = \psi^1|0\rangle - \psi^1\psi^2|2\rangle = \psi^1|\psi\rangle. \tag{9.136}$$

Here, we have used $\hat{\psi}^1|1\rangle = |0\rangle$, $\hat{\psi}^1|2\rangle = 0$, and $\hat{\psi}^1|12\rangle = \hat{\psi}^1\hat{\psi}^{2\dagger}\hat{\psi}^{1\dagger}|0\rangle = -\hat{\psi}^{2\dagger}|0\rangle = -|2\rangle$. Similarly, one can confirm that $|\psi\rangle$ is an eigenstate of $\hat{\psi}^2$. In addition (unlike for bosonic creation operators), we construct coherent Grassmann eigenstates of the creation operators

$$\langle\bar{\psi}|\hat{\psi}^{a\dagger} = \langle\bar{\psi}|\bar{\psi}^a, \quad \langle\bar{\psi}| = \langle0| - \langle1|\bar{\psi}^1 - \langle2|\bar{\psi}^2 + \langle12|\bar{\psi}^2\bar{\psi}^1. \tag{9.137}$$

Quick Question 9.10.1   Eigenstates of fermion field operators
Show that $|\psi\rangle$ is an eigenstate of $\hat{\psi}^2$, and that the states of eq. (9.137) are indeed eigenstates of $\hat{\psi}^{a\dagger}$.

The scalar product of two coherent Grassmann states is given by

$$\langle\bar{\psi}|\psi\rangle = \langle0|0\rangle + \langle1|1\rangle\bar{\psi}^1\psi^1 + \langle2|2\rangle\bar{\psi}^2\psi^2 + \langle12|12\rangle\bar{\psi}^2\bar{\psi}^1\psi^1\psi^2 = \exp\left(\bar{\psi}^1\psi^1 + \bar{\psi}^2\psi^2\right). \tag{9.138}$$

Similarly, the completeness relation takes the form

$$\int d\bar{\psi}^1 d\psi^1 d\bar{\psi}^2 d\psi^2 \, |\psi\rangle\langle\bar{\psi}| \exp\left(-\bar{\psi}^1\psi^1 - \bar{\psi}^2\psi^2\right) = |0\rangle\langle0| + |1\rangle\langle1| + |2\rangle\langle2| + |12\rangle\langle12|$$

$$= \mathbb{1}. \tag{9.139}$$

The trace of an operator $\hat{A}$ is obtained as

$$\text{Tr}\,\hat{A} = \int d\bar{\psi}^1 d\psi^1 d\bar{\psi}^2 d\psi^2 \, \exp\left(-\bar{\psi}^1\psi^1 - \bar{\psi}^2\psi^2\right) \langle\bar{\psi}|\hat{A}| - \psi\rangle. \tag{9.140}$$

The *negative* sign in $|-\psi\rangle$ is the reason for the *anti-periodic* boundary conditions of Grassmann fields in Euclidean time. The state $|-\psi\rangle$ is obtained from $|\psi\rangle$ by changing the sign of the Grassmann number coefficients in eq. (9.135), *i.e.* $|-\psi\rangle = |0\rangle + \psi^1|1\rangle + \psi^2|2\rangle + \psi^1\psi^2|12\rangle$.

Finally, we consider a Hermitian matrix Λ (in this case a 2×2 matrix, not to be confused with a Lorentz transformation) which defines a particular operator $\hat{A}$ that has the following matrix elements between Grassmann coherent states

$$\hat{A} = \exp\left(\hat{\psi}^\dagger \Lambda \hat{\psi}\right) \quad \Rightarrow \quad \langle\bar{\psi}|\hat{A}|\psi\rangle = \exp\left(\bar{\psi}e^\Lambda\psi\right). \tag{9.141}$$

Quick Question 9.10.2   Properties of the operator $\hat{A}$
Convince yourself of the orthogonality, completeness, and trace relations, as well as of the formula for the operator $\hat{A}$, *i.e.* of eqs. (9.138) – (9.141). Corresponding relations are valid in larger Grassmann algebras as well.

As it stands, in the continuum the fermionic path integral of eq. (9.132) is a rather formal expression that needs to be properly regularized. In particular, *a priori* it is not clear how to interpret a derivative like $\partial_4 \psi(x_4)$. Since a Grassmann number does not even take any particular values, one will not be able to decide whether or not it is a differentiable function of x_4.

On the other hand, we know that even in a bosonic quantum mechanical path integral, the paths that contribute significantly are not differentiable either. As we already did for bosonic path integrals in Section 1.3, we again introduce a *Euclidean time lattice* with spacing a and extent $Na = \beta$ in order to properly define the fermionic path integral. The lattice regularization is extremely powerful, particularly in quantum field theory. Compared to perturbative schemes, such as dimensional regularization, it has the advantage that it regularizes the entire theory at once (*i.e.* beyond perturbation theory), rather than regularizing individual Feynman diagrams in a perturbative expansion. The lattice regularization is applicable both within and beyond perturbation theory. However, for purely perturbative calculations other regularizations are easier to handle. For our present purpose, namely, to give the formal expression of eq. (9.132) a well-defined mathematical meaning, the lattice regularization is ideally suited.

First of all, on the lattice the integration measure is regularized – and thus well-defined – as

$$\mathcal{D}\bar{\psi}\mathcal{D}\psi = \prod_{n=1}^{N} \prod_{a=1,2} d\bar{\psi}_n^a d\psi_n^a, \tag{9.142}$$

where $\bar{\psi}_n^1, \psi_n^1, \bar{\psi}_n^2, \psi_n^2$ form a set of four independent Grassmann numbers associated with each point $x_4 = na$ (with $n \in \{1, 2, \ldots, N\}$) on the Euclidean-time lattice. The action is regularized as

$$S[\bar{\psi}, \psi] = a \sum_n \left(\frac{1}{2a} \Big[\bar{\psi}_n \left(\psi_{n+1} - \psi_n \right) - \left(\bar{\psi}_{n+1} - \bar{\psi}_n \right) \psi_{n+1} \Big] - \bar{\psi}_n \left(\vec{\sigma} \cdot \vec{p} + \mu \right) \psi_n \right). \tag{9.143}$$

As in Chapter 1, we have again implemented a manifestly time-reversal-invariant regularization of the derivative term. Here, however, this relation should not be viewed as a finite-difference approximation of the continuum action of eq. (9.132), which is, in fact, just a formal expression; *a priori* it is mathematically ill-defined. The lattice regularization instead provides a proper *definition* of this expression.

We will now construct a Euclidean transfer matrix $\hat{T}$ that approaches $\exp(-a(\hat{H} - \mu\hat{F}))$ in the Euclidean time continuum limit $a \to 0$, *i.e.*

$$Z(\vec{p}) = \int \mathcal{D}\bar{\psi}\mathcal{D}\psi \, \exp\left(-S[\bar{\psi}, \psi]\right) = \text{Tr} \, \hat{T}^N, \quad -\lim_{a \to 0} \frac{1}{a} \log(\hat{T}) = \hat{H} - \mu\hat{F}. \tag{9.144}$$

Using eq. (9.140), the trace is expressed as

$$\text{Tr } \hat{T}^N = \int d\bar{\psi}_N^1 d\psi_1^1 d\bar{\psi}_N^2 d\psi_1^2 \, \exp\left(-\bar{\psi}_N \psi_1\right) \langle \bar{\psi}_N | \hat{T}^N | -\psi_1 \rangle. \qquad (9.145)$$

We associate one factor of $\hat{T}$ with each instant of discrete Euclidean time n. We then insert complete sets of coherent Grassmann states between adjacent factors of $\hat{T}$ associated with the discrete times n and $n+1$, for $n \in \{1, 2, \ldots, N-1\}$,

$$\int d\bar{\psi}_n^1 d\psi_{n+1}^1 d\bar{\psi}_n^2 d\psi_{n+1}^2 |\psi_{n+1}\rangle \langle \bar{\psi}_n | \exp\left(-\bar{\psi}_n \psi_{n+1}\right) = \mathbb{1}. \qquad (9.146)$$

The Grassmann integrations provide the functional integral measure in eq. (9.144). The factor $\exp(-\bar{\psi}_n \psi_{n+1})$ corresponds to those discrete derivative contributions to the action of eq. (9.143) that couple $\bar{\psi}_n$ to ψ_{n+1}. The other contributions to the action are associated with single instants of time and give rise to the transfer matrix elements

$$\langle \bar{\psi}_n | \hat{T} | \psi_n \rangle = \exp\left(\bar{\psi}_n \psi_n + a\bar{\psi}_n \left(\vec{\sigma} \cdot \vec{p} + \mu\right) \psi_n\right). \qquad (9.147)$$

This expression has the form of eq. (9.141) with $e^\Lambda = \mathbb{1} + a(\vec{\sigma} \cdot \vec{p} + \mu)$, such that the transfer matrix is identified as

$$\hat{T} = \exp\left(\hat{\psi}^\dagger \Lambda \hat{\psi}\right) = \exp\left(\hat{\psi}^\dagger \log\left(\mathbb{1} + a\left(\vec{\sigma} \cdot \vec{p} + \mu\right)\right) \hat{\psi}\right) \quad \Rightarrow$$

$$\lim_{a \to 0} \hat{T} = \exp\left(a\hat{\psi}^\dagger \left(\vec{\sigma} \cdot \vec{p} + \mu\right) \hat{\psi}\right) = \exp\left(-a(\hat{H} - \mu\hat{F})\right). \qquad (9.148)$$

In the continuum limit, it is indeed consistent with the correct Hamilton and fermion number operator. Hence, we have convinced ourselves that the Hamiltonian formulation and the Euclidean functional integral lead to the same physical results.

The derivation presented here is inspired by Martin Lüscher's construction of the transfer matrix for Wilson's lattice gauge theory (Lüscher, 1977). It should be pointed out that, unlike in some other transfer matrix considerations which discard certain "small" terms, here no terms have been neglected. This is important because for Grassmann numbers there is no notion of large or small. Lüscher's construction simply provides the exact answer at finite lattice spacing and is well-behaved in the continuum limit.

Exercises

9.1 Grassmann integration by parts

Discuss the question, in which sense functions of Grassmann variables can be integrated by parts.

9.2 The Grassmannian δ-function

How would you formulate a "δ-function" under Grassmannian integrals? Explain a sensible criterion and its solution.

9.3 Fermion determinant

Derive eq. (9.16) for arbitrary N, without relying on the Jacobian for the Grassmann integration.

9.4   Fermionic 2-point function

Derive eqs. (9.28) and (9.31) for general square matrices A and M.

9.5   Discrete transformations of fermion fields

Explore whether or not the relations (9.114) and (9.115) also hold if one chooses the Dirac basis or the Majorana basis for the γ-matrices. In Minkowski space, the corresponding matrices take the form

Dirac basis:

$$\gamma^0 = \begin{pmatrix} \mathbb{1} & 0 \\ 0 & -\mathbb{1} \end{pmatrix},\ \gamma^1 = \begin{pmatrix} 0 & \mathbb{1} \\ -\mathbb{1} & 0 \end{pmatrix},\ \gamma^2 = \begin{pmatrix} 0 & \sigma^2 \\ -\sigma^2 & 0 \end{pmatrix},\ \gamma^3 = \begin{pmatrix} 0 & \sigma^3 \\ -\sigma^3 & 0 \end{pmatrix},$$

Majorana basis (purely imaginary):

$$\gamma^0 = \begin{pmatrix} 0 & \sigma^2 \\ \sigma^2 & 0 \end{pmatrix},\ \gamma^1 = \begin{pmatrix} i\sigma^3 & 0 \\ 0 & i\sigma^3 \end{pmatrix},\ \gamma^2 = \begin{pmatrix} 0 & -\sigma^2 \\ \sigma^2 & 0 \end{pmatrix},\ \gamma^3 = \begin{pmatrix} -i\sigma^1 & 0 \\ 0 & -i\sigma^1 \end{pmatrix}.$$

9.6   Implementing the Grassmann Majorana constraint

Apply the Grassmann Majorana constraint (9.64) to show that, alternatively to eq. (9.83), we can write

$$Z = \int \mathcal{D}\bar{\psi}_L \mathcal{D}\psi_L \exp\left(-\frac{1}{2}(\psi_L^\mathsf{T}, \bar{\psi}_L) A_L \begin{pmatrix} \psi_L \\ \bar{\psi}_L^\mathsf{T} \end{pmatrix}\right) = \mathrm{Pf}(A_L),$$

$$A_L = \begin{pmatrix} 0 & -i\sigma^2 \\ -i\sigma^2 & 0 \end{pmatrix} A_R \begin{pmatrix} 0 & i\sigma^2 \\ i\sigma^2 & 0 \end{pmatrix} = \begin{pmatrix} -i\sigma^2 m\delta(x-y) & W_L \\ -W_L^\mathsf{T} & i\sigma^2 m\delta(x-y) \end{pmatrix}.$$

9.7   Generating rotations in Euclidean space

a) Show that Λ, as defined in the upper line of eq. (9.91), and O according to the upper line in eq. (9.92), are matrices in the groups $SO(4)$ and $SO(3)$, respectively.

b) Derive the relations in the lower lines of these two equations.

9.8   Fermionic lattice path integral

Consider Grassmann variables ψ_x, $\bar{\psi}_x$, ψ_y, and $\bar{\psi}_y$ on a lattice with only two sites x and y. The fermionic action is given by

$$S[\bar{\psi}, \psi] = \bar{\psi}_x \psi_y + \bar{\psi}_y \psi_x + m(\bar{\psi}_x \psi_x + \bar{\psi}_y \psi_y).$$

Evaluate the partition function

$$Z = \int \mathcal{D}\bar{\psi}\mathcal{D}\psi\ \exp(-S[\bar{\psi}, \psi]),$$

using the rules for Grassmann integration. Determine the values of the 2-point function $\langle \bar{\psi}_x \psi_x \rangle$ and $\langle \bar{\psi}_x \psi_y \rangle$.

10 Chiral Symmetry in the Continuum and on the Lattice

In this chapter, we discuss the free fermion field – in the Euclidean functional integral formulation – in the continuum and on the lattice.

A 1-flavor, free, massless Dirac fermion field has a global $U(1)_L \times U(1)_R$ symmetry, known as *chiral symmetry*. A mass term explicitly breaks this symmetry down to its *vector subgroup* $U(1)_{L=R}$. In a variant for N_f quark flavors, an approximate, global chiral $SU(N_f)_L \times SU(N_f)_R$ symmetry plays an important role in Quantum Chromodynamics – the $SU(3)_c$ gauge theory of the strong interaction between quarks and gluons. The electroweak $SU(2)_L \times U(1)_Y$ gauge symmetry of the Standard Model is even a local, and hence exact, chiral symmetry. Maintaining chiral symmetry at the regularized level is non-trivial, but highly desirable, in particular, in chiral gauge theories including the Standard Model.

This may seem straightforward in the continuum, but the lattice regularization of chiral fermions is a non-trivial issue, even in the free case. When the Dirac Lagrangian is naively discretized, we are confronted with the famous *lattice fermion doubling problem*. Doubling can be avoided by the *Wilson fermion* formulation, which, however, breaks chiral symmetry explicitly. In view of the extension to the interacting case, which will be addressed in Chapter 18, it is of major interest to search for a way to preserve some form of chiral symmetry on the lattice, *i.e.* at the regularized level, in a regularization scheme that applies beyond perturbation theory.

Via block variable transformations of the fermion fields, the renormalization group leads the way to the *Ginsparg–Wilson relation,* which provides the key to the understanding of chiral symmetry on the lattice. We present two types of solutions to this relation, which are denoted as *perfect fermions* and *overlap fermions*. They are invariant under a lattice-modified form of chiral symmetry, as a direct consequence of the Ginsparg–Wilson relation.

10.1 Chiral Symmetry in the Continuum

Let us consider a free Dirac fermion field with the Euclidean Lagrangian

$$
\begin{aligned}
\mathcal{L}(\bar{\psi}, \psi) &= \bar{\psi}(x) \left(\gamma_\mu \partial_\mu + m \right) \psi(x) \\
&= \bar{\psi}_R(x) \sigma_\mu \partial_\mu \psi_R(x) + \bar{\psi}_L(x) \bar{\sigma}_\mu \partial_\mu \psi_L(x) + m \Big[\bar{\psi}_R(x) \psi_L(x) + \bar{\psi}_L(x) \psi_R(x) \Big].
\end{aligned}
\tag{10.1}
$$

We recall that in eq. (9.78) we have defined $\sigma_\mu = (-i\vec{\sigma}, \mathbb{1})$, $\bar{\sigma}_\mu = (i\vec{\sigma}, \mathbb{1})$. In the chiral limit, $m = 0$, the Lagrangian is invariant under global chiral transformations, which correspond to separate left- and right-handed fermion number symmetries, $U(1)_L \times U(1)_R$,

$$\psi_L(x) \to \exp(i\chi_L)\,\psi_L(x), \quad \bar{\psi}_L(x) \to \bar{\psi}_L(x)\exp(-i\chi_L),$$
$$\psi_R(x) \to \exp(i\chi_R)\,\psi_R(x), \quad \bar{\psi}_R(x) \to \bar{\psi}_R(x)\exp(-i\chi_R). \tag{10.2}$$

In the presence of a mass term, the symmetry is reduced to the vector subgroup $U(1)_{L=R}$, which is restricted to the transformations with $\exp(i\chi_L) = \exp(i\chi_R)$. This can be illustrated by rewriting the transformations (10.2) in terms of the Dirac spinor field and introducing a vectorial and an axial phase, χ_v and χ_a,

$$\psi(x) \to \exp(i\chi_v)\,\psi(x), \quad \bar{\psi}(x) \to \bar{\psi}(x)\exp(-i\chi_v),$$
$$\psi(x) \to \exp(i\chi_a\gamma_5)\,\psi(x), \quad \bar{\psi}(x) \to \bar{\psi}(x)\exp(i\chi_a\gamma_5). \tag{10.3}$$

If we include a mass term, the axial symmetry breaks explicitly, whereas the vectorial symmetry persists.

In theories with N_f fermion flavors, the chiral symmetry is enhanced to $U(N_f)_L \times U(N_f)_R$, but here we restrict ourselves to the case of a single Dirac fermion flavor, $N_f = 1$.

At the level of the equations of motion, chiral symmetry gives rise to two Noether currents. By referring to the transformations (10.3), they are denoted as the *vector current, j_μ,* and the *axial vector current, j_μ^5,*

$$j_\mu(x) = \bar{\psi}(x)\gamma_\mu\psi(x), \quad j_\mu^5(x) = \bar{\psi}(x)\gamma_\mu\gamma_5\psi(x). \tag{10.4}$$

The vector current is conserved, even if we include a fermion mass (we encountered its Minkowskian version in eq. (9.53)). The corresponding charge is associated with the total fermion number

$$F(x_4) = \int d^3x\, j_4(\vec{x}, x_4), \quad \partial_\mu j_\mu(x) = 0. \tag{10.5}$$

The continuity equation implies that fermion number is conserved. On the other hand, the conservation of the axial current is explicitly broken by a mass term,

$$\partial_\mu j_\mu^5(x) = 2m\bar{\psi}(x)\gamma_5\psi(x). \tag{10.6}$$

Quick Question 10.1.1　Divergence of the axial current
　　Show that eq. (10.6) can be derived by just inserting the equations of motion, *i.e.* the Dirac equation and its adjoint (cf. Section 9.2).

10.2　Lattice Fermion Doubling Problem

In the continuum, the Euclidean action of a free Dirac fermion in d space–time dimensions is given by

$$S[\bar{\psi}, \psi] = \int d^d x\, \bar{\psi}\,(\gamma_\mu\partial_\mu + m)\psi, \tag{10.7}$$

and the functional integral takes the form

$$Z = \int \mathcal{D}\bar{\psi}\mathcal{D}\psi \ \exp(-S[\bar{\psi},\psi]),$$ (10.8)

where $\bar{\psi}(x), \psi(x)$ are independent, Grassmann-valued fermion fields.

In the lattice regularization, they are replaced by fields $\bar{\Psi}_x$, Ψ_x, which reside at the sites x of a d-dimensional hypercubic Euclidean space–time lattice. Hence, the continuum derivative has to be discretized by a finite difference. This may seem like a straightforward variant of the lattice formulation of scalar fields that we discussed in Chapter 1. We are going to see, however, that it is not as simple as that.

The most obvious way to do so leads to the so-called *naive lattice fermion action*

$$S[\bar{\Psi},\Psi] = a^d \sum_{x,\mu} \frac{1}{2a} \left(\bar{\Psi}_x \gamma_\mu \Psi_{x+\hat{\mu}} - \bar{\Psi}_{x+\hat{\mu}} \gamma_\mu \Psi_x \right) + a^d \sum_x m \bar{\Psi}_x \Psi_x.$$ (10.9)

As in Section 1.7, $\hat{\mu}$ is a vector of length a in the μ-direction; *i.e.* it connects two nearest-neighbor lattice sites. In the continuum limit, $a \to 0$, the lattice sum $a^d \sum_x$ turns into the continuum space–time integral $\int d^d x$. The corresponding lattice Dirac operator D_{xy}, which is a matrix in the Dirac and space–time indices, appears as

$$S[\bar{\Psi},\Psi] = \sum_{xy} \bar{\Psi}_x D_{xy} \Psi_y \equiv \bar{\Psi} D \Psi,$$

$$\frac{1}{a^d} D_{xy} = \frac{1}{2a} \sum_\mu \gamma_\mu (\delta_{x+\hat{\mu},y} - \delta_{x-\hat{\mu},y}) + m \, \delta_{xy},$$ (10.10)

where the last term contains an implicit unit-matrix in spinor space. The lattice functional integral now takes the form

$$Z = \int \prod_x d\bar{\Psi}_x d\Psi_x \ \exp(-S[\bar{\Psi},\Psi]) = \int \mathcal{D}\bar{\Psi}\mathcal{D}\Psi \ \exp(-\bar{\Psi}D\Psi).$$ (10.11)

On the lattice, the fermionic Grassmann integration measure is completely regularized.

The momentum space of the lattice theory is a d-dimensional torus – the Brillouin zone $B =]-\pi/a, \pi/a]^d$ – which has periodic boundary conditions. Fourier transforming the naive fermion action of eq. (10.9) leads to the lattice fermion propagator

$$\left\langle \bar{\Psi}(-p)\Psi(p) \right\rangle = \left[i \sum_\mu \gamma_\mu \frac{1}{a} \sin(p_\mu a) + m \right]^{-1}.$$ (10.12)

We keep following the lines of Section 1.7 and perform an inverse Fourier transform in the Euclidean energy p_d to obtain the fermion 2-point function

$$\left\langle \bar{\Psi}(-\vec{p},0)\Psi(\vec{p},x_d) \right\rangle = \frac{1}{2\pi} \int_{-\pi/a}^{\pi/a} dp_d \ \left\langle \bar{\Psi}(-p)\Psi(p) \right\rangle \exp(ip_d x_d) \sim \exp\left(-E(\vec{p})x_d\right).$$

(10.13)

At large Euclidean time separation x_d, the 2-point function decays exponentially with the energy $E(\vec{p})$ of a fermion with spatial momentum $\vec{p}$. In close analogy to eq. (1.80), for the naive fermion action, the lattice dispersion relation reads

$$\sinh^2(E(\vec{p})a) = \sum_i \sin^2(p_i a) + (ma)^2.$$ (10.14)

The continuum limit $a \rightarrow 0$ successfully recovers the familiar continuum dispersion relation $E(\vec{p})^2 = \vec{p}\,^2 + m^2$.

However, this solution is *ambiguous*: In particular, $E(\vec{p}) = m$ is achieved not only by $\vec{p} = \vec{0}$, but additionally by all 3-momenta $\vec{p}$ which are located at the corners of the Brillouin zone. Generally, each momentum component has the choice between the values $p_i = 0$ and π/a, such that $\sin(p_i a) = 0$. As a result, the lattice dispersion relation leads to additional states in the spectrum, which are absent in the continuum theory, but which do not disappear when we approach the continuum limit. Consequently, the naive lattice fermion action gives rise to more than one physical fermion flavor in the continuum limit.

In total, the lattice fermion propagator of eq. (10.12) has 2^d poles instead of just one as in the continuum theory. The additional states that appear in the lattice dispersion relation manifest themselves as additional particles – the so-called *doubler fermions*. Thus, the fermion doubling problem entails a multiplication of fermion species. The doubler fermions pose a severe problem in lattice field theory. Without removing them, we cannot describe the fermions that we observe in Nature. In Chapter 18, we will substantiate this property by referring to the so-called axial anomaly.

10.3 Nielsen–Ninomiya No-Go Theorem

Before we eliminate the doubler fermions, let us demonstrate a general theorem due to Holger Nielsen and Masao Ninomiya (Nielsen and Ninomiya, 1981a,b):

> *A chirally invariant (free) fermion lattice action, which is local, lattice translation-invariant, and real, is necessarily plagued by fermion doubling.*

Below we will clarify what we mean by "locality of a lattice action". At this point, the term "chirally invariant" is understood such that the lattice Dirac operator D anti-commutes with γ_5, as in the continuum at $m = 0$. Indeed, the property $\{D, \gamma_5\} = 0$ implies that axial symmetry, *i.e.* invariance under the axial rotation described in eq. (10.3), holds. In Section 10.6, we will see how the chirality condition can be relaxed in order to overcome the fermion doubling problem.

The Nielsen-Ninomiya no-go theorem is based on the topology of the Brillouin zone B. It holds because the lattice momentum space has the structure of a torus, as we mentioned before. A quite general ansatz[1] for a chirally symmetric and translation-invariant lattice action for free fermions is given by

$$S[\bar{\Psi}, \Psi] = a^d \sum_{x,y,\mu} \bar{\Psi}_x \gamma_\mu \rho_\mu (x - y) \Psi_y. \tag{10.15}$$

The function $\rho_\mu (x - y)$ determines the strength of the coupling between the fermion field values $\bar{\Psi}_x$ and Ψ_y at two lattice sites x and y (which may be separated by an arbitrarily large distance). *Locality* of the lattice action means that $|\rho_\mu (x - y)|$ decays at least exponentially

---

[1] In principle, one could add terms with any product of an odd number of different γ-matrices, but additional terms of this kind do not help against the no-go theorem.

at large separations $|x - y|$. This criterion extends to the interacting case; it is relevant for the correct continuum limit. The stronger condition for $\rho_\mu(x-y)$ to have a compact support (*i.e.* to vanish when $|x - y|$ exceeds some finite number of lattice spacings) is denoted as *ultra-locality*. While ultra-locality is desirable in numerical lattice simulations, locality is sufficient for conceptual questions.

In momentum space, locality implies that the Fourier transform $\rho_\mu(p)$ is a smooth function: Since $\rho_\mu(x-y)$ decays faster than any polynomial at large $|x-y|$, $\rho_\mu(p)$ has an infinite number of continuous derivatives. In particular, $\rho_\mu(p)$ does not have any singularities or discontinuities. The corresponding lattice fermion propagator takes the form

$$\langle \bar{\Psi}(-p)\Psi(p)\rangle = \left[i \sum_\mu \gamma_\mu \rho_\mu(p) \right]^{-1}. \tag{10.16}$$

Reality and translation invariance of the lattice action imply that $\rho_\mu(p)$ is a real-valued, periodic function over the Brillouin zone.

The poles of the propagator manifest themselves as fermions, be they physical or doublers. They correspond to zeros of $\rho_\mu(p)$, *i.e.* to points p with $\rho_\mu(p) = 0$ for all μ. The Nielsen-Ninomiya theorem states that a regular, real-valued, and periodic function $\rho_\mu(p)$ necessarily vanishes at more than just the physical point in the Brillouin zone.

This is easy to show for $d = 1$. In that case, there is a single regular periodic function $\rho_1(p)$ which must have at least one zero that represents the physical fermion. Let us call the corresponding momentum p_0, and the correct continuum limit requires $\partial_p\rho_1(p_0) = 1$. Hence, the function is positive on one side of the zero and negative on the other side. Therefore, it must pass through zero again in order to be periodic. This necessarily leads to an additional doubler pole in the lattice fermion propagator. It actually implies an odd number of additional zeros of $\rho_1(p)$ (if we count them with their multiplicity).

In higher dimensions, the proof is similar. For example, for $d = 2$ there are two functions $\rho_1(p)$ and $\rho_2(p)$. The zeros of $\rho_1(p)$ are located on a closed curve in the two-dimensional Brillouin zone. This curve may also be closed via the periodic boundary conditions. However, it cannot be a single curve which crosses the Brillouin zone, say in one direction, because then $\rho_1(p)$ is not periodic in the other direction. The zeros of $\rho_2(p)$ are located on another closed curve that intersects the first one in the pole position of the physical fermion. The curves cannot just touch each other because this would lead to a quadratic rather than a linear dispersion relation for the physical fermion. Hence, as illustrated in Figure 10.1, due to the periodic boundary conditions of the Brillouin zone, the two curves must necessarily also intersect somewhere else.

In d dimensions, the zeros of $\rho_\mu(p)$ (with $\mu = 1, 2, \ldots, d$) lie on d closed $(d - 1)$-dimensional hyper-surfaces. Again, those cannot intersect in just one point. If they intersect once, they must necessarily also intersect somewhere else. This demonstrates lattice fermion doubling for a chirally symmetric, translation-invariant, real-valued, local lattice action. It should be noted that the theorem does not specify the number of doubler fermions, but it just implies that this number must be odd. It is indeed possible to reduce the number of doublers from $2^d - 1$ to 1, as discussed, *e.g.*, by Creutz (2011), but it is impossible to eliminate the doubler fermions completely.

One may try to evade the theorem by violating one of its basic assumptions. Giving up discrete translation invariance (*e.g.*, by employing a random lattice instead of a regular one) or the reality of the action (*e.g.*, by using a one-sided discrete derivative instead of

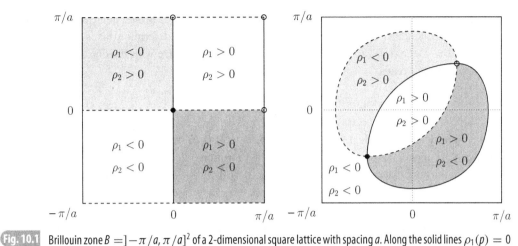

Fig. 10.1 Brillouin zone $B =]-\pi/a, \pi/a]^2$ of a 2-dimensional square lattice with spacing a. Along the solid lines $\rho_1(p) = 0$ and along the dashed lines $\rho_2(p) = 0$. The intersections of solid and dashed lines correspond to fermion poles. Left: The standard lattice fermion action gives rise to a physical fermion at $p = (0, 0)$ (filled circle) and to three doubler fermions at $(\pi/a, 0)$, $(0, \pi/a)$, and $(\pi/a, \pi/a)$ (open circles). Right: A situation with just one doubler fermion. Unlike for the standard action, in this case, the lines $\rho_1(p) = 0$ and $\rho_2(p) = 0$ do not close around the periodic boundary of B.

the forward-backward symmetric one in eq. (10.9)) has not led to acceptable solutions of the fermion doubling problem. The "staggered fermion" formulation, in its original form (Kogut and Susskind, 1975), tries to live with some doublers and treat them as a kind of "flavors" or "tastes"; in $d = 4$, this leads to four degenerate "tastes". A later variant takes the fourth root of the fermion determinant to construct a single flavor. This is sometimes used in lattice QCD simulations, although – due to this fourth root – locality is potentially compromised.

From a conceptual perspective, giving up locality is a major problem indeed in quantum field theory. For instance, there was an early attempt to solve the fermion doubling problem using so-called SLAC fermions, which set $\rho_\mu(p) = p_\mu$ (as in the continuum) inside the Brillouin zone (Drell *et al.*, 1976). Then, periodicity requires jumps at its edges. It turned out to be unacceptable because perturbation theory with gauge interactions revealed that it breaks both Lorentz invariance and locality even in the continuum limit (Karsten and Smit, 1978, 1979).

10.4 Absence of Neutrinos on a Lattice

Nielsen and Ninomiya entitled their original paper "Absence of neutrinos on a lattice" (Nielsen and Ninomiya, 1981a). They referred to the fact that in the Standard Model neutrinos exist only with left-handed chirality. According to the Nielsen-Ninomiya theorem, a single Weyl neutrino cannot be formulated on the lattice, because fermion doubling means that every left-handed fermion is accompanied by a right-handed doubler fermion partner. We illustrate this property by concentrating on the chirality of the doubler fermions

Let us try to put a single left-handed fermion (like a Standard Model neutrino) on the lattice. The Euclidean continuum action for a free left-handed fermion takes the form $S[\bar{\psi}_L, \psi_L] = \int d^4x \, \bar{\psi}_L \bar{\sigma}_\mu \partial_\mu \psi_L$. When it is discretized in the naive way, the corresponding lattice propagator reads

$$\langle \bar{\Psi}_L(-p) \Psi_L(p) \rangle = \left[i \sum_\mu \bar{\sigma}_\mu \frac{1}{a} \sin(p_\mu a) \right]^{-1}. \tag{10.17}$$

In the vicinity of the physical fermion pole at $p = (0, 0, 0, 0)$, we encounter the continuum momentum dependence $\bar{\sigma}_\mu p_\mu$. One of the doubler fermion poles is located at $(\pi/a, 0, 0, 0)$. In the vicinity of that pole, we have $p = (\pi/a, 0, 0, 0) + \delta p$ with $\sin(p_1 a) \approx -\delta p_1 a$. Hence, we encounter the momentum dependence $-\bar{\sigma}_1 \delta p_1 + \bar{\sigma}_2 \delta p_2 + \bar{\sigma}_3 \delta p_3 + \bar{\sigma}_4 \delta p_4$, which implies that, compared to the physical fermion pole, $\bar{\sigma}_1$ effectively changes sign. While the left-handed physical fermion is characterized by $\bar{\sigma}_\mu = (i\vec{\sigma}, \mathbb{1})$, this doubler fermion is characterized by

$$(-i\sigma^1, i\sigma^2, i\sigma^3, \mathbb{1}) = (i\sigma^1)(-i\sigma^1, -i\sigma^2, -i\sigma^3, \mathbb{1})(-i\sigma^1) = (i\sigma^1)\sigma_\mu(i\sigma^1)^\dagger. \tag{10.18}$$

This is equivalent to $\sigma_\mu = (-i\vec{\sigma}, \mathbb{1})$, which characterizes a right-handed fermion, by the unitary transformation $(i\sigma^1)$. Hence, the corresponding doubler fermion is, in fact, right-handed.

Similarly, the doubler fermion that resides near $(\pi/a, \pi/a, 0, 0)$ is characterized by

$$(-i\sigma^1, -i\sigma^2, i\sigma^3, \mathbb{1}) = (i\sigma^3)(i\sigma^1, i\sigma^2, i\sigma^3, \mathbb{1})(-i\sigma^3) = (i\sigma^3)\bar{\sigma}_\mu(i\sigma^3)^\dagger, \tag{10.19}$$

which is equivalent to $\bar{\sigma}_\mu$ by the unitary transformation $(i\sigma^3)$, so that doubler fermion is left-handed.

If there is an even (odd) number of momentum components π/a, the corresponding fermion is left- (right-)handed, respectively. As a result, half of the fermions are left- and the other half are right-handed.

One may wonder whether it is possible to simply interpret the doubler fermions as additional "flavors". However, this is problematical because anomalies, which are subtle quantum effects that can break chiral symmetries, are not correctly represented by the doubler fermions. We will address anomalies, in the continuum and also in the context of the lattice regularization, in Chapter 19.

10.5 Wilson Fermions

In his pioneering work on lattice field theory, Kenneth Wilson removed the doubler fermions – or moved them to high energy – in a direct and radical manner by breaking chiral symmetry explicitly (Wilson, 1974). Wilson added a term to the naive fermion action, which is proportional to a discretized Laplacian,

$$S[\bar{\Psi}, \Psi] = \sum_{x,y} \bar{\Psi}_x D_{W,xy} \Psi_y = a^d \sum_{x,\mu} \frac{1}{2a} \left(\bar{\Psi}_x \gamma_\mu \Psi_{x+\hat{\mu}} - \bar{\Psi}_{x+\hat{\mu}} \gamma_\mu \Psi_x \right) + a^d \sum_x m \bar{\Psi}_x \Psi_x$$

$$+ a^d \sum_{x,\mu} \frac{1}{2a} \left(2\bar{\Psi}_x \Psi_x - \bar{\Psi}_x \Psi_{x+\hat{\mu}} - \bar{\Psi}_{x+\hat{\mu}} \Psi_x \right), \tag{10.20}$$

and the resulting lattice propagator in momentum space reads

$$\langle \bar\Psi(-p)\Psi(p)\rangle = \left[\, \mathrm{i} \sum_\mu \gamma_\mu \frac{1}{a}\sin(p_\mu a) + m + \sum_\mu \frac{2}{a}\sin^2\left(\frac{p_\mu a}{2}\right)\right]^{-1} = D_{\mathrm{W}}(p)^{-1}. \quad (10.21)$$

Dimensional analysis shows that the Wilson term[2] is divided by a, not by a^2 as the discrete Laplacian. Hence, near the continuum limit, this term is $\mathcal{O}(a)$-suppressed, as we see explicitly in momentum space.[3]

The Nielsen-Ninomiya theorem does not apply to the Wilson–Dirac operator D_{W}, because it contains a Clifford scalar term (proportional to a unit-matrix), which implies $\{D_{\mathrm{W}}, \gamma_5\} \neq 0$, even at $m = 0$. Chiral symmetry is recovered only in the continuum limit, and in the interacting case this requires a delicate fine-tuning.

For small momenta, the Wilson term vanishes quadratically, and hence it does not affect the linear dispersion of the physical fermion, at least in the continuum limit. For the doubler fermions, on the other hand, the Wilson term is non-zero; it elevates them to a momentum-dependent energy of the order of the cut-off $1/a$. In the continuum limit, the doubler fermions are therefore eliminated from the physical particle spectrum.

For free fermions, this may look like a satisfactory solution to the doubling problem, but in interacting theories the Wilson term leads to several complications. When chiral symmetry is explicitly violated at the cut-off scale, it can no longer protect the fermion from additive mass renormalization. Recovering chiral symmetry in the continuum limit requires unnatural fine-tuning of the bare fermion mass. As a result, beyond perturbation theory there is a hierarchy problem not only for scalar fields but even for fermions. This problem will be discussed in Section 18.7.

At first glance, one might associate the fermion doubling problem just with peculiarities of the lattice regularization. It is, however, the lattice manifestation of a general, fundamental problem of the regularization of chiral fermions, which affects other regularization schemes as well. The axial symmetry, which is associated with γ_5, is a subtle issue in general. This is also manifest in the context of dimensional regularization, because there is no analytic continuation of $\gamma_5 = -\gamma_1\gamma_2 \ldots \gamma_d$ to non-integer dimensions d. We take the point of view that the hierarchy problem for fermions may be more physical than is generally appreciated. On the lattice, the problem is most obvious and its solution – to be discussed in the following sections – provides deep insights into the nature of chiral symmetry.

10.6 Perfect Lattice Fermions and the Ginsparg–Wilson Relation

In this section, we relate the continuum theory of free fermions to a corresponding lattice theory by an exact renormalization group transformation. This is achieved by defining lattice fermion fields as block averages of continuum fields integrated over hypercubes. We

---

[2] In front of this term, one can also put a real-valued coefficient, known as the *Wilson parameter*. It is of practical relevance for interacting Wilson fermions, but in the current context we can set it to 1.

[3] For a scalar field on the lattice, only a discrete Laplacian is required, and we see from the $\sin^2$-term that it does not suffer from a doubling problem. This problem arises for fermions due to the *first-order derivative* in the Dirac operator.

present this procedure following the lines and notation of Bietenholz and Wiese (1996a,b). The resulting lattice theory is in all physical respects equivalent to the underlying continuum theory; *i.e.* it is completely free of lattice artifacts. In particular, it has the same energy–momentum dispersion relation as the continuum theory. Lattice actions with these properties are known as *perfect actions* (Hasenfratz and Niedermayer, 1994).

In analogy to the renormalized trajectory obtained for scalar lattice field theory in Section 5.5, we now derive a perfect fermion action by blocking from the continuum. For this purpose, we average the continuum fermion fields $\psi(y)$ and $\bar{\psi}(y)$ over hypercubes c_x of size a^d, centered at the sites x of a d-dimensional Euclidean lattice,

$$\Psi_x = \frac{1}{a^d} \int_{c_x} d^d y \, \psi(y), \quad \bar{\Psi}_x = \frac{1}{a^d} \int_{c_x} d^d y \, \bar{\psi}(y), \tag{10.22}$$

which, in momentum space, corresponds to

$$\Psi(p) = \sum_{l \in \mathbb{Z}^d} \psi(p + 2\pi l/a) \Pi(p + 2\pi l/a),$$

$$\bar{\Psi}(-p) = \sum_{n \in \mathbb{Z}^d} \bar{\psi}(-p - 2\pi n/a) \Pi(p + 2\pi n/a). \tag{10.23}$$

The momentum p of the lattice fermion fields is defined in the Brillouin zone B, and these fields are periodic over B, while the momenta of the continuum fields take values all over $\mathbb{R}^d$, without periodicity. The blocking kernel in momentum space is given by

$$\Pi(p) = \prod_{\mu=1}^{d} \frac{2 \sin(p_\mu a/2)}{p_\mu a}. \tag{10.24}$$

Quick Question 10.6.1   Fermion fields blocked from the continuum to a lattice
Derive the relations (10.23) and (10.24).

The lattice fermion propagator is related to the continuum propagator by

$$\langle \bar{\Psi}(-p) \Psi(p) \rangle = \sum_{l \in \mathbb{Z}^d} \langle \bar{\psi}(-p - 2\pi l/a) \psi(p + 2\pi l/a) \rangle \, \Pi(p + 2\pi l/a)^2$$

$$= \sum_{l \in \mathbb{Z}^d} \frac{\Pi(p + 2\pi l/a)^2}{i\gamma_\mu(p_\mu + 2\pi l_\mu/a) + m} = D_{\text{perfect}}(p)^{-1}. \tag{10.25}$$

For $m = 0$, the lattice propagator corresponds to a lattice action of the form (10.15),

$$S[\bar{\Psi}, \Psi] = a^d \sum_{x,y} \bar{\Psi}_x \gamma_\mu \rho_\mu(x - y) \Psi_y = \sum_{x,y} \bar{\Psi}_x D_{\text{perfect},xy} \Psi_y, \tag{10.26}$$

with couplings $\rho_\mu(x - y)$ that can be (numerically) computed via summation over l and an inverse Fourier transform. (In view of the blocking kernel (10.24), there is no doubt that this series converges in any dimension.) This lattice action is perfect by construction; *i.e.* its energy spectrum is indistinguishable from that of the continuum theory. Hence, there can be no fermion doubling. Moreover, the action is chirally invariant, $\{D_{\text{perfect}}, \gamma_5\} = 0$. How can this be compatible with the Nielsen-Ninomiya no-go theorem?

The answer is that this action turns out to be *non-local* (Wiese, 1993). Its couplings $\rho_\mu(x - y)$ do not decay exponentially at large distances. Instead, for $d \geq 2$, their decay is only power-like (Bietenholz and Wiese, 1996b),

$$|\rho_\mu(x - y)| \propto |x - y|^{1-d}. \tag{10.27}$$

As a consequence, in momentum space $\rho_\mu(p)$ is not regular: It has poles, and therefore, the topological arguments that underlie the Nielsen-Ninomiya theorem do not apply.

The non-locality is most extreme in $d = 1$, where it can be analytically investigated by setting $\gamma_1 = 1$ and performing the summation in eq. (10.25) explicitly. The resulting massless propagator takes the form

$$\left\langle \bar{\Psi}(-p)\Psi(p) \right\rangle = \sum_{l \in \mathbb{Z}} \frac{\Pi(p + 2\pi l/a)^2}{i(p + 2\pi l/a)} = \frac{a}{2i} \cot\left(\frac{pa}{2}\right), \tag{10.28}$$

which implies

$$D_{\text{perfect}}(p) = \rho_1(p) = \frac{2}{a} \tan\left(\frac{pa}{2}\right). \tag{10.29}$$

This is a function with poles at the edge of the Brillouin zone ($p = \pm\pi/a$). The corresponding couplings in coordinate space,

$$\rho_1(x - y) = \frac{1}{a}(-1)^{(x-y)/a}, \tag{10.30}$$

do not decay at all at increasing distances $x - y$. This describes an extremely non-local action, which shows that relation (10.27) even holds in $d = 1$. We repeat that for $d \geq 2$ the chirally symmetric perfect action remains non-local with a power-law decay of the couplings at large distances.

The Nielsen-Ninomiya theorem implies that a local, perfect action can arise only if one explicitly breaks chiral symmetry, at least in its continuum form, $\{D, \gamma_5\} = 0$. Since the continuum theory of massless fermions is chirally invariant, this suggests to break chiral symmetry via the renormalization group blocking transformation. In fact, the "δ-function blocking" in eq. (10.22) was just an example, which can be generalized, for instance, to a "Gaussian-type" blocking. This leads to a generalized perfect lattice action $S[\bar{\Psi}, \Psi]$ given by

$$\exp(-S[\bar{\Psi}, \Psi]) = \int \mathcal{D}\bar{\psi}\mathcal{D}\psi \exp\left\{-\frac{1}{(2\pi)^d} \int d^d p\, \bar{\psi}(-p)\left(i\gamma_\mu p_\mu + m\right)\psi(p)\right\}$$

$$\times \exp\left\{-\frac{1}{\alpha}\frac{1}{(2\pi)^d} \int_B d^d p \left[\bar{\Psi}(-p) - \sum_{n \in \mathbb{Z}^d} \bar{\psi}(-p - 2\pi n/a)\Pi(p + 2\pi n/a)\right]\right.$$

$$\left. \times \left[\Psi(p) - \sum_{l \in \mathbb{Z}^d} \psi(p + 2\pi l/a)\Pi(p + 2\pi l/a)\right]\right\}. \tag{10.31}$$

The denominator α is a source of explicit chiral symmetry breaking, which enters the theory via the renormalization group transformation that maps the continuum theory to the lattice theory. For $\alpha \to 0$, we recover the "δ-function blocking", and, therefore, the chirally invariant but non-local perfect lattice action which corresponds to the propagator (10.25),

Its generalization reads

$$\langle \bar{\Psi}(-p)\Psi(p) \rangle = \sum_{l \in \mathbb{Z}^d} \frac{\Pi (p + 2\pi l/a)^2}{i\gamma_\mu(p_\mu + 2\pi l_\mu/a) + m} + \alpha, \tag{10.32}$$

which corresponds to a *local* perfect action, if $\alpha \neq 0$ (Bietenholz and Wiese, 1996b; Hasenfratz, 1998; Hasenfratz *et al.*, 1998).[4] Let us vary α in order to optimize the locality of the perfect action. For this purpose, we again consider $d = 1$. Then, the sum in eq. (10.32) can be performed analytically and the fermion propagator takes the form

$$\langle \bar{\Psi}(-p)\Psi(p) \rangle = \frac{1}{m} - \frac{2}{m^2 a} \left[\coth \left(\frac{ma}{2} \right) - i \cot \left(\frac{pa}{2} \right) \right]^{-1} + \alpha. \tag{10.33}$$

When we choose

$$\alpha = \frac{\exp(ma) - 1 - ma}{m^2 a}, \tag{10.34}$$

the propagator reduces to

$$\langle \bar{\Psi}(-p)\Psi(p) \rangle = \left(\frac{\exp(ma) - 1}{ma} \right)^2 \left[\frac{i}{a} \sin(pa) + \frac{\exp(ma) - 1}{a} + \frac{2}{a} \sin^2 \left(\frac{pa}{2} \right) \right]^{-1}. \tag{10.35}$$

This corresponds to the standard Wilson fermion action, except that the mass m is now replaced by $(\exp(ma) - 1)/a$. Hence, for the above choice of α, the 1-d perfect action is ultra-local, since it has only nearest-neighbor interactions. In the massless limit $m = 0$, the optimal choice for locality is $\alpha = a/2$, as we see from eq. (10.34). In higher dimensions, this is still a good choice: The action remains local, but is no longer ultra-local; the couplings between $\bar{\Psi}_x$ and Ψ_y decay exponentially at large distances $|x - y|$.

Let us now derive the energy–momentum dispersion relation of perfect lattice fermions. The fermion 2-point function takes the form

$$\begin{aligned}
\langle \bar{\Psi}(-\vec{p}, 0)\Psi(\vec{p}, x_d) \rangle &= \frac{1}{2\pi} \int_{-\pi/a}^{\pi/a} dp_d \, \langle \bar{\Psi}(-p)\Psi(p) \rangle \exp(ip_d x_d) \\
&= \frac{1}{2\pi} \int_{-\pi/a}^{\pi/a} dp_d \left[\sum_{l \in \mathbb{Z}^d} \frac{\Pi (p + 2\pi l/a)^2}{i\gamma_\mu(p_\mu + 2\pi l_\mu/a) + m} + \alpha \right] \exp(ip_d x_d) \\
&= \frac{1}{2\pi} \int_{-\infty}^{\infty} dp_d \sum_{\vec{l} \in \mathbb{Z}^{d-1}} \frac{m}{\left(\vec{p} + 2\pi \vec{l}/a \right)^2 + p_d^2 + m^2} \\
&\quad \times \prod_{i=1}^{d-1} \left(\frac{2 \sin(p_i a/2)}{p_i a + 2\pi l_i} \right)^2 \left(\frac{2 \sin(p_d a/2)}{p_d a} \right)^2 \exp(ip_d x_d) + \alpha \, \delta_{x_d, 0} \\
&= \sum_{\vec{l} \in \mathbb{Z}^{d-1}} C(\vec{p} + 2\pi \vec{l}/a) \exp \left(-E(\vec{p} + 2\pi \vec{l}/a) x_d \right) + \alpha \, \delta_{x_d, 0}, \tag{10.36}
\end{aligned}$$

where the last line defines the function C. In analogy to Section 5.5, the sum over l_d has been combined with the integral of p_d over the interval $]-\pi/a, \pi/a]$ to an integral over

---

[4] The constant α can be further generalized to some function $\alpha(p)$, which could be multiplied by an even number of γ-matrices. As long as this function is local in coordinate space, the locality of the perfect fermion action is preserved.

the entire momentum space $\mathbb{R}$ of the continuum theory. The sum over the spatial vectors $\vec{l} \in \mathbb{Z}^{d-1}$ leads to infinitely many poles of the integrand and hence to infinitely many states that contribute an exponential decay to the 2-point function. The energies of these states are determined by the locations of the poles, $E(\vec{p} + 2\pi \vec{l}/a) = -ip_d$, with

$$E(\vec{p} + 2\pi \vec{l}/a)^2 = -p_d^2 = \left(\vec{p} + 2\pi \vec{l}/a\right)^2 + m^2. \qquad (10.37)$$

Hence, the energy–momentum dispersion relation of perfect lattice fermions is *exactly the same as in the continuum*. As a consequence of the exact renormalization group blocking from the continuum, there are no lattice artifacts. In particular, the parameter α of the renormalization group blocking transformation (which makes the perfect action local) has no effect on the physical spectrum, although it explicitly breaks chiral symmetry – in the form that the Nielsen-Ninomiya theorem assumes.

In the case of the Wilson term (or of a mass), a local Clifford scalar term is added to the Dirac operator, which alters the pole structure of the propagator. The term proportional to α, however, is a local Clifford scalar addition *to the propagator*; hence, it does not affect its pole. (This property holds not only for Clifford scalar terms, but also for general, local terms that do not anti-commute with γ_5 (Ginsparg and Wilson, 1982).)

The explicit chiral symmetry breaking term α that we considered just leads to a contact term $\alpha \, \delta_{x_d,0}$ in the fermion propagator (or 2-point function) (10.36). As such, it has no effect on the spectrum, which is extracted from the 2-point function at large Euclidean time separations x_d. Remarkably, the spectrum of the lattice theory obeys continuous Poincaré invariance, despite the fact that the lattice action only has discrete lattice symmetries.

Chiral symmetry is present in a similar way. Due to the parameter α that enters the renormalization group blocking transformation, the perfect lattice action does not seem to be chirally invariant, not even for $m = 0$. Still, all physical consequences of chiral symmetry are correctly reproduced by the perfect action.

The crucial property that guarantees these features is the famous *Ginsparg–Wilson relation* (Ginsparg and Wilson, 1982; Hasenfratz, 1998),

$$\left\{D^{-1}, \gamma_5\right\} = 2\alpha\gamma_5, \qquad (10.38)$$

which the massless propagator D^{-1}_{perfect} fulfills.[5] A simple choice, which is particularly favorable for the locality, is $\alpha = a/2$; we saw that at least in the 1-d case it is manifestly optimal. This leads to the most commonly used form of the Ginsparg–Wilson relation,

$$\{D, \gamma_5\} = aD\gamma_5 D, \qquad (10.39)$$

where the right-hand side represents a product in momentum space, or a convolution in coordinate space, $a \sum_z D_{xz} \gamma_5 D_{zy}$.

The Ginsparg–Wilson relation turned out to be the key to the understanding of chiral symmetry on the lattice. In the continuum limit $a \to 0$, it turns into the standard form of chiral symmetry. At finite lattice spacing, however, it expresses invariance under a lattice-modified form of the chiral transformations (10.3) (Lüscher, 1998). It can be expressed in its infinitesimal form, to leading-order in the chiral rotation parameter ε,

---

[5] For a generalized term α, the right-hand side reads $\{\alpha, \gamma_5\}$.

$$\Psi \to \left(1 + i\gamma_5\varepsilon\left[1 - \frac{a}{2}D\right]\right)\Psi, \quad \bar{\Psi} \to \bar{\Psi}\left(1 + i\left[1 - \frac{a}{2}D\right]\varepsilon\gamma_5\right). \tag{10.40}$$

Let us consider the condition for invariance under this modified chiral transformation. Since we are dealing with a continuous chiral rotation, it is sufficient to keep considering $\mathcal{O}(\varepsilon)$. We stay with the short-hand notation for products in momentum space,

$$\mathcal{L} = \bar{\Psi}D\Psi \to \bar{\Psi}\left(1 + i\left[1 - \frac{a}{2}D\right]\varepsilon\gamma_5\right)D\left(1 + i\gamma_5\varepsilon\left[1 - \frac{a}{2}D\right]\right)\Psi$$

$$= \bar{\Psi}D\Psi + i\varepsilon\bar{\Psi}\left(\left[1 - \frac{a}{2}D\right]\gamma_5 D + D\gamma_5\left[1 - \frac{a}{2}D\right]\right)\Psi + \mathcal{O}(\varepsilon^2). \tag{10.41}$$

The condition for the term of $\mathcal{O}(\varepsilon)$ to vanish is exactly the Ginsparg–Wilson relation (10.39). Hence, if this relation holds, we do have exact chiral symmetry; it just takes a lattice modified form. In the interacting case, this is sufficient to avoid additive mass renormalization.

So we can circumvent the Nielsen-Ninomiya theorem by modifying the transformation with an additional term of $\mathcal{O}(a)$, such that the continuum limit reproduces the standard chiral transformation (Lüscher, 1998). This additional ingredient must be multiplied by a term of dimension mass, and the lattice Dirac operator takes this role.

In Chapter 18, we will return to the Ginsparg–Wilson relation in the framework of QCD, where it manifestly solves the problem of formulating chiral symmetry on the lattice.

10.7 Overlap Fermions

We have identified the Ginsparg–Wilson relation as the crucial criterion for chiral symmetry on the lattice. We already discussed the perfect lattice fermion as a solution to this relation, which is, however, somewhat complicated, even if the corresponding renormalization group transformation is simple and natural. Let us now construct a more obvious solution, by starting from the Wilson operator D_W, which we reviewed in Section 10.5.

This operator obeys γ_5-*Hermiticity*,

$$D_W^\dagger = \gamma_5 D_W \gamma_5. \tag{10.42}$$

For a γ_5-Hermitian lattice Dirac operator D, the Ginsparg–Wilson relation simplifies to

$$D + D^\dagger = aD^\dagger D, \tag{10.43}$$

but we know that D_W is not a solution.[6] Let us define the shifted operator $A = aD - \mathbb{1}$ and consider

$$A^\dagger A = a^2 D^\dagger D - aD - aD^\dagger + \mathbb{1}. \tag{10.44}$$

The Ginsparg–Wilson relation (10.39) is equivalent to the condition for A to be unitary. The operator $A_W = aD_W - 1$ is not unitary, but we can enforce unitarity by constructing

$$A_{\text{overlap}} = A_W\left(A_W^\dagger A_W\right)^{-1/2} \quad \Rightarrow \quad A_{\text{overlap}}^\dagger A_{\text{overlap}} = \mathbb{1}. \tag{10.45}$$

---

[6] If we insert D_W into the general Ginsparg–Wilson relation (10.38) and solve it for the term α, the latter is nonlocal. As anticipated in footnote 4, this is not an acceptable solution, because then α affects the pole structure, and also the transformation (10.40) would take a non-local form.

Thus, we readily obtain the so-called *overlap Dirac operator*

$$D_{\text{overlap}} = \frac{1}{a}(A_{\text{overlap}} + \mathbb{1}) = \left(D_{\text{W}} - \mathbb{1}/a\right)\left[(aD_{\text{W}}^{\dagger} - \mathbb{1})(aD_{\text{W}} - \mathbb{1})\right]^{-1/2} + \frac{\mathbb{1}}{a} \quad (10.46)$$

as a solution to the Ginsparg–Wilson relation (10.39) (Neuberger, 1998). This construction also shows that the spectrum of D_{overlap} – and of other solutions to eq. (10.39) – is located on a circle in the complex plane, which has the center $1/a$ and the radius $1/a$, such that it passes through the origin. It is there to stay under interaction, if eq. (10.39) still holds, which explicitly confirms the absence of additive mass renormalization. The continuum limit turns this circle into the imaginary axis.

The (non-analytic) square root operation in eq. (10.45) looks worrisome regarding locality. However, the shift in the definition of A_{W} allows for an expansion in powers of aD_{W}, which decays exponentially, so the free overlap Dirac operator is indeed local. In the interacting case, locality persists as long as the field fluctuations are not too violent (Hernández *et al.*, 1999); see Chapter 18.

We may replace the kernel D_{W} by another γ_5-Hermitian lattice Dirac operator and construct the corresponding overlap Dirac operator. In particular, if we insert a truncated version of D_{perfect}, the transition to D_{overlap} is only a minor modification, which leads to better locality and scaling behavior than the use of the Wilson kernel D_{W}. This was demonstrated for free fermions (Bietenholz, 1999), for the 2-flavor Schwinger model (Bietenholz and Hip, 2000), as well as for QCD (Bietenholz, 2002).

Originally, the overlap Dirac operator D_{overlap} was derived by Rajamani Narayanan and Herbert Neuberger in a more sophisticated manner, which also addresses chiral gauge theory (Narayanan and Neuberger, 1993a,b, 1994, 1995) and which is related to the so-called domain wall fermions (Kaplan, 1992), to be discussed in Section 18.8.

Exercises

10.1 Fixed point action in one dimension
Use eqs. (10.32) and (10.34) to derive eqs. (10.33) and (10.35)

10.2 Generalized Ginsparg–Wilson relation
Use the generalized Ginsparg–Wilson relation $\{D^{-1}, \gamma_5\} = \{\alpha, \gamma_5\}$ in order to derive the corresponding generalizations of eqs. (10.39), (10.40), and (10.41).

11 Non-Abelian Gauge Fields

Non-Abelian gauge symmetry provides us with a remarkable mathematical structure which is embodied by the $SU(2)_L$ and $SU(3)_c$ gauge fields of the Standard Model, where they mediate the electroweak and strong interaction, respectively. Just as Abelian gauge fields, non-Abelian gauge fields are associated with a redundancy in the mathematical description of the physics. Interestingly, non-Abelian gauge symmetries are not limited to fundamental interactions. They can also be used to summarize the leading relativistic effects in the Pauli equation, which describes fermions quantum mechanically, in an external electromagnetic background field (Fröhlich and Studer, 1993), they naturally occur in the form of Berry gauge fields in adiabatic processes in quantum systems with degeneracies (Wilczek and Zee, 1984), and they even arise in the description of the motion of deformable bodies including falling cats and amoebae swimming in a fluid at low Reynolds number (Shapere and Wilczek, 1987, 1989).

Historically, non-Abelian gauge theory was first suggested in an unpublished letter by Wolfgang Pauli to Abraham Pais in 1953 (Straumann, 2002). The first paper introducing SU(2) gauge theories was the groundbreaking work of Chen-Ning Yang and Robert Mills (Yang and Mills, 1954). At that time, the dynamical role of Yang–Mills fields was not clear. In particular, first it seemed that, just like Abelian gauge fields, they should give rise to massless particle excitations only. It took until 1964 before it became clear that gauge particles can pick up a mass via the Higgs mechanism. This is indeed how the W- and Z-bosons – quanta of the $SU(2)_L \times U(1)_Y$ gauge fields in the Standard Model – obtain their masses.

It took even longer to realize that, in the absence of the Higgs mechanism, due to their self-interaction, non-Abelian gauge fields generate mass non-perturbatively. Indeed, together with quarks, the gluon quanta of the $SU(3)_c$ non-Abelian gauge field in the Standard Model are permanently confined inside protons, neutrons, and other strongly interacting particles known as hadrons. The strong interaction energy between quarks and gluons manifests itself as the hadron mass. This is the origin of almost all the mass of any macroscopic object.

In this chapter, we introduce non-Abelian gauge fields, first at the classical level in the continuum. We then quantize the theory perturbatively, using the functional integral approach. As in the case of Abelian gauge theories, this requires gauge fixing, which is, however, more complicated in the non-Abelian case. In particular, this led Ludvig Faddeev and Victor Popov to the introduction of unphysical *ghost fields* (Faddeev and Popov, 1967). The ghost fields have spin zero but fermionic statistics and thus *violate the spin statistics theorem*. Their purpose is to cancel contributions of the unphysical, longitudinal polarization degrees of freedom of the gauge field. In his approach to quantize gravity, Richard Feynman was first to realize that the introduction of auxiliary degrees of freedom with fermionic statistics is necessary in order to correctly represent non-linear gauge fixing constraints (Feynman, 1963).

It is challenging to show that the resulting theory is *unitary*; *i.e.* unphysical degrees of freedom cannot be generated in the course of the time evolution. A groundbreaking discovery that guarantees this important fact is due to Carlo Becchi, Alain Rouet, Raymond Stora, and independently Igor Tyutin, who introduced *BRST symmetry* (Becchi *et al.*, 1974, 1975, 1976; Tyutin, 1975), a global fermionic symmetry (with a Grassmann-valued transformation parameter) of the gauge-fixed action, which holds to all orders of perturbation theory and is crucial for proving unitarity. In particular, BRST symmetry ensures that ghosts and anti-ghosts, as well as the longitudinal components of gauge bosons, lie in an unphysical sector of the Hilbert space; thus, they are clearly separated from the physical, transversely polarized components of the gauge bosons.

Finally, we also quantize the theory *non-perturbatively* using the lattice regularization. First, we apply canonical quantization on a spatial lattice, and then, we proceed to the functional integral on a Euclidean space–time lattice. Before we address non-Abelian lattice gauge theories, we use the Abelian compact U(1) lattice gauge theory as a warm-up exercise.

11.1 Non-Abelian Gauge Fields at the Classical Level

Non-Abelian gauge fields lead to a very specific, highly constrained type of fundamental interactions, which dominate the Standard Model of particle physics. These interactions manifest themselves at microscopic scales and lead, for instance, to the confinement of quarks and gluons. The resulting hadrons experience only a screened, short-ranged remnant of the strong interaction, which is, for example, responsible for the binding of atomic nuclei. Due to the Higgs mechanism, the W- and Z-bosons, which are mediators of the weak interaction, obtain a large mass and are unstable against the decay into other particles, including electrons and neutrinos. As a result, also the weak interaction is short-ranged. Consequently, non-Abelian gauge interactions manifest themselves mostly at the quantum level. In particular, unlike electromagnetism, the weak and strong interactions do not give rise to macroscopic, classical gauge field configurations. Still, we first introduce non-Abelian gauge fields at the classical level, to provide a starting point for their quantization.

In the continuum, non-Abelian gauge fields are described by an anti-Hermitian, Lie-algebra-valued vector potential

$$G_\mu(x) = i g_s G_\mu^a(x) T^a. \tag{11.1}$$

Here, T^a are the Hermitian generators of a Lie algebra with the commutation and normalization relations

$$[T^a, T^b] = i f_{abc} T^c, \quad \mathrm{Tr}[T^a T^b] = \frac{1}{2}\delta_{ab}. \tag{11.2}$$

The *structure constants* $f_{abc} \in \mathbb{R}$ are totally anti-symmetric and characterize the algebra. The Lie algebra SU(2) has three generators, $T^a = \tau^a/2$, which can be given in terms of the Pauli matrices τ^a. In that case, the structure constants correspond to the Levi-Civita symbol, $f_{abc} = \epsilon_{abc}$. The Lie algebra SU(3) has eight generators, $T^a = \lambda^a/2$, which can be given in terms of the Gell-Mann matrices λ^a. The mathematics of Lie algebras is discussed in Appendix F, which also displays the Gell-Mann matrices explicitly. There are as many vector potentials $G_\mu^a(x) \in \mathbb{R}$ as there are generators of the corresponding Lie algebra. For example, in the Standard Model the SU(2)$_L$ gauge field that mediates the weak interaction is associated with three vector potentials $W_\mu^a(x)$ (with $a \in \{1, 2, 3\}$), which multiply the three Pauli matrices. Similarly, the gluon field that mediates the strong interaction is described by eight vector potentials $G_\mu^a(x)$ (with $a \in \{1, 2, \ldots, 8\}$) that multiply the Gell-Mann matrices. It is convenient to absorb the coupling constant g_s, which for SU(3)$_c$ describes the strength of the strong interaction, in the definition of the matrix-valued field $G_\mu(x)$, as it is done in eq. (11.1).

Non-Abelian gauge fields can be transformed by a group-valued gauge transformation function $\Omega(x) = \exp(i\omega^a(x)T^a)$ with $\omega^a(x) \in \mathbb{R}$, such that

$$G_\mu'(x) = \Omega(x)\left(G_\mu(x) + \partial_\mu\right)\Omega(x)^\dagger. \tag{11.3}$$

In contrast to Abelian gauge fields, the non-Abelian field strength tensor

$$G_{\mu\nu}(x) = \partial_\mu G_\nu(x) - \partial_\nu G_\mu(x) + [G_\mu(x), G_\nu(x)], \tag{11.4}$$

is not gauge-invariant but transforms as

$$G_{\mu\nu}'(x) = \Omega(x)G_{\mu\nu}(x)\Omega(x)^\dagger. \tag{11.5}$$

The Lagrangian of a non-Abelian gauge theory

$$\mathcal{L}(G) = -\frac{1}{4}G_{\mu\nu}^a G^{a\mu\nu} = \frac{1}{2g_s^2}\mathrm{Tr}\left[G_{\mu\nu}G^{\mu\nu}\right], \quad G_{\mu\nu}(x) = ig_s G_{\mu\nu}^a(x)T^a, \tag{11.6}$$

where we used the trace in eq. (11.2), is gauge-invariant due to cyclic permutability under the trace. The commutator term in eq. (11.4) introduces a non-linearity and renders non-Abelian gauge fields *self-interacting*. In particular, in contrast to Abelian gauge fields, non-Abelian gauge fields are "charged" in the sense that they themselves carry non-Abelian gauge charge. As a consequence of the commutator term, the Lagrangian (11.6) contains self-interaction terms that are cubic or quartic in the vector potential. As we will see, the resulting gauge theory is strongly interacting and highly non-trivial.

Let us construct the classical Hamilton density of the theory. For this purpose, we first fix the *temporal gauge* $G^0(x) = 0$ and obtain the non-Abelian "electric" field $E_i(x) = ig_s E_i^a(x)T^a$ with

$$E_i(x) = G^{i0}(x) = \partial^i G^0(x) - \partial^0 G^i(x) + [G^i(x), G^0(x)] = -\partial^0 G^i(x). \tag{11.7}$$

The momentum field, which is canonically conjugate to the non-Abelian vector potential, results as

$$\Pi_i^a(x) = \frac{\delta \mathcal{L}}{\delta \partial^0 G^{ai}(x)} = \partial^0 G^i(x) = -E_i^a(x). \qquad (11.8)$$

The classical Hamilton density thus takes the form

$$\mathcal{H}(G^{ai}, \Pi_i^a) = \Pi_i^a(x) \partial^0 G^{ai}(x) - \mathcal{L} = \frac{1}{2} \left(E_i^a(x) E_i^a(x) + B_i^a(x) B_i^a(x) \right). \qquad (11.9)$$

Here, we have also introduced the non-Abelian "magnetic" field

$$B_i(x) = -\frac{1}{2} \epsilon_{ijk} G^{jk}(x) = \mathrm{i} g_s B_i^a(x) T^a. \qquad (11.10)$$

We could now try to proceed and apply the canonical quantization procedure to the continuum theory. However, this turns out to be more complicated than the functional integral approach. We will return to canonical quantization in Section 11.8, when we formulate the theory beyond perturbation theory using the lattice regularization.

11.2 Gauge Fixing and Faddeev–Popov Ghosts

In order to quantize non-Abelian gauge fields at the perturbative level, one needs to fix the gauge in order to eliminate an infinite factor in the Euclidean functional integral

$$Z = \int \mathcal{D}G \, \exp\left(-S[G]\right), \quad S[G] = -\int d^d x \, \frac{1}{2g_s^2} \mathrm{Tr}\left[G_{\mu\nu} G_{\mu\nu}\right], \qquad (11.11)$$

that is due to an integration over all redundant gauge copies. The first steps in the gauge fixing procedure are the same as in the Abelian case that we discussed in Section 6.9. We begin by considering an infinitesimal non-Abelian gauge transformation

$$\Omega(x) = \exp\left(\mathrm{i}\omega^a(x)T^a\right) \approx \mathbb{1} + \omega(x), \quad \omega(x) = \mathrm{i}\omega^a(x)T^a, \qquad (11.12)$$

and apply it to the vector potential

$$\begin{aligned}
{}^\omega G_\mu(x) &= \Omega(x) \left(G_\mu(x) + \partial_\mu\right) \Omega(x)^\dagger \approx \left(\mathbb{1} + \omega(x)\right)\left(G_\mu(x) + \partial_\mu\right)\left(\mathbb{1} - \omega(x)\right) \\
&= G_\mu(x) + \left[\omega(x), G_\mu(x)\right] - \partial_\mu\omega(x) + \mathcal{O}\left(\omega^2\right) \\
&= G_\mu(x) - D_\mu\omega(x) + \mathcal{O}\left(\omega^2\right).
\end{aligned} \qquad (11.13)$$

Here, we have identified the *covariant derivative* in the adjoint representation

$$D_\mu\omega(x) = \partial_\mu\omega(x) + \left[G_\mu(x), \omega(x)\right]. \qquad (11.14)$$

Next, we define

$$C^a(x) = \partial_\mu {}^\omega G_\mu^a(x), \quad C(x) = g_s C^a(x) T^a = -\mathrm{i}\partial_\mu {}^\omega G_\mu(x), \qquad (11.15)$$

and we multiply the functional integral by the irrelevant, ξ-dependent constant

$$\int \mathcal{D}C \, \exp\left(-S_{\mathrm{gf}}[C]\right) = \int \mathcal{D}C \, \exp\left(-\int d^d x \, \frac{1}{2\xi} C^a C^a\right), \quad \xi > 0. \qquad (11.16)$$

which cancels in expectation values. The *gauge fixing action* takes the form

$$S_{\text{gf}}[C] = \int d^d x \, \frac{1}{2\xi} C^a C^a = \int d^d x \, \frac{1}{\xi g_s^2} \text{Tr}\left[C^2\right] = -\int d^d x \, \frac{1}{\xi g_s^2} \text{Tr}\left[\left(\partial_\mu{}^\omega G_\mu\right)^2\right] = S_{\text{gf}}[G].$$

$$(11.17)$$

As in the Abelian case, for a given gauge field $G_\mu(x)$, we view $C(x)$ as a function of the gauge transformation $\omega(x)$, and we replace the functional integration $\mathcal{D}C$ by $\mathcal{D}\omega$, taking into account the corresponding *functional Jacobi determinant*

$$\mathcal{D}C = \mathcal{D}\omega \, \det\left(\frac{\delta C}{\delta \omega}\right) = \mathcal{D}\omega \, \det\left(-i \frac{\delta \partial_\mu{}^\omega G_\mu}{\delta \omega}\right).$$

$$(11.18)$$

Here, we have implicitly assumed the relation between $C(x)$ and $\omega(x)$ to be unique; *i.e.* $\partial_\mu{}^\omega G_\mu(x) = \partial_\mu{}^{\omega'} G_\mu(x)$ implies $\omega'(x) = \omega(x)$. While this is true in the Abelian case, in the non-Abelian case it only holds for the infinitesimal gauge transformations that are relevant in the perturbative regime. Beyond perturbation theory, the non-Abelian Lorenz–Landau gauge fixing condition is afflicted by *Gribov ambiguities* (Gribov, 1978). Fortunately, as we will see in Section 11.10, in the non-perturbative lattice regularization gauge fixing is not even necessary.

The variation of $C(x)$ under an infinitesimal gauge transformation is given by

$$\delta C(x) = -i\delta \, \partial_\mu{}^\omega G_\mu(x) = -i \left(\partial_\mu{}^\omega G_\mu(x) - \partial_\mu G_\mu(x)\right) = -\partial_\mu D_\mu \omega(x). \quad (11.19)$$

This leads to the functional Jacobi determinant

$$\det\left(\frac{\delta C}{\delta \omega}\right) = \det\left(-i \frac{\delta \partial_\mu{}^\omega G_\mu}{\delta \omega}\right) = \det(-\partial_\mu D_\mu). \quad (11.20)$$

In contrast to the Abelian case, the Jacobi determinant $\det(-\partial_\mu D_\mu)$ now is a functional that depends on the gauge field. This implies that it is no longer a trivial constant that can be dropped.

Since the Jacobi determinant can no longer be dropped, Faddeev and Popov decided to cast it into a particularly convenient form by expressing it as a functional integral (Faddeev and Popov, 1967)

$$\det(-\partial_\mu D_\mu) = \int \mathcal{D}\bar{c}\mathcal{D}c \, \exp\left(-S_{\text{gh}}\left[\bar{c}, c, G\right]\right),$$

$$S_{\text{gh}}\left[\bar{c}, c, G\right] = \int d^d x \, \text{Tr}\left[\bar{c}\partial_\mu D_\mu c\right], \quad D_\mu c(x) = \partial_\mu c(x) + \left[G_\mu(x), c(x)\right], \quad (11.21)$$

of anti-commuting *ghost fields*

$$c(x) = c^a(x)T^a, \quad \bar{c}(x) = \bar{c}^a(x)T^a. \quad (11.22)$$

Like the gauge field, the ghost fields belong to the adjoint representation of the gauge group. However, they transform as scalar fields under Lorentz transformations, despite the fact that they are anti-commuting Grassmann fields. As a consequence, they violate the spin–statistics theorem and should hence be unphysical. As we will see in the next sections, this is indeed the case.

The non-Abelian functional integral then takes the form

$$
\begin{aligned}
Z &= \int \mathcal{D}C \int \mathcal{D}G \, \exp\left(-S[G] - S_{\mathrm{gf}}[C]\right) \\
&= \int \mathcal{D}\omega \int \mathcal{D}G \, \det(-\partial_\mu D_\mu) \exp\left(-S[G] - S_{\mathrm{gf}}[{}^\omega G]\right) \\
&= \int \mathcal{D}\omega \int \mathcal{D}\,{}^\omega G \, \det(-\partial_\mu D_\mu) \exp\left(-S[{}^\omega G] - S_{\mathrm{gf}}[{}^\omega G]\right) \\
&= \int \mathcal{D}\omega \int \mathcal{D}G \, \det(-\partial_\mu D_\mu) \exp\left(-S[G] - S_{\mathrm{gf}}[G]\right) \\
&= \int \mathcal{D}G \mathcal{D}\bar{c}\mathcal{D}c \, \exp\left(-S[G] - S_{\mathrm{gf}}[G] - S_{\mathrm{gh}}[\bar{c}, c, G]\right).
\end{aligned}
\tag{11.23}
$$

Here, we have used the fact that the Yang–Mills action and the measure are gauge-invariant, *i.e.* $S[G] = S[{}^\omega G]$ and $\mathcal{D}G = \mathcal{D}\,{}^\omega G$, and we have then renamed the functional integration variable ${}^\omega G$ by G. Finally, we have replaced the functional Jacobi determinant by a functional integration over the Faddeev–Popov ghost fields, and we have dropped the irrelevant integration over ω on which nothing depends any longer.

While neither the gauge fixing action $S_{\mathrm{gf}}[G]$ nor the ghost action $S_{\mathrm{gh}}[\bar{c}, c, G]$ is gauge-invariant, both are at least Lorentz-invariant. In addition, the action conserves the *ghost number*, which is 1 for the ghost c, -1 for the anti-ghost $\bar{c}$, and 0 for the gauge field. Even if one starts in the sector of vanishing ghost number, at this stage it is not obvious that unphysical ghost–anti-ghost pairs cannot be generated in scattering processes.

11.3 Becchi–Rouet–Stora–Tyutin Symmetry

Obviously, gauge fixing explicitly breaks manifest gauge invariance. Before we often emphasized that gauge invariance should be strictly respected in order to prevent gauge-dependent redundant variables to contaminate the physics. The *BRST symmetry* that was discovered by Becchi, Rouet, Stora, and Tyutin (Becchi *et al.*, 1974, 1975, 1976; Tyutin, 1975) will enable us to understand that the Faddeev–Popov ghosts cancel the contributions of the unphysical, longitudinally polarized gauge bosons. Furthermore, BRST invariance guarantees that the resulting theory is *unitary* in the physical sector; *i.e.* unphysical degrees of freedom, namely, longitudinally polarized gauge bosons or Faddeev–Popov ghosts, cannot be created in scattering processes of physical, transversely polarized gauge bosons.

A BRST transformation takes the form of a ghost-field-dependent non-Abelian gauge transformation with a global anti-commuting Grassmann number parameter $\bar{\varepsilon}$, $\theta(x) = \bar{\varepsilon}c(x)$, such that

$$
\begin{aligned}
\exp\left(i\theta(x)\right) &= \exp\left(i\bar{\varepsilon}c(x)\right) = \mathbb{1} + i\bar{\varepsilon}c(x) = \mathbb{1} + i\bar{\varepsilon}c^a(x)T^a, \\
{}^\theta G_\mu(x) &= \exp\left(i\theta(x)\right)\left(G_\mu(x) + \partial_\mu\right)\exp\left(-i\theta(x)\right) \\
&= \left(\mathbb{1} + i\theta(x)\right)\left(G_\mu(x) + \partial_\mu\right)\left(\mathbb{1} - i\theta(x)\right) \\
&= G_\mu(x) + i\left[\theta(x), G_\mu(x)\right] - i\partial_\mu\theta(x).
\end{aligned}
\tag{11.24}
$$

Here, we have used $\bar{\varepsilon}^2 = 0$. The BRST variation of the gauge field is thus given by

$$\delta G_\mu(x) = {}^\theta G_\mu(x) - G_\mu(x) = \mathrm{i}\left[\theta(x), G_\mu(x)\right] - \mathrm{i}\partial_\mu\theta(x)$$
$$= \mathrm{i}\bar{\varepsilon}\left[c(x), G_\mu(x)\right] - \mathrm{i}\bar{\varepsilon}\partial_\mu c(x). \qquad (11.25)$$

The BRST variation of the ghost field is defined by

$$\delta c(x) = \mathrm{i}\bar{\varepsilon}\, c(x)c(x), \qquad (11.26)$$

and the BRST variation of the anti-ghost is set to

$$\delta\bar{c}(x) = \frac{\mathrm{i}}{\xi g_{\mathrm{s}}^2}\bar{\varepsilon}\,\partial_\mu G_\mu(x). \qquad (11.27)$$

Since $c(x)$ and $\bar{c}(x)$ are independent Grassmann fields, it is no problem that they transform differently under BRST transformations. The BRST transformation leaves the ghost number unchanged, provided that we assign ghost number -1 to the parameter $\bar{\varepsilon}$.

Let us now consider the BRST variation of the various terms in the action. First of all, the ordinary Yang–Mills action $S[G]$ is BRST-invariant because it is gauge-invariant and the BRST transformation takes the form of a gauge transformation, despite the fact that it depends on the product of the Grassmann number $\bar{\varepsilon}$ and the ghost field $c(x)$.

The BRST variation of the gauge fixing term $\mathcal{L}_{\mathrm{gf}} = -\mathrm{Tr}[\partial_\mu G_\mu \partial_\nu G_\nu]/(2\xi g_{\mathrm{s}}^2)$ is given by

$$\delta\mathcal{L}_{\mathrm{gf}} = -\frac{1}{\xi g_{\mathrm{s}}^2}\mathrm{Tr}\left[\partial_\mu\delta G_\mu \partial_\nu G_\nu\right] = -\frac{\mathrm{i}}{\xi g_{\mathrm{s}}^2}\mathrm{Tr}\left[\bar{\varepsilon}\partial_\mu\left(\left[c, G_\mu\right] - \partial_\mu c\right)\partial_\nu G_\nu\right]. \qquad (11.28)$$

The BRST variation of the ghost Lagrangian $\mathcal{L}_{\mathrm{gh}} = -\mathrm{Tr}[\bar{c}\partial_\mu(\partial_\mu c + [G_\mu, c])]$ takes the form

$$\delta\mathcal{L}_{\mathrm{gh}} = -\mathrm{Tr}\left\{\delta\bar{c}\,\partial_\mu\left(\partial_\mu c + \left[G_\mu, c\right]\right) + \bar{c}\,\partial_\mu\left(\partial_\mu\delta c + \left[\delta G_\mu, c\right] + \left[G_\mu, \delta c\right]\right)\right\}$$
$$= -\mathrm{Tr}\left\{\frac{\mathrm{i}}{\xi g_{\mathrm{s}}^2}\bar{\varepsilon}\partial_\nu G_\nu\partial_\mu\left(\partial_\mu c + \left[G_\mu, c\right]\right)\right.$$
$$\left. + \mathrm{i}\bar{c}\,\partial_\mu\left(\bar{\varepsilon}\partial_\mu cc + \bar{\varepsilon}c\partial_\mu c + \left[\bar{\varepsilon}\left[c, G_\mu\right], c\right] - \left[\bar{\varepsilon}\partial_\mu c, c\right] + \left[G_\mu, \bar{\varepsilon}cc\right]\right)\right\}. \qquad (11.29)$$

The terms proportional to $1/\xi$ exactly cancel the BRST variation of the gauge fixing term $\delta\mathcal{L}_{\mathrm{gh}}$. Some of the remaining terms do and others do not contain the gauge field. In Problem 11.4, one can convince oneself that both sets of terms cancel separately. This shows that the sum of the gauge fixing and the ghost Lagrangian, and thus the entire action, is BRST-invariant

$$\delta\left(\mathcal{L}_{\mathrm{gf}} + \mathcal{L}_{\mathrm{gh}}\right) = 0. \qquad (11.30)$$

As a result, we have traded the original Yang–Mills Lagrangian, which has an exact local symmetry but contains unphysical redundant degrees of freedom, for a gauge-fixed Lagrangian endowed with additional unphysical ghost fields and an exact, global, ghost-field-dependent BRST symmetry. As we emphasized earlier, gauge symmetries result from a redundant description, while global symmetries usually lead to distinct superselection sectors corresponding to different values of a conserved global charge. As we will discuss in the next section, BRST symmetry indeed allows us to identify the physical sector of the theory and to distinguish it from unphysical sectors, which are inaccessible to the actual dynamics.

11.4 Nilpotency and BRST Cohomology

Let us define an *anti-commuting* operation Q that describes the BRST variation of all fields

$$
\begin{aligned}
\delta G_\mu(x) = \bar\varepsilon\, {}^Q G_\mu(x) &\quad\Rightarrow\quad {}^Q G_\mu(x) = \mathrm{i}\left[c(x), G_\mu(x)\right] - \mathrm{i}\,\partial_\mu c(x), \\
\delta c(x) = \bar\varepsilon\, {}^Q c(x) &\quad\Rightarrow\quad {}^Q c(x) = \mathrm{i}\, c(x) c(x), \\
\delta \bar c(x) = \bar\varepsilon\, {}^Q \bar c(x) &\quad\Rightarrow\quad {}^Q \bar c(x) = \frac{\mathrm{i}}{\xi g_{\mathrm s}^2}\partial_\mu G_\mu(x).
\end{aligned}
\tag{11.31}
$$

We now want to show that Q is a *nilpotent* operation, *i.e.* $Q^2 = 0$, and we begin by operating twice with Q on the gauge field

$$
\begin{aligned}
{}^{Q^2} G_\mu(x) &= {}^Q \left\{ \mathrm{i}\left[c(x), G_\mu(x)\right] - \mathrm{i}\,\partial_\mu c(x) \right\} \\
&= \mathrm{i}\left\{ {}^Q c(x) G_\mu(x) - c(x)\, {}^Q G_\mu(x) - {}^Q G_\mu(x) c(x) - G_\mu(x)\, {}^Q c(x) - \partial_\mu\, {}^Q c(x) \right\} \\
&= 0.
\end{aligned}
\tag{11.32}
$$

Here, we have used the fact that Q itself is an anti-commuting object which leads to a change of sign when it is pushed through a Grassmann-valued field. In Problem 11.5, one can convince oneself that indeed ${}^{Q^2} G_\mu(x) = 0$. Next, we apply Q^2 to the ghost field

$$
\begin{aligned}
{}^{Q^2} c(x) &= {}^Q \{\mathrm{i}\, c(x) c(x)\} = \mathrm{i}\, {}^Q c(x) c(x) - \mathrm{i}\, c(x)\, {}^Q c(x) = -c(x) c(x) c(x) + c(x) c(x) c(x) \\
&= 0,
\end{aligned}
\tag{11.33}
$$

as well as to the anti-ghost field

$$
{}^{Q^2} \bar c(x) = {}^Q \left\{ \frac{1}{\xi g_{\mathrm s}^2}\partial_\mu G_\mu(x) \right\} = \frac{\mathrm{i}}{\xi g_{\mathrm s}^2}\partial_\mu \left\{ \left[c(x), G_\mu(x)\right] - \partial_\mu c(x) \right\} = 0.
\tag{11.34}
$$

This expression vanishes if we rely on the equation of motion, $\partial_\mu\left(\partial_\mu c(x) + [G_\mu(x), c(x)]\right) = 0$, which results when one varies the ghost action with respect to the anti-ghost field $\bar c(x)$.

In the Hamiltonian formulation, the nilpotent operation turns into an operator $\hat Q$ that obeys $\hat Q^2 = 0$ and commutes with the Hamilton operator $\hat H$, *i.e.* $[\hat Q, \hat H] = 0$. Related to the fact that the ghost fields violate the spin–statistics theorem, the corresponding states have an indefinite (*i.e.* not necessarily positive) norm, and thus they do not reside in the physical Hilbert space. Since the operator $\hat Q$ acts in an indefinite norm space $\mathcal{H}$, it is not Hermitian but known as "pseudo-Hermitian".

First of all, states can be distinguished by whether they are annihilated by $\hat Q$ or not. We associate the states $|\psi\rangle$ that are not annihilated by $\hat Q$ (*i.e.* with $\hat Q|\psi\rangle \neq 0$) with the subspace $\mathcal{H}_> \subset \mathcal{H}$.

States that are annihilated by $\hat Q$, *i.e.* $\hat Q|\psi\rangle = 0$, are called *BRST-closed*. They belong to the kernel $\ker(\hat Q)$. Some BRST-closed states $|\psi\rangle$ are expressible as

$$
|\psi\rangle = \hat Q|\chi\rangle \quad\Rightarrow\quad \hat Q|\psi\rangle = \hat Q^2|\chi\rangle = 0.
\tag{11.35}
$$

Such states are called *BRST-exact*. They belong to the image $\text{im}(\hat{Q})$. We associate these states with another subspace $\mathcal{H}_< = \text{im}(\hat{Q}) \subset \mathcal{H}$. Due to the nilpotency of $\hat{Q}$, all states in $\text{im}(\hat{Q})$ have vanishing norm

$$|\psi\rangle = \hat{Q}|\chi\rangle \quad \Rightarrow \quad \langle\psi|\psi\rangle = \langle\chi|\hat{Q}^2|\chi\rangle = 0. \tag{11.36}$$

In addition, any pair of BRST-exact states, $|\psi\rangle, |\psi'\rangle \in \mathcal{H}_<$, has a vanishing scalar product

$$\langle\psi|\psi'\rangle = \langle\chi|\hat{Q}^2|\chi'\rangle = 0. \tag{11.37}$$

It is obvious that BRST-exact states are automatically BRST-closed. It is particularly interesting to consider states that are BRST-closed but *not* BRST-exact. Such states are said to belong to the *cohomology space* of $\hat{Q}$, $\text{cohom}(\hat{Q})$, which forms yet another subspace $\mathcal{H}_0 = \text{cohom}(\hat{Q}) \subset \mathcal{H}$ of the indefinite norm space

$$\mathcal{H} = \mathcal{H}_> + \mathcal{H}_< + \mathcal{H}_0,$$
$$\mathcal{H}_> = \mathcal{H} \setminus \ker(\hat{Q}), \quad \mathcal{H}_< = \text{im}(\hat{Q}), \quad \mathcal{H}_0 = \text{cohom}(\hat{Q}) = \ker(\hat{Q}) \setminus \text{im}(\hat{Q}). \tag{11.38}$$

It turns out that $\mathcal{H}_0 = \text{cohom}(\hat{Q})$ corresponds to the *physical* sector of the Hilbert space. However, there is a certain redundancy within $\text{cohom}(\hat{Q})$. Let us consider a state $|\psi\rangle \in \text{cohom}(\hat{Q})$, which means that $\hat{Q}|\psi\rangle = 0$ but $|\psi\rangle \neq \hat{Q}|\chi\rangle$ for any state $|\chi\rangle$. Furthermore, let us consider another state $|\psi'\rangle \in \text{cohom}(\hat{Q})$ such that

$$|\psi'\rangle = |\psi\rangle - \hat{Q}|\eta\rangle \; ; \tag{11.39}$$

i.e. the difference between the two states is BRST-exact. Such states are identified with each other and describe the same physical state. In particular, both states have the same norm

$$\langle\psi'|\psi'\rangle = \langle\psi|\psi\rangle - \langle\eta|\hat{Q}|\psi\rangle - \langle\psi|\hat{Q}|\eta\rangle + \langle\eta|\hat{Q}^2|\eta\rangle = \langle\psi|\psi\rangle. \tag{11.40}$$

The physical Hilbert space therefore is the quotient space $\text{cohom}(\hat{Q})/\text{im}(\hat{Q})$.

One can show that physical states $|\psi\rangle$ of transversely polarized gauge bosons are indeed annihilated by $\hat{Q}$, *i.e.* $\hat{Q}|\psi\rangle = 0$, while $|\psi\rangle \neq \hat{Q}|\chi\rangle$, which means that they belong to $\text{cohom}(\hat{Q}) = \mathcal{H}_0$. Since $\hat{Q}$ commutes with the Hamilton operator, such a state remains within $\text{cohom}(\hat{Q})$ during its time evolution.

When one adds a positively longitudinally polarized, unphysical gauge boson to a physical state, then $\hat{Q}$ no longer annihilates it. As a result, it is no longer closed and hence no longer part of the cohomology space. Instead, it belongs to the sector $\mathcal{H}_>$. Similarly, one can show that unphysical gauge boson states with negative longitudinal polarization are BRST-closed and thus belong to $\mathcal{H}_<$.

11.5 Aharonov–Bohm Effect as an Analogue of BRST Cohomology

Let us try to illustrate these rather abstract algebraic considerations by a comparison with a mathematical analogue that a typical physicist may be more familiar with. We consider the quantum mechanical system of a charged particle in the field of a thin solenoid that is threaded by a magnetic flux. To a good approximation, there is no magnetic field outside the solenoid. Hence, the particle, which is shielded from the interior of the solenoid, moves in a space with vanishing magnetic field $\vec{B} = 0$. It is well understood that the particle, as it moves around the solenoid, picks up an Aharonov–Bohm phase, which is described by a Wegner–Wilson loop of the electromagnetic vector potential $\vec{A}$ (Aharonov and Bohm, 1959, 1961). What does the Aharonov–Bohm effect have to do with BRST cohomology? While it is physically completely unrelated, it bears close *mathematical* similarities.

First of all, the relation $\vec{B} = \vec{\nabla} \times \vec{A}$ can be expressed as $^\star B = dA$ in the language of differential forms. The vector field $\vec{A}$ is equivalent to a 1-form A. The vector field $\vec{B}$ is a component of the relativistic anti-symmetric field strength tensor $F_{\mu\nu}$. This is reflected by the fact that the dual field $^\star B$ is a 2-form. The differential d increases the form degree by 1. A gauge transformation $\vec{A}' = \vec{A} - \vec{\nabla}\alpha$ leaves the magnetic field unchanged

$$\vec{B}' = \vec{\nabla} \times \vec{A}' = \vec{\nabla} \times \vec{A} - \vec{\nabla} \times \vec{\nabla}\alpha = \vec{B}, \tag{11.41}$$

because $\vec{\nabla} \times \vec{\nabla} = 0$. In the notation of differential forms, the gauge transformation α, which is described by a scalar field, is a 0-form, and the gauge transformation takes the form

$$A' = A - d\alpha \quad \Rightarrow \quad {}^\star B' = dA' = dA - d^2\alpha = {}^\star B. \tag{11.42}$$

Here, d turns the 0-form α into the 1-form $d\alpha$. Just like the BRST operator $\hat{Q}$, the differential d is nilpotent, *i.e.* $d^2 = 0$, such that $d^2\alpha = 0$. The nilpotency of d also implies

$$d^\star B = d^2 A = 0. \tag{11.43}$$

In ordinary vector analysis notation, this corresponds to $\vec{\nabla} \cdot \vec{B} = 0$, which follows from $\vec{\nabla} \cdot (\vec{\nabla} \times \vec{A}) = 0$. This is one of the homogeneous Maxwell equations (a Bianchi identity), which encodes the absence of magnetic monopoles.

In the mathematical analogue with BRST cohomology, a state $|\psi\rangle \in \mathcal{H}_>$, which obeys $\hat{Q}|\psi\rangle \neq 0$, corresponds to a vector potential A with a non-zero magnetic field strength $^\star B = dA \neq 0$ ($\vec{B} = \vec{\nabla} \times \vec{A} \neq 0$). Such field configurations play no role in the Aharonov–Bohm effect. There we are concerned with vector potentials that lead to a vanishing magnetic field $^\star B = dA = 0$. Those belong to the kernel of d, *i.e.* $A \in \ker(d)$.

Among these are vector potentials which represent a pure gauge; *i.e.* they are a gauge transformation of a zero field, $A = d\alpha$, and thus belong to the image of d, $A \in \text{im}(d)$. Such trivial vector potentials do not yield a non-zero Aharonov–Bohm phase because the corresponding Wegner–Wilson loop simply vanishes. Those are analogous to the BRST-exact states $|\psi\rangle = \hat{Q}|\chi\rangle \in \text{im}(\hat{Q}) = \mathcal{H}_<$.

Finally, the vector potentials that give rise to a non-vanishing Aharonov–Bohm effect have vanishing magnetic field (outside the solenoid) but are not a pure gauge, *i.e.* $^\star B = dA = 0$ while $A \neq d\alpha$. Such fields belong to the kernel but not to the image of d; *i.e.* they belong to the cohomology space $\text{cohom}(d) = \ker(d) \setminus \text{im}(d)$. Obviously, they are analogous to the physical states $|\psi\rangle \in \text{cohom}(\hat{Q}) = \mathcal{H}_0$.

Just as in the BRST case, there is an ambiguity also in its mathematical Aharonov–Bohm analogue, which is just the ordinary gauge ambiguity. Indeed, two vector potentials A and $A' = A - d\alpha$, which differ just by a gauge transformation $d\alpha \in \text{im}(d)$, yield the same Aharonov–Bohm phase and are physically indistinguishable. The physical space of magnetic field configurations that lead to a non-trivial Aharonov–Bohm effect hence belongs to the quotient space $\text{cohom}(d)/\text{im}(d)$, which results from identifying gauge copies with one another. This is the mathematical analogue of the physical Hilbert space in the non-Abelian

Table 11.1 A dictionary that relates BRST cohomology to its Aharonov–Bohm mathematical analogue.

BRST Cohomology	Aharonov–Bohm Effect
state $|\psi\rangle$	vector potential A
BRST operator $\hat{Q}$	differential operator d
action of BRST operator $\hat{Q}|\psi\rangle$	magnetic field $^\star B = dA$
$|\psi\rangle \in \mathcal{H}_>,\ \hat{Q}|\psi\rangle \neq 0$	non-zero magnetic field $^\star B \neq 0$
BRST-closed state $\hat{Q}|\psi\rangle = 0,\ |\psi\rangle \in \ker(\hat{Q})$	zero magnetic field $^\star B = dA = 0,\ A \in \ker(d)$
BRST-exact state $|\psi\rangle = \hat{Q}|\chi\rangle,\ |\psi\rangle \in \mathrm{im}(\hat{Q}) = \mathcal{H}_<$	pure gauge $A = d\alpha,\ A \in \mathrm{im}(d)$
$\hat{Q}|\psi\rangle = 0,\ |\psi\rangle \neq \hat{Q}|\chi\rangle$ $|\psi\rangle \in \ker(\hat{Q}) \setminus \mathrm{im}(\hat{Q}) = \mathrm{cohom}(\hat{Q}) = \mathcal{H}_0$	$^\star B = dA = 0,\ A \neq d\alpha$ $A \in \ker(d) \setminus \mathrm{im}(d) = \mathrm{cohom}(d)$
equivalent states $|\psi'\rangle = |\psi\rangle - \hat{Q}|\eta\rangle$	gauge equivalence $A' = A - d\alpha$

gauge theory. In both of these cases, the interesting physics happens in the cohomology space (modulo the image) of the corresponding nilpotent operator, d for the Aharonov–Bohm effect and $\hat{Q}$ for a properly gauge-fixed non-Abelian gauge theory. The mathematical analogies between BRST cohomology and the Aharonov–Bohm effect are summarized in Table 11.1.

11.6 Lattice Gauge Theory

In the context of classical statistical mechanics, lattice gauge theories were first constructed by Franz Wegner (1971) for the discrete Abelian gauge group $\mathbb{Z}(2)$. They were generalized by Kenneth Wilson to arbitrary Abelian or non-Abelian gauge groups and employed in the non-perturbative functional integral formulation of gauge theories in particle physics (Wilson, 1974).[1] The continuum Lagrangian forms the basis of perturbation theory. However, the perturbative series is known to be only asymptotic and not summable to all orders. In contrast, lattice gauge theory provides a solid *definition* of the theory.

While there is no rigorous mathematical proof that its continuum limit indeed exists, there is overwhelming numerical evidence. Furthermore, there is a rigorous proof by Thomas Reisz to all orders that lattice perturbation theory defines the same continuum limit as dimensional regularization applied to the continuum Lagrangian (Reisz, 1988a,b, 1989).

In contrast to purely perturbative approaches, however, lattice gauge theory can also address physics at finite and even strong coupling. In particular, when we will discuss the strong interaction, in Chapters 14 and 18, we will make use of the lattice regularization. Lattice QCD has become a quantitative tool that allows us to compute the properties of

---

[1] For an account of the early history of lattice gauge theory, we refer to Wilson (2005). We add that Jan Smit formulated lattice quantum field theory independently around 1972, but his work was not published. Independent early activities along these lines in the former Soviet Union are reviewed by Polyakov (2005).

strongly interacting fields using Monte Carlo simulations (an overview of low-energy lattice world data is given in Aoki *et al.* (2021)).

The manifolds for non-Abelian Lie groups are compact spaces. For example, the SU(2) group manifold is a sphere S^3 with a 3-dimensional surface (which can be embedded in $\mathbb{R}^4$). In lattice gauge theory, the vector potentials of the continuum theory are replaced by group-valued parallel transporter matrices which reside on the links that connect neighboring lattice sites. Due to the compact nature of the gauge group, in contrast to the perturbative continuum formulation, the non-perturbative lattice formulation *does not require gauge fixing*. Without gauge fixing, in the functional integral one integrates over all gauge copies. On the lattice with compact link variables, this leads to a trivial, finite, multiplicative constant. In the continuum theory, on the other hand, without gauge fixing the functional integral over the non-compact vector potentials gives rise to a non-trivial, divergent factor, which needs to be eliminated by an appropriate integration over the Faddeev–Popov ghost fields, as we discussed in Section 11.2. While this procedure is of great importance for perturbative calculations (even on the lattice), it is not, in itself, a feature of non-Abelian gauge theories. We now use the lattice regularization as the theoretical basis for non-Abelian gauge theory, and we quantize the theory without any need for ghost fields.

11.7 Canonical Quantization of Compact U(1) Lattice Gauge Theory

As a warm-up exercise for the quantization of non-Abelian lattice gauge theories, we first consider the Abelian case. The group manifold for the free electromagnetic field is the non-compact real line $\mathbb{R}$. On the lattice, one can alternatively formulate an Abelian gauge theory with the compact gauge group U(1). Due to the compact nature of the group manifold U(1), which is just a circle S^1, this theory shares some features with non-Abelian gauge theories. In particular, it is self-interacting and even confining, at least at strong coupling.

Let us consider a spatial cubic lattice with lattice spacing a. In order to motivate the concept of a group-valued *parallel transporter* $U_{x,i} \in U(1)$ connecting neighboring lattice sites x and $x + \hat{i}$ (where $\hat{i}$ is a vector of length a in the i-direction), we start out with an Abelian vector potential $A_i(x) \in \mathbb{R}$ in the continuum, which transforms under gauge transformations $\alpha(x) \in \mathbb{R}$ as $A_i'(x) = A_i(x) - \partial_i \alpha(x)$. In order to maintain exact gauge invariance on the lattice, we integrate the vector potential along the link (x, i),

$$U_{x,i} = \exp \left(\mathrm{i}e \int_x^{x+\hat{i}} dy_i \, A_i(y) \right) \in U(1). \tag{11.44}$$

Here, e is the fundamental electric charge unit. Under gauge transformations of the continuum gauge field $A_i(x)$, the parallel transporter transforms as

$$U_{x,i}' = \exp \left(\mathrm{i}e \int_x^{x+\hat{i}} dy_i \, A_i'(y) \right) = \exp \left(\mathrm{i}e \int_x^{x+\hat{i}} dy_i \left[A_i(y) - \partial_i \alpha(y) \right] \right)$$

$$= \exp \left(\mathrm{i}e \left[\int_x^{x+\hat{i}} dy_i \, A_i(y) + \alpha(x) - \alpha(x + \hat{i}) \right] \right) = \Omega_x U_{x,i} \Omega_{x+\hat{i}}^{\dagger},$$

$$\Omega_x = \exp \left(\mathrm{i}e\alpha(x) \right) \in U(1). \tag{11.45}$$

In lattice gauge theory, the parallel transporters are the fundamental degrees of freedom. In particular, they are not derived from an underlying continuum vector potential. The link variables $U = \exp(i\varphi)$ of a compact Abelian lattice gauge theory are given in terms of complex phases which take values in the group space circle $U(1) = S^1$. Hence, every link variable is analogous to a quantum mechanical "particle" on a circle. The angular momentum operator of such a "particle", together with its eigenstates, is given by

$$\hat{J} = -i\partial_\varphi, \quad \hat{J}|m\rangle = m|m\rangle, \quad m \in \mathbb{Z}, \quad \langle\varphi|m\rangle = \frac{1}{\sqrt{2\pi}}\exp(im\varphi). \tag{11.46}$$

In lattice gauge theory, the operators $\hat{J}_{x,i}$ are canonically conjugate to the link operators $\hat{U}_{x,i}$. Since the link variables in compact $U(1)$ lattice gauge theory are angular variables, their canonically conjugate "momenta" are actually angular momenta. The electric field operators, which are again associated with the links, are defined as $\hat{E}_{x,i} = e\hat{J}_{x,i}/a^2$. The corresponding canonical commutation relations take the form

$$[\hat{J}_{x,i}, \hat{U}_{y,j}] = \delta_{xy}\delta_{ij}\hat{U}_{x,i}, \quad [\hat{J}_{x,i}, \hat{U}_{y,j}^\dagger] = -\delta_{xy}\delta_{ij}\hat{U}_{x,i}^\dagger,$$
$$[\hat{J}_{x,i}, \hat{J}_{y,j}] = [\hat{U}_{x,i}, \hat{U}_{y,j}] = [\hat{U}_{x,i}^\dagger, \hat{U}_{y,j}^\dagger] = [\hat{U}_{x,i}, \hat{U}_{y,j}^\dagger] = 0. \tag{11.47}$$

The algebraic structure of lattice gauge theory is link-based. Operators that reside on different links commute with each other. The Hilbert space of the theory is the direct product of local link Hilbert spaces. Already the individual link Hilbert spaces are infinite-dimensional because they resemble the Hilbert space of a particle on a circle.

The electric contribution to the Hamilton operator corresponds to the kinetic energy of the analogous "particles" and is given as a sum over all links. The magnetic contribution, on the other hand, is a sum over elementary plaquettes

$$a\hat{H} = \frac{e^2}{2}\sum_{x,i}\hat{J}_{x,i}^2 + \frac{1}{2e^2}\sum_{x,i>j}\left(2 - \hat{U}_{x,i}\hat{U}_{x+\hat{i},j}\hat{U}_{x+\hat{j},i}^\dagger\hat{U}_{x,j}^\dagger - \hat{U}_{x,j}\hat{U}_{x+\hat{j},i}\hat{U}_{x+\hat{i},j}^\dagger\hat{U}_{x,i}^\dagger\right). \tag{11.48}$$

In the *classical* continuum limit, $a \to 0$, one identifies $U_{x,i} = \exp\left(ieaA_i(x + \hat{i}/2)\right)$ such that

$$U_{x,i}U_{x+\hat{i},j}U_{x+\hat{j},i}^*U_{x,j}^* =$$
$$\exp\left(iea\left[A_i(x+\hat{i}/2) + A_j(x+\hat{i}+\hat{j}/2) - A_i(x+\hat{j}+\hat{i}/2) - A_j(x+\hat{j}/2)\right]\right) \to$$
$$\exp\left(iea^2F_{ij}(x+\hat{i}/2+\hat{j}/2)\right) \quad \Rightarrow$$
$$\frac{1}{a^4}\left(2 - U_{x,i}U_{x+\hat{i},j}U_{x+\hat{j},i}^*U_{x,j}^* - U_{x,j}U_{x+\hat{j},i}U_{x+\hat{i},j}^*U_{x,i}^*\right) =$$
$$\frac{2}{a^4}\left[1 - \cos\left(ea^2F_{ij}(x+\hat{i}/2+\hat{j}/2)\right)\right] \to e^2F_{ij}(x+\hat{i}/2+\hat{j}/2)^2 \quad \Rightarrow$$
$$H \to \int d^3x\, \frac{1}{2}\left(\vec{E}^2 + \vec{B}^2\right). \tag{11.49}$$

Quick Question 11.7.1   Classical continuum limit of the lattice Hamilton operator
Justify the steps in eq. (11.49).

The Hamilton operator commutes with the infinitesimal generators of gauge transformations associated with the lattice sites x

$$\hat{G}_x = \sum_i (\hat{J}_{x,i} - \hat{J}_{x-\hat{i},i}) = \frac{a^2}{e} \sum_i (\hat{E}_{x,i} - \hat{E}_{x-\hat{i},i}), \quad [\hat{G}_x, \hat{G}_y] = 0, \quad [\hat{H}, \hat{G}_x] = 0. \quad (11.50)$$

In the classical continuum limit, one obtains $eG_x/a^3 \rightarrow \vec{\nabla} \cdot \vec{E}(x)$. A general finite gauge transformation $\Omega_x = \exp(ie\alpha_x)$ is represented by a unitary operator $\hat{V}$ that acts as

$$\hat{V}\hat{U}_{x,i}\hat{V}^\dagger = \Omega_x \hat{U}_{x,i} \Omega_{x+\hat{i}}^\dagger, \quad \hat{V} = \prod_x \exp\left(ie\alpha_x \hat{G}_x\right). \quad (11.51)$$

Let us point out again that gauge symmetries reflect a redundancy in the description of the physics. The energy eigenstates $|\psi, Q\rangle$, with $Q = \{Q_x\}$, can be characterized by the eigenvalues $Q_x \in \mathbb{Z}$ of all gauge generators,

$$\hat{G}_x |\psi, Q\rangle = Q_x |\psi, Q\rangle. \quad (11.52)$$

These states represent a system in the presence of external static charges $Q_x \in \mathbb{Z}$. Due to the compact nature of the gauge group U(1), the charges are quantized in integer units. In the absence of external charges, physical states $|\psi\rangle$ must be gauge-invariant, $\hat{G}_x |\psi\rangle = 0$, which is just the *Gauss law*.

The canonical, quantum statistical partition function of a gauge theory in a fixed background of charges Q is

$$Z_Q = \mathrm{Tr}\left[\exp\left(-\beta\hat{H}\right)\hat{P}_Q\right]. \quad (11.53)$$

Here, $\hat{P}_Q$ is a projection operator on the appropriate charge sector. Let us consider the theory in the presence of two opposite external charges $Q_x = 1$ and $Q_y = -1$, located at different lattice sites x and y. The static potential $V(x - y)$ between the charges is then given by

$$\frac{Z_Q}{Z} = \exp\left(-\beta V(x - y)\right), \quad V(x - y) \sim \sigma |x - y|. \quad (11.54)$$

Here, Z is the partition function in the absence of external charges. In the limit of strong gauge coupling e and at low temperature (large β), lattice pure gauge theories with a compact gauge group (and with a non-trivial center, cf. Appendix F) are confining with a linearly rising charge–anti-charge potential that is characterized at asymptotic distances by the *string tension* σ.

11.8 Canonical Quantization of Non-Abelian Lattice Gauge Theory

Let us now consider non-Abelian lattice gauge theories. First of all, we again start from a vector potential $G_i(x)$ in the continuum and we construct the parallel transporter

$$
\begin{aligned}
U_{x,i} &= \mathcal{P}\exp\left(\int_x^{x+\hat{i}} dy_i\, G_i(y)\right) \\
&= \lim_{N\to\infty} \exp\left(\frac{a}{N}G_i(x+\hat{i}/2N)\right)\exp\left(\frac{a}{N}G_i(x+3\hat{i}/2N)\right)\exp\left(\frac{a}{N}G_i(x+5\hat{i}/2N)\right)\cdots \\
&\quad \cdots \exp\left(\frac{a}{N}G_i(x+(2N-3)\hat{i}/2N)\right)\exp\left(\frac{a}{N}G_i(x+(2N-1)\hat{i}/2N)\right). \quad (11.55)
\end{aligned}
$$

Here, $\mathcal{P}$ denotes path ordering. Under a gauge transformation $G_i'(x) = \Omega(x)(G_i(x) + \partial_i)\Omega(x)^\dagger$ with the group-valued function $\Omega(x) = \exp(i\omega^a(x)T^a)$, the parallel transporter transforms as

$$U_{x,i}' = \Omega_x U_{x,i} \Omega_{x+\hat{i}}^\dagger. \tag{11.56}$$

In non-Abelian lattice gauge theory, the parallel transporters $U_{x,i}$ are again the fundamental degrees of freedom, without assuming an underlying continuum gauge field.

Let us consider the quantum mechanical, analogous "particle" that describes a link variable in an SU(2) lattice gauge theory. The corresponding group manifold is the sphere S^3. The position of the analogous "particle" on S^3 is described in terms of three angles, α, θ, and φ, by the SU(2) matrix ($\vec{\tau}$ are still the Pauli matrices)

$$\hat{U} = \cos\alpha \mathbb{1} + i\sin\alpha \vec{e}_\alpha \cdot \vec{\tau}, \quad \vec{e}_\alpha = (\sin\theta\cos\varphi, \sin\theta\sin\varphi, \cos\theta). \tag{11.57}$$

Since SU(2) is non-Abelian, we need to distinguish transformations that multiply $\hat{U}$ from the left and from the right. The corresponding $SU(2)_L \times SU(2)_R$ algebra is generated by

$$\hat{R}^a = \frac{1}{2}(\hat{J}^a + \hat{K}^a), \quad \hat{L}^a = \frac{1}{2}(\hat{J}^a - \hat{K}^a), \quad \hat{J}^\pm = \exp(\pm i\varphi)\left(\pm\,\partial_\theta + i\cot\theta\,\partial_\varphi\right), \quad \hat{J}^3 = -i\partial_\varphi,$$

$$\hat{K}^\pm = \exp(\pm i\varphi)\left(i\sin\theta\,\partial_\alpha + i\cot\alpha\cos\theta\,\partial_\theta \mp \frac{\cot\alpha}{\sin\theta}\partial_\varphi\right),$$

$$\hat{K}^3 = i\left(\cos\theta\,\partial_\alpha - \cot\alpha\sin\theta\,\partial_\theta\right). \tag{11.58}$$

The operators $\hat{R}^a$ and $\hat{L}^a$ represent the angular momenta that are canonically conjugate to the angles α, θ, and φ, which define the link operator $\hat{U}$. The corresponding canonical commutation relations take the form

$$[\hat{R}^a, \hat{U}] = \hat{U}\frac{\tau^a}{2}, \quad [\hat{L}^a, \hat{U}] = -\frac{\tau^a}{2}\hat{U},$$

$$[\hat{R}^a, \hat{R}^b] = i\epsilon_{abc}\hat{R}^c, \quad [\hat{L}^a, \hat{L}^b] = i\epsilon_{abc}\hat{L}^c, \quad [\hat{R}^a, \hat{L}^b] = 0. \tag{11.59}$$

The kinetic energy of the analogous "particle" represents the "electric" contribution to the Hamilton operator and corresponds to the Laplacian on the group manifold

$$\hat{R}^a\hat{R}^a + \hat{L}^a\hat{L}^a = \frac{1}{2}(\hat{J}^a\hat{J}^a + \hat{K}^a\hat{K}^a) = \frac{a^4}{2g_s^2}\hat{E}^a\hat{E}^a. \tag{11.60}$$

The corresponding eigenvalues are $j_L(j_L + 1) + j_R(j_R + 1) = l(l + 2)/2$ with $j_L = j_R$ and $l = j_L + j_R$, $j_L, j_R \in \{0, 1, 2, \ldots\}$. Each eigenstate is $(2j_L + 1)(2j_R + 1) = (l + 1)^2$-fold degenerate.

The link-based algebraic structure of SU(2) lattice gauge theory generalizes to an arbitrary gauge group G with Lie algebra generators T^a and structure constants f_{abc} in a straightforward manner

$$[\hat{R}_{x,i}^a, \hat{U}_{y,j}] = \delta_{xy}\delta_{ij}\hat{U}_{x,i}T^a, \quad [\hat{L}_{x,i}^a, \hat{U}_{y,j}] = -\delta_{xy}\delta_{ij}T^a\hat{U}_{x,i}, \quad [\hat{U}_{x,i}, \hat{U}_{y,j}] = 0,$$

$$[\hat{R}_{x,i}^a, \hat{R}_{y,j}^b] = i\delta_{xy}\delta_{ij}f_{abc}\hat{R}_{x,i}^c, \quad [\hat{L}_{x,i}^a, \hat{L}_{y,j}^b] = i\delta_{xy}\delta_{ij}f_{abc}\hat{L}_{x,i}^c, \quad [\hat{R}_{x,i}^a, \hat{L}_{y,j}^b] = 0. \tag{11.61}$$

The Hamilton operator, which was first constructed by John Kogut and Leonard Susskind (Kogut and Susskind, 1975), again has an "electric" link and a "magnetic" plaquette contribution,

$$
a\hat{H} = g_s^2 \sum_{x,i} \left(\hat{R}_{x,i}^a \hat{R}_{x,i}^a + \hat{L}_{x,i}^a \hat{L}_{x,i}^a \right)
$$
$$
+ \frac{2}{g_s^2} \sum_{x,i>j} \mathrm{Tr} \left(\mathbb{1} - \frac{1}{2} \hat{U}_{x,i} \hat{U}_{x+\hat{i},j} \hat{U}_{x+\hat{j},i}^\dagger \hat{U}_{x,j}^\dagger - \frac{1}{2} \hat{U}_{x,j} \hat{U}_{x+\hat{j},i} \hat{U}_{x+\hat{i},j}^\dagger \hat{U}_{x,i}^\dagger \right), \quad (11.62)
$$

as in the Abelian case, eq. (11.48). In one of the exercises, one can convince oneself that this expression reduces to the one in eq. (11.9) in the classical continuum limit. In the non-Abelian case, the infinitesimal generators of gauge transformations associated with the lattice sites x are given by

$$
\hat{G}_x^a = \sum_i \left(\hat{L}_{x,i}^a + \hat{R}_{x-\hat{i},i}^a \right), \quad [\hat{H}, \hat{G}_x^a] = 0, \quad [\hat{G}_x^a, \hat{G}_y^b] = \mathrm{i} \delta_{xy} f_{abc} \hat{G}_x^c. \quad (11.63)
$$

In Hilbert space, the most general gauge transformation $\Omega_x = \exp(\mathrm{i}\omega_x^a T^a)$ is represented by a unitary operator

$$
\hat{V} = \prod_x \exp \left(\mathrm{i}\omega_x^a \hat{G}_x^a \right), \quad \hat{V}\hat{U}_{x,i}\hat{V}^\dagger = \Omega_x \hat{U}_{x,i} \Omega_{x+\hat{i}}^\dagger, \quad (11.64)
$$

in analogy to eq. (11.51).

The non-Abelian Gauss law takes the form $\hat{G}_x^a|\Psi\rangle = 0$. Local violations of Gauss' law manifest themselves as external static non-Abelian gauge charges, which are characterized by some representation of the gauge group. For example, for SU(2) the representations are given by $Q_x \in \{0, 1/2, 1, 3/2, \ldots\}$ with $2Q_x + 1$ states that are distinguished by $Q_x^3 \in \{-Q_x, -Q_x + 1, \ldots, Q_x - 1, Q_x\}$, such that

$$
\hat{G}_x^a \hat{G}_x^a |\Psi, Q, Q^3\rangle = Q_x(Q_x+1)|\Psi, Q, Q^3\rangle, \quad \hat{G}_x^3 |\Psi, Q, Q^3\rangle = Q_x^3 |\Psi, Q, Q^3\rangle. \quad (11.65)
$$

As in the Abelian case, the potential $V(x-y)$ between external charges is given by

$$
\frac{Z_Q}{Z} = \exp\left(-\beta V(x-y)\right), \quad Z_Q = \mathrm{Tr}\left[\exp(-\beta\hat{H})\hat{P}_Q\right], \quad (11.66)
$$

where $\hat{P}_Q$ again projects on the corresponding external charge sector. Unlike in the Abelian case of compact U(1) gauge theory at strong coupling, non-Abelian external charges are not always confined by a linearly rising potential with a non-zero string tension σ. Linear confinement arises only for external charges that transform non-trivially under the center of the gauge group (cf. Appendix F). For example, for SU(2) those correspond to the representations with half-odd-integer values of $Q_x \in \{1/2, 3/2, 5/2, \ldots\}$.

11.9 Functional Integral for Compact U(1) Lattice Gauge Theory

We now proceed from the Hamiltonian formulation to the functional integral for U(1) lattice gauge theory. For this purpose, we express the quantum statistical partition function $Z_Q = \mathrm{Tr}[\exp(-\beta\hat{H})\hat{P}_Q]$ as a path integral over link variables on a 4-dimensional Euclidean space–time lattice. In particular, we divide the Euclidean time extent $\beta = aN$ into N intervals of size a. For simplicity, we choose the lattice spacing a in Euclidean time to coincide with the one in space. We now introduce the transfer matrix $\hat{\mathcal{T}} = \exp(-a\hat{H})$ and write $Z_Q = \mathrm{Tr}[\hat{\mathcal{T}}^N \hat{P}_Q]$. Next, the Hamilton operator is decomposed into its "electric" and "magnetic" contributions, $\hat{H} = \hat{H}_E + \hat{H}_B$. Using the Baker–Campbell–Hausdorff formula (cf. Chapter 1), for small a the transfer matrix is approximated as

$$\hat{\mathcal{T}} \approx \exp\left(-\frac{1}{2}a\hat{H}_B\right) \exp\left(-a\hat{H}_E\right) \exp\left(-\frac{1}{2}a\hat{H}_B\right). \tag{11.67}$$

This is correct up to terms of order a^2, which are proportional to the commutator $[\hat{H}_E, \hat{H}_B]$. In this context, it is important that this commutator increases only linearly with the spatial volume, despite the fact that $\hat{H}_E$ and $\hat{H}_B$ are both extensive quantities individually.

It should be noted that it is possible to define the functional integral directly in the Euclidean-time continuum, if one sets it up in the electric flux basis of the eigenstates $|m_{x,i}\rangle$ of $\hat{J}_{x,i}$. Here, we follow the more standard route and use the link basis of eigenstates $|U_{x,i}\rangle$ of the $\hat{U}_{x,i}$, with $\hat{U}_{x,i}|U_{x,i}\rangle = U_{x,i}|U_{x,i}\rangle$. This approach works only in discrete Euclidean time. At each discrete time-slice, we insert a complete set of states

$$\mathbb{1} = \int \mathcal{D}U \, |U\rangle\langle U| = \prod_{x,i} \int_{U(1)} dU_{x,i} \, |U_{x,i}\rangle\langle U_{x,i}| = \prod_{x,i} \int_{-\pi}^{\pi} d\varphi_{x,i} \, |\varphi_{x,i}\rangle\langle\varphi_{x,i}|. \tag{11.68}$$

Two factors $\exp(-a\hat{H}_B/2)$ associated with adjacent time steps give a factor $\exp(-a\hat{H}_B)$. These factors are diagonal in the link basis and result in

$$\langle U| \exp(-a\hat{H}_B)|U\rangle = \prod_{x,i>j} \exp\left(-\frac{1}{2e^2}\left(2 - U_{x,i}U_{x+\hat{i},j}U^*_{x+\hat{j},i}U^*_{x,j} - U_{x,j}U_{x+\hat{j},i}U^*_{x+\hat{i},j}U^*_{x,i}\right)\right)$$

$$= \prod_{x,i>j} \exp\left(-\frac{1}{e^2}\left(1 - \cos(\varphi_{x,i} + \varphi_{x+\hat{i},j} - \varphi_{x+\hat{j},i} - \varphi_{x,j})\right)\right). \tag{11.69}$$

These terms manifest themselves as spatial plaquette contributions to the action in all time-slices. The factor $\exp(-a\hat{H}_E)$, on the other hand, is not diagonal in the link basis such that

$$\langle U| \exp(-a\hat{H}_E)|U'\rangle = \prod_{x,i} \langle U_{x,i}| \exp\left(-\frac{e^2}{2}\hat{J}^2_{x,i}\right) |U'_{x,i}\rangle$$

$$= \prod_{x,i} \sum_{m_{x,i}\in\mathbb{Z}} \langle\varphi_{x,i}|m_{x,i}\rangle \exp\left(-\frac{e^2}{2}m^2_{x,i}\right) \langle m_{x,i}|\varphi'_{x,i}\rangle$$

$$= \prod_{x,i} \sum_{m_{x,i}\in\mathbb{Z}} \exp\left(-\frac{e^2}{2}m^2_{x,i}\right) \frac{1}{2\pi} \exp\left(\mathrm{i}m_{x,i}(\varphi_{x,i} - \varphi'_{x,i})\right)$$

$$= \prod_{x,i} \frac{1}{\sqrt{2\pi e}} \sum_{n_{x,i}\in\mathbb{Z}} \exp\left(-\frac{1}{2e^2}(\varphi_{x,i} - \varphi'_{x,i} + 2\pi n_{x,i})^2\right). \tag{11.70}$$

In the last step, we have used the Poisson resummation formula that we encountered before in Problem 1.6. The result can be identified as a space–time plaquette contribution to the action. The only non-zero link variables that contribute to this plaquette are the spatial link variables $\varphi_{x,i}$ and $\varphi'_{x,i}$ in two adjacent time-slices. There are no non-trivial temporal links, because we set up the underlying Hamiltonian formulation in the temporal gauge $U_{x,4} = 1$.

It should be noted that, other than for the spatial plaquettes, the space–time plaquette action is not the standard Wilson action but the so-called *Villain action*. This action is periodic in the link variables due to an explicit summation over the auxiliary integer-valued field $n_{x,i} \in \mathbb{Z}$ that is associated with the corresponding space–time plaquette.

Finally, we consider the matrix elements of the projection operator $\hat{P}_Q$ that enforces the Gauss law

$$\frac{1}{a}\sum_i \left(\hat{E}_{x,i} - \hat{E}_{x-\hat{i},i}\right)|\Psi,Q\rangle = \frac{e}{a^3}\sum_i \left(\hat{J}_{x,i} - \hat{J}_{x-\hat{i},i}\right)|\Psi,Q\rangle = \frac{e}{a^3}Q_x|\Psi,Q\rangle = \rho_x|\Psi,Q\rangle.$$

(11.71)

Here, $\rho_x = eQ_x/a^3$ is the charge density associated with the charge $Q_x \in \mathbb{Z}$. The projector is diagonal in the flux basis where it represents the constraint $\sum_i(m_{x,i} - m_{x-\hat{i},i}) = Q_x$. This constraint can be enforced by an integration over a Lagrange multiplier field φ_x

$$\langle m|\hat{P}_Q|m\rangle = \prod_x \frac{1}{2\pi}\int_{-\pi}^{\pi} d\varphi_x \exp\left(i\Big(Q_x - \sum_i (m_{x,i} - m_{x-\hat{i},i})\Big)\varphi_x\right).$$

(11.72)

The object $\Phi_{Q_x} = \exp(iQ_x\varphi_x) \in$ U(1), which was introduced by Alexander Polyakov, is known as a *Polyakov loop* (Polyakov, 1975). It represents a static, external charge Q_x.

The projector $\hat{P}_Q$ commutes with both $\hat{H}_E$ and $\hat{H}_B$. It is convenient to lump it together with the factor $\exp(-a\hat{H}_E)$ in the last time step

$$\langle U|\exp\left(-a\hat{H}_E\right)\hat{P}_Q|U'\rangle =$$

$$\prod_x \frac{1}{2\pi}\int_{-\pi}^{\pi} d\varphi_x \exp(iQ_x\varphi_x) \prod_{x,i}\sum_{m_{x,i}\in\mathbb{Z}} \exp\left(-\frac{e^2}{2}m_{x,i}^2\right)\frac{1}{2\pi}\exp\left(im_{x,i}(\varphi_{x,i}-\varphi'_{x,i}-\varphi_x+\varphi_{x+\hat{i}})\right)$$

$$= \prod_x \frac{1}{2\pi}\int_{-\pi}^{\pi} d\varphi_x \Phi_{Q_x} \prod_{x,i}\frac{1}{\sqrt{2\pi e}}\sum_{n_{x,i}\in\mathbb{Z}} \exp\left(-\frac{1}{2e^2}(\varphi_{x,i}-\varphi'_{x,i}-\varphi_x+\varphi_{x+\hat{i}}+2\pi n_{x,i})^2\right).$$

(11.73)

This represents the Villain action of space–time plaquettes in the last time step, which also contain temporal parallel transporters $U_{x,4} = \exp(i\varphi_x) \in$ U(1). The Polyakov loop can then be expressed as $\Phi_{Q_x} = U_{x,4}^{Q_x}$.

Since we worked in the temporal gauge, $A_{x,4} = 0$, only the space–time plaquettes in the last time step involve non-trivial temporal links. Consequently, both the Polyakov loop and the action of the space–time plaquettes are invariant only under time-independent gauge transformations. It is convenient to return to an unfixed gauge, *e.g.*, by performing random gauge transformations at all space–time points x, $U'_{x,\mu} = \Omega_x U_{x,\mu}\Omega^{\dagger}_{x+\hat{\mu}}$. In this way, all temporal links (not only the ones in the last time-step) become non-trivial and the Polyakov loop turns into a product of N temporal links

$$\Phi_{Q_x} = \prod_{n=0}^{N-1} U_{x+n\hat{4},4}^{Q_x},$$

(11.74)

which is invariant against general gauge transformations that are periodic in Euclidean time. Collecting all spatial plaquette terms from the individual time slices, as well as all space–time plaquette contributions from the various time steps, one arrives at the final expression for the functional integral

$$Z_Q = \int \mathcal{D}U \exp(-S[U]) \prod_x \Phi_{Q_x} = \prod_{x,\mu} \int_{U(1)} dU_{x,\mu} \exp(-S[U]) \prod_x \Phi_{Q_x}. \quad (11.75)$$

The Boltzmann weight $\exp(-S[U])$ of a lattice gauge field configuration $[U]$ receives contributions from the standard Wilson action on the spatial plaquettes in all time slices as well as from the Villain action on the space–time plaquettes in all time steps,

$$\exp(-S[U]) = \prod_{x,i>j} \exp\left(-\frac{1}{e^2}\left(1 - \cos(\varphi_{x,i} + \varphi_{x+\hat{i},j} - \varphi_{x+\hat{j},i} - \varphi_{x,j})\right)\right)$$

$$\times \prod_{x,i} \sum_{n_{x,i}\in\mathbb{Z}} \exp\left(-\frac{1}{2e^2}(\varphi_{x,i} - \varphi'_{x,i} - \varphi_x + \varphi_{x+\hat{i}} + 2\pi n_{x,i})^2\right). \quad (11.76)$$

11.10 Functional Integral for Non-Abelian Lattice Gauge Theory

Finally, we consider the functional integral for non-Abelian lattice gauge theories as it was first constructed by Wilson

$$Z_Q = \int \mathcal{D}U \exp(-S[U]) \prod_x \Phi_{Q_x} = \prod_{x,\mu} \int_G dU_{x,\mu} \exp(-S[U]) \prod_x \Phi_{Q_x}. \quad (11.77)$$

Here, the integration of the link variables $U_{x,\mu} \in G$ over the group manifold of the gauge group G is performed with the *Haar measure* $dU_{x,\mu}$, which is invariant under left and right gauge transformations. The Polyakov loop is still given by eq. (11.74), except that $U^{Q_x}_{x+n\hat{4},4}$ now denotes the corresponding link variable in the representation Q_x. The standard Wilson action is now used for all plaquettes,

$$S[U] = \frac{2}{g_s^2} \sum_{x,\mu>\nu} \mathrm{Tr}\left(\mathbb{1} - \frac{1}{2}U_{x,\mu}U_{x+\hat{\mu},\nu}U^{\dagger}_{x+\hat{\nu},\mu}U^{\dagger}_{x,\nu} - \frac{1}{2}U_{x,\nu}U_{x+\hat{\nu},\mu}U^{\dagger}_{x+\hat{\mu},\nu}U^{\dagger}_{x,\mu}\right). \quad (11.78)$$

This has the advantage that it is invariant against discrete lattice rotations between space and Euclidean time.

The Wilson theory is not exactly equivalent to the Kogut–Susskind Hamiltonian formulation of Section 11.8: The Hamilton operator of eq. (11.62) propagates the system in continuous time, whereas the Wilson plaquette action refers to discrete time. However, it is still possible to construct a transfer matrix $\hat{\mathcal{T}}$ such that $Z_Q = \mathrm{Tr}(\hat{\mathcal{T}}^N P_Q)$, as it was first done independently by Michael Creutz (1977) and Martin Lüscher (1977). They considered the theory also including fermion fields. The transfer matrix is again constructed in temporal gauge, such that there are only spatial link variables (except in the last time step). The matrix elements of the transfer matrix between states of spatial links $U_{x,i}$ and $U'_{x,i}$ in two adjacent time slices receive contributions from the square roots of the Boltzmann factors associated with the spatial plaquettes in the two time slices, as well as from the Boltzmann factors of the temporal plaquettes connecting the two time slices,

$$\langle U|\hat{\mathcal{T}}|U'\rangle = \exp\left(-\frac{1}{g_s^2}\sum_{x,i>j}\mathrm{Tr}\left(\mathbb{1} - \frac{1}{2}U_{x,i}U_{x+\hat{i},j}U^\dagger_{x+\hat{j},i}U^\dagger_{x,j} - \frac{1}{2}U_{x,j}U_{x+\hat{j},i}U^\dagger_{x+\hat{i},j}U^\dagger_{x,i}\right)\right)$$

$$\times \exp\left(-\frac{2}{g_s^2}\sum_{x,i}\mathrm{Tr}\left(\mathbb{1} - \frac{1}{2}U_{x,i}U'^\dagger_{x,i} + \frac{1}{2}U'_{x,i}U^\dagger_{x,i}\right)\right)$$

$$\times \exp\left(-\frac{1}{g_s^2}\sum_{x,i>j}\mathrm{Tr}\left(\mathbb{1} - \frac{1}{2}U'_{x,i}U'_{x+\hat{i},j}U'^\dagger_{x+\hat{j},i}U'^\dagger_{x,j} - \frac{1}{2}U'_{x,j}U'_{x+\hat{j},i}U'^\dagger_{x+\hat{i},j}U'^\dagger_{x,i}\right)\right).$$

(11.79)

It was proved by Creutz (1977) and Lüscher (1977) that $\hat{\mathcal{T}}$ has positive eigenvalues. Hence, one can define a Hamilton operator $\hat{H}_a = -(1/a)\log\hat{\mathcal{T}}$ whose spectrum determines the energies of the states of Wilson's lattice gauge theory at arbitrary values of the lattice spacing a.

Lattice field theory provides us with an elegant and powerful non-perturbative formulation of non-Abelian gauge theories, which is of great practical and conceptual virtue. We will make use of it in numerous places in this book, and we address Monte Carlo simulation techniques in Appendix H.

Exercises

11.1 From the group to the algebra
Assume that $\Omega(x)$ is an SU(N) gauge transformation, and prove that $i\Omega(x)\partial_\mu\Omega(x)^\dagger$ is in the SU(N) algebra.

11.2 Gauge covariance of a parallel transporter
Consider a parallel transporter $U_C = \mathcal{P}\exp\left(\int_C dx_\mu G_\mu(x)\right)$ constructed from a non-Abelian SU(N) gauge field $G_\mu(x)$ along an open curve C from x_1 to x_2. Show that the gauge transformation $G'_\mu(x) = \Omega(x)(G_\mu(x) + \partial_\mu)\Omega(x)^\dagger$ induces the transformation $U'_C = \Omega(x_1)U_C\Omega(x_2)^\dagger$.

11.3 Gauge transformation of a non-Abelian field strength
Show that the non-Abelian field strength

$$G_{\mu\nu}(x) = \partial_\mu G_\nu(x) - \partial_\nu G_\mu(x) + [G_\mu(x), G_\nu(x)]$$

transforms as $G'_{\mu\nu}(x) = \Omega(x)G_{\mu\nu}(x)\Omega(x)^\dagger$ under non-Abelian gauge transformations $G'_\mu(x) = \Omega(x)(G_\mu(x) + \partial_\mu)\Omega(x)^\dagger$, cf. eq. (11.5).

11.4 BRST invariance of ghost plus gauge fixing terms
Convince yourself that $\delta(\mathcal{L}_{gf} + \mathcal{L}_{gh}) = 0$.

11.5 Nilpotency of the anti-commuting BRST operation Q
Convince yourself that $Q^2 G_\mu(x) = 0$.

11.6 Inventing "new" mathematics

As Dirac has demonstrated with his δ-function, physicists can invent their own mathematical tools, which may later make their way into mathematics itself. Inventing new mathematics is a creative process, which is difficult to teach. Still, this exercise aims to practicing exactly this.

As we know following Riemann, the integral $\int f(x)\,dx$ can be viewed as a continuous analogue of a discrete sum $\sum_x f_x$. Imagine that, in order to develop a new idea in theoretical physics, one needs to generalize the product $\prod_x f_x$ in a similar way. In other words, we are in need of a mathematical tool that allows us to build the "continuous product" of all values of a non-negative, real-valued function $f(x)$ in the interval $x \in [a, b]$. We already have a notation for the desired "product integral"; we denote it by $\mathcal{P}_a^b f(x)^{dx}$, which should have the following properties, in analogy to ordinary integration as a proper limit of a discrete sum:

$$\int_a^b f(x)\,dx + \int_b^c f(x)\,dx = \int_a^c f(x)\,dx \quad\leftrightarrow\quad \mathcal{P}_a^b f(x)^{dx} \cdot \mathcal{P}_b^c f(x)^{dx} = \mathcal{P}_a^c f(x)^{dx},$$

$$\int_b^a f(x)\,dx = -\int_a^b f(x)\,dx \quad\leftrightarrow\quad \mathcal{P}_b^a f(x)^{dx} = \left[\mathcal{P}_a^b f(x)^{dx}\right]^{-1},$$

$$F(b) - F(a) = \int_a^b f(x)\,dx, \quad \lim_{dx\to 0} \frac{F(x+dx) - F(x)}{dx} = f(x) \quad\leftrightarrow$$

$$\frac{F(b)}{F(a)} = \mathcal{P}_a^b f(x)^{dx}, \quad \lim_{dx\to 0} \left[\frac{F(x+dx)}{F(x)}\right]^{1/dx} = f(x).$$

a) Explicitly construct a "product integral" $\mathcal{P}_a^b f(x)^{dx}$ by expressing it in terms of existing mathematical structures, and show that it indeed has the desired properties. Mathematical rigor is not required, just the right ideas.

b) Use the newly invented mathematical tool to demonstrate the analogy

$$\int_a^b \left[f(x) + g(x)\right] dx = \int_a^b f(x)dx + \int_a^b g(x)dx \quad\leftrightarrow$$
$$\mathcal{P}_a^b [f(x)g(x)]^{dx} = \mathcal{P}_a^b f(x)^{dx} \cdot \mathcal{P}_a^b g(x)^{dx}.$$

c) Show that the analogue of $\int_{-a}^a f(x)dx = 0$ for an odd function $f(-x) = -f(x)$ is the relation $\mathcal{P}_{-a}^a f(x)^{dx} = 1$ for a "self-reciprocal" function obeying $f(-x) = f(x)^{-1}$.

d) Evaluate $\mathcal{P}_0^1 x^{dx}$; in other words, take the "continuous product" of all real numbers between 0 and 1.

e) Compare the "product integral" with the path-ordered matrix product introduced in eq. (11.55).

11.7 Commutation relations of SU(2) lattice gauge theory

Verify the commutation relations of eq. (11.59) based on eq. (11.58).

11.8 Classical continuum limit of the Kogut–Susskind Hamilton operator

Convince yourself that the Kogut–Susskind Hamilton operator of eq. (11.62) reduces to the Hamilton function of eq. (11.9) in the classical continuum limit.

PART II

CONSTRUCTION OF THE STANDARD MODEL

Intermezzo: Concepts of the Standard Model

In this Intermezzo, we anticipate fundamental concepts and basic principles of the Standard Model, in order to prepare the reader for the systematic exposition of the subject in the continuation. As in the case of the Ouverture, understanding the Intermezzo is by no means necessary for proceeding further with the book. We recommend consulting the Intermezzo repeatedly while working through the rest of the book.

The Standard Model: A Non-Abelian Chiral Gauge Theory

The Standard Model is an $SU(3)_c \times SU(2)_L \times U(1)_Y$ gauge theory. The non-Abelian gauge group $SU(3)_c$ governs the strong interaction that is mediated by the gluon field, while the non-Abelian gauge group $SU(2)_L$ and the Abelian gauge group $U(1)_Y$ are responsible for the electroweak interactions.

The fields of the Standard Model all have specific properties under gauge transformations. They also transform appropriately under the space–time transformations of the Poincaré group. The Lagrangian of the Standard Model (Glashow, 1961; Weinberg, 1967; Salam, 1968) comprises all renormalizable terms that are gauge- and Poincaré-invariant products of the fields involved.

The scalar Higgs field takes a non-zero vacuum expectation value. As a result, via the Higgs mechanism the $SU(2)_L \times U(1)_Y$ symmetry "breaks spontaneously", and the $U(1)_{em}$ gauge group of electromagnetism emerges as the unbroken subgroup. In this manner, the gauge bosons $W^\pm$ and Z pick up a mass while the photon remains massless.[1]

Besides the bosonic Higgs and gauge fields, the Standard Model contains fermionic matter fields structured in three generations. The corresponding members of the different generations have the same quantum numbers but different masses. Each generation contains quark and lepton fields. While quarks participate in all gauge interactions, leptons do not interact strongly. The quarks of the first generation have the flavors up (u) and down (d), those of the second generation charm (c) and strange (s), and the ones of the third generation top (t) and bottom (b). Correspondingly, the leptons of the three generations are electrons e, muons μ, and tauons τ, along with their associated neutrinos ν_e, ν_μ, and ν_τ.

The Standard Model is a *chiral gauge theory*; *i.e.* left- and right-handed fermions couple differently to the gauge fields. In particular, all left-handed fermions form $SU(2)_L$ doublets, while the right-handed fermions are $SU(2)_L$ singlets and thus do not couple to the $SU(2)_L$ gauge field. Also, the values of the weak hypercharges, which determine the couplings

---

[1] When a gauge symmetry is concerned, we write "spontaneous symmetry breaking" in inverted commas: A gauge symmetry reflects a redundancy which cannot break, unlike a global symmetry.

of the fermions to the Abelian $U(1)_Y$ gauge field, are different for left- and right-handed fermions. In the Standard Model, right-handed neutrino fields do not exist. The fundamental differences between left and right manifest themselves by the explicit breaking of the discrete symmetries of parity P and charge conjugation C. Still, the gauge interactions of the Standard Model are invariant under the combined symmetry CP.

Interestingly, left- and right-handed quark fields couple to the $SU(3)_c$ gluon gauge field in the same manner: They are both color triplets. Consequently, the theory of the strong force between quarks and gluons – Quantum Chromodynamics (QCD) – is a *vector-like* gauge theory, which makes no fundamental difference between left and right chirality. Indeed, the strong interaction respects both P and C invariance separately.

Since left- and right-handed fermions couple differently to the electroweak gauge fields, explicit fermion mass terms are forbidden in the Standard Model Lagrangian because they violate gauge invariance. However, the fermions do not only couple to the gauge fields but also to the Higgs field. When the Higgs field picks up a non-zero vacuum expectation value, the corresponding Yukawa couplings give rise to effective fermion masses. Since there are no right-handed neutrino fields, neutrinos remain massless in the Standard Model, at least when one limits oneself to renormalizable terms in the Lagrangian.

The Yukawa couplings between the fermion fields and the Higgs field also give rise to mixing between the various generations. In this way, CP violation arises, provided that there are at least three fermion generations.

Starting in Chapter 12, we will build the Standard Model step by step, by adding one field after the other, and discussing the corresponding dynamics. We will begin with the Higgs field, then add the gauge fields, and finally the lepton and quark fields, beginning with the first generation and then proceeding to the others. At some point, we treat the Standard Model as a low-energy effective theory, including also the leading non-renormalizable term.

Renormalizability of Non-Abelian Gauge Theories

The Standard Model is *perturbatively renormalizable*; *i.e.* a finite number of terms in the Lagrangian is sufficient to absorb the UV divergences to all orders. Terms with coupling constants of negative mass dimension are *irrelevant*; *i.e.* they vanish in the UV limit; those are not included in the minimal version of the Standard Model Lagrangian. Proving the renormalizability of the Standard Model is an intricate mathematical problem that will not be addressed in detail in this book. Still, we like to mention the most important steps that were taken in order to achieve this goal.

First of all, one must distinguish gauge-invariant regularizations from gauge-variant ones. For example, a sharp momentum cut-off affects locality at ultra-short distances and explicitly breaks gauge invariance, while the lattice regularization respects gauge invariance all the way down to the lattice spacing scale. Dimensional regularization does not explicitly introduce a UV cut-off and respects gauge invariance, at least in vector-like gauge theories, in which left- and right-handed fermions couple to the gauge field in the same way. On the other hand, chiral gauge theories, including the Standard Model, are characterized by the property that left- and right-handed fermions reside in different representations of the gauge group. In that case, the matrix γ_5, which has no analytic continuation to non-integer dimensions, explicitly enters the Lagrangian. As a result, in a chiral gauge theory, dimensional regularization explicitly breaks chiral gauge invariance, already at tree-level, and thus, it falls in the category of gauge-variant regularizations.

In order to renormalize a gauge theory in a gauge-variant regularization, counterterms that explicitly break the gauge symmetry must be introduced to absorb some of the UV divergences. In the context of Quantum Electrodynamics (QED), *i.e.* of a vector-like Abelian gauge theory, this has been achieved by a BPHZ construction which employs a gauge-variant momentum cut-off. Nikolai Bogoliubov and Ostap Parasiuk were first to understand overlapping divergences in nested loop Feynman diagrams (Bogoliubov and Parasiuk, 1957). Klaus Hepp extended this work to an algebraically highly non-trivial proof of renormalizability to all orders of perturbation theory (Hepp, 1966), which was later cast into a more accessible, elegant form by Wolfhart Zimmermann (1969).

A necessary step in the perturbative treatment of a gauge theory is *gauge fixing*. Otherwise, the functional integral diverges in an unphysical manner because one integrates over infinitely many redundant gauge copies. In an Abelian gauge theory, the Lorenz gauge fixing condition, $\partial_\mu {}^\alpha A_\mu = 0$ with ${}^\alpha A_\mu = A_\mu - \partial_\mu \alpha$ implies a linear constraint on the gauge transformation function $\alpha(x)$. In a non-Abelian gauge theory, on the other hand, $\partial_\mu {}^\Omega G_\mu = 0$ with ${}^\Omega G_\mu = \Omega(G_\mu + \partial_\mu)\Omega^\dagger$, implies a non-linear constraint on $\Omega(x)$. As we discussed in Section 11.2, in order to take into account the corresponding Jacobi determinant, Ludvig Faddeev and Victor Popov integrated unphysical anti-commuting scalar ghost fields into the functional integral of non-Abelian gauge theories (Faddeev and Popov, 1967). The ghost fields cancel the contributions of the unphysical, longitudinally polarized components of the gauge field.

First indications for the renormalizability of non-Abelian gauge theories were obtained by Gerard 't Hooft and Martinus Veltman, in their investigations of 1- and 2-loop Feynman diagrams ('t Hooft, 1971a,b; 't Hooft and Veltman, 1972). Around the same time, important insights resulted from the work of Benjamin Lee and Jean Zinn-Justin (Lee and Zinn-Justin, 1972a,b,c, 1973) based upon *Slavnov–Taylor identities* ('t Hooft, 1971a; Taylor, 1971; Slavnov, 1972) – the non-Abelian variant of Ward–Takahashi identities (Ward, 1950; Takahashi, 1957), *i.e.* symmetry relations between different correlation functions.

A crucial question concerning the renormalizability of non-Abelian gauge theories is related to unitarity: Are the unphysical Faddeev–Popov ghost fields (which violate the spin–statistics theorem) protected from being produced in scattering processes of physical particles? As we discussed in Section 11.3, this is guaranteed by the global Grassmannian BRST symmetry discovered by Carlo Becchi, Alain Rouet, and Raymond Stora (Becchi *et al.*, 1974, 1975, 1976), and independently by Igor Tyutin (1975), which holds to all orders of perturbation theory. BRST cohomology ensures that unphysical degrees of freedom, including ghosts, anti-ghosts, and the longitudinal components of gauge bosons, remain outside the physical sector of the Hilbert space, which contains the transversely polarized components of the gauge boson fields. A general proof of the renormalizability of non-Abelian gauge theories is based on a master equation, known as the *Zinn-Justin equation* (Zinn-Justin, 1975, 2009).

Another crucial aspect of renormalizability concerns anomalies, *i.e.* explicit symmetry violations due to quantum effects. Gauge anomalies, *i.e.* violations of local symmetries – which are actually redundancies – are unacceptable. They lead to mathematical and physical inconsistencies because unphysical, redundant gauge degrees of freedom then contaminate physical results. All possible anomalies are solutions of the *Wess–Zumino consistency condition* (Wess and Zumino, 1971), which corresponds to a cohomology problem. The general solution of the BRST cohomology problem, which amounts to a complete classification of all possible anomalies in Yang–Mills gauge theories, was achieved by Friedemann Brandt, Norbert Dragon, and Maximilian Kreuzer (Brandt *et al.* (1989, 1990a); for a review, see Brandt *et al.* (2000)).

Triviality and Incorporation of Gravity

Its renormalizability suggests that the Standard Model could, at least in principle, be valid up to arbitrarily high energy scales. However, there is a caveat: the issue of *triviality*. For a long time, there was overwhelming evidence (albeit no rigorous proof) that the Higgs sector of the Standard Model becomes non-interacting – and thus "trivial" – when one moves the UV cut-off all the way to infinity (Aizenman, 1981; Fröhlich, 1982; Callaway, 1988; Fernández *et al.*, 1992). In 2022, a Fields medal was awarded to Hugo Duminil-Copin who, together with Michael Aizenman, proved rigorously that the $\lambda\phi^4$ theory is trivial in four space–time dimensions (Aizenman and Duminil-Copin, 2021).

While renormalizability implies that the Standard Model physics is insensitive to the UV cut-off, one may question whether it is physically meaningful to send the cut-off to infinity. In particular, one would expect that, at some energy scale, either near or well above the regime of $\mathcal{O}(10)$ TeV, which is accessible in current accelerator experiments, *new physics* beyond the Standard Model could manifest itself. In that case, the scale Λ at which new physics arises would represent a physical cut-off for the Standard Model, which would no longer provide an accurate description of the physics around or above that energy scale. The Standard Model would then still remain a consistent effective field theory at low and moderate energy. However, as one proceeds to higher and higher energies approaching Λ, more and more non-renormalizable terms with negative mass dimension (suppressed by inverse powers of Λ) would have to be added to the effective Lagrangian.

In view of the relatively light, observed Higgs boson mass of 125 GeV, the triviality of the Standard Model is a somewhat academic issue: It is significant only beyond the Planck scale, which already provides a finite (yet extremely high) energy scale at which the Standard Model must be replaced by a more complete theory that should include non-perturbative *quantum gravity*.

While gravity is usually not considered as part of the Standard Model, it can be incorporated perturbatively as a low-energy effective theory, provided that Lorentz invariance is maintained as an exact symmetry. This is required for a unification with Einstein's theory of gravity (Einstein, 1915; Hilbert, 1915; Einstein, 1916), *i.e.* General Relativity, where global Lorentz invariance is promoted to a (necessarily exact) gauge symmetry. In contrast to some claims in the literature, it is not true that gravity resists quantization in the context of quantum field theory, in the following sense: While Einstein's gravity is not renormalizable, *i.e.* at higher and higher energies more and more terms enter the Lagrangian, it can be quantized as a weakly coupled low-energy effective field theory (Buchbinder *et al.*, 1992; Donoghue, 2012; Donoghue and Holstein, 2015; Buchbinder and Shapiro, 2021).

This effective theory is expected to break down when black holes form or at the Planck scale, where gravity is strongly coupled. A fully established *non-perturbative* theory of quantum gravity is still missing.

Fundamental Standard Model Parameters

The Standard Model contains a number of free parameters, whose values can only be determined by comparison with experiments. Remarkably, in the minimal version of the Standard Model Lagrangian, there is only *one* parameter endowed with a non-zero dimension, which determines the vacuum value of the Higgs field, $v \simeq 246$ GeV. The masses of the heavy $W^\pm$ and Z gauge bosons, which mediate the weak interaction, are given by $M_W = \frac{1}{2}gv$ and $M_Z = \frac{1}{2}\sqrt{g^2 + g'^2}\,v$, where g and g' are dimensionless gauge couplings, associated with the Standard Model gauge groups $SU(2)_L$ and $U(1)_Y$, respectively. The Higgs boson mass is given by $M_H = \sqrt{\lambda/3}\,v$, with λ being the dimensionless quartic self-coupling of the Higgs field.

The strong interaction between quarks and gluons is described by *Quantum Chromodynamics* (QCD) which was pioneered by Harald Fritzsch, Murray Gell-Mann, Heinrich Leutwyler (Fritzsch *et al.*, 1973), and others,[2] an $SU(3)_c$ color gauge theory, which is another integral part of the Standard Model. Here, the classical scale invariance (at vanishing quark masses) is explicitly broken by quantum effects. Hence, *dimensional transmutation* trades the dimensionless $SU(3)_c$ gauge coupling g_s for the dimensioned QCD scale $\Lambda_{\text{QCD}} = 332(14)$ MeV (with three flavors, in the modified minimal subtraction renormalization scheme, Bruno *et al.* (2016)). Strongly interacting particles, in particular the light hadrons, receive the dominant portion of their mass from the strong interaction energy of quarks and gluons, which is proportional to Λ_{QCD}; only a minor (percent-level) fraction of their masses is due to the quark masses, *i.e.* due to the Higgs mechanism (Aoki *et al.*, 2020).

The masses of quarks, $m_q = f_q v$, and leptons, $m_l = f_l v$, are products of v with the dimensionless *Yukawa couplings* f_q and f_l, *i.e.* couplings between fermion fields and the Higgs field. Quarks and leptons exist in *three generations* with the same quantum numbers, but with different masses. The mixing parameters between the electroweak eigenstates and the mass eigenstates for the quark fields of the three generations are incorporated in the *Cabibbo–Kobayashi–Maskawa matrix* (Cabibbo, 1963; Kobayashi and Maskawa, 1973) and represent additional fundamental Standard Model parameters, whose values can only be determined experimentally.

In the minimal, renormalizable version of the Standard Model, the neutrinos are massless, because only left-handed neutrinos and only renormalizable terms are included. Since the discovery of *neutrino oscillations* (Fukuda *et al.*, 1998; Ahmad *et al.*, 2001), it is clear that neutrinos do have a non-zero mass. Neutrino masses naturally arise from non-renormalizable interactions when the Standard Model is considered as a low-energy effective theory. That leads to a matrix of lepton mixing parameters, the *Pontecorvo–Maki–*

---

[2] For earlier suggestions to assign three states to the quark flavors, see, *e.g.*, Bogoliubov *et al.* (1964); Han and Nambu (1965); Miyamoto (1965), and for a historical account, Leutwyler (2014).

Nakagawa–Sakata matrix (Pontecorvo, 1957; Maki *et al.*, 1962), in analogy to the quark mixing matrix. One may even extend the field content of the Standard Model by introducing right-handed neutrino fields. In contrast to their left-handed counterpart, the right-handed neutrinos are *sterile*, which means that they do not couple to any Standard Model gauge field. As a consequence of their inclusion, further dimensioned parameters, forming the Majorana mass matrix for the right-handed neutrinos of the three fermion generations, enter the extended Standard Model Lagrangian. A mass-by-mixing mechanism, first described by Peter Minkowski (Minkowski, 1977) – also known as the *seesaw mechanism* – leads to small neutrino masses, provided that the Majorana masses are much larger than $v \simeq 246$ GeV (Gell-Mann *et al.*, 1979; Yanagida, 1980; Glashow, 1980; Mohapatra and Senjanovic, 1980; Schechter and Valle, 1980).

The Majorana masses could set the scale Λ at which new physics, beyond the Standard Model, shows up. The low-energy effects of new physics – in particular, the non-zero neutrino masses – give rise to non-renormalizable corrections to the minimal Standard Model Lagrangian, which are suppressed by the inverse energy scale $1/\Lambda$. An effective dimension-5 term is generated if we introduce a heavy right-handed neutrino field and integrate it out.

In view of the considerable number of 20 or 29 free parameters (without or with neutrino masses and lepton mixing angles), one might expect that there could be an even more fundamental structure underlying the Standard Model, which would allow us to understand the origin of its free parameters and to find relations between them. On the other hand, in light of the impressive number of high-precision predictions of the Standard Model, the number of its free parameters appears actually modest. Ultimately, the Standard Model will break down at the Planck scale (if not earlier), when non-perturbative quantum gravity becomes significant.

Possible further extensions of the minimal Standard Model, in addition to the neutrino mass terms, but still below the Planck scale, might include *technicolor* (Weinberg, 1976, 1979b; Susskind, 1979), axions (Peccei and Quinn, 1977a,b; Weinberg, 1978; Wilczek, 1978), *supersymmetry* (Gervais and Sakita, 1971; Golfand and Likhtman, 1971; Volkov and Akulov, 1972; Wess and Zumino, 1974), *Grand Unified Theories* (GUT), in particular with the unified gauge group SU(5) (Georgi and Glashow, 1974; Georgi *et al.*, 1974; Buras *et al.*, 1978) or SO(10) (Fritzsch and Minkowski, 1975; Georgi, 1975), or other theories that have – or have not – been subject of intense theoretical investigation. These approaches will also be discussed in this book, with the exception of supersymmetry, where we refer to the extensive literature.

At the time of writing this book, in 2022, there is no conclusive experimental evidence for physics beyond the Standard Model, with the exception of neutrino masses. There is evidence indeed for *dark matter* (Zwicky, 1933; Rubin and Ford, 1970; Smoot *et al.*, 1992; Clowe *et al.*, 2006), but it could again be related to sterile neutrinos. The concept of *cosmic inflation* suggests that there could be an *inflaton field* (Guth, 1981). Moreover, there is evidence for *dark energy* (Riess *et al.*, 1998; Perlmutter *et al.*, 1999) – *i.e. vacuum energy* – which is likely to emerge either as a static *cosmological constant* Λ_c or as dynamical *quintessence* (Wetterich, 1988; Ratra and Peebles, 1988), cf. Section 2.7. The former is a free low-energy parameter of Einstein's gravity (see eq. (2.42)), in addition to Newton's constant G, which determines the Planck scale $M_{\text{Planck}} = \sqrt{\hbar c/G}$. When we include perturbative quantum gravity, as well as right-handed neutrino fields, we can currently not exclude that the Standard Model – extended in this way – could, in principle, be valid all the way up to the Planck scale; *i.e.* it could even be a "Theory of Almost Everything".

Hierarchies of Scales and Approximate Global Symmetries

In the minimal Standard Model, extended by perturbative quantum gravity, we encounter *four* dimensioned parameters: the Planck scale $M_{\text{Planck}} \simeq 10^{19}$ GeV, which determines the strength of gravity, the vacuum expectation value of the Higgs field, $v \approx 10^{-17} M_{\text{Planck}}$, the QCD scale $\Lambda_{\text{QCD}} \approx 10^{-20} M_{\text{Planck}}$, and the cosmological constant Λ_c, with $\Lambda_c^{1/2} \approx 10^{-60} M_{\text{Planck}}$. Why are these scales so vastly different? What is the origin of these hierarchies of energy scales?

Since, according to our present understanding, these scales are free parameters, answering these questions requires to step beyond the Standard Model or perturbative quantum gravity. Within that framework, we may still ask whether these hierarchies arise "naturally". At first glance, it may seem unnatural that, *e.g.,* the QCD scale is so much below the Planck scale. However, the QCD property of *asymptotic freedom* (Gross and Wilczek, 1973; Politzer, 1973) provides an explanation for this hierarchy: Without unnatural fine-tuning of the strong coupling constant g_s, Λ_{QCD} is exponentially suppressed with respect to the UV cut-off, which we may identify with M_{Planck}.

This explanation does not apply to the hierarchy between the electroweak scale and the Planck scale. Understanding why $v \ll M_{\text{Planck}}$ is known as the *hierarchy problem,* which does not have any natural solution within the Standard Model, because the self-coupling of the Higgs field is *not* asymptotically free. Conceivable solutions to the hierarchy problem could be associated with new physics, such as supersymmetry or technicolor theories. However, despite intensive investigations, until 2022 there is no experimental evidence for any such theory. Beyond the perturbative expansion, in particular on the lattice, supersymmetry with light scalar fields appears intrinsically unnatural, because its construction itself seems to require fine-tuning.

The relation $\Lambda_c^{1/2} \ll M_{\text{Planck}}$ constitutes the *cosmological constant problem,* the most severe hierarchy problem in physics. If the correct theory of non-perturbative quantum gravity would have a property similar to asymptotic freedom, then this problem could perhaps be solved naturally.

Without knowing such a theory, some people invoke the *anthropic principle*. In contrast to generic considerations, it inserts the fact of our existence as an additional input, which drastically constrains the possible values of Λ_c. Steven Weinberg has argued that alternative Universes – with significantly larger positive or negative cosmological constants – would either expand or collapse very rapidly, respectively, so that the evolution of intelligent life seems unlikely (Weinberg, 1987). This idea is supported by the hypothesis of eternal cosmic inflation, which could provide a huge number of different Universes, forming a very large multiverse. If such a multiverse indeed exists – which is a matter of speculation and hardly falsifiable – we could only have evolved in a pocket Universe with hierarchies of energy scales that are hospitable to our existence.

In addition to this speculative assumption, it is not surprising that the extra input of our existence leaves fewer possibilities. Since it narrows the issue by extended assumptions, it is questionable whether this should be considered an acceptable "explanation". Therefore, the anthropic principle should at most be invoked as a last resort, when all other approaches fail. It should certainly not prevent us from thinking hard about explanations which do not depend on this principle.

The Standard Model provides us with even more hierarchy puzzles. While the dimensionless Yukawa coupling f_t of the top quark (the heaviest elementary particle that we know) is of order 1, such that the top quark mass, $m_t = f_t v \simeq 174\,\mathrm{GeV}$, is near the electroweak scale v, the Yukawa couplings of the lightest quarks, up and down, are much smaller, $f_u, f_d \simeq \mathcal{O}(10^{-5})$, such that m_u, $m_d \ll \Lambda_{\mathrm{QCD}}$. This hierarchy between the masses of the light quarks and the QCD scale entails an approximate global $\mathrm{SU}(2)_L \times \mathrm{SU}(2)_R$ chiral symmetry. Its $\mathrm{SU}(2)_{L-R}$ *isospin* subgroup manifests itself in the hadron spectrum; for instance, it can be considered as an explanation why the proton and the neutron have almost the same mass. In this sense, the hierarchy m_u, $m_d \ll \Lambda_{\mathrm{QCD}}$ explains numerous features of the light hadron spectrum. However, understanding the origin of the quark mass hierarchy itself, and thus a deeper reason for the approximate chiral symmetry, is yet another unsolved hierarchy problem. So far, we recognize this symmetry and utilize it to simplify our theoretical approaches.

Local and Global Symmetries

As we pointed out before, local symmetries – *i.e.* gauge symmetries – must be exact in order to prevent unphysical effects of the redundant gauge variables. This includes Lorentz symmetry, when it is promoted to a gauge symmetry in the context of General Relativity. Gauge invariance is very restrictive and, in combination with renormalizability, it implies strong predictive power, with only one free parameter – the (dimensionless) gauge coupling constant associated with the corresponding gauge group. Non-gauge-mediated interactions, in particular the Yukawa couplings between the Higgs field and quark and lepton fields, lead to an extended number of free parameters and thus limit the predictive power of the theory.

In contrast to gauge symmetries, global symmetries (such as isospin) tend to be only approximate and to result from a hierarchy of energy scales. For example, the discrete symmetries of *charge conjugation* C and *parity* P are broken by the weak interaction (Lee and Yang, 1956b; Wu *et al.*, 1957; Ioffe *et al.*, 1957), but not by the electromagnetic and strong interactions. Due to the hierarchy $\Lambda_{\mathrm{QCD}} \ll v$, which is the reason for the weakness of the W- and Z-boson-mediated interaction, C- and P-violating processes are suppressed. In the Standard Model, the origin of C and P violation is the chiral nature of the theory – the fact that left- and right-handed quark or lepton fields transform differently under $\mathrm{SU}(2)_L \times \mathrm{U}(1)_Y$ gauge transformations. As we mentioned earlier in this Intermezzo, this is characteristic of a *chiral gauge theory,* in contrast to a *vector theory* like QCD, where left- and right-handed quark fields occur with the same color charges.

While interactions between gauge and chiral matter fields break C and P invariance, they leave the combined symmetry CP intact (Landau, 1957). The *CP-violating processes* in the minimal Standard Model arise solely from the (modest) mixing among the three quark generations (through the aforementioned Cabibbo–Kobayashi–Maskawa matrix); hence, they are even more suppressed. When extended by right-handed neutrino fields, additional CP-violating effects enter the Standard Model (by means of the Pontecorvo–Maki–Nakagawa–Sakata matrix). These sources of CP violation are not sufficiently strong to explain the *baryon asymmetry* between matter and anti-matter in our Universe.

It is a puzzle – known as the *strong CP-problem* – why the interactions of the gluons respect CP symmetry, *i.e.* why the experimental value of the QCD vacuum-angle θ is compatible with zero ($|\theta| \lesssim \mathcal{O}(10^{-10})$) (Baker *et al.*, 2006; Pendlebury *et al.*, 2015; Abel *et al.*, 2020)). A potential explanation beyond the Standard Model – which, however, lacks experimental confirmation – is related to an approximate global $U(1)_{PQ}$ Peccei–Quinn symmetry (Peccei and Quinn, 1977a,b), which would be associated with an additional light particle, the *axion* (Weinberg, 1978; Wilczek, 1978).

Remarkably, the *CPT theorem* (Lüders, 1954, 1957; Pauli, 1957; Jost, 1957) states that the combination of CP with time reversal T is an exact symmetry of any relativistic (local) field theory. If we treat Lorentz symmetry as a gauge symmetry of General Relativity, then it must be exact, which automatically ensures exact CPT invariance.

Further exact global symmetries are, however, suspicious: By default, they should either be gauged or explicitly broken. In the minimal renormalizable version of the Standard Model, without right-handed neutrinos, the difference between baryon and lepton numbers, $B - L$, is an exact global symmetry. In the SO(10) GUT extension of the Standard Model (Fritzsch and Minkowski, 1975; Georgi, 1975), $U(1)_{B-L}$ is indeed gauged and appears as a subgroup of the SO(10) unified gauge group. On the other hand, in the extended Standard Model with additional right-handed neutrinos, the global $U(1)_{B-L}$ symmetry is explicitly broken if we include Majorana mass terms. Fermion number conservation modulo 2 then still remains as an exact global symmetry. However, just as CPT, this symmetry automatically follows from Lorentz invariance.

Explicit versus Spontaneous Symmetry Breaking

As we emphasized before, gauge symmetries must be exact, whereas global symmetries are in general only approximate. A simple source of *explicit* global symmetry breaking are non-invariant terms in the Lagrangian. A typical example is the $SU(2)_L \times SU(2)_R$ chiral symmetry of QCD, which is explicitly broken, *e.g.,* by distinct Yukawa couplings between the (light) up and down quarks, and the Higgs field.

QCD with *massless u-* and *d*-quarks, however, has an exact chiral symmetry. Interestingly, this symmetry would not manifest itself directly in the QCD spectrum, because it is *spontaneously broken*. This means that, although the Hamilton operator of massless 2-flavor QCD is invariant under $SU(2)_L \times SU(2)_R$ chiral symmetry transformations, its ground states are not. There is a continuous set of degenerate vacuum states of massless QCD, which are related by chiral rotations. The process of *spontaneous symmetry breaking* means that only one of these ground states is actually realized. This state is still invariant against transformations in the unbroken $SU(2)_{L=R}$ isospin subgroup of $SU(2)_L \times SU(2)_R$. Small fluctuations around the spontaneously chosen vacuum state take energy in pro-

Table I.1 Properties of local and global continuous symmetries, depending on their dynamical realization

Dynamical Realization	Local Symmetry	Global Symmetry
exact and unbroken	Coulomb phase: massless gauge bosons	"unnatural" idealization, may imply exact degeneracies in the energy spectrum
exact but spontaneously broken	Higgs phase: massive gauge bosons	"unnatural" idealization, exactly massless Nambu–Goldstone bosons
approximate and not spontaneously broken	mathematically and physically inconsistent	realistic, arises due to large hierarchies of scales, almost degenerate spectrum
approximate and spontaneously broken	mathematically and physically inconsistent	realistic, arises due to large hierarchies of scales, almost massless pseudo-Nambu–Goldstone bosons

portion to the magnitude of their momenta; thus, they manifest themselves as massless particles, the *Nambu–Goldstone bosons*. As a consequence of the spontaneous chiral symmetry breaking, there appear three massless Nambu–Goldstone bosons, the pion isospin triplet π^+, π^0, π^-. In the real world, with non-zero u- and d-quark masses, chiral symmetry is in addition explicitly broken, which turns the pions into light (but no longer massless) pseudo-Nambu–Goldstone bosons.

The pure Higgs sector of the Standard Model also has a global $SU(2)_L \times SU(2)_R$ symmetry. However, in contrast to QCD, its $SU(2)_L \times U(1)_Y$ subgroup is gauged and must therefore be an exact symmetry. Since gauge symmetry just reflects a redundancy in our theoretical description, it cannot break spontaneously in the same way as a global symmetry. When one gauges a spontaneously broken global symmetry, one induces the *Higgs mechanism* in which some gauge fields pick up a mass. The previously massless Nambu–Goldstone degrees of freedom are then transmuted into longitudinal degrees of freedom of the massive gauge bosons. This is indeed the way how the electroweak gauge bosons $W^\pm$ and Z acquire their masses. One says that the gauge bosons "eat" the Nambu–Goldstone bosons and become massive. General properties of local and global symmetries, depending on their realization, are summarized in Table I.1, while concrete examples of important symmetries in particle physics are listed in Table I.2.

Technicolor extensions of the Standard Model mimic QCD at the electroweak scale (Weinberg, 1976, 1979b; Susskind, 1979). In this scenario, electroweak symmetry breaking results from techni-quark–techni-anti-quark condensation, and the $W^\pm$- and Z-bosons become massive because they "eat" the massless techni-pions $\Pi^\pm$ and Π^0. This would indeed solve the hierarchy problem, $v \ll M_{\text{Planck}}$, because $\Lambda_{\text{TC}} \simeq v$ is naturally small thanks to asymptotic freedom. The original idea of technicolor – as an alternative to the Higgs mechanism – is ruled out now, but technicolor is still being investigated as a scenario of a composite Higgs particle. However, this "extended technicolor" approach has a hard time avoiding the prediction of flavor-changing neutral currents, with a strength which is again ruled out by experiment.

Table I.2 Examples of exact local and approximate global symmetries with some physical consequences for different symmetry realizations

Dynamical Realization	Exact Local Symmetries	Approximate Global Symmetries
not spontaneously broken	electromagnetism $U(1)_{em}$: massless photon γ strong interaction $SU(3)_c$: massless confined gluons	isospin $SU(2)_I$: proton and neutron with almost equal masses: $M_p \approx M_n$
spontaneously broken	electroweak interaction $SU(2)_L \times U(1)_Y$: massive gauge bosons $W^{\pm}$ and Z	chiral symmetry in QCD $SU(2)_L \times SU(2)_R$: almost massless pseudo-Nambu–Goldstone pions

Anomalies in Local and Global Symmetries

A more subtle form of explicit symmetry breaking does not manifest itself in the Lagrangian: It occurs only at the quantum level, but not in the classical theory. Whenever quantum effects break a symmetry that is exact at the classical level, one speaks of an *anomaly*.

Theories affected by *gauge anomalies* are mathematically and physically inconsistent, because unphysical (redundant) gauge variables contaminate physical observables via quantum effects. Gauge anomalies must therefore be canceled. Gauge *anomaly cancellation* imposes stringent constraints on chiral gauge theories, including the Standard Model. For example, as a consequence of the cancellation of Witten's so-called *global anomaly* (Witten, 1982), which otherwise would break the $SU(2)_L$ gauge symmetry of the Standard Model, the number of quark colors N_c (which is 3 in Nature) must be odd. In addition, anomaly cancellation determines the electric charge quantization, which will also be discussed in Part II.

In contrast to gauge anomalies, the anomalies in global symmetries do not need to be canceled, and they lead to observable effects. A striking example is *scale invariance*. In the absence of quark masses, the QCD Lagrangian contains only one parameter – the dimensionless gauge coupling g_s. Hence, the action of massless QCD is exactly scale-invariant. This global symmetry is broken by an anomaly, because the quantization of the theory requires the introduction of a dimensioned cut-off – such as the inverse lattice spacing, π/a, in the lattice regularization. Remarkably, even when the cut-off is removed, *i.e.* in the continuum limit $a \to 0$, the fundamental scale Λ_{QCD} emerges as a natural constant via the process of *dimensional transmutation*, which can be traced back to logarithmically divergent 1-loop corrections.

Another anomaly breaks the flavor-singlet axial $U(1)_A$ symmetry of the QCD Lagrangian with massless quarks (Adler, 1969; Bell and Jackiw, 1969). This quantum effect generates a large mass for the η'-meson. Only in the limit $N_c \to \infty$, which suppresses the $U(1)_A$ anomaly, the η'-meson would become a (massless) Nambu–Goldstone boson ('t Hooft, 1976a,b; Crewther, 1977; Witten, 1979b; Veneziano, 1979; 't Hooft, 1986). The fact that the η'-meson is amazingly heavy in Nature (heavier than a nucleon) is then explained as a topological effect in the $1/N_c$-expansion.

Yet another anomaly of electromagnetic origin breaks the discrete global *G-parity* invariance of QCD, which conserves the number of pions modulo 2. As a consequence, a single neutral pion can decay into two photons, $\pi^0 \to \gamma\gamma$ (Adler, 1969; Bell and Jackiw, 1969). This quantum effect changes the number of pions from one to zero and thus breaks G-parity anomalously. In contrast to many textbooks, we will point out that, in a gauge-anomaly-free Standard Model with N_c quark colors, the width of the neutral pion, associated with the decay into two photons, is not proportional to N_c^2, but is actually N_c-independent (Gerard and Lahna, 1995; Abbas, 2000; Bär and Wiese, 2001).

Examples of how some important global symmetries in particle physics are broken, either at the classical or at the quantum level, are provided in Table I.3. There we distinguish two types of symmetry breaking at the quantum level: anomalous breaking due to the functional integral and breaking through a "quantum" parameter (such as the vacuum-angle θ) in the Lagrangian. The corresponding terms in the Lagrangian are suppressed relative to the "classical" terms by a factor of $\hbar$. As a consequence, they do not contribute to the classical equations of motion and manifest themselves only at the quantum level. It is interesting to note that one and the same phenomenon, *e.g.*, the violation of the $U(1)_B$ baryon number symmetry at the quantum level – via so-called electroweak instantons – manifests itself in two ways: either in the fermionic measure of the Standard Model or directly in the low-energy effective Lagrangian via a topological term (known as the Wess–Zumino–Novikov–Witten term). In the latter case, the corresponding "quantum" parameter is again the number N_c of quark colors.

Power of Lattice Field Theory

Many aspects of quantum field theory and of the Standard Model in particular can be understood in the framework of the perturbative expansion, which is most elegantly realized using dimensional regularization. However, there are also important non-perturbative phenomena, in particular associated with the strong interaction at low energies. These include confinement, *i.e.* the absence of color non-singlet states, such as individual quarks or gluons, as well as spontaneous chiral symmetry breaking.

Lattice field theory, in particular *lattice Quantum Chromodynamics* – the lattice gauge theory of quark and gluon fields – provides us with a powerful tool to quantitatively address the non-perturbative QCD dynamics from first principles. In this approach, continuous Euclidean space–time is replaced by a regular 4-dimensional grid of lattice points. The quark fields then reside at the lattice sites, while the gluon vector fields are naturally associated with the links connecting neighboring lattice sites.

It should be stressed that the Lagrangian of the continuum theory is not sufficient to define QCD at the non-perturbative level. In particular, the perturbative expansion in terms of Feynman diagrams is known to be a divergent, asymptotic series. Hence, even perturbation theory to all orders does not define the theory non-perturbatively. Lattice field theory, on the other hand, does provide us with a non-perturbative definition of the theory. Consequently,

Table I.3 Preview on some phenomena related to explicit global symmetry breaking: examples of approximate, continuous or discrete global symmetries and some of their physical consequences, with a classical or quantum origin of the symmetry breaking

Nature of Breaking	Continuous Symmetries	Discrete Symmetries
explicitly broken at the Lagrangian level by a "classical" parameter	isospin $SU(2)_I$ in QCD coupled to QED broken by unequal quark masses $\Rightarrow$ neutron-proton mass difference $M_n > M_p$; also broken by unequal quark charges $\Rightarrow$ different pion masses $M_{\pi^\pm} > M_{\pi^0}$ (to be discussed in Chapter 23)	$\mathbb{Z}(3)$ center symmetry of pure gluon theory broken by quark fields in QCD $\Rightarrow$ string breaking by quark–anti-quark creation (to be discussed in Chapter 14)
anomalously broken at the quantum level via the measure of the fermionic functional integral	axial $U(1)_A$ broken in $N_c = 3$ QCD $\Rightarrow$ heavy η'-meson, $M_{\eta'} \gg M_\pi, M_K, M_\eta$ (to be discussed in Chapter 20) baryon number $U(1)_B$ in Standard Model broken by electroweak instantons $\Rightarrow$ proton decay to pion-positron $p \rightarrow \pi^0 + e^+$ (to be discussed in Chapter 15)	$\mathbb{Z}(2)$ G-parity in QCD coupled to QED broken by electromagnetism $\Rightarrow$ pion decay to two photons $\pi^0 \rightarrow \gamma\gamma$ (to be discussed in Chapter 24)
explicitly broken at the Lagrangian level by a "quantum" parameter	baryon number $U(1)_B$ in effective theory broken by topological WZNW-term $\Rightarrow$ skyrmion decay by electroweak instantons (to be discussed in Chapter 24)	$\mathbb{Z}(2)$ symmetries P and CP in QCD broken by vacuum angle $\theta \neq 0 \Rightarrow$ non-zero neutron electric dipole moment (to be discussed in Chapter 25)

lattice QCD should not be viewed as an approximation to any pre-existing continuum theory, but rather as a non-perturbative *definition* of the theory, from which the continuum theory can be defined by extrapolating the lattice spacing $a \rightarrow 0$.

Although the existence of the continuum limit has not been established rigorously, there is overwhelming numerical evidence from large-scale numerical Monte Carlo simulations that it indeed exists and agrees with experimental results. Lattice field theory is the only known, rigorous method that allows us to quantitatively access the low-energy QCD dynamics from first principles. Numerous important results, ranging from the hadron spectrum of baryons, mesons, and glueballs, to the properties of QCD at high temperatures have been obtained in this way. For instance, the light hadron masses have now been computed by lattice simulations up to percent-level uncertainties, and the results are consistent with the experimental values. Just as the Particle Data Group (PDG) (Zyla *et al.*, 2020) provides regular summaries of particle properties obtained from experiment, the Flavor Lattice Averaging Group, FLAG (Aoki *et al.*, 2020), compiles and reanalyzes numerical results obtained worldwide with lattice QCD.

Lattice QCD is sometimes associated only with its computational aspects, which are technically non-trivial and require large computational resources. However, while Monte Carlo simulations are of enormous practical importance, first and foremost lattice field theory provides a solid conceptual basis for quantum field theory. For example, at large values of the bare strong coupling constant, lattice QCD enables analytic understanding of confinement (Wilson, 1974). Whether confinement persists in the continuum, which is approached in the limit of weak bare coupling, is a very challenging, mathematical question. Again, there is overwhelming numerical evidence that this is indeed the case.

For some time, lattice field theory had a bad reputation, because it had problems with regularizing the chiral symmetries associated with massless left- and right-handed fermions. In fact, when the Dirac theory is naively discretized on a lattice, there is a multiplication of fermion species known as the *fermion doubling problem* (cf. Chapter 10). For a long time, it seemed that chiral fermions, such as neutrinos, cannot exist on the lattice. In fact, the Nielsen–Ninomiya no-go theorem (Nielsen and Ninomiya, 1981a), which is based on a few plausible assumptions, suggested that lattice field theory cannot regularize fermions with an exact chiral symmetry. While this is not disastrous for QCD, which has only an approximate global chiral symmetry, it is a very serious problem for chiral gauge theories including the Standard Model, for which chiral symmetry is an exact gauge symmetry.

In his original formulation of lattice QCD, using what is now known as Wilson fermions, Kenneth Wilson eliminated the doubler fermions by breaking chiral symmetry explicitly (Wilson, 1977). Later, together with Paul Ginsparg, he derived a relation that allows the construction of exact chiral symmetries on the lattice (Ginsparg and Wilson, 1982). However, the full power of this relation was realized only much later.

In the meantime, in 1992 David Kaplan made a groundbreaking discovery (Kaplan, 1992). Inspired by earlier work of Valery Rubakov and Mikhail Shaposhnikov and of Curtis Callan and Jeffrey Harvey in the continuum (Rubakov and Shaposhnikov, 1983; Callan and Harvey, 1985), he realized that massless 4-dimensional chiral fermions naturally emerge at a domain wall. The domain wall separates regions in a 5-dimensional space–time with different values of a scalar field, to which the fermion field is coupled by a Yukawa coupling. Remarkably, the localization of massless chiral fermions on a domain wall defect in a higher-dimensional space–time is a robust topological effect that persists in the lattice regularization.

Later it became clear that a limit of domain wall fermions – the so-called overlap fermions (Narayanan and Neuberger, 1993a,b) – as well as "perfect" lattice fermions (Bietenholz and Wiese, 1996b) obey the Ginsparg–Wilson relation, which lay dormant until then. In this way, the power of the Ginsparg–Wilson relation was realized and fully unleashed by Hasenfratz (1998); Hasenfratz *et al.* (1998); Neuberger (1998); and Lüscher (1998), when it became clear that one assumption underlying the Nielsen–Ninomiya no-go theorem could be evaded by modifying the form of lattice chiral symmetry. This opened the floodgates to a revolution of our non-perturbative understanding of chiral symmetry, both as a global approximate and as a local exact symmetry.

Any bad reputation that the lattice regularization may have had in the past due to its problems with chiral symmetry is certainly no longer justified. Thanks to Martin Lüscher's construction of chiral gauge theories on the lattice (Lüscher, 1999, 2002), which indeed represents a spectacular breakthrough, the non-perturbative understanding even extends to local chiral symmetry. The presentation of Lüscher's construction is beyond the scope of this book, and we refer to his papers. At the moment, a complete non-perturbative construction exists for Abelian chiral gauge theories. Due to complicated topological issues, related to the cohomology of the space of admissible lattice gauge field configurations, the construction of non-Abelian chiral gauge theories, including the Standard Model, is at the moment restricted to all orders in lattice perturbation theory (Lüscher, 2000). In other words, the cohomology problem that was solved for BRST continuum theories (Brandt *et al.*, 1990b) is still open for non-Abelian chiral lattice gauge theories. Trying to solve this very important open problem should be a high priority for the mathematical physics community.

Since the Dirac matrix γ_5 does not have any analytic continuation to general complex values of the space–time dimension, in the context of dimensional regularization there is no manifestly gauge-invariant formulation of chiral gauge theories (for which the fermionic chiral symmetry is local). In the continuum, all that exist are *ad hoc* prescriptions concerning γ_5 that restore gauge invariance at the 1- and in some cases at the 2-loop level. Lüscher's lattice construction of chiral gauge theories, on the other hand, is manifestly gauge-invariant and defines chiral gauge theories to all orders of perturbation theory. As a result, after understanding the deep problems related to chiral symmetry, which are manifest on the lattice, the lattice regularization of chiral gauge theories is superior to continuum regularization schemes. It substantially deepens our understanding of chiral gauge theories, for non-Abelian theories to all orders of the perturbative expansion and for Abelian gauge theories even at the non-perturbative level.

The investigation of chiral symmetry on the lattice reveals an under-appreciated, non-perturbative hierarchy problem associated with chiral symmetry. In fact, the original Wilson fermion required unnatural fine-tuning to reach a chirally invariant continuum limit. Realizing the Ginsparg–Wilson relation in four space–time dimensions may also require fine-tuning. However, this relation emerges naturally in Kaplan's construction of 5-dimensional domain wall fermions. We may wonder how Nature manages to produce chiral fermions, and domain wall lattice fermions may perhaps inspire us to contemplate the physical reality of an additional spatial dimension, because this could explain why fermions can be light without fine-tuning (although an extra dimension is not a necessary prerequisite to construct solutions to the Ginsparg–Wilson relation).

In the rest of this book, we will refer to the lattice regularization not only for important Monte Carlo results, but whenever it seems appropriate, in particular, because it provides us with deep understanding or because it simplifies the presentation. Still, most sections use the continuum field theory formulation.

12

Spontaneous Breakdown of Global Symmetries: From Condensed Matter to Higgs Bosons

In this chapter, we introduce the *scalar sector of the Standard Model*. Even without gauge fields or fermions (leptons and quarks) involved, there is interesting physics of the *Higgs field* alone. In the Higgs sector of the Standard Model, a global $SU(2)_L \times SU(2)_R \cong O(4)$ symmetry *breaks spontaneously* down to the subgroup $SU(2)_{L=R} \cong O(3)$.[1] According to the *Goldstone theorem*, this gives rise to three massless *Nambu–Goldstone bosons*. Once electroweak gauge fields are included in the Standard Model (which will be done in Chapter 13), the gauge bosons $W^{\pm}$ and Z become massive due to the *Higgs mechanism* by incorporating the Nambu–Goldstone bosons as longitudinal polarization degrees of freedom. The photon, on the other hand, remains massless, as a consequence of the *unbroken* $U(1)_{em}$ gauge symmetry of electromagnetism.

The analogy between the Higgs mechanism and the physics of superconductors was explained long ago by Philip Anderson (Anderson, 1963). In particular, the scalar field describing the *Cooper pairs* of electrons in a superconductor is a condensed matter analogue of the Higgs field in particle physics. When Cooper pairs condense, even the $U(1)_{em}$ gauge symmetry "breaks spontaneously" (as before, we use inverted commas to indicate that a gauge symmetry cannot really break) and, consequently, the photon then also becomes massive. However, in this chapter we do not yet include gauge fields. Instead, we concentrate on the dynamics of the scalar fields alone.

The spontaneous breakdown of continuous *global* symmetries also plays an important role in condensed matter physics. For example, in ultra-cold *Bose–Einstein condensates*, as well as in *superfluids*, a global $U(1)$ symmetry, which is responsible for particle number conservation, is spontaneously broken. In superfluid ^{3}He, Cooper pairs of neutral ^{3}He atoms form. Their condensation gives rise to superfluidity.

Obtaining an elementary Higgs particle with a renormalized mass much smaller than the Planck scale requires an unnatural fine-tuning of the bare mass. This is known as the *hierarchy problem*. A similar problem does not arise for composite Cooper pairs in condensed matter physics. Hence, one may wonder whether the physical Higgs particle could also be composite. This scenario has been suggested in theoretical approaches beyond the Standard Model, in particular in technicolor models, to be addressed in Section 15.12.

Another feature of the scalar sector of the Standard Model is its *triviality*. If one insists on completely removing the cut-off in the scalar quantum field theory describing the Higgs sector of the Standard Model, then the theory becomes non-interacting. As a result, the Standard Model can only be a *low-energy effective field theory*. However, the term "low energy" does not exclude that it could, at least in principle, be valid all the way up to the

---

[1] The symbol $\cong$ denotes a *local isomorphism* between two manifolds, which may still differ in their global topology.

Planck scale (where it necessarily breaks down because there quantum gravity is expected to be strongly coupled and can no longer be neglected).

12.1 Effective Scalar Fields for Cold Condensed Matter

In addition to the Higgs field in particle physics, scalar fields appear in low-energy effective theories of condensed matter. Here, we sketch how one arrives at such theories, which focus on collective phenomena, rather than keeping track of microscopic details.

Let us consider scalar particles (with spin 0). For example, we can think of the *Cooper pairs* of electrons in a *superconductor*. According to the BCS theory developed by John Bardeen, Leon Cooper, and Robert Schrieffer (Bardeen *et al.* (1957), see also Bogoliubov *et al.* (1958); de Gennes (1999)), electrons of opposite momenta near the Fermi surface[2] form bound pairs with a large spatial extent (up to $\approx 1000\,\text{Å}$). In ordinary superconductors, at temperatures up to a few Kelvin (K), the Coulomb repulsion between electrons is overcome by an attractive interaction mediated by phonon exchange (*i.e.* by couplings to the vibrations of the crystal lattice of ions). The resulting Cooper pairs form in the s-wave channel and have spin 0.[3] Hence, at energy scales well below the binding energy of a Cooper pair (*i.e.* below the energy gap of the superconductor), they can effectively be described by a scalar field.

Since we do not yet couple the scalar field to a gauge field, strictly speaking, the condensed matter systems that are analogous to the pure Higgs sector of the Standard Model are *superfluids*, with a spontaneously broken, *global* U(1) symmetry, rather than superconductors with a "spontaneously broken" U(1)$_{em}$ gauge symmetry of electromagnetism. In superfluid ^{4}He, the relevant atomic objects are bosons, consisting of an α-particle (two protons and two neutrons forming the atomic nucleus) and two electrons. When these bosons condense, at temperatures below 2.17 K, the global U(1) symmetry that describes the conserved boson number breaks spontaneously. Then, a macroscopic number of bosonic ^{4}He atoms occupy the same quantum state.

Superfluid ^{3}He, on the other hand, consists of fermionic atoms: a helium nucleus with two protons but only one neutron and two electrons. Before the fermionic ^{3}He atoms can condense, they must form bosonic bound states which are also denoted as Cooper pairs.[4] This happens at very low temperatures, below 2.6 mK. The Cooper pairs of superfluid ^{3}He form a spin triplet in the p-wave channel. Unlike in a superconductor, the Cooper pairs in superfluid ^{3}He are electrically neutral. Hence, their Bose condensation again leads to the spontaneous breaking of the global U(1) symmetry, corresponding to particle number conservation, but not of the local U(1)$_{em}$ symmetry of electromagnetism.

Another example for spontaneous symmetry breaking results from the *Bose–Einstein condensation* (BEC) of ultra-cold atomic gases, at temperatures in the nK range. Condensed

---

[2] The Fermi surface is a surface of constant energy in the momentum space of the electrons in a solid. All states with lower energies are occupied at zero temperature.

[3] The mechanism for binding Cooper pairs in *high-temperature superconductors*, which also have spin 0 but form in the d-wave channel, is not yet understood.

[4] Unpaired fermions cannot condense because, according to the Pauli principle, at most one fermion can occupy a given quantum state.

bosonic atoms can also be described by a complex scalar field, whose phase is affected by global U(1) symmetry transformations that are associated with the conserved particle number. Fermionic atomic gases, on the other hand, can condense only when two fermionic atoms form a bosonic molecule (of small spatial extent) or a large Cooper pair. By manipulating the scattering length of the fermionic atoms using Feshbach resonances (Feshbach, 1958), the BEC-BCS crossover from Bose–Einstein condensation to Cooper pair formation has been studied in detail (Chin *et al.*, 2010). When considered at sufficiently long distance scales, the bosonic pairs of fermionic atoms can again be described by a complex scalar field.

While the aforementioned condensed matter systems are non-relativistic, in the following we are going to consider relativistic scalar field dynamics (except for some remarks in Section 12.4).

12.2   Vacua in the $\lambda |\Phi|^4$ Model

As we saw in Section 7.1, a charged spinless particle is described by a complex scalar field $\Phi(x) = \phi_1(x) + i\phi_2(x) \in \mathbb{C}$ ($\phi_i(x) \in \mathbb{R}$), with the covariant Lagrangian (in Euclidean space–time)

$$\mathcal{L}(\Phi) = \frac{1}{2}\partial_\mu \Phi^* \partial_\mu \Phi + V(\Phi) = \frac{1}{2}\partial_\mu \phi_1 \partial_\mu \phi_1 + \frac{1}{2}\partial_\mu \phi_2 \partial_\mu \phi_2 + V(\phi_1, \phi_2),$$

$$V(\Phi) = \frac{m^2}{2}|\Phi|^2 + \frac{\lambda}{4!}|\Phi|^4, \quad |\Phi|^2 = \Phi^* \Phi = \phi_1^2 + \phi_2^2. \tag{12.1}$$

The Lagrangian of eq. (12.1) has a global symmetry: It is invariant under U(1) transformations

$$\Phi'(x) = \exp(\mathrm{i}Qe\alpha)\,\Phi(x) \quad \Leftrightarrow \quad \Phi'(x)^* = \exp(-\mathrm{i}Qe\alpha)\,\Phi^*(x), \tag{12.2}$$

where $\alpha \in \mathbb{R}$ is a space–time-independent phase.[5]

We proceed with a discussion of spontaneous symmetry breaking, which is essentially classical; it does not necessarily reveal the true nature of the quantum ground state.

To make sure that the potential is bounded from below, we assume the coupling λ to be strictly positive, $\lambda > 0$. We can, however, choose $m^2 \geq 0$ or $m^2 < 0$, and we distinguish the two cases for its sign:

- For $m^2 \geq 0$, the potential has a single minimum at $\Phi = 0$. The configuration that solves the classical Euclidean field equation of least action, the *classical vacuum*, is simply the constant field $\Phi(x) \equiv 0$. This vacuum configuration is invariant against

---

[5] At this point, it is not yet obvious why the phase comes with a coefficient $Qe \in \mathbb{R}$. The reason is that once the U(1) symmetry is gauged, the parameter e will be identified as the fundamental electric charge unit of the field Φ with charge $Q \in \mathbb{Z}$.

the U(1) transformations (12.2). Hence, in this case the U(1) symmetry is preserved; it is not spontaneously broken.

- For $m^2 < 0$, the trivial configuration $\Phi(x) = 0$ is unstable because it corresponds to a (local) maximum of the potential. The condition for a minimum now reads

$$\frac{\partial V}{\partial \Phi} = m^2 \Phi + \frac{\lambda}{3!} |\Phi|^2 \Phi = 0, \quad \Phi \neq 0 \quad \Rightarrow \quad |\Phi|^2 = -\frac{6m^2}{\lambda}. \tag{12.3}$$

In this case, the vacuum is no longer unique. Instead, there is a whole class of degenerate vacua

$$\Phi(x) = v \exp(i\varphi), \quad v = \sqrt{-\frac{6m^2}{\lambda}}, \tag{12.4}$$

parameterized by a phase[6] $\varphi \in] - \pi, \pi]$. The quantity v is the *vacuum expectation value*, or just vacuum value, of the field Φ.[7]

Let us choose the vacuum state with $\varphi = 0$. Of course, such a choice breaks the U(1) symmetry. Hence, in this case the global symmetry is *spontaneously broken*. There is no non-trivial subgroup of the U(1) symmetry that leaves the vacuum configuration invariant. Hence, the U(1) symmetry is spontaneously broken down to the trivial subgroup {1} (which only consists of the unit-element). Expanding around the spontaneously selected minimum of the potential, one obtains

$$\Phi(x) = v + \sigma(x) + i\pi(x) \quad \Rightarrow \quad \Phi(x)^* = v + \sigma(x) - i\pi(x),$$
$$|\Phi(x)|^2 = [v + \sigma(x)]^2 + \pi(x)^2,$$
$$\partial_\mu \Phi(x) = \partial_\mu \sigma(x) + i\partial_\mu \pi(x) \quad \Rightarrow \quad \partial_\mu \Phi(x)^* = \partial_\mu \sigma(x) - i\partial_\mu \pi(x). \tag{12.5}$$

We now want to express the Lagrangian in terms of the new, real-valued fields σ and π, which describe fluctuations around the vacuum configuration $\Phi(x) = v$ that we selected. We capture the low-energy physics – *i.e.* the dominant contributions to the functional integral – by expanding up to second order in σ and π,

$$\frac{1}{2} \partial_\mu \Phi^* \partial_\mu \Phi = \frac{1}{2} \partial_\mu \sigma \partial_\mu \sigma + \frac{1}{2} \partial_\mu \pi \partial_\mu \pi,$$

$$V(\Phi) = \frac{m^2}{2} (v + \sigma)^2 + \frac{m^2}{2} \pi^2 + \frac{\lambda}{4!} \left[(v + \sigma)^2 + \pi^2 \right]^2$$

$$\approx \frac{m^2}{2} v^2 + m^2 v\sigma + \frac{m^2}{2} \sigma^2 + \frac{m^2}{2} \pi^2 + \frac{\lambda}{4!} \left(v^4 + 4v^3 \sigma + 6v^2 \sigma^2 + 2v^2 \pi^2 \right)$$

$$= \frac{1}{2} \left(m^2 + \frac{\lambda}{2} v^2 \right) \sigma^2 + C, \tag{12.6}$$

where C is an irrelevant additive constant, and we assume $|\sigma(x)|, |\pi(x)| \ll v$. We interpret the term proportional to σ^2 as a *mass term* for the σ-field. Hence, the corresponding σ-*particle* has a mass squared

$$m_\sigma^2 = m^2 + \frac{\lambda}{2} v^2 = \frac{\lambda}{3} v^2 = -2m^2 > 0. \tag{12.7}$$

---

[6] As before, the notation implies that the interval includes the boundary value $\varphi = \pi$, but excludes $\varphi = -\pi$.
[7] After renormalization, this distinction of cases refers to the renormalized parameters m_r^2 and λ_r.

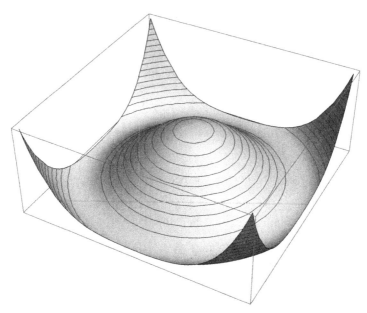

Fig. 12.1   Illustration of the potential $V(\Phi)$ of the $\lambda|\Phi|^4$ model in the case $m^2 < 0$. The set of degenerate minima form a circle of radius $|\Phi| = v = \sqrt{-6m^2/\lambda}$. The tangential fluctuations along the circle of vacua correspond to the massless Nambu–Goldstone boson, while radial excitations correspond to the massive σ-particle.

Since there is no term proportional to π^2, the corresponding π-*particle* is *massless*, $m_\pi = 0$. This massless particle is a *Nambu–Goldstone boson* (Nambu and Jona-Lasinio, 1961a,b; Goldstone, 1961). The situation is sketched in Figure 12.1.

The emergence of a Nambu–Goldstone boson is characteristic for the spontaneous breaking of a continuous, global symmetry. The Goldstone theorem, which determines the number of Nambu–Goldstone bosons (one in case of a spontaneously broken U(1) symmetry), will be discussed in Section 12.4. Once the U(1) symmetry is gauged, which will be done in Chapter 13, the Nambu–Goldstone boson turns into the longitudinal polarization state of a gauge boson, which then becomes massive. For example, in a superconductor the "spontaneously broken" symmetry is the U(1)$_{\text{em}}$ gauge symmetry of electromagnetism. In that case, the photon becomes massive.

12.3   Higgs Doublet Model

We now proceed to the Higgs sector of the Standard Model. Again a scalar field Φ plays a central role. However, the field Φ now has two complex components; it is a *complex doublet*. Thus, we deal with the *Higgs field*[8]

$$\Phi(x) = \begin{pmatrix} \Phi^+(x) \\ \Phi^0(x) \end{pmatrix}, \quad \Phi^+(x), \Phi^0(x) \in \mathbb{C}. \tag{12.8}$$

---

[8] The superscripts $+$ and 0 will later turn out to correspond to electric charges.

We follow the structure of the previous section and discuss a $\lambda|\Phi|^4$ model with a *global* symmetry,

$$\mathcal{L}(\Phi) = \frac{1}{2}\partial_\mu\Phi^\dagger\partial_\mu\Phi + V(\Phi),$$

$$V(\Phi) = \frac{m^2}{2}|\Phi|^2 + \frac{\lambda}{4!}|\Phi|^4, \quad |\Phi|^2 = \Phi^\dagger\Phi = \Phi^{+*}\Phi^+ + \Phi^{0*}\Phi^0. \quad (12.9)$$

This Lagrangian is invariant under a class of SU(2) transformations, which we denote as SU(2)$_L$,

$$\Phi'(x) = L\Phi(x), \quad L \in \text{SU}(2)_L. \quad (12.10)$$

We recall that SU(2) is the group of unitary 2×2 matrices with determinant 1,

$$L^\dagger = L^{-1}, \quad \det L = 1. \quad (12.11)$$

A general SU(2) matrix L can be written in terms of complex numbers z_1 and z_2, with $|z_1|^2 + |z_2|^2 = 1$,

$$L = \begin{pmatrix} z_1 & -z_2^* \\ z_2 & z_1^* \end{pmatrix} \quad \Rightarrow$$

$$L^\dagger = \begin{pmatrix} z_1^* & z_2^* \\ -z_2 & z_1 \end{pmatrix}, \quad L^\dagger L = \mathbb{1}, \quad \det L = |z_1|^2 + |z_2|^2 = 1. \quad (12.12)$$

This representation shows that the space of SU(2) matrices is isomorphic to the 3-dimensional unit-sphere S^3. The global SU(2)$_L$ invariance of the Lagrangian is readily confirmed,

$$|\Phi'(x)|^2 = \Phi'(x)^\dagger\Phi'(x) = [L\Phi(x)]^\dagger L\Phi(x) = \Phi(x)^\dagger L^\dagger L\Phi(x) = |\Phi(x)|^2,$$

$$\partial_\mu\Phi'(x)^\dagger\partial_\mu\Phi'(x) = \partial_\mu\Phi(x)^\dagger L^\dagger L\,\partial_\mu\Phi(x) = \partial_\mu\Phi(x)^\dagger\partial_\mu\Phi(x). \quad (12.13)$$

In addition to the SU(2)$_L$ symmetry, there is the so-called U(1)$_Y$ hypercharge symmetry, which acts as

$$\Phi'(x) = \exp\left(i\frac{1}{2}g'\varphi\right)\Phi(x). \quad (12.14)$$

Just like the SU(2)$_L$ symmetry, at this point we treat the U(1)$_Y$ hypercharge as a global symmetry. Once these symmetries will be gauged, *i.e.* made local, the parameter g' will be identified as the gauge coupling strength of the U(1)$_Y$ gauge field, and the prefactor $1/2 = Y_\Phi$ will be identified as the weak hypercharge of the Higgs field.

One might feel tempted to interpret the U(1)$_Y$ symmetry as a determinant of $L \in \text{U}(2)_L$, but the following interpretation is favorable. Actually, the global symmetry is even larger than the group SU(2)$_L \times$ U(1)$_Y$ identified so far: The action is indeed invariant under an extended group SU(2)$_L \times$ SU(2)$_R$, with U(1)$_Y$ being an Abelian subgroup of SU(2)$_R$. In order to make the additional $SU(2)_R$ symmetry evident, we introduce another notation for the same Higgs field by rewriting it as a matrix,

$$\mathbf{\Phi}(x) = \begin{pmatrix} \Phi^0(x)^* & \Phi^+(x) \\ -\Phi^+(x)^* & \Phi^0(x) \end{pmatrix}. \tag{12.15}$$

We see that the matrix field $\mathbf{\Phi}$ belongs to $SU(2)$, up to a scale factor $\sqrt{\det\mathbf{\Phi}}$ (provided that $\mathbf{\Phi}$ is non-zero). In this notation, the Lagrangian (12.9) takes the form

$$\mathcal{L}(\mathbf{\Phi}) = \frac{1}{4}\mathrm{Tr}\left[\partial_\mu\mathbf{\Phi}^\dagger\partial_\mu\mathbf{\Phi}\right] + \frac{m^2}{4}\mathrm{Tr}\left[\mathbf{\Phi}^\dagger\mathbf{\Phi}\right] + \frac{\lambda}{4!}\left(\frac{1}{2}\mathrm{Tr}\left[\mathbf{\Phi}^\dagger\mathbf{\Phi}\right]\right)^2, \tag{12.16}$$

which is manifestly invariant under the global transformations

$$\mathbf{\Phi}(x)' = L\mathbf{\Phi}(x)R^\dagger, \quad L \in SU(2)_L, \quad R \in SU(2)_R. \tag{12.17}$$

The $SU(2)_R$ symmetry is known as the *custodial symmetry*, a term which will be explained in Section 13.4. The special case

$$R = \begin{pmatrix} \exp\left(\mathrm{i}g'\varphi/2\right) & 0 \\ 0 & \exp\left(-\mathrm{i}g'\varphi/2\right) \end{pmatrix} \tag{12.18}$$

identifies $U(1)_Y$ as a subgroup of $SU(2)_R$.

As a third alternative, we introduce the notation with four real components,

$$\vec{\phi}(x) = (\phi_1(x), \phi_2(x), \phi_3(x), \phi_4(x)) \in \mathbb{R}^4,$$
$$\Phi^+(x) = \phi_2(x) + \mathrm{i}\,\phi_1(x), \quad \Phi^0(x) = \phi_4(x) - \mathrm{i}\,\phi_3(x),$$
$$\mathbf{\Phi}(x) = \phi_4(x)\,\mathbb{1} + \mathrm{i}\left[\phi_1(x)\tau^1 + \phi_2(x)\tau^2 + \phi_3(x)\tau^3\right], \tag{12.19}$$

where τ^1, τ^2, and τ^3 are the *Pauli matrices*[9]

$$\tau^1 = \begin{pmatrix} 0 & 1 \\ 1 & 0 \end{pmatrix}, \quad \tau^2 = \begin{pmatrix} 0 & -\mathrm{i} \\ \mathrm{i} & 0 \end{pmatrix}, \quad \tau^3 = \begin{pmatrix} 1 & 0 \\ 0 & -1 \end{pmatrix}. \tag{12.20}$$

They are Hermitian, $\tau^{a\dagger} = \tau^a$, and they obey the commutation and anti-commutation relations

$$[\tau^a, \tau^b] = 2\mathrm{i}\,\epsilon_{abc}\tau^c, \quad \{\tau^a, \tau^b\} = 2\delta_{ab}\mathbb{1}. \tag{12.21}$$

In the vector notation, the Lagrangian takes the form

$$\mathcal{L}(\vec{\phi}) = \frac{1}{2}\partial_\mu\vec{\phi}\cdot\partial_\mu\vec{\phi} + \frac{m^2}{2}\vec{\phi}\cdot\vec{\phi} + \frac{\lambda}{4!}\left(\vec{\phi}\cdot\vec{\phi}\right)^2. \tag{12.22}$$

This Lagrangian is manifestly O(4)-invariant under global rotations of the 4-component vector $\vec{\phi}$. This is precisely in agreement with the local isomorphism $SU(2)_L \times SU(2)_R \cong$ O(4). We also see now from two perspectives that the global symmetry group has, in total, six generators – no generator has been forgotten or doubly counted.

---

[9] For the Pauli matrices, we use the notation σ^a when they refer to the fermionic spin and τ^a otherwise.

As before, we distinguish between the symmetric and the spontaneously broken phase:

- At the classical level, for $m^2 \geq 0$ there is a unique vacuum field configuration

$$\Phi(x) = \begin{pmatrix} 0 \\ 0 \end{pmatrix}. \tag{12.23}$$

In this case, the $SU(2)_L \times SU(2)_R \cong O(4)$ symmetry is unbroken.
- For $m^2 < 0$, the vacuum is degenerate. It obeys the condition

$$|\Phi|^2 \equiv |\vec{\phi}|^2 \equiv \frac{1}{2} \mathrm{Tr} \left[\Phi^\dagger \Phi \right] = -\frac{6m^2}{\lambda}, \tag{12.24}$$

in analogy to eq. (12.3). Here, we choose

$$\Phi(x) = \begin{pmatrix} 0 \\ v \end{pmatrix}, \quad v = \sqrt{-\frac{6m^2}{\lambda}} \in \mathbb{R}_+, \tag{12.25}$$

which corresponds to

$$\Phi(x) = v\, \mathbb{1}, \quad \vec{\phi}(x) = (0, 0, 0, v). \tag{12.26}$$

This vacuum configuration is not invariant under general $SU(2)_L \times SU(2)_R$ transformations. It *is*, however, invariant under those transformations that obey $L = R$; they belong to the so-called vector subgroup $SU(2)_{L=R}$. Hence, in this case the $SU(2)_L \times SU(2)_R \cong O(4)$ symmetry is spontaneously broken down to the diagonal vector subgroup $SU(2)_{L=R} \cong O(3)$ which remains unbroken. (The $U(1)_{em}$ symmetry of electromagnetism will later be identified as a subgroup of $SU(2)_{L=R}$.)

Let us again expand around the selected vacuum configuration,

$$\Phi(x) = \begin{pmatrix} \pi_1(x) + i\pi_2(x) \\ v + \sigma(x) + i\pi_3(x) \end{pmatrix}, \quad \sigma(x), \pi_i(x) \in \mathbb{R}. \tag{12.27}$$

To second order in the fluctuation fields σ and $\vec{\pi} = (\pi_1, \pi_2, \pi_3)$, we obtain

$$\frac{1}{2}\partial_\mu \Phi^\dagger \partial_\mu \Phi = \frac{1}{2}\partial_\mu \sigma \partial_\mu \sigma + \frac{1}{2}\partial_\mu \vec{\pi} \cdot \partial_\mu \vec{\pi},$$

$$V(\Phi) = \frac{m^2}{2} \left[(v + \sigma)^2 + \vec{\pi}^{\,2} \right] + \frac{\lambda}{4!} \left[(v + \sigma)^2 + \vec{\pi}^{\,2} \right]^2$$

$$\approx \frac{m^2}{2} \left[v^2 + 2v\sigma + \sigma^2 + \vec{\pi}^{\,2} \right] + \frac{\lambda}{4!} \left[v^4 + 4v^3\sigma + 6v^2\sigma^2 + 2v^2\vec{\pi}^{\,2} \right]$$

$$= \frac{1}{2}\left(m^2 + \frac{\lambda}{2}v^2 \right) \sigma^2 + C, \tag{12.28}$$

where C is another irrelevant additive constant. Once more, we find a massive σ-particle; here, we denote its mass as m_H,

$$m_H^2 = m^2 + \frac{\lambda}{2}v^2 = \frac{\lambda}{3}v^2, \tag{12.29}$$

and in this case *three* massless Nambu–Goldstone bosons π_1, π_2, and π_3. The massive σ-particle – a quantized fluctuation of the σ field – is known as the *Higgs particle*. While the

Higgs particle is a singlet, the three Nambu–Goldstone bosons transform as a triplet under the unbroken $SU(2)_{L=R} \cong O(3)$ symmetry.

In 2012, the experiments ATLAS and CMS at the Large Hadron Collider (LHC) at CERN discovered the much-anticipated Higgs particle and its mass amounts to (Zyla *et al.*, 2020)

$$m_H = 125.18(16) \, \text{GeV}. \tag{12.30}$$

If physics were restricted to the scalar sector discussed in this chapter, one would expect a broad Higgs resonance. Extending the expansion in eq. (12.28) to third order yields a term $\propto \sigma \, \vec{\pi}^{\,2}$, which corresponds to the vertex of a decay $\sigma \to \pi\pi$; *i.e.* the heavy Higgs particle would rapidly decay into two massless Nambu–Goldstone bosons.

However, in Chapter 13 we will see that these Nambu–Goldstone bosons are strongly affected by the Higgs mechanism; they are absorbed in (or, as one says, "eaten" by) the massive gauge bosons $W^{\pm}$ and Z, and become their longitudinal polarization components. In this way, the previously massless Nambu–Goldstone bosons obtain masses, which are so heavy that the Higgs particle cannot decay into two of them (M_W, $M_Z > m_H/2$). Therefore, the observed decay of the Higgs particle is relatively "slow", with a lifetime of $\approx 1.6 \times 10^{-22}$ sec, and the Higgs resonance is narrow compared to the ungauged scenario $\sigma \to 2\pi$.

12.4 Goldstone Theorem

The presentation in this section is partly inspired by the corresponding section in the book by Michael Peskin and Daniel Schroeder (Peskin and Schroeder, 1997).

In this chapter, we have encountered several Nambu–Goldstone bosons. Let us now take a more general point of view and discuss a theorem due to Jeffrey Goldstone, which predicts the number of these massless field excitations for a general pattern of spontaneous breakdown of a continuous global symmetry (Goldstone, 1961).

As a prototype model, we consider an N-component real scalar field

$$\vec{\phi}(x) = (\phi_1(x), \phi_2(x), \dots, \phi_N(x)) \tag{12.31}$$

with a derivative term and a potential $V(\vec{\phi})$ that are invariant under global transformations of some Lie group G. This symmetry group has n_G generators T^a, with $a \in \{1, 2, \dots, n_G\}$, which are purely imaginary, anti-symmetric (and therefore traceless) $N \times N$ matrices. A general, infinitesimal symmetry transformation of the field takes the form[10]

$$\vec{\phi}'(x) = \exp(\mathrm{i}\omega_a T^a) \, \vec{\phi}(x) \approx (\mathbb{1} + \mathrm{i}\omega_a T^a) \, \vec{\phi}(x), \tag{12.32}$$

for small angles ω_a. Let us assume the potential to have a set of degenerate minima, corresponding to classical vacuum configurations. We pick one spontaneously, $\vec{\phi} = \vec{v}$, and ask about the masses of the fluctuation fields around this vacuum.

First of all, since $\vec{v}$ is a minimum of the potential, we know that

$$\left. \frac{\partial V}{\partial \phi_i} \right|_{\vec{\phi} = \vec{v}} = 0, \qquad i \in \{1, 2, \dots, N\}. \tag{12.33}$$

---

[10] As in Section 2.2, it would be more obvious to deal with anti-symmetric real generators $T'^a = \mathrm{i}T^a$, without introducing imaginary parts. Our notation is chosen for the sake of consistency with forthcoming chapters.

The matrix of second derivatives of the potential,

$$M_{ij} = \left.\frac{\partial^2 V}{\partial\phi_i \partial\phi_j}\right|_{\vec{\phi}=\vec{v}}, \qquad (12.34)$$

defines the masses: The eigenvalues of the matrix M are the squared masses of the physical particle fluctuations around the vacuum $\vec{v}$.

Let us now assume the vacuum to be invariant under the transformations in a subgroup H of G, $H \subset G$, which is generated by T^b with $b \in \{1, 2, \ldots, n_H\}$ and $n_H \leq n_G$, i.e.

$$\left(1 + i\omega_b T^b\right)\vec{v} = \vec{v} \quad \Rightarrow \quad T^b \vec{v} = 0. \qquad (12.35)$$

Invariance of the potential under the transformation group G implies for any vector $\vec{\phi}$

$$0 = V(\vec{\phi}') - V(\vec{\phi}) = i\frac{\partial V}{\partial\phi_i}\,\omega_a T^a_{ij}\,\phi_j. \qquad (12.36)$$

We differentiate this equation with respect to ϕ_k and evaluate it at $\vec{\phi} = \vec{v}$,

$$0 = \left.\frac{\partial^2 V}{\partial\phi_k\partial\phi_i}\right|_{\vec{\phi}=\vec{v}}\omega_a T^a_{ij}\, v_j + \left.\frac{\partial V}{\partial\phi_i}\right|_{\vec{\phi}=\vec{v}}\omega_a T^a_{ki} \quad \Rightarrow \quad M_{ki}\,(T^a\vec{v})_i = 0, \qquad (12.37)$$

where we have used eqs. (12.33) and (12.34). For the unbroken subgroup H, i.e. for $a \leq n_H$, this is trivially satisfied because $T^a\vec{v} = \vec{0}$, according to eq. (12.35). For the remaining generators with $a \in \{n_H + 1, n_H + 2, \ldots, n_G\}$, however, the equation implies that $T^a\vec{v}$ must be an eigenvector of the matrix M with eigenvalue zero. Hence, the difference $n_G - n_H$ is the degeneracy[11] of the eigenvalue 0, i.e. the dimension of the manifold of vacuum configurations. Therefore, there are $n_G - n_H$ *massless fluctuation modes* around a vacuum $\vec{v}$. Upon quantization, in a relativistic quantum field theory in more than two space–time dimensions (cf. Section 12.5), these modes turn into $n_G - n_H$ massless Nambu–Goldstone bosons.

For example, when Cooper pairs condense, the $G = U(1)$ symmetry, which breaks spontaneously down to the trivial subgroup $H = \{1\}$, gives rise to $n_G - n_H = 1 - 0 = 1$ Nambu–Goldstone boson. In the Higgs sector of the Standard Model, the symmetry $G = SU(2)_L \times SU(2)_R \cong O(4)$ breaks spontaneously to the subgroup $H = SU(2)_{L=R} \cong O(3)$. Hence, in this case there are $n_G - n_H = 6 - 3 = 3$ Nambu–Goldstone bosons. In general, when a symmetry $G = O(N)$ breaks to $H = O(N-1)$, the number of broken generators is

$$n_G - n_H = \frac{1}{2}N(N-1) - \frac{1}{2}(N-1)(N-2) = N - 1, \qquad (12.38)$$

such that there are $N - 1$ Nambu–Goldstone bosons.

In *non-relativistic* theories, the number of massless particles does not necessarily coincide with the number of Nambu–Goldstone boson fields. Holger Nielsen and Sudhir Chadha (Nielsen and Chadha, 1976) distinguish Nambu–Goldstone bosons of *types I and*

---

[11] We are referring to the *geometric multiplicity*, i.e. to the dimension of the eigenspace corresponding to eigenvalue 0.

II, with energies proportional to an odd or even power of the momentum, respectively. Under mild assumptions, the sum of type I bosons, plus twice the number of type II bosons, is at least equal to the number of spontaneously broken symmetry generators.

For example, a quantum ferromagnet with a global $G = O(3)$ symmetry that is spontaneously broken down to the subgroup $H = O(2)$ has $n_G - n_H = 3 - 1 = 2$ massless modes, but only one Nambu–Goldstone particle – a magnetic spinwave or *magnon*.[12] Indeed, this Nambu–Goldstone boson is of type II, since ferromagnetic magnons have a non-relativistic dispersion relation $E \propto |\vec{p}\,|^2$.[13] This results from the fact that the order parameter of the ferromagnet – the uniform magnetization, *i.e.* the total spin – is a conserved quantity.

Quantum anti-ferromagnets also have an $O(3)$ symmetry that is spontaneously broken down to $O(2)$. However, other than in a ferromagnet, the staggered magnetization order parameter of an anti-ferromagnet is not a conserved quantity. As a consequence, anti-ferromagnetic magnons have a *relativistic* dispersion relation, $E \propto |\vec{p}\,|$, and in this case there are indeed two massless Nambu–Goldstone particles, in agreement with the relativistic form of the Goldstone theorem in eq. (12.38).

12.5 Mermin–Wagner–Hohenberg–Coleman Theorem

Nambu–Goldstone bosons can only exist in more than two space–time dimensions. In the context of condensed matter physics, David Mermin and Herbert Wagner, as well as Pierre Hohenberg, were first to prove that the spontaneous breakdown of a continuous global symmetry cannot occur in space–time dimensions $d \leq 2$ (Mermin and Wagner, 1966; Hohenberg, 1967). In the condensed matter literature, this property is known as the Mermin–Wagner theorem. In the context of relativistic quantum field theories, a corresponding theorem was proved by Sidney Coleman; it is often referred to as Coleman's theorem (Coleman, 1973). The *Mermin–Wagner–Hohenberg–Coleman theorem* states that, in two space–time dimensions, an order parameter corresponding to a continuous global symmetry necessarily has a vanishing vacuum expectation value. Therefore, massless Nambu–Goldstone bosons – which appear as a consequence of spontaneous symmetry breaking – can only exist in space–time dimensions $d \geq 3$.

For a simple plausibility argument, let us consider again an $O(N)$ model ($N \geq 2$) with an N-component scalar field $\vec{\phi}_x \in \mathbb{R}^N$, associated with the sites x of a hypercubic lattice of spacing a and space–time volume $L^{d_s}\beta$ with periodic boundary conditions. Here, $d_s = d - 1$ is the spatial dimension, and $\beta = 1/T$ is the inverse temperature. For simplicity, we restrict the scalar field to a fixed length $|\vec{\phi}_x|^2 = v^2$ and we write $\vec{\phi}_x = v\vec{s}_x$ with $|\vec{s}_x| = 1$. The standard form of the Euclidean lattice action then reads

$$S[\vec{s}] = a^d \sum_x \sum_{\mu=1}^{d} \frac{v^2}{2a^2} \left(\vec{s}_{x+\hat{\mu}} - \vec{s}_x \right)^2 = \kappa \sum_{x,\mu} \left(1 - \vec{s}_x \cdot \vec{s}_{x+\hat{\mu}} \right), \qquad (12.39)$$

where $\kappa = v^2 a^{d-2}$ is the coupling constant and $\hat{\mu}$ is a vector of length a in the μ-direction.

---

[12] The literature often denotes objects like magnons as "quasi-particles". The difference from elementary particles is that quasi-particles require the presence of some medium, in which the wavicle forms.

[13] Although these particles are massless, they do not obey the relativistic dispersion relation $E \propto |\vec{p}\,|$.

We are interested in a large spatial size, $L \gg a$, while (at this point) β is moderate. We focus on the configurations of low action, with the maximal Boltzmann factor contributions to the partition function and to the expectation values. In particular, let us consider the static plane-wave configurations

$$\vec{s}_{\vec{x},x_d}(\vec{n}) = (\cos(\vec{p} \cdot \vec{x}), \sin(\vec{p} \cdot \vec{x}), 0, \ldots, 0),$$

$$\vec{p} \cdot \vec{x} = \sum_{i=1}^{d_s} p_i x_i, \quad \vec{p} = \frac{2\pi}{L}\vec{n}, \quad \vec{n} \in \mathbb{Z}^{d_s}. \tag{12.40}$$

We compare the contribution of the constant configuration in this set, which has $\vec{n} = \vec{0}$, to the contributions of a configuration with $\vec{n} \neq \vec{0}$. The ratio of their statistical weights takes the form

$$\frac{\exp\left(-S[\vec{s}\,(\vec{n} \neq \vec{0})]\right)}{\exp\left(-S[\vec{s}\,(\vec{n} = \vec{0})]\right)} = \exp\left(-2\pi^2 n^2 \kappa \beta L^{d_s-2}\left[1 + \mathcal{O}(1/L^2)\right]\right). \tag{12.41}$$

Quick Question 12.5.1   Action of revolving configurations
   Derive eq. (12.41).

As we send $L \to \infty$, the fate of this ratio depends on the spatial dimension d_s:

- For $d_s > 2$, the non-zero-modes are suppressed. Then, the dominance of the zero-mode, and therefore the spontaneous breaking of the O(N) symmetry, is possible in the thermodynamic limit $L \to \infty$.
- For $d_s \leq 2$, these modes are not suppressed; hence, spontaneous symmetry breaking cannot occur at finite temperature.

Finally, we extend the consideration to the limit $T \to 0$, which is taken such that $\beta \propto L$. Then, the transition between these two scenarios is shifted down to $d_s = 1$; *i.e.* at zero temperature, spontaneous symmetry breaking may happen even at $d_s = 2$.

This behavior is due to infrared fluctuations (they are interpreted as quantum or thermal fluctuations, depending on the physical context), which are particularly strong in lower dimensions. In more than two space–time dimensions, quantum fluctuations are more restricted because the field variables (*e.g.*, on a cubic lattice) are then coupled to a larger number of neighboring sites.

As we have seen, the Mermin–Wagner–Hohenberg–Coleman theorem even applies to theories in $(2+1)$ dimensions, at non-zero temperature $T > 0$. In that case, the extent $\beta = 1/T$ of the Euclidean time dimension is finite, and thus, there are only two infinitely extended dimensions, in which truly infrared fluctuations can develop. Consequently, in $(2+1)$ dimensions, a continuous global symmetry can break spontaneously only at zero temperature, $T = 0$.

This has interesting consequences, for example, for the $(2+1)$-dimensional *anti-ferromagnetic* precursors of high-temperature superconductors, which were investigated by Sudip Chakravarty, Bertrand Halperin, and David Nelson (Chakravarty *et al.*, 1988, 1989), as well as by Peter Hasenfratz and Ferenc Niedermayer (Hasenfratz and Niedermayer,

1993). In this case, at small non-zero temperatures $T > 0$, non-perturbative effects entail a finite correlation length of the Nambu–Goldstone modes. It is exponentially large in the inverse temperature, but it diverges only in the $T \to 0$ limit.

It should be stressed that the Mermin–Wagner–Hohenberg–Coleman theorem does *not* exclude the existence of massless bosons in one spatial dimension, $d_s = 1$; it just states that such particles cannot result from spontaneous symmetry breaking. For example, as first understood by Vadim Berezinskiĭ (Berezinskiĭ, 1970, 1971) and by Michael Kosterlitz and David Thouless (Kosterlitz and Thouless, 1972, 1973; Kosterlitz, 1974), at sufficiently weak (but still non-zero) coupling, the $(1 + 1)$-dimensional XY model – which they interpreted in the context of classical statistical mechanics, cf. Chapters 1 and 5 – contains a massless boson. This boson exists, despite the fact that the Abelian continuous global O(2) symmetry of the model is not spontaneously broken. In the XY model, the weak- and the strong-coupling phases are separated by the *Berezinskiĭ–Kosterlitz–Thouless phase transition* (an "essential" phase transition, of infinite order).[14]

Massless bosons can even arise in $(1 + 1)$-dimensional systems with a non-Abelian O(3) symmetry. For example, as was first derived by Hans Bethe, anti-ferromagnetic quantum spin chains with spin 1/2 are gapless (Bethe, 1931). Duncan Haldane pointed out that the low-energy effective field theory, which describes spin chains with half-integer spin, is a 2-dimensional O(3) model with a non-trivial vacuum-angle $\theta = \pi$ (Haldane, 1983).[15] It was understood by Ian Affleck that the low-energy physics of this model is a *conformal field theory* – the $k = 1$ *Wess–Zumino–Novikov–Witten model* (Novikov, 1981, 1982; Witten, 1984a) – which contains massless bosons, despite the fact that the O(3) symmetry is not spontaneously broken (Affleck, 1985, 1988). Quantum spin chains with integer spins, on the other hand, correspond to $\theta = 0$ (Haldane, 1983). Those systems do have a finite energy gap, $E_1 - E_0 > 0$; thus, they do not give rise to massless (quasi-)particles.

The Mermin–Wagner–Hohenberg–Coleman theorem does not exclude either the spontaneous breakdown of a *discrete* symmetry in two dimensions. As an example from classical statistical mechanics, at sufficiently low temperature the *2-dimensional Ising model* has a spontaneously broken discrete $\mathbb{Z}(2)$ symmetry (Peierls, 1936; Onsager, 1944), cf. Appendix I. Just as continuous symmetries cannot break spontaneously in two dimensions, discrete symmetries cannot break spontaneously in one dimension. In a single space–time dimension, there is just time, so quantum field theory formally reduces to quantum mechanics of a finite number of degrees of freedom. Since spontaneous symmetry breaking is a collective phenomenon that necessarily involves an infinite number of degrees of freedom, it cannot arise in quantum mechanics. For instance, the 1-d XY model of classical statistical mechanics, which is equivalent to a quantum rotor (a quantum mechanical particle on a circle that we discussed in Problem 1.6), does not have a critical coupling. The 1-d Ising model does not exhibit a phase transition either; it spontaneously magnetizes only at strictly zero temperature (Ising, 1925).

---

[14] Traditionally, one classifies a phase transition as being of order n when the nth derivative of the free energy F is discontinuous; this is Paul Ehrenfest's classification scheme (Ehrenfest, 1933). Therefore, the case of an essential singularity of F corresponds to infinite order, cf. Section 1.4. (A modern classification only distinguishes first order and continuous phase transitions; see, *e.g.*, Yeomans (1992).)

[15] In the framework of QCD, the vacuum-angle θ will be discussed in detail in Chapters 19 and 25.

12.6 Low-Energy Effective Field Theory

Since they are massless, Nambu–Goldstone bosons dominate the low-energy physics of any system with a spontaneously broken, continuous, global symmetry. There is a general *effective Lagrangian technique* that describes the low-energy dynamics of the Nambu–Goldstone bosons. This approach was pioneered by Steven Weinberg (Weinberg, 1979c) and extended to a systematic method by Jürg Gasser and Heinrich Leutwyler for pions and other light mesons – the Nambu–Goldstone bosons of the spontaneously broken chiral symmetry of QCD (Gasser and Leutwyler, 1984, 1985). Therefore, this method is known as *chiral perturbation theory*, to be discussed in Chapter 23. It is, however, generally applicable to any system of Nambu–Goldstone bosons.

We will now illustrate this technique for the Nambu–Goldstone bosons that arise in the Higgs sector of the Standard Model (before gauging the symmetry). As we have seen, the global symmetry of the Higgs sector is $G = \mathrm{SU}(2)_L \times \mathrm{SU}(2)_R \cong \mathrm{O}(4)$, which then breaks spontaneously down to a single $H = \mathrm{SU}(2)_{L=R} \cong \mathrm{O}(3)$ symmetry.

As was first pointed out by Curtis Callan, Sidney Coleman, Julius Wess, and Bruno Zumino, in general, in a low-energy effective theory, the Nambu–Goldstone bosons are described by fields in the *coset space G/H* – the manifold of the group G with points being identified if they are related by a symmetry transformation in the unbroken subgroup H (Coleman *et al.*, 1969; Callan *et al.*, 1969). We saw that the dimension of the coset space, $n_G - n_H$, corresponds to the number of Nambu–Goldstone boson fields. In the Higgs sector of the Standard Model, this coset space is

$$G/H = \mathrm{SU}(2)_L \times \mathrm{SU}(2)_R/\mathrm{SU}(2)_{L=R} = \mathrm{SU}(2) = S^3, \qquad (12.42)$$

or equivalently $G/H = \mathrm{O}(4)/\mathrm{O}(3) = \mathrm{O}(3) \cong S^3$. Hence, the three Nambu–Goldstone bosons can be described by a matrix-valued field $U(x) \in \mathrm{SU}(2)$. As we pointed out in Section 12.3, the SU(2) group manifold is isomorphic to a 3-dimensional sphere S^3. One may think of the field $U(x)$ as the "angular" or "tangential" degree of freedom of the Higgs field matrix of eq. (12.15), *i.e.*

$$\boldsymbol{\Phi}(x) = \begin{pmatrix} \Phi^0(x)^* & \Phi^+(x) \\ -\Phi^+(x)^* & \Phi^0(x) \end{pmatrix} = |\Psi(x)|\, U(x), \quad U(x) \in \mathrm{SU}(2),$$

$$|\Phi(x)|^2 = |\Phi^+(x)|^2 + |\Phi^0(x)|^2 = \det \boldsymbol{\Phi}(x). \qquad (12.43)$$

Indeed, we have seen that the "radial" fluctuations of the magnitude $|\Phi|$ lead to the massive Higgs particle, while the "tangential" fluctuations along the vacuum manifold give rise to three massless Nambu–Goldstone bosons (in Section 12.3, we denoted the fields of massive and of massless fluctuations by σ and $\vec{\pi}$, respectively). From the $\mathrm{SU}(2)_L \times \mathrm{SU}(2)_R$ transformation rule of the Higgs field $\boldsymbol{\Phi}(x)$, eq. (12.17), one obtains

$$U'(x) = L U(x) R^\dagger, \quad L \in \mathrm{SU}(2)_L, \quad R \in \mathrm{SU}(2)_R. \qquad (12.44)$$

The effective field theory is formulated as a systematic low-energy expansion. The low-energy physics of the Nambu–Goldstone bosons is dominated by those terms in the effective Lagrangian that contain a small number of derivatives. In Fourier space, a spatial derivative corresponds to a momentum and a temporal derivative to an energy. Therefore, higher-derivative terms are suppressed at low energies.

All terms of the effective Lagrangian must be invariant under all symmetries of the underlying microscopic system – in this case, of the Higgs sector of the Standard Model. In particular, the effective Lagrangian must be invariant under Lorentz transformations and under the $SU(2)_L \times SU(2)_R$ transformations (12.44).

The effective Lagrangian is constructed, order by order in the number of derivatives, by writing down all terms that are consistent with the symmetries. In the systematic derivative expansion, one starts with terms containing no derivatives. For example, the term $\mathrm{Tr}[U^\dagger U]$ is both Lorentz-invariant and $SU(2)_L \times SU(2)_R$-invariant. However, since $U \in SU(2)$ implies $U^\dagger U = \mathbb{1}$, this term is a trivial constant.

Indeed, there are no non-trivial terms without derivatives. Furthermore, there are no terms with just a single derivative (or any odd number of space–time derivatives), because its uncontracted Lorentz index would violate Lorentz invariance. The leading term of the effective Lagrangian therefore has two derivatives and is given by

$$\mathcal{L}(U) = \frac{F^2}{4} \mathrm{Tr}\left[\partial_\mu U^\dagger \partial_\mu U\right]. \tag{12.45}$$

Higher-order terms with four or six derivatives, etc., contribute less at low energies. Each term appears with a coefficient, like F^2 in eq. (12.45). These coefficients are known as *low-energy constants*. They enter the effective theory as free parameters; their values cannot be deduced from symmetry considerations. Hence a theoretical prediction for them must be based on the underlying, fundamental theory. Comparing eq. (12.45) with eq. (12.16), using eq. (12.43), and setting $|\Phi|^2$ to its vacuum value v^2, one obtains a classical estimate of the low-energy parameter $F = v$. (Of course, in order to properly identify the correct value of F based on the parameters of the Standard Model, one must take quantum effects into account.)

In principle, the string of terms in the effective Lagrangian – and the number of low-energy parameters – is infinite, but the restriction to low energies justifies its truncation.

An N-component $\lambda|\vec{\phi}|^4$ model ($N \geq 2$) is known as a *linear σ-model*. The Higgs sector of the Standard Model is an example with $N = 4$. A linear σ-model is characterized by the Lagrangian

$$\mathcal{L}(\vec{\phi}) = \frac{1}{2}\partial_\mu\vec{\phi}\cdot\partial_\mu\vec{\phi} + \frac{m^2}{2}|\vec{\phi}|^2 + \frac{\lambda}{4!}|\vec{\phi}|^4 = \frac{1}{2}\partial_\mu\vec{\phi}\cdot\partial_\mu\vec{\phi} + \frac{\lambda}{4!}\left(|\vec{\phi}|^2 - v^2\right)^2 + C, \tag{12.46}$$

where we substituted $v^2 = -6m^2/\lambda$ and C is yet another irrelevant constant. In the limit $\lambda \to \infty$, the action diverges, unless $|\vec{\phi}(x)| = v$. Introducing the N-component unit-vector field $\vec{s}(x) = \vec{\phi}(x)/v$, this limit takes us to the *non-linear σ-model* or $O(N)$ model (cf. Chapters 1 and 5, as well as Section 12.4)[16]

$$\mathcal{L}(\vec{s}) = \frac{v^2}{2}\partial_\mu\vec{s}\cdot\partial_\mu\vec{s}, \quad \vec{s}(x) = (s_1(x), s_2(x), s_3(x), s_4(x)), \quad |\vec{s}(x)| = 1. \tag{12.47}$$

The non-linear constraint, $|\vec{s}(x)| = 1$, can be implemented in the measure of the functional integral; see eq. (1.41). Identifying

---

[16] For the special case $N = 1$, *i.e.* for a one-component $\lambda\phi^4$ model on the lattice, this limit leads to the Ising model.

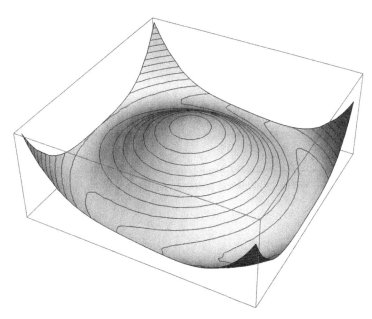

Fig. 12.2 A potential, which is modified compared to Figure 12.1: Here, a small explicit symmetry breaking is added, such that the circle of minima in Figure 12.1 is slightly tilted. This implies a unique minimum. Now the tangential fluctuations describe pseudo-Nambu–Goldstone bosons, which are massive, but still much lighter than the σ-particle.

$$U(x) = s_4(x)\mathbb{1} + i\left[s_1(x)\tau^1 + s_2(x)\tau^2 + s_3(x)\tau^3\right] \tag{12.48}$$

for $N = 4$, eq. (12.47) has just the structure of $\mathcal{L}(U)$ in eq. (12.45). Hence, the low-energy effective theory for the Higgs sector is in fact a non-linear σ-model. It takes high energies to activate the fluctuations around $|\vec{\phi}(x)| = \mathrm{v}$.

It is also interesting to anticipate that the very same effective Lagrangian describes the low-energy dynamics of the *pions in QCD* with two massless quark flavors (the pion is the lightest quark–anti-quark bound state, *i.e.* the lightest hadron). Only the value of the coupling F, which then corresponds to the *pion decay constant*, is different. The structure of the effective Lagrangian $\mathcal{L}(U)$ solely depends on the pattern of spontaneous symmetry breaking. Indeed, the chiral symmetry of two-flavor QCD is again $SU(2)_L \times SU(2)_R$, which breaks spontaneously to $SU(2)_{L=R}$; this will be addressed in Chapters 18 and 23.

It is an important feature of the effective Lagrangian technique that it is still applicable if the spontaneous symmetry breaking is supplemented by a small amount of *explicit* breaking. If one adds a symmetry breaking term like $\vec{h} \cdot \vec{\phi}$ (where $\vec{h}$ is constant and $|\vec{h}|$ is small) to the underlying microscopic Lagrangian of eq. (12.46), the Nambu–Goldstone bosons pick up a small mass and turn into *pseudo-Nambu–Goldstone bosons*. The settings without and with a small explicit symmetry breaking are illustrated in Figures 12.1 and 12.2.

In the case of explicit symmetry breaking, also the effective Lagrangian $\mathcal{L}(U)$ contains additional terms. It is then expanded in powers of the momenta *and* of the symmetry breaking parameters, according to a suitable power counting rule. In the description of a ferromagnet, an explicit symmetry breaking term represents a modest, external magnetic field. In two-flavor QCD, it corresponds to the small masses of the quark flavors u and d, which lead to a small but non-zero mass for the pions.

We anticipate that, in the Higgs sector of the Standard Model, the $SU(2)_L \times U(1)_Y$ subgroup of $SU(2)_L \times SU(2)_R$ must not be explicitly broken, because it will be gauged (see Chapter 13).

12.7 Hierarchy Problem

In the Higgs sector of the Standard Model, m^2 is the only dimensioned parameter, since λ is dimensionless.[17] We saw that these parameters determine the vacuum expectation value of the Higgs field,

$$v = \sqrt{-\frac{6m^2}{\lambda}}, \qquad (12.49)$$

which sets the energy scale for the breaking of the $SU(2)_L \times SU(2)_R$ symmetry. In fact, even after gauge and fermion fields have been added to the (minimal) Standard Model, its Lagrangian still contains only this single, dimensioned parameter. One could then simply take the point of view that v is a truly fundamental energy scale in units of which all other dimensioned physical quantities can be expressed. The experimental value $v \simeq 246\,\text{GeV}$ has been derived from the observed masses of the W- and Z-bosons and the strength of the electroweak interaction.

At the quantum level, however, the situation is more complicated. As we have seen in Chapter 4, quantum field theories must be regularized and renormalized. Indeed, the UV cut-off Λ represents another energy scale that enters the quantum theory through the process of regularization. When one renormalizes the theory, one attempts to move the cut-off to infinity, while keeping the physical masses and thus v fixed. As we will discuss in Section 12.10, this is tricky in the Standard Model, because the Higgs sector is "trivial"; *i.e.* it becomes a non-interacting (*i.e.* free) theory in the limit $\Lambda \to \infty$.

We know that the Standard Model cannot be the "Theory of Everything" because it does not include non-perturbative quantum gravity. The natural energy scale of gravity is the Planck scale that we encountered before in Section 2.7,

$$M_{\text{Planck}} = \frac{1}{\sqrt{G}} \simeq 1.22 \times 10^{19}\,\text{GeV}, \qquad (12.50)$$

where G is Newton's constant. Even if we would assume the Standard Model to describe the physics correctly all the way up to the Planck scale, it would necessarily have to break down at that scale. In this sense, we can think of M_{Planck} as an ultimate UV cut-off of the Standard Model.

Once we have appreciated the existence of the two fundamental scales v of the Standard Model and M_{Planck} of gravity with

$$\frac{v}{M_{\text{Planck}}} = \mathcal{O}(10^{-17}), \qquad (12.51)$$

we are confronted with the notorious

---

[17] In d space–time dimensions, the self-coupling λ has the mass dimension $d - 4$. Hence, $d = 4$ is a special case with respect to its dimension.

Box 12.1   **Hierarchy problem**

Why is the ratio between the electroweak scale and the Planck scale so tiny?

Although this is not a discrepancy by more than 30 orders of magnitude, this hierarchy problem represents a similar puzzle as the cosmological constant problem[18] that we addressed in Chapter 2. One wonders whether there may be a dynamical mechanism that arranges for v to be naturally very much smaller than M_{Planck} or any other relevant UV cut-off scale Λ.

Let us discuss the hierarchy problem in the context of the lattice regularized scalar field theory. Nature must have found a concrete way to regularize the Higgs physics at ultra-short distances. Due to renormalizability and universality, only the symmetries, but not the details of this regularization should matter at low energies. We will use the regularization on a space–time lattice with spacing a as a simplified model of Nature at ultra-short distances. In other words, in this context we identify the lattice cut-off $\Lambda = \pi/a$ with the Planck scale M_{Planck}. This does not mean that we claim that a rigid Planck-scale lattice actually exists in Nature; we rather mimic the unknown space–time structure there in this simple manner.[19] In the lattice regularization, the scalar field theory is characterized by the functional integral

$$Z = \int \mathcal{D}\vec{\phi} \, \exp(-S[\vec{\phi}]) = \prod_x \left(\frac{a}{2\pi}\right)^2 \int_{\mathbb{R}^4} d\vec{\phi}_x \, \exp(-S[\vec{\phi}]), \qquad (12.52)$$

with the Euclidean lattice action

$$S[\vec{\phi}] = a^4 \sum_x \left[\frac{1}{2} \sum_\mu \left(\frac{\vec{\phi}_{x+\hat{\mu}} - \vec{\phi}_x}{a} \right)^2 + V(\vec{\phi}_x) \right], \quad \vec{\phi}_x \in \mathbb{R}^4. \qquad (12.53)$$

Here, $\vec{\phi}_x$ is the 4-component scalar field at the lattice site x, and $\hat{\mu}$ is a vector of length a in the μ-direction. The first term is a finite difference regularization of the continuum expression $\partial_\mu \vec{\phi} \cdot \partial_\mu \vec{\phi}/2$, and the potential

$$V(\vec{\phi}_x) = \frac{m^2}{2} |\vec{\phi}_x|^2 + \frac{\lambda}{4!} |\vec{\phi}_x|^4 \qquad (12.54)$$

is the same as in the continuum theory. Just as the continuum theory – which is defined only perturbatively – the lattice theory exists in two phases, one with and one without spontaneous symmetry breaking. Its phase diagram is symbolically sketched in Figure 12.3. The two phases are separated by a phase transition line $\kappa_c(\lambda)$. For $\kappa = v^2 a^2 > \kappa_c(\lambda)$, the model is in the *broken phase* with three massless Nambu–Goldstone bosons, whereas for $\kappa < \kappa_c(\lambda)$ it is in the massive, *symmetric phase*. Extensive lattice studies have shown that the phase transition is of second order. This implies that the correlation length in units of

---

[18] Note that a fair comparison refers to the ratio of quantities of dimension mass, $(\rho_{naive}/\rho_{observation})^{1/4}$, where ρ is the vacuum energy density, cf. Section 2.7.

[19] A more subtle approach uses non-commutative geometry (Snyder, 1947), with a relation like $[\hat{x}_\mu, \hat{x}_\nu] \sim 1/\Lambda^2$ (for $\mu \neq \nu$), but quantum field theories on such spaces suffer from conceptual problems, like the mixing of UV and IR singularities (Minwalla *et al.*, 2000).

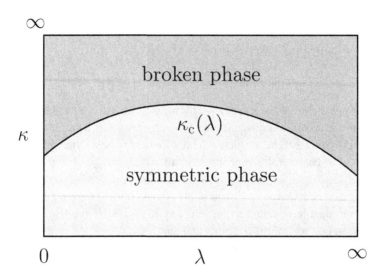

Schematic picture of the phase diagram of the 4-component $\lambda |\vec{\phi}|^4$ model. For the parameter $\kappa = v^2 a^2$ above its critical value $\kappa_c(\lambda)$, the O(4) symmetry is spontaneously broken, while for $\kappa < \kappa_c$ the symmetry remains intact.

the lattice spacing a, $\xi_H/a = 1/(m_H a)$, corresponding to the inverse Higgs particle mass in units of the lattice spacing, diverges at the phase transition line. Ignoring logarithmic corrections, the Higgs mass (and thus the vacuum value v of the scalar field) behaves as

$$m_H a = m_H \frac{\pi}{\Lambda} \propto \left| \frac{\kappa - \kappa_c(\lambda)}{\kappa_c(\lambda)} \right|^\nu . \tag{12.55}$$

Here, ν is a *critical exponent*, which takes the value $\nu = 1/2$. If we want to identify the lattice cut-off Λ with the Planck scale M_{Planck}, we must realize the hierarchy $m_H/M_{\text{Planck}} = \mathcal{O}(10^{-17})$. To achieve this, for a given value of λ, one has to fine-tune the bare parameter κ to the critical value $\kappa_c(\lambda)$ to a large number of decimal places. While fine-tuning to a critical point happens on a daily basis in condensed matter experiments, it appears unnatural in the fundamental description of Nature. Explaining the hierarchy between the electroweak and the Planck scale without a need for fine-tuning is the challenge of the hierarchy problem.

For gauge fields, a similar hierarchy problem does not exist. For example, the photon is naturally massless as a consequence of the unbroken gauge symmetry in the Coulomb phase of QED. Non-Abelian gauge fields also naturally exist at low energy scales, as a consequence of the property of asymptotic freedom.

From a perturbative point of view, there is no hierarchy problem for fermions either, because fermion mass terms are forbidden by chiral symmetry. However, as we will see in Chapter 18, when considered *beyond* perturbation theory, fermions do suffer from a severe hierarchy problem. In particular, as we have seen in Chapter 10, when regularized naively on a 4-dimensional space–time lattice, fermions suffer from species doubling. When the fermion doublers are removed by breaking chiral symmetry explicitly, without unnatural fine-tuning of the bare fermion mass, the renormalized mass resides at the cut-off scale. Remarkably, the non-perturbative hierarchy problem of fermions has been solved elegantly by formulating the theory with an additional spatial dimension of finite extent. In five dimensions fermions may get localized on a 4-dimensional domain wall. Indeed, as we will also see in Chapter 18, domain wall (Kaplan, 1992) as well as overlap fermions (Narayanan and Neuberger, 1993b) are naturally light without fine-tuning.

12.8 Solving the Hierarchy Problem with Supersymmetry?

There have been many attempts to solve the hierarchy problem of the Higgs sector of the Standard Model by postulating *supersymmetry* – a symmetry between bosons and fermions (see, *e.g.,* Wess and Bagger (1992)). In supersymmetric extensions of the Standard Model, each known boson has a fermionic superpartner, while each known fermion has a bosonic superpartner. For example, the electron has a scalar superpartner – the so-called selectron – and the photon has a fermionic superpartner – the photino. Similarly, the Higgs particle has a fermionic Higgsino partner. By supersymmetry, the Higgs mass (and thus the scale v) is then tied to the Higgsino mass, which would be protected from running to the cut-off scale by chiral symmetry, at least in the framework of perturbation theory. Hence, with super-symmetry, it is often argued that elementary scalar particles – such as the Higgs particle – can be light without unnatural fine-tuning.

The hierarchy problem is sometimes discussed separately for each order of perturbation theory, as if one would face the problem again and again, after completing another step of the calculation. While this seems to be true at a technical level, it is not compelling to think in this way. In particular, if one addresses the problem non-perturbatively, a single adjustment of the final result suffices to arrange for the hierarchy. In fact, supersymmetric theories do not explain the hierarchy between the electroweak and the Planck scale, but must also impose it by hand. However, once this is done at tree-level, supersymmetry stabilizes the hierarchy against radiative corrections.

If one thinks non-perturbatively, the hierarchy must be adjusted only once, both with and without supersymmetry. Still, in a supersymmetric theory one could place the electroweak scale at any desired value, without fine-tuning a bare parameter. However, even this apparent advantage might turn out to be a perturbative illusion. In particular, a satisfactory non-perturbative lattice definition of supersymmetric theories with naturally light scalars has not yet been found. Since supersymmetry is intimately related to infinitesimal space–time translations, it is not surprising that discretizing space–time breaks supersymmetry explicitly. Therefore, at a non-perturbative level, supersymmetry has the same status as chiral symmetry had before the successful construction of Ginsparg–Wilson fermions. In particular, achieving supersymmetry in a non-perturbative lattice context seems to require unnatural fine-tuning by itself and could then not be used to solve the naturalness problem.

It remains to be seen whether supersymmetry can be put on theoretically more solid grounds beyond perturbation theory. The success story behind lattice chiral symmetry, to be told in Section 18.6, which began with domain wall and overlap fermions, and culminated in Lüscher's construction of chiral gauge theories on the lattice, may encourage us to be optimistic that a similar story may repeat itself for lattice supersymmetry. Promising steps in this direction have already been taken, for example, by Kaplan *et al.* (2003) and Catterall *et al.* (2009). Irrespective of this, supersymmetry provides us with an interesting theoretical laboratory in which valuable insights into some specific quantum field theories can be gained. For example, regarding the impact of supersymmetry on the cosmological constant, we refer to Section 8.7. However, we do not know whether such theories can be realized naturally or whether their construction beyond perturbation theory itself requires unnatural fine-tuning, at least in a fully controlled non-perturbative framework such as the lattice regularization.

Until now, no superpartners have been observed. Hence, we do not know whether supersymmetry is realized in Nature. While the status of supersymmetry is yet unclear, hopefully the ongoing LHC experiments at CERN will decide whether or not supersymmetry exists at the TeV scale – there is no hint for its existence until 2022, despite intensive search at energies up to about 13 TeV. If supersymmetry is realized in Nature, it must be explicitly or spontaneously broken; otherwise, superpartners should be degenerate with the known particles and should have long been detected. Supersymmetry can make the electroweak scale appear natural only if it is around the corner at the 10 TeV scale. In view of the absence of superpartners at that scale, supersymmetry seems less natural than it may have appeared previously.

A non-perturbative approach to the hierarchy problem is *technicolor* (see Hill and Simmons (2003) for a review). In analogy to Cooper pair condensation, in technicolor models the electroweak symmetry is spontaneously broken by the condensation of fermion pairs. Just as quarks are bound by strong color forces, the so-called techni-fermions are bound by even stronger technicolor forces. In technicolor models, the Higgs particle is composed of two tightly bound *techni-fermions* (cf. Sections 15.12, 23.7, and 24.11). While technicolor suffers from its own severe problems, it remains to be seen whether the LHC will find evidence for this intriguing idea; as of 2022, there is no experimental support for this scenario either.

12.9   Is Nature Natural?

The concept of *naturalness* has played an important role in the development of particle physics. Fine-tuning of fundamental parameters to a large number of decimal places appears unnatural. Unlike in a condensed matter experiment, where a clever experimentalist may fine-tune the temperature to a critical value, who should have fine-tuned the electroweak scale v to lie 17 orders of magnitude below gravity's Planck scale? And how does the adjustment of the cosmological constant come about, around 120 orders of magnitude below what may appear as its "natural" value related to the Planck scale? At this point, one might feel tempted to invoke the anthropic principle.

Why do we not live on the surface of the sun? Obviously, the conditions there are not suitable for the development of life, which requires, for instance, liquid water. Hence, we are not surprised that life developed at a more hospitable location in the Universe, which indeed provides a vast variety of possible environments. As Weinberg has argued, if the cosmological constant would differ even slightly from its actual value, the Universe would no longer be hospitable for life as we know it (Weinberg, 1987). In particular, the time scale of the expansion of the Universe then seems too short for life to develop.

Similar anthropic arguments have been presented to "explain" the order of magnitude of the observed value of v or other fundamental parameters of the Standard Model. For example, a diverse population of stable nuclei, which forms the basis for the rich biochemistry that underlies life, seems to require finely tuned values of the quark masses. In particular, the Hoyle state of the ^{12}C nucleus, whose existence is vital for the abundant production of carbon inside stars, appears to be delicately fine-tuned (Hoyle, 1954). Just as life "decided" to develop on earth rather than on the surface of the sun, could it have developed anywhere else, not only on other planets but even beyond our Universe? The idea of a multiverse, of which our Universe would only be a very small part, indeed provides a viable framework for anthropic considerations. The landscape of string theory vacua (Susskind, 2006), as well as the possibly eternal process of cosmic inflation (Guth, 1981; Albrecht and Steinhardt, 1982; Linde, 1986), indeed allows us to contemplate the existence of other Universes with other values of fundamental physical parameters or even with different laws of physics.

If our Universe would be part of a very diverse, much larger multiverse, Nature, as we experience it locally, might indeed appear "unnatural" from a global perspective. Can the multiverse be taken seriously as a truly scientific idea that can, at least in principle, be falsified, or should it be considered a mere speculation? While it is impossible to directly probe other parts of the multiverse (assuming that it does exist), it is not excluded that one might eventually be able to establish string theory or cosmic inflation by ultra-high energy observations or by very detailed cosmological observations. The existence of a multiverse might then appear as a natural consequence of an established theory. While this is possible, at least in principle, we are currently far from this situation.

While it might be unfair to ban the anthropic principle as being non-scientific, we should resort to it only if all alternative explanations fail – or we simply admit that we do not yet have a satisfactory explanation. Imagine a 19th-century physicist, trying to explain the physical properties of liquid water, such as its boiling and freezing temperatures, which may have been viewed as "fundamental parameters" important for life. One could have already invoked the anthropic principle and said that water is just the way it is, because otherwise life could not exist. If one would be satisfied with such argumentation, one could have easily missed the deep understanding of chemistry and material science that resulted from quantum mechanics in the early 20th century. As one learned from Charles Darwin in the middle of the 19th century, as life evolves, it adapts to its environment, taking advantage of the properties of liquid water or the carbon abundance thanks to the existence of the Hoyle state. If a multiverse indeed exists, its different corners could easily be populated by totally different intelligent creatures that are well adapted to their local environment. The less curious species may be happy to "explain" their "fundamental parameters" based on an "alien-tropic" principle.

Fortunately, human curiosity has prevented us from viewing the boiling and freezing temperatures of water as "fundamental parameters" to be explained by the anthropic principle. Let us be optimistic and hope that at least some of the free parameters of the Standard Model will eventually find a natural explanation. There should be plenty of room for the curiosity-driven development of new concepts that will allow us to push the boundaries of current knowledge further into the unknown.

12.10 Triviality of the Standard Model

As we have seen, the Higgs sector of the Standard Model is a 4-component $\lambda|\vec{\phi}|^4$ model, a linear σ-model. The Lagrangian contains the dimensioned parameter m^2, as well as the dimensionless scalar self-coupling λ. These parameters determine the Higgs boson mass, as well as the vacuum expectation value of the scalar field

$$m_H = \sqrt{-2m^2} = \sqrt{\frac{\lambda}{3}}\, v, \quad v = \sqrt{-\frac{6m^2}{\lambda}}. \tag{12.56}$$

Hence, at fixed v a *heavy* Higgs particle would require a *strongly* coupled scalar field (with a large value of λ, such that the potential strongly increases when $|\vec{\phi}|$ deviates from v). We have obtained these results essentially by considering the model just at the level of classical field theory. When the theory is quantized using perturbation theory, the bare parameters are renormalized, but remain free parameters, as we have seen for the 1-component $\lambda\phi^4$ model in Chapter 4. Thus, eq. (12.56) actually refers to the renormalized self-coupling λ_r, but the Higgs mass m_H still appears arbitrary.

However, when a $\lambda|\vec{\phi}|^4$ model is addressed beyond perturbation theory, a new feature arises. Since the 1980s, in the lattice regularization there has been convincing evidence that the $\lambda|\vec{\phi}|^4$ model – and hence the Standard Model – is *trivial* in $d = 4$ (Lüscher and Weisz, 1987, 1988). Triviality has been rigorously proved in $d \geq 5$ a long time ago (Aizenman, 1981; Fröhlich, 1982; Callaway, 1988). It still took a long time until Michael Aizenman and Hugo Duminil-Copin presented a proof of triviality for the 1-component $\lambda\phi^4$ model in $d = 4$ (Aizenman and Duminil-Copin, 2021). This means that the renormalized self-coupling λ_r goes to zero if one insists on sending the UV cut-off to infinity. In other words, the continuum limit $a \to 0$ of a lattice $\lambda\phi^4$ model is just a free field theory. How can we then use it to define the Standard Model as an interacting field theory beyond perturbation theory?

Indeed, one *should not* insist on completely removing the UV cut-off. This means that the Standard Model cannot possibly make sense at arbitrarily high energies (beyond the finite cut-off). Hence, it represents a low-energy effective theory, which must necessarily be replaced by something more fundamental at very high energies. In other words, the Standard Model could not even in principle be the "Theory of Everything". This is actually a remarkable property of the Standard Model: It kindly informs us about its own limitations and tells us that it will eventually break down. Non-trivial theories (like QCD), on the other hand, remain interacting even when the cut-off is removed completely, as we described in Chapter 5. These theories could, at least in principle, be valid at arbitrarily high energy scales. However, as we will discuss in Chapter 18, Nature's QCD does not stand on its own, but is embedded in the Standard Model. As a result, QCD itself plays the role of a low-energy effective theory within the Standard Model, and so does QED.

The triviality of the Standard Model leads to an estimate for an upper bound on the Higgs boson mass. We saw that a heavy Higgs boson corresponds to a large value of λ_r. On the other hand, we just pointed out that we end up with a free theory, $\lambda_r = 0$, when we remove the cut-off completely. Only when we leave the cut-off finite, we obtain a heavy Higgs particle. However, the theory would clearly not be sensible if it led to a Higgs mass similar to – or even larger than – the UV cut-off.

Although the Higgs boson has been discovered, and its mass is determined as $m_H \simeq$ 125 GeV (cf. eq. (12.30)), it is still instructive to review in which range its mass could have been on conceptual grounds. As a quantitative example, a (non-perturbatively renormalized) Higgs mass of $m_H \approx 100$ GeV translates into a cut-off $\Lambda \simeq \mathcal{O}(10^{36})$ GeV = $\mathcal{O}(10^{17}) M_{Planck}$ (Smit, 2002). This is a regime where physics is completely unknown; hence, a cut-off of this magnitude is unproblematical. However, when we move the Higgs mass up to $m_H \approx 600$ GeV, we find the corresponding cut-off around 6 TeV, *i.e.* at $\Lambda \approx$ $10 m_H$. An even heavier Higgs particle would come too close to the cut-off scale to make any sense. Therefore, 600 GeV appears as a reasonable magnitude for the upper bound. It should be noted that the upper bound is not universal; *e.g.*, it depends on the short-distance details of the lattice action. In practice, based on different regularizations that seem reasonable, this ambiguity implies an $\mathcal{O}(10)$ percent uncertainty on the upper bound. Therefore, the lattice regularization sets an upper limit on the Higgs mass of around 600 to 700 GeV.

Long before the Higgs particle was discovered (in 2011/12), fits to processes which include virtual Higgs particle contributions suggested a much lower mass, below 150 GeV, which is well compatible with the observed value of eq. (12.30). There is also extensive literature about estimates of a *lower* bound for the Higgs mass, due to the "condition of vacuum stability". They depend, however, on subtleties in the assumptions; see, *e.g.*, Fodor *et al.* (2008), which we do not review here.

12.11 Electroweak Symmetry Restoration at High Temperature

Since the (hot) Big Bang, which happened about 1.38×10^{10} years ago, the Universe has undergone a dramatic evolution. The Big Bang itself represents a mathematical singularity in the solutions of classical General Relativity, hinting at our incomplete understanding of gravity. In the very early Universe, the energy density was at the Planck scale, where quantum effects of gravity are strong, and one might suspect that the classical singularity was eliminated by quantum fluctuations of the space–time metric. Since we do not have an established theory of non-perturbative quantum gravity, we can only speculate about the times at and immediately after the Big Bang. However, since the Universe is expanding and thus cooling down, it soon reached the energy scales of the Standard Model. Only about 10^{-14} sec after the Big Bang, the Universe had cooled down to temperatures in the TeV range, and its further evolution can be understood based on the Standard Model, combined with classical General Relativity.[20]

To a good approximation, the early Universe undergoes an adiabatic expansion (*i.e.* the total entropy is conserved); thus, it maintains its thermal equilibrium. When the Universe had a temperature around 1 TeV, it contained an extremely hot gas of quarks, leptons, gauge bosons, Higgs particles, and perhaps other yet undetected particles, *e.g.*, those forming the dark matter component of the Universe. At temperatures $T \gg v \simeq 246$ GeV, thermal fluctuations are so violent that the very early Universe was in an unbroken, symmetric

---

[20] Classical General Relativity also has to deal with singularities in the centers of black holes, which are, however, not of central importance for the cosmic evolution as a whole.

phase.[21] Just as the magnetization of a ferromagnet is destroyed at high temperatures (cf. Appendix I), the spontaneous order of the Higgs field cannot be maintained in the presence of strong thermal fluctuations. One sometimes says that the $SU(2)_L \times SU(2)_R$ symmetry was "restored" in the early Universe. Of course, it would be more appropriate to say that it was not yet spontaneously broken.

The high-temperature $SU(2)_L \times SU(2)_R$ symmetric phase that was realized in the early Universe should not be confused with the unbroken vacuum state (with $v = 0$) that exists at zero temperature for $m^2 > 0$.

As the Universe expanded and cooled down, it eventually ended up in the broken symmetry vacuum (with $v > 0$) that we populate today. It is interesting to ask how the expectation value $v(T)$ of the Higgs field depends on the temperature T. In particular, one may expect a phase transition at some critical temperature T_c. For temperatures $T > T_c$, the early Universe is in the symmetric phase, with $v(T) = 0$. When it expands and cools, it passes through the phase transition and enters the broken phase, with $v(T) > 0$. The order of this so-called *electroweak phase transition* has an impact on the dynamics of the early Universe. In particular, a *strong first-order phase transition* would have drastic consequences. Just as boiling water forms expanding bubbles of steam inside the liquid phase, a first-order electroweak phase transition would also proceed via *bubble nucleation*. In this case, bubbles of the broken phase would form inside the early symmetric phase. Since the bubble walls cost surface free energy, the energy consumed by their formation leads to cooling below the critical temperature. This "supercooling" delays the completion of the phase transition.

In such a hypothetical scenario, once the Universe had cooled sufficiently, bubbles of the broken phase would suddenly nucleate and expand quickly, soon filling all of the Universe. The violent dynamics of a first-order phase transition takes the system out of thermal equilibrium. As discussed by Andrei Sakharov, besides C and CP violation and the existence of baryon number violating processes, the deviation from thermal equilibrium is a necessary prerequisite for dynamically generating the *baryon asymmetry* – the observed surplus of matter over anti-matter (Sakharov, 1967). We will see in Chapters 15 and 17 that the Standard Model indeed violates both C and CP, as well as baryon number conservation (at sufficiently high temperatures). In order to understand how the Standard Model physics affects the baryon asymmetry, it is hence vital to understand the nature of the electroweak phase transition.

This is a non-perturbative question that has been addressed with Monte Carlo simulations applied to lattice field theory. In the pure Higgs sector of the Standard Model, one finds a *second* rather than first-order phase transition. At the phase transition temperature T_c, the thermal expectation value of the Higgs field – which acts as an order parameter – vanishes, *i.e.* $v(T_c) = 0$. As a consequence, the low-temperature broken and the high-temperature symmetric phase become indistinguishable at the phase transition. In particular, a second-order phase transition does not proceed by the nucleation of bubbles separating distinct phases; hence, it does not take the system far out of thermal equilibrium. Therefore, the effects of a second-order phase transition on the baryon asymmetry are much less drastic.

---

[21] For comparison, the energy v corresponds to a temperature of 2.85×10^{15} K, while the present temperature of the cosmic microwave background amounts to 2.73 K.

In the presence of electroweak gauge fields (which will be added in Chapter 13), even the second-order phase transition is washed out to a (smooth) *crossover*. Not even the t-quark, which is most strongly coupled to the Higgs field (cf. Chapter 17), alters this scenario qualitatively. As we will discuss in Section 26.5, topologically non-trivial Standard Model processes near T_c, known as sphaleron events, partially wash out a baryon asymmetry generated at higher energy scales, but maintain the baryon-minus-lepton number asymmetry.

12.12 Extended Model with Two Higgs Doublets

This section discusses physics beyond the Standard Model and may be skipped in a first reading.

Some theories beyond the Standard Model have an extended Higgs sector. For example, the minimal supersymmetric extension of the Standard Model (the so-called MSSM) contains two Higgs doublets.

A two Higgs doublet extension of the Standard Model (without supersymmetry) was introduced by Roberto Peccei and Helen Quinn (Peccei and Quinn, 1977a,b), in an attempt to solve the so-called *strong CP-problem,* which we will discuss in more detail in Chapter 25. One year later, Steven Weinberg and Frank Wilczek realized independently that the spontaneous breakdown of the $U(1)_{PQ}$ symmetry, a so-called *Peccei–Quinn symmetry* (see below), which plays an important role in the proposed solution of the strong CP-problem, gives rise to a pseudo-Nambu–Goldstone boson – the *axion* (Weinberg, 1978; Wilczek, 1978, 1987).

As of 2022, there are no experimental data which prompt us to go beyond the Standard Model assumption of a minimal Higgs sector, with just a single Higgs doublet, and one observable Higgs particle. Nevertheless, it is still conceivable that there could be several Higgs particles.

In this section, we consider an extension of the Standard Model by adding a second Higgs doublet $\widetilde{\Phi}$. We parameterize the two complex Higgs doublets as

$$\Phi(x) = \left(\begin{array}{c} \Phi^{|}(x) \\ \Phi^0(x) \end{array} \right), \quad \widetilde{\Phi}(x) = \left(\begin{array}{c} \widetilde{\Phi}^0(x) \\ \widetilde{\Phi}^-(x) \end{array} \right). \tag{12.57}$$

Under the $SU(2)_L$ symmetry, they transform as

$$\Phi'(x) = L\Phi(x), \quad \widetilde{\Phi}'(x) = L\widetilde{\Phi}(x), \quad L \in SU(2)_L, \tag{12.58}$$

and under $U(1)_Y$ as

$$\Phi'(x) = \exp\left(i\frac{g'}{2}\varphi \right) \Phi(x), \quad \widetilde{\Phi}'(x) = \exp\left(-i\frac{g'}{2}\varphi \right) \widetilde{\Phi}(x). \tag{12.59}$$

In addition to these symmetries, the extension of the Standard Model that we consider here has the aforementioned global $U(1)_{PQ}$ symmetry, which acts as

$$\Phi'(x) = \exp\left(i\gamma \right) \Phi(x), \quad \widetilde{\Phi}'(x) = \exp\left(i\gamma \right) \widetilde{\Phi}(x). \tag{12.60}$$

The corresponding Lagrangian of the two Higgs doublet model takes the form

$$\mathcal{L}(\Phi, \widetilde{\Phi}) = \frac{1}{2}\partial_\mu \Phi^\dagger \partial_\mu \Phi + \frac{1}{2}\partial_\mu \widetilde{\Phi}^\dagger \partial_\mu \widetilde{\Phi} + V(\Phi, \widetilde{\Phi}). \tag{12.61}$$

There is no need to include derivative terms that mix the two scalar fields. If such terms were present, one could eliminate them by a field redefinition (this would amount to a diagonalization of the corresponding matrix). In the potential $V(\Phi, \widetilde{\Phi})$, however, mixed terms may be present. The most general, renormalizable potential, which is invariant under $SU(2)_L$, $U(1)_Y$, as well as $U(1)_{PQ}$, is given by

$$V(\Phi, \widetilde{\Phi}) = \frac{m^2}{2}|\Phi|^2 + \frac{\lambda}{4!}|\Phi|^4 + \frac{\widetilde{m}^2}{2}|\widetilde{\Phi}|^2 + \frac{\widetilde{\lambda}}{4!}|\widetilde{\Phi}|^4 + \frac{\varkappa}{2}|\Phi|^2|\widetilde{\Phi}|^2 + \frac{\varkappa'}{2}|\Phi^\dagger\widetilde{\Phi}|^2,$$
$$|\Phi|^2 = \Phi^{+*}\Phi^+ + \Phi^{0*}\Phi^0, \quad |\widetilde{\Phi}|^2 = \widetilde{\Phi}^{0*}\widetilde{\Phi}^0 + \widetilde{\Phi}^{-*}\widetilde{\Phi}^-. \tag{12.62}$$

In contrast to the Standard Model, the extended model with $\varkappa' \neq 0$ does not have an additional $SU(2)_R$ symmetry. This symmetry would act by matrix multiplication from the right on a 2×2 matrix that is built by combining the column vectors $\widetilde{\Phi}$ and Φ (similar to the Φ that we used before). For $\varkappa' > 0$, the classical vacuum configurations obey $\Phi^\dagger\widetilde{\Phi} = 0$. For $m^2 < 0$, $\widetilde{m}^2 < 0$, one possible choice is

$$\Phi(x) = \begin{pmatrix} 0 \\ v \end{pmatrix}, \quad v = \sqrt{\frac{\widetilde{\lambda}m^2 - 6\varkappa\widetilde{m}^2}{6\varkappa^2 - \lambda\widetilde{\lambda}/6}},$$

$$\widetilde{\Phi}(x) = \begin{pmatrix} \widetilde{v} \\ 0 \end{pmatrix}, \quad \widetilde{v} = \sqrt{\frac{\lambda\widetilde{m}^2 - 6\varkappa m^2}{6\varkappa^2 - \lambda\widetilde{\lambda}/6}}. \tag{12.63}$$

This vacuum configuration is not invariant under either $SU(2)_L$, $U(1)_Y$, or $U(1)_{PQ}$. However, it is invariant against the $U(1)_{em}$ subgroup of $SU(2)_L \times U(1)_Y$, which acts as

$$\Phi'(x) = \begin{pmatrix} \exp(ie\alpha) & 0 \\ 0 & 0 \end{pmatrix} \Phi(x), \quad \widetilde{\Phi}'(x) = \begin{pmatrix} 0 & 0 \\ 0 & \exp(-ie\alpha) \end{pmatrix} \widetilde{\Phi}(x), \tag{12.64}$$

and will soon be identified as the symmetry of electromagnetism. Hence, the symmetry group $G = SU(2)_L \times U(1)_Y \times U(1)_{PQ}$ is spontaneously broken down to the subgroup $H = U(1)_{em}$. According to the Goldstone theorem (see Section 12.4), in this case, there appear $n_G - n_H = 3 + 1 + 1 - 1 = 4$ Nambu–Goldstone bosons. The additional fourth Nambu–Goldstone boson, which results from the spontaneous breakdown of the $U(1)_{PQ}$ Peccei–Quinn symmetry, is the axion.

Let us again expand around the vacuum configuration by generalizing the ansatz (12.27),

$$\Phi(x) = \begin{pmatrix} \pi_1(x) + i\pi_2(x) \\ v + \sigma(x) + i\pi_3(x) \end{pmatrix}, \quad \widetilde{\Phi}(x) = \begin{pmatrix} \widetilde{v} + \widetilde{\sigma}(x) - i\widetilde{\pi}_3(x) \\ -\widetilde{\pi}_1(x) - i\widetilde{\pi}_2(x) \end{pmatrix}. \tag{12.65}$$

Up to quadratic order in the fluctuations, the potential takes the form

$$\begin{aligned} V(\Phi, \widetilde{\Phi}) &= \frac{m^2}{2}\left[(v + \sigma)^2 + \pi_1^2 + \pi_2^2 + \pi_3^2\right] + \frac{\lambda}{4!}\left[(v + \sigma)^2 + \pi_1^2 + \pi_2^2 + \pi_3^2\right]^2 \\ &\quad + \frac{\widetilde{m}^2}{2}\left[(\widetilde{v} + \widetilde{\sigma})^2 + \widetilde{\pi}_1^2 + \widetilde{\pi}_2^2 + \widetilde{\pi}_3^2\right] + \frac{\widetilde{\lambda}}{4!}\left[(\widetilde{v} + \widetilde{\sigma})^2 + \widetilde{\pi}_1^2 + \widetilde{\pi}_2^2 + \widetilde{\pi}_3^2\right]^2 \\ &\quad + \frac{\varkappa}{2}\left[(v + \sigma)^2 + \pi_1^2 + \pi_2^2 + \pi_3^2\right]\left[(\widetilde{v} + \widetilde{\sigma})^2 + \widetilde{\pi}_1^2 + \widetilde{\pi}_2^2 + \widetilde{\pi}_3^2\right] \\ &\quad + \frac{\varkappa'}{2}\left|(\pi_1 - i\pi_2)(\widetilde{v} + \widetilde{\sigma} - i\widetilde{\pi}_3) + (v + \sigma - i\pi_3)(-\widetilde{\pi}_1 + i\widetilde{\pi}_2)\right|^2 \end{aligned}$$

$$\approx \frac{1}{2}\left(m^2 + \frac{\lambda}{2}v^2 + \varkappa\tilde{v}^2\right)\sigma^2 + \frac{1}{2}\left(\tilde{m}^2 + \frac{\tilde{\lambda}}{2}\tilde{v}^2 + \varkappa v^2\right)\tilde{\sigma}^2$$

$$+ \frac{\varkappa'}{2}(v^2 + \tilde{v}^2)\left[\left(\frac{\tilde{v}\pi_1 - v\tilde{\pi}_1}{\sqrt{v^2 + \tilde{v}^2}}\right)^2 + \left(\frac{\tilde{v}\pi_2 - v\tilde{\pi}_2}{\sqrt{v^2 + \tilde{v}^2}}\right)^2\right] + C, \qquad (12.66)$$

where C is yet another irrelevant constant. Now there are *four* massive modes, *i.e.* four Higgs-type particles,[22]

$$\sigma(x), \quad \tilde{\sigma}(x), \quad \rho_1(x) = \frac{\tilde{v}\pi_1(x) - v\tilde{\pi}_1(x)}{\sqrt{v^2 + \tilde{v}^2}}, \quad \rho_2(x) = \frac{\tilde{v}\pi_2(x) - v\tilde{\pi}_2(x)}{\sqrt{v^2 + \tilde{v}^2}}, \qquad (12.67)$$

with the corresponding mass squares

$$m_\sigma^2 = \frac{\lambda}{3}v^2, \quad m_{\tilde{\sigma}}^2 = \frac{\tilde{\lambda}}{3}\tilde{v}^2, \quad m_{\rho_1}^2 = m_{\rho_2}^2 = \varkappa'(v^2 + \tilde{v}^2), \qquad (12.68)$$

as well as four massless Nambu–Goldstone modes,

$$\pi_3(x) \quad \tilde{\pi}_3(x), \quad \zeta_1(x) = \frac{v\pi_1(x) + \tilde{v}\tilde{\pi}_1(x)}{\sqrt{v^2 + \tilde{v}^2}}, \quad \zeta_2(x) = \frac{v\pi_2(x) + \tilde{v}\tilde{\pi}_2(x)}{\sqrt{v^2 + \tilde{v}^2}}. \qquad (12.69)$$

The modes σ, $\tilde{\sigma}$, π_3, and $\tilde{\pi}_3$ are neutral, whereas the modes $\rho^\pm = (\rho_1 \pm i\rho_2)/\sqrt{2}$ and $\zeta^\pm = (\zeta_1 \pm i\zeta_2)\sqrt{2}$ are charged under the unbroken subgroup $H = U(1)_{em}$ of the symmetry $G = SU(2)_L \times U(1)_Y \times U(1)_{PQ}$. As we will see in Chapter 13, via the Higgs mechanism three of the Nambu–Goldstone bosons will be incorporated in the massive gauge bosons $W^\pm$ and Z. In particular, the charged modes $\zeta^\pm$ will become longitudinal polarization modes of the charged $W^\pm$-bosons. Similarly, a linear combination of the neutral modes π_3 and $\tilde{\pi}_3$ will become the longitudinal polarization mode of the massive neutral Z-boson. The orthogonal linear combination of π_3 and $\tilde{\pi}_3$, on the other hand, remains massless and represents the *neutral* axion.

Finally, let us construct the leading terms in the low-energy effective theory for the two Higgs doublet model. Following the general scheme, the fields describing the Nambu–Goldstone bosons parameterize the coset space

$$G/H = SU(2)_L \times U(1)_Y \times U(1)_{PQ}/U(1)_{em} = SU(2) \times U(1), \qquad (12.70)$$

and hence take the form $V(x) \in SU(2)$ and $\exp(ia(x)/\tilde{F}) \in U(1)$. The leading terms of the effective Lagrangian are given by

$$\mathcal{L}(V, a) = \frac{F^2}{4}\mathrm{Tr}\left[\partial_\mu V^\dagger \partial_\mu V\right] + K\mathrm{Tr}\left[\partial_\mu V^\dagger \partial_\mu V \tau^3\right] + \frac{1}{2}\partial_\mu a \partial_\mu a$$

$$= \frac{F^2}{4}\mathrm{Tr}\left[\partial_\mu V^\dagger \partial_\mu V\right] + \frac{1}{2}\partial_\mu a \partial_\mu a. \qquad (12.71)$$

Here, we identified the *axion field* $a(x) \in \mathbb{R}$.

The term proportional to the constant K seems to explicitly break the $SU(2)_R$ symmetry down to $U(1)_Y$. However, this term simply vanishes as we will see in Problem 12.4. Consequently, despite the fact that there is no $SU(2)_R$ symmetry in the underlying two Higgs doublet model, to leading-order the Lagrangian of the corresponding low-energy effective theory is still compatible with an $SU(2)_R$ custodial symmetry. At higher-order, on the other

---

[22] The factors v and $\tilde{v}$ are arranged such that all four Higgs-type fields have the dimension of mass.

hand, $SU(2)_R$ breaking terms do arise. Hence, the custodial symmetry is an *accidental*, approximate, global symmetry: It arises only because no symmetry breaking terms exist in the leading low-energy Lagrangian.

Experimental axion searches have thus far been unsuccessful (Zyla *et al.*, 2020). Hence, there is no experimental support for the hypothesis of a two Higgs doublet extension of the Standard Model. Still, a relatively light axion is often considered as a promising dark matter candidate. Axions will be discussed further in Chapter 25.

Exercises

12.1 The group SU(2) and the Higgs field

a) Start from a general complex 2×2 matrix

$$M = \begin{pmatrix} a & b \\ c & d \end{pmatrix}, \qquad a, b, c, d \in \mathbb{C}.$$

Now assume $M \in SU(2)$ and show that this implies the properties

$$d = a^*, \quad c = -b^*, \quad |a|^2 + |b|^2 = 1.$$

b) The Higgs field in its matrix form has the structure

$$\Phi = \begin{pmatrix} \Phi^{0*} & \Phi^+ \\ -\Phi^{+*} & \Phi^0 \end{pmatrix}, \qquad \Phi^+, \ \Phi^0 \in \mathbb{C}.$$

Show that it maintains this structure under the transformations

$$\Phi \to L \Phi \quad \text{and} \quad \Phi \to \Phi R^\dagger,$$

if $L, R \in SU(2)$.

12.2 Symmetries under compact Lie groups

a) Count the number of generators for the following groups: $U(n)$, $SU(n)$, $O(n)$, $SO(n)$. Determine also the *rank* of these groups.

b) In low-energy hadron physics, the following symmetry breaking plays an essential role,

$$SU(N_f)_L \times SU(N_f)_R \to SU(N_f).$$

Here, we refer to a global symmetry of the Lagrangian (the chiral flavor symmetry), which is realized in the limit of massless quarks. If it undergoes spontaneous symmetry breaking, how many Nambu–Goldstone bosons emerge?

In one case, this symmetry breaking pattern can be described isomorphically by orthogonal groups as $O(N) \to O(n)$. For which numbers N_f, N, and n is this so? Show that this case is unique.

12.3 Global O(N) symmetry breaking

Consider an N-component real scalar field $\vec{\phi} = (\phi_1, \phi_2, \ldots, \phi_N)$ ($\phi_i \in \mathbb{R}$) with the Lagrangian

$$\mathcal{L}(\vec{\phi}) = \frac{1}{2} \partial_\mu \vec{\phi} \cdot \partial_\mu \vec{\phi} + \frac{m^2}{2} \vec{\phi} \cdot \vec{\phi} + \frac{\lambda}{4!} (\vec{\phi} \cdot \vec{\phi})^2.$$

Assume $m^2 < 0$ and determine the unbroken subgroup H that remains intact after spontaneous breakdown of the global O(N) symmetry. How many Nambu–Goldstone bosons do arise? What kind of fields can be used to describe the Nambu–Goldstone bosons in a low-energy effective theory? Write down the leading-order term of the corresponding low-energy effective Lagrangian.

12.4 Extended Standard Model with two Higgs doublets

Consider the extension of the Standard Model with two Higgs doublets, as in Section 12.12,

$$\Phi(x) = \begin{pmatrix} \Phi^+(x) \\ \Phi^0(x) \end{pmatrix}, \quad \widetilde{\Phi}(x) = \begin{pmatrix} \widetilde{\Phi}^0(x) \\ \widetilde{\Phi}^-(x) \end{pmatrix}.$$

The Lagrangian is given by

$$\mathcal{L}(\Phi, \partial_\mu \Phi, \widetilde{\Phi}, \partial_\mu \widetilde{\Phi}) = \frac{1}{2} \partial_\mu \Phi^\dagger \partial_\mu \Phi + \frac{1}{2} \partial_\mu \widetilde{\Phi}^\dagger \partial_\mu \widetilde{\Phi} + V(\Phi, \widetilde{\Phi}),$$

with a potential which does not include the $\varkappa'$-term of eq. (12.62),

$$V(\Phi, \widetilde{\Phi}) = \frac{m^2}{2} |\Phi|^2 + \frac{\lambda}{4!} |\Phi|^4 + \frac{\widetilde{m}^2}{2} |\widetilde{\Phi}|^2 + \frac{\widetilde{\lambda}}{4!} |\widetilde{\Phi}|^4 + \frac{\varkappa}{2} |\Phi|^2 |\widetilde{\Phi}|^2,$$

$$|\Phi|^2 = \Phi^{+*} \Phi^+ + \Phi^{0*} \Phi^0, \quad |\widetilde{\Phi}|^2 = \widetilde{\Phi}^{0*} \widetilde{\Phi}^0 + \widetilde{\Phi}^{-*} \widetilde{\Phi}^-.$$

a) Find all internal symmetries of the Lagrangian, and discuss how they act on the fields.

b) Assume $\varkappa > 0$ and $m^2 < 0$, and determine one of the degenerate vacuum configurations. What is the corresponding unbroken subgroup? How many massless particles does the Goldstone theorem predict in this case?

c) Expand the field around the vacuum configuration, and diagonalize the mass matrix, in order to explicitly verify the number of Nambu–Goldstone bosons. What are the masses of the Higgs particles?

d) Show that the term proportional to K in the effective Lagrangian of eq. (12.71) vanishes.

12.5 Adjoint Higgs field in SU(3)

Consider the Higgs field

$$\Xi(x) = \Xi_a(x) \lambda^a, \quad \Xi_a(x) \in \mathbb{R}, \quad a \in \{1, 2, \dots, 8\},$$

in the adjoint representation of SU(3). Here, λ^a are the eight Gell-Mann matrices that generate the SU(3) algebra. Under global transformations $\Upsilon \in$ SU(3), the scalar field transforms as

$$\Xi'(x) = \Upsilon \, \Xi(x) \, \Upsilon^\dagger.$$

We consider the potential

$$V(\Xi) = \frac{M^2}{2} \text{Tr} \left[\Xi^2 \right] + \frac{\Lambda}{4!} \text{Tr} \left[\Xi^4 \right].$$

a) Show that this potential has an SO(8) symmetry.

b) Assume $m^2 < 0$ and construct a vacuum configuration. What is the corresponding unbroken subgroup? How many Nambu–Goldstone bosons arise?

c) Construct a renormalizable term that explicitly breaks the SO(8) symmetry down to SU(3).

d) Consider possible symmetry breaking patterns in the presence of this term. How many Nambu–Goldstone bosons are now obtained?

12.6 Metropolis algorithm for the Ising model

Consult Appendix H to implement a Markov chain Monte Carlo simulation of the 1-d Ising model with periodic boundary conditions, using a Metropolis algorithm.

The method allows us to use different initial configurations, *e.g.,* a completely ordered one with all spins parallel, a staggered one with neighboring spins pointing in opposite directions, or a random one.

The program should contain a subroutine in which one Metropolis sweep through the lattice is performed. In one sweep, each spin is subjected to one Metropolis update; *i.e.* its neighbors are examined, and the change of the energy under spin flip is calculated. Then, the spin is flipped or not, according to the rules of the Metropolis algorithm. If the spin should be flipped with probability p, one can pick a uniformly distributed random number $r \in [0, 1]$ and flip the spin if $r < p$.

There should also be a subroutine which measures the energy of a given configuration. This value is then used in the computation of the specific heat.

An important part of each Monte Carlo routine is the error analysis. When the Monte Carlo data are written into a file, after the thermalization and with sufficiently separated measurements, the error analysis can be performed with a separate program. One should then compare Monte Carlo and analytic results (obtained, *e.g.,* in Problem 1.7) in order to decide whether they agree within the statistical errors.

It is straightforward to extend the simulations to the 2-d Ising model which exhibits a second-order phase transition. Lattices of moderate sizes around 16^2, 32^2, or 48^2 already reveal the transition at the critical temperature $T_c = 2J/\log(1 + \sqrt{2})$. A reliable indicator is the peak of the specific heat (which diverges at T_c in infinite volume).

13

Local Symmetry and the Higgs Mechanism: From Superconductivity to Electroweak Gauge Bosons

In this chapter, we discuss the *gauge fields* mediating the *electromagnetic* and *weak interactions*. The weak interaction is responsible, for example, for radioactive decays.

In order to describe these interactions, two symmetry groups that we know from Chapter 12 are revisited. In particular, when *electroweak gauge fields* are included, both the $SU(2)_L$ symmetry and the $U(1)_Y$ subgroup of $SU(2)_R$ are promoted to *local* symmetries. This electroweak $SU(2)_L \times U(1)_Y$ gauge symmetry "breaks spontaneously" down to $U(1)_{em}$ – the gauge group of electromagnetism.

Due to the *Higgs mechanism,* the gauge bosons W^+, W^-, and Z become *massive*. The additional longitudinal polarization states of these three massive vector bosons are provided by three Nambu–Goldstone modes. One says that "the gauge bosons eat the Nambu–Goldstone bosons" and thus pick up a mass. The photon, on the other hand, remains massless, as a consequence of the *unbroken* $U(1)_{em}$ gauge symmetry of electromagnetism.

The full gauge symmetry group of the Standard Model is $SU(3)_c \times SU(2)_L \times U(1)_Y$, where the color gauge group $SU(3)_c$ is associated with the strong interaction between quarks, which is mediated by gluons. Before adding quarks (and neglecting gravity), the gluons do not interact with Higgs bosons, W- and Z-bosons, or photons. Hence, we will add the gluons only in Chapter 14, when we address the strong interaction.

To illustrate the basic ideas behind the Higgs mechanism, we first turn to a simpler model motivated by the condensed matter physics of *superconductors:* electrodynamics with a charged scalar field representing Cooper pairs. We encountered Cooper pairs before (in Chapter 12) as neutral scalar fields in a superfluid with a global symmetry. Now we are going to discuss the role of the electric charge of Cooper pairs in a superconductor, *i.e.* their coupling to the $U(1)_{em}$ gauge field of electromagnetism. When Cooper pairs condense inside a superconductor, the $U(1)_{em}$ gauge symmetry undergoes the Higgs mechanism and the photon becomes *massive*.

The Higgs mechanism was discovered by Francois Englert and Robert Brout, by Peter Higgs, as well as by Gerald Guralnik, Carl Hagen, and Tom Kibble (Englert and Brout, 1964; Higgs, 1964a,b; Guralnik *et al.*, 1964), with important early contributions in the condensed matter context by Philip Anderson (1963). The prediction of an observable, scalar particle was explicitly expressed by Peter Higgs.

13.1 Higgs Mechanism in Scalar Electrodynamics

Let us return to scalar electrodynamics, which we introduced in Section 7.2. Its Euclidean Lagrangian,

$$\mathcal{L}(\Phi, A) = \frac{1}{2}(D_\mu \Phi)^* D_\mu \Phi + V(\Phi) + \frac{1}{4} F_{\mu\nu} F_{\mu\nu}, \quad V(\Phi) = \frac{m^2}{2}|\Phi|^2 + \frac{\lambda}{4!}|\Phi|^2, \quad (13.1)$$

is invariant against $U(1)_{em}$ gauge transformations

$$\Phi'(x) = \exp(iQe\alpha(x))\,\Phi(x), \quad A'_\mu(x) = A_\mu(x) - \partial_\mu\alpha(x). \quad (13.2)$$

We have generalized the scalar field $\Phi(x) \in \mathbb{C}$ to have an arbitrary quantized electric charge $Q \in \mathbb{Z}$ such that the covariant derivative now takes the form

$$D_\mu \Phi(x) = \left[\partial_\mu + iQeA_\mu(x)\right]\Phi(x). \quad (13.3)$$

For example, the effective, charged scalar field that describes a Cooper pair in a superconductor has charge $Q = -2$, because it represents two negatively charged electrons.

As we already pointed out in Section 7.2, one cannot add an explicit photon mass term $M_\gamma^2 A_\mu A_\mu / 2$ to the Lagrangian of eq. (13.1), because it is not gauge-invariant. We will now see that an Abelian gauge field can still pick up a mass via the Higgs mechanism, provided the scalar field Φ assumes a non-zero vacuum expectation value.

As in the case of the global $U(1)$ symmetry, we distinguish two phases:

- For $m^2 \geq 0$, the symmetry is unbroken, and the system is in its *Coulomb phase* with scalar infraparticles (cf. Section 7.3) of charge Q and massless photons. In such a phase, the electric charge is a conserved quantity. Indeed, in the vacuum of QED – and even of the full Standard Model – the $U(1)_{em}$ symmetry of electromagnetism is realized in a Coulomb phase.
- Again, the broken phase (which, at the classical level, corresponds to $m^2 < 0$) is particularly interesting. Inside a superconductor, the $U(1)_{em}$ symmetry is "spontaneously broken" as a consequence of Cooper pair condensation, which corresponds to a non-vanishing ground state expectation value of the charged scalar field. Since Cooper pairs are charged, their condensation implies that the ground state of a superconductor itself contains an uncertain number of charges. As a consequence, the ground state is not a charge eigenstate.[1]

Once more, there are degenerate vacuum configurations with $|\Phi| = v = \sqrt{-m^2/\lambda} > 0$, but they are now *related by gauge transformations*, as we see from eq. (13.2). Therefore, they represent *the same physical state*. As a result – in contrast to systems with a spontaneously broken global symmetry – in a gauge theory "spontaneous symmetry breaking" does *not* lead to vacuum degeneracy. In fact, gauge symmetries cannot break. Strictly speaking, they are not even symmetries of the physical world but merely redundancies in our theoretical description. Still, we adopt the common practice to speak of "spontaneous gauge symmetry breaking". However, in order to remind the reader of the subtleties related to this notion, we will keep on writing this expression in inverted commas.

---

[1] Of course, any real superconductor is a finite piece of material embedded in the Coulomb phase of the QED vacuum. Hence, the total charge of the entire superconductor still remains conserved.

To take a closer look at the resulting so-called *Higgs phase*, it is helpful to fix the gauge, so that we obtain a reference point for an expansion. We choose the *unitary gauge*, which means (referring to the notation of Section 12.2)

$$\text{Re } \Phi(x) = \phi_1(x) \geq 0, \quad \text{Im } \Phi(x) = \phi_2(x) = 0. \tag{13.4}$$

As in Section 12.2, we investigate the fluctuations around the vacuum configuration $\phi_1(x) = \mathrm{v}$. Due to unitary gauge fixing, in this case we only allow for real-valued fluctuations,

$$\Phi(x) = \mathrm{v} + \sigma(x); \tag{13.5}$$

thus, there is no π-excitation. To $\mathcal{O}(\sigma^2)$, we obtain

$$
\begin{aligned}
V(\Phi) &= \frac{m^2}{2}(\mathrm{v} + \sigma)^2 + \frac{\lambda}{4!}(\mathrm{v} + \sigma)^4 \\
&\approx \frac{m^2}{2}\mathrm{v}^2 + m^2\mathrm{v}\sigma + \frac{m^2}{2}\sigma^2 + \frac{\lambda}{4!}(\mathrm{v}^4 + 4\mathrm{v}^3\sigma + 6\mathrm{v}^2\sigma^2) \\
&= \frac{1}{2}(m^2 + \frac{\lambda}{2}\mathrm{v}^2)\sigma^2 + C = \frac{\lambda}{6}\mathrm{v}^2\sigma^2 + C = \frac{1}{2}m_\sigma^2 + C,
\end{aligned}
\tag{13.6}
$$

where we have used eq. (12.7). There is again a σ-particle with the same mass as in the case of the spontaneously broken global symmetry. However, the massless Nambu–Goldstone boson π has disappeared, since – as we just mentioned – the degeneracy of vacua is no longer physical.

What happened to the π degree of freedom? Let us consider the covariant derivative term and expand it to second order in σ and A_μ,

$$
\begin{aligned}
\frac{1}{2}(D_\mu\Phi)^* D_\mu\Phi &= \frac{1}{2}\Big[(\partial_\mu - \mathrm{i}QeA_\mu)(\mathrm{v} + \sigma)\Big]\Big[(\partial_\mu + \mathrm{i}QeA_\mu)(\mathrm{v} + \sigma)\Big] \\
&= \frac{1}{2}(\partial_\mu\sigma - \mathrm{i}QeA_\mu\mathrm{v} - \mathrm{i}QeA_\mu\sigma)(\partial_\mu\sigma + \mathrm{i}QeA_\mu\mathrm{v} + \mathrm{i}QeA_\mu\sigma) \\
&\approx \frac{1}{2}\partial_\mu\sigma\partial_\mu\sigma + \frac{1}{2}Q^2e^2\mathrm{v}^2 A_\mu A_\mu.
\end{aligned}
\tag{13.7}
$$

Amazingly, we do obtain a *massive photon* with

$$M_\gamma = |Q|\, e\, \mathrm{v}. \tag{13.8}$$

Therefore, the missing degree of freedom (which was formerly identified as the π-particle) has turned into an additional, longitudinal polarization state of the photon.

This mechanism of mass generation is known as the *Higgs mechanism*. It is based on the "spontaneous breakdown" of a gauge symmetry. A phase in which the gauge symmetry is "spontaneously broken", so that the gauge bosons are massive, is called a *Higgs phase*. While the QED vacuum is in a Coulomb phase, inside a superconductor the $U(1)_{\text{em}}$ gauge symmetry *is* "spontaneously broken", and the photon becomes massive.

This mass can be *measured,* because it is related to the penetration depth of magnetic fields in the superconductor. This penetration was described by Fritz and Heinz London for type I superconductors (London and London, 1935).[2] It falls off exponentially, with a factor $\exp(-M_\gamma r)$. One then identifies $1/M_\gamma$ as the *range* of the electromagnetic interaction.

---

[2] In contrast, type II superconductors can be pierced by strings of magnetic flux tubes.

This resembles the form of a Yukawa potential, $V_{\text{Yukawa}}(r) \propto \exp(-Mr)/r$, which Hideki Yukawa postulated for the nuclear force (Yukawa, 1935). In contrast, in the *Coulomb phase* the potential follows a power decay, $V(r) \propto 1/r$, so its decay is slower than exponential, which means that the electromagnetic interaction has an infinite range, as a consequence of the fact that, in the ordinary vacuum (but not inside a superconductor), the photon is massless.

13.2 Higgs Mechanism in the Electroweak Theory

Let us proceed to the *electroweak gauge interactions* in the Standard Model. Here, we promote the $SU(2)_L$ symmetry as well as the $U(1)_Y$ subgroup of $SU(2)_R$ to local symmetries.

To begin, we turn $SU(2)_L$ into a gauge symmetry; *i.e.* we demand invariance of the Lagrangian against the gauge transformation

$$\Phi'(x) = L(x)\,\Phi(x), \quad \Phi(x) = \begin{pmatrix} \Phi^+(x) \\ \Phi^0(x) \end{pmatrix} \in \mathbb{C}^2. \tag{13.9}$$

The potential $V(\Phi)$ of eq. (12.9) is already invariant, but the derivative term is not, because

$$\partial_\mu \Phi'(x) = L(x)\partial_\mu \Phi(x) + \partial_\mu L(x)\Phi(x) = L(x)\left[\partial_\mu + L(x)^\dagger \partial_\mu L(x)\right]\Phi(x). \tag{13.10}$$

As in Section 7.2, we want to cancel the additional term. For this purpose, we introduce a gauge field $W_\mu(x)$ and construct a covariant derivative of the form

$$D_\mu \Phi(x) = \left[\partial_\mu + W_\mu(x)\right]\Phi(x). \tag{13.11}$$

Each component of W_μ is a complex 2×2 matrix field. In the derivative term of the Lagrangian, the above covariant derivative is multiplied by

$$(D_\mu \Phi(x))^\dagger = \partial_\mu \Phi(x)^\dagger + \Phi(x)^\dagger W_\mu(x)^\dagger = \partial_\mu \Phi(x)^\dagger - \Phi(x)^\dagger W_\mu(x). \tag{13.12}$$

This ensures that the derivative term in the Lagrangian is real-valued. In the last step, we have assumed W_μ to be *anti-Hermitian*,

$$W_\mu(x)^\dagger = -W_\mu(x). \tag{13.13}$$

In this form, W_μ is a natural generalization of the term $iQeA_\mu$, which entered the covariant derivative in the gauging of a single complex scalar field theory (in Section 7.2), and $\Phi^\dagger$ is the anti-field of Φ.

Hence, this gauge field can be written as

$$W_\mu(x) = igW_\mu^a(x)\frac{\tau^a}{2}, \quad a = 1, 2, 3, \tag{13.14}$$

where τ^a are the (Hermitian) Pauli matrices given in eq. (12.20) and the factor $1/2$ is a convention.[3] The parameter g is the *gauge coupling;* it characterizes the universal strength of the weak interaction, in this case between the Higgs field and the gauge field W_μ.

As we already discussed in Section 11.1, for the behavior of this matrix-valued gauge field under a non-Abelian gauge transformation, we write

$$W_\mu'(x) = L(x)\left[W_\mu(x) + \partial_\mu\right]L(x)^\dagger, \quad L(x) \in \mathrm{SU(2)}_L. \tag{13.15}$$

This leads to the simple relation

$$
\begin{aligned}
D_\mu'\Phi'(x) &= \left[\partial_\mu + W_\mu'(x)\right]\Phi'(x) \\
&= L(x)\left[\partial_\mu\Phi(x) + L(x)^\dagger\partial_\mu L(x)\Phi(x) + W_\mu(x)L(x)^\dagger L(x)\Phi(x) + \partial_\mu L(x)^\dagger L(x)\Phi(x)\right] \\
&= L(x)\left[\partial_\mu + W_\mu(x)\right]\Phi(x) = L(x)D_\mu\Phi(x).
\end{aligned}
\tag{13.16}
$$

We have applied the identity $\partial_\mu[L(x)^\dagger L(x)] = \partial_\mu\mathbb{1} = 0$, which is often useful to simplify expressions that involve unitary matrix fields. Consequently, we obtain

$$\left(D_\mu\Phi'(x)\right)^\dagger = \left(D_\mu\Phi(x)\right)^\dagger L(x)^\dagger, \tag{13.17}$$

and we arrive at the desired gauge-invariant Lagrangian

$$\mathcal{L}(\Phi, W) = \frac{1}{2}(D_\mu\Phi)^\dagger D_\mu\Phi + V(\Phi). \tag{13.18}$$

So far, the gauge field was treated as an external field. We still have to add its own derivative term. As we discussed in Section 11.1, the field strength tensor of a non-Abelian gauge field is given by

$$
\begin{aligned}
W_{\mu\nu}(x) &= \left[D_\mu, D_\nu\right] = \left[\partial_\mu + W_\nu(x), \partial_\nu + W_\nu(x)\right] \\
&= \partial_\mu W_\nu(x) - \partial_\nu W_\mu(x) + \left[W_\mu(x), W_\nu(x)\right],
\end{aligned}
\tag{13.19}
$$

and it transforms as

$$W_{\mu\nu}'(x) = L(x)W_{\mu\nu}(x)L(x)^\dagger. \tag{13.20}$$

It is necessary to include the commutator term in $W_{\mu\nu}$ in order to obtain a meaningful transformation behavior. Moreover, it is consistent to use the covariant derivative also for the formulation of the field strength. Hence, we consider eq. (13.19) as the general form of a field strength. The case of a $U(1)$ gauge field of eq. (7.14) was just the special situation where the commutator vanishes.

In analogy to the Abelian gauge theory, we write

$$\mathcal{L}(W) = \frac{1}{4}\,W_{\mu\nu}^a W_{\mu\nu}^a = -\frac{1}{2g^2}\,\mathrm{Tr}\left[W_{\mu\nu}W_{\mu\nu}\right], \quad W_{\mu\nu}(x) = \mathrm{i}gW_{\mu\nu}^a(x)\frac{\tau^a}{2}, \tag{13.21}$$

which is indeed gauge-invariant, as eq. (13.20) shows.

---

[3] We write τ^a for Pauli matrices that generate an internal SU(2) symmetry, in contrast to σ^a which generate the spatial spin rotation symmetry $\mathrm{SU(2)}_S$. We use again Einstein's convention of an implicit sum over the repeated index a.

Thus far, we have limited the gauging to the $SU(2)_L$ transformations and therefore to transformations with determinant 1. Now we also want to gauge the additional $U(1)$ transformations related to the determinant. The group of these transformations is $U(1)_Y$. The Higgs field then transforms as

$$\Phi'(x) = \exp\left(\frac{1}{2}ig'\varphi(x)\right)\Phi(x).\tag{13.22}$$

Here, g' is another coupling constant – the *weak hypercharge* (and the factor $1/2$ is just conventional, as in eq. (13.14)). As we discussed in Section 12.3, the $U(1)_Y$ symmetry is actually a subgroup of $SU(2)_R$ with

$$R(x) = \begin{pmatrix} \exp\left(ig'\varphi(x)/2\right) & 0 \\ 0 & \exp\left(-ig'\varphi(x)/2\right) \end{pmatrix}.\tag{13.23}$$

We emphasize once more that only the $U(1)_Y$ subgroup, but not the whole $SU(2)_R$ symmetry, is promoted to a local symmetry. Gauging solely the $U(1)_Y$ subgroup implies an explicit breaking of the global $SU(2)_R$ symmetry. Therefore, at this point we are not going to consider the remaining two generators of $SU(2)_R$.[4]

We now introduce an Abelian $U(1)_Y$ gauge field that transforms as

$$B'_\mu(x) = B_\mu(x) - \partial_\mu\varphi(x).\tag{13.24}$$

This additional gauge field contributes another term to the covariant derivative,

$$D_\mu\Phi(x) = \left[\partial_\mu + W_\mu(x) + i\frac{g'}{2}B_\mu(x)\right]\Phi(x)$$
$$= \left[\partial_\mu + igW_\mu^a(x)\frac{\tau^a}{2} + i\frac{g'}{2}B_\mu(x)\right]\begin{pmatrix}\Phi^+(x)\\\Phi^0(x)\end{pmatrix},\tag{13.25}$$

which is anti-Hermitian as well. (Note that ∂_μ and B_μ actually mean $\partial_\mu\mathbb{1}$ and $B_\mu\mathbb{1}$.) We employ again the Abelian gauge-invariant field strength of eq. (7.14),

$$B_{\mu\nu}(x) = \partial_\mu B_\nu(x) - \partial_\nu B_\mu(x),\tag{13.26}$$

to add another gauge term, and we arrive at the complete Lagrangian

$$\mathcal{L}(\Phi, W, B) = \frac{1}{2}\left(D_\mu\Phi\right)^\dagger D_\mu\Phi + V(\Phi) - \frac{1}{2g^2}\text{Tr}\left[W_{\mu\nu}W_{\mu\nu}\right] + \frac{1}{4}B_{\mu\nu}B_{\mu\nu}.\tag{13.27}$$

Let us consider the symmetry breaking case $m^2 < 0$, again in the unitary gauge, generalizing eq. (13.4). In the non-Abelian case, it fixes $\Phi^+(x) = 0$, $\Phi^0(x) \in \mathbb{R}_+$; hence, it selects the classical vacuum configuration of the Higgs field as

$$\Phi(x) = \begin{pmatrix}0\\v\end{pmatrix},\quad v \in \mathbb{R}_+.\tag{13.28}$$

Here, we observe a feature which is qualitatively new compared to Section 13.1. The vacuum configuration (13.28) is *invariant* under $U(1)$ gauge transformations of the type

---

[4] The group $SU(2)_R$ is known as the "custodial symmetry", which Section 13.4 is devoted to.

$$\Phi'(x) = \begin{pmatrix} \exp\left(i e \alpha(x)\right) & 0 \\ 0 & 1 \end{pmatrix} \Phi(x). \tag{13.29}$$

which have a U(1)$_Y$ hypercharge part, along with a diagonal SU(2)$_L$ part,

$$\begin{pmatrix} \exp\left(i e \alpha(x)\right) & 0 \\ 0 & 1 \end{pmatrix} =$$

$$\begin{pmatrix} \exp\left(i e \alpha(x)/2\right) & 0 \\ 0 & \exp\left(i e \alpha(x)/2\right) \end{pmatrix} \begin{pmatrix} \exp\left(i e \alpha(x)/2\right) & 0 \\ 0 & \exp\left(-i e \alpha(x)/2\right) \end{pmatrix}. \tag{13.30}$$

Hence, the choice of the vacuum state does not "break" the SU(2)$_L \times$ U(1)$_Y$ gauge symmetry completely. Instead, there remains a local U(1) symmetry, which we denote as U(1)$_{\text{em}}$, because we will soon identify it with the electromagnetic gauge group. Since that symmetry remains *unbroken*, despite the Higgs mechanism, there will be one massless gauge boson – the *photon*.

All other gauge bosons "eat" a Nambu–Goldstone boson and become massive. To see this, we consider again the fluctuations within the unitary gauge,

$$\Phi(x) = \begin{pmatrix} 0 \\ v + \sigma(x) \end{pmatrix}, \quad \sigma(x) \in \mathbb{R}. \tag{13.31}$$

Expanding in powers of the field $\sigma(x)$, the usual potential term yields

$$V(\Phi) = \frac{m^2}{2}(v + \sigma)^2 + \frac{\lambda}{4!}(v + \sigma)^4 = -m^2 \sigma^2 + C + \mathcal{O}(\sigma^4), \quad v^2 = -\frac{6m^2}{\lambda}, \tag{13.32}$$

so there is once again a massive σ-particle, namely, the *Higgs particle,* with

$$M_{\text{H}}^2 = 2m^2 > 0. \tag{13.33}$$

We anticipated in Chapter 12 that its mass was measured as $M_{\text{H}} \simeq 125\,\text{GeV}$, cf. eq. (12.30). Similarly, from the covariant derivative term in the Lagrangian we obtain

$$\frac{1}{2}(D_\mu \Phi)^\dagger D_\mu \Phi = \frac{1}{2} \left| \left(\partial_\mu + i g W_\mu^a \frac{\tau^a}{2} + i \frac{g'}{2} B_\mu \right) \begin{pmatrix} 0 \\ v + \sigma \end{pmatrix} \right|^2$$

$$= \frac{1}{2} \partial_\mu \sigma \partial_\mu \sigma + \frac{(v + \sigma)^2}{2} (0, 1) \left[\left(g W_\mu^a \frac{\tau^a}{2} + \frac{g'}{2} B_\mu \right) \left(g W_\mu^b \frac{\tau^b}{2} + \frac{g'}{2} B_\mu \right) \right] \begin{pmatrix} 0 \\ 1 \end{pmatrix}$$

$$= \frac{1}{2} \partial_\mu \sigma \partial_\mu \sigma + \frac{1}{8}(v + \sigma)^2 \left[g^2 W_\mu^1 W_\mu^1 + g^2 W_\mu^2 W_\mu^2 + \left(g W_\mu^3 - g' B_\mu \right) \left(g W_\mu^3 - g' B_\mu \right) \right]. \tag{13.34}$$

We infer that there are two *W-bosons* of mass

$$M_W = \frac{1}{2} g v. \tag{13.35}$$

Furthermore, the linear combination[5]

$$Z_\mu(x) = \frac{g W_\mu^3(x) - g' B_\mu(x)}{\sqrt{g^2 + g'^2}},$$ (13.36)

represents the *Z-boson* with the mass

$$M_Z = \frac{1}{2}\sqrt{g^2 + g'^2} \, v.$$ (13.37)

The orthonormal linear combination

$$A_\mu(x) = \frac{g' W_\mu^3(x) + g B_\mu(x)}{\sqrt{g^2 + g'^2}}$$ (13.38)

remains massless and describes the *photon*.

In order to write these linear combinations in a transparent form, we introduce the *Weinberg angle* (or weak mixing angle) θ_W,

$$\begin{pmatrix} A_\mu(x) \\ Z_\mu(x) \end{pmatrix} = \begin{pmatrix} \cos\theta_W & \sin\theta_W \\ -\sin\theta_W & \cos\theta_W \end{pmatrix} \begin{pmatrix} B_\mu(x) \\ W_\mu^3(x) \end{pmatrix},$$ (13.39)

such that

$$\frac{g}{\sqrt{g^2 + g'^2}} = \cos\theta_W, \qquad \frac{g'}{\sqrt{g^2 + g'^2}} = \sin\theta_W,$$ (13.40)

and therefore

$$\frac{M_W}{M_Z} = \frac{g}{\sqrt{g^2 + g'^2}} = \cos\theta_W.$$ (13.41)

The W- and Z-bosons have first been produced and directly detected in high-energy experiments at CERN in 1983. The experimental values of the masses are (Workman *et al.*, 2022)

$$M_W = 80.377(12) \text{ GeV}, \quad M_Z = 91.1876(21) \text{ GeV}.$$ (13.42)

Their ratio determines the absolute value of the Weinberg angle. Along with further ways of its determination, the observed value (at energy M_Z, in the so-called $\overline{\text{MS}}$ renormalization scheme) amounts to (Workman *et al.*, 2022)

$$\sin^2\theta_W = 0.23119(14).$$ (13.43)

In fact, there are a number of ways to measure $\sin^2\theta_W$ in high-energy experiments, and the results based on different methods are consistent with the value obtained from the ratio

---

[5] In the space spanned by the basis vectors W_μ^3 and B_μ, we recognize in eq. (13.34) the vector $(g, -g')$, which we normalize. This implies the canonical normalization of a vector field.

M_W/M_Z. This is a beautiful confirmation of the consistency of the Standard Model. On the downside, θ_W is one of the parameters which are *free* in the Standard Model – a prediction for its value would require a more fundamental theory, beyond the Standard Model. Grand Unified Theories (GUTs), which will be discussed in Chapter 26 and in a simplified form in Section 13.8, indeed provide predictions for θ_W.

13.3 Identification of the Electric Charge

The coupling constant to the photon field is the elementary charge e. The covariant derivative of the Higgs field reads

$$
\begin{aligned}
D_\mu \Phi &= \left[\partial_\mu + i g W_\mu^a \frac{\tau^a}{2} + i \frac{g'}{2} B_\mu \right] \begin{pmatrix} \Phi^+ \\ \Phi^0 \end{pmatrix} \\
&= \left[\partial_\mu + i g W_\mu^1 \frac{\tau^1}{2} + i g W_\mu^2 \frac{\tau^2}{2} + \frac{i}{2} \begin{pmatrix} g W_\mu^3 + g' B_\mu & 0 \\ 0 & -g W_\mu^3 + g' B_\mu \end{pmatrix} \right] \begin{pmatrix} \Phi^+ \\ \Phi^0 \end{pmatrix} \\
&= \left[\partial_\mu + i g W_\mu^1 \frac{\tau^1}{2} + i g W_\mu^2 \frac{\tau^2}{2} \right. \\
&\quad \left. + i \begin{pmatrix} \left[\frac{1}{2}(g^2 - g'^2) Z_\mu + g g' A_\mu \right] / \sqrt{g^2 + g'^2} & 0 \\ 0 & -\frac{1}{2}\sqrt{g^2 + g'^2}\, Z_\mu \end{pmatrix} \right] \begin{pmatrix} \Phi^+ \\ \Phi^0 \end{pmatrix}.
\end{aligned}
$$
(13.44)

We see that indeed only the component Φ^+ couples to the electromagnetic field A_μ, and we can read off its electric charge

$$
e = \frac{g g'}{\sqrt{g^2 + g'^2}}.
$$
(13.45)

We further confirm that the component Φ^0 is electrically neutral (as we anticipated in Chapter 12, footnote 8), and we see that the Z-boson is neutral as well; hence, it is often denoted as Z^0 (although Z already stands for "zero").

The charged W-bosons manifest themselves in the linear combinations

$$
W_\mu^\pm(x) = \frac{1}{\sqrt{2}} \left(W_\mu^1(x) \mp i W_\mu^2(x) \right).
$$
(13.46)

These are the fields which couple the Higgs field components Φ^+ and Φ^0. The corresponding vertices are shown in Figure 13.1: Due to the conservation of the electric charge, also the weak gauge bosons W^+ and W^- carry the electric charge which is indicated by their superscript.

Under the electromagnetic $U(1)_{\text{em}}$ gauge transformations of eq. (13.29), the various gauge fields transform as

$$
A_\mu'(x) = A_\mu(x) - \partial_\mu \alpha(x), \quad Z_\mu'(x) = Z_\mu(x), \quad W_\mu^{\pm\prime}(x) = \exp\left(\pm i e \alpha(x) \right) W_\mu^\pm(x). \tag{13.47}
$$

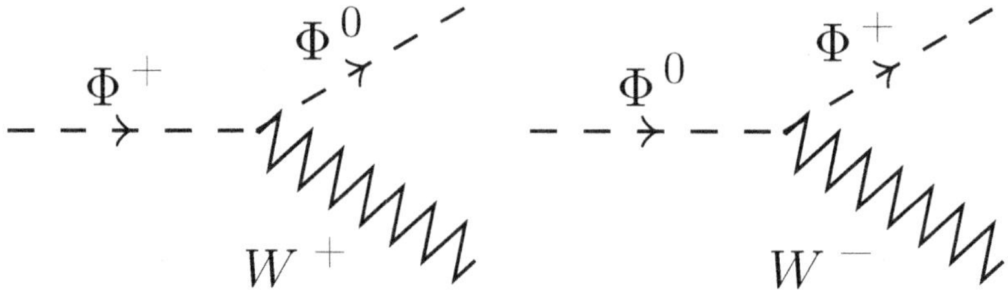

Fig. 13.1 The vertices for the transitions $\Phi^+ \rightarrow \Phi^0$ and $\Phi^0 \rightarrow \Phi^+$, which are mediated by the charged weak gauge bosons W^+ and W^-. In both cases, the electric charge is conserved.

At this point, let us take an overview over the fundamental Standard Model parameters that we have introduced until now. In Chapter 12, the Lagrangian of the Higgs field involved the constants m^2 and λ, which lead to the electroweak scale v and to the Higgs boson mass M_H (albeit the latter is still affected by gauge interactions). In this chapter, the W and B gauge fields come along with the coupling constants g and g'. They fix the electric charge unit e as well as the Weinberg angle θ_W, and (together with v) the gauge boson masses M_W and M_Z. Thus v, λ, g, and g' represent the full set of independent parameters that we have encountered so far; v has the dimension of a mass; λ, g, and g' are dimensionless. Hence, it does not come as a surprise that the particle masses M_H, M_W, and M_Z are all proportional to v, whereas the formulae for the dimensionless quantities e and θ_W only involve g and g'.

An overview over all free, fundamental parameters in the Standard Model will be provided in Section 17.8.

13.4 Accidental Custodial Symmetry

Let us again consider the matrix representation of the Higgs field

$$\Phi(x) = \begin{pmatrix} \Phi^0(x)^* & \Phi^+(x) \\ -\Phi^+(x)^* & \Phi^0(x) \end{pmatrix}, \tag{13.48}$$

which transforms as $\Phi'(x) = L\Phi(x)R^\dagger$ under the global $SU(2)_L \times SU(2)_R$ symmetry. Since only the $U(1)_Y$ subgroup of $SU(2)_R$ is gauged in the Standard Model, under $SU(2)_L \times U(1)_Y$ gauge transformations this field transforms as

$$\Phi'(x) = L(x)\Phi(x)R(x)^\dagger, \quad L(x) \in SU(2)_L,$$
$$R(x) = \begin{pmatrix} \exp\left(i g'\varphi(x)/2\right) & 0 \\ 0 & \exp\left(-i g'\varphi(x)/2\right) \end{pmatrix} \in U(1)_Y. \tag{13.49}$$

The corresponding covariant derivative takes the form

$$D_\mu \Phi(x) = \partial_\mu \Phi(x) + W_\mu(x)\Phi(x) - \Phi(x) i g' B_\mu(x)\frac{\tau^3}{2}. \tag{13.50}$$

Before the $U(1)_Y$ subgroup of $SU(2)_R$ is gauged (or equivalently when one sets $g' = 0$), $SU(2)_R$ is an exact global symmetry, known as the *custodial symmetry*. Once $U(1)_Y$ is gauged, the custodial symmetry is explicitly broken, but it keeps a status as an approximate global symmetry, since the coupling g' is weak. With g' included, we have obtained in eq. (13.41) the tree-level mass ratio

$$\rho = \frac{M_W^2}{M_Z^2 \cos^2 \theta_W} = 1. \tag{13.51}$$

This ratio is still modified by quantum corrections, due to gauge and Higgs boson loops. However, at weak g' these corrections are small; they only set in at $\mathcal{O}(g'^2) = \mathcal{O}(\theta_W^2)$. In fact, the experimental values given in eqs. (13.42), (13.43) confirm that $\rho \approx 1$ is valid in Nature. This *protection* from large quantum corrections is based on the approximate custodial symmetry, and it motivates this expression. For a pedagogical review and further details, we refer to Willenbrock (2004) and references therein.

Generally, gauge theories contain redundant (*i.e.* unphysical) degrees of freedom which do not affect the physics due to gauge invariance. Hence, in order to maintain only the physically relevant degrees of freedom, gauge symmetries must not be broken. Global symmetries, on the other hand, are usually only approximate and arise due to some hierarchies of energy scales, whose origin may or may not be understood.

Let us now ask whether we understand the origin of the approximate custodial symmetry. In particular, we ask whether there are other sources of custodial symmetry breaking besides the weak $U(1)_Y$ gauge interactions. While such symmetry breaking terms can always be constructed using sufficiently many derivatives or field values, here we limit ourselves to renormalizable interactions, which are the ones that dominate the physics at low (or moderate) energies. Since

$$\mathbf{\Phi}(x)^\dagger \mathbf{\Phi}(x) = \mathbf{\Phi}(x)\mathbf{\Phi}(x)^\dagger = \left(|\Phi^0(x)|^2 + |\Phi^+(x)|^2 \right) \mathbb{1} \tag{13.52}$$

is proportional to the unit-matrix, one cannot construct any $SU(2)_L \times U(1)_Y$-invariant terms without derivatives that explicitly break the custodial $SU(2)_R$ symmetry. For example, the terms $\text{Tr}\left[\mathbf{\Phi}^\dagger \mathbf{\Phi}\tau^3\right]$ and $\text{Tr}\left[\mathbf{\Phi}^\dagger \mathbf{\Phi}\mathbf{\Phi}^\dagger \mathbf{\Phi}\tau^3\right]$ simply vanish, and

$$\text{Tr}\left[\mathbf{\Phi}(x)^\dagger \mathbf{\Phi}(x)\tau^3 \mathbf{\Phi}(x)^\dagger \mathbf{\Phi}(x)\tau^3\right] = 2\left(|\Phi^0(x)|^2 + |\Phi^+(x)|^2 \right)^2 \tag{13.53}$$

just reduces to the standard $SU(2)_L \times SU(2)_R$-invariant quartic self-coupling.

By involving two covariant derivatives, we can further construct the additional term $\text{Tr}\left[(D_\mu \mathbf{\Phi})^\dagger D_\mu \mathbf{\Phi}\,\tau^3\right]$, which may seem to explicitly break $SU(2)_L \times SU(2)_R$ down to $SU(2)_L \times U(1)_Y$. If this were the case, this term should also be included in the Standard Model Lagrangian with an adjustable prefactor. If this prefactor would not be unnaturally small, the custodial symmetry should suffer from strong explicit breaking, and it would not even remain as a useful approximate symmetry. Only if the prefactor of such a term would be small (perhaps due to some not yet understood hierarchy of energy scales), the symmetry would remain only weakly broken. It would then be puzzling why the symmetry is not more strongly broken.

However, such a puzzle does not exist for the custodial symmetry of the Standard Model: It turns out that $\text{Tr}\left[(D_\mu \mathbf{\Phi})^\dagger D_\mu \mathbf{\Phi}\,\tau^3\right]$ vanishes as well. This can be seen by inserting the parameterization (12.19) of $\mathbf{\Phi}$ and the electroweak covariant derivative of eq. (13.44)

Indeed, besides the gauge coupling g', in the gauge–Higgs sector there is no other renormalizable interaction that explicitly breaks the custodial symmetry.[6] One says that the custodial symmetry is *accidental*. It simply arises because g' is relatively small and no other renormalizable symmetry breaking terms exist.

13.5 Variants of the Standard Model with Modified Gauge Symmetry

In this section, we explore hypothetical variants of the Standard Model by modifying its gauge symmetry $SU(2)_L \times U(1)_Y$, in order to understand how this would affect the physics.

First, let us imagine that we gauge not only $U(1)_Y$ but the full symmetry $SU(2)_R$. In that case, the corresponding covariant derivative would take the form

$$D_\mu \Phi(x) = \partial_\mu \Phi(x) + W_\mu(x)\Phi(x) - \Phi(x)X_\mu(x), \qquad (13.54)$$

where $X_\mu(x) = \mathrm{i}g'X_\mu^a(x)\tau^a/2$ is a hypothetical, non-Abelian gauge field that transforms as

$$X'_\mu(x) = R(x)\left[X_\mu(x) + \partial_\mu\right]R(x)^\dagger \qquad (13.55)$$

under $SU(2)_R$ gauge transformations. Since only the $U(1)_Y$ subgroup of $SU(2)_R$ is gauged in the Standard Model, the hypothetical non-Abelian gauge field X_μ then reduces to the Abelian Standard Model gauge field B_μ, *i.e.*

$$X_\mu(x) = \mathrm{i}g'B_\mu(x)\frac{\tau^3}{2}. \qquad (13.56)$$

When $SU(2)_L \times SU(2)_R$ is gauged, the global custodial symmetry disappears because it is turned into a gauge redundancy.

As before, as long as the $SU(2)_L \times SU(2)_R$ symmetry is global, it breaks spontaneously to $SU(2)_{L=R}$, which gives rise to three massless Nambu–Goldstone bosons. However, since we now have six gauge field components (instead of the four in the Standard Model), now three gauge fields A_μ^a (and not only the photon field $A_\mu = A_\mu^3$) remain unaffected by the Higgs mechanism, while three others, Z_μ^a, "eat" the Nambu–Goldstone bosons and become massive

$$Z_\mu^a(x) = \frac{gW_\mu^a(x) - g'X_\mu^a(x)}{\sqrt{g^2 + g'^2}}, \quad A_\mu^a(x) = \frac{g'W_\mu^a(x) + gX_\mu^a(x)}{\sqrt{g^2 + g'^2}}. \qquad (13.57)$$

One might expect the three "photons" $A_\mu^a(x)$ to be massless, because they are unaffected by the Higgs mechanism. However, this is not the case. The $A_\mu^a(x)$ are non-Abelian $SU(2)_{L=R}$ gauge fields, which are confined and would thus form what one might call "photon balls" (in analogy to the glueballs that are formed by confined gluons, as we will discuss in Chapter 14.)

Finally, let us consider a variant of the Standard Model in which we gauge only the unbroken $SU(2)_{L=R}$ subgroup of $SU(2)_L \times SU(2)_R$. In that case, the left and right gauge

---

[6] In Chapter 15, we will couple the Higgs field to leptons and quarks, which gives rise to additional terms that do break the custodial symmetry.

fields are the same, such that $g'X_\mu^a(x) = gW_\mu^a(x) = \sqrt{g^2 + g'^2}A_\mu^a(x)/2$ and $Z_\mu^a(x) = 0$. Then, the three previously massless Nambu–Goldstone bosons remain uneaten. However, just as the gauging of the $U(1)_Y$ subgroup explicitly breaks the custodial symmetry $SU(2)_R$ in the Standard Model, here the gauging of the $SU(2)_{L=R}$ subgroup explicitly breaks $SU(2)_L \times SU(2)_R$. As a result, first of all, the uneaten Nambu–Goldstone bosons pick up a non-zero mass via loop effects and thus become pseudo-Nambu–Goldstone bosons. Even more importantly, since the pseudo-Nambu–Goldstone bosons form a triplet under $SU(2)_{L=R}$ and are hence "charged" under the gauge group of the non-Abelian "photons" A_μ^a, they are confined and thus also pick up a mass non-perturbatively. For example, the triplet of pseudo-Nambu–Goldstone bosons can form a confined $SU(2)_{L=R}$ singlet together with the triplet of non-Abelian "photons".

13.6 Scalar Electrodynamics on the Lattice

The previous discussion of gauge theories was essentially at a classical level. The quantization of gauge theories is a non-trivial issue. In Chapter 6, the simplest gauge theory – an Abelian theory of free photons – has been quantized canonically. In Section 11.9, we have quantized compact $U(1)$ lattice gauge theory using the functional integral. In this section, we do the same for non-compact $U(1)$ lattice gauge theory coupled to a complex scalar field, which is nothing but the lattice formulation of scalar electrodynamics.

As before, we consider a 4-dimensional hypercubic Euclidean space–time lattice with spacing a. A complex scalar field $\Phi_x \in \mathbb{C}$ is associated with the lattice sites x, and its self-interaction potential $V(\Phi_x) = \lambda(|\Phi_x|^2 - v^2)^2/4!$ takes the same form as in the continuum. The corresponding term in the lattice action is given by

$$S[\Phi] = a^4 \sum_x V(\Phi_x) = a^4 \sum_x \frac{\lambda}{4!} \left(|\Phi_x|^2 - v^2\right)^2. \tag{13.58}$$

Under $U(1)$ lattice gauge transformations α_x, which are naturally associated with the lattice sites, the scalar field transforms just as in the continuum,

$$\Phi'_x = \exp\left(iQe\alpha_x\right)\Phi_x. \tag{13.59}$$

The lattice variants of the covariant derivative of the continuum theory, which is given by $D_\mu \Phi(x) = \left(\partial_\mu + iQeA_\mu(x)\right)\Phi(x)$, are naturally associated with the links connecting neighboring lattice sites x and $x + \hat{\mu}$. For this purpose, we first associate a real-valued non-compact vector potential, $A_{x,\mu} \in \mathbb{R}$, with the links. Under lattice gauge transformations, it transforms as

$$A'_{x,\mu} = A_{x,\mu} - \frac{1}{a}\left(\alpha_{x+\hat{\mu}} - \alpha_x\right), \tag{13.60}$$

which is the finite-difference variant of the continuum relation $A'_\mu(x) = A_\mu(x) - \partial_\mu \alpha(x)$. We also introduce a compact link parallel transporter for the charge Q as

$$U_{x,\mu}^Q = \exp\left(iQeaA_{x,\mu}\right) \quad \Rightarrow \quad U_{x,\mu}^{Q\,'} = \exp\left(iQe\alpha_x\right)U_{x,\mu}^Q \exp\left(-iQe\alpha_{x+\hat{\mu}}\right). \tag{13.61}$$

On the lattice, it is natural to distinguish forward and backward covariant derivatives

$$\left(D^{\mathrm{f}}\Phi\right)_{x,\mu} = \frac{1}{a}\left(U^{Q}_{x,\mu}\Phi_{x+\hat{\mu}} - \Phi_x\right) \quad\Rightarrow\quad \left(D^{\mathrm{f}}\Phi\right)'_{x,\mu} = \exp\left(\mathrm{i}Qe\alpha_x\right)\left(D^{\mathrm{f}}\Phi\right)_{x,\mu},$$

$$\left(D^{\mathrm{b}}\Phi\right)_{x,\mu} = \frac{1}{a}\left(\Phi_{x+\hat{\mu}} - U^{Q*}_{x,\mu}\Phi_x\right) \quad\Rightarrow\quad \left(D^{\mathrm{b}}\Phi\right)'_{x,\mu} = \exp\left(\mathrm{i}Qe\alpha_{x+\hat{\mu}}\right)\left(D^{\mathrm{b}}\Phi\right)_{x,\mu}. \quad (13.62)$$

The lattice variant of the contribution $|D_\mu\Phi(x)|^2/2$ to the Lagrangian of the continuum theory then takes the form

$$\frac{1}{2a^2}\left|\left(D^{\mathrm{f}}\Phi\right)_{x,\mu}\right|^2 = \frac{1}{2a^2}\left|\left(D^{\mathrm{b}}\Phi\right)_{x,\mu}\right|^2 = \frac{1}{2a^2}\left(|\Phi_x|^2 + |\Phi_{x+\hat{\mu}}|^2 - \Phi_x^* U^{Q}_{x,\mu}\Phi_{x+\hat{\mu}} - \Phi_{x+\hat{\mu}}^* U^{Q*}_{x,\mu}\Phi_x\right). \quad (13.63)$$

The corresponding contribution to the lattice action is given by

$$S[\Phi,A] = a^4 \sum_{x,\mu} \frac{1}{2a^2}\left(|\Phi_x|^2 + |\Phi_{x+\hat{\mu}}|^2 - \Phi_x^* U^{Q}_{x,\mu}\Phi_{x+\hat{\mu}} - \Phi_{x+\hat{\mu}}^* U^{Q*}_{x,\mu}\Phi_x\right). \quad (13.64)$$

It should be noted that, although $S[\Phi,A]$ depends only on the compact link variables U^Q, those depend on the more fundamental, non-compact vector potential A.

The contribution to the lattice action that resembles $F_{\mu\nu}F_{\mu\nu}/4$ in the continuum theory is based on the gauge-invariant non-compact Abelian lattice field strength

$$F_{x,\mu\nu} = \frac{1}{a}\left(A_{x+\hat{\mu},\nu} - A_{x,\nu} - A_{x+\hat{\nu},\mu} + A_{x,\mu}\right), \quad (13.65)$$

which is the finite-difference variant of the continuum expression $F_{\mu\nu}(x) = \partial_\mu A_\nu(x) - \partial_\nu A_\mu(x)$. The corresponding contribution to the lattice action reads

$$S[A] = a^4 \sum_{x,\mu,\nu} \frac{1}{4}F_{x,\mu\nu}^2 = a^4 \sum_{x,\mu,\nu} \frac{1}{4a^2}\left(A_{x,\mu} + A_{x+\hat{\mu},\nu} - A_{x+\hat{\nu},\mu} - A_{x,\nu}\right)^2. \quad (13.66)$$

In the last equality, we have reordered the link variables in the order of the parallel transport around the plaquette. While this is not necessary in an Abelian theory, it would be inevitable in the non-Abelian case (in which one works with parallel transporters).

The measure of the functional integral for lattice scalar QED takes the form

$$\int \mathcal{D}\Phi\mathcal{D}A = \prod_x \frac{a}{2\pi}\int_{\mathbb{C}} d\Phi_x \prod_{y,\mu} \sqrt{\frac{a}{2\pi}}\int_{\mathbb{R}} dA_{y,\mu}, \quad (13.67)$$

which corresponds to independent integrations over all degrees of freedom over their respective integration ranges $\mathbb{C}$ and $\mathbb{R}$. The gauge orbits of the scalar field are circles around the origin in the complex plane $\mathbb{C}$. The angular integration over all gauge copies of Φ_x just contributes a finite factor 2π, which cancels in all expectation values.

In contrast to the gauge orbits of the compact link parallel transporters $U^{Q}_{x,\mu}$, the gauge orbits of the non-compact vector potential $A_{x,\mu}$ are infinitely large. This implies that, in contrast to compact U(1) or to non-Abelian lattice gauge theories, non-compact U(1) lattice gauge theory requires gauge fixing, in order to make the measure of the functional integral mathematically well-defined.

In lattice QED, gauge fixing can be realized by adding a term

$$S_{\mathrm{gf}}[A] = a^4 \sum_x \frac{1}{2\xi}\left[\sum_\mu \frac{1}{a}\left(A_{x+\hat{\mu},\mu} - A_{x,\mu}\right)\right]^2 \quad (13.68)$$

to the lattice action. This is a finite-difference variant of the expression $\left[\partial_\mu A_\mu(x)\right]^2/2\xi$, which can be added to the continuum Lagrangian. In the limit $\xi \to 0$, it enforces the Lorenz–Landau gauge condition, as we saw in Section 6.9.

The functional integral of scalar QED on the lattice finally takes the form

$$Z = \int \mathcal{D}\Phi \mathcal{D}A \exp\left(-S[\Phi,A] - S[\Phi] - S[A] - S_{\text{gf}}[A]\right). \tag{13.69}$$

As we discussed in Section 1.7, this resembles a system of classical statistical mechanics, which can be treated numerically with Monte Carlo simulation techniques. Using the method of importance sampling, one then generates a Markov chain of field configurations, such that the probability to generate a configuration $[\Phi,A]$ is proportional to its Boltzmann weight (cf. Appendix H). Observables $O[\Phi,A]$ are then evaluated on the generated field configurations and averaged. This yields non-perturbative results, which are, however, affected by statistical errors due to the finite length of the Markov chain. When efficient Monte Carlo algorithms are available, very accurate numerical results can be obtained. By carefully analyzing effects of the finite lattice volume, using, for example, the so-called finite-size scaling method (Fisher and Barber, 1972; Privman and Fisher, 1984; Landau and Binder, 2014), one can obtain precise results on the phase structure of a lattice model. Scalar lattice QED exists in two phases: a Coulomb phase with massless photons and a Higgs phase in which the photon is massive.

Let us return to the issue of *triviality*. While there is even a rigorous proof (Aizenman and Duminil-Copin, 2021), already for a long time, detailed lattice calculations provided overwhelming evidence that 4-d $\lambda|\Phi|^4$ theory is "trivial", *i.e.* when one insists on completely removing the cut-off, the theory becomes non-interacting. The same is expected for scalar QED. However, when one stays with a large (but not infinite) cut-off, one can still obtain meaningful physical results. Triviality implies that the scalar self-coupling λ is irrelevant in the renormalization group sense (cf. Chapter 5). Taking this into account, one can set the bare lattice coupling to $\lambda = \infty$ and thus enforce a fixed length, $|\Phi_x| = v$, of the scalar field. This reduces the relevant parameter space and thus significantly simplifies the numerical investigations. Rescaling the fields as

$$\widetilde{\Phi}_x = \frac{\Phi_x}{v}, \quad \widetilde{A}_{x,\mu} = ea A_{x,\mu}, \tag{13.70}$$

up to an irrelevant additive constant, the contribution of the scalar field to the lattice action then reduces to

$$S[\Phi,A] + S[\Phi] \to S[\widetilde{\Phi},\widetilde{A}] = -\sum_{x,\mu} \kappa \, \text{Re}\left[\widetilde{\Phi}_x^* \, U_{x,\mu}^Q \, \widetilde{\Phi}_{x+\hat{\mu}}\right], \quad U_{x,\mu}^Q = \exp\left(iQ\widetilde{A}_{x,\mu}\right). \tag{13.71}$$

Here, we have introduced the *hopping parameter* $\kappa = v^2 a^2$. Similarly, the action of the pure gauge theory takes the form

$$S[A] \to S[\widetilde{A}] = \sum_{x,\mu,\nu} \frac{1}{4e^2} \left(\widetilde{A}_{x,\mu} + \widetilde{A}_{x+\hat{\mu},\nu} - \widetilde{A}_{x+\hat{\nu},\mu} - \widetilde{A}_{x,\nu}\right)^2. \tag{13.72}$$

As a consequence, in the limit $\lambda \to \infty$, after rescaling the fields, besides the fixed charge $Q \in \mathbb{Z}$, the physics depends only on the two coupling constants e^2 and κ.

Figure 13.2 shows a sketch of the corresponding phase diagram of lattice scalar QED, which has been obtained using a combination of analytic and numerical methods (Borgs

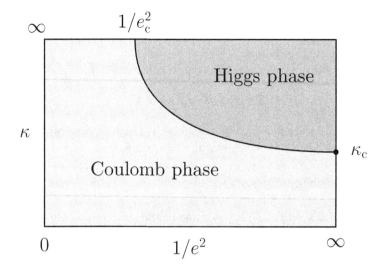

Fig. 13.2 The schematic phase diagram of non-compact lattice scalar QED with a complex scalar field of charge $Q = 1$ at quartic self-coupling $\lambda = \infty$ in eq. (13.58). The system is in the Coulomb phase with a massless photon, except for the Higgs phase at small gauge coupling e and large vacuum expectation value v, which corresponds to large $\kappa = v^2 a^2$. The phase transition line $\kappa_c(e^2)$ is expected to be of first order, except at $e^2 = 0$ where it ends in a second-order critical point (indicated by a dot).

and Nill, 1987). At $\kappa = 0$, the scalar field decouples and the model reduces to a free theory of massless photons, forming a Coulomb phase. At $e^2 \to 0$, the field strength $F_{x,\mu\nu}$ is forced to zero, such that the field $A_{x,\mu}$ can be gauge transformed to zero. The model then reduces to a $\lambda|\Phi|^4$ theory with just a global U(1) symmetry. For small $\kappa < \kappa_c$, the U(1) symmetry is unbroken, while for $\kappa > \kappa_c$ it is spontaneously broken, thus giving rise to a single Nambu–Goldstone boson. It turns out that, at the critical coupling κ_c, the symmetric and the broken phases are separated by a second-order phase transition. By approaching this *critical point* (cf. Appendix I), one can take a continuum limit of scalar QED, which is likely to be a trivial (*i.e.* non-interacting) theory.

The phase transition at $\kappa = \kappa_c$ and $e^2 = 0$ is the endpoint of a line $\kappa_c(e^2)$ of first-order phase transitions, which extends throughout the phase diagram up to $\kappa = \infty$. This line separates the Coulomb phase with massless photons and massive charged scalar infraparticles (cf. Section 7.3) from the Higgs phase with a massive photon that has "eaten" the Nambu–Goldstone boson. In addition, this phase contains a massive scalar Higgs particle.

13.7 SU(2)$_L$ Gauge–Higgs Model on the Lattice

In this section, we consider a non-Abelian lattice gauge theory of the weak interaction by gauging the SU(2)$_L$ symmetry. For simplicity, we do not gauge the U(1)$_Y$ symmetry, thus leaving SU(2)$_R$ as an exact custodial symmetry. It is convenient to use the matrix representation of the complex doublet Higgs field

$$\Phi_x = \begin{pmatrix} \Phi_x^{0*} & \Phi_x^+ \\ -\Phi_x^{+*} & \Phi_x^0 \end{pmatrix}, \quad \Phi_x' = L_x \Phi_x R^\dagger. \tag{13.73}$$

The Higgs field $\mathbf{\Phi}_x$, as well as the gauge transformations $L_x \in SU(2)_L$, are naturally associated with the lattice sites x, while $SU(2)_R$ is just a global symmetry that acts uniformly everywhere in space–time.

The quartic self-interaction of the Higgs field is described by

$$S[\mathbf{\Phi}] = a^4 \sum_x V(\mathbf{\Phi}_x) = a^4 \sum_x \frac{\lambda}{4!} \left(\left(\frac{1}{2} \mathrm{Tr} \left[\mathbf{\Phi}_x^\dagger \mathbf{\Phi}_x \right] \right)^2 - v^2 \right)^2,$$

$$\frac{1}{2} \mathrm{Tr} \left[\mathbf{\Phi}_x^\dagger \mathbf{\Phi}_x \right] = |\Phi_x^+|^2 + |\Phi_x^0|^2. \tag{13.74}$$

Following Section 11.10, next we introduce a link-based $SU(2)_L$ lattice gauge field $U_{x,\mu} \in SU(2)_L$ that transforms as

$$U'_{x,\mu} = L_x U_{x,\mu} L_{x+\hat\mu}^\dagger. \tag{13.75}$$

Its self-interaction can be described by the standard Wilson plaquette action

$$S[U] = \frac{2}{g^2} \sum_{x,\mu > \nu} \mathrm{Tr} \left[\mathbb{1} - \frac{1}{2} U_{x,\mu} U_{x+\hat\mu,\nu} U_{x+\hat\nu,\mu}^\dagger U_{x,\nu}^\dagger - \frac{1}{2} U_{x,\nu} U_{x+\hat\nu,\mu} U_{x+\hat\mu,\nu}^\dagger U_{x,\mu}^\dagger \right]. \tag{13.76}$$

Again it is natural to distinguish a forward from a backward variant of the covariant derivative of the continuum theory, $D_\mu \mathbf{\Phi}(x) = \left(\partial_\mu + W_\mu(x) \right) \mathbf{\Phi}(x)$, such that

$$\left(D^{\mathrm{f}} \mathbf{\Phi} \right)_{x,\mu} = \frac{1}{a} \left(U_{x,\mu} \mathbf{\Phi}_{x+\hat\mu} - \mathbf{\Phi}_x \right) \quad \Rightarrow \quad \left(D^{\mathrm{f}} \mathbf{\Phi} \right)'_{x,\mu} = L_x \left(D^{\mathrm{f}} \mathbf{\Phi} \right)_{x,\mu} R^\dagger,$$

$$\left(D^{\mathrm{b}} \mathbf{\Phi} \right)_{x,\mu} = \frac{1}{a} \left(\mathbf{\Phi}_{x+\hat\mu} - U_{x,\mu}^\dagger \mathbf{\Phi}_x \right) \quad \Rightarrow \quad \left(D^{\mathrm{b}} \mathbf{\Phi} \right)'_{x,\mu} = L_{x+\hat\mu} \left(D^{\mathrm{b}} \mathbf{\Phi} \right)_{x,\mu} R^\dagger. \tag{13.77}$$

The lattice variant of the contribution $\mathrm{Tr} \left[(D_\mu \mathbf{\Phi})^\dagger D_\mu \mathbf{\Phi} \right]/4$ to the Lagrangian of the continuum theory is then given by

$$\frac{1}{4a^2} \mathrm{Tr} \left[\left(D^{\mathrm{f}} \mathbf{\Phi} \right)_{x,\mu}^\dagger \left(D^{\mathrm{f}} \mathbf{\Phi} \right)_{x,\mu} \right] = \frac{1}{4a^2} \mathrm{Tr} \left[\left(D^{\mathrm{b}} \mathbf{\Phi} \right)_{x,\mu}^\dagger \left(D^{\mathrm{b}} \mathbf{\Phi} \right)_{x,\mu} \right]$$

$$= \frac{1}{4a^2} \mathrm{Tr} \left[\mathbf{\Phi}_x^\dagger \mathbf{\Phi}_x + \mathbf{\Phi}_{x+\hat\mu}^\dagger \mathbf{\Phi}_{x+\hat\mu} - \mathbf{\Phi}_x^\dagger U_{x,\mu} \mathbf{\Phi}_{x+\hat\mu} - \mathbf{\Phi}_{x+\hat\mu}^\dagger U_{x,\mu}^\dagger \mathbf{\Phi}_x \right]. \tag{13.78}$$

The corresponding contribution to the lattice action thus takes the form

$$S[\mathbf{\Phi}, U] = a^4 \sum_{x,\mu} \frac{1}{4a^2} \mathrm{Tr} \left[\mathbf{\Phi}_x^\dagger \mathbf{\Phi}_x + \mathbf{\Phi}_{x+\hat\mu}^\dagger \mathbf{\Phi}_{x+\hat\mu} - \mathbf{\Phi}_x^\dagger U_{x,\mu} \mathbf{\Phi}_{x+\hat\mu} - \mathbf{\Phi}_{x+\hat\mu}^\dagger U_{x,\mu}^\dagger \mathbf{\Phi}_x \right]. \tag{13.79}$$

The measure for the functional integration over the $SU(2)_L$ lattice gauge field is finite and fully regularized, such that, unlike for non-compact $U(1)$ lattice gauge theory, gauge fixing is not necessary. The functional integral for the $SU(2)_L$ gauge–Higgs model hence takes the form

$$Z = \int \mathcal{D}\mathbf{\Phi} \mathcal{D}U \exp \left(-S[\mathbf{\Phi}, U] - S[\mathbf{\Phi}] - S[U] \right)$$

$$= \prod_x \left(\frac{a}{2\pi} \right)^2 \int_{\mathbb{C}^2} d\mathbf{\Phi}_x \prod_{y,\mu} \int_{SU(2)} dU_{y,\mu} \exp \left(-S[\mathbf{\Phi}, U] - S[\mathbf{\Phi}] - S[U] \right). \tag{13.80}$$

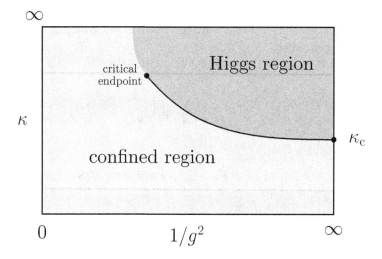

Fig. 13.3 The schematic phase diagram for the SU(2)$_L$ lattice gauge–Higgs model with a complex doublet scalar Higgs field at quartic self-coupling $\lambda = \infty$. At $\kappa = 0$, the Higgs field decouples and the W-bosons of the resulting SU(2)$_L$ Yang–Mills theory are confined inside W-balls. At small gauge coupling g and large vacuum expectation value v (corresponding to large values of $\kappa = v^2 a^2$), the system is in the Higgs region with massive W-bosons. The line $\kappa_c(g^2)$ of first-order phase transitions terminates at two critical endpoints (indicated by two dots). As a result, the confined and the Higgs region are continuously connected, without any order parameter that could distinguish them.

Again, numerical simulations have been used to analyze the phase structure of this model (Lang *et al.*, 1981; Jersák *et al.*, 1985; Montvay, 1985).

For simplicity, one again sets $\lambda = \infty$ which enforces at each lattice site $\Phi_x^\dagger \Phi_x = \Phi_x \Phi_x^\dagger = v^2 \mathbb{1}$. Rescaling the Higgs field to $\widetilde{\Phi}_x = \Phi_x/v$ and again ignoring a trivial additive constant, the Higgs field contribution to the action then simplifies to

$$S[\Phi, U] + S[\Phi] \rightarrow S[\widetilde{\Phi}, U] = - \sum_{x,\mu} \frac{\kappa}{2} \mathrm{Re}\, \mathrm{Tr} \left[\widetilde{\Phi}_x^\dagger U_{x,\mu} \widetilde{\Phi}_{x+\hat{\mu}} \right]. \tag{13.81}$$

As a result, for $\lambda \rightarrow \infty$, after field rescaling, the physics depends only on g^2 and $\kappa = v^2 a^2$.

Figure 13.3 shows a sketch of the phase diagram of the SU(2)$_L$ gauge–Higgs model. At hopping parameter $\kappa = 0$, the Higgs field again decouples. However, unlike the non-compact Abelian pure photon theory, the non-Abelian SU(2)$_L$ Yang–Mills theory is not a free field theory. As we will discuss in Chapter 14, this theory is actually confining. Despite the fact that, for $\kappa = 0$, no Higgs mechanism operates, the W-bosons then pick up a non-perturbatively generated mass from their self-interaction and form what one might want to call "W-balls" again in analogy to the glueballs formed by confined gluons (cf. Section 14.12). At $g^2 = 0$, the non-Abelian field strength again vanishes and the gauge field can be removed by a gauge transformation. The model then reduces to a scalar field theory with a global SU(2)$_L$ × SU(2)$_R$ symmetry and with a critical coupling κ_c. For $\kappa > \kappa_c$, this symmetry is spontaneously broken to SU(2)$_{L=R}$, which gives rise to three massless Nambu–Goldstone bosons. Once $g^2 \neq 0$, the Nambu–Goldstone bosons are "eaten" and become the longitudinal polarization states of three massive W-bosons, which are degenerate due to the exact global SU(2)$_R$ custodial symmetry.

For $g^2 = 0$, the transition that separates the broken phase from the symmetric phase at $\kappa < \kappa_c$ again turns out to be of second order. This critical point is again an endpoint of a first-order phase transition line given by $\kappa_c(g^2)$. However, in contrast to the non-compact Abelian model, this line terminates inside the phase diagram and does not extend to $\kappa = \infty$. In fact, at $\kappa = \infty$ the model is trivial without any phase transition. This is easy to see in the unitary gauge, $\widetilde{\Phi}_x = \mathbb{1}$, because then $\kappa = \infty$ enforces the trivial gauge field $U_{x,\mu} = \mathbb{1}$. Since the phase transition line terminates in another critical endpoint, the Higgs region and the confined region are continuously connected (Fradkin and Shenker, 1979). This implies that a state of a massive W-boson can be deformed continuously into a W-ball state. In 1981, Larry Abbott and Edward Farhi asked the provocative question: "Are the weak interactions strong?" (Abbott and Farhi, 1981). While this seems not to be the case in Nature, in the confined region of the phase diagram of the $SU(2)_L$ gauge–Higgs model this is indeed so.

The continuity between the Higgs and the confined regions of the phase diagram also implies that there is no order parameter that can qualitatively distinguish the Higgs from the confined phase. As we will discuss in Section 14.13, this is due to the fact that the $\mathbb{Z}(2)$ center symmetry of the $SU(2)_L$ gauge symmetry is explicitly broken by the Higgs field. Still, it is possible to distinguish a Higgs confinement phase (with massive W-bosons or W-balls) from a Coulomb phase (with a massless gauge boson). In Section 14.11, we will investigate the so-called Fredenhagen–Marcu operator that achieves exactly that.

13.8 Small Electroweak Unification

This section discusses physics beyond the Standard Model and may be skipped in a first reading. At this point, we return to continuum notation.

As we have seen, the gauge groups $SU(2)_L$ and $U(1)_Y$ come along with two distinct gauge couplings g and g'. Hence, in the Standard Model the electroweak interactions are *not* truly unified. In the next chapter, we will also include the strong interaction with the gauge group $SU(3)_c$, which is associated with yet another gauge coupling g_s. Hence, the full gauge group of the Standard Model, $SU(3)_c \times SU(2)_L \times U(1)_Y$, has three gauge couplings. In the framework of Grand Unified Theories (GUTs), which are extensions of the Standard Model, one embeds the electroweak and strong interactions in one single, simple gauge group, *e.g.*, $SU(5)$ or $Spin(10)$ (the universal covering group of $SO(10)$), or the exceptional Lie group $E(6)$, which relates g, g', and g_s.

The gauge symmetries $SU(5)$ or $SO(10)$ are too large to be manifest at low temperatures. They must be "spontaneously broken" down to the $SU(3)_c \times SU(2)_L \times U(1)_Y$ symmetry of the Standard Model. In GUTs, this happens at temperatures around 10^{16} GeV, which were realized in the Universe about 10^{-34} sec after the Big Bang. At present (and in the foreseeable future), these energy scales cannot be probed experimentally.[7] Hence, GUTs

---

[7] While accelerator experiments are far from this energy scale, cosmic rays hit our atmosphere with energies up to $\mathcal{O}(10^{11})$ GeV, which triggers extensive air showers. Although their flux is very low and their energy is significantly reduced in the center-of-mass frame, this might possibly open a window for observations which is not that far below the energy scale that we are referring to.

rely on theoretical arguments and sometimes on speculation. GUTs may naturally explain light neutrino masses and are consistent with a relatively light Higgs boson. In addition, GUTs imply subtle effects beyond Standard Model physics, like proton decay, which can be tested experimentally. We will discuss GUTs in detail in Chapter 26.

To illustrate the ideas behind GUTs with a simple example, let us unify only the electroweak gauge interactions by embedding $SU(2)_L \times U(1)_Y$ into one single gauge group. Since the group $SU(2)_L \times U(1)_Y$ has two commuting generators, *i.e.* its rank is $1 + 1 = 2$, the embedding unified group must also have a rank of at least 2. There are three so-called simple Lie groups of rank 2 – the special unitary group $SU(3)$ (of dimension 8), the symplectic group $Sp(2)$ (of dimension 10), and the exceptional group $G(2)$ (of dimension 14), which contains $SU(3)$ as a subgroup. Hence, the minimal unifying group that contains $SU(2)_L \times U(1)_Y$ is $SU(3)$ (not to be confused with the color gauge group $SU(3)_c$, although their group theory is the same). We will return to small unification within the groups $SU(3)$, $G(2)$, $SU(4) = Spin(6)$, and $Spin(7)$ in Section 26.13.

When the electroweak interactions of the Standard Model are embedded in the gauge group $SU(3)$, half of the gauge bosons can be identified with known particles: $SU(2)_L$ has three W-bosons, and $U(1)_Y$ has one B-boson which, together with W^3, forms the Z-boson and the photon, as we have seen. The remaining 4 gauge bosons of $SU(3)$ are new hypothetical particles, which we call X^1, X^2 and Y^1, Y^2. In order to make these particles heavy, the $SU(3)$ symmetry must be "spontaneously broken" to $SU(2)_L \times U(1)_Y$. This can be achieved with the Higgs mechanism, in this case using an 8-component scalar field transforming in the adjoint representation $\{8\}$ of $SU(3)$. We write

$$\Xi(x) = \Xi_a(x)\lambda^a, \quad \Xi_a(x) \in \mathbb{R}, \quad a = 1, 2, \ldots, 8, \tag{13.82}$$

where λ^a are the eight Gell-Mann matrices – generators of $SU(3)$ – given in Appendix F. They are Hermitian, traceless 3×3 matrices, analogous to the three Pauli matrices which generate $SU(2)$. Under gauge transformations, the scalar field transforms as

$$\Xi'(x) = \Upsilon(x)\Xi(x)\Upsilon(x)^\dagger, \quad \Upsilon(x) \in SU(3). \tag{13.83}$$

We introduce a potential of the form

$$V(\Xi) = \frac{M^2}{2}\mathrm{Tr}\left[\Xi^2\right] + \frac{\kappa}{3!}\mathrm{Tr}\left[\Xi^3\right] + \frac{\Lambda}{4!}\mathrm{Tr}\left[\Xi^4\right]. \tag{13.84}$$

This potential is gauge-invariant due to the cyclicity of the trace. We can choose a unitary gauge, in which the scalar field is diagonal (one uses the unitary transformation $\Upsilon(x)$ to diagonalize the Hermitian matrix $\Xi(x)$)

$$\Xi(x) = \begin{pmatrix} \xi_1(x) & 0 & 0 \\ 0 & \xi_2(x) & 0 \\ 0 & 0 & -\xi_1(x) - \xi_2(x) \end{pmatrix}, \tag{13.85}$$

so the potential then takes the form

$$V(\Xi) = \frac{M^2}{2}\left(\xi_1^2 + \xi_2^2 + (\xi_1 + \xi_2)^2\right)$$
$$+ \frac{\kappa}{3!}\left(\xi_1^3 + \xi_2^3 - (\xi_1 + \xi_2)^3\right) + \frac{\Lambda}{4!}\left(\xi_1^4 + \xi_2^4 + (\xi_1 + \xi_2)^4\right)$$

$$= M^2 \left(\xi_1^2 + \xi_2^2 + \xi_1 \xi_2 \right) - \frac{\kappa}{2} \xi_1 \xi_2 (\xi_1 + \xi_2)$$

$$+ \frac{\Lambda}{4!} \left(2\xi_1^4 + 4\xi_1^3 \xi_2 + 6\xi_1^2 \xi_2^2 + 4\xi_1 \xi_2^3 + 2\xi_2^4 \right). \tag{13.86}$$

We are assuming $\Lambda > 0$ for the potential to be bounded from below. It is important to include all terms that are gauge-invariant and renormalizable. In particular, one also needs to consider the term

$$\det(\Xi) = -\xi_1 \xi_2 (\xi_1 + \xi_2) = \frac{1}{3} \text{Tr} \left[\Xi^3 \right]. \tag{13.87}$$

Since this term is proportional to $\text{Tr} \left[\Xi^3 \right]$, it does not need to be added separately. One should also investigate the term

$$\left(\text{Tr} \left[\Xi^2 \right] \right)^2 = 4 \left(\xi_1^2 + \xi_2^2 + \xi_1 \xi_2 \right)^2$$

$$= 4 \left(\xi_1^4 + 2\xi_1^3 \xi_2 + 3\xi_1^2 \xi_2^2 + 2\xi_1 \xi_2^3 + \xi_2^4 \right) = 2\text{Tr} \left[\Xi^4 \right]. \tag{13.88}$$

Since that term is proportional to $\text{Tr} \left[\Xi^4 \right]$, it is already included in the potential (13.84).

The absolute minima of the potential determine the symmetry breaking pattern. For $M^2 > 0$, there is a local minimum at $\xi_1 = \xi_2 = 0$ with $V(0) = 0$. If this is the absolute minimum, the symmetry remains unbroken. For $M^2 < 0$, on the other hand, $\xi_1 = \xi_2 = 0$ is a local maximum, and the symmetry breaks spontaneously. If $\xi_1 = \xi_2$, the SU(3) gauge symmetry indeed "breaks spontaneously" to the Standard Model gauge group SU(2)$_L$ × U(1)$_Y$, while $\xi_1 \neq \xi_2$ would generically correspond to the breaking pattern SU(3) → U(1) × U(1).

The conditions for a stationary point take the form

$$\frac{\partial V(\Xi)}{\partial \xi_1} = M^2 (2\xi_1 + \xi_2) - \frac{\kappa}{2} \left(2\xi_1 \xi_2 + \xi_2^2 \right) + \frac{\Lambda}{3!} \left(\xi_1^2 + \xi_2^2 + \xi_1 \xi_2 \right) (2\xi_1 + \xi_2) = 0,$$

$$\frac{\partial V(\Xi)}{\partial \xi_2} = M^2 (2\xi_2 + \xi_1) - \frac{\kappa}{2} \left(2\xi_1 \xi_2 + \xi_1^2 \right) + \frac{\Lambda}{3!} \left(\xi_1^2 + \xi_2^2 + \xi_1 \xi_2 \right) (2\xi_2 + \xi_1) = 0. \tag{13.89}$$

Adding the two equations, one obtains

$$3M^2 (\xi_1 + \xi_2) - \frac{\kappa}{2} \left(\xi_1^2 + \xi_2^2 + 4\xi_1 \xi_2 \right) + \frac{\Lambda}{2} \left(\xi_1^2 + \xi_2^2 + \xi_1 \xi_2 \right) (\xi_1 + \xi_2) = 0. \tag{13.90}$$

Subtracting the two equations results in

$$M^2 (\xi_1 - \xi_2) + \frac{\kappa}{2} \left(\xi_1^2 - \xi_2^2 \right) + \frac{\Lambda}{3!} \left(\xi_1^2 + \xi_2^2 + \xi_1 \xi_2 \right) (\xi_1 - \xi_2) = 0, \tag{13.91}$$

which implies $\xi_1 = \xi_2$, or (if $\xi_1 \neq \xi_2$)

$$M^2 + \frac{\kappa}{2} (\xi_1 + \xi_2) + \frac{\Lambda}{3!} \left(\xi_1^2 + \xi_2^2 + \xi_1 \xi_2 \right) = 0. \tag{13.92}$$

Inserting this result in eq. (13.90), one obtains

$$(\xi_1 + \xi_2)^2 + \frac{1}{2} \xi_1 \xi_2 = 0. \tag{13.93}$$

This quadratic equation has two solutions: $\xi_2 = -2\xi_1$ and $\xi_2 = -\frac{1}{2}\xi_1$, which imply $-\xi_1 - \xi_2 = \xi_1$ and $-\xi_1 - \xi_2 = \xi_2$, respectively. In both cases, two of the three eigenvalues of Ξ are degenerate, such that (despite the fact that $\xi_1 \neq \xi_2$) the symmetry breaking pattern still is SU(3) → SU(2)$_L$ × U(1)$_Y$. This shows that this is indeed the only symmetry breaking pattern that results from this unified theory. In fact, the two minima at $-\xi_1 - \xi_2 = \xi_1$ and at $-\xi_1 - \xi_2 = \xi_2$ are gauge equivalent to the minimum at $\xi_1 = \xi_2$, by a simple permutation of the eigenvalues.

We can hence limit ourselves to $\xi_1 = \xi_2 = \xi_\pm$ for which eq. (13.90) implies

$$\xi_\pm = \frac{1}{2\Lambda}\left(\kappa \pm \sqrt{\kappa^2 - 8M^2\Lambda}\right). \tag{13.94}$$

The corresponding value of the potential is given by

$$V(\Xi) = \xi_\pm^2\left(\frac{3M^2}{2} - \frac{\kappa\xi_\pm}{4}\right). \tag{13.95}$$

Keeping in mind that $M^2 < 0$, for $\kappa > 0$ the absolute minimum of the potential corresponds to

$$\xi_+ = \frac{1}{2\Lambda}\left(\kappa + \sqrt{\kappa^2 - 8M^2\Lambda}\right) = \frac{\mathcal{V}}{\sqrt{3}}, \tag{13.96}$$

while for $\kappa < 0$ it corresponds to ξ_-. Here, we have introduced the GUT scale $\mathcal{V}$ which determines the vacuum value of the adjoint scalar field as

$$\Xi(x) = \Xi_0 = \frac{\mathcal{V}}{\sqrt{3}}\begin{pmatrix} 1 & 0 & 0 \\ 0 & 1 & 0 \\ 0 & 0 & -2 \end{pmatrix} = \mathcal{V}\lambda^8. \tag{13.97}$$

As usual, here we have normalized the SU(3) generator λ^8 that multiplies $\mathcal{V}$ such that $\mathrm{Tr}[\lambda^8\lambda^8] = 2$.

In the SU(5) GUT that unifies all gauge interactions of the Standard Model, $\mathcal{V}$ is of the order of 10^{16} GeV. The GUT scale is significantly below the Planck scale of 10^{19} GeV, which justifies neglecting quantum gravity in the above considerations.

Let us now consider the SU(3) unified gauge field with the coupling g_3

$$V_\mu(x) = ig_3 V_\mu^a(x)\frac{\lambda^a}{2}. \tag{13.98}$$

Under non-Abelian gauge transformations, it transforms as

$$V_\mu'(x) = \Upsilon(x)\left(V_\mu(x) + \partial_\mu\right)\Upsilon(x)^\dagger, \quad \Upsilon(x) \in \mathrm{SU}(3). \tag{13.99}$$

The W- and B-bosons of the Standard Model are associated with the Gell-Mann matrices $\lambda^1, \lambda^2, \lambda^3$, and λ^8, which generate the $\mathrm{SU}(2)_L \times \mathrm{U}(1)_Y$ subgroup of SU(3). In addition, we introduce X- and Y-bosons that are associated with the remaining generators $\lambda^4, \lambda^5, \lambda^6, \lambda^7$, such that

$$W_\mu^1(x) = V_\mu^1(x), \quad W_\mu^2(x) = V_\mu^2(x), \quad W_\mu^3(x) = V_\mu^3(x), \quad B_\mu(x) = V_\mu^8(x),$$
$$X_\mu^1(x) = V_\mu^4(x), \quad X_\mu^2(x) = V_\mu^5(x), \quad Y_\mu^1(x) = V_\mu^6(x), \quad Y_\mu^2(x) = V_\mu^7(x). \tag{13.100}$$

Introducing

$$X_\mu(x) = X_\mu^1(x) + iX_\mu^2(x), \quad \overline{X}_\mu(x) = X_\mu^1(x) - iX_\mu^2(x),$$
$$Y_\mu(x) = Y_\mu^1(x) + iY_\mu^2(x), \quad \overline{Y}_\mu(x) = Y_\mu^1(x) - iY_\mu^2(x), \tag{13.101}$$

one obtains

$$V_\mu(x) = i\frac{g_3}{2}\begin{pmatrix} W_\mu^3(x) + B_\mu(x)/\sqrt{3} & W_\mu^-(x) & \overline{X}_\mu(x) \\ W_\mu^+(x) & -W_\mu^3(x) + B_\mu(x)/\sqrt{3} & \overline{Y}_\mu(x) \\ X_\mu(x) & Y_\mu(x) & -2B_\mu(x)/\sqrt{3} \end{pmatrix}.$$

(13.102)

Under embedded Standard Model gauge transformations in $SU(2)_L \times U(1)_Y \subset SU(3)$,

$$\Upsilon(x) = \begin{pmatrix} \exp\left(ig'\varphi(x)/2\right)L(x) & 0_{2\times 1} \\ 0_{1\times 2} & \exp(-ig'\varphi(x)) \end{pmatrix} \in SU(3), \quad L(x) \in SU(2)_L,$$

(13.103)

the various fields transform as

$$W_\mu'(x) = L(x)(W_\mu(x) + \partial_\mu)L(x)^\dagger, \quad B_\mu'(x) = B_\mu(x) - \partial_\mu\varphi(x),$$

$$\left(X_\mu'(x), Y_\mu'(x)\right) = \exp\left(-ig'\frac{3}{2}\varphi(x)\right)\left(X_\mu'(x), Y_\mu'(x)\right)L(x)^\dagger,$$

$$\begin{pmatrix} \overline{X}_\mu'(x) \\ \overline{Y}_\mu'(x) \end{pmatrix} = \exp\left(ig'\frac{3}{2}\varphi(x)\right)L(x)\begin{pmatrix} \overline{X}_\mu(x) \\ \overline{Y}_\mu(x) \end{pmatrix}.$$

(13.104)

We see that the X- and Y-bosons form a (charge conjugate) electroweak doublet with weak hypercharge $Y_X = Y_Y = -3/2$, while the $\overline{X}$- and $\overline{Y}$-bosons are an electroweak doublet with weak hypercharge $Y_{\overline{X}} = Y_{\overline{Y}} = 3/2$. This corresponds to the electric charges $Q = T_L^3 + Y$, such that

$$Q_X = -\frac{1}{2} - \frac{3}{2} = -2, \quad Q_Y = \frac{1}{2} - \frac{3}{2} = -1, \quad Q_{\overline{X}} = 2, \quad Q_{\overline{Y}} = 1.$$

(13.105)

As illustrated in Figure F.2, under the $SU(2)_L \times U(1)_Y$ subgroup the $SU(3)$ octet decomposes as

$$\{8\} = \{3\}_0 + \{1\}_0 + \{2\}_{2/3} + \{2\}_{-2/3}.$$

(13.106)

Here, the subscripts refer to the weak hypercharges of the corresponding $SU(2)_L$ representations.

From eq. (13.102), we read off the Standard Model gauge couplings

$$g = g_3, \quad g' = \frac{1}{\sqrt{3}}g_3 \quad \Rightarrow \quad e = \frac{gg'}{\sqrt{g^2 + g'^2}} = \frac{g_3}{2}, \quad \sin^2\theta_W = \frac{g'^2}{g^2 + g'^2} = \frac{1}{4}.$$

(13.107)

Unlike in the Standard Model, after genuine electroweak unification, g and g' are no longer independent parameters, and we obtain a prediction for the Weinberg angle $|\theta_W|$. Somewhat surprisingly, this prediction is quite close to the experimentally determined value, which amounts to $\sin^2\theta_W = 0.23119(14)$. As we will see in Chapter 26, the $SU(5)$ GUT which unifies not only the electroweak but also the strong interactions predicts $\sin^2\theta_W = 3/8$, which seems to agree less well with the experimental value. However, as we have seen in Section 5.10, in quantum field theory the couplings are "running"; *i.e.* they depend on the energy scale. The above relation is valid at the GUT scale $\mathcal{V}$ and not at the Standard Model scale of $v = 246$ GeV. As mentioned in Section 5.11, the non-Abelian $SU(2)_L$ gauge coupling g is *asymptotically free* and thus decreases with increasing energy scale. The Abelian gauge coupling g', on the other hand, increases with the energy scale. As a result, $\sin^2\theta_W$ also increases with the energy scale and eventually reaches the values resulting from unification at the scale $\mathcal{V}$. Since $1/4 < 3/8$, the SU(3) unification scale $\mathcal{V}$ would be below the SU(5) GUT scale of about 10^{16} GeV.

As we will see next, the X- and Y-bosons become massive after the "spontaneous breakdown" of SU(3) to SU(2)$_L$ × U(1)$_Y$. For the adjoint scalar field, the covariant derivative takes the form

$$D_\mu \Xi(x) = \partial_\mu \Xi(x) + [V_\mu(x), \Xi(x)]. \tag{13.108}$$

With the field strength tensor

$$V_{\mu\nu}(x) = \partial_\mu V_\nu(x) - \partial_\nu V_\mu(x) + [V_\mu(x), V_\nu(x)], \tag{13.109}$$

the bosonic part of the SU(3) unified Lagrangian is then given by

$$\mathcal{L}(\Xi, V) = \frac{1}{4} \mathrm{Tr} \left[D_\mu \Xi D_\mu \Xi \right] + V(\Xi) - \frac{1}{2g_3^2} \mathrm{Tr} \left[V_{\mu\nu} V_{\mu\nu} \right]. \tag{13.110}$$

We now insert the vacuum value of the scalar field, *i.e.* $\Xi(x) = \Xi_0$, to obtain the mass terms for the gauge field

$$\frac{1}{4} \mathrm{Tr} \left[D_\mu \Xi D_\mu \Xi \right] = \frac{1}{4} \mathrm{Tr} \left[[V_\mu, \Xi_0][V_\mu, \Xi_0] \right] = \frac{3}{8} g_3^2 \mathcal{V}^2 (\overline{X}_\mu X_\mu + \overline{Y}_\mu Y_\mu), \tag{13.111}$$

because one obtains

$$[V_\mu, \Xi_0] = \mathrm{i} \frac{\sqrt{3}}{2} g_3 \mathcal{V} \begin{pmatrix} 0 & 0 & -\overline{X}_\mu \\ 0 & 0 & -\overline{Y}_\mu \\ X_\mu & Y_\mu & 0 \end{pmatrix}. \tag{13.112}$$

Thus, the X- and Y-bosons both pick up the mass

$$M_X = M_Y = \frac{\sqrt{3}}{2} g_3 \mathcal{V}. \tag{13.113}$$

These 4 gauge bosons become massive by "eating" 4 Nambu–Goldstone bosons. Indeed, when the "small" unified group $G =$ SU(3) "breaks spontaneously" down to the subgroup $H =$ SU(2)$_L$ × U(1)$_Y$, according to the Goldstone theorem, there are $8 - (3 + 1) = 4$ Nambu–Goldstone bosons.

Quick Question 13.8.1   SU(5) Grand Unified Theory
Howard Georgi and Sheldon Glashow proposed a Grand Unified Theory based on an extended gauge group SU(5), which incorporates the gauge group of the Standard Model, SU(3)$_c$ × SU(2)$_L$ × U(1)$_Y$. How many additional gauge bosons did they introduce?

13.9 Electroweak Symmetry Breaking in an SU(3) Unified Theory

While the X- and Y-bosons pick up a mass at the scale $\mathcal{V}$, at this level the Standard Model gauge bosons are still massless.[8] Just as in the Standard Model, in order to give mass to the W- and Z-boson, the $SU(2)_L \times U(1)_Y$ gauge symmetry should break further to $U(1)_{em}$. In the unified theory, however, this cannot be achieved with just the standard complex Higgs doublet, simply because SU(3) has no doublet representations. Instead, it is natural to introduce a complex Higgs field in the fundamental triplet representation $\{3\}$

$$\Phi(x) = \begin{pmatrix} \Phi^+(x) \\ \Phi^0(x) \\ \phi^-(x) \end{pmatrix} \in \mathbb{C}^3, \tag{13.114}$$

which transforms as $\Phi'(x) = \Upsilon(x)\Phi(x)$ and decomposes into the standard $SU(2)_L$ Higgs doublet and an additional complex $SU(2)_L$ singlet ϕ^-, i.e.

$$\{3\} = \{2\}_{1/2} + \{1\}_{-1}. \tag{13.115}$$

This is again illustrated in Figure F.2. The additional complex singlet has the weak hypercharge $Y_\phi = -1$, which implies the electric charge $Q_\phi = -1$.

Let us construct the most general, SU(3)-invariant, renormalizable potential that depends on both the fundamental Higgs field Φ and the adjoint scalar field Ξ,

$$V(\Phi, \Xi) = \frac{m'^2}{2}\Phi^\dagger\Phi + \frac{\kappa'}{3!}\Phi^\dagger\Xi\Phi + \frac{\lambda'}{4!}\Phi^\dagger\Xi^2\Phi + \frac{\lambda}{4!}(\Phi^\dagger\Phi)^2. \tag{13.116}$$

Since Ξ picks up a vacuum value $\Xi_0 = \mathcal{V}\lambda^8$ at the large unification scale $\mathcal{V}$, while Φ should pick up a vacuum value at the much smaller electroweak scale $v \ll \mathcal{V}$, we can replace Ξ by Ξ_0 and consider

$$V(\Phi, \Xi_0) = \frac{m'^2}{2}(\Phi^{+*}\Phi^+ + \Phi^{0*}\Phi^0 + \phi^{-*}\phi^-) + \frac{\lambda}{4!}(\Phi^{+*}\Phi^+ + \Phi^{0*}\Phi^0 + \phi^{-*}\phi^-)^2$$
$$+ \frac{\kappa'}{3!}\frac{\mathcal{V}}{\sqrt{3}}(\Phi^{+*}\Phi^+ + \Phi^{0*}\Phi^0 - 2\phi^{-*}\phi^-) + \frac{\lambda'}{4!}\frac{\mathcal{V}^2}{3}(\Phi^{+*}\Phi^+ + \Phi^{0*}\Phi^0 + 4\phi^{-*}\phi^-). \tag{13.117}$$

The extrema of this potential are determined by

$$\frac{\partial V(\Phi, \Xi_0)}{\partial \Phi^{+*}} = \left(\frac{m'^2}{2} + \frac{\kappa'}{3!}\frac{\mathcal{V}}{\sqrt{3}} + \frac{\lambda'}{4!}\frac{\mathcal{V}^2}{3}\right)\Phi^+ + \frac{2\lambda}{4!}(\Phi^{+*}\Phi^+ + \Phi^{0*}\Phi^0 + \phi^{-*}\phi^-)\Phi^+ = 0,$$

$$\frac{\partial V(\Phi, \Xi_0)}{\partial \Phi^{0*}} = \left(\frac{m'^2}{2} + \frac{\kappa'}{3!}\frac{\mathcal{V}}{\sqrt{3}} + \frac{\lambda'}{4!}\frac{\mathcal{V}^2}{3}\right)\Phi^0 + \frac{2\lambda}{4!}(\Phi^{+*}\Phi^+ + \Phi^{0*}\Phi^0 + \phi^{-*}\phi^-)\Phi^0 = 0,$$

$$\frac{\partial V(\Phi, \Xi_0)}{\partial \phi^{-*}} = \left(\frac{m'^2}{2} - \frac{2\kappa'}{3!}\frac{\mathcal{V}}{\sqrt{3}} + \frac{\lambda'}{3!}\frac{\mathcal{V}^2}{3}\right)\phi^- + \frac{2\lambda}{4!}(\Phi^{+*}\Phi^+ + \Phi^{0*}\Phi^0 + \phi^{-*}\phi^-)\phi^- = 0. \tag{13.118}$$

---

[8] As we will see in Chapter 14, SU(2) gauge bosons that do not pick up mass via the Higgs mechanism are not strictly massless. Instead, they are confined and pick up a small mass non-perturbatively.

Just as in the Standard Model, using the $SU(2)_L$ gauge freedom we put the vacuum value of the standard Higgs field to $\Phi^+ = 0$ and $\Phi^0 = v$. If the electrically charged *Higgs singlet* would pick up a vacuum value as well, $U(1)_{em}$ would "break spontaneously" and, in contrast to observation, the vacuum would turn into a superconductor. Hence, we are interested in situations in which the absolute minimum of $V(\Phi, \Xi)$ occurs at $\phi^- = 0$. This leaves us with the equation

$$\frac{m'^2}{2} + \frac{\kappa'}{3!} \frac{\mathcal{V}}{\sqrt{3}} + \frac{\lambda'}{4!} \frac{\mathcal{V}^2}{3} + \frac{2\lambda}{4!} v^2 = 0. \tag{13.119}$$

Achieving a small value $v \ll \mathcal{V}$ requires a delicate fine-tuning of the parameters m', κ', or λ'. This is how the *hierarchy problem* manifests itself in unified theories.

By expanding around the minimum of $V(\Phi, \Xi_0)$, we obtain the mass squared of the Standard Model Higgs particle

$$M_H^2 = m'^2 + \frac{\kappa'}{3\sqrt{3}}\mathcal{V} + \frac{\lambda'}{36}\mathcal{V}^2 + \frac{\lambda}{2}v^2 = \frac{\lambda}{3}v^2. \tag{13.120}$$

This is indeed the familiar Standard Model result, which we recovered here by using eq. (13.119), without additional fine-tuning of m', κ', or λ'. Again using eq. (13.119), the mass squared of the Higgs singlet results as

$$M_\phi^2 = m'^2 - \frac{2\kappa'}{3\sqrt{3}}\mathcal{V} + \frac{\lambda'}{9}\mathcal{V}^2 + \frac{\lambda}{6}v^2 = -\frac{\kappa'}{\sqrt{3}}\mathcal{V} + \frac{\lambda'}{12}\mathcal{V}^2, \tag{13.121}$$

which naturally remains at the scale $\mathcal{V}$. Splitting the Higgs doublet from the Higgs singlet relies on the fine-tuning that is required to arrange for $v \ll \mathcal{V}$.

In the SU(5) GUT, the corresponding problem is known as the *doublet–triplet splitting problem*, which is again a manifestation of the hierarchy problem. In that case, the Standard Model Higgs doublet is embedded in the fundamental {5} representation of SU(5), which also contains an additional color-triplet Higgs {3} that transforms under the $SU(3)_c$ gauge symmetry of the strong interaction.

The Lagrangian of the fundamental Higgs triplet takes the form

$$\mathcal{L}(\Phi, \Xi, V) = \frac{1}{2}(D_\mu \Phi)^\dagger D_\mu \Phi + V(\Phi, \Xi), \tag{13.122}$$

with the covariant derivative given by

$$D_\mu \Phi(x) = \left(\partial_\mu + V_\mu(x)\right) \Phi(x). \tag{13.123}$$

Just as in the Standard Model, this gives rise to the usual masses of the W- and Z-boson and leaves the photon massless.

GUTs extend the Standard Model to much higher energies, which are not directly testable in collider experiments. Hence, they remain a matter of speculation. We will return to the discussion of GUTs in Chapter 26, where we will concentrate on their effects on the fermion dynamics, which includes proton decay.

Exercises

13.1 Non-Abelian gauge theory and matter

We consider an SU(N) gauge theory ($N \geq 2$). A gauge transformation is given by a matrix of the form

$$\Omega(x) = \exp(i\omega_a(x)T^a) \in \text{SU}(N).$$

Here, $\omega_a(x)$ are real, differentiable functions, and the T^a form a complete set of generators of SU(N).

Hence, the gauge field $G_\mu(x) = igG_\mu^a(x)T^a$ (where $g \in \mathbb{R}$ is a coupling constant) transforms as

$$G_\mu(x) \rightarrow \Omega(x)\left(G_\mu(x) + \partial_\mu\right)\Omega(x)^\dagger. \tag{13.124}$$

We now add a charged matter field

$$\Phi(x) = \begin{pmatrix} \Phi_1(x) \\ \Phi_2(x) \\ . \\ \Phi_N(x) \end{pmatrix} \in \mathbb{C}^N.$$

Under a gauge transformation (13.124), the matter field transforms as

$$\Phi(x) \rightarrow \Omega(x)\Phi(x).$$

How do we have to define the covariant derivative D_μ, if we want the Lagrangian

$$\mathcal{L}(\Phi, G) = \frac{1}{2}(D_\mu\Phi)^\dagger D_\mu\Phi + \frac{m^2}{2}\Phi^\dagger\Phi + \frac{\lambda}{4!}(\Phi^\dagger\Phi)^2$$

to be gauge-invariant? Start from the ansatz

$$D_\mu = \partial_\mu + cG_\mu(x)$$

and derive the suitable constant c.

13.2 Symmetry breaking in the presence of a gauge field

Consider again an N-component complex field $\Phi \in \mathbb{C}^N$, now minimally coupled to an Abelian gauge field A_μ with field strength $F_{\mu\nu}$, and the Lagrangian

$$\mathcal{L}(\Phi, A) = \frac{1}{2}D_\mu\Phi^\dagger D_\mu\Phi + \frac{m^2}{2}\Phi^\dagger\Phi + \frac{\lambda}{4!}(\Phi^\dagger\Phi)^2 + \frac{1}{4}F_{\mu\nu}F_{\mu\nu}.$$

The covariant derivative is given by $D_\mu\Phi = (\partial_\mu + ieA_\mu)\Phi$. What is the global symmetry group G of the Lagrangian? Assume that $m^2 < 0$ and determine the unbroken subgroup H that remains intact after spontaneous breakdown of the global symmetry G. How many Nambu–Goldstone bosons arise? What kind of fields can be used to describe the Nambu–Goldstone bosons in a low-energy effective theory?

13.3 Invariance of the gauge–Higgs sector Lagrangian

Show explicitly that the Lagrangian of the gauge–Higgs sector of the Standard Model is gauge-invariant. Use the gauge transformation properties of Higgs and gauge fields only. For example, do not use the transformation property of a non-Abelian field strength, but derive it.

13.4 Couplings of gauge bosons

Consider the Lagrangian $\mathcal{L}(\Phi, W, B)$ of the gauge–Higgs sector of the Standard Model. Work out the interaction terms that are cubic or quartic in the gauge fields, and express them in terms of the physical fields $W_\mu^\pm$, Z_μ, and A_μ. Do the fields $W_\mu^\pm$ and Z_μ couple to the photon field with the appropriate strengths?

13.5 Accidental custodial symmetry

Show that the custodial symmetry breaking terms $\mathrm{Tr}\left[\Phi^\dagger \Phi \tau^3\right]$, $\mathrm{Tr}\left[\Phi^\dagger \Phi \Phi^\dagger \Phi \tau^3\right]$, as well as $\mathrm{Tr}\left[D_\mu \Phi^\dagger D_\mu \Phi \tau^3\right]$ vanish in the Standard Model. Also show that the term $\mathrm{Tr}\left[\Phi^\dagger \Phi \tau^3 \Phi^\dagger \Phi \tau^3\right]$ does not violate the custodial symmetry.

13.6 Higgs mechanism in variants of the Standard Model

In the Standard Model, only the $SU(2)_L \times U(1)_Y$ subgroup of the global $SU(2)_L \times SU(2)_R$ symmetry of the Higgs sector is gauged. As a consequence, the W- and Z-bosons pick up different masses, while the photon remains massless. Now consider the following two variants of the Standard Model: In the first variant, only $SU(2)_L$ is gauged (gauge coupling g), while in the second variant both $SU(2)_L$ and $SU(2)_R$ are gauged (gauge couplings g and g'). For both cases, work out the gauge boson mass terms that appear in the Lagrangian. Do these mass parameters represent the masses of physical or of confined particles?

13.7 SU(3) electroweak unification

Consider an $SU(3)$ gauge–Higgs theory with the gauge field $V_\mu(x) = \mathrm{i} g_3 V_\mu^a(x) \lambda^a / 2$ and an adjoint Higgs field $\Xi(x) = \Xi_a(x) \lambda^a$, $a \in \{1, 2, \ldots, 8\}$, with the Lagrangian

$$\mathcal{L}(\Xi, V) = \frac{1}{4} \mathrm{Tr}\left[D_\mu \Xi D_\mu \Xi\right] + V(\Xi) - \frac{1}{2g_3^2} \mathrm{Tr}\left[V_{\mu\nu} V_{\mu\nu}\right],$$

and the scalar potential

$$V(\Xi) = \frac{M^2}{2} \mathrm{Tr}\left[\Xi^2\right] + \frac{\Lambda}{4!} \mathrm{Tr}\left[\Xi^4\right].$$

a) What is the global symmetry of the potential and how is it spontaneously broken for $M^2 < 0$? How many Nambu–Goldstone bosons emerge for $g_3 = 0$?

b) Determine the mass acquired by the gauge bosons via the Higgs mechanism.

c) How many gauge bosons remain massless? How many massless particles persist after spontaneous symmetry breaking?

13.8 SU(3) electroweak unification with Higgs triplet

Add a complex Higgs triplet field

$$\Phi(x) = \begin{pmatrix} \Phi^+(x) \\ \Phi^0(x) \\ \phi^-(x) \end{pmatrix} \in \mathbb{C}^3,$$

to the $SU(3)$ electroweak unification model.

a) Construct the most general, $SU(3)$-invariant, renormalizable potential $V(\Xi, \Phi)$. Which patterns of symmetry breaking can be achieved?

b) Assume that the symmetry breaking proceeds in two steps, ending in the same way as in the Standard Model. Determine the masses that the various gauge bosons pick up via the Higgs mechanism.

14 Gluons: From Confinement to Deconfinement

In this chapter, we introduce the *non-Abelian* $SU(3)_c$ *color gauge field* that gives rise to the *strong interaction*. The fields that we encountered so far, the Higgs field as well as the $SU(2)_L$ and $U(1)_Y$ gauge fields, are color-neutral. Quarks will be added to the theory only in Chapter 15; in their absence, the electroweak gauge–Higgs sector is decoupled from the $SU(3)_c$ gauge field.[1]

Remarkably, *gluons* – the quanta of the color gauge field – do not appear as observable particles in the spectrum of the theory. Instead, gluons are *confined* inside color-neutral *glueballs* (unless they appear in a deconfined gluon plasma at high temperature). As a consequence of confinement, there are no color-charged states in the physical spectrum.

Since quarks transform as color triplets, they do not exist as isolated physical particles either, but are confined inside color-neutral particles known as *hadrons*. Hadrons containing an equal number of quarks and anti-quarks are denoted as *mesons*, while hadrons with an excess of three quarks over anti-quarks are known as *baryons*.

The glueballs of the pure $SU(3)_c$ Yang–Mills gauge theory are (hypothetical) hadrons consisting of gluons only. Once quarks will be added to the theory, the glueballs mix with meson states. Since quarks also couple to $SU(2)_L$ and $U(1)_Y$ gauge fields, the gluons will then indirectly affect the electroweak sector. However, in this chapter, the gluons form a strongly interacting world of their own.

Although we are not yet introducing dynamical quarks, we will already consider "quarks" as infinitely heavy static color sources. A *static quark–anti-quark pair* is then connected by a *confining color flux string,* with a tension that measures its energy per length. The Yang–Mills string is an interesting dynamical object that supports massless transverse fluctuations.

At high temperature, gluons undergo a first-order phase transition that separates the low-temperature glueball phase from a high-temperature gluon plasma. The $\mathbb{Z}(3)$ *center symmetry* of the $SU(3)_c$ gauge group plays a "central" role for these dynamics.

14.1 Gluons in the Continuum and on the Lattice

We introduce the gluon field via an algebra-valued vector potential

$$G_\mu(x) = \mathrm{i}g_s G_\mu^a(x)T^a, \quad T^a = \frac{1}{2}\lambda^a, \tag{14.1}$$

---

[1] In the absence of quarks and leptons, only gravity (which one usually does not couple to the Standard Model) would induce interactions between Higgs or electroweak gauge bosons and the $SU(3)_c$ color gauge field.

where the index $a = 1, 2, \ldots, 8$ is summed over, following the Einstein summation convention. Here, g_s is the *strong coupling constant* and λ^a are the 3×3 Gell-Mann matrices that we encountered before in Section 13.8, when we discussed the unification of $SU(2)_L$ and $U(1)_Y$ gauge theory (the Gell-Mann matrices are written down in Appendix F.7). Like the Pauli matrices, they are Hermitian, such that $G_\mu^\dagger = -G_\mu$. Although gluons are described by the same mathematical tools, they are physical objects that should certainly not be confused with the hypothetical fields X_μ and Y_μ discussed in Section 13.8. In particular, the gluons do not pick up mass through a Higgs mechanism.

Using the structure of non-Abelian gauge fields that was introduced in Chapter 11, we postulate the usual behavior under gauge transformations,

$$G'_\mu(x) = \Omega(x)\left(G_\mu(x) + \partial_\mu\right)\Omega(x)^\dagger, \quad \Omega(x) \in SU(3)_c. \tag{14.2}$$

The gluon field strength tensor and its gauge transformation behavior are given by

$$G_{\mu\nu}(x) = \partial_\mu G_\nu(x) - \partial_\nu G_\mu(x) + [G_\mu(x), G_\nu(x)], \quad G'_{\mu\nu}(x) = \Omega(x)G_{\mu\nu}(x)\Omega(x)^\dagger. \tag{14.3}$$

The latter follows from the derivation in Problem 11.3. The Lagrangian of the $SU(3)_c$ *Yang–Mills theory* takes the gauge- and Lorentz-invariant form

$$\mathcal{L}(G) = -\frac{1}{2g_s^2}\text{Tr}\left[G_{\mu\nu}G_{\mu\nu}\right] + \frac{i\theta}{32\pi^2}\epsilon_{\mu\nu\rho\sigma}\text{Tr}\left[G_{\mu\nu}G_{\rho\sigma}\right]. \tag{14.4}$$

It is remarkable that the Lagrangian of 4-dimensional Yang–Mills theory has just *two* parameters, which are both dimensionless: the *strong gauge coupling* g_s and the *vacuum-angle* θ. The experimental value of the vacuum-angle is indistinguishable from zero. Since the θ-term breaks CP symmetry (it breaks P but leaves C invariant), the related puzzle (why is $\theta = 0$?) is known as the "strong CP-problem", to which we will return in Chapter 25. For the moment, we simply set $\theta = 0$.

Since the theory does not contain any dimensioned parameter, at the classical level it is *scale-invariant*. Consequently, there is no preselected energy that could set the scale of the strong interaction. However, scale invariance is explicitly broken at the quantum level; *i.e.* it is affected by an *anomaly*. Indeed, a cut-off that is introduced when one regularizes a quantum field theory is a dimensioned quantity that explicitly breaks scale invariance. Even when the cut-off is removed to infinity, in Yang–Mills theory scale invariance remains broken in the continuum limit, and a dynamically generated scale emerges via the mechanism of *dimensional transmutation*.

We remark that there are also *conformal field theories*, where scale invariance re-emerges when the cut-off is removed (see, *e.g.*, Di Francesco *et al.* (1997)). These include the supersymmetric Yang–Mills theory with $\mathcal{N} = 4$ supercharges, which, besides gauge fields, also contains scalar and fermion fields. Pure Yang–Mills theory, on the other hand, is scale-invariant only at the classical but not at the quantum level.

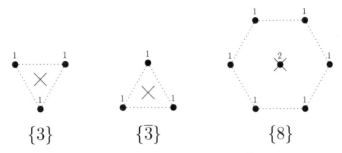

$\{3\}$ $\{\overline{3}\}$ $\{8\}$

Fig. 14.1 The weight diagrams of the irreducible SU(3) representations {3} (quarks), {$\overline{3}$} (anti-quarks), and {8} (gluons). The labels indicate the multiplicities of the states.

Since gluons – as well as quarks – are confined (unless they form a quark–gluon plasma at high temperatures), the number of colors is not (easily) directly observable; because we cannot simply count the number of color-charged particle states: 8 gluons or 3 quarks for each flavor. In later chapters, we will explore the dependence of physical quantities on the number of colors N_c, which we often keep as a free, integer-valued parameter. The Standard Model then has an $SU(N_c) \times SU(2)_L \times U(1)_Y$ gauge symmetry. As will be discussed in Sections 21.8 and 22.2, the number $N_c = 3$ in Nature can be deduced from appropriate experiments. The corresponding $SU(3)_c$ weight diagrams for color triplets (quarks), color anti-triplets (anti-quarks), and color octets (gluons) are illustrated in Figure 14.1. The group theory of SU(3) is discussed in Appendix F, especially in Sections F.6 and F.7. The representations of SU(3) fall in three classes, characterized by their triality, which affects the coupling rules of irreducible representations. Triplet and anti-triplet have opposite, non-trivial triality, while the octet has trivial triality.

Since gluons are strongly interacting particles, their low-energy physics cannot be understood using perturbation theory. In this case, the use of the non-perturbative lattice regularization is indispensable. The formulation of an $SU(N_c)$ Yang–Mills theory on a lattice has been discussed in Section 11.10.

The fundamental degrees of freedom describing the gluon field on the lattice are *parallel transporter link variables* $U_{x,\mu} \in SU(N_c)$. In the classical continuum limit, $a \rightarrow 0$, where one naively assumes smooth field configurations, they are (asymptotically) related to a continuum vector potential $G_\mu(x)$ according to

$$U_{x,\mu} = \exp(aG_\mu(x)) = \exp\left(i a g_s G_\mu^a(x) T^a\right), \qquad (14.5)$$

and they transform as

$$U'_{x,\mu} = \Omega_x U_{x,\mu} \Omega_{x+\hat{\mu}}^\dagger \qquad (14.6)$$

under non-Abelian gauge transformations with $\Omega_x \in SU(N_c)$. On the lattice, the parallel transport along a contour $\mathcal{C}$ of adjacent links (x, μ) that connect the lattice sites x and y is given by

$$U_\mathcal{C} = \mathcal{P} \prod_{(x,\mu)\in\mathcal{C}} U_{x,\mu}, \quad U'_\mathcal{C} = \Omega_x U_\mathcal{C} \Omega_y^\dagger. \qquad (14.7)$$

The symbol $\mathcal{P}$ indicates a "path-ordered product", which consecutively runs over all link variables along the contour in an oriented manner. Traces of parallel transporters around closed contours are gauge-invariant.

The standard Wilson action of an $SU(N_c)$ lattice Yang–Mills theory takes the form

$$S[U] = \frac{1}{g_s^2} \sum_{x,\,\mu>\nu} \text{Tr} \left[2 \, \mathbb{1} - U_{x,\mu\nu} - U_{x,\mu\nu}^\dagger \right]. \tag{14.8}$$

The sum extends over elementary lattice plaquettes with the parallel transporter around a plaquette given by

$$U_{x,\mu\nu} = U_{x,\mu} U_{x+\hat{\mu},\nu} U_{x+\hat{\nu},\mu}^\dagger U_{x,\nu}^\dagger, \tag{14.9}$$

thus forming the simplest (non-trivial) closed contour. The corresponding functional integral reads

$$Z = \prod_{x,\,\mu} \int_{SU(N_c)} dU_{x,\mu} \, \exp(-S[U]), \tag{14.10}$$

where we integrate all link variables independently over the $SU(N_c)$ group manifold.

Quick Question 14.1.1   Naive continuum limit of the Wilson action
> Use eq. (14.5) to show that, in the naive continuum limit $a \to 0$, the Wilson action of eq. (14.8) reproduces the continuum action of eq. (14.4) at $\theta = 0$.

14.2 Quark Confinement and the Wegner–Wilson Loop

To analyze the phase structure of a gauge theory, one needs to study order parameters, similar to the magnetization $\langle s_x \rangle$ in the Ising model (see Appendix I), or other spin models (discussed in Section 1.4). A simple, local order parameter such as $\langle U_{x,\mu} \rangle$ does not make sense in a gauge theory. This follows from *Elitzur's theorem*, which states that gauge-*variant* observables vanish in lattice gauge theory (formulated with parallel transporters) (Elitzur, 1975). This theorem is trivial to prove using eq. (14.18) below. An order parameter in a gauge theory must be gauge-invariant and, in addition, non-local. An appropriate order parameter in the pure gluon theory is the *Wegner–Wilson loop*

$$\langle W_\mathcal{C} \rangle = \langle \text{Tr} U_\mathcal{C} \rangle = \frac{1}{Z} \prod_{x,\mu} \int_{SU(N_c)} dU_{x,\mu} \, \text{Tr} \left[\mathcal{P} \prod_{(x,\mu)\in\mathcal{C}} U_{x,\mu} \right] \exp(-S[U]) \,, \tag{14.11}$$

for a closed (lattice) contour $\mathcal{C}$. If this contour is rectangular, with side lengths R and T (in a spatial and the temporal direction), the Wegner–Wilson loop describes the instantaneous creation and annihilation of a static, external charge–anti-charge pair (a source and sink of color flux) at distance R that exists for a Euclidean time T. As we have seen in Chapter 11, a straight line of links in the time direction, which form a *Polyakov loop* that is closed over the periodic Euclidean-time direction (see Section 11.9), represents a static, external color charge. The temporal edges of the Wegner–Wilson loop play the same role. The spatial edges, on the other hand, represent instantaneous creation and annihilation processes of the charge–anti-charge pair that are separated by the time T. The Wegner–Wilson loop is related to the static charge–anti-charge potential $V(R)$,[2]

$$\lim_{T \to \infty} \langle W_C \rangle \sim \exp(-V(R)T). \tag{14.12}$$

In 4-d non-Abelian gauge theories, there are no isolated color-charged states. Instead color-charged particles are *confined*. Indeed, the flux that emanates from a color charge in the fundamental $\{N_c\}$ representation of $SU(N_c)$ and terminates in an anti-charge $\{\overline{N_c}\}$ is constricted to a thin *flux tube* – a confining string. This flux tube has a *string tension* σ that leads to a static charge–anti-charge potential, which rises linearly at large separation R,

$$V(R) \sim \sigma R. \tag{14.13}$$

In a *confinement phase*, the Wegner–Wilson loop therefore exhibits an *area law*,

$$- \lim_{R,T \to \infty} \log \langle W_C \rangle \sim \sigma A, \quad A = RT. \tag{14.14}$$

This is a proportionality relation, which holds asymptotically for large T and R.

When confinement is lost by passing through a phase transition, the Wegner–Wilson loop no longer shows an area law behavior. A massless *Coulomb phase*, for example, is characterized by a *perimeter law* with a constant γ

$$\lim_{R,T \to \infty} \langle W_C \rangle \sim \exp\left(- \gamma (R + T) \right). \tag{14.15}$$

14.3 Character Expansion and Group Integration

This section is mathematical. For a first quick reading of this chapter, the reader might opt to skip it and trust the group integration formulae to be used in the continuation.

The title of this section may sound as if it belonged to a guide book for personal development. However, this is true only as far as it shall empower us to perform calculations in the strong coupling limit of lattice gauge theory.

Integration of the $SU(N)$ group manifold applies the *Haar measure dU* (Haar, 1933), which is invariant against group multiplication both from the left and from the right. In particular, one obtains

$$\int_{SU(N)} dU\, f(\Omega_L U) = \int_{SU(N)} dU\, f(U \Omega_R^\dagger) = \int_{SU(N)} dU\, f(U), \tag{14.16}$$

---

[2] At large T, the behavior is dominated by this exponential, although there may be a prefactor which is polynomial in R and T.

for any function f and for any SU(N) matrices Ω_L and Ω_R. It is convenient to normalize the measure such that

$$\int_{\text{SU}(N)} dU = 1. \tag{14.17}$$

The Haar measure is unique, and it has the following additional properties

$$\int_{\text{SU}(N)} dU\, U^{ij} = 0, \quad \int_{\text{SU}(N)} dU\, U^{ij}(U^\dagger)^{kl} = \frac{1}{N}\delta_{jk}\delta_{il}. \tag{14.18}$$

Here, ij and kl determine specific matrix elements, *i.e.* $i, j, k, l \in \{1, 2, \ldots, N\}$. In particular, the case $j = k$ is consistent with the normalization (14.17).

Lie algebras as well as Lie groups have different irreducible representations (cf. Appendix F). At the level of the SU(N) algebra, a d_Γ-dimensional *representation* Γ is defined by a set of $N^2 - 1$ Hermitian $d_\Gamma \times d_\Gamma$ matrices Γ^a that satisfy the SU(N) commutation relations $[\Gamma^a, \Gamma^b] = \mathrm{i}f_{abc}\Gamma^c$, where f_{abc} are the structure constants. For example, the fundamental representation $\{3\}$ of the SU(3) algebra is given by the eight Hermitian 3×3 Gell-Mann matrices $\lambda^a/2$, while the adjoint representation $\{8\}$ is given in terms of eight 8×8 Hermitian matrices that satisfy the same commutation relations.

The representations U_Γ of the SU(N) group are obtained by exponentiating the elements of the corresponding representation of the algebra, $U_\Gamma = \exp(\mathrm{i}\omega^a\Gamma^a)$, $\omega^a \in \mathbb{R}$. As a result, for example, the matrix elements U_Γ^{ab} ($a, b \in \{1, 2, \ldots, N^2 - 1\}$) of the *adjoint group representation* $\Gamma = \{N^2 - 1\}$ are given in terms of the fundamental $N \times N$ group representation matrices U as

$$U_{\Gamma=\{N^2-1\}}^{ab} = \frac{1}{2}\mathrm{Tr}\left[\lambda^a U \lambda^b U^\dagger\right]. \tag{14.19}$$

The $N \times N$ unit-matrix $U = \mathbb{1}$ maps to the $d_\Gamma \times d_\Gamma$ unit-matrix $U_\Gamma = \mathbb{1}$. Indeed for the adjoint representation, using $U = \mathbb{1}$, one obtains

$$U_{\Gamma=\{N^2-1\}}^{ab} = \frac{1}{2}\mathrm{Tr}\left[\lambda^a \mathbb{1}\lambda^b \mathbb{1}\right] = \delta_{ab}. \tag{14.20}$$

In general, the inverse $U^\dagger$ of U maps to $U_\Gamma^\dagger$. For the adjoint representation, $U^\dagger$ maps to $U_{\Gamma=\{N^2-1\}}^\mathsf{T}$ because

$$\frac{1}{2}\mathrm{Tr}\left[\lambda^a U^\dagger \lambda^b U\right] = \frac{1}{2}\mathrm{Tr}\left[\lambda^b U \lambda^a U^\dagger\right] = U_{\Gamma=\{N^2-1\}}^{ba}. \tag{14.21}$$

In fact, the adjoint representation is real, such that $U_{\Gamma=\{N^2-1\}}^\mathsf{T} = U_{\Gamma=\{N^2-1\}}^\dagger$.

Quick Question 14.3.1   Reality of the adjoint representation
Show that $U_{\Gamma=\{N^2-1\}}^{ab} \in \mathbb{R}$.

One can also show that the group multiplication rule is inherited for any representation,

$$UV = W \implies U_\Gamma V_\Gamma = W_\Gamma. \tag{14.22}$$

Group elements can be classified into equivalence classes known as *conjugacy classes*. Two group elements U and U' are *equivalent* if there exists another group element Ω such that $U' = \Omega U \Omega^\dagger$. Obviously, equivalent group elements have the same trace, $\mathrm{Tr}U' = \mathrm{Tr}U$. Hence, the trace is constant over a conjugacy class and is thus known as a *class function*. The *character* $\chi_\Gamma(U)$ of a group element U in the representation Γ is defined as the corresponding trace

$$\chi_\Gamma(U) = \mathrm{Tr}U_\Gamma. \tag{14.23}$$

For example, based on eq. (14.19), one obtains

$$\chi_{\Gamma=\{N^2-1\}}(U) = \mathrm{Tr}U\,\mathrm{Tr}U^\dagger - 1 = |\mathrm{Tr}U|^2 - 1. \tag{14.24}$$

The representations Γ of a group can be viewed as "conjugate momenta" of the group elements U. If we formally represent the group elements by "position eigenstates" $|U\rangle$, the representations play the role of "momentum eigenstates" $|\Gamma, ab\rangle$ with $a, b \in \{1, 2, \ldots, d_\Gamma\}$, such that

$$\langle U|\Gamma, ab\rangle = \sqrt{d_\Gamma} U_\Gamma^{ab} \quad \Rightarrow \quad \chi_\Gamma(U) = \mathrm{Tr}[U_\Gamma] = U_\Gamma^{aa} = \frac{1}{\sqrt{d_\Gamma}} \sum_{a=1}^{d_\Gamma} \langle U|\Gamma, aa\rangle. \tag{14.25}$$

In the Hamiltonian formulation of lattice gauge theory (which we discussed in Chapter 11, but which we are not using in this section), the states $|\Gamma, ab\rangle$ are eigenstates of the electric flux operator associated with a link. In this section, where we use the Euclidean functional integral, the states $|\Gamma, ab\rangle$ are just employed to derive useful relations between the characters. The completeness relation

$$\int_{\mathrm{SU}(N)} dU\, |U\rangle\langle U| = \mathbb{1} \tag{14.26}$$

gives rise to an orthogonality relation between the characters

$$\delta_{\Gamma,\Gamma'}\delta_{aa'}\delta_{bb'} = \langle\Gamma, ab|\Gamma', a'b'\rangle = \int_{\mathrm{SU}(N)} dU\, \langle\Gamma, ab|U\rangle\langle U|\Gamma', a'b'\rangle = \int_{\mathrm{SU}(N)} dU\sqrt{d_\Gamma d_{\Gamma'}} U_\Gamma^{ab*} U_{\Gamma'}^{a'b'}$$

$$\Rightarrow \quad \int_{\mathrm{SU}(N)} dU\, \chi_\Gamma(U)^* \chi_{\Gamma'}(U) = \int_{\mathrm{SU}(N)} dU\, U_\Gamma^{aa*} U_{\Gamma'}^{a'a'} = \frac{1}{\sqrt{d_\Gamma d_{\Gamma'}}} \delta_{\Gamma,\Gamma'}\delta_{aa'}\delta_{aa'} = \delta_{\Gamma,\Gamma'}. \tag{14.27}$$

Similarly, the completeness relation

$$\sum_{\Gamma,ab} |\Gamma, ab\rangle\langle\Gamma, ab| = \mathbb{1} \tag{14.28}$$

implies

$$\langle U|V\rangle = \sum_{\Gamma,ab} \langle U|\Gamma, ab\rangle\langle\Gamma, ab|V\rangle = \sum_\Gamma d_\Gamma U_\Gamma^{ab} V_\Gamma^{ab*} = \sum_\Gamma d_\Gamma U_\Gamma^{ab} (V_\Gamma^\dagger)^{ba}$$

$$= \sum_\Gamma d_\Gamma (UV^\dagger)_\Gamma^{aa} = \sum_\Gamma d_\Gamma \chi_\Gamma(UV^\dagger). \tag{14.29}$$

Another useful relation is the *fusion rule*

$$\int_{\mathrm{SU}(N)} dU\, \chi_\Gamma(VU)\, \chi_{\Gamma'}(U^\dagger W) = \frac{1}{d_\Gamma} \chi_\Gamma(VW)\delta_{\Gamma,\Gamma'}. \tag{14.30}$$

Any state $|f\rangle$ can be written as a superposition of "momentum eigenstates"

$$|f\rangle = \sum_{\Gamma,ab} |\Gamma,ab\rangle\langle\Gamma,ab|f\rangle \quad \Rightarrow$$

$$\langle U|f\rangle = \sum_{\Gamma,ab}\langle U|\Gamma,ab\rangle\langle\Gamma,ab|f\rangle - \sum_{\Gamma,ab} U_\Gamma^{ab}\langle\Gamma,ab|f\rangle, \qquad (14.31)$$

with the expansion coefficients given by

$$\langle\Gamma,ab|f\rangle = \int_{SU(N)} dU \,\langle\Gamma,ab|U\rangle\langle U|f\rangle = \int_{SU(N)} dU \, U_\Gamma^{ab*}\langle U|f\rangle. \qquad (14.32)$$

The matrix element $\langle U|f\rangle = f(U)$ defines a function over the group manifold, and $\langle\Gamma,ab|f\rangle$ plays the role of its "Fourier transform". A class function obeys the property $f(\Omega U\Omega^\dagger) = f(U)$, which implies $\langle\Gamma,ab|f\rangle = f_\Gamma\delta_{ab}$, such that

$$f_\Gamma = \sum_{a=1}^{d_\Gamma}\langle\Gamma,aa|f\rangle = \int_{SU(N)} dU \, U_\Gamma^{aa*}\langle U|f\rangle = \int_{SU(N)} dU \, \chi_\Gamma(U)^* f(U),$$

$$f(U) = \sum_{\Gamma,ab} U_\Gamma^{ab}\langle\Gamma,ab|f\rangle = \sum_\Gamma U_\Gamma^{ab} f_\Gamma\delta_{ab} = \sum_\Gamma U_\Gamma^{aa} f_\Gamma = \sum_\Gamma \chi_\Gamma(U) f_\Gamma. \quad (14.33)$$

The last equation defines the *character expansion* of the function $f(U)$.

14.4 Strong Coupling Limit of Lattice Yang–Mills Theory

In lattice Yang–Mills theory, it is straightforward to prove confinement for sufficiently large values of the bare gauge coupling g_s (Wilson, 1974). However, in the strong coupling region the correlation length is short and one cannot take the continuum limit. Attaining the continuum limit requires taking $g_s \to 0$. Assuming that there is no phase transition that separates the strong from the weak-coupling regime, the proof of confinement in the strong coupling regime carries over to the continuum limit. However, this assumption is not necessarily justified; while there is no analytic proof, there is convincing numerical evidence that this is indeed the case. In the strong coupling regime, one may expand the Boltzmann factor $\exp(-S[U])$ in powers of $1/g_s^2$. Since the action is a class function, one can perform a character expansion of the plaquette Boltzmann factor

$$\exp\left(\frac{1}{g_s^2}\mathrm{Tr}U_{x,\mu\nu}\right) = \sum_\Gamma \chi_\Gamma(U_{x,\mu\nu})f_\Gamma, \quad f_\Gamma = \int_{SU(N_c)} dU \, \chi_\Gamma(U)^* \exp\left(\frac{1}{g_s^2}\mathrm{Tr}U_{x,\mu\nu}\right).$$

$$(14.34)$$

Let us again consider a rectangular loop $\mathcal{C}$ of size $R \times T$ in the $\mu\nu$-plane, oriented in a clockwise manner. Applying the group integration rules of eq. (14.18), the leading term with $\exp(-S[U]) = 1$ (for $1/g_s^2 = 0$) then leads to

$$\langle W_\mathcal{C}\rangle = \frac{1}{Z}\prod_{x,\mu}\int_{SU(N_c)} dU_{x,\mu} \,\mathrm{Tr}\left[\mathcal{P}\prod_{(x,\mu)\in\mathcal{C}} U_{x,\mu}\right] = 0, \qquad (14.35)$$

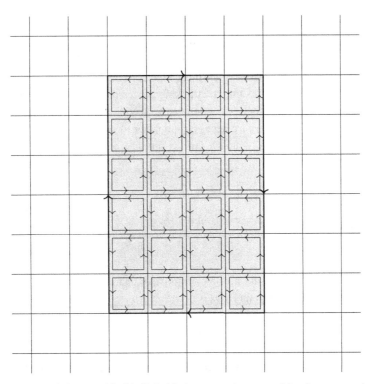

Fig. 14.2 A Wegner–Wilson loop (thick external line) is tiled with elementary plaquettes, giving rise to an area law. The integrations over the link variables do not vanish, because each link enclosed by the loop appears twice, with opposite orientations.

because the integration over the links that belong to the Wegner–Wilson loop gives zero. The second formula of eq. (14.18) suggests that, in order to obtain a non-zero result, one needs a product of two matrix elements for each link variable along the Wegner–Wilson loop.

When we expand to the next order for a plaquette that touches the loop along a link (x, μ), the integration over that link indeed provides a non-zero result, because both $U_{x,\mu}$ and $U_{x,\mu}^\dagger$ contribute to the integrand. However, as long as there remain some links along the loop that are not touched by plaquettes, the integration over them still gives zero. Furthermore, the plaquettes that result from expanding the Boltzmann factor bring in additional links that must again be paired. Consequently, the leading non-zero result is obtained when the Boltzmann factors for all plaquettes enclosed by the planar Wegner–Wilson loop are expanded to first order, such that the interior of the loop is completely tiled with plaquettes. Since there are $R/a \times T/a = A/a^2$ tiling plaquettes, the expansion gives rise to a factor $1/(g_s^2)^{A/a^2}$. This situation is illustrated in Figure 14.2.

Let us now perform the corresponding link integrations. First of all, the integration over all links that do not belong to the tiling plaquettes, or to the loop contour, provides a trivial factor of 1. The four links of each tiling plaquette occur twice in the integrand (once as $U_{x,\mu}$ and once as $U_{x,\mu}^\dagger$), because they are shared either with an adjacent plaquette or with the loop contour. When we integrate out a link that belongs to the loop contour, according to the fusion rule of eq. (14.30), the loop gets deformed around the adjacent plaquette, and a factor of $1/N_c$ is generated. By integrating out more links, the original loop gets deformed further and further until it ultimately shrinks to a point. In this process, it is important to be aware of the following subtlety. The deformed loop may contain pieces on which it backtracks along itself. Such pieces cancel because $U_{x,\mu} U_{x,\mu}^\dagger = U_{x,\mu}^\dagger U_{x,\mu} = \mathbb{1}$. The integrations over the links on backtracking pieces (which trivially "deform" the loop by cutting out such pieces) therefore contribute a trivial factor of 1 instead of $1/N_c$. Since A/a^2 non-trivial deformations are necessary in order to shrink the original loop to a point, from the link integrations one obtains an overall factor of $1/N_c^{A/a^2}$. Altogether, we thus obtain

$$\langle W_{\mathcal{C}} \rangle = \frac{1}{N_c^{A/a^2}} \frac{1}{(g_s^2)^{A/a^2}} = \exp\left(-A \log\left(N_c g_s^2\right)/a^2\right) \quad \Rightarrow \quad \sigma = \frac{1}{a^2} \log\left(N_c g_s^2\right).$$

(14.36)

This leading contribution in the strong coupling expansion indeed gives rise to an area law, as characterized by the relation (14.14).

It is worth noting that $N_c g_s^2$ is the so-called 't Hooft coupling, which is kept constant in the large-N_c limit, such that σ remains finite.

We add that the particular result (14.36) only holds for $N_c > 2$. This is because SU(2) is pseudo-real (see Appendix F) such that the orientation of a loop loses its meaning. Still, SU(2) lattice gauge theory follows an area law as well. In that case, the leading-order of the strong coupling expansion yields the string tension $\sigma = \log\left(g_s^2\right)/a^2$; see, e.g., Smit (2002).

14.5 Asymptotic Freedom and Natural Continuum Limit

Besides the vacuum-angle $\theta \in]-\pi, \pi]$ (which we have put to 0 in agreement with experiment), classical Yang–Mills theory has only one parameter, g_s, which is dimensionless; hence, the model is scale-invariant. In the quantum theory, on the other hand, scale invariance is explicitly broken by the UV cut-off. Since scale invariance is anomalous, it remains broken even after the cut-off is removed to infinity. As a result, in the process of renormalization, by means of *dimensional transmutation*, the dimensionless bare coupling g_s is traded for a single, dimensioned scale. In the context of the so-called modified, minimal subtraction scheme, which is popular in perturbation theory, this scale is denoted as $\Lambda_{\overline{\text{MS}}}$.

On the lattice, *a priori*, the most natural scale is the lattice spacing a. The continuum limit is approached when the correlation length $\xi = 1/M$ (which corresponds to the inverse mass of the lightest glueball) diverges in units of the lattice spacing, $\xi/a \to \infty$ (cf. Chapter 1). Therefore, the long-distance, low-energy physics becomes insensitive to the short-distance lattice details. How can one reach this limit? The functional integral of lattice Yang–Mills theory contains a single parameter – the bare gauge coupling g_s. When one wants to reach the continuum limit, one must search for a value of g_s for which the correlation length ξ diverges. As usual, we are looking for a *second-order phase transition*, so we are searching for the *critical value* of g_s.

According to Section 12.7, as a manifestation of the hierarchy problem, approaching the continuum limit in scalar field theory requires an unnatural fine-tuning of the bare parameters in the lattice action. Fortunately, a similar hierarchy problem does not exist for 4-d non-Abelian gauge fields. This is due to their remarkable property of *asymptotic freedom*, which we mentioned in Chapter 5 and which will be discussed in more detail in Chapter 18. On the lattice, asymptotic freedom implies that

$$\xi/a \sim \left(\beta_0 g_s^2\right)^{\beta_1/(2\beta_0^2)} \exp\left(\frac{1}{2\beta_0 g_s^2}\right), \quad \beta_0 = \frac{11 N_c}{3(4\pi)^2}, \quad \beta_1 = \frac{34 N_c^2}{3(4\pi)^4}. \quad (14.37)$$

Here, β_0 and β_1 are the 1- and 2-loop coefficients of the β-function, which we anticipated in Section 5.11 (Vanyashin and Terentev, 1965; Khriplovich, 1969; Gross and Wilczek, 1973; Politzer, 1973; Caswell, 1974; Jones, 1974). As a consequence of eq. (14.37), the correlation length diverges, *i.e.* the continuum limit of lattice Yang–Mills theory is reached, in the limit $g_s \to 0$. Hence, without any fine-tuning, a very long correlation length, $\xi \gg a$, can be obtained simply by sending g_s to a sufficiently (but not unnaturally) small value. The glueball mass M is thus naturally much smaller than the lattice momentum cut-off π/a. When we again identify the lattice spacing a with the Planck length, we find that, unlike Higgs particles, glueballs are naturally very much lighter than this ultimate high-energy scale.

By matching calculations in lattice and continuum perturbation theory, one can relate $\Lambda_{\overline{MS}}$ to Λ_{lattice}, which is a scale defined in lattice perturbation theory. Such a study was presented by Anna and Peter Hasenfratz (Hasenfratz and Hasenfratz, 1980): By matching 2- and 3-point functions to 1 loop in the continuum and on the lattice, they obtained a large conversion factor for SU(3)$_c$ Yang–Mills theory, $\Lambda_{\overline{MS}} \simeq 83.5 \, \Lambda_{\text{lattice}}$. In this way, one can convert a non-perturbative lattice result $\xi \Lambda_{\text{lattice}}$ into a result $\xi \Lambda_{\overline{MS}} = C$ in the continuum $\overline{MS}$-scheme. Here, C is a dimensionless, universal, non-perturbative prediction of Yang–Mills theory, which is independent of the UV cut-off. Obviously, it depends on the choice of the modified, minimal subtraction renormalization scheme, which is convenient in perturbative calculations. This way of matching lattice and continuum scales would work only if the lattice calculation is performed very close to the continuum limit, in the so-called asymptotic scaling regime, which is difficult to reach in practice.

14.6 How Strong is the Strong Force?

As we will discuss in Section 14.7, in lattice Yang–Mills theory the radius of convergence of the strong coupling expansion is limited by a roughening transition of the string world-sheet that is bounded by the Wegner–Wilson loop. Consequently, the strong coupling expansion cannot reach the continuum limit that lies in the weak-coupling regime. Numerical simulation techniques, on the other hand, are sufficiently powerful to enter the continuum regime. The results of a Monte Carlo simulation of the potential between external static charges in the fundamental $\{3\}$ and $\{\bar{3}\}$ representations of SU(3)$_c$ are shown in Figure 14.3. A potential of the form

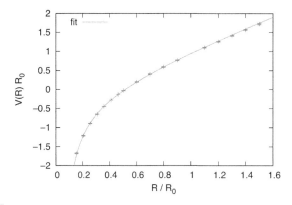

Fig. 14.3 The confining potential $V(R)$ for a static quark–anti-quark pair at a distance R in $SU(3)_c$ Yang–Mills theory, expressed in units of the Sommer scale $R_0 \simeq 0.470(4)$ fm, using Monte Carlo data from Necco and Sommer (2002). We show the dimensionless term $V(R)\,R_0$, with $V(R) = \sigma R + \mu - \pi/(12R)$, cf. eq. (14.53), where σ and μ are treated as fit parameters. The resulting string tension $\sigma \simeq 1.39/R_0^2 \simeq (0.495(5)\ \text{GeV})^2$ has a similar value as the one of a centimeter-thick steel cable, shown on the right-hand side. However, the Yang–Mills string is about 13 orders of magnitude thinner.

$$V(R) = \sigma R + \mu - \frac{\pi}{12R}, \qquad (14.38)$$

with a non-zero string tension σ, is obtained with high statistical accuracy.

From a non-perturbative point of view, the correlation length ξ itself provides a natural, physical unit. Yang–Mills theory makes predictions only for dimensionless ratios of physical quantities, such as, *e.g.*, the string tension expressed in units of ξ, $\sigma\xi^2 = C'$, where C' is another universal, dimensionless number. Lattice Monte Carlo calculations are performed in a finite space–time volume, usually with periodic boundary conditions (for the sake of (discrete) translation invariance). From a practical point of view, neither the correlation length ξ nor the string tension σ are among the most easily accessible quantities, in particular, because they are affected by strong finite-size effects and by severe statistical fluctuations.

Rainer Sommer defined a more easily and accurately accessible length scale R_0, known as the *Sommer scale* (Sommer, 1994). It is implicitly defined via the force $F(R) = - \,dV(R)/dR$ derived from the static quark–anti-quark potential $V(R)$ at the specific distance $R = R_0$, such that

$$R_0^2\,|F(R_0)| = 1.65. \qquad (14.39)$$

While this definition of a length scale may seem quite arbitrary, it is very well chosen, so that it can be determined accurately in practical lattice calculations. In particular, it is much more easily accessible than the string tension σ, which requires the accurate computation of the potential $V(R) \sim \sigma R$ at asymptotically large distances. In the meantime, alternative theoretically defined scales related to the so-called gradient flow have been defined (Lüscher, 2010). These scales are very useful because they can be computed extremely accurately, although they are not directly related to an experimentally measurable physical quantity.

In the real world, there are many more scales than just that of the strong interaction. However, at least at the present stage of our construction of the Standard Model, gluons do not couple to electroweak gauge or Higgs fields and they thus form a "world" of their own. Still gravity, which is usually ignored in the Standard Model, would also affect the gluons of SU(3)$_c$ Yang–Mills theory. The scale of gravity is set by Newton's constant $G = 1/M_{Planck}^2$, which defines the Planck scale $M_{Planck} \approx 2.2 \cdot 10^{-8}$ kg. The corresponding length scale is the Planck length $l_{Planck} \approx 1.6 \cdot 10^{-35}$ m. The Sommer scale R_0, on the other hand, is of the magnitude of 1 fm $= 10^{-15}$ m, the typical size of atomic nuclei. As we pointed out before, the hierarchy of length scales $R_0/l_{Planck} \approx 10^{20}$ does *not* give rise to a hierarchy problem, because, thanks to asymptotic freedom, R_0 is naturally (*i.e.* without fine-tuning of g_s) much larger than l_{Planck} (which we may again identify with the lattice spacing a in the present context).

As we discuss in Appendix B, the kilogram and the meter are man-made units whose origin is historical and not a matter of physics. Expressing the Sommer scale R_0 in meters (or better fm) is a non-trivial exercise. First of all, R_0 refers to the force between infinitely heavy static color charges which do not exist in Nature. Hence, a direct experimental measurement of R_0 is not possible. From a theoretical point of view, this is not a serious issue, because static sources are a legitimate theoretical probe, not only in Yang–Mills theory but also in the full Standard Model.

It is important to note that, in contrast to R_0, the status of the string tension σ is very different in Yang–Mills theory and in the Standard Model: σ vanishes in the presence of dynamical quarks, because the confining string, connecting an external static quark to a static anti-quark, can then break due to the creation of a dynamical quark–anti-quark pair.

It is another important feature of R_0 that it is not too sensitive to other Standard Model parameters, including the quark masses. Its *proper determination* in fm proceeds by performing a lattice calculation of a dimensionless combination $M_p R_0$, where M_p could, for example, be the proton mass, and then inserting the experimental value of M_p. A less rigorous procedure is to fit the spacings of energy levels of heavy, non-relativistic quark–anti-quark bound states (such as $c\text{-}\bar{c}$ and $b\text{-}\bar{b}$) with a phenomenological Cornell potential $V_C(R) = \sigma R + \mu + C/R$ at intermediate distances and to extract R_0 from $V_C(R)$. The most accurate lattice determinations yield $R_0 = 0.470(4)$ fm (Aoki *et al.*, 2020).

In order to gain a more intuitive understanding for the strength of the strong force, let us perform a *Gedankenexperiment* by asking which weight a single Yang–Mills string can support, against the gravitational attraction on planet earth. Lattice calculations result in the string tension $\sigma \approx 1.39\, \hbar c/R_0^2$. The gravitational force exerted on a mass M near earth's surface is $F = Mg$, where $g \approx 9.81$ m/sec^2 is earth's gravitational acceleration (not to be confused with the electroweak SU(2)$_L$ gauge coupling). Equating this with the confining force $F = -dV(R)/dR = -\sigma$ at asymptotically large distances, $R \to \infty$, determines the mass that can be supported by the tension of a single Yang–Mills string

$$\sigma = Mg \quad \Rightarrow \quad M = \frac{\sigma}{g} \approx 1.39 \frac{\hbar c}{g R_0^2} \approx 2 \cdot 10^4 \, \text{kg}. \tag{14.40}$$

This is of the order of the weight that a centimeter-thick steel cable can support. Remarkably, a Yang–Mills string, which is able to support the same weight, is about 13 orders of magnitude thinner, cf. Figures 14.3 and 14.4,

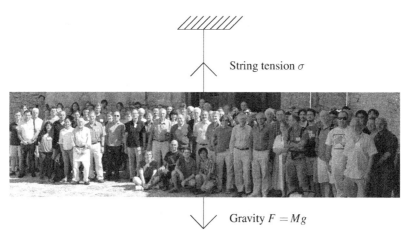

String tension σ

Gravity $F = Mg$

Fig. 14.4 A single Yang–Mills string is strong enough to support more than 100 people. A "Yang–Mills elevator" was installed by one of the authors as a Gedankenexperiment at the Erice workshop "From Quarks and Gluons to Hadrons and Nuclei" in 2011, in order to "elevate" our intuitive understanding for the strength of the strong force (Wiese, 2012).

> **Quick Question 14.6.1 Overloaded Yang–Mills elevator**
> What happens when about 1000 (adult) people enter the Yang–Mills elevator?

14.7 Roughening Transition

The string tension has been computed to high orders of the strong coupling expansion. The resulting contributions are associated with plaquette surfaces that are deformations of the minimal tiling surface of plaquettes, which are still bounded by the rectangular Wegner–Wilson loop contour. For example, the next-to-leading contribution comes from a deformation along the surface of an elementary cube that touches the minimal surface. Higher-order corrections result from surfaces with more complicated step-like shapes, which can be interpreted as Euclidean world-sheets of the fluctuating confining string.

In condensed matter physics, piecewise constant step-like surfaces are known as *rigid interfaces*. For example, let us consider the two broken phases that exist in the 3-d Ising model (of classical statistical mechanics) at temperatures $T < T_{\rm c}$, *i.e.* below the critical point. At very low temperatures $T < T_{\rm r}$, the interfaces that separate the two broken phases are rigid. At the roughening temperature $T_{\rm r} < T_{\rm c}$, the interfaces themselves – but not yet the bulk of the system – undergo a phase transition known as the *roughening transition*. At temperatures above the roughening transition, but still below $T_{\rm c}$ (*i.e.* for $T_{\rm r} < T < T_{\rm c}$), the interfaces are no longer step-like (and thus rigid), but become rough because they develop soft modes. The corresponding massless excitations – which are known as *capillary waves* – are associated with the transverse translational symmetry that is spontaneously broken by the position of the interface.

The position of the interface above a 2-d base plane is described by an integer-valued height variable, $h_x \in \mathbb{Z}$, where x is a lattice point in a 2-d plane and h_x determines the height of the interface above that plane in units of the lattice spacing a. The resulting effective theory is a 2-d *height model* with a discrete translation symmetry $\mathbb{Z}$, with the classical Hamilton function and partition function

$$\mathcal{H}[h] = \frac{\kappa}{2} \sum_{x,i=1,2} \left(h_x - h_{x+\hat{i}}\right)^2, \quad Z = \prod_x \sum_{h_x \in \mathbb{Z}} \exp\left(-\beta\mathcal{H}[h]\right), \quad \kappa > 0. \qquad (14.41)$$

One might not expect the spontaneous breakdown of a discrete symmetry to give rise to a massless boson. However, the 2-d height model with the discrete $\mathbb{Z}$ symmetry is connected to the 2-d XY model with a continuous, global O(2) symmetry by an exact duality transformation. The high-temperature phase of the height model, in which the interface is rough, is then mapped to the low-temperature phase of the XY model which is indeed massless. Despite the fact that, due to the Mermin–Wagner–Hohenberg–Coleman theorem, in $d = 2$ the O(2) symmetry cannot break spontaneously, below the phase transition temperature, there is still an exactly massless mode. Indeed, the roughening transition in the 2-d height model is dual to the Berezinskiĭ–Kosterlitz–Thouless transition of the 2-d XY model. The roughening transition in the 3-d Ising model (which only affects the 2-d interfaces below T_c but not the bulk of the system) falls in the same universality class. We address the massless transverse modes of the string world-sheet in Yang–Mills theory in the next section.

14.8 Systematic Low-Energy Effective String Theory

In lattice Yang–Mills theory, the roughening transition occurs at a critical value $g_{s,r}$ of the bare gauge coupling. At strong coupling $g_s > g_{s,r}$, the world-sheet of the confining string that connects two external, fundamental charges forms a rigid step-like surface. In the weak-coupling regime, $g_s < g_{s,r}$, which is reached when one approaches the continuum limit $g_s \to 0$, on the other hand, the world-sheet is rough and supports massless excitations. These excitations, which correspond to the capillary waves in condensed matter physics, play the role of Nambu–Goldstone bosons associated with the spontaneous breakdown of translation invariance in the spatial directions transverse to the world-sheet. Indeed, the location of a free-floating world-sheet breaks translation invariance spontaneously. The world-sheet of a string that connects external static charges, however, is constrained to end at the location of the charges, such that translation invariance is, in addition, explicitly broken.

The massless, transverse Nambu–Goldstone excitations of the Yang–Mills string are described by a systematic low-energy effective continuum field theory, which, at the same time, turns out to be a *string theory*. When string theory is considered as a candidate for a fundamental theory of Nature, for consistency it must be formulated in $d = 26$ space–time dimensions for bosonic strings and in $d = 10$ space–time dimensions for supersymmetric strings. The string theory that describes the Yang–Mills string, on the other hand, is a systematic low-energy effective theory that breaks down at energies around the glueball mass $M = 1/\xi$, but is consistent in any space–time dimension for which Yang–Mills strings are rough. Let us perform the calculation not just in 4, but generally in $d \geq 3$ Euclidean space–time dimensions. The most relevant cases are $d = 3$ and 4. In higher dimensions, $d > 5$, lattice Yang–Mills theories still confine at strong coupling, but not in the continuum limit. As long as the strong coupling phase contains a regime in which the string world-sheets are rough, the calculation still applies to Yang–Mills theories with $d \geq 5$.

We consider a string that connects two external static color charges at the spatial positions $\vec{x} = (0, 0, \ldots, 0)$ and $\vec{y} = (0, 0, \ldots, R)$, separated by a distance R in the spatial $(d - 1)$-direction. The location of the string world-sheet is then described by a "height field"

$$X^a(x_{d-1}, x_d) = \left(X^1(x_{d-1}, x_d), X^2(x_{d-1}, x_d), \ldots, X^{d-2}(x_{d-1}, x_d) \right) , \quad a \in \{1, 2, \ldots, d-2\},$$
(14.42)

which represents the vector that points from the base point (x_{d-1}, x_d) to the string. For $d = 3$, it corresponds to the height variable $h_x \in \mathbb{Z}$ in eq. (14.41). The string is described by an *effective 2-d field theory* for the $(d - 2)$-component scalar field $X^a(x_{d-1}, x_d)$ over a spatial interval $x_{d-1} \in [0, R]$. Since the string must end in the charges, the field X obeys the Dirichlet boundary conditions

$$X^a(0, x_d) = X^a(R, x_d) = 0.$$
(14.43)

The leading terms in the Euclidean effective action for the string at inverse temperature β take the form

$$S_0[X] = \int_0^\beta dx_d \int_0^R dx_{d-1} \, \sigma \left(1 + \frac{1}{2} \partial_\nu X^a \partial_\nu X^a \right) + \int_0^\beta dx_d \, \mu$$

$$= \sigma R \beta + \mu \beta + \sigma \int_0^\beta dx_d \int_0^R dx_{d-1} \, \frac{1}{2} \partial_\nu X^a \partial_\nu X^a,$$
(14.44)

where $\nu \in \{d-1, d\}$, σ represents the string tension and μ is the energy associated with the two endpoints of the string. As usual, at finite temperature, the scalar field obeys periodic boundary conditions in Euclidean-time

$$X^a(x_{d-1}, x_d + \beta) = X^a(x_{d-1}, x_d).$$
(14.45)

The corresponding Hamilton operator is given by

$$\hat{H} = \mu + \int_0^R dx_{d-1} \left(\sigma + \frac{1}{2\sigma} \hat{\Pi}^a \hat{\Pi}^a + \frac{\sigma}{2} \partial_{d-1} \hat{X}^a \partial_{d-1} \hat{X}^a \right).$$
(14.46)

Here, $\hat{X}^a(x_{d-1})$ and $\hat{\Pi}^a(x_{d-1})$ are scalar field operators and their canonically conjugate momenta, which obey the commutation relations

$$\left[\hat{X}^a(x_{d-1}), \hat{\Pi}^b(y_{d-1}) \right] = i \delta_{ab} \, \delta(x_{d-1} - y_{d-1}),$$

$$\left[\hat{X}^a(x_{d-1}), \hat{X}^b(y_{d-1}) \right] = \left[\hat{\Pi}^a(x_{d-1}), \hat{\Pi}^b(y_{d-1}) \right] = 0.$$
(14.47)

Performing a Fourier analysis,

$$\hat{X}^a(x_{d-1}) = \frac{2}{R} \sum_{n=1}^{\infty} \hat{X}_n^a \sin(\pi n x_{d-1}/R), \quad \hat{X}_n^a = \int_0^R dx_{d-1}\, \hat{X}^a(x_{d-1}) \sin(\pi n x_{d-1}/R),$$

$$\hat{\Pi}^a(x_{d-1}) = \frac{2}{R} \sum_{n=1}^{\infty} \hat{\Pi}_n^a \sin(\pi n x_{d-1}/R), \quad \hat{\Pi}_n^a = \int_0^R dx_{d-1}\, \hat{\Pi}^a(x_{d-1}) \sin(\pi n x_{d-1}/R),$$

$$(14.48)$$

results in the commutation relations

$$\left[\hat{X}_n^a, \hat{\Pi}_m^b\right] = \frac{1}{2} i R \delta_{ab} \delta_{nm}, \quad \left[\hat{X}_n^a, \hat{X}_m^b\right] = \left[\hat{\Pi}_n^a, \hat{\Pi}_m^b\right] = 0, \quad (14.49)$$

and the Hamilton operator takes the form

$$\hat{H} = \sigma R + \mu + \frac{1}{R} \sum_{n=1}^{\infty} \left[\frac{1}{\sigma} \hat{\Pi}_n^a \hat{\Pi}_n^a + \sigma \left(\frac{\pi n}{R}\right)^2 \hat{X}_n^a \hat{X}_n^a \right]. \quad (14.50)$$

We now introduce the annihilation and creation operators

$$\hat{a}_n^a = \frac{1}{\sqrt{R}} \left(\sqrt{\frac{\pi n \sigma}{R}} \hat{X}_n^a + i \sqrt{\frac{R}{\pi n \sigma}} \hat{\Pi}_n^a \right), \quad \hat{a}_n^{a\dagger} = \frac{1}{\sqrt{R}} \left(\sqrt{\frac{\pi n \sigma}{R}} \hat{X}_n^a - i \sqrt{\frac{R}{\pi n \sigma}} \hat{\Pi}_n^a \right),$$

$$(14.51)$$

and finally obtain the Hamilton operator and the commutation relation

$$\hat{H} = \sigma R + \mu + \sum_{n=1}^{\infty} \frac{\pi n}{R} \left(\hat{a}_n^{a\dagger} \hat{a}_n^a + \frac{d-2}{2} \right), \quad \left[\hat{a}_n^a, \hat{a}_m^{b\dagger}\right] = \delta_{ab} \delta_{nm}. \quad (14.52)$$

14.9 Lüscher Term as a Casimir Effect

Following Lüscher (1981), we now consider the leading correction to the ground state energy – which corresponds to the static quark–anti-quark potential – by summing up the zero-point energies of all the string modes n,

$$V(R) = \sigma R + \mu + \sum_{n=1}^{\infty} \frac{\pi n}{R} \frac{d-2}{2} \hat{=} \sigma R + \mu - \frac{\pi(d-2)}{24R}. \quad (14.53)$$

The symbol $\hat{=}$ means "corresponds to" in the sense of analytic continuation or renormalization, as we are going to discuss at the end of this section.

The Lüscher term $-\pi(d-2)/(24R)$ provides a universal, long-distance correction to $V(R)$, which corresponds to the *Casimir effect* (Casimir and Polder, 1948; Casimir, 1948),[3]

---

[3] The Casimir effect has been experimentally observed for the quasi-vacuum between metal surfaces, first between a plate and a sphere (Lamoreaux, 1997; Mohideen and Roy, 1998), and later also between parallel plates (Bressi *et al.*, 2002). In this form, the effect is more complicated because it is affected by the interaction of the electromagnetic field with the electric charges and currents inside the plates.

i.e. to a finite-size effect of the vacuum energy due to the Dirichlet boundary conditions of eq. (14.43). The corresponding attractive force corrects the string tension to

$$F(R) = -\frac{dV(R)}{dR} = -\sigma - \frac{\pi(d-2)}{24R^2}.$$ (14.54)

The line in Figure 14.3 is a fit to eq. (14.53) with $d = 4$. Surprisingly, it works all the way from short to long distances, including at the Sommer scale $R = R_0$, which implies

$$\sigma R_0^2 + \frac{\pi}{12} \approx 1.65 \quad \Rightarrow \quad \sigma R_0^2 \approx 1.39 \quad \Rightarrow \quad \sigma = (0.495(5)\ \text{GeV})^2.$$ (14.55)

Actually, at these moderate distance scales the excited states of the string observed in Monte Carlo simulations do not yet seem to follow the effective string theory, which predicts an energy difference to the ground state of π/R (Juge *et al.*, 2003). However, pushing the numerical calculations to even larger distance scales and controlling their continuum limit is quite difficult.

For completeness, let us derive the Lüscher term, which is based on the relation

$$\sum_{n=1}^{\infty} n \overset{\wedge}{=} -\frac{1}{12}.$$ (14.56)

This relation, which may look strange at first glance, follows naturally from the analytically continued ζ-function. At $\text{Re}\, s > 1$, it is defined by the convergent series

$$\zeta(s) = \sum_{n=1}^{\infty} n^{-s}.$$ (14.57)

Bernhard Riemann analyzed its analytic continuation to $\mathbb{C} - \{1\}$: There is a simple pole at $s = 1$ (Riemann, 1859). His result can be condensed into the formula

$$\zeta(s) = \frac{(2\pi)^s}{\pi} \sin\left(\frac{\pi s}{2}\right) \Gamma(1-s)\,\zeta(1-s),$$ (14.58)

which is valid all over $\mathbb{C}$. We read off $\zeta(-1) = (2\pi^2)^{-1}(-1)\pi^2/6 = -1/12$, where we inserted Leonhard Euler's "Basel formula" $\zeta(2) = \sum_{n\geq 1} 1/n^2 = \pi^2/6$. This justifies the relation (14.56), which was rediscovered by Srinivasa Ramanujan, and it also justifies eq. (14.53).

A perhaps more physical regularization results when we consider the momenta $p_n = \pi n/R$, $n \in \mathbb{N}$, associated with the finite interval $[0, R]$ and compare the sum $(\pi/R) \sum_{n=1}^{\infty} p_n$ with the momentum integral $\int_0^{\infty} dp\, p$, which plays the role of a counter-term. Obviously, both the sum and the integral are UV-divergent. Let us apply the *heat kernel regularization*, which inserts a convergence-generating factor $\exp(-tp)$, $t > 0$,

$$\int_0^{\infty} dp\, p \exp(-tp) = -\frac{d}{dt} \int_0^{\infty} dp \exp(-tp) = -\frac{d}{dt}\frac{1}{t} = \frac{1}{t^2}.$$ (14.59)

It still diverges in the limit $t \to 0$. Next, we regularize the sum in the same manner

$$\frac{\pi}{R} \sum_{n=1}^{\infty} p_n \exp(-tp_n) = -\frac{\pi}{R}\frac{d}{dt} \sum_{n=1}^{\infty} \exp(-t\pi n/R) = -\frac{\pi}{R}\frac{d}{dt}\left(\frac{1}{1-\exp(-t\pi/R)} - 1\right)$$

$$= \frac{\pi^2}{R^2}\frac{\exp(-t\pi/R)}{\left[1-\exp(-t\pi/R)\right]^2} \overset{t\to 0}{\longrightarrow} \frac{1}{t^2} - \frac{\pi^2}{12R^2} + \dots.$$ (14.60)

The difference between the sum and the integral (times R/π) hence takes the form

$$\sum_{n=1}^{\infty} p_n \exp(-tp_n) - \frac{R}{\pi} \int_0^{\infty} dp\, p \exp(-tp) \xrightarrow[t \to 0]{} \frac{R}{\pi} \left(\frac{1}{t^2} - \frac{\pi^2}{12R^2} - \frac{1}{t^2} \right) = -\frac{\pi}{12R}, \quad (14.61)$$

which is again just the Lüscher term. The ground state energy at finite R is smaller than its value in the $R \to \infty$ limit, by a UV-independent amount.

Riemann's analytic continuation subtracts the same divergent term from the divergent series $\sum_{n \geq 1} n$.

For a more general class of regularizations, this difference between the sum and the corresponding integral (the counter-term to be subtracted) can be expanded by means of the Euler–Maclaurin formula (Casimir and Polder, 1948). The result unambiguously agrees with eq. (14.61), which confirms that it does not depend on the choice of the regularizing function, as long as it fulfills suitable conditions (they are discussed for instance in Bietenholz (2021)).

14.10 Cosmological Constant Problem on the String World-Sheet

As we discussed in Section 2.7, the vacuum energy of a scalar quantum field theory receives a UV-divergent additive term, to be removed by renormalization. The cosmological constant problem is a hierarchy problem that arises because the observed vacuum energy density is about 120 orders of magnitude smaller than its naively expected value (employing a cut-off at the Planck scale), as we discussed in Section 2.7. It seems to require an enormous amount of fine-tuning to arrange for the observed value. This was motivation enough for Weinberg to seek an anthropic explanation for the cosmological constant (Weinberg, 1987): Were its magnitude just a bit larger than the observed value, the Universe may no longer be hospitable for life as we know it.

The world-sheet of a Yang–Mills string is a $(1 + 1)$-dimensional "world" of its own. In fact, since Yang–Mills theory has a mass gap, this is the only place where low-energy excitations can exist. It is inhabited by massless "creatures", the transverse Nambu–Goldstone excitations, that constantly travel from one end of their world to the other at the speed of light. If they would find time to do physics (which is obviously impossible), just like we, they might get puzzled about their cosmological constant problem. When they perform calculations in the low-energy effective string theory, which we may view as their "standard model", they encounter the string tension σ as their vacuum energy density, which is again UV-divergent. On the other hand, observation may tell them that σ has a specific finite value, way below any ultimate UV cut-off.

As their technology advances, the inhabitants of the string world-sheet may discover higher-order terms in their "standard model" Lagrangian. Since their world has two ends, besides μ, there are potentially additional boundary terms with two derivatives, which contribute to the Euclidean action at sub-leading-order

$$S_1[X] = \frac{b}{4} \int_0^{\beta} dx_d \left(\partial_{d-1} X^a \partial_{d-1} X^a |_{x_{d-1}=0} + \partial_{d-1} X^a \partial_{d-1} X^a |_{x_{d-1}=R} \right). \quad (14.62)$$

In addition, there may be two sub-leading bulk terms with four derivatives

$$S_2[X] = \frac{c_2}{4} \int_0^\beta dx_d \int_0^R dx_{d-1} \, \partial_\nu X^a \partial_\nu X^a \partial_\rho X^b \partial_\rho X^b,$$

$$S_3[X] = \frac{c_3}{4} \int_0^\beta dx_d \int_0^R dx_{d-1} \, \partial_\nu X^a \partial_\nu X^b \partial_\rho X^a \partial_\rho X^b. \tag{14.63}$$

It was pointed out by Martin Lüscher and Peter Weisz (Lüscher and Weisz, 2004) that open-closed string duality, which interchanges β with $2R$ and is manifest in the underlying Yang–Mills theory, implies

$$b = 0, \quad (d-2)c_2 + c_3 = \frac{d-4}{2} \sigma. \tag{14.64}$$

Once the $(1 + 1)$-d physicists' measurements provide $b = 0$ and $2c_2 + c_3 = 0$, it may dawn on them that they actually live on a string embedded in a 4-d space–time. This also explains their long-standing puzzle why there are two distinct massless modes, $a \in \{1, 2\}$: They reflect the fluctuations of their world in the two transverse extra dimensions.

After intense theoretical studies of 4-d theories, they may eventually encounter Yang–Mills theory along with its remarkable property of asymptotic freedom. After further tremendous advances of their collider technology, which necessarily involves linear accelerators, the 2-d physicists may accomplish an enormous concentration of energy at some point on the string, eventually leading to the emission of a glueball into the 4-d bulk. This may convince them that Yang–Mills theory should indeed be their "Theory of Everything". In this way, they will learn that the string tension σ, which plays the role of their cosmological constant, is protected from additive renormalization. In fact, thanks to asymptotic freedom of the underlying Yang–Mills theory, without fine-tuning, it is naturally much smaller than the ultimate UV cut-off. This seems to solve their cosmological constant problem and to obviate the need for an "alien-tropic" explanation.

One should not forget that Yang–Mills theory has its own vacuum energy problem. The vacuum energy density associated with the 4-d bulk is still UV-divergent, but the 2-d world residing on the string is affected by this only via the divergent self-energy μ of the static quark–anti-quark pair. The 2-d vacuum energy σR, which is proportional to the size R of the string "universe", is indeed UV-protected.

An important aspect of the cosmological constant problem, which we have ignored in our discussion, concerns the incorporation of *gravity*. If gravity was affected by the 4-d cosmological constant, this might still indirectly affect the 2-d string world.

We have presented the story in this way in order to encourage the reader to think hard about our own cosmological constant problem, which is obviously much more difficult, but perhaps not impossible to solve without resorting to anthropic arguments either. Could our own Universe work in a similar way as the string world-sheet? If so, we might live on a 3-brane, a 3-d space-filling membrane that sweeps out a $(3 + 1)$-d manifold embedded in a higher-dimensional space–time (Rubakov and Shaposhnikov, 1983). The Standard Model fields might then provide some insight into the nature of the extra dimensions, and the cosmological constant problem might be resolved due to some special feature of the higher-dimensional theory, perhaps similar to asymptotic freedom. String theory may or may not have something to say about a scenario like this.

We bring up these highly speculative ideas mostly to encourage the reader to remain curious and open-minded, and not to be satisfied with our present, poor understanding of the cosmological constant.

14.11   Gluon Confinement and the Fredenhagen–Marcu Operator

As we have seen, in an $SU(N_c)$ Yang–Mills theory external static color charges in the fundamental $\{N_c\}$ and in the anti-fundamental $\{\overline{N_c}\}$ representation are confined to one another by a linearly rising potential (at distances beyond about 0.5 fm). Dynamical gluons, on the other hand, transform in the adjoint $\{N_c^2 - 1\}$ representation. When the gluon theory is exposed to a pair of external color charges in the *adjoint representation*, those are not confined by a non-zero string tension. This is because a color flux string that connects the external static charges can break by the creation of a pair of dynamical gluons, which screen the external color charges.

To understand this more quantitatively, let us again employ the strong coupling expansion. A pair of external static charges in the adjoint representation is again described by a Wegner–Wilson loop. However, the parallel transporters along the loop are now taken in the adjoint representation. Using the character expansion and tiling the Wegner–Wilson loop with plaquettes in the adjoint representation, one again obtains a contribution which alone would give rise to an area law (cf. Figure 14.5). However, there is another contribution, which arises from a tube of plaquettes in the fundamental representation, that dominates at large R and T and gives rise to a perimeter law. As a result, external adjoint charges are not confined by an unbreakable string.

The same is true for dynamical quarks in QCD. External fundamental charges can then be screened by the creation of a dynamical quark–anti-quark pair, which again leads to *string breaking*. This certainly does not mean that dynamical quarks in QCD are not confined. It just means that their confinement is more subtle and cannot be characterized by a string tension related to the area law of a Wegner–Wilson loop. In order to distinguish confined dynamical quarks in QCD from unconfined free charges in a Coulomb phase (like in QED), Klaus Fredenhagen and Mihail Marcu have constructed a *vacuum overlap operator* (Fredenhagen and Marcu, 1986). This order parameter can also be used to characterize the confinement of gluons in Yang–Mills theory.

The *Fredenhagen–Marcu operator* is a ratio of two expectation values. The numerator consists of parallel transporters along a staple-shaped path connecting a source and a sink of color flux, while the denominator is the square root of a Wegner–Wilson loop,

$$\rho(R, T) = \frac{\left\langle \, \boxed{} \, \right\rangle}{\left\langle \, \boxed{} \, \right\rangle^{1/2}}. \tag{14.65}$$

The object in the numerator stands for

$$\boxed{} = \mathrm{Tr}\left[U_{x,ij}\lambda^a\right]\mathrm{Tr}\left[\lambda^a U_{\mathcal{C}_{xy}}\lambda^b U_{\mathcal{C}_{xy}}^\dagger\right]\mathrm{Tr}\left[U_{y,ij}^\dagger\lambda^b\right], \quad U_{\mathcal{C}_{xy}} = \mathcal{P}\prod_{(z,\mu)\in\mathcal{C}_{xy}} U_{z,\mu}. \tag{14.66}$$

Here, $U_{\mathcal{C}_{xy}}$ is a path-ordered product along the staple-shaped path $\mathcal{C}_{xy}$ of time extent T connecting two points x and y that are spatially separated by a distance R. At these points, dynamical gluons are generated by plaquette operators $U_{x,ij}$ and $U_{y,ij}$. The Gell-Mann matrices

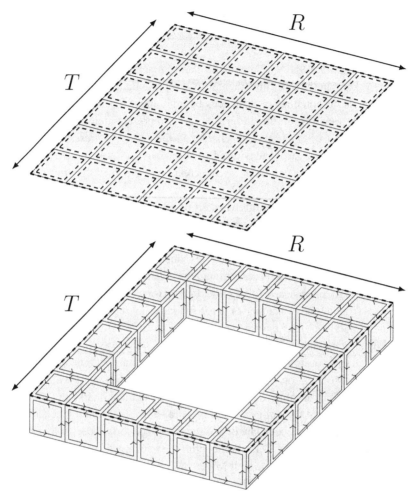

A Wegner–Wilson loop in the adjoint representation (external dashed line) is tiled with elementary plaquettes in the adjoint representation, which alone would give rise to an area law (top). However, for large R and T this term is dominated by a contribution resulting from a tube of fundamental plaquettes that gives rise to a perimeter law (bottom).

λ^a and λ^b reflect the fact that gluons transform in the adjoint representation. The object in the denominator is a Wegner–Wilson loop of size $R \times 2T$ in the adjoint representation.

The Fredenhagen–Marcu order parameter describes the creation of a pair of gluons that propagate for a time T, and it measures their overlap with the vacuum in the limit $R, T \to \infty$. This can be understood as follows: In a Coulomb phase, states of free, unconfined charges exist in the physical spectrum, which are orthogonal to the vacuum state. Consequently, the Fredenhagen–Marcu vacuum overlap order parameter then vanishes. In a confined phase, on the other hand, the dynamical charges are screened and $\rho(R, T)$ approaches a non-zero constant at large R and T.

As illustrated in Figure 14.6, in the strong coupling limit the leading contribution to the numerator of the Fredenhagen–Marcu operator is a tube of plaquettes emanating from the source plaquette (x, ij), following the staple-shaped path, and ending at the sink plaquette

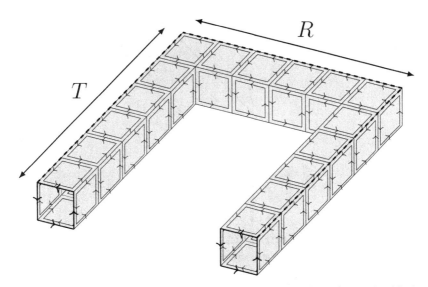

Fig. 14.6 The Fredenhagen–Marcu operator (thick dashed and solid lines) receives a contribution from a tube of fundamental plaquettes. This implies that gluons are still confined, albeit not by an unbreakable string. In particular, they do not exist as color-charged objects in an unconfined, free-charge, non-Abelian Coulomb phase.

(y, ij). This implies a perimeter law for the numerator. Being a Wegner–Wilson loop in the adjoint representation, the denominator also obeys a perimeter law. Due to the square root and the doubled temporal extent, the perimeter behavior in the numerator and in the denominator cancels exactly and one is left with $\rho(R, T) \neq 0$. This shows that gluons are indeed confined (without any string tension, however, with color charge screening). In particular, they do not exist in a non-Abelian Coulomb phase of massless, unconfined gluons. One expects this result to persist in the continuum limit.

Strictly speaking, both the Wegner–Wilson loop and the Fredenhagen–Marcu order parameter provide valid confinement criteria only at zero temperature, because they require taking the limit $T \rightarrow \infty$ (which is impossible when Euclidean-time has a finite extent corresponding to the inverse temperature β).

14.12 Glueball Spectrum

As we have seen, in a Yang–Mills theory well separated, external static charges in the fundamental representations $\{N_c\}$ and $\{\overline{N}_c\}$ experience a linearly rising confining potential. Dynamical gluons in the $(N_c^2 - 1)$-dimensional representation of SU(N_c) are also confined; *i.e.* there are no states of massless color-charged gluons in the physical spectrum. Indeed, in SU(N_c) Yang–Mills theory, gluons are confined inside massive, color-neutral glueballs. The mass M of the lightest glueball is a non-perturbatively generated physical quantity that sets the energy scale of the strong interaction.

When one uses the lattice regularization, the dimensioned cut-off parameter that explicitly breaks scale invariance is the lattice spacing a. In Yang–Mills theory on the lattice, the mass M of a glueball state can be determined from the asymptotic exponential decay of an

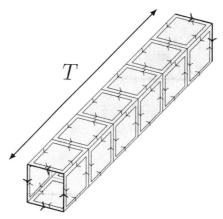

Fig. 14.7 The connected correlation function of two plaquettes (thick lines) receives a contribution from a tube of fundamental plaquettes. This implies that glueballs are massive objects.

appropriate correlation function

$$\langle \mathcal{O}(0)\mathcal{O}(x_4) \rangle = \frac{1}{Z} \prod_{x,\mu} \int_{SU(N_c)} dU_{x,\mu} \, \mathcal{O}(0)\mathcal{O}(x_4) \exp(-S[U])$$

$$\sim |\langle 0|\mathcal{O}|0\rangle|^2 + \left|\langle J^{PC}|\mathcal{O}|0\rangle\right|^2 \exp(-Mx_4), \qquad (14.67)$$

(at large x_4), where $\mathcal{O}$ represents an operator that creates glueball states $|J^{PC}\rangle$, with appropriate angular momentum J, parity P, and charge conjugation parity C, from the vacuum state $|0\rangle$. The lowest mass glueball has angular momentum $J = 0$ and is both parity and charge conjugation even; *i.e.* it is a $J^{PC} = 0^{++}$ glueball. An operator that creates this state from the vacuum is given by

$$\mathcal{O}(x_4) = \sum_{\vec{x},i,j} \mathrm{Tr} \, U_{ij,\vec{x},x_4}. \qquad (14.68)$$

Here, i and j determine the two directions of a spatial plaquette. The summation over i and j ensures that the generated glueball states are lattice rotation-invariant and have positive parity $P = +$. Since the summation includes both orientations of the plaquette, it also implies $C = +$. The summation over the spatial position $\vec{x}$ of the plaquette ensures that the state has vanishing spatial momentum. Taking the trace of the plaquette parallel transporter $U_{ij,\vec{x},x_4}$ guarantees gauge invariance. The masses of glueball states with other quantum numbers can be determined by investigating the correlation functions of other operators. As illustrated in Figure 14.7, in the strong coupling limit the connected glueball correlation function

$$\langle \mathcal{O}(0)\mathcal{O}(x_4) \rangle_c = \langle \mathcal{O}(0)\mathcal{O}(x_4) \rangle - |\langle \mathcal{O} \rangle|^2 = \frac{1}{(N_c g_s^2)^{4x_4/a}} \sim \exp(-Mx_4), \qquad (14.69)$$

is dominated by a tube of plaquettes connecting the plaquettes of the operators $\mathcal{O}(0)$ and $\mathcal{O}(x_4)$. One hence reads off the mass of the lightest 0^{++} glueball

$$M = \frac{4}{a} \log(N_c g_s^2). \qquad (14.70)$$

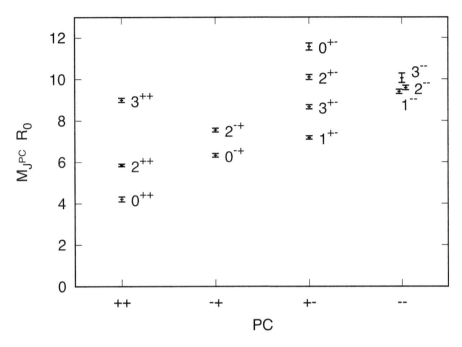

Fig. 14.8 The mass spectrum of the lightest glueballs in $SU(3)_c$ Yang–Mills theory. The masses are given in terms of the Sommer scale R_0. Using $R_0 = 0.470(4)$ fm, the lightest glueball state has a mass of 1.77(5) GeV.

As usual, the mass gap is given by the inverse correlation length, $M = 1/\xi$ (cf. Chapter 1). In this case, M is the mass of the lightest glueball. As we discussed before, when this correlation length becomes much larger than the lattice spacing, $\xi \gg a$, the theory approaches its continuum limit. Using Monte Carlo simulations of lattice Yang–Mills theory, Colin Morningstar and Michael Peardon have computed the spectrum of low-lying glueball states in $SU(3)_c$ Yang–Mills theory (Morningstar and Peardon, 1999). Their results, which have been extrapolated to the continuum limit, are illustrated in Figure 14.8.

Obviously, any numerical Monte Carlo result does not have the same status as a rigorous mathematical derivation. Rigorously proving the existence of a non-zero mass gap in 4-dimensional Yang–Mills theory is one of the millennium problems for which the Clay Institute for Mathematics awards one million US dollars.

It should be pointed out that the spectrum changes qualitatively once dynamical quarks are added (which will be done only in the next chapter). Then, glueballs mix with quark–anti-quark states and become unstable against the decay into mesons. In the pure Yang–Mills theory (without quarks), on the other hand, the lightest glueballs are absolutely stable particles (there is nothing that they could decay to).

14.13 Polyakov Loop and Center Symmetry

Let us now consider Yang–Mills theory at finite temperature $1/\beta$, which manifests itself as a periodic Euclidean-time dimension of finite extent β. An important physical quantity is the *Polyakov loop*,

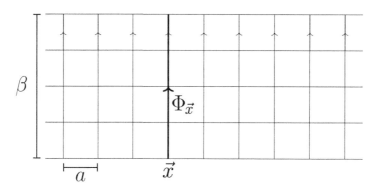

Fig. 14.9 A Polyakov loop $\Phi_{\vec{x}}$ (thick line) wraps around the periodic Euclidean-time dimension whose extent β determines the inverse temperature. In a center transformation, all time-like link variables in the last time-slice (marked by an arrow) are multiplied with a center element $z \in \mathbb{Z}(N_c)$. This leaves the plaquette action invariant but changes the Polyakov loop to $\Phi_{\vec{x}}' = z\Phi_{\vec{x}}$.

$$\Phi_{\vec{x}} = \mathrm{Tr}\left[\mathcal{P}\prod_{x_4} U_{4,\vec{x},x_4}\right], \tag{14.71}$$

a Wegner–Wilson loop closed over the periodic Euclidean-time dimension, cf. Figure 14.9 and Section 11.9.

The Polyakov loop represents a *single static quark,* which is used here to probe the dynamics of gluons in the absence of dynamical quarks. The thermal expectation value of the Polyakov loop,

$$\langle\Phi_{\vec{x}}\rangle = \exp(-\beta F), \tag{14.72}$$

determines the *free energy F of the static quark.* When the free energy is infinite, *i.e.* when $\langle\Phi_{\vec{x}}\rangle = 0$, the quark is confined. When the free energy is finite, *i.e.* when $\langle\Phi_{\vec{x}}\rangle \neq 0$, on the other hand, the quark is deconfined.

The *center* of a group consists of those group elements that commute with all other elements. Typically, these are multiples of the unit-matrix. The center $\mathbb{Z}(N_c)$ of $SU(N_c)$ contains the elements $z\mathbb{1}$, where

$$z = \exp\left(2\pi\,in/N_c\right) \in \mathbb{Z}(N_c), \quad n \in \{1, 2, \ldots, N_c\}. \tag{14.73}$$

Indeed, the matrices $z\mathbb{1}$ belong to $SU(N_c)$, because they all have the determinant $z^{N_c} = 1$.

The Euclidean functional integral of an $SU(N_c)$ Yang–Mills theory at finite temperature has a *global $\mathbb{Z}(N_c)$ center symmetry.* This symmetry is actually not a symmetry of the Hamilton operator or of the Euclidean-time transfer matrix of the lattice theory, but of the transfer matrix that transports the system in a spatial direction. As such, it is a symmetry that the theory enjoys only in thermal equilibrium at finite temperature (Holland and Wiese, 2001). As illustrated in Figure 14.9, a center transformation multiplies all time-like link variables in the last time-slice with a center element $z \in \mathbb{Z}(N_c)$. Since each plaquette contains either no or two oppositely oriented links of this type, the plaquette action is invariant against center transformations.

Each Polyakov loop, on the other hand, contains exactly one such link, and hence, it changes to $\Phi'_{\vec{x}} = z\,\Phi_{\vec{x}}$. As a consequence, the thermal expectation value of the Polyakov loop acts as an order parameter for the spontaneous breakdown of the center symmetry.

14.14 Deconfinement at High Temperatures

Despite extended searches, at particle accelerators and in all kinds of other places, isolated quarks or gluons have never been observed. This is in agreement with the evidence for confinement from lattice gauge theory.

Moreover, there is numerical evidence for *deconfinement* at high temperatures. The Polyakov loop picks up a non-zero expectation value at sufficiently high temperatures, indicating that the free energy of a static quark is then finite. The *spontaneous breakdown of the center symmetry* at high temperature hence signals deconfinement.

At high temperatures, the extent of periodic Euclidean-time is short and the Polyakov loop thus extends over a small interval. The parallel transport that it describes hence does not deviate much from the unit-matrix or from one of its center symmetry copies. In this setting, the Polyakov loop has a non-zero thermal expectation value, indicating the spontaneous breakdown of the center symmetry that is characteristic for a deconfined phase. $SU(N_c)$ gluons at high temperature exist in N_c distinct deconfined phases with $\langle\Phi\rangle = \Phi_0 z$.

When the temperature is lowered, the Euclidean-time extent increases and the parallel transport corresponding to the Polyakov loop extends over a larger periodic interval. Large fluctuations then wash out its expectation value, and one enters the confined phase that is characterized by $\langle\Phi\rangle = 0$. At some critical temperature T_c, there is a phase transition that separates the low-temperature confined glueball phase from N_c high-temperature deconfined *gluon plasma phases*.

With an increasing number of colors, there is a large difference between the N_c-independent number of glueball states and the number $N_c^2 - 1$ of deconfined gluons. The large mismatch between the number of active degrees of freedom in the low- and high-temperature phases then leads to a *first-order* deconfinement phase transition.

Monte Carlo simulations of $SU(N_c)$ Yang–Mills theory confirm that the phase transition is of first order for $N_c \geq 3$; see for instance Kogut *et al.* (1983); Brown *et al.* (1988), and Fukugita *et al.* (1989). This implies that the expectation value of the Polyakov loop remains non-zero as one approaches the phase transition from the deconfined phase. Consequently, just as gas and liquid in boiling water (see Figure I.1), the confined and deconfined phases remain distinguishable and coexist at the transition temperature.

The completion of a first-order phase transition from the deconfined gluon plasma at high temperature to the confined glueball phase at low temperature is delayed by *supercooling*. This is because the phase transition proceeds by the nucleation of bubbles of the low-temperature phase within the high-temperature bulk. The bubble walls, which separate the confined and deconfined phases, have some *interface tension* α_{cd}. The required interface free energy for bubble nucleation becomes available only at temperatures below T_c, when the bulk confined phase is already sufficiently favored over the deconfined phase. Otherwise small bubbles of the confined phase shrink again and disappear. Only after sufficient supercooling, small bubbles can grow and transform the gluon plasma into the bulk confined phase. Bubble nucleation is a rather violent process that is characteristic for any strong first-order phase transition.

Interestingly, in $SU(3)_c$ Yang–Mills theory the transition from the deconfined high-temperature to the confined low-temperature phase may also proceed in a different manner. Thanks to the $\mathbb{Z}(3)$ center symmetry, there are *three* distinct deconfined phases. If different deconfined phases are realized in different regions of space, they are separated by deconfined–deconfined interfaces. The free energy of such an interface is given by its area times the interface tension α_{dd}. Benjamin Widom's inequality, $\alpha_{dd} \leq 2\alpha_{cd}$ (Widom, 1975), expresses the fact that a deconfined–deconfined interface cannot possibly cost more free energy than a pair of confined–deconfined interfaces. Otherwise a deconfined–deconfined interface would simply split into a pair of confined–deconfined interfaces. This is exactly what happens when Widom's inequality is saturated, $\alpha_{dd} = 2\alpha_{cd}$. This is indeed the case in $SU(3)_c$ Yang–Mills theory; see Holland and Wiese (2001) and references therein. When a deconfined–deconfined interface is cooled down to the phase transition temperature, without any supercooling the confined phase already appears above T_c, because the deconfined–deconfined interface splits into a pair of confined–deconfined interfaces. The confined phase then forms a *complete wetting* layer that separates the two deconfined phases. Complete wetting is a universal behavior of interfaces that is well-known in condensed matter physics. It even happens in the *human eye:* The body of the eye is completely covered by a wetting layer of liquid of tears that separates it from the surrounding air.

Benjamin Svetitsky and Laurence Yaffe have conjectured that a second-order deconfinement phase transition of a $(d_s + 1)$-dimensional $SU(N_c)$ Yang–Mills theory should fall in the universality class of d_s-dimensional $\mathbb{Z}(N_c)$-symmetric classical spin models (Svetitsky and Yaffe, 1982).

For 4-d $SU(2)$ Yang–Mills theory, the mismatch between the number of active degrees of freedom in the two phases is small, and one finds a second-order deconfinement phase transition, which indeed falls in the universality class of the 3-d Ising model (Kuti *et al.*, 1981; Engels *et al.*, 1982; Kogut *et al.*, 1983). In that case, the thermal expectation value of the Polyakov loop vanishes as one approaches the phase transition from the deconfined side. As a result, at the transition temperature the two deconfined phases become indistinguishable from the confined phase. A second-order phase transition proceeds without bubble nucleation and is thus much smoother than a first-order transition.

Immediately after the Big Bang, during the first 10^{-6} sec in the history of the Universe, quarks and gluons were not yet confined. In full QCD including light quarks, there is just a *crossover* (smoother than an actual phase transition) between the confined and the deconfined regimes, which takes place at temperatures around 156 MeV (Borsanyi *et al.*, 2010; Bhattacharya *et al.*, 2014; Bazavov *et al.*, 2019).[4] A smooth crossover does not proceed via bubble nucleation either and thus does not have a large impact on the cosmic evolution. In particular, it does not lead to spatial inhomogeneities, which could otherwise have an impact on the primordial synthesis of light nuclei that happened during the first few minutes after the Big Bang. Convincing experimental evidence for the existence of a deconfined quark–gluon plasma at high temperatures has been obtained in detailed heavy-ion collision experiments (for a review, see Cunqueiro and Sickles (2022)).

---

[4] In contrast to a phase transition, the crossover temperature depends somewhat on the criterion that one considers. This explains why some collaborations report a higher crossover temperature.

14.15 Exceptional Confinement and Deconfinement in G(2) Yang–Mills Theory

This section addresses physics with the exceptional gauge group G(2), which is not realized in Nature. It can be skipped at a first reading.

The group G(2) is one of five exceptional Lie groups (also including F(4), E(6), E(7), and E(8)),[5] which do not fall into the main sequences SU(n), SO(n) (or more precisely Spin(n)), and Sp(n). The groups G(2), F(4), and E(8) are remarkable because they only have a *trivial center* that consists of the unit-element. Therefore, confinement and deconfinement in the corresponding Yang–Mills theories are exceptional. G(2) is the smallest exceptional group. It has rank 2, 14 generators, and contains SU(3) as a subgroup. G(2) has a 7-dimensional fundamental representation and is a subgroup of SO(7). The 7×7 real orthogonal matrices Ω of SO(7) have determinant 1 and obey the condition

$$\Omega_{ab}\Omega_{bc}^{\mathsf{T}} = \Omega_{ab}\Omega_{cb} = \delta_{ac}. \tag{14.74}$$

The elements of the G(2) subgroup, in addition, satisfy the cubic constraint

$$T_{abc} = T_{def}\Omega_{da}\Omega_{eb}\Omega_{fc}, \tag{14.75}$$

where T is a totally anti-symmetric tensor with the following non-vanishing elements

$$T_{127} = T_{154} = T_{163} = T_{235} = T_{264} = T_{374} = T_{576} = 1. \tag{14.76}$$

The group theory of G(2) is addressed in more detail in Appendix F.14.

Under its SU(3) subgroup, the fundamental representation {7} and the adjoint representation {14} of G(2), whose weight diagrams are illustrated in Figure 14.10, decompose as

$$\{7\} = \{1\} + \{3\} + \{\bar{3}\}, \quad \{14\} = \{3\} + \{\bar{3\}} + \{8\}. \tag{14.77}$$

Consequently, the 14 gauge bosons of G(2) Yang–Mills theory can be interpreted as 8 gluons plus 6 additional gauge bosons with the color quantum numbers of quarks and anti-quarks. Similarly, an external static charge in the {7} representation of G(2) incorporates a quark, an anti-quark, as well as a color singlet.

Since G(2) contains SU(3) as a subgroup, its center cannot be larger than $\mathbb{Z}(3)$. In fact, it turns out that G(2) has only a trivial center that consists of the 7×7 unit-matrix. This has drastic consequences for confinement and deconfinement (Holland *et al.*, 2003). In particular, in G(2) Yang–Mills theory, the string connecting two external static charges in the fundamental {7} representation can break by the pair creation of dynamical gauge bosons. Due to the absence of a non-trivial center, for G(2) there is no concept like triality

[5] The indices 2, 4, 6, 7, 8 correspond to the rank.

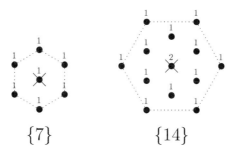

$$\{7\} \qquad\qquad \{14\}$$

Fig. 14.10 The weight diagrams of the irreducible G(2) representations {7} and {14}. The multiplicities of the states are indicated by small labels.

in SU(3). This implies that an external charge in the representation {7} can be screened by a sufficiently large number of G(2) gauge bosons (Holland *et al.*, 2003). From the tensor product reduction

$$\{14\} \times \{14\} \times \{14\} = \{1\} + \{7\} + 5\{14\} + 3\{27\} + 2\{64\} + 4\{77\}$$
$$+ 3\{77'\} + \{182\} + 3\{189\} + \{273\} + 2\{448\}, \quad (14.78)$$

we conclude that three G(2) gauge bosons in the adjoint representation {14} can indeed screen an external {7} charge. As a result, the confining string connecting two static fundamental {7} charges can break by the creation of three dynamical gauge boson pairs. Hence, confinement in G(2) Yang–Mills theory differs qualitatively from Yang–Mills theories with a non-trivial center. In fact, since six of the G(2) gauge bosons have the same color charges as quarks and anti-quarks, confinement in G(2) Yang–Mills theory is similar to quark confinement in QCD and is hence not that exceptional after all.

In order to verify that G(2) gauge bosons are *still confined* – albeit without any string tension – we again employ the Fredenhagen–Marcu operator. The operator of eq. (14.66) in the adjoint representation describes the creation of a pair of dynamical gauge bosons and can be considered for any gauge group. However, for G(2) there is an additional Fredenhagen–Marcu operator which has no analogue for gauge groups with a non-trivial center

$$\boxed{}\ \ \boxed{} = (U^{ab}_{x,ij} T_{abc}) U^{cd}_{\mathcal{C}_{xy}} (T_{def} U^{ef}_{y,ij}). \quad (14.79)$$

Much like the original operator which Fredenhagen and Marcu constructed for QCD, this operator describes the creation of a pair of fundamental charges. Due to its trivial center, the group G(2) has the special feature that a fundamental {7} charge can be composed of adjoint {14} gauge bosons. In this case, the Wegner–Wilson loop in the denominator of eq. (14.65) is taken in the fundamental representation. At strong coupling, the corresponding order parameter is again non-zero, which confirms that G(2) Yang–Mills theory indeed confines, at least in the strong coupling limit.

Finally, let us consider G(2) Yang–Mills theory at finite temperature. Since G(2) only has a trivial center, there is no symmetry reason that would imply the existence of a deconfinement phase transition. In particular, there is no center symmetry that could break spontaneously at high temperatures. Hence, one might have expected just a crossover between the low- and the high-temperature regimes. However, Monte Carlo simulations of G(2) Yang–Mills theory indeed show a rather strong first-order phase transition (Pepe and

Wiese, 2007), stronger than the one in $SU(3)_c$ Yang–Mills theory. Due to the 14 G(2) gauge bosons compared to the 8 $SU(3)_c$ gluons, this can again be attributed to the large mismatch between the number of active confined versus deconfined degrees of freedom.

Exercises

14.1 String tension of 2-d U(1) lattice gauge theory at strong coupling
Derive the basic rules for performing integrations over the link variables of a U(1) lattice gauge theory, and use them to derive the value of the string tension in the strong coupling limit.

14.2 Fredenhagen–Marcu operator at strong coupling
Evaluate the Fredenhagen–Marcu operator in the strong coupling limit of 4-d SU(3) Yang–Mills theory in order to show that gluons (which transform in the adjoint representation) are confined.

14.3 String breaking in Yang–Mills theory at strong coupling
Use the strong coupling expansion of 4-d SU(3) Yang–Mills theory in order to determine the potential between two external, static charges in the adjoint representation. Short distances are characterized by an area law, while large distances are associated with a perimeter law. Determine the string-breaking distance that separates the two regimes.

14.4 Monte Carlo simulation of 2-d U(1) lattice gauge theory
Implement a Metropolis algorithm for the numerical simulation of 2-d U(1) lattice gauge theory on a square lattice with periodic boundary conditions. Consult Appendix H for the basics of the Monte Carlo method. Propose a change of an individual link variable in a symmetric interval around its current value. Accept or reject the proposal, using a Metropolis step, by comparing the corresponding action values before and after the proposed change. Determine the expectation value of a 1×1 Wegner–Wilson loop (an elementary plaquette) at various values of the coupling e, and compare it with an analytic strong coupling result.

14.5 Tensor product reduction in G(2)
Use the graphical method described in Appendix F.15 to verify the G(2) tensor product reduction $\{14\} \times \{14\} \times \{14\}$ of eq. (14.78).

15

One Generation of Leptons and Quarks

In this chapter, we add the *fermions* to the Lagrangian of the Standard Model. The fermions are *leptons* and *quarks*. The leptons participate in the electroweak gauge interactions only, whereas the quarks are engaged in both electroweak and strong interactions.

It is interesting that we need to add leptons *and* quarks at the same time; a simplification of the Standard Model without quarks would be mathematically inconsistent. This is because the quarks cancel anomalies, which would explicitly break the gauge symmetry in a purely leptonic model at the quantum level. *Cancellation of anomalies in gauge symmetries is mandatory,* both perturbatively and beyond perturbation theory.

Anomalies in global symmetries, on the other hand, are a perfectly acceptable form of explicit symmetry breaking. In fact, they are necessary to correctly describe some aspects of the particle phenomenology. A prominent example is the decay of the neutral pion into two photons, $\pi^0 \to \gamma\gamma$, which is due to the anomalous breakdown of the global G-parity transformation (cf. Section 24.4). As we will see in this chapter, the global symmetries associated with the separate conservation of the *lepton number L* and the *baryon number B* are both *anomalous,* whereas the conservation of the *difference $B - L$* corresponds to an *exact* global symmetry of the Standard Model.

This chapter will be limited to *a single generation* of fermions. In this minimal version, the Standard Model includes only two quark flavors with both chiralities, a left-handed electron and neutrino, as well as a right-handed electron, but no right-handed neutrino. This does not allow for the construction of a renormalizable neutrino mass term. However, by now we know that neutrinos do have small but non-zero masses. This motivates an extension by a non-renormalizable term or by a right-handed neutrino field. Still, we will follow our strategy of adding fields step by step, and so we will first work with a left-handed neutrino only.

15.1 Electron and Left-Handed Neutrino

The leptons of the *first generation* are the electron and its neutrino. We start with a left-handed neutrino only and denote the spinor fields of these leptons as $\nu_L(x)$, $\bar{\nu}_L(x)$, $e_L(x)$, $\bar{e}_L(x)$, $e_R(x)$, and $\bar{e}_R(x)$. Right-handed neutrino fields are not part of the minimal version of the Standard Model. Before we introduce a right-handed neutrino field as an extension of the Standard Model, the neutrinos are massless, while the electrons will pick up a mass through the Higgs mechanism. However, before we introduce the so-called Yukawa couplings between the lepton fields and the Higgs field (in Chapter 16), even the electron is massless.

At this point – without mass terms – the free lepton Lagrangian

$$\mathcal{L}_0(\bar{v}, v, \bar{e}, e) = \bar{v}_L \bar{\sigma}_\mu \partial_\mu v_L + \bar{e}_L \bar{\sigma}_\mu \partial_\mu e_L + \bar{e}_R \sigma_\mu \partial_\mu e_R \qquad (15.1)$$

has several global symmetries (σ_μ and $\bar{\sigma}_\mu$ are defined in eq. (9.78)). First, all lepton fields can be multiplied by the same (space–time independent) phase factor $\exp(i\chi)$, $\chi \in \mathbb{R}$, while the anti-leptons pick up the complex conjugated phase factor $\exp(-i\chi)$,

$$\begin{aligned}
v_L'(x) &= \exp(i\chi)v_L(x), & \bar{v}_L'(x) &= \bar{v}_L(x)\exp(-i\chi), \\
e_L'(x) &= \exp(i\chi)e_L(x), & \bar{e}_L'(x) &= \bar{e}_L(x)\exp(-i\chi), \\
e_R'(x) &= \exp(i\chi)e_R(x), & \bar{e}_R'(x) &= \bar{e}_R(x)\exp(-i\chi).
\end{aligned} \qquad (15.2)$$

The corresponding global symmetry $U(1)_L$ is associated with *lepton number conservation*.[1] This symmetry is *vector-like*, which means that it affects left- and right-handed lepton fields in the same way (except that we have not yet introduced a right-handed neutrino field).

The left-handed neutrino and electron fields form an $SU(2)_L$ *doublet*

$$l_L(x) = \begin{pmatrix} v_L(x) \\ e_L(x) \end{pmatrix}, \qquad \bar{l}_L(x) = (\bar{v}_L(x), \bar{e}_L(x)), \qquad (15.3)$$

which enters the free lepton Lagrangian as

$$\mathcal{L}_0(\bar{v}, v, \bar{e}, e) = \bar{l}_L \bar{\sigma}_\mu \partial_\mu l_L + \bar{e}_R \sigma_\mu \partial_\mu e_R. \qquad (15.4)$$

Another global symmetry of the free Lagrangian rotates the left-handed neutrino and electron fields into each other. In the Standard Model, this symmetry is promoted to a local one,

$$\begin{aligned}
l_L'(x) &= \begin{pmatrix} v_L'(x) \\ e_L'(x) \end{pmatrix} = L(x) \begin{pmatrix} v_L(x) \\ e_L(x) \end{pmatrix} = L(x)l_L(x), \\
\bar{l}_L'(x) &= (\bar{v}_L'(x), \bar{e}_L'(x)) = (\bar{v}_L(x), \bar{e}_L(x))L(x)^\dagger = \bar{l}_L(x)L(x)^\dagger, \quad L(x) \in SU(2)_L. \quad (15.5)
\end{aligned}$$

The right-handed component of the electron field, $e_R(x)$, on the other hand, is an $SU(2)_L$ *singlet*; *i.e.* it remains invariant under $SU(2)_L$ transformations,

$$e_R'(x) = e_R(x). \qquad (15.6)$$

Since left- and right-handed fields transform differently under $SU(2)_L$, we are confronted with a *chiral gauge theory*.

There is yet another global Abelian symmetry of the free Lagrangian (15.4) that acts differently in the left-handed and in the right-handed sector. In the Standard Model, it is promoted to the local $U(1)_Y$ symmetry

$$\begin{aligned}
v_L'(x) &= \exp\left(iY_{l_L}g'\varphi(x)\right)v_L(x), & \bar{v}_L'(x) &= \bar{v}_L(x)\exp\left(-iY_{l_L}g'\varphi(x)\right), \\
e_L'(x) &= \exp\left(iY_{l_L}g'\varphi(x)\right)e_L(x), & \bar{e}_L'(x) &= \bar{e}_L(x)\exp\left(-iY_{l_L}g'\varphi(x)\right), \\
e_R'(x) &= \exp\left(iY_{e_R}g'\varphi(x)\right)e_R(x), & \bar{e}_R'(x) &= \bar{e}_R(x)\exp\left(-iY_{e_R}g'\varphi(x)\right).
\end{aligned} \qquad (15.7)$$

---

[1] It should be noted that the subscript L in v_L and e_L means "left-handed", whereas in $U(1)_L$ it refers to the lepton number L. We add that the indices of the groups $SU(2)_L$ and $SU(2)_R$ refer to matrices $L, R \in SU(2)$, which multiply the matrix form Φ of the Higgs field from the left and from the right, respectively.

Here, we assign *different weak hypercharges* Y_{l_L} and Y_{e_R} to the left-handed leptons and to the right-handed electron, respectively. Hence, the $U(1)_Y$ gauge symmetry is *chiral* as well. It is necessary, however, to assign the same hypercharge Y_{l_L} to both members of the $SU(2)_L$ doublet (ν_L and e_L), because otherwise the $SU(2)_L$ gauge symmetry would be broken explicitly.

In analogy to spin, one introduces a *"weak isospin"* which acts on the left-handed doublet as $T_L^3 = \tau^3/2$. The leptons have the following 3-components of the weak isospin

$$T_{L\nu_L}^3 = \frac{1}{2}, \quad T_{Le_L}^3 = -\frac{1}{2}, \quad T_{Le_R}^3 = 0, \tag{15.8}$$

as displayed in Table 15.1. This corresponds to the weak isodoublet l_L and the singlet e_R.

Analogously, we introduce a generator T_R^3 which takes the values

$$T_{R\nu_L}^3 = 0, \quad T_{Re_L}^3 = 0, \quad T_{Re_R}^3 = -\frac{1}{2}. \tag{15.9}$$

This operator generates an Abelian subgroup of $SU(2)_R$. In Section 15.10, we will also introduce a right-handed neutrino field $\nu_R(x)$ for which

$$T_{L\nu_R}^3 = 0, \quad T_{R\nu_R}^3 = \frac{1}{2}. \tag{15.10}$$

In the Standard Model, the $SU(2)_L$ and $U(1)_Y$ (but not the full $SU(2)_R$) symmetries are promoted to gauge symmetries. Just as in the gauge–Higgs Lagrangian of the Standard Model (discussed in Chapter 13), this is achieved by substituting ordinary derivatives ∂_μ by covariant derivatives D_μ. For the left-handed lepton doublet, the covariant derivative takes the form

$$D_\mu \begin{pmatrix} \nu_L(x) \\ e_L(x) \end{pmatrix} = \left[\partial_\mu + iY_{l_L} g' B_\mu(x) + ig W_\mu^a(x) \frac{\tau^a}{2} \right] \begin{pmatrix} \nu_L(x) \\ e_L(x) \end{pmatrix}. \tag{15.11}$$

As in Chapter 13, both the derivative ∂_μ and the gauge field B_μ act as unit 2×2 matrices in the flavor space.[2] Using

$$W_\mu(x) = ig W_\mu^a(x) \frac{\tau^a}{2}, \tag{15.12}$$

(with an implicit sum over $a = 1, 2, 3$), eq. (15.11) can be rewritten in a compact form,

$$D_\mu l_L(x) = \left[\partial_\mu + iY_{l_L} g' B_\mu(x) + W_\mu(x) \right] l_L(x). \tag{15.13}$$

For the right-handed electron singlet, the covariant derivative reads

$$D_\mu e_R(x) = \left[\partial_\mu + iY_{e_R} g' B_\mu(x) \right] e_R(x). \tag{15.14}$$

Hence, the covariant derivative D_μ corresponds to eq. (15.13) or to eq. (15.14), depending on whether it acts on l_L or on e_R.

The Lagrangian describing the propagation of the leptons, as well as their interactions with the $U(1)_Y$ and $SU(2)_L$ gauge fields, takes the form

---

[2] To be pedantic, we could write $\partial_\mu \mathbb{1}$ and $B_\mu \mathbb{1}$ each time, but for the sake of a better legibility of the formulae, we will suppress the factor $\mathbb{1}$.

$$\mathcal{L}(\bar{\nu}, \nu, \bar{e}, e, W, B) = \bar{l}_L \bar{\sigma}_\mu D_\mu l_L + \bar{e}_R \sigma_\mu D_\mu e_R$$

$$= (\bar{\nu}_L, \bar{e}_L) \bar{\sigma}_\mu D_\mu \begin{pmatrix} \nu_L \\ e_L \end{pmatrix} + \bar{e}_R \sigma_\mu D_\mu e_R. \tag{15.15}$$

In order to ensure $SU(2)_L$ gauge invariance, the gauge coupling g has to take the same universal value as in the gauge–Higgs sector, *i.e.* as in Chapter 13.

It is important that a direct mass term $m_e(\bar{e}_L e_R + \bar{e}_R e_L)$ is *not* gauge-invariant, because the left- and right-handed electron fields transform differently under both $SU(2)_L$ and $U(1)_Y$ gauge transformations. Therefore, direct mass terms are forbidden in the Standard Model Lagrangian. This is a nice feature of chiral gauge theories, because it protects the fermions from additive mass renormalization. Thus, in contrast to the scalar Higgs field, there is no hierarchy problem for chiral fermions, at least at the level of perturbation theory.

In Chapter 16, we will construct Yukawa interaction terms between the fermions and the Higgs field. After spontaneous symmetry breaking, *i.e.* when the Higgs field picks up a non-zero vacuum expectation value v, such terms give rise to dynamically generated fermion masses. In this way, in the minimal Standard Model (with massless neutrinos) all fermion masses are tied to the electroweak symmetry breaking scale v.

15.2 CP and T Invariance of Gauge Interactions

As we have seen, the left-handed electron and neutrino form an $SU(2)_L$ doublet, while the right-handed electron is a singlet. Consequently, left- and right-handed particles have different physical properties, which makes the Standard Model a *chiral gauge theory*.

As a result of this asymmetric treatment of left- and right-handed degrees of freedom, parity P and charge conjugation C are explicitly broken in the Standard Model. Parity violation was predicted by Tsung-Dao Lee and Chen-Ning Yang in 1956 (Lee and Yang, 1956b) and indeed observed in the β-decay by Madame Chien-Shiung Wu (Wu *et al.*, 1957) (cf. Appendix A). As we will see next, the gauge interactions still respect the combined discrete symmetry CP as well as the time reversal T.[3]

Let us introduce the transformation behavior of the gauge fields under the discrete symmetries C, P, and T. The Abelian gauge field B_μ transforms as

$$
\begin{aligned}
{}^C B_\mu(x) &= -B_\mu(x), \\
{}^P B_i(\vec{x}, x_4) &= -B_i(-\vec{x}, x_4), \quad {}^P B_4(\vec{x}, x_4) = B_4(-\vec{x}, x_4), \\
{}^T B_i(\vec{x}, x_4) &= -B_i(\vec{x}, -x_4), \quad {}^T B_4(\vec{x}, x_4) = B_4(\vec{x}, -x_4).
\end{aligned}
\tag{15.16}
$$

---

[3] As we will discuss in Chapter 17, with three or more generations of fermions, it is natural that the fermion–Higgs couplings explicitly violate CP and T invariance. Furthermore, Chapter 19 deals with the QCD vacuum-angle θ as another possible source of explicit CP and T breaking. However, in Nature this parameter is consistent with zero, $|\theta| \leq \mathcal{O}(10^{-10})$.

Consequently, the combined transformations CP and CPT take the form

$$^{\mathrm{CP}}B_i(\vec{x}, x_4) = B_i(-\vec{x}, x_4), \quad ^{\mathrm{CP}}B_4(\vec{x}, x_4) = -B_4(-\vec{x}, x_4),$$
$$^{\mathrm{CPT}}B_\mu(x) = -B_\mu(-x). \tag{15.17}$$

Similarly, the non-Abelian gauge field W_μ transforms as

$$^{\mathrm{C}}W_\mu(x) = W_\mu(x)^*,$$
$$^{\mathrm{P}}W_i(\vec{x}, x_4) = -W_i(-\vec{x}, x_4), \quad ^{\mathrm{P}}W_4(\vec{x}, x_4) = W_4(-\vec{x}, x_4),$$
$$^{\mathrm{T}}W_i(\vec{x}, x_4) = W_i(\vec{x}, -x_4)^*, \quad ^{\mathrm{T}}W_4(\vec{x}, x_4) = -W_4(\vec{x}, -x_4)^*, \tag{15.18}$$

which implies

$$^{\mathrm{CP}}W_i(\vec{x}, x_4) = -W_i(-\vec{x}, x_4)^*, \quad ^{\mathrm{CP}}W_4(\vec{x}, x_4) = W_4(-\vec{x}, x_4)^*,$$
$$^{\mathrm{CPT}}W_\mu(x) = -W_\mu(-x). \tag{15.19}$$

The difference between eqs. (15.17) and (15.19) is due to the factor i in the definition of W_μ; see eq. (15.12).

Let us now investigate the CP transformation properties of the interaction terms that couple the right-handed electron to the $U(1)_Y$ gauge field. Using the CP transformation rules for the fermions of eq. (9.106), we obtain

$$\begin{aligned}
S[^{\mathrm{CP}}\bar{e}_{\mathrm{R}}, {}^{\mathrm{CP}}e_{\mathrm{R}}, {}^{\mathrm{CP}}B] &= \int d^4x \, ^{\mathrm{CP}}\bar{e}_{\mathrm{R}}(\vec{x}, x_4)\sigma_\mu i Y_{l_{\mathrm{L}}} g' \, ^{\mathrm{CP}}B_\mu(\vec{x}, x_4)^{\mathrm{CP}}e_{\mathrm{R}}(\vec{x}, x_4) \\
&= -\int d^4x \, e_{\mathrm{R}}(-\vec{x}, x_4)^{\mathsf{T}} P^{\mathsf{T}} C^{-1} \left[-\sigma_i i Y_{l_{\mathrm{L}}} g' B_i(-\vec{x}, x_4) \right. \\
&\quad \left. + \sigma_4 i Y_{l_{\mathrm{L}}} g' B_4(-\vec{x}, x_4) \right] CP\bar{e}_{\mathrm{R}}(-\vec{x}, x_4)^{\mathsf{T}} \\
&= \int d^4x \, e_{\mathrm{R}}(-\vec{x}, x_4)^{\mathsf{T}} \left[\sigma_i^{\mathsf{T}} i Y_{l_{\mathrm{L}}} g' B_i(-\vec{x}, x_4) \right. \\
&\quad \left. + \sigma_4^{\mathsf{T}} i Y_{l_{\mathrm{L}}} g' B_4(-\vec{x}, x_4) \right] \bar{e}_{\mathrm{R}}(-\vec{x}, x_4)^{\mathsf{T}} \\
&= \int d^4x \, \bar{e}_{\mathrm{R}}(-\vec{x}, x_4)\sigma_\mu i Y_{l_{\mathrm{L}}} g' B_\mu(-\vec{x}, x_4)e_{\mathrm{R}}(-\vec{x}, x_4) \\
&= S[\bar{e}_{\mathrm{R}}, e_{\mathrm{R}}, B], \tag{15.20}
\end{aligned}$$

where we have used the Grassmann rules. In the same manner, one can show that the couplings of the left-handed leptons are CP-invariant as well. The *CP theorem* (Lüders, 1954; Bell, 1955; Pauli, 1957; Lüders, 1957; Jost, 1957) assures that the observables in any local, Lorentz-invariant quantum field theory are CPT-invariant. Therefore, the interaction terms under consideration are automatically CPT- and thus (due to CP invariance) also T-invariant.

Finally, we list the C, P, and T transformation properties of the Higgs field

$$^{\mathrm{C}}\Phi(x) = \Phi(x)^*,$$
$$^{\mathrm{P}}\Phi(\vec{x}, x_4) = \Phi(-\vec{x}, x_4),$$
$$^{\mathrm{T}}\Phi(\vec{x}, x_4) = \Phi(\vec{x}, -x_4)^*. \tag{15.21}$$

which then implies

$$^{\mathrm{CP}}\Phi(\vec{x}, x_4) = \Phi(-\vec{x}, x_4)^*,$$
$$^{\mathrm{CPT}}\Phi(x) = \Phi(-x). \tag{15.22}$$

Using these transformation rules, it is straightforward to return to the gauge–Higgs sector and address the

---

Quick Question 15.2.1   C, P, and T invariance of the gauge–Higgs action
Show that the action (without fermions)

$$S[\Phi, W, B] = \int d^4x \left[\frac{1}{2} D_\mu \Phi^\dagger D_\mu \Phi + V(\Phi) - \frac{1}{2g^2} \mathrm{Tr}(W_{\mu\nu} W_{\mu\nu}) + \frac{1}{4} B_{\mu\nu} B_{\mu\nu} \right] \tag{15.23}$$

is *invariant separately under C, P, and T.*

15.3 Fixing the Lepton Weak Hypercharges

We know that the electron carries electric charge $-e$, while the neutrino is electrically neutral. In the Lagrangian (15.15), we recognize off-diagonal terms that couple leptons of different electric charge, associated with W^1 and W^2. In order to preserve the electric charge under interactions, these gauge bosons must be charged themselves. In particular, in Chapter 13 we already identified a positive and a negative W-boson given by

$$W_\mu^\pm(x) = \frac{1}{\sqrt{2}} \left(W_\mu^1(x) \mp i W_\mu^2(x) \right). \tag{15.24}$$

This implies

$$W_\mu^1(x) \frac{\tau^1}{2} + W_\mu^2(x) \frac{\tau^2}{2} = \frac{1}{\sqrt{2}} \begin{pmatrix} 0 & W_\mu^+(x) \\ W_\mu^-(x) & 0 \end{pmatrix} = \frac{1}{\sqrt{2}} \left(W_\mu^+(x)\tau^+ + W_\mu^-(x)\tau^- \right), \tag{15.25}$$

where $\tau^\pm = (\tau^1 \pm i\tau^2)/2$. Unlike τ^a, these matrices are not Hermitian, but they obey $(\tau^+)^\dagger = \tau^-$. The electrically charged gauge bosons $W^\pm$ can decay into leptons, for instance, following the pattern of Figure 15.1.

We observed in Section 13.2 that the electrically neutral gauge fields, *i.e.* the flavor diagonal fields, split physically into a massless photon and a massive Z-boson

$$A_\mu(x) = \frac{g' W_\mu^3(x) + g B_\mu(x)}{\sqrt{g^2 + g'^2}}, \quad Z_\mu(x) = \frac{g W_\mu^3(x) - g' B_\mu(x)}{\sqrt{g^2 + g'^2}}. \tag{15.26}$$

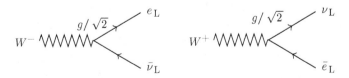

Fig. 15.1 Illustration of a decay of the gauge boson W^- into an electron and an anti-neutrino (on the left), and of a gauge boson W^+ into a positron and a neutrino (on the right). Time evolves from left to right, and the vertices correspond to the off-diagonal matrix elements in eq. (15.28), which describe the coupling of $W^\pm$ to lepton currents. The arrows indicate the flow of lepton number.

They are natural to consider after spontaneous symmetry breaking. Inserting the inverse relations

$$W_\mu^3(x) = \frac{g'A_\mu(x) + gZ_\mu(x)}{\sqrt{g^2 + g'^2}}, \quad B_\mu(x) = \frac{gA_\mu(x) - g'Z_\mu(x)}{\sqrt{g^2 + g'^2}}, \tag{15.27}$$

we write the lepton–gauge coupling terms in the Lagrangian (15.15) as

$$\mathcal{L}(\bar{\nu}, \nu, \bar{e}, e, A, Z) = (\bar{\nu}_L, \bar{e}_L)\bar{\sigma}_\mu \left[\partial_\mu + i \left(\begin{array}{cc} X_\mu^1 & gW_\mu^+/\sqrt{2} \\ gW_\mu^-/\sqrt{2} & X_\mu^2 \end{array} \right) \right] \left(\begin{array}{c} \nu_L \\ e_L \end{array} \right)$$

$$+ \bar{e}_R\sigma_\mu \left[\partial_\mu + i \frac{Y_{e_R}g'}{\sqrt{g^2 + g'^2}} \left(gA_\mu - g'Z_\mu \right) \right] e_R, \tag{15.28}$$

where

$$X_\mu^1(x) = \frac{1}{\sqrt{g^2 + g'^2}} \left[gg' \left(1/2 + Y_{l_L} \right) A_\mu(x) + \left(g^2/2 - Y_{l_L}g'^2 \right) Z_\mu(x) \right],$$

$$X_\mu^2(x) = \frac{1}{\sqrt{g^2 + g'^2}} \left[gg' \left(-1/2 + Y_{l_L} \right) A_\mu(x) + \left(-g^2/2 - Y_{l_L}g'^2 \right) Z_\mu(x) \right]. \tag{15.29}$$

Since the (electrically neutral) neutrino does not couple to the photon field A_μ, the term X_μ^1 must not contain a contribution from A_μ. This implies $Y_{l_L} = -1/2$, and therefore,

$$X_\mu^1(x) = \frac{\sqrt{g^2 + g'^2}}{2} Z_\mu(x),$$

$$X_\mu^2(x) = \frac{1}{\sqrt{g^2 + g'^2}} \left[\frac{g'^2 - g^2}{2} Z_\mu(x) - gg'A_\mu(x) \right]$$

$$= -\frac{\sqrt{g^2 + g'^2}}{2} \left[\cos(2\theta_W)Z_\mu(x) + \sin(2\theta_W)A_\mu(x) \right], \tag{15.30}$$

where θ_W is the Weinberg angle introduced in eq. (13.40). Again, we identify

$$e = \frac{gg'}{\sqrt{g^2 + g'^2}} \tag{15.31}$$

as the unit of electric charge, in exact agreement with eq. (13.45). Indeed, $-e$ is the correct electric charge of the left-handed electron. In order to obtain the same value $-e$ also for the right-handed electron, we adjust its weak hypercharge to $Y_{e_R} = -1$. We now see that Y_{l_L} and Y_{e_R} are different. This confirms that not only the SU(2)$_L$, but also the U(1)$_Y$ gauge couplings are *chiral*.

The corresponding vertex enables transitions like the decay of a Z-boson into an electron–positron or a neutrino–anti-neutrino pair, or electron–positron pair creation mediated by a photon, as illustrated in Figure 15.2.

At this point, we observe a simple relation between the weak hypercharge Y (*i.e.* the coupling to B_μ in units of g'), the generator T_R^3, which was introduced in eq. (15.9), and the lepton number L,

$$Y = T_R^3 - \frac{1}{2}L. \tag{15.32}$$

This can be readily verified in each case from Table 15.1.

Furthermore, the electric charge Q (in units of e) is related to Y and the third component of the weak isospin T_L^3 (which was introduced in eq. (15.8)) as

Table 15.1 The leptons of the first fermion generation, with their quantum numbers, which are relevant in this section (including the lepton number L). An extended version follows in Table 15.2.

	L	Y	T_L^3	T_R^3	Q
ν_L	1	$-1/2$	$1/2$	0	0
e_L	1	$-1/2$	$-1/2$	0	-1
e_R	1	-1	0	$-1/2$	-1

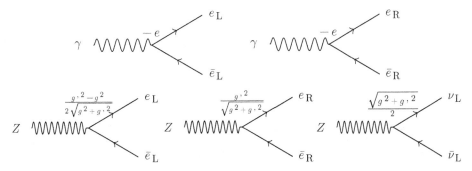

Fig. 15.2 Illustration of the vertices of the electrically neutral gauge bosons γ and Z coupled to left- or right-handed electrons or neutrinos, together with the corresponding coupling strength.

$$Q = T_L^3 + Y = T_L^3 + T_R^3 - \frac{1}{2}L. \qquad (15.33)$$

In the following, the relation (15.33) will be given a prominent status. It can be checked in each case from Table 15.1. We remark here that its validity is also a consequence of the parameter choice $Y_{l_L} = -1/2$ that we made in order to decouple the neutrino from A_μ.

We can also interpret these expressions in terms of gauge couplings to *fermionic currents*. Generally, currents are 4-vectors $j_\mu(x)$ obeying the *continuity equation* $\partial_\mu j_\mu(x) = 0$, at least at the classical level. From the Lagrangian of a free fermion, we obtain the Noether current $\bar{\psi}\gamma_\mu\psi$. Its conservation can also be derived from the free Dirac equation (9.40).

The electromagnetic current of the electron (with charge $-e$) takes the form

$$j_\mu^{em}(x) = j_{\mu,L}^{em}(x) + j_{\mu,R}^{em}(x) = -e\left(\bar{e}_L(x)\bar{\sigma}_\mu e_L(x) + \bar{e}_R(x)\sigma_\mu e_R(x)\right). \qquad (15.34)$$

In these terms, the photon field couples to the electromagnetic current via the term $iA_\mu j_\mu^{em}$ in the Lagrangian. It should be noted that this current is *neutral*. This means that there is no change in the charge between the initial and the final fermion states, *i.e.* the state before and after the scattering on the photon. Clearly, the charge cannot change because the photon is neutral.

Besides the couplings to the photon, we also have *weak current interactions*. The *weak neutral current* j_μ^0 couples to the Z-boson field,

$$j_\mu^0(x) = \frac{\sqrt{g^2 + g'^2}}{2} \bar{v}_L(x) \bar{\sigma}_\mu v_L(x) + \frac{g'^2 - g^2}{2\sqrt{g^2 + g'^2}} \bar{e}_L(x) \bar{\sigma}_\mu e_L(x) + \frac{g'^2}{\sqrt{g^2 + g'^2}} \bar{e}_R(x) \sigma_\mu e_R(x)$$

$$= \frac{\sqrt{g^2 + g'^2}}{2} \left(\bar{v}_L(x) \bar{\sigma}_\mu v_L(x) - \cos(2\theta_W) \bar{e}_L(x) \bar{\sigma}_\mu e_L(x) + 2 \sin^2 \theta_W \bar{e}_R(x) \sigma_\mu e_R(x) \right).$$

(15.35)

The *charged currents*, on the other hand, couple to the charged gauge bosons $W^\pm$ and take the form

$$j_\mu^+(x) = \frac{g}{\sqrt{2}} \bar{v}_L(x) \bar{\sigma}_\mu e_L(x), \quad j_\mu^-(x) = \frac{g}{\sqrt{2}} \bar{e}_L(x) \bar{\sigma}_\mu v_L(x).$$

(15.36)

For the current j_μ^+ (or j_μ^-), the charge increases (decreases) in the transition from the initial to the final state. Note that the neutral currents contain both left- and right-handed contributions, while the charged currents must be purely left-handed in order to couple to $W^\pm$. These gauge–current vertices are diagrammatically illustrated in Figure 15.1.

This set of charged and neutral currents enables a number of physical transitions, such as the decays $W^- \to e_L + \bar{v}_L$, $Z \to v_L + \bar{v}_L$, or $Z \to e_L + \bar{e}_L$. As a general principle, particles tend to *decay* into lighter particles if there is no conservation law preventing such a decay. Nevertheless, for instance the decay $Z \to e_L + \bar{e}_L$ can also be *inverted:* An electron–positron pair collides at very high energy and generates a Z-boson, which very soon again decays into leptons. In the electron–positron scattering amplitude, this channel is then visible as a *resonance* at the energy that is needed to provide the Z-mass. The coupling of Z_μ to the weak neutral current j_μ^0 also describes the scattering of a neutrino or an electron off a Z-boson.

The Large Electron–Positron Collider (LEP) at CERN has studied in great detail the Z decay channels. This included all three fermion generations, which will be introduced in Chapter 17. To high precision, no decays were missing, which suggests that it is unlikely that more fermion generations exist. Otherwise, the Z-decay into the neutrino–anti-neutrino pair of a fourth generation should occur (unless that neutrino is extremely heavy, $m_v > M_Z/2 \simeq 45.6$ GeV).

15.4 Triangle Gauge Anomalies in the Lepton Sector

As it stands, the Standard Model with just electrons and neutrinos is *inconsistent* because it suffers from *anomalies* in its gauge interactions. Anomalies represent a form of explicit symmetry breaking due to quantum effects, while at the classical level the corresponding symmetry is exact. Anomalies usually arise from a non-invariance of the functional measure, while the action of the theory is invariant.

Gauge anomalies represent an explicit violation of gauge invariance. Since gauge invariance is vital for eliminating redundant gauge-dependent degrees of freedom, theories with an explicitly broken gauge symmetry are mathematically and physically inconsistent. In order to render the Standard Model consistent, the gauge anomalies of the leptons must be canceled by other fields. The quarks, which participate in both the electroweak and the strong interaction, serve this purpose. As we will see later, in contrast to gauge anomalies, anomalies in global symmetries are perfectly acceptable and even necessary to describe the physics correctly (Adler, 1969; Bell and Jackiw, 1969).

In the Standard Model, there are different types of gauge anomalies that must be canceled. First, there is a triangle anomaly in the $U(1)_Y$ gauge interaction which manifests itself already within (but also beyond) perturbation theory. One considers the interaction with a triangle built from fermion propagators; each corner is a vertex with a coupling to an external gauge field B_μ. The interaction at each vertex is proportional to the weak hypercharge Y of the fermion that propagates around the triangle (it cannot change at the vertices). Hence, the total contribution is proportional to Y^3. It turns out that, at the quantum level, gauge invariance is maintained only if the anomaly,

$$A = \sum_{L} Y^3 - \sum_{R} Y^3, \tag{15.37}$$

vanishes. The sums extend over the left- and right-handed degrees of freedom, respectively.[4] Since the left-handed neutrino and electron carry the weak hypercharge $Y_{l_L} = -1/2$, while the right-handed electron has $Y_{e_R} = -1$, cf. Table 15.1, the $U(1)_Y$ triangle anomaly in the lepton sector is given by

$$A_l = 2Y_{l_L}^3 - Y_{e_R}^3 = 2\left(-\frac{1}{2}\right)^3 - (-1)^3 = \frac{3}{4} \neq 0. \tag{15.38}$$

This shows that the lepton sector of the Standard Model by itself does not constitute a consistent quantum field theory. As we will see, this non-zero anomaly in the lepton sector will be canceled by a corresponding anomaly in the quark sector.

The general expression for $SU(2)_L \times U(1)_Y$ triangle anomalies is given by

$$A^{abc} = \text{Tr}_L \left[(T^a T^b + T^b T^a) T^c \right] - \text{Tr}_R \left[(T^a T^b + T^b T^a) T^c \right]. \tag{15.39}$$

The T^a with $a \in \{1, 2, 3\}$ refer to the generators of $SU(2)_L$, $T^a = \tau^a/2$, and $T^4 = Y\mathbb{1}$. If all three indices a, b, and c are equal to 4, we recover the $U(1)_Y$ anomaly discussed above, $i.e.$ $A^{444} = A_l$. If one index belongs to $\{1, 2, 3\}$ and the other two are equal to 4, the tracelessness of the $SU(2)_L$ generators leads to a vanishing anomaly. Similarly, if all three indices belong to $\{1, 2, 3\}$, the Pauli matrix identity,

$$\text{Tr} \left[(\tau^a \tau^b + \tau^b \tau^a) \tau^c \right] = 2\delta_{ab} \text{Tr}\, \tau^c = 0, \tag{15.40}$$

again leads to a vanishing triangle anomaly. However, if two indices belong to $\{1, 2, 3\}$ while the third one, say c, is equal to 4, the anomaly takes the form

$$A_l^{ab4} = \text{Tr}_L \left[\frac{1}{4} (\tau^a \tau^b + \tau^b \tau^a) Y \right] = \frac{1}{2} \delta_{ab} \text{Tr}_L Y \mathbb{1}. \tag{15.41}$$

---

[4] We do not present an explicit evaluation of this triangle diagram. Pedagogical explanations of this calculation can be found, $e.g.$, in Ryder (1996) and in Peskin and Schroeder (1997). We will consider anomalies in more detail using the lattice regularization in Chapter 19.

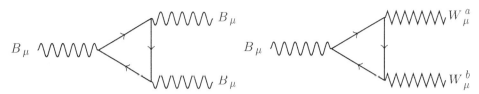

Fig. 15.3 Illustration of the triangle diagrams that contribute to the gauge anomalies $A^{444} = A_l$ on the left and A^{ab4} with $a, b \in \{1, 2, 3\}$ on the right.

Here, we have used the fact that in the Standard Model the left-handed fermions are $SU(2)_L$ doublets, while the right-handed fermions are $SU(2)_L$ singlets ($T_R^a = 0$). In the lepton sector, the corresponding anomaly is given by

$$A_l^{ab4} = \delta_{ab} Y_{l_L} = -\frac{1}{2}\delta_{ab} \neq 0, \tag{15.42}$$

which thus gives rise to another inconsistency. Both types of contributing triangle diagrams are illustrated in Figure 15.3.

It should be mentioned that anomalies that show up in perturbation theory (but not global anomalies) are entirely captured by triangle diagrams and do not receive higher-order corrections (Adler and Bardeen, 1969).

15.5 Witten's Global SU(2)$_L$ Gauge Anomaly in the Lepton Sector

In addition to the triangle anomalies, there is a so-called *global anomaly* in the $SU(2)_L$ gauge interactions, which was discovered by Edward Witten (1982). It should be stressed that here "global" does not refer to a global symmetry. Instead, it refers to the global topological properties of $SU(2)_L$ gauge transformations. Two gauge transformations are considered topologically equivalent if they can be deformed continuously into one another. The corresponding equivalence classes are known as homotopy classes. In four space–time dimensions, the homotopy group of $SU(2)_L$ gauge transformations is

$$\Pi_4[SU(2)] = \Pi_4[S^3] = \mathbb{Z}(2) ; \tag{15.43}$$

i.e. these gauge transformations fall into two distinct topological classes.[5] The topologically trivial class contains all gauge transformations that can be continuously deformed into the gauge transformation $L(x) = \mathbb{1}$, while the topologically non-trivial class contains all other gauge transformations. As Witten first realized, the fermionic measure of each doublet of Weyl fermions in an $SU(2)_L$ gauge theory changes sign under topologically non-trivial $SU(2)_L$ gauge transformations. Hence, in order to be gauge-invariant, the theory must contain an *even number of doublets*.

---

[5] The homotopy group $\Pi_d[\mathcal{M}]$ contains the winding numbers of maps from a d-dimensional sphere S^d into the target manifold $\mathcal{M}$. In this case, S^d represents the compactified 4-d Euclidean space–time, and the target manifold $\mathcal{M}$ is the group space of $SU(2)$, namely, the 3-dimensional sphere S^3. Homotopy groups are discussed in Appendix G and applied extensively in Chapters 19 and 24.

Since the lepton sector of the first generation of the Standard Model contains a single $SU(2)_L$ doublet (consisting of the left-handed electron and neutrino), it suffers from Witten's global gauge anomaly. In order to cancel this anomaly, we must add an *odd number of $SU(2)_L$ doublets*. Since it is associated with "large" gauge transformations, which are not located in the neighborhood of the identity, the global gauge anomaly manifests itself only beyond perturbation theory. In particular, it is not visible in perturbative triangle diagrams.

We add that both the perturbative triangle anomalies and the non-perturbative global anomaly are correctly represented in the lattice regularization (Bär and Campos, 2000; Bär, 2003), provided the lattice chiral symmetry is introduced carefully, cf. Section 18.6.

15.6 Up and Down Quarks

In the first fermion generation, there are also the *up* and *down quarks, u* and *d*, which will come to our rescue and cancel both, the triangle anomalies and the global gauge anomaly. The quarks are massive and thus require the introduction of *left- and right-handed fields*.[6] In addition to the electroweak interaction, the quarks participate in the strong interaction and thus they carry an $SU(N_c)$ *color charge*. The color index on a quark field then takes values $c \in \{1, 2, \ldots, N_c\}$. In the real world, the number of colors is $N_c = 3$. However, as we will see, a consistent variant of the Standard Model can be formulated with any odd number of colors. There are some misconceptions about this issue in large parts of the textbook literature. In order to illuminate this point, we keep N_c general in this and some other chapters.

The left-handed u- and d-quark fields (with color index c) then form N_c distinct $SU(2)_L$ doublets

$$q_L^c(x) = \begin{pmatrix} u_L^c(x) \\ d_L^c(x) \end{pmatrix}, \quad \bar{q}_L^c(x) = \left(\bar{u}_L^c(x), \bar{d}_L^c(x) \right), \quad c \in \{1, 2, \ldots, N_c\}, \tag{15.44}$$

or two $SU(N_c)$ color N_c-plets. The right-handed quarks u_R^c and d_R^c also form two $SU(N_c)$ color N_c-plets, but they are $SU(2)_L$ singlets. In analogy to the lepton sector, in the quark sector the generators T_L^3 and T_R^3 take the values

$$T_{Lu_L}^3 = \frac{1}{2}, \ T_{Ld_L}^3 = -\frac{1}{2}, \ T_{Lu_R}^3 = 0, \ T_{Ld_R}^3 = 0,$$

$$T_{Ru_L}^3 = 0, \ T_{Rd_L}^3 = 0, \ T_{Ru_R}^3 = \frac{1}{2}, \ T_{Rd_R}^3 = -\frac{1}{2}. \tag{15.45}$$

Using the Einstein summation convention for the color index c, the Lagrangian for free massless quarks can be written as

$$\mathcal{L}_0(\bar{u}, u, \bar{d}, d) = \bar{u}_L^c \bar{\sigma}_\mu \partial_\mu u_L^c + \bar{u}_R^c \sigma_\mu \partial_\mu u_R^c + \bar{d}_L^c \bar{\sigma}_\mu \partial_\mu d_L^c + \bar{d}_R^c \sigma_\mu \partial_\mu d_R^c. \tag{15.46}$$

---

[6] For some time, there was a controversy whether the u-quark mass might vanish. However, this turned out to be inconsistent with phenomenology. Still, even if some quarks were massless, one would need to introduce both, left- and right-handed quark fields, in order to achieve anomaly cancellation.

It has a global $U(1)_B$ symmetry, which means that $\mathcal{L}_0$ is invariant when we multiply all quark fields by the same phase,

$$
\begin{aligned}
u_L^{c\,\prime}(x) &= \exp\left(i\rho/N_c\right) u_L^c(x), & \bar{u}_L^{c\,\prime}(x) &= \bar{u}_L^c(x) \exp\left(-i\rho/N_c\right), \\
u_R^{c\,\prime}(x) &= \exp\left(i\rho/N_c\right) u_R^c(x), & \bar{u}_R^{c\,\prime}(x) &= \bar{u}_R^c(x) \exp\left(-i\rho/N_c\right), \\
d_L^{c\,\prime}(x) &= \exp\left(i\rho/N_c\right) d_L^c(x), & \bar{d}_L^{c\,\prime}(x) &= \bar{d}_L^c(x) \exp\left(-i\rho/N_c\right), \\
d_R^{c\,\prime}(x) &= \exp\left(i\rho/N_c\right) d_R^c(x), & \bar{d}_R^{c\,\prime}(x) &= \bar{d}_R^c(x) \exp\left(-i\rho/N_c\right).
\end{aligned}
\tag{15.47}
$$

In analogy to the lepton number, the corresponding conserved charge is the *quark number* or equivalently the *baryon number B*. Each baryon contains an excess of N_c confined quarks over anti-quarks. Hence, the baryon number of a quark is $B = 1/N_c$, while an anti-quark has $B = -1/N_c$.

We still need to assign weak hypercharges to the quark fields. $SU(2)_L$ gauge invariance requires that the left-handed u- and d-quarks carry the same charge Y_{q_L}. On the other hand, since the right-handed quarks are $SU(2)_L$ singlets, u and d may carry different hypercharges Y_{u_R} and Y_{d_R}. The $U(1)_Y$ gauge transformations act as

$$
\begin{aligned}
u_L^{c\,\prime}(x) &= \exp\left(i Y_{q_L} g'\varphi(x)\right) u_L^c(x), & \bar{u}_L^{c\,\prime}(x) &= \bar{u}_L^c(x)\, \exp\left(-i Y_{q_L} g'\varphi(x)\right), \\
u_R^{c\,\prime}(x) &= \exp\left(i Y_{u_R} g'\varphi(x)\right) u_R^c(x), & \bar{u}_R^{c\,\prime}(x) &= \bar{u}_R^c(x) \exp\left(-i Y_{u_R} g'\varphi(x)\right), \\
d_L^{c\,\prime}(x) &= \exp\left(i Y_{q_L} g'\varphi(x)\right) d_L^c(x), & \bar{d}_L^{c\,\prime}(x) &= \bar{d}_L^c(x) \exp\left(-i Y_{q_L} g'\varphi(x)\right), \\
d_R^{c\,\prime}(x) &= \exp\left(i Y_{d_R} g'\varphi(x)\right) d_R^c(x), & \bar{d}_R^{c\,\prime}(x) &= \bar{d}_R^c(x) \exp\left(-i Y_{d_R} g'\varphi(x)\right).
\end{aligned}
\tag{15.48}
$$

Under $SU(2)_L$ gauge transformations, the quark fields transform as

$$
\begin{aligned}
q_L^{c\,\prime}(x) &= \begin{pmatrix} u_L^{c\,\prime}(x) \\ d_L^{c\,\prime}(x) \end{pmatrix} = L(x) \begin{pmatrix} u_L^c(x) \\ d_L^c(x) \end{pmatrix} = L(x) q_L^c(x), \\
\bar{q}_L^{c\,\prime}(x) &= \left(\bar{u}_L^{c\,\prime}(x),\, \bar{d}_L^{c\,\prime}(x)\right) = \left(\bar{u}_L^c(x),\, \bar{d}_L^c(x)\right) L(x)^\dagger = \bar{q}_L^c(x) L(x)^\dagger, \\
u_R^{c\,\prime}(x) &= u_R^c(x), \quad d_R^{c\,\prime}(x) = d_R^c(x), \quad \bar{u}_R^{c\,\prime}(x) = \bar{u}_R^c(x), \quad \bar{d}_R^{c\,\prime}(x) = \bar{d}_R^c(x).
\end{aligned}
\tag{15.49}
$$

Before the quarks are coupled to the gluons, they do not yet participate in the strong interaction. The gluons are then still strongly interacting among each other and are confined inside glueballs, but they decouple from the other fields.[7] Without quark–gluon couplings, the quarks are not confined inside hadrons but represent physical states. Such a world may be considered a "theorist's paradise", because the quark physics would be mostly perturbative and thus analytically calculable.

We will now switch on the quark–gluon coupling. While the real world, in which quarks are confined, is much more interesting than the "theorist's paradise", it will also turn out to be much more difficult to understand. In particular, since strong non-perturbative effects then dominate at low energies, perturbation theory breaks down, and quantitative results can often be obtained only by means of numerical Monte Carlo calculations of lattice regularized systems.

---

[7] In the absence of quark–gluon couplings, only gravity would establish some interaction between gluons and the rest of the world.

Suppressing color indices, under $SU(N_c)$ gauge transformations $\Omega(x)$ the quark fields transform as

$$
\begin{aligned}
q_L{}'(x) &= \Omega(x)q_L(x), & \bar{q}_L{}'(x) &= \bar{q}_L(x)\Omega(x)^\dagger, \\
u_R{}'(x) &= \Omega(x)u_R(x), & \bar{u}_R{}'(x) &= \bar{u}_R(x)\Omega(x)^\dagger, \\
d_R{}'(x) &= \Omega(x)d_R(x), & \bar{d}_R{}'(x) &= \bar{d}_R(x)\Omega(x)^\dagger, & \Omega(x) \in SU(N_c).
\end{aligned}
\tag{15.50}
$$

In the mathematical jargon, this means that the quark fields q transform in the *fundamental* $\{N_c\}$ representation of $SU(N_c)$, while the anti-quarks $\bar{q}$ transform in the *anti-fundamental* $\{\overline{N_c}\}$ representation. Since both left- and right-handed quarks transform in the same way under color gauge transformations, the $SU(N_c)$ symmetry is not chiral but vector-like. Making the color indices c explicit, for the left-handed quark doublet the covariant derivative takes the form

$$
D_\mu q_L^c(x) = \left[\left(\partial_\mu + i Y_{q_L} g' B_\mu(x) + i g W_\mu^a(x) \frac{\tau^a}{2} \right) \delta_{cc'} + i g_s G_\mu^a(x) \frac{\lambda^a_{cc'}}{2} \right] q_L^{c'}(x). \tag{15.51}
$$

Here, λ^a with $a \in \{1, 2, \dots, N_c^2 - 1\}$ are the generators of $SU(N_c)$, which obey $\mathrm{Tr}[\lambda^a \lambda^b] = 2\delta_{ab}$. For $N_c = 3$, we can choose the eight Gell-Mann matrices displayed in Appendix F.7. Inserting $W_\mu(x) = i g W_\mu^a(x) \tau^a/2$ and $G_\mu(x) = i g_s G_\mu^a(x) \lambda^a/2$, one arrives at the short-hand notation

$$
D_\mu q_L(x) = \left[\partial_\mu + i Y_{q_L} g' B_\mu(x) + W_\mu(x) + G_\mu(x) \right] q_L(x). \tag{15.52}
$$

For the right-handed quark singlets, the covariant derivatives are given by

$$
\begin{aligned}
D_\mu u_R^c(x) &= \left[\left(\partial_\mu + i Y_{u_R} g' B_\mu(x) \right) \delta_{cc'} + i g_s G_\mu^a(x) \frac{\lambda^a_{cc'}}{2} \right] u_R^{c'}(x), \\
D_\mu d_R^c(x) &= \left[\left(\partial_\mu + i Y_{d_R} g' B_\mu(x) \right) \delta_{cc'} + i g_s G_\mu^a(x) \frac{\lambda^u_{cc'}}{2} \right] d_R^{c'}(x),
\end{aligned}
\tag{15.53}
$$

or alternatively, suppressing the color indices,

$$
\begin{aligned}
D_\mu u_R(x) &= \left[\partial_\mu + i Y_{u_R} g' B_\mu(x) + G_\mu(x) \right] u_R(x), \\
D_\mu d_R(x) &= \left[\partial_\mu + i Y_{d_R} g' B_\mu(x) + G_\mu(x) \right] d_R(x).
\end{aligned}
\tag{15.54}
$$

The Lagrangian describing the propagation of the quarks, as well as their interactions with the $U(1)_Y$, $SU(2)_L$, and $SU(N_c)$ gauge fields, then takes the form

$$
\begin{aligned}
\mathcal{L}(\bar{u}, u, \bar{d}, d, G, W, B) &= \bar{q}_L \bar{\sigma}_\mu D_\mu q_L + \bar{u}_R \sigma_\mu D_\mu u_R + \bar{d}_R \sigma_\mu D_\mu d_R \\
&= (\bar{u}_L, \bar{d}_L) \bar{\sigma}_\mu D_\mu \begin{pmatrix} u_L \\ d_L \end{pmatrix} + \bar{u}_R \sigma_\mu D_\mu u_R + \bar{d}_R \sigma_\mu D_\mu d_R.
\end{aligned}
\tag{15.55}
$$

Among other processes, the coupling of quarks to W-bosons gives rise to the β-decay of the neutron, as illustrated in Figure 15.4.

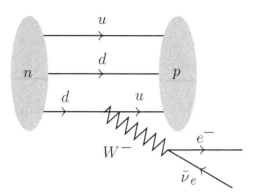

Fig. 15.4 β-decay of the neutron into a proton, an electron, and an anti-neutrino. A d-quark inside the neutron converts into a u-quark by emitting a W^--boson, which soon decays into leptons.

15.7 Anomaly Cancellation between Leptons and Quarks

As we have seen, quarks carry a color index $c \in \{1, 2, \dots, N_c\}$. Hence, there are N_c left-handed $SU(2)_L$ quark doublets. Together with the left-handed lepton doublet, these are $(N_c + 1)$ left-handed $SU(2)_L$ doublets. Thus, in order to cancel the global $SU(2)_L$ anomaly of the lepton sector, the number of colors N_c must be *odd* in the Standard Model.[8]

In analogy to the leptons, the quarks also contribute to the various triangle anomalies. First of all, the quark triangle diagram with external $U(1)_Y$ bosons attached to all three vertices contributes

$$A_q^{444} = N_c \left(2Y_{q_L}^3 - Y_{u_R}^3 - Y_{d_R}^3 \right). \tag{15.56}$$

For the same reasons as in the lepton sector, the triangle diagrams with one or three external $SU(2)_L$ gauge bosons vanish. The diagram with two external $SU(2)_L$ and one external $U(1)_Y$ boson, on the other hand, is non-zero and contributes

$$A_q^{ab4} = \delta_{ab} N_c Y_{q_L}, \quad a, b \in \{1, 2, 3\}. \tag{15.57}$$

In order to cancel the triangle anomalies in the lepton sector, we now demand

$$A_l^{ab4} + A_q^{ab4} = 0 \quad \Rightarrow \quad Y_{q_L} = \frac{1}{2N_c},$$

$$A_l^{444} + A_q^{444} = 0 \quad \Rightarrow \quad 2Y_{q_L}^3 - Y_{u_R}^3 - Y_{d_R}^3 = -\frac{3}{4N_c} \quad \Rightarrow$$

$$Y_{u_R}^3 + Y_{d_R}^3 = \frac{1}{4N_c^3} + \frac{3}{4N_c}. \tag{15.58}$$

Since the quarks also couple to the gluons, there are additional triangle anomalies which are absent in the lepton sector. In particular, the range of the indices a, b, c now extends from 1, 2, 3 for $SU(2)_L$ and 4 for $U(1)_Y$ to $a - 4, b - 4, c - 4 \in \{1, 2, \dots, N_c^2 - 1\}$.

[8] Here, we assume that the anomalies are canceled within a single generation of fermions. If the number of generations would be even, the global gauge anomaly would also cancel for even N_c. However, since baryons (which – in a naive constituent quark picture – consist of N_c quarks, cf. Chapter 21) would then be bosons, the resulting physics would be drastically different from the real world.

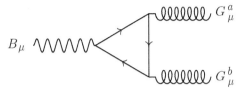

Fig. 15.5 Triangle diagram that contributes to the gauge anomaly A^{ab4} with $a - 4, b - 4 \in \{1, 2, \ldots, N_c^2 - 1\}$. The spiraling lines denote gluon propagators.

Since there is the same number of left- and right-handed color N_c-plets, the pure QCD part of the Standard Model is a non-chiral *vector-like theory*, in which the corresponding pure SU(N_c) anomaly cancels trivially. As a consequence, the triangle diagrams with three external gluons vanish. Triangle diagrams with a single external gluon vanish due to the tracelessness of λ^a, while those with two external gluons and one external SU(2)$_L$ gauge boson vanish due to the tracelessness of τ^a. The triangle diagram with two external gluons and one external U(1)$_Y$ boson, which is illustrated in Figure 15.5, on the other hand, is proportional to

$$A_q^{ab4} = \delta_{ab} \left(2Y_{q_L} - Y_{u_R} - Y_{d_R}\right), \quad a - 4, b - 4 \in \{1, 2, \ldots, N_c^2 - 1\}. \tag{15.59}$$

The cancellation of this anomaly, which does not receive any contribution from the lepton sector, thus requires

$$Y_{u_R} + Y_{d_R} = 2Y_{q_L} = \frac{1}{N_c}, \tag{15.60}$$

where we used eq. (15.58). This relation further implies

$$Y_{q_L} = \frac{1}{2N_c}, \quad Y_{u_R} = \frac{1}{2}\left(\frac{1}{N_c} + 1\right), \quad Y_{d_R} = \frac{1}{2}\left(\frac{1}{N_c} - 1\right) ; \tag{15.61}$$

i.e. anomaly cancellation completely fixes the weak hypercharges of the quarks.

Interestingly, the resulting values are related to the generator T_R^3 and the baryon number $B = 1/N_c$ by

$$Y = T_R^3 + \frac{1}{2}B. \tag{15.62}$$

The quantum numbers of the u- and d-quarks – for a general number of colors N_c – are summarized in Table 15.2, together with those of the leptons.

In the real world with $N_c = 3$, the baryon number of a quark is $B = 1/3$ and the weak hypercharges are given by

$$Y_{q_L} = \frac{1}{6}, \quad Y_{u_R} = \frac{2}{3}, \quad Y_{d_R} = -\frac{1}{3}. \tag{15.63}$$

It is often argued that in the Standard Model the number of colors must be exactly $N_c = 3$ in order to achieve anomaly cancellation. In contrast to this claim, we have just seen that the Standard Model is indeed *consistent for any odd number N_c* (Rudaz, 1990; Abbas, 1990; Bär and Wiese, 2001). As we will discuss in Sections 21.8 and 22.2, there is sufficient experimental evidence to single out $N_c = 3$. However, we would like to point out that the reasons for this are more subtle than it is often assumed. In particular, $N_c = 3$

does *not* follow from the requirement of mathematical consistency (*i.e.* gauge anomaly cancellation) of the Standard Model.

It is worth noting that no additional anomalies arise when the Standard Model is coupled to perturbative quantum gravity; *i.e.* gravitational anomalies and mixed gauge–gravitational anomalies are automatically canceled as well (Alvarez-Gaumé and Witten, 1984). This is essential because otherwise the Standard Model would be mathematically inconsistent with General Relativity – the established theory of the gravitational interaction.

15.8 Electric Charges of Quarks and Baryons

Again in analogy to the lepton sector, one identifies the electric charge of the quarks as

$$Q = T_{\rm L}^3 + Y = T_{\rm L}^3 + T_{\rm R}^3 + \frac{1}{2}B. \tag{15.64}$$

For the left-handed u- and d-quark, this equation takes the form

$$Q_{u_{\rm L}} = T_{{\rm L}u_{\rm L}}^3 + Y_{q_{\rm L}} = \frac{1}{2} + \frac{1}{2N_{\rm c}} = \frac{1}{2}\left(\frac{1}{N_{\rm c}} + 1\right),$$
$$Q_{d_{\rm L}} = T_{{\rm L}d_{\rm L}}^3 + Y_{q_{\rm L}} = -\frac{1}{2} + \frac{1}{2N_{\rm c}} = \frac{1}{2}\left(\frac{1}{N_{\rm c}} - 1\right). \tag{15.65}$$

For the right-handed quark fields, one finds the same values of the electric charges,

$$Q_{u_{\rm R}} = T_{{\rm L}u_{\rm R}}^3 + Y_{u_{\rm R}} = 0 + \frac{1}{2}\left(\frac{1}{N_{\rm c}} + 1\right),$$
$$Q_{d_{\rm R}} = T_{{\rm L}d_{\rm R}}^3 + Y_{d_{\rm R}} = 0 + \frac{1}{2}\left(\frac{1}{N_{\rm c}} - 1\right). \tag{15.66}$$

In the real world with $N_{\rm c} = 3$, the electric charges of the quarks are thus given by

$$Q_u = \frac{1}{2}\left(\frac{1}{N_{\rm c}} + 1\right) \stackrel{N_{\rm c}=3}{=} \frac{2}{3}, \quad Q_d = \frac{1}{2}\left(\frac{1}{N_{\rm c}} - 1\right) \stackrel{N_{\rm c}=3}{=} -\frac{1}{3}. \tag{15.67}$$

Since quarks have lepton number $L = 0$, and leptons have baryon number $B = 0$, the electric charges of the fermionic matter fields in the Standard Model are

$$Q = T_{\rm L}^3 + Y = T_{\rm L}^3 + T_{\rm R}^3 + \frac{1}{2}(B - L). \tag{15.68}$$

As we will see in Section 15.11, the difference between baryon and lepton number, $B - L$, generates an exact global symmetry of the (minimal) Standard Model, while B and L individually are explicitly broken by anomalies. Once we will extend the Standard Model by right-handed neutrino fields and add Majorana mass terms for the neutrinos, also the $B - L$ symmetry will be explicitly broken.

As we will discuss in Chapter 21, just like gluons, quarks are confined inside *hadrons*. Hadrons containing N_c more quarks than anti-quarks are known as *baryons* (with baryon number $B = 1$) and *vice versa* for anti-baryons. In the real world, baryons have an excess of three more quarks than anti-quarks. Of major importance are the *nucleons, i.e.* the *proton* and the *neutron,* which are the lightest baryons. In a naive constituent quark picture, the proton has the quark content (*uud*), and the neutron consists of (*udd*). Actually, baryons contain in addition a fluctuating number of gluons and quark–anti-quark pairs. Since those are electrically neutral, the electric charges, which result from the naive picture,

$$Q_p = 2Q_u + Q_d = 2\frac{2}{3} - \frac{1}{3} = 1,$$

$$Q_n = Q_u + 2Q_d = \frac{2}{3} - 2\frac{1}{3} = 0, \qquad (15.69)$$

are the familiar charges of the proton and the neutron; they are integer multiples of the electron charge $-e$. Despite numerous experimental studies, including Millikan-type experiments, fundamental fractional electric charges have never been observed in Nature.[9] This is a consequence of quark confinement combined with anomaly cancellation.

In a hypothetical, but mathematically fully consistent world with an arbitrary odd number of colors N_c, there would still be protons and neutrons. However, a proton would then (naively) contain $(N_c + 1)/2$ *u*-quarks and $(N_c - 1)/2$ *d*-quarks, while a neutron would contain $(N_c - 1)/2$ *u*-quarks and $(N_c + 1)/2$ *d*-quarks. Hence, just as in the real world, we would still obtain

$$Q_p = \frac{N_c + 1}{2}Q_u + \frac{N_c - 1}{2}Q_d = \frac{N_c + 1}{4}\left(\frac{1}{N_c} + 1\right) + \frac{N_c - 1}{4}\left(\frac{1}{N_c} - 1\right) = 1,$$

$$Q_n = \frac{N_c - 1}{2}Q_u + \frac{N_c + 1}{2}Q_d = \frac{N_c - 1}{4}\left(\frac{1}{N_c} + 1\right) + \frac{N_c + 1}{4}\left(\frac{1}{N_c} - 1\right) = 0. \quad (15.70)$$

Consequently, confinement combined with anomaly cancellation is responsible for charge quantization in integer units even for an arbitrary odd number N_c of colors.[10]

15.9 Anomaly Matching

Gerard 't Hooft's *anomaly matching condition* states that gauge anomaly cancellation should also work out separately with the degrees of freedom which are relevant at low energy scales ('t Hooft, 1980). In particular, this implies that anomaly cancellation should take place even if one considers only the *low-energy regime* of a given theory. Anomalies must therefore be canceled properly also in a low-energy effective theory for the Standard Model. This condition puts non-trivial constraints on the possible dynamics of such theories.

---

[9] Fractional charges carried by Laughlin quasi-particles emerge as a collective phenomenon in the condensed matter physics of the fractional quantum Hall effect. Quarks may also occur as deconfined objects, in a quark–gluon plasma, which does, however, not enable the direct detection of fractional charges either.

[10] When the number of fermion generations was even, N_c could as well be even, cf. footnote 8. In that case, baryons would be bosons with half-integer electric charges. This would drastically change the physics.

For example, at low energies quarks are confined inside protons and neutrons (the nucleons), and so the anomalies should also cancel between leptons and nucleons. To convince ourselves that this is indeed the case, let us reconsider the first generation now expressed in terms of nucleon (rather than quark) degrees of freedom,

$$l_{\mathrm{L}} = \begin{pmatrix} \nu_{\mathrm{L}} \\ e_{\mathrm{L}} \end{pmatrix}, \ e_{\mathrm{R}}, \quad N_{\mathrm{L}} = \begin{pmatrix} p_{\mathrm{L}} \\ n_{\mathrm{L}} \end{pmatrix}, \ p_{\mathrm{R}}, \ n_{\mathrm{R}}. \tag{15.71}$$

Indeed, the global gauge anomaly is still canceled because the left-handed nucleons form one $SU(2)_L$ doublet. The weak hypercharge assignments for the nucleons are

$$Y_{N_{\mathrm{L}}} = \frac{1}{2}, \quad Y_{p_{\mathrm{R}}} = 1, \quad Y_{n_{\mathrm{R}}} = 0. \tag{15.72}$$

Here, the index N refers to a nucleon. The electric charges of the left-handed nucleons then take the well-known values

$$Q_{p_{\mathrm{L}}} = T^3_{\mathrm{L}p_{\mathrm{L}}} + Y_{N_{\mathrm{L}}} = \frac{1}{2} + \frac{1}{2} = 1,$$
$$Q_{n_{\mathrm{L}}} = T^3_{\mathrm{L}n_{\mathrm{L}}} + Y_{N_{\mathrm{L}}} = -\frac{1}{2} + \frac{1}{2} = 0. \tag{15.73}$$

Similarly, for the right-handed proton and neutron we obtain the same values, which confirms the consistency,

$$Q_{p_{\mathrm{R}}} = T^3_{\mathrm{L}p_{\mathrm{R}}} + Y_{p_{\mathrm{R}}} = 0 + 1 = 1,$$
$$Q_{n_{\mathrm{R}}} = T^3_{\mathrm{L}n_{\mathrm{R}}} + Y_{n_{\mathrm{R}}} = 0 + 0 = 0. \tag{15.74}$$

The corresponding contributions to the $SU(2)_L \times U(1)_Y$ triangle anomalies in the nucleon sector are then given by

$$A_N^{444} = 2Y_{N_{\mathrm{L}}}^3 - Y_{p_{\mathrm{R}}}^3 - Y_{n_{\mathrm{R}}}^3 = 2\left(\frac{1}{2}\right)^3 - 1^3 - 0^3 = -\frac{3}{4},$$
$$A_N^{ab4} = \delta_{ab} Y_{N_{\mathrm{L}}} = \frac{1}{2}\delta_{ab}, \tag{15.75}$$

which again cancel the anomalies A_l^{444} and A_l^{ab4} of the leptons.

In view of our analysis for a general odd number N_{c}, this nucleon consideration corresponds exactly to the case $N_{\mathrm{c}} = 1$, provided we identify the proton with the u-"quark" and the neutron with the d-"quark". In this respect, the discussion of a possible generalization is not just academic.

In fact, as early as 1949 Hiroshi Fukuda and Yoneji Miyamoto, as well as Jack Steinberger, were first to calculate a triangle diagram with nucleons propagating around the loop (Fukuda and Miyamoto, 1949; Steinberger, 1949).[11] It is sometimes stated that they accidentally arrived at the right answer although they neglected the quark content of protons and neutrons and thus the color factor N_{c}. Of course, in 1949 Fukuda, Miyamoto, and Steinberger did not know about quarks or color, but they were still using a consistent low-energy picture, so their result was not accidental. Indeed, thanks to anomaly matching, their result is the correct answer irrespective of the value of N_{c}.

---

[11] Their calculations referred to the triangle diagram that is relevant for the neutral pion decay $\pi^0 \to \gamma\gamma$. This is different from the triangle diagrams discussed here, but the qualitative low-energy argument is generic, so it applies there as well.

15.10 Right-Handed Neutrinos

The minimal version of the Standard Model, which we have presented so far, does not contain right-handed neutrino fields. If one insists on perturbative renormalizability, the absence of right-handed neutrino fields implies that neutrinos are massless, which may thus be viewed as a prediction of the Standard Model. However, since the observation of neutrino oscillations in 1998 in the Super-Kamioka Neutrino Detection Experiment (Super-Kamiokande) (Fukuda *et al.*, 1998) and in 2001 by the Sudbury Neutrino Observatory (SNO) (Ahmad *et al.*, 2001), it is known that neutrinos must have mass.[12] In this regard, the minimal Standard Model is in conflict with experiment and must therefore be extended. One may do this in at least two alternative ways.

- First, one may view the Standard Model as an effective theory, formulated only in terms of the relevant low-energy degrees of freedom. The leading terms in the effective Lagrangian are indeed the renormalizable interactions that we have considered until now. However, in an effective theory framework there are additional higher-order corrections to the effective Lagrangian which need not be renormalizable. As we will discuss in more detail in Section 16.3, one can indeed construct non-renormalizable neutrino mass terms by using just the left-handed neutrino fields introduced until now.
- An alternative way to proceed, which reflects a drastically different point of view, is to assume that the Standard Model is an integral part of a renormalizable theory with a larger field content that may extend to much higher energies beyond the TeV range. This approach is pursued, for example, in the framework of Grand Unified Theories (GUTs), which will be discussed in Chapter 26. If one insists on perturbative renormalizability, the incorporation of neutrino mass terms requires the introduction of right-chirality neutrino fields ν_R and $\bar{\nu}_R$. It is conceivable that there are even relatively light right-handed neutrinos that could have evaded detection. As we will now discuss, the Standard Model can be extended by right-handed neutrinos in a straightforward manner.

Right-handed neutrinos are leptons (with lepton number $L = 1$); *i.e.* under global $U(1)_L$ transformations they transform as

$$\nu_R'(x) = \exp(i\chi)\nu_R(x), \quad \bar{\nu}_R'(x) = \bar{\nu}_R(x)\exp(-i\chi), \tag{15.76}$$

with the same phase χ as in eq. (15.2). Just like right-handed electrons, right-handed neutrinos are both $SU(N_c)$ color and $SU(2)_L$ singlets, with the weak isospin 3-components

$$T_{L\nu_R}^3 = 0, \quad T_{R\nu_R}^3 = \frac{1}{2}. \tag{15.77}$$

Using the relations $Y = T_R^3 - L/2$ and $Q = T_L^3 + Y$ of Section 15.3, we then obtain

$$Y_{\nu_R} = T_{R\nu_R}^3 - \frac{1}{2} = 0, \quad Q_{\nu_R} = T_{L\nu_R}^3 + Y_{\nu_R} = 0. \tag{15.78}$$

---

[12] As of 2022, the values of the neutrino masses are not known experimentally. Neutrino oscillations only provide values for the neutrino mass differences squared, leaving uncertainties in the individual values and even in their hierarchy. Decay experiments further provide upper bounds, which are given in Table B.2. Other bounds and estimates are obtained from cosmology, but they rely on further assumptions.

Table 15.2 The leptons and quarks of the first fermion generation, with their quantum numbers, which are relevant in this chapter. Here we refer to the extended version of the Standard Model, which includes a right-handed neutrino. All these quantum numbers are identical in the second and third fermion generation, which will be added in Chapter 17. In the quark sector, we start with the expression for a general number of colors, N_c. The arrows point to the values for $N_c = 3$, which are realized in Nature.

	L	B	Y	T_L^3	T_R^3	Q	SU(N_c)
ν_L	1	0	$-1/2$	$1/2$	0	0	$\{1\}$
e_L	1	0	$-1/2$	$-1/2$	0	-1	$\{1\}$
ν_R	1	0	0	0	$1/2$	0	$\{1\}$
e_R	1	0	-1	0	$-1/2$	-1	$\{1\}$
u_L	0	$1/N_c \to 1/3$	$1/2N_c \to 1/6$	$1/2$	0	$(1/N_c+1)/2 \to 2/3$	$\{N_c\} \to \{3\}$
d_L	0	$1/N_c \to 1/3$	$1/2N_c \to 1/6$	$-1/2$	0	$(1/N_c-1)/2 \to -1/3$	$\{N_c\} \to \{3\}$
u_R	0	$1/N_c \to 1/3$	$(1/N_c+1)/2 \to 2/3$	0	$1/2$	$(1/N_c+1)/2 \to 2/3$	$\{N_c\} \to \{3\}$
d_R	0	$1/N_c \to 1/3$	$(1/N_c-1)/2 \to -1/3$	0	$-1/2$	$(1/N_c-1)/2 \to -1/3$	$\{N_c\} \to \{3\}$

This implies that the right-handed neutrino is neutral, not only electrically, but under all gauge interactions in the Standard Model. Consequently, right-handed neutrinos are *"sterile"*; *i.e.* they do not participate in the electromagnetic, weak, or strong interaction. Since right-handed neutrinos do not couple to the gauge fields of the Standard Model, they do not contribute to the gauge anomalies. Hence, these anomalies remain properly canceled.

Table 15.2 gives an overview over the quantum numbers of the leptons (including ν_R) and quarks of the first fermion generation. In the higher generations – to be supplemented in Chapter 17 – all these quantum numbers are repeated identically.

As we will see in Chapter 16, left- and right-handed neutrino fields can be combined in a so-called Yukawa term coupled to the Higgs field. When the Higgs field picks up a vacuum value v, this term gives rise to a non-zero neutrino mass proportional to v. We will also see that right-handed neutrino fields alone can be used to form additional Majorana mass terms, which are not tied to the electroweak symmetry breaking scale v. In fact, besides v, the Majorana masses will appear as further dimensioned parameters in the extended Standard Model Lagrangian.

15.11 Lepton and Baryon Number Anomalies

As we discussed before, the lepton–gauge-field Lagrangian $\mathcal{L}(\bar{\nu}, \nu, \bar{e}, e, W, B)$ of eq. (15.15) as well as the quark–gauge-field Lagrangian $\mathcal{L}(\bar{u}, u, \bar{d}, d, G, W, B)$ of eq. (15.55) is invariant against global U(1)$_L$ lepton number and U(1)$_B$ baryon number transformations, respectively. Hence, at the classical level, lepton and baryon number are conserved quantities.

As we discussed in the Intermezzo, usually global symmetries are only approximate, while exact symmetries are local. Would it be possible to gauge the U(1)$_L$ and U(1)$_B$ symmetries in the Standard Model? As we will see, this is not possible, because both symmetries are explicitly broken by anomalies and are thus only approximate. Still, we will find that the combination $B - L$ is conserved exactly.

Let us imagine a hypothetical $U(1)_L$ gauge boson that couples to the lepton number. Such a particle would mediate a *fifth gauge force*, beyond gravity, electromagnetism, as well as the weak and strong interactions. There is no experimental evidence for such a force, and we will now see that gauging $U(1)_L$ is, in fact, impossible in the Standard Model because this symmetry suffers from triangle anomalies. After the introduction of right-handed neutrinos, $U(1)_L$ is a vector-like symmetry; *i.e.* both left- and right-handed leptons carry the same lepton number $L = 1$. As a consequence, the pure $U(1)_L$ triangle anomaly with three external (hypothetical) $U(1)_L$ gauge bosons vanishes trivially. Still, there may be mixed anomalies. First of all, triangle diagrams containing external gluons vanish because leptons do not participate in the strong interaction. Triangle diagrams with two external $U(1)_L$ and one external $SU(2)_L$ gauge boson vanish due to the tracelessness of τ^a. The mixed anomaly with one $U(1)_Y$ and two $U(1)_L$ gauge bosons is proportional to

$$A^{4LL} = 2\left[\mathrm{Tr}_L L^2 Y - \mathrm{Tr}_R L^2 Y\right] = 2\left[2Y_{l_L} - Y_{\nu_R} - Y_{e_R}\right] = 2\left[2\left(-\frac{1}{2}\right) - 0 - (-1)\right] = 0,$$
(15.79)

and thus vanishes as well.

We still need to consider the triangle diagrams with just one external $U(1)_L$ gauge boson. The diagram with two external $U(1)_Y$ gauge bosons contributes

$$A^{44L} = 2\left[\mathrm{Tr}_L Y^2 L - \mathrm{Tr}_R Y^2 L\right] = 2\left[2Y_{l_L}^2 - Y_{\nu_R}^2 - Y_{e_R}^2\right] = 2\left[2\left(-\frac{1}{2}\right)^2 - 0^2 - (-1)^2\right]$$
$$= -1 ;$$
(15.80)

thus, it leads to an inconsistency when the $U(1)_L$ lepton number symmetry is gauged. The diagram with one external $U(1)_Y$ and one external $SU(2)_L$ gauge boson vanishes due to the tracelessness of τ^a. On the other hand, the triangle diagram with two external $SU(2)_L$ gauge bosons (with $a, b \in \{1, 2, 3\}$) contributes

$$A^{abL} = \mathrm{Tr}_L\left[(T^a T^b + T^b T^a)L\right] - \mathrm{Tr}_R\left[(T^a T^b + T^b T^a)L\right]$$
$$= \mathrm{Tr}_L\left[\frac{1}{4}(\tau^a\tau^b + \tau^b\tau^a)L\right] = \frac{1}{2}\delta_{ab}\mathrm{Tr}_L L = \delta_{ab},$$
(15.81)

and thus gives rise to another anomaly.

Let us now investigate potential anomalies in the $U(1)_B$ baryon number symmetry. In that case, only quarks propagate around the triangle diagrams. In analogy to the lepton case, one may convince oneself that the only non-vanishing anomalies are

$$A^{44B} = 2\left[\mathrm{Tr}_L Y^2 B - \mathrm{Tr}_R Y^2 B\right] = 2N_c\left[2Y_{q_L}^2 - Y_{u_R}^2 - Y_{d_R}^2\right]\frac{1}{N_c}$$
$$= 2\left[2\left(\frac{1}{2N_c}\right)^2 - \frac{1}{4}\left(\frac{1}{N_c}+1\right)^2 - \frac{1}{4}\left(\frac{1}{N_c}-1\right)^2\right] = -1,$$
$$A^{abB} = \mathrm{Tr}_L\left[(T^a T^b + T^b T^a)B\right] - \mathrm{Tr}_R\left[(T^a T^b + T^b T^a)B\right]$$
$$= \mathrm{Tr}_L\left[\frac{1}{4}(\tau^a\tau^b + \tau^b\tau^a)B\right] = \frac{1}{2}\delta_{ab}\mathrm{Tr}_L B = N_c\delta_{ab}\frac{1}{N_c} = \delta_{ab}.$$
(15.82)

Remarkably, for any number of colors, we end up with

$$A^{44B} = A^{44L}, \quad A^{abB} = A^{abL}, \quad a, b \in \{1, 2, 3\},$$
(15.83)

such that the anomalies cancel in the combination $B - L$. Hence, although B and L are individually broken at the quantum level, the difference between baryon and lepton number is an exactly conserved quantum number in the Standard Model.

Since $U(1)_{B-L}$ is an exact global symmetry of the Standard Model, the question arises why it is not gauged. Interestingly, the addition of a $U(1)_{B-L}$ gauge field to the minimal Standard Model alone would not allow this. Cancellation of the triangle anomaly with three external $U(1)_{B-L}$ bosons requires the additional introduction of an (otherwise sterile) right-handed neutrino field. This is a rather unusual situation in which a global symmetry, although it is exact, cannot be gauged in a straightforward manner. The reason is that the fermionic measure is invariant only against global, but not against local symmetry transformations. Once a right-handed neutrino field is introduced, the anomaly with three external $U(1)_{B-L}$ bosons is canceled and the fermionic measure becomes locally invariant.

Indeed, there are GUT extensions of the Standard Model with an $SO(10)$ gauge group which contains the $U(1)_{B-L}$ subgroup as a local symmetry, as we will see in Chapter 26. Alternatively, when $U(1)_{B-L}$ remains a global symmetry, Majorana mass terms involving the right-handed neutrino field ν_R explicitly break L invariance even at the Lagrangian level and thus turn $U(1)_{B-L}$ into an approximate symmetry. Similarly, if one views the Standard Model as a low-energy effective theory, $U(1)_{B-L}$ is an *accidental* global symmetry, which will be violated by non-renormalizable higher-order corrections to the Lagrangian.

Finally, we stress that – although both the numbers B and L are anomalously broken – processes that violate B or L are strongly suppressed at low temperature. However, B or L violating processes should become noticeable at very high temperatures in the $100\,\text{GeV}$ regime; see Section 26.5. Experimentally, the baryon and lepton number have always been observed to be separately conserved. In particular, as of 2022, proton decay has never been detected, despite extensive searches, which implies a proton life-time longer than 10^{34} years, cf. Appendix B.

15.12 Gauge Anomaly-Free Technicolor Model

This section addresses physics beyond the Standard Model, so it can be skipped at a first reading.

Why is the electroweak scale $v \simeq 246\,\text{GeV}$ so much smaller than the Planck scale $M_{\text{Planck}} \simeq 10^{19}\,\text{GeV}$? This is the hierarchy problem that we have discussed in Section 12.7. One approach to this problem is based on the concept of *"technicolor"* – a hypothetical gauge interaction even stronger than the strong force. It (supposedly) confines new fundamental fermions – the so-called *techni-quarks* – to form the Higgs particle as a tightly bound composite object.

This bears some analogy to the binding of electrons that form Cooper pairs in a superconductor. In that case, the condensation of Cooper pairs leads to the "spontaneous breaking" of $U(1)_{\text{em}}$. Similarly, in technicolor models the condensation of techni-quark–techni-anti-quark pairs leads to the "spontaneous breaking" of $SU(2)_L \times U(1)_Y$ down to $U(1)_{\text{em}}$. Thanks to the property of *asymptotic freedom*, which technicolor models share with QCD, one can explain the large hierarchy between v and M_{Planck} in a natural manner, *i.e.* without fine-tuning parameters. In fact, technicolor models mimic the dynamics of QCD at the electroweak scale. Since we will address the QCD dynamics only in Chapter 18, we will postpone the discussion of the technicolor dynamics. However, in this section we already introduce the basic ingredients of a minimal technicolor extension of the Standard Model, and we show that the extended model is still gauge anomaly-free. It should be mentioned that, at present, there is no experimental evidence supporting the concept of technicolor models. Instead, in these models there are severe problems due to flavor-changing neutral currents as well as due to fermion mass generation. Hence, altogether it seems that technicolor is not a very promising approach for going beyond the Standard Model.

Still, let us construct a concrete technicolor model, in particular, to show explicitly that such constructions are at all possible. In addition to the Standard Model fermions, we want to add a *techni-up* and a *techni-down quark*, U and D, whose left-handed components form an $SU(2)_L$ doublet and whose right-handed components are $SU(2)_L$ singlets. The *technicolor gauge group* is chosen to be $SU(N_t)$, and both the left- and the right-handed techni-quarks transform in the fundamental representation of $SU(N_t)$. All the Standard Model fermions are assumed to be technicolor singlets. We will not introduce any techni-leptons. Hence, anomaly cancellation works differently from the Standard Model. For simplicity, we choose the techni-quarks to be $SU(N_c)$ color singlets. Nevertheless, they are still confined by technicolor interactions. Let us denote the $U(1)_Y$ quantum numbers of the techni-quarks by Y_{Q_L}, Y_{U_R}, and Y_{D_R}. These parameters will be determined by anomaly cancellation conditions.

The gauge group of this technicolor model is given by $SU(N_t) \times SU(N_c) \times SU(2)_L \times U(1)_Y$. Let us now demand anomaly cancellation. Since, like $SU(N_c)$ color, the technicolor gauge group $SU(N_t)$ is a vector-like symmetry, the triangle diagram with three external techni-gluons automatically vanishes. Triangle diagrams with only external techni-gluons and external (QCD) gluons vanish as well. Triangle diagrams with only one external techni-gluon vanish because the generators of $SU(N_t)$ are traceless. The triangle diagram with two techni-gluons and one $SU(2)_L$ boson vanishes due to the tracelessness of τ^a. The triangle diagram with two external techni-gluons and one external $U(1)_Y$ boson vanishes only if

$$2Y_{Q_L} = Y_{U_R} + Y_{D_R}. \tag{15.84}$$

The techni-quarks also contribute to the anomalies of the Standard Model gauge symmetries. For example, the triangle diagram with two $SU(2)_L$ bosons and one $U(1)_Y$ boson still vanishes only if

$$Y_{Q_L} = 0, \tag{15.85}$$

while the diagram with three external $U(1)_Y$ bosons vanishes only if

$$2Y_{Q_L}^3 = Y_{U_R}^3 + Y_{D_R}^3. \tag{15.86}$$

Anomaly cancellation hence implies

$$Y_{Q_L} = 0, \quad Y_{U_R} + Y_{D_R} = 0. \tag{15.87}$$

We still want to be able to couple the new theory to gravity, which is possible only if we cancel gravitational as well as mixed gauge-gravitational anomalies. This again requires eq. (15.84), which is hence already satisfied.

In order to reproduce the physics of the Standard Model, we must maintain $U(1)_{\text{em}}$ as an unbroken gauge symmetry. This requires the electric charges of the left- and right-handed techni-quarks to be equal.[13] Based on $Q = T_L^3 + Y$, we obtain

$$Q_{U_L} = \frac{1}{2}, \quad Q_{D_L} = -\frac{1}{2}, \quad Q_{U_R} = Y_{U_R}, \quad Q_{D_R} = Y_{D_R}. \tag{15.88}$$

Hence, in order to have equal charges for left- and right-handed techni-quarks, we request

$$Y_{U_R} = -Y_{D_R} = \frac{1}{2}. \tag{15.89}$$

As we will discuss in Chapter 24, this implies that the neutral techni-pion would not decay anomalously into two photons.

In order to also cancel Witten's global gauge anomaly, the total number of $SU(2)_L$ doublets must be even; hence, N_t must be *even*.

The simplest choice $N_t = 2$ is not analogous to QCD. Due to the pseudo-real nature of $SU(2)$,[14] techni-quarks and techni-anti-quarks would then be indistinguishable and the chiral symmetry would be $Sp(4)$ instead of $SU(2)_L \times SU(2)_R$. Therefore, the simplest *valid* choice is $N_t = 4$, such that the gauge symmetry of the Standard Model, extended by this simple version of technicolor, takes the form $SU(4)_t \times SU(3)_c \times SU(2)_L \times U(1)_Y$.

Exercises

15.1 Electric charges of W- and Z-bosons

Examine the couplings of photons with W- and Z-bosons resulting from the gauge–Higgs sector of the Standard Model Lagrangian, and verify that the W-bosons have electric charges ± 1 while the Z-boson is electrically neutral.

15.2 Physics of an $SU(2)_L \times SU(2)_R$ -symmetric model

In the Standard Model, only the $SU(2)_L \times U(1)_Y$ subgroup of the global $SU(2)_L \times SU(2)_R$ symmetry of the Higgs sector is gauged. Now consider again a variant of the Standard Model in which both $SU(2)_L$ and $SU(2)_R$ are gauged (gauge couplings g and g'). Consider one generation of Standard Model fermions including a right-handed neutrino, and assume the $SU(2)_L$ and $SU(3)_c$ transformation properties to be the usual ones. Further assume that under $SU(2)_R$ the left-handed fields are singlets, while the right-handed fields form one lepton and one quark $SU(2)_R$ doublet. Check whether this model is free of gauge anomalies.

15.3 Anomaly cancellation in a Standard Model with two generations

Consider a Standard Model with two generations. The members of the second generation have the same quantum numbers as those of the first generation. Instead of

---

[13] If the left- and right-handed techni-quarks have the same electric charges, the breaking of the techni-chiral $SU(2)_L \times SU(2)_R$ symmetry leaves $U(1)_{\text{em}}$ intact. Otherwise, the techni-chiral condensate $\langle \bar{U}_L U_R + \bar{U}_L U_R + \bar{D}_L D_R + \bar{D}_L D_R \rangle$ would carry an electric charge and would turn the vacuum into a superconductor.

[14] This means that its representations are not necessarily real, but still unitarily equivalent to their complex conjugates.

canceling the global anomaly within a single generation, explore the alternative option to cancel it between the two generations. The cancellation of triangle anomalies works as before. What consequences does this have for the allowed values for the number of colors N_c? What are the corresponding electric charges of the quarks? A baryon is (naively speaking) formed of N_c quarks. What are the electric charges of the baryons and what are their spin and statistics?

15.4 Hypothetical world with 1 + 1/2 generations

Consider a hypothetical world described by a non-standard model with gauge group $SU(N_c) \times SU(2)_L \times U(1)_Y$ where the number of colors N_c is not necessarily equal to 3. The leptons in the hypothetical world are just the leptons of our world's first and second generation, with the usual charge assignments. In the hypothetical world, there are just u- and d-quarks. As usual, the left-handed u- and d-quarks form an $SU(2)_L$ doublet with weak hypercharge Y_L, while the right-handed quarks are $SU(2)_L$ singlets with weak hypercharges Y_{u_R} and Y_{d_R}. Note that the values of Y_L, Y_{u_R}, and Y_{d_R} may differ from the ones in our world. Impose the cancellation of all gauge anomalies in this non-standard model in order to determine Y_L, Y_{u_R}, and Y_{d_R}, as well as the values of the quark's electric charges Q_u and Q_d for general N_c.

15.5 Quarks in unusual representations

Consider another hypothetical world with the usual $SU(3)_c \times SU(2)_L \times U(1)_Y$ gauge group and with one generation of leptons with the same charge assignments as in the Standard Model. Now take the left-handed quarks of all three generations (u_L, d_L, c_L, s_L, t_L, b_L) and put them in a sextet of $SU(2)_L$. Still, each quark comes in three colors so there are six left-handed $SU(3)_c$ triplets. As in the Standard Model, the right-handed quarks are $SU(2)_L$ singlets and $SU(3)_c$ triplets. However, the weak hypercharges of the quarks may be different from the Standard Model assignments. Impose gauge anomaly cancellation and request that left- and right-handed quarks carry the same electric charge. Use the fact that an $SU(2)_L$ sextet contributes to the global $SU(2)_L$ anomaly in the same way as an $SU(2)_L$ doublet. What are the electric charges of the baryons in this theory?

15.6 Hypothetical world with two colors

Consider a world with an $SU(2)_c \times SU(2)_L \times U(1)_Y$ gauge group and with one generation of leptons and quarks. The leptons have the same charge assignments as in the Standard Model (except that they are now $SU(2)_c$ instead of $SU(3)_c$ singlets). The u- and d-quarks are now put in $SU(2)_c$ triplets, and, as usual, they form $SU(2)_L$ doublets. The weak hypercharges of the quarks may be different from the Standard Model assignments. Again, impose anomaly cancellation and request that left- and right-handed quarks carry the same electric charge. Are there states with fractional electric charge in the spectrum of this theory?

16

Fermion Masses

At this point, we have introduced all fields of the Standard Model with one generation of fermions. Lorentz invariance and locality, combined with gauge invariance, which requires gauge anomaly cancellation, have led to stringent restrictions on the terms that are allowed to enter the Lagrangian. Altogether, until now, we have introduced *five* adjustable fundamental parameters: only one dimensioned parameter – the vacuum value v of the Higgs field – as well as the dimensionless Higgs self-coupling λ and the three dimensionless gauge couplings g, g' (or alternatively e and θ_W), and g_s (the latter gives rise to the mass scale Λ_{QCD} by means of dimensional transmutation, cf. Chapters 14 and 18). In addition, we have made appropriate choices for the fermion representations. For example, in the way we have presented the Standard Model, one may consider the number of colors N_c as another (odd integer-valued) parameter to be determined by experiment. In any case, the number of parameters is still modest at this stage.

Usually, it is emphasized that the Standard Model describes the electromagnetic, weak, and strong interaction, and that there are four fundamental forces, including gravity. However, the self-interaction of the Higgs field can be viewed as a fundamental force as well. In this chapter, we will see that, in addition, the Standard Model contains *Yukawa interactions* between the Higgs field and the fermions. When the Higgs field picks up its vacuum value v, the Yukawa interactions lead to fermion masses as well as to mixing between different fermion generations. With one generation of fermions, we will now have the Dirac masses of the up and down quark, and of the electron, which are proportional to corresponding Yukawa couplings whose values are adjusted to the experimentally observed fermion masses. Although they do not involve gauge fields, the Yukawa interactions are also constrained by gauge invariance. When one limits oneself to renormalizable interactions, the neutrino remains massless. However, if we view the Standard Model as a low-energy effective theory of some yet unknown, more fundamental structure, it is natural to include higher-order non-renormalizable corrections. As we will see, the leading correction, which is associated with a dimension-5 operator, endows the neutrino with mass.

In the next chapter, we will add two more fermion generations which will significantly increase the number of parameters. While it is possible to determine the values of the Standard Model parameters by comparison with experiment, their theoretical understanding is a very challenging, unsolved problem beyond the Standard Model.

One may hope for the third run of the Large Hadron Collider (LHC) at CERN – or some future accelerator, which will attain even higher energy – to shed light on the origin of the Higgs phenomenon and thus of the dynamical mechanism responsible for electroweak symmetry breaking. It is possible that an extended version of the Standard Model will replace the Higgs field and the parameters associated with its Yukawa interactions by a more fundamental and more constrained dynamics, perhaps driven by yet unknown gauge forces. On the other hand, different ideas beyond the Standard Model, *e.g.*, those relying on supersymmetry – a symmetry that pairs fermions with bosons – could increase the number of adjustable parameters even further.

16.1 Electron and Down Quark Masses

We have not yet introduced any mass terms for the fermions. An electron mass term in the Lagrangian would have the form

$$\mathcal{L}(\bar{e}_\mathrm{L}, e_\mathrm{L}, \bar{e}_\mathrm{R}, e_\mathrm{R}) = m_e \left[\bar{e}_\mathrm{R} e_\mathrm{L} + \bar{e}_\mathrm{L} e_\mathrm{R} \right]. \tag{16.1}$$

As we mentioned before, since left- and right-handed fermions transform differently under $SU(2)_L$ and $U(1)_Y$ gauge transformations, this term violates the chiral gauge symmetry and is thus *forbidden*. To be explicit, in the term (16.1) $SU(2)_L$ solely transforms e_L, while $U(1)_Y$ transforms both e_L and e_R, but not in the same manner, because their hypercharges are different.

We remember that we encountered a similar situation before for the weak gauge bosons: We know experimentally that they are massive, but explicit mass terms for them are forbidden by gauge invariance. The way out was the *Higgs mechanism*: by picking up a vacuum value, the Higgs field Φ gives mass to the gauge bosons via "spontaneous gauge symmetry breaking". Similarly, Φ can also give mass to fermions. Let us write down a Yukawa interaction term[1] with the *Yukawa coupling* f_e

$$\mathcal{L}(\bar{l}_\mathrm{L}, l_\mathrm{L}, \bar{e}_\mathrm{R}, e_\mathrm{R}, \Phi) = f_e\, \bar{l}_\mathrm{L} \Phi e_\mathrm{R} + f_e^*\, \bar{e}_\mathrm{R} \Phi^\dagger l_\mathrm{L}$$

$$= f_e \left(\bar{\nu}_\mathrm{L}, \bar{e}_\mathrm{L} \right) \begin{pmatrix} \Phi^+ \\ \Phi^0 \end{pmatrix} e_\mathrm{R} + f_e^*\, \bar{e}_\mathrm{R} \left(\Phi^{+*}, \Phi^{0*} \right) \begin{pmatrix} \nu_\mathrm{L} \\ e_\mathrm{L} \end{pmatrix}. \tag{16.2}$$

The second term is multiplied by f_e^*, in order to ensure Hermiticity of the corresponding Hamilton operator.[2] The above Lagrangian is $SU(2)_L$ gauge-invariant because the left-handed leptons and the Higgs field are both $SU(2)_L$ doublets, while the right-handed electron is an $SU(2)_L$ singlet. Moreover, the Lagrangian is also $U(1)_Y$-invariant. To see this, we sum up the hypercharges of the fields in the first term in eq. (16.2),

$$-Y_{l_\mathrm{L}} + Y_\Phi + Y_{e_\mathrm{R}} = \frac{1}{2} + \frac{1}{2} - 1 = 0. \tag{16.3}$$

Since the hypercharges add up to zero, the corresponding $U(1)_Y$ gauge transformation factors $\exp(\mathrm{i} Y g' \varphi(x))$ cancel each other, and the term is thus $U(1)_Y$ gauge-invariant. (In

---

[1] Generally, a Yukawa interaction term has the structure $\bar{\psi}(x)\phi(x)\psi(x)$, where ϕ is a scalar field and $\bar{\psi}$ and ψ are fermion fields.

[2] Note that choosing a Yukawa coupling like f_e to be complex does not yield any problem for the convergence of the functional integral. This is in contrast to the scalar self-coupling λ, which must be real-positive, $\lambda \in \mathbb{R}_0^+$. The integrals $\int \mathcal{D}\bar{l}_\mathrm{L} \mathcal{D}l_\mathrm{L} \mathcal{D}\bar{e}_\mathrm{R} \mathcal{D}e_\mathrm{R}$, however, converge in any case because they are Grassmannian

the second term, the signs of all hypercharges are flipped, and hence its total hypercharge vanishes as well.)

Since charge conjugation, as well as parity, turns left- into right-handed neutrinos, and since there are no right-handed neutrino fields in the Standard Model, it is clear that the Lagrangian of eq. (16.2) explicitly breaks P and C invariance. This breaking was predicted by Lee and Yang (1956b) and first observed in the β-decay by Wu *et al.* (1957). Let us now perform a combined CP transformation in the corresponding action,

$$
\begin{aligned}
S[^{CP}\bar{l}_L, {}^{CP}l_L, {}^{CP}\bar{e}_R, {}^{CP}e_R, {}^{CP}\Phi] \\
= \int d^4x \left[-f_e \, l_L(-\vec{x}, x_4)^\top P^\top C^{-1} \Phi(-\vec{x}, x_4)^* CP\bar{e}_R(-\vec{x}, x_4)^\top \right. \\
\left. -f_e^* \, e_R(-\vec{x}, x_4)^\top P^\top C^{-1} \Phi(-\vec{x}, x_4)^\top CP\bar{l}_L(-\vec{x}, x_4)^\top \right] \\
= \int d^4x \left[f_e \, \bar{e}_R(-\vec{x}, x_4) \Phi(-\vec{x}, x_4)^\dagger l_L(-\vec{x}, x_4) + f_e^* \, \bar{l}_L(-\vec{x}, x_4) \Phi(-\vec{x}, x_4) e_R(-\vec{x}, x_4) \right] \\
= \int d^4x \left[f_e \, \bar{e}_R(\vec{x}, x_4) \Phi(\vec{x}, x_4)^\dagger l_L(\vec{x}, x_4) + f_e^* \, \bar{l}_L(\vec{x}, x_4) \Phi(\vec{x}, x_4) e_R(\vec{x}, x_4) \right]. \quad (16.4)
\end{aligned}
$$

Hence, it seems that the action is CP-invariant only if the Yukawa coupling f_e is real. However, as we will now discuss, f_e can always be made real by a field redefinition. Let us write $f_e = |f_e| \exp(i\theta)$. Next we redefine

$$
e'_R(x) = e_R(x) \exp(i\theta), \quad \bar{e}'_R(x) = \bar{e}_R(x) \exp(-i\theta), \quad (16.5)
$$

which absorbs the complex phase $\exp(i\theta)$ into the right-handed electron field. Expressed in terms of the redefined fields, the Lagrangian then contains the real-positive Yukawa coupling $|f_e|$. It is important that the field redefinition leaves the gauge–fermion terms of the Lagrangian invariant. On the other hand, as we will discuss in Section 18.6, such field redefinitions may have subtle effects on the fermionic integration measure; for an extensive discussion, we refer to Fujikawa and Suzuki (2004). In any case, from now on we may assume f_e to be real-valued and positive, $f_e > 0$.

Inserting again the vacuum configuration of the Higgs field that we selected before, we obtain

$$
\mathcal{L}(\bar{l}_L, l_L, \bar{e}_R, e_R, v) = f_e \left[(\bar{v}_L, \bar{e}_L) \begin{pmatrix} 0 \\ v \end{pmatrix} e_R + \bar{e}_R (0, v) \begin{pmatrix} v_L \\ e_L \end{pmatrix} \right] = f_e v \left[\bar{e}_L e_R + \bar{e}_R e_L \right]. \quad (16.6)
$$

Indeed, we have arrived at a mass term for the electron with the mass

$$
m_e = f_e v, \quad (16.7)
$$

while the neutrino remains massless. Via the Yukawa coupling f_e, we have just introduced another free parameter into the theory which determines the value of the electron mass. The Standard Model itself does not make any prediction about this parameter. If we want to understand the value of the electron mass, we need to go beyond the Standard Model. At present, nobody understands why the electron has its particular mass of $m_e \simeq 0.511$ MeV. As we continue to add mass terms, the number of adjustable parameters in the Standard Model increases accordingly.

We see that the Standard Model contains more than just electroweak and strong interactions. Every Yukawa coupling parameterizes a *fundamental force* that is not often emphasized at the same level as the gauge forces. There is reason to believe that the Yukawa couplings are not as fundamental as the gauge interactions. For example, in a future theory beyond the Standard Model, the Yukawa couplings might ultimately be replaced by some gauge force of a new kind. In this way, we would perhaps gain predictive power and finally understand the value of the electron mass. This underscores that the true origin of mass may not yet be sufficiently well understood. The celebrated Higgs mechanism leaves this fundamental question unanswered.

Since the d-quark appears in the same position of an $SU(2)_L$ doublet as the electron, and since

$$-Y_{q_L} + Y_\Phi + Y_{d_R} = -\frac{1}{2N_c} + \frac{1}{2} + \frac{1}{2}\left(\frac{1}{N_c} - 1\right) = 0, \qquad (16.8)$$

we can give the d-quark a mass $m_d = f_d v$ by adding a further term

$$\mathcal{L}(\bar{q}_L, q_L, \bar{d}_R, d_R, \Phi) = f_d \left[\bar{q}_L \Phi d_R + \bar{d}_R \Phi^\dagger q_L\right]$$

$$= f_d \left[(\bar{u}_L, \bar{d}_L)\begin{pmatrix}\Phi^+ \\ \Phi^0\end{pmatrix} d_R + \bar{d}_R (\Phi^{+*}, \Phi^{0*})\begin{pmatrix}u_L \\ d_L\end{pmatrix}\right] \qquad (16.9)$$

to the Standard Model Lagrangian. On the other hand, we cannot give mass to the u-quark in the same way, just as we did not obtain a neutrino mass.[3]

16.2 Up Quark Mass

We could easily construct a mass term for the u-quark if we had an additional Higgs field at hand; let us call it

$$\widetilde{\Phi}(x) = \begin{pmatrix}\widetilde{\Phi}^0(x) \\ \widetilde{\Phi}^-(x)\end{pmatrix}, \qquad (16.10)$$

which would be an $SU(2)_L$ doublet that takes a vacuum value in its *upper* component

$$\widetilde{\Phi}(x) = \begin{pmatrix}\widetilde{v} \\ 0\end{pmatrix}. \qquad (16.11)$$

---

[3] For some time, it was not clear whether the u-quark could be massless (Leutwyler, 1990), which is by now excluded (Zyla *et al.*, 2020). However, even if we were ready to accept $m_u = 0$, we still had to find a way to give mass to an upper $SU(2)_L$ doublet partner. In the next chapter, we will add two generations of heavier fermions, and the charm and top quarks – which take the position of the up quark in the second and third generation – clearly have large masses.

Then, we could just add another Yukawa term

$$
\begin{aligned}
\mathcal{L}(\bar{q}_{\mathrm{L}}, q_{\mathrm{L}}, \bar{u}_{\mathrm{R}}, u_{\mathrm{R}}, \widetilde{\Phi}) &= f_u \left[\bar{q}_{\mathrm{L}} \widetilde{\Phi} u_{\mathrm{R}} + \bar{u}_{\mathrm{R}} \widetilde{\Phi}^\dagger q_{\mathrm{L}} \right] \\
&= f_u \left[(\bar{u}_{\mathrm{L}}, \bar{d}_{\mathrm{L}}) \begin{pmatrix} \widetilde{\Phi}^0 \\ \widetilde{\Phi}^- \end{pmatrix} u_{\mathrm{R}} + \bar{u}_{\mathrm{R}} (\widetilde{\Phi}^{0*}, \widetilde{\Phi}^{-*}) \begin{pmatrix} u_{\mathrm{L}} \\ d_{\mathrm{L}} \end{pmatrix} \right].
\end{aligned} \quad (16.12)
$$

To render this term gauge-invariant, the weak hypercharge of the field $\widetilde{\Phi}$ must obey

$$
-Y_{q_{\mathrm{L}}} + Y_{\widetilde{\Phi}} + Y_{u_{\mathrm{R}}} = \frac{1}{2N_{\mathrm{c}}} + Y_{\widetilde{\Phi}} + \frac{1}{2}\left(\frac{1}{N_{\mathrm{c}}} + 1\right) = 0 \quad \Rightarrow \quad Y_{\widetilde{\Phi}} = -\frac{1}{2}. \quad (16.13)
$$

At this point, we could just add a new Higgs field $\widetilde{\Phi}$ with the desired features. In fact, as we have discussed in Section 12.12, this is exactly what Roberto Peccei and Helen Quinn have proposed in order to solve the strong CP-problem (Peccei and Quinn, 1977a,b), which will be addressed in Chapter 25.

However, here we focus on the Standard Model, which does not proceed in this manner: It follows a more economic procedure by "recycling" the Higgs field introduced previously. It may come as a surprise that a field $\widetilde{\Phi}$ with the desired properties can be constructed directly from the known Higgs field Φ as

$$
\widetilde{\Phi}(x) = \begin{pmatrix} \widetilde{\Phi}^0(x) \\ \widetilde{\Phi}^-(x) \end{pmatrix} = \begin{pmatrix} \Phi^0(x)^* \\ -\Phi^+(x)^* \end{pmatrix} = i\tau^2 \Phi(x)^*. \quad (16.14)
$$

While it is clear that this field indeed has $Y_{\widetilde{\Phi}} = -1/2$, it is less obvious that it also transforms as an $SU(2)_L$ doublet. To see this, it is useful to return to the matrix form

$$
\mathbf{\Phi}(x) = \begin{pmatrix} \Phi^0(x)^* & \Phi^+(x) \\ -\Phi^+(x)^* & \Phi^0(x) \end{pmatrix}. \quad (16.15)
$$

As we have seen in Section 13.2, under $SU(2)_L$ gauge transformations $L(x)$ the matrix field transforms as $\mathbf{\Phi}'(x) = L(x)\mathbf{\Phi}(x)$. Since the field $\widetilde{\Phi}$ is nothing but the first column vector of the matrix $\mathbf{\Phi}$, it is clear that it transforms indeed as an $SU(2)_L$ doublet.

Using the matrix field $\mathbf{\Phi}$, the quark Yukawa coupling terms can be written as

$$
\mathcal{L}(\bar{q}_{\mathrm{L}}, q_{\mathrm{L}}, \bar{u}_{\mathrm{R}}, u_{\mathrm{R}}, \bar{d}_{\mathrm{R}}, d_{\mathrm{R}}, \mathbf{\Phi}) = (\bar{u}_{\mathrm{L}}, \bar{d}_{\mathrm{L}}) \mathbf{\Phi} \mathcal{F} \begin{pmatrix} u_{\mathrm{R}} \\ d_{\mathrm{R}} \end{pmatrix} + (\bar{u}_{\mathrm{R}}, \bar{d}_{\mathrm{R}}) \mathcal{F}^\dagger \mathbf{\Phi}^\dagger \begin{pmatrix} u_{\mathrm{L}} \\ d_{\mathrm{L}} \end{pmatrix}, \quad (16.16)
$$

where the Yukawa couplings are contained in the diagonal matrix

$$
\mathcal{F} = \begin{pmatrix} f_u & 0 \\ 0 & f_d \end{pmatrix} = \mathcal{F}^\dagger. \quad (16.17)
$$

The above construction implies $\widetilde{v} = v^* = v$; hence, the u-quark mass is given by $m_u = f_u v$. Inserting the vacuum value of the Higgs field, the quark mass matrix then results as

$$
\mathcal{M} = \mathbf{\Phi} \mathcal{F} = \begin{pmatrix} v & 0 \\ 0 & v \end{pmatrix} \begin{pmatrix} f_u & 0 \\ 0 & f_d \end{pmatrix} = \begin{pmatrix} m_u & 0 \\ 0 & m_d \end{pmatrix}. \quad (16.18)
$$

16.3 Neutrino Mass from a Dimension-5 Operator

Oscillations between different pairs of neutrino flavors were observed by the Super Kamio-kande Experiment in Japan (Fukuda *et al.*, 1998) and by the Sudbury Neutrino Observatory in Canada (Ahmad *et al.*, 2001). This implies that at least two of the three neutrinos must be massive. Since the Standard Model does not contain right-handed neutrino fields, one cannot even write down a neutrino mass term, at least as long as one restricts oneself to renormalizable interactions. As we have discussed in Section 12.10, already due to its triviality, the Standard Model is actually a low-energy effective theory, which cannot be valid up to arbitrarily high energy scales.

If one accepts that the Standard Model is a low-energy effective theory, there is no good reason to exclude non-renormalizable interactions. Instead, those should be added as higher-order corrections to the leading renormalizable Standard Model interactions. The leading term of this type is a dimension-5 operator that gives rise to the Lagrangian contribution (Weinberg, 1979a), where we introduce ${}^c l_{\rm L} = {\rm i} \sigma^2 \bar{l}_{\rm L}^{\rm T}$, ${}^c \bar{l}_{\rm L} = -l_{\rm L}^{\rm T} {\rm i} \sigma^2$

$$\mathcal{L}(\bar{l}_{\rm L}, l_{\rm L}, \tilde{\Phi}) = \frac{1}{2\Lambda} \left[\left({}^c \bar{l}_{\rm L} \tilde{\Phi}^* \right) \left(\tilde{\Phi}^\dagger l_{\rm L} \right) + \left(\bar{l}_{\rm L} \tilde{\Phi} \right) \left(\tilde{\Phi}^{\rm T} {}^c l_{\rm L} \right) \right]. \tag{16.19}$$

In contrast to C, c operates in the absence of a field $l_{\rm R}$ of opposite chirality. Since a fermion field has canonical mass dimension $3/2$, while a scalar field has dimension 1, the terms in square brackets have dimension 5. The prefactor $1/\Lambda$ hence has mass dimension -1, and the corresponding interaction is therefore *non-renormalizable*. This is no problem for an effective theory that will break down at some high energy scale Λ. The higher-dimensional, non-renormalizable terms just parameterize small effects that new high-energy degrees of freedom – residing at the scale Λ – have on the low-energy physics. The field products $\bar{l}_{\rm L} \tilde{\Phi}$, $\tilde{\Phi}^{\rm T} {}^c l_{\rm L}$, ${}^c \bar{l}_{\rm L} \tilde{\Phi}^*$, and $\tilde{\Phi}^\dagger l_{\rm L}$ all are SU(2)$_L$ singlets. In addition, they are U(1)$_Y$ gauge-invariant because the weak hypercharges of the left-handed leptons and the Higgs field cancel each other. When the Higgs field assumes its vacuum value v, the dimension-5 operator takes the form of a (left-handed) *Majorana mass term*

$$\mathcal{L}(\bar{l}_{\rm L}, l_{\rm L}, {\rm v}) = \frac{{\rm v}^2}{2\Lambda} \left[{}^c \bar{\nu}_{\rm L} \nu_{\rm L} + \bar{\nu}_{\rm L} {}^c \nu_{\rm L} \right]. \tag{16.20}$$

We see that the resulting *neutrino mass*

$$m_\nu = \frac{{\rm v}^2}{\Lambda} \tag{16.21}$$

is suppressed by the high energy scale Λ.

It is interesting to convince oneself that, besides the terms of eq. (16.19), there are no other gauge- and Lorentz-invariant dimension-5 terms that can be formed from the Standard Model fields. For example, while the field products $\bar{l}_{\rm L} \Phi$, $\Phi^{\rm T} {}^c l_{\rm L}$, ${}^c \bar{l}_{\rm L} \Phi^*$, and $\Phi^\dagger l_{\rm L}$ are also SU(2)$_L$ singlets, they are not U(1)$_Y$-invariant, but transform with a total $Y = 1, 1, -1,$

and -1, respectively. As a result, these terms cannot be combined in a gauge- and Lorentz-invariant manner.

One may also ask whether one can form dimension-5 operators with a pair of fermion fields combined with gauge fields. Since each fermion field has dimension $3/2$, the gauge field contribution should then have dimension 2, such that $2 \cdot \frac{3}{2} + 2 = 5$. This is the case, for example, for the Lorentz scalars $\sigma_{\mu\nu} G_{\mu\nu}$, $\sigma_{\mu\nu} W_{\mu\nu}$, and $\sigma_{\mu\nu} B_{\mu\nu}$, where $\sigma_{\mu\nu} = [\gamma_\mu, \gamma_\nu]/(2\mathrm{i})$. These terms transform in the adjoint representations of $SU(3)_c$ or $SU(2)_L$, and are gauge-invariant under $U(1)_Y$, respectively. In order to obey Lorentz invariance, these terms could be combined with fermion fields $\bar{\psi}_R \psi_L$ or $\bar{\psi}_L \psi_R$, but none of these combinatons is gauge-invariant. Similarly, one can convince oneself that there are no purely bosonic dimension-5 operators either.

16.4 Mass Hierarchies of Fermions

The masses of the u- and d-quarks and of the electron are proportional to the vacuum value of the Higgs field

$$m_u = f_u \mathrm{v} \approx 2\,\mathrm{MeV}, \quad m_d = f_d \mathrm{v} \approx 5\,\mathrm{MeV}, \quad m_e = f_e \mathrm{v} \approx 0.5\,\mathrm{MeV}. \tag{16.22}$$

Since quarks are confined, their masses are not directly measurable. In fact, the masses are *running* parameters that depend on the energy scale. The approximate values quoted above correspond to an energy scale in the GeV range.[4] The masses of the electron and of the u- and d-quark are about a factor of 10^5 smaller than the electroweak symmetry breaking scale $\mathrm{v} \simeq 246\,\mathrm{GeV}$. This means that the corresponding dimensionless Yukawa couplings f_u, f_d, and f_e are around 10^{-5}. The masses of the quarks and the charged leptons of the other two generations are larger; in particular, the top quark, the heaviest known elementary particle, has a mass of $m_t = f_t \mathrm{v} \approx 173\,\mathrm{GeV}$, such that its Yukawa coupling f_t is of order 1. The Standard Model offers no clue to why the quark masses take their particular values. Although they are all tied to the electroweak symmetry breaking scale v, the quark masses are vastly different because the corresponding Yukawa couplings vary over 5 orders of magnitude. Are dimensionless numbers as small as 10^{-5} "unnatural"? In any case, Nature has "chosen" these numbers, which appear as freely adjustable parameters in the theoretical framework of the Standard Model. A yet unknown, more fundamental theory would be required to explain the hierarchy of fermion masses.

In contrast to the masses of the other fermions, the neutrino mass (16.21) is not proportional to v, but we saw that it amounts to v^2/Λ. In other words, neutrinos are "naturally" a factor of v/Λ lighter than the top quark. While the values of the neutrino masses themselves are still experimentally undetermined, the current upper bound on neutrino masses is in the eV range; in particular, the KATRIN Collaboration measured $m_\nu < 0.8\,\mathrm{eV}$ (with 90 percent confidence level) (Aker *et al.*, 2019), by high-precision monitoring of the electron energy spectrum in β-decays. This suggests $\mathrm{v}/\Lambda \lesssim 100\,\mathrm{GeV}/1\mathrm{eV} = 10^{11}$, which hints

---

[4] More precise values of the quark and lepton masses are listed in Appendix B, Zyla *et al.* (2020), or the latest edition of the Particle Data Book.

at $\Lambda \gtrsim 10^{13}$ GeV as a rough order-of-magnitude estimate of the scale where new physics beyond the Standard Model, which is responsible for neutrino masses, is residing.

In Section 16.9, we will discuss the so-called *seesaw mechanism,* which generates light neutrino masses by mixing with a heavy sterile Majorana neutrino, whose mass is of order Λ. This is a concrete extension of the Standard Model which indeed gives rise to the dimension-5 operator (16.19). However, we do not even need concrete assumptions about the physics beyond the Standard Model in order to arrive "naturally" at very light neutrinos. When treated as an effective theory, via the dimension-5 operator, the Standard Model itself predicts that neutrino masses are smaller than the other fermion masses, by a factor v/Λ. From this point of view, the fact that the renormalizable Standard Model predicts massless neutrinos is not a deficit at all. Instead, it should be viewed as an attractive feature, because it provides a very good leading-order result. Once the values of the neutrino masses will be determined experimentally, by means of its dimension-5 operator the Standard Model even kindly informs us about a scale Λ where it will necessarily break down.

16.5 Neutrino Mass Term and Reconsideration of CP

As we have seen in Section 8.7, a Majorana mass term, while being C-invariant, explicitly breaks the standard parity transformation P, and it thus also breaks CP. At the same time, it breaks the fermion number symmetry $U(1)_F$ down to $\mathbb{Z}(2)_F$. Still, the Majorana mass term remains invariant against the modified parity symmetry P' of eq. (8.78), which combines P with a fermion number transformation $\pm i$. As a result, in contrast to a Dirac mass term, which is invariant separately against CP and $U(1)_F$, a Majorana mass term is invariant only against the modified transformation CP' and against $\mathbb{Z}(2)_F$.

In the renormalizable Standard Model, there are only Dirac mass terms. For a single generation of quarks and leptons, the theory is then invariant against CP (at least when the QCD vacuum-angle θ is set to zero, in agreement with experiment). While baryon and lepton number are broken by anomalies, the combination $B - L$ remains an exact global symmetry, $U(1)_{B-L}$, which plays the same role as the fermion number in Section 8.7. Once the non-renormalizable dimension-5 operator, which endows the neutrinos with Majorana mass, is added to the Standard Model, CP and $U(1)_{B-L}$ are both explicitly broken, but again the modified transformation CP' and $\mathbb{Z}(2)_{B-L}$ remain as exact global symmetries.

Since CP' is a combination of CP with the $U(1)_{B-L}$ transformations $\pm i$, which are exact symmetries of the renormalizable Standard Model, CP' is a symmetry of both the renormalizable terms and the non-renormalizable dimension-5 term in the Standard Model Lagrangian. Indeed, in the presence of the dimension-5 operator, CP' takes over the role of CP. It is then useful to consider P' instead of P and hence CP' instead of CP. As a result, despite the fact that it breaks CP, the Majorana mass term that results from the dimension-5 operator remains *invariant against CP'*.

As we will see in Section 17.3, CP (and also CP') are explicitly broken once there are more than two generations of quarks and leptons. What was interpreted as CP violation in the renormalizable Standard Model with unbroken $U(1)_{B-L}$ symmetry can be reinterpreted as a physically equivalent CP' breaking when $U(1)_{B-L}$ is explicitly broken to $\mathbb{Z}(2)_{B-L}$. In other words, in the renormalizable Standard Model, the definition of CP is *ambiguous,* because it can be modified by $U(1)_{B-L}$ transformations. In the presence of the dimension-5 operator, this ambiguity is reduced because only CP' remains a symmetry. Some ambiguity remains, because CP' could still be combined with $\mathbb{Z}(2)_{B-L}$.

16.6 Lepton and Baryon Number Violation by Higher-Dimensional Operators

Within the renormalizable Standard Model, B and L are already explicitly broken by an anomaly, while $B - L$ remains as an *accidental* global symmetry, because no renormalizable gauge-invariant term breaks it. Non-renormalizable higher-dimensional operators, however, are in general able to explicitly break accidental global symmetries. In particular, the dimension-5 operator of eq. (16.19) explicitly breaks $U(1)_L$ down to $\mathbb{Z}(2)_L$. This also breaks $U(1)_{B-L}$ to $\mathbb{Z}(2)_{B-L}$ (which is equivalent to $\mathbb{Z}(2)_{B+L}$). As a result, only *fermion number modulo 2* remains a global symmetry of the Standard Model. This is a simple consequence of Lorentz invariance, which requires an even number of fermion fields (ψ or $\bar{\psi}$) to appear in the Lagrangian.

In the context of the Standard Model, Lorentz invariance is maintained as an exact global symmetry, because it is a gauge symmetry of General Relativity and should hence not be violated explicitly. Consequently, the dimension-5 operator finally breaks the last, truly global symmetry of the Standard Model. This confirms that all global symmetries are, in fact, accidental. They are approximate symmetries that arise in the presence of a hierarchy of different energy scales. When one expands the Lagrangian in inverse powers of a high scale Λ, terms that eventually break the accidental symmetry explicitly will arise at sufficiently high order.

The fact that the dimension-5 operator of eq. (16.19) changes the lepton number by two units has physical consequences that should, in principle, be measurable. First of all, there are intense experimental searches for a *neutrinoless double β-decay* (for a review, see Giuliani and Poves (2012)). There are some atomic nuclei for which ordinary single β-decay is energetically forbidden, while the simultaneous double β-decay of two neutrons is allowed, cf. Figure 16.1. Such a decay is usually associated with the emission of two anti-neutrinos, which ensures that lepton number is conserved. If, due to the dimension-5 operator, the lepton number can be violated by two units, the double β-decay should also be possible without any anti-neutrinos (Furry, 1939). If this decay exists, then it is extremely rare; as of 2022, there is no independently confirmed experimental evidence for it.

In addition, there are non-renormalizable dimension-6 operators that explicitly break baryon number conservation, which were first constructed by Steven Weinberg and independently by Frank Wilczek and Anthony Zee (Weinberg, 1979a; Wilczek and Zee, 1979). These are 4-fermion operators consisting of three quark fields and one lepton field, which each have dimension $3/2$ and $4 \cdot 3/2 = 6$. The three quark fields are coupled to an $SU(3)_c$ singlet which – together with the lepton field – couples to an $SU(2)_L$ singlet with vanishing weak hypercharge, such that the corresponding terms are $SU(3)_c \times SU(2)_L \times U(1)_Y$ gauge-invariant. It turns out that there are four independent operators of this kind within the first generation,

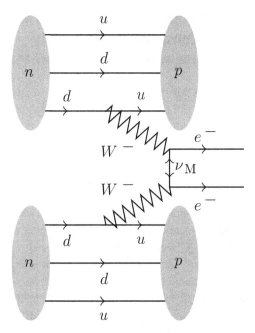

Fig. 16.1 Neutrinoless double-β-decay: Two neutrons in an atomic nucleus undergo a simultaneous β-decay into two protons. Ordinary β-decay of two neutrons would lead to the emission of two electrons e^- and two anti-neutrinos $\bar{\nu}_e$. In a neutrinoless double-β-decay, via a Majorana mass term, the two anti-neutrinos can annihilate each other, thus violating lepton number conservation by two units (indicated by the two oppositely oriented arrows on the neutrino propagator).

$$\mathcal{O}_1 = \epsilon_{abc}(d_R^a{}^c \bar{u}_R^b)(u_R^c{}^c \bar{e}_R), \qquad \mathcal{O}_2 = \epsilon_{abc}(d_R^a{}^c \bar{u}_R^b)(q_L^c i\tau^2{}^c \bar{l}_L),$$
$$\mathcal{O}_3 = \epsilon_{abc}(q_L^a i\tau^2{}^c \bar{q}_L^b)(u_R^c{}^c \bar{e}_R), \qquad \mathcal{O}_4 = \epsilon_{abc}(q_L^a i\tau^2{}^c \bar{q}_L^b)(q_L^c i\tau^2{}^c \bar{l}_L). \qquad (16.23)$$

If all these dimension-6 operators would originate from the same new physics as the dimension-5 operator, they would be suppressed by a factor $1/\Lambda^2$, thus leading to a tiny proton decay rate. The operators of eq. (16.23) are appropriate only for $N_c = 3$. In general, the baryon number violating operators involve N_c quark fields and one lepton field; thus, they have dimension $3(N_c + 1)/2$. This would make them even more suppressed for $N_c > 3$. On the other hand, as we will discuss in Section 16.11, for $N_c = 1$ there are even renormalizable baryon number violating operators of dimension 3, which would lead to unsuppressed baryon decay. Then, due to a lack of strong interactions, there is no physical distinction between baryons and leptons, and B and L would cease to be useful concepts.

Interestingly, while all four operators of eq. (16.23) break B and L invariance individually, they still conserve the difference $B - L$. This implies that no additional $B - L$ breaking (besides the one that results from the dimension-5 operator) arises at this higher-order of the effective field theory expansion. Even with all higher-order terms included, the quantity $B - L$ modulo 2 remains an exact symmetry that is protected by Lorentz invariance.

16.7 Charge Quantization, Fermion Masses, and Consistency with Gravity

As recognized in experiments which were significantly advanced by Robert Millikan and Harvey Fletcher (Millikan, 1913), electric charge is quantized in integer multiples of a basic unit e. In particular, the negative electric charge $-e$ of an electron exactly compensates the positive charge e of a proton, such that, *e.g.*, a hydrogen atom is electrically neutral. Can we understand why electric charge is quantized?

In quantum field theory, all particles of a given type are indistinguishable quantized excitations of the same field. Therefore, it is not mysterious that all electrons carry the same charge $-e$. As a consequence of gauge and Lorentz invariance, which implies CPT symmetry, all positrons carry the opposite charge e. However, a theory of just the electromagnetic and strong interactions, *i.e.* QCD combined with QED, with electrons and positrons of charge $\pm e$, and of protons and anti-protons of a different charge $\pm e' \neq \pm e$, would be mathematically consistent, although inconsistent with experiments.

Abelian gauge charges are, *a priori,* parameters which could be varied continuously. The quantization of non-Abelian charges, on the other hand, is a consequence of group theory, because the irreducible representations of Lie algebras are discrete objects. As a consequence, when an Abelian U(1) gauge symmetry is embedded in a non-Abelian gauge group, electric charge quantization follows automatically. This is the case, *e.g.*, in Grand Unified Theories (GUT) which will be discussed in Chapter 26. Besides electric charges, GUTs also contain magnetic monopoles, *i.e.* magnetic charges. Indeed, as Dirac realized, the existence of magnetic monopoles would automatically imply electric charge quantization (Dirac, 1931). Since there is no experimental evidence for the existence of magnetic monopoles, but overwhelming evidence for very precise charge quantization, it is obvious to ask whether the latter may follow even in the absence of magnetic monopoles.

What is the origin of charge quantization in the Standard Model? This is a subtle question that is usually not addressed in the textbook literature. As we have seen in Chapter 15, charge quantization indeed follows from the condition of gauge anomaly cancellation, once the electric charges of the leptons have been fixed to their experimental values. In this section, we will leave the electric charges of the leptons as free parameters. In particular, we will not assume *a priori* that neutrinos must be electrically neutral, and we will ask under which circumstances anomaly cancellation still implies electric charge quantization.

Again, we consider a single generation of fermions and we treat the weak hypercharges of leptons and quarks as *continuous, free parameters*. By convention, we fix the weak hypercharge of the Higgs field to $Y_\Phi = 1/2$.[5] As usual, the Higgs mechanism then leads to the "spontaneous breakdown" of the gauge group $SU(2)_L \times U(1)_Y$ to $U(1)_{em}$, and the photon emerges as a massless gauge boson that couples to the electric charge $Q = T_L^3 + Y$. The remaining free parameters are the weak hypercharges of the left- and right-handed leptons, Y_{l_L}, Y_{e_R}, and of the left- and right-handed quarks, Y_{q_L}, Y_{u_R}, and Y_{d_R}.

The Standard Model is formulated in flat space–time. Hence, it does not contain gravity. Still, respecting Lorentz or Euclidean space–time rotation invariance as an exact symmetry is an important principle in the construction of the Standard Model Lagrangian. This

---

[5] Changing this convention would amount to a redefinition of the $U(1)_Y$ gauge coupling g'.

is justified because local Lorentz invariance is a gauge symmetry of General Relativity. Although string theory may seem promising, at present we have no established theory of strongly coupled quantum gravity. In any case, direct experimental support for such a theory would require systematically probing Planck-scale energies, which will remain far beyond reach in the foreseeable future. Still, based on General Relativity, which has been very accurately tested in a large variety of experiments (as reviewed by Will (2014)), we have a good understanding of gravity at low energies. Since Newton's constant has negative mass dimension, namely, -2, General Relativity is not perturbatively renormalizable. Nevertheless, it can be quantized consistently as a systematic low-energy effective field theory (Donoghue, 2012; Donoghue and Holstein, 2015).

We have already encountered several reasons why the Standard Model is not the "Theory of Everything" but "just" a low-energy effective theory that will for sure break down at the Planck scale, if not below it. It is important to investigate whether the Standard Model and General Relativity are consistent with each other, when they are both treated as low-energy effective theories. In particular, we want to know whether the chiral gauge symmetry of the Standard Model can be maintained in a curved space–time background and whether general coordinate invariance (or local space–time rotation and translation symmetry) can be maintained in the presence of the fundamental fields of the Standard Model. If not, our two most fundamental descriptions of Nature, despite their internal mathematical consistency, would be inconsistent with each other.

Fortunately, this is not the case: In four space–time dimensions, there are no purely gravitational anomalies and the mixed gauge-gravitational anomalies cancel among the chiral fermions of the Standard Model (Alvarez-Gaumé and Witten, 1984). The condition for the cancellation of this anomaly, which is associated with a triangle diagram with two external gravitons and one external B-boson, takes the form

$$2Y_{l_{\mathrm{L}}} - Y_{e_{\mathrm{R}}} + N_c \left(2Y_{q_{\mathrm{L}}} - Y_{u_{\mathrm{R}}} - Y_{d_{\mathrm{R}}}\right) = 0. \tag{16.24}$$

The triangle diagram with two external gluons and one external B-boson (see Figure 15.5) gives rise to the anomaly cancellation condition

$$N_c \left(2Y_{q_{\mathrm{L}}} - Y_{u_{\mathrm{R}}} - Y_{d_{\mathrm{R}}}\right) = 0, \tag{16.25}$$

such that the combination of both conditions implies

$$2Y_{l_{\mathrm{L}}} = Y_{e_{\mathrm{R}}}. \tag{16.26}$$

In Section 15.7, we had adjusted the weak hypercharges of the leptons to their experimentally observed values,

$$Y_{l_{\mathrm{L}}} = -\frac{1}{2}, \quad Y_{e_{\mathrm{R}}} = -1. \tag{16.27}$$

We see that eq. (16.26) is then indeed satisfied, which means that the mixed gauge-gravitational anomaly is canceled in the Standard Model. Hence, gravitational fields do not spoil the gauge invariance of the Standard Model, from which we conclude that General Relativity and the Standard Model happily coexist as mutually consistent low-energy descriptions of our world.[6]

---

[6] The term "low energy" refers to possibly different energy scales in these two cases. While perturbative quantum gravity is usually expected to break down only at the Planck scale, the Standard Model may or may not break down much earlier.

When we ignore the experimental input on the weak hypercharges of the leptons and treat Y_{l_L} and Y_{e_R} as free parameters, the gauge-gravitational anomaly cancellation condition (16.26) is obviously not sufficient to fix these two parameters. However, there is a second condition, which results from the Yukawa coupling that gives mass to the electron. In order to endow the electron with non-zero mass, this coupling must be non-zero and the corresponding term must be gauge-invariant, which implies

$$m_e \neq 0 \quad \Rightarrow \quad Y_{l_L} - \frac{1}{2} = Y_{e_R}. \tag{16.28}$$

Here, we have inserted the value of the weak hypercharge of the Higgs field $Y_\Phi = 1/2$. Together, the two conditions indeed lead to the experimentally observed values cited in eq. (16.27). As we have seen in Section 15.7, the gauge anomaly cancellation conditions then fix the quark hypercharges to

$$Y_{q_L} = \frac{1}{2N_c}, \quad Y_{u_R} = Q_u = \frac{1}{2}\left(\frac{1}{N_c} + 1\right), \quad Y_{d_R} = Q_d = \frac{1}{2}\left(\frac{1}{N_c} - 1\right), \tag{16.29}$$

which implies that electric charge is quantized, and $Q_u - Q_d = 1$ is generic.

It is remarkable that the combination of gauge-gravitational anomaly cancellation with a non-zero electron mass automatically opens the door to non-zero masses of the other fermions as well. A non-vanishing u- or d-quark mass requires gauge invariance of the corresponding Yukawa interactions, which are obeyed only if

$$Y_{q_L} + \frac{1}{2} = Y_{u_R}, \quad Y_{q_L} - \frac{1}{2} = Y_{d_R}. \tag{16.30}$$

Indeed, according to eq. (16.29), both relations are satisfied, and hence non-zero quark masses are then automatically consistent with gauge invariance.

It is worth noting that it would be possible to trade the gauge-gravitational anomaly cancellation condition against a condition for a gauge-invariant Yukawa coupling of the d-quark. Although neutrinos are massless in the renormalizable Standard Model, they can pick up a mass from the non-renormalizable dimension-5 operator, as we saw in Section 16.3. This operator is gauge-invariant only if $Y_{l_L} = -Y_\Phi = -1/2$.

We conclude that the consistency of the Standard Model as a quantum field theory, as well as its consistency with General Relativity as a low-energy effective theory, which manifests itself by the cancellation of the gauge and the mixed gauge-gravitational anomalies, combined with the existence of a massive electron, guarantees that the Yukawa couplings of the other fermions are automatically also gauge-invariant. This includes the neutrino whose mass results from the dimension-5 operator which is also automatically gauge-invariant. As a consequence, electric charge is then automatically quantized in integer units of e.[7]

We stress that we have naturally obtained electric charge quantization in the Standard Model without embedding its Abelian gauge groups $U(1)_Y$ or $U(1)_{em}$ in a grand unified non-Abelian group, provided that the electron is massive. As we will see in Chapter 26, the resulting Standard Model fits naturally into the GUT framework. In that case, the mixed gauge-gravitational anomaly cancellation condition (16.24) arises automatically from embedding $U(1)_Y$ in a non-Abelian group. As a result, GUTs give rise to very heavy magnetic monopoles, which, despite substantial experimental efforts, have not been observed.

---

[7] While the electric charges of the quarks are fractional, the hadrons in which they are confined carry integer charges.

This may be explained by cosmic inflation, which leads to an exponential dilution of pre-existing monopoles. However, GUTs or magnetic monopoles are not necessary (although sufficient) to explain charge quantization. As we have seen, electric charge quantization already follows naturally within the Standard Model, as soon as we demand its consistency with gravity and endow the electron with mass.

16.8 Dirac and Majorana Masses from Right-Handed Neutrino Fields

The experimental observation of neutrino oscillations implies that (at least two) neutrinos have mass. The Standard Model, viewed as an effective theory, incorporates this fact via its dimension-5 operator (16.19), which implies that the Standard Model will necessarily break down at the scale Λ, which was introduced in eq. (16.21). What kind of renormalizable theory might replace the Standard Model at those high energy scales? In order to construct renormalizable terms that give rise to non-zero neutrino masses, one must explicitly introduce *right-handed neutrino fields*. As we have seen in Section 15.10, right-handed neutrinos are sterile; *i.e.* they do not participate in the gauge interactions of the Standard Model.

Once we extend the Standard Model by introducing a right-handed neutrino field ν_R and its partner $\bar{\nu}_R$, we can give a Dirac mass to the neutrino in the same way as we gave mass to the u-quark. In addition, since the right-handed neutrino field is sterile (and thus gauge-invariant), we can construct a gauge-invariant dimension-3 Majorana mass term without invoking the Higgs field. This term has a prefactor M with mass dimension 1, the so-called *Majorana mass,* which plays the role of the high energy scale Λ. Besides v, which is the only dimensioned parameter in the Standard Model Lagrangian, M enters the extended model as a second dimensioned parameter. Introducing ${}^c\nu_R = -i\sigma^2 \bar{\nu}_R^{\mathsf{T}}, {}^c\bar{\nu}_R = \nu_R^{\mathsf{T}} i\sigma^2$, the renormalizable neutrino mass terms then take the form

$$\mathcal{L}(\bar{l}_L, l_L, \bar{\nu}_R, \nu_R, \widetilde{\Phi}) = f_\nu \, \bar{l}_L \widetilde{\Phi} \nu_R + f_\nu^* \, \bar{\nu}_R \widetilde{\Phi} l_L + \frac{1}{2}\left(M\, {}^c\bar{\nu}_R \nu_R + M^*\bar{\nu}_R{}^c\nu_R\right)$$

$$= f_\nu \, (\bar{\nu}_L, \bar{e}_L)\begin{pmatrix} \widetilde{\Phi}^0 \\ \widetilde{\Phi}^- \end{pmatrix}\nu_R + f_\nu^* \, \bar{\nu}_R \left(\widetilde{\Phi}^{0*}, \widetilde{\Phi}^{-*}\right)\begin{pmatrix} \nu_L \\ e_L \end{pmatrix}$$

$$+ \frac{1}{2}\left(M\, {}^c\bar{\nu}_R \nu_R + M^*\bar{\nu}_R{}^c\nu_R\right). \tag{16.31}$$

When the Higgs field assumes its vacuum value, this term reduces to

$$\mathcal{L}(\bar{l}_L, l_L, \bar{\nu}_R, \nu_R, v) = \frac{1}{2}\left(\bar{\nu}_L, {}^c\bar{\nu}_R\right)\begin{pmatrix} 0 & f_\nu v \\ f_\nu v & M \end{pmatrix}\begin{pmatrix} {}^c\nu_L \\ \nu_R \end{pmatrix}$$

$$+ \frac{1}{2}\left({}^c\bar{\nu}_L, \bar{\nu}_R\right)\begin{pmatrix} 0 & f_\nu^* v \\ f_\nu^* v & M^* \end{pmatrix}\begin{pmatrix} \nu_L \\ {}^c\nu_R \end{pmatrix}$$

$$= \frac{1}{2}\left(\bar{\nu}_L, {}^c\bar{\nu}_R\right)\mathcal{M}_\nu\begin{pmatrix} {}^c\nu_L \\ \nu_R \end{pmatrix} + \frac{1}{2}\left({}^c\bar{\nu}_L, \bar{\nu}_R\right)\mathcal{M}_\nu^*\begin{pmatrix} \nu_L \\ {}^c\nu_R \end{pmatrix}, \tag{16.32}$$

with the symmetric neutrino mass matrix

$$\mathcal{M}_\nu = \begin{pmatrix} 0 & f_\nu v \\ f_\nu v & M \end{pmatrix}. \tag{16.33}$$

It can be diagonalized by a unitary transformation ($N \in U(2)$),

$$\begin{pmatrix} \nu'_L \\ {}^c\nu'_R \end{pmatrix} = N \begin{pmatrix} \nu_L \\ {}^c\nu_R \end{pmatrix}, \quad (\bar{\nu}'_L, {}^c\bar{\nu}'_R) = (\bar{\nu}_L, {}^c\bar{\nu}_R) N^\dagger,$$

$$\begin{pmatrix} {}^c\nu'_L \\ \nu'_R \end{pmatrix} = N^* \begin{pmatrix} {}^c\nu_L \\ \nu_R \end{pmatrix}, \quad ({}^c\bar{\nu}'_L, \bar{\nu}'_R) = ({}^c\bar{\nu}_L, \bar{\nu}_R) N^\mathsf{T},$$

$$N \mathcal{M}_\nu N^\mathsf{T} = N^* \mathcal{M}_\nu^* N^\dagger = \begin{pmatrix} m_\nu & 0 \\ 0 & M_\nu \end{pmatrix}. \tag{16.34}$$

It should be pointed out that the neutrino mass matrix is in general not real but *symmetric*. (This property persists in case of more than one generation, which will be discussed in Chapter 17.) Only if $\mathcal{M}_\nu$ is *real symmetric,* it can be diagonalized by an orthogonal basis transformation. When $\mathcal{M}_\nu$ is complex symmetric (and not Hermitian), it cannot be diagonalized by an ordinary basis transformation. The above unitary transformation with $N \in U(2)$ is not an ordinary basis transformation, but, as we will see explicitly in the next section, it allows us to diagonalize $\mathcal{M}_\nu$ with non-negative real-valued mass eigenvalues m_ν and M_ν.

16.9 Seesaw Mass-by-Mixing Mechanism

As was first pointed out by Peter Minkowski (1977) and independently by Gell-Mann *et al.* (1979); Yanagida (1980); Glashow (1980); Mohapatra and Senjanovic (1980); Schechter and Valle (1980), in the presence of a large Majorana mass, the mixing of active (weakly interacting, left-handed) and sterile (right-handed) neutrinos naturally leads to very light weakly interacting neutrinos, in addition to a very heavy Majorana neutrino. The mass-by-mixing mechanism by which one neutrino becomes heavy, while the other one only picks up a very small mass, is known as the *seesaw mechanism.*

We can parameterize the matrix $N \in U(2)$ of eq. (16.34) as

$$N = \begin{pmatrix} A \exp(i\varphi) & B \exp(i\varphi) \\ -B^* & A^* \end{pmatrix}$$

$$= \begin{pmatrix} \exp(i(\alpha + \beta + \varphi)) & 0 \\ 0 & 1 \end{pmatrix} \begin{pmatrix} |A| & |B| \\ -|B| & |A| \end{pmatrix} \begin{pmatrix} \exp(-i\beta) & 0 \\ 0 & \exp(-i\alpha) \end{pmatrix}, \tag{16.35}$$

with $A = |A| \exp(i\alpha)$, $B = |B| \exp(i\beta)$, and $|A|^2 + |B|^2 = 1$. We write $f_\nu = |f_\nu| \exp(i\chi)$ and $M = |M| \exp(i\theta)$, and fix α and β such that

$$\exp(i\chi) = \exp(i(\alpha + \beta)), \quad \exp(i\theta) = \exp(2i\alpha). \tag{16.36}$$

This turns the neutrino mass matrix $\mathcal{M}_\nu$ into the real symmetric matrix

$$\begin{pmatrix} \exp(-i\beta) & 0 \\ 0 & \exp(-i\alpha) \end{pmatrix} \begin{pmatrix} 0 & f_\nu v \\ f_\nu v & M \end{pmatrix} \begin{pmatrix} \exp(-i\beta) & 0 \\ 0 & \exp(-i\alpha) \end{pmatrix}^\mathsf{T} = \begin{pmatrix} 0 & |f_\nu| v \\ |f_\nu| v & |M| \end{pmatrix}, \tag{16.37}$$

which can be diagonalized by the orthogonal transformation

$$\begin{pmatrix} |A| & |B| \\ -|B| & |A| \end{pmatrix} \begin{pmatrix} 0 & |f_\nu| v \\ |f_\nu| v & |M| \end{pmatrix} \begin{pmatrix} |A| & |B| \\ -|B| & |A| \end{pmatrix}^\mathsf{T} = \begin{pmatrix} -m_\nu & 0 \\ 0 & M_\nu \end{pmatrix}, \tag{16.38}$$

with the neutrino masses

$$M_\nu = \frac{1}{2}\left(\sqrt{|M|^2 + 4|f_\nu|^2 v^2} + |M|\right), \quad m_\nu = \frac{1}{2}\left(\sqrt{|M|^2 + 4|f_\nu|^2 v^2} - |M|\right). \quad (16.39)$$

We see that the negative sign in front of m_ν in eq. (16.38) is required in order to obtain $m_\nu > 0$. The corresponding eigenvectors are determined by the mixing parameters

$$|A|^2 = \frac{1}{2}\left(1 + \frac{|M|}{\sqrt{|M|^2 + 4|f_\nu|^2 v^2}}\right), \quad |B|^2 = \frac{1}{2}\left(1 - \frac{|M|}{\sqrt{|M|^2 + 4|f_\nu|^2 v^2}}\right). \quad (16.40)$$

Finally, to justify the positive neutrino mass m_ν, we fix φ such that $\exp(2\mathrm{i}(\alpha + \beta + \varphi)) = -1$, which implies

$$\begin{pmatrix} \exp(\mathrm{i}(\alpha + \beta + \varphi)) & 0 \\ 0 & 1 \end{pmatrix} \begin{pmatrix} -m_\nu & 0 \\ 0 & M_\nu \end{pmatrix} \begin{pmatrix} \exp(\mathrm{i}(\alpha + \beta + \varphi)) & 0 \\ 0 & 1 \end{pmatrix}^{\mathsf{T}} = \begin{pmatrix} m_\nu & 0 \\ 0 & M_\nu \end{pmatrix}. \quad (16.41)$$

As we have just seen, the transformation of eq. (16.34) allows us to diagonalize the complex symmetric neutrino mass mixing matrix $\mathcal{M}_\nu$ to a matrix with real-positive entries M_ν and m_ν (note that these are not the eigenvalues of $\mathcal{M}_\nu$, which are the zeros of its characteristic polynomial).

Interestingly, the four continuous parameters of the U(2) matrix N are all determined in this process. This is different from a unitary transformation that diagonalizes a Hermitian matrix. In that case, the two complex phases of the eigenvectors remain undetermined in the diagonalization process. That will be important in the next chapter, when we count CP-violating phases in the Pontecorvo–Maki–Nakagawa–Sakata (PMNS) matrix that describes neutrino mixing for several generations. In the present case of a single generation, after diagonalization of the neutrino mixing matrix which has two complex parameters, $f_\nu, M \in \mathbb{C}$, we encounter *two physical, real-positive neutrino masses, m_ν and M_ν.*

When we assume the scale hierarchy $|M| \gg |f_\nu| v$, the expressions for the neutrino masses of eq. (16.39) simplify to

$$M_\nu \simeq M, \quad m_\nu \simeq \frac{|f_\nu|^2 v^2}{|M|}, \quad (16.42)$$

and the mixing parameters of eq. (16.40) reduce to

$$|A|^2 \simeq 1 - \frac{|f_\nu|^2 v^2}{|M|^2}, \quad |B|^2 \simeq \frac{|f_\nu|^2 v^2}{|M|^2}. \quad (16.43)$$

In the *seesaw mechanism*, by mixing, a heavy sterile neutrino endows an active neutrino that participates in the weak interaction with a small mass. This mechanism, which results from diagonalizing a simple 2×2 matrix, leads to neutrino masses that are naturally a factor $v/|M|$ smaller than the masses of the other fermions.

Remarkably, via its non-renormalizable dimension-5 operator, the Standard Model *anticipates* the seesaw mechanism, because it leads to the neutrino mass given in eq. (16.21), $m_\nu = v^2/\Lambda$. When we extend the Standard Model by a heavy right-handed sterile neutrino, we can relate the scale Λ to the Majorana mass parameter. However, we should not forget that eq. (16.42) for the light neutrino mass also involves the factor $|f_\nu|^2$, *i.e.* $|M| = |f_\nu|^2\Lambda$. Since the Yukawa couplings for the other fermions vary over 5 orders of magnitude, one can imagine that the different neutrino masses even vary over 10 orders of magnitude, due to $|f_\nu|$.

At present, we simply do not understand the deep origin of the Yukawa couplings. Nor do we know the value of the scale Λ, let alone the nature of the physics that takes place at that scale. However, we do know the Standard Model as a very precise effective theory, which kindly informs us that it will eventually break down at some scale Λ where the deep origin of non-zero neutrino masses is hiding.

16.10 Right-Handed Neutrinos and Electric Charge Quantization

Let us return to the question of electric charge quantization, this time in the presence of a right-handed neutrino field. We will not assume *a priori* the right-handed neutrino to be sterile, *i.e.* like the hypercharges of the other fermions; we will treat Y_{ν_R} as a continuous, free parameter.

As discussed in Section 15.7, gauge anomaly cancellation leads to the constraints

$$2Y_{l_L}^3 - Y_{\nu_R}^3 - Y_{e_R}^3 + N_c(2Y_{q_L}^3 - Y_{u_R}^3 - Y_{d_R}^3) = 0,$$

$$2Y_{l_L} + 2N_c Y_{q_L} = 0 \quad \Rightarrow \quad Y_{l_L} = -N_c Y_{q_L},$$

$$N_c(2Y_{q_L} - Y_{u_R} - Y_{d_R}) = 0 \quad \Rightarrow \quad Y_{q_L} = \frac{1}{2}(Y_{u_R} + Y_{d_R}). \tag{16.44}$$

Using the latter two relations, the first one can be rewritten as

$$(Y_{u_R} - Y_{d_R})^2 = \frac{2}{3Y_{l_L}}(Y_{\nu_R}^3 + Y_{e_R}^3 - 2Y_{l_L}^3). \tag{16.45}$$

The mixed gauge-gravitational anomaly cancellation condition takes the form

$$2Y_{l_L} - Y_{\nu_R} - Y_{e_R} + N_c\left(2Y_{q_L} - Y_{u_R} - Y_{d_R}\right) = 0 \quad \Rightarrow \quad 2Y_{l_L} - Y_{\nu_R} - Y_{e_R} = 0. \tag{16.46}$$

Altogether, there are four independent relations for six unknowns.

As in the Standard Model without right-handed neutrino fields, let us also demand the Yukawa coupling for the electron to be gauge-invariant,

$$m_e \neq 0 \quad \Rightarrow \quad Y_{l_L} - \frac{1}{2} = Y_{e_R}. \tag{16.47}$$

Together with eq. (16.46), this results in

$$Y_{l_L} + \frac{1}{2} = Y_{\nu_R}, \tag{16.48}$$

which implies that a Yukawa coupling that gives rise to a Dirac mass term for the neutrino is gauge-invariant as well. Using these relations, eq. (16.45) can be rewritten as

$$
\begin{aligned}
(Y_{u_R} - Y_{d_R})^2 &= \frac{2}{3Y_{l_L}}\left(Y_{\nu_R}^3 + Y_{e_R}^3 - 2Y_{l_L}^3\right) \\
&= \frac{2}{3Y_{l_L}}\left((Y_{\nu_R} + Y_{e_R})^3 - 3Y_{\nu_R}Y_{e_R}(Y_{\nu_R} + Y_{e_R}) - 2Y_{l_L}^3\right) \\
&= \frac{2}{3Y_{l_L}}\left(8Y_{l_L}^3 - 6Y_{\nu_R}Y_{e_R}Y_{l_L} - 2Y_{l_L}^3\right) = 4Y_{l_L}^2 - 4Y_{\nu_R}Y_{e_R} \\
&= (Y_{\nu_R} + Y_{e_R})^2 - 4Y_{\nu_R}Y_{e_R} = (Y_{\nu_R} - Y_{e_R})^2 = 1.
\end{aligned}
\tag{16.49}
$$

Adopting the natural convention that $Y_{u_R} > Y_{d_R}$, we obtain $Y_{u_R} - Y_{d_R} = 1$. Combining this with the last equality of eq. (16.44), we obtain $Y_{l_L} + 1/2 = Y_{u_R}$ and $Y_{l_L} - 1/2 = Y_{d_R}$, which implies that the Yukawa couplings for the quarks are also automatically gauge-invariant, thus allowing non-zero Dirac masses for the quarks as well. As in the Standard Model (without ν_R), one could again trade the gauge-gravitational anomaly cancellation condition for the condition of a gauge-invariant Yukawa coupling of the d-quark. Until now, we have introduced only five constraints for six unknowns; hence, there is a continuous family of solutions, and electric charge is *not* quantized.

In a hypothetical world of this kind, neutrinos would generically be charged particles with an electric charge

$$Q_\nu = T^3_{L\nu_L} + Y_{l_L} = \frac{1}{2} + Y_{l_L} = Y_{\nu_R}, \qquad (16.50)$$

where we inserted eq. (16.48). This automatically implies that the neutrino cannot pick up a Majorana mass term, because that would not be gauge-invariant. Such a Dirac neutrino could not participate in the seesaw mechanism and would hence not be exceptionally light for a natural reason.

By using the constraints that arise from anomaly cancellation as well as from the non-vanishing electron mass, one further obtains

$$Q_e = -\frac{1}{2} + Y_{l_L} = -1 + Q_\nu, \quad Q_u = \frac{1}{2}\left(\frac{1}{N_c} + 1\right) - \frac{Q_\nu}{N_c}, \quad Q_d = \frac{1}{2}\left(\frac{1}{N_c} - 1\right) - \frac{Q_\nu}{N_c}. \qquad (16.51)$$

As a result, the pions π^+, π^0, π^- as well as other mesons, which contain a valence quark–anti-quark pair, would keep their usual integer-valued electric charges. On the other hand, the electric charges of the baryons, which contain N_c valence quarks, would now differ from their standard values by $-Q_\nu$. For example, proton and neutron would now have the electric charges

$$Q_p = 1 - Q_\nu = -Q_e, \quad Q_n = -Q_\nu. \qquad (16.52)$$

In particular, just as the neutrino, the neutron would no longer be electrically neutral. Remarkably, although charge would not be quantized in integer units, the electric charges of proton and electron would still exactly cancel each other, and the hydrogen atom would hence remain neutral. Deuterium, with an additional neutron in the atomic nucleus, on the other hand would carry the total charge $Q_p + Q_n + Q_e = -Q_\nu$. Such an exotic variant of heavy hydrogen would form a neutral atom only if it would capture a neutrino, which would indeed be attracted by the Coulomb force (regardless of the sign of Q_ν). The same would be true for heavier neutral atoms. They would contain as many neutrinos as there are neutrons in the atomic nucleus.

The β-decay of the neutron into a proton, electron, and anti-neutrino would still be consistent with charge conservation because $Q_n = Q_p + Q_e - Q_\nu$. In particular, the W^--boson that mediates this decay would still have its usual integer-valued electric charge $-e$. Also the weak decay of the charged pions into a lepton and its neutrino, as well as the electromagnetic decay of the neutral pion into two photons, would still proceed as usual.

Why do we not live in an exotic world like this? As we will now argue, the reason seems to be that neutrinos are naturally much lighter than the other fermions (which is indeed automatically the case in the Standard Model). In extensions of the Standard Model with an explicit right-handed neutrino field, this manifests itself via the seesaw mass-by-mixing mechanism which involves a large Majorana mass. Gauge invariance of the Majorana mass term requires

$$M \neq 0 \quad \Rightarrow \quad Y_{\nu_R} = 0, \tag{16.53}$$

which finally implies that neutrinos are electrically neutral and that charge is quantized.

We may conclude that charge quantization relies on the Majorana nature of the right-handed neutrino. This is naturally incorporated in the Standard Model via the non-renormalizable dimension-5 operator. In extensions of the Standard Model with a right-handed neutrino field, it requires $Y_{\nu_R} = 0$. As we will see in Chapter 26, this is naturally realized in GUTs, because hypercharge is then embedded in a unifying non-Abelian gauge group, such as SU(5) or SO(10).

16.11 Lepton–Baryon Mixing for $N_c = 1$

This section deals with questions related to the $N_c = 1$ variant of the Standard Model, which is not realized in Nature. It can be skipped at a first reading.

As we have seen in Chapter 15, the Standard Model is gauge anomaly-free for any odd number of colors, including $N_c = 1$. The models with odd $N_c \geq 5$ are quite similar to our real world, at least at a qualitative level. As we will discuss now, the $N_c = 1$ variant of the Standard Model is qualitatively different, because it lacks the strong interaction. As a result, there is no physical distinction between quarks and leptons, and baryon and lepton number are no longer useful concepts. In particular, there are Yukawa couplings that mix leptons and baryons. Furthermore, just like the right-handed neutrino, for $N_c = 1$ the right-handed neutron is also a gauge singlet, thus allowing for a Majorana mass term. (We will return to the $N_c = 1$ model in Chapter 26, where we embed it in the small unification scheme that was introduced in Section 13.8.)

In the $N_c = 1$ variant of the Standard Model, the gauge group is just $\mathrm{SU}(2)_L \times \mathrm{U}(1)_Y$. Hence, there is no strong interaction and the up and down "quarks" are just the proton and the neutron. The fermion content of a single generation of fermions then is

$$q_L = \begin{pmatrix} p_L \\ n_L \end{pmatrix}, \quad p_R, \quad n_R, \quad l_L = \begin{pmatrix} \nu_L \\ e_L \end{pmatrix}, \quad e_R. \tag{16.54}$$

In the absence of the strong interaction, the right-handed neutron is a sterile particle that does not participate in the electroweak gauge interactions. As a consequence, one can construct a Majorana mass term ${}^c\bar{n}_R n_R$, which introduces a second mass scale beyond v. Since the Standard Model with $N_c = 1$ has no strong interaction, we repeat that there is nothing that physically distinguishes baryons from leptons. As a result, electron and anti-proton, as well as neutrino and anti-neutron, can mix via additional Majorana terms ${}^c\bar{p}_R e_R$ and ${}^c\bar{q}_L \left(i\tau^2 \right) l_L$, which give rise to two additional independent Majorana mixing scales.

These terms are the $N_c = 1$ analogues of the baryon number violating 4-fermion operators listed in eq. (16.23) for $N_c = 3$. For general odd N_c, the corresponding (N_c+1)-fermion operators contain N_c quark fields and one lepton field, and thus have dimension $3(N_c+1)/2$. Hence, they are renormalizable only for $N_c = 1$. This results in strong lepton–baryon mixing, such that (unlike for $N_c \geq 3$) baryon and lepton number are not even separately conserved at the classical level.

Interestingly, $U(1)_{B-L}$ remains an exact global symmetry. In the Standard Model with odd $N_c \geq 3$, renormalizable lepton–quark mixing terms are forbidden because they are not $SU(N_c)$-invariant. As a consequence, the Standard Model has only one mass scale v, while its $N_c = 1$ variant has, in addition, three independent Majorana mass or mixing scales, associated with the terms

$$^c\bar{n}_R n_R, \quad {}^c\bar{p}_R e_R, \quad {}^c\bar{q}_L \left(i\tau^2 \right) l_L, \tag{16.55}$$

which are unrelated to spontaneous symmetry breaking. The gauge-invariant mass terms as well as the Yukawa couplings are given by

$$\begin{aligned}
\mathcal{L}(\bar{e}, e, \bar{v}, v, \bar{p}, p, \bar{n}, n, \Phi) = & f_e \bar{l}_L \Phi e_R + f_e^* \bar{e}_R \Phi^\dagger l_L + f_p \bar{q}_L \tilde{\Phi} p_R + f_p^* \bar{p}_R \tilde{\Phi}^\dagger q_L \\
& + M_R{}^c\bar{p}_R e_R + M_R^* \bar{e}_R{}^c p_R + M_L \bar{l}_L \left(i\tau^2 \right)^\dagger {}^c q_L + M_L^{*c}\bar{q}_L \left(i\tau^2 \right) l_L \\
& + f_n \bar{q}_L \Phi n_R + f_n^* \bar{n}_R \Phi^\dagger q_L + \frac{M_n}{2} \left({}^c\bar{n}_R n_R + \bar{n}_R{}^c n_R \right).
\end{aligned} \tag{16.56}$$

As before, the Higgs field takes the form

$$\Phi(x) = \begin{pmatrix} \Phi^+(x) \\ \Phi^0(x) \end{pmatrix}, \quad \tilde{\Phi}(x) = i\tau^2 \Phi(x)^* = \begin{pmatrix} \Phi^0(x)^* \\ -\Phi^+(x)^* \end{pmatrix}. \tag{16.57}$$

When it is replaced by its vacuum value $\Phi^0(x) \to v$, $\Phi^+(x) \to 0$, the Lagrangian of eq. (16.56) turns into

$$\begin{aligned}
\mathcal{L}(\bar{e}, e, \bar{v}, v, \bar{p}, p, \bar{n}, n, v) = & \left(\bar{e}_L, {}^c\bar{p}_R \right) \begin{pmatrix} f_e v & M_L \\ M_R & f_p v \end{pmatrix} \begin{pmatrix} e_R \\ {}^c p_L \end{pmatrix} \\
& + \left(\bar{e}_R, {}^c\bar{p}_L \right) \begin{pmatrix} f_e^* v & M_R^* \\ M_L^* & f_p^* v \end{pmatrix} \begin{pmatrix} e_L \\ {}^c p_R \end{pmatrix} \\
& + \frac{1}{2} \left(\bar{v}_L, \bar{n}_L, {}^c\bar{n}_R \right) \begin{pmatrix} 0 & -M_L & 0 \\ -M_L & 0 & f_n v \\ 0 & f_n v & M_n \end{pmatrix} \begin{pmatrix} {}^c v_L \\ {}^c n_L \\ n_R \end{pmatrix} \\
& + \frac{1}{2} \left({}^c\bar{v}_L, {}^c\bar{n}_L, \bar{n}_R \right) \begin{pmatrix} 0 & -M_L^* & 0 \\ -M_L^* & 0 & f_n^* v \\ 0 & f_n^* v & M_n \end{pmatrix} \begin{pmatrix} v_L \\ n_L \\ {}^c n_R \end{pmatrix} \\
= & \left(\bar{e}_L, {}^c\bar{p}_R \right) \mathcal{M}_C \begin{pmatrix} e_R \\ {}^c p_L \end{pmatrix} + \left(\bar{e}_R, {}^c\bar{p}_L \right) \mathcal{M}_C^\dagger \begin{pmatrix} e_L \\ {}^c p_R \end{pmatrix} \\
& + \frac{1}{2} \left(\bar{v}_L, \bar{n}_L, {}^c\bar{n}_R \right) \mathcal{M}_N \begin{pmatrix} {}^c v_L \\ {}^c n_L \\ n_R \end{pmatrix} + \frac{1}{2} \left({}^c\bar{v}_L, {}^c\bar{n}_L, \bar{n}_R \right) \mathcal{M}_N^* \begin{pmatrix} v_L \\ n_L \\ {}^c n_R \end{pmatrix}.
\end{aligned} \tag{16.58}$$

Here, the mass and mixing matrices of the charged and the neutral fermions are given by

$$\mathcal{M}_C = \begin{pmatrix} f_e v & M_L \\ M_R & f_p v \end{pmatrix}, \quad \mathcal{M}_N = \begin{pmatrix} 0 & -M_L & 0 \\ -M_L & 0 & f_n v \\ 0 & f_n v & M_n \end{pmatrix}. \tag{16.59}$$

They can be diagonalized by the bi-unitary transformations

$$
\begin{pmatrix} e'_{\mathrm{R}} \\ {}^c p'_{\mathrm{L}} \end{pmatrix} = C_{\mathrm{R}} \begin{pmatrix} e_{\mathrm{R}} \\ {}^c p_{\mathrm{L}} \end{pmatrix}, \quad
\begin{pmatrix} e'_{\mathrm{L}} \\ {}^c p'_{\mathrm{R}} \end{pmatrix} = C_{\mathrm{L}} \begin{pmatrix} e_{\mathrm{L}} \\ {}^c p_{\mathrm{R}} \end{pmatrix}, \quad
C_{\mathrm{L}} \mathcal{M}_C C_{\mathrm{R}}^{\dagger} = \begin{pmatrix} m_e & 0 \\ 0 & m_p \end{pmatrix},
$$

$$
\begin{pmatrix} {}^c v'_{\mathrm{L}} \\ {}^c n'_{\mathrm{L}} \\ n'_{\mathrm{R}} \end{pmatrix} = N_{\mathrm{R}} \begin{pmatrix} {}^c v_{\mathrm{L}} \\ {}^c n_{\mathrm{L}} \\ n_{\mathrm{R}} \end{pmatrix}, \quad
\begin{pmatrix} v'_{\mathrm{L}} \\ n'_{\mathrm{L}} \\ {}^c n'_{\mathrm{R}} \end{pmatrix} = N_{\mathrm{L}} \begin{pmatrix} v_{\mathrm{L}} \\ n_{\mathrm{L}} \\ {}^c n_{\mathrm{R}} \end{pmatrix}, \quad
N_{\mathrm{R}} \mathcal{M}_N N_{\mathrm{L}}^{\dagger} = \begin{pmatrix} m_v & 0 & 0 \\ 0 & m_n & 0 \\ 0 & 0 & M \end{pmatrix}.
$$

$$(16.60)$$

It is straightforward (but not very instructive) to work out the masses explicitly. It is interesting to note that – due to its Majorana mass term – via lepton–baryon mixing the right-handed neutron field provides mass also to the neutrino.

As we will discuss in Chapter 17, in the odd $N_c \geq 3$ Standard Model with massless neutrinos, the lepton numbers of the different fermion generations are separately conserved (ignoring the $B + L$ anomaly). On the other hand, the quarks of the different fermion generations mix with each other. As a result, only their overall baryon number (summed over all quark flavors) is conserved.

In the $N_c = 1$ variant of the Standard Model, positron and proton mix in a similar way as the quark generations in the $N_c \geq 3$ Standard Model, as if baryons and anti-leptons would represent something like two different "generations", with identical gauge quantum numbers. Just as the baryon numbers of the different quark generations are not separately conserved in the $N_c \geq 3$ Standard Model, in its $N_c = 1$ variant baryon and lepton number are not separately conserved either. For $N_c = 1$, the analogue of the overall baryon number is $B - L$, simply because B characterizes the first (baryon) "generation", while $-L$ characterizes the second (lepton) "generation".

To summarize, due to the absence of the strong interaction, the $N_c = 1$ variant of the Standard Model is qualitatively different from the models with odd $N_c \geq 3$, which differ from each other quantitatively but not qualitatively. While the Lagrangian of the Standard Model with $N_c \geq 3$ contains only one dimensioned parameter v, which sets the scale at the classical level, the variant of the model with $N_c = 1$ possesses *three additional mass and mixing scales* (we expressed them by M_{L}, M_{R}, and M_n) that are unrelated to spontaneous symmetry breaking and are each afflicted by their own hierarchy problem. The absence of such scales can be considered an attractive feature of the Standard Model with odd $N_c \geq 3$. Of course, in that case, the strong interaction generates its own dimensioned scale $\Lambda_{\overline{\mathrm{MS}}}$ by dimensional transmutation. Thanks to asymptotic freedom, there is no hierarchy problem associated with that scale.

Exercises

16.1 Gauge invariance of Yukawa couplings

Show that the Yukawa terms of eqs. (16.2), (16.9), and (16.12) are invariant under all gauge symmetries of the Standard Model.

16.2 Fermion masses in an $SU(2)_L \times SU(2)_R$-symmetric model

Investigate once more the modified Standard Model with an $SU(2)_L \times SU(2)_R$ gauge symmetry, with gauge couplings g and g'. Again, consider one generation of Standard Model fermions including a right-handed neutrino and assume that the $SU(2)_L$ and $SU(3)_c$ transformation properties are the usual ones. Further assume that under $SU(2)_R$ the left-handed fields are singlets, while the right-handed fields form one lepton and one quark $SU(2)_R$ doublet. Construct Yukawa terms that give mass to leptons and quarks and calculate the resulting mass parameters. Would these mass parameters represent the masses of physical particles?

16.3 Dimension-5 operators

Show that the dimension-5 operator of eq. (16.19) is Lorentz-invariant and invariant under all gauge symmetries of the Standard Model. Convince yourself that there are no other allowed dimension-5 operators.

There are two additional copies of the first generation of quarks and leptons, with the same quantum numbers but with different masses. The first particle of the second generation to be discovered was the *muon*, a heavy cousin of the electron. On that occasion, at a dinner in a Chinese restaurant, Isidor Rabi asked "Who ordered that?". Indeed, we do not understand why the second and third generation exist. Gauge anomalies cancel within each generation separately, and thus, there is no mathematical reason for a specific number of generations.

The muon comes with its own *muon-neutrino* with which it forms an $SU(2)_L$ doublet. The quarks of the second generation are known as *charm* (*c*) and *strange* (*s*), which have the same quantum numbers as the up and down quarks of the first generation. Finally, the whole structure repeats itself again in the third generation, which consists of the *τ-lepton* (or *tauon*) and its neutrino, as well as of the quarks *top* (*t*) and *bottom* (*b*) (or *truth* and *beauty*). The phenomena and puzzles associated with the existence of several generations define the area of *flavor physics*.

Since the quarks and leptons of the different generations have the same quantum numbers, they can mix with each other. We hence distinguish between *electroweak eigenstates*, which are produced pairwise in the weak charged current interactions, and *mass eigenstates*, which result from a unitary mixing matrix applied to the electroweak eigenstates. The mixing of *u*-, *c*-, and *t*-quarks, as well as of *d*-, *s*-, and *b*-quarks, leads to the 3×3 *Cabibbo–Kobayashi–Maskawa (CKM) matrix*, which depends on three real-valued Euler angles and one complex *CP-violating phase*. With just two generations, there is only one angle – the Cabibbo angle – and no CP violation.

Because the Standard Model is a chiral gauge theory, charge conjugation C and parity P are individually violated. However, the combination CP is respected by all gauge interactions (at least if we assume a vanishing θ-vacuum angle, whose experimental value is, in any case, consistent with zero). In 1964, James Cronin, Val Fitch *et al.* observed *CP violation in the decays of neutral kaons* (meson bound states with down and anti-strange ($d\bar{s}$) or strange and anti-down ($s\bar{d}$) quarks). This led Makoto Kobayashi and Toshihide Maskawa to predict the existence of a third generation of quarks (Kobayashi and Maskawa, 1973).

In the beginning of the 1970s, there was no direct experimental evidence for any member of the third generation, and even the experimental discovery of the second generation was not yet complete: Only the three lightest quark flavors *u*, *d*, and *s* were known at that time. Also, the need for gauge anomaly cancellation had not yet been fully appreciated. Still, in 1970 Sheldon Glashow, John Iliopoulos, and Luciano Maiani predicted the *c*-quark, based on what is now known as the *GIM mechanism*. In the absence of the charm quark, large *flavor-changing neutral currents* would conflict with the observed mixing in the neutral kaon system.

In the absence of neutrino masses, there would be no mixing among the leptons of the different generations and thus no neutrino oscillations. In that case, the lepton numbers of the different generations would be separately conserved (when we ignore electroweak instanton effects). Since *neutrino oscillations* have now been observed, we know that (at least two) neutrinos do have mass. As we have seen in Section 16.3, this can be incorporated in the Standard Model via a non-renormalizable dimension-5 term. When extended to three generations, this operator induces lepton mixing, which is characterized by the *Pontecorvo–Maki–Nakagawa–Sakata (PMNS) matrix*. Like the CKM quark mixing matrix, the PMNS matrix again has three real-valued Euler angles, but it has in addition three (and not just one) complex CP-violating phases. Even with just two generations, the PMNS matrix already has one CP-violating phase.

17.1 Electroweak versus Mass Eigenstates

Let us now add the remaining two generations of fermions. We first return to the scenario with massless neutrinos. In particular, we do not introduce fields for the right-handed neutrinos. For the first generation, we have

$$\begin{pmatrix} \nu_{e\mathrm{L}}(x) \\ e_{\mathrm{L}}(x) \end{pmatrix}, \; e_{\mathrm{R}}(x), \quad \begin{pmatrix} u_{\mathrm{L}}(x) \\ d_{\mathrm{L}}(x) \end{pmatrix}, \; u_{\mathrm{R}}(x), \; d_{\mathrm{R}}(x). \tag{17.1}$$

The neutrino that we dealt with so far is now denoted as the *electron-neutrino* ν_e, in order to distinguish it from the other neutrinos that we are about to introduce.

In the second generation, we have the *muon* μ (as a heavier copy of the electron) and its neutrino ν_μ, as well as charm c- and strange s-quarks (as the heavier copies of up and down quarks). The lepton and quark doublets and singlets of the second generation then take the form

$$\begin{pmatrix} \nu_{\mu\mathrm{L}}(x) \\ \mu_{\mathrm{L}}(x) \end{pmatrix}, \; \mu_{\mathrm{R}}(x), \quad \begin{pmatrix} c_{\mathrm{L}}(x) \\ s_{\mathrm{L}}(x) \end{pmatrix}, \; c_{\mathrm{R}}(x), \; s_{\mathrm{R}}(x). \tag{17.2}$$

The charge assignments (T_{L}^3, Y, and thus $Q = T_{\mathrm{L}}^3 + Y$) are exactly the same as for the first generation. The heavy fermions tend to decay into lighter ones. The heavier they are, the faster this happens, and it is more difficult to generate such particles at all. In light of the concept of one generation, "strange" effects were sometimes observed and related to the s-quark, which was then completed to a generation by the subsequent discovery of the c-quark in 1974.

Later on, yet another generation was discovered step by step; see Appendix A. In the third generation, we have the tauon τ, its neutrino ν_τ, as well as the top and bottom quarks (t and b)

$$\begin{pmatrix} \nu_{\tau\mathrm{L}}(x) \\ \tau_{\mathrm{L}}(x) \end{pmatrix}, \; \tau_{\mathrm{R}}(x), \quad \begin{pmatrix} t_{\mathrm{L}}(x) \\ b_{\mathrm{L}}(x) \end{pmatrix}, \; t_{\mathrm{R}}(x), \; b_{\mathrm{R}}(x). \tag{17.3}$$

As a last ingredient, the t-quark was discovered experimentally at the Tevatron proton–antiproton collider at Fermilab (near Chicago) in 1995. Its existence had been expected long before on theoretical grounds. Since the gauge anomalies must be canceled, the Standard Model only works if generations are complete, and the b-quark had been observed already in 1977. However, like the masses of other elementary particles, the Standard Model does not predict the value of the t-quark mass, which was found experimentally at 173.3(8) GeV.

We now introduce spinor fields that contain the *mass eigenstates* with the same quantum numbers from the various generations

$$N_{\mathrm{L}}(x) = \begin{pmatrix} v_{e\mathrm{L}}(x) \\ v_{\mu\mathrm{L}}(x) \\ v_{\tau\mathrm{L}}(x) \end{pmatrix}, \quad E_{\mathrm{L,R}}(x) = \begin{pmatrix} e_{\mathrm{L,R}}(x) \\ \mu_{\mathrm{L,R}}(x) \\ \tau_{\mathrm{L,R}}(x) \end{pmatrix},$$

$$U_{\mathrm{L,R}}(x) = \begin{pmatrix} u_{\mathrm{L,R}}(x) \\ c_{\mathrm{L,R}}(x) \\ t_{\mathrm{L,R}}(x) \end{pmatrix}, \quad D_{\mathrm{L,R}}(x) = \begin{pmatrix} d_{\mathrm{L,R}}(x) \\ s_{\mathrm{L,R}}(x) \\ b_{\mathrm{L,R}}(x) \end{pmatrix}, \tag{17.4}$$

and we form the corresponding left-handed lepton and quark $\mathrm{SU}(2)_L$ doublets

$$L_{\mathrm{L}}(x) = \begin{pmatrix} N_{\mathrm{L}}(x) \\ E_{\mathrm{L}}(x) \end{pmatrix}, \quad Q_{\mathrm{L}}(x) = \begin{pmatrix} U_{\mathrm{L}}(x) \\ D_{\mathrm{L}}(x) \end{pmatrix}. \tag{17.5}$$

In general, the mass eigenstates differ from the electroweak eigenstates, to which they are related by global unitary transformations $U^{N_\mathrm{L}}, U^{E_\mathrm{L,R}}, U^{U_\mathrm{L,R}}, U^{D_\mathrm{L,R}} \in \mathrm{U}(N_\mathrm{g})$, where N_g is the number of generations ($N_\mathrm{g} = 3$ in Nature),

$$N_{\mathrm{L}}'(x) = U^{N_\mathrm{L}} N_{\mathrm{L}}(x), \quad E_{\mathrm{L,R}}'(x) = U^{E_\mathrm{L,R}} E_{\mathrm{L,R}}(x),$$

$$U_{\mathrm{L,R}}'(x) = U^{U_\mathrm{L,R}} U_{\mathrm{L,R}}(x), \quad D_{\mathrm{L,R}}'(x) = U^{D_\mathrm{L,R}} D_{\mathrm{L,R}}(x). \tag{17.6}$$

The gauge interactions of the leptons are then described by the Lagrangian

$$\mathcal{L}(\bar{L}_{\mathrm{L}}', L_{\mathrm{L}}', \bar{E}_{\mathrm{R}}', E_{\mathrm{R}}', B, W) = \bar{L}_{\mathrm{L}}' \bar{\sigma}_\mu D_\mu L_{\mathrm{L}}' + \bar{E}_{\mathrm{R}}' \sigma_\mu D_\mu E_{\mathrm{R}}', \tag{17.7}$$

with the covariant derivatives given by

$$D_\mu L_{\mathrm{L}}'(x) = \left(\partial_\mu - i\frac{g'}{2} B_\mu(x) + W_\mu(x) \right) L_{\mathrm{L}}'(x), \quad D_\mu E_{\mathrm{R}}'(x) = \left(\partial_\mu - ig' B_\mu(x) \right) E_{\mathrm{R}}'(x). \tag{17.8}$$

By definition, the gauge interactions are diagonal in the *basis of electroweak eigenstates*. However, the Yukawa couplings are not diagonal in this basis; they take the form

$$\mathcal{L}(\bar{L}_{\mathrm{L}}', L_{\mathrm{L}}', \bar{E}_{\mathrm{R}}', E_{\mathrm{R}}', \Phi) = \bar{L}_{\mathrm{L}}' \Phi \mathcal{F}_E E_{\mathrm{R}}' + \bar{E}_{\mathrm{R}}' \mathcal{F}_E^\dagger \Phi^\dagger L_{\mathrm{L}}'. \tag{17.9}$$

Here, $\mathcal{F}_E$ is a general complex $N_\mathrm{g} \times N_\mathrm{g}$ matrix of Yukawa couplings that describes the Higgs-induced generation mixing between the electroweak eigenstates. On the other hand, again by definition, when the Higgs field assumes its vacuum value v, the Yukawa couplings are diagonal in the basis of the mass eigenstates, *i.e.*

$$\mathcal{L}(\bar{L}_{\mathrm{L}}, L_{\mathrm{L}}, \bar{E}_{\mathrm{R}}, E_{\mathrm{R}}, \mathrm{v}) = \bar{E}_{\mathrm{L}} \mathcal{M}_E E_{\mathrm{R}} + \bar{E}_{\mathrm{R}} \mathcal{M}_E E_{\mathrm{L}}, \tag{17.10}$$

where $\mathcal{M}_E = \mathrm{diag}(m_e, m_\mu, m_\tau)$ is the diagonal mass matrix of the charged leptons, with positive (real-valued) lepton masses m_e, m_μ, and m_τ. Consequently, the matrix $\mathcal{F}_E$ is diagonalized by the bi-unitary transformations that relate the electroweak and the mass eigenstates

$$U^{E_L\dagger} v \, \mathcal{F}_E U^{E_R} = \mathcal{M}_E. \tag{17.11}$$

This relation constrains the unitary transformations U^{E_L} and U^{E_R}.

Since, at this stage, the neutrinos are massless, there is no similar relation for U^{N_L}, which we can hence fix as $U^{N_L} = U^{E_L}$, such that $L'_L(x) = U^{E_L} L_L(x)$. This ensures that $N_L(x)$ and $E_L(x)$ remain partners in $SU(2)_L$ doublets even in the basis of the mass eigenstates. The gauge interactions of the leptons then take the form

$$\mathcal{L}(\bar{L}_L, L_L, \bar{E}_R, E_R, B, W) = \bar{L}_L U^{E_L\dagger} \bar{\sigma}_\mu D_\mu U^{E_L} L_L + \bar{E}_R U^{E_R\dagger} \sigma_\mu D_\mu U^{E_R} E_R$$
$$= \bar{L}_L \bar{\sigma}_\mu D_\mu L_L + \bar{E}_R \sigma_\mu D_\mu E_R. \tag{17.12}$$

We conclude that, as long as the neutrinos are massless, the gauge interactions of the leptons are diagonal even in the basis of mass eigenstates; hence, there is no mixing between the leptons of different generations. As we will see in Section 17.3, since quarks are massive, for them there is generation mixing via the charged current interactions. In Section 17.6, we will encounter generation mixing even for the leptons, once we take into account non-zero neutrino masses.

17.2 Generation-Specific Lepton Numbers and Lepton Universality

Since there is no lepton mixing in the absence of neutrino masses, the individual generation-specific lepton numbers L_e, L_μ, and L_τ are *separately conserved*. For (anti-)leptons, this number is 1 (-1) in the corresponding generation and zero otherwise. A typical example is the decay $\mu \rightarrow e + \bar{\nu}_e + \nu_\mu$, which occurs after a muon life-time of $2.2 \cdot 10^{-6}$ sec; see the left panel of Figure 17.1. This process preserves $L_\mu = 1$ and $L_e = 0$. The process $\mu \rightarrow e + \gamma$, on the other hand, is *forbidden*. Although it leaves the total lepton number $L = L_e + L_\mu + L_\tau$ invariant, it violates both L_e and L_μ individually. The tauon is still significantly heavier than the muon ($m_\tau \simeq 1.8$ GeV versus $m_\mu \simeq 106$ MeV). Hence, its life-time is much shorter, about $2.9 \cdot 10^{-13}$ sec. It can decay either into $e + \bar{\nu}_e + \nu_\tau$ or into $\mu + \bar{\nu}_\mu + \nu_\tau$ (where the muon will soon decay further). Both processes preserve $L_\tau = 1$ and $L_e = L_\mu = 0$. Since the mass difference between electron and muon is small compared to the mass of the tauon, the phase space for both decays is more or less the same, and indeed both have essentially the same branching ratio (17.52 percent versus 17.39 percent). This universal behavior of leptons, irrespective of what generation they belong to, is a general phenomenon as long as the differences in the lepton masses are negligible. This is known as *lepton universality*. Unlike the muon, the tauon is so heavy that it can even decay into pions. The electron is the lightest electrically charged particle and is hence absolutely stable.

The conservation of lepton number, separately in each generation, also restricts the possible Z-decays, in addition to charge conservation. Therefore, the leptonic decay of the Z-boson can only lead to a lepton and its own anti-lepton (cf. right panel of Figure 17.1).

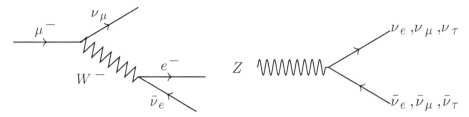

Fig. 17.1 Left: Weak decay of a muon μ^- into a muon-neutrino ν_μ, an electron e^-, and an electron–anti-neutrino $\bar{\nu}_e$. The generation-specific lepton numbers $L_e = 0, L_\mu = 1$, and $L_\tau = 0$ are conserved throughout this process. Right: Z-boson decay into neutrino–anti-neutrino pairs of the different generations. The arrows describe the flow of the lepton numbers.

The decay width of Z allows us to sum up the leptonic decay channels and thus to identify the number of generations. The result – found in particular in LEP experiments at CERN – implies that there are no further leptons beyond the known three generations. A conceivable exception is a generation whose leptons are so heavy that the Z-boson cannot decay into any of its members. However, given the sequence of masses found so far, this scenario seems unlikely; it would require a new type of neutrino with a mass $m_\nu > m_Z/2 \simeq 45.6$ GeV (as we mentioned before, in Section 15.3).

17.3 Cabibbo–Kobayashi–Maskawa Quark Mixing Matrix

Let us return to the discussion of electroweak versus mass eigenstates, and let us now focus on the quarks. In the basis of electroweak eigenstates, by definition, the gauge interactions take the form

$$\mathcal{L}(\bar{Q}'_\mathrm{L}, Q'_\mathrm{L}, \bar{U}'_\mathrm{R}, U'_\mathrm{R}, \bar{D}'_\mathrm{R}, D'_\mathrm{R}, B, W, G) = \bar{Q}'_\mathrm{L}\bar{\sigma}_\mu D_\mu Q'_\mathrm{L} + \bar{U}'_\mathrm{R}\sigma_\mu D_\mu U'_\mathrm{R} + \bar{D}'_\mathrm{R}\sigma_\mu D_\mu D'_\mathrm{R}, \tag{17.13}$$

with the covariant derivatives now given by

$$D_\mu Q'_\mathrm{L}(x) = \left(\partial_\mu + \mathrm{i}\frac{g'}{2N_\mathrm{c}}B_\mu(x) + W_\mu(x) + G_\mu(x)\right)Q'_\mathrm{L}(x),$$

$$D_\mu U'_\mathrm{R}(x) = \left(\partial_\mu + \mathrm{i}\frac{g'}{2}\left(\frac{1}{N_\mathrm{c}}+1\right)B_\mu(x) + G_\mu(x)\right)U'_\mathrm{R}(x),$$

$$D_\mu D'_\mathrm{R}(x) = \left(\partial_\mu + \mathrm{i}\frac{g'}{2}\left(\frac{1}{N_\mathrm{c}}-1\right)B_\mu(x) + G_\mu(x)\right)D'_\mathrm{R}(x). \tag{17.14}$$

The Yukawa couplings are again off-diagonal in the basis of electroweak eigenstates and take the form

$$\mathcal{L}(\bar{Q}'_\mathrm{L}, Q'_\mathrm{L}, \bar{U}'_\mathrm{R}, U'_\mathrm{R}, \bar{D}'_\mathrm{R}, D'_\mathrm{R}, \Phi) = \bar{Q}'_\mathrm{L}\widetilde{\Phi}\mathcal{F}_U U'_\mathrm{R} + \bar{U}'_\mathrm{R}\mathcal{F}_U^\dagger\widetilde{\Phi}^\dagger Q'_\mathrm{L} + \bar{Q}'_\mathrm{L}\Phi\mathcal{F}_D D'_\mathrm{R} + \bar{D}'_\mathrm{R}\mathcal{F}_D^\dagger\Phi^\dagger Q'_\mathrm{L}. \tag{17.15}$$

Here, $\mathcal{F}_U$ and $\mathcal{F}_D$ are *a priori* general complex $N_g \times N_g$ matrices of Yukawa couplings. When the Higgs field assumes its vacuum value v, by definition the Yukawa couplings are diagonal in the basis of the mass eigenstates, *i.e.*

$$\mathcal{L}(\bar{Q}_L, Q_L, \bar{U}_R, U_R, \bar{D}_R, D_R, v) = \bar{U}_L \mathcal{M}_U U_R + \bar{U}_R \mathcal{M}_U U_L + \bar{D}_L \mathcal{M}_D D_R + \bar{D}_R \mathcal{M}_D D_L.$$

(17.16)

Here, $\mathcal{M}_U = \mathrm{diag}(m_u, m_c, m_t)$ and $\mathcal{M}_D = \mathrm{diag}(m_d, m_s, m_b)$ are the diagonal quark mass matrices with positive (real-valued) quark masses. As for the leptons, the matrices $\mathcal{F}_U$ and $\mathcal{F}_D$ are diagonalized by the bi-unitary transformations that relate the electroweak and the mass eigenstates

$$U^{U_L \dagger} v \mathcal{F}_U U^{U_R} = \mathcal{M}_U, \quad U^{D_L \dagger} v \mathcal{F}_D U^{D_R} = \mathcal{M}_D.$$

(17.17)

These relations constrain the unitary transformations $U^{U_{L,R}}$ and $U^{D_{L,R}}$.

The $SU(3)_c \times U(1)_Y$ gauge interactions of the quarks are associated with electrically neutral, flavor-conserving currents with the contributions

$$\bar{U}'_{L,R}(x) \gamma_\mu U'_{L,R}(x) = \bar{U}_{L,R}(x) U^{U_{L,R} \dagger} \gamma_\mu U^{U_{L,R}} U_{L,R}(x) = \bar{U}_{L,R}(x) \gamma_\mu U_{L,R}(x),$$

$$\bar{D}'_{L,R}(x) \gamma_\mu D'_{L,R}(x) = \bar{D}_{L,R}(x) U^{D_{L,R} \dagger} \gamma_\mu U^{D_{L,R}} D_{L,R}(x) = \bar{D}_{L,R}(x) \gamma_\mu D_{L,R}(x). \quad (17.18)$$

When we rotate these terms into the basis of mass eigenstates, we can simply drop the primes, because $U^{U_{L,R} \dagger} U^{U_{L,R}} = \mathbb{1}$, $U^{D_{L,R} \dagger} U^{D_{L,R}} = \mathbb{1}$. Hence, the neutral current interactions do not lead to changes among different quark flavors: The Standard Model is *free of flavor-changing neutral currents*, at least at tree level.

The $SU(2)_L$ gauge interactions contain both neutral and charged current interactions, which are mediated by W^3 and $W^\pm$, respectively. While the neutral currents are again flavor-conserving, the charged current interactions are flavor-changing. Explicitly, the charged currents are given by

$$\sqrt{2} j_\mu^+(x)/g = \bar{U}'_L(x) \bar{\sigma}_\mu D'_L(x) = \bar{U}_L(x) U^{U_L \dagger} \bar{\sigma}_\mu U^{D_L} D_L(x) = \bar{U}_L(x) \bar{\sigma}_\mu V D_L(x),$$

$$\sqrt{2} j_\mu^-(x)/g = \bar{D}'_L(x) \bar{\sigma}_\mu U'_L(x) = \bar{D}_L(x) U^{D_L \dagger} \bar{\sigma}_\mu U^{U_L} U_L(x) = \bar{D}_L(x) \bar{\sigma}_\mu V^\dagger U_L(x). \quad (17.19)$$

Here, we have introduced the *Cabibbo–Kobayashi–Maskawa (CKM) quark mixing matrix*

$$V = U^{U_L \dagger} U^{D_L} \in U(N_g).$$

(17.20)

This matrix describes the extent of flavor-changing in the charged current interactions of the Standard Model.

Let us count the number of physical parameters in the mixing matrix V for the case of N_g generations. We proceed in three steps:

- Since V is unitary, one might naively expect N_g^2 real parameters.

- However, one can change the matrices U^{U_L} and U^{D_L} by multiplying them with diagonal unitary matrices – we call them D^{U_L} and D^{D_L} – from the right,

$$U^{U_L\,'} = U^{U_L} D^{U_L}, \quad U^{D_L\,'} = U^{D_L} D^{D_L}. \tag{17.21}$$

This still leaves the resulting mass matrices diagonal, and it turns the matrix V into

$$V' = U^{U_L\,'\dagger} U^{D_L\,'} = D^{U_L\dagger} V D^{D_L}. \tag{17.22}$$

Our requirement was the diagonalization of the mass matrices; now we see that this can be achieved in different ways. The corresponding ambiguity should be subtracted from the set of physical parameters. To be more explicit: the matrices D^{U_L} and D^{D_L} together have $2N_g$ parameters (the complex phases on their diagonals), which should not be counted as physical parameters in V.

- However, an overall phase factor common to both, D^{U_L} and D^{D_L}, would not affect V' at all. This means that in the preceding step we wanted to subtract one parameter, which does not actually exist. The correct number of parameters to be subtracted is therefore $2N_g - 1$.

Hence, the proper counting of physical parameters in the CKM matrix is

$$N_g^2 - (2N_g - 1) = (N_g - 1)^2. \tag{17.23}$$

With a single generation, there is no mixing and hence no free parameter.

With two generations, there is one physical parameter. This situation was assumed for some time, and the allowed quark mixing was described by the so-called *Cabibbo angle* θ_C – named after Nicola Cabibbo. It is instructive to take a closer look at the case of $N_g = 2$ generations: A general unitary 2×2 matrix can be written as

$$V = \begin{pmatrix} A\exp(i\varphi) & B\exp(i\varphi) \\ -B^* & A^* \end{pmatrix} = \begin{pmatrix} |A|\exp(i(\alpha+\varphi)) & |B|\exp(i(\beta+\varphi)) \\ -|B|\exp(-i\beta) & |A|\exp(-i\alpha) \end{pmatrix}, \tag{17.24}$$

where A and B are arbitrary complex numbers with phases α and β, and $|A|^2 + |B|^2 = 1$ (φ is the phase of $\det V$, which generalizes our former representation of SU(2) matrices). We now use the freedom to introduce D^{U_L} and D^{D_L} in order to change any unitary matrix V to

$$V' = D^{U_L\dagger} V D^{D_L} = \begin{pmatrix} \cos\theta_C & \sin\theta_C \\ -\sin\theta_C & \cos\theta_C \end{pmatrix}. \tag{17.25}$$

Choosing

$$D^{U_L} = \mathrm{diag}(\exp(i(\varphi+\beta)), \exp(-i\alpha)), \quad D^{D_L} = \mathrm{diag}(\exp(i(\beta-\alpha)), 1), \tag{17.26}$$

one indeed turns the general U(2) matrix V (which seems to have four parameters) into the special form $V' \in$ SO(2), which only depends on the Cabibbo angle. Experiments led to a Cabibbo angle of $\theta_C \simeq 13°$. Since it is non-zero, flavor-changing weak decays do occur, but their transition rate is small due to the modest mixing angle.

As we have seen, for a general number of generations N_g, the number of physical parameters in the matrix V is $(N_g - 1)^2$. For $N_g \geq 3$, the matrix V will in general not belong to SO(N_g) (which has only $N_g(N_g - 1)/2$ parameters) because

$$(N_g - 1)^2 - \frac{N_g(N_g - 1)}{2} = \frac{(N_g - 1)(N_g - 2)}{2} > 0. \tag{17.27}$$

For example, for the physical case of $N_g = 3$ generations, the CKM matrix contains $(N_g - 1)(N_g - 2)/2 = 1$ *complex phase*, in addition to $N_g(N_g - 1)/2 = 3$ Cabibbo-type Euler angles.

In the quark sector alone, the Yukawa couplings give rise to ten free parameters of the Standard Model – the six quark masses, three mixing angles, and one complex phase. Experimentalists are working hard on the even more precise identification of these parameters; there are constraints for them based on a variety of processes.

The complex phase is a source of explicit CP symmetry breaking which manifests itself, for example, in the neutral kaon system. For instance, one considers the interaction between two currents through a gauge boson $W^{\pm}$. Such a scattering amplitude is proportional to the product of two CKM matrix elements – one for each flavor-changing transition in the two currents. If we invert the directions of the scattering and replace the particles involved by their anti-particles (and vice versa), we obtain the complex conjugate of the above product of matrix elements, cf. eq. (17.19). In total, this means that a CP transformation changes V to V^*. Therefore, the complex phase in the CKM matrix implies an *explicit violation of CP invariance* (if this phase is non-zero).

The breaking of CP invariance was in fact observed already in 1964 in the decay of neutral kaons. For a long period, this remained the only process where CP violation could be detected. Since the year 2001, this has also been achieved in decays of neutral mesons which include the b-quark (Aubert *et al.*, 2001; Abe *et al.*, 2001).

17.4 Flavor-Changing Neutral Currents and the GIM Mechanism

As we will see in Chapter 21, quarks and anti-quarks are confined inside bound states known as mesons. The lightest mesons are the pions π^+, π^-, and π^0, the eta-meson η, as well as the kaons K^+, K^-, K^0, and $\overline{K^0}$. As we will discuss in Chapters 18 and 23, these eight pseudo-scalar particles are the pseudo-Nambu–Goldstone bosons of the spontaneously broken, approximate SU(3)$_L$ × SU(3)$_R$ chiral symmetry of QCD. In the (rather naive) constituent quark model, the K^0 consists of a down quark and an anti-strange quark ($d\bar{s}$), while its anti-particle $\overline{K^0}$ consists of a strange quark and an anti-down quark ($s\bar{d}$). Because gluons are flavor-blind and couple only to a quark's color index, the strong interaction leaves the quark flavor invariant. As a result, K^0 and $\overline{K^0}$ cannot mix under the strong interaction. The weak neutral current, which couples to the Z-boson, is not associated with a change of the quark flavor either. As a result, the Standard Model is free of flavor-changing neutral current interactions, at least at tree level.

On the other hand, as we discussed in the previous section, the weak charged current interactions (mediated by the $W^{\pm}$-bosons) mix the flavors, with the Cabibbo angle θ_C quantifying the amount of mixing between the flavors d and s. Since this process is associated with the weak interaction, it proceeds much more slowly (on subatomic scales) than processes of the strong or electromagnetic interactions. Figure 17.2 shows the two leading

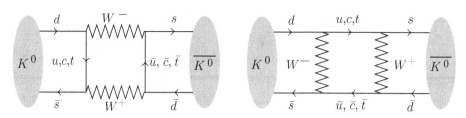

Fig. 17.2 The mixing between the mesons K^0 and $\overline{K^0}$ amounts to a flavor-changing neutral current effect at the 1-loop level. The GIM mechanism leads to a suppression of this effect due to the presence of the c-quark.

Feynman box diagrams that lead to the mixing of K^0 and $\overline{K^0}$. The intermediate particles can be u-, c-, or t-quarks, and their anti-quarks. Since the neutral mesons K^0 and $\overline{K^0}$ have different flavors, their mixing amounts to a *flavor-changing neutral current* interaction, which arises here at the 1-loop level.

At the beginning of the 1970s, only u-, d-, and s-quarks were known. Hence, only u-quarks were expected to appear in the loops. The corresponding amplitudes pick up two factors of $V_{ud} = \cos\theta_C$ and two factors of $V_{us} = \sin\theta_C$ at the four vertices, such that the amplitudes are proportional to $V_{du}V_{ud}V_{su}V_{us} = \cos^2\theta_C \sin^2\theta_C$. This would give rise to a much larger flavor-changing neutral current effect than the one that is actually observed. In order to resolve this puzzle, in 1970, Glashow, Iliopoulos, and Maiani (GIM) postulated the existence of a fourth quark flavor, now known as *charm*, c. The GIM mechanism relies on the unitarity of the CKM matrix or actually at that time on the orthogonality of the 2×2 Cabibbo quark mixing matrix

$$V = \begin{pmatrix} V_{ud} & V_{us} \\ V_{cd} & V_{cs} \end{pmatrix} = \begin{pmatrix} \cos\theta_C & \sin\theta_C \\ -\sin\theta_C & \cos\theta_C \end{pmatrix}. \tag{17.28}$$

When both virtual u- and c-quarks contribute to the two box diagrams, ignoring the different values of the quark masses, the total amplitude is proportional to

$$V_{du}V_{ud}V_{su}V_{us} + V_{du}V_{ud}V_{sc}V_{cs} + V_{dc}V_{cd}V_{su}V_{us} + V_{dc}V_{cd}V_{sc}V_{cs} = 0 ; \tag{17.29}$$

i.e. the flavor-changing neutral current effect would be completely absent. Taking account of the different quark masses, the effect turns out to be proportional to $(m_c^2 - m_u^2)/M_W^2$, which is not zero but much more suppressed than in the absence of the c-quark. Actually, the comparison with the observed value led to a rough estimate of the c-quark mass of $m_c \approx 3$ GeV. In 1974, the c-quark was indeed observed independently by the experimental collaborations of Samuel Ting and Burton Richter in the form of the J/Ψ meson – a very narrow $c\bar{c}$ resonance. The mass of the c-quark is $m_c = 1.27(3)$ GeV.

17.5   CP Violation with Neutral Kaons and *B*-Mesons

The strong interaction respects charge conjugation C and parity P (if we ignore the negligibly small θ-vacuum angle which explicitly breaks P and hence CP). Like the pions and the η-meson, the kaons are pseudo-scalar particles; *i.e.* they are parity odd and thus have intrinsically P $= -1$. The neutral kaon K^0 and its anti-particle $\overline{K^0}$ are related by C. When the strong interaction eigenstates K^0 and $\overline{K^0}$ mix via the weak interaction, in the absence of the third generation they would form two linear combinations

$$|K_L\rangle = \frac{1}{\sqrt{2}}\left[|K^0\rangle + |\overline{K^0}\rangle\right], \quad |K_S\rangle = \frac{1}{\sqrt{2}}\left[|K^0\rangle - |\overline{K^0}\rangle\right]. \tag{17.30}$$

The even combination K_L (known as K-long) has C $= 1$ and hence CP $= -1$. If we assume CP symmetry, it can decay only into three pions (via the weak interaction). Since the phase space for this decay is small, K_L is relatively long-lived ($\tau_{K_L} \approx 5.18 \times 10^{-8}$ sec). The odd combination K_S (K-short) has C $= -1$ and CP $= 1$; hence, it can decay into two pions. Since the phase space for this two-particle decay is bigger, K_S has a much shorter life-time ($\tau_{K_S} \approx 0.59 \times 10^{-10}$ sec).

In the presence of the third generation, the situation is more complicated because CP is then explicitly broken by the complex phase in the CKM matrix. The amount of CP violation in the mixing is quantified by a small parameter ε

$$|K_L\rangle = \frac{1}{\sqrt{2(1+|\varepsilon|^2)}}\left[(1+\varepsilon)|K^0\rangle + (1-\varepsilon)|\overline{K^0}\rangle\right],$$

$$|K_S\rangle = \frac{1}{\sqrt{2(1+|\varepsilon|^2)}}\left[(1-\varepsilon)|K^0\rangle - (1+\varepsilon)|\overline{K^0}\rangle\right]. \tag{17.31}$$

Since $|K_L\rangle$ now contains a small CP $= 1$ component, it may sometimes also decay into two pions (with a branching ratio of about 1 permille). This type of CP violation via K^0-$\overline{K^0}$ mixing, followed by a CP respecting decay, is known as *indirect CP violation*.

There is also *direct CP violation*, which happens when CP is violated in the decay process itself, for example, when a CP $= -1$ state decays into two pions (which would be forbidden by CP symmetry). Direct CP violation is quantified by another small parameter ε'. First evidence for direct CP violation was obtained in the NA31 experiment at CERN in 1988.

The story around K^0-$\overline{K^0}$ mixing, which started with the experimental observation of CP violation by Cronin, Fitch *et al.* in 1964, continued in the 1990s with the heavier b-quark. Its physics was intensively investigated at the B-factories Belle at KEK (Abe *et al.*, 2001) and BaBar at SLAC (Aubert *et al.*, 2001). Still referring to a naive picture, the neutral meson B^0 consists of a down quark and an anti-bottom quark ($d\bar{b}$), while its anti-particle $\overline{B^0}$ consists of a bottom quark and an anti-down quark ($b\bar{d}$). At the B-factories, CP violation has been observed in B^0-$\overline{B^0}$ mixing and the subsequent decays, which proceed via a large number of decay channels. The LHCb collaboration at CERN observed CP violation also in the B_s^0 meson, which consists of a strange quark and an anti-bottom quark ($s\bar{b}$), and in the D^0 meson, which consists of a charm quark and an anti-up quark ($c\bar{u}$) (Aaij *et al.*, 2013, 2019). There is a very diverse and interesting phenomenology behind the various heavy quark decays. Some of them are very rare in the Standard Model, and their accurate experimental investigation is promising in the search for new physics beyond the Standard Model. All this is very well worth studying in much more detail, but this is beyond the scope of this book.

17.6 Pontecorvo–Maki–Nakagawa–Sakata Lepton Mixing Matrix

As we have seen in Section 17.2, in the renormalizable Standard Model with massless neutrinos there is no lepton mixing. We will now again extend the Standard Model by including non-renormalizable dimension-5 terms that give rise to non-vanishing neutrino masses, as in Section 16.3. As a result, there will then be mixing also in the lepton sector.

The non-renormalizable dimension-5 operators that couple the electroweak eigenstates of the left-handed leptons to the Higgs field $\widetilde{\Phi} = i\tau^2\Phi^*$ take the form

$$\mathcal{L}(\bar{L}'_L, L'_L, \widetilde{\Phi}) = \frac{1}{2\Lambda}\left[\left({}^c\bar{L}'_L\widetilde{\Phi}^*\right)\mathcal{G}_N\left(\widetilde{\Phi}^\dagger L'_L\right) + \left(\bar{L}'_L\widetilde{\Phi}\right)\mathcal{G}_N^*\left(\widetilde{\Phi}^{\mathsf{T}\,c}L'_L\right)\right]. \qquad (17.32)$$

Here, $\mathcal{G}_N$ is a complex $N_g \times N_g$ matrix. In order to guarantee that the corresponding Hamilton operator is Hermitian, the matrix must be symmetric, *i.e.* $\mathcal{G}_N = \mathcal{G}_N^{\mathsf{T}}$, so $\mathcal{G}_N^* = \mathcal{G}_N^\dagger$. When the Higgs field is replaced by its vacuum value v, the above expression assumes the form of a Majorana mass term for the neutrinos

$$\mathcal{L}(\bar{L}'_L, L'_L, \mathrm{v}) = \frac{\mathrm{v}^2}{2\Lambda}\left[{}^c\bar{N}'_L\mathcal{G}_N N'_L + \bar{N}'_L\mathcal{G}_N^{*\,c}N'_L\right]. \qquad (17.33)$$

We now express the Lagrangian in terms of mass eigenstates N_L, which are related to the electroweak eigenstates N'_L by the unitary transformation

$$N'_L(x) = U^{N_L}N_L(x), \quad \bar{N}'_L(x) = \bar{N}_L(x)U^{N_L\,\dagger},$$
$${}^cN'_L(x) = i\sigma^2\bar{N}'_L(x)^{\mathsf{T}} = i\sigma^2[\bar{N}_L(x)U^{N_L\,\dagger}]^{\mathsf{T}} = U^{N_L\,*}i\sigma^2\bar{N}_L(x)^{\mathsf{T}} = U^{N_L\,*\,c}N_L(x),$$
$${}^c\bar{N}'_L(x) = -N'_L(x)^{\mathsf{T}}i\sigma^2 = -[U^{N_L}N_L(x)]^{\mathsf{T}}i\sigma^2 = -N_L(x)^{\mathsf{T}}i\sigma^2U^{N_L\,\mathsf{T}} = {}^c\bar{N}_L(x)U^{N_L\,\mathsf{T}}.$$
$$(17.34)$$

The transformed Lagrangian then takes the form

$$\mathcal{L}(\bar{L}_L, L_L, \mathrm{v}) = \frac{\mathrm{v}^2}{2\Lambda}\left[{}^c\bar{N}_L U^{N_L\,\mathsf{T}}\mathcal{G}_N U^{N_L}N_L + \bar{N}_L U^{N_L\,\dagger}\mathcal{G}_N^* U^{N_L\,*\,c}N_L\right]. \qquad (17.35)$$

Under the unitary transformation, the matrix $\mathcal{G}_N$ hence transforms into

$$\mathcal{G}'_N = U^{N_L\,\mathsf{T}}\mathcal{G}_N U^{N_L}. \qquad (17.36)$$

Since $\mathcal{G}_N$ is complex rather than real symmetric, this is not an ordinary orthogonal transformation. It is, however, the natural transformation that maintains the symmetry of the matrix, *i.e.*

$$\mathcal{G}'_N{}^{\mathsf{T}} = U^{N_L\,\mathsf{T}}\mathcal{G}_N^{\mathsf{T}}U^{N_L} = \mathcal{G}'_N. \qquad (17.37)$$

In order to obtain the mass eigenstates for the neutrinos, we need to diagonalize the matrix $\mathcal{G}_N$, such that

$$U^{N_L\,\mathsf{T}}\frac{\mathrm{v}^2}{\Lambda}\mathcal{G}_N U^{N_L} = \mathcal{M}_N = \mathrm{diag}(m_{\nu_1}, m_{\nu_2}, m_{\nu_3}), \qquad (17.38)$$

where the neutrino mass matrix $\mathcal{M}_N$ has real, non-negative entries. Note that the neutrino masses m_{ν_i} are not associated with the electron-, muon-, or tau-neutrino. Those are the electroweak eigenstates which have no definite mass. For example, it is not meaningful to ask about the mass of the electron-neutrino ν_e, because this state is a superposition of three mass eigenstates. Hence, in contrast to the quark flavors (up, down, strange, etc.) which, by definition, are mass eigenstates, the neutrino "flavors" e, μ, and τ denote electroweak eigenstates.

For a single generation ($N_g = 1$), $U^{N_L} = \exp(-i\varphi/2) \in U(1)$ can be used to absorb the complex phase of $v^2 \mathcal{G}_N / \Lambda = m_\nu \exp(i\varphi)$ with the resulting neutrino mass m_ν being real and positive.

For two generations, the diagonalization of the matrix $\mathcal{G}_N$ implies four conditions. First, the two diagonal elements of the matrix $\mathcal{G}'_N$ should be real and positive, which represents two conditions. Second, the off-diagonal elements (which are the same because $\mathcal{G}'_N$ is symmetric) must vanish. This represents two more conditions because the off-diagonal elements have both a real and an imaginary part. The unitary matrix $U^{N_L} \in U(2)$ has four continuous parameters, which are hence completely fixed by the four conditions that must be satisfied in order to diagonalize $\mathcal{G}_N$. (This is in contrast to diagonalizing a Hermitian matrix which automatically has real eigenvalues, such that two parameters of the diagonalizing unitary transformation – namely, the complex phases of the corresponding eigenvectors – remain undetermined. This is analogous to the transformation in Section 16.9.)

Similarly, for N_g generations, the diagonalization of $\mathcal{G}_N$ implies N_g^2 conditions; N_g for ensuring that the diagonal elements of $\mathcal{G}'_N$ are real and positive (their complex phases are zero), and $N_g(N_g-1)$ to ensure that both the real and the imaginary parts of the $N_g(N_g-1)/2$ independent off-diagonal elements vanish. Hence, diagonalizing $\mathcal{G}_N$ again determines all the N_g^2 continuous parameters of $U^{N_L} \in U(N_g)$.

Just as for the quarks, the charged currents of the leptons are given by

$$\sqrt{2}\, j_\mu^+(x)/g = \bar{N}'_L(x)\bar{\sigma}_\mu E'_L(x) = \bar{N}_L(x)U^{N_L\dagger}\bar{\sigma}_\mu U^{E_L}E_L(x) = \bar{N}_L(x)\bar{\sigma}_\mu U E_L(x),$$
$$\sqrt{2}\, j_\mu^-(x)/g = \bar{E}'_L(x)\bar{\sigma}_\mu N'_L(x) = \bar{E}_L(x)U^{E_L\dagger}\bar{\sigma}_\mu U^{N_L}N_L(x) = \bar{E}_L(x)\bar{\sigma}_\mu U^\dagger N_L(x). \quad (17.39)$$

In analogy to the CKM quark mixing matrix $V = U^{U_L\dagger}U^{D_L}$, here we have introduced the *Pontecorvo–Maki–Nakagawa–Sakata (PMNS) lepton-mixing matrix*

$$U = U^{N_L\dagger}U^{E_L} \in U(N_g). \quad (17.40)$$

For massless neutrinos, we have chosen $U^{N_L} = U^{E_L}$ in order to ensure that $N_L(x)$ and $E_L(x)$ remain partners in an $SU(2)_L$ doublet in the basis of mass eigenstates for the charged leptons. In order to maintain this feature even in the presence of non-zero neutrino masses, we still define

$$\begin{pmatrix} \nu_{eL}(x) \\ \nu_{\mu L}(x) \\ \nu_{\tau L}(x) \end{pmatrix} = U^{E_L\dagger}N'_L(x). \quad (17.41)$$

These are the *electroweak neutrino eigenstates* that appear in charged current interactions, together with the corresponding charged leptons. The *mass eigenstates* of the neutrinos are denoted as

$$\begin{pmatrix} \nu_{1L}(x) \\ \nu_{2L}(x) \\ \nu_{3L}(x) \end{pmatrix} = N_L(x) = U^{N_L\dagger}N'_L(x) = U^{N_L\dagger}U^{E_L}\begin{pmatrix} \nu_{eL}(x) \\ \nu_{\mu L}(x) \\ \nu_{\tau L}(x) \end{pmatrix} = U\begin{pmatrix} \nu_{eL}(x) \\ \nu_{\mu L}(x) \\ \nu_{\tau L}(x) \end{pmatrix}.$$
$$(17.42)$$

We conclude that the PMNS matrix U rotates the electroweak eigenstates of the neutrinos into their mass eigenstates.

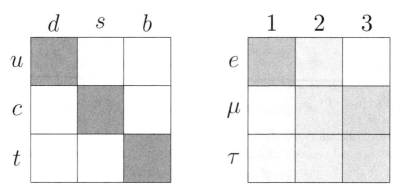

Fig. 17.3 Illustration of the Cabibbo–Kobayashi–Maskawa (left) and Pontecorvo–Maki–Nakagawa–Sakata (right) matrices V and U. The gray scale corresponds to the absolute value squared of the matrix elements. We see that the quark mixing matrix is almost diagonal, in contrast to the lepton-mixing matrix. Note that all elements of V and U are non-zero. As a last ingredient, this was established for U_{e3} (An *et al.*, 2012; Ahn *et al.*, 2012; Abe *et al.*, 2012).

As we have seen in Section 17.3, for the quarks U^{U_L} and U^{D_L} can be multiplied on the right with diagonal unitary matrices D^{U_L} and D^{D_L}, respectively. This has no impact on the diagonalization of the mass matrices $\mathcal{M}_U$ and $\mathcal{M}_D$ of the up- and down-type quarks, but leads to a reduction of the number of physical parameters of the CKM matrix. The unitary transformation associated with the charged leptons U^{E_L} can again be multiplied with a diagonal matrix D^{E_L}, *i.e.*

$$U^{E_L\,'} = U^{E_L} D^{E_L}, \tag{17.43}$$

without compromising the diagonalization of the mass matrix $\mathcal{M}_E$. However, as we have just seen, the unitary transformation U^{N_L} associated with the neutrino fields does not allow for such a modification. As a result, by a field redefinition the PMNS matrix can be turned into

$$U' = U^{N_L\dagger} U^{E_L\,'} = U D^{E_L}. \tag{17.44}$$

Let us now count the number of physical parameters of the matrix $U \in U(N_g)$, which naively has N_g^2 parameters. The multiplication with D^{E_L} can be used to absorb N_g complex phases by a field redefinition, which leaves us with $N_g(N_g - 1)$ physical parameters. Just as for the CKM matrix, $N_g(N_g - 1)/2$ of them are real-valued Euler-type angles, but in the leptonic case there remain $N_g(N_g - 1)/2$ complex CP-violating phases; *i.e.* there are just as many complex phases as real rotation angles. Recall that, in the quark case the CKM matrix had only $(N_g - 1)(N_g - 2)/2$ CP-violating phases. We conclude that in the lepton sector with massive neutrinos, a CP-violating phase already exists for $N_g = 2$, and there are three CP-violating phases in the physical case of three generations.

The magnitudes squared of the elements of the CKM and PMNS matrices are illustrated in Figure 17.3.

17.7 Neutrino Oscillations

Neutrinos are typically produced as electroweak eigenstates in a process of the weak interaction. For example, *solar neutrinos* originate from nuclear reactions inside the sun, which

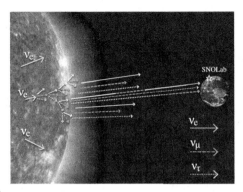 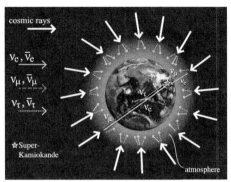

Fig. 17.4 Left: The standard solar model predicts the generation of about $2 \cdot 10^{38}$ electron neutrinos ν_e per second, which leads to an expected flux on earth around $6 \cdot 10^{10} \nu_e / (\text{cm}^2 \text{sec})$. Due to neutrino oscillations, only 1/3 of them arrive as ν_e, as the SNO Laboratory in Canada demonstrated. It uses heavy water, D_2O as detection material, and the orientation of the Cherenkov cone singles out solar neutrino events. A modified β-decay (with an incoming ν_e instead of an outgoing $\bar{\nu}_e$) measures the solar ν_e flux. In addition, deuterium dissolution processes, $D + \nu_f \rightarrow n + p + \nu_f$, $f \in \{e, \mu, \tau\}$ measure the total solar neutrino flux, summed over all flavors. Right: Illustration of the Japanese Super-Kamiokande experiment on atmospheric neutrinos. Cosmic rays generate air showers with up to 10^{11} secondary particles, including electron- and muon-neutrinos or anti-neutrinos, often through pion or muon decay. An underground water tank monitors high-energy neutrino events (up to several GeV), again by Cherenkov radiation. ν and $\bar{\nu}$ are not distinguished, but the cone profile does distinguish ν_e from ν_μ events. Their flux from above matches accurately the expectation, but the ν_μ flux from below (after crossing the earth) is significantly smaller: The main reason is that these ν_μ have sufficient time to oscillate into ν_τ (artistic illustrations first published by Aguilar-Arévalo and Bietenholz, 2015).

produce electron-neutrinos; the crucial process is the transition $4p \rightarrow \,^4\text{He} + 2\bar{e} + 2\nu_e$. Since these are superpositions of different mass eigenstates, which are associated with different energies, the time evolution of this quantum state gives rise to oscillations into other neutrino flavors, namely, muon- or tau-neutrinos. In order to detect the neutrinos on earth, they must interact with some detector material that contains electrons. This usually filters out the electron-neutrinos, while neutrinos that have oscillated into other flavors remain undetected. Indeed, in the late 1960s in his experiments in the Homestake Mine in South Dakota, Ray Davis observed only about one-third of the neutrinos that were predicted by the standard solar model (Davis *et al.*, 1968). This gave rise to the *solar neutrino puzzle*. After decades of controversial discussions about the correctness of the standard solar model, it turned out that the model works correctly. On their way through the sun and to the earth, the missing neutrinos indeed oscillate into other neutrino flavors. In this context, it is worth mentioning the Mikheyev–Smirnov–Wolfenstein (MSW) effect, which affects the oscillations while the neutrinos travel through the electron-rich matter inside the sun (Wolfenstein, 1978; Mikheyev and Smirnov, 1985). The charged current interactions between the electrons and the neutrinos affect their propagation in matter and boost the oscillations. Taking the MSW effect into account was important in order to correctly interpret the observations of high-energy solar neutrinos at the Sudbury Neutrino Observatory (SNO) in Ontario (Canada), cf. Figure 17.4 (left).

In 1998, first evidence for oscillations of *atmospheric neutrinos* was provided by the Super-Kamioka neutrino detector (Super-Kamiokande) in Japan. Atmospheric neutrinos are generated when cosmic rays (*e.g.*, containing high-energy protons) hit the earth's atmosphere, and produce electron- or muon-neutrinos and anti-neutrinos, often by creating light mesons that decay into leptons. Super-Kamiokande is equipped with Cherenkov detectors that can measure the direction of the incoming neutrinos. In this way, it was able to distinguish between neutrinos that have traveled different distances through the earth (sometimes hitting the detector from below after traveling through the entire earth and sometimes arriving directly from the atmosphere above), which implies different oscillation lengths, cf. Figure 17.4 (right). This allowed the determination of some parameters of the PMNS matrix.

Particle accelerators can produce well-controlled *neutrino beams* and direct them toward a neutrino detector that may be several hundred kilometers away. In this way, a ν_μ-beam from CERN near Geneva (Switzerland) has been directed to the OPERA detector deep inside the Gran Sasso mountain in Italy, 730 km away, which led to ν_τ-appearance via oscillations. Using the same technique, a neutrino beam from J-PARC in Japan was sent to the Super-Kamiokande detector and led to the measurement of a mixing parameter in the PMNS matrix.

Yet another method detects neutrinos in a short distance of about 2 km from a nuclear reactor. This led to the evidence that even the smallest element U_{e3} of the PMNS matrix is non-zero (An *et al.*, 2012; Ahn *et al.*, 2012; Abe *et al.*, 2012).

In this section, we illustrate the basic principle behind neutrino oscillations, without going into any details of the rich corresponding phenomenology. In practice, neutrinos that are produced in particle decays have very high energy compared to their tiny rest mass and are hence ultra-relativistic (in the lab frame). Then, one can treat the mass as a small perturbation and derive formulae for the probabilities of neutrino oscillations that are very useful in practical situations. Here, we proceed differently by simply considering neutrinos at zero momentum, *i.e.* in their rest frame. This situation may apply to neutrinos in the cosmic background of neutrinos that decoupled from the other particles about 1 second after the Big Bang. The initially very large momenta of these neutrinos have been red-shifted due to the expansion of the Universe, such that today some should be non-relativistic, with estimated kinetic energies around 10^{-4} to 10^{-6} eV. Unfortunately, the cosmic background neutrinos are undetectable with presently available technology.

Let us consider a neutrino at rest, *i.e.* with momentum $\vec{p} = \vec{0}$. If it is initially (at time $t = 0$) in one of the mass eigenstates $|\nu_i\rangle$ (and hence has the energy $E_i = \sqrt{\vec{p}^{\,2} + m_{\nu_i}^2} = m_{\nu_i}$), it remains in that eigenstate and undergoes the trivial time evolution

$$|\nu_i(t)\rangle = \exp(-\mathrm{i}E_i t)|\nu_i\rangle = \exp(-\mathrm{i}m_{\nu_i} t)|\nu_i\rangle. \qquad (17.45)$$

However, if the neutrino is produced in a weak decay, it is initially in an electroweak "flavor" eigenstate $f \in \{e, \mu, \tau\}$, which is a superposition of mass eigenstates

$$|\nu_f\rangle = \sum_{i=1}^{N_g} U_{fi}^\dagger |\nu_i\rangle = \sum_{i=1}^{N_g} U_{if}^* |\nu_i\rangle, \qquad (17.46)$$

whose time evolution is given by

$$|\Psi(t)\rangle = \sum_{i=1}^{N_g} U_{if}^* |\nu_i(t)\rangle = \sum_{i=1}^{N_g} U_{if}^* \exp(-\mathrm{i}m_{\nu_i} t)|\nu_i\rangle. \qquad (17.47)$$

The probability amplitude for detecting the neutrino at time t in an electroweak eigenstate with flavor f' is then given by

$$\langle v_{f'}|\Psi(t)\rangle = \sum_{i=1}^{N_g} U_{if}^* U_{if'} \exp(-\mathrm{i}m_{v_i}t). \tag{17.48}$$

For simplicity, let us consider just two generations ($N_g = 2$). The oscillation probability to detect an initial v_e (with $f = e$) at the later time t as v_μ (with $f' = \mu$) is then given by

$$\begin{aligned}
|\langle v_\mu|\Psi(t)\rangle|^2 &= \left|\sum_{i=1}^{2} U_{ie}^* U_{i\mu} \exp(-\mathrm{i}m_{v_i}t)\right|^2 \\
&= |U_{1e}|^2|U_{1\mu}|^2 + |U_{2e}|^2|U_{2\mu}|^2 \\
&\quad + 2\mathrm{Re}\left[U_{1e}U_{1\mu}^* U_{2e}^* U_{2\mu} \exp(\mathrm{i}(m_{v_1} - m_{v_2})t)\right].
\end{aligned} \tag{17.49}$$

For $N_g = 2$, the PMNS matrix can be parameterized as

$$U = \begin{pmatrix} U_{1e} & U_{1\mu} \\ U_{2e} & U_{2\mu} \end{pmatrix} = \begin{pmatrix} \cos\theta & \sin\theta\,\exp(\mathrm{i}\phi) \\ -\sin\theta\,\exp(-\mathrm{i}\phi) & \cos\theta \end{pmatrix} \in \mathrm{SU}(2). \tag{17.50}$$

It contains one real Euler angle θ and one complex phase ϕ. Inserting this parametrization in eq. (17.49), one obtains

$$|\langle v_\mu|\Psi(t)\rangle|^2 = \sin^2(2\theta)\sin^2\left(\frac{1}{2}(m_{v_1} - m_{v_2})t\right). \tag{17.51}$$

By measuring the time dependence of the oscillation probability, one can determine the absolute value of the neutrino mass difference $|m_{v_1} - m_{v_2}|$ (but not the individual masses, nor the sign of this difference) as well as the mixing angle $|\theta|$. It is also interesting that the CP-violating phase ϕ is not measurable in this way. It can be detected as an interference effect in a setup that is not time reversal-invariant.

A similar situation arises in the realistic case of three generations, in which there are three Euler angles $\theta_{ff'}$ and three CP-violating phases $\phi_{ff'}$ associated with the mixing among the different pairs of generations f and f'. The experimentally determined values of the neutrino mixing angles $\theta_{ff'}$ are much larger than the corresponding angles in the CKM quark mixing matrix. This is reflected in the plots of Figure 17.3.

17.8 Overview of Fundamental Standard Model Parameters

At this point, we have completed the construction of the Standard Model. Let us provide an overview over its fundamental parameters. While their values can be determined experimentally, they are not understood on theoretical grounds. In fact, achieving a deeper understanding of these (apparently) free parameters would require the discovery of even more fundamental structures underlying the Standard Model.

First, we concentrate on the *minimal Standard Model* with massless neutrinos and we consider its parameters in the order in which they appear in the book. Since the Standard Model is renormalizable, only a finite number of terms enter its Lagrangian (in particular, terms of dimension 5 or higher are excluded at this stage). Consequently, the number of fundamental Standard Model parameters is finite. (We refer to renormalized parameters throughout the following discussion.)

Chapter 12 addressed the Higgs sector of the Standard Model, whose Lagrangian contains two fundamental parameters – v and λ – which determine the quartic potential $V(\Phi) = \lambda(\Phi^\dagger\Phi - v^2)^2/4!$. The vacuum value $v \simeq 246\,\text{GeV}$ of the Higgs field is the only dimensioned parameter in the Lagrangian of the minimal Standard Model. Together with the dimensionless scalar self-coupling λ, it determines the Higgs boson mass $m_H = \sqrt{\lambda/3}\,v$.

Chapter 13 extended the Higgs sector by gauging its $SU(2)_L \times U(1)_Y$ symmetry. In this way, two additional fundamental parameters – the gauge couplings g and g' – appear, which (together with v) determine the masses of the W- and Z-bosons, $M_W = gv/2$ and $M_Z = \sqrt{g^2 + g'^2}\,v/2$, as well as the Weinberg angle $\cos\theta_W = g/\sqrt{g^2 + g'^2}$ and the electric charge $e = gg'/\sqrt{g^2 + g'^2}$.

In Chapter 14, the gluon field was added as the $SU(3)_c$ gauge field mediating the strong interaction, with the corresponding gauge coupling g_s as another fundamental Standard Model parameter. By dimensional transmutation, the dimensionless coupling g_s is traded for the dimensioned parameter Λ_{QCD} of eq. (B.16) that sets the energy scale below which the running gauge coupling becomes strong. The Higgs field expectation value v (which enters the theory via the Lagrangian) and Λ_{QCD} (which appears in the process of renormalization) are the only dimensioned, fundamental parameters of the minimal Standard Model.

One could argue that the weak hypercharges of the leptons and quarks, $Y_{l_L}, Y_{e_R}, Y_{q_L}, Y_{u_R}$, and Y_{d_R} should be counted as additional parameters (after setting $Y_\Phi = 1/2$ by convention). While this would indeed be correct at the classical level, as we have seen in Chapter 15, at the quantum level these parameters are fixed by the condition of gauge anomaly cancellation and should therefore not be counted as free parameters.

Several parameters arise from the Yukawa couplings between the Higgs field and the lepton and quark fields, which were addressed in Chapters 16 and 17. In the minimal Standard Model with massless neutrinos, there are three dimensionless Yukawa couplings f_e, f_μ, and f_τ, which determine the masses of the charged leptons $m_e = f_e v$, $m_\mu = f_\mu v$, and $m_\tau = f_\tau v$. Similarly, there are six Yukawa couplings f_u, f_d, f_s, f_c, f_b, and f_t, for the different quark flavors. There are four more parameters which determine the Cabibbo–Kobayashi–Maskawa quark mixing matrix: three angles (including the Cabibbo angle) and one CP-violating phase. Hence, including the Higgs self-coupling λ as well as the vacuum value v, altogether there are $2 + 3 + 6 + 4 = 15$ fundamental Standard Model parameters associated with the non-gauge interactions between Higgs and fermion fields. For N_g generations, there are N_g charged lepton masses and $2N_g$ quark masses, as well as $(N_g - 1)^2$ mixing parameters (including $(N_g - 1)(N_g - 2)/2$ complex CP-violating phases). Hence, the total number of parameters in the Higgs–fermion sector then is

$$2 + N_g + 2N_g + (N_g - 1)^2 = N_g(N_g + 1) + 3. \tag{17.52}$$

Besides the three gauge couplings g, g', and g_s, the gauge interactions also give rise to the strong and electromagnetic vacuum angles θ and θ_{QED}, to be discussed in Chapters 19 and 25. The weak interaction vacuum angle, on the other hand, is unphysical and can be absorbed by a field redefinition. The experimentally determined QCD vacuum angle $|\theta| \leq \mathcal{O}(10^{-10})$ (Zyla et al., 2020) leads to the *strong CP-problem*: Why is θ consistent with zero? Like understanding the value of any other fundamental Standard Model parameter, solving this problem requires physics beyond the Standard Model. The value of the electromagnetic vacuum angle θ_{QED} is not known. In fact, this parameter is often ignored, but it deserves more attention than it received in the past.

Altogether, the gauge and non-gauge interactions of the minimal Standard Model give rise to

$$3 + 2 + N_g(N_g + 1) + 3 = N_g(N_g + 1) + 8 \tag{17.53}$$

fundamental parameters, which amounts to 20 parameters for the physical case of $N_g = 3$ generations. Interestingly, for general odd $N_c \geq 3$, the number of parameters is independent of the number of colors. As discussed in Section 16.11, for $N_c = 1$ the number of parameters is larger, although g_s and θ are missing, due to the lack of the constraining $\mathrm{SU}(N_c)$ gauge symmetry of the strong interaction. In general, the $\mathrm{SU}(N_c) \times \mathrm{SU}(2)_L \times \mathrm{U}(1)_Y$ gauge symmetry substantially reduces the number of parameters, because all allowed terms in the Lagrangian must be gauge-invariant.

Often the Higgs field, which appears less natural than the gauge fields from a theoretical point of view, is blamed for the large number of Standard Model parameters. Indeed, in the absence of the Higgs field, the number of fundamental parameters would be much smaller. However, the physics would then be qualitatively different. In particular, ignoring a small QCD effect to be discussed in Section 23.6, the electroweak symmetry would remain unbroken and quarks and leptons would remain massless. Given the diverse dynamical roles that the Higgs field is playing, the number of parameters associated with its interactions may no longer seem excessive. In fact, one could as well blame the fermion fields for the proliferation of fundamental parameters. Indeed, if the fermions would not exist in three generations with identical quantum numbers, their mixing would be prevented by gauge invariance. With only one generation ($N_g = 1$), only three fundamental parameters would be associated with the Yukawa couplings, while for $N_g = 3$ there are thirteen parameters.

Since the discovery of neutrino oscillations, it is clear that neutrinos have mass. As discussed in Chapter 16, this can be accounted for by adding non-renormalizable dimension-5 terms to the minimal Standard Model, thus treating it as a low-energy effective theory. In this way, three neutrino mass parameters m_{ν_1}, m_{ν_2}, and m_{ν_3} enter the extended Lagrangian. Furthermore, as an analogue of the CKM matrix, the Pontecorvo–Maki–Nakagawa–Sakata (PMNS) lepton-mixing matrix arises. As discussed in Section 17.6, it contains three mixing angles as well as three CP-violating phases, so it contributes six additional fundamental parameters. Hence, in the non-renormalizable extension of the Standard Model, there are $3 + 6 = 9$ additional fundamental parameters at the level of dimension-5 terms. For a general number of generations, there are N_g neutrino masses as well as $N_g(N_g - 1)$ parameters in the PMNS lepton mixing matrix (half of them being complex CP-violating phases), such that there are N_g^2 additional parameters at the level of dimension-5 terms. When non-renormalizable interactions of dimension 6 and higher are included, the number of parameters increases rapidly.

Perturbative quantum gravity can actually be incorporated in the Standard Model, at least as a low-energy effective theory. To leading order, the corresponding Lagrangian is the one of classical General Relativity, with Newton's constant G (or equivalently the Planck mass M_{Planck}) and the cosmological constant Λ_c as two more dimensioned parameters.

Up to now, we have counted those fundamental Standard Model parameters that can take continuous values. In addition, there are many less obvious, *discrete* parameters, such as the number of fermion generations N_g or the number of quark colors N_c, which does play a prominent role in this book. Other discrete parameters are associated with the number of fundamental fields and their representations under the various gauge groups. What one considers a discrete parameter is a matter of choice, and thus, counting them is ambiguous. There are many deep questions beyond the Standard Model related to its discrete parameters, such as: Why are there three fermion generations? Why is the gauge group $SU(3)_c \times SU(2)_L \times U(1)_Y$? Why do quarks transform in the fundamental representation of $SU(3)_c$? Why are there three space and one time dimension? This list could easily be extended, and we have no clue how to answer any of these deep questions.

While the total number of continuous fundamental parameters may seem quite large, we should not forget that *all* other physical quantities (and, in fact, all other known natural phenomena, with the exception of dark matter and dark energy), at least in principle, originate from those parameters. An impressive, long list of successful Standard Model predictions can be found in the Particle Data Book. In addition, while many of the parameters – including the masses and mixing angles of the heavy quarks or the neutrinos – are relevant in particle physics, their exact values have hardly any direct impact on the rest of physics. Only a few parameters – namely, the masses of the light quarks and the electron, the Cabibbo angle, and the three gauge couplings, as well as v, determine – at least in principle – all of nuclear and atomic physics.

The question whether or not the number of parameters in some theory should be considered as "large" has to be decided based on the number of its successful predictions. As we just argued, applying this criterion to the Standard Model (or just to its low-energy sector) leads to the clear-cut conclusion that the set of its free parameters is actually *not* large.

Moreover, the Standard Model predictions are not only incredibly numerous, but they also attain a *precision* that science has not seen before. For instance, at present (in 2022) the magnetic moment of the muon is under intense debate: It is an open question whether its value exactly agrees with the Standard Model. However, even the possible deviation would only be of the relative magnitude of 10^{-10}. This is as if the Standard Model had predicted the distance between two points in Mexico City and Bern (Switzerland), which is about 10^4 km, and *possibly* made a mistake in the mm regime.

The concept of the Standard Model is simple: Introduce the quantum fields necessary for the known elementary particles, and assemble the Lagrangian with all terms compatible with locality, power-counting renormalizability, as well as Lorentz and gauge invariance. This involves some free parameters, but it leads to a huge number of predictions, which are confirmed with unprecedented precision. Therefore, regardless of possible extensions, the Standard Model certainly deserves our profound respect.

17.9   Low-Energy Theory Perspective on the Standard Model Physics

Let us look back at some of the effects that we have discussed within the Standard Model and reconsider them from a low-energy perspective. For this purpose, we may imagine a hypothetical, alternative historical development of particle physics, in which QCD would have been discovered already around 1960, when first experiments were able to reach energies around 10 GeV. Let us assume that the electromagnetic and strong interaction could already then have been described in terms of QED and QCD, while the weak interaction would not yet be understood at the gauge theory level. In this situation, the "standard model" would comprise QCD and QED with the gauge group $SU(3)_c \times U(1)_{em}$. However, the fact that $U(1)_{em}$ results from the "spontaneous breakdown" of the larger gauge group $SU(2)_L \times U(1)_Y$ would not yet have been discovered in this scenario. Instead, the weak interaction would manifest itself in the form of higher-dimensional operators.

This is, in fact, not so different from the actual historical development, in which Enrico Fermi's 4-fermion operators of dimension 6 (which he introduced to describe the β-decay of the neutron (Fermi, 1934)) provided an effective low-energy description of the weak interaction. The 4-fermion coupling G_F is related to the weak coupling g and M_W in a simple manner, $G_F = 2^{-5/2} g^2 / M_W^2$. In the actual historical development, QCD was formulated in the early 1970s, shortly after the rest of the Standard Model.

The hypothetical QCD-QED "standard model" is just the low-energy effective theory that emerges when one integrates out the W- and Z-boson fields as well as the Higgs field and the t-quark from the actual Standard Model. In the resulting scenario, the left-handed neutrinos would be "sterile" because they do not participate in the strong or electromagnetic gauge interactions, which are vector-like. This "standard model" would leave many questions unanswered:

- Why is electric charge quantized?
- How many quark flavors can we expect?
- Why are parity and charge conjugation violated?
- Why do quarks and electrons interact weakly with the same strength?
- Why are neutrinos so light?

All these questions are understood within the actual Standard Model, but not within the low-energy effective theory of QCD and QED.

- As we have seen in Section 16.7, electric charge quantization follows from anomaly cancellation, which is not an issue in vector-like theories such as QCD and QED. This would also prevent the prediction of the t-quark because it would no longer be required for anomaly cancellation within the third generation.
- The Standard Model determines the number of quark flavors based on gauge anomaly cancellation in a single generation.
- The explicit breaking of parity and charge conjugation finds a natural explanation in the transformation behavior of quark and lepton fields under the chiral gauge group $SU(2)_L \times U(1)_Y$.
- Left-handed quarks and electrons couple to neutrinos with the same strength, because there is only one universal coupling g associated with the gauge group $SU(2)_L$.
- Neutrinos are massless in the renormalizable Standard Model. As we discussed in Sections 16.3 and 16.9, they are naturally light due to the mass-by-mixing seesaw mechanism, which manifests itself by a highly suppressed dimension-5 term when the Standard Model is considered as an effective theory. In the QCD-QED "standard model", on the other hand, even the left-handed neutrino field is "sterile" and can thus be endowed with a dimension-3 Majorana mass term, which would not be small without unnatural fine-tuning.

Here, we see the great predictive power of $SU(2)_L \times U(1)_Y$ gauge theory at work. It would be wonderful if the higher-dimensional operators that correct the Standard Model would find a similar explanation by some future theory that extends it to even higher energies. Since 1960, within half a century, accelerator technology has advanced from 10 GeV to energies around 10 TeV. Can we perhaps today repeat the theoretical triumph of the Standard Model, which was constructed around 1970, about 40 years before it was fully experimentally tested by the discovery of the Higgs particle in 2012? We know the values of all parameters of the renormalizable Standard Model quite precisely, and we have already begun to gain insight into neutrino mixing, which can be associated with non-renormalizable dimension-5 corrections. How much do we need to learn before we can take the next leap toward an even deeper understanding of Nature at ultra-high energies? While experimental guidance is of utmost importance, new theoretical ideas are more than welcome as well.

Exercises

17.1 Gauging the generation symmetry?

Consider the usual three generations of quarks and leptons in the Standard Model, each extended by a right-handed neutrino field. Putting all Yukawa couplings to zero, there is a global $SU(3)_g$ symmetry that rotates the generations into one another. For example, the electron, muon, and tau-lepton form a triplet of $SU(3)_g$, and so do their neutrinos. The u-, c-, and t-quarks form another triplet, and finally the quarks d, s, and b are members of yet another $SU(3)_g$ triplet. Can the $SU(3)_g$ global symmetry be consistently promoted to a gauge symmetry at the quantum level? Does the situation change in the absence of the right-handed neutrino fields?

PART III

STRONG INTERACTION

18   Quantum Chromodynamics

This chapter is devoted to the theory of the strong interaction – *Quantum Chromodynamics* (QCD) – the $SU(3)_c$ gauge theory of quark and gluon fields. At energies in the GeV range, which are much smaller than the electroweak symmetry breaking scale $v \simeq 246$ GeV, the strong interaction is much stronger than the other fundamental forces. It is therefore a good approximation to neglect the electromagnetic and weak gauge interactions. In addition, the Higgs field can safely be replaced by its vacuum expectation value v, such that the Yukawa couplings just reduce to quark mass terms. Since the top quark is very heavy, it also has negligible effects on the strong interactions at low energies. The light quarks up, down, and strange have masses below the characteristic QCD energy scale $\Lambda_{QCD} \approx 0.33$ GeV and thus dominate the low-energy physics.

Single quarks have never been observed despite numerous experimental efforts. In fact, the strong interaction is so strong that quarks are confined. The confining force is mediated by the gluons. As a consequence of confinement, colored quarks and gluons form color-neutral hadrons which have integer electric charges. Hadrons include baryons (states with 3 more quarks than anti-quarks), anti-baryons (with 3 more anti-quarks than quarks), mesons (with the same number of quarks and anti-quarks), or more exotic creatures like glueballs (which are flavor-singlet states dominated by gluons). *Confinement* is a complicated dynamical phenomenon that is still not fully understood. Fortunately, lattice QCD provides a non-perturbative formulation of the theory that leads to quantitative results for the properties of hadrons from first principles. Monte Carlo calculations in the framework of lattice QCD face numerous challenges in order to reach the continuum limit in a large volume at physical values of the quark masses and thus require large-scale numerical resources. After decades of intense research, lattice QCD is now in a position to provide reliable results for a variety of physical quantities.

Usually, quark mass terms do not play a dominant role in the QCD dynamics. In particular, the masses of hadrons are not at all the sum of the masses of the quarks within them. Even with exactly massless quarks, due to confinement the hadrons (except for the Nambu–Goldstone bosons among them) would still have masses of the order of Λ_{QCD}. The strong interaction energy of quarks and gluons manifests itself as the mass of hadrons. Sometimes, one can read that the origin of mass in the Universe is the Higgs mechanism, and indeed we have seen that the quark masses would be zero if the Higgs potential would not have the Mexican hat shape. However, the dominant contribution to the mass of the ordinary matter that surrounds us is due to protons and neutrons, and thus due to QCD interaction energy.

Despite the fact that quarks do not exist as free particles, there is a lot of indirect experimental evidence for quarks (Bloom *et al.*, 1969; Breidenbach *et al.*, 1969), thanks to another fundamental property of the strong interaction. As discovered independently by David Gross and Frank Wilczek as well as by David Politzer, QCD is *asymptotically free* (Gross and Wilczek, 1973; Politzer, 1973). With increasing energy, quarks and gluons behave more and more like free particles, as observed in deep-inelastic lepton–hadron scattering processes (see Chapter 22). The high-energy physics of QCD is accessible to perturbative calculations, which predict the running of the strong coupling g_s depending on the energy scale.

When the mass terms of the light u- and d-quarks are neglected, the QCD Lagrangian has a global $SU(2)_L \times SU(2)_R$ chiral symmetry. Hence, one might expect corresponding degeneracies in the hadron spectrum. Since this is not what is actually observed, one concludes that chiral symmetry is spontaneously broken. Under spontaneous symmetry breaking, only the $SU(2)_{L=R}$ isospin subgroup of chiral symmetry remains unbroken. When a global, continuous symmetry breaks spontaneously, the Nambu–Goldstone phenomenon gives rise to a number of massless particles. In QCD with $N_f = 2$ quark flavors, the Nambu–Goldstone bosons are the three pions π^+, π^0, and π^-. Due to the small but non-zero quark masses, chiral symmetry is not purely spontaneously but also explicitly broken, and the pions are not exactly massless. Chiral symmetry leads to interesting predictions about the low-energy dynamics of QCD. A systematic method to investigate this is provided by *chiral perturbation theory*, which is based upon a low-energy pion effective Lagrangian (cf. Section 12.6 and Chapter 23).

Constructing a regularization of QCD beyond perturbation theory, which respects chiral symmetry, has been a long-standing problem. The lattice regularization gives rise to the fermion doubling problem, and the Nielsen–Ninomiya no-go theorem suggested that chiral symmetry cannot be maintained on the lattice. However, as we discussed in Chapter 10, the Ginsparg–Wilson relation implements chiral symmetry in a lattice-modified form. This led to a breakthrough in our understanding of chiral symmetry in general. In particular, domain wall fermions (Kaplan, 1992), which naturally possess a chiral symmetry beyond perturbation theory without fine-tuning, solve an under-appreciated fermionic hierarchy problem. This may even be a concrete hint to the physical reality of extra dimensions.

18.1 Deconstructing the Standard Model

In Part II, we have constructed the Standard Model step by step, starting from the bosonic Higgs and gauge sectors, and then adding the leptons and quarks of the various fermion generations. In order to arrive at QCD alone, we need to "deconstruct" the Standard Model, removing all degrees of freedom that are not strongly interacting at low energies. Mathematically, we can achieve this by switching off the electroweak gauge interactions, *i.e.* by putting $g = g' = 0$, and by freezing the Higgs field to its vacuum value v. The other fields then decouple from the quark and gluon fields, which remain strongly coupled to each other via the strong gauge coupling g_s.

We recall that the gluons are described by an $SU(3)_c$ algebra-valued gauge potential

$$G_\mu(x) = i g_s G_\mu^a(x) \frac{\lambda^a}{2}, \quad a \in \{1, 2, \ldots, 8\}, \tag{18.1}$$

which transforms as

$$G'_\mu(x) = \Omega(x) \left(G_\mu(x) + \partial_\mu \right) \Omega(x)^\dagger, \tag{18.2}$$

under gauge transformations $\Omega(x) \in SU(3)$. The gluon field strength tensor

$$G_{\mu\nu}(x) = \partial_\mu G_\nu(x) - \partial_\nu G_\mu(x) + \left[G_\mu(x), G_\nu(x)\right], \tag{18.3}$$

transforms as

$$G'_{\mu\nu}(x) = \Omega(x)G_{\mu\nu}(x)\Omega(x)^\dagger, \tag{18.4}$$

as we saw in Problem 11.3. The Euclidean QCD Lagrangian takes the gauge-invariant form

$$\mathcal{L}_{QCD}(\bar{q}, q, G) = \sum_f \bar{q}^f \left[\gamma_\mu \left(\partial_\mu + G_\mu\right) + m_f\right] q^f - \frac{1}{2g_s^2} \text{Tr}\left[G_{\mu\nu}G_{\mu\nu}\right]. \tag{18.5}$$

The quark field $q^f(x) = q_L^f(x) + q_R^f(x)$ with the flavor index $f \in \{u, d, s, c, b, t\}$ is just a collection of the quark fields we have introduced before. For example, $q^u(x) = u_L(x) + u_R(x)$.

Although left- and right-handed quarks couple differently to the weak $SU(2)_L$ gauge field (which makes the Standard Model a chiral gauge theory), gluons do not distinguish between left- and right-handed quarks, which are both color triplets. Similarly, left- and right-handed anti-quarks are color anti-triplets. As a result, QCD is a *vector-like gauge theory*; the gluon field is chirality-blind. As a consequence, QCD does not suffer from gauge anomalies and is thus mathematically consistent. In particular, unlike the full Standard Model, it does not require gauge anomaly cancellation between quarks and leptons.

The QCD Lagrangian is renormalizable, which allows us to completely remove the ultraviolet cut-off. Unlike the full Standard Model, which is affected by the triviality of the Higgs sector, QCD alone could, at least in principle, be valid at arbitrarily high energy scales, provided that the quark masses are inserted by hand, rather than being realized via Yukawa couplings. In reality, QCD is embedded into the Standard Model, of which it is a strongly coupled low-energy effective theory, and it is indeed affected by the Higgs sector as soon as one reaches energies around the electroweak scale $v \simeq 246$ GeV.

18.2 Asymptotic Freedom

Like the couplings in other quantum field theories, in QCD the strong coupling g_s varies with the energy scale. Remarkably, with increasing energy scale, this coupling becomes weaker and weaker, thus leading to free quarks and gluons in the ultimate, asymptotic high-energy limit. *Asymptotic freedom* results from a crucial difference between Abelian and non-Abelian gauge theories, namely, that non-Abelian gauge fields are themselves charged with respect to their gauge group. The non-Abelian charge of the gluons leads to a self-interaction that is not present for the (Abelian) photons.

The objects in the QCD Lagrangian do not directly correspond to observable quantities. Both fields and couplings get renormalized. In particular, the formal expression

$$Z = \int \mathcal{D}\bar{q}\mathcal{D}q\mathcal{D}G \, \exp\left(-\int d^4x \, \mathcal{L}_{QCD}(\bar{q}, q, G)\right), \tag{18.6}$$

for the QCD functional integral is, *a priori*, mathematically undefined, *i.e.* divergent, until it is regularized and appropriately renormalized. Perturbative regularization schemes operate at the level of individual Feynman diagrams. The gluon self-interaction results from the commutator term in the gluon field strength in eq. (18.3). It gives rise to *3- and 4-gluon vertices* in the QCD Feynman rules. We will not derive these Feynman rules, but discuss them only at a qualitative level. The terms in the Lagrangian that are quadratic in G_μ give rise to the *gluon propagator*. In addition, there is a *quark propagator* and a *quark–gluon vertex*.

In gauge theories, it is essential that gauge invariance is preserved in the regularized theory. However, as we saw in Chapter 11, the perturbative quantization of a non-Abelian gauge theory requires gauge fixing. Otherwise, the functional integral over the gluon field would contribute an uncontrolled, field-dependent, infinite factor to the functional integral. The Lorenz–Landau gauge constraint, $\partial_\mu G_\mu(x) = 0$, is then enforced by integrating Faddeev–Popov ghost fields (Faddeev and Popov, 1967) into the functional integral. In QCD, the ghosts are color octets. Correspondingly, there is a ghost propagator and a ghost-gluon vertex. In contrast to quarks and gluons, ghosts arise only in the loops of a Feynman diagram, not in external legs. As we saw in Section 11.4, their purpose is to cancel the effects of unphysical longitudinal gauge degrees of freedom. Ghosts are mathematical integration variables in the functional integral that leave no trace in the particle spectrum of the theory. In some sense, one could say the same about quarks and gluons, because ultimately, due to confinement, they cannot exist as asymptotic states either. However, as we will discuss in Chapter 22, their physical "reality" (which is not meant in a philosophical, ontological sense) has been verified indirectly in detail by deep-inelastic lepton–hadron scattering experiments (Bloom *et al.*, 1969; Breidenbach *et al.*, 1969).

In Section 18.5, we will consider lattice QCD which defines the theory beyond perturbation theory. Lattice QCD is formulated in terms of quark and gluon fields only, without the need to introduce ghost fields. As we have already seen in Chapter 14, ghosts are not needed in the non-perturbative formulation of the pure gluon theory, because the link variables that represent the lattice gluon field are integrated over the compact group manifold. Without gauge fixing, the lattice functional integral extends over all gauge copies. This contributes a factor, proportional to the volume of the gauge group manifold, for each lattice point. This factor is irrelevant because it is field-independent and cancels in all expectation values. Still, ghost fields appear when one engages in lattice perturbation theory calculations.

The loop integrations in Feynman diagrams can be divergent in four dimensions. As we discussed in detail in Chapter 4, in dimensional regularization one formally works in $d \in \mathbb{C}$ dimensions (by analytic continuation in d) and one performs the limit $\varepsilon = 4 - d \to 0$, such that the physics remains constant. In order to absorb the divergences, quark and gluon fields are renormalized

$$q(x) = Z_q(\varepsilon)^{1/2} q^{\mathrm{r}}(x), \quad G_\mu(x) = Z_G(\varepsilon)^{1/2} G_\mu^{\mathrm{r}}(x), \tag{18.7}$$

and also the coupling is renormalized via

$$g_{\mathrm{s}} = \frac{Z(\varepsilon)}{Z_q(\varepsilon) Z_G(\varepsilon)^{1/2}} g_{\mathrm{s}}^{\mathrm{r}}. \tag{18.8}$$

Here, the unrenormalized quantities as well as the Z-factors are divergent, but the renormalized quantities are finite in the limit $\varepsilon \to 0$. Correspondingly, one renormalizes the n-point functions and the resulting vertex functions as

$$\Gamma^{r}_{n_q,n_G}(k_i,p_j) = \lim_{\varepsilon \to 0} Z_q(\varepsilon)^{n_q/2} Z_G(\varepsilon)^{n_G/2} \Gamma_{n_q,n_G}(k_i,p_j,\varepsilon). \tag{18.9}$$

Here, n_q and n_G denote the numbers of quark and gluon fields in the n-point function, while k_i and p_j are the corresponding momenta. Demanding convergence of the renormalized vertex function fixes the divergent part of the Z-factors. In order to fix the finite part as well, one must specify renormalization conditions. In QCD, this can be done using the vertex functions $\Gamma_{0,2}$, $\Gamma_{2,0}$, and $\Gamma_{2,1}$, $i.e.$ the inverse gluon and quark propagators as well as the quark–gluon vertex. As opposed to QED, where mass and charge of the electron are directly observable, in QCD one uses an arbitrarily chosen energy scale μ (not to be confused with a space–time index) to formulate the renormalization conditions

$$\Gamma^{r}_{0,2}(p,-p)^{ab}_{\nu\rho}|_{p^2=\mu^2} = (p_\nu p_\rho - \delta_{\nu\rho}p^2)\delta_{ab},$$
$$\Gamma^{r}_{2,0}(k,k)|_{k^2=\mu^2} = \gamma_\nu k_\nu,$$
$$\Gamma^{r}_{2,1}(k,k,k)^{a}_{\nu}|_{k^2=\mu^2} = i g^{r}_s \gamma_\nu \frac{\lambda^a}{2}. \tag{18.10}$$

The renormalized vertex functions are functions of the renormalized coupling g^{r}_s and of the renormalization scale μ, while the unrenormalized vertex functions depend on the bare coupling g_s and on the regularization parameter ε (which replaces a cut-off in dimensional regularization). Hence, as we discussed for scalar field theory in Chapter 5, there is an implicit relation

$$g^{r}_s = g^{r}_s(g_s,\varepsilon,\mu), \tag{18.11}$$

which defines the β-function

$$\beta\left(g^{r}_s\right) = \lim_{\varepsilon \to 0} \mu \frac{\partial}{\partial \mu} g^{r}_s(g_s,\varepsilon,\mu). \tag{18.12}$$

The β-function can be computed in QCD perturbation theory. To leading order in the coupling, one obtains

$$\beta\left(g^{r}_s\right) = -\frac{\left(g^{r}_s\right)^3}{16\pi^2}\left(11 - \frac{2}{3}N_f\right). \tag{18.13}$$

Here, N_f is the number of quark flavors. As we discussed in Chapter 5, fixed points g^{r*}_s of the renormalization group are of special interest. They are invariant under a change of the arbitrarily chosen renormalization scale μ; and hence, they correspond to zeros of the β-function. In QCD, there is a single Gaussian fixed point at $g^{r*}_s = 0$. For

$$11 - \frac{2}{3}N_f > 0 \quad \Rightarrow \quad N_f \le 16, \tag{18.14}$$

$i.e.$ for not too many quark flavors, the β-function is $negative$ close to the fixed point. This feature is responsible for asymptotic freedom. It is typical for non-Abelian gauge theories in four space–time dimensions, as long as there are not too many fermions or scalars involved. Asymptotic freedom is due to the self-interaction of the gauge field (which gives rise to the contribution 11) that is not present in an Abelian theory. We now use

$$\beta\left(g_s^r\right) = \mu\frac{\partial}{\partial\mu}g_s^r = -\frac{\left(g_s^r\right)^3}{16\pi^2}\left(11 - \frac{2}{3}N_f\right) \quad \Rightarrow$$

$$\frac{\partial g_s^r}{\partial\mu}\frac{1}{\left(g_s^r\right)^3} = \frac{1}{2}\frac{\partial\left(g_s^r\right)^2}{\partial\mu}\frac{1}{\left(g_s^r\right)^4} = -\frac{11 - \frac{2}{3}N_f}{16\pi^2}\frac{1}{\mu} \quad \Rightarrow$$

$$\frac{\partial\left(g_s^r\right)^2}{\left(g_s^r\right)^4} = -\frac{33 - 2N_f}{24\pi^2}\frac{\partial\mu}{\mu} \quad \Rightarrow \quad \frac{1}{\left(g_s^r\right)^2} = \frac{33 - 2N_f}{24\pi^2}\log\frac{\mu}{\Lambda_{QCD}}. \qquad (18.15)$$

Here, Λ_{QCD} enters as an integration constant. Introducing the renormalized strong "fine-structure constant" $\alpha_s^r = \left(g_s^r\right)^2/4\pi$, one obtains

$$\alpha_s^r(\mu) = \frac{6\pi}{33 - 2N_f}\frac{1}{\log\left(\mu/\Lambda_{QCD}\right)}. \qquad (18.16)$$

At high energy scales μ, the renormalized coupling slowly (*i.e.* logarithmically) approaches zero. Hence, quarks and gluons then behave like free particles. This behavior indeed matches experimental data. Lattice QCD studies revealed $\Lambda_{QCD} = 0.332(14)$ GeV, using $N_f = 3$ and the $\overline{MS}$ renormalization scheme (Bruno *et al.*, 2016). The great importance of asymptotic freedom was first understood and successfully elaborated by Gross and Wilczek and independently by Politzer (Gross and Wilczek, 1973; Politzer, 1973).

The QCD Lagrangian with massless fermions has no dimensioned parameter. Hence, the classical theory is scale-invariant; *i.e.* each classical solution with energy E can be mapped onto other solutions with scaled energies λE, for any arbitrary scale parameter $\lambda > 0$. Scale invariance, however, is *anomalous*: It does not survive the quantization of the theory. This is the reason why there is a proton with a very specific mass M_p, but no scaled version of it with mass λM_p, and the same for all other hadrons. We can now understand better why this is the case. In the process of quantization, the dimensioned scale μ (and related to it Λ_{QCD}) emerges, thus leading to an explicit breaking of the scale invariance of the classical theory. As we already discussed in Chapter 14, the process of trading the dimensionless coupling g_s of the classical theory for the dimensioned scale Λ_{QCD} in the quantum theory is known as *dimensional transmutation*. As a result, scale transformations are not a symmetry of QCD, not even when all quarks are massless.

18.3 Structure of Chiral Symmetry

Chiral symmetry is an approximate global symmetry of the QCD Lagrangian which results from the fact that the u- and d-quark masses – and to some extent even the mass of the heavier s-quark – are small compared to the QCD scale Λ_{QCD}. Neglecting the u- and d-quark masses, the QCD Lagrangian is invariant against separate U(2) transformations of the left- and right-handed quarks, such that we have a U(2)$_L$ × U(2)$_R$ global symmetry. We can decompose each U(2) symmetry into an SU(2) and a U(1) part, and hence, we obtain SU(2)$_L$ × SU(2)$_R$ × U(1)$_L$ × U(1)$_R$. The U(1)$_B$ = U(1)$_{L=R}$ symmetry describes the simultaneous rotations of left- and right-handed quarks and corresponds to baryon number conservation. The remaining so-called axial $U(1)_A = U(1)_{L=R^*}$ symmetry is broken by the Adler–Bell–Jackiw anomaly (cf. Chapter 19). Thus, it is explicitly broken by quantum effects, so it is not a symmetry of QCD. In Chapter 20, we will return to the $U(1)_A$-problem related to this symmetry.

Let us consider the quark contribution to the QCD Lagrangian with N_f flavors,

$$\mathcal{L}(\bar{q}, q, G) = \bar{q} \left[\gamma_\mu \left(\partial_\mu + G_\mu \right) + \mathcal{M} \right] q, \qquad (18.17)$$

where $\mathcal{M} = \mathrm{diag}(m_u, m_d, m_s, \ldots, m_{N_f})$ is the diagonal mass matrix of the quarks. Decomposing the quark fields into left- and right-handed components

$$q_L(x) = \frac{1 - \gamma_5}{2} q(x), \quad q_R(x) = \frac{1 + \gamma_5}{2} q(x), \quad q(x) = q_L(x) + q_R(x),$$

$$\bar{q}_L(x) = \bar{q}(x) \frac{1 + \gamma_5}{2}, \quad \bar{q}_R(x) = \bar{q}(x) \frac{1 - \gamma_5}{2}, \quad \bar{q}(x) = \bar{q}_L(x) + \bar{q}_R(x), \quad (18.18)$$

turns the quark contribution to the Lagrangian into

$$\mathcal{L}(\bar{q}_L, q_L, \bar{q}_R, q_R, G) = \bar{q}_L \bar{\sigma}_\mu \left(\partial_\mu + G_\mu \right) q_L + \bar{q}_R \sigma_\mu \left(\partial_\mu + G_\mu \right) q_R$$
$$+ \bar{q}_L \mathcal{M} q_R + \bar{q}_R \mathcal{M} q_L. \qquad (18.19)$$

The derivative term decomposes into two decoupled contributions from left- and right-handed quarks. This part of the Lagrangian is hence invariant against separate $U(N_f)$ transformations of the left- and right-handed components in flavor space

$$q'_L(x) = L q_L(x), \quad \bar{q}'_L(x) = \bar{q}_L(x) L^\dagger, \quad L \in U(N_f)_L,$$
$$q'_R(x) = R q_R(x), \quad \bar{q}'_R(x) = \bar{q}_R(x) R^\dagger, \quad R \in U(N_f)_R. \qquad (18.20)$$

Without the mass term, the QCD Lagrangian thus has a $U(N_f)_L \times U(N_f)_R$ symmetry. As mentioned before, and as we will discuss in detail in Chapter 19, there is an anomaly in the axial $U(1)_A = U(1)_{L=R^*}$ symmetry and hence the chiral symmetry of the quantum theory is reduced to

$$SU(N_f)_L \times SU(N_f)_R \times U(1)_{L=R} = SU(N_f)_L \times SU(N_f)_R \times U(1)_B. \qquad (18.21)$$

Still, chiral symmetry is only approximate, because the mass term couples right- and left-handed fermions. In addition, the mass matrix does not commute with R or L. Only if all quarks had the same mass, i.e. if $\mathcal{M} = m\mathbb{1}$, one would have

$$\bar{q}'_L(x) \mathcal{M} q'_R(x) = \bar{q}_L(x) R^\dagger m \mathbb{1} L q_R(x) = \bar{q}_L(x) R^\dagger L \mathcal{M} q_R(x). \qquad (18.22)$$

In that case, the mass term would be invariant against simultaneous transformations $R = L$, such that $R^\dagger L = R^\dagger R = \mathbb{1}$. Hence, for equal quark masses, chiral symmetry is explicitly broken to

$$SU(N_f)_{L=R} \times U(1)_{L=R} = SU(N_f)_f \times U(1)_B. \qquad (18.23)$$

This corresponds to the flavor and baryon number symmetry.

In the real world, the quark masses are different, and the symmetry is, in fact, explicitly broken to

$$\prod_f U(1)_f = U(1)_u \times U(1)_d \times U(1)_s \times \cdots \times U(1)_{N_f}. \qquad (18.24)$$

While the masses of the u- and d-quarks turn out to be quite different (cf. Section 23.4), it is most important that they are both small compared to Λ_{QCD} and can hence almost be neglected. Therefore, the $SU(2)_L \times SU(2)_R \times U(1)_B \times U(1)_s$ symmetry is hardly broken explicitly. Since the s-quark is heavier, $SU(3)_L \times SU(3)_R \times U(1)_B$ is a more approximate chiral symmetry, because it is more strongly, explicitly broken. It should be pointed out that, in the context of QCD, the $SU(2)_L$ symmetry (which is a gauge symmetry in the full Standard Model) is reduced to an approximate global symmetry, as a consequence of replacing the Higgs field by its vacuum expectation value.

18.4 Dynamical Realization of Chiral Symmetry

Let us first discuss the $N_f = 2$ case of u- and d-quarks only. Since the masses of these quarks are very small, the $SU(2)_L \times SU(2)_R \times U(1)_B$ chiral symmetry is almost exact. Then, the chiral symmetry is explicitly broken only very weakly, so one might expect corresponding degeneracies in the QCD spectrum. As we will discuss in detail in Chapter 21, hadrons can be classified as isospin multiplets. The isospin transformations are $SU(2)_I$ rotations that act on left- and right-handed fermions simultaneously, *i.e.* $SU(2)_I = SU(2)_{L=R}$. The symmetry that is manifest in the spectrum is indeed $SU(2)_I \times U(1)_B$, but not the full chiral symmetry $SU(2)_L \times SU(2)_R \times U(1)_B$. One concludes that chiral symmetry must be spontaneously broken.

Let us neglect the quark masses, such that chiral symmetry is not broken explicitly. As a consequence, there are exactly conserved currents

$$j_\mu^{La}(x) = \bar{q}_L(x)\bar{\sigma}_\mu \frac{\tau^a}{2} q_L(x), \quad j_\mu^{Ra}(x) = \bar{q}_R(x)\sigma_\mu \frac{\tau^a}{2} q_R(x), \qquad (18.25)$$

where $a \in \{1, 2, 3\}$ is a flavor (isospin) index and τ^a are the Pauli matrices for isospin. From the left- and right-handed currents, we construct a vector and an axial-vector current

$$j_\mu^a(x) = j_\mu^{Ra}(x) + j_\mu^{La}(x) = \bar{q}(x)\gamma_\mu \frac{\tau^a}{2} q(x),$$

$$j_\mu^{5a}(x) = j_\mu^{Ra}(x) - j_\mu^{La}(x) = \bar{q}(x)\gamma_\mu \gamma_5 \frac{\tau^a}{2} q(x). \qquad (18.26)$$

Let us consider an $SU(2)_L \times SU(2)_R$ invariant state $|\Phi\rangle$ as a candidate for the QCD vacuum. One then obtains

$$\langle\Phi|j_\mu^{La}(x)j_\nu^{Rb}(y)|\Phi\rangle = \langle\Phi|j_\mu^{Ra}(x)j_\nu^{Lb}(y)|\Phi\rangle = 0 \quad \Rightarrow$$

$$\langle\Phi|j_\mu^a(x)j_\nu^b(y)|\Phi\rangle = \langle\Phi|j_\mu^{5a}(x)j_\nu^{5b}(y)|\Phi\rangle. \qquad (18.27)$$

On both sides of the last equation, one can insert complete sets of states between the currents. On the left-hand side, states with quantum numbers $J^P = 0^+, 1^-, \ldots$ contribute, while on the right-hand side the non-vanishing contributions come from states $0^-, 1^+, \ldots$. The two expressions can be equal only if the corresponding parity partners are energetically degenerate.

The observed particle spectrum consists of hadrons which are confined bound states of quarks and gluons. Baryons are states containing three more quarks than anti-quarks, while anti-baryons contain three more anti-quarks than quarks. Meson states, on the other hand, contain the same number of quarks and anti-quarks. We will discuss the hadron spectrum in more detail in Chapter 21. In the observed hadron spectrum, there is no degeneracy of particles with even and odd parity, not even approximately. We conclude that the $SU(2)_L \times SU(2)_R$ invariant state $|\Phi\rangle$ is *not* the actual QCD vacuum. The true QCD vacuum $|0\rangle$ cannot be chirally invariant. The same is true for all other eigenstates of the QCD Hamilton operator. This means that chiral symmetry must be spontaneously broken.

Let us now briefly switch to the Hamiltonian formulation and construct the operators that describe the conserved flavor-non-singlet vector and axial charges

$$\hat{Q}^a = \int d^3x\, \hat{q}(\vec{x})^\dagger \frac{\tau^a}{2} \hat{q}(\vec{x}), \quad \hat{Q}^{5a} = \int d^3x\, \hat{q}(\vec{x})^\dagger \gamma^5 \frac{\tau^a}{2} \hat{q}(\vec{x}). \tag{18.28}$$

We now consider the states $\hat{Q}^a|0\rangle$ and $\hat{Q}^{5a}|0\rangle$, constructed from the vacuum by acting with the vector and axial charges. If the vacuum were chirally symmetric, we would have

$$\hat{Q}^a|\Phi\rangle = \hat{Q}^{5a}|\Phi\rangle = 0. \tag{18.29}$$

The true QCD vacuum $|0\rangle$ is not chirally invariant and, in fact,

$$\hat{Q}^{5a}|0\rangle \neq 0. \tag{18.30}$$

Since for massless quarks the axial current is conserved, we have

$$[\hat{H}_{QCD}, \hat{Q}^{5a}] = 0 \quad \Rightarrow \quad \hat{H}_{QCD}\hat{Q}^{5a}|0\rangle = \hat{Q}^{5a}\hat{H}_{QCD}|0\rangle = 0. \tag{18.31}$$

Hence, the state $\hat{Q}^{5a}|0\rangle$ is yet another eigenstate of the QCD Hamilton operator with zero energy. Such a state corresponds to a massless Nambu–Goldstone boson with quantum numbers $J^P = 0^-$ in the limit of vanishing momentum. For two flavors ($N_f = 2$), these pseudo-scalar particles are identified with the pions π^+, π^0, and π^-.

If one would also have $\hat{Q}^a|0\rangle \neq 0$, the vector flavor symmetry, which for $N_f = 2$ corresponds to isospin $SU(2)_I = SU(2)_{L=R}$, would also be spontaneously broken, and there would be another set of scalar Nambu–Goldstone bosons with $J^P = 0^+$. Such particles do not exist in the hadron spectrum, and we conclude that the isospin symmetry is not spontaneously broken. Up to small effects due to explicit isospin breaking terms, the $SU(2)_I$ symmetry is indeed manifest in the QCD spectrum.

As we discussed in Section 12.4, when a continuous global symmetry breaks spontaneously, massless particles – the Nambu–Goldstone bosons – appear in the spectrum. According to the Goldstone theorem, the number of massless bosons is the difference of the number of generators of the full symmetry group and the subgroup that remains intact after spontaneous breaking. When the chiral symmetry $SU(2)_L \times SU(2)_R \times U(1)_B$ breaks spontaneously to $SU(2)_I \times U(1)_B$, we hence expect $3 + 3 + 1 - 3 - 1 = 3$ Nambu–Goldstone bosons. In QCD, these are identified as the pions π^+, π^0, and π^-. Indeed, the pions are light – they are by far the lightest hadrons – but they are not massless. This is due to the explicit chiral symmetry breaking induced by the non-zero masses of the u- and d-quarks. Hence, since chiral symmetry is only approximate, the pions are light *pseudo-Nambu–Goldstone bosons*.

When the still relatively light s-quark is included, the chiral symmetry is extended to $SU(3)_L \times SU(3)_R \times U(1)_B$, which then breaks spontaneously to $SU(3)_f \times U(1)_B$. In that case, one expects $8 + 8 + 1 - 8 - 1 = 8$ Nambu–Goldstone bosons. The five additional bosons are identified as the four kaons K^+, K^0, $\overline{K^0}$, K^-, and the η-meson. Since the s-quark is heavier than the u- and d-quarks, these pseudo-Nambu–Goldstone bosons are heavier than the pions.

The pseudo-Nambu–Goldstone bosons are the lightest particles in QCD. Therefore, they determine the dynamics at low energies. One can construct effective theories that are applicable in the low-energy regime, and that are formulated entirely in terms of Nambu–Goldstone boson fields, as we sketched in Section 12.6. At low energies, the Nambu–Goldstone bosons interact only weakly and can hence be treated perturbatively. This is done systematically in chiral perturbation theory, which will be discussed in Chapter 23.

One can detect spontaneous chiral symmetry breaking by investigating the chiral order parameter

$$\langle \bar{q}q \rangle = \langle 0|\bar{q}(x)q(x)|0\rangle = \langle 0|\Big(\bar{q}_{\mathrm{L}}(x)q_{\mathrm{R}}(x) + \bar{q}_{\mathrm{R}}(x)q_{\mathrm{L}}(x)\Big)|0\rangle. \qquad (18.32)$$

The order parameter is invariant against simultaneous transformations $R = L$, but not against general chiral rotations. If chiral symmetry would be intact, the chiral condensate would vanish. When the symmetry is spontaneously broken, on the other hand, $\langle \bar{q}q \rangle \neq 0$.

18.5 Lattice QCD

In Chapter 10, we have considered free lattice fermions, and in Chapter 14, we have applied the lattice regularization to the Yang–Mills theory of the pure gluon field. Here, we introduce lattice QCD as a non-perturbative regularization of the theory of the strong interaction. It should be stressed that the QCD Lagrangian endowed with Feynman rules and dimensional regularization defines the theory only order by order in perturbation theory. This is known to yield a non-convergent, asymptotic series, which is, however, still extremely useful at high energies.

Lattice QCD should hence not be viewed as an approximation to an *a priori* well-defined continuum theory, but rather as a non-perturbative *definition* of QCD. As was proved rigorously by Thomas Reisz: At the level of perturbation theory, lattice QCD defines the same continuum theory as dimensional regularization (Reisz, 1988a,b, 1989). Since the lattice explicitly breaks Poincaré invariance, additional vertices enter perturbative calculations. Therefore, lattice perturbation theory is quite a bit more tedious than continuum perturbation theory. Hence, lattice gauge theory is usually not the method of choice for perturbative calculations, but is primarily used beyond perturbation theory.

In practice, non-perturbative calculations are possible because lattice QCD can be treated with numerical Monte Carlo calculations. Thanks to numerous theoretical and algorithmic breakthroughs, over the decades numerical lattice QCD has reached a stage at which it provides quantitative, percent-level results for the strong interaction dynamics from QCD first principles. This requires controlling a number of systematic errors, which are due to effects of a finite space–time volume, artifacts due to a non-zero lattice spacing, as well as deviations from the physical values of the quark masses. Using lattice QCD, a wide variety of physical quantities, ranging from a multitude of hadron properties to QCD at finite temperature, is being computed reliably with high accuracy.

Lattice QCD is a very interesting and broad field of its own, which has been described in several textbooks (Creutz, 1983; Rothe, 1992; Montvay and Münster, 1994; Smit, 2002; DeGrand and DeTar, 2006; Gattringer and Lang, 2009; Knechtli *et al.*, 2017). Here, we restrict ourselves to the main theoretical structures underlying lattice QCD. In particular, we do not discuss QCD-specific algorithmic techniques and will just refer to a few numerical results. Generalities of the Monte Carlo method are summarized in Appendix H.

We recall that the gluon field is described by parallel transporters $U_{x,\mu} \in SU(3)_c$ associated with the links that connect neighboring lattice sites x and $x + \hat{\mu}$ of a 4-dimensional Euclidean space–time lattice with spacing a. Here, $\hat{\mu}$ is a vector of length a pointing in the μ-direction. As we discussed for free fermions in Section 10.5, we start out by describing quarks as Wilson fermions. The lattice quark fields, which are associated with the lattice sites x, are represented by two independent Grassmann fields Q_x and $\bar{Q}_x$. In order to couple the quarks to the gluons, we gauge $SU(3)_c$ in the action of free Wilson fermions of eq. (10.20),

$$
\begin{aligned}
S_{\text{QCD}}[\bar{Q}, Q, U] &= a^4 \sum_{x,\mu} \frac{1}{2a} (\bar{Q}_x \gamma_\mu U_{x,\mu} Q_{x+\hat{\mu}} - \bar{Q}_{x+\hat{\mu}} \gamma_\mu U_{x,\mu}^\dagger Q_x) + a^4 \sum_x \bar{Q}_x \mathcal{M} Q_x \\
&+ a^4 \sum_{x,\mu} \frac{1}{2a} (2\bar{Q}_x Q_x - \bar{Q}_x U_{x,\mu} Q_{x+\hat{\mu}} - \bar{Q}_{x+\hat{\mu}} U_{x,\mu}^\dagger Q_x) \\
&- a^4 \sum_{x,\mu,\nu} \frac{1}{g_s^2 a^4} \text{Tr} \left[U_{x,\mu} U_{x+\hat{\mu},\nu} U_{x+\hat{\nu},\mu}^\dagger U_{x,\nu}^\dagger + U_{x,\nu} U_{x+\hat{\nu},\mu} U_{x+\hat{\mu},\nu}^\dagger U_{x,\mu}^\dagger \right].
\end{aligned}
$$

(18.33)

In order to eliminate the doubler fermions, we have included the Wilson term which breaks chiral symmetry explicitly. The lattice-regularized QCD functional integral then takes the form

$$
Z = \prod_x \int d\bar{Q}_x dQ_x \prod_{y,\mu} \int_{SU(3)} dU_{y,\mu} \exp\left(-S_{\text{QCD}}[\bar{Q}, Q, U]\right).
$$

(18.34)

In contrast to the continuum, due to the compact nature of the gauge group, no gauge fixing is necessary on the lattice. Hence, in this case, one need not introduce Faddeev–Popov ghost fields, at least as long as one does not engage in lattice perturbation theory calculations.

The lattice QCD functional integral depends on the quark mass matrix $\mathcal{M}$ as well as on the bare gauge coupling g_s. Due to asymptotic freedom, in order to reach the continuum limit, one must take the limit $g_s \to 0$. When one puts $\mathcal{M} = 0$ for free Wilson fermions, one reaches the chiral limit of massless quarks. In the interacting theory, on the other hand, this is no longer the case. In particular, since chiral symmetry is explicitly broken by the Wilson term, the bare quark masses must be fine-tuned in order to reach a massless limit. This fine-tuning is unnatural from a theoretical point of view. In particular, if one imagines Wilson's lattice QCD as an oversimplified model for the short-distance physics at the ultimate cut-off scale, one would not understand why light fermions exist in Nature. For Wilson fermions, an exact chiral symmetry that can break spontaneously does not exist at finite lattice spacing. It emerges only in the continuum limit after a delicate fine-tuning of the bare quark masses to appropriate negative values.

18.6 Ginsparg–Wilson Relation and Lüscher's Lattice Chiral Symmetry

In the discussion of the perfect free fermion action in Section 10.6, we have encountered the Ginsparg–Wilson relation (10.32). Following Lüscher (1998), we now use this relation in order to construct a version of chiral symmetry that is natural for a lattice theory and reduces to the usual one in the continuum limit. This generalizes the free fermion considerations of Chapter 10. For this purpose, we consider a lattice fermion action

$$S[\bar{Q}, Q, U] = a^4 \, \bar{Q} D[U] Q = a^4 \sum_{x,y} \bar{Q}_x D[U]_{xy} Q_y, \qquad (18.35)$$

that is defined in terms of the lattice Dirac operator $D[U]$. This operator should be local (*i.e.* it should decay exponentially at large distances $|x - y|$) but not necessarily ultra-local (*i.e.* limited to a finite range of neighboring sites). The lattice Dirac operator obeys the Ginsparg–Wilson relation if the corresponding fermion propagator $D[U]^{-1}$ satisfies

$$\{D[U]^{-1}, \gamma_5\} = D[U]^{-1}\gamma_5 + \gamma_5 D[U]^{-1} = a\gamma_5. \qquad (18.36)$$

Alternatively, the Ginsparg–Wilson relation takes the form

$$\gamma_5 D[U] + D[U]\gamma_5 = aD[U]\gamma_5 D[U]. \qquad (18.37)$$

It is non-trivial to construct lattice actions that obey this relation. In Sections 10.6 and 10.7, we have seen that the perfect action for massless free fermions as well as overlap fermions indeed satisfies it. It is straightforward to convince oneself that overlap fermions obey the Ginsparg–Wilson relation even in the interacting case. For this purpose, one uses the Wilson–Dirac operator $D_W[U]$ to define $A_W[U] = aD_W[U] - \mathbb{1}$. As for the free theory, one then constructs the unitary operator (Neuberger, 1998)

$$A_{\text{overlap}}[U] = A_W[U]/\sqrt{A_W[U]^\dagger A_W[U]},$$
$$D_{\text{overlap}}[U] = \frac{1}{a}\left(A_{\text{overlap}}[U] + \mathbb{1}\right)$$
$$= (D_W[U] - \mathbb{1}/a)/\sqrt{\left(aD_W[U]^\dagger - \mathbb{1}\right)\left(aD_W[U] - \mathbb{1}\right)} + \frac{1}{a}\mathbb{1}. \qquad (18.38)$$

For the moment, we do not worry about the concrete form of $D[U]$ and just assume that it obeys eq. (18.37).

Let us first consider an infinitesimal chiral rotation of the form, which is familiar from the continuum theory,

$$Q' = Q + \delta Q = \left(\mathbb{1} + i\varepsilon^a T^a \gamma_5\right) Q,$$
$$\bar{Q}' = \bar{Q} + \delta\bar{Q} = \bar{Q}\left(\mathbb{1} + i\varepsilon^a T^a \gamma_5\right). \qquad (18.39)$$

Here, T^a (with $a \in \{1, 2, \ldots, N_f^2 - 1\}$) are the generators of $\mathrm{SU}(N_f)$ and ε^a are small parameters. In order to discuss flavor-singlet axial transformations with an infinitesimal parameter ε^0, we also define $T^0 = \mathbb{1}$. If the lattice action is local and has no fermion doubling, the Nielsen–Ninomiya theorem implies that it cannot be invariant under the above chiral rotations. On the other hand, the lattice fermion measure is invariant under the full chiral symmetry $\mathrm{U}(N_f)_L \times \mathrm{U}(N_f)_R$. This is quite different from massless QCD in the continuum. In the continuum, the action is invariant under $\mathrm{U}(N_f)_L \times \mathrm{U}(N_f)_R$ chiral transformations, while the measure is invariant only under $\mathrm{SU}(N_f)_L \times \mathrm{SU}(N_f)_R \times \mathrm{U}(1)_{L=R}$. As we will discuss in Chapter 19, due to the axial anomaly, the measure of the continuum theory is not invariant under flavor-singlet axial transformations, whereas the measure of the lattice theory is invariant.

Next we consider Lüscher's modification of the standard chiral transformation (Lüscher, 1998)

$$Q' = Q + \delta Q = \left(\mathbb{1} + i\varepsilon^a T^a \gamma_5 \left(1 - \frac{a}{2}D[U]\right)\right) Q,$$
$$\bar{Q}' = \bar{Q} + \delta\bar{Q} = \bar{Q}\left(\mathbb{1} + i\varepsilon^a T^a \left(1 - \frac{a}{2}D[U]\right)\gamma_5\right). \tag{18.40}$$

In contrast to chiral transformations in the continuum, via $D[U]$, Lüscher's lattice-modified chiral transformation depends on the gluon field. Still, in the continuum limit $a \to 0$ it reduces to the standard chiral symmetry of the continuum theory. It is remarkable that eq. (18.40) is a symmetry of any lattice action that obeys the Ginsparg–Wilson relation eq. (18.37). This follows from

$$\bar{Q}'D[U]Q' = \bar{Q}\left(\mathbb{1} + i\varepsilon^a T^a \left(1 - \frac{a}{2}D[U]\right)\gamma_5\right) D[U]\left(\mathbb{1} + i\varepsilon^a T^a \gamma_5 \left(1 - \frac{a}{2}D[U]\right)\right) Q$$
$$= \bar{Q}D[U]Q + \bar{Q}\left(i\varepsilon^a T^a\left[\gamma_5 D[U] + D[U]\gamma_5 - aD[U]\gamma_5 D[U]\right]\right)Q + \mathcal{O}(\varepsilon^2)$$
$$= \bar{Q}D[U]Q + \mathcal{O}(\varepsilon^2). \tag{18.41}$$

Let us also consider the transformation of the fermionic measure

$$\mathcal{D}\bar{Q}'\mathcal{D}Q' = \mathcal{D}\bar{Q}\left(\mathbb{1} + i\varepsilon^a T^a \left(1 - \frac{a}{2}D[U]\right)\gamma_5\right)\left(\mathbb{1} + i\varepsilon^a T^a \gamma_5 \left(1 - \frac{a}{2}D[U]\right)\right)\mathcal{D}Q$$
$$= \mathcal{D}\bar{Q}\mathcal{D}Q\left(\mathbb{1} + i\varepsilon^a \mathrm{Tr}\left[T^a \gamma_5 \left(2\mathbb{1} - aD[U]\right)\right] + \mathcal{O}(\varepsilon^2)\right)$$
$$= \mathcal{D}\bar{Q}\mathcal{D}Q\left(\mathbb{1} - i\varepsilon^0 a\mathrm{Tr}\left[\gamma_5 D[U]\right] + \mathcal{O}(\varepsilon^2)\right). \tag{18.42}$$

Here, the trace extends over color, flavor, and Dirac indices, as well as over the space–time points. While any lattice fermion action that obeys the Ginsparg–Wilson relation is invariant against Lüscher's lattice chiral transformations, the lattice fermion measure is not. It changes under flavor-singlet axial transformations, while it is invariant under $SU(N_f)_L \times SU(N_f)_R \times U(1)_{L=R}$. This is exactly as in the continuum theory. In Chapter 19, we will see that the non-invariance of the fermionic measure under flavor-singlet axial transformations indeed gives rise to the correct axial anomaly.

Based on the Ginsparg–Wilson relation, Lüscher has even been able to *regularize chiral gauge theories* on the lattice (Lüscher, 1999, 2000). His construction is complete for Abelian chiral gauge theories, but not yet for non-Abelian chiral gauge theories including the Standard Model, for which some subtle cohomology problems have remained unsolved. However, even in the non-Abelian case, his construction works to all orders of lattice perturbation theory. This is in contrast to continuum regularization schemes like dimensional regularization, where recipes for dealing with γ_5 exist only up to a few orders in perturbation theory. We may conclude that the deeper understanding of lattice chiral symmetry, which resulted from the Ginsparg–Wilson relation, represents a breakthrough in our understanding of chiral symmetry in general, even within perturbation theory.

18.7 Under-Appreciated Fermionic Hierarchy Problem

Why do fermions exist so far below the Planck scale? In contrast to the well-appreciated hierarchy problem for scalar fields, this is an under-appreciated hierarchy problem for fermions. A common point of view is that fermions are naturally light, because – in contrast to scalar fields – they are protected from additive mass renormalization by their chiral symmetry. The point of view to be taken here is that chiral symmetry is non-trivial to maintain in the presence of an ultraviolet regulator, and the question arises whether such a regulator is just a mathematical trick (like in dimensional regularization) or if it is deeply rooted in physical reality.

The Planck scale, where strongly coupled quantum gravity is expected to emerge, serves as an ultimate ultraviolet regulator for the physics of the Standard Model. For the purpose of the present discussion, we mimic the Planck scale by a lattice with a spacing a at the order of the Planck length. This provides a concrete realization of a physical ultraviolet cut-off beyond perturbation theory. It is a matter of speculation whether or not some kind of discrete space–time structure might be physical reality.

Box 18.1	Deep question
The lattice provides the only fully satisfactory non-perturbative regularization scheme in quantum field theory that we know. Could this be a hint that space–time in Nature indeed has some kind of discrete structure?	

Once an ultraviolet cut-off scale is introduced, it may not be obvious why physics can take place very much below this scale. The hierarchy problem for scalar fields arises because we have no natural explanation for the fact that the Higgs field, which drives electroweak symmetry breaking in the Standard Model, exists so far below the Planck scale.

Besides the scalar Higgs field, the Standard Model contains gauge fields and fermion fields. Abelian gauge fields may exist in a Coulomb phase with naturally massless photons. Non-Abelian gauge fields are asymptotically free, which implies that their physics naturally takes place way below the ultraviolet cut-off. Hence, there is no gauge theory hierarchy problem. If one takes chiral symmetry for granted, one concludes the same for fermions. However, in particular beyond perturbation theory, it is far from obvious how to preserve chiral symmetry at the ultraviolet cut-off scale.

Hence, we should not take chiral symmetry for granted easily. Instead, we should be puzzled how fermions can naturally be light and how chiral symmetry emerges. This is not idle thought about unphysical regularization procedures, but a deep question whose answer may provide concrete hints to some physics that lies beyond the Standard Model.

18.8 Domain Wall Fermions and a Fifth Dimension of Space–Time

Remarkably, the under-appreciated fermion hierarchy problem has been solved – for vector-like theories – by David Kaplan (1992), who introduced lattice domain wall fermions that reside at the boundaries of a $(4 + 1)$-d space–time with an additional fourth spatial dimension. Domain wall fermions have a lower-dimensional condensed matter analogue: the $(1 + 1)$-d edge states of a $(2 + 1)$-d quantum Hall sample. The fermion hierarchy puzzle hence suggests that our $(3 + 1)$-d world might be the edge of a gigantic $(4 + 1)$-d "quantum Hall sample". This way of thinking is more productive than taking chiral symmetry for granted and blaming the lattice regularization for its problems to respect it.[1]

---

[1] Similar issues exist for supersymmetric theories with naturally massless scalars, which also wait to be put on equally solid grounds beyond perturbation theory.

Related to earlier work in the continuum by Valery Rubakov and Mikhail Shaposhnikov (Rubakov and Shaposhnikov, 1983), as well as by Curtis Callen and Jeffrey Harvey (Callan and Harvey, 1985), which implies that massless chiral fermions can be localized on a domain wall, Kaplan considered a Wilson–Dirac operator in five Euclidean dimensions with a domain wall mass term that is a function of the fifth direction. An appealing variant replaces a domain-wall–anti-domain-wall pair by an extra-dimensional interval of finite extent L_5 (Shamir, 1993). A left-handed chiral fermion then gets localized on the left, and a right-handed chiral fermion gets localized on the right 4-d boundary of the 5-d space–time slab. For a finite value of L_5, tunneling of fermions from one wall to the other induces a small fermion mass, which vanishes exponentially in the $L_5 \to \infty$ limit.

While Kaplan proposed domain wall fermions, Rajamani Narayanan and Herbert Neuberger developed overlap fermions, by using an infinite number of regulator "flavor" fields in order to preserve chiral symmetry (Narayanan and Neuberger, 1993a,b). This is closely related to Kaplan's construction, as well as to a proposal with an infinite set of the Pauli–Villars fields (Frolov and Slavnov, 1994). Thus, the "fifth dimension" can also be interpreted as a "flavor" space.[2] In the overlap fermion formalism, the determinant of a chiral fermion in the background of a gauge field is equivalent to the overlap of two many-body fermionic ground states. In the limit $L_5 \to \infty$, domain wall fermions turn into overlap fermions.

Thanks to the additional fifth dimension, domain wall fermions naturally possess a chiral symmetry without requiring any fine-tuning (which is necessary in Wilson's original approach). We conclude that the existence of a fifth dimension could explain why fermions can naturally be light and possess an (approximate) chiral symmetry, beyond perturbation theory.

Box 18.2	Deep question

Why do chiral fermions exist in Nature? Could this be a concrete hint for the existence of a fifth space–time dimension?

---

[2] It indeed appears as a fifth dimension for free fermions, but if gauge interactions are included, a gauge field component in the fifth direction is not required.

Exercises

18.1 Symmetries of the QCD Lagrangian

Reinstall explicit color, flavor, and Dirac indices in the quark contribution to the QCD Lagrangian of eq. (18.17). Convince yourself that it is Lorentz-invariant as well as invariant under $SU(3)_c$ gauge transformations. Also show that, for $\mathcal{M} = 0$, it is chirally invariant.

18.2 Form and realization of chiral symmetry

Consider $N_f = 3$ quark flavors and $N_c = 3$ colors. Assume that all quark masses are much smaller than Λ_{QCD}. Distinguish the following cases:

a) $0 = m_u = m_d = m_s$,
b) $0 = m_u = m_d < m_s$,
c) $0 = m_u < m_d < m_s$,
d) $0 = m_u < m_d = m_s$,
e) $0 < m_u = m_d < m_s$,
f) $0 < m_u < m_d = m_s$,
g) $0 < m_u = m_d = m_s$,
h) $0 < m_u < m_d < m_s$.

What is the corresponding symmetry, both of the Lagrangian and of the full quantum theory, in each of these cases? How many Nambu–Goldstone and how many pseudo-Nambu–Goldstone bosons arise in the different cases? How are the masses of the pseudo-Nambu–Goldstone bosons ordered and how many have the same mass?

19

Topology of Gauge Fields

In this chapter, we investigate the *topological structure of non-Abelian gauge fields*. In the Standard Model, the non-trivial topology of SU(2)$_L$ gauge fields gives rise to *baryon number violating* processes. Similarly, in QCD a non-trivial topology of the gluon field leads to an *explicit breaking of the flavor-singlet axial symmetry*. This provides a solution to the famous U(1)$_A$-*problem in QCD* – the question why the η'-meson is not a pseudo-Nambu–Goldstone boson. In addition, the gauge field topology gives rise to an additional parameter in QCD – the vacuum-angle θ. That confronts us with the *strong CP-problem:* Why is $|\theta|$ tiny and consistent with zero in Nature? We are going to discuss the U(1)$_A$-problem in Chapter 20, and we will address the strong CP-problem in Chapter 25.

19.1 Adler–Bell–Jackiw Anomaly

Let us consider the flavor-singlet *axial current* in QCD,

$$j_\mu^5(x) = \sum_{\text{f}} \bar{q}^{\text{f}}(x)\gamma_\mu\gamma_5 q^{\text{f}}(x) = \sum_{\text{f}} \left(\bar{q}_{\text{R}}^{\text{f}}(x)\sigma_\mu q_{\text{R}}^{\text{f}}(x) - \bar{q}_{\text{L}}^{\text{f}}(x)\bar{\sigma}_\mu q_{\text{L}}^{\text{f}}(x) \right), \qquad (19.1)$$

where q^{f} is the quark field of flavor $\text{f} \in \{u, d, s, c, b, t\}$. The QCD Lagrangian with massless quarks is invariant under global U(1)$_A$ transformations, and hence, the current j_μ^5 is conserved at the classical level, $\partial_\mu j_\mu^5 = 0$. At the *quantum* level, however, this symmetry cannot be maintained; it is explicitly violated by the *Adler–Bell–Jackiw anomaly* (Adler, 1969; Bell and Jackiw, 1969)

$$\langle \partial_\mu j_\mu^5(x) \rangle = 2N_{\text{f}}P(x), \qquad (19.2)$$

where N_{f} is the number of quark flavors and

$$P(x) = -\frac{1}{32\pi^2}\epsilon_{\mu\nu\rho\sigma}\text{Tr}\left[G_{\mu\nu}(x)G_{\rho\sigma}(x)\right] \qquad (19.3)$$

is the *Chern–Pontryagin density* of the gluon field. The factor of 2 in eq. (19.2) arises because both left- and right-handed quark fields couple to the gluon field. The expectation value in eq. (19.2) refers to an integration over the fermionic Grassmann fields in a fixed background gauge field G_μ. In the continuum, the anomaly equation (19.2) can be derived in perturbation theory, where it follows from a triangle diagram, after a somewhat tedious calculation; see, for instance, the book by Peskin and Schroeder (1997). On the lattice, on the other hand, it follows directly from the Ginsparg–Wilson relation, beyond perturbation theory. In eq. (18.42), we have seen that the fermionic measure of lattice QCD is not invariant against Lüscher's flavor-singlet axial transformations. As we will see at the end of this chapter, the variation of the fermionic measure of the lattice theory, $-a \, \text{tr}[\gamma_5 D_{xx}]/2$ (where the trace now extends only over Dirac, flavor, and color, but no longer also over space–time indices), indeed corresponds to the right-hand side of eq. (19.2).

Let us also consider the *baryon number current* in the Standard Model

$$j_\mu^B(x) = \sum_f \bar{q}^f(x)\gamma_\mu q^f(x) = \sum_f \left(\bar{q}_R^f(x)\sigma_\mu q_R^f(x) + \bar{q}_L^f(x)\bar{\sigma}_\mu q_L^f(x) \right). \tag{19.4}$$

The Standard Model Lagrangian is invariant under global $U(1)_B$ baryon number transformations. Hence, the corresponding Noether current j_μ^B is conserved at the classical level, $\partial_\mu j_\mu^B = 0$. However, at the quantum level also that symmetry is explicitly broken by an anomaly

$$\langle \partial_\mu j_\mu^B(x) \rangle = N_g P(x). \tag{19.5}$$

Here, N_g is the number of fermion generations (with $N_g = 3$ in the real world), and

$$P(x) = -\frac{1}{32\pi^2} \epsilon_{\mu\nu\rho\sigma} \text{Tr}\left[W_{\mu\nu}(x) W_{\rho\sigma}(x) \right] \tag{19.6}$$

is now the Chern–Pontryagin density of the $SU(2)_L$ gauge field. As before, $W_{\mu\nu}$ is the field strength tensor of the $SU(2)_L$ gauge field. In eq. (19.5), no factor of 2 arises, because only the left-handed quarks couple to the $SU(2)_L$ gauge field.

In the following, we consider the topology of a *general* $SU(N)$ *vector potential* G_μ. The Chern–Pontryagin density can be expressed as a total divergence

$$P(x) = \partial_\mu \Omega_\mu^{(0)}(x), \tag{19.7}$$

where $\Omega_\mu^{(0)}(x)$ is the so-called *Chern–Simons density* or *0-cochain*, which is given by

$$\Omega_\mu^{(0)}(x) = -\frac{1}{8\pi^2} \epsilon_{\mu\nu\rho\sigma} \text{Tr}\left[G_\nu(x) \left(\partial_\rho G_\sigma(x) + \frac{2}{3} G_\rho(x) G_\sigma(x) \right) \right]. \tag{19.8}$$

We could now formally construct a conserved current

$$\tilde{j}_\mu^5(x) = j_\mu^5(x) - 2N_f \Omega_\mu^{(0)}(x), \tag{19.9}$$

because

$$\partial_\mu \tilde{j}_\mu^5(x) = \partial_\mu j_\mu^5(x) - 2N_f P(x) = 0. \tag{19.10}$$

One might thus think that we have found a modified $U(1)$ symmetry, which is free of the anomaly. This is, however, not the case, because the current $\tilde{j}_\mu^5$ contains the term $\Omega_\mu^{(0)}$, which is not gauge-invariant. Therefore, although the gauge-variant current is formally conserved, it has no physical meaning.

19.2 Topological Charge

Let us consider the topological charge as the integral of the Chern–Pontryagin density over Euclidean space–time,

$$Q = -\frac{1}{32\pi^2} \int d^4x\, \epsilon_{\mu\nu\rho\sigma} \text{Tr}[G_{\mu\nu} G_{\rho\sigma}] = \int d^4x\, P = \int d^4x\, \partial_\mu \Omega_\mu^{(0)} = \int_{S^3} d^3\sigma_\mu\, \Omega_\mu^{(0)}.$$

$$(19.11)$$

In the last step, we have used Gauss' theorem to reduce the integral over Euclidean space–time to an integral over its boundary, which is topologically a (large) 3-sphere S^3.

For now, we will restrict our attention to gauge field configurations with a *finite* action. Their field strength vanishes at infinity, and consequently, the gauge potential is a *pure gauge configuration* (a gauge transformation of a zero field)

$$G_\mu(x) = \Omega(x) \partial_\mu \Omega(x)^\dagger, \quad \Omega(x) \in \text{SU}(N). \quad (19.12)$$

Of course, this expression is only valid at space–time infinity. Inserting it into the expression (19.8) for the Chern–Simons density, we obtain

$$\begin{aligned}
Q &= -\frac{1}{8\pi^2} \int_{S^3} d^3\sigma_\mu\, \epsilon_{\mu\nu\rho\sigma} \text{Tr}\Big[\big(\Omega\partial_\nu\Omega^\dagger\big) \big\{ \partial_\rho\big(\Omega\partial_\sigma\Omega^\dagger\big) + \frac{2}{3}\big(\Omega\partial_\rho\Omega^\dagger\big)\big(\Omega\partial_\sigma\Omega^\dagger\big) \big\} \Big] \\
&= -\frac{1}{8\pi^2} \int_{S^3} d^3\sigma_\mu\, \epsilon_{\mu\nu\rho\sigma} \text{Tr}\Big[-\big(\Omega\partial_\nu\Omega^\dagger\big)\big(\Omega\partial_\rho\Omega^\dagger\big)\big(\Omega\partial_\sigma\Omega^\dagger\big) \\
&\quad + \frac{2}{3}\big(\Omega\partial_\nu\Omega^\dagger\big)\big(\Omega\partial_\rho\Omega^\dagger\big)\big(\Omega\partial_\sigma\Omega^\dagger\big) \Big] \\
&= \frac{1}{24\pi^2} \int_{S^3} d^3\sigma_\mu\, \epsilon_{\mu\nu\rho\sigma} \text{Tr}\Big[\big(\Omega\partial_\nu\Omega^\dagger\big)\big(\Omega\partial_\rho\Omega^\dagger\big)\big(\Omega\partial_\sigma\Omega^\dagger\big) \Big].
\end{aligned}$$

$$(19.13)$$

The gauge transformation $\Omega(x)$ defines a map of the sphere S^3 at space–time infinity into the gauge group SU(N),

$$\Omega : S^3 \to \text{SU}(N). \quad (19.14)$$

Such maps have topological properties. They fall into equivalence classes – known as *homotopy classes* – which represent topologically distinct sectors. Two maps are *equivalent* if they can be deformed continuously into one another. The homotopy properties are described by so-called *homotopy groups*. In our case, the relevant homotopy group is

$$\Pi_3[\text{SU}(N)] = \mathbb{Z}, \quad (19.15)$$

as we are going to see. General aspects of homotopy groups are addressed in Appendix G.

The index 3 in eq. (19.15) indicates that we consider maps of the 3-dimensional sphere S^3. The third homotopy group of SU(N) is given by the integers. This means that, for each integer $Q \in \mathbb{Z}$, there is a class of maps that can be continuously deformed into one another, whereas maps with different Q are topologically distinct. The integer Q that characterizes the map topologically is the *topological charge*. Now we want to show that the above expression for Q is exactly that integer.

For this purpose, we decompose $\Omega(x)$ as follows:

$$
\Omega = VW, \qquad W = \begin{pmatrix}
1 & 0 & 0 & \cdots & 0 \\
0 & \widetilde{\Omega}_{11} & \widetilde{\Omega}_{12} & \cdots & \widetilde{\Omega}_{1,N-1} \\
0 & \widetilde{\Omega}_{21} & \widetilde{\Omega}_{22} & \cdots & \widetilde{\Omega}_{1,N-1} \\
\cdot & \cdot & \cdot & & \cdot \\
\cdot & \cdot & \cdot & & \cdot \\
\cdot & \cdot & \cdot & & \cdot \\
0 & \widetilde{\Omega}_{N-1,1} & \widetilde{\Omega}_{N-1,2} & \cdots & \widetilde{\Omega}_{N-1,N-1}
\end{pmatrix}, \tag{19.16}
$$

with the *embedded matrix* $\widetilde{\Omega} \in \mathrm{SU}(N-1)$. It is (indirectly) defined by

$$
V = \begin{pmatrix}
\Omega_{11} & -\Omega_{21}^* & -\dfrac{\Omega_{31}^*(1+\Omega_{11})}{1+\Omega_{11}^*} & \cdots & -\dfrac{\Omega_{N1}^*(1+\Omega_{11})}{1+\Omega_{11}^*} \\[2mm]
\Omega_{21} & \dfrac{1+\Omega_{11}^* - |\Omega_{21}|^2}{1+\Omega_{11}} & -\dfrac{\Omega_{31}^*\Omega_{21}}{1+\Omega_{11}^*} & \cdots & -\dfrac{\Omega_{N1}^*\Omega_{21}}{1+\Omega_{11}^*} \\[2mm]
\Omega_{31} & -\dfrac{\Omega_{21}^*\Omega_{31}}{1+\Omega_{11}} & \dfrac{1+\Omega_{11}^* - |\Omega_{31}|^2}{1+\Omega_{11}^*} & \cdots & -\dfrac{\Omega_{N1}^*\Omega_{31}}{1+\Omega_{11}^*} \\[2mm]
\cdot & \cdot & \cdot & & \cdot \\
\cdot & \cdot & \cdot & & \cdot \\
\Omega_{N1} & -\dfrac{\Omega_{21}^*\Omega_{N1}}{1+\Omega_{11}} & -\dfrac{\Omega_{31}^*\Omega_{N1}}{1+\Omega_{11}^*} & \cdots & \dfrac{1+\Omega_{11}^* - |\Omega_{N1}|^2}{1+\Omega_{11}^*}
\end{pmatrix} \in \mathrm{SU}(N).
$$
$$\tag{19.17}$$

The matrix V is constructed entirely from the elements $\Omega_{11}, \Omega_{21}, \ldots, \Omega_{N1}$ of the first column of the matrix Ω.

The idea is now to reduce the expression for the topological charge from $\mathrm{SU}(N)$ to $\mathrm{SU}(N-1)$ by using the identity

$$
\epsilon_{\mu\nu\rho\sigma}\mathrm{Tr}\Big[(VW)\partial_\nu(VW)^\dagger(VW)\partial_\rho(VW)^\dagger(VW)\partial_\sigma(VW)^\dagger\Big] = \epsilon_{\mu\nu\rho\sigma}\mathrm{Tr}\Big[(V\partial_\nu V^\dagger)(V\partial_\rho V^\dagger)(V\partial_\sigma V^\dagger)\Big]
$$
$$
+\epsilon_{\mu\nu\rho\sigma}\mathrm{Tr}\Big[(W\partial_\nu W^\dagger)(W\partial_\rho W^\dagger)(W\partial_\sigma W^\dagger)\Big] + 3\epsilon_{\mu\nu\rho\sigma}\partial_\nu\mathrm{Tr}\Big[(V\partial_\rho V^\dagger)(W\partial_\sigma W^\dagger)\Big]. \tag{19.18}
$$

The application of this formula (whose verification is left as an exercise) to the expression for the topological charge, substituting $\Omega = VW$, yields

$$
Q = \frac{1}{24\pi^2}\int_{S^3} d^3\sigma_\mu\, \epsilon_{\mu\nu\rho\sigma}\mathrm{Tr}\Big[\big(\Omega\partial_\nu\Omega^\dagger\big)\big(\Omega\partial_\rho\Omega^\dagger\big)\big(\Omega\partial_\sigma\Omega^\dagger\big)\Big]
$$
$$
= \frac{1}{24\pi^2}\int_{S^3} d^3\sigma_\mu\, \epsilon_{\mu\nu\rho\sigma}\mathrm{Tr}\Big[\big(V\partial_\nu V^\dagger\big)\big(V\partial_\rho V^\dagger\big)\big(V\partial_\sigma V^\dagger\big)
$$
$$
+ \big(W\partial_\nu W^\dagger\big)\big(W\partial_\rho W^\dagger\big)\big(W\partial_\sigma W^\dagger\big)\Big]. \tag{19.19}
$$

The last term in eq. (19.18) drops out using Gauss' theorem, together with the fact that S^3 has no boundary. We conclude that the topological charge of a product of two gauge transformations V and W is the sum of the topological charges of V and W. Since V only depends on $\Omega_{11}, \Omega_{21}, \ldots, \Omega_{N1}$, it can be viewed as a map of S^3 into the sphere S^{2N-1},

$$
V: S^3 \to S^{2N-1}, \tag{19.20}
$$

because $|\Omega_{11}|^2 + |\Omega_{21}|^2 + \cdots + |\Omega_{N1}|^2 = 1$ (we recall that the rows and columns of a unitary matrix are orthonormal). The corresponding homotopy group is trivial for $N \geq 3$, *i.e.*

$$
\Pi_3[S^{2N-1}] = \{0\}. \tag{19.21}
$$

	$m < n$	$m = n$	$m > n$
Table 19.1 The quantity $\Pi_m[S^n]$ for different dimensions m and n.			
$\Pi_m[S^n]$	$\{0\}$	$\mathbb{Z}$	not always trivial
			e.g. $\Pi_4[S^3] = \mathbb{Z}(2)$

In other words: All maps of S^3 into the higher-dimensional sphere S^{2N-1} are topologically equivalent (they can be continuously deformed into each other).

This can be understood more easily in the lower-dimensional example

$$\Pi_1[S^2] = \{0\}. \tag{19.22}$$

Each closed curve on an ordinary sphere can be continuously constricted to a point, and hence, it is topologically trivial. The same holds for higher-dimensional spheres, $\Pi_1[S^n] = \{0\}$, for $n > 1$. Regarding the quantity $\Pi_m[S^n]$ in general dimensions m and n, Table 19.1 gives an overview.

Since the map V is topologically trivial for $N \geq 3$, its contribution to the topological charge vanishes. The remaining W-term reduces to the SU$(N-1)$ contribution

$$Q = \frac{1}{24\pi^2} \int_{S^3} d^3\sigma_\mu \, \epsilon_{\mu\nu\rho\sigma} \text{Tr}\left[\left(\widetilde{\Omega}\partial_\nu\widetilde{\Omega}^\dagger\right)\left(\widetilde{\Omega}\partial_\rho\widetilde{\Omega}^\dagger\right)\left(\widetilde{\Omega}\partial_\sigma\widetilde{\Omega}^\dagger\right) \right]. \tag{19.23}$$

The separation of the V contribution only works if the decomposition of Ω into V and $\widetilde{\Omega}$ is non-singular. In fact, the expression for V is singular for $\Omega_{11} = -1$. This corresponds to an $[(N-1)^2 - 1]$-dimensional subspace of the $(N^2 - 1)$-dimensional SU(N) group manifold. The map Ω itself covers only a 3-d subspace of SU(N). Since for $N \geq 3$

$$(N^2 - 1) - [(N-1)^2 - 1] = 2N - 1 > 3, \tag{19.24}$$

hitting a singularity is of zero measure. Thus we have reduced the SU(N) topological charge to the SU$(N-1)$ case. This can be iterated step by step, all the way down to SU(2).

It remains to be shown that the SU(2) expression is actually an integer. First of all

$$\widetilde{\Omega} : S^3 \rightarrow \text{SU}(2) = S^3, \tag{19.25}$$

and indeed

$$\Pi_3[\text{SU}(2)] = \Pi_3[S^3] = \mathbb{Z}. \tag{19.26}$$

The topological charge specifies how often the SU(2) group space (which is isomorphic to the 3-sphere) is covered by $\widetilde{\Omega}$ as we move along the boundary of Euclidean space–time (which is also topologically S^3).

Again, it is useful to first consider a lower-dimensional example: maps from the circle S^1 to the group U(1), which is topologically a circle too,

$$\Omega = \exp(i\alpha) : S^1 \rightarrow \text{U}(1) = S^1. \tag{19.27}$$

The relevant homotopy group is

$$\Pi_1[\text{U}(1)] = \Pi_1[S^1] = \mathbb{Z}. \tag{19.28}$$

For each integer, there is an equivalence class of maps that can be continuously deformed into one another. Moving over the circle S^1, the map may cover the group space U(1) any integer number of times. In U(1), the expression for the topological charge is similar to the one in SU(N),

$$Q = -\frac{1}{2\pi \mathrm{i}} \int_{S^1} d\sigma_\mu \, \epsilon_{\mu\nu} \, \Omega \, \partial_\nu \Omega^\dagger = \frac{1}{2\pi} \int_{S^1} d\sigma_\mu \, \epsilon_{\mu\nu} \, \partial_\nu \alpha = \frac{1}{2\pi} \Big(\alpha(2\pi) - \alpha(0) \Big). \quad (19.29)$$

If $\Omega(x) = \exp(\mathrm{i}\alpha(x))$ is continuous over the circle, $\alpha(2\pi)$ and $\alpha(0)$ differ by 2π times an integer. That integer is the topological charge. It counts how many times the map Ω covers the group space U(1), as we move along the circle S^1.

We are looking for an analogous expression in SU(2). For this purpose, we parameterize the map $\widetilde{\Omega}$ as

$$\widetilde{\Omega}(x) = \exp(\mathrm{i}\vec{\alpha}(x) \cdot \vec{\tau}) = \cos\alpha(x) + \mathrm{i}\sin\alpha(x) \, \vec{e}_\alpha(x) \cdot \vec{\tau},$$
$$\vec{\alpha}(x) = \alpha(x) \, \vec{e}_\alpha(x), \quad \vec{e}_\alpha(x) = (\sin\theta(x)\sin\varphi(x), \sin\theta(x)\cos\varphi(x), \cos\theta(x)). \quad (19.30)$$

In an exercise, we will verify the identity

$$\epsilon_{\mu\nu\rho\sigma} \mathrm{Tr}\Big[\Big(\widetilde{\Omega}\partial_\nu\widetilde{\Omega}^\dagger \Big) \Big(\widetilde{\Omega}\partial_\rho\widetilde{\Omega}^\dagger \Big) \Big(\widetilde{\Omega}\partial_\sigma\widetilde{\Omega}^\dagger \Big) \Big] = 12\sin^2\alpha \sin\theta \, \epsilon_{\mu\nu\rho\sigma} \, \partial_\nu\alpha \, \partial_\rho\theta \, \partial_\sigma\varphi. \quad (19.31)$$

This is exactly the volume element of a 3-sphere and hence of the SU(2) group space. Thus, we can now write

$$Q = \frac{1}{2\pi^2} \int_{S^3} d^3\sigma_\mu \sin^2\alpha \sin\theta \, \epsilon_{\mu\nu\rho\sigma} \, \partial_\nu\alpha \, \partial_\rho\theta \, \partial_\sigma\varphi = \frac{1}{2\pi^2} \int_{S^3} d\widetilde{\Omega}. \quad (19.32)$$

The volume of the 3-sphere S^3 amounts to $2\pi^2$. When the map $\widetilde{\Omega}$ covers the sphere Q times, the integral gives Q times the volume of S^3. This finally explains why the prefactor $1/32\pi^2$ was introduced in expression (19.11) for the topological charge (eqs. (19.13) and (19.31) are intermediate steps).

Quick Question 19.2.1   Volume of the unit-sphere S^{d-1}
Consider the d-dimensional Gauss integral to derive the volume of the sphere S^{d-1} by using the definition of the Γ-function.

19.3 Topology of a Gauge Field on a Compact Manifold

Imagine that space–time was compact and hence free of boundaries. Our previous discussion – for which the value of the gauge field at the boundary was essential – would suggest that in a compact space–time the topology is trivial. On the other hand, we expect that topology has *local consequences*. For example, baryon number conservation is violated because the topological charge does not vanish. It is hardy conceivable that a laboratory experiment could be sensitive to the nature of the "boundary" of the Universe. To resolve

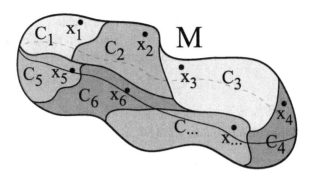

Fig. 19.1 A symbolic illustration of the manifold M and its division into sets c_i, which contain the singular points x_i.

this apparent paradox, we will now discuss the topology of a gauge field on a *compact* Euclidean space–time manifold M, and we will see that non-trivial topology indeed persists.

Let us again consider the topological charge

$$Q = \int_M d^4x\, P. \tag{19.33}$$

Writing the Chern–Pontryagin density as the total divergence of the 0-cochain

$$P(x) = \partial_\mu \Omega_\mu^{(0)}(x), \tag{19.34}$$

and employing Gauss' theorem, we obtain

$$Q = \int_M d^4x\, \partial_\mu \Omega_\mu^{(0)} = \int_{\partial M} d^3\sigma_\mu \Omega_\mu^{(0)} = 0. \tag{19.35}$$

In the last step, we have used the property that M has no boundary; *i.e.* ∂M is an empty set. A gauge field whose Chern–Pontryagin density can globally be written as a total divergence is indeed topologically trivial on a compact manifold.

The relevant objection is that eq. (19.34) may be valid only locally. In other words, *gauge singularities* may prevent us from applying the Gauss theorem (as we did above). In general, it turns out to be impossible to work in a gauge that makes the gauge field non-singular all over the space–time manifold. Instead, we have to subdivide space–time into local patches in which the gauge field is smooth and glue the patches together by non-trivial gauge transformations, which form a *fiber bundle* of transition functions.

A topologically non-trivial gauge field contains gauge singularities at some points $x_i \in M$. We cover the manifold M by closed sets c_i, $M = \cup_i c_i$ with $c_i \cap c_j = \partial c_i \cap \partial c_j$, such that $x_i \in c_i \backslash \partial c_i$; *i.e.* each singularity lies in the interior of one set c_i. This situation is illustrated in Figure 19.1.

The next step is to remove the gauge singularities at x_i by performing gauge transformations Ω_i within each local patch

$$G_\mu^i(x) = \Omega_i(x)\left[G_\mu(x) + \partial_\mu\right]\Omega_i(x)^\dagger. \tag{19.36}$$

After the gauge transformation, the gauge potential $G_\mu^i(x)$ is regular in the patch c_i. Hence, we can now use Gauss' theorem and obtain

$$Q = \sum_i \int_{c_i} d^4x\, P = \sum_i \int_{\partial c_i} d^3\sigma_\mu\, \Omega_{\mu,i}^{(0)} = \frac{1}{2}\sum_{ij}\int_{c_i\cap c_j} d^3\sigma_\mu \left[\Omega_{\mu,i}^{(0)} - \Omega_{\mu,j}^{(0)}\right]. \tag{19.37}$$

The index i of the 0-cochain indicates that we are in the patch c_i. At the intersection of two sets, $c_i \cap c_j$, the gauge field G_μ^i differs from G_μ^j, although the original gauge field G_μ was continuous there. In fact, the two gauge fields are related by a gauge transformation v_{ij}

$$G_\mu^i(x) = v_{ij}(x)\left[G_\mu^j(x) + \partial_\mu\right]v_{ij}(x)^\dagger, \tag{19.38}$$

which is defined only on the common boundary $c_i \cap c_j$. The gauge transformations v_{ij} form a *fiber bundle* of transition functions given by

$$v_{ij}(x) = \Omega_i(x)\Omega_j(x)^\dagger. \tag{19.39}$$

This relation immediately implies a consistency equation: This *cocycle condition* relates the transition functions in the intersection $c_i \cap c_j \cap c_k$ of three patches,

$$v_{ik}(x) = v_{ij}(x)\, v_{jk}(x). \tag{19.40}$$

The above difference of two 0-cochains in different gauges is given by the so-called *coboundary operator* Δ

$$\Delta\Omega_{\mu,ij}^{(0)} = \Omega_{\mu,i}^{(0)} - \Omega_{\mu,j}^{(0)}. \tag{19.41}$$

In an exercise, we are going to show that

$$\Delta\Omega_{\mu,ij}^{(0)} = -\frac{1}{24\pi^2}\epsilon_{\mu\nu\rho\sigma}\,\mathrm{Tr}\left[\left(v_{ij}\partial_\nu v_{ij}^\dagger\right)\left(v_{ij}\partial_\rho v_{ij}^\dagger\right)\left(v_{ij}\partial_\sigma v_{ij}^\dagger\right)\right]$$
$$-\frac{1}{8\pi^2}\epsilon_{\mu\nu\rho\sigma}\,\partial_\nu\mathrm{Tr}\left[\partial_\rho v_{ij}^\dagger v_{ij}G_\sigma^i\right]. \tag{19.42}$$

Eq. (19.37) for the topological charge then takes the form

$$Q = -\frac{1}{48\pi^2}\sum_{ij}\int_{c_i\cap c_j}d^3\sigma_\mu\,\epsilon_{\mu\nu\rho\sigma}\,\mathrm{Tr}\left[\left(v_{ij}\partial_\nu v_{ij}^\dagger\right)\left(v_{ij}\partial_\rho v_{ij}^\dagger\right)\left(v_{ij}\partial_\sigma v_{ij}^\dagger\right)\right]$$
$$-\frac{1}{16\pi^2}\sum_{ij}\int_{\partial(c_i\cap c_j)}d^2\sigma_{\mu\nu}\,\epsilon_{\mu\nu\rho\sigma}\,\mathrm{Tr}\left[\partial_\rho v_{ij}^\dagger v_{ij}G_\sigma^i\right]. \tag{19.43}$$

Using eq. (19.38) and the emerging cocycle condition, this term can be rewritten as

$$Q = -\frac{1}{48\pi^2}\sum_{ij}\int_{c_i\cap c_j}d^3\sigma_\mu\,\epsilon_{\mu\nu\rho\sigma}\,\mathrm{Tr}\left[\left(v_{ij}\partial_\nu v_{ij}^\dagger\right)\left(v_{ij}\partial_\rho v_{ij}^\dagger\right)\left(v_{ij}\partial_\sigma v_{ij}^\dagger\right)\right]$$
$$-\frac{1}{48\pi^2}\sum_{ijk}\int_{c_i\cap c_j\cap c_k}d^2\sigma_{\mu\nu}\,\epsilon_{\mu\nu\rho\sigma}\,\mathrm{Tr}\left[\left(v_{ij}\partial_\rho v_{ij}^\dagger\right)\left(v_{jk}\partial_\sigma v_{jk}^\dagger\right)\right]. \tag{19.44}$$

This shows that the topology of the fiber bundle is entirely encoded in the transition functions.

In the mathematical terminology, the gauge transformations Ω_i form a *section of the fiber bundle*. Using formula (19.18) together with eq. (19.39), one can infer that the topological charge is expressed in terms of the sections as

$$Q = \sum_i Q_i = \frac{1}{24\pi^2} \sum_i \int_{\partial c_i} d^3\sigma_\mu \, \epsilon_{\mu\nu\rho\sigma} \, \mathrm{Tr}\left[(\Omega_i \partial_\nu \Omega_i^\dagger)(\Omega_i \partial_\rho \Omega_i^\dagger)(\Omega_i \partial_\sigma \Omega_i^\dagger)\right]. \quad (19.45)$$

We recognize the integer winding number Q_i that characterizes the map Ω_i topologically. In fact, the boundary ∂c_i is topologically a 3-sphere, such that

$$\Omega_i : \partial c_i \to \mathrm{SU}(N), \quad (19.46)$$

and hence

$$Q_i \in \Pi_3[\mathrm{SU}(N)] = \mathbb{Z}. \quad (19.47)$$

We conclude that the topological charge Q is the sum of the local winding numbers $Q_i \in \mathbb{Z}$, which are associated with the regions c_i, $Q = \sum_i Q_i$.

In general, the Q_i are not gauge-invariant. Hence, individually they have no physical meaning. However, the total charge – as the sum of all Q_i – *is* gauge-invariant.

19.4 SU(2) Instanton

We have argued that 4-d SU(N) gauge field configurations fall into topologically distinct classes. Now we want to construct concrete examples of topologically non-trivial field configurations. Here, we consider *instantons,* which have $Q = 1$ and are solutions of the Euclidean classical field equations.

An instanton occurs at a given "instant in Euclidean time". Since these solutions do not exist in Minkowski space–time, they have no direct interpretation in terms of real-time events. It is not entirely clear which role they play in the fully non-perturbative quantum field theory. Certainly, not only instantons but also other gluon field configurations with non-zero topological charge contribute significantly to the functional integral – the instantons actually form a subset of configurations of measure zero. Instantons describe tunneling processes between degenerate classical vacuum states semi-classically. Their existence gives rise to the θ-vacuum structure of non-Abelian gauge theories.

Here, we concentrate on SU(2). This is sufficient, because we have seen that the SU(N) topological charge, for $N \geq 3$, can be successively reduced to the SU(2) case. In this section, we go back to an infinite space with a boundary sphere S^3, and we demand the gauge field to have finite action, which is the setting of Section 19.2. Then, as we saw before, the gauge potential is in a pure gauge at space–time infinity,

$$G_\mu(x)|_{|x|\to\infty} = \Omega(x)\partial_\mu\Omega(x)^\dagger, \quad |x| = \sqrt{\vec{x}^{\,2} + x_4^2}. \quad (19.48)$$

Provided the gauge field is otherwise smooth, the topology resides entirely in the map Ω. We want to construct a field configuration with topological charge $Q = 1$, *i.e.* one in which the map Ω covers the group space SU(2) $= S^3$ just *once* as we integrate over the boundary sphere S^3. The simplest map of this sort is the identity; *i.e.* each point at the boundary of space–time is mapped to the corresponding point in the group space,

$$\Omega(x) = \frac{x_4 + i\vec{x} \cdot \vec{\tau}}{|x|}. \tag{19.49}$$

Next, we want to extend the gauge field to the interior of space–time, without introducing singularities. We cannot simply maintain the form of eq. (19.48) because $\Omega(x)$ is singular at $x = 0$. To avoid this singularity, we make the ansatz

$$G_\mu(x) = f(|x|)\,\Omega(x)\,\partial_\mu\Omega(x)^\dagger, \quad f(\infty) = 1,\ f(0) = 0. \tag{19.50}$$

For any smooth function f with these properties, the above gluon field configuration has $Q = 1$. However, this does not yet mean that we have constructed an instanton.

Single or multiple *instantons* or *anti-instantons* are field configurations with $Q > 0$ or $Q < 0$, respectively, that are, in addition, solutions of the Euclidean classical equations of motion; *i.e.* they are (local) minima of the Euclidean action

$$S[G] = -\frac{1}{2g_s^2} \int d^4x \, \mathrm{Tr}\left[G_{\mu\nu}G_{\mu\nu}\right]. \tag{19.51}$$

Note that, despite the minus sign, $S[G] \geq 0$ because $G_{\mu\nu}$ is anti-Hermitian. Let us now consider the following integral

$$-\int d^4x \, \mathrm{Tr}\left[\left(G_{\mu\nu} \pm \frac{1}{2}\epsilon_{\mu\nu\rho\sigma}G_{\rho\sigma}\right)\left(G_{\mu\nu} \pm \frac{1}{2}\epsilon_{\mu\nu\kappa\lambda}G_{\kappa\lambda}\right)\right] =$$

$$-\int d^4x \, \mathrm{Tr}\left[G_{\mu\nu}G_{\mu\nu} \pm \epsilon_{\mu\nu\rho\sigma}G_{\mu\nu}G_{\rho\sigma} + G_{\mu\nu}G_{\mu\nu}\right] = 4g_s^2 S[G] \pm 32\pi^2 Q[G]. \tag{19.52}$$

Since we have integrated a square, we obtain the *Schwarz inequality*

$$S[G] \pm \frac{8\pi^2}{g_s^2}Q[G] \geq 0 \quad \Rightarrow \quad S[G] \geq \frac{8\pi^2}{g_s^2}|Q[G]|\,; \tag{19.53}$$

i.e. a topologically non-trivial field configuration costs at least a minimal action proportional to the topological charge. Single or multiple instantons (or anti-instantons) are configurations with an action that is minimal within a fixed non-trivial topological charge sector; *i.e.* they obey

$$S[G] = \frac{8\pi^2}{g_s^2}|Q[G]|. \tag{19.54}$$

From eq. (19.52), it is clear that a minimum action configuration arises solely if

$$G_{\mu\nu}(x) = \pm\frac{1}{2}\epsilon_{\mu\nu\rho\sigma}G_{\rho\sigma}(x). \tag{19.55}$$

Configurations that obey this equation with a plus (minus) sign are called *self-dual (anti–self-dual)*.

Solving the self-duality equation, the instanton configuration takes the form (Belavin *et al.*, 1975)

$$G_\mu(x) = \frac{|x|^2}{|x|^2 + \rho^2}\,\Omega(x)\partial_\mu\Omega(x)^\dagger. \tag{19.56}$$

There is a whole family of instantons with different radii $\rho > 0$. As a consequence of classical scale invariance, they all have the same action value $S[G] = 8\pi^2/g_s^2$.

19.5 θ-Vacuum States

The existence of topologically non-trivial gauge transformations has drastic consequences for non-Abelian gauge theories. In fact, there is not just one classical vacuum state, but there is one for each integer topological winding number. Instantons describe tunneling processes between topologically distinct vacua. Due to tunneling, the degeneracy of the classical vacuum states is lifted, and the true quantum vacua turn out to be θ-vacuum states, in which configurations with different winding numbers are superimposed. This was first understood by Roman Jackiw and Claudio Rebbi (Jackiw and Rebbi, 1976) and independently by Curtis Callan, Roger Dashen, and David Gross (Callan *et al.*, 1976).

In the following, we apply temporal gauge fixing to set $G_4(x) = 0$ (though this does not fix the gauge completely), and we consider space to be compactified from $\mathbb{R}^3$ to S^3.[1] The classical vacuum solutions of such a theory are the pure gauge fields

$$G_i(\vec{x}) = \Omega(\vec{x})\partial_i\Omega(\vec{x})^\dagger. \tag{19.57}$$

Since we have compactified space, the classical vacua can be classified by their winding number

$$n[\Omega] \in \Pi_3[SU(N)] = \mathbb{Z}, \tag{19.58}$$

which is given by

$$n[\Omega] = \frac{1}{24\pi^2} \int_{S^3} d^3x \, \epsilon_{ijk} \mathrm{Tr}\left[\left(\Omega\partial_i\Omega^\dagger\right)\left(\Omega\partial_j\Omega^\dagger\right)\left(\Omega\partial_k\Omega^\dagger\right)\right]. \tag{19.59}$$

One might think that one could construct a *quantum vacuum* $|n\rangle$ just by considering small fluctuations around a classical vacuum with given n. Quantum tunneling, however, induces transitions between the different classical vacua.

Assume the system to be in a classical vacuum state with winding number m at early Euclidean times, $x_4 \to -\infty$. Then, it changes continuously (now deviating from a pure gauge), and finally, at late times $x_4 \to \infty$, it returns to a classical vacuum state with a possibly different winding number n. The Euclidean time evolution is described by a concrete configuration in the functional integral. The corresponding gauge field interpolates between the initial and final classical vacua.

When we calculate its topological charge, we can use the Gauss theorem. This yields an integral of the Chern–Simons density over the space–time boundary, which consists of the spatial spheres S^3 at $x_4 = -\infty$ and at $x_4 = \infty$. At both boundary spheres, the gauge field is in a pure gauge, and the integral yields the corresponding winding number such that

---

[1] This is just a mathematical trick which simplifies the discussion. Using transition functions, one could choose any other compactification, *e.g.*, on a torus T^3, or one could choose appropriate boundary conditions on $\mathbb{R}^3$ itself.

$$Q = n - m. \tag{19.60}$$

Hence, a configuration with topological charge Q induces a transition from a classical vacuum with winding number m to one with winding number $n = m + Q$. In other words, the functional integral that describes the amplitude for transitions from one classical vacuum to another is restricted to field configurations in the topological sector $Q = n - m$,

$$\langle n | \hat{U}(\infty, -\infty) | m \rangle = \int \mathcal{D} G_\mu^{(n-m)} \exp(-S[G_\mu]). \tag{19.61}$$

Here, $G_\mu^{(Q)}$ denotes a gauge field with topological charge Q and $\hat{U}$ is the (Euclidean) time evolution operator.

Note that, in contrast to Q, the winding number n is *not gauge-invariant*. We emphasize at this point that Gauss' law just demands that physical states be annihilated by the infinitesimal generators of gauge transformations. In addition, there are *large gauge transformations* with integer winding numbers, which give rise to a shift symmetry $\mathbb{Z}$, under which physical states may transform non-trivially.[2]

In particular, if we perform a gauge transformation with winding number 1, the winding number of the pure gauge field increases to $n + 1$. In the canonical formalism, such a gauge transformation Ω is implemented by a unitary operator $\hat{T}$ that acts on wave functionals $\Psi[G]$ by gauge transforming the spatial components of the gauge field G_i,

$$\hat{T} \Psi[G] = \Psi[^\Omega G], \quad {}^\Omega G_i(\vec{x}) = \Omega \Big(G_i(\vec{x}) + \partial_i \Big) \Omega(\vec{x})^\dagger. \tag{19.62}$$

Acting on a state $|n\rangle$ that describes small fluctuations around a classical vacuum with winding number n, one obtains

$$\hat{T} |n\rangle = |n + 1\rangle ; \tag{19.63}$$

i.e. $\hat{T}$ acts as a *shift operator*. Since it implements a gauge transformation, it commutes with the Hamilton operator $\hat{H}$, because the theory is gauge-invariant. This means that $\hat{H}$ and $\hat{T}$ can be diagonalized simultaneously, so they have common eigenstates. Since $\hat{T}$ is unitary, its eigenvalues are complex phases $\exp(i\theta)$, with $\theta \in]-\pi, \pi]$. Hence, an eigenstate – in particular a vacuum – can be written as $|\theta\rangle$, with

$$\hat{T} |\theta\rangle = \exp(i\theta) |\theta\rangle. \tag{19.64}$$

There are many different large gauge transformations Ω which all have the same winding number $n[\Omega] = 1$. Each of them has its own associated shift operator $\hat{T}$. Since all maps Ω in the same homotopy class can be deformed into each other, the various $\hat{T}$ operators differ only by topologically trivial, *i.e.* "small", gauge transformations. Since Gauss' law implies that physical states must be invariant under those small gauge transformations, the $\hat{T}$ operators that are associated with different maps Ω are physically equivalent.

We can construct a θ-*vacuum* as a linear combination

$$|\theta\rangle = \sum_{n \in \mathbb{Z}} c_n |n\rangle. \tag{19.65}$$

---

[2] The question of invariance under large gauge transformations will be clarified at the end of this section.

Using

$$\hat{T}|\theta\rangle = \sum_n c_n \hat{T}|n\rangle = \sum_n c_n |n+1\rangle = \sum_n c_{n-1}|n\rangle = \exp(i\theta) \sum_n c_n |n\rangle, \qquad (19.66)$$

we obtain $c_{n-1} = \exp(i\theta)c_n$. With the choice $c_0 = 1/\sqrt{2\pi}$ (which is unique up to a trivial overall complex phase), this implies $c_n = \exp(-in\theta)/\sqrt{2\pi}$, and therefore,

$$|\theta\rangle = \frac{1}{\sqrt{2\pi}} \sum_n \exp(-in\theta)\,|n\rangle. \qquad (19.67)$$

The true vacuum of a non-Abelian gauge theory is hence a superposition of fluctuations around classical vacuum states of different winding numbers. For each value of θ, there is a corresponding (true) quantum vacuum state. Note that the θ-vacuum states are orthonormal,

$$\langle\theta|\theta'\rangle = \frac{1}{2\pi} \sum_{n,m} \exp(i(n\theta - m\theta'))\langle n|m\rangle = \frac{1}{2\pi} \sum_m \exp(im(\theta - \theta')) = \delta_{2\pi}(\theta - \theta'). \qquad (19.68)$$

Here, $\delta_{2\pi}$ is the 2π-periodic δ-function; we used the orthonormality relation $\langle n|m\rangle = \delta_{nm}$, along with Poisson's summation formula.

By construction, θ-vacuum states are eigenstates of the unitary operator $\hat{T}$, which implements a topologically non-trivial gauge transformation Ω with winding number $n[\Omega] = 1$. Since the corresponding eigenvalue $\exp(i\theta)$ is in general not equal to 1, θ-vacuum states are not completely gauge-invariant. This may seem to violate Gauss' law, according to which physical states must be gauge-invariant. However, this does not apply to topologically non-trivial Ω, which are known as "large gauge transformations", as we explained before.

Let us now consider the quantum transition amplitude between different θ-vacua

$$\langle\theta|\hat{U}(\infty, -\infty)|\theta'\rangle = \sum_{n,m} \exp(in\theta)\,\langle n|\hat{U}(\infty, -\infty)|m\rangle\,\exp(-im\theta')$$

$$= \sum_{m,Q=n-m} \exp\left(im(\theta - \theta') + i\theta Q\right) \int \mathcal{D}G_\mu^{(Q)} \exp(-S[G])$$

$$= \delta_{2\pi}(\theta - \theta') \sum_Q \int \mathcal{D}G_\mu^{(Q)} \exp(-S[G] + i\theta Q[G])$$

$$= \delta_{2\pi}(\theta - \theta') \int \mathcal{D}G_\mu \exp(-S_\theta[G]). \qquad (19.69)$$

There is *no transition between different θ-vacua,* which confirms that they are energy eigenstates. In the exponent, we have identified the action, in the presence of a vacuum-angle θ, as

$$S_\theta[G] = S[G] - i\theta Q[G]. \qquad (19.70)$$

It is interesting that the topological charge enters the Euclidean action with a prefactor i. This is characteristic for topological terms in Euclidean quantum field theory.

When one reintroduces $\hbar$ (which we have set to 1), it divides $S[G]$ but not $Q[G]$, such that $S_\theta[G]/\hbar = S[G]/\hbar - \mathrm{i}\theta Q[G]$. This shows that θ *only enters the quantum theory*, but it has no effect in the classical limit $\hbar \to 0$.

19.6 Analogy with Energy Bands in a Periodic Crystal

In this section, we write $\hbar$ explicitly. The mathematics of θ-vacua is analogous to the one used to describe the energy bands in a 1-dimensional periodic crystal with lattice spacing a. In that case, the quantum mechanical single-particle Hamilton operator takes the form

$$\hat{H} = \frac{\hat{p}^2}{2M} + V(x), \quad V(x+a) = V(x). \tag{19.71}$$

Here, $\hat{p} = -\mathrm{i}\hbar\partial_x$ is the momentum operator and $V(x)$ is a periodic potential for a single electron. Let us assume the potential – which is generated by the ions in a crystal – to have its (classically degenerate) minima at the positions $x_n = na$, with $n \in \mathbb{Z}$. Due to the periodicity of the potential, $\hat{H}$ commutes with the unitary shift operator $\hat{T} = \exp(\mathrm{i}\hat{p}a/\hbar)$ that acts on the single-particle wave function $\Psi(x)$ as

$$\hat{T}\Psi(x) = \Psi(x+a). \tag{19.72}$$

This equation is analogous to eqs. (19.62) and (19.63). The gauge field G_i is analogous to the particle coordinate x, and the large gauge transformation Ω with winding number $n[\Omega] = 1$ is analogous to a shift by one lattice spacing. The classical configurations of minimal energy, where the particle is at rest at some position x_n, are analogous to the classical vacua of the gauge theory, which are characterized by a pure gauge configuration in the winding number sector n. For the periodic crystal, one can construct states $|n\rangle$ that are localized around the minimum at position x_n, which are analogous to the states $|n\rangle$ in the gauge theory. Again, due to quantum tunneling between different classical minima, the single-particle energy eigenstates in the periodic crystal are *Bloch waves, i.e.* superpositions

$$|\theta\rangle = \frac{1}{\sqrt{2\pi}} \sum_{n\in\mathbb{Z}} \exp(-\mathrm{i}n\theta)\,|n\rangle, \tag{19.73}$$

which are eigenstates of the shift operator $\hat{T}|\theta\rangle = \exp(\mathrm{i}\theta)|\theta\rangle$. In particular, the wave function $\Psi_\theta(x) = \langle x|\theta\rangle$ obeys the condition

$$\Psi_\theta(x+a) = \hat{T}\Psi_\theta(x) = \exp(\mathrm{i}\theta)\Psi_\theta(x). \tag{19.74}$$

Hence, the vacuum-angle θ is analogous to the *Bloch momentum*. Just as the Bloch momentum labels the states in an energy band of a periodic crystal, the vacuum-angle θ characterizes physically distinct eigenstates of the gauge theory Hamilton operator.

The analogy with the periodic crystal raises the question whether the vacuum-angle should be viewed as a quantum number that distinguishes different possible energy eigenstates or as a fundamental parameter of the Standard Model Hamilton operator (or Lagrangian).

We discuss this question by again considering the periodic crystal. The way we have presented this so far, θ labels a wave function $\Psi_\theta(x)$ that satisfies eq. (19.74), while the

Hamilton operator of eq. (19.71) is θ-independent. As we will see next, the explicit θ-dependence (which manifests itself in eq. (19.74)) can be moved from the wave function to the Hamilton operator by the unitary transformation $U_\theta = \exp(-i\theta x/a)$. Acting with U_θ on a wave function $\Psi_\theta(x)$, one obtains the new wave function

$$\widetilde{\Psi}(x) = U_\theta \Psi_\theta(x) = \exp(-i\theta x/a)\Psi_\theta(x), \qquad (19.75)$$

which now obeys the θ-independent condition

$$\widetilde{\Psi}(x+a) = \exp(-i\theta(x+a)/a)\Psi_\theta(x+a) = \exp(-i\theta(x+a)/a)\exp(i\theta)\Psi_\theta(x)$$
$$= \exp(-i\theta x/a)\Psi_\theta(x) = \widetilde{\Psi}(x), \qquad (19.76)$$

and is thus strictly periodic. When we apply the unitary transformation to the momentum operator, we obtain the new operator

$$\hat{p}_\theta = U_\theta \hat{p} U_\theta^\dagger = \exp(-i\theta x/a)\,(-i\hbar\partial_x)\,\exp(i\theta x/a) = -i\hbar\partial_x + \hbar\theta/a = \hat{p} + \hbar\theta/a, \quad (19.77)$$

which is explicitly θ-dependent. The same is true for the unitarily transformed Hamilton operator

$$\hat{H}_\theta = U_\theta \hat{H} U_\theta^\dagger = \frac{(\hat{p} + \hbar\theta/a)^2}{2M} + V(x). \qquad (19.78)$$

Hence, we see that the explicit θ-dependence has been moved from the wave function to the Hamilton operator. We also confirm that the θ-dependence disappears in the classical limit $\hbar \to 0$. It should be noted that the eigenstates of $\hat{H}_\theta$ still implicitly depend on θ (which is now a parameter in the Hamilton operator), although θ no longer enters the condition (19.76) explicitly.

Of course, the θ-dependent Hamilton operator still obeys a shift symmetry, which is now represented by

$$\hat{T}_\theta = U_\theta \hat{T} U_\theta^\dagger = \exp(i\hat{p}_\theta a/\hbar) = \exp(i\hat{p}a/\hbar)\exp(i\theta) = \exp(i\theta)\hat{T}. \qquad (19.79)$$

Indeed, the θ-independent wave functions $\widetilde{\Psi}(x)$ are still eigenstates of the (now θ-dependent) shift operator

$$\hat{T}_\theta \widetilde{\Psi}(x) = \exp(i\theta)\hat{T}\widetilde{\Psi}(x) = \exp(i\theta)\widetilde{\Psi}(x+a) = \exp(i\theta)\widetilde{\Psi}(x). \qquad (19.80)$$

In the gauge theory, one can proceed analogously and move the explicit θ-dependence from the states to the Hamilton operator. The wave functionals $\widetilde{\Psi}[G]$ are then invariant even under large gauge transformations

$$\hat{T}\widetilde{\Psi}[G] = \widetilde{\Psi}[^\Omega G] = \widetilde{\Psi}[G]. \qquad (19.81)$$

This suggests that the most appropriate place of the vacuum-angle is in the Hamilton operator (or in the Lagrangian), not in the state. Then, we arrive at a consistent picture of invariance under all gauge transformations. The necessity of arriving at this picture is most transparent if we start from the lattice formulation: Here, any gauge transformation at each site is allowed, and the notion of differentiable gauge transformations, which seems to be a requisite in the continuum, is not on solid grounds: The distinction between "small" (Gauss law restricted) and "large" gauge transformations fades away.

19.7   Some Questions Related to θ

The topological properties of gauge fields have local consequences. For example, a value of the QCD vacuum-angle $\theta \neq 0$ or π would break the parity symmetry P (while charge conjugation C remains unbroken). As a result, the combined symmetry CP would then be broken by the strong interaction, while other gauge interactions in the Standard Model break both P and C, but keep CP intact. A strong CP violation would manifest itself by a non-zero electric dipole moment of the neutron, which would be proportional to θ (at least for small θ) (Crewther *et al.*, 1979). The experimental bound on the neutron's dipole moment – which is so far immeasurably small – implies the upper bound $|\theta| < 10^{-10}$ (Baker *et al.*, 2006; Pendlebury *et al.*, 2015; Abel *et al.*, 2020), which may suggest that $\theta = 0$. The *strong CP-problem*, which will be discussed in Chapter 25, then confronts us with the puzzle why θ is so incredibly small (if not zero).

As we have seen in eq. (19.70), in the functional integral, θ manifests itself as the prefactor of the topological charge contribution to the Euclidean action. Correspondingly, the Lagrangian receives a topological contribution $-\mathrm{i}\theta P(x)$ from the Chern–Pontryagin density. From this point of view, which is completely justified, θ appears as yet another *fundamental parameter in the Standard Model*. However, from our previous discussion we know that the explicit θ-dependence can also be removed from the Lagrangian or Hamilton operator. Then, θ appears explicitly as a parameter that characterizes the wave functionals $\Psi_\theta[G]$, by their transformation properties under large gauge transformations Ω with $n[\Omega] = 1$, *i.e.*

$$\hat{T}\Psi_\theta[G] = \exp(\mathrm{i}\theta)\Psi_\theta[G]). \tag{19.82}$$

This relation follows immediately from eq. (19.81) when one identifies $\Psi_\theta[G] = \langle G_i|\theta\rangle$. From this equivalent point of view, θ has mutated from a parameter in the Lagrangian to a *conserved quantum number* that characterizes a state.

Remarkably, large gauge transformations give rise to an exact "global" shift symmetry $\mathbb{Z}$, and the corresponding conserved quantum number is $\exp(\mathrm{i}\theta)$. This actually seems to contradict the Intermezzo, where we argued that global symmetries are usually only approximate, while gauge symmetries just reflect redundancies in our theoretical description. Simply demanding that physical states must be invariant against all gauge transformations – including large ones – does not resolve the problem, because then θ can still exist as a parameter in the Hamilton operator $\hat{H}_\theta$, which still commutes with $\hat{T}_\theta$. It seems appropriate to think of θ as a manifestation of an exact "global" shift symmetry that is protected from explicit breaking because it is embedded in a gauge symmetry.

Since θ characterizes energy eigenstates, one may wonder whether the "wave function of the Universe" may be a superposition of different θ-states. Already from an abstract mathematical point of view, this is problematical, because the different θ-states exist in different domains of the Hilbert space. Defining the Hamilton operator as a self-adjoint (rather than just Hermitian) operator requires specifying the domain of the Hilbert space in which it acts. From a mathematical point of view, θ is a self-adjoint extension parameter that specifies that domain. Hence, superimposing different θ-states is not mathematically meaningful. The same can be concluded on physical grounds when one moves the θ-dependence into the Hamilton operator. It makes no sense in physics to superimpose states that evolve according to different Hamilton operators.

If θ is a strictly conserved quantity, its physical value can only be "explained" by the initial conditions. Let us compare the situation with the one for another conserved quantity: the total electric charge Q. Are we puzzled by the overall charge neutrality of the Universe? If not, why not? A Universe that carries a net electric charge per unit volume would probably just explode (much faster than the observed accelerated cosmic expansion), due to its enormous Coulomb energy. Obviously, we would not be able to exist in such a Universe. Hence, it seems tempting to invoke the anthropic principle to explain why the net electric charge of the Universe is $Q = 0$. However, other potential explanations, although not necessarily easily falsifiable either, are possible as well. For example, a compact Universe without spatial boundary, *e.g.*, a very large torus, would automatically have $Q = 0$, as a consequence of Gauss' law.

Can $\theta = 0$ perhaps be explained in a similar manner? Indeed, as we will discuss in the next section, the existence of an exactly massless quark, say the u-quark, would eliminate all θ effects. While this "solution" seems not phenomenologically viable, it would in any case immediately confront us with another problem. We would just trade the strong CP-problem (why is $\theta = 0$?) for the "u-quark problem" (why is $m_u = 0$?).

Could there be an anthropic explanation for $\theta = 0$? A non-zero electric dipole moment of the neutron may well be consistent with the existence of intelligent life forms, which could be capable of asking "why does θ take this value?". In this context, one may also wonder about the contribution of θ to the vacuum energy, which in turn contributes to the net cosmological constant. At present, we have no clue how to solve the cosmological constant problem, and we may thus accept the anthropic principle as a possible explanation. Certainly, as Weinberg has argued, a Universe with a large (positive or negative) cosmological constant would not be hospitable to life. Could we argue that vacuum energy associated with $\theta \neq 0$ would yield an additional large contribution to the cosmological constant, which would be inconsistent with our existence? This is, in fact, not convincing, because there are already many non-vanishing contributions to the vacuum energy, *e.g.*, due to the QCD chiral condensate $\langle \bar{q}q \rangle$ or due to the vacuum value v of the Higgs field. Additional θ-vacuum energy would not change this situation. A solution of the cosmological constant problem that does not rely on the anthropic principle should explain why the sum of all these forms of vacuum energy seems to have no gravitational effect. When we invoke the anthropic principle, we just assume that in our part of a "multiverse" all forms of vacuum energy add up to an almost vanishing cosmological constant. This argument would remain unchanged in the presence of additional θ-vacuum energy, and thus, there seems to be no compelling anthropic argument for $\theta = 0$.

One may also ask whether classical gravity in an expanding Universe may have an impact on θ. Would θ remain conserved or could it be driven to zero as a function of time? Since θ multiplies a topological term, which is insensitive to the metric, such speculations do not seem very promising either.

A naive hope to avoid the strong CP-problem might be that gluon field configurations with non-vanishing topological charge are negligible in the QCD functional integral (due to their action requirement, which is at least proportional to $|Q|$). Chapter 20 addresses the $U(1)_A$-problem and explains why this simple assumption is wrong. Topologically non-trivial gluon field configurations indeed contribute significantly to the functional integral. As a result, the flavor-singlet axial $U(1)_A$ symmetry is strongly, anomalously broken. Hence, unlike the flavor-non-singlet chiral symmetry, it cannot break spontaneously. This solves the $U(1)_A$-problem and explains why the η'-meson is not a light pseudo-Nambu–Goldstone boson.

As we will discuss in Chapter 25, $\theta = 0$ may find a natural explanation in new physics beyond the Standard Model. In particular, when θ is promoted from a fundamental parameter to a dynamical field, it may naturally evolve toward $\theta(x) = 0$. This potential solution of the strong CP-problem is associated with a spontaneously broken approximate global $U(1)_{PQ}$ *Peccei–Quinn symmetry*, and a corresponding pseudo-Nambu–Goldstone boson – the hypothetical *axion* – which has until now resisted experimental detection.

19.8 Atiyah–Singer Index Theorem

Let us consider a *massless fermion* in a fixed background gauge field G_μ, thus leading to the massless Dirac operator

$$D[G] = \gamma_\mu [\partial_\mu + G_\mu(x)]. \tag{19.83}$$

An *index theorem* due to Michael Atiyah and Isadore Singer relates the topological charge of the gauge field to the zero-modes of the Dirac operator (Atiyah and Singer, 1963). They studied the eigenvalue problem

$$D[G]\psi_\lambda(x) = \lambda\psi_\lambda(x). \tag{19.84}$$

Here, $\psi_\lambda(x)$ is a 4-component Dirac spinor that transforms in the fundamental representation of the gauge group and takes values in the complex numbers (*i.e.* it is not a Grassmann field). Since $D[G]$ is anti-Hermitian, it has purely imaginary eigenvalues $\lambda = -\lambda^*$. It is important that $D[G]$ anti-commutes with γ_5,

$$\{D[G], \gamma_5\} = D[G]\gamma_5 + \gamma_5 D[G] = 0. \tag{19.85}$$

This implies that the non-zero eigenvalues $\lambda \neq 0$ come in complex conjugate pairs, because

$$D[G]\gamma_5\psi_\lambda(x) = -\gamma_5 D[G]\psi_\lambda(x) = -\lambda\gamma_5\psi_\lambda(x). \tag{19.86}$$

This shows that $\gamma_5\psi_\lambda(x)$ is also an eigenvector of $D[G]$, with the eigenvalue $-\lambda = \lambda^*$, *i.e.*

$$\psi_{-\lambda}(x) = \mathcal{N}_\lambda \gamma_5 \psi_\lambda(x). \tag{19.87}$$

Here, $\mathcal{N}_\lambda \in \mathbb{C}$ is a normalization factor which reduces to a complex phase, because ψ_λ and $\gamma_5\psi_\lambda$ have the same norm. By fixing the arbitrary phase in $\psi_{-\lambda}$ accordingly, one can set $\mathcal{N}_\lambda = 1$.

A special situation arises for eigenvectors with zero eigenvalue

$$D[G]\psi_0(x) = 0. \tag{19.88}$$

These eigenvalues are generically not paired. Instead, the corresponding eigenvectors have a definite chirality, because for them

$$\gamma_5\psi_0(x) = \pm\psi_0(x) \quad \Rightarrow \quad (1 \mp \gamma_5)\psi_0(x) = 0. \tag{19.89}$$

In general, there is some number n_R of zero eigenvalues with a right-handed eigenvector that obeys $(1 - \gamma_5)\psi_0(x) = 0$ and n_L zero eigenvalues with a left-handed eigenvector that obeys $(1 + \gamma_5)\psi_0(x) = 0$. The left- and right-handed zero-modes are eigenstates of γ_5 with eigenvalues -1 and 1, respectively.

The Atiyah–Singer index theorem states that the difference between the numbers of left- and right-handed zero-modes of the Dirac operator $D[G]$ is given by the topological charge of the gauge field $Q[G]$, *i.e.*

$$n_R - n_L = Q[G]. \tag{19.90}$$

Hence, a topologically non-trivial gauge field configuration necessarily has at least one zero-mode.[3] The proof of the Atiyah–Singer index theorem is relatively straightforward on the lattice, which will be discussed in Section 19.10.

Let us now consider the functional integral over a non-Abelian gauge field G_μ and a massless Dirac fermion described by the Grassmann field ψ

$$\begin{aligned}
Z &= \int \mathcal{D}G \mathcal{D}\bar{\psi} \mathcal{D}\psi \ \exp(-S_\theta[G]) \exp(-\bar{\psi} D[G]\psi) \\
&= \int \mathcal{D}G \mathcal{D}\bar{\psi} \mathcal{D}\psi \ \exp(-S[G] + i\theta Q[G]) \exp(-\bar{\psi} D[G]\psi) \\
&= \sum_{Q \in \mathbb{Z}} \int \mathcal{D}G^{(Q)} \exp(-S[G] + i\theta Q) \det D[G] \\
&= \int \mathcal{D}G^{(0)} \exp(-S[G]) \det D[G]. \tag{19.91}
\end{aligned}$$

Here, we have divided the gluon functional integral into topological sectors containing gauge fields $\mathcal{D}G^{(Q)}$ with fixed topological charge Q, and we have applied the Atiyah–Singer index theorem to conclude that

$$\det D[G] = \delta_{Q[G],0} \det D[G]. \tag{19.92}$$

Any zero-mode of the Dirac operator makes the fermion determinant vanish; hence, it eliminates topologically non-trivial field configurations from theories with massless fermions. Then, $Q[G] = 0$ holds for all configurations that contribute to the functional integral. In that case, the θ-term in the action has no effect, and all θ-vacua become physically equivalent. Non-trivial topological sectors still contribute to certain observables, including the chiral condensate.

This scenario had been suggested as a possible solution to the strong CP-problem. If the lightest quark – the u-quark – would be massless, then θ could not generate an electric dipole moment for the neutron.

A priori, it is not obvious that this scenario can be excluded. For example, to leading-order in the quark masses, the pion mass depends only on the sum $m_u + m_d$, so $m_d > 0$ is sufficient to explain the non-zero pion mass. One thus needs to understand more subtle effects. Based on lattice QCD calculations, according to the Particle Data Book, the

---

[3] In principle, the value of Q could emerge from a cancellation between some left-handed and some right-handed zero-modes. This happens in models which assume a "gas" or "liquid" of instantons and anti-instantons. However, in the set of QCD configurations, the subset which provides such a cancellation is of measure zero. This is observed in lattice simulations with chiral quarks: For typical configurations, one always obtains $n_L = 0$ or $n_R = 0$ (or both).

u-quark mass is incompatible with zero: It quotes a value of $m_u = 2.16^{+0.49}_{-0.26}$ MeV in the $\overline{\text{MS}}$ scheme at a scale of 2 GeV (Workman *et al.*, 2022).

19.9 Zero-Mode of the SU(2) Instanton

Let us illustrate the index theorem by explicitly constructing the *zero-mode* of the SU(2) instanton. For this purpose, we use the notation ('t Hooft, 1976b)

$$\tau_\mu = (-i\vec{\tau}, \mathbb{1}), \quad \bar{\tau}_\mu = (i\vec{\tau}, \mathbb{1}), \quad \bar{\tau}_\mu \tau_\nu = \delta_{\mu\nu}\mathbb{1} + \eta_{\mu\nu}, \quad \tau_\mu \bar{\tau}_\nu = \delta_{\mu\nu}\mathbb{1} + \bar{\eta}_{\mu\nu}. \quad (19.93)$$

The anti-symmetric *'t Hooft symbols* $\eta_{\mu\nu} = -\eta_{\nu\mu}$ and $\bar{\eta}_{\mu\nu} = -\bar{\eta}_{\nu\mu}$ are (anti-)self-dual and obey

$$\widetilde{\eta}_{\mu\nu} = \frac{1}{2}\epsilon_{\mu\nu\rho\sigma}\eta_{\rho\sigma} = \eta_{\mu\nu}, \quad \widetilde{\bar{\eta}}_{\mu\nu} = \frac{1}{2}\epsilon_{\mu\nu\rho\sigma}\bar{\eta}_{\rho\sigma} = -\bar{\eta}_{\mu\nu},$$

$$[\eta_{\mu\nu}, \eta_{\rho\sigma}] = 2\left(\delta_{\mu\sigma}\eta_{\nu\rho} - \delta_{\mu\rho}\eta_{\nu\sigma} - \delta_{\nu\sigma}\eta_{\mu\rho} + \delta_{\nu\rho}\eta_{\mu\sigma}\right), \quad \text{Tr}(\eta_{\mu\nu}\eta_{\mu\nu}) = -24. \quad (19.94)$$

In an exercise, we are going to show that the SU(2) instanton of eq. (19.56) and its field strength can be expressed as

$$G_\mu(x) = \frac{x_\nu}{x^2 + \rho^2}\eta_{\nu\mu}, \quad G_{\mu\nu}(x) = \frac{2\rho^2}{(x^2 + \rho^2)^2}\eta_{\mu\nu}. \quad (19.95)$$

Next, we extend the definition of the 't Hooft tensor from SU(2)$_c$ color to SU(2)$_S$ spin indices

$$\sigma_\mu = (-i\vec{\sigma}, \mathbb{1}), \; ^\bullet\bar{\sigma}_\mu = (i\vec{\sigma}, \mathbb{1}), \quad \bar{\sigma}_\mu\sigma_\nu = \delta_{\mu\nu}\mathbb{1} + \zeta_{\mu\nu}, \quad \sigma_\mu\bar{\sigma}_\nu = \delta_{\mu\nu}\mathbb{1} + \bar{\zeta}_{\mu\nu}, \quad (19.96)$$

which implies

$$\frac{1}{2}[\gamma_\mu, \gamma_\nu] = \frac{1}{2}\left[\begin{pmatrix} 0 & \sigma_\mu \\ \bar{\sigma}_\mu & 0 \end{pmatrix}, \begin{pmatrix} 0 & \sigma_\nu \\ \bar{\sigma}_\nu & 0 \end{pmatrix}\right] = \begin{pmatrix} \bar{\zeta}_{\mu\nu} & 0 \\ 0 & \zeta_{\mu\nu} \end{pmatrix}, \quad (19.97)$$

and we consider the Dirac operator in the instanton background

$$D[G] = \gamma_\mu\left(\partial_\mu + G_\mu(x)\right) = \begin{pmatrix} 0 & \sigma_\mu \\ \bar{\sigma}_\mu & 0 \end{pmatrix}\left(\partial_\mu + G_\mu(x)\right) \quad \Rightarrow$$

$$D[G]^2 = \begin{pmatrix} \sigma_\mu\bar{\sigma}_\nu & 0 \\ 0 & \bar{\sigma}_\mu\sigma_\nu \end{pmatrix}\left(\partial_\mu + G_\mu(x)\right)\left(\partial_\nu + G_\nu(x)\right)$$

$$= \left(\partial_\mu + G_\mu(x)\right)\left(\partial_\mu + G_\mu(x)\right) + \frac{1}{2}\begin{pmatrix} \bar{\zeta}_{\mu\nu} & 0 \\ 0 & \zeta_{\mu\nu} \end{pmatrix}[\partial_\mu + G_\mu(x), \partial_\nu + G_\nu(x)]$$

$$= \left(\partial_\mu + G_\mu(x)\right)\left(\partial_\mu + G_\mu(x)\right) + \frac{1}{2}\begin{pmatrix} \bar{\zeta}_{\mu\nu} & 0 \\ 0 & \zeta_{\mu\nu} \end{pmatrix}G_{\mu\nu}(x). \quad (19.98)$$

The instanton field strength $G_{\mu\nu}(x)$ contains the 't Hooft tensor $\eta_{\mu\nu}$ in color space, which is now contracted with the corresponding tensors $\zeta_{\mu\nu}$ and $\bar{\zeta}_{\mu\nu}$ in spin space. Remarkably, $\bar{\zeta}_{\mu\nu}\eta_{\mu\nu} = 0$, such that

$$D[G]^2 = \left(\partial_\mu + G_\mu(x)\right)\left(\partial_\mu + G_\mu(x)\right) + \frac{1}{2}\begin{pmatrix} 0 & 0 \\ 0 & \zeta_{\mu\nu} \end{pmatrix} G_{\mu\nu}(x) \quad \Rightarrow$$

$$D[G]^2 \psi_{\mathrm{OR}}(x) = \left\{ \left(\partial_\mu + G_\mu(x)\right)\left(\partial_\mu + G_\mu(x)\right) + \frac{1}{2}\zeta_{\mu\nu}G_{\mu\nu}(x) \right\}\psi_{\mathrm{OR}}(x) = 0,$$

$$D[G]^2 \psi_{\mathrm{OL}}(x) = \left(\partial_\mu + G_\mu(x)\right)\left(\partial_\mu + G_\mu(x)\right) \psi_{\mathrm{OL}}(x) = 0. \tag{19.99}$$

The left- and right-handed components of the zero-mode $\psi_0(x) = \begin{pmatrix} \psi_{\mathrm{OL}}(x) \\ \psi_{\mathrm{OR}}(x) \end{pmatrix}$, which

satisfies $\gamma_\mu\left(\partial_\mu + G_\mu(x)\right)\psi_0(x) = 0$, obey two qualitatively different equations. In particular, the operator $\left(\partial_\mu + G_\mu(x)\right)\left(\partial_\mu + G_\mu(x)\right)$, which acts on $\psi_{\mathrm{OL}}(x)$, is negative definite,
because it is the square of the anti-Hermitian operator $\left(\partial_\mu + G_\mu(x)\right)$. Consequently, it can-
not have a normalizable zero-mode, and hence $\psi_{\mathrm{OL}}(x) = 0$. As a result, the zero-mode is
purely right-handed and obeys the Weyl equation

$$\sigma_\mu\left(\partial_\mu + G_\mu(x)\right)\psi_{\mathrm{OR}}(x) = 0 \quad \Rightarrow \quad (x^2 + \rho^2)\sigma_\nu\partial_\nu\psi_{\mathrm{OR}}(x) = x_\nu\sigma_\mu\eta_{\mu\nu}\psi_{\mathrm{OR}}(x). \tag{19.100}$$

The right-handed zero-mode $\psi_{\mathrm{OR}}(x)$ has a spin index $s \in \{1,2\}$ and a color index $c \in \{1,2\}$.
When one makes the ansatz $\psi_{\mathrm{OR}}^{sc}(x) = \psi(x)\epsilon_{sc}$, where the 2-d Levi-Civita symbol ϵ_{sc} anti-
symmetrizes spin and color indices, one can convince oneself that

$$\sigma_\mu\eta_{\mu\nu}\psi_{\mathrm{OR}}^{sc}(x) = -3\sigma_\nu\psi_{\mathrm{OR}}^{sc}(x) \quad \Rightarrow \quad (x^2 + \rho^2)\partial_\nu\psi(x) = -3x_\nu\psi(x) \quad \Rightarrow$$

$$\psi(x) = \frac{A}{(x^2 + \rho^2)^{3/2}}. \tag{19.101}$$

Finally, the normalized right-handed zero-mode for the SU(2) instanton takes the form

$$\psi_{\mathrm{OR}}^{sc}(x) = \frac{\rho}{\pi}\frac{1}{(x^2 + \rho^2)^{3/2}}\epsilon_{sc}. \tag{19.102}$$

This is consistent with the index theorem, $Q[G] = n_{\mathrm{R}} - n_{\mathrm{L}}$, since the instanton has $Q[G] = 1$, $n_{\mathrm{R}} = 1$, and $n_{\mathrm{L}} = 0$.

19.10 Index Theorem on the Lattice

For a long time, it was not understood how to naturally (without fine-tuning) recover chiral
symmetry in the lattice regularization. Indeed, for some time the lattice had a bad reputation because it seemed to mutilate the chiral properties of fermions. As we discussed in
Section 18.6, the situation changed when it was realized that the Ginsparg–Wilson relation is the key to understanding chiral symmetry beyond perturbation theory. This insight
also affected the non-perturbative status of the index theorem. Very remarkably, based on
the Ginsparg–Wilson relation, Peter Hasenfratz, Victor Laliena, and Ferenc Niedermayer
were able to prove the index theorem on the lattice in an elegant manner (Hasenfratz et al.,
1998). Their proof simplifies when one takes advantage of Martin Lüscher's definition of
lattice chiral symmetry, which is again based upon the Ginsparg–Wilson relation (Lüscher,
1998) that we already discussed in Section 18.6.

Let us hence consider a lattice Dirac operator $D[U]$ that obeys the *Ginsparg–Wilson
relation*

$$\{D[U], \gamma_5\} = D[U]\gamma_5 + \gamma_5 D[U] = aD[U]\gamma_5 D[U]. \tag{19.103}$$

In addition, we assume $D[U]$ to be γ_5-*Hermitian*

$$D[U]^\dagger = \gamma_5 D[U]\gamma_5. \qquad (19.104)$$

That property holds for the commonly used lattice Dirac operators. This assumption is not really necessary, but it somewhat simplifies the proof. For a γ_5-Hermitian Dirac operator, the Ginsparg–Wilson relation reduces to

$$D[U] + D[U]^\dagger = aD[U]^\dagger D[U] = aD[U]D[U]^\dagger. \qquad (19.105)$$

As in Section 10.7, we now define $A[U] = aD[U] - \mathbb{1}$, such that

$$A[U]^\dagger A[U] = a^2 D[U]^\dagger D[U] - aD[U]^\dagger - aD[U] + \mathbb{1} = \mathbb{1},$$
$$A[U]A[U]^\dagger = a^2 D[U]D[U]^\dagger - aD[U] - aD[U]^\dagger + \mathbb{1} = \mathbb{1}. \qquad (19.106)$$

Hence, $A[U]$ is unitary with eigenvalues $\exp(i\omega) \in U(1)$. The eigenvectors $\Psi_{\omega,x}$ of the lattice Dirac operator thus obey

$$D[U]\Psi_{\omega,x} = \frac{1}{a}[1 + \exp(i\omega)]\Psi_{\omega,x}. \qquad (19.107)$$

For $\omega = \pi - ia\lambda$, the eigenvalue of $D[U]$ reduces to the eigenvalue $\lambda \in i\mathbb{R}$ of $D[G]$ in the continuum limit.

Quick Question 19.10.1   Commutator of $D[U]$ with $D[U]^\dagger$
Use the Ginsparg–Wilson relation along with γ_5-Hermiticity to show that $[D[U], D[U]^\dagger] = 0$.

Based on this commutation property, we obtain

$$D[U]^\dagger \gamma_5 \Psi_{\omega,x} = \frac{1}{a}[1 + \exp(i\omega)]\gamma_5 \Psi_{\omega,x}. \qquad (19.108)$$

This implies that non-real eigenvalues $\exp(i\omega)$ and $\exp(-i\omega)$ are paired. With an appropriate choice of a complex phase of $\Psi_{-\omega,x}$, we obtain

$$\Psi_{-\omega,x} = \gamma_5 \Psi_{\omega,x}. \qquad (19.109)$$

Real-valued eigenvalues $\exp(i\omega) = \pm 1$ (with $\omega = 0$ or π), on the other hand, are generically not paired. Their eigenmodes are simultaneous eigenvectors of γ_5 with eigenvalues ± 1

$$\gamma_5 \Psi_{0,x} = \pm\Psi_{0,x}, \quad \gamma_5 \Psi_{\pi,x} = \pm\Psi_{\pi,x}. \qquad (19.110)$$

We distinguish the zero-modes of $D[U]$ with $\omega = \pi$ from the modes with $\omega = 0$, which have the eigenvalue $2/a$ of $D[U]$. As in the continuum, we denote the number of left- and right-handed zero-modes, which have γ_5-eigenvalues -1 and 1, as n_L and n_R, respectively. Similarly, we denote the left- and right-handed modes with eigenvalue $2/a$ of $D[U]$ as n'_L and n'_R.

Let us first consider $\text{Tr}(\gamma_5) = 0$, where the trace extends not only over Dirac but also over color and lattice site indices. In the basis of eigenvectors of $D[U]$, due to eq. (19.109),

γ_5 has non-vanishing diagonal elements only for the modes with $\exp(i\omega) = \pm 1$. As a result, we obtain

$$\text{Tr}(\gamma_5) = n_R - n_L + n'_R - n'_L = 0 \quad \Rightarrow \quad n_R - n_L = n'_L - n'_R. \tag{19.111}$$

Next, we consider $\text{Tr}(\gamma_5 D[U])$. Due to eq. (19.109), this trace again receives contributions only from the modes with $\exp(i\omega) = \pm 1$. Since the eigenvectors with $\exp(i\omega) = -1$ are zero-modes of $D[U]$, in this case, only the modes with $\exp(i\omega) = 1$ (which have $D[U]$-eigenvalue $2/a$) contribute and we obtain

$$\text{Tr}(\gamma_5 D[U]) = \frac{2}{a}(n'_R - n'_L) \quad \Rightarrow \quad n_R - n_L = -\frac{a}{2}\text{Tr}(\gamma_5 D[U]). \tag{19.112}$$

The right-hand side of this expression coincides with the gauge variation of the fermionic measure under Lüscher's lattice chiral transformation of eq. (18.42). We will now identify this term with the topological charge $Q[U]$ of the lattice gauge configuration $[U]$.

A priori, a lattice gauge field does not have well-defined topological properties. In particular, the field configurations that contribute significantly to the functional integral are not differentiable and not even continuous. By extending the trace (now written as tr) only over Dirac and color indices (but not also over the lattice sites), we *define* the *topological charge density* as

$$q_x[U] = -\frac{a}{2}\text{tr}(\gamma_5 D[U]_{xx}) \quad \Rightarrow \quad Q[U] = \sum_x q_x[U], \tag{19.113}$$

such that the index theorem extends exactly to the fully non-perturbative framework of lattice gauge theory with strongly fluctuating quantum gauge fields,

$$n_R - n_L = Q[U]. \tag{19.114}$$

Since we *defined* $q_x[U]$ accordingly, the index theorem on the lattice may appear as a tautology. However, this is not the case. First of all, Hasenfratz, Laliena, and Niedermayer showed that $Q[U]$ agrees with the topological charge that is associated with the corresponding fixed point of the renormalization group (Hasenfratz *et al.*, 1998). Furthermore, it is straightforward to inspect $q_x[U]$ in the classical continuum limit of smooth lattice gauge fields. By construction, $q_x[U]$ is gauge-invariant, invariant under charge conjugation, and it is odd under parity and time reversal. Since its sum over the 4-d space–time lattice yields the integer $n_R - n_L$, $q_x[U]$ has dimension 4. The only continuum expression with these features is indeed $-1/32\pi^2 \epsilon_{\mu\nu\rho\sigma}\text{Tr}[G_{\mu\nu}(x)G_{\rho\sigma}(x)]$. Finally, $Q[U]$ is a topological quantity even on the lattice; *i.e.* it is invariant against (sufficiently small) continuous deformations of the lattice gauge field configuration. It inherits this feature from the exact zero-modes of the lattice Dirac operator which are also topologically protected. We will return to the discussion of the lattice topological charge in Section 20.4.

This simple derivation of the index theorem underscores the power of lattice gauge theory. It deepens our understanding of chiral symmetry by extending it beyond perturbation theory. Any bad reputation that the lattice regularization may have had in the past, due to its problems with chiral symmetry, is certainly no longer justified.

Exercises

19.1 Divergence of the Chern–Simons density
Verify that the Chern–Simons density (19.8) is consistent with eqs. (19.3) and (19.7).

19.2 Derivation of a useful formula
Demonstrate that the formula (19.18) is correct.

19.3 Coset decomposition of SU(3)
Perform the coset decomposition of an SU(3) matrix $\Omega = VW$, where W is an embedded SU(2) matrix. Show that, for $N = 3$, the matrix V defined in eq. (19.17) is indeed in SU(3).

19.4 Volume element of the group SU(2)
Take the SU(2) matrix $\tilde{\Omega} = \cos\alpha + i\sin\alpha\, \vec{e}_\alpha \cdot \vec{\tau}$ of eq. (19.30) and verify eq. (19.31).

19.5 Gauge invariance of the topological charge
Demonstrate explicitly the gauge invariance of $Q = \sum_i Q_i$ in eq. (19.45), by performing a gauge transformation on the original gauge field

$$G'_\mu(x) = \Omega(x)\left(G_\mu(x) + \partial_\mu\right)\Omega(x)^\dagger. \tag{19.115}$$

Based on the gauge transformation properties of the section and using formula (19.18), this is feasible.

19.6 Coboundary of the 0-cochain
Show that eq. (19.42) holds.

19.7 1-cochain for SU(2)
Consider the integral

$$I[U] = \frac{1}{24\pi^2}\int_{H^3} d^3x\, \epsilon_{ijk} \mathrm{Tr}\left[\left(U\partial_i U^\dagger\right)\left(U\partial_j U^\dagger\right)\left(U\partial_k U^\dagger\right)\right],$$

where H^3 is a 3-dimensional hemisphere and $U = \cos\alpha + i\sin\alpha\,\vec{e}_\alpha \cdot \vec{\tau} \in \mathrm{SU}(2)$.
a) Show that

$$\frac{1}{24\pi^2}\epsilon_{ijk}\mathrm{Tr}\left[\left(U\partial_i U^\dagger\right)\left(U\partial_j U^\dagger\right)\left(U\partial_k U^\dagger\right)\right] = \partial_i\Omega_i^{(1)},$$

where the 1-cochain is given by

$$\Omega_i^{(1)} = -\frac{1}{8\pi^2}\left(\alpha - \sin\alpha\cos\alpha\right)\epsilon_{ijk}\,\vec{e}_\alpha \cdot \left(\partial_j\vec{e}_\alpha \times \partial_k\vec{e}_\alpha\right).$$

b) Show that, besides contributions from the 2-dimensional boundary $\partial H^3 = S^2$, $I[U]$ receives integer-valued contributions from the bulk of H^3.

19.8 SU(2) instanton
Convince yourself that the gluon field (19.49), (19.50) with

$$f(|x|) = \frac{|x|^2}{|x|^2 + \rho^2} \tag{19.116}$$

is indeed an instanton for any value of the scale parameter $\rho > 0$.

19.9 Specific form of the SU(2) instanton

Show that the SU(2) instanton of eq. (19.56) and its field strength can be expressed as eq. (19.95).

19.10 Instanton zero-mode

Convince yourself that the instanton zero-mode of eq. (19.102) indeed obeys the eigenvalue eq. (19.84) with $\lambda = 0$.

20

U(1)$_A$-Problem

The U(1)$_A$-*problem* in QCD is to explain why the η'-meson has a large mass (slightly heavier than the nucleon) and hence cannot be interpreted as a pseudo-Nambu–Goldstone boson. This is *qualitatively* understood based on the Adler–Bell–Jackiw anomaly – the *explicit* breaking of the axial U(1)$_A$ symmetry of QCD, as a quantum effect, cf. eq. (19.2). However, the question remains how strong this breaking really is and how far up it drives the η'-meson mass.

As we will discuss below, the *quantitative* solution to the U(1)$_A$-problem – *i.e.* an explanation of the large η'-meson mass – requires gluon field configurations with non-zero topological charge to contribute significantly to the functional integral. This is confirmed by lattice calculations, and it offers a convincing solution to the U(1)$_A$-problem. On the other hand, when we understand that topologically non-trivial gluon field configurations explain the U(1)$_A$-problem, we cannot discard these configurations when we face the still unsolved strong CP-problem, which will be discussed in Chapter 25.

20.1 Nature of the Problem

Let us describe the U(1)$_A$-problem more explicitly. The (global) chiral symmetry of the QCD Lagrangian – with N_f massless quark flavors – is U(N_f)$_L$ × U(N_f)$_R$, while in the spectrum only the flavor and baryon number symmetries SU(N_f)$_{L=R}$ × U(1)$_{L=R}$ = U(N_f)$_{L=R}$ are manifest. According to the Goldstone theorem, one might thus expect $N_f^2 + N_f^2 - N_f^2 = N_f^2$ Nambu–Goldstone bosons, but one finds only $N_f^2 - 1$ of them. The missing Nambu–Goldstone boson should be a pseudo-scalar, flavor-singlet particle. The lightest particle with these quantum numbers is the η'-meson. However, its mass of $M_{\eta'} = 0.958$ GeV is too heavy for a pseudo-Nambu–Goldstone boson. Hence, the U(1)$_A$-problem starts with the question why the axial U(1)$_A$ symmetry is not spontaneously broken, although it is not manifest in the spectrum.

We have already pointed out, in particular in Section 19.1, that the axial U(1)$_A$ symmetry is not really a symmetry of QCD. Although this symmetry is present in the massless QCD Lagrangian, it cannot be maintained under quantization because of its anomaly. This explains qualitatively why the η'-meson is not a Nambu–Goldstone boson. To understand the problem more quantitatively, one must consider the extent of the quantum symmetry breaking. Gerard 't Hooft realized that topologically non-trivial configurations of the gluon field – for example, instantons – give mass to the η'-meson.[1]

---

[1] What matters is a sufficient fraction of topologically non-trivial gluon field configurations. Strictly speaking,

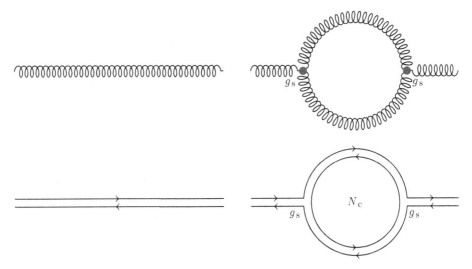

Fig. 20.1   Feynman diagrams (top) versus 't Hooft double-line diagrams (bottom): gluon propagator (left) and gluon loop diagram (right).

20.2 QCD in the Large-N_c Limit

Again we begin with a preview: To understand the $U(1)_A$-problem, we consider QCD with a large number of colors N_c. If the color symmetry is extended in this way, the explicit axial $U(1)_A$ breaking via the anomaly disappears in the limit $N_c \to \infty$. Then, $U(1)_A$ becomes an exact symmetry even at the quantum level. Its spontaneous breakdown implies that the η'-meson then indeed represents a massless Nambu–Goldstone boson. For large but finite N_c, the η'-meson picks up a mass proportional to the *topological susceptibility* $\chi_t = \langle Q^2 \rangle / V$ – the vacuum value of the topological charge Q squared per space–time volume V – evaluated in the pure gluon theory.

QCD simplifies in the large-N_c limit, but is still too complicated to be solved analytically.

Now let us be more explicit and classify the subset of Feynman diagrams that contribute in the large-N_c limit. An essential observation is that, for many colors, the distinction between $SU(N_c)$ and $U(N_c)$ becomes irrelevant. In particular, the color quantum numbers of a gluon can then be factorized to those of a quark and an anti-quark (*i.e.* to a product of a fundamental and anti-fundamental color representation). As illustrated in Figure 20.1 (left), each gluon propagator in a Feynman diagram may then be replaced by the "color flow" of a quark–anti-quark pair, thus resulting in a *double-line representation*. In this way, any large-N_c QCD diagram can be represented as a quark double-line diagram. The gluon self-energy diagram illustrated in Figure 20.1 (right), for example, has two 3-gluon vertices (each contributing a factor g_s) attached to an internal gluon loop. In addition to the color flow associated with the external lines, there is a single loop whose color flow is not associated with the external lines. Summing over the corresponding color index, one obtains a factor N_c, such that the diagram contributes in proportion to $g_s^2 N_c$. Taking the

the instantons (as specific configurations that minimize the Euclidean action within a topological sector) are of zero measure in the functional integral.

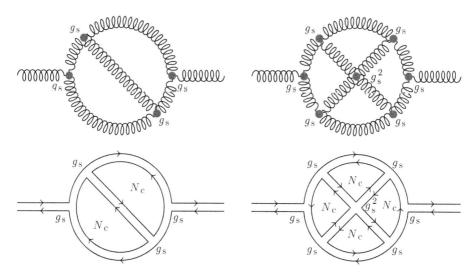

Fig. 20.2 Feynman diagrams (top) versus 't Hooft double-line diagrams (bottom): a 2-loop diagram (left) and a planar 4-loop diagram (right).

$N_c \to \infty$ limit naively, this diagram would thus diverge. Following 't Hooft, we absorb this divergence in a redefinition of the strong coupling as

$$g_{tH}^2 = g_s^2 N_c. \tag{20.1}$$

We take the large-N_c limit such that g_s goes to zero, while keeping g_{tH} fixed.

Another 1-loop diagram, which contributes to the gluon self-energy, involves a single 4-gluon vertex that contributes a factor g_s^2, as well as one internal color line that again contributes a factor N_c, such that this diagram is also proportional to $g_s^2 N_c = g_{tH}^2$.

Next, we consider the 2-loop diagram illustrated in Figure 20.2 (left). There are two internal loops, which together yield a factor N_c^2. In addition, there are four 3-gluon vertices contributing a factor g_s^4, such that this diagram is proportional to $g_s^4 N_c^2 = g_{tH}^4$; hence, it remains finite in the $N_c \to \infty$ limit.

At last, we consider the *4-loop diagram* illustrated in Figure 20.2 (right). The loops give rise to a factor N_c^4, while the six 3-gluon vertices contribute g_s^6, and the 4-gluon vertex contributes g_s^2. Altogether, this diagram is hence proportional to $g_s^8 N_c^4 = g_{tH}^8$; it is again finite in the large-N_c limit.

All diagrams considered until now are *planar; i.e.* they can be drawn in a plane, without any quark lines crossing each other. Next, we consider the *non-planar* 3-loop diagram shown in Figure 20.3 (left). It again has six 3-gluon vertices that contribute g_s^6, but the 4-gluon vertex is now missing. In addition, the color flow in the interior of the diagram is described by just a single color loop, which contributes a factor N_c. Hence, this diagram is proportional to $g_s^6 N_c = g_{tH}^6/N_c^2$; it is thus *suppressed* in the large-N_c limit.

One can convince oneself that only planar diagrams contribute to the large-N_c limit. In particular, if we add another gluon propagator to a planar diagram such that it remains planar, we add two 3-gluon vertices and hence a factor g_s^2, and we generate an additional

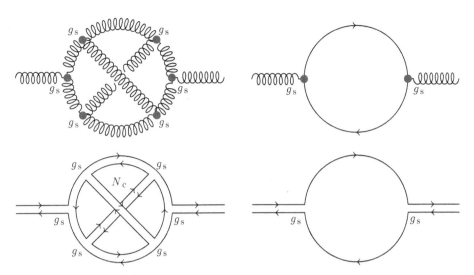

Fig. 20.3 Feynman diagrams (top) versus 't Hooft double-line diagrams (bottom): non-planar gluon diagram (left) and quark loop diagram (right). These diagrams are suppressed in the large-N_c limit.

loop, thus introducing an extra loop color factor N_c. The total weight remains of order 1. Non-planar gluon diagrams, on the other hand, are suppressed by powers of $1/N_c^2$.

Let us also consider the quark contribution to the gluon propagator illustrated in Figure 20.3 (right). There is no color factor N_c for this diagram, and still, there are two quark–gluon vertices contributing a factor $g_s^2 = g_{tH}^2/N_c$. We see that diagrams with internal quark loops are suppressed by powers of $1/N_c$. They do not contribute in the large-N_c limit either.

Even though this eliminates a huge class of diagrams, the remaining planar gluon diagrams are still too complicated to be summed up analytically. Nevertheless, the above N_c-counting allows us to understand some aspects of the QCD dynamics. In the large-N_c limit, QCD reduces to a theory of non-interacting, *stable mesons and glueballs*, while baryons have a mass proportional to N_c and are thus removed from the meson and glueball spectrum. This can be understood qualitatively in a simple constituent quark model. For $SU(N_c)$, a color singlet baryon consists of N_c quarks, each contributing the constituent quark mass to the total baryon mass. Mesons, on the other hand, still consist of one quark–anti-quark pair, so the mass that we attribute to the constituent quark remains finite for $N_c \to \infty$. Although the baryon mass diverges $\propto N_c$, since baryons reside in a superselection sector (characterized by the conserved baryon number), their physics can still be addressed in the large-N_c limit.

The topology of the gluon field is also affected by the large-N_c limit. In Section 19.4, we have derived the instanton action bound

$$S[G] \geq \frac{8\pi^2}{g_s^2} |Q[G]| = \frac{8\pi^2 N_c}{g_{tH}^2} |Q[G]| , \tag{20.2}$$

which is valid for all $SU(N_c)$ gauge groups ($N_c \geq 2$). In the large-N_c limit, the action of an instanton diverges, and topologically non-trivial field configurations may appear to be suppressed in the functional integral. At first glance, this seems to suggest that the source of symmetry breaking via the anomaly disappears, and the η'-meson indeed becomes a Nambu–Goldstone boson in the $N_c \to \infty$ limit. While this conclusion is correct, the argument to reach this conclusion is too naive. In particular, the Yang–Mills action itself takes the form

$$S[G] = -\frac{1}{2g_s^2} \int d^4x \, \mathrm{Tr}\left[G_{\mu\nu}G_{\mu\nu}\right] = -\frac{N_c}{2g_{tH}^2} \int d^4x \, \mathrm{Tr}\left[G_{\mu\nu}G_{\mu\nu}\right] , \qquad (20.3)$$

so that all field configurations are affected by the prefactor N_c.

20.3 Witten–Veneziano Formula for the η'-Meson Mass

In the large-N_c setting, one can derive a mass formula for the η'-meson. As we will see in Chapter 23, the same is true for the pion, even for arbitrary values of N_c. The pion mass results from an explicit chiral symmetry breaking due to non-zero quark masses. Similarly, the η'-mass results from an explicit axial U(1)$_A$ symmetry breaking via the anomaly due to finite N_c. As we will now see, this can be expressed as a $1/N_c$-effect.

Let us consider the topological susceptibility as the integrated correlation function of two Chern–Pontryagin densities,

$$\chi_t = \int_V d^4x \, _{YM}\langle 0|P(0)P(x)|0\rangle_{YM} = \frac{\langle Q^2\rangle}{V} , \qquad (20.4)$$

in the pure gluon theory (without quarks). The state $|0\rangle_{YM}$ is the vacuum (of the pure gluon theory), and V is the volume of space–time. When we add massless quarks, the Atiyah–Singer index theorem implies that the topological charge – and hence χ_t – vanishes, because the zero-modes of the Dirac operator eliminate topologically non-trivial field configurations. Therefore, in full QCD – with massless dynamical quarks – one expects

$$\int d^4x \, \langle 0|P(0)P(x)|0\rangle = 0 , \qquad (20.5)$$

where $|0\rangle$ is the vacuum of full QCD (in the chiral limit). However, as we argued before, the effects of quarks are $1/N_c$-suppressed. So, it is not obvious how they can eliminate the topological susceptibility of the pure gluon theory at large N_c.

As we discussed, in the large-N_c limit the quark effects manifest themselves entirely in terms of mesons. To leading order in the large-N_c expansion, one obtains (Witten, 1979b)

$$\chi_t - \sum_m \frac{\langle 0|P|m\rangle\langle m|P|0\rangle}{M_m^2} = 0. \qquad (20.6)$$

The sum extends over all pseudo-scalar, flavor-singlet meson states $|m\rangle$ (whose quantum numbers match those of P), and M_m are the corresponding meson masses. While only the glueball states contribute to χ_t in the pure Yang–Mills theory, the mesons cancel their contribution to the topological susceptibility in full QCD.

Using large-N_c techniques, one can show that $|\langle 0|P|m\rangle|^2$ is of $\mathcal{O}(1/N_c)$, while χ_t is of $\mathcal{O}(1)$ (Witten, 1979b; Veneziano, 1979). If also the masses of all pseudo-scalar, flavor-singlet mesons were of $\mathcal{O}(1)$, there would be a contradiction. The puzzle gets resolved when one assumes that the lightest flavor-singlet, pseudo-scalar meson – the η'-meson – has a mass squared of order $1/N_c$, such that

$$\chi_t = \frac{|\langle 0|P|\eta'\rangle|^2}{M_{\eta'}^2}. \tag{20.7}$$

Using the anomaly equation (19.2), one obtains

$$\langle 0|P|\eta'\rangle = \frac{1}{2N_f}\langle 0|\partial_\mu j_\mu^5|\eta'\rangle = \frac{1}{\sqrt{2N_f}}M_{\eta'}^2 F_{\eta'}. \tag{20.8}$$

It turns out that, in the large-N_c limit, the pion and η' decay constants coincide, $F_{\eta'} = F_\pi$, and we arrive at the celebrated *Witten–Veneziano formula* (Witten, 1979b; Veneziano, 1979)

$$\chi_t = \frac{F_\pi^2 M_{\eta'}^2}{2N_f}. \tag{20.9}$$

In this equation, χ_t is of $\mathcal{O}(1)$, F_π^2 is of $\mathcal{O}(N_c)$, and $M_{\eta'}^2$ is of $\mathcal{O}(1/N_c)$. Based on the Ginsparg–Wilson relation, a controlled derivation of the Witten–Veneziano formula has been carried out using the lattice regularization in Giusti *et al.* (2002).

The Witten–Veneziano formula implies that, in an infinitely colorful world, the η'-meson is indeed a Nambu–Goldstone boson. At finite N_c, the anomaly arises, leading to an explicit axial $U(1)_A$ symmetry breaking proportional to $1/N_c$. The pseudo-Nambu–Goldstone boson mass squared is hence proportional to $1/N_c$.

So far, we have assumed all quarks to be massless. When a non-zero s-quark mass is taken into account, for $N_f = 3$ formula (20.9) is modified to

$$\chi_t = \frac{F_\pi^2}{6}\left(M_{\eta'}^2 + M_\eta^2 - 2M_K^2\right) \simeq (0.180\,\text{GeV})^4. \tag{20.10}$$

In an exercise to Chapter 25, this equation will be derived.

20.4 Topological Susceptibility from Lattice Gauge Theory

If a continuum field theory has topological sectors, its topological charge Q is defined on differentiable configurations, as we saw in Chapter 19 for the case of 4-d SU(N) Yang–Mills gauge fields. However, when we consider the functional integral on finer and finer lattices, it turns out that it is dominated by configurations which do not extrapolate to a differentiable – or at least continuous – form; their fluctuations are too violent. Still, thanks to asymptotic freedom, a certain degree of "smoothness" of the lattice gauge field arises on fine lattices. However, when we want to address a topological observable, such as the topological susceptibility χ_t, it is not obvious what we are actually referring to.

On the lattice, one might refer to a naively discretized version of the Chern–Pontryagin density, which does, however, not lead to integer values of Q. Moreover, naive discretizations give rise to devastating short-distance power-law divergences, proportional to a^{-4}, which prevent any controlled continuum limit of χ_t. When combined with smearing, which smoothens the gauge field, one can still obtain meaningful results (Dürr *et al.*, 2007).

Integer values $Q \in \mathbb{Z}$ are achieved by geometric definitions of the lattice topological charge. These values are invariant against continuous deformations of the gauge field, thus maintaining its topological nature (Lüscher, 1982; Philipps and Stone, 1986; Göckeler *et al.*, 1986). However, the segmentation of the lattice configuration space into topological sectors depends on the details of the geometric definition and on the lattice spacing, at least to some extent. It is difficult to decide analytically whether or not this leads to an unambiguous continuum limit of quantities like χ_t.

Fortunately, based on a deeper understanding of the Ginsparg–Wilson relation, different sound definitions of the lattice topological charge have been discovered, and the corresponding topological susceptibilities agree with each other in the continuum limit.

As we have seen in Section 19.10, the Atiyah–Singer index theorem provides a relation between the zero-modes of the Dirac operator in a background gauge field and its topological charge. Remarkably, this relation is maintained exactly on the lattice, provided that the fermion action obeys the Ginsparg–Wilson relation (Hasenfratz *et al.*, 1998). Leonardo Giusti, Giancarlo Rossi, and Massimo Testa were able to show that power-divergent short-distance contributions cancel in the topological susceptibility that results from the index theorem definition of the topological charge (Giusti *et al.*, 2004). Again based on the Ginsparg–Wilson relation, Martin Lüscher constructed a universal, so-called density–chain definition of the topological charge, which is provably unaffected by short-distance ambiguities (Lüscher, 2004). In addition, he constructed another definition of the lattice topological charge, based on the so-called *gradient flow* (Lüscher, 2010), which naturally leads to a segmentation of the lattice configuration space into topological sectors. The resulting topological susceptibility agrees with the ones obtained from definitions of Q that are based on the Ginsparg–Wilson relation. Finally, in the framework of dimensional regularization, Martin Lüscher and Peter Weisz were able to prove to all orders in perturbation theory that the Chern–Pontryagin density $P(x)$ is unaffected by multiplicative renormalization (Lüscher and Weisz, 2021).

As a result, we are now in the comfortable situation to have a theoretically well-founded, deep understanding of the topological charge with several robust lattice constructions that yield unambiguous results in the continuum limit. This goes way beyond semi-classical considerations based on instantons.

Lattice definitions of Q that are based on the Ginsparg–Wilson relation are tedious to evaluate in numerical Monte Carlo calculations. Still, χ_t has been determined (Del Debbio *et al.*, 2005). The gradient flow definition is numerically more easily accessible and leads to consistent, but more accurate results (Cè *et al.*, 2015). Expressed in units of the Sommer scale R_0, the topological susceptibility of SU(3)$_c$ Yang–Mills theory takes the value

$$\chi_t = (0.483(2)/R_0)^4 = (0.203(2)\,\text{GeV})^4\,. \tag{20.11}$$

Here, we have again inserted $R_0 = 0.470(4)$ fm. Scale setting in physical units is a nontrivial issue in lattice field theory (Sommer, 2014), which we discussed in Section 14.6 and which dominates the error of χ_t in GeV. Ultimately, assigning physical units makes sense only if one simulates the real world with realistic quark masses, ideally including electromagnetic effects, which is not easy to handle. The error of χ_t in units of the Sommer scale R_0 is smaller than the error in units of GeV, due to the uncertainty of R_0 itself.

Given the fact that eq. (20.10) was derived at large N_c, it is remarkable that it agrees so well with the $N_c = 3$ result of eq. (20.11). Using Monte Carlo simulations of SU(N_c) lattice Yang–Mills theory at increasing values of N_c, the topological susceptibility (which shows only a mild N_c-dependence) has also been extrapolated to the $N_c \to \infty$ limit (Cè *et al.*, 2016). The lattice calculations show that the topological charge distribution is not exactly Gaussian. It also differs significantly from the distribution that characterizes a dilute gas of instantons. We thus conclude that, not unexpectedly, topological charge carriers other than dilute instantons dominate the non-perturbative functional integral.

Exercises

20.1 Non-planar diagrams

Convince yourself with some examples that *non-planar* diagrams vanish in the large-N_c limit.

21

Spectrum of Light Baryons and Mesons

In this chapter, we consider *hadrons* – *i.e. baryons* and *mesons* – the low-energy bound states of quarks and gluons (for glueballs we refer to Section 14.12). Baryons contain three more quarks than anti-quarks, while mesons contain the same number of both. As first realized by Murray Gell-Mann (1964) and George Zweig (1964), most hadron states that exist in the QCD spectrum follow a simple group theoretical classification scheme in which one constructs a wave function from its spin, flavor, and color components. This rather simplistic procedure, which underlies the non-relativistic *constituent quark* model (Isgur and Karl, 1977, 1979), should not be taken too seriously at a dynamical level. While it provides a good overview of the states that exist, the hadron masses can be calculated from first principles only by using lattice QCD.

It should be noted that most hadrons are unstable against the decay into lighter particles. However, many of these decays are due to the electroweak interaction and thus proceed at rather slow rates. When we concentrate on QCD and neglect the electroweak gauge interactions, several hadrons, including the pseudo-Nambu–Goldstone bosons, become stable particles, because their decay would violate quark flavor conservation. Still, many other hadrons are unstable even under the strong interaction and can decay into lighter hadrons, in particular, by emitting pseudo-Nambu–Goldstone bosons. Even if most hadrons are extremely short-lived resonances, they manifest themselves as pronounced signals in the scattering cross sections of their decay products.

As we have seen, for reasons that we do not understand, the u- and d-quarks have masses much smaller than the characteristic QCD energy scale $\Lambda_{QCD} \simeq 0.33$ GeV. As a result, for $N_f = 2$ quark flavors the chiral symmetry $SU(2)_L \times SU(2)_R$ is a rather good, approximate global symmetry of QCD. Its $SU(2)_{L=R} = SU(2)_I$ subgroup, which persists under spontaneous chiral symmetry breaking, is known as *isospin*. It provides a powerful group theoretical classification of hadron states.

In the presence of the heavier strange quark s, isospin is replaced by the less accurate *flavor symmetry* $SU(3)_{L=R} = SU(3)_f$. Here, we concentrate on the hadrons that predominantly consist of the light quarks u, d, and s, with masses below Λ_{QCD}. It should be pointed out that hadrons containing the heavier quarks c and b are very interesting physical systems as well. Their discussion is, however, beyond the scope of this book. The extremely heavy t-quark does not form experimentally identifiable hadrons because it decays too rapidly.

21.1 Isospin Symmetry

Proton and *neutron* have almost the same masses (they differ by about 0.1 percent)

$$M_p \simeq 0.9383 \, \text{GeV}, \quad M_n \simeq 0.9396 \, \text{GeV}. \tag{21.1}$$

While the proton seems to be absolutely stable, a free neutron decays radioactively into a proton, an electron, and an electron–anti-neutrino, $n \rightarrow p + e + \bar{\nu}_e$. This is the β-decay which is illustrated in Figure 15.4.

Protons and neutrons (the nucleons) are the constituents of atomic nuclei. Originally, Hideki Yukawa postulated a light particle mediating the interaction between protons and neutrons. The π-meson or *pion* is a boson with spin 0, which exists in three charge states π^+, π^0, and π^-. The corresponding masses are similar to the mass of the particle that Yukawa predicted,

$$M_{\pi^+} = M_{\pi^-} \simeq 0.140\,\text{GeV}, \quad M_{\pi^0} \simeq 0.135\,\text{GeV}. \tag{21.2}$$

In pion–nucleon scattering, a resonance occurs in the total cross section as a function of the pion–nucleon center-of-mass energy. The resonance energy is interpreted as the mass of an unstable particle – the so-called Δ-*isobar*. One may view the Δ-particle as an excited state of the nucleon. It exists in four charge states Δ^{++}, Δ^+, Δ^0, and Δ^- with masses

$$M_{\Delta^{++}} \simeq M_{\Delta^+} \simeq M_{\Delta^0} \simeq M_{\Delta^-} \simeq 1.230\,\text{GeV}. \tag{21.3}$$

Similar to pion–nucleon scattering, there is also a resonance in pion–pion scattering. This so-called ρ-*meson* comes in three charge states ρ^+, ρ^0, and ρ^- with masses

$$M_{\rho^+} \simeq M_{\rho^0} \simeq M_{\rho^-} \simeq 0.770\,\text{GeV}. \tag{21.4}$$

In these multiplets, particles with different electric charges have (almost) degenerate masses, and it is natural to associate this with an (approximate) symmetry. This so-called *isospin symmetry* was introduced by Werner Heisenberg (1932). Isospin is mathematically equivalent to the ordinary $SU(2)_S$ spin rotational symmetry. Isospin is an intrinsic symmetry, not related to space–time transformations. As we know, each total spin $S = 0, 1/2, 1, 3/2, \ldots$ is associated with an irreducible representation of the $SU(2)_S$ rotation group containing $2S + 1$ states that are distinguished by their spin projection

$$S_z = -S, -S + 1, \ldots, S - 1, S. \tag{21.5}$$

In complete analogy, the representations of the $SU(2)_I$ isospin symmetry group are characterized by their total isospin $I = 0, 1/2, 1, 3/2, \ldots$. The states of an isospin representation are distinguished by their isospin projection

$$I_3 = -I, -I + 1, \ldots, I - 1, I. \tag{21.6}$$

A representation with isospin I contains $2I + 1$ states and is denoted by $\{2I + 1\}$. We can classify the hadrons by their isospin, as it is done in Table 21.1. For the baryons in this table (nucleon and Δ), isospin projection and electric charge are related by $Q = I_3 + 1/2$, and for the mesons (π and ρ) by $Q = I_3$.

Isospin is an approximate, global symmetry of QCD. For example, the proton–pion scattering reaction $p + \pi \rightarrow \Delta$ is consistent with isospin symmetry because the coupling of the isospin representations of nucleon and pion

$$\{2\} \times \{3\} = \{2\} + \{4\}, \tag{21.7}$$

does indeed contain the quadruplet isospin 3/2 representation $\{4\}$ of the Δ-isobar. The isospin symmetry of the hadron spectrum indicates that the strong interaction is electric-charge-independent. This is no surprise because the electric charge Q is responsible for the electromagnetic but not for the strong interaction.

Hadron	Representation	I	I_3	Q	S
Table 21.1 The isospin classification of light hadrons.					
p, n	$\{2\}$	$1/2$	$1/2, -1/2$	$1, 0$	$1/2$
$\Delta^{++}, \Delta^{+}, \Delta^{0}, \Delta^{-}$	$\{4\}$	$3/2$	$3/2, 1/2, -1/2, -3/2$	$2, 1, 0, -1$	$3/2$
$\pi^{+}, \pi^{0}, \pi^{-}$	$\{3\}$	1	$1, 0, -1$	$1, 0, -1$	0
$\rho^{+}, \rho^{0}, \rho^{-}$	$\{3\}$	1	$1, 0, -1$	$1, 0, -1$	1

21.2 Nucleon and Δ-Isobar

Let us consider the question of the hadronic constituents by investigating various symmetries. First we consider isospin. Since the hadrons form isospin multiplets, the same should be true for their constituents. The only SU(2) representation from which we can generate all others is the fundamental representation – the isospin doublet $\{2\}$ with $I = 1/2$ and $I_3 = \pm 1/2$. We identify the two states of this multiplet with the *"constituent quarks"* u ($I_3 = 1/2$) and d ($I_3 = -1/2$). A constituent quark is not a completely well-defined physical concept. One may view a constituent quark as a heavy quasi-particle that consists of a fundamental light "current" quark, dressed non-perturbatively with a cloud of gluons and quark–anti-quark pairs. Although it is not fundamental, the constituent quark concept is useful, because it leads to a successful group theoretical classification scheme for hadron states.

Since the Δ-isobar has isospin $I = 3/2$, it contains at least three constituent quarks. We couple

$$\{2\} \times \{2\} \times \{2\} = (\{1\} + \{3\}) \times \{2\} = \{2\} + \{2\} + \{4\}, \tag{21.8}$$

so we do indeed encounter a quadruplet. For the electric charges of the baryons, we obtain

$$Q = I_3 + \frac{1}{2} = \sum_{q=1}^{3}\left(I_{3q} + \frac{1}{6}\right) = \sum_{q=1}^{3} Q_q, \tag{21.9}$$

and hence for the charges of the constituent quarks we arrive at

$$Q_q = I_{3q} + \frac{1}{6} \quad \Rightarrow \quad Q_u = \frac{1}{2} + \frac{1}{6} = \frac{2}{3}, \quad Q_d = -\frac{1}{2} + \frac{1}{6} = -\frac{1}{3}. \tag{21.10}$$

The constituent quarks thus indeed have the usual fractional electric charges. Using Young tableaux (cf. Appendix F) as well as Clebsch–Gordan coefficients of SU(2), one finds

$$\boxed{1\,2\,3}_{3/2} = uuu \equiv \Delta^{++},$$

$$\boxed{1\,2\,3}_{1/2} = \frac{1}{\sqrt{3}}\,(uud + udu + duu) \equiv \Delta^{+},$$

$$\boxed{1\,2\,3}_{-1/2} = \frac{1}{\sqrt{3}}\,(udd + dud + ddu) \equiv \Delta^{0},$$

$$\boxed{1\,2\,3}_{-3/2} = ddd \equiv \Delta^{-}. \tag{21.11}$$

These isospin states are completely symmetric under permutations of the constituent quarks. We write the general coupling of the three constituent quarks as

$$
\boxed{1} \times \boxed{2} \times \boxed{3} = \boxed{1\,2\,3} + \boxed{\begin{array}{cc} 1 & 2 \\ \hline 3 & \end{array}} + \boxed{\begin{array}{cc} 1 & 3 \\ \hline 2 & \end{array}} + \boxed{\begin{array}{c} 1 \\ \hline 2 \\ \hline 3 \end{array}}. \tag{21.12}
$$

Translated into SU(2) language, this Young tableaux equation reads

$$
\{2\} \times \{2\} \times \{2\} = \{4\} + \{2\} + \{2\} + \{0\}. \tag{21.13}
$$

Here, $\{0\}$ denotes an empty representation – one that cannot be realized in SU(2) because the corresponding Young tableau has more than two rows. We identify the totally symmetric representation as the four charge states of the Δ-isobar.

Before we can characterize the state of the Δ-isobar in more detail, we must consider the other symmetries of the problem. The Δ-isobar is a resonance in the scattering of spin-1/2 nucleons and spin-0 pions. The experimentally observed spin of the resonance is 3/2. In order to account for this, we associate a spin 1/2 with the constituent quarks. Then, in complete analogy to isospin, we construct a totally symmetric spin representation for the Δ-particle

$$
\begin{aligned}
\boxed{1\,2\,3}_{3/2} &= \uparrow\uparrow\uparrow, \\
\boxed{1\,2\,3}_{1/2} &= \frac{1}{\sqrt{3}} (\uparrow\uparrow\downarrow + \uparrow\downarrow\uparrow + \downarrow\uparrow\uparrow), \\
\boxed{1\,2\,3}_{-1/2} &= \frac{1}{\sqrt{3}} (\uparrow\downarrow\downarrow + \downarrow\uparrow\downarrow + \downarrow\downarrow\uparrow), \\
\boxed{1\,2\,3}_{-3/2} &= \downarrow\downarrow\downarrow.
\end{aligned} \tag{21.14}
$$

The isospin–spin part of the Δ-isobar state hence takes the form

$$
|\Delta I_3 S_z\rangle = \boxed{1\,2\,3}_{I_3} \, \boxed{1\,2\,3}_{S_z}. \tag{21.15}
$$

This state is symmetric with respect to both isospin and spin. Consequently, it is symmetric under simultaneous isospin–spin permutations. For illustration, we write down the state for a Δ^+ particle with spin projection $S_z = 1/2$ as an example,

$$
\begin{aligned}
|\Delta \tfrac{1}{2}\tfrac{1}{2}\rangle = \frac{1}{3} (&u\uparrow u\uparrow d\downarrow + u\uparrow u\downarrow d\uparrow + u\downarrow u\uparrow d\uparrow \\
&+ u\uparrow d\uparrow u\downarrow + u\uparrow d\downarrow u\uparrow + u\downarrow d\uparrow u\uparrow \\
&+ d\uparrow u\uparrow u\downarrow + d\uparrow u\downarrow u\uparrow + d\downarrow u\uparrow u\uparrow).
\end{aligned} \tag{21.16}
$$

It is explicitly manifest that this state is totally symmetric.

As we have seen, the Young tableau $\boxed{}$ is associated with the iso-doublet $\{2\}$. Hence, it is natural to expect that the nucleon state can be constructed from it. Now we have two possibilities, $\boxed{\begin{array}{c} 1\,2 \\ 3 \end{array}}$ and $\boxed{\begin{array}{c} 1\,3 \\ 2 \end{array}}$, corresponding to symmetric or anti-symmetric couplings of the quarks 1 and 2. Using Clebsch–Gordan coefficients, one finds

$$\young(12,3)_{1/2} = \frac{1}{\sqrt{6}}\,(2uud - udu - duu)\,, \qquad \young(12,3)_{-1/2} = \frac{1}{\sqrt{6}}\,(udd + dud - 2ddu)\,,$$

$$\young(13,2)_{1/2} = \frac{1}{\sqrt{2}}\,(udu - duu)\,, \qquad \young(13,2)_{-1/2} = \frac{1}{\sqrt{2}}\,(udd - dud)\,. \qquad (21.17)$$

Proton and neutron have spin 1/2. Hence, we again have two possible coupling schemes $\young(12,3)$ and $\young(13,2)$ in spin space. We now combine the mixed isospin and spin permutation symmetries to an isospin–spin representation of definite permutation symmetry. This requires to reduce the inner product

$$\young(\ \ ,\ )_{I_3} \times \young(\ \ ,\ )_{S_z} = \young(\ \ \ )_{I_3 S_z} + \young(\ \ ,\ )_{I_3 S_z} + \young(\ ,\ ,\ )_{I_3 S_z} \qquad (21.18)$$

in the permutation group S_3. The two isospin and spin representations can be coupled to a symmetric, mixed, or anti-symmetric isospin–spin representation. As in the case of the Δ-isobar, we want to couple isospin and spin symmetrically. To do this explicitly, we need the Clebsch–Gordan coefficients of the group S_3. One obtains

$$|N I_3 S_z\rangle = \frac{1}{\sqrt{2}}\left(\young(12,3)_{I_3}\young(12,3)_{S_z} + \young(13,2)_{I_3}\young(13,2)_{S_z}\right)\,. \qquad (21.19)$$

In this construction, we have implicitly assumed the orbital angular momentum of the constituent quarks inside a hadron to vanish. Then, the orbital state is completely symmetric in the coordinates of the quarks.[1] The orbital part of the baryon wave function is therefore described by the Young tableau $\young(\ \ \ )$. Since also the isospin–spin part is totally symmetric, the baryon wave function seems completely symmetric under permutations of the quarks. Since we have treated constituent quarks as spin-1/2 fermions, this contradicts the *Pauli principle* which requires a totally anti-symmetric fermion wave function and hence the Young tableau $\young(\ ,\ ,\ )$. To satisfy the Pauli principle, the color symmetry comes to the rescue. In $SU(3)_c$, $\young(\ ,\ ,\ )$ corresponds to a singlet representation, which means that baryons are *color-neutral*. Since we have three colors, we can completely anti-symmetrize three quarks

$$\young(\ ,\ ,\ ) = \frac{1}{\sqrt{6}}(rgb - rbg + gbr - grb + brg - bgr)\,. \qquad (21.20)$$

The color symmetry is the key to the fundamental understanding of the strong interaction. As opposed to isospin, as we have seen, color is an exact gauge symmetry.

---

[1] Here and in the continuation of this chapter, we often write "quark" for simplicity, although we actually mean a constituent quark. Similarly, we sometimes just write "charge" when referring to the electric charge.

21.3 Anti-Quarks and Mesons

We have seen that the baryons (nucleon and Δ) consist of three constituent quarks, which are isospin doublets, spin doublets, and color triplets. Now we want to construct the meson iso-triplets (π and ρ) in a similar manner. Since these particles have spin 0 and 1, respectively, they must contain an even number of constituent quarks. When we use two quarks, *i.e.* when we construct states like uu, ud, or dd, the resulting electric charges are $4/3$, $1/3$, and $-2/3$ in contradiction to experiment. Also, the coupling of two color triplets

$$\square \times \square = \square\square + \begin{array}{c}\square\\\square\end{array}$$

$$\{3\} \times \{3\} = \{6\} + \{\bar{3}\}, \tag{21.21}$$

does not contain a singlet as desired by the confinement hypothesis.

A representation together with its anti-representation can always be coupled to a singlet. In SU(3), this corresponds to

$$\square \times \begin{array}{c}\square\\\square\end{array} = \begin{array}{c}\square\\\square\\\square\end{array} + \begin{array}{c}\square\square\\\square\end{array}$$

$$\{3\} \times \{\bar{3}\} = \{1\} + \{8\}, \tag{21.22}$$

Hence, it is natural to work with *anti-quarks*. Anti-quarks are isospin doublets, spin doublets, and color anti-triplets. The anti-quarks $\bar{u}$ and $\bar{d}$ have electric charges $Q_{\bar{u}} = -2/3$ and $Q_{\bar{d}} = 1/3$. Now we consider combinations of quark and anti-quark $u\bar{d}$, $u\bar{u}$, $d\bar{d}$, and $d\bar{u}$, which have charges 1, 0, 0, and -1, just as we need them for the mesons. First, we couple the isospin wave function

$$\square \times \square = \square\square + \begin{array}{c}\square\\\square\end{array}$$

$$\{2\} \times \{2\} = \{3\} + \{1\}, \tag{21.23}$$

and we obtain

$$\square\square_{1} = u\bar{d}, \quad \square\square_{0} = \frac{1}{\sqrt{2}}(u\bar{u} - d\bar{d}), \quad \square\square_{-1} = d\bar{u},$$

$$\begin{array}{c}\square\\\square\end{array}_{0} = \frac{1}{\sqrt{2}}(u\bar{u} + d\bar{d}). \tag{21.24}$$

We proceed analogously with the spin and obtain

$$|\pi I_3 S_z\rangle = \square\square_{I_3} \begin{array}{c}\square\\\square\end{array}_{S_z}, \quad |\rho I_3 S_z\rangle = \square\square_{I_3} \square\square_{S_z}. \tag{21.25}$$

Since quarks and anti-quarks are distinguishable particles (e.g., they have opposite charges), we need not worry about the Pauli principle in this case. As opposed to the baryons, here the coupling to color singlets follows from confinement.

Of course, we can combine isospin and spin wave functions also in different ways

$$|\omega I_3 S_z\rangle = \boxed{}_{I_3} \;\; \boxed{}_{S_z}, \quad |\eta' I_3 S_z\rangle = \boxed{}_{I_3} \;\; \boxed{}_{S_z}. \tag{21.26}$$

Indeed, one observes mesons with these quantum numbers, with masses $M_\omega \simeq 0.782\,\text{GeV}$ and $M_{\eta'} \simeq 0.958\,\text{GeV}$.

21.4 Strange Hadrons

Up to now, we have considered hadrons that are built of u- and d-quarks and their anti-particles. However, one also observes hadrons containing *strange quarks*. The masses of the pseudo-scalar mesons (spin $S = 0$, parity $P = -1$), cf. Section 18.4, are given by

$$M_\pi \simeq 0.138\,\text{GeV}, \quad M_K \simeq 0.496\,\text{GeV}, \quad M_\eta \simeq 0.549\,\text{GeV}, \quad M_{\eta'} \simeq 0.958\,\text{GeV}, \tag{21.27}$$

while the masses of the vector mesons ($S = 1$, $P = -1$) are

$$M_\rho \simeq 0.770\,\text{GeV}, \quad M_\omega \simeq 0.783\,\text{GeV}, \quad M_{K^*} \simeq 0.892\,\text{GeV}, \quad M_\varphi \simeq 1.020\,\text{GeV}. \tag{21.28}$$

Altogether, we have nine scalar and nine vector mesons. In each group, we have so far classified four members (π^+, π^0, π^-, η' and ρ^+, ρ^0, ρ^-, ω). The number four resulted from the SU(2)$_I$ isospin relation

$$\{2\} \times \{2\} = \{1\} + \{3\}. \tag{21.29}$$

The number nine then suggests to consider the corresponding SU(3)$_f$ identity

$$\{3\} \times \{\bar{3}\} = \{1\} + \{8\}. \tag{21.30}$$

Indeed, we obtain nine mesons if we generalize isospin to a larger symmetry SU(3)$_f$. This so-called flavor group has nothing to do with the color symmetry SU(3)$_c$. It is only an approximate symmetry of QCD, with SU(2)$_I$ as a subgroup. In SU(3)$_f$, we have another quark flavor s – the strange quark.

The generators of SU(3) can be chosen as the Gell-Mann matrices

$$\lambda^1 = \begin{pmatrix} 0 & 1 & 0 \\ 1 & 0 & 0 \\ 0 & 0 & 0 \end{pmatrix}, \quad \lambda^2 = \begin{pmatrix} 0 & -i & 0 \\ i & 0 & 0 \\ 0 & 0 & 0 \end{pmatrix}, \quad \lambda^3 = \begin{pmatrix} 1 & 0 & 0 \\ 0 & -1 & 0 \\ 0 & 0 & 0 \end{pmatrix},$$

$$\lambda^4 = \begin{pmatrix} 0 & 0 & 1 \\ 0 & 0 & 0 \\ 1 & 0 & 0 \end{pmatrix}, \quad \lambda^5 = \begin{pmatrix} 0 & 0 & -i \\ 0 & 0 & 0 \\ i & 0 & 0 \end{pmatrix},$$

$$\lambda^6 = \begin{pmatrix} 0 & 0 & 0 \\ 0 & 0 & 1 \\ 0 & 1 & 0 \end{pmatrix}, \quad \lambda^7 = \begin{pmatrix} 0 & 0 & 0 \\ 0 & 0 & -i \\ 0 & i & 0 \end{pmatrix}, \quad \lambda^8 = \frac{1}{\sqrt{3}} \begin{pmatrix} 1 & 0 & 0 \\ 0 & 1 & 0 \\ 0 & 0 & -2 \end{pmatrix}. \tag{21.31}$$

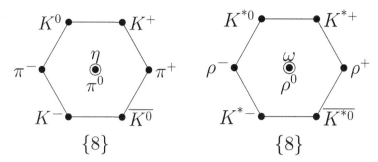

Fig. 21.1 SU(3)$_f$ flavor octets of pseudo-scalar mesons with spin $S = 0$ and negative parity (left) and of vector mesons with spin $S = 1$ and negative parity (right). The x-direction corresponds to the isospin $I_3 = \lambda^3/2$, and the y-direction corresponds to the strong "hypercharge" $Y = \lambda^8/\sqrt{3}$.

SU(3) has rank 2, which means that two of the generators commute with each other, namely, $[\lambda^3, \lambda^8] = 0$. We identify the generators of the isospin subgroup SU(2)$_I$ as

$$I_1 = \frac{1}{2}\lambda^1, \quad I_2 = \frac{1}{2}\lambda^2, \quad I_3 = \frac{1}{2}\lambda^3. \tag{21.32}$$

It is also convenient to introduce the so-called strong "hypercharge"

$$Y = \frac{1}{\sqrt{3}}\lambda^8, \tag{21.33}$$

(not to be confused with the generator of U(1)$_Y$ gauge transformations in the Standard Model). Then, I^2, I_3, and Y commute with each other, and we can characterize the states of an SU(3)$_f$ multiplet by their isospin quantum numbers and by their strong "hypercharge". Starting with the SU(3)$_f$ triplet, we have

$$I^2 u = \frac{1}{2}\left(\frac{1}{2}+1\right)u = \frac{3}{4}u, \quad I_3 u = \frac{1}{2}u, \quad Yu = \frac{1}{3}u,$$

$$I^2 d = \frac{1}{2}\left(\frac{1}{2}+1\right)d = \frac{3}{4}d, \quad I_3 d = -\frac{1}{2}d, \quad Yd = \frac{1}{3}d,$$

$$I^2 s = 0, \quad I_3 s = 0, \quad Ys = -\frac{2}{3}s. \tag{21.34}$$

The electric charge is now given by

$$Q = I_3 + \frac{1}{2}Y, \tag{21.35}$$

such that

$$Q_u = \frac{2}{3}, \quad Q_d = -\frac{1}{3}, \quad Q_s = -\frac{1}{3}\,; \tag{21.36}$$

i.e. as expected, the charge of the constituent s-quark is the same as the one of the d-quark. If SU(3)$_f$ would be a symmetry as good as SU(2)$_I$, the states in an SU(3)$_f$ multiplet should be almost degenerate. This is, however, not quite the case, since SU(3)$_f$ is a less accurate symmetry of QCD than SU(2)$_I$. The SU(3)$_f$ multiplets of the scalar and vector meson states are illustrated in Figure 21.1.

Of course, the s-quark also appears in baryons. There, we have

$$\{3\} \times \{3\} \times \{3\} = \{10\} + 2\{8\} + \{1\} \tag{21.37}$$

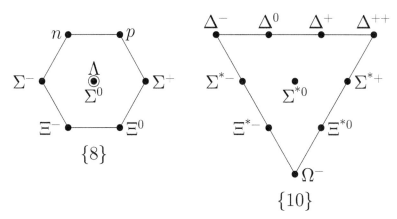

Fig. 21.2 SU(3)$_f$ flavor octet of baryons with spin $S = 1/2$ (left) and flavor decuplet of baryons with spin $S = 3/2$ (right). The x-direction corresponds to the isospin $I_3 = \lambda^3/2$, and the y-direction corresponds to the strong "hypercharge" $Y = \lambda^8/\sqrt{3}$.

which replaces the previous SU(2)$_I$ relation

$$\{2\} \times \{2\} \times \{2\} = \{4\} + 2\{2\} + \{0\}. \tag{21.38}$$

Indeed, one observes more baryons than just the nucleon and the Δ-isobar.

The masses for the spin-1/2 baryons are

$$M_N \simeq 0.939\,\text{GeV}, \quad M_\Lambda \simeq 1.116\,\text{GeV}, \quad M_\Sigma \simeq 1.193\,\text{GeV}, \quad M_\Xi \simeq 1.318\,\text{GeV}, \tag{21.39}$$

while the spin-3/2 baryon masses are

$$M_\Delta \simeq 1.232\,\text{GeV}, \quad M_{\Sigma^*} \simeq 1.385\text{GeV}, \quad M_{\Xi^*} \simeq 1.530\,\text{GeV}, \quad M_\Omega \simeq 1.672\,\text{GeV}. \tag{21.40}$$

The SU(3)$_f$ multiplets of the $S = 1/2$ and $S = 3/2$ baryon states are illustrated in Figure 21.2. Proton and neutron are part of an octet: ⊞ is $\{2\}$ in SU(2)$_I$ and $\{8\}$ in SU(3)$_f$. The Δ-isobar is part of a decuplet: ⬜⬜⬜ is $\{4\}$ in SU(2)$_I$ and $\{10\}$ in SU(3)$_f$. One does not find an SU(3)$_f$ singlet. This is because a spatially symmetric color singlet wave function is totally anti-symmetric. In order to obtain a totally anti-symmetric wave function, also the spin part should be totally anti-symmetric in all three quarks: ⬛. For SU(2)$_S$, this is impossible.

We know that the SU(3)$_f$ symmetry is explicitly broken because the s-quark is significantly heavier than the u- and d-quarks. Based on the quark content, one would expect

$$M_{\Sigma^*} - M_\Delta \simeq M_{\Xi^*} - M_{\Sigma^*} \simeq M_\Omega - M_{\Xi^*}. \tag{21.41}$$

Indeed, one finds experimentally

$$M_{\Sigma^*} - M_\Delta \simeq 0.153\,\text{GeV}, \quad M_{\Xi^*} - M_{\Sigma^*} \simeq 0.145\,\text{GeV}, \quad M_\Omega - M_{\Xi^*} \simeq 0.142\,\text{GeV}. \tag{21.42}$$

In fact, based on the quark model, the existence of the Ω-baryon was predicted near its observed mass before it was detected in experiments.

21.5　Gell-Mann–Okubo Baryon Mass Formula

We have seen that the constituent quark model leads to a successful classification of hadron states in terms of flavor symmetry. The results about the hadron dynamics are, however, of more qualitative nature, and the assumption that a hadron is essentially a collection of a few constituent quarks is certainly too naive. The fundamental theory of the strong interaction is QCD. In the following, we combine basic QCD physics with group theory to describe patterns in the hadron spectrum. The interaction between quarks and gluons is flavor-independent and therefore $SU(3)_f$ symmetric. Also, the gluon self-interaction is flavor-symmetric because the gluons are flavor singlets. A violation of the $SU(3)_f$ flavor symmetry results only from the quark mass matrix

$$\mathcal{M} = \begin{pmatrix} m_u & 0 & 0 \\ 0 & m_d & 0 \\ 0 & 0 & m_s \end{pmatrix}. \tag{21.43}$$

Here, we refer to the actual quark masses, no longer to constituent quarks. Let us replace the u- and d-quark masses by their average $m_q = (m_u + m_d)/2$, which is much smaller than the s-quark mass ($m_q \ll m_s$). The quark mass matrix is then decomposed as

$$\mathcal{M} = \frac{2m_q + m_s}{3} \begin{pmatrix} 1 & 0 & 0 \\ 0 & 1 & 0 \\ 0 & 0 & 1 \end{pmatrix} + \frac{m_q - m_s}{3} \begin{pmatrix} 1 & 0 & 0 \\ 0 & 1 & 0 \\ 0 & 0 & -2 \end{pmatrix}$$

$$= \frac{2m_q + m_s}{3} \mathbb{1} + \frac{m_q - m_s}{\sqrt{3}} \lambda^8. \tag{21.44}$$

The mass matrix contains an $SU(3)_f$ singlet as well as an octet piece. Correspondingly, the QCD Hamilton operator can be written as

$$\hat{H}_{QCD} = \hat{H}_1 + \hat{H}_8. \tag{21.45}$$

We assume $\hat{H}_8$ to be small, so it can be treated as a perturbation. This is justified if the mass difference $m_s - m_q$ is small. First we consider $\hat{H}_1$ alone. Since it is $SU(3)_f$ symmetric, we expect degenerate states in $SU(3)_f$ multiplets – the hadron octets and decuplets. Here, we assume the flavor symmetry not to be spontaneously broken, which is indeed correct for QCD.

Let us start with the baryons. The eigenstates of $\hat{H}_1$ are denoted by $|B_1 Y II_3\rangle$

$$\hat{H}_1 |B_1 Y II_3\rangle = M_{B_1} |B_1 Y II_3\rangle. \tag{21.46}$$

We use degenerate perturbation theory to first order in $\hat{H}_8$ and obtain

$$M_B = M_{B_1} + \langle B_1 Y II_3 | \hat{H}_8 | B_1 Y II_3 \rangle. \tag{21.47}$$

A diagonalization in the space of degenerate states is not necessary, since $\hat{H}_8$ transforms as the λ^8 component of an octet and can therefore not change Y, I, and I_3.

Next we will compute the required matrix elements using group theory. Starting with the baryon decuplet, we obtain a non-zero value only if $\{8\}$ and $\{10\}$ can couple to $\{10\}$. Indeed, the decuplet appears in the reduction. Using the *Wigner–Eckart theorem*, we obtain

$$\langle B_1 Y II_3 | \hat{H}_8 | B_1 Y II_3 \rangle = \langle B_1 || \hat{H}_8 || B_1 \rangle \langle \{10\} Y II_3 | \{8\}000\{10\} Y II_3 \rangle, \tag{21.48}$$

where $\langle B_1||\hat{H}_8||B_1 \rangle$ is a reduced matrix element, and the second factor is an $SU(3)_f$ Clebsch–Gordan coefficient given by

$$\langle \{10\}YII_3|\{8\}000\{10\}YII_3 \rangle = Y/\sqrt{8}. \tag{21.49}$$

Then, we obtain for the baryon masses in the decuplet

$$M_B = M_{B_1} + \langle B_1||\hat{H}_8||B_1 \rangle Y/\sqrt{8}, \tag{21.50}$$

and hence

$$M_{\Sigma^*} - M_\Delta = M_{\Xi^*} - M_{\Sigma^*} = M_\Omega - M_{\Xi^*} = -\langle B_1||\hat{H}_8||B_1 \rangle/\sqrt{8}. \tag{21.51}$$

Indeed, as we saw in relation (21.42), the experimental values of the three mass differences are similar. In view of the fact that we have just used first-order perturbation theory, this is quite remarkable.

Next we consider the mass splittings in the baryon octet. Here, we ask whether $\{8\}$ and $\{8\}$ can couple to $\{8\}$. One finds

$$\{8\} \times \{8\} = \{27\} + \{10\} + \{\overline{10}\} + 2\{8\} + \{1\}. \tag{21.52}$$

Hence, there are even two ways to couple two octets to an octet. One is symmetric, and the other is anti-symmetric against the exchange of the two octets. We write

$$\langle B_1 YII_3|\hat{H}_8|B_1 YII_3 \rangle = \langle B_1||\hat{H}_8||B_1 \rangle_s \langle \{8\}YII_3|\{8\}000\{8\}YII_3 \rangle_s$$
$$+ \langle B_1||\hat{H}_8||B_1 \rangle_a \langle \{8\}YII_3|\{8\}000\{8\}YII_3 \rangle_a. \tag{21.53}$$

The Clebsch–Gordan coefficients are given by

$$\langle \{8\}YII_3|\{8\}000\{8\}YII_3 \rangle_s = \left(I(I+1) - Y^2/4 - 1 \right)/\sqrt{5},$$
$$\langle \{8\}YII_3|\{8\}000\{8\}YII_3 \rangle_a = \sqrt{3/4}\, Y, \tag{21.54}$$

and we obtain for the baryon octet

$$M_B = M_{B_1} + \langle B_1||\hat{H}_8||B_1 \rangle_s \left(I(I+1) - Y^2/4 - 1 \right)/\sqrt{5} + \langle B_1||\hat{H}_8||B_1 \rangle_a \sqrt{3/4}\, Y. \tag{21.55}$$

These formulae for the baryon masses were first derived by Murray Gell-Mann (Gell-Mann, 1961) and by Susumu Okubo (Okubo, 1962a,b). From the octet formula, one infers

$$2M_N + 2M_\Xi = 4M_{B_1} + \langle B_1||\hat{H}_8||B_1 \rangle_s 4\,(3/4 - 1/4 - 1)/\sqrt{5},$$
$$M_\Sigma + 3M_\Lambda = 4M_{B_1} + \langle B_1||\hat{H}_8||B_1 \rangle_s\,[(2-1) + 3(-1)]/\sqrt{5}, \tag{21.56}$$

which implies the *Gell-Mann–Okubo formula*

$$2M_N + 2M_\Xi = M_\Sigma + 3M_\Lambda. \tag{21.57}$$

Experimentally, the two sides of this equation give 1.129 GeV and 1.135 GeV, in remarkable agreement with the theory.

21.6 Meson Mixing

Similar to the baryons, the explicit $SU(3)_f$ symmetry breaking due to the larger s-quark mass leads to mass splittings also for the mesons. There, however, one obtains, in addition, a mixing between flavor octet and flavor singlet states. (For the baryons, a mixing between octet and decuplet is excluded because they have different spins.) First, we again consider eigenstates of $\hat{H}_1$

$$\hat{H}_1|M_1 YII_3\rangle = M_{M_1}|M_1 YII_3\rangle. \tag{21.58}$$

The following analysis applies both to scalar and to vector mesons. In both cases, we have an $SU(3)_f$ octet and a singlet. In perturbation theory, we have to diagonalize a 9×9 matrix. Like in the baryon case, the matrix is, however, already almost diagonal. Let us first consider the seven meson states with $Y, I, I_3 \neq 0, 0, 0$. These are π and K for the scalar and ρ and K^* for the vector mesons, with

$$M_M = M_{M_1} + \langle M_1 YII_3|\hat{H}_8|M_1 YII_3\rangle. \tag{21.59}$$

In complete analogy to the baryon octet, we obtain

$$M_M = M_{M_1} + \langle M_1||\hat{H}_8||M_1\rangle_s \left(I(I+1) - Y^2/4 - 1\right)/\sqrt{5} + \langle M_1||\hat{H}_8||M_1\rangle_a\sqrt{3/4}Y. \tag{21.60}$$

As opposed to the baryons, the mesons and their anti-particles belong to the same multiplet. For example, we have

$$M_{K^+} = M_{M_1} + \langle M_1||\hat{H}_8||M_1\rangle_s \left(3/4 - 1/4 - 1\right)/\sqrt{5} + \langle M_1||\hat{H}_8||M_1\rangle_a\sqrt{3/4},$$
$$M_{K^-} = M_{M_1} + \langle M_1||\hat{H}_8||M_1\rangle_s \left(3/4 - 1/4 - 1\right)/\sqrt{5} - \langle M_1||\hat{H}_8||M_1\rangle_a\sqrt{3/4}. \tag{21.61}$$

According to the CPT theorem, particles and anti-particles have exactly the same masses in a relativistic quantum field theory, which implies

$$\langle M_1||\hat{H}_8||M_1\rangle_a = 0. \tag{21.62}$$

Now we come to the issue of *mixing* between the mesons η_1 and η_8 and between ω_1 and ω_8. We concentrate on the vector mesons. Then, we need the following matrix elements

$$\langle\omega_1|\hat{H}_8|\omega_1\rangle = 0,$$
$$\langle\omega_8|\hat{H}_8|\omega_8\rangle = \langle M_1||\hat{H}_8||M_1\rangle_s\langle\{8\}000|\{8\}000\{8\}000\rangle_s$$
$$= \langle M_1||\hat{H}_8||M_1\rangle_s \left(-1/\sqrt{5}\right). \tag{21.63}$$

The actual meson masses are the eigenvalues of the Hermitian matrix

$$\mathcal{M}_M = \begin{pmatrix} M_{\omega_1} & \langle\omega_1|\hat{H}_8|\omega_8\rangle \\ \langle\omega_8|\hat{H}_8|\omega_1\rangle & M_{\omega_8} - \langle M_1||\hat{H}_8||M_1\rangle_s/\sqrt{5} \end{pmatrix}. \tag{21.64}$$

The particles φ and ω that one observes correspond to the eigenstates

$$\begin{pmatrix} |\varphi\rangle \\ |\omega\rangle \end{pmatrix} = \begin{pmatrix} \cos\theta & -\sin\theta \\ \sin\theta & \cos\theta \end{pmatrix} \begin{pmatrix} |\omega_1\rangle \\ |\omega_8\rangle \end{pmatrix}. \tag{21.65}$$

Here, θ is a *meson mixing angle*. One obtains

$$M_\varphi + M_\omega = M_{\omega_1} + M_{\omega_8} - \langle M_1||\hat{H}_8||M_1\rangle_s/\sqrt{5},$$
$$M_\varphi M_\omega = M_{\omega_1}\left(M_{\omega_8} - \langle M_1||\hat{H}_8||M_1\rangle_s\sqrt{5}\right) - |\langle\omega_1|\hat{H}_8|\omega_8\rangle|^2. \tag{21.66}$$

In addition, we have

$$M_\rho = M_{\omega_8} + \langle M_1 || \hat{H}_8 || M_1 \rangle_s (2-1)/\sqrt{5},$$
$$M_{K^*} = M_{\omega_8} + \langle M_1 || \hat{H}_8 || M_1 \rangle_s \, (3/4 - 1/4 - 1) \,/\sqrt{5}, \qquad (21.67)$$

and hence

$$4M_{K^*}/3 - M_\rho/3 = M_{\omega_8} + \langle M_1 || \hat{H}_8 || M_1 \rangle_s \, (4/3\,(-1/2) - 1/3) \,/\sqrt{5}$$
$$= M_{\omega_8} - \langle M_1 || \hat{H}_8 || M_1 \rangle_s /\sqrt{5}, \qquad (21.68)$$

such that

$$M_{\omega_1} = M_\varphi + M_\omega - 4M_{K^*}/3 + M_\rho/3 = 0.870\,\text{GeV},$$
$$|\langle \omega_1 | \hat{H}_8 | \omega_8 \rangle|^2 = M_{\omega_1} \left(4M_{K^*}/3 - M_\rho/3 \right) - M_\varphi M_\omega \simeq (0.113\,\text{GeV})^2. \quad (21.69)$$

The mixing angle is now determined from

$$M_{\omega_1} \cos\theta - \langle \omega_1 | \hat{H}_8 | \omega_8 \rangle \sin\theta = M_\varphi \cos\theta,$$
$$\langle \omega_8 | \hat{H}_8 | \omega_1 \rangle \cos\theta - \left(M_{\omega_8} - \langle M_1 || \hat{H}_8 || M_1 \rangle_s /\sqrt{5} \right) \sin\theta = -M_\varphi \sin\theta, \quad (21.70)$$

and we arrive at

$$\left(M_{\omega_1} + M_{\omega_8} - \langle M_1 || \hat{H}_8 || M_1 \rangle_s /\sqrt{5} \right) \sin\theta \cos\theta - \langle \omega_1 | \hat{H}_8 | \omega_8 \rangle = 2M_\varphi \sin\theta \cos\theta \quad \Rightarrow$$

$$\frac{1}{2} \sin(2\theta) = \pm \frac{\sqrt{\left(M_\varphi + M_\omega - 4M_{K^*}/3 + M_\rho/3 \right) \left(4M_{K^*}/3 - M_\rho/3 \right) - M_\varphi M_\omega}}{M_\varphi - M_\omega}. \quad (21.71)$$

Numerically, one obtains $\theta = \pm 52.6°$ and therefore $\cos\theta \approx 1/\sqrt{3}$, $\sin\theta \approx \pm\sqrt{2/3}$, such that

$$|\varphi\rangle \approx s\bar{s} \qquad \text{or} \qquad \frac{1}{3} \left(2u\bar{u} + 2d\bar{d} - s\bar{s} \right),$$
$$|\omega\rangle \approx \frac{1}{\sqrt{2}} \left(u\bar{u} + d\bar{d} \right) \qquad \text{or} \qquad -\frac{1}{\sqrt{18}} \left(u\bar{u} + d\bar{d} + 4s\bar{s} \right). \quad (21.72)$$

Experimentally, the φ meson decays in 84 percent of all cases into kaons ($\varphi \to K^+ + K^-, K^0 + \overline{K^0}$) and only in 16 percent of all cases into pions ($\varphi \to \pi^+ + \pi^0 + \pi^-$). Hence, one concludes that the φ-meson is dominated by s-quarks, such that one obtains almost ideal mixing

$$|\varphi\rangle \approx s\bar{s}, \quad |\omega\rangle \approx \frac{1}{\sqrt{2}} \left(u\bar{u} + d\bar{d} \right). \qquad (21.73)$$

One can repeat the calculation of meson mixing for the scalar mesons η and η'. The η-meson is a pseudo-Nambu–Goldstone boson of the spontaneously broken approximate $SU(3)_L \times SU(3)_R$ chiral symmetry of QCD. As we have seen in Chapter 20, the η'-meson also becomes a Nambu–Goldstone boson in the large-N_c limit, but for $N_c = 3$ it is heavier than a nucleon. In the constituent quark model, one obtains

$$\eta_1 = \frac{1}{\sqrt{3}} \left(u\bar{u} + d\bar{d} + s\bar{s} \right), \quad \eta_8 = \frac{1}{\sqrt{6}} \left(u\bar{u} + d\bar{d} - 2s\bar{s} \right),$$
$$\begin{pmatrix} \eta \\ \eta' \end{pmatrix} = \begin{pmatrix} \cos\bar{\theta} & -\sin\bar{\theta} \\ \sin\bar{\theta} & \cos\bar{\theta} \end{pmatrix} \begin{pmatrix} \eta_8 \\ \eta_1 \end{pmatrix}. \qquad (21.74)$$

In this case, the mixing angle results as $\bar{\theta} = -11.5°$, such that $\eta \approx \eta_8$ and $\eta' \approx \eta_1$.

The rather naive non-relativistic constituent model is not appropriate to address the physics of light pseudo-Nambu–Goldstone bosons. We will address their masses in Chapter 23 in the framework of chiral perturbation theory.

21.7 Hadron Spectrum from Lattice QCD

In contrast to the constituent quark model, lattice QCD provides a method to address the hadron spectrum from first principles. Via dimensional transmutation, the gauge coupling g_s is traded for the QCD scale Λ_{QCD} in the continuum limit. The bare quark masses enter lattice calculations as input parameters that must be adjusted to match experimental data for hadron masses.

Lattice QCD calculations are Monte Carlo simulations that consume a large amount of numerical resources and are typically pursued in relatively large, international collaborations. Different collaborations use different lattice actions. In particular, quarks can be regularized, *e.g.*, with Wilson fermions and variants thereof, twisted-mass fermions, domain wall or overlap fermions, or with staggered fermions. The details of the lattice action also vary, related to different improvement schemes that aim at reducing lattice cut-off artifacts. At the end, after taking the continuum limit, the details of the lattice action become irrelevant (in the renormalization group sense), and the various approaches should give the same universal results (as long as locality is respected). In order to obtain reliable results, several statistical and systematic errors must be controlled carefully.

Monte Carlo calculations generate a Markov chain of field configurations using appropriate simulation algorithms that are supposed to satisfy the conditions of ergodicity and detailed balance, as discussed in Appendix H. These conditions are sufficient to guarantee convergence to the desired probability distribution. However, the simulation algorithms suffer from autocorrelations; *i.e.* subsequent configurations in the Markov chain are not statistically independent. In fact, the autocorrelation times increase as the chiral limit – or the continuum limit – is approached. The development of efficient algorithms has been vital for reaching the current status of lattice QCD and continues to be highly desirable. Due to autocorrelations, the analysis of statistical errors is complicated and requires special care.

Lattice QCD simulations also suffer from a variety of systematic errors. Approaching the continuum limit requires a sequence of simulations performed with finer and finer lattice spacings. Fortunately, the way in which the continuum limit is approached as a function of the lattice spacing can be understood analytically (Symanzik, 1983a,b; Lüscher and Weisz, 1985). Also, the space–time volume of the lattice is necessarily finite and must be extrapolated to the infinite volume limit. The way in which this limit is approached can also be understood analytically (Lüscher, 1986a,b), thus making reliable extrapolations possible. In addition, unless one works directly with the physical values of the renormalized quark masses, one must extrapolate to the physical point. Sufficiently close to the chiral limit, analytic results from chiral perturbation theory can facilitate this task.

As illustrated in Figure 21.3, in this way reliable results have been obtained for the light hadron spectrum, for instance, in the case of u- and d-quarks of the same mass m_q and an s-quark of mass m_s. It is a highly non-trivial, non-perturbative test of QCD that the

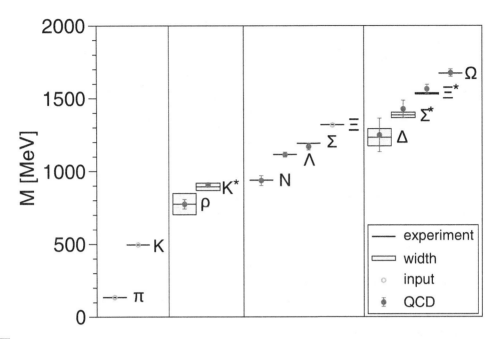

Fig. 21.3 Hadron spectrum obtained from Monte Carlo simulations of lattice QCD with u- and d-quarks of degenerate mass $m_q = (m_u + m_d)/2$ and a heavier s-quark. The quark masses have been adjusted to the experimental values of the pion and kaon masses, and the mass of the Ξ-baryon has been used to convert lattice units into MeV. All other hadron masses are first principle QCD results without any further adjustable parameters. The agreement with experiment is at the few percent level (courtesy BMW Collaboration (Dürr *et al.*, 2008)).

theory agrees very well (*i.e.* at percent level) with experiment. With increasing accuracy, including other effects is vital for comparison with experimental data. For example, virtual c-$\bar{c}$ sea quark pairs or the electromagnetic interaction have small but non-negligible effects on the light hadron spectrum. To account for such effects, refined simulations are being performed with additional c-quarks or by coupling lattice QCD to lattice QED.

Over the decades, lattice QCD has matured to a reliable numerical tool for calculating properties of the strong interaction. This goes far beyond the hadron spectrum and includes hadronic matrix elements, scattering lengths, as well as the width of unstable hadronic resonances. In addition, QCD has been investigated in thermal equilibrium and properties of the quark–gluon plasma at high temperatures have been studied in detail.

Lattice QCD is also an indispensable tool for determining fundamental Standard Model parameters, such as the running strong coupling, the values of the quark masses, or the elements of the CKM quark mixing matrix, by comparison with experiments. The results of particle physics experiments are summarized on a regular basis by the *Particle Data Group*, which averages results of different experiments, taking into account their statistical as well as systematic errors. In some sense, lattice QCD calculations can be viewed as virtual numerical "experiments", which also are affected by statistical and systematic errors. The *Flavor Lattice Averaging Group (FLAG)* averages the results obtained by different collaborations, again taking into account the corresponding errors (see Aoki *et al.* (2021)). This important enterprise helps to ensure that reliable lattice QCD results are made available to the scientific community at large.

21.8 Hadrons for $N_c = 5$

This section describes hadrons in a hypothetical world with $N_c = 5$ quark colors. It can be skipped at a first reading.

As we have seen in Chapter 15, the Standard Model is gauge anomaly free, and thus mathematically consistent, for any odd number of colors N_c. The electric quark charges are generally given by

$$Q_u = \frac{1}{2}\left(\frac{1}{N_c} + 1\right), \quad Q_d = Q_s = \frac{1}{2}\left(\frac{1}{N_c} - 1\right). \tag{21.75}$$

In particular, for $N_c = 5$ they are $Q_u = 3/5$ and $Q_d = Q_s = -2/5$.

We will use the group theoretical classification of the constituent quark model in order to investigate how N_c affects the hadron spectrum. As we will see, the effects are much more pronounced for three than for two flavors.

Let us first consider mesons for an arbitrary number of colors, which in the naive constituent quark model consist of a quark–anti-quark pair. Since a fundamental quark representation $\{N_c\}$ can always be combined with an anti-fundamental representation $\{\overline{N_c}\}$ to form a color singlet, the same spin–flavor meson states can be formed for any value of N_c. While their masses and decay widths are expected to depend on the value of N_c, their symmetry properties in terms of spin and flavor are N_c-independent.

This is very different for baryons, whose number of quark constituents is given by N_c. First of all, as we have seen in Section 15.7, in order to cancel Witten's global $\mathrm{SU}(2)_L$ anomaly, N_c must be odd, which implies that baryons are fermions. In a world with $N_c = 1$, there would still be a nucleon, but no Δ-isobar, while for any odd $N_c \geq 3$, there is a Δ-isobar with the usual electric charges

$$\begin{aligned}
Q_{\Delta^{++}} &= \frac{N_c + 3}{2}Q_u + \frac{N_c - 3}{2}Q_d = 2, \\
Q_{\Delta^{+}} &= \frac{N_c + 1}{2}Q_u + \frac{N_c - 1}{2}Q_d = 1, \\
Q_{\Delta^{0}} &= \frac{N_c - 1}{2}Q_u + \frac{N_c + 1}{2}Q_d = 0, \\
Q_{\Delta^{-}} &= \frac{N_c - 3}{2}Q_u + \frac{N_c + 3}{2}Q_d = -1.
\end{aligned} \tag{21.76}$$

Hence, the existence of the Δ-isobar is evidence for $N_c \geq 3$.

In general, in a world with just two light quark flavors and an arbitrary odd number of colors, baryons have equal half-integer valued spin S and isospin I, in the range $1/2 \leq S = I \leq N_c/2$. This follows from the group theory underlying the constituent quark model. Since the color singlet wave function of N_c quarks is totally anti-symmetric, the spin–isospin wave function must be totally symmetric. This is possible only if $S = I$. For $N_c \geq 5$, there would be an additional baryon resonance Δ', beyond the Δ-isobar of eq. (21.76), with $S = I = 5/2$. For arbitrary odd N_c, the highest of these additional resonances has $S = I = N_c/2$. The absence of such resonances thus constitutes experimental evidence for $N_c = 3$. However, if we limit ourselves to the lowest energies in the baryon number $B = 1$ sector (*i.e.* below the Δ-isobar), for $N_f = 2$ the number of baryon states is N_c-independent.

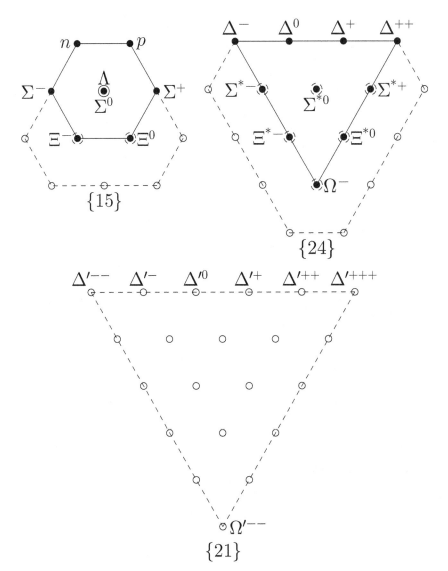

Fig. 21.4 Baryon multiplets for $N_f = 3$ and $N_c = 5$. The SU(3)$_f$ representation $\{15\}$ has spin $S = 1/2$ and contains the usual baryon octet, while the representation $\{24\}$ has $S = 3/2$ and contains the usual flavor decuplet. The representation $\{21\}$ has spin $S = 5/2$ and cannot be realized with $N_c = 3$. It contains a resonance Δ' with isospin $I = 5/2$ and a state Ω'^{--} that consists of five s-quarks. States that cannot be realized with $N_c = 3$ are shown as empty circles or as dashed circles for doubly occupied states. As in Figure 21.1, the axes represent $I_3 = \lambda^3/2$ (horizontal) and $Y = \lambda^8/\sqrt{3}$ (vertical).

Let us now consider the baryon spectrum with three light flavors u, d, and s for $N_c = 5$. Since the color wave function of five constituent quarks coupled to a color singlet is totally anti-symmetric, their spin–flavor wave function must again be totally symmetric. This is possible only if spin and flavor have the same permutation properties, *i.e.* if they belong to the same Young tableau of the permutation group S_5. Both the SU(2)$_S$ spin and SU(3)$_f$ flavor representation, must hence correspond to the same of the following three Young tableaux

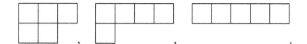

For SU(2)$_S$, these correspond to $S = 1/2, 3/2$, and $5/2$, respectively, while for SU(3)$_f$ they correspond to {15}, {24}, and {21}, as illustrated in Figure 21.4. As we have seen, for $N_f = 2$ and $N_c > 3$ there are no additional, low-energy $S = 1/2$ baryon states besides the nucleon. As we see now, for $N_f = 3$ and $N_c = 5$, the actual $S = 1/2$ baryon octet is enlarged to a 15-plet. As Figure 21.4 shows, there are 7 additional baryon states with isospin $I = 3/2$ (containing two s-quarks) and with $I = 1$ (containing three s-quarks). The fact that these additional states are not observed provides strong experimental evidence for $N_c = 3$.

Similarly, for $N_c = 5$ the ordinary $S = 3/2$ baryon decuplet is contained in a 24-plet. The additional 14 baryon states correspond to isospin $I = 2$ containing one, $I = 3/2$ containing two, $I = 1$ containing three, and $I = 1/2$ containing four s-quarks. Finally, there is a 21-plet that contains baryons with spin $S = 5/2$. This includes the $I = 5/2$ states Δ' of the two-flavor theory, with Δ'^{+++} consisting of five u-quarks and Δ'^{--} consisting of five d-quarks. Besides many other states, there is also an isospin-singlet Ω'^{--}, which consists of five s-quarks. This underscores that $N_c > 3$ has much stronger effects on the baryon spectrum for $N_f = 3$ than for $N_f = 2$. In any case, the fact that one does not observe any of the additional baryon states underscores that in Nature $N_c = 3$.

Exercises

21.1 Young tableaux for S_5

Consult Appendix F in order to construct all Young tableaux for the permutation group S_5. Determine their dimensions by the box enumeration method and independently by the dimension formula eq. (F.56). Check the results by performing the dimension test of eq. (F.55).

21.2 Adjoint representation of SU(n)

Use eq. (F.60) to prove that the dimension of the adjoint representation of SU(n) is $n^2 - 1$.

21.3 External product of two Young tableaux

Couple two S_3 representations of mixed symmetry by forming the external product, *i.e.* construct the resulting Young tableaux of S_6. Then, translate the equation into SU(2), SU(3), and SU(4) language, and check the results by a dimensions test.

21.4 Coupling of the fundamental representation and its anti-partner in SU(n)

Use the Young tableaux technique to couple the fundamental representation of SU(n) to its anti-partner, and show that one obtains the adjoint representation as well as the singlet.

21.5 Coupling of octets and decouplets in SU(3)

Use the Young tableaux method to reduce the following products in SU(3) into irreducible representations: {8} × {8}, {8} × {10}, {10} × {10}.

21.6 Baryon mass formulae for QCD with two flavors

Consider QCD with up and down quarks only ($N_f = 2$). Use degenerate perturbation theory and the fact that isospin is a good approximate symmetry ($m_d - m_u \ll \Lambda_{QCD}$) to derive baryon mass formulae for the nucleon and the Δ-isobar. Use group theoretical methods similar to those used to derive the Gell-Mann–Okubo formula in the three-flavor case. Take into account the effect of different u- and d-quark masses, but ignore effects due to electromagnetism.

21.7 Magnetic moments of baryons

Use the constituent quark model wave function of eq. (21.19) to determine the ratio of the magnetic moments of proton and neutron. The magnetic moment of a baryon is given by

$$\mu_{B_{I_3}} = \langle BI_3 S | \mu_{B_z} | BI_3 S \rangle.$$

Here, S is the maximal possible spin projection S_z for a baryon of total spin S. The magnetic moment operator is given by

$$\vec{\mu}_B = \mu_0 \sum_{q=1}^{3} Q_q \vec{\sigma}_q,$$

where Q_q is the electric charge of a quark and $\vec{\sigma}_q/2$ is its spin vector. Now use the constituent quark model wave function of eq. (21.15) to determine the magnetic moments of the various charge states of the Δ-isobar.

21.8 Clebsch–Gordan coefficients for SU(3)

Verify eq. (21.54) regarding the Clebsch–Gordan coefficients for coupling two SU(3) octets to another octet $\langle \{8\} Y I I_3 | \{8\} 000 \{8\} Y I I_3 \rangle_{a,s}$ in a symmetric and in an antisymmetric manner.

21.9 Meson mixing of η and η'

In close analogy to Section 21.6, determine the amount of mixing leading to the pseudo-scalar mesons η and η'.

22 Partons and Hard Processes

The presentation of this chapter is partly inspired by the textbook by Peter Becher, Manfred Böhm, and Hans Joos (Becher et al., 1983).

As we have seen, the constituent quark model provides a group theoretical classification of hadrons. Constituent quarks can be viewed as non-relativistic quasi-particles, which are composed of fundamental quarks, anti-quarks, and gluons in a non-transparent manner.

In this chapter, we investigate the dynamics of hadrons in the *high-energy regime,* in which non-relativistic concepts break down. Historically, this regime first became accessible via the *parton model,* which was proposed by Richard Feynman (1969). However, unlike the constituent quark model, the parton model has found a solid theoretical basis as the leading-order of a systematic perturbative expansion of QCD at high energies where, due to asymptotic freedom, perturbation theory applies. In fact, it has been proved that the high-energy QCD physics of hadrons can be *factorized* into *hard* parton processes, to be treated in QCD perturbation theory, and *soft*, non-perturbative physics which, for example, manifests itself via so-called *parton distribution functions* (PDFs). A detailed discussion of these important topics is beyond the scope of this book; we refer the interested reader to the extensive literature (Collins *et al.*, 1985, 1989; Campbell *et al.*, 2018; Brock *et al.*, 1995).

We concentrate our discussion mostly on the leading-order parton "model", but we stress again that it is not just a model, but follows from QCD first principles. As illustrated in Figure 22.1 (left), the parton model describes a fast moving hadron as a collection of parallel moving partons, which were soon identified as quarks, anti-quarks, and gluons that are sharing the total momentum of the hadron. The parton structure of hadrons is determined by the strong interaction and is characterized by so-called *structure functions.* Some typical *hard processes* – i.e. high-energy processes – are electron–positron annihilation into hadrons ($e^+ + e^- \rightarrow X$) and *deep-inelastic lepton–nucleon scattering* ($e^- + N \rightarrow e^- + X$).

At first, it may be counter-intuitive that, on the one hand, quarks and gluons cannot escape from the hadrons as free particles (they are confined), and that, on the other hand, they appear as quasi-free partons at high energies. However, the latter behavior is consistently explained by asymptotic freedom; see Section 18.2. As opposed to QED, the color charge is not screened at large distances, and it even gets enhanced. At short distances, on the other hand, the strong interaction becomes weaker, and quarks and gluons behave more and more like free particles. Only at large distances, the coupling grows and quarks and gluons are confined.

The internal structure of hadrons can be studied by electromagnetically or weakly interacting probes, namely, electrons or neutrinos, which exchange a photon, a $W^{\pm}$-, or a Z-boson with one parton inside the hadron. Various processes of this kind are illustrated in Figure 22.2. The elementary parton processes can be computed perturbatively in the Standard Model. Then, one coherently sums over all parton processes.

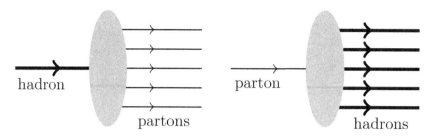

Fig. 22.1 A hadron as a bunch of partons (left) and a parton converted into a jet of hadrons (right).

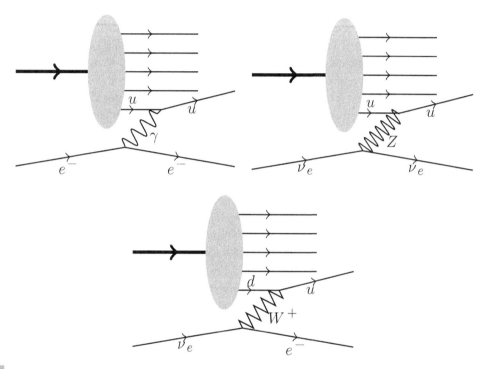

Fig. 22.2 Various deep-inelastic scattering processes involving the exchange of a photon, a Z-boson, or a W-boson.

In the final state of a hadronic reaction, due to confinement, one never finds single partons but only entire hadrons (besides leptons and photons). Each parton therefore hadronizes into a *jet*, which represents a bunch of hadrons. This is illustrated in Figure 22.1 (right). We will now study some hard processes in more detail, in order to relate the parton properties to the observed properties of hadrons.

22.1 Electron–Positron Annihilation into Hadrons

Electron–positron annihilation into hadrons has been studied in detail at e^+e^-–colliders like PETRA at DESY in Hamburg, PEP at SLAC in Stanford, SuperKEKB in Tsukuba,

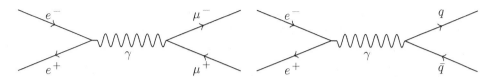

Feynman diagrams for electron–positron annihilation into a muon–anti-muon and into a quark–anti-quark pair.

BEPC II in Beijing, DAFNE in Frascati, VEPP in Novosibirsk and – last but not least – LEP at CERN near Geneva. There are plans for future e^+e^-–colliders, which should be well suited to study the Higgs boson in great detail, in particular the International Linear Collider (ILC) and the Compact Linear Collider (CLIC), whose tentative locations are currently uncertain, or the Circular Electron Positron Collider (CEPC) in China.[1]

The energy released in an e^+e^- annihilation materializes again as particles. Of course, not only strongly interacting hadrons are created but also lepton pairs, including muons. The elementary processes that turn an electron–positron pair into a muon–anti-muon or a quark–anti-quark pair are described by the Feynman diagrams in Figure 22.3. The quark–anti-quark pair then hadronizes in a complicated manner, involving non-perturbative strong interaction processes.

We now discuss the so-called *R-ratio* of the total cross sections for electron–positron annihilation into hadrons versus muon–anti-muon pairs at high energies,

$$R = \frac{\sigma(e^+e^- \to \text{hadrons})}{\sigma(e^+e^- \to \mu^+\mu^-)} = \sum_{\text{partons } q} \frac{\sigma(e^+e^- \to q\bar{q})}{\sigma(e^+e^- \to \mu^+\mu^-)}. \tag{22.1}$$

The cross sections σ measure the number of muon–anti-muon pairs or hadrons that are created per unit of time, normalized to the incoming flux of electron–positron pairs. In this case, the partons that we sum over do not include the gluons, because the interaction is electromagnetic, and gluons are electrically neutral.

Let us consider the kinematics of the elementary processes. Before and after the annihilation, we have the following Minkowskian 4-momenta (in the center-of-mass frame)

$$p_{e^-} = \left(\sqrt{m_e^2 + |\vec{p}\,|^2}, \vec{p}\right), \quad p_{e^+} = \left(\sqrt{m_e^2 + |\vec{p}\,|^2}, -\vec{p}\right),$$

$$p_{\mu^-} = \left(\sqrt{m_\mu^2 + |\vec{p}\,'|^2}, \vec{p}\,'\right), \quad p_{\mu^+} = \left(\sqrt{m_\mu^2 + |\vec{p}\,'|^2}, -\vec{p}\,'\right). \tag{22.2}$$

Energy conservation implies

$$\sqrt{m_e^2 + |\vec{p}\,|^2} = \sqrt{m_\mu^2 + |\vec{p}\,'|^2}. \tag{22.3}$$

The square of the invariant mass of the system is denoted by the Mandelstam variable s, which we encountered before in Section 4.9,

$$s = (p_{e^-} + p_{e^+})^2 = (p_{\mu^-} + p_{\mu^+})^2 = 4\left(m_e^2 + |\vec{p}\,|^2\right) = 4\left(m_\mu^2 + |\vec{p}\,'|^2\right). \tag{22.4}$$

---

[1] Hadron colliders attain higher energies, but lepton colliders have an advantage regarding the cleaner analysis of the generated particles.

The dimension of a cross section is length squared. The leading contributions to the cross section can be computed from the S-matrix elements corresponding to the Feynman diagrams of Figure 22.3. Here, we will not derive the result completely. In any case, we will only need its most elementary features, which we are going to make plausible. We restrict ourselves to high energies, *i.e.* $s \gg m_e^2, m_\mu^2$, and obtain

$$\sigma(e^+e^- \to \mu^+\mu^-) = \frac{4\pi\alpha^2}{3s}, \tag{22.5}$$

where $\alpha = e^2/(4\pi) \simeq 1/137$ is the fine-structure constant. It is easy to understand the essential parts of this result.

- The two vertices in the Feynman diagram each contribute a factor e to the S-matrix element. The cross section is proportional to the S-matrix element squared and hence proportional to e^4.
- In the high-energy limit, $\vec{p}^{\,2} = \vec{p}^{\,\prime 2} = s/4$ is the only dimensioned quantity. Hence, σ must be proportional to $1/s$.
- In the following, we will be interested only in the ratio of cross sections, so we do not derive the coefficient $4\pi/3$, which drops out anyhow.

The relevant parton process is also illustrated in Figure 22.3. Again we consider the high-energy limit $s \gg m_e^2, m_q^2$. One of the two vertices is still given by the charge of the electron. The other vertex, however, contributes the electric charge Q_q of the parton. The charged partons – quarks or anti-quarks – are elementary fermions, just like the muon. Thus, we obtain

$$\sigma(e^+e^- \to q\bar{q}) - \frac{4\pi\alpha^2}{3s}Q_q^2, \tag{22.6}$$

and therefore

$$R = \sum_{\text{partons } q} \frac{\sigma(e^+e^- \to q\bar{q})}{\sigma(e^+e^- \to \mu^+\mu^-)} = \sum_{\text{partons } q} Q_q^2. \tag{22.7}$$

Experimental results for R over a large energy range are illustrated in Figure 22.4.

22.2   R-Ratio as Evidence for $N_c = 3$

The R-ratio data at intermediate energies are compared to the leading-order theoretical prediction in Figure 22.5. First, one observes that, in the range $2\,\text{GeV} \lesssim \sqrt{s} \lesssim 3\,\text{GeV}$, the R-ratio is essentially energy-independent and close to the value 2. Taking into account the three light quark flavors u, d, and s with their electric charges

$$Q_u = \frac{1}{2}\left(\frac{1}{N_c} + 1\right), \quad Q_d = Q_s = \frac{1}{2}\left(\frac{1}{N_c} - 1\right), \tag{22.8}$$

and summing over these three flavors, and over the N_c colors, we arrive at

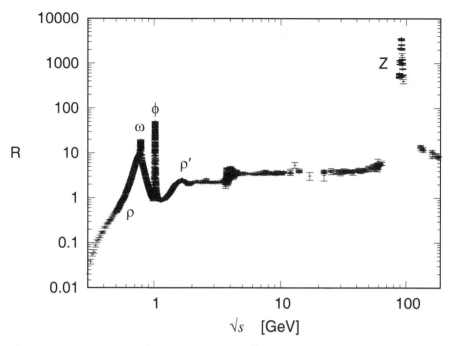

Fig. 22.4 Data for the *R*-ratio as a function of the center-of-mass energy $\sqrt{s}$ over a large energy range extending from the light vector meson resonances ρ, ω, ϕ, and ρ' all the way up to the *Z*-boson.

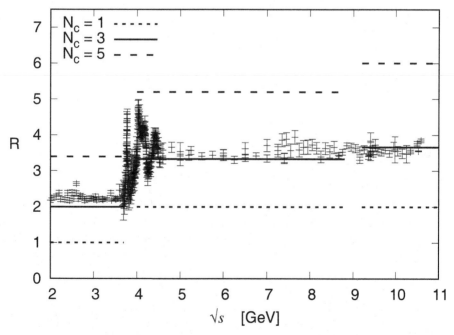

Fig. 22.5 Comparison of intermediate energy data for the *R*-ratio with leading-order theoretical predictions for different odd numbers of color $N_c = 1, 3$, and 5. The data are consistent only with $N_c = 3$.

$$\sum_{\text{partons } q} Q_q^2 = N_c \sum_{f=u,d,s} Q_f^2 = N_c \left[\frac{1}{4} \left(\frac{1}{N_c} + 1 \right)^2 + \frac{1}{2} \left(\frac{1}{N_c} - 1 \right)^2 \right]$$

$$= \frac{3N_c}{4} - \frac{1}{2} + \frac{3}{4N_c} \xrightarrow{N_c=3} 2. \tag{22.9}$$

In the literature, the N_c-dependence of the quark charges is usually ignored in these considerations. As we discussed in Section 15.8, this is inconsistent. In any case, only for $N_c = 3$ one obtains $\sum_{\text{partons } q} Q_q^2 = 2$ (if we require $N_c \in \mathbb{N}$), in agreement with experiments for $2\,\text{GeV} \lesssim \sqrt{s} \lesssim 3\,\text{GeV}$. This is strong experimental evidence that there are indeed *three* quark colors in Nature.

From the experimental data, we further see that for $\sqrt{s} \gtrsim 3\,\text{GeV}$ the R-ratio increases beyond 2. This is due to the (heavier) c-quark with a mass around $1.3\,\text{GeV}$, which cannot be pair-created at lower energies. The c-quark leads to the J/Ψ resonance (a $c\bar{c}$ bound state, with a mass close to $3\,\text{GeV}$) and has the same electric charge as the u-quark, $Q_c = Q_u$. In the higher energy range, we thus obtain

$$\sum_{\text{partons } q} Q_q^2 = N_c \sum_{f=u,d,s,c} Q_f^2 = N_c \left[\frac{1}{2} \left(\frac{1}{N_c} + 1 \right)^2 + \frac{1}{2} \left(\frac{1}{N_c} - 1 \right)^2 \right]$$

$$= N_c + \frac{1}{N_c} \xrightarrow{N_c=3} \frac{10}{3}. \tag{22.10}$$

The same phenomenon occurs again when we reach the b-quark threshold: The b-quark has a mass around $4.2\,\text{GeV}$, it gives rise to the Υ resonance (a $b\bar{b}$ bound state, with mass $\approx 9.5\,\text{GeV}$), and it has the same charge as the s- and the d-quark, $Q_b = Q_s = Q_d$. Hence, we now obtain

$$\sum_{\text{partons } q} Q_q^2 = N_c \sum_{f=u,d,s,c,b} Q_f^2 = N_c \left[\frac{1}{2} \left(\frac{1}{N_c} + 1 \right)^2 + \frac{3}{4} \left(\frac{1}{N_c} - 1 \right)^2 \right]$$

$$= \frac{5N_c}{4} - \frac{1}{2} + \frac{5}{4N_c} \xrightarrow{N_c=3} \frac{11}{3}, \tag{22.11}$$

in good agreement with the experiments. The remaining small deviations can be understood from QCD perturbation theory.

So far, the R-ratio has been measured up to energies around $\sqrt{s} \approx 150\,\text{GeV}$. If future experiments proceed to much higher energies in the TeV regime – which corresponds to the plans for future linear e^+e^-–accelerators – one is tempted to consider the heaviest quark (and the heaviest elementary particle) in the Standard Model: the t-quark with a mass around $173\,\text{GeV}$ and with $Q_t = Q_c = Q_u$. Naively, one might then expect for $\sqrt{s} \gtrsim 360\,\text{GeV}$ the ratio

$$\sum_{\text{partons } q} Q_q^2 = N_c \sum_{f=u,d,s,c,b,t} Q_f^2 = N_c \left[\frac{3}{4} \left(\frac{1}{N_c} + 1 \right)^2 + \frac{3}{4} \left(\frac{1}{N_c} - 1 \right)^2 \right]$$

$$= \frac{3N_c}{2} + \frac{3}{2N_c} \xrightarrow{N_c=3} 5. \tag{22.12}$$

This is not realistic, however, because below the t-quark threshold there is the Z-boson, which opens another annihilation channel, beyond just the photon, which is independent

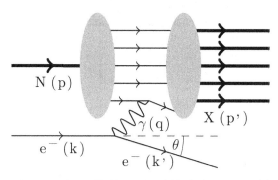

Fig. 22.6 Deep-inelastic electron–nucleon scattering and its kinematic variables in the parton model.

of the electric charge. Moreover, due to its large Yukawa coupling, the top quark couples very strongly to the Higgs field, such that the above purely electromagnetic determination of R certainly fails at these very high energies.

22.3 Deep-Inelastic Electron–Nucleon Scattering

To study the interior of hadrons, one has bombarded them with electromagnetic or weakly interacting probes like electrons or neutrinos. Let us first discuss electron–nucleon scattering

$$e + N \rightarrow e + X$$

at high energies. To leading-order in α, the corresponding diagram in the parton model is shown in Figure 22.6.

We are interested in reactions at high momentum transfer $q = k' - k$. In a hard reaction, the nucleon will in general break up into fragments, so we will find a variety of hadrons X in the final state. Among those, there will certainly be another baryon, because the strong interaction conserves the total baryon number. Since other hadrons are created as well, the reaction is *inelastic*. The experiments have been performed such that the scattered electron is analyzed for scattering angle and energy, while the emerging hadrons are usually left undetected.

Let us first consider the *invariant mass* W of the hadronic final state

$$W^2 = (p+q)^2 = p^2 + 2pq + q^2 = M_N^2 + 2M_N\nu - Q^2. \qquad (22.13)$$

We have inserted the nucleon mass $M_N^2 = p^2$, the energy transfer in the rest frame of the nucleon ν, as well as the 4-momentum transfer squared Q^2, which are defined by

$$\nu = \frac{pq}{M_N}, \quad Q = \sqrt{-q^2}. \qquad (22.14)$$

In the rest frame of the nucleon, we further have

$$p = \left(M_N, \vec{0}\right), \quad k = \left(E, \vec{k}\right), \quad k' = \left(E', \vec{k}'\right), \qquad (22.15)$$

such that $pk = M_N E$, $pk' = M_N E'$ and hence

$$\nu = E - E' = \frac{pq}{M_N} = \frac{p}{M_N}(k - k'). \qquad (22.16)$$

The scattering angle θ of the electron (still in the nucleon rest frame) results from

$$\cos\theta = \frac{\vec{k}\cdot\vec{k}'}{|\vec{k}|\,|\vec{k}'|}. \tag{22.17}$$

At high energies, we can neglect the electron mass, *i.e.*

$$E \simeq |\vec{k}|, \quad E' \simeq |\vec{k}'| \quad \Rightarrow$$
$$\cos\theta = 1 - 2\sin^2\frac{\theta}{2} \simeq \frac{EE' - kk'}{EE'}, \quad kk' \simeq 2EE'\sin^2\frac{\theta}{2}. \tag{22.18}$$

Hence, we obtain

$$Q^2 = -q^2 = -\left(k - k'\right)^2 = 2kk' = 4EE'\sin^2\frac{\theta}{2}. \tag{22.19}$$

We already mentioned that the hadronic final state contains at least one baryon, due to baryon number conservation. Therefore, the invariant mass must be at least as large as the nucleon mass

$$W^2 = M_N^2 + 2M_N\nu - Q^2 \geq M_N^2, \tag{22.20}$$

such that the *Bjorken variable x* (Bjorken, 1969) – named after James Bjorken – is at most 1,

$$x = \frac{Q^2}{2M_N\nu} \leq 1. \tag{22.21}$$

Since $E \geq E'$ (*i.e.* $\nu \geq 0$), we also have $x \geq 0$. In the *Bjorken limit*, both the energy transfer ν and the momentum transfer Q^2 go to infinity, in such a way that their ratio x remains constant.

The S-matrix element for the corresponding transition takes the form

$$\langle e(k')X(p')|S|e(k)N(p)\rangle =$$
$$-\mathrm{i}(2\pi)^4\delta\left(p' + k' - p - k\right) 4\pi\alpha\,\bar{e}(k')\gamma^\mu e(k)\frac{1}{q^2}\langle X(p')|j_\mu|N(p)\rangle. \tag{22.22}$$

Here, the δ-function ensures 4-momentum conservation, $\bar{e}(k')\gamma^\mu e(k)$ is the electromagnetic current of the electron, $1/q^2$ is the photon propagator in the Feynman gauge, and j_μ is the hadronic electromagnetic current. In QCD, the current j^μ is expressed in terms of quark fields

$$j^\mu(x) = \sum_{\mathrm{f}=u,d,s,c,b,t} Q_{\mathrm{f}}e\,\bar{q}^{\mathrm{f}}(x)\gamma^\mu q^{\mathrm{f}}(x). \tag{22.23}$$

To evaluate the differential cross section, one squares the absolute value of the S-matrix element (except for the δ-function, of course)

$$\frac{d^2\sigma}{dE'd\Omega} = \frac{\alpha^2}{Q^4}\frac{E'}{E}\frac{1}{2}\sum_{\text{spins}}\left[\bar{e}(k')\gamma^\mu e(k)\right]\left[\bar{e}(k')\gamma^\nu e(k)\right]^*$$

$$\times \sum_X \frac{1}{M_N}(2\pi)^3\delta\left(p'-p-q\right)\frac{1}{2}\sum_{\text{spins}}\langle N(p)|j_\mu|X(p')\rangle\,\langle X(p')|j_\nu|N(p)\rangle$$

$$= \frac{\alpha^2}{Q^4}\frac{E'}{E}L^{\mu\nu}W_{\mu\nu}. \tag{22.24}$$

By means of the factors $1/2$, one averages over the unobserved initial spin states. Summing over the unobserved electron spin polarizations, one obtains for the *leptonic tensor*

$$L^{\mu\nu} = \frac{1}{2}\sum_{\text{spins}}\left[\bar{e}(k')\gamma^\mu e(k)\right]\left[\bar{e}(k')\gamma^\nu e(k)\right]^* = 2\left(k^\mu k'^\nu + k^\nu k'^\mu - kk'g^{\mu\nu}\right). \tag{22.25}$$

In eq. (22.24), we have also introduced a *hadronic tensor*

$$W_{\mu\nu} = \sum_X \frac{1}{M_N}(2\pi)^3\delta\left(p'-p-q\right)\frac{1}{2}\sum_{\text{spins}}\langle N(p)|j_\mu|X(p')\rangle\,\langle X(p')|j_\nu|N(p)\rangle. \tag{22.26}$$

It contains the complicated dynamics of the strong interaction, whereas the leptonic tensor is simple. Still, symmetries restrict the form of the hadronic tensor. First of all, $W_{\mu\nu}$ only depends on the 4-vectors p_μ and q_μ. From those, one can build three Lorentz scalars

$$p^2 = p^\mu p_\mu = M_N^2, \quad pq = p^\mu q_\mu = \nu M_N, \quad q^2 = q^\mu q_\mu = -Q^2. \tag{22.27}$$

For the two independent scalars, we choose ν and Q^2.

Next we ask which tensors we can form. Obvious options are $p_\mu p_\nu$, $p_\mu q_\nu$, $q_\mu p_\nu$, and $q_\mu q_\nu$. In addition, there are the metric tensor $g_{\mu\nu}$ and the totally anti-symmetric Levi-Civita symbol $\epsilon_{\mu\nu\rho\sigma}$ from which we can construct $\epsilon_{\mu\nu\rho\sigma}p^\rho q^\sigma$. For symmetry reasons, we can hence write

$$W_{\mu\nu} = Ap_\mu p_\nu + Bp_\mu q_\nu + Cq_\mu p_\nu + Dq_\mu q_\nu + Eg_{\mu\nu} + F\epsilon_{\mu\nu\rho\sigma}p^\rho q^\sigma. \tag{22.28}$$

Here, $A, B, \dots, F$ are scalar functions of the variables ν and Q^2. We restrict $W_{\mu\nu}$ further by using additional properties of the electromagnetic current j^μ:

- The current j^μ is Hermitian such that $W_{\nu\mu} = W_{\mu\nu}^*$. Hence, A, D, and E are real, F is purely imaginary, and $B^* = C$.
- In contrast to the weak interaction, the electromagnetic interaction conserves parity. Therefore, the parity odd $\epsilon_{\mu\nu\rho\sigma}$-term vanishes, and $F = 0$.
- Finally, the electromagnetic current is conserved, *i.e.*

$$\partial^\mu j_\mu = 0 \quad \Rightarrow \quad q^\mu W_{\mu\nu} = q^\nu W_{\mu\nu} = 0. \tag{22.29}$$

For arbitrary p and q, we then obtain

$$q^\mu W_{\mu\nu} = A\nu M_N p_\nu + B\nu M_N q_\nu - CQ^2 p_\nu - DQ^2 q_\nu + Eq_\nu = 0,$$
$$q^\nu W_{\mu\nu} = A\nu M_N p_\mu - BQ^2 p_\mu + C\nu M_N q_\mu - DQ^2 q_\mu + Eq_\mu = 0 \quad \Rightarrow$$
$$A\nu M_N - CQ^2 = 0, \quad B\nu M_N - DQ^2 + E = 0,$$
$$A\nu M_N - BQ^2 = 0, \quad C\nu M_N - DQ^2 + E = 0 \quad \Rightarrow \quad B = C, \tag{22.30}$$

such that $B = C$ is also real. Thus, we infer

$$B = \frac{\nu M_N}{Q^2} A, \quad E = DQ^2 - B\nu M_N = DQ^2 - \frac{M_N^2 \nu^2}{Q^2} A,$$

$$W_{\mu\nu} = A \left(p_\mu p_\nu + \frac{M_N \nu}{Q^2} \left(p_\mu q_\nu + p_\nu q_\mu - M_N \nu g_{\mu\nu} \right) \right) + D \left(q_\mu q_\nu + Q^2 g_{\mu\nu} \right)$$

$$= A \left(p_\mu + \frac{M_N \nu}{Q^2} q_\mu \right) \left(p_\nu + \frac{M_N \nu}{Q^2} q_\nu \right) + \left(D - A \frac{\nu^2 M_N^2}{Q^4} \right) \left(q_\mu q_\nu + Q^2 g_{\mu\nu} \right). \quad (22.31)$$

It is common to replace A and D by the *structure functions* W_1^{eN} and W_2^{eN},

$$W_1^{eN}(\nu, Q^2) = \frac{\nu^2 M_N^2}{Q^2} A(\nu, Q^2) - Q^2 D(\nu, Q^2),$$

$$W_2^{eN}(\nu, Q^2) = M_N^2 A(\nu, Q^2). \quad (22.32)$$

They encode the information about the non-perturbative dynamics of the strong interaction, but they cannot be determined from symmetry considerations.

By inserting the leptonic and hadronic tensors, after some calculus one finds

$$\frac{d^2\sigma}{dE' d\Omega} = \frac{\alpha^2}{Q^4} \frac{E'}{E} L^{\mu\nu} W_{\mu\nu} = \frac{\alpha^2}{Q^4} \frac{E'}{E} 2 \left(k^\mu k'^\nu + k^\nu k'^\mu - kk' g^{\mu\nu} \right)$$

$$\times \left[-W_1^{eN}(\nu, Q^2) \frac{1}{Q^2} \left(q_\mu q_\nu + Q^2 g_{\mu\nu} \right) \right.$$

$$\left. + W_2^{eN}(\nu, Q^2) \frac{1}{M_N^2} \left(p_\mu + \frac{M_N \nu}{Q^2} q_\mu \right) \left(p_\nu + \frac{M_N \nu}{Q^2} q_\nu \right) \right]$$

$$= 4 \frac{\alpha^2 E'^2}{Q^4} \left[2 W_1^{eN}(\nu, Q^2) \sin^2 \frac{\theta}{2} + W_2^{eN}(\nu, Q^2) \cos^2 \frac{\theta}{2} \right]. \quad (22.33)$$

Up to this point, we have just taken into account kinematics. Next we will describe *deep-inelastic scattering* in the parton model, which corresponds to the leading order in a systematic perturbative treatment of QCD at high energy. For this purpose, we leave the nucleon rest frame and move to the *Breit frame*, which is characterized by the property that there is only momentum exchange but no energy transfer, $q_0 = 0$. Furthermore, in this frame the momentum transfer $\vec{q}$ is aligned with the negative z-axis, such that

$$q = (0, \vec{q}) = (0, 0, 0, -Q), \quad p = \left(\sqrt{P^2 + M_N^2}, 0, 0, P \right),$$

$$q^2 = -Q^2, \quad M_N \nu = pq = PQ \quad \Rightarrow \quad P = \frac{M_N \nu}{Q} = \frac{Q}{2x}. \quad (22.34)$$

The Bjorken limit, $Q^2, \nu \to \infty$ with $x = Q^2/2M_N\nu$ fixed, drives $P \to \infty$. We see that in the Breit frame the nucleon moves very fast.

In the parton model, a nucleon (proton or neutron) $N \in \{p, n\}$ is characterized by a set of *parton distribution functions* (PDFs) $N_t(\xi)$ that describe the probability to find a parton of type t inside N that carries the momentum fraction ξ. The possible types $t \in \{f, \bar{f}, g\}$ of partons are either quarks or anti-quarks of some flavor f or gluons g. For example, the probability to find an anti-u-quark with momentum fraction ξ inside a neutron is $n_{\bar{u}}(\xi)$, while the probability to find a gluon with momentum fraction ξ inside a proton is $p_g(\xi)$. In the Breit frame, the total nucleon momentum is shared by parallel moving partons with momenta

$$p_t = (\xi P, 0, 0, \xi P), \quad \xi \in [0, 1]. \tag{22.35}$$

We assume all partons to move *parallel* to each other (not anti-parallel). A momentum sum rule guarantees that the parton momenta add up to the total momentum of the nucleon

$$\sum_{t \in \{f, \bar{f}, g\}} \int_0^1 d\xi \; \xi \, N_t(\xi) = 1. \tag{22.36}$$

Since ξ appears as an integration variable, it need not be indexed explicitly by the parton type t.

Let us consider an elementary electron scattering reaction on a single parton with electric charge $\pm Q_f$ (a quark or anti-quark, *i.e.* $t \in \{f, \bar{f}\}$). Electrically neutral partons, *i.e.* gluons, do not contribute to electron scattering. For a quark of flavor f, the reaction is described by the partonic tensor[2]

$$P_{\mu\nu}^f = \frac{Q_f^2}{p_f^0} \, \delta(2p_f q + q^2) \frac{1}{4} \sum_{\text{spins}} \left[\bar{q}^f(p_f') g_{\mu\rho} \gamma^{\,\rho} q^f(p_f) \right] \left[\bar{q}^f(p_f') g_{\nu\sigma} \gamma^{\,\sigma} q^f(p_f) \right]^*$$

$$= \frac{Q_f^2}{p_f^0} \, \delta(2p_f q + q^2) \left(p_{f\mu} p_{f\nu}' + p_{f\nu} p_{f\mu}' - p_f p_f' g_{\mu\nu} \right). \tag{22.37}$$

We insert $p_f' = p_f + q$ in the last term,

$$p_{f\mu} p_{f\nu}' + p_{f\nu} p_{f\mu}' - p_f p_f' g_{\mu\nu} = \left(g_{\mu\nu} - \frac{q_\mu q_\nu}{q^2} \right) \frac{q^2}{2} + 2 \left(p_{f\mu} - \frac{p_f q}{q^2} q_\mu \right) \left(p_{f\nu} - \frac{p_f q}{q^2} q_\nu \right), \tag{22.38}$$

and obtain the partonic tensor

$$P_{\mu\nu}^f = \frac{Q_f^2}{p_f^0} \, \delta(2p_f q + q^2) \left[\left(-g_{\mu\nu} + \frac{q_\mu q_\nu}{q^2} \right) \left(-\frac{q^2}{2} \right) + 2 \left(p_{f\mu} - \frac{p_f q}{q^2} q_\mu \right) \left(p_{f\nu} - \frac{p_f q}{q^2} q_\nu \right) \right]. \tag{22.39}$$

Since we are at high energy, we have neglected the nucleon mass. The δ-function implements momentum conservation

$$2p_f q + q^2 = 0 \quad \Rightarrow \quad p_f^2 = p_f'^2 = (p_f + q)^2 = 0. \tag{22.40}$$

Using $p = (P, 0, 0, P)$, we arrive at

$$p_f = \xi p \quad \Rightarrow \quad \xi = \frac{p_f q}{pq} = \frac{-q^2}{2M_N \nu} = x. \tag{22.41}$$

This consideration reveals a new interpretation of Bjorken's scaling variable x. It characterizes the *momentum fraction of the scattering parton before it is hit by the hard virtual photon.*

---

[2] We (somewhat artificially) use the metric tensor to raise the indices on the γ-matrices, just to comply with our convention to use upper indices for all γ-matrices in Minkowski space–time.

In order to compare with the hadronic tensor, in the next step we sum over the different types of partons and integrate over the parton momenta,

$$
\sum_{t\in\{f,\bar{f}\}} \int_0^1 d\xi \, N_t(\xi) P_{\mu\nu}^t = \sum_{t\in\{f,f\}} \int_0^1 d\xi \, N_t(\xi) \frac{Q_t^2}{\xi P} \delta(2p_f q + q^2)
$$

$$
\times \left[\left(-g_{\mu\nu} + \frac{q_\mu q_\nu}{q^2} \right) \left(-\frac{q^2}{2} \right) + 2\xi^2 \left(p_\mu - \frac{pq}{q^2} q_\mu \right) \left(p_\nu - \frac{pq}{q^2} q_\nu \right) \right] =
$$

$$
\sum_{t\in\{f,\bar{f}\}} \int_0^1 d\xi \, N_t(\xi) \frac{Q_t^2}{\xi P} \frac{1}{2pq} \delta(\xi - x)
$$

$$
\times \left[\left(-g_{\mu\nu} + \frac{q_\mu q_\nu}{q^2} \right) \left(-\frac{q^2}{2} \right) + 2\xi^2 \left(p_\mu - \frac{pq}{q^2} q_\mu \right) \left(p_\nu - \frac{pq}{q^2} q_\nu \right) \right] =
$$

$$
\sum_{t\in\{f,\bar{f}\}} N_t(x) Q_t^2 \left[\left(-g_{\mu\nu} + \frac{q_\mu q_\nu}{q^2} \right) \frac{Q^2}{4xPpq} + \left(p_\mu - \frac{pq}{q^2} q_\mu \right) \left(p_\nu - \frac{pq}{q^2} q_\nu \right) \frac{2x^2}{2xPpq} \right] =
$$

$$
\sum_{t\in\{f,\bar{f}\}} N_t(x) Q_t^2 \left[\left(-g_{\mu\nu} + \frac{q_\mu q_\nu}{q^2} \right) \frac{Q}{2M_N \nu} + \left(p_\mu - \frac{pq}{q^2} q_\mu \right) \left(p_\nu - \frac{pq}{q^2} q_\nu \right) \frac{xQ}{(M_N \nu)^2} \right].
$$

$$(22.42)$$

Comparing with the hadronic tensor

$$
\frac{M_N}{p^0} W_{\mu\nu} = \left(-g_{\mu\nu} + \frac{q_\mu q_\nu}{q^2} \right) \frac{Q}{\nu} W_1^{eN}(\nu, Q^2) + \left(p_\mu - \frac{pq}{q^2} q_\mu \right) \left(p_\nu - \frac{pq}{q^2} q_\nu \right) \frac{Q}{\nu M_N^2} W_2^{eN}(\nu, Q^2),
$$

$$(22.43)$$

we identify

$$
W_1^{eN}(\nu, Q^2) = \frac{1}{2M_N} \sum_{t\in\{f,\bar{f}\}} N_t(x) Q_t^2,
$$

$$
W_2^{eN}(\nu, Q^2) = \frac{x}{\nu} \sum_{t\in\{f,\bar{f}\}} N_t(x) Q_t^2.
$$

$$(22.44)$$

It is common to define the *scaling functions*

$$
F_1^{eN}(x) = \lim_{\nu, Q^2 \to \infty} 2M_N W_1^{eN}(\nu, Q^2) = \sum_{t\in\{f,\bar{f}\}} N_t(x) Q_t^2,
$$

$$
F_2^{eN}(x) = \lim_{\nu, Q^2 \to \infty} \nu W_2^{eN}(\nu, Q^2) = x \sum_{t\in\{f,\bar{f}\}} N_t(x) Q_t^2,
$$

$$(22.45)$$

which do not depend on ν and Q^2 individually, but only on their ratio, *i.e.* on the Bjorken variable $x = Q^2/(2M_N \nu)$, that we defined in eq. (22.21). This behavior is known as *Bjorken scaling*. Indeed, Bjorken scaling is observed in the experimentally determined structure functions, at least for not too small x, as illustrated in Figure 22.7. Deviations from scaling are understood in terms of QCD perturbation theory.

Eqs. (22.45) imply the *Callan–Gross relation* (Callan and Gross, 1969)

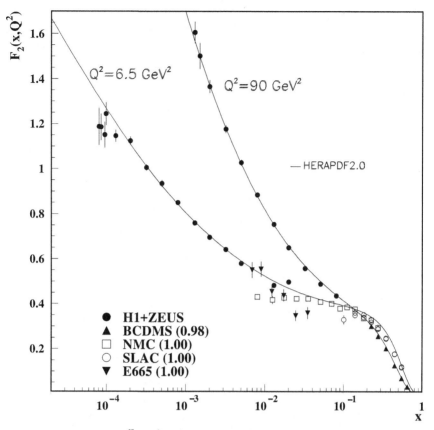

Fig. 22.7 Data for the structure function $\nu W_2^{ep}(\nu, Q^2)$ as a function of the Bjorken scaling variable x for two values of Q^2 (Workman *et al*., 2022, by permission of Oxford University Press). For $x > 0.1$, the data show approximate Q^2-independence, *i.e.* Bjorken scaling. At smaller values of x, there are deviations from scaling that result, *e.g.*, from the radiation of hard gluons, which is described by perturbative QCD corrections.

$$F_2^{eN}(x) = x F_1^{eN}(x). \tag{22.46}$$

Averaging over x one finds experimentally

$$\left\langle \frac{F_2^{eN}(x) - x F_1^{eN}(x)}{F_2^{eN}(x)} \right\rangle_x = -0.01 \pm 0.11. \tag{22.47}$$

The Callan–Gross relation is satisfied solely if the electrically charged partons are spin-1/2 particles. Hence, the experimental data confirm that the charged partons of the nucleon are indeed elementary fermions.

At large but finite momentum transfer Q^2, there are deviations from Bjorken scaling that cannot be understood in the naive parton model. However, thanks to asymptotic freedom, perturbative QCD provides a systematic expansion about the free parton limit and successfully predicts the observed scaling violations, for example, in terms of the radiation of hard gluons. The deviations from scaling in the structure functions are described by the

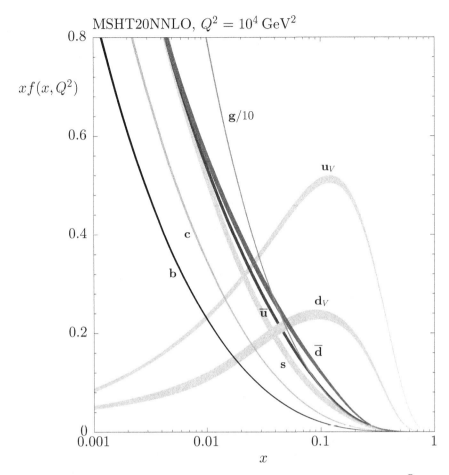

Fig. 22.8 The parton distribution functions (PDFs) $xp_t(x) = xf(x, Q^2)$ of the different types of partons t $\in \{f, \bar{f}, g\}$ inside a proton, as a function of the momentum fraction x, as obtained by the MSHT Collaboration from global fits of experimental data. In agreement with QCD, the PDFs also depend on Q^2. For $x \in [0.2, 1]$, the valence quarks, here denoted by u_V and d_V, dominate, while for $x \in [0, 0.2]$ also gluons (t = g) as well as so-called sea quarks (t $\in \{\bar{u}, \bar{d}, s, c, b\}$) contribute significantly (courtesy of Lucian Harland-Lang).

so-called *DGLAP evolution equations*, which were worked out by Vladimir Gribov and Lev Lipatov (Gribov and Lipatov, 1972), Yuri Dokshitzer (Dokshitzer, 1977) as well as by Guido Altarelli and Giorgio Parisi (Altarelli and Parisi, 1977).

In 2020, the MSHT Collaboration released updated results for PDFs of the various partons inside the proton (Bailey *et al.*, 2021), which are illustrated in Figure 22.8. The PDFs are extracted from global fits to a large variety of experimental data, taking into account next-to-next-to-leading-order (NNLO) QCD effects.

22.4 Deep-Inelastic Neutrino–Nucleon Scattering

In neutrino-nucleon scattering, we are most interested in processes that have an easily detectable charged lepton in the final state and that hence proceed via charged currents,

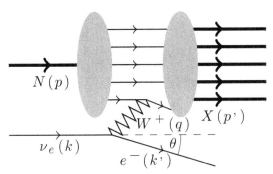

Deep-inelastic neutrino–nucleon scattering and its kinematic variables in the parton model.

i.e. via $W^{\pm}$-bosons. In the parton model, such a process is illustrated in Figure 22.9. The fact that the weak interaction violates parity is reflected by the weak, charged, light quark current, which only involves left-handed quarks,

$$j^{+\mu} = \frac{g}{\sqrt{2}} \left(\bar{u}_L \, \bar{\sigma}^{\mu} \cos \theta_C \, d_L + \bar{u}_L \, \bar{\sigma}^{\mu} \sin \theta_C \, s_L \right). \tag{22.48}$$

The Cabibbo angle $\theta_C \simeq 13°$ describes the mixing of d- and s-quarks under the weak interaction.[3] Due to parity violation, there is now an additional term $F\epsilon_{\mu\nu\rho\sigma}p^{\rho}q^{\sigma}$ in the hadronic tensor. Correspondingly, there is a third structure function W_3 involved. For the differential cross section one then obtains

$$\frac{d^2\sigma}{dE'd\Omega} = \frac{G_F^2 E'^2}{2\pi^2} \left[2W_1^{\nu N}(\nu, Q^2) \sin^2 \frac{\theta}{2} + W_2^{\nu N}(\nu, Q^2) \cos^2 \frac{\theta}{2} - W_3^{\nu N}(\nu, Q^2) \frac{E+E'}{M_N} \sin^2 \frac{\theta}{2} \right]. \tag{22.49}$$

Here, $G_F = 1.16 \times 10^{-5} \, \text{GeV}^{-2}$ is the Fermi coupling constant of the weak interaction. It refers to a low-energy effective theory where the weak interaction is reduced to a 4-fermion vertex (cf. Section 17.9).

The sign of the W_3-term changes under a replacement of neutrino and anti-neutrino. In analogy to electron–nucleon scattering, one obtains in the Bjorken limit

$$F_1^{\nu N}(x) = \lim_{\nu, Q^2 \to \infty} 2M_N W_1^{\nu N}(\nu, Q^2) = 2\cos^2\theta_C N_d(x) + 2\sin^2\theta_C N_s(x) + 2N_{\bar{u}}(x),$$

$$F_2^{\nu N}(x) = \lim_{\nu, Q^2 \to \infty} \nu W_2^{\nu N}(\nu, Q^2) = x F_1^{\nu N}(x),$$

$$F_3^{\nu N}(x) = \lim_{\nu, Q^2 \to \infty} \nu W_3^{\nu N}(\nu, Q^2) = -2\cos^2\theta_C N_d(x) - 2\sin^2\theta_C N_s(x) + 2N_{\bar{u}}(x). \tag{22.50}$$

In experiments at CERN, liquid freon – which contains an equal number of protons and neutrons – has been investigated by scattering electrons as well as neutrinos. For protons and neutrons, one expects a linear combination of probabilities,

---

[3] Here, we neglect the contributions of c-, b-, and t-quarks. The mixing among all three generations is described by the CKM matrix, which θ_C is part of, that we first encountered in Section 17.3.

$$F_1^{ep}(x) = \frac{4}{9}p_u(x) + \frac{1}{9}p_d(x) + \frac{1}{9}p_s(x) + \frac{4}{9}p_{\bar{u}}(x) + \frac{1}{9}p_{\bar{d}}(x) + \frac{1}{9}p_{\bar{s}}(x),$$

$$F_1^{en}(x) = \frac{4}{9}n_u(x) + \frac{1}{9}n_d(x) + \frac{1}{9}n_s(x) + \frac{4}{9}n_{\bar{u}}(x) + \frac{1}{9}n_{\bar{d}}(x) + \frac{1}{9}n_{\bar{s}}(x),$$

$$F_1^{\nu p}(x) = 2\cos^2\theta_C\, p_d(x) + 2\sin^2\theta_C\, p_s(x) + 2p_{\bar{u}}(x),$$

$$F_1^{\nu n}(x) = 2\cos^2\theta_C\, n_d(x) + 2\sin^2\theta_C\, n_s(x) + 2n_{\bar{u}}(x). \tag{22.51}$$

Let us assume the probability for finding s-quarks in the nucleon to be small, *i.e.* p_s, $p_{\bar{s}}$, n_s, $n_{\bar{s}} \approx 0$. In the naive constituent quark model, this would be correct. At high energy, however, it is an approximation. Assuming isospin symmetry, we relate

$$p_u(x) = n_d(x), \quad p_d(x) = n_u(x), \quad p_{\bar{u}}(x) = n_{\bar{d}}(x), \quad p_{\bar{d}}(x) = n_{\bar{u}}(x). \tag{22.52}$$

Summing over proton and neutron, we then obtain

$$F_1^{ep}(x) + F_1^{en}(x) = \frac{5}{9}\Big(p_u(x) + p_d(x) + p_{\bar{u}}(x) + p_{\bar{d}}(x)\Big),$$

$$F_1^{\nu p}(x) + F_1^{\nu n}(x) = 2\cos^2\theta_C\Big(p_d(x) + p_u(x)\Big) + 2\Big(p_{\bar{u}}(x) + p_{\bar{d}}(x)\Big). \tag{22.53}$$

Approximating $\cos^2\theta_C = 0.95 \approx 1$, we arrive at

$$\frac{F_2^{\nu p}(x) + F_2^{\nu n}(x)}{F_2^{ep}(x) + F_2^{en}(x)} = \frac{F_1^{\nu p}(x) + F_1^{\nu n}(x)}{F_1^{ep}(x) + F_1^{en}(x)} \approx \frac{18}{5}. \tag{22.54}$$

This result is in good agreement with experiment, again confirming the established fractional electric charges of the quarks.

22.5 Sum Rules

One can use conserved quantities to derive sum rules for the parton distributions. For example, the electric charge of the nucleon, in units of e, is given by

$$Q_N = \int_0^1 dx \sum_{f=u,d,s} \Big(Q_f N_f(x) - Q_f N_{\bar{f}}(x)\Big), \tag{22.55}$$

and the baryon number imposes the constraint

$$1 = \int_0^1 dx \sum_{f=u,d,s} \left(\frac{1}{3}N_f(x) - \frac{1}{3}N_{\bar{f}}(x)\right). \tag{22.56}$$

Both constraints are in agreement with experiment; they are known as *sum rules*.

One can also formulate a sum rule for the total momentum

$$1 = \int_0^1 dx\, x \sum_i N_i(x). \tag{22.57}$$

Here, we have to sum over all partons of the nucleon. The contribution of quarks plus anti-quarks amounts to

$$I_q = \int_0^1 dx\, x \sum_{f=u,d,s} \Big(N_f(x) + N_{\bar{f}}(x)\Big). \tag{22.58}$$

Again neglecting the Cabibbo angle, *i.e.* by inserting $\cos^2 \theta_c \approx 1$, $\sin^2 \theta_C \approx 0$, we obtain

$$F_1^{eN}(x) = \frac{4}{9}N_u(x) + \frac{1}{9}N_d(x) + \frac{1}{9}N_s(x) + \frac{4}{9}N_{\bar{u}}(x) + \frac{1}{9}N_{\bar{d}}(x) + \frac{1}{9}N_{\bar{s}}(x),$$

$$F_1^{\nu N}(x) = 2N_d(x) + 2N_{\bar{u}}(x). \tag{22.59}$$

For the proton and neutron, this takes the form

$$F_1^{ep}(x) = \frac{4}{9}p_u(x) + \frac{1}{9}p_d(x) + \frac{1}{9}p_s(x) + \frac{4}{9}p_{\bar{u}}(x) + \frac{1}{9}p_{\bar{d}}(x) + \frac{1}{9}p_{\bar{s}}(x),$$

$$F_1^{en}(x) = \frac{4}{9}n_u(x) + \frac{1}{9}n_d(x) + \frac{1}{9}n_s(x) + \frac{4}{9}n_{\bar{u}}(x) + \frac{1}{9}n_{\bar{d}}(x) + \frac{1}{9}n_{\bar{s}}(x)$$

$$= \frac{4}{9}p_d(x) + \frac{1}{9}p_u(x) + \frac{1}{9}p_s(x) + \frac{4}{9}p_{\bar{d}}(x) + \frac{1}{9}p_{\bar{u}}(x) + \frac{1}{9}p_{\bar{s}}(x),$$

$$F_1^{\nu p}(x) = 2p_d(x) + 2p_{\bar{u}}(x),$$

$$F_1^{\nu n}(x) = 2n_d(x) + 2n_{\bar{u}}(x) = 2p_u(x) + 2p_{\bar{d}}(x). \tag{22.60}$$

Next we compute

$$\frac{9}{2}\left(F_1^{ep}(x) + F_1^{en}(x)\right) - \frac{3}{4}\left(F_1^{\nu p}(x) + F_1^{\nu n}(x)\right) =$$

$$\frac{5}{2}p_u(x) + \frac{5}{2}p_d(x) + p_s(x) + \frac{5}{2}p_{\bar{u}}(x) + \frac{5}{2}p_{\bar{d}}(x) + p_{\bar{s}}(x)$$

$$- \frac{3}{2}p_u(x) - \frac{3}{2}p_d(x) - \frac{3}{2}p_{\bar{u}}(x) - \frac{3}{2}p_{\bar{d}}(x) =$$

$$p_u(x) + p_d(x) + p_s(x) + p_{\bar{u}}(x) + p_{\bar{d}}(x) + p_{\bar{s}}(x), \tag{22.61}$$

which yields a neat expression for the term in eq. (22.58),

$$I_q = \int_0^1 dx \left[\frac{9}{2}\left(F_1^{ep}(x) + F_1^{en}(x)\right) - \frac{3}{4}\left(F_1^{\nu p}(x) + F_1^{\nu n}(x)\right)\right]. \tag{22.62}$$

Experimentally, one finds $I_q \approx 0.5$, not 1. *The charged partons (quarks and anti-quarks) only carry about one half of the momentum of the nucleon.* This implies that there are additional, electrically neutral partons carrying the other half of the momentum. In QCD, these partons are naturally identified with the *gluons,* which we discussed in Chapter 14.

We add that the gluons also carry a substantial fraction of the *nucleon spin*, as reviewed by Ji *et al.* (2021). In particular, in 1987 experiments by the European Muon Collaboration at CERN shocked the community with the observation that the valence quarks hardly contribute to the proton spin (Ashman *et al.*, 1988). This gave rise to the so-called proton spin crisis (Jaffe, 1995).

Exercises

22.1 Scaling functions of the cascade baryons Ξ^0 and Ξ^-

Ignore the fact that lepton scattering experiments on the unstable cascade baryons Ξ^0 and Ξ^- are impossible in practice and consider their scaling functions.

a) Construct the scaling functions $F_1^{e\Xi^0}(x)$, $F_1^{e\Xi^-}(x)$, $F_1^{\nu\Xi^0}(x)$, and $F_1^{\nu\Xi^-}(x)$ in analogy to eq. (22.51).

b) Consider a "substance" that consists of an equal mixture of Ξ^0 and Ξ^- (a strange variant of freon). Make the same approximation for the Cabibbo angle ($\cos\theta_C \approx 1$) and derive an expression for $\left(F_1^{\nu\Xi^0}(x) + F_1^{\nu\Xi^-}(x)\right) / \left(F_1^{e\Xi^0}(x) + F_1^{e\Xi^-}(x)\right)$ in analogy to eq. (22.54).

22.2 Scaling functions of "freon" for general N_c

Consider the Standard Model for a general, odd number of colors N_c. Make the same assumptions about isospin and the same approximation for the Cabibbo angle ($\cos\theta_C \approx 1$), and derive the analogue of eq. (22.54) for general N_c.

23 Chiral Perturbation Theory

Being almost massless, the pseudo-Nambu–Goldstone bosons are the lightest particles in QCD. Therefore, they dominate the dynamics of the strong interaction at low energies, and it is possible to switch to a low-energy effective description that only involves the Nambu–Goldstone boson fields. This is the case not only for QCD but also for any other microscopic theory with a continuous, global symmetry that breaks spontaneously to a subgroup. If there are other massless particles besides the Nambu–Goldstone bosons, they must also be included in the low-energy effective theory. It is important that, at low energies, Nambu–Goldstone bosons interact only weakly and can thus be treated perturbatively. In a systematic derivative expansion of the effective Lagrangian – known as *chiral perturbation theory* – terms with a small number of derivatives dominate the physics at low momenta and thus at low energies. Chiral perturbation theory has been used to address a large variety of low-energy properties of the Standard Model, in a systematic multi-loop expansion. For reviews of this very powerful approach, see Pich (1995); Bernard and Meißner (2007); Scherer (2003); Leutwyler (2012).

23.1 Effective Theory for Pions, Kaons, and the η-Meson

If a global, continuous symmetry G breaks spontaneously down to a subgroup $H \subset G$, the emerging Nambu–Goldstone bosons are described by fields in the coset space G/H, in which points are identified if they are connected by symmetry transformations in the unbroken subgroup H (Coleman *et al.*, 1969; Callan *et al.*, 1969). For the spontaneously broken chiral symmetry of QCD, we have

$$G = SU(N_f)_L \times SU(N_f)_R \times U(1)_B,$$
$$H = SU(N_f)_{L=R} \times U(1)_B, \tag{23.1}$$

and the corresponding coset space is

$$G/H = SU(N_f). \tag{23.2}$$

Hence, the Nambu–Goldstone boson fields are represented by special unitary matrices $U(x) \in SU(N_f)$. Under chiral rotations, they transform as

$$U'(x) = LU(x)R^\dagger, \quad L, R \in SU(N_f). \tag{23.3}$$

Let us construct an effective low-energy theory that involves only the Nambu–Goldstone boson fields. As was first pointed out by Steven Weinberg (1979c), such a theory must simply respect the basic principles of quantum field theory, including locality and unitarity, as well as the symmetries of the underlying microscopic theory. All terms that satisfy these criteria contribute to the effective Lagrangian. In particular, they must be Lorentz-invariant and respect the symmetries of the underlying QCD theory. As pioneered by Jürg Gasser and Heinrich Leutwyler, the effective theory is constructed step by step in a systematic low-energy expansion, known as chiral perturbation theory (Gasser and Leutwyler, 1984, 1985). The leading terms contain the smallest possible number of derivatives. The only allowed term without derivatives, which respects the chiral symmetry of massless QCD, $\text{Tr}[U^\dagger U]$, is a trivial constant. Next, we consider the derivative

$$\partial_\mu U'(x) = L \partial_\mu U(x) R^\dagger. \tag{23.4}$$

A term with a single derivative would have an uncontracted index μ and would thus not be Lorentz-invariant. The leading term in the effective, Euclidean Lagrangian has two derivatives forming a chirally invariant Lorentz scalar

$$\mathcal{L}(U) = \frac{F_\pi^2}{4} \text{Tr} \left[\partial_\mu U^\dagger \partial_\mu U \right]. \tag{23.5}$$

The coupling constant F_π has the dimension of a mass and determines the strength of the interaction between the Nambu–Goldstone bosons. It also determines the weak decay of the charged pion and is therefore known as the *pion decay constant*, whose physical value is about 92 MeV. The value of F_π in eq. (23.5) refers to the chiral limit of massless quarks; it is a bit below the physical value.

The above Lagrangian is chirally invariant because

$$\mathcal{L}(U') = \frac{F_\pi^2}{4} \text{Tr} \left[\partial_\mu U'^\dagger \partial_\mu U' \right] = \frac{F_\pi^2}{4} \text{Tr} \left[R \partial_\mu U^\dagger L^\dagger L \partial_\mu U R^\dagger \right] = \mathcal{L}(U). \tag{23.6}$$

There are several sub-leading higher-order terms with four derivatives, whose prefactors are low-energy constants known as Gasser–Leutwyler coefficients (Gasser and Leutwyler, 1984).

We also want to include chiral symmetry breaking effects due to non-zero quark masses m_f, which enter the underlying QCD Lagrangian via the diagonal quark mass matrix $\mathcal{M} = \text{diag}(m_u, m_d, m_s)$, such that

$$\mathcal{L}(\bar{q}, q, G) = \bar{q}_L \bar{\sigma}_\mu \left(\partial_\mu + G_\mu \right) q_L + \bar{q}_R \sigma_\mu \left(\partial_\mu + G_\mu \right) q_R$$
$$+ \bar{q}_L \mathcal{M} q_R + \bar{q}_R \mathcal{M}^\dagger q_L - \frac{1}{2g_s^2} \text{Tr} \left[G_{\mu\nu} G_{\mu\nu} \right]. \tag{23.7}$$

The quark–gluon interaction terms are invariant against chiral transformations, which for $N_f = 3$ take the form

$$q_L'(x) = \begin{pmatrix} u_L'(x) \\ d_L'(x) \\ s_L'(x) \end{pmatrix} = L \begin{pmatrix} u_L(x) \\ d_L(x) \\ s_L(x) \end{pmatrix} = L q_L(x), \quad \bar{q}_L'(x) = \bar{q}_L(x) L^\dagger, \quad L \in \text{SU(3)}_L,$$

$$q_R'(x) = \begin{pmatrix} u_R'(x) \\ d_R'(x) \\ s_R'(x) \end{pmatrix} = R \begin{pmatrix} u_R(x) \\ d_R(x) \\ s_R(x) \end{pmatrix} = R q_R(x), \quad \bar{q}_R'(x) = \bar{q}_R(x) R^\dagger, \quad R \in \text{SU(3)}_R, \tag{23.8}$$

while the mass terms explicitly break the chiral symmetry. Interestingly, we can make even the mass terms chirally invariant if we replace the constant mass matrix by a *"spurion"* $\mathcal{M}$ that itself transforms non-trivially as

$$\mathcal{M}' = L\mathcal{M}R^\dagger. \tag{23.9}$$

This transformation behavior, which characterizes the way in which the quark masses break chiral symmetry, also determines the way how the mass matrix enters the effective Lagrangian

$$\mathcal{L}(U) = \frac{F_\pi^2}{4}\text{Tr}\left[\partial_\mu U^\dagger \partial_\mu U\right] - \frac{\Sigma}{2}\text{Tr}\left[\mathcal{M}^\dagger U + U^\dagger \mathcal{M}\right]. \tag{23.10}$$

The mass term is also invariant if $\mathcal{M}$ transforms as a spurion. On the other hand, when it is put to its constant physical value $\mathcal{M} = \mathcal{M}^\dagger = \text{diag}(m_u, m_d, m_s)$, it breaks chiral symmetry explicitly

$$\text{Tr}\left[\mathcal{M}^\dagger U' + U'^\dagger \mathcal{M}\right] = \text{Tr}\left[\mathcal{M}\left(LUR^\dagger + RU^\dagger L^\dagger\right)\right]. \tag{23.11}$$

If all quark masses are equal, *i.e.* for $\mathcal{M} = m\mathbb{1}$, the Lagrangian is still invariant against $\text{SU}(N_f)_{L=R}$ flavor transformations for which $L = R$. For a general, diagonal mass matrix, the symmetry is reduced to $\prod_f \text{U}(1)_f$; *i.e.* the individual flavors are separately conserved. The baryon number symmetry $\text{U}(1)_B$ assigns the same $\text{U}(1)$ transformation to all flavors and is thus the diagonal subgroup of $\prod_f \text{U}(1)_f$.

We still need to determine the prefactor Σ of the mass term in eq. (23.10). For this purpose, we determine the vacuum expectation value of $\partial_{m_f}\mathcal{L}|_{\mathcal{M}=0}$, first in QCD,

$$\langle 0|\partial_{m_f}\mathcal{L}|_{\mathcal{M}=0}|0\rangle = \langle 0|\left(\bar{q}_L^f q_R^f + \bar{q}_R^f q_L^f\right)|0\rangle = \frac{1}{N_f}\langle\bar{q}q\rangle. \tag{23.12}$$

Here, $\langle\bar{q}q\rangle$ is the *chiral condensate* in the chiral limit, *i.e.* the order parameter that signals spontaneous chiral symmetry breaking. The condensate $\langle\bar{q}q\rangle$ receives additive contributions from all flavors. This result for the underlying QCD theory should be matched by the corresponding low-energy effective theory.

The classical vacuum of the effective theory with $\mathcal{M} = 0$ corresponds to a constant field $U(x) = \mathbb{1}$. Hence, in the effective theory, we obtain

$$\langle 0|\partial_{m_f}\mathcal{L}|_{\mathcal{M}=0}|0\rangle = -\frac{\Sigma}{2}\text{Tr}\left[\text{diag}(1,0,\ldots,0)(\mathbb{1}+\mathbb{1})\right] = -\Sigma \quad \Rightarrow \quad \Sigma = -\frac{\langle\bar{q}q\rangle}{N_f}. \tag{23.13}$$

It should be noted that $\langle\bar{q}q\rangle < 0$. The negative sign results from eq. (9.31). The parameters F_π and Σ determine the leading low-energy dynamics of QCD. Their values can be obtained either by comparison with experiment or from QCD itself, by means of numerical lattice QCD calculations. F_π is directly experimentally accessible via the weak decay of the charged pions. On the other hand, the experimental determination of Σ is rather indirect and relies more heavily on theoretical input. Up to these two parameters, the Nambu–Goldstone boson dynamics at the lowest energies are completely determined by chiral symmetry (and the other symmetries of the theory).

At higher energies, higher-order terms (with more derivatives or more insertions of the mass matrix) contribute to the effective Lagrangian. Again, these terms are restricted by symmetry considerations, and each is associated with an additional low-energy parameter. Chiral perturbation theory provides a systematic low-energy expansion, whose predictive power at higher energies is, unfortunately, limited, due to the proliferation of *a priori* unknown low-energy parameters. Here, we restrict ourselves to the lowest order and thus to the Lagrangian of eq. (23.10).

It is interesting to note that, from the point of view of the Standard Model, the spurion transforming as $\mathcal{M}' = L\mathcal{M}R^\dagger$ is not as spurious as it may seem. In fact, in its matrix form the Higgs field transforms as $\mathbf{\Phi}'(x) = L\mathbf{\Phi}(x)R^\dagger$ under global $SU(2)_L \times SU(2)_R$ transformations. The quark masses $m_{\rm f} = f_{\rm f}v$ indeed arise from the Yukawa couplings $f_{\rm f}$ and the vacuum value v of the Higgs field via the Lagrangian

$$\mathcal{L}(\bar{q}_{\rm L}, q_{\rm L}, \bar{q}_{\rm R}, q_{\rm R}, \mathbf{\Phi}) = \bar{q}_{\rm L}\mathbf{\Phi}\mathcal{F}q_{\rm R} + \bar{q}_{\rm R}\mathcal{F}^\dagger\mathbf{\Phi}^\dagger q_{\rm L}, \tag{23.14}$$

where $\mathcal{F} = \mathrm{diag}(f_u, f_d)$ is the diagonal matrix of Yukawa couplings. Hence, the spurion resembles features of the dynamical Higgs field.

23.2 Masses of Pseudo-Nambu–Goldstone Bosons

Chiral perturbation theory is an expansion around the classical vacuum solution $U(x) = \mathbb{1}$. We write

$$U(x) = \exp\left(\mathrm{i}\pi^a(x)\lambda^a/F_\pi\right), \quad a \in \{1, 2, \ldots, N_{\rm f}^2 - 1\}, \tag{23.15}$$

where λ^a are the generators of $SU(N_{\rm f})$, and we expand in powers of π^a/F_π. To lowest order, we thus obtain

$$U(x) \simeq 1 + \mathrm{i}\pi^a(x)\lambda^a/F_\pi, \quad \partial_\mu U(x) \simeq \mathrm{i}\partial_\mu\pi^a(x)\lambda^a/F_\pi, \tag{23.16}$$

and hence

$$\mathcal{L}(\pi) \simeq \frac{1}{4}\mathrm{Tr}\left[\partial_\mu\pi^a\lambda^a\partial_\mu\pi^b\lambda^b\right] - \frac{\Sigma}{2}\mathrm{Tr}\left[\mathcal{M}(\mathbb{1}+\mathbb{1})\right] = \frac{1}{2}\partial_\mu\pi^a\partial_\mu\pi^a - \Sigma\,\mathrm{Tr}\mathcal{M}. \tag{23.17}$$

The introduction of F_π in eq. (23.15) ensures the generic normalization of the first term on the right-hand side. The last term is a constant that is irrelevant in this context. Next, we expand consistently to $\mathcal{O}\left(\pi^2/F_\pi^2\right)$,

$$U(x) \simeq 1 + \mathrm{i}\pi^a(x)\lambda^a/F_\pi + \frac{1}{2}\left(\mathrm{i}\pi^a(x)\lambda^a/F_\pi\right)^2, \tag{23.18}$$

such that for quarks with equal masses, *i.e.* for $\mathcal{M} = m\mathbb{1}$,

$$\mathrm{Tr}\left[\mathcal{M}(U + U^\dagger)\right] = m\mathrm{Tr}\left[U + U^\dagger\right]$$
$$\simeq 2mN_{\rm f} - m\frac{1}{F_\pi^2}\pi^a\pi^b\mathrm{Tr}\left[\lambda^a\lambda^b\right] = 2mN_{\rm f} - 2m\frac{1}{F_\pi^2}\pi^a\pi^a. \tag{23.19}$$

Altogether we obtain

$$\mathcal{L}(\pi) \simeq \frac{1}{2}\partial_\mu\pi^a\partial_\mu\pi^a - \Sigma\left(mN_{\rm f} - m\frac{1}{F_\pi^2}\pi^a\pi^a\right). \tag{23.20}$$

This is the leading-order Lagrangian of free particles with mass

$$M_\pi^2 = \frac{2m\Sigma}{F_\pi^2}. \tag{23.21}$$

This behavior is generic for the mass squared of a pseudo-Nambu–Goldstone boson: It is proportional to the explicit symmetry breaking parameter (m in this case).

It is straightforward to derive a corresponding mass formula for a general non-degenerate mass matrix. For $N_f = 2$, with the light quark masses m_u and m_d, at leading order one obtains the *Gell-Mann–Oakes–Renner* relation (Gell-Mann *et al.*, 1968)

$$M_{\pi^+}^2 = M_{\pi^0}^2 = M_{\pi^-}^2 = \frac{(m_u + m_d)\Sigma}{F_\pi^2}. \tag{23.22}$$

Interestingly, despite the fact that the quark masses break the isospin symmetry $SU(2)_{L=R}$ explicitly to $U(1)_u \times U(1)_d/U(1)_B$, the masses of the three charge states of the pion are the same, at least to this leading order. The mass difference $m_d - m_u$ enters only at higher order. As we will see in Section 23.4, the masses of the charged pions $\pi^\pm$ receive additional electromagnetic corrections, which make them slightly heavier than the neutral pion π^0.

For $N_f = 3$, with an additional s-quark of mass m_s, we parametrize the Nambu–Goldstone boson field as $U(x) = \exp\left(i\pi^a(x)\lambda^a/F_\pi\right)$, where λ^a are the eight Gell-Mann matrices that generate the flavor algebra $SU(3)_f$,

$$\pi^a \lambda^a = \sqrt{2} \begin{pmatrix} \pi^3(x)/\sqrt{2} + \pi^8(x)/\sqrt{6} & \pi^+(x) & K^+(x) \\ \pi^-(x) & -\pi^3(x)/\sqrt{2} + \pi^8(x)/\sqrt{6} & K^0(x) \\ K^-(x) & \overline{K^0}(x) & -2\pi^8(x)/\sqrt{6} \end{pmatrix},$$

$$\pi^\pm(x) = \frac{1}{\sqrt{2}}\left(\pi^1(x) \mp i\pi^2(x)\right), \quad K^\pm(x) = \frac{1}{\sqrt{2}}\left(\pi^4(x) \mp i\pi^5(x)\right),$$

$$K^0(x) = \frac{1}{\sqrt{2}}\left(\pi^6(x) - i\pi^7(x)\right), \quad \overline{K^0}(x) = \frac{1}{\sqrt{2}}\left(\pi^6(x) + i\pi^7(x)\right). \tag{23.23}$$

One then obtains

$$\text{Tr}\left[\mathcal{M}U^\dagger + U\mathcal{M}^\dagger\right] = \text{Tr}\left[\mathcal{M}\left(U + U^\dagger\right)\right]$$

$$= 2\text{Tr}[\mathcal{M}] - \frac{1}{F_\pi^2}\text{Tr}\left[\mathcal{M}\left(\pi^a\lambda^a\right)^2\right] + \mathcal{O}\left(\left(\frac{\pi^a\lambda^a}{F_\pi}\right)^4\right),$$

$$\frac{1}{2}\text{Tr}\left[\mathcal{M}\left(\pi^a\lambda^a\right)^2\right] = (m_u + m_d)\pi^+\pi^- + (m_u + m_s)K^+K^- + (m_d + m_s)K^0\overline{K^0}$$

$$+ (m_u + m_d)\frac{(\pi^3)^2}{2} + \frac{1}{\sqrt{3}}(m_u - m_d)\pi^3\pi^8 + \frac{1}{3}(m_u + m_d + 4m_s)\frac{(\pi^8)^2}{2}. \tag{23.24}$$

We immediately read off the masses of the charged pions and the kaons to this leading order

$$M_{\pi^+}^2 = M_{\pi^-}^2 = \frac{(m_u + m_d)\Sigma}{F_\pi^2},$$

$$M_{K^+}^2 = M_{K^-}^2 = \frac{(m_u + m_s)\Sigma}{F_\pi^2}, \quad M_{K^0}^2 = M_{\overline{K}^0}^2 = \frac{(m_d + m_s)\Sigma}{F_\pi^2}. \tag{23.25}$$

As a result of isospin symmetry breaking due to different u- and d-quark masses, there is $\pi^3 - \pi^8$ mixing proportional to $m_d - m_u$. After diagonalizing the mixing matrix, we obtain the masses of the neutral pion and the η-meson as

$$M^2_{\pi^0} = \frac{\Sigma}{F^2_\pi} \left[\frac{2}{3}(m_u + m_d + m_s) - \frac{1}{3}\sqrt{(2m_s - m_u - m_d)^2 + 3(m_u - m_d)^2} \right],$$

$$M^2_\eta = \frac{\Sigma}{F^2_\pi} \left[\frac{2}{3}(m_u + m_d + m_s) + \frac{1}{3}\sqrt{(2m_s - m_u - m_d)^2 + 3(m_u - m_d)^2} \right]. \qquad (23.26)$$

Using the fact that $m_s \gg m_u, m_d$ and neglecting terms of order $\mathcal{O}\left((m_{u,d}/m_s)^2\right)$, these results simplify to

$$M^2_{\pi^0} = \frac{(m_u + m_d)\Sigma}{F^2_\pi}, \qquad M^2_\eta = \frac{(m_u + m_d + 4m_s)\Sigma}{3F^2_\pi}. \qquad (23.27)$$

Hence, to this order, the masses of the pions and the η-meson are unaffected by the isospin breaking due to $m_u \neq m_d$. However, this breaking does lead to a difference between the masses of the charged and neutral kaons. As we will see in the next section, further isospin breaking effects arise due to electromagnetic corrections, because $Q_u \neq Q_d = Q_s$.

The η-meson is also affected by mixing with the lightest flavor-singlet meson, which is a pseudo-Nambu–Goldstone boson only in the large-N_c limit, but not in the real world with $N_c = 3$. As we discussed in Chapter 20, the axial $U(1)_A$ anomaly gives a large mass to the η'-meson.

In analogy to the Gell-Mann–Okubo mass formula for the baryon octet, eq. (21.57), to leading order of chiral perturbation theory, one obtains the mass relation

$$M^2_\pi + 3M^2_\eta = 2M^2_{K^\pm} + 2M^2_{K^0}, \qquad (23.28)$$

Before one compares this with experimental values, it is useful to also consider electromagnetic corrections to the pseudo-Nambu–Goldstone boson masses.

23.3 Low-Energy Effective Theory for Nambu–Goldstone Bosons and Photons

Let us now consider chiral perturbation theory as a low-energy effective theory not only for the strong but also for the electromagnetic interaction of quarks with gluons and photons. The Lagrangian of the underlying microscopic theory then takes the form

$$\mathcal{L}(\bar{q}, q, G_\mu, A_\mu) = \bar{q}_L \bar{\sigma}_\mu \left(\partial_\mu + G_\mu + ieQ_L A_\mu\right) q_L + \bar{q}_R \sigma_\mu \left(\partial_\mu + G_\mu + ieQ_R A_\mu\right) q_R$$

$$+ \bar{q}_L \mathcal{M} q_R + \bar{q}_R \mathcal{M}^\dagger q_L - \frac{1}{2g^2_s}\text{Tr}\left[G_{\mu\nu}G_{\mu\nu}\right] + \frac{1}{4}F_{\mu\nu}F_{\mu\nu}$$

$$Q_L = Q_R = Q = \text{diag}(Q_u, Q_d, Q_s) = \text{diag}\left(\frac{2}{3}, -\frac{1}{3}, -\frac{1}{3}\right). \qquad (23.29)$$

Here, Q_L and Q_R are the charge matrices of the left- and right-handed quark fields. Since QED is a vector-like theory, the physical values of both charge matrices are the same. Since the charge matrix Q is not proportional to the unit-matrix, the electromagnetic interaction breaks chiral symmetry explicitly. However, just as for the mass matrix, we can introduce appropriate spurions which in this case transform as

$$Q'_L = LQ_L L^\dagger, \quad Q'_R = RQ_L R^\dagger. \tag{23.30}$$

This formal transformation leaves even the quark–photon interaction terms chirally invariant.

At the level of the low-energy effective theory, the spurions Q_L and Q_R play a similar role as $\mathcal{M}$, by characterizing the way in which electromagnetism breaks chiral symmetry. In the presence of the electromagnetic interaction, to lowest order the chiral Lagrangian takes the form

$$\mathcal{L}(U) = \frac{F_\pi^2}{4} \text{Tr}\left[D_\mu U^\dagger D_\mu U\right] - \frac{\Sigma}{2} \text{Tr}\left[\mathcal{M}^\dagger U + U^\dagger \mathcal{M}\right]$$
$$- Ce^2 \text{Tr}\left[Q_L U Q_R U^\dagger\right] + \frac{1}{4} F_{\mu\nu} F_{\mu\nu}, \tag{23.31}$$

where C is another low-energy parameter. Since – in contrast to gluons – photons are not confined and are, in fact, massless, they explicitly enter the chiral Lagrangian via the covariant derivative

$$D_\mu U(x) = \partial_\mu U(x) + ieA_\mu(x)Q_L U(x) - ieA_\mu(x)U(x)Q_R. \tag{23.32}$$

At the end, the spurions are replaced by their constant physical values $Q_L = Q_R = Q$. It is important that, at the level of the low-energy effective theory, in order to incorporate electromagnetism it is not sufficient to include the photon field by its minimal coupling via the covariant derivative. In particular, the field strength $F_{\mu\nu}$ may also enter directly, and the spurions may enter even without A_μ.

In order to better understand the structure of the electromagnetic covariant derivative, let us recall how electromagnetism emerges in the Standard Model from the "spontaneous breaking" of $SU(2)_L \times U(1)_Y$ down to $U(1)_{em}$. For this purpose, we consider $N_f = 2$ and we gauge $SU(2)_L$ as well as the $U(1)_Y$ subgroup of $SU(2)_R$ at the level of the effective theory. We consider the $SU(2)_L$ gauge field $W_\mu(x) = igW_\mu^a(x)\tau^a/2$, and we embed the $U(1)_Y$ gauge field $B_\mu(x)$ in a hypothetical $SU(2)_R$ gauge field $X_\mu(x) = ig'X_\mu^a(x)\tau^a/2$, by identifying $X_\mu^3(x) = B_\mu(x)$. Under $SU(2)_L \times SU(2)_R$ gauge transformations, the fields then transform as

$$W'_\mu(x) = L(x)\left(W_\mu(x) + \partial_\mu\right)L(x)^\dagger, \quad X'_\mu(x) = R(x)\left(X_\mu(x) + \partial_\mu\right)R(x)^\dagger,$$
$$U'(x) = L(x)U(x)R(x)^\dagger \quad \Rightarrow$$
$$D'_\mu U'(x) = \partial_\mu U'(x) + W'_\mu(x)U'(x) - U'(x)X'_\mu(x)$$
$$= L(x)\left[\partial_\mu U(x) + W_\mu(x)U(x) - U(x)X_\mu(x)\right]R(x)^\dagger$$
$$= L(x)D_\mu U(x)R(x)^\dagger. \tag{23.33}$$

The electromagnetic gauge transformations are embedded in $SU(2)_{L=R}$ and take the form

$$L(x) = R(x) = \text{diag}\left(\exp\left(ig'\varphi(x)/2\right), \exp\left(-ig'\varphi(x)/2\right)\right),$$
$$B'_\mu(x) = B_\mu - \partial_\mu\varphi(x). \tag{23.34}$$

The photon field results from

$$W_\mu^3(x) = \frac{g' A_\mu(x) + g Z_\mu(x)}{\sqrt{g^2 + g'^2}}, \quad B_\mu(x) = \frac{g A_\mu(x) - g' Z_\mu(x)}{\sqrt{g^2 + g'^2}} \quad \Rightarrow$$

$$A_\mu(x) = \frac{g' W_\mu^3(x) + g B_\mu(x)}{\sqrt{g^2 + g'^2}}, \quad A_\mu'(x) = A_\mu(x) - \partial_\mu \alpha(x),$$

$$e\alpha(x) = g' \varphi(x), \quad e = \frac{g g'}{\sqrt{g^2 + g'^2}}. \tag{23.35}$$

Setting $Z_\mu(x) = 0$ and inserting $W_\mu(x) = i g W_\mu^3(x) \tau^3 / 2$ and $X_\mu(x) = i g' B_\mu(x) \tau^3 / 2$ in the covariant derivative of eq. (23.33), we arrive at the electromagnetic covariant derivative of eq. (23.32)

$$D_\mu U(x) = \partial_\mu U(x) + i \frac{g g' A_\mu(x)}{\sqrt{g^2 + g'^2}} \frac{\tau^3}{2} U(x) - U(x) i \frac{g' g A_\mu(x)}{\sqrt{g^2 + g'^2}} \frac{\tau^3}{2}$$

$$= \partial_\mu U(x) + \left[i e A_\mu(x) Q', U(x) \right] = \partial_\mu U(x) + \left[i e A_\mu(x) Q, U(x) \right]. \tag{23.36}$$

Here, we have introduced the traceless part $Q' = \tau^3 / 2$ of the quark charge matrix Q via

$$Q = \begin{pmatrix} 2/3 & 0 \\ 0 & -1/3 \end{pmatrix} = \begin{pmatrix} 1/2 & 0 \\ 0 & -1/2 \end{pmatrix} + \frac{1}{6} \begin{pmatrix} 1 & 0 \\ 0 & 1 \end{pmatrix} = Q' + \frac{B}{2} \mathbb{1}. \tag{23.37}$$

The trace of the charge matrix is given in terms of the baryon number $B = 1/3$ (for $N_c = 3$). A subtle effect, namely, the decay of the neutral pion into two photons, is due to the baryon number contribution to the electric charge. This effect, which is based on the axial anomaly, is not yet included in the chiral Lagrangian in the form of eq. (23.31). It will be discussed in detail in Chapter 24.

In the theory with $N_f = 3$ and $N_c = 3$, the quark charge matrix Q of eq. (23.29) is traceless and directly enters the covariant derivative of eq. (23.32). Applying the electromagnetic gauge transformation eq. (23.33), $U'(x) = L(x) U(x) R(x)^\dagger$, with

$$L(x) = R(x) = \mathrm{diag}\left(\exp\left(i 2 e \alpha(x) / 3 \right), \exp\left(-i e \alpha(x) / 3 \right), \exp\left(-i e \alpha(x) / 3 \right) \right), \tag{23.38}$$

to $\pi^a(x) \lambda^a$ of eq. (23.23), one obtains the expected gauge transformation behavior of the various Nambu–Goldstone boson fields

$$\pi^{\pm'}(x) = \exp\left(\pm i e \alpha(x) \right) \pi^\pm(x), \quad \pi^{0'}(x) = \pi^0(x), \quad \eta'(x) = \eta(x),$$

$$K^{\pm'}(x) = \exp\left(\pm i e \alpha(x) \right) K^\pm(x), \quad K^{0'}(x) = K^0(x), \quad \overline{K^0}'(x) = \overline{K^0}(x). \tag{23.39}$$

Here, we have used the fact that the fields $\pi^0(x)$ and $\eta(x)$ result from the mixing of the neutral fields $\pi^3(x)$ and $\pi^8(x)$.

23.4 Electromagnetic Corrections to the Nambu–Goldstone Boson Masses

The pseudo-Nambu–Goldstone boson masses receive additional contributions from the term

$$\mathrm{Tr}\left[Q_L U Q_R U^\dagger\right] = \mathrm{Tr}\left[QUQU^\dagger\right]$$

$$= \mathrm{Tr}[Q^2] + \frac{1}{F_\pi^2}\mathrm{Tr}\left[Q\pi^a\lambda^a[Q,\pi^b\lambda^b]\right] + \mathcal{O}\left(\left(\frac{\pi^a\lambda^a}{F_\pi}\right)^3\right). \quad (23.40)$$

The term that contributes to the mass squares of the Nambu–Goldstone bosons results in

$$\mathrm{Tr}\left[Q\pi^a\lambda^a[Q,\pi^b\lambda^b]\right] = -2(\pi^+\pi^- + K^+K^-). \quad (23.41)$$

Quick Question 23.4.1   Trace evaluation
Reproduce eqs. (23.40) and (23.41).

These results correct the masses of the charged bosons to

$$M_{\pi^\pm}^2 = \frac{(m_u + m_d)\Sigma}{F_\pi^2} + \frac{2Ce^2}{F_\pi^2}, \quad M_{K^\pm}^2 = \frac{(m_u + m_s)\Sigma}{F_\pi^2} + \frac{2Ce^2}{F_\pi^2}. \quad (23.42)$$

The fact that the mass squares of the charged pions and kaons receive the same electromagnetic correction is known as *Dashen's theorem* (Dashen, 1969). It should be noted that photon loops also contribute to the order $\mathcal{O}(e^2)$. However, they are sub-leading in the systematic derivative expansion of the low-energy effective theory. As a consequence, in the effective theory the leading-order effects arise already at tree-level, whereas they are entirely due to loop diagrams in the underlying microscopic QCD theory coupled to QED.

In the presence of leading-order electromagnetic corrections, the Gell-Mann–Okubo-type relation can be maintained in the form

$$M_{\pi^+}^2 + M_{\pi^-}^2 - M_{\pi^0}^2 + 3M_\eta^2 = 2(m_u + m_d + 2m_s)\frac{\Sigma}{F_\pi^2} + \frac{4Ce^2}{F_\pi^2} = M_{K^+}^2 + M_{K^-}^2 + M_{K^0}^2 + M_{\overline{K^0}}^2. \quad (23.43)$$

Like Σ and F_π, the value of the low-energy parameter C is not predicted by the effective theory, but can be extracted from experiment or from lattice calculations. The experimental values of the pseudo-Nambu–Goldstone boson masses are given by

$$M_{\pi^\pm} = 139.6\,\mathrm{MeV}, \quad M_{\pi^0} = 135.0\,\mathrm{MeV}, \quad M_\eta = 547.9\,\mathrm{MeV},$$
$$M_{K^\pm} = 493.7\,\mathrm{MeV}, \quad M_{K^0} = M_{\overline{K^0}} = 497.6\,\mathrm{MeV}. \quad (23.44)$$

Inserting these values, the left- and right-hand sides of eq. (23.43) result in $0.92\,\mathrm{GeV}^2$ and $0.98\,\mathrm{GeV}^2$, respectively. Their difference of 6 percent is remarkably small for a (leading-order) tree-level calculation.

The mass of the charged pions is larger than that of the neutral pion. At leading order, the difference is entirely due to electromagnetic effects. The charged kaons, on the other hand, are lighter than the neutral ones, because the isospin breaking effect due to $m_u \neq m_d$ counteracts the electromagnetic corrections.

This is similar for proton and neutron. The electromagnetic contribution to the proton mass is overwhelmed by the difference of the u- and d-quark masses, such that the neutron is heavier than the proton. Otherwise, the proton and not the neutron would undergo radioactive decay. A careful analysis of the QCD and QED contributions to the neutron–proton mass difference (Gasser *et al.*, 2021) shows that

$$\Delta M = M_n - M_p = 939.56\,\mathrm{MeV} - 938.27\,\mathrm{MeV} = 1.29\,\mathrm{MeV} = \Delta M_{\mathrm{QCD}} + \Delta M_{\mathrm{QED}},$$

$$\Delta M_{\mathrm{QCD}} = 1.87(16)\,\mathrm{MeV}, \quad \Delta M_{\mathrm{QED}} = -0.58(16)\,\mathrm{MeV}. \tag{23.45}$$

Since the quark masses m_u, m_d, and m_s always arise in products with Σ/F_π^2, using the experimental values of the meson masses, one can only determine quark mass *ratios*. In particular, one obtains the following relations

$$
\begin{aligned}
&\frac{2m_s}{m_u + m_d} = \frac{M_{K^0}^2 + M_{K^\pm}^2 - M_{\pi^\pm}^2}{M_{\pi^0}^2} = 25.9, \quad
\frac{m_s - m_d}{m_u + m_d} = \frac{M_{K^\pm}^2 - M_{\pi^\pm}^2}{M_{\pi^0}^2} = 12.3, \\[2mm]
&\frac{m_s - m_u}{m_u + m_d} = \frac{M_{K^0}^2 - M_{\pi^0}^2}{M_{\pi^0}^2} = 12.6, \quad
\frac{m_d - m_u}{m_u + m_d} = \frac{(M_{\pi^\pm}^2 - M_{\pi^0}^2) - (M_{K^\pm}^2 - M_{K^0}^2)}{M_{\pi^0}^2} = 0.28, \\[2mm]
&\frac{m_u}{m_d} = \frac{2M_{\pi^0}^2 - M_{\pi^\pm}^2 + M_{K^\pm}^2 - M_{K^0}^2}{M_{\pi^\pm}^2 - M_{K^\pm}^2 + M_{K^0}^2} = 0.56, \quad
\frac{m_s}{m_d} = \frac{M_{K^\pm}^2 - M_{\pi^\pm}^2 + M_{K^0}^2}{M_{\pi^\pm}^2 - M_{K^\pm}^2 + M_{K^0}^2} = 19.2.
\end{aligned}
\tag{23.46}
$$

A determination of the quark masses themselves requires lattice calculations in the underlying QCD theory compared with experiment. Like the strong coupling g_s, the quark masses are "running" parameters; *i.e.* they depend on the renormalization scheme and scale. In the $\overline{\mathrm{MS}}$ renormalization scheme at a scale of 2 GeV, the light quark masses are obtained as (Workman *et al.*, 2022)

$$m_u = 2.16^{+0.49}_{-0.26}\,\mathrm{MeV}, \quad m_d = 4.67^{+0.48}_{-0.17}\,\mathrm{MeV}, \quad m_s = 93.4^{+8.6}_{-3.4}\,\mathrm{MeV}. \tag{23.47}$$

This yields

$$\frac{m_u}{m_d} = 0.474^{+0.056}_{-0.074}, \quad \frac{m_s}{m_d} = 19.5^{+2.5}_{-2.5}, \quad \frac{2m_s}{m_u + m_d} = 27.33^{+0.67}_{-0.77}, \tag{23.48}$$

in remarkable agreement with the leading-order results of chiral perturbation theory.

23.5 Effective Theory for Nucleons and Pions

Chiral perturbation theory can be extended to superselection sectors with non-zero baryon number B (Gasser *et al.*, 1988; Jenkins and Manohar, 1991; Bernard *et al.*, 1992; Becher and Leutwyler, 1999). Although states in the $B = 1$ sector necessarily have energies larger than (or equal to) the nucleon mass, they are still accessible to a systematic low-energy expansion. This is a consequence of baryon number conservation, which implies that protons are stable and thus cannot release their rest energy. As a result, one can perform a systematic low-energy expansion in the small excitation energies above the large nucleon mass.

For $N_f = 2$, nucleons enter the low-energy effective theory in the form of Dirac spinor fields $\psi(x)$ and $\bar{\psi}(x)$ that transform as an $SU(2)_I$ isospin doublet. The upper and lower isospin components of $\psi(x)$ represent the proton and neutron field. Global chiral rotations $L, R \in SU(2)_L \times SU(2)_R$ are *non-linearly* realized on this field by means of the transformations

$$\psi'(x) = V(x)\psi(x), \quad \bar{\psi}'(x) = \bar{\psi}(x)V(x)^{\dagger}. \tag{23.49}$$

The field $V(x)$ depends on L and R as well as on the field $U(x)$ and is expressed as

$$V(x) = R[R^{\dagger}LU(x)]^{1/2}[U(x)^{1/2}]^{\dagger} = L[L^{\dagger}RU(x)^{\dagger}]^{1/2}U(x)^{1/2}. \tag{23.50}$$

Here, the square root $U(x)^{1/2}$ resides in the middle of the shortest geodesic that connects $U(x)$ with the unit-matrix $\mathbb{1}$ in the $SU(2)$ group manifold (which is the sphere S^3). For transformations in the unbroken isospin vector subgroup $SU(2)_I = SU(2)_{L=R}$, the field $V(x)$ reduces to the global flavor transformation $V(x) = L = R$. The local transformation $V(x)$ provides a non-linear representation of chiral symmetry, which takes the form of a local $SU(2)_I$ "gauge" transformation, despite the fact that it represents just the global $SU(2)_L \times SU(2)_R$ symmetry.

In order to construct a chirally invariant (*i.e.* $SU(2)_I$ "gauge"-invariant) Lagrangian, one needs an $SU(2)_I$ isospin "gauge" field. For this purpose, one constructs a field $u(x) \in SU(2)$ from the pion field $U(x)$ as

$$u(x) = U(x)^{1/2}. \tag{23.51}$$

Under chiral rotations, the field $u(x)$ transforms as

$$u(x)' = Lu(x)V(x)^{\dagger} = V(x)u(x)R^{\dagger}. \tag{23.52}$$

The anti-Hermitian composite field

$$v_{\mu}(x) = \frac{1}{2}\left[u(x)^{\dagger}\partial_{\mu}u(x) + u(x)\partial_{\mu}u(x)^{\dagger}\right] \tag{23.53}$$

indeed transforms as an isospin "gauge" field

$$\begin{aligned} v'_{\mu}(x) &= \frac{1}{2}\left[V(x)u(x)^{\dagger}L^{\dagger}\partial_{\mu}\left(Lu(x)V(x)^{\dagger}\right) + V(x)u(x)R^{\dagger}\partial_{\mu}\left(Ru(x)^{\dagger}V(x)^{\dagger}\right)\right] \\ &= V(x)\left(v_{\mu}(x) + \partial_{\mu}\right)V(x)^{\dagger}. \end{aligned} \tag{23.54}$$

In addition, a Hermitian composite field is constructed as

$$a_{\mu}(x) = \frac{i}{2}\left[u(x)^{\dagger}\partial_{\mu}u(x) - u(x)\partial_{\mu}u(x)^{\dagger}\right]. \tag{23.55}$$

It transforms as a "charged" vector field

$$\begin{aligned} a'_{\mu}(x) &= \frac{1}{2}\left[V(x)u(x)^{\dagger}L^{\dagger}\partial_{\mu}\left(Lu(x)V(x)^{\dagger}\right) - V(x)u(x)R^{\dagger}\partial_{\mu}\left(Ru(x)^{\dagger}V(x)^{\dagger}\right)\right] \\ &= V(x)a_{\mu}(x)V(x)^{\dagger}. \end{aligned} \tag{23.56}$$

The leading terms in the Lagrangian of a low-energy effective theory for nucleons and pions take the form

$$\mathcal{L}(U, \bar{\psi}, \psi) = M\bar{\psi}\psi + \bar{\psi}\gamma_\mu(\partial_\mu + v_\mu)\psi + ig_A\bar{\psi}\gamma_\mu\gamma_5 a_\mu\psi$$
$$+ \frac{F_\pi^2}{4}\text{Tr}\left[\partial_\mu U^\dagger \partial_\mu U\right] - \frac{\Sigma}{2}\text{Tr}\left[\mathcal{M}\left(U^\dagger + U^\dagger\right)\right]. \tag{23.57}$$

Here, M is the nucleon mass in the chiral limit (which results entirely from spontaneous chiral symmetry breaking) and g_A is the coupling to the isovector axial current. It is remarkable that – thanks to the non-linear realization of chiral symmetry – the fermion mass term is chirally invariant. This makes sense, because the mass M arises even in the chiral limit from the spontaneous breakdown of chiral symmetry. A simple consequence of the effective theory is the *Goldberger–Treiman relation*

$$g_{\pi NN}F_\pi = g_A M, \tag{23.58}$$

which relates the pion–nucleon coupling constant $g_{\pi NN}$ to the axial vector coupling g_A of the nucleon that determines the weak decay rate of the neutron (Goldberger and Treiman, 1958).

It is remarkable that fermions with a non-linearly realized chiral symmetry do not contribute to anomalies (Manohar and Georgi, 1984; Manohar and Moore, 1984). As we will see in Chapter 24, in the low-energy effective theory anomalies enter through Wess–Zumino–Novikov–Witten (Wess and Zumino, 1971; Novikov, 1981, 1982; Witten, 1983a,b) and Goldstone–Wilczek terms (Goldstone and Wilczek, 1981).

23.6   QCD Contributions to the *W*- and *Z*-Boson Masses

As we have seen in Chapter 13, the *W*- and *Z*-bosons receive their masses from the Higgs mechanism. As a result, the masses of the electroweak gauge bosons are proportional to the vacuum value v of the Higgs field. While this is by far the dominant contribution to the *W*- and *Z*-boson masses, as we will see now, they also receive a small additional contribution from chiral symmetry breaking in QCD. This observation will serve as a motivation to continue our discussion of technicolor models in the next section. In those models, electroweak symmetry breaking results from chiral symmetry breaking induced by a new strong gauge force between techni-quarks and techni-gluons.

For the moment, we restrict ourselves to Standard Model physics. Let us consider the low-energy effective theory with two massless quark flavors. We now gauge $SU(2)_L$ by replacing ordinary derivatives $\partial_\mu U(x)$ by covariant derivatives,

$$D_\mu U(x) = \left(\partial_\mu + W_\mu(x)\right) U(x), \tag{23.59}$$

where $W_\mu(x) = igW_\mu^a(x)\tau^a/2$ is the gauge field that describes the *W*-bosons. Under $SU(2)_L$ gauge transformations, the fields transform as

$$U'(x) = L(x)U(x), \quad W'_\mu(x) = L(x)\left(W_\mu(x) + \partial_\mu\right)L(x)^\dagger. \tag{23.60}$$

In the chiral limit of massless quarks, the gauge-invariant Lagrangian, which also includes the self-interaction of the $SU(2)_L$ gauge field, then takes the form

$$\mathcal{L}(U, W) = \frac{F_\pi^2}{4} \text{Tr} \left[D_\mu U^\dagger D_\mu U \right] - \frac{1}{2g^2} \text{Tr} \left[W_{\mu\nu} W_{\mu\nu} \right]. \tag{23.61}$$

In Section 13.2, we have calculated the W- and Z-boson masses resulting from the Higgs mechanism. In that case, we have chosen the unitary gauge for the Higgs field. In the same way, we now choose the *unitary gauge for the pion field*; *i.e.* we arrange $L(x)$ such that $U(x) = \mathbb{1}$. Then, the action turns into

$$\mathcal{L}(\mathbb{1}, W) = \frac{F_\pi^2}{4} \text{Tr} \left[W_\mu^\dagger W_\mu \right] - \frac{1}{2g^2} \text{Tr} \left[W_{\mu\nu} W_{\mu\nu} \right]. \tag{23.62}$$

From this expression, we read off the W-mass

$$M_W = \frac{1}{2} g F_\pi, \tag{23.63}$$

which, in this case, results just from chiral symmetry breaking in QCD. Of course, numerically $F_\pi = 0.092$ GeV is negligible compared to the electroweak scale v $= 246$ GeV, which provides by far the dominant contribution to the physical W-mass, gv$/2$. Still, it is interesting that QCD alone "breaks" the electroweak symmetry dynamically. In technicolor models, QCD is replaced by another confining gauge theory that operates at much higher energy scales and has a value of F_π that corresponds to the electroweak scale v of the Standard Model.

Let us also gauge $U(1)_Y$ as a subgroup of $SU(2)_R$ in order to see what happens to the Z-boson. Then, the covariant derivative again takes the form

$$D_\mu U(x) = \partial_\mu U(x) + W_\mu(x) U(x) - U(x) X_\mu(x), \tag{23.64}$$

where $X_\mu(x) = i g' B_\mu(x) \tau^3 / 2$, and the gauge-invariant Lagrangian now reads

$$\mathcal{L}(U, W, X) = \frac{F_\pi^2}{4} \text{Tr}[D_\mu U^\dagger D_\mu U] - \frac{1}{2g^2} \text{Tr} \left[W_{\mu\nu} W_{\mu\nu} \right] - \frac{1}{2g'^2} \text{Tr} \left[X_{\mu\nu} X_{\mu\nu} \right]. \tag{23.65}$$

In the unitary gauge $U(x) = \mathbb{1}$, this expression takes the form

$$\mathcal{L}(\mathbb{1}, W, B) = \frac{F_\pi^2}{4} \text{Tr} \left[(W_\mu - X_\mu)(W_\mu - X_\mu) \right] - \frac{1}{2g^2} \text{Tr} \left[W_{\mu\nu} W_{\mu\nu} \right] - \frac{1}{2g'^2} \text{Tr} \left[X_{\mu\nu} X_{\mu\nu} \right]. \tag{23.66}$$

Using eq. (23.35), one obtains that the Z-boson picks up a mass

$$M_Z = \frac{1}{2} \sqrt{g^2 + g'^2} F_\pi, \tag{23.67}$$

while the photon remains massless. Again, this contribution to the Z-boson mass, which is due to chiral symmetry breaking in QCD, is negligible compared to the mass proportional to v that is generated via the Higgs mechanism, $\sqrt{g^2 + g'^2} \, \text{v}/2$.

Quick Question 23.6.1   QCD contribution to the Z-boson mass
Convince yourself that one indeed arrives at eq. (23.67).

It is interesting that for both M_W and M_Z the leading contribution of Chapter 13 is corrected by substituting $v \rightarrow v + F_\pi$. Therefore, the relation

$$\frac{M_W}{M_Z} = \frac{g}{\sqrt{g^2 + g'^2}} = \cos\theta_W \qquad (23.68)$$

remains unchanged. This implies that, in complete analogy to QCD, a technicolor model would indeed be capable of replacing the Higgs mechanism to give mass to the W- and Z-bosons.

23.7 Remarks about Technicolor

The last two sections of this chapter address physics beyond the Standard Model and can be skipped at a first reading.

As we already mentioned in Section 15.12, technicolor models mimic QCD at a much higher energy scale, such that the analogue of F_π for the technicolor interaction takes its value at the electroweak scale v. The main motivation behind the technicolor idea is to solve the hierarchy problem (cf. Section 12.7). While in the Standard Model the stabilization of the electroweak scale v against radiative corrections requires unnatural fine-tuning, the asymptotic freedom of a technicolor gauge theory would protect the analogue of F_π from similar effects.

For concreteness, let us consider the anomaly-free $SU(N_t) \times SU(N_c) \times SU(2)_L \times U(1)_Y$ technicolor extension of the Standard Model (*e.g.*, with $N_t = 4$) that we discussed in Section 15.12. This model has two flavors, U and D, of exactly massless techni-quarks, whose left-handed components U_L and D_L form an $SU(2)_L$ doublet, while the right-handed components U_R and D_R are $SU(2)_L$ singlets. The strong $SU(N_t)$ technicolor gauge interaction then breaks the $SU(2)_L \times SU(2)_R$ chiral symmetry of the techni-U- and techni-D-quarks spontaneously. The corresponding Nambu–Goldstone bosons are three massless techni-pions. When $SU(2)_L$ is gauged, these are "eaten" and become the longitudinal polarization states of the massive W- and Z-bosons. This is exactly what we just discussed in the previous section, except that the physics happens at the technicolor scale and not at the QCD scale.

If the Higgs sector of the Standard Model is replaced by a strongly interacting technicolor gauge interaction, one does not only obtain techni-pions, but – just as in QCD – also massive techni-mesons, including a techni-ρ-meson, as well as techni-baryons, with masses in the TeV range. At the time of writing of this book (in 2022), the Large Hadron Collider (LHC) at CERN has not provided experimental evidence for the existence of such states. The Higgs boson itself was discovered at the LHC in 2012 at a mass of 125 GeV. In technicolor models, the Higgs particle arises as the techni-σ-meson. In view of the electroweak scale v = 246 GeV, it is a challenge for technicolor models to provide a state as light as the observed Higgs particle. To perhaps achieve this, models near a *conformal window*, with almost massless excitations and with a slowly running (*i.e.* "walking") coupling, have been investigated (Holdom, 1981; Banks and Zaks, 1982).

While technicolor naturally provides masses for the W- and Z-bosons without fine-tuning, it works much less well when it comes to giving masses to the fermions. Since technicolor models get rid of the elementary Higgs field, they can no longer generate fermion masses from Yukawa couplings. Since the physical Higgs particle would correspond to a techni-σ-meson (*i.e.* techni-quark–techni-anti-quark bound state), the Yukawa couplings are now replaced by four-fermion couplings between two techni-quarks and two Standard Model fermions. It is very difficult to achieve realistic fermion masses and mixing parameters in this way. In particular, technicolor models have a severe problem with *flavor-changing neutral currents*, which are absent in the Standard Model, at least at tree-level. Experimental bounds on flavor-changing neutral currents have ruled out most technicolor models. Therefore, these models no longer seem promising. Still, they are quite interesting theoretically, and one can learn a lot, even from an unsuccessful attempt to find a natural explanation of electroweak symmetry breaking.

23.8 Hypothesis of Minimal Flavor Violation

In the absence of Yukawa couplings, all elementary fermions are massless (while confinement still gives masses of order Λ_{QCD} to the hadrons). The Standard Model Lagrangian then has a large global symmetry $U(N_g)^5 = U(3)^5$, which rotates the members of the $N_g = 3$ generations that share the same quantum numbers,

$$Q_L(x) = \left(\begin{array}{c} U_L(x) \\ D_L(x) \end{array} \right), \quad U_L(x) = \left(\begin{array}{c} u_L(x) \\ c_L(x) \\ t_L(x) \end{array} \right), \quad D_L(x) = \left(\begin{array}{c} d_L(x) \\ s_L(x) \\ b_L(x) \end{array} \right),$$

$$U_R(x) = \left(\begin{array}{c} u_R(x) \\ c_R(x) \\ t_R(x) \end{array} \right), \quad D_R(x) = \left(\begin{array}{c} d_R(x) \\ s_R(x) \\ b_R(x) \end{array} \right),$$

$$L_L(x) = \left(\begin{array}{c} N_L(x) \\ E_L(x) \end{array} \right), \quad N_L(x) = \left(\begin{array}{c} \nu_{eL}(x) \\ \nu_{\mu L}(x) \\ \nu_{\tau_L}(x) \end{array} \right), \quad E_L(x) = \left(\begin{array}{c} e_L(x) \\ \mu_L(x) \\ \tau_L(x) \end{array} \right),$$

$$E_R(x) = \left(\begin{array}{c} e_R(x) \\ \mu_R(x) \\ \tau_R(x) \end{array} \right), \tag{23.69}$$

into each other

$$Q_L'(x) = V^{Q_L} Q_L(x), \quad U_R'(x) = V^{U_R} U_R(x), \quad D_R'(x) = V^{D_R} D_R(x),$$
$$L_L'(x) = V^{L_L} L_L(x), \quad E_R'(x) = V^{E_R} E_R(x). \tag{23.70}$$

The Yukawa couplings of the Standard Model take the form

$$\mathcal{L}(\bar{L}_L', L_L', \bar{E}_R', E_R', \bar{Q}_L', Q_L', \bar{U}_R', U_R', \bar{D}_R', D_R', \Phi) = \bar{L}_L' \Phi \mathcal{F}_E E_R' + \bar{E}_R' \mathcal{F}_E^\dagger \Phi^\dagger L_L'$$
$$+ \bar{Q}_L' \Phi \mathcal{F}_D D_R' + \bar{D}_R' \mathcal{F}_D^\dagger \Phi^\dagger Q_L' + \bar{Q}_L' \tilde{\Phi} \mathcal{F}_U U_R' + \bar{U}_R' \mathcal{F}_U^\dagger \tilde{\Phi}^\dagger Q_L'. \tag{23.71}$$

Here, the primed fermion fields represent the electroweak eigenstates and $\mathcal{F}_E$, $\mathcal{F}_D$, and $\mathcal{F}_U$ are general complex $N_g \times N_g$ matrices that parameterize the Yukawa couplings of leptons and quarks to the Higgs field. They explicitly break the global $U(N_g)^5$ symmetry that arises in the massless case. Just like the quark mass matrix $\mathcal{M}$ in QCD, the Yukawa matrices $\mathcal{F}_E$, $\mathcal{F}_D$, and $\mathcal{F}_U$ can be extended to spurions that transform as

$$\mathcal{F}_E' = V^{L_L} \mathcal{F}_E V^{E_R \, \dagger}, \quad \mathcal{F}_D' = V^{Q_L} \mathcal{F}_D V^{D_R \, \dagger}, \quad \mathcal{F}_U' = V^{Q_L} \mathcal{F}_U V^{U_R \, \dagger}. \tag{23.72}$$

Under these transformations, even the Yukawa couplings become invariant under the global $U(3)^5$ symmetry.

A model-independent way of characterizing new physics beyond the renormalizable Standard Model is to systematically investigate non-renormalizable higher-dimensional terms which arise when the Standard Model is considered as a low-energy effective theory for the new physics. In Section 16.3, we have investigated the dimension-5 operators that give rise to non-zero neutrino masses. We have also discussed some dimension-6 operators that give rise to baryon number violating processes. The complete list of dimension-6 operators contains more than 3500 items (Buchmüller and Wyler, 1986; Grzadkowski *et al.*, 2010; Jenkins *et al.*, 2018), which reflects the vast variety of possible extensions of the Standard Model at higher energies.

As we have seen in Section 17.4, in the Standard Model flavor-changing neutral currents are absent in the Lagrangian (*i.e.* at tree-level) and are even suppressed by the GIM mechanism at the 1-loop level. Potential extensions of the Standard Model are strongly constrained by the observed large suppression of flavor-changing neutral currents, which arises naturally (*i.e.* without fine-tuning) in the Standard Model. This suggests that flavor-changing new physics beyond the Standard Model should arise only at a scale around 1000 TeV. On the other hand, in order to resolve the hierarchy problem in a natural way, one would expect new physics already around the TeV scale.

This conundrum would be resolved in models that satisfy the assumption of *"minimal flavor violation"* (D'Ambrosio *et al.*, 2002). By definition, models that obey the hypothesis of minimal flavor violation break the $U(N_g)^5 = U(3)^5$ generation symmetry in the same way as the Standard Model. This means that the breaking is characterized by the spurion fields of eq. (23.72). As a result, such models are protected from new flavor-breaking effects beyond the Standard Model that would already arise at the TeV scale. The hypothesis of minimal flavor violation also leads to relations between flavor-changing neutral current processes in the B-meson and kaon systems. While there is no guarantee that the correct high-energy extension of the Standard Model obeys this hypothesis, the corresponding models have attractive features. Even if minimal flavor violation would not be realized in the physics beyond the Standard Model, the hypothesis provides a useful classification scheme of flavor violation patterns.

Exercises

23.1 Additional leading term in the $N_f = 2$ chiral Lagrangian?

Investigate the term $\det[U^\dagger \partial_\mu U]$ as an additional candidate for the leading-order chiral Lagrangian for $N_f = 2$.

23.2 Sub-leading terms in the chiral Lagrangian

Construct all independent, chirally- and Lorentz-invariant terms in the chiral Lagrangian that contain exactly four derivatives. Do not include mass terms. Ignore terms that contain the Levi-Civita symbol $\epsilon_{\mu\nu\rho\sigma}$. First, consider the cases $N_f = 2$ and 3. Are there additional terms for $N_f = 4$?

23.3 Masses of pseudo-Nambu–Goldstone mesons

Use an $SU(3)_L \times SU(3)_R$ invariant chiral Lagrangian with an additional explicit symmetry breaking mass matrix $\mathcal{M} = \text{diag}(m_u, m_d, m_s)$ to derive mass formulae for the pseudo-Nambu–Goldstone mesons π, K, and η. Ignore mass effects due to electromagnetic interactions. Note that the various charge states may still have different masses because $m_u \neq m_d$.

23.4 Transformation behavior of $u(x)$, $v_\mu(x)$, and $a_\mu(x)$

Verify the transformation behavior of the fields $u(x)$, $v_\mu(x)$, and $a_\mu(x)$ given in eqs. (23.52), (23.54), and (23.56).

As we have seen in Chapter 23, the low-energy physics of QCD is dominated by light pseudo-Nambu–Goldstone bosons – the pions, kaons, and the η-meson. Their dynamics at low energies are described by an effective chiral Lagrangian. In the theory with N_f light-quark flavors, the pseudo-Nambu–Goldstone boson fields take values in the coset space $SU(N_f)_L \times SU(N_f)_R / SU(N_f)_{L=R} = SU(N_f)$. For $N_f \geq 2$, there is the non-trivial homotopy group $\Pi_3[SU(N_f)] = \mathbb{Z}$ (cf. Appendix G), which gives rise to a conserved topological winding number that can be identified with the *baryon number*. The topological excitations that carry this winding number were first studied by Tony Skyrme (1961, 1962) and are known as *skyrmions*. It is quite remarkable that meson fields, which do not carry any baryon number themselves, can nevertheless represent baryons in this manner.

Depending on the number of colors N_c, the baryons of QCD are either fermions (for odd N_c) or bosons (for even N_c). At the level of the effective theory, the statistics of the skyrmions is again determined by a topological term. In a theory with $N_f = 2$ flavors of just u- and d-quarks, the non-trivial homotopy group $\Pi_4[SU(2)] = \mathbb{Z}(2)$ provides a topological number, $\text{Sign}[U] = \pm 1$, which enters the functional integral as a factor $\text{Sign}[U]^{N_c}$. When N_c is odd and the skyrmions need to be quantized as fermions, this factor implements the Pauli principle, by correctly incorporating the fermion permutation sign for skyrmions in the functional integral.

In a theory with $N_f \geq 3$ light flavors (which also includes the s-quark), the homotopy group is trivial, $\Pi_4[SU(N_f)] = \{0\}$. Remarkably, this enables the construction of yet another topological term, the so-called *Wess–Zumino–Novikov–Witten (WZNW) term* (Wess and Zumino, 1971; Novikov, 1981, 1982; Witten, 1983a,b, 1984a), which contains four physical space–time derivatives and is thus of higher order in the low-energy expansion. For this term to be well-defined, due to the non-trivial homotopy group $\Pi_5[SU(N_f)] = \mathbb{Z}$ for $N_f \geq 3$, it must have an integer-valued quantized prefactor. Interestingly, this factor turns out to be the number of colors N_c. For $N_f \geq 3$, among other properties, the WZNW term determines the spin and statistics of skyrmions. In particular, for a Nambu–Goldstone boson field that takes its values in an $SU(2)$ subgroup of $SU(N_f)$, the WZNW term reduces to $\text{Sign}[U]^{N_c}$.

The leading terms of the chiral Lagrangian, which have up to two space–time derivatives, have more symmetries than the underlying QCD Lagrangian itself. In particular, they are restricted to processes that conserve the number of Nambu–Goldstone bosons modulo 2, while in QCD there are processes that can change this number from even to odd, at least for $N_f \geq 3$. These effects are correctly represented by the WZNW term. In the $N_f = 2$ case, on the other hand, there is no WZNW term, and the number of pions is indeed conserved modulo 2. The corresponding symmetry is known as *G-parity* (Lee and Yang, 1956a).

Additional interesting effects arise when the electroweak $SU(2)_L \times U(1)_Y$ gauge interactions are included in the low-energy effective description. As we saw in Section 15.7, in the Standard Model, gauge anomalies are canceled between quark and lepton fields. The gauged chiral Lagrangian, which does not include the lepton fields, should hence have un-canceled gauge anomalies. Indeed, these are also accounted for by the WZNW term, thus realizing 't Hooft's anomaly matching mechanism.

In the presence of electromagnetic interactions, G-parity is explicitly broken, because the neutral pion can then decay into two photons (thus changing the number of pions from 1 to 0). In the underlying Standard Model, this is a subtle quantum effect due to the *Adler–Bell–Jackiw anomaly* (Adler, 1969; Bell and Jackiw, 1969). The anomaly is not present in the Standard Model Lagrangian, but manifests itself in the fermionic measure of the functional integral. Generally, when we include electroweak interactions in the effective low-energy theory, anomalies are represented by topological terms in the chiral Lagrangian and no longer in the (then entirely bosonic) measure of the functional integral. For example, for $N_f = 3$ and $N_c = 3$, the decay $\pi^0 \to \gamma\gamma$ is described entirely by the WZNW term. For $N_f = 2$, however, the π^0 decay is accounted for by yet another topological term. This term contains a *Goldstone–Wilczek current* (Goldstone and Wilczek, 1981), which is the gauge-covariant extension of the topologically conserved Skyrme current.

While the skyrmions themselves are not accessible to a systematic low-energy expansion, their baryon number current still affects the electromagnetic interactions of pions. The reason is that, according to eqs. (15.64), (15.65), and (15.66), the electric charges of the quarks receive a contribution that is proportional to their baryon number $1/N_c$. In the gauge-anomaly-free Standard Model with general odd N_c, for $N_f = 3$ the chiral Lagrangian contains both the WZNW term (with prefactor N_c) and the Goldstone–Wilczek term (with prefactor $(1 - N_c/3)$). As a result, in contradiction to most textbook literature, this implies that the decay width of $\pi^0 \to \gamma\gamma$ is N_c-independent.

24.1 Skyrmions

As we have stated in Chapter 23, the Nambu–Goldstone bosons of the strong interaction are described by a matrix-valued field

$$U(x) \in SU(N_f)_L \times SU(N_f)_R / SU(N_f)_{L=R} = SU(N_f). \tag{24.1}$$

At each instant in time, the Nambu–Goldstone boson field can be viewed as a map of 3-d space into the group $SU(N_f)$. The non-trivial homotopy group $\Pi_3[SU(N_f)] = \mathbb{Z}$ implies that such maps fall in topologically distinct equivalence classes characterized by an integer winding number

$$B = \frac{1}{24\pi^2} \int d^3x \, \epsilon_{ijk} \text{Tr}\left[\left(U^\dagger \partial_i U\right)\left(U^\dagger \partial_j U\right)\left(U^\dagger \partial_k U\right)\right] \in \Pi_3\left[SU(N_f)\right] = \mathbb{Z}. \tag{24.2}$$

Continuous maps with the same winding number can be deformed into each other and are thus topologically equivalent. Remarkably, the winding number can be identified with the *baryon number*; the topologically non-trivial Nambu–Goldstone boson configurations hence describe baryons. It is far from obvious that B is indeed the baryon number, but we will confirm this later.

While B is defined at each instant of time, it actually does not change with time – in the absence of the weak interaction, baryon number is conserved. At the level of the effective theory, this is a consequence of topological current conservation. Skyrme's baryon number current

$$j_\mu^S(x) = \frac{1}{24\pi^2}\epsilon_{\mu\nu\rho\sigma} \text{Tr}\left[\left(U(x)^\dagger \partial_\nu U(x)\right)\left(U(x)^\dagger \partial_\rho U(x)\right)\left(U(x)^\dagger \partial_\sigma U(x)\right)\right] \qquad (24.3)$$

is conserved, *i.e.*

$$\partial_\mu j_\mu^S(x) = 0, \qquad (24.4)$$

independently of the equations of motion. Consequently, the baryon number

$$B = \int d^3x\, j_4^S \qquad (24.5)$$

does not change with time (in this case, with Euclidean time), *i.e.*

$$\partial_4 B = \int d^3x\, \partial_4 j_4^S = -\int d^3x\, \partial_i j_i^S = -\int d^2\sigma_i\, j_i^S = 0, \qquad (24.6)$$

provided the baryon number current vanishes at spatial infinity.

Depending on the higher-order terms in the effective action, the corresponding classical equations may or may not give rise to stable solutions with $B = 1$. In order to stabilize baryonic solutions, Skyrme added a specific higher-order term with four derivatives to the leading-order Lagrangian (Skyrme, 1961, 1962). The emerging topological soliton is known as a *skyrmion*. For $N_f = 2$, it takes the form

$$U(\vec{x}) = \exp\left(\mathrm{i}f(|\vec{x}|)\frac{\vec{x}}{|\vec{x}|}\cdot\vec{\tau}\right), \qquad (24.7)$$

where $\vec{\tau}$ are the isospin Pauli matrices and the function $f(|\vec{x}|)$ describes the radial profile of the soliton. The properties $f(\infty) = \pi$ and $f(0) = 0$ ensure that the soliton has $B = 1$.

The Skyrme model has been used to describe baryons phenomenologically (Adkins *et al.*, 1983). This is certainly not as rigorous as QCD, but it works reasonably well. In fact, Witten has argued that, in the large-N_c limit, baryons turn into heavy topological solitons (Witten, 1979a). As mentioned before, skyrmions are not accessible to a systematic low-energy expansion. As we have seen in Chapter 23, a consistent low-energy effective theory in the baryon number $B = 1$ sector is possible using explicit baryon fields. Here, we will not address any detailed properties of skyrmions, because those are necessarily model-dependent. Instead, we concentrate on the model-independent topological aspects of their dynamics.

First of all, let us consider the case $N_f = 2$. Due to the non-trivial homotopy group $\Pi_4[SU(2)] = \mathbb{Z}(2)$, the space–time-dependent Nambu–Goldstone field "histories" then fall into two distinct classes. We can hence assign a winding number

$$\text{Sign}[U] = \pm 1 \qquad (24.8)$$

to each field configuration. Configurations that can be deformed continuously into the trivial configuration $U(x) = \mathbb{1}$ have $\text{Sign}[U] = 1$, while the others have $\text{Sign}[U] = -1$.

For subtle reasons, the $\Pi_4[SU(2)]$ winding number appears in the functional integral of the low-energy effective theory as

$$Z = \int \mathcal{D}U \, \exp(-S[U]) \, \text{Sign}[U]^{N_c}. \tag{24.9}$$

This was rigorously analyzed by Eric D'Hoker and Edward Farhi (D'Hoker and Farhi, 1984a). When the number of colors N_c is even, the baryons are bosons and the winding number drops out of the functional integral. When N_c is odd, on the other hand, the baryons are fermions and $\text{Sign}[U]$ enters the functional integral. Indeed, $\text{Sign}[U]$ emerges as a fermion permutation sign that encodes the Pauli principle for skyrmions in the pion effective theory. One can show this using various space–time-dependent field configurations U.

For example, for fermionic baryons, a configuration with $B = 2$, in which two skyrmions interchange their positions as time evolves, has $\text{Sign}[U] = -1$ in accordance with the Pauli principle. In addition, a configuration with $B = 1$, in which a single skyrmion rotates by 2π as time evolves, has $\text{Sign}[U] = -1$ as well. This reveals that skyrmions have half-integer spin. In other words, the $\Pi_4[SU(2)] = \mathbb{Z}(2)$ winding number $\text{Sign}[U] = \pm 1$ ensures that skyrmions behave properly as fermions that obey the spin–statistics theorem.

It is interesting that, in four space–time dimensions, particles can *only* be quantized as either bosons or fermions, because $\Pi_4[SU(2)] = \Pi_4[S^3] = \mathbb{Z}(2)$. This is in contrast to theories in three space–time dimensions, where particles can exist with any spin or statistics – *anyons* – because $\Pi_3[S^2] = \mathbb{Z}$ (Wilczek, 1982).

24.2 Anomaly Matching for $N_f = 2$

The term $\text{Sign}[U]^{N_c}$ also ensures the proper cancellation of Witten's global anomaly at the level of the effective theory. To see this, let us gauge $SU(2)_L$ in the chiral Lagrangian. This is easy to do, because the global chiral transformation $U'(x) = LU(x)$ with $L \in SU(2)_L$ is just a symmetry that is gauged in the Standard Model, as we saw in Part II. To obtain a low-energy effective theory of the entire Standard Model, the pion Lagrangian should then be coupled to external Higgs fields, W-bosons, as well as three generations of leptons. The global anomaly of the left-handed $SU(2)_L$ lepton doublets is canceled by an odd number N_c of left-handed quark doublets, cf. Chapter 15. At the level of the low-energy effective theory, the fundamental quark fields of the Standard Model are replaced by pion fields, so the question arises how the global anomaly is canceled now.

When we perform an $SU(2)_L$ gauge transformation $L(x)$ with winding number $\text{Sign}[L] = -1$, the fermionic measure of the lepton fields is not invariant, but changes by a minus sign. In the Standard Model, this minus sign is canceled by N_c minus signs, spit out by the fermionic measure of the quarks, so that the full theory is gauge-invariant. In the effective theory, there are no quarks, but the functional integral now contains the factor $\text{Sign}[U]^{N_c}$, which transforms as

$$\text{Sign}[U']^{N_c} = \text{Sign}[LU]^{N_c} = \text{Sign}[L]^{N_c} \text{Sign}[U]^{N_c}. \tag{24.10}$$

This gauge variation is exactly what we need in order to cancel the global anomaly of the leptons.

In Section 15.11, we discussed that baryon number is anomalous in the Standard Model. As we saw in Chapter 19, topologically non-trivial configurations of the W-boson field – so-called *electroweak instantons* – violate the conservation of the baryon current $j^B_\mu(x)$ via the Adler–Bell–Jackiw anomaly

$$\partial_\mu j^B_\mu(x) = -\frac{1}{32\pi^2}\epsilon_{\mu\nu\rho\sigma}\mathrm{Tr}\left[W_{\mu\nu}(x)W_{\rho\sigma}(x)\right]. \tag{24.11}$$

Here, $W_{\mu\nu}(x) = \partial_\mu W_\nu(x) - \partial_\nu W_\mu(x) + [W_\mu(x), W_\nu(x)]$ is the usual field strength of the W-boson field.

If the Skyrme current indeed represents baryon number, after gauging the $SU(2)_L$ symmetry, we should obtain the same result in the effective theory. When we gauge $SU(2)_L$, we need to replace ordinary derivatives $\partial_\mu U(x)$ by covariant derivatives,

$$D_\mu U(x) = \left(\partial_\mu + W_\mu(x)\right)U(x), \quad (D_\mu U(x))^\dagger = \partial_\mu U(x)^\dagger - U(x)^\dagger W_\mu(x). \tag{24.12}$$

When we only replace ordinary derivatives by covariant derivatives in eq. (24.3), we do not obtain eq. (24.11). Instead one should construct the *Goldstone–Wilczek current* (Goldstone and Wilczek, 1981), which represents the baryon number current in the presence of $SU(2)_L$ gauge fields

$$j^{\mathrm{GW}}_\mu(x) = \frac{1}{24\pi^2}\epsilon_{\mu\nu\rho\sigma}\mathrm{Tr}\left[\left(U(x)^\dagger D_\nu U(x)\right)\left(U(x)^\dagger D_\rho U(x)\right)\left(U(x)^\dagger D_\sigma U(x)\right)\right]$$
$$- \frac{1}{16\pi^2}\epsilon_{\mu\nu\rho\sigma}\mathrm{Tr}\left[W_{\nu\rho}(x)\left(D_\sigma U(x)U(x)^\dagger\right)\right], \tag{24.13}$$

whose divergence is indeed given by

$$\partial_\mu j^{\mathrm{GW}}_\mu(x) = -\frac{1}{32\pi^2}\epsilon_{\mu\nu\rho\sigma}\mathrm{Tr}\left[W_{\mu\nu}(x)W_{\rho\sigma}(x)\right]. \tag{24.14}$$

Obviously, in the absence of $SU(2)_L$ gauge fields, the Goldstone–Wilczek current reduces to Skyrme's baryon number current of eq. (24.3). Since eq. (24.11) is now satisfied at the level of the effective theory, just like the Standard Model, it gives rise to baryon number violating processes. In particular, if we consider a skyrmion propagating in an electroweak instanton background (a W-field configuration with topological charge $Q = 1$, cf. Chapter 19), the baryon number changes by one unit

$$B(x_4 = \infty) - B(x_4 = -\infty) = \int d^3x\, j^{\mathrm{GW}}_4(x_4 = \infty) - \int d^3x\, j^{\mathrm{GW}}_4(x_4 = -\infty)$$
$$= \int d^4x\, \partial_\mu j^{\mathrm{GW}}_\mu = -\frac{1}{32\pi^2}\int d^4x\, \epsilon_{\mu\nu\rho\sigma}\mathrm{Tr}\left[W_{\mu\nu}W_{\rho\sigma}\right]$$
$$= Q[W] = 1. \tag{24.15}$$

In the same process, the lepton number L also changes by one unit, such that $B - L$ remains conserved.

Of course, in the Standard Model we also gauge the symmetry $U(1)_Y$. We saw in Chapter 15 that the electroweak hypercharge of the left-handed quarks is $Y_{q_L} = 1/(2N_c)$, while the hypercharges of the right-handed quarks coincide with their electric charges, $Y_{u_R} = (1/N_c + 1)/2$ and $Y_{d_R} = (1/N_c - 1)/2$. Hence, at the level of the effective theory, the corresponding transformations act both on the left and on the right, *i.e.* $U(x)' = \tilde{L}(x)U(x)$ $\tilde{R}(x)^\dagger$, with

$$
\tilde{L}(x) = \begin{pmatrix} \exp\left(\mathrm{i}g'Y_{q_L}\varphi(x)\right) & 0 \\ 0 & \exp\left(\mathrm{i}g'Y_{q_L}\varphi(x)\right) \end{pmatrix},
$$

$$
\tilde{R}(x) = \begin{pmatrix} \exp\left(\mathrm{i}g'Y_{u_R}\varphi(x)\right) & 0 \\ 0 & \exp\left(\mathrm{i}g'Y_{d_R}\varphi(x)\right) \end{pmatrix}. \tag{24.16}
$$

Note that $\tilde{L}(x)$ and $\tilde{R}(x)$, written in this form, are not elements of SU(2), but only of U(2). However, since $\tilde{L}(x)$ is proportional to the unit-matrix, it commutes with any matrix $U(x)$, so we can pull it over to the right-hand side and combine it with $\tilde{R}(x)$ to form a new SU(2) matrix that multiplies $U(x)$ only from the right. We denote the new SU(2)$_R$ matrix as $R(x)$,

$$
U'(x) = U(x)R(x)^\dagger, \quad R(x) = \begin{pmatrix} \exp(\mathrm{i}g'\varphi(x)/2) & 0 \\ 0 & \exp(-\mathrm{i}g'\varphi(x)/2) \end{pmatrix}. \tag{24.17}
$$

In this way, we have identified U(1)$_Y$ as an Abelian subgroup of SU(2)$_R$. The total covariant derivative takes the form

$$
D_\mu U(x) = \partial_\mu U(x) + W_\mu(x)U(x) - U(x)X_\mu(x)
$$

$$
= \partial_\mu U(x) + \mathrm{i}g W_\mu^a(x)\frac{\tau^a}{2}U(x) - \mathrm{i}g'U(x)B_\mu(x)\frac{\tau^3}{2}. \tag{24.18}
$$

Here, we have again introduced a hypothetical fully fledged SU(2)$_R$ gauge field $X_\mu(x) = \mathrm{i}g'X_\mu^a(x)\tau^a/2$. Then, we have identified $X_\mu^3(x) = B_\mu(x)$ as the gauge field associated with the Abelian subgroup U(1)$_Y \subset$ SU(2)$_R$, and we have put the non-Abelian components $X_\mu^1(x) = X_\mu^2(x) = 0$. The field strength $B_{\mu\nu} = \partial_\mu B_\nu - \partial_\nu B_\mu$ of the U(1)$_Y$ gauge field is embedded correspondingly, $X_{\mu\nu}(x) = \mathrm{i}g'B_{\mu\nu}(x)\tau^3/2$. With U(1)$_Y$ also being gauged, we extend the definition of the *Goldstone–Wilczek current* to (Goldstone and Wilczek, 1981)

$$
j_\mu^{\mathrm{GW}}(x) = \frac{1}{24\pi^2}\epsilon_{\mu\nu\rho\sigma}\mathrm{Tr}\left[\left(U(x)^\dagger D_\nu U(x)\right)\left(U(x)^\dagger D_\rho U(x)\right)\left(U(x)^\dagger D_\sigma U(x)\right)\right]
$$

$$
- \frac{1}{16\pi^2}\epsilon_{\mu\nu\rho\sigma}\mathrm{Tr}\left[W_{\nu\rho}(x)\left(D_\sigma U(x)U(x)^\dagger\right)\right]
$$

$$
- \frac{1}{16\pi^2}\epsilon_{\mu\nu\rho\sigma}\mathrm{Tr}\left[X_{\nu\rho}(x)\left(U(x)^\dagger D_\sigma U(x)\right)\right]. \tag{24.19}
$$

Now the divergence of the baryon number current takes the form

$$
\partial_\mu j_\mu^{\mathrm{GW}}(x) = -\frac{1}{32\pi^2}\epsilon_{\mu\nu\rho\sigma}\mathrm{Tr}\left[W_{\mu\nu}(x)W_{\rho\sigma}(x)\right] + \frac{1}{32\pi^2}\epsilon_{\mu\nu\rho\sigma}\mathrm{Tr}\left[X_{\mu\nu}(x)X_{\rho\sigma}(x)\right]
$$

$$
= -\frac{1}{32\pi^2}\epsilon_{\mu\nu\rho\sigma}\mathrm{Tr}\left[W_{\mu\nu}(x)W_{\rho\sigma}(x)\right] - \frac{g'^2}{64\pi^2}\epsilon_{\mu\nu\rho\sigma}B_{\mu\nu}(x)B_{\rho\sigma}(x). \tag{24.20}
$$

Even though the preceding arguments have allowed us to identify the correct expressions for the covariant derivatives and the Goldstone–Wilczek current, we have not yet coupled the U(1)$_Y$ gauge field in the correct form. From our previous discussion, we know that the full U(1)$_Y$ gauge transformations of the Standard Model are described by embedded SU(2)$_L$ × SU(2)$_R$ × U(1)$_{L=R}$ matrices. At the level of the pion effective theory, these are indistinguishable from equivalent SU(2)$_L$ × SU(2)$_R$ transformations, because one can commute a multiple of the unit-matrix from the left to the right (as we did when proceeding to eq. (24.17)). Based on this observation, it might seem that the low-energy effective theory of the Standard Model with $Y_{q_L} = 1/(2N_c)$, $Y_{u_R} = (1/N_c + 1)/2$, and $Y_{d_R} = (1/N_c - 1)/2$ would be the same as the one of a theory with $Y_{q_L} = 0$, $Y_{u_R} = 1/2$, and $Y_{d_R} = -1/2$, in which the baryon number contributions to the weak hypercharges (which are proportional to the baryon number $1/N_c$ of a quark) have been dropped.

Since the actual quark hypercharges receive the contribution $1/(2N_c)$ proportional to the baryon number, the $U(1)_Y$ gauge field of the Standard Model indeed also couples to the baryon number current. At the level of the effective pion theory, this coupling enters the action via the Goldstone–Wilczek current, which carries the baryon number. Indeed, a Goldstone–Wilczek term must be added to the effective pion Lagrangian. When we include electroweak gauge fields, the corresponding contribution to the action takes the form

$$S_{GW}[U, W, B] = \frac{g'}{2} \int d^4x \, B_\mu j_\mu^{GW}. \tag{24.21}$$

It is important that (just like $\text{Sign}[U]$) this term alone is not gauge-invariant. As we discussed before, $\text{Sign}[U]$ changes sign under topologically non-trivial $SU(2)_L$ gauge transformations. This gauge variation is exactly compensated by the global anomaly in the lepton sector. Similarly, while j_μ^{GW} is both $SU(2)_L$ and $U(1)_Y$ gauge-invariant, the action $S_{GW}[U, W, B]$ is only $SU(2)_L$ gauge-invariant, but it varies under $U(1)_Y$ gauge transformations $B_\mu(x)' = B_\mu(x) - \partial_\mu \varphi(x)$. Its violation of gauge invariance is given by

$$S_{GW}[U', W, B'] - S_{GW}[U, W, B] = -\frac{g'}{2} \int d^4x \, \partial_\mu \varphi j_\mu^{GW} = \frac{g'}{2} \int d^4x \, \varphi \, \partial_\mu j_\mu^{GW}$$

$$= -\frac{g'}{2} \int d^4x \, \varphi \left\{ \frac{1}{32\pi^2} \epsilon_{\mu\nu\rho\sigma} \text{Tr}\left[W_{\mu\nu} W_{\rho\sigma} \right] + \frac{g'^2}{64\pi^2} \epsilon_{\mu\nu\rho\sigma} B_{\mu\nu} B_{\rho\sigma} \right\}. \tag{24.22}$$

This gauge variation is exactly what is needed in order to cancel the triangle anomalies in the lepton sector and render the entire theory gauge-invariant.

Replacing ordinary derivatives by the covariant derivatives of eq. (24.18), the leading contribution to the pion effective action takes the form

$$S[U, W, B] = \int d^4x \, \frac{F_\pi^2}{4} \text{Tr}\left[D_\mu U^\dagger D_\mu U \right]. \tag{24.23}$$

Assuming an odd value for the number of colors N_c in order to cancel Witten's global anomaly in the lepton sector, we obtain $\text{Sign}[U]^{N_c} = \text{Sign}[U]$. The functional integral for pion fields coupled to external electroweak background gauge fields then finally takes the form

$$Z[W, B] = \int \mathcal{D}U \exp(-S[U, W, B]) \, \text{Sign}[U] \exp(iS_{GW}[U, W, B]). \tag{24.24}$$

Remarkably, N_c does not appear explicitly in this expression. It only enters implicitly in $S[U, W, B]$ because the value of the pion decay constant, F_π, which is determined by the non-perturbative QCD dynamics, changes with the number of colors. The explicit N_c-independence of the above functional integral implies that the structure of the low-energy physics of pions is insensitive to the number of quark colors. As we will see in the next section, in contrast to standard textbook knowledge, this implies that the width of the anomalous decay of the neutral pion into two photons is N_c-independent.

In Chapter 23, we already addressed the electromagnetic interactions of the Nambu–Goldstone bosons for $N_c = 3$. Let us now identify the $U(1)_{em}$ gauge symmetry of electromagnetism for an arbitrary number of colors. The electric charge of the left- and right-handed u-quarks is $Q_u = (1/N_c + 1)/2$, while that of the d-quarks is $Q_d = (1/N_c - 1)/2$. Hence, we write $U'(x) = \bar{L}(x)U(x)\bar{R}(x)^\dagger$, now with

$$\bar{L}(x) = \bar{R}(x) = \begin{pmatrix} \exp\left(ieQ_u\alpha(x)\right) & 0 \\ 0 & \exp\left(ieQ_d\alpha(x)\right) \end{pmatrix}. \tag{24.25}$$

As before, $\bar{L}(x)$ and $\bar{R}(x)$ are elements of U(2), not of SU(2). However, by commuting through the factor $\exp\left(ie\alpha(x)/(2N_c)\right)\mathbb{1}$, we can replace them by a new matrix

$$V(x) = \begin{pmatrix} \exp\left(ie\alpha(x)/2\right) & 0 \\ 0 & \exp\left(-ie\alpha(x)/2\right) \end{pmatrix}, \tag{24.26}$$

which is in the vector subgroup $SU(2)_{L=R}$. Hence, at the level of the low-energy effective theory, for the covariant derivative of electromagnetism we again obtain

$$D_\mu U = \partial_\mu U(x) + ieA_\mu(x)\left[\frac{\tau^3}{2}, U(x)\right]. \tag{24.27}$$

In the presence of just electromagnetic gauge fields, the Goldstone–Wilczek current simplifies to

$$\begin{aligned}
j_\mu^{GW}(x) = \frac{1}{24\pi^2}\epsilon_{\mu\nu\rho\sigma}\text{Tr}\left[\left(U(x)^\dagger D_\nu U(x)\right)\left(U(x)^\dagger D_\rho U(x)\right)\left(U(x)^\dagger D_\sigma U(x)\right)\right] \\
- \frac{ie}{16\pi^2}\epsilon_{\mu\nu\rho\sigma}F_{\nu\rho}\text{Tr}\left[\frac{\tau^3}{2}\left(D_\sigma U(x)U(x)^\dagger\right)\right],
\end{aligned} \tag{24.28}$$

where $F_{\mu\nu}(x) = \partial_\mu A_\nu(x) - \partial_\nu A_\mu(x)$ is the field strength of the electromagnetic field. Since QCD and QED are vector-like theories, the baryon current is now conserved, *i.e.* $\partial_\mu j_\mu^{GW}(x) = 0$, and the corresponding Goldstone–Wilczek term

$$S_{GW}[U, A] = \frac{e}{2}\int d^4x\, A_\mu j_\mu^{GW} \tag{24.29}$$

is $U(1)_{em}$ gauge-invariant. Again, there is no explicit N_c-dependence.

24.3 G-Parity and its Explicit Breaking

Tsung-Dao Lee and Chen-Ning Yang introduced *G-parity* as a combination of charge conjugation with the $SU(2)_{L=R} = SU(2)_I$ isospin transformation $i\tau^2$ for $N_f = 2$ (Lee and Yang, 1956a). At the level of the low-energy effective theory, charge conjugation takes the form

$${}^C U(x) = U(x)^\dagger, \quad U(x) = \exp\left(i\pi^a(x)\tau^a/F_\pi\right), \quad \pi^a(x)\tau^a = \begin{pmatrix} \pi^0(x) & \sqrt{2}\pi^+(x) \\ \sqrt{2}\pi^-(x) & -\pi^0(x) \end{pmatrix} \Rightarrow$$

$${}^C\pi^\pm(x) = \pi^\mp(x), \quad {}^C\pi^0(x) = \pi^0(x). \tag{24.30}$$

As a consequence, G-parity acts as

$${}^G U(x) = i\tau^2\, {}^C U(x)\left(i\tau^2\right)^\dagger = U(x)^{*\top} = U(x)^\dagger \quad \Rightarrow \quad {}^G\pi^a(x) = -\pi^a(x). \tag{24.31}$$

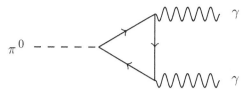

Diagram of neutral pion decay into two photons.

This discrete symmetry is equivalent to the intrinsic parity P_0, which changes the sign of the pseudo-scalar pion field without performing a spatial inversion. For $N_f = 2$, G-parity is a symmetry of QCD even in the presence of explicit isospin breaking due to different u- and d-quark masses. It implies that, for $N_f = 2$, no strong interaction process can change the number of pions from even to odd, or vice versa. Even the action $S[U, W, B]$ of eq. (24.23) is still G-invariant. However, the Goldstone–Wilczek action $S_{GW}[U, W, B]$ of eq. (24.21) is not, such that G-parity is explicitly broken by the electroweak interactions.

24.4 Electromagnetic Decay of the Neutral Pion

Let us identify a vertex for the decay of a neutral pion into two photons from the Goldstone–Wilczek term. For this purpose, we use the field

$$U(x) = \exp\left(i\pi^0(x)\tau^3/F_\pi\right) \approx 1 + i\pi^0(x)\tau^3/F_\pi. \tag{24.32}$$

Then, the electromagnetic covariant derivative takes the form

$$D_\mu U(x) = \partial_\mu U(x) + ieA_\mu(x)\left[\frac{\tau^3}{2}, U(x)\right] \approx i\partial_\mu\pi^0(x)\tau^3/F_\pi. \tag{24.33}$$

Inserting this leading contribution in the Goldstone–Wilczek term (24.29), one obtains the vertex

$$\mathcal{L}_{\pi^0\gamma\gamma} = -i\frac{e^2}{32\pi^2 F_\pi}\pi^0\epsilon_{\mu\nu\rho\sigma}F_{\mu\nu}F_{\rho\sigma}, \tag{24.34}$$

which is not explicitly N_c-dependent.

According to standard textbook knowledge, the width of the electromagnetic neutral pion decay is proportional to N_c^2. This is concluded from the correct expression

$$\mathcal{L}_{\pi^0\gamma\gamma} = -iN_c\left(Q_u^2 - Q_d^2\right)\frac{e^2}{32\pi^2 F_\pi}\pi^0\epsilon_{\mu\nu\rho\sigma}F_{\mu\nu}F_{\rho\sigma}, \tag{24.35}$$

which follows from a Standard Model triangle diagram with a quark loop attached to two external photon lines and one external isovector axial current (which represents the pion), as illustrated in Figure 24.1. Since the decay width follows from squaring the decay amplitude described by $\mathcal{L}_{\pi^0\gamma\gamma}$, keeping track of all constants, one obtains

$$\Gamma(\pi^0 \to \gamma\gamma) = \left[N_c\left(Q_u^2 - Q_d^2\right)\right]^2 \frac{e^4 M_\pi^3}{1024\pi^5 F_\pi^2}. \tag{24.36}$$

Ironically, the widespread argument assumes N_c to be varied, while the quark charges are fixed to their physical values $2/3$ and $-1/3$. Obviously, this assumption does not make much sense, because in a world with N_c colors, baryons consist (in a simple constituent quark picture) not of three but of N_c quarks. In particular, for general odd N_c, the proton is still a fermion and contains $(N_c + 1)/2$ u-quarks and $(N_c - 1)/2$ d-quarks, while the neutron contains $(N_c - 1)/2$ u-quarks and $(N_c + 1)/2$ d-quarks. Demanding the correct electric charges for proton and neutron, namely,

$$Q_p = \frac{N_c + 1}{2} Q_u + \frac{N_c - 1}{2} Q_d = 1,$$
$$Q_n = \frac{N_c - 1}{2} Q_u + \frac{N_c + 1}{2} Q_d = 0, \tag{24.37}$$

immediately implies the consistent N_c-dependent quark charges

$$Q_u = \frac{1}{2}\left(\frac{1}{N_c} + 1\right), \quad Q_d = \frac{1}{2}\left(\frac{1}{N_c} - 1\right). \tag{24.38}$$

Remarkably, these are exactly the same charges that follow from anomaly cancellation in the N_c-dependent Standard Model, as we saw in Chapter 15. Using the consistent electric charges of the quarks, the factor that enters the neutral pion decay width is

$$N_c \left(Q_u^2 - Q_d^2\right) = \frac{N_c}{4}\left[\left(\frac{1}{N_c} + 1\right)^2 - \left(\frac{1}{N_c} - 1\right)^2\right] = 1, \tag{24.39}$$

which is N_c-independent.

24.5 Evidence for $N_c = 3$ from $\pi^0 \to \gamma\gamma$?

The simple argument from above was first presented by Serge Rudaz (1990) and by Afsar Abbas (1990), and further elaborated by Jean-Marc Gérard and Touria Lahna (Gerard and Lahna, 1995) and by Oliver Bär and Uwe-Jens Wiese (Bär and Wiese, 2001). Unfortunately, it has until now not penetrated into the textbook literature. We therefore like to point out again that the decay width of the neutral pion does *not* depend explicitly on N_c. It still depends on N_c implicitly, because the pion mass M_π and the pion decay constant F_π (which enter the expression (24.36) for $\Gamma(\pi^0 \to \gamma\gamma)$) depend on the number of colors. When one takes the values of M_π and F_π from experiment, the implicit dependence is irrelevant and the observed width does *not* imply $N_c = 3$. On the other hand, if one would compute M_π and F_π using lattice QCD for different values of N_c – which is a non-trivial task – one could still infer $N_c = 3$ by comparison with experiment. However, using that method, one can deduce the number of colors from any QCD observable, not only from the decay width of the neutral pion.

Since the 1970s, the misleading statement that the $\pi^0 \to \gamma\gamma$ decay width is proportional to N_c^2 has been used to lend support to the conclusion that $N_c = 3$. While the conclusion itself is correct, this way of reaching it is just wrong. In this context, it is sometimes stated that Hiroshi Fukuda and Yoneji Miyamoto, as well as Jack Steinberger, who obtained almost the correct width from a nucleon triangle diagram as early as 1949 (Fukuda and

Miyamoto, 1949; Steinberger, 1949), "accidentally" got the right answer from an incorrect theory. This is wrong as well. Their result is indeed the right answer that one obtains from the low-energy effective theory of the Standard Model for any odd value of N_c. It corresponds to $N_c = 1$, *i.e.* to a colorless world without quarks. In 1949, nobody knew about quarks or color, but Fukuda, Miyamoto, and Steinberger were still using a consistent low-energy description of our world. As we saw in Chapter 15, the Standard Model with $N_c = 1$, and hence without strong interactions, indeed is also gauge anomaly-free. According to eq. (24.38), the "quark" charges are then equal to $Q_u = 1$ and $Q_d = 0$, and the u- and d-"quarks" are, in fact, just the proton and the neutron. A Standard Model with the usual Higgs and electroweak sector, but with $N_c = 1$, has a nucleon, but no strong interactions, no spontaneous symmetry breaking, and thus no pions. However, one can add the pions by hand in the form of a linear σ-model, without spoiling renormalizability or anomaly cancellation. In such a model, which is equivalent to what was used in 1949, one still obtains eq. (24.36), despite the fact that there are no colors.

To summarize, neither anomaly cancellation in the N_c-dependent Standard Model nor the width of the decay $\pi^0 \to \gamma\gamma$ imply $N_c = 3$. What are then good pieces of evidence for the number of colors?

- First of all, the fermionic nature of baryons implies that N_c is odd, which is also necessary in order to cancel Witten's global $SU(2)_L$ anomaly.
- Second, as we have seen in Section 21.8, the existence of the Δ-isobar is evidence for $N_c \geq 3$. For $N_c \geq 5$, there would be additional baryon resonances beyond the observed Δ-isobar. The absence of such resonances constitutes experimental evidence for $N_c = 3$.
- At last, as we discussed in Chapter 22, the R-ratio between the electron–positron annihilation cross sections into hadrons versus muon–anti-muon pairs provides further, compelling experimental evidence for $N_c = 3$. Both the high end of the baryon spectrum and the R-ratio refer to energy scales beyond the characteristic QCD scale Λ_{QCD}.

In this sense, at least with only two light-quark flavors, evidence for $N_c = 3$ comes only from high energies. This already follows from the fact that the low-energy effective theory for a single generation of quarks in the Standard Model with arbitrary odd N_c does not depend on N_c explicitly (cf. eq. (24.24)). As we will see in the next section, one can also deduce the value of N_c from physical processes at low energies, provided that one includes more than two light-quark flavors.

24.6　Wess–Zumino–Novikov–Witten Term

In this section, we proceed to the theory with $N_f \geq 3$ light-quark flavors, and the chiral symmetry $SU(N_f)_L \times SU(N_f)_R$. Because $\Pi_3[SU(N_f)] = \mathbb{Z}$ for all $N_f \geq 2$, skyrmions still exist and the Goldstone–Wilczek current still keeps the same form. However, as we already mentioned, for $N_f \geq 3$ the topological properties of the Nambu–Goldstone boson fields are qualitatively different from the two-flavor case. In particular, the homotopy group $\Pi_4[SU(N_f)] = \{0\}$ is trivial for $N_f \geq 3$. This implies that the winding number $\text{Sign}[U] \in \mathbb{Z}(2)$ that we introduced in Section 24.1 no longer exists for more than two light flavors.

One might think that this implies that skyrmions are always quantized as bosons. However, the situation is much more interesting. In fact, in contrast to the two-flavor case, $\Pi_4[\mathrm{SU}(N_\mathrm{f} \geq 3)] = \{0\}$ implies that any 4-d space–time-dependent Nambu–Goldstone boson field $U(x) \in \mathrm{SU}(N_\mathrm{f})$ can be continuously deformed into the trivial field $U(x) = \mathbb{1}$. This implies that there is no topological obstruction against interpolating the 4-d field into a *mathematical fifth dimension*. This allows us to construct the *Wess–Zumino–Novikov–Witten (WZNW) term* (Wess and Zumino, 1971; Novikov, 1981, 1982; Witten, 1983a,b, 1984a). By introducing a fifth coordinate $x_5 \in [0, 1]$, which plays the role of a mathematical deformation parameter, one can continuously interpolate the physical 4-d field $U(x)$ to a field $U(x, x_5)$ on the 5-d hemisphere H^5. The boundary of this hemisphere (at $x_5 = 0$), $\partial H^5 = S^4$, is the 4-d space–time (compactified to the sphere S^4), such that $U(x, 1) = \mathbb{1}$ and $U(x, 0) = U(x)$. The WZNW action takes the form

$$S_{\mathrm{WZNW}}[U] = \frac{1}{240\pi^2 \mathrm{i}} \int_{H^5} d^5x \, \epsilon_{\mu\nu\rho\sigma\lambda} \mathrm{Tr}\left[\left(U^\dagger \partial_\mu U\right)\left(U^\dagger \partial_\nu U\right)\left(U^\dagger \partial_\rho U\right)\left(U^\dagger \partial_\sigma U\right)\left(U^\dagger \partial_\lambda U\right)\right].$$
(24.40)

The factor i is necessary in order to obtain a real-valued term.

Quick Question 24.6.1   Reality of $S_{\mathrm{WZNW}}[U]$
Show that $S_{\mathrm{WZNW}}[U]$ is real-valued. Why is no explicit factor i needed in eq. (24.2) in order to make the baryon number B real? What happens for analogous expressions in an arbitrary, odd number of dimensions?

Of course, in order to be well-defined in terms of the physical 4-d field $U(x)$, the physics should be independent of how the field $U(x, x_5)$ is deformed into the mathematical fifth dimension. This is indeed possible because the integrand in eq. (24.40) is a total divergence. Since for $N_\mathrm{f} \geq 3$ the homotopy group $\Pi_5[\mathrm{SU}(N_\mathrm{f})] = \mathbb{Z}$ is non-trivial, integrating the same integrand over an entire 5-d sphere S^5 (without boundary), instead of just the hemisphere H^5, yields 2π times an integer-valued winding number, *i.e.*

$$w[U] = \frac{1}{480\pi^3 \mathrm{i}} \int_{S^5} d^5x \, \epsilon_{\mu\nu\rho\sigma\lambda} \mathrm{Tr}\left[\left(U^\dagger \partial_\mu U\right)\left(U^\dagger \partial_\nu U\right)\left(U^\dagger \partial_\rho U\right)\left(U^\dagger \partial_\sigma U\right)\left(U^\dagger \partial_\lambda U\right)\right]$$
$$\in \Pi_5[\mathrm{SU}(N_\mathrm{f})] = \mathbb{Z}.$$
(24.41)

Let us investigate the ambiguity of the WZNW term by comparing two different interpolations $U^{(1)}(x, x_5)$ and $U^{(2)}(x, x_5)$, which both agree in the physical 4-d space–time (at $x_5 = 0$), *i.e.* $U^{(1)}(x, 0) = U^{(2)}(x, 0)$. The difference between the two corresponding WZNW terms is then given by

$$S_{\mathrm{WZNW}}\left[U^{(1)}\right] - S_{\mathrm{WZNW}}\left[U^{(2)}\right] = 2\pi w[U].$$
(24.42)

Here, the 5-d field $U(x, x_5)$ (now with $x_5 \in [-1, 1]$) maps the entire sphere S^5 (and not only just H^5) to $\mathrm{SU}(N_\mathrm{f})$. On the "northern" hemisphere H^5 (*i.e.* for $x_5 \in [0, 1]$), the field U agrees with $U^{(1)}$, *i.e.* $U(x, x_5) = U^{(1)}(x, x_5)$. On the "southern" hemisphere (*i.e.* for $x_5 \in [-1, 0]$), on the other hand, it agrees with $U^{(2)}$, *i.e.* $U(x, x_5) = U^{(2)}(x, -x_5)$. Hence,

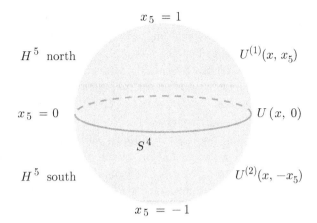

$x_5 = 1$

H^5 north

$U^{(1)}(x, x_5)$

$x_5 = 0$

$U(x, 0)$

S^4

H^5 south

$U^{(2)}(x, -x_5)$

$x_5 = -1$

Fig. 24.2 Extended 5-d manifold consisting of a northern and a southern hemisphere, separated by the equator S^4 which represents the compactified, physical 4-d space–time. The physical Nambu–Goldstone boson field $U(x, 0)$ is extrapolated into the mathematical fifth dimension in two different ways. The extrapolations $U^{(1)}(x, x_5)$ and $U^{(2)}(x, -x_5)$ extend over the northern and southern hemisphere, respectively. The extrapolation ambiguity of $S_{\text{WZNW}}[U]$ corresponds to 2π times the winding number $\Pi_5[SU(N_f)] = \mathbb{Z}$, which is associated with the entire sphere S^5.

along the "equator" (the 4-d physical space–time (compactified to S^4) at $x_5 = 0$), the interpolated 5-d field $U(x, x_5)$ agrees with the physical 4-d Nambu–Goldstone boson field $U(x)$, *i.e.* $U(x, 0) = U^{(1)}(x, 0) = U^{(2)}(x, 0) = U(x)$. This is symbolically sketched in Figure 24.2.

Equation (24.42) indeed implies that the WZNW action is affected by an interpolation ambiguity. However, the ambiguity is restricted to 2π times the integer winding number $w[U] \in \Pi_5[SU(N_f \geq 3)] = \mathbb{Z}$. Of course, one must still ensure that this ambiguity does not affect the 4-d physics. For this purpose, we consider the functional integral of the effective theory, to which the WZNW action contributes with a prefactor $(\mathrm{i}n)$

$$Z = \int \mathcal{D}U \, \exp(-S[U]) \exp(\mathrm{i}n S_{\text{WZNW}}[U]). \qquad (24.43)$$

The ambiguity of the WZNW action affects the factor that enters the functional integral as

$$\exp\left(\mathrm{i}n S_{\text{WZNW}}\left[U^{(1)}\right]\right) = \exp\left(\mathrm{i}n S_{\text{WZNW}}\left[U^{(2)}\right]\right) \exp(2\pi \, \mathrm{i} \, n \, w[U]). \qquad (24.44)$$

The 4-d physics thus remains independent of the details of the interpolation to the fifth mathematical dimension, provided that

$$\exp(2\pi \, \mathrm{i} \, n \, w[U]) = 1 \quad \Rightarrow \quad n \in \mathbb{Z}. \qquad (24.45)$$

The non-trivial homotopy group $\Pi_5[SU(N_f \geq 3)]$ hence implies that the prefactor n of the WZNW action is quantized in integer units. This is quite remarkable, because it means that one of the QCD low-energy parameters (which include quantities like F_π and $\Sigma = -\langle \bar{q}q \rangle / N_f$) – namely, n – is actually restricted to integer values.

One may wonder what happens to this quantization condition in the classical limit $\hbar \to 0$. This is not obvious, because we are using natural units in which $\hbar = 1$. In fact, when we reintroduce $\hbar$, the term $\exp(-S[U])$ is replaced by $\exp(-S[U]/\hbar)$, which is the generic form introduced in Chapter 1. However, since its exponent is dimension-less, the topological WZNW factor $\exp(inS_{\text{WZNW}}[U])$ does not involve $\hbar$ and thus remains unchanged. This shows that the WZNW term *solely causes quantum effects*, without any impact on classical physics.

The question is obvious: In the low-energy theory for QCD with $N_f \geq 3$ light-quark flavors, what is the correct integer value for the low-energy parameter n? In this context, it is instructive to consider the $N_f = 2$ case, embedded into the $N_f \geq 3$ effective theory. One can show that for an embedded SU(2) matrix $\widetilde{U}(x)$,

$$U(x) = \begin{pmatrix} \widetilde{U}(x) & 0 \\ 0 & \mathbb{1} \end{pmatrix}, \tag{24.46}$$

(where $\mathbb{1}$ is the $(N_f - 2) \times (N_f - 2)$ unit-matrix), one obtains

$$\exp(inS_{\text{WZNW}}[U]) = \text{Sign}[\widetilde{U}]^n. \tag{24.47}$$

Here, the argument of the WZNW term is a 5-d Nambu–Goldstone boson field $U(x, x_5) \in$ SU(N_f) which reduces to a 4-d embedded SU(2) field $\widetilde{U}(x)$ at the physical space–time boundary of H^5. The argument of the sign factor, on the other hand, is just the 4-d field $\widetilde{U}(x) \in$ SU(2). Indeed, in the $N_f \geq 3$ theory the WZNW term incorporates those effects that were described by $\text{Sign}[U]$ in the $N_f = 2$ theory. In particular, for odd n, the WZNW term ensures that the skyrmion is again quantized as a fermion with half-integer spin. We will soon convince ourselves that $n = N_c$. Based on the spin and statistics of skyrmions alone, we can only conclude that n must be odd when N_c is odd and even otherwise.

24.7 Intrinsic Parity and Its Anomalous Breaking

The effective action $S[U]$ (which does not include $S_{\text{WZNW}}[U]$) has additional symmetries compared to QCD itself. In particular, it is invariant under the *intrinsic parity* P_0, while QCD is invariant only with respect to the full parity transformation P. Being pseudo-scalar particles, the Nambu–Goldstone boson fields transform under P as

$$^P\pi^a(\vec{x}, x_4) = -\pi^a(-\vec{x}, x_4), \tag{24.48}$$

whereas the intrinsic parity P_0 only transforms them as

$$^{P_0}\pi^a(\vec{x}, x_4) = -\pi^a(\vec{x}, x_4). \tag{24.49}$$

With respect to the field $U(x) = \exp(i\pi^a(x)\lambda^a/F_\pi)$ (where λ^a are the generators of the SU(N_f) algebra), these transformations take the form

$$^PU(\vec{x}, x_4) = U(-\vec{x}, x_4)^\dagger, \qquad ^{P_0}U(\vec{x}, x_4) = U(\vec{x}, x_4)^\dagger. \tag{24.50}$$

The effective action $S[U]$ is invariant under both P and P_0, in particular, $S[^{P_0}U] = S[U^\dagger] = S[U]$. If P_0 were a symmetry of QCD, the number of Nambu–Goldstone bosons could never change from even to odd. This is indeed the case for $N_f = 2$, and the corresponding discrete symmetry (which is then equivalent to P_0) is just G-parity. For $N_f \geq 3$, on the other hand, there are QCD processes which violate the intrinsic symmetry P_0 (while respecting the full parity P) (Witten, 1983a). For instance, a φ-meson (a vector meson that we discussed in Section 21.6) can decay both into two kaons or into three pions. Equivalently, two-kaon scattering can lead to three-pion final states (thus changing the number of pseudo-Nambu–Goldstone bosons from even to odd). Such effects are indeed also described by the WZNW term. This is not surprising, because the WZNW term explicitly breaks the intrinsic parity P_0,

$$S_{\text{WZNW}}[^{P_0}U] = S_{\text{WZNW}}[U^\dagger] = -S_{\text{WZNW}}[U], \tag{24.51}$$

while it leaves the full parity symmetry intact, *i.e.* $S_{\text{WZNW}}[^{P}U] = S_{\text{WZNW}}[U]$. The WZNW term thus reduces the symmetries of the low-energy effective theory to those of QCD. In the $N_{\text{f}} = 2$ case, the WZNW term reduces to $\text{Sign}[\widetilde{U}]$, which indeed respects G-parity (*i.e.* P_0) because

$$\text{Sign}[^{P_0}\widetilde{U}] = \text{Sign}[\widetilde{U}^\dagger] = \text{Sign}[\widetilde{U}]. \tag{24.52}$$

The amplitude for scattering two kaons into a final three-pion state is determined by the WZNW term and is thus proportional to the integer n of eq. (24.44). The corresponding cross section (which results from squaring the amplitude) is thus proportional to n^2. Unfortunately, such scattering processes are impractical to observe in experiments. However, at least from a theoretical point of view, one can use the corresponding cross section to extract the value of n from a physical low-energy process that only involves the strong interaction.

24.8 Electromagnetic Interactions of Pions, Kaons, and η-Mesons

As we have seen, in the $N_{\text{f}} = 2$ theory, the factor $\text{Sign}[U]^{N_{\text{c}}}$ that enters the functional integral represents Witten's global anomaly at the level of the effective theory. Therefore, we expect the WZNW term to play a similar role for $N_{\text{f}} = 3$. Indeed, in the low-energy effective theory for the real world with three light-quark flavors u, d, and s, and with $N_{\text{c}} = 3$ colors, all gauge anomalies of the quarks in the underlying Standard Model are represented by the WZNW term. However, as we will see, if the trace of the light-quark charge matrix is non-zero, a Goldstone–Wilczek term arises in addition.

In order to determine the correct integer value of the low-energy parameter n (the quantized prefactor of the WZNW term), Witten considered QCD with $N_{\text{f}} = 3$ light flavors and a general number of colors N_{c}, coupled to QED with the physical quark charge matrix (Witten, 1983a)

$$Q' = \text{diag}\left(Q'_u, Q'_d, Q'_s\right) = \text{diag}\left(\frac{2}{3}, -\frac{1}{3}, -\frac{1}{3}\right). \tag{24.53}$$

The $U(1)_{\text{em}}$-gauged WZNW term then takes the form

$$
\begin{aligned}
S_{\text{WZNW}}[U, A] = {} & S_{\text{WZNW}}[U] + \frac{e}{48\pi^2} \int d^4x \, \epsilon_{\mu\nu\rho\sigma} A_\mu \\
& \times \text{Tr}\left[Q' \left(\partial_\nu U U^\dagger\right) \left(\partial_\rho U U^\dagger\right) \left(\partial_\sigma U U^\dagger\right) + Q' \left(U^\dagger \partial_\nu U\right) \left(U^\dagger \partial_\rho U\right) \left(U^\dagger \partial_\sigma U\right) \right] \\
& - \frac{i e^2}{48\pi^2} \int d^4x \, \epsilon_{\mu\nu\rho\sigma} A_\mu F_{\nu\rho} \text{Tr}\left[Q' \left(\partial_\sigma U U^\dagger\right) \left(Q' + \frac{1}{2} U Q' U^\dagger\right)\right. \\
& \left. + Q' \left(U^\dagger \partial_\sigma U\right) \left(Q' + \frac{1}{2} U^\dagger Q' U\right) \right].
\end{aligned}
\tag{24.54}
$$

It should be noted that all derivatives in this expression can be replaced by covariant derivatives without changing the result. Witten then showed that the WZNW term contribution to the vertex $\mathcal{L}_{\pi^0\gamma\gamma}$ is proportional to n, from which he correctly concluded that

$$n = N_c. \tag{24.55}$$

As we argued in Section 24.4, varying the number of colors while fixing the quark charges to their physical values (corresponding to $N_c = 3$) is not very natural, because, e.g., proton and neutron (which would then contain N_c instead of three quarks) do not obtain their correct electric charges. In the Standard Model with general N_c, the above charge assignment would even lead to a mathematically inconsistent theory, due to un-canceled gauge anomalies. The consistent N_c-dependent charge matrix of the light u-, d-, and s-quarks, which guarantees anomaly cancellation and yields the correct charges for proton and neutron, takes the form

$$Q = \text{diag}\,(Q_u, Q_d, Q_s) = \text{diag}\left(\frac{1}{2}\left(\frac{1}{N_c}+1\right), \frac{1}{2}\left(\frac{1}{N_c}-1\right), \frac{1}{2}\left(\frac{1}{N_c}-1\right)\right)$$

$$= Q' + \left(1 - \frac{N_c}{3}\right)\frac{1}{2}B. \tag{24.56}$$

Hence, for $N_c \neq 3$ the quark charge matrix is not just a traceless generator of $SU(3)_{L=R}$, but it also receives a contribution from the baryon number. This implies that, just as in the $N_f = 2$ case, in order to account for the baryon number contribution, we must include a Goldstone–Wilczek term in the effective theory.

In the $N_f = 2$ effective theory, the consistent quark charge matrix is

$$\text{diag}\,(Q_u, Q_d) = \text{diag}\left(\frac{1}{2}\left(\frac{1}{N_c}+1\right), \frac{1}{2}\left(\frac{1}{N_c}-1\right)\right) = \text{diag}\left(\frac{1}{2}, -\frac{1}{2}\right) + \frac{1}{2}B. \tag{24.57}$$

This implies that, compared to $N_f = 2$, the Goldstone–Wilczek term for $N_f = 3$, which takes the form

$$S_{\text{GW}}[U, A] = \frac{e}{48\pi^2}\int d^4x\, \epsilon_{\mu\nu\rho\sigma} A_\mu \text{Tr}\left[\left(U^\dagger \partial_\nu U\right)\left(U^\dagger \partial_\rho U\right)\left(U^\dagger \partial_\sigma U\right)\right]$$

$$- \frac{ie^2}{32\pi^2}\int d^4x\, \epsilon_{\mu\nu\rho\sigma} A_\mu F_{\nu\rho} \text{Tr}\left[Q'\left(\partial_\sigma U U^\dagger + U^\dagger \partial_\sigma U\right)\right], \tag{24.58}$$

enters the functional integral with an additional prefactor $(1 - N_c/3)$, such that

$$Z[A] = \int \mathcal{D}U \exp(-S[U, A])\exp(iN_c S_{\text{WZNW}}[U, A])\exp(i(1 - N_c/3)\,S_{\text{GW}}[U, A]). \tag{24.59}$$

Let us again consider the decay $\pi^0 \to \gamma\gamma$, now from the perspective of the $N_f = 3$ theory. For $N_f = 2$, i.e. in the absence of the WZNW term, the Goldstone–Wilczek term alone contributes the correct N_c-independent vertex $\mathcal{L}_{\pi^0\gamma\gamma}$. For $N_f = 3$, this contribution is modified by the factor $(1 - N_c/3)$. However, there is also the contribution of the WZNW term, which is proportional to the factor

$$N_c\left(Q_u'^2 - Q_d'^2\right) = N_c\left[\left(\frac{2}{3}\right)^2 - \left(-\frac{1}{3}\right)^2\right] = \frac{N_c}{3}. \tag{24.60}$$

As a result, the N_c-dependent part of the Goldstone–Wilczek term completely cancels the contribution from the WZNW term, and the N_c-independent part of the Goldstone–Wilczek term provides the vertex $\mathcal{L}_{\pi^0\gamma\gamma}$ with the correct N_c-independent strength. Finally, as we already concluded in the $N_f = 2$ case, the width of the neutral pion decay into two photons is indeed N_c-independent.

24.9   Electromagnetic Interactions of Nambu–Goldstone Bosons for $N_f \geq 3$

Let us finally consider the general case of $N_f \geq 3$, assuming that there are N_u light u-type quarks (like up, charm, and top) with charge $Q_u = (1/N_c + 1)/2$ and N_d light d-type quarks (like down, strange, and bottom) with charge $Q_d = (1/N_c - 1)/2$, with $N_u + N_d = N_f$. While the physical situation in the real world corresponds to $N_u = 1, N_d = 2$, and $N_f = 3$, it is still instructive to also consider the general case. In the context of the Standard Model, it is easy to realize the general case, by adjusting the quark Yukawa couplings in order to make the desirable number of quark flavors light.

For N_u light flavors with charge Q_u and N_d light flavors with charge Q_d, the charge matrix of the quarks takes the form

$$Q = Q' + \left(1 - N_c \frac{N_d - N_u}{N_f}\right) \frac{1}{2} B. \tag{24.61}$$

Here, Q' is a traceless diagonal generator of $SU(N_f)_{L=R}$.

> **Quick Question 24.9.1**   **Quark charge matrix for general N_c, N_u, and N_d**
> Show that the general quark charge matrix takes the form of eq. (24.61).

The functional integral of the low-energy effective theory then reads

$$Z[A] = \int \mathcal{D}U \exp(-S[U, A]) \exp(\mathrm{i} N_c S_{\mathrm{WZNW}}[U, A])$$
$$\times \exp\left(\mathrm{i}\left(1 - N_c \frac{N_d - N_u}{N_f}\right) S_{\mathrm{GW}}[U, A]\right). \tag{24.62}$$

Let us again consider the decay $\pi^0 \to \gamma\gamma$. The Goldstone–Wilczek term now contributes $(1 - N_c(N_d - N_u)/N_f)$ times the correct vertex $\mathcal{L}_{\pi^0\gamma\gamma}$. The WZNW term, on the other hand, contains the traceless $SU(N_f)$ generator Q', which has diagonal elements Q'_u and Q'_d given by

$$Q'_u = Q_u - \frac{1}{2}\left(\frac{1}{N_c} - \frac{N_d - N_u}{N_f}\right) = \frac{N_d}{N_f},$$
$$Q'_d = Q_d - \frac{1}{2}\left(\frac{1}{N_c} - \frac{N_d - N_u}{N_f}\right) = -\frac{N_u}{N_f}. \tag{24.63}$$

The strength with which the WZNW term contributes to the vertex $\mathcal{L}_{\pi^0\gamma\gamma}$ is given by

$$N_c \left(Q_u'^{\,2} - Q_d'^{\,2} \right) = N_c \frac{N_d^2 - N_u^2}{N_f^2} = N_c \frac{N_d - N_u}{N_f}. \tag{24.64}$$

Hence, as before, the N_c-dependent part of the Goldstone–Wilczek term completely cancels the contribution of the WZNW term, while the N_c-independent part of the Goldstone–Wilczek term alone determines the correct vertex $\mathcal{L}_{\pi^0\gamma\gamma}$.

It is straightforward to convince oneself that the treatment for general N_f is completely consistent with what we found in the $N_f = 2$ case. For this purpose, we again embed a pion SU(2) matrix $\tilde{U}(x)$ into SU(N_f)

$$U(x) = \begin{pmatrix} \tilde{U}(x) & 0 \\ 0 & \mathbb{1} \end{pmatrix}. \tag{24.65}$$

Here, $\mathbb{1}$ is again the $(N_f - 2) \times (N_f - 2)$ unit-matrix. Then, only a 2×2 submatrix $\tilde{Q}'$ of the $N_f \times N_f$ matrix Q' enters the calculations. Using

$$\tilde{Q}' = \begin{pmatrix} Q_u' & 0 \\ 0 & Q_d' \end{pmatrix} = \frac{N_d - N_u}{2N_f} + \frac{1}{2}\tau^3,$$

$$\tilde{Q}'^2 = \begin{pmatrix} Q_u'^{\,2} & 0 \\ 0 & Q_d'^{\,2} \end{pmatrix} = \frac{N_d^2 + N_u^2}{2N_f^2} + \frac{N_d - N_u}{2N_f}\tau^3, \tag{24.66}$$

one can show that

$$N_c(S_{\text{WZNW}}[U,A] - S_{\text{WZNW}}[U]) + \left(1 - N_c \frac{N_d - N_u}{N_f}\right) S_{\text{GW}}[U,A] = S_{\text{GW}}[\tilde{U},A]. \tag{24.67}$$

This shows explicitly that the general N_f case is completely consistent with the $N_f = 2$ result of eq. (24.24). In particular, N_c does not enter explicitly in the low-energy effective theory of pions and photons. Hence, it is impossible to literally *see* the number of colors, by detecting the photons emitted in some low-energy process involving only pions and photons.

24.10 Can One See the Number of Colors?

This is the title of the paper (Bär and Wiese, 2001) that this section is based on. The question is whether the number of colors appears explicitly in the low-energy effective theory of Nambu–Goldstone bosons and photons, and whether there are physical processes in which photons are emitted at a rate that depends on N_c. As we have argued before, in contrast to the standard textbook knowledge, this is not the case for the neutral pion decay into two photons, or for any other low-energy process with $N_f = 2$. We have also seen that, for $N_f \geq 3$, there are strong interaction processes (described by the WZNW term) that change the number of Nambu–Goldstone bosons from even to odd, with an amplitude proportional to N_c. Unfortunately, these processes are difficult to detect in experiments and are therefore impractical for determining N_c at low energy.

We now refine the question and ask: Can one see the number of colors in the $N_f = 3$ case, *i.e.* in low-energy processes of pions, kaons, η-mesons (denoted by $\pi^1, \pi^2, \ldots, \pi^8$), and photons? First, let us consider processes involving η-mesons and photons only. We ignore mixing of π^8 with π^3 and with the flavor singlet state η^1, and thus identify $\eta^8 = \pi^8$. We then write

$$U(x) = \exp\left(i\eta^8(x)\lambda^8/F_\pi\right), \tag{24.68}$$

and we use $Q' = \lambda^3 + \lambda^8/\sqrt{3}$, where λ^3 and λ^8 are the two diagonal Gell-Mann generators of $SU(3)_{L=R}$. It is straightforward to show that

$$N_c(S_{\mathrm{WZNW}}[U,A] - S_{\mathrm{WZNW}}[U]) + (1 - N_c/3)S_{\mathrm{GW}}[U,A]$$
$$= \frac{e^2}{32\sqrt{3}\pi^2 F_\pi} \int d^4x \, \eta^8 \epsilon_{\mu\nu\rho\sigma} F_{\mu\nu} F_{\rho\sigma}, \tag{24.69}$$

which is again N_c-independent.

As a next step, let us consider processes involving η-mesons, pions, and photons. One can convince oneself that the vertex for the decay $\eta \to \pi^+\pi^-\gamma$ receives a contribution only from the WZNW term (and not also from the Goldstone–Wilczek term),

$$\mathcal{L}_{\eta^8\pi^+\pi^-\gamma} = \frac{eN_c}{4\sqrt{3}\pi^2 F_\pi^3} \epsilon_{\mu\nu\rho\sigma} A_\mu \partial_\nu \eta^8 \partial_\rho \pi^+ \partial_\sigma \pi^-. \tag{24.70}$$

Hence, at least to leading order in the low-energy expansion, the width of the corresponding decay is indeed proportional to N_c. Unfortunately, since the η-meson is rather heavy, chiral perturbation theory converges less well than in the $N_f = 2$ case. As a consequence, there are important sub-leading corrections which obscure the leading N_c-dependence, and in practice, it seems impossible to extract the value of N_c from experimental data for this decay (Borasoy and Lipartia, 2005).

While the vertex for the process $\eta^8 \to \gamma\gamma$ is N_c-independent, the one for the corresponding flavor singlet process is not. The quark triangle diagram that describes the decay $\eta^1 \to \gamma\gamma$ in the underlying Standard Model is proportional to

$$\mathrm{Tr}\left(\frac{1}{\sqrt{6}}Q^2\right) = \frac{N_c}{\sqrt{6}}\left(Q_u^2 + 2Q_d^2\right) = \frac{3}{4\sqrt{6}}\left(N_c + \frac{1}{N_c} - \frac{2}{3}\right), \tag{24.71}$$

and is thus N_c-dependent. As we have seen in Chapter 21, the physical η-meson is a mixture of the flavor octet η^8 and the flavor singlet η^1. The linear combination of η^1 and η^8 that is orthogonal to the η-meson is the η'-meson. As we have seen in Chapter 20, the η'-meson is not a Nambu–Goldstone boson, because the axial $U(1)_A$ symmetry is strongly, explicitly broken by the Adler–Bell–Jackiw anomaly, at least in the real world with $N_c = 3$. However, this anomaly disappears in the large-N_c limit, in which the η'-meson becomes a light pseudo-Nambu–Goldstone boson. For large N_c, one can incorporate both η^1 and η^8 (and hence the mixed states η and η') in the low-energy chiral Lagrangian (Bijnens, 1993; Kaiser and Leutwyler, 2000). Then, one can address the issues of η-meson mixing and the N_c-dependence of the anomalous decays $\eta, \eta' \to \gamma\gamma$ in the framework of the low-energy effective theory. Since in the real world the η'-meson is much heavier than the η-meson, again sub-leading corrections play an important role. When these corrections are taken into account, one concludes that one can indeed determine the value $N_c = 3$ from a comparison with experimental data for these decays (Borasoy, 2004).

24.11 Techni-Baryons, Techni-Skyrmions, and Topological Dark Matter

As we discussed in Section 15.12, in technicolor models the low-energy physics of techni-pions is described by the same mathematical formalism as the one for the pions in QCD, except that the relevant energy scale changes from the QCD scale F_π to the electroweak scale v. Let us now discuss skyrmions in the context of technicolor models – so-called *techni-skyrmions* (D'Hoker and Farhi, 1984b). To be specific, we again consider the gauge-anomaly-free $SU(N_t) \times SU(3)_c \times SU(2)_L \times U(1)_Y$ technicolor model (*e.g.*, with $N_t = 4$) that we discussed in Section 15.12. In this model, techni-baryons are bosons that consist of four techni-quarks. As we have seen, at the level of the low-energy pion effective theory for QCD, baryons manifest themselves as skyrmions. Similarly, in the low-energy description of technicolor models, techni-baryons manifest themselves as techni-skyrmions. As before, a Skyrme current describes the techni-baryon number, which is related to the homotopy group $\Pi_3[SU(2)] = \mathbb{Z}$. While baryon number is conserved at the classical level, at the quantum level it is affected by an anomaly due to electroweak instantons. The same is true for the techni-baryon number. Since in our specific technicolor model $N_t = 4$ is even, the techni-skyrmion must be quantized as a boson, which implies that the sign factor $\mathrm{Sign}[U] \in \Pi_4[SU(2)] = \mathbb{Z}(2)$ enters the techni-pion functional integral in the trivial form $\mathrm{Sign}[U]^{N_t} = 1$.

When we gauged $U(1)_Y$ in the chiral Lagrangian for the Standard Model with $N_f = 2$, we introduced the Goldstone–Wilczek term in order to account for the decay $\pi^0 \to \gamma\gamma$. When the neutral techni-pion is "eaten", it becomes the longitudinal component of the Z-boson, which might hence also undergo this anomalous decay. However, the width of the physical Z-boson is well described by the Standard Model, in which this process is absent. Hence, in the construction of technicolor models we should make sure that the techni-pion cannot decay into two photons. This implies that the electric charges of the techni-U- and techni-D-quarks should be equal with opposite signs. This is indeed the case in the concrete technicolor model considered here.

The techni-baryons in this model are bosons with equal spin S and techni-isospin I_t. The lightest techni-baryon is expected to have $S = I_t = 0$. It consists of two techni-U- and two techni-D-quarks and is thus electrically neutral. According to a naive constituent techni-quark model, one would also expect more massive excited techni-baryon resonances (analogous to the Δ-isobar in QCD) with $S = I_t = 1$ and 2, with electric charges $0, \pm 1$ and $0, \pm 1, \pm 2$, respectively. The conservation of the techni-baryon number (up to anomalous processes involving electroweak instantons) guarantees the stability of the lightest techni-baryon. Since this object is an electrically neutral color singlet, it could be a promising candidate for the *dark matter* in the Universe, which is about four times more abundant than the ordinary, luminous baryonic matter. In the low-energy effective theory of a technicolor model, techni-baryons appear as techni-skyrmions, which would then be a topological form of dark matter (Murayama and Shu, 2010). This takes us to another

---

Box 24.1   **Deep question**

What is the nature of the dark matter in the Universe?

At least to leading order, the low-energy effective theory for technicolor models is the same as for the Standard Model. Hence, one may wonder whether techni-skyrmion-like excitations can also arise in the Higgs sector of the Standard Model. If so, they would even provide a dark matter candidate within the Standard Model itself. Since the Higgs sector of the Standard Model represents a linear (rather than a non-linear) σ-model, a techni-skyrmion-like excitation in the angular part $U(x) \in SU(2)$ of the Higgs field matrix $\Phi(x) = |\Phi(x)|U(x)$ can unwind itself and is thus not topologically stable. Unwinding can happen when the magnitude of the Higgs field goes to $|\Phi(x)| = 0$ at some space–time point x. While this is energetically disfavored by the Mexican hat potential $V(\Phi) = (\lambda/4!)(|\Phi|^2 - v^2)^2$, it is not impossible. In addition, in the Standard Model (just like in a non-linear σ-model without a stabilizing Skyrme term) a techni-skyrmion-like excitation of the Higgs field is unstable against shrinking to a point. A non-renormalizable stabilizing term may possibly be induced by some physics beyond the Standard Model, such as technicolor.

Exercises

24.1 Baryon number of the skyrmion

Show that the baryon number of a skyrmion, as parameterized in eq. (24.7), is indeed equal to one.

24.2 Topological conservation law

Prove that the Skyrme current

$$j_\mu^S(x) = \frac{1}{24\pi^2}\epsilon_{\mu\nu\rho\sigma}\mathrm{Tr}\left[\left(U(x)^\dagger \partial_\nu U(x)\right)\left(U(x)^\dagger \partial_\rho U(x)\right)\left(U(x)^\dagger \partial_\sigma U(x)\right)\right]$$

is conserved, *i.e.* $\partial_\mu j_\mu^S(x) = 0$, independent of the classical equations of motion of the low-energy pion theory.

24.3 Discrete symmetries of pion field and Skyrme current

Consider the discrete symmetries of charge conjugation C, parity P, and intrinsic parity P_0 (G-parity for $N_f = 2$). They act on the pion field as

$$^C\pi^\pm(\vec{x}, x_4) = \pi^\mp(\vec{x}, x_4), \ ^C\pi^0(\vec{x}, x_4) = \pi^0(\vec{x}, x_4),$$
$$^P\pi^a(\vec{x}, x_4) = -\pi^a(-\vec{x}, x_4), \ ^{P_0}\pi^a(\vec{x}, x_4) = -\pi^a(\vec{x}, x_4).$$

How do these symmetries act on the Nambu–Goldstone boson field $U(\vec{x}, x_4) \in SU(2)$ of the low-energy effective theory? How does the Skyrme current transform under these symmetries? Is G-parity an exact symmetry of two-flavor QCD?

24.4 Coefficient in the Goldstone–Wilczek current

Replace the ordinary derivatives in the Skyrme current with covariant derivatives by gauging $SU(2)_L$. Modify this current to the Goldstone–Wilczek current

$$j_\mu^{\mathrm{GW}}(x) = \frac{1}{24\pi^2}\epsilon_{\mu\nu\rho\sigma}\mathrm{Tr}\left[\left(U(x)^\dagger D_\nu U(x)\right)\left(U(x)^\dagger D_\rho U(x)\right)\left(U(x)^\dagger D_\sigma U(x)\right)\right]$$
$$+ C\epsilon_{\mu\nu\rho\sigma}\mathrm{Tr}\left[W_{\nu\rho}(x)\left(D_\sigma U(x)U(x)^\dagger\right)\right].$$

Fix the constant C in order to reproduce the anomaly in the baryon number

$$\partial_\mu j_\mu^{\mathrm{GW}}(x) = -\frac{1}{32\pi^2}\epsilon_{\mu\nu\rho\sigma}\mathrm{Tr}\left[W_{\mu\nu}(x)W_{\rho\sigma}(x)\right].$$

24.5 Goldstone–Wilczek current and the photon
Construct the baryon number Goldstone–Wilczek current in the presence of an electromagnetic gauge field. Show that there are no baryon number violating processes induced by photons. Consider the low-energy effective action of pions coupled to photons. Which of the discrete symmetries C, P, and P_0 are exact symmetries of two-flavor QCD coupled to QED?

24.6 Symmetries of the Wess–Zumino–Novikov–Witten term
Consider the discrete symmetries C, P, and P_0 in the case of $N_f \geq 3$ flavors. Which of these are exact symmetries of QCD alone and of QCD and QED combined?

24.7 Consistency of the $N_f \geq 3$ Wess–Zumino–Novikov–Witten term with $N_f = 2$
Verify eq. (24.67) in order to convince yourself that the $N_f \geq 3$ Wess–Zumino–Novikov–Witten term is consistent with the $N_f = 2$ Goldstone–Wilczek term.

24.8 N_c-independence of the decay $\eta^8 \to \gamma\gamma$
Verify eq. (24.69) in order to convince yourself that the width of the decay $\eta^8 \to \gamma\gamma$ is independent of the number of colors.

24.9 A useful formula in five dimensions
Generalize eq. (19.18) to five dimensions; *i.e.* show that for $U(x) = V(x)W(x)$

$$\epsilon_{\mu\nu\rho\sigma\lambda}\mathrm{Tr}\left[\left(U^\dagger\partial_\mu U\right)\left(U^\dagger\partial_\nu U\right)\left(U^\dagger\partial_\rho U\right)\left(U^\dagger\partial_\sigma U\right)\left(U^\dagger\partial_\lambda U\right)\right] =$$
$$\epsilon_{\mu\nu\rho\sigma\lambda}\mathrm{Tr}\left[\left(V^\dagger\partial_\mu V\right)\left(V(x)^\dagger\partial_\nu V\right)\left(V^\dagger\partial_\rho V\right)\left(V^\dagger\partial_\sigma V\right)\left(V^\dagger\partial_\lambda V\right)\right] +$$
$$\epsilon_{\mu\nu\rho\sigma\lambda}\mathrm{Tr}\left[\left(W^\dagger\partial_\mu W\right)\left(W^\dagger\partial_\nu W\right)\left(W^\dagger\partial_\rho W\right)\left(W^\dagger\partial_\sigma W\right)\left(W^\dagger\partial_\lambda W\right)\right] + \partial_\mu K_\mu,$$

and determine $K_\mu(x)$ explicitly.

PART IV

SELECTED TOPICS BEYOND
THE STANDARD MODEL

25

Strong CP-Problem

We have seen in Chapter 19 that SU(N) gauge fields have non-trivial topological structures. In particular, classical vacua (*i.e.* pure gauge configurations) are characterized by an integer winding number in the homotopy group $\Pi_3[\mathrm{SU}(N)] = \mathbb{Z}$. Instantons are examples of Euclidean field configurations with topological charge $Q = 1$ that describe tunneling processes between topologically distinct classical vacua. Due to tunneling, the quantum vacuum is a linear superposition of classical vacua characterized by a vacuum-angle $\theta \in]-\pi, \pi]$. In the Euclidean action, the vacuum-angle manifests itself as an additional term $-i\theta Q$. For $\theta \neq 0$ or π, this term explicitly breaks the CP symmetry. As a consequence, the neutron would have an electric dipole moment which (for small θ) is proportional to θ (Crewther *et al.*, 1979); without CP violation, the dipole moment vanishes. Indeed, the observed electric dipole moment of the neutron is indistinguishable from zero. This puts a stringent bound on the vacuum-angle $|\theta| \lesssim 10^{-10}$ (Baker *et al.*, 2006; Pendlebury *et al.*, 2015; Abel *et al.*, 2020). The question arises why in Nature $\theta = 0$ holds to such a high accuracy. This puzzle is the *strong CP-problem*.

In the Standard Model, the Yukawa couplings lead to CP violation which is indeed observed in the neutral kaon and B-meson decays, as we discussed in Section 17.5. This effect is rather subtle and requires the presence of at least three fermion generations. If there were significant *CP violation in the strong interaction*, it would give rise to more drastic effects.

One might hope to solve the strong CP-problem by the assumption that gluon field configurations with $Q \neq 0$ are very much suppressed in the functional integral. However, this is not the case. In particular, the quantitative solution to the U(1)$_A$-problem relies on the fact that gluon field configurations with $Q \neq 0$ do appear frequently in pure gauge theory (cf. Chapter 20), which is confirmed by non-perturbative lattice calculations of SU(3)$_c$ Yang–Mills theory.

Of course, this need not necessarily be the case in full QCD with dynamical quarks. Indeed, if the u-quark were massless, the Atiyah–Singer index theorem would imply that fermionic zero-modes of the Dirac operator completely eliminate gluon field configurations with $Q \neq 0$ from the functional integral. In that case, the θ-vacuum term would vanish and all θ-vacua would be physically equivalent and CP-conserving. For some time, it was a controversial issue whether or not the u-quark could indeed be massless. However, by now the accuracy of chiral perturbation theory as well as lattice QCD calculations has increased to the point where this possibility is excluded. In any case, if the u-quark would indeed be massless, and the strong CP-problem would resolve itself in that way, we would immediately face an "m_u-problem": Why should the u-quark be massless?

In Section 19.1, we have seen that the Chern–Pontryagin topological charge density is intimately connected with the divergence of the flavor-singlet axial current. This implies that the vacuum-angle can be rotated using an axial U(1)$_A$ transformation. However, it cannot be eliminated in this manner, because it just gets transformed into a complex phase of the determinant of the quark mass matrix. Still, this transformation is useful: For example, it enables the investigation of the θ-vacuum dynamics using chiral Lagrangians. For unequal u- and d-quark masses, one finds a phase transition at $\theta = \pi$ at which CP is spontaneously broken (Dashen, 1971). Hence, despite the fact that $\theta = \pi$ does not break CP explicitly, the CP symmetry would then be broken dynamically. However, observation implies that θ cannot be equal to π in Nature and must indeed be very close to zero.

The chiral Lagrangian method also allows us to study θ-vacuum effects in the large-N_c limit, keeping the 't Hooft coupling $g_s^2 N_c$ fixed. In this limit, the axial $U(1)_A$ anomaly vanishes and the η'-meson becomes a massless Nambu–Goldstone boson. In fact, the η'-meson couples to the complex phase of the quark mass matrix – and thus to θ – and can indeed be used to rotate θ away. Hence, there is no strong CP-problem at $N_c = \infty$. Of course, we know that in our world $N_c = 3$, so we do face the strong CP-problem.

A promising solution to the strong CP-problem was proposed by Roberto Peccei and Helen Quinn (Peccei and Quinn, 1977a,b). They suggested an extension of the Standard Model with two Higgs doublets (cf. Section 12.12). The presence of the second Higgs field entails an additional $U(1)_{PQ}$ symmetry – known as a *Peccei–Quinn symmetry* – which allows one to absorb θ in a new dynamical field, even at finite N_c. The $U(1)_{PQ}$ Peccei–Quinn symmetry is assumed to break spontaneously below a very high energy scale Λ_{PQ}. It was pointed out independently by Steven Weinberg (1978) and by Frank Wilczek (1978) that this leads to a pseudo-Nambu–Goldstone boson – the so-called *axion*. However, despite numerous experimental efforts, up to 2022, no axion has been detected, and it is an open question whether this is indeed the correct solution of the strong CP-problem. Although the original Peccei–Quinn model was soon ruled out by experiments, the symmetry breaking scale of the model can be shifted to higher energy scales making the axion more weakly coupled, such that it becomes very hard to detect, *i.e.* almost invisible.

If they exist, axions are interesting players in the Universe: They couple only weakly to ordinary matter, but they still induce interesting effects. First of all, they could shorten the life-time of stars. In addition, they would provide a component of the dark matter in the Universe. Axions could be generated in the early Universe in different ways. For example, they could just be thermally produced. Then, they could be generated by a disoriented $U(1)_{PQ}$ condensate. Furthermore, the spontaneous breakdown of a U(1) symmetry is generically accompanied by the generation of topological line defects – so-called *cosmic strings* – that could extend throughout the Universe. Indeed, if axions exist, axionic cosmic strings should exist as well. A network of such fluctuating strings could radiate energy by emitting the corresponding pseudo-Nambu–Goldstone bosons – namely, axions – and thus generate them as a dark matter component.

It is also interesting to consider the vacuum-angles θ_2 and θ_1 that are associated with the $SU(2)_L$ and $U(1)_Y$ gauge fields, respectively. As we will see in this chapter, θ_2 can be eliminated by an appropriate anomalous baryon number transformation of the quark fields in the background of a topologically non-trivial $SU(2)_L$ gauge field. The Abelian vacuum-angle θ_1 is associated with several open questions, and it is interesting to ask whether it gives rise to yet another CP-problem, now associated with the $U(1)_Y$ gauge symmetry.

25.1 Rotating θ into the Mass Matrix

Let us assume that there is a θ-vacuum term $\exp(i\theta Q[G])$ in the Euclidean functional integral of QCD with

$$Q[G] = -\frac{1}{32\pi^2} \int d^4x\, \epsilon_{\mu\nu\rho\sigma} \text{Tr}\left[G_{\mu\nu} G_{\rho\sigma}\right]. \tag{25.1}$$

We have already seen that such a term is intimately connected with the flavor-singlet axial $U(1)_A$ symmetry. Indeed, due to the non-invariance of the fermionic measure under axial transformations, the axial current is anomalous. Let us discuss this in the theory with N_f quark flavors. Under a flavor-singlet axial $U(1)_A$ transformation

$$\begin{aligned}
q_L^{f'}(x) &= \exp\left(i\theta/2N_f\right) q_L^f(x), \\
q_R^{f'}(x) &= \exp\left(-i\theta/2N_f\right) q_R^f(x),
\end{aligned} \tag{25.2}$$

the fermion determinant in the background of a gluon field with topological charge $Q[G]$ is not invariant: It changes by $\exp(i\,\delta_A\theta\, Q[G])$ with

$$\delta_A\theta = -N_f\frac{\theta}{2N_f} - N_f\frac{\theta}{2N_f} = -\theta \quad \Rightarrow \quad \theta' = \theta + \delta_A\theta = 0. \tag{25.3}$$

Hence, the above axial transformation can be used to eliminate the pre-existing θ-vacuum-angle. Of course, the transformation must be applied consistently everywhere. It cancels out in the quark–gluon gauge interactions, which are chirally invariant, but not in the mass terms. In fact, the mass matrix $\mathcal{M} = \text{diag}(m_u, m_d, \ldots, m_{N_f})$ then turns into

$$\mathcal{M}' = \text{diag}\left(m_u \exp(i\theta/N_f), m_d \exp(i\theta/N_f), \ldots, m_{N_f} \exp(i\theta/N_f)\right); \tag{25.4}$$

i.e. θ turns into the complex phase of the determinant of the quark mass matrix. If one of the quarks is massless, the determinant vanishes and its phase becomes irrelevant. Interestingly, strong CP violation would manifest itself by a complex phase in the quark mass matrix, while the CP violation due to the Yukawa couplings leads to the complex phase in the Cabibbo–Kobayashi–Maskawa quark mixing matrix. The latter requires at least three fermion generations, whereas the strong CP violation due to a non-vanishing θ-angle applies to any number of generations.

25.2 θ-Angle in Chiral Perturbation Theory

Let us now discuss how the vacuum-angle affects the low-energy QCD dynamics. Since we know how the quark mass matrix enters the chiral Lagrangian, and since θ is just the complex phase of the determinant of that matrix, it is clear how to include θ in chiral perturbation theory. To lowest order, the chiral perturbation theory action then takes the form (cf. eq. (23.10))

$$S[U] = \int d^4x \left\{\frac{F_\pi^2}{4}\text{Tr}\left[\partial_\mu U^\dagger \partial_\mu U\right] - \frac{\Sigma}{2}\text{Tr}\left[\mathcal{M}'^\dagger U + U^\dagger \mathcal{M}'\right]\right\}, \tag{25.5}$$

where $\mathcal{M}'$ is the θ-dependent quark mass matrix of eq. (25.4). The above action is not 2π-periodic in θ: It is only $2\pi N_f$-periodic. In an exercise, we will verify that the resulting functional integral

$$Z(\theta) = \int \mathcal{D}U \, \exp(-S[U]) \tag{25.6}$$

is indeed 2π-periodic in θ. For this purpose, we add 2π to θ, which amounts to a multiplication of $\mathcal{M}'$ with the diagonal matrix

$$z = \exp(2\pi \, \mathrm{i}/N_f)\mathbb{1} \in \mathbb{Z}(N_f), \tag{25.7}$$

which belongs to the center $\mathbb{Z}(N_f)$ of the flavor group $SU(N_f)$. Redefining $U'(x) = z^\dagger U(x)$ and using the invariance of the measure, $\mathcal{D}U' = \mathcal{D}U$, one confirms the 2π-periodicity $Z(\theta + 2\pi) = Z(\theta)$.

The situation in QCD itself is similar. While the contribution $-\mathrm{i}\theta Q$ to the action is not periodic in θ, it enters the functional integral through the 2π-periodic Boltzmann factor $\exp(\mathrm{i}\theta Q)$. Hence, the functional integral is periodic, while the action itself is not.

Let us confirm that a non-zero vacuum-angle indeed breaks CP. On the level of the chiral Lagrangian, charge conjugation takes the form $^C U(x) = U(x)^\mathsf{T}$, while parity corresponds to $^P U(\vec{x}, x_4) = U(-\vec{x}, x_4)^\dagger$. The action (25.5) breaks P but is C-invariant; hence, it indeed violates CP.

Let us now examine the effect of θ on the vacuum of the pion theory in the $N_f = 2$ case. Then, the mass matrix takes the form

$$\mathcal{M}' = \mathrm{diag}\left(m_u \exp(\mathrm{i}\theta/2), m_d \exp(\mathrm{i}\theta/2)\right). \tag{25.8}$$

In order to find the vacuum configuration, we minimize the potential energy; hence, we maximize

$$\mathrm{Tr}\left[\mathcal{M}'^\dagger U + U^\dagger \mathcal{M}'\right] = m_u \cos\left(\frac{\theta}{2} + \varphi\right) + m_d \cos\left(\frac{\theta}{2} - \varphi\right). \tag{25.9}$$

Here, we have parameterized $U = \mathrm{diag}\left(\exp(\mathrm{i}\varphi), \exp(-\mathrm{i}\varphi)\right)$. The minimum energy configuration has

$$\tan \varphi = \frac{m_d - m_u}{m_u + m_d} \tan \frac{\theta}{2}. \tag{25.10}$$

As expected, for $\theta = 0$ one obtains $\varphi = 0$ and hence $U(x) = \mathbb{1}$. It is interesting that the θ-angle affects the pion vacuum only for non-degenerate quark masses, at least to leading order. At $\theta = \pi$, the pion vacuum configurations have $\varphi = \pm\pi/2$, i.e. $U(x) = \pm\mathrm{diag}(\mathrm{i}, -\mathrm{i})$. These two vacua are not CP-invariant. Instead, they are CP partners of one another. This confirms that, at $\theta = \pi$, the CP symmetry is spontaneously broken (Dashen, 1971). Hence, despite the fact that for $\theta = \pi$ there is no explicit strong CP violation, the symmetry is not intact either. Observations are indeed compatible with $\theta = 0$, but by no means with $\theta = \pi$.

25.3 θ-Angle at Large N_c

In Chapter 20, we have seen that the $U(1)_A$-problem can be understood quantitatively in the limit of many colors N_c. At $N_c = \infty$, the anomalous axial $U(1)_A$ symmetry is restored and the η'-meson becomes a massless Nambu–Goldstone boson. As we discussed in Chapter 20, at large but finite N_c, the η'-meson is a pseudo-Nambu–Goldstone boson with a mass squared (in the chiral limit)

$$M_{\eta'}^2 = \frac{2N_f \chi_t}{F_\pi^2} \qquad (25.11)$$

that is proportional to $1/N_c$ (note that F_π^2 is of order N_c) (Witten, 1979b; Veneziano, 1979). Here, $\chi_t = \langle Q^2 \rangle / V$ is the topological susceptibility of the pure gluon theory which is of order 1 with respect to the large-N_c limit.

Since for large N_c the η'-meson becomes light, it must be included in the low-energy chiral Lagrangian (Witten, 1979b; Kaiser and Leutwyler, 2000). In the limit $N_c \rightarrow \infty$, the axial $U(1)_A$ symmetry is restored in the action and breaks spontaneously. Hence, the chiral symmetry is then $U(N_f)_L \times U(N_f)_R$, which is broken to $U(N_f)_{L=R}$. Consequently, the effective Nambu–Goldstone boson fields now reside in the coset space

$$U(N_f)_L \times U(N_f)_R / U(N_f)_{L=R} = U(N_f), \qquad (25.12)$$

so there are N_f^2 Nambu–Goldstone bosons. The additional η' Nambu–Goldstone boson is described by the complex phase of the determinant of a unitary matrix $\widetilde{U} \in U(N_f)$, which would have determinant 1 if the η'-meson were heavy

$$\det \widetilde{U}(x) = \exp\left(i\sqrt{2N_f}\eta'(x)/F_\pi \right). \qquad (25.13)$$

For large N_c, the leading terms in the chiral perturbation theory action take the form

$$S[\widetilde{U}] = \int d^4x \left\{ \frac{F_\pi^2}{4}\mathrm{Tr}\left[\partial_\mu \widetilde{U}^\dagger \partial_\mu \widetilde{U}\right] - \frac{\Sigma}{2}\mathrm{Tr}\left[\mathcal{M}'^\dagger \widetilde{U} + \widetilde{U}^\dagger \mathcal{M}'\right] + \frac{\chi_t}{2}\left(i \log \det \widetilde{U}\right)^2 \right\}. \qquad (25.14)$$

In an exercise, we will convince ourselves that this action implies the η'-mass of eq. (25.11).

If there is a non-zero θ-angle in the quark mass matrix $\mathcal{M}'$, this angle can now be absorbed into the η'-meson field, i.e. in the complex phase of the determinant of the Nambu–Goldstone boson field $\widetilde{U}$. Then, the action turns into

$$S[\widetilde{U}] = \int d^4x \left\{ \frac{F_\pi^2}{4}\mathrm{Tr}\left[\partial_\mu \widetilde{U}^\dagger \partial_\mu \widetilde{U}\right] - \frac{\Sigma}{2}\mathrm{Tr}\left[\mathcal{M}^\dagger \widetilde{U} + \widetilde{U}^\dagger \mathcal{M}\right] + \frac{\chi_t}{2}\left(i \log \det \widetilde{U} - \theta\right)^2 \right\}. \qquad (25.15)$$

In the large-N_c limit, the last term, which is of order 1, can be neglected compared to the other terms which are of order N_c. Hence, at $N_c = \infty$, the vacuum-angle drops out of the theory, and all θ-vacua become physically equivalent. We see that, for infinitely many colors, there is no strong CP-problem. Essentially, after its anomaly is removed, the $U(1)_A$ symmetry allows us to eliminate θ, despite the fact that this symmetry is still explicitly broken by the quark masses.

25.4 Peccei–Quinn Symmetry

At finite N_c and in the absence of an exactly massless quark flavor, the $U(1)_A$ symmetry is inevitably broken by the axial anomaly. Therefore, we are not able to eliminate θ by using that symmetry. The idea of Peccei and Quinn (Peccei and Quinn, 1977a,b) was to introduce an additional $U(1)_{PQ}$ symmetry – now known as the *Peccei–Quinn symmetry*. This symmetry will allow us to absorb θ in an additional dynamical field (to be called $a(x)$), which will reach its minimal energy at $a(x) = 0$, thus solving the strong CP-problem.

We will discuss the Peccei–Quinn scenario in the context of the single-generation Standard Model. The generalization to more generations is straightforward. It should again be noted that with fewer than three generations, there is no CP-violating phase in the quark mixing matrix and θ would be the only source of CP violation.

Let us recall how the u- and d-quarks obtain their masses. As we have seen in Section 16.1, the mass of the d-quark, $m_d = f_d v$, is due to the Yukawa coupling of eq. (16.9),

$$\mathcal{L}(\bar{q}_L, q_L, \bar{d}_R, d_R, \Phi) = f_d \left[(\bar{u}_L, \bar{d}_L) \begin{pmatrix} \Phi^+ \\ \Phi^0 \end{pmatrix} d_R + \bar{d}_R (\Phi^{+*}, \Phi^{0*}) \begin{pmatrix} u_L \\ d_L \end{pmatrix} \right], \quad (25.16)$$

while the mass of the u-quark (cf. Section 16.2, eq. (16.12)), $m_u = f_u v$, is due to the term

$$\mathcal{L}(\bar{q}_L, q_L, \bar{u}_R, u_R, \widetilde{\Phi}) = f_u \left[(\bar{u}_L, \bar{d}_L) \begin{pmatrix} \widetilde{\Phi}^0 \\ \widetilde{\Phi}^- \end{pmatrix} u_R + \bar{u}_R (\widetilde{\Phi}^{0*}, \widetilde{\Phi}^{-*}) \begin{pmatrix} u_L \\ d_L \end{pmatrix} \right]. \quad (25.17)$$

In the Standard Model, the Higgs field $\widetilde{\Phi}$ is constructed out of the Higgs field Φ as

$$\widetilde{\Phi}(x) = \begin{pmatrix} \widetilde{\Phi}^0(x) \\ \widetilde{\Phi}^-(x) \end{pmatrix} = \begin{pmatrix} \Phi_0(x)^* \\ -\Phi^+(x)^* \end{pmatrix}. \quad (25.18)$$

When we write the Higgs field as the matrix

$$\boldsymbol{\Phi}(x) = \begin{pmatrix} \Phi^0(x)^* & \Phi^+(x) \\ -\Phi^+(x)^* & \Phi^0(x) \end{pmatrix} = \left(\widetilde{\Phi}(x), \Phi(x) \right), \quad (25.19)$$

both Yukawa couplings can be combined into one expression (cf. eq. (16.16))

$$\mathcal{L}(\bar{q}_L, q_L, \bar{u}_R, u_R, \bar{d}_R, d_R, \boldsymbol{\Phi}) = (\bar{u}_L, \bar{d}_L) \boldsymbol{\Phi} \mathcal{F} \begin{pmatrix} u_R \\ d_R \end{pmatrix} + (\bar{u}_R, \bar{d}_R) \mathcal{F}^\dagger \boldsymbol{\Phi}^\dagger \begin{pmatrix} u_L \\ d_L \end{pmatrix}, \quad (25.20)$$

where $\mathcal{F} = \mathrm{diag}(f_u, f_d)$ is the diagonal matrix of Yukawa couplings. When a θ-term is present in the QCD Lagrangian, it can be rotated into the matrix of Yukawa couplings by the transformation

$$\begin{pmatrix} u_L'(x) \\ d_L'(x) \end{pmatrix} = \exp(i\theta/4) \begin{pmatrix} u_L(x) \\ d_L(x) \end{pmatrix}, \quad \begin{pmatrix} u_R'(x) \\ d_R'(x) \end{pmatrix} = \exp(-i\theta/4) \begin{pmatrix} u_R(x) \\ d_R(x) \end{pmatrix}. \quad (25.21)$$

This turns the matrix of Yukawa couplings into

$$\mathcal{F}' = \mathrm{diag}\left(f_u \exp(i\theta/2), f_d \exp(i\theta/2) \right). \quad (25.22)$$

The Higgs field matrix $\boldsymbol{\Phi}$ takes the form of an $SU(2)$ matrix (with determinant 1), multiplied by a positive real number. Therefore, the complex phase $\exp(i\theta)$ of the matrix of Yukawa couplings cannot be absorbed into it. Hence, θ cannot be eliminated. Here, we have assumed f_u and f_d to be real. Otherwise, the effective vacuum-angle would still be the complex phase of the determinant of $\mathcal{F}'$.

It is instructive to include the Yukawa couplings in the chiral perturbation theory action

$$S[U, \Phi] = \int d^4x \left\{ \frac{F_\pi^2}{4} \mathrm{Tr}\left[\partial_\mu U^\dagger \partial_\mu U \right] - \frac{\Sigma}{2} \mathrm{Tr}\left[\mathcal{F}'^\dagger \Phi^\dagger U + U^\dagger \Phi \mathcal{F}' \right] \right\}. \tag{25.23}$$

Again, the complex phase in $\mathcal{F}'$ cannot be absorbed into the Higgs field matrix Φ because it is an SU(2) matrix times a positive real number. The Nambu–Goldstone boson matrix U is also an SU(2) matrix, and hence, θ cannot be eliminated. As we have seen, θ can actually be rotated away if the Nambu–Goldstone boson matrix is in U(2) and contains the η'-meson field as a complex phase of its determinant. This, however, is the case only for $N_c = \infty$.

The basic idea of Peccei and Quinn can be boiled down to extending the Standard Model Higgs field to a matrix proportional to an element of U(2) – not just of SU(2). The additional angular degree of freedom effectively turns θ from a constant into a dynamical field $a(x)$. The extra U(1)$_\mathrm{PQ}$ Peccei–Quinn symmetry then allows for global shifts of $a(x)$.

The actual proposal of Peccei and Quinn does more than just this. It introduces two independent Higgs doublets Φ and $\widetilde{\Phi}$ which can be combined to form a GL(2, $\mathbb{C}$) matrix. Working with GL(2, $\mathbb{C}$) rather than with U(2) matrices ensures that the Higgs sector is described by a renormalizable, linear σ-model, instead of a perturbatively non-renormalizable, nonlinear σ-model. For our purposes, however, this is irrelevant.

For simplicity, we will not follow Peccei and Quinn all the way and introduce two Higgs fields. Instead, we will just extend the standard Higgs field to a matrix Φ' that is given by a U(2) matrix multiplied by a positive real number. This means that we introduce just one additional degree of freedom, while Peccei and Quinn introduced four. The complex phase in $\mathcal{F}'$ can then be absorbed by a redefinition of Φ' and one obtains

$$S[U, \Phi'] = \int d^4x \left\{ \frac{F_\pi^2}{4} \mathrm{Tr}\left[\partial_\mu U^\dagger \partial_\mu U \right] - \frac{\Sigma}{2} \mathrm{Tr}\left[\mathcal{F}^\dagger \Phi'^\dagger U + U^\dagger \Phi' \mathcal{F} \right] \right\}. \tag{25.24}$$

Since this expression now contains the original, real Yukawa coupling matrix $\mathcal{F}$, all manifestations of the constant vacuum-angle θ have completely disappeared from the theory. Instead, the dynamical complex phase $\exp\left(\mathrm{i}a(x)/\Lambda_\mathrm{PQ}\right)$ of $\Phi'(x)$ now takes the role of θ. Here, $a(x)$ is the *axion field*. In particular, $a(x)/\Lambda_\mathrm{PQ}$ behaves like a space–time-dependent θ-vacuum-angle. Here, Λ_PQ is a potentially very high Peccei–Quinn energy scale.

Irrespective of its origin at the scale Λ_PQ, at Standard Model energy scales the axion amounts to an extension of the field content. It should be pointed out that the axion is "sterile" in the sense that it transforms trivially under all Standard Model gauge transformations. Two important axion contributions to the Lagrangian then take the form

$$\mathcal{L}(a, G) = \frac{1}{2} \partial_\mu a \partial_\mu a - \mathrm{i} \frac{a}{\Lambda_\mathrm{PQ}} \frac{1}{32\pi^2} \epsilon_{\mu\nu\rho\sigma} \mathrm{Tr}\left[G_{\mu\nu} G_{\rho\sigma} \right]. \tag{25.25}$$

Since, like other scalar fields, the axion field has mass dimension 1, on dimensional grounds one might also expect a quartic potential $V(a)$ as well as a mixed term $a^2 \Phi^\dagger \Phi$ with a dimensionless prefactor. However, since $a(x)/\Lambda_\mathrm{PQ}$ represents an angle with 2π-periodicity, one can argue that these terms are forbidden. Besides its free propagation described by the kinetic dimension-4 term, the axion interacts with the Standard Model fields via the gluon topological charge density. This interaction represents a dimension-5 operator that is suppressed by the Peccei–Quinn scale Λ_{PQ}. While several other terms are also forbidden because they are not 2π-periodic in $a(x)/\Lambda_{PQ}$, there are further allowed dimension-5 terms that couple $\partial_\mu a$ to gauge-invariant fermionic currents.

Quick Question 25.4.1   Axion–fermion coupling
Construct some dimension-5 operators that couple the axion to fermions.

25.5   U(1)$_{PQ}$ Symmetry Breaking and the Axion

The scalar potential $V(\Phi')$ in the extension of the Standard Model is invariant against $SU(2)_L \times SU(2)_R \times U(1)_{PQ}$ transformations. While the $U(1)_{PQ}$ symmetry is spontaneously broken below the scale Λ_{PQ}, the Higgs field takes the vacuum value $\Phi = \text{diag}(v, v)$. At low energy, the full symmetry is spontaneously broken down to $SU(2)_{L=R}$. Hence, there are $3 + 3 + 1 - 3 = 4$ massless Nambu–Goldstone bosons. As usual, we then gauge $SU(2)_L$ as well as the $U(1)_Y$ subgroup of $SU(2)_R$, which amounts to a partial explicit breaking of $SU(2)_R$. The unbroken subgroup of $SU(2)_{L=R}$ is just $U(1)_{em}$. Via the Higgs mechanism, three of the four Nambu–Goldstone bosons are "eaten" by the gauge bosons and become the longitudinal components of the electroweak bosons $W^\pm$ and Z. Since $U(1)_{PQ}$ remains a global symmetry, the fourth Nambu–Goldstone boson does not get "eaten". This Nambu–Goldstone boson is the *axion*, whose emergence from the Peccei–Quinn scenario was realized independently by Weinberg (1978) and Wilczek (1978).

Let us now construct a low-energy effective theory that contains all Nambu–Goldstone bosons of the extended Standard Model, namely, the pions and the axion. This is easy to do, because we have already included Φ' in the chiral Lagrangian. After spontaneous symmetry breaking at the electroweak scale v, we write

$$\Phi'(x) = v\, \text{diag}\left(\exp(i a(x)/\Lambda_{PQ}), \exp(i a(x)/\Lambda_{PQ})\right), \tag{25.26}$$

where $a(x)$ still parameterizes the pseudo-scalar axion field. Similarly, we write

$$U(x) = \text{diag}\left(\exp\left(i\pi^0(x)/F_\pi\right), \exp\left(-i\pi^0(x)/F_\pi\right)\right). \tag{25.27}$$

Of course, the field $U(x)$ also contains the charged pions. However, at this point, we are interested in *axion–pion mixing*. Since the axion is electrically neutral, it cannot mix with the charged pions and we thus ignore them. Let us first search for the vacuum of the axion–pion system. Minimizing the energy implies maximizing

$$\text{Tr}\left[\mathcal{F}^\dagger \Phi'^\dagger U + U^\dagger \Phi' \mathcal{F}\right] = m_u \cos\left(a/\Lambda_{PQ} + \pi^0/F_\pi\right) + m_d \cos\left(a/\Lambda_{PQ} - \pi^0/F_\pi\right). \tag{25.28}$$

Obviously, this expression is maximal for $a(x) = \pi^0(x) = 0$. Next we expand around this vacuum to second order in the fields. The resulting mass-squared matrix takes the form

$$M^2 = \Sigma \begin{pmatrix} (m_u + m_d)/F_\pi^2 & (m_u - m_d)/(F_\pi \Lambda_{PQ}) \\ (m_u - m_d)/(F_\pi \Lambda_{PQ}) & (m_u + m_d)/\Lambda_{PQ}^2 \end{pmatrix}. \tag{25.29}$$

In the limit $\Lambda_{\text{PQ}}/F_\pi \to \infty$, this matrix is reduced to

$$M^2 = \Sigma \begin{pmatrix} (m_u + m_d)/F_\pi^2 & 0 \\ 0 & 0 \end{pmatrix}, \tag{25.30}$$

from which we read off the familiar Gell-Mann–Oakes–Renner relation for the mass-squared of the pion,

$$M_\pi^2 = \frac{(m_u + m_d)\Sigma}{F_\pi^2}. \tag{25.31}$$

In this limit, the axion remains massless and there is no axion–pion mixing.

Next, we keep $\Lambda_{\text{PQ}}/F_\pi$ finite, but we still assume $\Lambda_{\text{PQ}} \gg F_\pi$. Then, there is a small amount of mixing between the axion and the pion, but to leading order the pion mass remains unaffected. The determinant of the mass-squared matrix M^2 is given by

$$\Sigma^2 \left[\frac{(m_u + m_d)^2}{F_\pi^2 \Lambda_{\text{PQ}}^2} - \frac{(m_u - m_d)^2}{F_\pi^2 \Lambda_{\text{PQ}}^2} \right] = \Sigma^2 \frac{4 m_u m_d}{F_\pi^2 \Lambda_{\text{PQ}}^2} = M_\pi^2 M_a^2. \tag{25.32}$$

Hence, the axion mass squared is given by

$$M_a^2 = \frac{4 m_u m_d \Sigma}{(m_u + m_d) \Lambda_{\text{PQ}}^2}. \tag{25.33}$$

It vanishes in the chiral limit, even if just one of the quark masses is zero. The ratio of the axion and pion mass squares is

$$\frac{M_a^2}{M_\pi^2} = \frac{4 m_u m_d F_\pi^2}{(m_u + m_d)^2 \Lambda_{\text{PQ}}^2}. \tag{25.34}$$

In the original proposal by Peccei and Quinn, the scale of U(1)$_{\text{PQ}}$ breaking was related to the electroweak scale, $\Lambda_{\text{PQ}} \approx$ v. Setting $m_u = m_d$ for simplicity, one then obtains the estimate

$$\frac{M_a}{M_\pi} = \frac{F_\pi}{\text{v}} \approx \frac{0.1\,\text{GeV}}{250\,\text{GeV}} = \frac{1}{2500} \quad \Rightarrow \quad M_a \approx \frac{0.14\,\text{GeV}}{2500} \approx 50\,\text{keV}. \tag{25.35}$$

This is an unusually light particle that would have observable effects. Indeed, there have been several experimental searches for this "standard" axion, but they did not find anything. As a consequence, the standard axion with a mass around 50 keV has been ruled out. Still, by pushing the U(1)$_{\text{PQ}}$ breaking scale Λ_{PQ} far above the electroweak scale v, one can make the axion lighter and even more weakly coupled. This makes it essentially invisible to all experiments performed until now (Kim, 1979; Shifman et al., 1980; Zhitnitsky, 1980; Dine et al., 1981). Still, searches continue attempting to detect the "invisible" axion. So far, we do not know whether Peccei and Quinn's theoretically appealing solution of the strong CP-problem is actually realized in Nature.

Box 25.1   **Deep question**

Does the axion field provide the right solution to the strong CP-problem?

25.6 Astrophysical and Cosmological Axion Effects

Invisible axions are light and interact only weakly, because the axion mass and its couplings are suppressed by $1/\Lambda_{PQ}$. Still, unless they are too light – and thus too weakly interacting – axions could *cool stars* very efficiently. Their low interaction cross section allows them to carry away energy more easily than the more strongly coupled photon. Stars live rather long, because they cannot get rid of their energy by radiation very quickly.

For example, a photon that is generated in a nuclear reaction in the center of the sun spends 10^7 years before it reaches the sun's surface (if we assume it to still be "the same" photon), simply because its electromagnetic cross section with the charged matter in the sun is large. An axion, on the other hand, interacts weakly and can thus get out much faster. Like neutrinos, axions can therefore act as a super-coolant for stars. The observed life-time of stars can thus be used to put astrophysical limits on the axion mass. In this way, invisible axions heavier than 1 eV have been ruled out. This implies that the Peccei–Quinn symmetry breaking scale must be above 10^7 GeV.

If they exist, axions would also affect the cooling of a neutron star that forms after a supernova explosion. There would be less energy taken away by neutrinos. The observed neutrino burst of the supernova SN 1987A would have consisted of fewer neutrinos if axions had also cooled the neutron star. This astrophysical observation excludes axions of masses between 1 keV and 20 keV – a range in which it can hardly be detected.

There are various mechanisms in the early Universe that can lead to the generation of axions (Kolb and Turner, 1990).

- The simplest is via thermal excitations. One can estimate that thermally generated axions must be rather heavy in order to contribute substantially to the energy density of the Universe. In fact, *thermally produced* axions are not a promising dark matter candidate, because their required mass is already ruled out by the astrophysical limits.
- Another interesting mechanism for axion production relies on the fact that the axion potential is not completely flat but has a unique minimum (modulo its periodicity). At high temperatures, the small axion mass is irrelevant, and the potential is practically flat and corresponds to a continuous family of degenerate minima related to each other by $U(1)_{PQ}$ symmetry transformations. Hence, there are several degenerate vacua labeled by different values of the axion field (and thus with different values of θ). Under such conditions, all values of θ are equally probable, and different regions of the hot early Universe must have been in different θ-sectors. When the temperature decreases as the Universe expands, $a(x) = 0$ is singled out as the unique minimum. In order to minimize its energy, the scalar field then "rolls" down to the minimum of the potential and oscillates around it. These oscillations are damped by axion production, and finally $a(x) = 0$ is reached everywhere in the Universe. The axions produced in this way could form a Bose–Einstein condensate that may contribute significantly to the energy density in the Universe, provided the axion mass is in the 10^{-5} eV range. This makes *coherently produced* axions an attractive candidate for dark matter (Preskill *et al.*, 1983; Abbott and Sikivie, 1983). This axion production mechanism via a disoriented Peccei–Quinn condensate is similar to the pion production mechanism via disorienting the chiral condensate in a heavy-ion collision that has been discussed by Krishna Rajagopal and Frank Wilczek (Rajagopal and Wilczek, 1993a,b).

- At temperatures far above the QCD scale, U(1)$_{PQ}$ is an almost exact global symmetry, which is spontaneously broken below Λ_{PQ}. If this happens after cosmic inflation, the *Kibble mechanism* (Kibble, 1976) leads to the generation of a network of *cosmic strings*. Such a string network can lower its energy by radiating axions. Once the axion mass becomes important, the string solutions become unstable, and the string network dissolves, again leading to axion emission. This production mechanism may also lead to sufficiently many axions as dark matter candidates.

25.7 Elimination of the Weak SU(2)$_L$ Vacuum-Angle

As we have seen, the θ-vacuum-angle associated with the gluon field can be rotated into the complex phase of the determinant of the quark mass matrix by an anomalous axial transformation. In this way, it appears in a different term of the action, but it still has the same physical effects. In particular (at least for finite N_c and in the absence of an exactly massless quark), θ cannot be removed from the theory. Thus, we are definitely confronted with the strong CP-problem.

A corresponding CP-violating term can also be constructed for the SU(2)$_L$ gauge field, which raises the question whether there is also a *weak CP-problem*. We will first address this question for the renormalizable minimal version of the Standard Model with massless neutrinos. However, we will see that its answer is affected by the presence of higher-dimensional operators, which can be attributed to physics beyond the Standard Model.

First of all, besides the topological charge of the SU(N_c) gluon field, one can define an integer-valued topological charge of the SU(2)$_L$ gauge field

$$Q[W] = -\frac{1}{32\pi^2} \int d^4x \, \epsilon_{\mu\nu\rho\sigma} \, \mathrm{Tr}\left[W_{\mu\nu} W_{\rho\sigma}\right] \in \Pi_3[\mathrm{SU}(2)] = \mathbb{Z}, \qquad (25.36)$$

and one can add a θ-vacuum term $-i\theta_2 Q[W]$ to the action of the Standard Model. As we discussed in Section 15.11, electroweak instantons with $Q[W] = 1$ give rise to baryon-number-violating Standard Model processes, irrespective of the value of θ_2.

Could the vacuum-angle θ_2, associated with SU(2)$_L$, be another source of CP-violation? Could this have observable consequences? If no corresponding CP violating effects are observed, would this confront us with a "weak CP-problem"? As we will see, in the renormalizable Standard Model, even when it is extended by non-renormalizable dimension-5 neutrino mass operators, the answer to these questions is *no*, because the weak vacuum-angle θ_2 can be eliminated by an appropriate redefinition of the quark fields. We will also see that, once baryon-number-violating dimension-6 operators are included, the angle θ_2 may have physical consequences, which, however, are very much suppressed at low energies.

Let us aim at eliminating θ_2 from the Standard Model, by performing a field transformation that leaves the action – but not the measure – invariant. The transformation can then be adjusted in such a way that the variation of the fermionic measure cancels a pre-existing topological term $\exp(i\theta_2 Q[W])$. There are different possible ways of achieving this. First of all, one can perform an anomalous lepton number transformation on the lepton fields. This works within the renormalizable Standard Model, because there are no lepton number violating terms in the Lagrangian, and the leptonic measure can indeed produce the desired variation. However, once the Standard Model is extended by the non-renormalizable lepton-number-violating dimension-5 operators discussed in Section 16.3, θ_2 would manifest itself as a complex phase of the neutrino mass matrix.

In order to avoid the reappearance of θ_2, we prefer to perform an anomalous baryon number transformation on the quark fields. This leaves the Lagrangian – even including the dimension-5 operators – invariant and still induces the desired variation via the fermionic measure of the quark fields. In particular, this does not regenerate the strong θ-term $\exp(\mathrm{i}\theta Q[G])$, because baryon number is anomalous only in the background of a topologically non-trivial $\mathrm{SU}(2)_L \times \mathrm{U}(1)_Y$ gauge field, but not in an $\mathrm{SU}(N_c)$ gauge background. When we perform a generation-independent baryon number transformation of the quark mass eigenstates

$$
\begin{aligned}
Q_L'(x) &= \exp\left(\mathrm{i}\rho/N_c\right) Q_L(x), \\
U_R'(x) &= \exp\left(\mathrm{i}\rho/N_c\right) U_R(x), \\
D_R'(x) &= \exp\left(\mathrm{i}\rho/N_c\right) D_R(x),
\end{aligned} \tag{25.37}
$$

the fermionic measure is not invariant. It varies by a factor $\exp(\mathrm{i}\delta_B\theta_2 Q[W])$ with

$$
\delta_B\theta_2 = -N_g N_c \rho/N_c = -N_g\,\rho, \tag{25.38}
$$

because there are $N_g N_c$ left-handed $\mathrm{SU}(2)_L$ quark doublets. Adjusting $\rho = \theta_2/N_g$ accordingly, this cancels the pre-existing weak topological term $\exp(\mathrm{i}\theta_2 Q[W])$ and yields the net vacuum-angle

$$
\theta_2' = \theta_2 + \delta_B\theta_2 = \theta_2 - N_g\,\rho = 0. \tag{25.39}
$$

This eliminates the weak vacuum-angle from the theory, even in the presence of neutrino mass terms. Hence, there is no weak CP-problem, at least not at this level.

Of course, there are also higher-order terms of dimension 6, which have been classified by Buchmüller and Wyler (1986) and by Grzadkowski *et al.* (2010). These include the baryon-number-violating operators $\mathcal{O}_i$, first constructed independently by Weinberg (1979a) and by Wilczek and Zee (1979) that we discussed in eq. (16.23). These 4-fermion operators contain three quark fields and one lepton field and are thus specific to the real-world case $N_c = 3$. Interestingly, although they break both B and L, they still conserve $B - L$. In any case, since they are B-violating, they change under the anomalous baryon number transformation of eq. (25.37) and pick up a phase $\exp(\mathrm{i}\rho) = \exp(\mathrm{i}\theta_2/N_g)$. In this way, the weak vacuum-angle θ_2 manifests itself in the complex phases of the baryon-number-violating operators $\mathcal{O}_i$, just as the QCD vacuum-angle θ manifests itself in the phase of the determinant of the quark mass matrix, which changes under anomalous axial $\mathrm{U}(1)_A$ transformations.

The dimension-6 operators $\mathcal{O}_i$ are suppressed by Λ^{-2}, where Λ is the scale from which baryon-number-violating processes beyond the Standard Model originate. Their effects on the low-energy physics, and thus possible effects of θ_2, are therefore expected to be extremely small at low energies.

Alternatively, we could have rotated θ_2 into the dimension-5 neutrino term by a transformation of the left-handed lepton fields. This would still leave the physics unchanged, but it would then not be obvious that the effects of θ_2 are suppressed by $1/\Lambda^2$.

25.8 Is there an Electromagnetic CP-Problem?

Let us also address the issue of a vacuum-angle θ_1 associated with the U(1)$_Y$ gauge field B_μ. For this purpose, we construct the term

$$Q[B] = -\frac{1}{32\pi^2} \int d^4x\, \epsilon_{\mu\nu\rho\sigma}\, \mathrm{Tr}\left[X_{\mu\nu}X_{\rho\sigma}\right] = \frac{g'^2}{64\pi^2} \int d^4x\, \epsilon_{\mu\nu\rho\sigma} B_{\mu\nu} B_{\rho\sigma}. \qquad (25.40)$$

Its prefactor arises from the expression for the SU(2) topological charge when one embeds the Abelian U(1)$_Y$ gauge field in SU(2)$_R$ as $X_\mu(x) = \mathrm{i}\,g' B_\mu(x)\tau^3/2$. Since $\Pi_3[\mathrm{U}(1)] = \{0\}$, the Abelian topological charge $Q[B]$ vanishes when one demands that the field strength approaches zero on a sphere S^3 at Euclidean space–time infinity, as one does for non-Abelian instantons. This is how it is often argued that θ_1 can be ignored.

However, the situation is more subtle. First of all, when one compactifies space–time to a torus, $Q[B]$ can take non-zero integer values. *(Anti-)selfdual solutions* of the Euclidean equations of motion then have a constant field strength that obeys $B_{\mu\nu} = \pm\frac{1}{2}\epsilon_{\mu\nu\rho\sigma} B_{\rho\sigma}$. Hence, finite-action field configurations with topological charge $Q[B] = \pm 1$ exist also in the Abelian case. However, they are not instantons; *i.e.* they are not localized around an instant of Euclidean time. In particular, unlike in the non-Abelian case, (anti-)selfdual Abelian solutions extend homogeneously throughout space–time and have no scale associated with them. In contrast to the QCD vacuum, which is filled with localized topological objects (including instantons), the vacuum associated with the U(1)$_Y$ gauge theory contains a negligible amount of topological charge, which, in addition, is smeared out homogeneously throughout space–time. This means that, unlike neutrons, which are sensitive to QCD instantons and other topological charge carriers, neutrinos or other leptons do not pick up a significant electric dipole moment, even if θ_1 would differ significantly from zero.

In this context, *'t Hooft–Polyakov monopoles* play an interesting role. They are classical solutions in Grand Unified Theories (GUTs), which were discovered independently by Alexander Polyakov (1974) and Gerard 't Hooft (1974). Their magnetic charge is associated with a topological winding number in $\Pi_2[\mathrm{SU}(2)/\mathrm{U}(1)] = \Pi_2[S^2] = \mathbb{Z}$ (cf. Appendix G). Classical magnetic monopole solutions with additional electric charge were discovered by Bernard Julia and Anthony Zee and are known as *Julia–Zee dyons* (Julia and Zee, 1975). Remarkably, as Edward Witten (1979c) was first to realize, when a dyon is considered at the quantum level in a theory with non-zero electromagnetic vacuum-angle θ_{QED}, its electric charge is given by $-e\theta_{\mathrm{QED}}/2\pi$. This sensitivity to θ_{QED} results from the fact that, unlike ordinary elementary particles, a dyon is a topologically non-trivial excitation of the vacuum. The electric charge of a dyon could hence reveal information about θ_{QED}, provided that such objects indeed exist. Even if they could exist in principle, observations of magnetic monopoles or dyons seem unlikely in the foreseeable future. We will consider monopoles and dyons in more detail in Sections 26.7 and 26.9.

We conclude that topologically non-trivial field configurations, such as dyons, may affect Abelian gauge theory as well and that θ_1 should thus not be ignored from the outset. If we take θ_1 seriously, it is natural to ask whether, like θ_2, it can be rotated away by a field transformation. Via the fermionic measure of the quark fields in the background of a topologically non-trivial U(1)$_Y$ gauge field, the previously performed baryon number transformation of eq. (25.37) indeed also generates a topological term $\exp(\mathrm{i}\,\delta_B\theta_1\, Q[B])$

with

$$\delta_B \theta_1 = -N_g N_c 2 \left[2Y_{Q_L}^2 - Y_{U_R}^2 - Y_{D_R}^2 \right] \rho/N_c$$

$$= -N_g 2 \left[2\frac{1}{4N_c^2} - \frac{1}{4}\left(\frac{1}{N_c} + 1\right)^2 - \frac{1}{4}\left(\frac{1}{N_c} - 1\right)^2 \right] \rho = N_g \rho. \quad (25.41)$$

Together with a pre-existing vacuum-angle θ_1, this yields

$$\theta_1' = \theta_1 + \delta_B \theta_1 = \theta_1 + N_g \rho = \theta_1 + \theta_2. \quad (25.42)$$

We conclude that, while θ_2 can be rotated to $\theta_2' = 0$, the sum of the two vacuum-angles,

$$\theta_1' + \theta_2' = \theta_1 + \theta_2, \quad (25.43)$$

remains invariant. This combination cannot be eliminated by the anomalous baryon number transformation (25.37). It constitutes an (under-appreciated) parameter of the Standard Model, whose potential physical effects deserve further investigation. First, it may play a role in the generation of large magnetic fields during the cosmic evolution (Turner and Widrow, 1988; Joyce and Shaposhnikov, 1997; Alekseev *et al.*, 1998; Fröhlich and Pedrini, 2000). Moreover, it determines the electric charge of dyons, which are potential relics from a GUT epoch in the early Universe.

Since we did not want θ_2 to enter the dimension-5 terms, we limited ourselves to the anomalous baryon number transformation of eq. (25.37) which leaves those terms invariant. Alternatively, one could perform an anomalous $B + L$ transformation

$$Q_L'(x) = \exp\left(i\rho/N_c\right) Q_L(x),$$
$$U_R'(x) = \exp\left(i\rho/N_c\right) U_R(x),$$
$$D_R'(x) = \exp\left(i\rho/N_c\right) D_R(x),$$
$$L_L'(x) = \exp\left(i\rho\right) L_L(x),$$
$$E_R'(x) = \exp\left(i\rho\right) E_R(x), \quad (25.44)$$

which leaves the action of the renormalizable Standard Model (but not the dimension-5 terms) invariant. The variation of the fermionic measure of the leptons then gives rise to the additional shifts

$$\delta_L \theta_1 = -N_g 2 \left[2Y_{L_L}^2 - Y_{E_R}^2 \right] \rho = -N_g 2 \left[2\left(-\frac{1}{2}\right)^2 - (-1)^2 \right] \rho = N_g \rho,$$

$$\delta_L \theta_2 = -N_g \rho, \quad (25.45)$$

such that now

$$\theta_1' = \theta_1 + \delta_B \theta_1 + \delta_L \theta_1 = \theta_1 + 2N_g \rho,$$
$$\theta_2' = \theta_2 + \delta_B \theta_2 + \delta_L \theta_2 = \theta_2 - 2N_g \rho \quad \Rightarrow$$
$$\theta_1' + \theta_2' = \theta_1 + \theta_2. \quad (25.46)$$

As we concluded before, the sum $(\theta_1 + \theta_2)$ remains invariant and cannot be rotated away.

In order to find an appropriate interpretation for this sum of the two electroweak vacuum-angles, it is instructive to consider the low-energy limit of electrodynamics in which we keep only the photon and ignore the $W^\pm$- and Z-boson fields. This implies $W_\mu^{1,2}(x) \to 0$ as well as

$$W_\mu^3(x) = \frac{g'A_\mu(x) + gZ_\mu(x)}{\sqrt{g^2 + g'^2}} \to \frac{g'A_\mu(x)}{\sqrt{g^2 + g'^2}},$$

$$B_\mu(x) = \frac{gA_\mu(x) - g'Z_\mu(x)}{\sqrt{g^2 + g'^2}} \to \frac{gA_\mu(x)}{\sqrt{g^2 + g'^2}}. \tag{25.47}$$

The electroweak topological terms then reduce to

$$i\theta_1 \frac{g'^2}{64\pi^2} \epsilon_{\mu\nu\rho\sigma} B_{\mu\nu} B_{\rho\sigma} - i\theta_2 \frac{1}{32\pi^2} \epsilon_{\mu\nu\rho\sigma} \mathrm{Tr}\left[W_{\mu\nu} W_{\rho\sigma}\right] \to$$

$$i(\theta_1 + \theta_2) \frac{g^2 g'^2}{g^2 + g'^2} \frac{1}{64\pi^2} \epsilon_{\mu\nu\rho\sigma} F_{\mu\nu} F_{\rho\sigma} = i\theta_{\mathrm{QED}} \frac{e^2}{64\pi^2} \epsilon_{\mu\nu\rho\sigma} F_{\mu\nu} F_{\rho\sigma}. \tag{25.48}$$

We have identified the electromagnetic vacuum-angle, as well as the electric charge unit

$$\theta_{\mathrm{QED}} = \theta_1 + \theta_2, \quad e = \frac{gg'}{\sqrt{g^2 + g'^2}}. \tag{25.49}$$

Since QED is a vector-like theory, its vacuum-angle can indeed *not* be eliminated by an anomalous baryon or lepton number transformation.

To summarize, by an appropriate $U(1)_A$ transformation we have rotated a pre-existing QCD vacuum-angle θ into the complex phase of the quark mass matrix. Similarly, by an anomalous baryon number transformation, we canceled a pre-existing $SU(2)_L$ term $\exp(i\theta_2 Q[W])$. This does not even have an impact on the lepton-number-violating dimension-5 terms, but only on the baryon-number-violating dimension-6 terms. However, θ_2 then reappears in the $U(1)_Y$ topological term as $\exp(i(\theta_1 + \theta_2)Q[B]) \to \exp(i\theta_{\mathrm{QED}} Q[A])$.

Consequently, θ_{QED} does enter the renormalizable Standard Model as another source of CP violation. As an overview, there are:

- θ as the phase of the determinant of the quark mass matrix
- $(N_g - 1)(N_g - 2)/2$ complex phases of the CKM quark mixing matrix
- the QED vacuum-angle $\theta_{\mathrm{QED}} = \theta_1 + \theta_2$
- $N_g(N_g - 1)/2$ complex phases of the PMNS lepton mixing matrix, at the level of dimension-5 operators.

Altogether, these are $(N_g - 1)^2 + 2$ CP-violating parameters, *i.e.* 6 for three fermion generations ($N_g = 3$).

Since θ_{QED} determines the electric charge of magnetic monopoles, the question arises whether there is also an electromagnetic CP-problem. Unfortunately, we do not know, since no reproducible observation of a magnetic monopole has been reported, and there is no experimental information about θ_{QED}. Let us imagine the unlikely event that somebody finds an electrically neutral magnetic monopole. This would indeed raise an electromagnetic CP-problem: Why is $\theta_{\mathrm{QED}} = 0$?

In Chapter 26, we will discuss Grand Unified Theories (GUTs), in which all Standard Model gauge fields are embedded in one unifying gauge group, such as SU(5) or SO(10). In that case, at very high energies there is only one vacuum-angle. Since θ does not get renormalized, GUTs indeed imply that $\theta_1 = \theta_2 = \theta$. Witten pointed out that O(32) superstring theory naturally gives rise to an axion field in the framework of GUTs (Witten,

1984b). Provided that only the strong interaction breaks the corresponding $U(1)_{PQ}$ symmetry, such a field would solve not only the strong but also the potential electromagnetic CP-problem. In other words, in such a scenario magnetic monopoles would automatically be electrically neutral.

The questions related to θ_{QED} are somewhat subtle and are not covered extensively in the literature. We encourage the reader to be open-minded and not to discard θ_{QED} from the outset. If there is also an electromagnetic CP-problem, one should address the

---

Box 25.2   **Deep question**

What are the solutions to the strong and to the electromagnetic CP-problem?

Exercises

25.1 Periodicity of $Z(\theta)$

Show that the partition function $Z(\theta)$ in eq. (25.6) is 2π-periodic. Extend this result to higher-order terms of the chiral Lagrangian.

25.2 Witten–Veneziano formula and the chiral Lagrangian

Show that the Witten–Veneziano formula (25.11) for the η'-mass is consistent with the chiral Lagrangian of eq. (25.14).

25.3 Witten–Veneziano formula for non-zero strange quark mass

Derive eq. (20.10) by inserting $m_u = m_d = 0$ and $m_s > 0$ in the quark mass matrix of eq. (25.14).

25.4 Axion mass for $N_f = 3$

Consider the standard axion in a theory with three light quark flavors. What is the axion mass as a function of m_u, m_d, and m_s? Show that the axion is massless if one of the quark masses vanishes. Also convince yourself that one recovers the $N_f = 2$ result in the limit of a very heavy strange quark.

25.5 The axion mass at large N_c

Generalize the chiral perturbation theory action of eq. (25.14) to large N_c in order to include the standard axion besides the η'-meson. Consider two flavors of quarks with general values of m_u and m_d, and determine the resulting masses of π^0, η', and the axion.

26

Grand Unified Theories

Although the Standard Model (augmented by neutrino masses) agrees very well with experiments – it is the most precise theory in the history of science – from a conceptual point of view it is not completely satisfactory. In particular, the considerable number of free parameters, like fermion masses (Yukawa couplings) or gauge boson masses (gauge couplings), cf. Section 17.8, suggests the existence of a more fundamental, underlying theory that might allow us to relate some of these parameters.

As we discussed in Section 13.8, electromagnetism and the weak force are not truly unified in the Standard Model, because there are two independent coupling constants g for $SU(2)_L$ and g' for $U(1)_Y$. The strong $SU(3)_c$ gauge interaction has yet another coupling constant g_s, which is unrelated to g and g'. In the framework of *Grand Unified Theories (GUTs)*, one embeds the electroweak and strong interactions in one single, superior gauge group, such as $SU(5)$ or $Spin(10)$ (the universal covering group of $SO(10)$), which relates g, g', and g_s. For a textbook and a review, we refer to Ross (1985) and Langacker (2012).

If we assume $SU(5)$ to be the unified gauge group, then – at low energy – it must be "spontaneously broken" to the $SU(3)_c \times SU(2)_L \times U(1)_Y$ gauge symmetry of the Standard Model. The corresponding phase transition is supposed to happen near the *GUT scale* of about 10^{16} GeV, which is far beyond the energies that are accessible to collider experiments today. The unification of the electroweak and strong interactions naturally leads to the decay of strongly interacting particles into particles which participate in the electroweak interactions only. In particular, quarks can decay into leptons. This inevitably leads to the decay of the proton and hence to a violation of baryon number conservation. The $SU(5)$ GUT predicts a proton life-time of about 10^{30} to 10^{31} years.

Experimentally, however, the proton is even longer lived, with a life-time exceeding 10^{32} years. This is a tremendous time span, for instance, compared to the age of the Universe, which is of the order of 10^{10} years. The experimental lower bound on the proton life-time was established by observing a well-shielded, large amount of water for a long time with photo-multiplier tubes, which would detect the Cherenkov light that could emerge when a proton in the water tank decays. Since no such decay has ever been observed, the simplest version of the $SU(5)$ GUT, which predicts proton decay at a faster rate, has already been ruled out. Still, we will discuss that model in some detail, because it is the simplest representative in a larger class of GUTs.

Other GUTs allow for longer life-times of the proton, which are not experimentally ruled out so far. Proton decay is an attractive feature of GUTs, because it may offer an explanation of the *baryon asymmetry*. This asymmetry refers to the fact that some small amount of matter – namely, all the matter that still exists in the Universe today – survived the mass extinction of matter and anti-matter very soon after the Big Bang.

26.1 Minimal SU(5) Model

This section provides a straightforward generalization of Section 13.8 from SU(3) to SU(5). Here, we will go into less detail than we did in Chapter 13. The SU(5) GUT is also known as the Georgi–Glashow model (Georgi and Glashow, 1974; Georgi *et al.*, 1974).

The group $SU(n)$ has rank $n - 1$; *i.e.* $n - 1$ of its $n^2 - 1$ generators commute with each other. For the group $U(n)$, the rank and the dimension (number of generators) both increase by one unit to n and n^2, respectively. Thus, the rank of the Standard Model gauge group $SU(3)_c \times SU(2)_L \times U(1)_Y$ is $2 + 1 + 1 = 4$. Hence, if we want to embed this group in a single Lie group, its rank must be at least 4. A simple Lie group with that property is SU(5), which has rank 4 and 24 generators. Consequently, in an SU(5) gauge theory there are 24 gauge bosons.

When the Standard Model is embedded in SU(5), half of the gauge bosons can be identified with known particles: $SU(3)_c$ has $3^2 - 1 = 8$ gluons, $SU(2)_L$ has $2^2 - 1 = 3$ W-bosons, and $U(1)_Y$ has one B-boson. The remaining 12 gauge bosons of SU(5) are new particles, which are called X and Y. To make these particles heavy, the SU(5) symmetry is "spontaneously broken" to $SU(3)_c \times SU(2)_L \times U(1)_Y$ by invoking the Higgs mechanism, using a scalar field $\Xi(x)$ that transforms in the 24-dimensional adjoint representation of SU(5). We write

$$\Xi(x) = \Xi_a(x)\eta^a, \quad a \in \{1, 2, \ldots, 24\}, \quad \mathrm{Tr}(\eta^a \eta^b) = 2\delta_{ab}. \tag{26.1}$$

The η^a are the 24 generators of SU(5), normalized with the condition on $\mathrm{Tr}(\eta^a \eta^b)$ that we apply to the generators of all $SU(n)$ groups. They are Hermitian, traceless 5×5 matrices, analogous to the 3 Pauli matrices of SU(2), and to the 8 Gell-Mann matrices of SU(3). Under a gauge transformation $\Upsilon(x) \in SU(5)$, the scalar field transforms as

$$\Xi'(x) = \Upsilon(x)\, \Xi(x)\, \Upsilon(x)^\dagger. \tag{26.2}$$

We introduce the most general gauge-invariant and renormalizable potential, which takes the form of the linear combination

$$V(\Xi) = \frac{M^2}{2}\mathrm{Tr}\left(\Xi^2\right) + \frac{\kappa}{3!}\mathrm{Tr}\left(\Xi^3\right) + \frac{\Lambda_1}{4!}\mathrm{Tr}\left(\Xi^4\right) + \frac{\Lambda_2}{4!}\left(\mathrm{Tr}\left(\Xi^2\right)\right)^2. \tag{26.3}$$

In contrast to the SU(3) case that we discussed in Section 13.8, in the SU(5) case the two quartic terms (with the coupling constants Λ_1 and Λ_2) are linearly independent and must hence both be included in the potential. A term $\det(\Xi)$, on the other hand, is now of fifth order; thus, it is not renormalizable and need not be included in the potential.

One can diagonalize the Hermitian matrix field $\Xi(x)$ by a gauge transformation $\Upsilon(x)$, thus fixing the *unitary gauge,* such that

$$\Xi(x) = \begin{pmatrix} \xi_1(x) & 0 & 0 & 0 & 0 \\ 0 & \xi_2(x) & 0 & 0 & 0 \\ 0 & 0 & \xi_3(x) & 0 & 0 \\ 0 & 0 & 0 & \xi_4(x) & 0 \\ 0 & 0 & 0 & 0 & \xi_5(x) \end{pmatrix}, \quad \xi_i(x) \in \mathbb{R}, \quad \sum_i \xi_i = 0. \tag{26.4}$$

Then, the potential simplifies to

$$V(\Xi) = \frac{M^2}{2} \sum_i \xi_i^2 + \frac{\kappa}{3!} \sum_i \xi_i^3 + \frac{\Lambda_1}{4!} \sum_i \xi_i^4 + \frac{\Lambda_2}{4!} \left(\sum_i \xi_i^2 \right)^2. \tag{26.5}$$

The minima of the potential are characterized by

$$\frac{\partial V(\Xi)}{\partial \xi_i} = M^2 \xi_i + \frac{\kappa}{2} \xi_i^2 + \frac{\Lambda_1}{3!} \xi_i^3 + \frac{\Lambda_2}{3!} \sum_j \xi_j^2 \xi_i = C, \tag{26.6}$$

where C is a Lagrange multiplier that implements the constraint $\mathrm{Tr}\,\Xi = \sum_i \xi_i = 0$. We are interested in minima with an unbroken $SU(3)_c \times SU(2)_L \times U(1)_Y$ symmetry, with the vacuum configuration

$$\Xi(x) = \Xi_0 = \mathcal{V}\eta^{24} = \mathcal{V}\sqrt{\frac{3}{5}} \begin{pmatrix} 1 & 0 & 0 & 0 & 0 \\ 0 & 1 & 0 & 0 & 0 \\ 0 & 0 & -2/3 & 0 & 0 \\ 0 & 0 & 0 & -2/3 & 0 \\ 0 & 0 & 0 & 0 & -2/3 \end{pmatrix}. \tag{26.7}$$

According to eq. (26.1), we have normalized the $SU(5)$ generator η^{24} such that $\mathrm{Tr}(\eta^{24}\eta^{24})$ $= 2$. Using eq. (26.6), it is then straightforward to derive an explicit expression for the GUT scale $\mathcal{V}$. For $M^2 < 0$, the $SU(5)$ symmetric phase at $\Xi = 0$ corresponds to a local maximum of the potential. In Problem 26.1, we identify the scale $\mathcal{V}$ and the parameter regime that leads to the $SU(5) \to SU(3)_c \times SU(2)_L \times U(1)_Y$ breaking pattern, rather than, e.g., $SU(5) \to SU(4) \times U(1)$.

Let us now consider the $SU(5)$ gauge field

$$V_\mu(x) = \mathrm{i}g_5 V_\mu^a(x) \frac{\eta^a}{2}. \tag{26.8}$$

Under general non-Abelian gauge transformations, it turns into

$$V_\mu'(x) = \Upsilon(x)(V_\mu(x) + \partial_\mu)\Upsilon(x)^\dagger. \tag{26.9}$$

At this point, we introduce the *X- and Y-bosons* via

$$V_\mu(x) = \mathrm{i}\frac{g_5}{2} \begin{pmatrix} W_\mu^a(x)\tau^a + \sqrt{3/5}B_\mu(x)\mathbb{1}_{2\times 2} & \begin{matrix} \bar{X}_\mu^1(x) & \bar{X}_\mu^2(x) & \bar{X}_\mu^3(x) \\ \bar{Y}_\mu^1(x) & \bar{Y}_\mu^2(x) & \bar{Y}_\mu^3(x) \end{matrix} \\ \begin{matrix} X_\mu^1(x) & Y_\mu^1(x) \\ X_\mu^2(x) & Y_\mu^2(x) \\ X_\mu^3(x) & Y_\mu^3(x) \end{matrix} & G_\mu^a(x)\lambda^a - 2/\sqrt{15}B_\mu(x)\mathbb{1}_{3\times 3} \end{pmatrix}. \tag{26.10}$$

Under embedded Standard Model $SU(3)_c \times SU(2)_L \times U(1)_Y \subset SU(5)$ gauge transformations,

$$\Upsilon(x) = \begin{pmatrix} \exp\left(ig'/2\varphi(x)\right) L(x) & 0_{2\times3} \\ 0_{3\times2} & \exp\left(-ig'/3\varphi(x)\right) \Omega(x) \end{pmatrix},$$

$$L(x) \in SU(2)_L,$$

$$\Omega(x) \in SU(3)_c,$$

(26.11)

the various gauge fields transform as

$$G'_\mu(x) = \Omega(x)(G_\mu(x) + \partial_\mu)\Omega(x)^\dagger,$$

$$W'_\mu(x) = L(x)(W_\mu(x) + \partial_\mu)L(x)^\dagger,$$

$$B'_\mu(x) = B_\mu(x) - \partial_\mu\varphi(x),$$

$$\begin{pmatrix} X_\mu^{1'}(x) & Y_\mu^{1'}(x) \\ X_\mu^{2'}(x) & Y_\mu^{2'}(x) \\ X_\mu^{3'}(x) & Y_\mu^{3'}(x) \end{pmatrix} = \exp\left(-ig'\frac{5}{6}\varphi(x)\right) \Omega(x) \begin{pmatrix} X_\mu^1(x) & Y_\mu^1(x) \\ X_\mu^2(x) & Y_\mu^2(x) \\ X_\mu^3(x) & Y_\mu^3(x) \end{pmatrix} L(x)^\dagger,$$

$$\begin{pmatrix} \bar{X}_\mu^{1'}(x) & \bar{X}_\mu^{2'}(x) & \bar{X}_\mu^{3'}(x) \\ \bar{Y}_\mu^{1'}(x) & \bar{Y}_\mu^{2'}(x) & \bar{Y}_\mu^{3'}(x) \end{pmatrix} = \exp\left(ig'\frac{5}{6}\varphi(x)\right) L(x) \begin{pmatrix} \bar{X}_\mu^1(x) & \bar{X}_\mu^2(x) & \bar{X}_\mu^3(x) \\ \bar{Y}_\mu^1(x) & \bar{Y}_\mu^2(x) & \bar{Y}_\mu^3(x) \end{pmatrix} \Omega(x)^\dagger.$$

(26.12)

Hence, the X- and Y-bosons are a color triplet and a (charge conjugate) electroweak doublet with weak hypercharge $Y_X = Y_Y = -5/6$, while the $\bar{X}$- and $\bar{Y}$-bosons form a color anti-triplet and electroweak doublet with weak hypercharge $Y_{\bar{X}} = Y_{\bar{Y}} = 5/6$. This corresponds to the electric charges $Q = T_L^3 + Y$, such that

$$Q_X = -\frac{1}{2} - \frac{5}{6} = -\frac{4}{3}, \quad Q_Y = \frac{1}{2} - \frac{5}{6} = -\frac{1}{3}, \quad Q_{\bar{X}} = \frac{4}{3}, \quad Q_{\bar{Y}} = \frac{1}{3}.$$

(26.13)

Just as the quarks, the X- and Y-bosons, which are both color triplets, carry electric charges that are quantized in integer units shifted by $-1/3$. Correspondingly, the electric charges of the $\bar{X}$- and $\bar{Y}$-bosons, which are color anti-triplets, are quantized in integer units shifted by $1/3$, like the anti-quarks.

As we will see, via the Higgs mechanism the X- and Y-bosons pick up masses at the GUT scale. They are very heavy, unstable, and extremely short-lived. Consequently, although they participate in the strong interaction and have the same color quantum numbers as the quarks, they do not form exotic hadrons. In any case, such hadrons would again have integer electric charges.

To summarize, under the $SU(3)_c \times SU(2)_L \times U(1)_Y$ subgroup, which is the point of departure of the Standard Model, the adjoint representation of $SU(5)$ decomposes as

$$\{24\} = \{8,1\}_0 + \{1,3\}_0 + \{1,1\}_0 + \{\bar{3},2\}_{5/6} + \{3,\bar{2}\}_{-5/6}.$$

(26.14)

This also follows from the $SU(2)$, $SU(3)$, and $SU(5)$ relations

$$\{2\} \times \{\bar{2}\} = \{2\} \times \{2\} = \{3\} + \{1\},$$

$$\{3\} \times \{\bar{3}\} = \{8\} + \{1\},$$

$$\{5\} \times \{\bar{5}\} = \{24\} + \{1\}.$$

(26.15)

When we use the fact that the fundamental representation $\{5\}$ of $SU(5)$ decomposes into an $SU(3)_c$ triplet and an $SU(2)_L$ doublet with corresponding $U(1)_Y$ quantum numbers,

$$\{5\} = \{3,1\}_{-1/3} + \{1,2\}_{1/2},$$

(26.16)

we indeed obtain

$$
\begin{aligned}
\{5\} \times \{\bar{5}\} &= \left(\{3,1\}_{-1/3} + \{1,2\}_{1/2}\right) \times \left(\{\bar{3},1\}_{1/3} + \{1,\bar{2}\}_{-1/2}\right) \\
&= \{3,1\}_{-1/3} \times \{\bar{3},1\}_{1/3} + \{1,2\}_{1/2} \times \{\bar{3},1\}_{1/3} \\
&\quad + \{3,1\}_{-1/3} \times \{1,\bar{2}\}_{-1/2} + \{1,2\}_{1/2} \times \{1,\bar{2}\}_{-1/2} \\
&= \{8,1\}_0 + \{1,1\}_0 + \{\bar{3},2\}_{5/6} + \{3,\bar{2}\}_{-5/6} + \{1,3\}_0 + \{1,1\}_0. \quad (26.17)
\end{aligned}
$$

From eq. (26.10), we read off the Standard Model gauge couplings

$$
g_{\mathrm{s}} = g = g_5, \quad g' = \sqrt{\frac{3}{5}}\, g_5 \quad \Rightarrow \quad e = \frac{gg'}{\sqrt{g^2 + g'^2}} = \sqrt{\frac{3}{8}}\, g_5, \quad \sin^2\theta_{\mathrm{W}} = \frac{3}{8}. \quad (26.18)
$$

These relations hold at the GUT scale $\mathcal{V}$. Using the renormalization group, one can evolve the running couplings down to the Standard Model scale v and compare them to their physical values (Georgi et al., 1974). If one assumes that there is a "desert" (not populated by additional fields) extending over the vast energy range separating the Standard Model scale v from the GUT scale $\mathcal{V}$, the unification of all three Standard Model gauge couplings g_{s}, g, and g' at approximately the same scale $\mathcal{V}$ works remarkably well, but not perfectly.

In order to facilitate more accurate unification at one scale, one may invoke supersymmetry (Dimopoulos et al., 1981). This brings along many additional fields: a fermionic superpartner for each boson as well as a bosonic superpartner for each fermion, which one would expect around the TeV scale. Supersymmetric GUTs have their unification scale not too far below the Planck scale. As a result, the Hubble time is then too short to reach a thermal equilibrium at the corresponding higher GUT scale. Despite intense searches, until 2022 the LHC has not provided any experimental evidence for supersymmetry. If no superpartners but some other exotic "beasts" populate the desert, one can still imagine the coupling constant unification to work out. In any case, if the GUT scale is quite close to the Planck scale, there seems to be no good reason for a precise unification of the gauge interactions excluding gravity.

The X- and Y-bosons become massive by the "spontaneous breakdown" of SU(5) to $\mathrm{SU}(3)_{\mathrm{c}} \times \mathrm{SU}(2)_L \times \mathrm{U}(1)_Y$ via a variant of the Higgs mechanism. For the adjoint scalar field, the covariant derivative takes the form

$$
D_\mu \Xi(x) = \partial_\mu \Xi(x) + [V_\mu(x), \Xi(x)]. \quad (26.19)
$$

Introducing the field strength tensor

$$
V_{\mu\nu}(x) = \partial_\mu V_\nu(x) - \partial_\nu V_\mu(x) + [V_\mu(x), V_\nu(x)], \quad (26.20)
$$

the bosonic part of the SU(5) GUT Lagrangian is given by

$$
\mathcal{L}(\Xi, V_\mu) = \frac{1}{4}\mathrm{Tr}(D_\mu \Xi D_\mu \Xi) + V(\Xi) - \frac{1}{2g_5^2}\mathrm{Tr}(V_{\mu\nu} V_{\mu\nu}). \quad (26.21)
$$

Inserting the vacuum value $\Xi(x) = \Xi_0$, we obtain the mass terms for the gauge fields X and Y,

$$
\frac{1}{4}\mathrm{Tr}\left(D_\mu \Xi D_\mu \Xi\right) = \frac{1}{4}\mathrm{Tr}\left([V_\mu, \Xi_0][V_\mu, \Xi_0]\right) = \frac{5}{24}g_5^2 \mathcal{V}^2 \left(\bar{X}_\mu X_\mu + \bar{Y}_\mu Y_\mu\right), \quad (26.22)
$$

because

$$[V_\mu, \Xi_0] = i\sqrt{\frac{5}{3}}\frac{g_5}{2}\mathcal{V}\begin{pmatrix} 0_{2\times2} & \begin{matrix} -\bar{X}^1_\mu & -\bar{X}^2_\mu & -\bar{X}^3_\mu \\ -\bar{Y}^1_\mu & -\bar{Y}^2_\mu & -\bar{Y}^3_\mu \end{matrix} \\ \begin{matrix} X^1_\mu & Y^1_\mu \\ X^2_\mu & Y^2_\mu \\ X^3_\mu & Y^3_\mu \end{matrix} & 0_{3\times3} \end{pmatrix}. \tag{26.23}$$

Hence, the X- and Y-bosons both pick up the mass

$$M_X = M_Y = \sqrt{\frac{5}{12}}g_5\mathcal{V}. \tag{26.24}$$

These 12 gauge bosons become massive by "eating" 12 Nambu–Goldstone bosons. When (before gauging) the grand unified group $G = \mathrm{SU}(5)$ breaks spontaneously down to the Standard Model subgroup $H = \mathrm{SU}(3)_c \times \mathrm{SU}(2)_L \times \mathrm{U}(1)_Y$, according to the Goldstone theorem, there are indeed $24 - (8 + 3 + 1) = 12$ Nambu–Goldstone bosons.

26.2 Fermion Multiplets

How can we accommodate quarks and leptons in representations of $\mathrm{SU}(5)$? Let us consider the Standard Model fermions of the first generation,

$$\begin{pmatrix} u^1_L \\ d^1_L \end{pmatrix}, u^1_R, d^1_R, \begin{pmatrix} u^2_L \\ d^2_L \end{pmatrix}, u^2_R, d^2_R, \begin{pmatrix} u^3_L \\ d^3_L \end{pmatrix}, u^3_R, d^3_R, \begin{pmatrix} \nu_L \\ e_L \end{pmatrix}, e_R. \tag{26.25}$$

These are 15 fermionic degrees of freedom. Let us couple two fundamental representations in $\mathrm{SU}(2)$, $\mathrm{SU}(3)$, and $\mathrm{SU}(5)$

$$\begin{aligned} \{2\} \times \{2\} &= \{3\} + \{1\}, \\ \{3\} \times \{3\} &= \{6\} + \{\bar{3}\}, \\ \{5\} \times \{5\} &= \{15\} + \{10\}. \end{aligned} \tag{26.26}$$

Indeed, there is a 15-dimensional representation. Can we host the fermions of one generation in that representation? Let us investigate the $\mathrm{SU}(3)$ and $\mathrm{SU}(2)$ content by using eq. (26.16),

$$\begin{aligned} \{5\} \times \{5\} &= \big(\{3,1\}_{-1/3} + \{1,2\}_{1/2}\big) \times \big(\{3,1\}_{-1/3} + \{1,2\}_{1/2}\big) \\ &= \{3,1\}_{-1/3} \times \{3,1\}_{-1/3} + \{1,2\}_{1/2} \times \{3,1\}_{-1/3} \\ &\quad + \{3,1\}_{-1/3} \times \{1,2\}_{1/2} + \{1,2\}_{1/2} \times \{1,2\}_{1/2} \\ &= \{6,1\}_{-2/3} + \{\bar{3},1\}_{-2/3} + \{3,2\}_{1/6} + \{3,2\}_{1/6} \\ &\quad + \{1,3\}_1 + \{1,1\}_1 = \{15\} + \{10\}. \end{aligned} \tag{26.27}$$

The symmetric combination $\{15\}$ and the anti-symmetric combination $\{10\}$ decompose as

$$\begin{aligned} \{15\} &= \{6,1\}_{-2/3} + \{3,2\}_{1/6} + \{1,3\}_1, \\ \{10\} &= \{\bar{3},1\}_{-2/3} + \{3,2\}_{1/6} + \{1,1\}_1. \end{aligned} \tag{26.28}$$

The Standard Model does not contain fermions in a sextet representation of $SU(3)_c$ (the quarks are triplets and the leptons are singlets). Therefore, the 15 fermions of a generation do *not* form a $\{15\}$-plet of $SU(5)$.

We do have

$$\left\{ \begin{pmatrix} u_L^1 \\ d_L^1 \end{pmatrix}, \begin{pmatrix} u_L^2 \\ d_L^2 \end{pmatrix}, \begin{pmatrix} u_L^3 \\ d_L^3 \end{pmatrix} \right\} = \{3,2\}_{1/6},$$

$$\{u_R^1, u_R^2, u_R^3\} = \{3,1\}_{2/3}, \quad \{d_R^1, d_R^2, d_R^3\} = \{3,1\}_{-1/3},$$

$$\left\{ \begin{pmatrix} \nu_L \\ e_L \end{pmatrix} \right\} = \{1,2\}_{-1/2}, \quad \{e_R\} = \{1,1\}_{-1}. \tag{26.29}$$

Both $\{3,2\}$ and $\{1,1\}$ are contained in $\{10\}$. However, there is also a $\{\bar{3},1\}$. In order to respect Lorentz invariance, we should not mix left- and right-handed fermions. To this end, using the operation c of eq. (16.31), we express everything with left-handed fields

$$\left\{ \begin{pmatrix} u_L^1 \\ d_L^1 \end{pmatrix}, \begin{pmatrix} u_L^2 \\ d_L^2 \end{pmatrix}, \begin{pmatrix} u_L^3 \\ d_L^3 \end{pmatrix} \right\} = \{3,2\}_{1/6},$$

$$\left\{ {}^c u_R^1, {}^c u_R^2, {}^c u_R^3 \right\} = \{\bar{3},1\}_{-2/3}, \quad \left\{ {}^c d_R^1, {}^c d_R^2, {}^c d_R^3 \right\} = \{\bar{3},1\}_{1/3},$$

$$\left\{ \begin{pmatrix} \nu_L \\ e_L \end{pmatrix} \right\} = \{1,2\}_{-1/2}, \quad \left\{ {}^c e_R \right\} = \{1,1\}_1. \tag{26.30}$$

As a result, we identify an $SU(5)$ decuplet

$$\left\{ \begin{pmatrix} u_L^1 \\ d_L^1 \end{pmatrix}, \begin{pmatrix} u_L^2 \\ d_L^2 \end{pmatrix}, \begin{pmatrix} u_L^3 \\ d_L^3 \end{pmatrix}, {}^c u_R^1, {}^c u_R^2, {}^c u_R^3, {}^c e_R \right\} = \{3,2\}_{1/6} + \{\bar{3},1\}_{-2/3} + \{1,1\}_1$$

$$= \{10\}. \tag{26.31}$$

The various quark and lepton fields fill an anti-symmetric matrix

$$\chi(x) = \begin{pmatrix} 0 & -{}^c e_R(x) & u_L^1(x) & u_L^2(x) & u_L^3(x) \\ {}^c e_R(x) & 0 & d_L^1(x) & d_L^2(x) & d_L^3(x) \\ -u_L^1(x) & -d_L^1(x) & 0 & {}^c u_R^3(x) & -{}^c u_R^2(x) \\ -u_L^2(x) & -d_L^2(x) & -{}^c u_R^3(x) & 0 & {}^c u_R^1(x) \\ -u_L^3(x) & -d_L^3(x) & {}^c u_R^2(x) & -{}^c u_R^1(x) & 0 \end{pmatrix} = -\chi(x)^\mathsf{T}, \tag{26.32}$$

where T denotes transpose, as usual. Under $SU(5)$ gauge transformations, this matrix field transforms as

$$\chi'(x) = \Upsilon(x)\,\chi(x)\,\Upsilon(x)^\mathsf{T}, \quad \Upsilon(x) \in SU(5), \tag{26.33}$$

such that the anti-symmetry is indeed maintained under gauge transformations,

$$\chi'(x)^\mathsf{T} = \left(\Upsilon(x)\chi(x)\Upsilon(x)^\mathsf{T}\right)^\mathsf{T} = \Upsilon(x)\chi(x)^\mathsf{T}\Upsilon(x)^\mathsf{T} = -\Upsilon(x)\chi(x)\Upsilon(x)^\mathsf{T} = -\chi'(x). \tag{26.34}$$

The corresponding covariant derivative takes the form

$$D_\mu \chi(x) = \partial_\mu \chi(x) + V_\mu(x)\chi(x) + \chi(x)V_\mu(x)^\mathsf{T}, \tag{26.35}$$

which implies

$$(D_\mu \chi(x))^\mathsf{T} = -D_\mu \chi(x), \quad D_\mu \chi'(x) = \Upsilon(x) D_\mu \chi(x) \Upsilon(x)^\mathsf{T}. \tag{26.36}$$

Let us investigate how the various building blocks of $\chi(x)$ transform under the embedded Standard Model gauge transformations of eq. (26.11). The 2×2 block, which contains the charge conjugate of the right-handed electron field, transforms as

$$-{}^c e'_\mathrm{R}(x) \mathrm{i}\tau^2 = \begin{pmatrix} 0 & -{}^c e'_\mathrm{R}(x) \\ {}^c e'_\mathrm{R}(x) & 0 \end{pmatrix}$$

$$= \exp\left(\mathrm{i}\frac{g'}{2}\varphi(x)\right) L(x) \begin{pmatrix} 0 & -{}^c e_\mathrm{R}(x) \\ {}^c e_\mathrm{R}(x) & 0 \end{pmatrix} \exp\left(\mathrm{i}\frac{g'}{2}\varphi(x)\right) L(x)^\mathsf{T}$$

$$= -\exp\left(\mathrm{i}g'\varphi(x)\right) {}^c e_\mathrm{R}(x) L(x) \mathrm{i}\tau^2 L(x)^\mathsf{T} = -\exp\left(\mathrm{i}g'\varphi(x)\right) {}^c e_\mathrm{R}(x) \mathrm{i}\tau^2,$$

$$\tag{26.37}$$

which reflects the fact that $Y_{e_\mathrm{R}} = -1$. Here, we have used that $\mathrm{i}\tau^2 L(x)^\mathsf{T} = L(x)^\dagger \mathrm{i}\tau^2$.

Quick Question 26.2.1   Identities for SU(2) matrices

Check the identities $U^* \tau^2 = \tau^2 U$ and $\tau^2 U^\mathsf{T} = U^\dagger \tau^2$ for any matrix $U \in \mathrm{SU}(2)$.

The 2×3 block transforms as

$$\begin{pmatrix} u_\mathrm{L}^{1\,'}(x) & u_\mathrm{L}^{2\,'}(x) & u_\mathrm{L}^{3\,'}(x) \\ d_\mathrm{L}^{1\,'}(x) & d_\mathrm{L}^{2\,'}(x) & d_\mathrm{L}^{3\,'}(x) \end{pmatrix} =$$

$$\exp\left(\mathrm{i}\frac{g'}{2}\varphi(x)\right) L(x) \begin{pmatrix} u_\mathrm{L}^1(x) & u_\mathrm{L}^2(x) & u_\mathrm{L}^3(x) \\ d_\mathrm{L}^1(x) & d_\mathrm{L}^2(x) & d_\mathrm{L}^3(x) \end{pmatrix} \exp\left(-\mathrm{i}\frac{g'}{3}\varphi(x)\right) \Omega(x)^\mathsf{T} =$$

$$\exp\left(\mathrm{i}\frac{g'}{6}\varphi(x)\right) L(x) \begin{pmatrix} u_\mathrm{L}^1(x) & u_\mathrm{L}^2(x) & u_\mathrm{L}^3(x) \\ d_\mathrm{L}^1(x) & d_\mathrm{L}^2(x) & d_\mathrm{L}^3(x) \end{pmatrix} \Omega(x)^\mathsf{T}. \tag{26.38}$$

This confirms that the left-handed quarks form an $\mathrm{SU}(2)_L$ doublet and an $\mathrm{SU}(3)_\mathrm{c}$ color triplet with the weak hypercharge $Y_{q_\mathrm{L}} = 1/6$. Finally, the 3×3 block that contains the charge conjugate of the right-handed up quark transforms as

$$\begin{pmatrix} 0 & {}^c u_\mathrm{R}^{3\,'}(x) & -{}^c u_\mathrm{R}^{2\,'}(x) \\ -{}^c u_\mathrm{R}^{3\,'}(x) & 0 & {}^c u_\mathrm{R}^{1\,'}(x) \\ {}^c u_\mathrm{R}^{2\,'}(x) & -{}^c u_\mathrm{R}^{1\,'}(x) & 0 \end{pmatrix} =$$

$$\exp\left(-\mathrm{i}g'\frac{2}{3}\varphi(x)\right) \Omega(x) \begin{pmatrix} 0 & {}^c u_\mathrm{R}^3(x) & -{}^c u_\mathrm{R}^2(x) \\ -{}^c u_\mathrm{R}^3(x) & 0 & {}^c u_\mathrm{R}^1(x) \\ {}^c u_\mathrm{R}^2(x) & -{}^c u_\mathrm{R}^1(x) & 0 \end{pmatrix} \Omega(x)^\mathsf{T}, \tag{26.39}$$

which is consistent with $Y_{u_\mathrm{R}} = 2/3$. The 3×3 block matrix is given by $\epsilon_{abc} {}^c u_\mathrm{R}^c(x)$. The charge conjugate of the right-handed up quark is a color anti-triplet that transforms under $\mathrm{SU}(3)_\mathrm{c}$ gauge transformations as ${}^c u'_\mathrm{R}(x) = \Omega(x)^* {}^c u_\mathrm{R}(x)$. This is indeed consistent because

$$\Omega_{ad}(x) \epsilon_{def} {}^c u_\mathrm{R}^f(x) \Omega_{be}(x) = \epsilon_{abc} \Omega_{cf}(x)^* {}^c u_\mathrm{R}^f(x) = \epsilon_{abc} {}^c u'_\mathrm{R}{}^c(x). \tag{26.40}$$

Here, we have used the identity

$$\epsilon_{def}\Omega_{ad}(x)\Omega_{be}(x) = \epsilon_{abc}\Omega_{cf}(x)^*, \tag{26.41}$$

which is satisfied by every SU(3) matrix $\Omega(x)$. It just means that the vector cross-product of two column vectors of an SU(3) matrix provides the complex conjugate of its third column vector.

The remaining five fermions are

$$\left\{{}^c d_R^1, {}^c d_R^2, {}^c d_R^3\right\} = \{\bar{3}, 1\}_{1/3}, \quad \left\{\begin{pmatrix} \nu_L \\ e_L \end{pmatrix}\right\} = \{1, 2\}_{-1/2}. \tag{26.42}$$

They naturally fit into an SU(5) anti-quintet

$$\{\bar{5}\} = \{\bar{3}, 1\}_{1/3} + \{1, \bar{2}\}_{-1/2}. \tag{26.43}$$

Since SU(2) is pseudo-real, the representation $\{\bar{2}\}$ is equivalent to $\{2\}$, from which it results by the unitary transformation $i\tau^2$, such that

$$\left\{{}^c d_R^1, {}^c d_R^2, {}^c d_R^3, i\tau^2 \begin{pmatrix} \nu_L \\ e_L \end{pmatrix}\right\} = \{\bar{3}, 1\}_{1/3} + \{1, \bar{2}\}_{-1/2} = \{\bar{5}\}. \tag{26.44}$$

The corresponding fermion field takes the form

$$\psi(x) = \begin{pmatrix} e_L(x) \\ -\nu_L(x) \\ {}^c d_R^1(x) \\ {}^c d_R^2(x) \\ {}^c d_R^3(x) \end{pmatrix}, \quad \psi'(x) = \Upsilon(x)^* \psi(x). \tag{26.45}$$

In this case, the covariant derivative is given by

$$D_\mu \psi(x) = \partial_\mu \psi(x) + V_\mu(x)^* \psi(x), \quad D_\mu \psi'(x) = \Upsilon(x)^* D_\mu \psi(x). \tag{26.46}$$

Under the embedded Standard Model gauge transformations of eq. (26.11), the two upper components of $\psi(x)$ transform as

$$\begin{pmatrix} e_L'(x) \\ -\nu_L'(x) \end{pmatrix} = \exp\left(-i\frac{g'}{2}\varphi(x)\right) L(x)^* \begin{pmatrix} e_L(x) \\ -\nu_L(x) \end{pmatrix} = \exp\left(-i\frac{g'}{2}\varphi(x)\right) L(x)^* i\tau^2 \begin{pmatrix} \nu_L(x) \\ e_L(x) \end{pmatrix}$$

$$= \exp\left(-i\frac{g'}{2}\varphi(x)\right) i\tau^2 L(x) \begin{pmatrix} \nu_L(x) \\ e_L(x) \end{pmatrix}, \tag{26.47}$$

which confirms that the left-handed leptons form an SU(2)$_L$ doublet with $Y_{l_L} = -1/2$. Similarly, the three lower components transform as

$$\begin{pmatrix} {}^c d_R^{1'}(x) \\ {}^c d_R^{2'}(x) \\ {}^c d_R^{3'}(x) \end{pmatrix} = \exp\left(i\frac{g'}{3}\varphi(x)\right) \Omega(x)^* \begin{pmatrix} {}^c d_R^1(x) \\ {}^c d_R^2(x) \\ {}^c d_R^3(x) \end{pmatrix}. \tag{26.48}$$

This is consistent with the facts that the charge conjugate of the right-handed down quark is a color anti-triplet and that $Y_{d_R} = -1/3$.

As we see, the fermions of one generation form a reducible 15-dimensional representation of SU(5), which decomposes into $\{10\}$ and $\{\bar{5}\}$. In the SU(5) GUT, quarks and leptons belong to the same irreducible representation. This immediately explains why proton and positron have the same electric charge. It also implies that baryon and lepton number are no longer conserved. However, $B - L$ invariance remains exact. This becomes obvious when we define the *quantum number F*,

$$F = B - L - \frac{4}{5}Y, \quad F_\chi = \frac{1}{5}, \quad F_\psi = -\frac{3}{5}. \tag{26.49}$$

We see that F coincides for all members of a fermion representation. As a consequence, the gauge interactions conserve F. Since $U(1)_Y$ is an exact gauge symmetry, $B - L$ remains conserved, while B and L are individually violated by GUT scale gauge interactions, thus leading to proton decay.

In GUTs, $U(1)_{em}$ is embedded in a compact non-Abelian group, so the electric charge is automatically quantized. GUTs contain superheavy magnetic monopoles. As Dirac pointed out almost a century ago, the existence of magnetic monopoles would explain the quantization of electric charge (Dirac, 1931). However, despite intensive experimental searches, magnetic monopoles have not been observed (in a reproducible manner).

The absence of monopoles does not imply that GUTs are incorrect, because an inflationary phase in the very early Universe may have led to an extreme dilution of these objects. As we have seen in Sections 15.8 and 15.10, electric charge quantization follows already in the Standard Model, if one takes into consideration that the electron is massive. In the minimal SU(5) GUT (without right-handed neutrino fields), neutrinos remain massless and $B - L$ is an exact global symmetry (unless one introduces a non-renormalizable dimension-5 term), and still electric charge remains quantized.

26.3 Lepton–Quark Transitions and Proton Decay

Let us consider the contribution to the GUT Lagrangian that describes the gauge interactions of the fermions

$$\mathcal{L}(\bar{\chi}, \chi, \bar{\psi}, \psi, V_\mu) = \text{Tr}\left(\bar{\chi}\,\bar{\sigma}_\mu D_\mu\,\chi \right) + \bar{\psi}\,\bar{\sigma}_\mu D_\mu\,\psi, \tag{26.50}$$

with the covariant derivatives given by eqs. (26.35) and (26.46). In Problem 26.4, we will check that this Lagrangian is gauge-invariant and that it reproduces all gauge interactions of the Standard Model if we set $X_\mu = Y_\mu = 0$. Let us now concentrate on those terms that involve the X- and Y-bosons and are thus not present in the Standard Model, by subtracting the Standard Model contribution. We obtain

$$\mathcal{L}(\bar{\chi}, \chi, \bar{\psi}, \psi, V_\mu) - \mathcal{L}(\bar{\chi}, \chi, \bar{\psi}, \psi, V_\mu)|_{X_\mu = Y_\mu = 0} =$$

$$\mathrm{i}g_5 \left\{ (\bar{u}_\mathrm{L}, \bar{d}_\mathrm{L}) \left(\mathrm{i}\tau^2 \right) \bar{\sigma}_\mu \begin{pmatrix} X_\mu \\ Y_\mu \end{pmatrix} {}^c e_\mathrm{R} + {}^c\bar{e}_\mathrm{R}\bar{\sigma}_\mu \left(\bar{X}_\mu, \bar{Y}_\mu \right) \left(-\mathrm{i}\tau^2 \right) \begin{pmatrix} u_\mathrm{L} \\ d_\mathrm{L} \end{pmatrix} \right\}$$

$$-\mathrm{i}g_5\epsilon_{abc} \left\{ (\bar{u}_\mathrm{L}^a, \bar{d}_\mathrm{L}^a) \bar{\sigma}_\mu \begin{pmatrix} \bar{X}_\mu^b \\ \bar{Y}_\mu^b \end{pmatrix} {}^c u_\mathrm{R}^c + {}^c\bar{u}_\mathrm{R}^a\bar{\sigma}_\mu \left(X_\mu^b, Y_\mu^b \right) \begin{pmatrix} u_\mathrm{L}^c \\ d_\mathrm{L}^c \end{pmatrix} \right\}$$

$$+\mathrm{i}\frac{g_5}{2} \left\{ (\bar{\nu}_\mathrm{L}, \bar{e}_\mathrm{L}) \left(\mathrm{i}\tau^2 \right) \bar{\sigma}_\mu \begin{pmatrix} X_\mu \\ Y_\mu \end{pmatrix} {}^c d_\mathrm{R} + {}^c\bar{d}_\mathrm{R}\bar{\sigma}_\mu \left(\bar{X}_\mu, \bar{Y}_\mu \right) \left(-\mathrm{i}\tau^2 \right) \begin{pmatrix} \nu_\mathrm{L} \\ e_\mathrm{L} \end{pmatrix} \right\}. \tag{26.51}$$

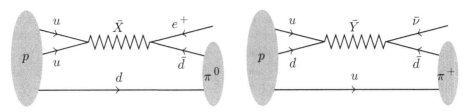

Fig. 26.1 Diagrams of two proton decay scenarios, according to the upper line (left) and the lower line (right) of relation (26.53).

> **Quick Question 26.3.1   Gauge invariance**
> Check that the terms in eq. (26.51) are invariant under Standard Model gauge transformations.

These terms violate both baryon and lepton number, but they keep the difference $B - L$ invariant. Quarks and leptons, or anti-quark pairs, can annihilate into X- or Y-bosons, $e.g.$,

$$d + e \to X, \quad u + e \to Y, \quad d + \nu \to Y, \quad \bar{u} + \bar{u} \to X, \quad \bar{u} + \bar{d} \to Y. \qquad (26.52)$$

Since the X- and Y-bosons induce lepton–quark transitions, they are also known as *leptoquarks*.

An important prediction of GUTs is the instability of the proton. The predicted proton decay proceeds via the X- or Y-boson channel. The following decays, which are illustrated in Figure 26.1, are dominant

$$p \sim uud \to \bar{X}d \to \bar{d}d + \bar{e} \to \pi^0 + \bar{e},$$
$$p \sim uud \to u\bar{Y} \to u\bar{d} + \bar{\nu} \sim \pi^+ + \bar{\nu}. \qquad (26.53)$$

Proton decay is suppressed by the large mass of the X- and Y-bosons. The resulting proton life-time is given by

$$\tau_p \propto \frac{M_X^4}{M_p^5} \approx 10^{31} \text{years}. \qquad (26.54)$$

Although this predicted life-time is much larger than the age of the Universe, the minimal SU(5) model has been ruled out experimentally, because the proton indeed lives longer than the model predicts. In particular, for the first decay in eqs. (26.53) the Super-Kamiokande Collaboration observed a lower bound of 1.6×10^{34} years (Abe *et al.*, 2017).

On the other hand, more sophisticated GUTs based on Spin(10) or E(6) gauge symmetries are not experimentally ruled out so far.

26.4 Baryon Asymmetry in the Universe

Why is there a baryon asymmetry in the Universe? In other words, why does the Universe consist of matter and not also of equal amounts of anti-matter?

In the early Universe, anti-matter was also present. In particular, at temperatures in the GeV range, about the same number of quarks and anti-quarks was around. When the

Universe expanded and cooled down, quarks and anti-quarks, or baryons and anti-baryons, annihilated each other almost completely, and all matter in the Universe today is the tiny fraction that survived this mass extinction.

Since the Universe is electrically neutral, it contains as many electrons as protons, but almost no positrons. Correspondingly, there is also a lepton asymmetry. However, since also the only weakly interacting neutrinos contribute to it, it is extremely hard to probe experimentally.

As was pointed out by Andrei Sakharov, in order to be able to explain the baryon asymmetry, besides baryon number conservation, also C and CP invariance must be violated, and the Universe must get out of thermal equilibrium (Sakharov, 1967).

A trivial condition for the explanation of the baryon asymmetry is the existence of baryon number violating processes. Only then, an initial state with unknown baryon number (in particular $B = 0$) may turn into the situation that we observe today. As we have seen, GUTs indeed give rise to such processes. At temperatures in the 10^{16} GeV range, which were realized in the very early Universe, baryon number violating GUT processes would have been in thermal equilibrium, such that the effects of even earlier initial conditions would have been washed out. In the SU(5) theory, for example, the following processes are possible

$$ u + u \to \bar{X} \to e^+ + \bar{d}, \quad u + d \to \bar{Y} \to \bar{\nu}_e + \bar{d}, \quad u + d \to \bar{Y} \to \bar{u} + e^+. \quad (26.55) $$

All these processes change the baryon number ($\Delta B = -1$) and the lepton number ($\Delta L = -1$), but $B - L$ remains unchanged. Indeed, $B - L$ is conserved in the SU(5) GUT, just as in the Standard Model (but this does not necessarily hold in other GUTs, as we will discuss below).

If C or CP were conserved, baryon number violating processes would generate anti-baryons at the same rate as baryons, and no net baryon asymmetry could emerge. The charge conjugate of a left-handed quark is a right-handed anti-quark; *i.e.* baryon number changes sign under charge conjugation. Similarly, under a CP transformation a left-handed quark is turned into a left-handed anti-quark (see Chapters 8 and 9), and again baryon number changes sign. Both the Standard Model and the SU(5) GUT are chiral gauge theories in which C is violated. In the Standard Model with three generations, CP symmetry is broken by the complex phase of the Cabibbo–Kobayashi–Maskawa matrix (cf. Section 17.3), which manifests itself, for example, in K^0-$\overline{K^0}$ mixing (see Section 17.5). However, the amount of CP violation in the Standard Model alone is insufficient to explain the baryon asymmetry quantitatively.

In thermal equilibrium, baryon number violating processes proceed in both directions; *i.e.* baryons are generated but also annihilated at the same rate. Hence, a net baryon asymmetry cannot be generated in thermal equilibrium. More formally, this results from CPT invariance (a symmetry of all relativistic quantum field theories). However, since the Universe expands and cools, baryon number violating processes eventually fall out of thermal equilibrium, such that indeed all necessary Sakharov conditions for the generation of a baryon asymmetry are satisfied in our Universe.

An attractive alternative to the generation of a baryon asymmetry at the GUT scale is *leptogenesis* (Buchmüller *et al.*, 2005a,b). Then, first a lepton asymmetry is generated, for example, by the out-of-equilibrium decay of a heavy right-handed neutrino. In a second step, the lepton asymmetry is then partly converted into a baryon asymmetry, by B and L violating (but $B - L$ conserving) Standard Model processes.

26.5 Thermal Baryon Number Violation in the Standard Model

The Lagrangian of the Standard Model does not contain baryon number violating interactions. However, as we discussed in Section 15.11, this does not imply that the Standard Model conserves baryon number after quantization. Indeed, due to the chiral couplings of the quarks, the baryon number current has an anomaly. Actually, the Lagrangian has a global $U(1)_B$ baryon number symmetry, but this symmetry is explicitly broken in the quantum theory. The same is true for the lepton number. The difference, $B - L$, on the other hand, remains conserved. The existence of baryon number violating processes at the electroweak scale may change a baryon asymmetry that has been generated at the GUT scale or partly convert a lepton asymmetry (generated via leptogenesis) into a baryon asymmetry; for a review, see Bödeker and Buchmüller (2021).

In analogy to Section 19.5, let us now consider the vacuum structure of the $SU(2)_L$ gauge–Higgs sector of the Standard Model. A classical (static) vacuum solution is

$$\Phi(\vec{x}) = \begin{pmatrix} \Phi^+(\vec{x}) \\ \Phi^0(\vec{x}) \end{pmatrix} = \begin{pmatrix} 0 \\ v \end{pmatrix}, \quad W_i(\vec{x}) = 0. \tag{26.56}$$

Of course, gauge transformations of this solution are vacua as well. However, states that are related by a gauge transformation are physically equivalent, and one should not consider the other solutions as additional vacua. Still, there is a subtlety, because there are gauge transformations with different topological properties.

First of all, there are the so-called *small gauge transformations,* which can be continuously deformed into the identity, and one should indeed not distinguish between states related by small gauge transformations. However, there are also *large gauge transformations* – those that cannot be deformed into a trivial gauge transformation – and they indeed give rise to additional vacuum states. The gauge transformations

$$L(\vec{x}) : \mathbb{R}^3 \to SU(2)_L \tag{26.57}$$

can be viewed as maps from coordinate space into the group space. When one identifies points at spatial infinity, $\mathbb{R}^3$ is compactified to S^3. On the other hand, the group space of $SU(2)_L$ is also S^3. Hence, the gauge transformations are maps

$$L(\vec{x}) : S^3 \to S^3. \tag{26.58}$$

Such maps are known to fall into topologically distinct classes characterized by a winding number

$$n[L] \in \Pi_3[SU(2)_L] = \mathbb{Z} \tag{26.59}$$

from the third homotopy group of the gauge group; see Appendix G. In this case, maps with any integer winding number are possible.

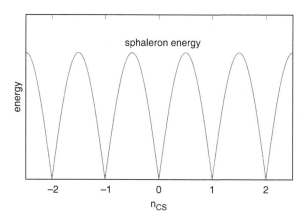

Fig. 26.2 The periodic potential $V(n_{CS})$ corresponds to the minimal value of the gauge–Higgs field energy for a given Chern–Simons number $n_{CS} \in \mathbb{R}$. A sphaleron configuration (which is "ready to decay" to a classical vacuum configuration with winding number $n_{CS} = n \in \mathbb{Z}$) resides at a maximum of the potential.

Denoting a map with winding number n by L_n, we can thus construct a set of topologically inequivalent vacuum configurations

$$\Phi^{(n)}(\vec{x}) = L_n(\vec{x}) \begin{pmatrix} 0 \\ v \end{pmatrix}, \quad W_i^{(n)}(\vec{x}) = L_n(\vec{x}) \partial_i L_n(\vec{x})^\dagger. \tag{26.60}$$

Topologically distinct vacua are separated by energy barriers, and thus, there is a periodic potential in the space of field configurations. To see this, we consider the Chern–Simons density (or 0-cochain), $\Omega_\mu^{(0)}$, that was introduced in eq. (19.8). The spatial integral of its temporal component associates a time-dependent Chern–Simons number

$$n_{CS}[W] = \int d^3x \, \Omega_4^{(0)} = -\frac{1}{8\pi^2} \int d^3x \, \epsilon_{ijk} \text{Tr}\left[W_i \left(\partial_j W_k + \frac{2}{3} W_j W_k \right) \right] \tag{26.61}$$

with any $SU(2)_L$ gauge field $W_\mu(x)$. For a vacuum configuration $W_i^{(n)}(\vec{x})$ of eq. (26.60), which is characterized by a pure gauge transformation, $L_n(\vec{x}) \in SU(2)_L$, with winding number

$$n[L_n] = \frac{1}{24\pi^2} \int d^3x \, \epsilon_{ijk} \text{Tr}\left[\left(L_n \partial_i L_n^\dagger \right) \left(L_n \partial_j L_n^\dagger \right) \left(L_n \partial_k L_n^\dagger \right) \right] = n \in \Pi_3 \left[SU(2)_L \right] = \mathbb{Z}, \tag{26.62}$$

the Chern–Simons number is simply given by $n_{CS}[W^{(n)}] = n \in \mathbb{Z}$. For non-vacuum configurations, on the other hand, the Chern–Simons number is a general real number. One can vary a static gauge–Higgs field configuration in order to minimize its energy, keeping the Chern–Simons number $n_{CS} \in \mathbb{R}$ fixed. The minimal value of the energy then defines a periodic potential $V(n_{CS})$, which is illustrated in Figure 26.2. Interestingly, this energy approaches the minima with non-zero derivative (Akiba *et al.*, 1988).

Classically, the system may be in one of the degenerate vacuum configurations. Quantum effects, however, can result in tunneling events that lead from one vacuum to another. It turns out that a transition from the vacuum characterized by the winding number m to the vacuum characterized by n is accompanied by a baryon number violating process with $\Delta B = N_g(n - m)$, where N_g is the number of generations of quarks and leptons. Also, the lepton number changes by $\Delta L = N_g(n - m)$, such that $B - L$ is conserved.

The tunneling amplitude – and hence the rate of baryon number violating processes – is controlled by the barrier height between adjacent classical vacua. The unstable field configurations at the top of a barrier are known as *sphalerons* (which means "ready to decay") (Kuzmin *et al.*, 1985; Arnold and McLerran, 1987). In the Standard Model, the height of the barrier (*i.e.* the sphaleron energy) is given by $4\pi v/g$ and the resulting tunneling rate is $\exp(-8\pi^2/g^2) \approx \exp(-200)$, which is negligible.

Hence, one might assume that baryon number violation in the Standard Model is only of academic interest. However, in the early Universe there was no need to tunnel through the barrier – the field could simply step over it due to large thermal fluctuations.

One concludes that in the TeV energy range, baryon number violating processes are unsuppressed in the Standard Model. This means that any pre-existing baryon or lepton asymmetry – created at the GUT scale or via leptogenesis – would be washed out, because baryon number violating processes reach a thermal equilibrium. Since electroweak symmetry breaking is not associated with a strong first-order phase transition, a sufficient baryon asymmetry is not regenerated at the electroweak scale.

Let us assume some initial net baryon number and lepton number asymmetries ΔB_i and ΔL_i. If sphaleron transitions are frequent, they lead to an equilibrium at the final time (today),

$$\Delta \left(B_f + L_f \right) = 0. \tag{26.63}$$

We saw, however, that they preserve the difference between these asymmetries, which implies

$$\Delta \left(B_f - L_f \right) = \Delta \left(B_i - L_i \right). \tag{26.64}$$

Hence, the present baryon and lepton asymmetries turn out to be

$$\Delta B_f = -\Delta L_f = \frac{1}{2} \Delta \left(B_i - L_i \right). \tag{26.65}$$

This again leads to a puzzle, because also the minimal SU(5) model conserves $B - L$. An asymmetry $\Delta(B_i - L_i)$ would then have to be due to processes in the even earlier Universe. Fortunately, there is a way out. Other GUTs, including those with gauge group Spin(10) or E(6), are not ruled out by proton decay experiments and indeed do not conserve $B - L$.

The reason for $B - L$ violation in these models is related to the existence of massive neutrinos. The seesaw mass-by-mixing mechanism that we discussed in Section 16.9 employs heavy neutrinos, possibly with a mass around the GUT scale, in order to obtain light neutrinos with masses in the eV range, which are identified with the neutrinos that are observed. As a result, we may be able to explain the baryon asymmetry using either GUTs or leptogenesis only if the neutrinos are massive. Since the observation of neutrino oscillations indeed implies that neutrinos do have mass (cf. Section 17.7), this encourages an explanation of the baryon asymmetry along these lines. However, until a satisfactory explanation has been firmly established, we are left with the

Box 26.1   **Deep question**

What is the origin of the baryon asymmetry in the Universe?

26.6 Topological Excitations as Cosmic Relics

GUTs give rise to a variety of topological excitations which might even persist as cosmic relics in the Universe today. In this section, we consider Φ as some generic scalar field (not necessarily the Standard model Higgs field).

When a discrete symmetry is spontaneously broken in the early Universe, it is natural to expect the generation of *domain walls* as 2-dimensional topological defects. This is because then there is a discrete set of degenerate vacua

$$\mathcal{M} = \{\Phi | \Phi \text{ is a global minimum of } V(\Phi)\}. \tag{26.66}$$

Let us consider two halves of the Universe (left $z < 0$, right $z > 0$). At $z = -\infty$ and at $z = \infty$, the field Φ must assume a vacuum value in order to have finite energy, *i.e.*

$$\Phi(z = -\infty), \Phi(z = \infty) \in \mathcal{M}. \tag{26.67}$$

Topologically, the boundaries $z = -\infty$ and $z = \infty$ correspond to a 0-dimensional "sphere" (consisting of just two points) $\{-\infty, \infty\} = S^0$, such that the field at spatial infinity can be viewed as a map

$$\Phi : S^0 \to \mathcal{M}. \tag{26.68}$$

Such maps are characterized by homotopy classes (cf. Appendix G). The relevant homotopy group is $\Pi_0[\mathcal{M}]$. It is non-trivial only if the vacuum manifold is not simply connected.

In the Standard Model, $\mathcal{M}$ is simply connected. Hence,

$$\Pi_0[\mathcal{M}] = \{0\}, \tag{26.69}$$

and no domain walls arise. This is fine, because domain walls would pose a problem for cosmology. Since their energy is proportional to their area, and since they could span the entire Universe, domain walls carry enormous energies. They could close the Universe, which would be in contradiction to cosmological observations.

When a symmetry group G breaks spontaneously to a subgroup H, the vacuum manifold is $\mathcal{M} = G/H$. In particular, when $G = U(1)$ breaks to $H = \{\mathbb{1}\}$, the vacuum manifold is a circle, $\mathcal{M} = S^1$. This naturally gives rise to *cosmic strings*, which are spatially 1-dimensional topological defects. The relevant homotopy group is

$$\Pi_1[\mathcal{M}] = \Pi_1[U(1)/\{\mathbb{1}\}] = \Pi_1[U(1)] = \Pi_1[S^1] = \mathbb{Z}. \tag{26.70}$$

In the Standard Model, $G = SU(3)_c \times SU(2)_L \times U(1)_Y$ and $H = SU(3)_c \times U(1)_{em}$, such that

$$\Pi_1[\mathcal{M}] = \Pi_1[SU(3) \times SU(2) \times U(1)/SU(3) \times U(1)] = \Pi_1[SU(2)] = \Pi_1[S^3] = \{0\}. \tag{26.71}$$

Hence, there are no stable cosmic string solutions in the Standard Model. In some GUTs, cosmic strings appear because a $U(1)$ symmetry breaks spontaneously. Then, a network of cosmic strings emerges that may extend throughout the entire Universe.

Finally, *magnetic monopoles* are point-like topological excitations, which arise when the second homotopy group of the vacuum manifold, $\Pi_2[\mathcal{M}]$, is non-trivial. For example, when $G = SU(2)$ breaks spontaneously to $H = U(1)$, one obtains

$$\Pi_2[\mathcal{M}] = \Pi_2[G/H] = \Pi_2[SU(2)/U(1)] = \Pi_2[S^2] = \mathbb{Z}. \tag{26.72}$$

In the Standard Model, there are no magnetic monopoles, because

$$\Pi_2[SU(3) \times SU(2) \times U(1)/SU(3) \times U(1)] = \Pi_2[SU(2)] = \Pi_2[S^3] = \{0\}. \quad (26.73)$$

In the SU(5) GUT, on the other hand,

$$\Pi_2[SU(5)/SU(3) \times SU(2) \times U(1)] = \Pi_1[SU(3) \times SU(2) \times U(1)]$$
$$= \Pi_1[U(1)] = \Pi_1[S^1] = \mathbb{Z}. \quad (26.74)$$

The same is true for other GUTs like SO(10) or E(6), since then again $\Pi_2[G] = \{0\}$, and the $U(1)_{em}$ symmetry of the Standard Model remains as an unbroken subgroup. Consequently, magnetic monopoles are unavoidable in GUTs.

Why have monopoles not yet been observed? Maxwell's theory with $\vec{\nabla} \cdot \vec{B} = 0$ agrees very well with observations. Would it be possible that GUTs have monopole solutions, but monopoles have simply not been created? The *Kibble mechanism* demonstrates that this is impossible (Kibble, 1976). Let us assume that, in the very early Universe, a GUT phase transition has taken place. In the high-temperature phase, some scalar field Φ fluctuates wildly. When the system undergoes a phase transition and enters the low-temperature phase, the field fluctuations get frozen, and Φ obtains its vacuum value. The originally random orientations of Φ then inevitably lead to the creation of topological defects. Depending on the symmetry breaking pattern, those could be domain walls, cosmic strings, or monopoles. However, if *cosmic inflation* happens after a GUT phase transition, the generated topological excitations may get diluted by the exponential expansion of space, so much that they can no longer be found in the Universe today.

26.7 't Hooft–Polyakov Monopole and Callan–Rubakov Effect

The existence of magnetic monopoles was contemplated by Paul Dirac as early as 1931 (Dirac, 1931). The Standard Model does not contain magnetically charged particles, and even Dirac did not believe in the existence of magnetic monopoles shortly before the end of his life.[1] Still, some extensions of the Standard Model – for example, the SU(5) GUT – contain very heavy *'t Hooft–Polyakov monopoles*, which look like Dirac monopoles from large distances. In the monopole core, the SU(5) symmetry is "unbroken" and quarks and leptons are indistinguishable there. As a consequence, baryons that enter the monopole core can reappear as leptons and thus the monopole itself can catalyze baryon decay. This effect – known as the *Callan–Rubakov effect* – was first described by Valery Rubakov (1981, 1982) and independently by Curtis Callan (1983). In the SU(5) GUT, $B - L$ is conserved and thus baryon and lepton number are violated by the same amount. As a result, SU(5) monopoles also catalyze lepton decay.

Magnetic monopoles are particle-like topological excitations, which arise naturally in GUTs. A simplified (and certainly unrealistic) GUT is the SU(2) analogue of the

---

[1] As a student, one of the authors attended a meeting of Nobel laureates in Lindau (Germany) in 1983. A year earlier, the observation of a magnetic monopole (which remained unconfirmed) had been reported. At a dinner, administered by Dirac's wife, each student at Dirac's table was allowed one yes-or-no question. UJW asked Dirac whether he believes in the existence of magnetic monopoles and his answer was *no*.

Georgi–Glashow model. In the SU(5) model, we have used a Higgs field in the adjoint representation to "break" the gauge symmetry spontaneously. We now do the same in the simplified SU(2) model, using a Higgs triplet

$$\Xi(x) = \Xi_a(x)\tau^a. \tag{26.75}$$

The Lagrangian in Minkowski space–time is then given by

$$\mathcal{L}(\Xi, V) = \frac{1}{4}\mathrm{Tr}\left[D_\mu \Xi D^\mu \Xi\right] - V(\Xi) + \frac{1}{2e^2}\mathrm{Tr}\left[V_{\mu\nu}V^{\mu\nu}\right], \tag{26.76}$$

and again

$$V_{\mu\nu}(x) = \partial_\mu V_\nu(x) - \partial_\nu V_\mu(x) + [V_\mu(x), V_\nu(x)], \quad V_\mu(x) = ieV_\mu^a(x)\frac{\tau^a}{2},$$
$$D_\mu \Xi(x) = \partial_\mu \Xi(x) + [V_\mu(x), \Xi(x)]. \tag{26.77}$$

At the classical level, we have identified the unified non-Abelian gauge coupling with the electric charge e (which is still affected by renormalization group running at the quantum level).

At spatial infinity, the scalar field must approach its vacuum value $\mathcal{V}$. Using the temporal gauge $V_0 = 0$, we make an ansatz for a static solution of the classical equations of motion

$$\Xi_a(r,\theta,\varphi) = \xi(r)\frac{x^a}{r}, \quad V_i^a(r,\theta,\varphi) = \mathrm{v}(r)\epsilon_{iab}\frac{x^b}{r^2}, \tag{26.78}$$

with $\xi(0) = 0, \xi(\infty) = \mathcal{V}, \mathrm{v}(0) = 0$. When the resulting equations of motion for $\xi(r)$ and $\mathrm{v}(r)$ are solved numerically, one obtains monotonically rising solutions for these functions.

Fixing to the unitary gauge, the Higgs field can be diagonalized by a unitary transformation, such that $\Xi(x) \to \Xi_3(x)\tau^3$. The unitary gauge leaves an Abelian subgroup of SU(2) unfixed

$$\Upsilon(x)\left(\Xi_3(x)\tau^3\right)\Upsilon(x)^\dagger = \Xi_3(x)\tau^3, \quad \Upsilon(x) = \begin{pmatrix} \exp(ie\alpha(x)/2) & 0 \\ 0 & \exp(-ie\alpha(x)/2) \end{pmatrix}. \tag{26.79}$$

We identify the unbroken subgroup with the U(1)$_{\mathrm{em}}$ gauge symmetry of electromagnetism. In the unitary gauge, the non-Abelian gauge field V_μ decomposes into the diagonal photon field A_μ and two electrically charged fields $X_\mu^\pm$

$$A_\mu(x) = V_\mu^3(x), \quad X^\pm(x) = \frac{1}{\sqrt{2}}\left(V_\mu^1(x) \mp iV_\mu^2(x)\right). \tag{26.80}$$

Under the remnant U(1)$_{\mathrm{em}}$ gauge transformations $\Upsilon(x)$, these fields transform as

$$A_\mu'(x) = A_\mu(x) - \partial_\mu\alpha(x), \quad X_\mu^{\pm\prime}(x) = \exp(\pm ie\alpha(x))X_\mu^\pm(x). \tag{26.81}$$

At large distances, the electromagnetic field strength $F_{\mu\nu} = \partial_\mu A_\nu - \partial_\nu A_\mu$ of the static solution of eq. (26.78) takes the form

$$B_i(\vec{x}) = \frac{1}{2}\epsilon_{ijk}F^{jk}(\vec{x}) = \frac{g}{4\pi}\frac{x_i}{r^3}, \quad g = \frac{2\pi}{e}. \tag{26.82}$$

This is the field of a magnetic charge g (not to be confused with the SU(2)$_L$ weak gauge coupling), which obeys the Dirac quantization condition, to be discussed further in Section 26.8.

Since the $U(1)_{em}$ symmetry remains after SU(2) "spontaneous symmetry breaking", the vacuum manifold is $\mathcal{M} = SU(2)/U(1)$. Spatial infinity is topologically a sphere S^2. The field configurations are therefore characterized by winding numbers in the homotopy group

$$\Pi_2[SU(2)/U(1)] = \Pi_2[S^3/S^1] = \Pi_1[S^1] = \mathbb{Z}. \tag{26.83}$$

The solution of eq. (26.78) has winding number 1 and thus corresponds to a monopole with one unit of magnetic charge g.

The monopole mass turns out to be $M = \mathcal{O}(g\mathcal{V})$; *i.e.* GUT monopoles have a mass of about 10^{16} GeV – a hypothetical elementary particle which would be as heavy as a bacterium.

Let us describe baryon decay catalyzed by magnetic monopoles. Following Curtis Callan and Edward Witten, we model baryons as Skyrme solitons (Callan and Witten, 1984). The magnetic current of a monopole is given by

$$m_\mu(x) = \frac{1}{2}\epsilon_{\mu\nu\rho\sigma}\partial^\nu F^{\rho\sigma}(x), \tag{26.84}$$

which measures the amount of violation of the Abelian *Bianchi identity*. In the presence of magnetic charge, the Goldstone–Wilczek current of eq. (24.13) is no longer conserved because

$$\partial^\mu j_\mu^{GW}(x) = -\frac{ie}{8\pi^2}m_\mu(x)\text{Tr}\left[T^3(D^\mu U(x)U(x)^\dagger + U(x)^\dagger D^\mu U(x))\right]. \tag{26.85}$$

For a magnetic monopole at rest at the origin, $\vec{x} = \vec{0}$, we have

$$m_0(\vec{x}, t) = g\delta(\vec{x}), \quad m_i(\vec{x}, t) = 0, \tag{26.86}$$

where g is again the magnetic charge. In spherical coordinates (r, θ, φ), a vector potential that describes this situation is given by

$$\vec{A}(\vec{x}) = g\frac{1 - \cos\theta}{r\sin\theta}\vec{e}_\varphi. \tag{26.87}$$

This potential is singular along the negative z-axis, due to a Dirac string, which constitutes a coordinate singularity. The location of the Dirac string, which extends from the monopole to infinity, is gauge-dependent and thus not physical.

Writing $U(x) = \exp(i\pi^0(x)\tau^3/F_\pi)$ and integrating eq. (26.85) over space, we obtain the rate of change of the baryon number as

$$\partial_t B(t) = \frac{eg}{\pi F_\pi}\partial_t\pi^0(\vec{0}, t). \tag{26.88}$$

Using the Dirac quantization condition of eq. (26.82), $eg = 2\pi$, one obtains

$$B(\infty) - B(-\infty) = \frac{1}{2\pi F_\pi}\left[\pi^0(\vec{0}, \infty) - \pi^0(\vec{0}, -\infty)\right]. \tag{26.89}$$

Hence, if the neutral pion field $\pi^0(\vec{0})/F_\pi$ at the location of the monopole rotates by $2\pi n$, baryon number is violated by n units.

If monopoles were still present in the Universe, they would accumulate inside neutron stars and catalyze baryon decay. This would lead to the emission of intense radiation, which is not observed. Hence, the question arises where the GUT monopoles went (if they indeed existed). A possible scenario contemplates monopole confinement by cosmic strings. One then assumes that the Universe has been superconducting for a while. Monopoles and anti-monopoles would then be connected by *Abrikosov cosmic strings*. The strings cost energy in proportion to their length and thus pull monopoles and anti-monopoles together until they annihilate. Cosmic inflation offers an alternative explanation for why no monopoles are observed today: An exponential expansion of space in the early Universe may have diluted them, provided that inflation took place after monopoles had been created in a GUT phase transition.

26.8 Dirac–Schwinger–Zwanziger Dyon Quantization Condition

As Dirac showed in 1931, a theory that contains both electric charges e and magnetic charges g is consistent at the quantum level only if the quantization condition

$$eg = 2\pi n \hbar c, \quad n \in \mathbb{Z} \tag{26.90}$$

is satisfied. In order to emphasize the quantum nature of the condition, we have explicitly written $\hbar$ (along with c). The condition results when one demands that the Dirac string, which emanates from a magnetic monopole, must not be observable. The Dirac string can be replaced by an infinitely thin solenoid that carries the magnetic flux of the monopole to infinity. This resembles the setup of the Aharonov–Bohm effect. A quantum mechanical point particle of charge e, which moves in the field of the solenoid, experiences an Aharonov–Bohm phase $eg/(\hbar c)$. Dirac's quantization condition implies that this phase is a multiple of 2π and thus not observable.

Daniel Zwanziger (1968a,b) and Julian Schwinger (1969) have generalized the Dirac quantization condition to particles that carry both electric and magnetic charge, known as *dyons*. In dyon electrodynamics, the Maxwell equations are extended to

$$\partial_\mu F^{\mu\nu}(x) = \frac{1}{c}j^\nu(x), \quad \partial_\mu \widetilde{F}^{\mu\nu}(x) = \frac{1}{c}m^\nu(x), \quad \widetilde{F}_{\mu\nu}(x) = \frac{1}{2}\epsilon_{\mu\nu\rho\sigma}F^{\rho\sigma}(x). \tag{26.91}$$

The generalized Maxwell equations are invariant against a field redefinition

$$\begin{pmatrix} F^{\mu\nu\prime}(x) \\ \widetilde{F}^{\mu\nu\prime}(x) \end{pmatrix} = \begin{pmatrix} \cos\gamma & \sin\gamma \\ -\sin\gamma & \cos\gamma \end{pmatrix} \begin{pmatrix} F^{\mu\nu}(x) \\ \widetilde{F}^{\mu\nu}(x) \end{pmatrix},$$
$$\begin{pmatrix} j^{\mu\prime}(x) \\ m^{\mu\prime}(x) \end{pmatrix} = \begin{pmatrix} \cos\gamma & \sin\gamma \\ -\sin\gamma & \cos\gamma \end{pmatrix} \begin{pmatrix} j^\mu(x) \\ m^\mu(x) \end{pmatrix}. \tag{26.92}$$

Let us assume that there is one type of point particles which carry electric charge e_1 and magnetic charge g_1, and another type of point particles with electric charge e_2 and magnetic charge g_2. By an appropriate field redefinition, one can then rotate the magnetic charge of the first particle type to $g_1' = 0$,

$$\begin{pmatrix} \cos\gamma & \sin\gamma \\ -\sin\gamma & \cos\gamma \end{pmatrix} \begin{pmatrix} e_1 \\ g_1 \end{pmatrix} = \begin{pmatrix} e_1' \\ 0 \end{pmatrix} \quad \Rightarrow \quad e_1 \sin\gamma = g_1 \cos\gamma. \tag{26.93}$$

The same rotation must be applied to the charges of the second particle type

$$
\begin{pmatrix} \cos \gamma & \sin \gamma \\ -\sin \gamma & \cos \gamma \end{pmatrix} \begin{pmatrix} e_2 \\ g_2 \end{pmatrix} = \begin{pmatrix} e_2 \cos \gamma + g_2 \sin \gamma \\ -e_2 \sin \gamma + g_2 \cos \gamma \end{pmatrix} = \begin{pmatrix} e_2' \\ g_2' \end{pmatrix}. \tag{26.94}
$$

Using $e_1 \sin \gamma = g_1 \cos \gamma$, Dirac's quantization condition (which was derived assuming that $g_1 = 0$) then generalizes to a relation due to Zwanziger and Schwinger

$$
\begin{aligned}
e_1' g_2' &= (e_1 \cos \gamma + g_1 \sin \gamma)(-e_2 \sin \gamma + g_2 \cos \gamma) \\
&= -e_1 e_2 \cos \gamma \sin \gamma + e_1 g_2 \cos^2 \gamma - g_1 e_2 \sin^2 \gamma + g_1 g_2 \sin \gamma \cos \gamma \\
&= -g_1 e_2 \cos^2 \gamma + e_1 g_2 \cos^2 \gamma - g_1 e_2 \sin^2 \gamma + e_1 g_2 \sin^2 \gamma \\
&= e_1 g_2 - e_2 g_1 = 2\pi n \hbar c, \quad n \in \mathbb{Z}. \tag{26.95}
\end{aligned}
$$

The electric and magnetic field of a pair of point-like dyons (one of type 1 and the other of type 2) at the positions $\vec{x}_1 = (0, 0, d/2)$ and $\vec{x}_2 = (0, 0, -d/2)$ (and thus at distance d) are given by

$$
\vec{E}(\vec{x}) = \frac{e_1}{4\pi} \frac{\vec{x} - \vec{x}_1}{|\vec{x} - \vec{x}_1|^3} + \frac{e_2}{4\pi} \frac{\vec{x} - \vec{x}_2}{|\vec{x} - \vec{x}_2|^3}, \quad \vec{B}(\vec{x}) = \frac{g_1}{4\pi} \frac{\vec{x} - \vec{x}_1}{|\vec{x} - \vec{x}_1|^3} + \frac{g_2}{4\pi} \frac{\vec{x} - \vec{x}_2}{|\vec{x} - \vec{x}_2|^3}. \tag{26.96}
$$

Interestingly, the angular momentum of the electromagnetic field of the two static dyons is non-zero

$$
\begin{aligned}
\vec{J} &= \frac{1}{c} \int d^3 x \, \vec{x} \times \left(\vec{E}(\vec{x}) \times \vec{B}(\vec{x}) \right) \\
&= \frac{e_1 g_2 - e_2 g_1}{16\pi^2 c} \int d^3 x \, \vec{x} \times \left(\frac{\vec{x} - \vec{x}_1}{|\vec{x} - \vec{x}_1|^3} \times \frac{\vec{x} - \vec{x}_2}{|\vec{x} - \vec{x}_2|^3} \right) \\
&= \frac{e_1 g_2 - e_2 g_1}{16\pi^2 c} \int d^3 x \, \frac{\vec{x} \times (\vec{x} \times \vec{d})}{|\vec{x} - \vec{x}_1|^3 |\vec{x} - \vec{x}_2|^3}. \tag{26.97}
\end{aligned}
$$

Next, we project the angular momentum on the direction $\vec{d} = \vec{x}_1 - \vec{x}_2$ that connects the two dyons, and some calculus leads to

$$
\vec{J} \cdot \frac{\vec{d}}{d} = -\frac{e_1 g_2 - e_2 g_1}{16\pi^2 c} \int d^3 x \, \frac{|\vec{x} \times \vec{d}|^2}{d |\vec{x} - \vec{x}_1|^3 |\vec{x} - \vec{x}_2|^3} = -\frac{e_1 g_2 - e_2 g_1}{4\pi c}. \tag{26.98}
$$

Imposing the condition that angular momentum is quantized in half-integer multiples of $\hbar$, one again obtains the Dirac–Schwinger–Zwanziger quantization condition eq. (26.95).

26.9 Julia–Zee Dyon and Witten Effect

As we pointed out in Section 25.8, as a result of the *Witten effect*, the electromagnetic vacuum-angle θ_{QED} determines the electric charge $-e\theta_{\text{QED}}/(2\pi)$ of magnetic monopoles (Witten, 1979c), which thus turn into dyons (if $\theta_{\text{QED}} \neq 0$). In a GUT with a single unified gauge symmetry, in contrast to the Standard Model, there is only one vacuum-angle θ. In the SU(2) Georgi–Glashow model, the vacuum-angle θ can thus be identified with θ_{QED}.

The static, classical monopole in the temporal gauge $V_0 = 0$ that we discussed in Section 26.7 does not carry electric charge. Allowing for $V_0 \neq 0$, Bernard Julia and Anthony Zee constructed a classical dyon solution with both magnetic and electric charge (Julia and Zee,

1975). At the classical level, these charges are not restricted by a quantization condition. By returning to the $V_0 = 0$ gauge, Terry Tomboulis and Gordon Woo turned the static Julia–Zee dyon into a gauge-equivalent, time-dependent, periodic classical solution. Applying semi-classical quantization, they showed that the electric charge of the dyon is quantized in integer units (Tomboulis and Woo, 1976).

Edward Witten extended these investigations by including a θ-vacuum contribution to the Lagrangian in Minkowski space–time

$$\mathcal{L}_\theta(V) = -\hbar \frac{\theta}{32\pi^2} \epsilon_{\mu\nu\rho\sigma} \operatorname{Tr}\left[V^{\mu\nu}V^{\rho\sigma}\right]. \tag{26.99}$$

It should be noted that $\mathcal{L}_\theta(V)$ is a quantum contribution to the Lagrangian, which contains $\hbar$ explicitly. Since the action enters the real-time functional integral via $\exp(iS/\hbar)$, the explicit factor of $\hbar$ is necessary already for dimensional reasons. In any case, the θ-term is unobservable at the classical level (even if $\hbar \neq 0$), because it is a total divergence that does not affect the Euler–Lagrange equations.

The static, classical monopole that we discussed in Section 26.7 does not carry electric charge. Still, the monopole core contains the electrically charged fields $X_\mu^\pm$. As we also discussed in Section 26.7, the unitary gauge (which turns the adjoint Higgs field into $\Xi_3\tau^3$) leaves the Abelian $U(1)_{\text{em}}$ gauge freedom unfixed. The 't Hooft–Polyakov monopole is a magnetically charged *infraparticle* whose magnetic "Coulomb" field extends to infinity. As we saw in Section 7.3, when we transform the electromagnetic field to the Coulomb gauge, $\partial_i A_i = 0$, we endow a bare electric charge with its surrounding Coulomb field. When one imposes the Coulomb gauge on the 't Hooft–Polyakov monopole (after fixing the unitary gauge), a global $U(1)_{\text{em}}$ gauge freedom still remains. This global symmetry transformation affects only the charged fields,

$$X_\mu^{\pm\prime}(x) = \exp(\pm ie\alpha)X_\mu^\pm(x). \tag{26.100}$$

Here, α parametrizes a family of degenerate classical, static monopole solutions: It plays the role of a zero-mode. In order to derive the Witten effect, we now quantize this zero-mode semi-classically, by turning α into a dynamical, time-dependent function $\alpha(t)$. We investigate the Lagrange function,

$$L(\alpha, \dot\alpha) = \int d^3x \left[\mathcal{L}(\Xi, V) + \mathcal{L}_\theta(V)\right] = \frac{I}{2}\dot\alpha^2 + \hbar\frac{\theta}{2\pi}e\dot\alpha, \tag{26.101}$$

as a function of $\alpha(t)$ and its time-derivative $\dot\alpha(t) = \partial_t\alpha(t)$ only. Here, I is an abstract "moment of inertia". Its value can be obtained from the classical 't Hooft–Polyakov monopole solution, but it is not important in the present context. The momentum canonically conjugate to α, which is actually an abstract "angular momentum", is given by

$$p_\alpha = \frac{\delta L(\alpha, \dot\alpha)}{\delta\dot\alpha} = I\dot\alpha + \hbar\frac{\theta}{2\pi}e. \tag{26.102}$$

The Lagrange function leads to the Hamilton function

$$H = p_\alpha\dot\alpha - L = \frac{I}{2}\dot\alpha^2 = \frac{1}{2I}\left(p_\alpha - \hbar\frac{\theta}{2\pi}e\right)^2. \tag{26.103}$$

Upon canonical quantization, p_α turns into the operator

$$\hat{p}_\alpha = -i\hbar\partial_\alpha, \tag{26.104}$$

which turns the Hamilton function H into the Hamilton operator

$$\hat{H} = \frac{\hbar^2}{2I} \left(-i\partial_\alpha - \frac{\theta}{2\pi} e \right)^2 = \frac{\hbar^2 \hat{q}^2}{2I}. \tag{26.105}$$

In the presence of a θ-vacuum term, the electric charge operator is given by

$$\hat{q} = -i\partial_\alpha - \frac{\theta}{2\pi} e. \tag{26.106}$$

Since $e\alpha$ is an angle, the wave function of the zero-mode takes the form $\Psi_n(\alpha) \propto \exp(i\, n\, e\, \alpha)$ with $n \in \mathbb{Z}$, such that

$$\hat{q}\,\Psi_n(\alpha) = \left(n - \frac{\theta}{2\pi} \right) e\, \Psi_n(\alpha). \tag{26.107}$$

As a result, in the presence of a non-zero vacuum-angle θ, the electric charge of the 't Hooft–Polyakov monopole is not an integer multiple of the electric charge unit e. Instead, it is shifted by $\theta/(2\pi)$. In particular, in a non-trivial θ-vacuum, all magnetic monopoles are influenced by the Witten effect – they turn into magnetically and electrically charged dyons.

26.10 Fermion Masses and the Hierarchy Problem

Let us return to the discussion of the SU(5) GUT. Up to this point, all fermions and all gauge bosons of the Standard Model have been considered massless. This means that we will need another scalar field, in particular, the $SU(2)_L$ complex Higgs doublet of the Standard Model, which is naturally contained in the fundamental representation of SU(5). Hence, we introduce a quintet of complex scalar fields

$$\Phi(x) = \begin{pmatrix} \Phi^+(x) \\ \Phi^0(x) \\ \phi^1(x) \\ \phi^2(x) \\ \phi^3(x) \end{pmatrix} \in \mathbb{C}^5, \quad \Phi'(x) = \Upsilon(x)\Phi(x), \tag{26.108}$$

which contains the Standard Model Higgs field in its first two components and a complex color triplet $\phi^c(x)$ with weak hypercharge $Y_\phi = -1/3$ in the remaining three components. Using the fact that the standard Higgs doublet has $Y_\Phi = 1/2$ and $(B - L)_\Phi = 0$, one identifies the quantum number

$$F = B - L - \frac{4}{5} Y, \quad F_\Phi = -\frac{2}{5}. \tag{26.109}$$

This implies that the color triplet ϕ carries $(B - L)_\phi = -2/3$, which corresponds to the value of a pair of anti-quarks (which may also form a color triplet). There is a cubic term $\Phi^\dagger \Xi \Phi$ in the scalar potential that couples the fundamental Higgs field $\{5\}$ to the adjoint scalar field $\{24\}$. It is F-invariant if we assign

$$F_\Xi = 0 \tag{26.110}$$

to the adjoint scalar field. Then, the cubic and quartic self-interaction terms of Ξ are also $U(1)_F$-invariant. Under the unbroken $SU(3)_c \times U(1)_{em}$ subgroup of the Standard Model, the $\{5\}$ representation decomposes as

$$\{5\} = \{3\}_{-1} + \{1\}_1 + \{1\}_0. \tag{26.111}$$

Here, the subscripts denote the electric charges associated with the various $SU(3)_c$ representations. We see that this decomposition contains an electrically neutral color singlet state $\{1\}_0$. Therefore, when Φ picks up a vacuum expectation value, the $SU(3)_c \times U(1)_{em}$ symmetry indeed remains unbroken.

As discussed in Section 13.9, splitting the Standard Model Higgs at the electroweak scale v from the color triplet at the GUT scale $\mathcal{V}$ requires fine-tuning. This so-called *doublet–triplet splitting problem* is just another variant of the hierarchy problem.

It has often been argued that supersymmetry can "solve" the hierarchy problem, because it stabilizes the hierarchy between the Standard Model scale and the GUT scale against radiative corrections, which cancel between bosonic and fermionic particles and their superpartners. A widespread argument is that non-supersymmetric theories require fine-tuning again and again, at each order of perturbation theory, while in supersymmetric theories one fixes the hierarchy once and for all at the classical level. However, these arguments are not very convincing:

- If one thinks non-perturbatively, there is no need to consider different orders of perturbation theory one after the other. Thus, there is only a single fine-tuning, at this time at the quantum level.
- In any case, the hierarchy as such is not explained by supersymmetry.
- In addition, supersymmetric theories that contain massless scalars may suffer from severe conceptual problems. Constructing such theories beyond perturbation theory, in particular, by using the lattice regularization, seems to require unnatural fine-tuning already.

Before supersymmetric theories with massless scalars will be constructed satisfactorily beyond perturbation theory, without relying on fine-tuning, they should not be considered as a convincing solution to the hierarchy problem.

Let us now address the Yukawa couplings that the fermions in the $\{10\}$ and $\{\bar{5}\}$ representation can participate in. The group theory of $SU(5)$ implies

$$\begin{aligned}
\{\bar{5}\} \times \{\bar{5}\} &= \{\overline{15}\} + \{\overline{10}\}, \\
\{\bar{5}\} \times \{10\} &= \{5\} + \{45\}, \\
\{10\} \times \{10\} &= \{\bar{5}\} + \{\overline{45}\} + \{50\}.
\end{aligned} \tag{26.112}$$

As a consequence, there is no gauge-invariant Yukawa interaction that couples two fermion fields in the $\{\bar{5}\}$ representation to the adjoint or the fundamental scalar, simply because neither $\{5\}$ nor $\{24\}$ appear in the tensor product reduction of $\{\bar{5}\} \times \{\bar{5}\}$. There are, however, two other Yukawa interactions, one that couples a $\{\bar{5}\}$ to a $\{10\}$ and one that couples two $\{10\}$ fermion fields. Both of these Yukawa terms involve the fundamental scalar field Φ (or its complex conjugate), because $\{5\}$ (or $\{\bar{5}\}$) occurs in the corresponding tensor product reductions. This implies that, together with Φ, the two fermion fields can be coupled in a gauge-invariant way (to form an SU(5) singlet).

If Φ takes its expectation value at the Standard Model scale v (which requires fine-tuning), the fermion masses that result from the corresponding Yukawa couplings are associated with the scale v. In particular, they are prevented from running up to the GUT scale $\mathcal{V}$. This is an attractive feature of this particular GUT, which results from the fact that the adjoint representation {24} does not arise on the right-hand side of the tensor product reductions of eq. (26.112).

Otherwise, one could construct additional gauge-invariant Yukawa couplings to the adjoint scalar field Ξ, which would provide the fermions with masses at the GUT scale $\mathcal{V}$. Since this is not the case here, no additional hierarchy problem arises that would require an unnatural fine-tuning of Yukawa couplings.

Let us now explicitly construct the Yukawa couplings between the fields χ, ψ, and Φ. The corresponding contribution to the Lagrangian of the SU(5) GUT takes the form

$$\mathcal{L}(\bar{\chi}, \chi, \bar{\psi}, \psi, \Phi) = -f_d \left({}^c\bar{\psi}_a \chi_{ab} \Phi_b^* + \Phi_a^{*c} \bar{\chi}_{ab} \psi_b \right) + f_u \frac{1}{4} \epsilon_{abcde} {}^c\bar{\chi}_{ab} \chi_{cd} \Phi_e. \quad (26.113)$$

Using an SU(5) analogue of the SU(3) matrix relation eq. (26.41),

$$\epsilon_{abcde} \Upsilon_{af} \Upsilon_{bg} \Upsilon_{ch} \Upsilon_{di} = \epsilon_{fghij} \Upsilon_{ej}^*, \quad (26.114)$$

we are going to show in Problem 26.5 that the Lagrangian terms in eq. (26.113) are indeed gauge-invariant. It also preserves F and thus $B - L$ invariance, because

$$F_\psi + F_\chi - F_\Phi = 0, \quad 2F_\chi + F_\Phi = 0. \quad (26.115)$$

When one replaces Φ, given in eq. (26.108), by its vacuum expectation value, $\Phi^0 = \text{v}$, $\Phi^+ = \phi^1 = \phi^2 = \phi^3 = 0$, the Lagrangian reduces to the fermion mass terms

$$\mathcal{L}(\bar{\chi}, \chi, \bar{\psi}, \psi, \text{v}) = f_d \text{v} \left(\bar{e}_L e_R + \bar{e}_R e_L + \bar{d}_L d_R + \bar{d}_R d_L \right) + f_u \text{v} \left(\bar{u}_L u_R + \bar{u}_R u_L \right). \quad (26.116)$$

We read off the quark and lepton masses

$$m_e = m_d = f_d \text{v}, \quad m_u = f_u \text{v}, \quad m_\nu = 0. \quad (26.117)$$

As a result of grand unification, the masses of the electron and of the down quark are the same at the electroweak scale v. One can use the renormalization group to run them down to lower energy scales. In this way, one obtains reasonable values for the phenomenological fermion masses.

GUTs do not shed any light on the puzzles related to the additional generations. Just as in the Standard Model, the story of the first generation repeats itself for the other two generations. In this way, one also obtains the tree-level relations $m_s = m_\mu$ and $m_b = m_\tau$, which need to be corrected by renormalization group running before they should be compared with experiments. Once this is taken into account, only the prediction $m_b = m_\tau$, but not $m_s = m_\mu$, is consistent with observation. As a result, several renormalizable GUTs are already ruled out for this reason. However, if one considers GUTs as low-energy effective theories that break down at the Planck scale, higher-dimensional terms, which are proportional to powers of $\mathcal{V}/M_{\text{Planck}}$, can remedy such discrepancies.

26.11 Spin(10) Structure

An attractive alternative to SU(5) is provided by the grand unified group Spin(10), as proposed by Harald Fritzsch and Peter Minkowski (Fritzsch and Minkowski, 1975) and by Howard Georgi (1975), which has rank 5 and 45 generators. This implies that it gives rise to even more, namely, $45 - 12 = 33$, ultra-heavy gauge bosons, and that it contains an additional $U(1)_F$ subgroup, which is related to $B - L$ invariance. Since F is a generator of Spin(10), it is traceless in any representation.

For example, the symmetric tensor $\{54\}$, the anti-symmetric tensor (adjoint) $\{45\}$, and the vector $\{10\}$ representations of Spin(10) decompose as follows under the $SU(5) \times U(1)_F$ subgroup

$$\{54\} = \{24\}_0 + \{15\}_{-4/5} + \{\overline{15}\}_{4/5} \quad \Rightarrow \quad \text{Tr}(F) = 15\left(-\frac{4}{5} + \frac{4}{5}\right) = 0,$$

$$\{45\} = \{24\}_0 + \{10\}_{-4/5} + \{\overline{10}\}_{4/5} + \{1\}_0 \quad \Rightarrow \quad \text{Tr}(F) = 10\left(-\frac{4}{5} + \frac{4}{5}\right) = 0,$$

$$\{10\} = \{5\}_{-2/5} + \{\bar{5}\}_{2/5} \quad \Rightarrow \quad \text{Tr}(F) = 5\left(-\frac{2}{5} + \frac{2}{5}\right) = 0. \tag{26.118}$$

Remarkably, Spin(10) also has a 16-dimensional complex spinor representation that decomposes under $SU(5) \times U(1)_F$ as

$$\{16\} = \{10\}_{1/5} + \{\bar{5}\}_{-3/5} + \{1\}_1 \quad \Rightarrow \quad \text{Tr}(F) = 10 \cdot \frac{1}{5} - 5 \cdot \frac{3}{5} + 1 = 0. \tag{26.119}$$

Hence, it incorporates all Standard Model fermions and, in addition, one SU(5) singlet, which is naturally associated with a sterile, right-handed neutrino.

There are several possible symmetry breaking patterns that take us from Spin(10) via $SU(3)_c \times SU(2)_L \times U(1)_Y$ to the unbroken subgroup $SU(3)_c \times U(1)_{em}$ of the Standard Model. One possible pattern proceeds via SU(5)

$$\text{Spin}(10) \underset{\{16\}}{\longrightarrow} \text{SU}(5) \underset{\{45\}}{\longrightarrow} \text{SU}(3)_c \times \text{SU}(2)_L \times \text{U}(1)_Y \underset{\{10\}}{\longrightarrow} \text{SU}(3)_c \times \text{U}(1)_{em}. \tag{26.120}$$

This symmetry breaking pattern is realized if three different scalar fields in the spinor $\{16\}$, adjoint $\{45\}$, and vector $\{10\}$ representations pick up non-zero vacuum expectation values at subsequent energy scales. Since the decomposition of $\{16\}$ in eq. (26.119) contains the SU(5) singlet $\{1\}_1$, SU(5) remains unbroken when the scalar field picks up a corresponding vacuum value. However, since this representation is associated with the non-zero value $F = 1$, the gauge symmetry $U(1)_F$ "breaks spontaneously". A scalar field in the adjoint Spin(10) representation $\{45\}$ would induce an alternative breaking pattern via $SU(5) \times U(1)_F$, thus leaving $U(1)_F$ unbroken. This is because $\{45\}$ contains the SU(5) singlet $\{1\}_0$ that has $F = 0$; hence, it is neutral with respect to $U(1)_F$.

When a scalar field in the representation $\{10\}$ or $\{54\}$ picks up a non-zero vacuum expectation value, SU(5) would not remain unbroken, because $\{10\}$ and $\{54\}$ do not contain SU(5) singlets. It is interesting to note the relation

$$\text{Spin}(10) \supset \text{Spin}(6) \times \text{Spin}(4) = \text{SU}(4) \times \text{SU}(2)_L \times \text{SU}(2)_R. \tag{26.121}$$

Here, the universal covering group of SO(6), Spin(6) = SU(4), contains $SU(3)_c$ and, in addition, incorporates *lepton number as a "fourth color"*. Under the $SU(4) \times SU(2)_L \times SU(2)_R$ subgroup, the various Spin(10) representations decompose as

$$\{54\} = \{20, 1, 1\} + \{6, 2, 2\} + \{1, 3, 3\} + \{1, 1, 1\},$$
$$\{45\} = \{15, 1, 1\} + \{6, 2, 2\} + \{1, 3, 1\} + \{1, 1, 3\},$$
$$\{16\} = \{4, 2, 1\} + \{\bar{4}, 1, 2\},$$
$$\{10\} = \{6, 1, 1\} + \{1, 2, 2\}. \tag{26.122}$$

Correspondingly, there is an alternative hierarchy of symmetry breaking patterns

$$\begin{aligned} \text{Spin}(10) &\xrightarrow[\{54\}]{} \text{Spin}(6) \times \text{Spin}(4) = SU(4) \times SU(2)_L \times SU(2)_R \\ &\xrightarrow[\{45\}]{} SU(3)_c \times U(1)_{B-L} \times SU(2)_L \times SU(2)_R \\ &\xrightarrow[\{16\}]{} SU(3)_c \times SU(2)_L \times U(1)_Y \\ &\xrightarrow[\{10\}]{} SU(3)_c \times U(1)_{em}, \end{aligned} \tag{26.123}$$

which proceeds via the left–right symmetric *Pati–Salam model* with an $SU(4) \times SU(2)_L \times SU(2)_R$ symmetry (Pati and Salam, 1974), and there are several further alternative breaking patterns. Since $\{54\}$ contains the $SU(4) \times SU(2)_L \times SU(2)_R$ singlet $\{1, 1, 1\}$, this subgroup remains unbroken when the corresponding scalar field picks up a vacuum expectation value at the GUT scale $\mathcal{V}$.

One can contemplate GUTs involving even larger unifying groups including the exceptional groups E(6), E(7), and E(8). Since SU(5) can be interpreted as E(4) and Spin(10) can be interpreted as E(5), there is the subgroup chain

$$SU(5) = E(4) \subset \text{Spin}(10) = E(5) \subset E(6) \subset E(7) \subset E(8). \tag{26.124}$$

The group E(6) has rank 6 and possesses complex representations. The fundamental representations of E(6) are $\{27\}$ and $\{\overline{27}\}$, with the following decomposition into Spin(10) representations

$$\{27\} = \{1\} + \{10\} + \{16\}. \tag{26.125}$$

This contains both the representation $\{16\}$ that hosts the Standard Model fermions and the representation $\{10\}$ that contains the Standard Model Higgs field.

Inspired by superstring theory, which operates in a 10-dimensional space–time, one can consider higher-dimensional supersymmetric GUTs (Kawamura, 2000, 2001; Asaka *et al.*, 2001), which can be dimensionally reduced to 4-dimensional space–time, for example, by compactifying the extra dimensions on an orbifold. An orbifold results when distinct points in a manifold are identified by a discrete symmetry operation. Compactification on an orbifold has several interesting features. First of all, orbifolding leads to gauge symmetry breaking and may thus obviate the need for employing high-dimensional Higgs representations, which indeed seem rather baroque. In addition, the discrete symmetry of the orbifold can lead to a natural solution of the doublet–triplet splitting problem. Furthermore, when applied to a supersymmetric E(6) GUT, orbifolding leads to interesting relations between the gauge and matter sectors (Raby, 2017).

26.12 Neutrino Masses in the Spin(10) GUT

The subject of this section was discussed by Buchmüller *et al.* (1991). We first note that the gauge-invariant Yukawa couplings can be deduced from

$$\{16\} \times \{16\} = \{10\} + \{120\} + \{126\}. \tag{26.126}$$

It is appealing that $\{54\}$, $\{45\}$, and $\{16\}$ do not appear in this decomposition. Otherwise, some Standard Model fermions would receive masses at the GUT scale $\mathcal{V}$. The representation $\{10\} = \{5\} + \{\bar{5}\}$ contains the fundamental representation $\{5\}$ of SU(5), which in turn contains the usual Higgs doublet. Hence, when the corresponding scalar field $\{10\}$ receives a vacuum value at the scale v (which requires fine-tuning as a consequence of the hierarchy problem), the Standard Model fermions obtain their usual masses at the electroweak scale without any need for additional fine-tuning of the Yukawa couplings. Since the right-handed neutrino resides in the same Spin(10) representation $\{16\}$ as the other fermions, we then obtain a neutrino with the Dirac mass $m_\nu = m_u$ at the GUT scale.

Can neutrinos in the Spin(10) GUT be naturally much lighter than the other fermions? In the corresponding extension of the Standard Model or of the SU(5) GUT, ν_R transforms as a sterile gauge singlet, which has its own Majorana mass scale that is unrelated to spontaneous symmetry breaking. If that scale is large, it leads to the seesaw mass-by-mixing mechanism that we discussed in Section 16.9, thus giving rise to very light non-sterile neutrinos. Can ν_R be naturally much heavier than the other fermions if it belongs to the same Spin(10) representation $\{16\}$?

This is a delicate question that is again intimately related to the hierarchy problem. Due to its more complicated symmetry breaking pattern involving several scalar fields, the Spin(10) GUT has more than just one ultra-high energy scale. All these scales mix at the quantum level and must be fine-tuned against each other. Via quantum effects, the various fermion masses thus receive contributions also from scales other than just v that is related with $\{10\}$. By fine-tuning the various parameters, it is conceivable to keep the ν_R mass near the GUT scale $\mathcal{V}$, thus facilitating the seesaw mechanism, while keeping the other fermion masses at the low scale v. Alternatively, one can add yet another scalar field in the representation $\{126\}$, which may directly endow ν_R with a Majorana mass at the GUT scale.

In any case, as a consequence of the hierarchy problem, stabilizing the various scales against each other remains a delicate issue of fine-tuning.

26.13 Small Unification with SU(3), G(2), Spin(6), or Spin(7)

Let us finally return to the small electroweak SU(3) unification, whose bosonic aspects we discussed in Sections 13.8 and 13.9. Now we will address the corresponding fermionic degrees of freedom. This setup corresponds to the gauge anomaly-free and thus mathematically consistent version of the Standard Model with $N_c = 1$, *i.e.* without strong interaction (cf. Section 15.9). In this case, the up and down "quarks" are just the proton and the neutron, and the fermion content of the first generation of fermions is

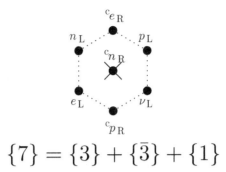

$$\{7\} = \{3\} + \{\bar{3}\} + \{1\}$$

Fig. 26.3 Illustration of the decomposition of colorless $N_c = 1$ fermions into an SU(3) triplet, a singlet, and an anti-triplet. Remarkably, these three SU(3) representations combine to form the fundamental $\{7\}$ representation of the exceptional group G(2).

$$q_L = \begin{pmatrix} p_L \\ n_L \end{pmatrix}, \ p_R, \ n_R, \quad l_L = \begin{pmatrix} \nu_L \\ e_L \end{pmatrix}, \ e_R. \tag{26.127}$$

Just as the right-handed neutrino in the Standard Model with $N_c \geq 3$, now the right-handed neutron is a sterile particle that does not participate in any Standard Model gauge interactions. As a consequence, one can construct a Majorana mass term $^c\bar{n}_R n_R$, which introduces a second mass scale beyond v. Since the Standard Model with $N_c = 1$ has no strong interaction, there is nothing that physically distinguishes baryons from leptons. Therefore, electron and anti-proton, as well as neutrino and anti-neutron, can mix via additional Majorana mass terms $^c\bar{p}_R e_R$ and $^c\bar{q}_L i\tau^2 l_L$, which give rise to two additional independent Majorana mixing scales. As a result of lepton–baryon mixing, baryon and lepton number are no longer useful concepts. Nevertheless, $U(1)_{B-L}$ remains an exact global symmetry.

In the Standard Model with $N_c \geq 3$, lepton–quark mixing terms are forbidden because they are not $SU(N_c)$ invariant. Therefore, the Standard Model has only one mass scale v, whereas its $N_c = 1$ variant has, in addition, three independent Majorana mass or mixing scales, which are not related to spontaneous symmetry breaking.

As illustrated in Figure 26.3, the fermions of eq. (26.127) naturally fit into an SU(3) triplet, singlet, and anti-triplet

$$\left\{ \begin{pmatrix} p_L \\ n_L \end{pmatrix}, {}^c p_R \right\} = \{2\}_{1/2} + \{1\}_{-1} = \{3\}, \quad \{n_R\} = \{1\}_0 = \{1\},$$

$$\left\{ i\tau^2 \begin{pmatrix} \nu_L \\ e_L \end{pmatrix}, {}^c e_R \right\} = \{\bar{2}\}_{-1/2} + \{1\}_1 = \{\bar{3}\}. \tag{26.128}$$

We introduce the fermion fields

$$\chi(x) = \begin{pmatrix} p_L(x) \\ n_L(x) \\ -{}^c p_R(x) \end{pmatrix}, \quad \chi'(x) = \Upsilon(x)\chi(x),$$

$$\psi(x) = \begin{pmatrix} e_L(x) \\ -\nu_L(x) \\ -{}^c e_R(x) \end{pmatrix}, \quad \psi'(x) = \Upsilon(x)^* \psi(x). \tag{26.129}$$

Similar to the case of SU(5), one can identify a conserved quantum number

$$F = B - L - \frac{4}{3}Y,$$

$$F_\chi = \frac{1}{3}, \quad F_{n_R} = 1, \quad F_\psi = -\frac{1}{3}, \quad F_\Phi = -\frac{2}{3}, \quad F_\Xi = 0. \qquad (26.130)$$

Since $U(1)_Y$ is an exact gauge symmetry, $U(1)_{B-L}$ remains as an exact global symmetry.

As we discussed in Sections 13.8 and 13.9, the $SU(3)$ model follows the symmetry breaking pattern

$$SU(3) \underset{\{8\}}{\longrightarrow} SU(2)_L \times U(1)_Y \underset{\{3\}}{\longrightarrow} U(1)_{em}. \qquad (26.131)$$

Under the $SU(2)_L \times U(1)_Y$ subgroup, the adjoint representation decomposes as

$$\{8\} = \{3\}_0 + \{2\}_1 + \{2\}_{-1} + \{1\}_0. \qquad (26.132)$$

Here, the subscripts refer to the weak hypercharge of the corresponding $SU(2)_L$ representation. When the adjoint scalar $\{8\}$ picks up a non-zero vacuum expectation value at the scale $\mathcal{V}$, the $SU(3)$ gauge symmetry "breaks spontaneously" to $SU(2)_L \times U(1)_Y$. This is because the singlet $\{1\}_0$ with $Y = 0$ appears in the decomposition of eq. (26.132). When the fundamental Higgs field $\{3\}$ obtains a vacuum value at the Standard Model scale v, the symmetry is further reduced to $U(1)_{em}$. This is reflected in the corresponding decomposition of the $SU(3)$ triplet with respect to the $U(1)_{em}$ subgroup of electromagnetism

$$\{3\} = \{1\}_1 + \{1\}_0 + \{1\}_{-1}, \qquad (26.133)$$

which includes the neutral object $\{1\}_0$ that corresponds to Φ^0. In this case, the subscripts refer to the electric charge $Q = T_L^3 + Y$.

What kinds of fermion mass terms can be constructed for this small electroweak $SU(3)$ unified theory of the $N_c = 1$ variant of the Standard Model? Since the right-handed neutron is a gauge singlet, it can again acquire a Majorana mass term that is unrelated to the vacuum values $\mathcal{V}$ of Ξ or v of Φ. Using the fundamental Higgs triplet, one can also construct a Dirac mass term for the neutron, $\bar{\chi}\Phi n_R$, which is associated with the scale v. The $SU(3)$ coupling rules

$$\{3\} \times \{3\} = \{\bar{3}\} + \{6\}, \quad \{\bar{3}\} \times \{\bar{3}\} = \{3\} + \{\bar{6}\}, \qquad (26.134)$$

allow the additional Yukawa couplings $f_p \epsilon_{abc}{}^c \bar{\chi}_a \chi_b \Phi_c$ and $f_e \epsilon_{abc}{}^c \bar{\psi}_a \psi_b \Phi_c^*$, which provide Dirac masses for proton and electron associated with the scale v. Finally, the $SU(3)$ coupling rule

$$\{3\} \times \{\bar{3}\} = \{1\} + \{8\} \qquad (26.135)$$

gives rise to two lepton–baryon mixing terms, ${}^c\bar{\psi}\chi$ and ${}^c\bar{\psi}\Xi\chi$, which are equivalent to the two mixing scales that are already present in the $N_c = 1$ Standard Model.

Interestingly, there is another simple Lie group of rank 2, the *exceptional group* $G(2)$, which has 14 generators and contains $SU(3)$ as a subgroup; see Section F.14. The adjoint representation of $G(2)$ decomposes as

$$\{14\} = \{8\} + \{3\} + \{\bar{3}\},$$
$$= \{3\}_0 + \{2\}_1 + \{2\}_{-1} + \{1\}_0 + \{2\}_{1/3} + \{1\}_{-2/3} + \{2\}_{-1/3} + \{1\}_{2/3}, \quad (26.136)$$

under the $SU(3)$ and $SU(2)_L \times U(1)_Y$ subgroups, respectively. When an adjoint scalar field $\{14\}$ obtains a non-zero vacuum expectation value at the scale $\mathcal{V}$, the following symmetry breaking pattern may arise

$$G(2) \xrightarrow[\{14\}]{} SU(2)_L \times U(1)_Y \xrightarrow[\{7\}]{} U(1)_{em}. \tag{26.137}$$

Since the decomposition of $\{14\}$ does not include an SU(3) singlet, SU(3) does not remain unbroken. However, since $\{14\}$ contains the $SU(2)_L \times U(1)_Y$ singlet $\{1\}_0$, $SU(2)_L \times U(1)_Y$ remains unbroken when the adjoint scalar field picks up a non-zero vacuum expectation value. The standard Higgs doublet can be embedded in the $\{7\}$ of G(2), which then leads to the final breaking down to $U(1)_{em}$.

Remarkably, under the SU(3) subgroup the 7-dimensional fundamental representation of G(2) decomposes as

$$\{7\} = \{3\} + \{\bar{3}\} + \{1\}, \tag{26.138}$$

such that one generation of $N_c = 1$ fermions fits in. For the group G(2), one obtains

$$\{7\} \times \{7\} = \{1\} + \{7\} + \{14\} + \{27\}. \tag{26.139}$$

This can be derived by using the Antoine–Speiser scheme described in Section F.15. In the absence of an additional scalar $\{27\}$, this relation implies that there are three different Yukawa couplings. The one associated with the adjoint scalar $\{14\}$ gives rise to lepton–baryon mixing at the scale $\mathcal{V}$, while the one associated with the fundamental Higgs gives rise to Dirac masses of the nucleons and the electron at the scale v. In addition, the coupling to the singlet $\{1\}$ provides Majorana masses for all fermions (not just for the right-handed neutron) at an additional scale that is unrelated to spontaneous symmetry breaking. This potentially undesirable feature is a consequence of the fact that G(2) has only real representations. We conclude that G(2) unification would not be such a grand idea.

Let us investigate yet another unifying gauge group, Spin(6) = SU(4), which has complex representations. The group SU(4) has rank 3 and thus possesses an additional Abelian subgroup, which turns out to be $U(1)_F$, just as in the Spin(10) GUT. As in that case, we explicitly introduce a right-handed neutrino field and find that the resulting eight fermion fields form a $\{4\}$ and $\{\bar{4}\}$ of SU(4),

$$\left\{ \begin{pmatrix} p_L \\ n_L \end{pmatrix}, {}^c p_R, {}^c n_R \right\} = \{3\}_{1/3} + \{1\}_{-1} = \{4\},$$

$$\left\{ i\tau^2 \begin{pmatrix} \nu_L \\ e_L \end{pmatrix}, {}^c e_R, {}^c \nu_R \right\} = \{\bar{3}\}_{-1/3} + \{1\}_1 = \{\bar{4}\}. \tag{26.140}$$

Here, the subscript refers to the F-value of the corresponding SU(3) representation.

The spinor $\{4\}$, vector $\{6\}$, adjoint $\{15\}$, and symmetric tensor $\{20\}$ representations of Spin(6) = SU(4) are analogous to the corresponding representations $\{16\}$, $\{10\}$, $\{45\}$, and $\{54\}$ of Spin(10), respectively. Under the $SU(3) \times U(1)_F$ subgroup of SU(4), they decompose as

$$\{20\} = \{8\}_0 + \{6\}_{-4/3} + \{\bar{6}\}_{4/3} \quad \Rightarrow \quad \text{Tr}(F) = 6\left(-\frac{4}{3} + \frac{4}{3}\right) = 0,$$

$$\{15\} = \{8\}_0 + \{\bar{3}\}_{-4/3} + \{3\}_{4/3} + \{1\}_0 \quad \Rightarrow \quad \text{Tr}(F) = 3\left(-\frac{4}{3} + \frac{4}{3}\right) = 0,$$

$$\{6\} = \{3\}_{-2/3} + \{\bar{3}\}_{2/3} \quad \Rightarrow \quad \text{Tr}(F) = 3\left(-\frac{2}{3} + \frac{2}{3}\right) = 0,$$

$$\{4\} = \{3\}_{1/3} + \{1\}_{-1} \quad \Rightarrow \quad \text{Tr}(F) = 3 \cdot \frac{1}{3} - 1 = 0. \tag{26.141}$$

Since $\{4\}$ contains the singlet $\{1\}_{-1}$ with non-zero $F = -1$, a corresponding scalar field induces symmetry breaking to SU(3). This is the first step in the symmetry breaking pattern

$$\text{Spin}(6) \xrightarrow[\{4\}]{} \text{SU}(3) \xrightarrow[\{15\}]{} \text{SU}(2)_L \times \text{U}(1)_Y \xrightarrow[\{6\}]{} \text{U}(1)_{\text{em}}. \qquad (26.142)$$

The last step is induced by a scalar field $\{6\}$ that contains the SU(3) triplet Φ, which in turn contains the standard Higgs doublet.

Let us comment on yet another interesting subgroup of Spin(6),

$$\text{Spin}(6) \supset \text{Spin}(4) \times \text{Spin}(2) = \text{SU}(2)_L \times \text{SU}(2)_R \times \text{U}(1)_{B-L}. \qquad (26.143)$$

It gives rise to the decomposition

$$\begin{aligned}
\{20\} &= \{3,3\}_0 + \{2,2\}_2 + \{2,2\}_{-2} + \{1,1\}_0 + \{1,1\}_4 + \{1,1\}_{-4}, \\
\{15\} &= \{2,2\}_2 + \{2,2\}_{-2} + \{3,1\}_0 + \{1,3\}_0 + \{1,1\}_0, \\
\{6\} &= \{2,2\}_0 + \{1,1\}_2 + \{1,1\}_{-2}, \\
\{4\} &= \{2,1\}_1 + \{1,2\}_{-1}.
\end{aligned} \qquad (26.144)$$

Since $\{20\}$ contains the neutral singlet $\{1,1\}_0$, it induces a breaking that leaves $\text{SU}(2)_L \times \text{SU}(2)_R \times \text{U}(1)_{B-L}$ invariant. This is the first step in the Spin(6) version of the Pati–Salam breaking pattern

$$\begin{aligned}
\text{Spin}(6) &\xrightarrow[\{20\}]{} \text{Spin}(4) \times \text{Spin}(2) = \text{SU}(2)_L \times \text{SU}(2)_R \times \text{U}(1)_{B-L} \\
&\xrightarrow[\{15\}]{} \text{SU}(2)_L \times \text{U}(1)_Y \xrightarrow[\{6\}]{} \text{U}(1)_{\text{em}}.
\end{aligned} \qquad (26.145)$$

The SU(4) coupling rules reveal the allowed Yukawa couplings

$$\{4\} \times \{\bar{4}\} = \{1\} + \{15\}, \quad \{4\} \times \{4\} = \{6\} + \{10\}, \quad \{\bar{4}\} \times \{\bar{4}\} = \{6\} + \{\overline{10}\}. \qquad (26.146)$$

As for SU(3), those couplings again give rise to lepton–baryon mixing, as well as to Dirac mass terms for proton and electron, associated with three distinct energy scales. However, similar to Spin(10), the right-handed neutrino now belongs to the same representation $\{\bar{4}\}$ as the other leptons and thus receives the same Dirac mass from the Higgs field $\{6\}$. Using fine-tuning, an additional scalar $\{\overline{10}\}$ (which is analogous to the $\{126\}$ of Spin(10)) could be used to provide ν_R with a large Majorana mass, thus inducing the seesaw mass-by-mixing mechanism.

As a further approach, let us consider the unifying group Spin(7), which again has rank 3 and contains Spin(6) = SU(4) as a subgroup. In contrast to Spin(6), Spin(7) has no complex representations. As a result, the two inequivalent complex spinor representations $\{4\}$ and $\{\bar{4}\}$ of Spin(6) combine to the single 8-dimensional spinor representation of Spin(7),

$$\{8\} = \{4\} + \{\bar{4}\}. \qquad (26.147)$$

Similar to Spin(10), now all fermions of the $N_c = 1$ variant of the Standard Model, together with the right-handed neutrino, fit into a single representation. The group theory of Spin(7) implies

$$\{8\} \times \{8\} = \{1\} + \{7\} + \{21\} + \{35\}. \qquad (26.148)$$

Unlike the complex spinor representation $\{16\}$ of Spin(10), the spinor representation $\{8\}$ of Spin(7) is real. Hence, the singlet $\{1\}$ arises in the above decomposition, and, just like for G(2), all fermions pick up an undesirable Majorana mass term at a scale unrelated to $\mathcal{V}$ or v. Hence, SO(7) unification would not be such a grand idea either. We conclude that one should obey the

Box 26.2	Condition for good GUTs

Satisfactory unification requires the fermions to reside in complex representations.

Finally, let us attempt a unification for the gauge anomaly-free Standard Model with odd $N_c \geq 5$, with an additional right-handed neutrino field. The gauge group $SU(N_c) \times SU(2)_L \times U(1)_Y$ then has rank $N_c - 1 + 1 + 1 = N_c + 1$. A conceivable unifying gauge group must have at least the same rank. One generation of fermions then consists of N_c left-handed quark doublets and one left-handed lepton $SU(2)_L$ doublet, as well as of $2N_c$ right-handed quark singlets and two right-handed lepton $SU(2)_L$ singlets. Thus, altogether, there are $4(N_c + 1)$ fermionic degrees of freedom.

- For $N_c = 1$, these are the 8 fermions that fill the two inequivalent complex spinor representations $\{4\}$ and $\{\bar{4}\}$ of $\text{Spin}(6) = SU(4)$.
- For $N_c = 3$, they are the 16 fermions that fill a complex spinor representation of $\text{Spin}(10)$.
- For $N_c = 5$, the rank of the Standard Model group is 6, which may suggest $SU(7)$ or the exceptional group $E(6)$ as the unification group, because both have rank 6 and possess complex representations. However, neither has a 24-dimensional representation.

The dimension of the fundamental spinor representations of $\text{Spin}(2n + 2)$ is 2^n. Hence, one might think that for $N_c = 7$, one generation of fermions could fit into the 32-dimensional spinor representation of $\text{Spin}(12)$. However, this is not the case, because $\text{Spin}(12)$ has rank 6, which is smaller than $N_c + 1 = 8$. In any case, $\text{Spin}(12)$ is not a promising unification group either, because it lacks complex representations.

The mismatch between the size of the spinor representation and the rank extends to larger values of N_c. We arrive at the conclusion:

Box 26.3	GUT condition on the number of colors N_c

Grand unification seems to work successfully only for $N_c = 3$.

To some extent, it also works for $N_c = 1$. However, unlike the Standard Model with $N_c \geq 3$, that model has lepton–baryon mixing and possesses multiple energy scales besides v, even before one considers unification.

26.14 Grand or not so Grand Unification?

After all, what should we think about the whole GUT approach? First of all, it might simply provide a correct extension of the Standard Model to extremely high energies. Of course, in the absence of experimental confirmation, this remains a matter of speculation based on some GUT feeling: It is certainly also possible that unification does not happen in Nature.

While directly accessing the GUT scale will remain impossible in the foreseeable future, indirect low-energy tests are conceivable. When the $U(1)_{em}$ gauge group of electromagnetism is embedded in a compact non-Abelian group, this leads to the possible existence of magnetic monopoles. This would automatically imply electric charge quantization (Dirac, 1931). However, as we have seen in Sections 15.8 and 15.10, charge quantization already follows within the Standard Model, provided that we take a non-zero electron mass into account. It is indeed quite satisfactory that charge quantization, which manifests itself at low energies, does not require a GUT scale explanation. Other topological excitations that may exist in GUTs include cosmic strings – spatially 1-dimensional objects that could extend throughout the Universe. Depending on whether possible cosmic inflation took place above or below the GUT scale, monopoles or cosmic strings might still exist in the Universe today. If they would be observed, they could reveal important hints to the physics at the GUT scale. The same is true for possible future observations of proton decay.

The Spin(10) GUT is particularly attractive in several respects. It hosts all Standard Model fermions, plus the right-handed sterile neutrino, in a single irreducible complex representation {16}. Based on the seesaw mass-by-mixing mechanism, after some fine-tuning, it provides very light neutrinos. This can be viewed as a successful GUT prediction of non-zero neutrino masses, long before this was confirmed by the observation of neutrino oscillations. Furthermore, the Spin(10) GUT eliminates $B-L$ as an exact global symmetry, simply by incorporating it as a "spontaneously broken" local symmetry.

GUTs relate not only the various Standard Model gauge couplings, but also the quark and lepton masses, and thus lead to a reduction in the number of fundamental parameters. On the other hand, via their scalar potential they introduce at least as many new parameters, which characterize the new physics predicted at the GUT scale. We emphasize that GUTs do not shed light on the puzzle why there are exactly three generations of fermions. They also involve generation mixing, and they lead to relations between the CKM and the PMNS matrix.

Like in the Standard Model, a major puzzle in GUTs is the hierarchy problem which requires fine-tuning that appears unnatural. Even on purely theoretical grounds, supersymmetry does not provide a solution to the hierarchy problem. Moreover, at the TeV scale supersymmetry lacks experimental support. Although superpartners could facilitate coupling constant unification, other exotic "beasts" may populate the "desert" that separates the Standard Model scale v from the GUT scale $\mathcal{V}$ and they may also do the job if necessary.

Finally, although the Standard Model is gauge anomaly-free, and thus mathematically consistent, for all odd values of N_c, grand unification seems to work well only for $N_c = 3$. In particular, as we have seen, for $N_c = 1$ small electroweak unification using SU(3), G(2), Spin(6), or Spin(7) would not be such a grand idea, because the fermion masses would end up at the GUT scale. Grand unification does not seem to work either for $N_c > 3$. One may hence speculate that in Nature $N_c = 3$ could be realized because grand unification works successfully just in that case.

In any case, GUTs are intellectually quite appealing and they may very well provide valuable hints to what awaits us at the ultimate high-energy frontier. This is worth emphasizing, although the powerful GUT community, which existed in the 1970s, seems not to be as active any longer. Since this is the last technical chapter, we end with the dictum in German: "Ende GUT, alles gut".

Exercises

26.1 Potential in the SU(5) GUT

a) Derive the GUT scale V in the potential of the SU(5) GUT, based on eqs. (26.6) and (26.7).

b) Now assume $M^2 < 0$ and discuss which pattern of SU(5) "gauge symmetry breaking" occurs in different parameter intervals.

26.2 Covariant derivative for adjoint Higgs field

Show that the covariant derivative for an adjoint Higgs field of eq. (26.19) transforms appropriately under SU(5) gauge transformations.

26.3 Covariant derivative of the anti-symmetric leptoquark matrix

Starting from definition (26.35), derive the relation (26.36).

26.4 Lagrangian term of the leptoquark transition

Consider the Lagrangian contribution (26.50).

a) Show that this term is gauge-invariant.

b) Now set $X_\mu = Y_\mu = 0$, and show that this term still contains all gauge interactions of the Standard Model.

26.5 Identities of unitary matrices and applications

a) Verify the identities (26.41) for SU(3) matrices and (26.114) for SU(5) matrices.

b) Apply the former identity to verify eq. (26.40) and the latter to check the gauge invariance of the terms in eq. (26.113).

26.6 Coupling of quintets and decouplets in SU(5)

Use the Young tableaux method described in Appendix F to reduce the following products in SU(5) into irreducible representations: $\{5\} \times \{5\}$, $\{\bar{5}\} \times \{10\}$, $\{10\} \times \{10\}$.

26.7 Number of degrees of freedom in the Standard Model and in the SU(5) GUT

Consider the Standard Model as well as the SU(5) GUT with an adjoint and a fundamental scalar field, each with three generations of quarks and leptons. Count the number of bosonic and fermionic, relativistic degrees of freedom.

26.8 Exceptional group G(2)

a) Show that the center of G(2) is trivial.

b) Verify the G(2) tensor product reduction of eq. (26.139) by using the graphical Antoine–Speiser scheme described in Appendix F.15.

Finale

We have reached the end of our multifaceted journey, which took us from the quantum mechanical path integral to quantum field theory, to the Standard Model of particle physics, and to some theories beyond. We hope that the reader has enjoyed this journey, and that it has not turned into an arduous Odyssey. The intention was to acquire a clear understanding of these challenging topics and be well prepared and motivated for further explorations.

Of course, there remains a lot to be learned about fundamental physics. Important next challenges are to study in more detail the perturbative quantization of non-Abelian gauge theories, the rich phenomenology of the Standard Model, a variety of further effective field theories that address the Standard Model physics at selected energy scales, and theories beyond the Standard Model which are continued to be tested in the LHC era, or string theory, which may not be experimentally testable in the foreseeable future, but addresses fundamental questions about the quantization of gravity. Fortunately, there is excellent textbook literature on all these topics.

While the Standard Model has a considerable number of free parameters, it is amazing that only very few of them – including the masses of the electron and the light quarks, as well as the gauge couplings – determine all of nuclear and atomic physics, and, at least in principle, condensed matter physics, chemistry, biology, and all other aspects of Nature as we experience them every day.

However, one should keep in mind that understanding the formation of emergent complex structures requires the identification of the relevant degrees of freedom and their dynamics, more than detailed knowledge of the underlying microscopic theory. In fact, just as there are true statements, *e.g.,* about integer numbers that cannot be derived from an axiomatic approach, as attempted by Alfred Whitehead and Bertrand Russell in their Principia Mathematica, similar Gödelian incompleteness may afflict theoretical physics as well. It may indeed be impossible – not only in practice, but even in principle – to rigorously derive some emergent collective phenomena – such as high-temperature superconductivity – from first principles of the Standard Model. The quest for a Theory of Everything, which can provide a complete explanation of all natural phenomena, could thus be doomed from the outset. In this case, even if a theory has a wide range of applications, we may ultimately have to be content with its *consistency* instead of urging for completeness.

Still, the search for an even more fundamental theory of Nature – whose predictive power captures phenomena at the Planck scale – is well motivated. While Kurt Gödel's proof demonstrated the incompleteness of Whitehead's and Russell's framework, it did not imply its inconsistency. Physicists have a deeply ingrained trust in the mathematical consistency of the laws of Nature, based on the overwhelming success of this concept.[1]

---

[1] The question *why* mathematics is so enormously successful in natural sciences is rather philosophical; Eugene Wigner wrote an interesting essay about it (Wigner, 1960). We should stress, however, that logical and mathe-

Thus, we expect that theories for coarse-grained phenomena, for example, in condensed matter physics – although possibly not rigorously derivable from first principles – must not be inconsistent with the underlying structures of the Standard Model. Hence, while particle physicists should not overestimate the predictive power of their theories as providing a complete description of Nature, physicists working on coarser scales should be aware of the fundamental structures underlying all physics and the other sciences.

The reader who has worked through all the subjects in this book in depth and has solved the exercises should now have a solid basis in quantum field theory and the Standard Model. This is a remarkable achievement which provides a sound point of departure for solving actual research problems. As we have seen, quantum field theory is at the basis of some of Nature's greatest puzzles. On our journey, we have encountered some deep questions of current research in fundamental physics. Let us reconsider a list of them.

Questions and challenges within the Standard Model:

- Can we rigorously prove confinement or the existence of a mass gap in Yang–Mills theory?
- How does the QCD phase diagram look at high baryon density?
- Can we define the electroweak chiral gauge theory non-perturbatively?
- Is the value of the anomalous magnetic moment of the muon compatible with the Standard Model?

Questions referring to Standard Model extensions:

- Up to which energy scale does the Standard Model apply, and what comes beyond it?
- Can we understand the values of the Standard Model parameters or relate them to each other?
- Why is the QCD vacuum-angle compatible with zero?
- Why do the fermions tend to be light?
- What is the origin of the baryon asymmetry?
- What is the deeper reason for electroweak symmetry breaking?
- How many Higgs particles are there and what are their properties?
- Why are there no magnetic monopoles?
- What is the nature of dark matter?

Questions far beyond the Standard Model:

- What is the origin of gauge invariance?
- Why is the Standard Model gauge symmetry $SU(3)_c \times SU(2)_L \times U(1)_Y$?
- Why is there a generation structure of quarks and leptons, and why is the number of generations (apparently) just three?
- Why is the electroweak scale so much below the Planck scale?
- What is the correct theory of non-perturbative quantum gravity?
- What is the nature of the vacuum energy, and why is the cosmological constant so incredibly small?
- Why do we live in $3 + 1$ space–time dimensions?
- What is the origin of space and time?

matical consistency is only a necessary, but not a sufficient condition for the scientific description of Nature – some theory may well be mathematically correct, but disconnected from reality. An example is the SU(5) GUT that we reviewed in Chapter 26.

Answering any one of those deep questions would be a great leap forward in our understanding of Nature. Even the beginning graduate student should have the courage to stare these questions in the face and think about them at a profound level. After all, it was our goal to introduce the Standard Model at an early stage, in order to be able to progress quickly to the questions that drive current research. If we can master the basics of quantum field theory and the Standard Model, nothing should stop us from exploring everything else that we urge to understand.

At this point, our "symphony" of field vibrations ends: It involved an Ouverture, an Intermezzo, and Four Parts – somewhat like Antonio Vivaldi's Four Seasons – and finally this Finale. It could be viewed as a modern variation of an ancient theme, inspired by Pythagoras' Music of Spheres. Hopefully, this book shows not only the impressive successes of the Standard Model, but also the beauty and harmony of our most fundamental theory of Nature. This reflects the beauty of both Nature herself and of the human mind. Still, as the previous list of open questions demonstrates, our understanding remains incomplete. This confronts us with great challenges, but it also offers wonderful opportunities to push the boundaries of our knowledge further into the unknown.

Appendix A Highlights in the Development of Particle Physics

In this appendix, we highlight the development of experimental and theoretical particle physics, with an emphasis on breakthrough discoveries and groundbreaking achievements which paved the way toward the Standard Model of particle physics. We limit our attention to theoretical developments within the Standard Model, excluding theories that go beyond it. We provide a (certainly incomplete) list of important events in more or less chronological order, but make no attempt to shed light on particle physics from a history of science point of view. An overview of the various developments in chronological order is provided in Table A.1.

A.1 Development of Experimental High-Energy Physics

Important breakthroughs that initiated particle physics were the discovery of the *electron* in cathode ray tubes by Joseph John Thomson in 1897 and of the *atomic nucleus* by Hans Geiger, Ernest Marsden, and Ernest Rutherford in 1911. In 1919, Rutherford further reported the discovery of the *proton*. Then, it took until 1932 when also the *neutron* was observed by James Chadwick, and in the same year, Carl Anderson detected the *positron*. In 1936, it was again Anderson, now together with Seth Neddermeyer, who discovered the *muon*.

The first meson, namely, the charged *pion,* was observed in 1947 by Cecil Powell, César Lattes, Giuseppe Occhialini *et al.* In the same year, also the *kaon* was discovered, by George Rochester and Clifford Butler. This was before the age of particle accelerators: The positron, muon, pion, and kaon were all first detected in cosmic rays.

When particle accelerators went into operation, soon the *anti-proton* was discovered by Emilio Segrè and Owen Chamberlain in 1955.

A very different but groundbreaking experiment was directed by Madame Chien-Shiung Wu in 1957: It showed that processes of the weak interaction are not invariant under spatial inversion; *i.e.* the *parity symmetry is broken.*

Further milestones on the way toward the Standard Model were the evidence for *quarks* obtained by Jerome Friedman, Henry Kendall, and Richard Taylor at the Stanford Linear Accelerator Center (SLAC) in a series of experiments from 1967–1973 and of *gluons* in the PLUTO detector at the Deutsches Elektronen-Synchrotron (DESY, in Hamburg) in the period 1978–1980.

The production and detection of *W- and Z-bosons* was achieved in the Super Proton Synchrotron (SPS) at the Conseil Européen pour la Recherche Nucléaire (CERN, near Geneva) in 1983, using advanced accelerator technology such as stochastic cooling. Indirect evidence for the existence of the Z-boson had already been obtained in 1974 in the form

584   Appendix A

Table A.1 Chronological development of experimental and theoretical particle physics

Experimental Discoveries	Theoretical Breakthroughs
	1864 Classical electrodynamics (Maxwell)
1897 Electron e^- (Thompson)	
	1905 Special relativity
1911 Atomic nucleus	(Einstein; Lorentz; Poincaré)
(Geiger, Marsden, Rutherford)	
1919 Proton p (Rutherford)	1920s Quantum mechanics
	(Bohr; Heisenberg; Schrödinger; Born; ...)
	1928 Relativistic quantum physics,
	positron prediction (Dirac)
1932 Neutron n (Chadwick)	1930 Neutrino prediction (Pauli)
Positron e^+ (Anderson)	
	1933/4 Theory of neutrino interactions (Fermi)
	1934 Scalar field quantization (Pauli, Weisskopf)
1936 Muon μ (Anderson, Neddermeyer)	
	1940s Quantum Electrodynamics (QED)
	(Dyson; Feynman; Schwinger; Tomonaga)
1947 Pion $\pi^\pm$ (Lattes, Occhialini, Powell)	
Kaon $K^\pm$ (Butler, Rochester)	1953/4 Renormalization group
	(Petermann, Stückelberg; Gell-Mann, Low)
	1954 Non-Abelian gauge theory (Mills, Yang)
1955 Anti-proton $\bar{p}$ (Chamberlain, Segrè)	1955/6 LSZ reduction
	(Lehmann, Symanzik, Zimmermann)
	Axiomatic quantum field theory (Wightman)
1956 Electron-neutrino ν_e (Cowan, Reyes)	1956 Prediction of P violation (Lee, Yang)
1957 P violation (Wu)	1957 CP symmetry postulated (Landau)
	CPT theorem (Pauli; Lüders; Jost)
	1960 Prediction of muon-neutrino (Pontecorvo)
	1960/1 Concept of spontaneous symmetry breaking
1962 Muon-neutrino ν_μ	(Nambu, Jona-Lasinio; Goldstone)
(Lederman, Schwartz, Steinberger)	1962–69 Prediction of neutrino oscillations
	(Maki, Nakagawa, Sakata; Gribov, Pontecorvo)
	1963 Flavor mixing (Cabbibo)
	1957–64 Higgs mechanism
	(Anderson; Englert, Brout; Higgs;
	Guralnik, Hagen, Kibble; Stückelberg)
1964 CP violation (Cronin, Fitch)	1964 Algebraic quantum field theory
	(Haag, Kastler)
	Higgs particle prediction (Higgs)
	Constituent quarks (Gell-Mann; Zweig)
	1967 Electroweak sector of the Standard Model
	(Glashow 1961; Salam; Weinberg 1967)
1967–73 Evidence for quarks (SLAC)	1968 Quantization of Yang–Mills theories
(Freedman, Kendall, Taylor)	(Faddeev, Popov)
	1969 Triangle anomalies (Adler; Bell, Jackiw)
	Effective theory for Nambu–Goldstone bosons
	(Callan, Coleman, Wess, Zumino)

(cont.)

Table A.1 (cont.)

Experimental Discoveries	Theoretical Breakthroughs
	1970 Prediction of c-quark, GIM mechanism (Glashow, Iliopoulos, Maiani) Scale invariance and its breaking (Callan; Symanzik)
	1971/2 Dimensional regularization and loop calculations in the Standard Model (Bollini, Giambiagi; 't Hooft, Veltman)
	1972 Anomaly cancellation in the Standard Model (Bouchiat, Iliopoulos, Meyer)
1973 Weak neutral currents (CERN)	1973 Prediction of third quark generation (Kobayashi, Maskawa) Asymptotic freedom (Gross, Wilczek; Politzer) Quantum Chromodynamics (QCD) (Fritzsch, Gell-Mann, Leutwyler; Nambu)
1974 c-quark (SLAC, BNL)	1970s Renormalization Group (Kadanoff; Fisher, Wilson)
	1975 Instanton gauge field configurations (Belavin, Polyakov, Schwarz, Tyupkin)
1975 τ-lepton (Perl, SLAC)	1976 θ-vacuum states (Jackiw, Rebbi; Callan, Dashen, Gross) Solution of the $U(1)_A$-problem ('t Hooft; Witten; Veneziano) Baryon and lepton number violation by electroweak instantons ('t Hooft)
1977 b-quark (FNAL)	1970s to 1980s Lattice regularization
1978–80 Evidence for gluons (DESY)	(Wegner; Wilson; Kogut, Susskind; Creutz; Hasenfratz; Lüscher; ...)
	1981/2 Triviality of the 4-d $\lambda\phi^4$ model (Aizenman; Fröhlich)
	1982 Global $SU(2)_L$ anomaly (Witten)
1983 $W^{\pm}$ and Z detection (CERN)	1984 Chiral perturbation theory (Gasser, Leutwyler)
	1992/3 Domain wall and overlap lattice fermions (Kaplan; Narayanan, Neuberger)
1995 t-quark (FNAL)	
1998–2001: neutrino oscillations (Super-Kamiokande; SNO-Lab)	1998 Exact lattice index theorem (Hasenfratz, Laliena, Niedermayer) Exact lattice chiral symmetry (Lüscher)
2000 τ-neutrino ν_τ (DONUT, FNAL)	1999–2000 Lattice chiral gauge theories (Lüscher)
2012 Neutrino mixing with $U_{e3} \neq 0$ (Daya Bay; RENO; Double Chooz)	
2012 Higgs particle (ATLAS, CMS, CERN)	
	2019 Rigorous triviality proof of 4-d $\lambda\phi^4$ (Aizenman, Duminil-Copin)

of neutral currents at the CERN Gargamelle bubble chamber. In the 1990s, the electroweak W- and Z-bosons were further investigated in great detail at the Large Electron-Positron (LEP) collider also at CERN, which provided evidence against more than three fermion generations.

Further triumphs of particle physics were the discoveries of the heavy *charm quark* independently by the groups of Burton Richter at SLAC and of Samuel Ting at the Brookhaven National Laboratory (BNL) in 1974, of the even heavier *bottom quark* by the E288 experiment in 1977, and of the superheavy *top quark* by the CDF and D0 collaborations in 1995, all three at the Fermi National Accelerator Laboratory (FNAL) near Chicago.

The heavy quarks are intimately related to *CP violation* (the explicit violation of the combined symmetries of parity and charge conjugation), which was discovered by a collaboration led by James Cronin and Val Fitch at BNL in 1964 and further investigated in detail by the BaBar collaboration at SLAC and by the Belle experiment at Kō Enerugī Kasokuki Kenkyū Kikō (KEK) in Japan.

In landmark experiments, the *electron-neutrino* was discovered by Clyde Cowan, Frederick Reines, and collaborators in 1956. Next, the *muon-neutrino* was detected by Leon Lederman, Melvin Schwartz, and Jack Steinberger in 1962. Regarding the third lepton generation, the τ-*lepton* was observed in 1977 by Martin Perl's group at SLAC. This suggested the existence of a τ-neutrino, and it stimulated further investigations of the fascinating, extremely elusive neutrinos. In fact, the τ-*neutrino* was detected in the year 2000 in the DONUT experiment at FNAL.

The detection of solar and cosmic neutrinos was achieved by the groups of Raymond Davis in the Homestake gold mine and of Masatoshi Koshiba using the gigantic Kamiokande detector in the underground Mozumi mine in Japan. In 2001, this line of research culminated in the solution of the solar neutrino problem, which was ultimately provided at the Sudbury Neutrino Observatory (SNO) in Ontario, Canada, led by Arthur McDonald. It is solved by *neutrino oscillation,* which had also been observed, since 1998, for atmospheric neutrinos in the Super-Kamiokande experiment, under the direction of Takaaki Kajita. At that time, the lepton mixing matrix (PMNS matrix) element U_{e3} seemed to be compatible with zero. However, in 2012 the reactor neutrino experiments Daya Bay in China, RENO in South Korea, and Double Chooz in France demonstrated that U_{e3} does not vanish.

The latest highlight so far was the discovery of the Higgs particle. Stringent evidence was presented by the ATLAS and CMS collaborations in 2012, soon after the Large Hadron Collider (LHC) went into operation at CERN. Therefore, *all* particles of the Standard Model have now been discovered, but *no* particle beyond has ever been detected (although the observation of neutrino oscillations, which implies non-zero neutrino masses, represents a step beyond the minimal renormalizable version of the Standard Model). There is strong evidence for cosmic dark matter, starting with the observations of Fritz Zwicky in 1933, but its nature remains unknown.

Since the middle of the twentieth century, the particle search was performed systematically with more and more powerful accelerators; until 2022, the LHC has attained collision energies up to 13.6 TeV. However, cosmic ray particles hit the atmosphere with energies reaching up to $\mathcal{O}(10^8)$ TeV (which causes extended air showers), though with a very low intensity. These *ultra-high-energy cosmic rays* have been detected in particular by the observatories AGASA (Japan), HighRes (USA), and Pierre Auger (Argentina), but even there no contradiction with the Standard Model has been found.

A.2 Development of Quantum Field Theory and the Standard Model

Besides the spectacular experimental achievements, the theoretical developments that have led to the formulation of the Standard Model are equally impressive. Ultimately, the Standard Model rests on the theoretical basis of *quantum physics* and *special relativity,* embedded into the framework of a covariant *quantum field theory.*

The first established field theory emerged in James Clerk Maxwell's classical theory of *electromagnetism* from 1864. In 1905, the physics of special relativity was fully understood by Albert Einstein; some of its mathematical properties had been explored before by Hendrik Lorentz and Henri Poincaré.

In the 1920s, *quantum mechanics* was elaborated by Niels Bohr, Max Born, Louis de Broglie, Werner Heisenberg, Wolfgang Pauli, Erwin Schrödinger, and others. In 1928, it was further developed by Paul Adrien Maurice Dirac to a *"relativistic quantum mechanics"* of the electron and positron. The theories of Maxwell and Dirac were merged to form *Quantum Electrodynamics* (QED) by Freeman Dyson, Richard Feynman, Julian Schwinger, and Sin-Itiro Tomonaga in the 1940s. In 1954, Chen-Ning Yang and Robert Mills proposed an extension of gauge theory to *non-Abelian gauge groups.* This concept had first been mentioned in an unpublished letter by Wolfgang Pauli to Abraham Pais in 1953.

Pauli also predicted the existence of a *neutrino* in 1930, and in 1933/4, Enrico Fermi proposed a theoretical description for its interaction, in particular in the β-decay. Another neutrino flavor, the *muon-neutrino,* was predicted by Bruno Pontecorvo in 1960, and the process of *neutrino oscillation* was proposed in the 1960s by Ziro Maki, Masami Nakagawa, and Shoichi Sakata as well as by Vladimir Gribov and Bruno Pontecorvo.

Symmetries and their breaking play a central role in particle physics. In particular, the explicit *violation of parity symmetry* was predicted by Tsung-Dao Lee and Chen-Ning Yang in 1956. Soon afterward, Boris Ioffe, Lev Okun, and Aleksei Rudik pointed out that parity violation in the β-decay also implies a violation of the invariance under charge conjugation, C. In 1957, Lev Landau suggested that the combination CP of C and parity P remains conserved. Later, the small *CP violation* (observed in 1964) inspired Makoto Kobayashi and Toshihide Maskawa in 1973 to anticipate the existence of the bottom and the top quark. They extended the concept of *quark flavor mixing,* which had been introduced for two flavors by Nicola Cabibbo in 1963. On the other hand, to the best of our knowledge, physics is invariant under the combined transformations CPT. The corresponding *CPT theorem* was demonstrated for any local and covariant field theory by Wolfgang Pauli and Gerhart Lüders in 1954, and most rigorously by Res Jost in 1957.

Spontaneous symmetry breaking, which was studied by Giovanni Jona-Lasinio and Yoichiro Nambu since 1960 and investigated further by Jeffrey Goldstone in 1961, plays a central role both in particle and in condensed matter physics. An elegant effective field theory description of massless Nambu–Goldstone bosons was provided by Curtis Callan, Sidney Coleman, Julius Wess, and Bruno Zumino in 1969. In the 1980s, inspired by earlier work by Steven Weinberg, *chiral perturbation theory* was constructed, as a systematic low-energy effective theory for light hadrons, in particular, by Jürg Gasser and Heinrich Leutwyler.

Although gauge symmetries are unbreakable, an effect which appears similar to spontaneous symmetry breaking endows the gauge bosons with mass. This *Higgs mechanism* was first understood by Philip Anderson, Robert Brout, Francois Englert, Gerald Guralnik, Carl Hagen, Peter Higgs, Tom Kibble, and Ernst Stückelberg in the early 1960s. In particular, Higgs predicted the existence of a scalar particle – now known as the *Higgs particle* – based on this mechanism in 1964.

In 1967, the Higgs mechanism was incorporated in what is now the electroweak sector of the *Standard Model* of particle physics, by Abdus Salam and by Steven Weinberg, revitalizing earlier work by Sheldon Glashow of 1961. By now, most aspects of the Standard Model – which also includes *Quantum Chromodynamics* (QCD) – have been tested so exquisitely well that it has long become *the* established theory of particle physics.

Anomalies were brought to prominence by Stephen Adler, John Bell, and Roman Jackiw in 1969. They are a form of explicit symmetry breaking upon quantization, which imposes stringent constraints on the consistency of the Standard Model. The importance of gauge triangle anomaly cancellation in the Standard Model was pointed out by Claude Bouchiat, John Iliopoulos, and Philippe Meyer in 1972. In this context, Edward Witten made an important contribution in 1982, when he discovered a so-called global anomaly in SU(2) gauge theories.

As was realized in the mid-1970s, anomalies are intimately related to the topology of quantum fields. Instantons are examples of topologically non-trivial field configurations, which provide a solution to the so-called $U(1)_A$-problem. They give rise to θ-vacuum states in the strong interaction, as well as to non-perturbative baryon number violating effects in the weak interaction. Leading figures in this field include Curtis Callan, Roger Dashen, David Gross, Gerard 't Hooft, Roman Jackiw, Sergei Novikov, Alexander Polyakov, Claudio Rebbi, Gabriele Veneziano, Edward Witten, and others.

While *quarks* were first introduced by Murray Gell-Mann and George Zweig in 1964 as group theoretical entities, they were soon established as physical constituents of protons and neutrons. The Standard Model contains a non-Abelian $SU(3)_c$ color gauge field. It gives rise to the gluons of QCD which was put forward in the early 1970s by Harald Fritzsch, Murray Gell-Mann, and Heinrich Leutwyler, as well as Yoichiro Nambu and others. The QCD theory for the strong interaction involves quarks as elementary particles, which are more fundamental than the previously suggested group theoretical hadronic building blocks (now denoted as "constituent quarks"), but share the same quantum numbers. The investigation of QCD at high energies was pioneered by David Gross, Frank Wilczek, and by David Politzer in 1973, who derived and correctly interpreted the property of *asymptotic freedom*.

In 1970, Sheldon Glashow, John Iliopoulos, and Luciano Maiani predicted the existence of a fourth quark flavor based on the so-called *GIM mechanism*. After an earlier conjecture by James Bjorken and Sheldon Glashow in 1964, the GIM mechanism provided solid indirect evidence for the existence of a charm quark before it was first detected in 1974.

The understanding of the mathematical structure of *quantum field theory* was pioneered by Murray Gell-Mann, Rudolf Haag, Daniel Kastler, Harry Lehmann, Francis Low, Ernst Stückelberg, Kurt Symanzik, Arthur Wightman, and Wolfhart Zimmermann, with numerous essential contributions made by Felix Berezin, Curtis Callan, Sidney Coleman, Jürg Fröhlich, James Glimm, Arthur Jaffe, Jean Zinn-Justin, and many others.

The consistent elimination of ultraviolet divergences is intimately connected with the *renormalization group* whose studies were pioneered by Nikolai Bogoliubov, Michael Fisher, Murray Gell-Mann, Leo Kadanoff, Francis Low, André Petermann, Ernst Stückelberg, and – last but not least – Kenneth Wilson.

The intricacies of the *quantization of non-Abelian gauge theories* were analyzed by Ludvig Faddeev and Victor Popov in 1967. The mathematical treatment of gauge field theory requires, in particular, a gauge-invariant regularization. In the framework of perturbation theory, this was provided by Wolfgang Pauli and Felix Villars in 1949 for Abelian gauge theory. In 1971/2, Carlos Bollini and Juan José Giambiagi – as well as Gerard 't Hooft and Martinus Veltman – proposed *dimensional regularization*, which applies also to non-Abelian gauge theory. The *renormalizability* of these theories was investigated by 't Hooft in 1971/2, thus providing a promising basis for the use of non-Abelian gauge theories as fundamental structures underlying particle physics. Around 1975, the BRST formalism of Carlo Becchi, Alain Rouet, Raymond Stora, and Igor Tyutin provided additional important mathematical consolidation of the perturbative quantization of gauge theories. The proof of their perturbative renormalizability rests on identities derived by John Ward and Yasushi Takahashi in the Abelian, and by Andrei Slavnov and John Taylor in the non-Abelian case, as well as on the so-called Zinn-Justin equation.

Quantum field theory in Euclidean space–time was pioneered by Julian Schwinger in 1958/9 and developed further by Kurt Symanzik since 1966, as well as by Konrad Osterwalder and Robert Schrader in 1973–5. In 1981/2, Michael Aizenman and Jürg Fröhlich independently presented strong arguments in favor of the *triviality* of scalar $\lambda\phi^4$ theory. Much later, in 2019, Michael Aizenman and Hugo Duminil-Copin provided a rigorous proof of this feature that affects the Higgs sector of the Standard Model.

Beyond perturbation theory, the *lattice regularization* provides a very powerful tool to address quantum field theory from first principles – this scheme was elaborated in particular by Kenneth Wilson in the 1970s; further key players include Michael Creutz, Peter Hasenfratz, John Kogut, Julius Kuti, Martin Lüscher, Herbert Neuberger, Alexander Polyakov, Jan Smit, Leonard Susskind, and Franz Wegner.

In the early 1990s, long-lasting problems associated with the non-perturbative regularization of theories with an exact global chiral symmetry were solved elegantly by David Kaplan and by Rajamani Narayanan and Herbert Neuberger, who constructed domain wall and overlap lattice fermions, respectively. The Ginsparg–Wilson relation, which had been derived by Paul Ginsparg and Kenneth Wilson in 1982, turned out to hold the key to understanding chiral symmetry beyond perturbation theory. In 1998, this relation led Peter Hasenfratz, Victor Laliena, and Ferenc Niedermayer to a proof of an exact lattice variant of the Atiyah–Singer index theorem, with far reaching physical consequences. Also based on the Ginsparg–Wilson relation, in the same year Martin Lüscher constructed an exact lattice variant of chiral symmetry. In 1999 and 2000, this culminated in a spectacular breakthrough: Lüscher's construction of lattice chiral gauge theories, both within and beyond perturbation theory. Even perturbative continuum chiral gauge theories benefit from the deeper understanding of chiral symmetry that emerges from this work in the very powerful framework of lattice field theory.

Appendix B Units, Hierarchies, and Fundamental Parameters

Physical units represent man-made conventions influenced by the historical development of physics. The magnitudes of the most commonly used units refer to quantities which are relevant in our daily life; thus, they are related to the measures of the human body.

In addition, there are also *natural units* which express physical quantities in terms of fundamental constants of Nature: Newton's gravitational constant G, the speed of light c, and Planck's action quantum h. In this appendix, we consider the units commonly used in particle physics, and we discuss energy scales and mass hierarchies, as well as the fundamental parameters of the Standard Model.

B.1 Man-Made versus Natural Units

The most basic physical quantities – length, time, and mass – are usually measured in units of *meters* (m), *seconds* (sec), and *kilograms* (kg). Obviously, these are man-made units, appropriate for the use at our human scales. For example, the length of a step is roughly one meter, the duration of a heartbeat is about one second, and one kilogram is the mass of typical objects that we handle manually, *e.g.*, a tool, a loaf of bread, or a bottle of wine.

Time is measured by counting periodic phenomena. An individual cesium atom is an extremely accurate clock. Nowadays, 1 second is defined as $9\,192\,631\,770$ periods of a particular microwave transition of the ^{133}Cs atom. The meter was originally defined by the length of the meter stick kept in the "Bureau International des Poids et Measures" in Paris. Its first version dates back to the wake of the French revolution, at the end of the eighteenth century. Nowadays, one defines the meter through the speed of light c and the second as

$$1\,\text{m} = 3.33564095 \times 10^{-9}\,c\,\text{sec}. \tag{B.1}$$

In other words, the measurement of distance is reduced to the measurement of time by invoking a natural constant. Together with the meter stick, a certain amount of platinum–iridium alloy was deposited in Paris, and its mass was defined to be one kilogram.

Expressed in those man-made units, Nature's most fundamental constants are the *speed of light*

$$c = 2.99792458 \times 10^{8}\,\text{m sec}^{-1}, \tag{B.2}$$

Planck's action quantum (divided by 2π)

$$\hbar = \frac{h}{2\pi} = 1.054571817 \times 10^{-34}\,\text{kg m}^{2}\text{sec}^{-1}, \tag{B.3}$$

and *Newton's gravitational constant*

$$G = 6.67430(15) \times 10^{-11}\, \mathrm{m}^3\,\mathrm{kg}^{-1}\mathrm{sec}^{-2}. \tag{B.4}$$

Appropriately combining these fundamental constants, Nature provides us with her own natural units (also known as *Planck units*): the Planck length

$$l_{\mathrm{Planck}} = \sqrt{\frac{G\hbar}{c^3}} = 1.61626(2) \times 10^{-35}\, \mathrm{m}, \tag{B.5}$$

and the Planck time

$$t_{\mathrm{Planck}} = \sqrt{\frac{G\hbar}{c^5}} = 5.39125(6) \times 10^{-44}\, \mathrm{sec}, \tag{B.6}$$

which represent the shortest distances and times of interest in physics. Today, we are very far from exploring such short length and time scales experimentally. It is even expected that our classical concepts of space and time may break down at the Planck scale.

One also defines the *Planck mass*

$$M_{\mathrm{Planck}} = \sqrt{\frac{\hbar c}{G}} = 2.17643(2) \times 10^{-8}\, \mathrm{kg}. \tag{B.7}$$

While this seems very light in our daily life, it is very heavy in the context of particle physics.

Planck units are not very practical in our everyday life. For example, a step has a length of about $10^{35}\, l_{\mathrm{Planck}}$, a heartbeat lasts roughly $10^{43}\, t_{\mathrm{Planck}}$, and the mass of our body is of order $10^9\, M_{\mathrm{Planck}}$. Still, l_{Planck}, t_{Planck}, and M_{Planck} are the most fundamental basic units that Nature provides us with. It is interesting to ask why we exist at scales so far away from the Planck scale. For example, why does a kilogram correspond to about $10^8\, M_{\mathrm{Planck}}$? In some sense, this is a "historical" question. The amount of platinum–iridium alloy deposited in Paris a long time ago, which defined the kilogram, obviously is an arbitrarily chosen man-made unit. Why was it chosen in this particular manner?

If we assume that the kilogram was chosen because it is a reasonable fraction of our body mass, we may rephrase the question as a biological one: Why do intelligent beings weigh 10^9 to $10^{10}\, M_{\mathrm{Planck}}$? If biology could explain the number of cells in our body and, with some help from chemistry, could also explain the number of atoms necessary to form a cell, we could reduce the question to a physics problem. Since atoms get their mass from the nucleons (protons and neutrons, which have approximately the same mass), we are led to ask: Why is the proton mass,

$$M_p = 1.67262 \times 10^{-27}\, \mathrm{kg} = 7.6852 \times 10^{-20}\, M_{\mathrm{Planck}}, \tag{B.8}$$

so light compared to the Planck mass? This hierarchy puzzle, which is addressed for instance in the Intermezzo and in Chapters 14 and 18, has been understood at least qualitatively using the property of asymptotic freedom of Quantum Chromodynamics – the quantum field theory of quarks and gluons whose interaction energy explains the mass of the proton.

Since the ratio $M_p/M_{\mathrm{Planck}} \approx 10^{-19}$ is so tiny, it is unpractical to use M_{Planck} as a basic unit of mass in particle physics. Instead, it is common to refer to the *electron Volt*, the energy that an electron (of charge $-e$) picks up when it is accelerated by a potential difference of one Volt,

$$1\,\mathrm{eV} \simeq 1.602176634 \times 10^{-19}\,\mathrm{kg}\,\mathrm{m}^2\,\mathrm{sec}^{-2}, \tag{B.9}$$

as a basic energy unit. Obviously, the Volt, and therefore also the eV, is again a man-made unit – as arbitrarily chosen as, 1 kg, 1 m, and 1 sec. In this unit, the rest energy of a proton is given by

$$M_p c^2 = 0.9382720882(3)\,\mathrm{GeV}. \tag{B.10}$$

This number, and practically all results which are relevant in particle physics, can be found in the *Review of Particle Physics,* which is regularly updated. In this book, and in particular in this appendix, we quote the values given in Workman *et al.* (2022).

   In particle physics, it is convenient and common practice to set $\hbar = c = 1$. Then, masses and momenta are measured in energy units, and lengths and times are measured in units of inverse energy. In particular, one has

$$\hbar c \simeq 3.1615 \times 10^{-26}\,\mathrm{kg}\,\mathrm{m}^3\,\mathrm{sec}^{-2}, \tag{B.11}$$

such that $\hbar c = 1$ implies

$$1\,\mathrm{fm} = 10^{-15}\,\mathrm{m} \simeq (0.1973\,\mathrm{GeV})^{-1}, \tag{B.12}$$

which is the scale of nuclear radii.

   The strength of the electromagnetic interaction is determined by the quantized charge unit e (the electric charge of a proton). In natural units, it gives rise to the experimentally determined *fine-structure constant*, which was introduced by Arnold Sommerfeld (Sommerfeld, 1916) and which takes (at zero momentum transfer) the value

$$\alpha = \frac{e^2}{4\pi\,\hbar c} = \frac{1}{137.03599908(2)}. \tag{B.13}$$

The strength of electromagnetism is determined by this dimensionless number which is completely independent of any man-made convention.

   As was pointed out by Frank Wilczek (2004), eq. (B.8) also explains why gravity is an extremely weak force. To understand this, let us compare the magnitude of the attractive gravitational force $F_{\mathrm{g}}(R)$ between two protons of charge e at a distance R with the magnitude of the repulsive electrostatic Coulomb force $F_{\mathrm{e}}(R)$

$$F_{\mathrm{g}} = G\frac{M_p^2}{R^2}, \quad F_{\mathrm{e}} = \frac{1}{4\pi}\frac{e^2}{R^2} \quad \Rightarrow \quad \frac{F_{\mathrm{g}}}{F_{\mathrm{e}}} = \frac{4\pi G M_p^2}{e^2} = \frac{1}{\alpha}\left(\frac{M_p}{M_{\mathrm{Planck}}}\right)^2 \approx 10^{-36}. \tag{B.14}$$

This is the reason why gravity effects are completely negligible in particle physics (at least at energy scales way below the Planck scale). Here and throughout, we are using Heaviside–Lorentz units.

   We have again used the fine-structure constant, whose value is not understood on theoretical grounds. It is an interesting question (that, *e.g.,* Wolfgang Pauli was fascinated by (Enz and Meyenn, 1994)) why α takes this particular value. Until now, we have no clue how to answer this question. Some people like to use the anthropic principle: If α would be significantly different, atomic physics and thus chemistry would work differently, and life as we know it would be impossible. Obviously, we can only exist in a Universe with a value of α that is hospitable to life. According to the anthropic principle, our existence may "explain" the approximate value of α. The authors prefer to be more optimistic and hope that some extension of the Standard Model will eventually explain the value of α.

B.2 Energy Scales and Particle Masses

Table B.1 lists the charges, masses, and life-times of some important particles. The photon is, as far as we can tell, exactly massless, in agreement with the unbroken gauge symmetry $U(1)_{em}$. Then, it cannot possibly decay into any lighter particles and is therefore stable. Just as the photon mediates the electromagnetic interaction, the heavy gauge bosons W^+, W^-, and Z mediate the weak interaction. Unlike the photon, the electroweak gauge bosons are unstable against the decay into other particles and live only for $\mathcal{O}(10^{-25})$ sec. Due to their large mass, there is a large phase space for the decay into light particles which causes the short life-times of the W- and Z-bosons. The inverse of their mass determines the very short range of the weak interaction, $M_{W,Z}^{-1} = \mathcal{O}(10^{-17})$ m.

The mass of ordinary matter is dominated by the masses of protons and neutrons, forming atomic nuclei, which are surrounded by an electron cloud. While the life-time of an isolated neutron is finite (about 15 minutes), because it decays into proton, electron, and anti-neutrino (β-decay), a neutron bound inside a stable atomic nucleus cannot decay. Despite numerous experimental searches, a proton has never been observed to decay. Still, as discussed in Chapter 15, the Standard Model does predict proton decay, however, at such a tiny rate that its experimental confirmation is practically impossible. Grand Unified Theories (GUTs) predict proton decays at higher and perhaps detectable rates (cf. Chapter 26). Such theories may eventually explain the baryon asymmetry in the Universe – the fact that some small amount of matter (all that is present in the Universe today) survived the annihilation of matter and anti-matter around 1 second after the Big Bang.

The pions π^+, π^0, and π^- are the lightest strongly interacting particles (hadrons). They are responsible for the large-distance contribution to the (still very short-ranged) nuclear force between protons and neutrons. The charged pions $\pi^\pm$ are relatively long-lived, because they decay only through processes of the weak interaction. The neutral pion π^0, on the other hand, lives much shorter, because (as discussed in Chapter 24) it can decay electromagnetically into two photons.

The minimal Standard Model Lagrangian contains only one dimensioned parameter – the vacuum expectation value v of the Higgs field. Its value is extracted from experimental data as

$$v \simeq 246\,\text{GeV} \simeq 2.02 \times 10^{-17} M_{\text{Planck}}. \tag{B.15}$$

We see a huge hierarchy separating the electroweak scale v from the Planck scale M_{Planck} set by the gravitational force. Since v is a free parameter of the Standard Model, we do not understand where the hierarchy originates from. Indeed, in order to adjust v to its experimental magnitude, the bare mass parameter of the Higgs field must be fine-tuned to a large number of decimal places. Many physicists consider this unnatural, which confronts us with the so-called *hierarchy problem*. Some theories beyond the Standard Model (*e.g.,* those based on technicolor) attempt to solve this hierarchy problem by explaining the ratio v/M_{Planck} without any need for fine-tuning.

The charges and masses of the leptons are listed in Table B.2. There are three generations of leptons including the electrically charged leptons – electron, muon, and tau – as well as the corresponding neutrinos. The masses of the charged leptons are experimentally known to high precision. In the Standard Model, the lepton masses are free parameters, resulting

Table B.1 Electric charges (in units of e), masses, and life-times of some important particles. (Actually, ν_e describes an electroweak flavor, not a mass eigenstate.)

Particle type	Particle	Electric charge	Mass [GeV]	Life-time
scalar boson	Higgs particle	0	125.25(17)	$\simeq 1.9(7) \times 10^{-22}$ sec
gauge bosons	photon γ	0	$< 10^{-27}$	stable
	$W^{\pm}$-bosons	± 1	80.377(12)	$3.16(6) \times 10^{-25}$ sec
	Z-boson	0	91.1876(21)	$2.638(2) \times 10^{-25}$ sec
leptons	neutrino ν_e	0	$< 0.8 \times 10^{-9}$	expected to be stable
	electron e	-1	$0.51099895000(15) \times 10^{-3}$	$> 6.6 \times 10^{28}$ years
baryons	proton p	1	0.93827208816(29)	$> 10^{31} \ldots 10^{34}$ years
	neutron n	0	0.9395654205(5)	878.4(5) sec
mesons	pion π^0	0	0.1349768(5)	$8.43(13) \times 10^{-17}$ sec
	pions $\pi^{\pm}$	± 1	0.13957039(18)	$2.6033(5) \times 10^{-8}$ sec

Table B.2 Electric charges (in units of e), masses, and life-times of the three generations of leptons. Neutrinos are assumed to be stable (although their flavors oscillate). The upper bounds on the masses originate from the direct measurement in the β-decay and in the decays $\pi^+ \rightarrow \mu^+ + \nu_\mu$ and $\tau \rightarrow \nu_\tau + 5\pi$, respectively. It should be noted that ν_e, ν_μ, and ν_τ refer to the electroweak and not to the mass eigenstates.

Lepton	Electric charge	Mass [GeV]	Life-time
electron-neutrino ν_e	0	$< 0.8 \times 10^{-9}$	
electron e	-1	$0.51099895000(15) \times 10^{-3}$	$> 6.6 \times 10^{28}$ years
muon-neutrino ν_μ	0	$< 0.19 \times 10^{-3}$	
muon μ	-1	0.1056583755(23)	$2.1969811(22) \times 10^{-6}$ sec
tau-neutrino ν_τ	0	$< 18.2 \times 10^{-3}$	
tau-lepton τ	-1	1.77686(12)	$2.903(5) \times 10^{-13}$ sec

from the Yukawa couplings to the Higgs field, multiplied by v. At present, we have no clue either why the lepton masses take their specific values. In particular, we do not understand why the masses of the electron and the tau-lepton differ by more than three orders of magnitude, or why the electron mass is more than five orders of magnitude smaller than the electroweak scale v. It is often considered "unnatural" that the corresponding Yukawa coupling is as small as $\mathcal{O}(10^{-5})$, despite the fact that Nature made exactly that "choice".

In the minimal Standard Model, the neutrinos are exactly massless. However, the observations of neutrino oscillations imply that neutrinos must have a non-zero mass, since we have data about their squared mass differences. This does not specify their mass hierarchy, and also for the masses themselves we only have upper bounds.

The so-called seesaw mass-by-mixing mechanism, which is discussed in Section 16.9, accounts for very small neutrino masses, at the expense of introducing a new, very high energy scale, which could be as large as $\mathcal{O}(10^{16})$ GeV, corresponding to the masses of extremely heavy Majorana neutrinos. This scenario can be incorporated in the Grand Unified

Table B.3 Electric charges (in units of e) and running masses (in the $\overline{\text{MS}}$ scheme at the respective mass scales) for the six quark flavors.

Generation	Quark flavor	Electric charge	Mass [GeV]
1.	up u	2/3	$0.00216^{+0.00049}_{-0.00026}$
	down d	−1/3	$0.00467^{+0.00048}_{-0.00017}$
2.	charm c	2/3	1.27(2)
	strange s	−1/3	$0.0934^{+0.0086}_{-0.0034}$
3.	top t	2/3	172.7(3)
	bottom b	−1/3	4.18(3)

Theories, which assume the unification of the gauge couplings at energies of a similar magnitude, besides introducing additional heavy gauge bosons.

Table B.3 summarizes the electric charges and masses of the quarks. They appear in three generations as well, with the up (u) and down (d) quark forming the first, the charm (c) and strange (s) quark the second, and the top (t) and bottom (b) quark the third generation. The electric charges of quarks are either 2/3 or −1/3 of the elementary charge e. However, since quarks do not exist in isolation but are confined inside hadrons, in agreement with Millikan-type experiments, at an observable level no fractionally charged particle states seem to occur in Nature.[1] Confinement also implies that quark masses do not represent the inertia of physical objects. Only the masses of the resulting hadrons are truly physical masses determining inertia and gravitational coupling strengths.

Like other quantities in quantum field theory, quark masses are *"running"; i.e.* they depend on the chosen renormalization scale and scheme. The quark masses in Table B.3 are defined in the so-called $\overline{\text{MS}}$ minimal subtraction renormalization scheme. The masses of the light quarks u, d, and s are quoted at a scale of 2 GeV, while the masses of the heavy quarks c, b, and t are quoted at the scale of the respective mass itself. Like the lepton masses, the quark masses are determined by the scale v – the only dimensioned parameter in the (minimal) Standard Model Lagrangian – multiplied by the respective Yukawa coupling to the Higgs field. Again, we do not understand why the quark masses take these specific values. In particular, we do not know why the masses of the u- and the t-quark differ by almost five orders of magnitude.

It is interesting that the masses of the proton and other hadrons are *not* proportional to v. In QCD, hadrons arise non-perturbatively as states containing confined quarks and gluons. Remarkably, the proton mass is still ≈ 0.9 GeV even when the quark masses are set to zero. For massless quarks (*i.e.* in the chiral limit), the QCD action contains no dimensioned parameter and is thus scale invariant. However, scale invariance is anomalous; *i.e.* although it is present in the classical theory, it is explicitly broken at the quantum level. Since quantum field theories must be regularized and renormalized, upon quantization a dimensioned parameter enters the theory. Even when the regularization is removed, a dimensioned scale is left behind. This phenomenon – which is known as *dimensional transmutation* – is manifest already in perturbation theory. In particular, in the $\overline{\text{MS}}$ renormalization scheme, the

---

[1] Quarks also exist in short-lived, deconfined quark–gluon plasmas, which, however, does not enable the direct observation of fractional charges. Charge fractionalization of electrons is known to occur as a *collective* phenomenon in the condensed matter physics of the fractional quantum Hall effect.

perturbatively defined scale Λ_{QCD} arises, whose value in the three-flavor theory (with u-, d-, and s-quarks) is given by Bruno *et al.* (2016)

$$\Lambda_{QCD} = 0.332(14)\,\text{GeV}. \tag{B.16}$$

In the chiral limit of massless quarks, Λ_{QCD} is the only scale of QCD, to which all hadron masses are proportional. For example, the dimensionless ratio M_p/Λ_{QCD} is predicted by non-perturbative QCD without any adjustable parameters.

While the proton mass is provided by Nature, the $\overline{\text{MS}}$ scheme is again man-made by theoretical physicists to ease perturbative calculations in QCD. Of course, in contrast to the kg, which was chosen at our human scale, Λ_{QCD} is defined at the relevant energy scale of the strong interaction. As a consequence of asymptotic freedom of QCD, it is natural for the proton mass M_p (and hence the QCD scale Λ_{QCD}) to be much smaller than the Planck scale M_{Planck}, as we mentioned before. On the other hand, since the Higgs sector of the Standard Model is not asymptotically free, the hierarchy problem (the question why $v \ll M_{\text{Planck}}$) arises. As long as this problem remains unsolved, we will not understand either why Λ_{QCD} is about three orders of magnitude smaller than v. In any case, this property allows us to study QCD separately, as a strongly interacting, effective low-energy sector of the Standard Model, where quark masses appear directly in the Lagrangian.

Just as the fundamental electric charge e determines the strength of the electromagnetic interaction between photons and electrons or other charged particles, the strong coupling g_s determines the strength of the strong interaction between quarks and gluons and among gluons themselves. Like the quark masses, the QCD analogue $\alpha_s = g_s^2/4\pi\hbar c$ of the fine-structure constant $\alpha = e^2/4\pi\hbar c$ also depends on the renormalization scale and scheme. Asymptotic freedom implies that α_s approaches zero in the high-energy limit; *i.e.* the strong interaction becomes weak at high momentum transfers. At the scale of the Z-boson mass, the quark–gluon coupling constant amounts to

$$\alpha_s(M_Z) = 0.1179(9). \tag{B.17}$$

The parameter Λ_{QCD} sets the energy scale at which α_s becomes strong.

The electroweak interactions are described by the gauge group $SU(2)_L \times U(1)_Y$ with two corresponding gauge coupling constants g and g'. At temperature scales below v, in particular in the vacuum, the $SU(2)_L \times U(1)_Y$ symmetry is "spontaneously broken" down to the $U(1)_{\text{em}}$ gauge group of electromagnetism with the elementary electric charge given by

$$e = \frac{gg'}{\sqrt{g^2 + g'^2}}. \tag{B.18}$$

The ratio of the W- and Z-boson masses defines the Weinberg angle θ_W

$$\frac{M_W}{M_Z} = \frac{g}{\sqrt{g^2 + g'^2}} = \cos\theta_W. \tag{B.19}$$

Its measured value (in the $\overline{\text{MS}}$ scheme at the scale M_Z) amounts to

$$\sin^2\theta_W = \frac{g'^2}{g^2 + g'^2} = 0.23121(4). \tag{B.20}$$

Eqs. (B.18), (B.19), and (B.20) are discussed in Chapter 13.

Just as the value of the fine-structure constant[2], $\alpha \approx 1/137$, is not understood theoretically, the values of the three gauge couplings $g_{\rm s}$, g, and g' associated with the Standard Model gauge group $SU(3)_{\rm c} \times SU(2)_L \times U(1)_Y$ are not understood either. When one uses the renormalization group to evolve the three gauge couplings – from the currently experimentally accessible energy scales all the way up to the GUT scale of about 10^{16} GeV – in a supersymmetric GUT extension of the Standard Model the couplings converge to one unified value. Hence, properly designed GUT theories are indeed able to relate the values of the gauge couplings $g_{\rm s}$, g, and g', or equivalently $\Lambda_{\rm QCD}$, $\sin^2 \theta_{\rm W}$, and α (cf. Chapter 26).

---

[2] Actually the value of α is "running", *i.e.* energy-scale-dependent, as well. As usual, we refer to its low-energy, or long-distance, limit.

Appendix C Structure of Minkowski Space–Time

Relativistic theories, such as Albert Einstein's special relativity (Einstein, 1905) or James Clerk Maxwell's electrodynamics (Clerk Maxwell, 1865), are invariant not only under spatial translations and rotations but also under Lorentz transformations that rotate spatial and temporal coordinates into one another. The resulting enlarged invariance is known as *Lorentz invariance* (against space–time rotations) or as *Poincaré invariance* against space–time rotations and translations.

C.1 Lorentz Transformations

After related work by Henri Poincaré, Hermann Minkowski was first to fully realize that in relativistic theories space and time (which are separate entities in Newtonian mechanics) are naturally united to space–time (Minkowski, 1908). A point in *Minkowski space–time* (which could specify an event) is described by four coordinates – one for time and three for space – which form a *4-vector*

$$x^\mu = \left(x^0, x^1, x^2, x^3\right) = (c\,t, \vec{x}), \quad x^0 = c\,t. \tag{C.1}$$

In particular, the time t (multiplied by the velocity of light c) plays the role of the 0-component of the 4-vector. Minkowski's space–time does not follow the rules of Euclidean geometry, and terms like "norm", "metric", "scalar product", and "distance" have an unconventional meaning, since they are not positive definite. Positivity does hold for the "distance" between events, which are causally connected.

In particular, the *norm* squared of the 4-vector x^μ is given by

$$s^2 = \left(x^0\right)^2 - \left(x^1\right)^2 - \left(x^2\right)^2 - \left(x^3\right)^2, \tag{C.2}$$

and may thus be negative. In addition to the *contra-variant* vector x^μ, it is useful to introduce the *co-variant* 4-vector

$$x_\mu = (x_0, x_1, x_2, x_3) = (c\,t, -\vec{x}). \tag{C.3}$$

Both the co- and the contra-variant 4-vectors contain the same physical information. Their components are simply related by

$$x_0 = x^0, \quad x_1 = -x^1, \quad x_2 = -x^2, \quad x_3 = -x^3. \tag{C.4}$$

The norm squared of the 4-vector can then be written as

$$\sum_{\mu=0}^{3} x_{\mu} x^{\mu} = x_{\mu} x^{\mu} = \left(x^0\right)^2 - \left(x^1\right)^2 - \left(x^2\right)^2 - \left(x^3\right)^2 = s^2. \tag{C.5}$$

Instead of frequently writing sums over space–time indices $\mu = 0, \ldots, 3$ explicitly, Einstein introduced a *summation convention* according to which repeated indices (one co- and one contra-variant index) will automatically be summed. In the second term of eq. (C.5), we have already used Einstein's summation convention.

The norm of a 4-vector induces a corresponding *metric* via

$$\left(x^0\right)^2 - \left(x^1\right)^2 - \left(x^2\right)^2 - \left(x^3\right)^2 = g_{\mu\nu} x^{\mu} x^{\nu}. \tag{C.6}$$

We have again used Einstein's summation convention by dropping the explicit sums over the repeated indices μ and ν. The metric tensor g, with its elements $g_{\mu\nu}$, is given by the 4×4 matrix

$$g = \begin{pmatrix} 1 & 0 & 0 & 0 \\ 0 & -1 & 0 & 0 \\ 0 & 0 & -1 & 0 \\ 0 & 0 & 0 & -1 \end{pmatrix}. \tag{C.7}$$

The metric tensor can also be used to relate co- and contra-variant 4-vectors by lowering a contra-variant index, *i.e.*,

$$x_{\mu} = g_{\mu\nu} x^{\nu}. \tag{C.8}$$

Again, the repeated index ν is summed over, whereas the unrepeated index μ is fixed. Let us also introduce the inverse metric g^{-1} with the components $g^{\mu\nu}$,

$$g^{-1} = \begin{pmatrix} 1 & 0 & 0 & 0 \\ 0 & -1 & 0 & 0 \\ 0 & 0 & -1 & 0 \\ 0 & 0 & 0 & -1 \end{pmatrix}, \tag{C.9}$$

which obeys

$$g g^{-1} = \mathbb{1}. \tag{C.10}$$

In components, this relation takes the form

$$g_{\mu\nu} g^{\nu\rho} = \delta_{\mu}^{\ \rho}, \tag{C.11}$$

where $\delta_{\mu}^{\ \rho}$ is the Kronecker symbol; *i.e.*, it represents the matrix elements of the unit-matrix $\mathbb{1}$. The inverse metric can now be used to raise co-variant indices, *e.g.*,

$$x^{\mu} = g^{\mu\nu} x_{\nu}. \tag{C.12}$$

Let us ask under what kind of rotations the norm squared of a 4-vector is invariant. The rotated 4-vector can be written as

$$x'^{\mu} = \Lambda^{\mu}_{\ \nu} x^{\nu}, \tag{C.13}$$

where Λ is a 4×4 space–time rotation matrix. Similarly, we obtain

$$x'_{\mu} = g_{\mu\nu} x'^{\nu} = g_{\mu\nu} \Lambda^{\nu}_{\ \rho} x^{\rho}. \tag{C.14}$$

The norm squared of x'^{μ} is then given by

$$s'^2 = x'_{\mu} x'^{\mu} = g_{\mu\nu} \Lambda^{\nu}_{\ \rho} x^{\rho} \Lambda^{\mu}_{\ \sigma} x^{\sigma}. \tag{C.15}$$

It is invariant under space–time rotations, *i.e.*, $s'^2 = s^2$, only if

$$g_{\mu\nu} \Lambda^{\nu}_{\ \rho} \Lambda^{\mu}_{\ \sigma} = g_{\rho\sigma}. \tag{C.16}$$

With a little algebra, this relation can be rewritten as

$$(\Lambda^{\mathsf{T}})^{\mu}_{\ \sigma} g_{\mu\nu} \Lambda^{\nu}_{\ \rho} = g^{\mathsf{T}}_{\sigma\rho}, \tag{C.17}$$

or equivalently as the matrix multiplication

$$\Lambda^{\mathsf{T}} g \Lambda = g^{\mathsf{T}} = g. \tag{C.18}$$

This condition is the Minkowski space–time analogue of the Euclidean space condition for orthogonal spatial rotations, $\Omega^{\mathsf{T}} \Omega = \mathbb{1}$. One now obtains

$$x'_{\mu} = g_{\mu\nu} \Lambda^{\nu}_{\ \rho} x^{\rho} = g_{\mu\nu} \Lambda^{\nu}_{\ \rho} g^{\rho\sigma} x_{\sigma} = \left[g \Lambda g^{-1} \right]^{\sigma}_{\ \mu} x_{\sigma} = x_{\sigma} \left(\left[g \Lambda g^{-1} \right]^{\mathsf{T}} \right)^{\sigma}_{\ \mu} = x_{\sigma} (\Lambda^{-1})^{\sigma}_{\ \mu}. \tag{C.19}$$

Here, we have used eq. (C.18), which leads to

$$\left[g \Lambda g^{-1} \right]^{\mathsf{T}} = g^{-1} \Lambda^{\mathsf{T}} g = \Lambda^{-1}. \tag{C.20}$$

Finally, as a consequence of eq. (C.19), we obtain

$$x_{\nu} = x'_{\mu} \Lambda^{\mu}_{\ \nu}. \tag{C.21}$$

Space–time rotations which obey eq. (C.18) are known as *Lorentz transformations*.

The distance squared in space–time between two 4-vectors x^{μ} and y^{μ} is given by

$$(\Delta s)^2 = \left(x^0 - y^0 \right)^2 - \left(x^1 - y^1 \right)^2 - \left(x^2 - y^2 \right)^2 - \left(x^3 - y^3 \right)^2, \tag{C.22}$$

and may again be negative. This distance is invariant under both Lorentz transformations Λ and space–time translations d^{μ}, *i.e.*, $(\Delta s')^2 = (\Delta s)^2$, with

$$x'^{\mu} = \Lambda^{\mu}_{\ \nu} x^{\nu} + d^{\mu}, \quad y'^{\mu} = \Lambda^{\mu}_{\ \nu} y^{\nu} + d^{\mu}. \tag{C.23}$$

Lorentz transformations and space–time translations form another group – the 10-dimensional *Poincaré group* – which contains the 6-dimensional Lorentz group as a subgroup (cf. Appendix E).

C.2 Gradient as a 4-Vector and d'Alembert Operator

In order to construct field theories in manifestly Lorentz-invariant form, we also need to combine temporal and spatial derivatives to a 4-vector. Let us introduce

$$\partial^\mu = \left(\partial^0, \partial^1, \partial^2, \partial^3\right) = \left(\frac{\partial}{\partial x_0}, \frac{\partial}{\partial x_1}, \frac{\partial}{\partial x_2}, \frac{\partial}{\partial x_3}\right) = \left(\frac{1}{c}\partial_t, -\vec{\nabla}\right). \tag{C.24}$$

How does this operator transform under Lorentz transformations? Using eq. (C.21), one obtains

$$\partial^{\mu'} = \frac{\partial}{\partial x'_\mu} = \frac{\partial x_\nu}{\partial x'_\mu}\frac{\partial}{\partial x_\nu} = \Lambda^\mu_{\ \nu}\partial^\nu. \tag{C.25}$$

This shows that ∂^μ indeed transforms as a contra-variant 4-vector. Similarly, one can define the co-variant 4-vector

$$\partial_\mu = \left(\partial_0, \partial_1, \partial_2, \partial_3\right) = \left(\frac{\partial}{\partial x^0}, \frac{\partial}{\partial x^1}, \frac{\partial}{\partial x^2}, \frac{\partial}{\partial x^3}\right) = \left(\frac{1}{c}\partial_t, \vec{\nabla}\right). \tag{C.26}$$

Now we form the *scalar product* of the co- and contra-variant derivative 4-vectors. In this way, we obtain a second-derivative operator which transforms as a space–time scalar; *i.e.*, it is invariant under Lorentz transformations. This Minkowski space–time analogue of the Laplace operator in Euclidean space is known as the *d'Alembert operator*,

$$\Box = \partial_\mu \partial^\mu = \frac{\partial^2}{\partial x_0^2} - \frac{\partial^2}{\partial x_1^2} - \frac{\partial^2}{\partial x_2^2} - \frac{\partial^2}{\partial x_3^2}. \tag{C.27}$$

For example, we can write the continuity equation of some 4-vector current $j^\mu = (j^0, \vec{j})$ as $\partial_\mu j^\mu = \partial^\mu j_\mu = 0$. Similarly, the exponent in a Minkowski space–time Fourier transform is proportional to $p^\mu x_\mu = p_\mu x^\mu$, with the 4-momentum $p^\mu = (E/c, \vec{p})$.

Appendix D   Relativistic Formulation of Classical Electrodynamics

The relativistic nature of Maxwell's equations is not evident in their original form. In this appendix, we formulate electrodynamics such that its invariance under Lorentz transformations – *i.e.* under rotations in Minkowski space–time – becomes evident. In order to better see the connection with the Maxwell equations in their original form, in this appendix we do write the speed of light, c, rather than setting $c = 1$.

D.1   Current and Vector Potential

In order to express electrodynamics in a manifestly Lorentz co-variant form, we proceed step by step and begin with the charge and current densities $\rho(\vec{x}, t)$ and $\vec{j}(\vec{x}, t)$. The corresponding *continuity equation*, which expresses charge conservation, takes the form

$$\partial_t \rho(\vec{x}, t) + \vec{\nabla} \cdot \vec{j}(\vec{x}, t) = 0. \tag{D.1}$$

Since temporal and spatial derivatives are combined to a gradient 4-vector, it is natural to combine $\rho(\vec{x}, t)$ and $\vec{j}(\vec{x}, t)$ to the current 4-vector

$$j^{\mu}(x) = \left(c\rho(\vec{x}, t), j_x(\vec{x}, t), j_y(\vec{x}, t), j_z(\vec{x}, t) \right). \tag{D.2}$$

Here, we have introduced $x = (ct, \vec{x})$ as a short-hand notation for a point in space–time. Of course, this should not be confused with the x-component of the spatial vector $\vec{x}$. In Lorentz-invariant form, the continuity equation thus takes the form

$$\partial_{\mu} j^{\mu}(x) = \frac{1}{c} \partial_t c\rho(\vec{x}, t) + \partial_x j_x(\vec{x}, t) + \partial_y j_y(\vec{x}, t) + \partial_z j_z(\vec{x}, t) = 0. \tag{D.3}$$

Here, we have combined the co-variant 4-vector ∂_{μ} and the contra-variant 4-vector $j^{\mu}(x)$ to the Lorentz scalar 0. The Lorentz invariance of the continuity equation implies that charge conservation is valid in any inertial frame, independent of the motion of an observer.

Of course, the charge and current densities themselves depend on the reference frame. If a general (non-uniform and non-static) charge and current density is transformed into another reference frame, one must also transform the space–time point x at which the density is evaluated, *i.e.*

$$j'^{\mu}(x') = \Lambda^{\mu}_{\ \nu} j^{\nu}(x) = \Lambda^{\mu}_{\ \nu} j^{\nu}(\Lambda^{-1} x'). \tag{D.4}$$

From the scalar potential $\phi(\vec{x}, t)$ and the vector potential $\vec{A}(\vec{x}, t)$, one can construct another 4-vector field

$$A^\mu(x) = (\phi(\vec{x}, t), A_x(\vec{x}, t), A_y(\vec{x}, t), A_z(\vec{x}, t)). \tag{D.5}$$

Under Lorentz transformations, it again transforms as

$$A'^\mu(x') = \Lambda^\mu_{\ \nu} A^\nu(\Lambda^{-1} x'). \tag{D.6}$$

Scalar and vector potentials change non-trivially under gauge transformations

$$^\alpha\phi(\vec{x}, t) = \phi(\vec{x}, t) - \frac{1}{c}\partial_t\alpha(\vec{x}, t), \quad {}^\alpha\vec{A}(\vec{x}, t) = \vec{A}(\vec{x}, t) + \vec{\nabla}\alpha(\vec{x}, t). \tag{D.7}$$

In 4-vector notation, this relation takes the form

$$^\alpha A^\mu(x) = A^\mu(x) - \partial^\mu\alpha(x). \tag{D.8}$$

Here, $\alpha(x)$ is an arbitrary (differentiable) space–time-dependent gauge transformation function. This function is a space–time scalar; *i.e.* under Lorentz transformations, it transforms as

$$\alpha'(x') = \alpha(\Lambda^{-1}x'). \tag{D.9}$$

The Lorenz gauge[1] or Landau gauge fixing condition

$$\frac{1}{c}\partial_t\phi(\vec{x}, t) + \vec{\nabla}\cdot\vec{A}(\vec{x}, t) = 0 \tag{D.10}$$

can hence be expressed as

$$\partial_\mu A^\mu(x) = 0. \tag{D.11}$$

In the Lorenz–Landau gauge, the wave equations take the form

$$\frac{1}{c^2}\partial_t^2\phi(\vec{x}, t) - \Delta\phi(\vec{x}, t) = \rho(\vec{x}, t), \quad \frac{1}{c^2}\partial_t^2\vec{A}(\vec{x}, t) - \Delta\vec{A}(\vec{x}, t) = \frac{1}{c}\vec{j}(\vec{x}, t), \tag{D.12}$$

which can be combined to

$$\Box A^\mu(x) = \frac{1}{c}j^\mu(x). \tag{D.13}$$

D.2 Field Strength Tensor

It may not be entirely obvious how to express the electric and magnetic field, $\vec{E}(\vec{x}, t)$ and $\vec{B}(\vec{x}, t)$, in a relativistic form. We need to use

$$\vec{E}(\vec{x}, t) = -\vec{\nabla}\phi(\vec{x}, t) - \frac{1}{c}\partial_t\vec{A}(\vec{x}, t), \quad \vec{B}(\vec{x}, t) = \vec{\nabla}\times\vec{A}(\vec{x}, t). \tag{D.14}$$

---

[1] This term is named after the Danish physicist Ludvig Lorenz (1829–1891). He should not be confused with the Dutch physicist Hendrik Lorentz (1853–1928); the Lorentz transformation and Lorentz group refer to his name.

It is now obvious that $\vec{E}(\vec{x}, t)$ and $\vec{B}(\vec{x}, t)$ are constructed form the 4-vectors ∂^μ and $A^\mu(x)$. The scalar product of these two 4-vectors

$$\partial_\mu A^\mu(x) = \frac{1}{c}\partial_t\phi(\vec{x}, t) + \vec{\nabla} \cdot \vec{A}(\vec{x}, t) \tag{D.15}$$

appears in the Lorenz–Landau gauge fixing condition, but it does not yield the electric or magnetic field. The 4-vectors ∂^μ and $A^\mu(x)$ can also be combined to the symmetric tensor

$$D^{\mu\nu}(x) = \partial^\mu A^\nu(x) + \partial^\nu A^\mu(x), \tag{D.16}$$

as well as to the anti-symmetric tensor

$$F^{\mu\nu}(x) = \partial^\mu A^\nu(x) - \partial^\nu A^\mu(x). \tag{D.17}$$

Under gauge transformations, these tensors transform as

$$
\begin{aligned}
{}^\alpha D^{\mu\nu}(x) &= \partial^\mu\,{}^\alpha A^\nu(x) + \partial^\nu\,{}^\alpha A^\mu(x) \\
&= \partial^\mu A^\nu(x) - \partial^\mu\partial^\nu\alpha(x) + \partial^\nu A^\mu(x) - \partial^\nu\partial^\mu\alpha(x) = D^{\mu\nu}(x) - 2\partial^\mu\partial^\nu\alpha(x), \\
{}^\alpha F^{\mu\nu}(x) &= \partial^\mu\,{}^\alpha A^\nu(x) - \partial^\nu\,{}^\alpha A^\mu(x) \\
&= \partial^\mu A^\nu(x) - \partial^\mu\partial^\nu\alpha(x) - \partial^\nu A^\mu(x) + \partial^\nu\partial^\mu\alpha(x) = F^{\mu\nu}(x), \tag{D.18}
\end{aligned}
$$

i.e. the anti-symmetric tensor $F^{\mu\nu}$ is gauge invariant, whereas the symmetric tensor $D^{\mu\nu}$ is not. As a consequence, $D^{\mu\nu}$ does not play any particular role in electrodynamics. Since the electromagnetic fields are gauge invariant, we expect them to be related to $F^{\mu\nu}$. Let us consider the components of this tensor

$$
\begin{aligned}
F^{01}(x) &= \partial^0 A^1(x) - \partial^1 A^0(x) = \frac{1}{c}\partial_t A_x(\vec{x}, t) + \partial_x\phi(\vec{x}, t) = -E_x(\vec{x}, t), \\
F^{02}(x) &= \partial^0 A^2(x) - \partial^2 A^0(x) = \frac{1}{c}\partial_t A_y(\vec{x}, t) + \partial_y\phi(\vec{x}, t) = -E_y(\vec{x}, t), \\
F^{03}(x) &= \partial^0 A^3(x) - \partial^3 A^0(x) = \frac{1}{c}\partial_t A_z(\vec{x}, t) + \partial_z\phi(\vec{x}, t) = -E_z(\vec{x}, t), \\
F^{12}(x) &= \partial^1 A^2(x) - \partial^2 A^1(x) = -\partial_x A_y(\vec{x}, t) + \partial_y A_x(\vec{x}, t) = -B_z(\vec{x}, t), \\
F^{23}(x) &= \partial^2 A^3(x) - \partial^3 A^2(x) = -\partial_y A_z(\vec{x}, t) + \partial_z A_y(\vec{x}, t) = -B_x(\vec{x}, t), \\
F^{31}(x) &= \partial^3 A^1(x) - \partial^1 A^3(x) = -\partial_z A_x(\vec{x}, t) + \partial_x A_z(\vec{x}, t) = -B_y(\vec{x}, t). \tag{D.19}
\end{aligned}
$$

Hence, the anti-symmetric *field strength tensor* indeed contains the electric and magnetic fields as

$$
F^{\mu\nu}(x) = \begin{pmatrix}
0 & -E_x(\vec{x}, t) & -E_y(\vec{x}, t) & -E_z(\vec{x}, t) \\
E_x(\vec{x}, t) & 0 & -B_z(\vec{x}, t) & B_y(\vec{x}, t) \\
E_y(\vec{x}, t) & B_z(\vec{x}, t) & 0 & -B_x(\vec{x}, t) \\
E_z(\vec{x}, t) & -B_y(\vec{x}, t) & B_x(\vec{x}, t) & 0
\end{pmatrix}. \tag{D.20}
$$

The co-variant components of this tensor are given by

$$
F_{\mu\nu}(x) = \begin{pmatrix}
0 & E_x(\vec{x},t) & E_y(\vec{x},t) & E_z(\vec{x},t) \\
-E_x(\vec{x},t) & 0 & -B_z(\vec{x},t) & B_y(\vec{x},t) \\
-E_y(\vec{x},t) & B_z(\vec{x},t) & 0 & -B_x(\vec{x},t) \\
-E_z(\vec{x},t) & -B_y(\vec{x},t) & B_x(\vec{x},t) & 0
\end{pmatrix}. \tag{D.21}
$$

D.3 Maxwell Equations

Let us first consider the *inhomogeneous* Maxwell equations

$$
\vec{\nabla} \cdot \vec{E}(\vec{x},t) = \rho(\vec{x},t), \quad \vec{\nabla} \times \vec{B}(\vec{x},t) - \frac{1}{c}\partial_t\vec{E}(\vec{x},t) = \frac{1}{c}\vec{j}(\vec{x},t). \tag{D.22}
$$

These are four equations with the components of the 4-vector current $j^\mu(x)$ on the right-hand side. Hence, on the left-hand side there must also be a 4-vector. The left-hand side consists of derivatives, *i.e.* of components of the gradient 4-vectors ∂_μ, and of the electromagnetic fields, *i.e.* of the components of the field strength tensor $F^{\mu\nu}$. Hence, the 4-vector ∂_μ and the tensor $F^{\mu\nu}$ on the left-hand side must be combined to another 4-vector. This can be achieved by forming $\partial_\mu F^{\mu\nu}(x)$ and thus by contracting (*i.e.* by summing) one co- and one contra-variant index. The components of this term take the form

$$
\begin{aligned}
\partial_\mu F^{\mu 0}(x) &= \partial_x E_x(\vec{x},t) + \partial_y E_y(\vec{x},t) + \partial_z E_z(\vec{x},t) \\
&= \vec{\nabla} \cdot \vec{E}(\vec{x},t) = \rho(\vec{x},t), \\
\partial_\mu F^{\mu 1}(x) &= -\frac{1}{c}\partial_t E_x(\vec{x},t) + \partial_y B_z(\vec{x},t) - \partial_z B_y(\vec{x},t) \\
&= \left[\vec{\nabla} \times \vec{B}\right]_x (\vec{x},t) - \frac{1}{c}\partial_t E_x(\vec{x},t) = \frac{1}{c}j_x(\vec{x},t), \\
\partial_\mu F^{\mu 2}(x) &= -\frac{1}{c}\partial_t E_y(\vec{x},t) - \partial_x B_z(\vec{x},t) + \partial_z B_x(\vec{x},t) \\
&= \left[\vec{\nabla} \times \vec{B}\right]_y (\vec{x},t) - \frac{1}{c}\partial_t E_y(\vec{x},t) = \frac{1}{c}j_y(\vec{x},t), \\
\partial_\mu F^{\mu 3}(x) &= -\frac{1}{c}\partial_t E_z(\vec{x},t) + \partial_x B_y(\vec{x},t) - \partial_y B_x(\vec{x},t) \\
&= \left[\vec{\nabla} \times \vec{B}\right]_z (\vec{x},t) - \frac{1}{c}\partial_t E_z(\vec{x},t) = \frac{1}{c}j_z(\vec{x},t).
\end{aligned} \tag{D.23}
$$

Indeed, these equations can be summarized as

$$
\partial_\mu F^{\mu\nu}(x) = \frac{1}{c}j^\nu(x). \tag{D.24}
$$

At this point, the usefulness of the compact 4-dimensional notation should be obvious. Inserting eq. (D.17) into the inhomogeneous Maxwell equations, we obtain

$$
\partial_\mu F^{\mu\nu}(x) = \partial_\mu\left(\partial^\mu A^\mu(x) - \partial^\nu A^\mu(x)\right) = \Box A^\nu(x) - \partial^\nu\partial_\mu A^\mu(x) = \frac{1}{c}j^\nu(x). \tag{D.25}
$$

If the 4-vector potential obeys the Lorenz–Landau gauge fixing condition $\partial_\mu A^\mu(x) = 0$, this is just the wave equation (D.13).

How can we express the *homogeneous* Maxwell equations

$$\vec{\nabla} \cdot \vec{B}(\vec{x}, t) = 0, \quad \vec{\nabla} \times \vec{E}(\vec{x}, t) + \frac{1}{c}\partial_t \vec{B}(\vec{x}, t) = 0, \tag{D.26}$$

in 4-dimensional form? Except for the vanishing right-hand side, they look very similar to the inhomogeneous equations. All we need to do is to substitute

$$\vec{E}(\vec{x}, t) \longrightarrow -\vec{B}(\vec{x}, t), \quad \vec{B}(\vec{x}, t) \longrightarrow \vec{E}(\vec{x}, t). \tag{D.27}$$

Such a substitution is known as a *duality transformation*. Under this operation, the field strength tensor turns into the *dual tensor*

$$\widetilde{F}^{\mu\nu}(x) = \begin{pmatrix} 0 & B_x(\vec{x}, t) & B_y(\vec{x}, t) & B_z(\vec{x}, t) \\ -B_x(\vec{x}, t) & 0 & -E_z(\vec{x}, t) & E_y(\vec{x}, t) \\ -B_y(\vec{x}, t) & E_z(\vec{x}, t) & 0 & -E_x(\vec{x}, t) \\ -B_z(\vec{x}, t) & -E_y(\vec{x}, t) & E_x(\vec{x}, t) & 0 \end{pmatrix}, \tag{D.28}$$

and the homogeneous Maxwell equations can thus be expressed as

$$\partial_\mu \widetilde{F}^{\mu\nu}(x) = 0. \tag{D.29}$$

The co-variant components of the dual field strength tensor take the form

$$\widetilde{F}_{\mu\nu}(x) = \begin{pmatrix} 0 & -B_x(\vec{x}, t) & -B_y(\vec{x}, t) & -B_z(\vec{x}, t) \\ B_x(\vec{x}, t) & 0 & -E_z(\vec{x}, t) & E_y(\vec{x}, t) \\ B_y(\vec{x}, t) & E_z(\vec{x}, t) & 0 & -E_x(\vec{x}, t) \\ B_z(\vec{x}, t) & -E_y(\vec{x}, t) & E_x(\vec{x}, t) & 0 \end{pmatrix}. \tag{D.30}$$

We see that the field strength tensor $F^{\mu\nu}(x)$ and its dual $\widetilde{F}_{\mu\nu}(x)$ consist of the same components, namely, of the electric and magnetic fields $\vec{E}(\vec{x}, t)$ and $\vec{B}(\vec{x}, t)$. Hence, there must be a relation between the two tensors. This relation reads

$$\widetilde{F}_{\mu\nu}(x) = \frac{1}{2}\epsilon_{\mu\nu\rho\sigma}F^{\rho\sigma}(x). \tag{D.31}$$

Here, $\epsilon_{\mu\nu\rho\sigma}$ is a totally anti-symmetric object with elements 0 or ± 1, which is known as the *Levi-Civita symbol*. If any of the indices μ, ν, ρ, and σ are equal, the value of $\epsilon_{\mu\nu\rho\sigma}$ is zero. Only if all indices are different, the value of $\epsilon_{\mu\nu\rho\sigma}$ is non-zero. The value is $\epsilon_{\mu\nu\rho\sigma} = 1$ if $\mu\nu\rho\sigma$ is an even permutation of 0123 (*i.e.* it requires an even number of index pair permutations to turn $\mu\nu\rho\sigma$ into 0123). Similarly, $\epsilon_{\mu\nu\rho\sigma} = -1$ if $\mu\nu\rho\sigma$ is an odd permutation of 0123. For example, we obtain

$$\widetilde{F}_{01}(x) = \frac{1}{2}\epsilon_{01\rho\sigma}F^{\rho\sigma}(x) = \frac{1}{2}\left(\epsilon_{0123}F^{23}(x) - \epsilon_{0132}F^{32}(x)\right) = F^{23}(x) = -B_x(\vec{x}, t), \tag{D.32}$$

as well as

$$\tilde{F}_{12}(x) = \frac{1}{2}\epsilon_{12\rho\sigma}F^{\rho\sigma}(x) = \frac{1}{2}\left(\epsilon_{1203}F^{03}(x) - \epsilon_{1230}F^{30}(x)\right) = F^{03}(x) = -E_z(\vec{x}, t), \quad (D.33)$$

Inserting eq. (D.31) into the homogeneous Maxwell equations (D.29), one obtains

$$\partial^{\mu}\tilde{F}_{\mu\nu}(x) = \frac{1}{2}\epsilon_{\mu\nu\rho\sigma}\partial^{\mu}F^{\rho\sigma}(x) = \frac{1}{2}\epsilon_{\mu\nu\rho\sigma}\partial^{\mu}\left(\partial^{\rho}A^{\sigma}(x) - \partial^{\sigma}A^{\rho}(x)\right) = 0. \quad (D.34)$$

Due to the anti-symmetry of $\epsilon_{\mu\nu\rho\sigma}$ and the commutativity of the derivatives ∂^{μ} and ∂^{ρ}, this equation is automatically satisfied. This is no surprise, because the original Maxwell equations were also automatically satisfied by the introduction of the scalar and vector potentials $\phi(\vec{x}, t)$ and $\vec{A}(\vec{x}, t)$.

The homogeneous Maxwell equations can alternatively be expressed as

$$\partial^{\mu}F^{\rho\sigma}(x) + \partial^{\rho}F^{\sigma\mu}(x) + \partial^{\sigma}F^{\mu\rho}(x) = 0, \quad (D.35)$$

which is a variant of the *Bianchi identity*. Indeed, multiplying this relation with $\epsilon_{\mu\nu\rho\sigma}$ and applying cyclic permutations to the indices μ, ρ, and σ, one again arrives at eq. (D.34).

D.4 Space–Time Scalars from Field Strength Tensors

Which scalar quantities can be formed by combining the field strength tensors $F^{\mu\nu}(x)$ and $\tilde{F}^{\mu\nu}(x)$? First, we construct the combination

$$\frac{1}{4}F_{\mu\nu}(x)F^{\mu\nu}(x) = \frac{1}{2}\left(\vec{B}(\vec{x}, t)^2 - \vec{E}(\vec{x}, t)^2\right), \quad (D.36)$$

which will later turn out to be the Lagrangian (or Lagrange density) of electrodynamics. Then, we can construct

$$\frac{1}{4}\tilde{F}_{\mu\nu}(x)\tilde{F}^{\mu\nu}(x) = \frac{1}{2}\left(\vec{E}(\vec{x}, t)^2 - \vec{B}(\vec{x}, t)^2\right), \quad (D.37)$$

which is thus the same up to a minus sign. While the electromagnetic fields themselves are obviously not Lorentz-invariant, the difference of their magnitudes squared is invariant.

One can also mix the field strength tensor with its dual,

$$\frac{1}{4}F_{\mu\nu}(x)\tilde{F}^{\mu\nu}(x) = \vec{E}(\vec{x}, t) \cdot \vec{B}(\vec{x}, t) = \frac{1}{4}\tilde{F}_{\mu\nu}(x)F^{\mu\nu}(x). \quad (D.38)$$

Interestingly, the projection of the electric on the magnetic field $\vec{E}(\vec{x}, t) \cdot \vec{B}(\vec{x}, t)$ is also Lorentz-invariant; *i.e.* it has the same value in all inertial frames.

D.5 Transformation of Electromagnetic Fields

We have seen that the electromagnetic fields $\vec{E}(\vec{x}, t)$ and $\vec{B}(\vec{x}, t)$ are not Lorentz-invariant; *i.e.* they depend on the motion of the observer, since they form the components of the field strength tensor $F^{\mu\nu}$. Under a Lorentz transformation, this tensor transforms as

$$F^{\mu\nu\prime}(x') = \Lambda^{\mu}{}_{\rho}\Lambda^{\nu}{}_{\sigma}F^{\rho\sigma}(x) = \Lambda^{\mu}{}_{\rho}\Lambda^{\nu}{}_{\sigma}F^{\rho\sigma}(\Lambda^{-1}x'). \quad (D.39)$$

This can be rewritten in matrix form as

$$F^{\mu\nu\prime}(x') = \Lambda^{\mu}_{\ \rho} F^{\rho\sigma}(\Lambda^{-1}x')\Lambda^{\mathsf{T}\ \nu}_{\ \ \sigma} \quad\Rightarrow\quad F'(x') = \Lambda F(\Lambda^{-1}x')\Lambda^{\mathsf{T}}. \qquad \text{(D.40)}$$

D.6 Action and Euler–Lagrange Equation

The Lagrangian for the electromagnetic field, interacting with a charge and current distribution j^{ν}, is given by

$$\mathcal{L}(A) = -\frac{1}{4}F_{\mu\nu}F^{\mu\nu} - \frac{1}{c}A_{\nu}j^{\nu}. \qquad \text{(D.41)}$$

The corresponding action is obtained by integrating the Lagrangian over space–time,

$$S[A] = \int dt\, d^3x\, \mathcal{L}(A) = \int d^4x\, \frac{1}{c}\left(-\frac{1}{4}F_{\mu\nu}F^{\mu\nu} - \frac{1}{c}A_{\nu}j^{\nu}\right). \qquad \text{(D.42)}$$

The action is a functional (*i.e.* a function that itself depends on a whole function, not just on its value at one point) of the electromagnetic 4-vector potential A^{μ}. The Euler–Lagrange equation of motion resulting from the principle of least action now takes the form

$$\partial_{\mu}\frac{\delta\mathcal{L}}{\delta\partial_{\mu}A_{\nu}(x)} - \frac{\delta\mathcal{L}}{\delta A_{\nu}(x)} = \partial_{\mu}F^{\mu\nu}(x) - \frac{1}{c}j^{\nu}(x) = 0. \qquad \text{(D.43)}$$

Indeed, this yields just the inhomogeneous Maxwell equations

$$\partial_{\mu}F^{\mu\nu}(x) = \frac{1}{c}j^{\nu}(x). \qquad \text{(D.44)}$$

The homogeneous Maxwell equations are automatically satisfied as a consequence of the definition $F^{\mu\nu} = \partial^{\mu}A^{\nu} - \partial^{\nu}A^{\mu}$.

D.7 Energy–Momentum Tensor

Let us consider the *energy–momentum tensor* of the free electromagnetic field (*i.e.* in the absence of charges and currents),

$$\mathcal{T}^{\mu\nu}(x) = -F^{\mu\rho}(x)F^{\nu}_{\ \rho}(x) - \mathcal{L}g^{\mu\nu}, \qquad \text{(D.45)}$$

which obeys the continuity equation

$$\partial_{\mu}\mathcal{T}^{\mu\nu}(x) = 0. \qquad \text{(D.46)}$$

The time–time component of the energy–momentum tensor is given by

$$\mathcal{T}^{00}(x) = -F^{0\rho}(x)F^0_{\ \rho}(x) + \frac{1}{4}F_{\rho\sigma}(x)F^{\rho\sigma}(x)g^{00}$$
$$= \vec{E}(\vec{x},t)^2 + \frac{1}{2}\left(\vec{B}(\vec{x},t)^2 - \vec{E}(\vec{x},t)^2\right) = \frac{1}{2}\left(\vec{E}(\vec{x},t)^2 + \vec{B}(\vec{x},t)^2\right), \quad \text{(D.47)}$$

which is the energy density of the electromagnetic field. The space–time components of the energy–momentum tensor take the form

$$\mathcal{T}^{i0}(x) = -F^{i\rho}(x)F^0_{\ \rho}(x) + \frac{1}{4}F_{\rho\sigma}(x)F^{\rho\sigma}(x)g^{i0}$$
$$= \epsilon_{ijk}E_j(\vec{x},t)B_k(\vec{x},t) = \left[\vec{E}(\vec{x},t) \times \vec{B}(\vec{x},t)\right]_i. \quad \text{(D.48)}$$

Here, ϵ_{ijk} is again a completely anti-symmetric Levi-Civita symbol, and $\vec{E} \times \vec{B}$ is the *Poynting vector* which is known to represent the momentum density of the electromagnetic field. The continuity equation that represents energy conservation takes the form

$$\partial_\mu \mathcal{T}^{\mu 0}(x) = \partial_0 \mathcal{T}^{00}(x) + \partial_i \mathcal{T}^{i0}(x) = 0. \quad \text{(D.49)}$$

Appendix E From the Galilei to the Poincaré Algebra

In this appendix, we contrast the non-relativistic Galilei algebra that describes the symmetries of space and time underlying Newtonian physics, with the relativistic Poincaré algebra that describes the symmetries of Minkowski space–time.

E.1 Galilei Algebra

In Newtonian physics, one assumes space to be homogeneous and isotropic, and time to be absolute. The symmetry of the resulting space–time is known as *Galilei invariance*. The *Galilei algebra* is generated by infinitesimal translations and rotations in space and time. While the Hamilton operator of a quantum system $\hat{H}$ generates infinitesimal translations in time, its momentum operator $\hat{\vec{P}}$ generates translations in space. Similarly, the angular momentum operator $\hat{\vec{J}}$ generates infinitesimal spatial rotations and the Galilean boost operator $\hat{\vec{G}}$ generates transitions between different inertial frames.

The Galilei algebra is characterized by the following commutation relations

$$\left[\hat{P}_i, \hat{H}\right] = 0, \quad \left[\hat{J}_i, \hat{H}\right] = 0, \quad \left[\hat{G}_i, \hat{H}\right] = i\hat{P}_i, \quad \left[\hat{P}_i, \hat{P}_j\right] = 0, \quad \left[\hat{J}_i, \hat{P}_j\right] = i\epsilon_{ijk}\hat{P}_k,$$

$$\left[\hat{G}_i, \hat{P}_j\right] = i\delta_{ij}\mathcal{M}, \quad \left[\hat{J}_i, \hat{J}_j\right] = i\epsilon_{ijk}\hat{J}_k, \quad \left[\hat{J}_i, \hat{G}_j\right] = i\epsilon_{ijk}\hat{G}_k, \quad \left[\hat{G}_i, \hat{G}_j\right] = 0. \quad \text{(E.1)}$$

Here, $\mathcal{M}$ denotes the total mass of the physical system, which is conserved at the non-relativistic level. The commutation relations that involve the angular momentum operator $\hat{\vec{J}}$ determine whether an object is a scalar or a vector under spatial rotations. The Hamilton operator $\hat{H}$ is a scalar and thus commutes with $\hat{\vec{J}}$. An arbitrary vector $\hat{\vec{V}}$, on the other hand, transforms as

$$\left[\hat{J}_i, \hat{V}_j\right] = i\epsilon_{ijk}\hat{V}_k. \quad \text{(E.2)}$$

Indeed, the vectors $\hat{\vec{P}}$, $\hat{\vec{J}}$, and $\hat{\vec{G}}$ have just this commutation behavior with the components of $\hat{\vec{J}}$.

E.2 Poincaré and Lorentz Algebras

In contrast to non-relativistic Newtonian physics, the physics of special relativity takes place in Minkowski space–time. The corresponding *Lorentz boost* operators $\hat{\vec{K}}$ are generators of infinitesimal space–time rotations, which act differently from the Galilei boosts $\overset{\scriptscriptstyle\triangle}{G}$. The *Poincaré algebra* generates space–time translations and rotations, with the latter forming the *Lorentz subalgebra*. The commutation relations of the Poincaré algebra take the form

$$\left[\hat{P}_i, \hat{H}\right] = 0, \quad \left[\hat{J}_i, \hat{H}\right] = 0, \quad \left[\hat{K}_i, \hat{H}\right] = \mathrm{i}\hat{P}_i, \quad \left[\hat{P}_i, \hat{P}_j\right] = 0, \quad \left[\hat{J}_i, \hat{P}_j\right] = \mathrm{i}\epsilon_{ijk}\hat{P}_k,$$

$$\left[\hat{K}_i, \hat{P}_j\right] = \mathrm{i}\delta_{ij}\hat{H}, \quad \left[\hat{J}_i, \hat{J}_j\right] = \mathrm{i}\epsilon_{ijk}\hat{J}_k, \quad \left[\hat{J}_i, \hat{K}_j\right] = \mathrm{i}\epsilon_{ijk}\hat{K}_k, \quad \left[\hat{K}_i, \hat{K}_j\right] = -\mathrm{i}\epsilon_{ijk}\hat{J}_k. \quad (\text{E.3})$$

We see that only the commutators $\left[\hat{K}_i, \hat{P}_j\right]$ and $\left[\hat{K}_i, \hat{K}_j\right]$ differ from the corresponding ones for the Galilei boosts $\overset{\scriptscriptstyle\triangle}{G}$. In particular, compared to $\left[\hat{G}_i, \hat{P}_j\right] = \mathrm{i}\delta_{ij}\mathcal{M}$, the mass $\mathcal{M}$ in the Galilean algebra is replaced by the Hamilton operator $\hat{H}$ in the Poincaré algebra. This has important physical consequences. For example, one can no longer add an arbitrary constant to the Hamilton operator (a kind of "cosmological constant"), without adjusting the boost $\overset{\scriptscriptstyle\triangle}{K}$ accordingly.

The Hamilton operator and the spatial momentum form the momentum 4-vector

$$\hat{P}_\mu = \left(\hat{H}, \hat{P}_1, \hat{P}_2, \hat{P}_3\right), \tag{E.4}$$

which generates infinitesimal space–time translations. Similarly, the angular momenta and boosts combine to the anti-symmetric tensor

$$\hat{M}_{\mu\nu} = \begin{pmatrix} 0 & \hat{K}_1 & \hat{K}_2 & \hat{K}_3 \\ -\hat{K}_1 & 0 & \hat{J}_3 & -\hat{J}_2 \\ -\hat{K}_2 & -\hat{J}_3 & 0 & \hat{J}_1 \\ -\hat{K}_3 & \hat{J}_2 & -\hat{J}_1 & 0 \end{pmatrix}, \tag{E.5}$$

which generates the infinitesimal space–time rotations that form the Lorentz subalgebra. A manifestly covariant form of the Poincaré algebra is given by

$$\left[\hat{P}_\mu, \hat{P}_\nu\right] = 0, \quad \left[\hat{M}_{\mu\nu}, \hat{P}_\rho\right] = \mathrm{i}\left(g_{\nu\rho}\hat{P}_\mu - g_{\mu\rho}\hat{P}_\nu\right),$$

$$\left[\hat{M}_{\mu\nu}, \hat{M}_{\rho\sigma}\right] = \mathrm{i}\left(g_{\mu\sigma}\hat{M}_{\nu\rho} - g_{\mu\rho}\hat{M}_{\nu\sigma} - g_{\nu\sigma}\hat{M}_{\mu\rho} + g_{\nu\rho}\hat{M}_{\mu\sigma}\right), \tag{E.6}$$

where $g_{\mu\nu} = \mathrm{diag}(1, -1, -1, -1)$ is the diagonal metric tensor of flat Minkowski space–time. The metric is used to raise or lower tensor indices, for example, $\hat{V}_\mu = g_{\mu\nu}\hat{V}^\nu$, and to define the scalar product $\hat{V}_\mu\hat{V}^\mu = g_{\mu\nu}\hat{V}^\nu\hat{V}^\mu$.

Let us investigate the *Casimir operators* of the Poincaré algebra: By definition, they commute with all ten generators and correspond to Lorentz-invariant physical quantities; *i.e.* they are the same in all inertial frames. It is easy to convince oneself that the operator that describes the squared rest mass[1]

---

[1] Of course, the "relativistic mass", which still floats around in the literature (in particular in high school text-books), is an inappropriate concept: The mass should be understood as a Lorentz scalar, as clarified by Lev Okun (1989).

$$\hat{\mathcal{M}}^2 = \hat{P}_\mu \hat{P}^\mu = \hat{H}^2 - \hat{\vec{P}}^2, \tag{E.7}$$

indeed commutes with all $\hat{P}_\mu$ and $\hat{M}_{\mu\nu}$. This simply reflects the fact that $\hat{P}_\mu \hat{P}^\mu$ is a Lorentz scalar. Two more Lorentz scalars are

$$\hat{M}_{\mu\nu}\hat{M}^{\mu\nu} = 2\left(\hat{\vec{J}}^2 - \hat{\vec{K}}^2\right), \quad \frac{1}{2}\epsilon_{\mu\nu\rho\sigma}\hat{M}^{\mu\nu}\hat{M}^{\rho\sigma} = 4\hat{\vec{J}} \cdot \hat{\vec{K}}. \tag{E.8}$$

It is also interesting to consider the *Pauli–Lubanski vector*

$$\hat{I}_\mu = \frac{1}{2}\epsilon_{\mu\nu\rho\sigma}\hat{M}^{\nu\rho}\hat{P}^\sigma, \tag{E.9}$$

whose components are given by

$$\hat{I}_0 = \hat{P}_i \hat{J}_i, \quad \hat{I}_i = \epsilon_{ijk}\hat{K}_j \hat{P}_k - \hat{H}\hat{J}_i, \tag{E.10}$$

and which indeed transforms as a 4-vector

$$\left[\hat{P}_\mu, \hat{I}_\nu\right] = 0, \quad \left[\hat{M}_{\mu\nu}, \hat{I}_\rho\right] = \mathrm{i}\left(g_{\nu\rho}\hat{I}_\mu - g_{\mu\rho}\hat{I}_\nu\right). \tag{E.11}$$

The components of the Pauli–Lubanski vector obey the commutation relations

$$\left[\hat{I}_\mu, \hat{I}_\nu\right] = \mathrm{i}\epsilon_{\mu\nu\rho\sigma}\hat{I}^\rho \hat{P}^\sigma. \tag{E.12}$$

By construction, the scalar product of the Pauli–Lubanski vector with the 4-momentum vanishes, $\hat{I}_\mu \hat{P}^\mu = 0$. However, one can still form the non-trivial Lorentz scalar

$$\hat{I}_\mu \hat{I}^\mu = \hat{I}_0^2 - \hat{I}_i^2 = -\hat{\mathcal{M}}^2 \hat{\vec{J}}^2, \tag{E.13}$$

which is another Casimir operator. It determines the total spin in the rest frame.

Thus, we have reviewed the complete set of Casimir operators of the Poincaré algebra.

Appendix F Lie Groups and Lie Algebras

In the early 1870s, the Norwegian mathematician Sophus Lie discovered the general concept of non-Abelian continuous symmetry groups and their associated algebras, now known as *Lie groups* and *Lie algebras*. The simplest continuous, non-Abelian (*i.e.* non-commuting) symmetry group is SO(3) – the rotation group in a 3-dimensional Euclidean space, consisting of real orthogonal 3×3 matrices with determinant 1. The associated so(3) algebra[1] is characterized by the angular momentum commutation relations among its three generators. This algebra is equivalent to the su(2) algebra, which generates the group SU(2) of complex unitary 2×2 matrices with determinant 1. The group manifold of SU(2) is the 3-dimensional sphere S^3, while the group manifold of SO(3) is the coset space $S^3/\mathbb{Z}(2)$. Here, $\mathbb{Z}(2) = \{1, -1\}$ is the *center* of the group SU(2), such that SO(3) = SU(2)/$\mathbb{Z}(2)$. SU(2) is the *universal covering group* of SO(3).

All so-called *simple compact Lie algebras* have been completely classified by Elie Cartan in 1894. They include the orthogonal algebras so(n) (with $n \geq 3$), the unitary algebras su(n) (with $n \geq 2$), the symplectic algebras sp(n) (with $n \geq 1$), as well as the exceptional algebras g(2), f(4), e(6), e(7), and e(8), which were discovered by Wilhelm Killing around 1890. These generate the Lie groups SO(n), SU(n), Sp(n), G(2), F(4), E(6), E(7), and E(8), respectively. The SO(n) groups have a universal covering group known as Spin(n).

F.1 Definition of a Lie Algebra

A Lie algebra is spanned by its generators T^a with $a \in \{1, 2, \ldots, n_G\}$, where n_G denotes the number of generators. A general element of the algebra is given by a linear combination

$$H = \sum_{a=1}^{n_G} \omega^a T^a = \omega^a T^a, \tag{F.1}$$

with parameters $\omega^a \in \mathbb{R}$. Here, we use Einstein's convention of summing over repeated indices. In this case, we do not distinguish between upper and lower indices (in the algebra there is no concept of co- and contra-variant). Different elements of the algebra can be added, *i.e.*

$$H_1 + H_2 = H, \quad H_1 = \omega_1^a T^a, \quad H_2 = \omega_2^a T^a, \quad \omega_1^a + \omega_2^a = \omega^a, \tag{F.2}$$

such that H again belongs to the algebra. The product $H_1 H_2$, on the other hand, does in general not belong to the algebra. Instead, the concept of "multiplication" in a Lie algebra is represented by the *commutator*

---

[1] In this appendix, but not in the rest of the book, we distinguish the algebra from the corresponding group by using lower-case instead of capital letters.

$$[H_1, H_2] = H_1 H_2 - H_2 H_1 = \omega_1^a \omega_2^b [T^a, T^b]. \tag{F.3}$$

The structure of a Lie algebra is characterized by its *structure constants* $f_{abc} \in \mathbb{R}$, which determine the commutation relations between the generators

$$[T^a, T^b] = \mathrm{i} f_{abc} T^c. \tag{F.4}$$

It is a non-trivial feature of Lie algebras that the commutator of two generators is again a linear combination of generators. This structure can be realized only with very specific sets of generators. In particular, $\mathrm{i}[T^a, T^b] = -f_{abc} T^c$ and thus $\mathrm{i}[H_1, H_2]$ is again a member of the algebra. The generators of a Lie algebra are Hermitian, *i.e.*

$$T^{a\dagger} = T^a, \tag{F.5}$$

such that the commutator is anti-Hermitian,

$$[T^a, T^b]^\dagger = (T^a T^b - T^b T^a)^\dagger = T^{b\dagger} T^{a\dagger} - T^{a\dagger} T^{b\dagger} = T^b T^a - T^a T^b = -[T^a, T^b], \quad \text{(F.6)}$$

and $\mathrm{i}[T^a, T^b]$ is indeed again Hermitian.

Due to the anti-symmetry of the commutator, the structure constants satisfy $f_{bac} = -f_{abc}$. The *Jacobi identity*,

$$[[T^a, T^b], T^c] + [[T^b, T^c], T^a] + [[T^c, T^a], T^b] = 0, \tag{F.7}$$

even implies that the f_{abc} are totally anti-symmetric against all permutations of the indices.

F.2 Simple and Semi-Simple Lie Algebras

Important internal symmetries in particle physics are associated with so-called simple or semi-simple Lie algebras. In order to define these notions, we first need to introduce the concept of a *subalgebra*. A subalgebra is generated by a subset S^a of generators (which are linear combinations of the T^a) that is closed under commutation, *i.e.*

$$[S^a, S^b] = \mathrm{i} f_{abc} S^c. \tag{F.8}$$

An *invariant subalgebra* obeys the additional requirement that

$$[T^a, S^b] = \mathrm{i} f_{abc} S^c, \tag{F.9}$$

for all generators T^a of the algebra and all generators S^b of the subalgebra. A Lie algebra is called *simple* if it has no invariant subalgebra. A subalgebra is *Abelian* if its generators commute with each other, *i.e.* $[S^a, S^b] = 0$. A Lie algebra is called *semi-simple* if it has no invariant Abelian subalgebra. A semi-simple Lie algebra may still have non-Abelian invariant subalgebras.

The number of generators of the maximal Abelian subalgebra of a semi-simple Lie algebra (which cannot be an invariant subalgebra) determines the *rank r* of the algebra. In the following, we will focus on compact Lie algebras. By definition, those are associated with Lie groups that have a compact group manifold. It turns out that compact Lie algebras are semi-simple.

F.3 Representations of Lie Algebras

A Lie algebra has many possible *representations* which can be viewed as concrete real-izations of its generators, which we have, until now, defined in an abstract and implicit manner, through their commutation relations. As we will see, the generators can, for ex-ample, be represented by finite-dimensional square matrices. In fact, every semi-simple Lie algebra has an n_G-dimensional representation, which is known as the *adjoint repre-sentation*. Here, n_G is again the number of generators. In the adjoint representation, the generators are realized as $n_G \times n_G$ matrices whose elements can be chosen as

$$T^a_{bc} = -\mathrm{i}f_{abc}, \tag{F.10}$$

such that

$$[T^a, T^b]_{de} = T^a_{df}T^b_{fe} - T^b_{df}T^a_{fe} = -f_{adf}f_{bfe} + f_{bdf}f_{afe}. \tag{F.11}$$

In order to convince oneself that this indeed defines a representation of the algebra, *i.e.* $[T^a, T^b]_{de} = \mathrm{i}f_{abc}T^c_{de} = f_{abc}f_{cde}$, one thus needs to show that

$$-f_{adf}f_{bfe} + f_{bdf}f_{afe} = f_{abc}f_{cde}. \tag{F.12}$$

This indeed follows directly from the Jacobi identity (F.7).

The *conjugate* $\overline{T}$ of a representation is defined as

$$\mathrm{i}\overline{T}^a = (\mathrm{i}T^a)^* = -\mathrm{i}(T^a)^\mathsf{T} \quad \Rightarrow \quad \overline{T}^a = -(T^a)^\mathsf{T}, \tag{F.13}$$

where $*$ denotes complex conjugation and T denotes transpose. The conjugate is again a representation of the algebra because

$$[\overline{T}^a, \overline{T}^b] = [(T^a)^\mathsf{T}, (T^b)^\mathsf{T}] = (T^a)^\mathsf{T}(T^b)^\mathsf{T} - (T^b)^\mathsf{T}(T^a)^\mathsf{T} = (T^bT^a - T^aT^b)^\mathsf{T}$$
$$= -[T^a, T^b]^\mathsf{T} = -\mathrm{i}f_{abc}(T^c)^\mathsf{T} = \mathrm{i}f_{abc}\overline{T}^c. \tag{F.14}$$

A representation is called *real*, if $\overline{T}^a = T^a$. It is instructive to convince oneself that the adjoint representation is real.

A representation is called *pseudo-real* if it is not real, but if $\overline{T}^a$ and T^a are unitarily equiv-alent, *i.e.* if there is a unitary transformation U such that $U\overline{T}^aU^\dagger = T^a$. Representations that are inequivalent to their conjugate representation are known as *complex* representa-tions. Among the compact semi-simple Lie algebras, only su(n) with $n \geq 3$, so($4n + 2$) with $n \geq 1$, and e(6) have complex representations.

F.4 Lie Algebra so(3) and its Representations

The simplest simple Lie algebra has three generators ($n_G = 3$) which obey the commuta-tion relations

$$[T^a, T^b] = \mathrm{i}\epsilon_{abc}T^c, \tag{F.15}$$

where ϵ_{abc} is the totally anti-symmetric Levi-Civita symbol. The interested reader may want to show that the consistency condition (F.12) is indeed satisfied.

The above algebra is well-known from quantum mechanics where the generators T^a are the three components of an angular momentum vector. Angular momentum is conserved as a consequence of the isotropy of 3-dimensional Euclidean coordinate space. The rotations in this space are described by real orthogonal 3×3 matrices with determinant 1. Under multiplication, these matrices form the group SO(3). The corresponding Lie algebra is known as so(3). It is instructive to convince oneself that so(3) is simple.

The quantum mechanical orbital angular momentum vector of a point particle

$$\hat{\vec{L}} = \hat{\vec{r}} \times \hat{\vec{p}} = -i\hbar\, \vec{r} \times \vec{\nabla} \tag{F.16}$$

provides a concrete realization of the commutation relations of the so(3) algebra when we identify T^a with $\hat{L}^a/\hbar$, $a \in \{1, 2, 3\}$. As we know, the quantum mechanical angular momentum operator acts in the Hilbert space of square integrable wave functions, which is infinite-dimensional. Hence, the representation of the so(3) generators by the operator $\hat{\vec{L}}$ is infinite-dimensional.

As we know from quantum mechanics, general wave functions can be decomposed into linear combinations of angular momentum eigenstates $|lm\rangle$ which obey

$$\hat{\vec{L}}^2|lm\rangle = \hbar^2 l(l+1)|lm\rangle, \quad \hat{L}^3|lm\rangle = \hbar m|lm\rangle, \quad l \in \mathbb{N}, \quad m \in \{-l, -l+1, \ldots, l\}. \tag{F.17}$$

The operator $\hat{\vec{L}}^2 = \hat{L}^a \hat{L}^a$ commutes with all components of the angular momentum vector, i.e. $[\hat{\vec{L}}^2, \hat{L}^a] = 0$. This is characteristic of a so-called *Casimir operator* of a Lie algebra, which consists of sums of products of generators and *commutes with all generators* T^a. The Casimir operators themselves do not belong to the Lie algebra. According to a general theorem due to Giulio Racah, the number of independent Casimir operators of a semi-simple Lie algebra is given by the rank. This includes all simple Lie algebras, in particular so(3). Since the different components of an angular momentum vector do not commute, so(3) has rank 1. Consequently, $\hat{\vec{L}}^2$ is its only independent Casimir operator. Obviously, any polynomial of $\hat{\vec{L}}^2$ also commutes with all the generators. However, those operators are not independent of $\hat{\vec{L}}^2$.

If we consider a particle moving on the surface of a 2-dimensional sphere S^2, in spherical coordinates its position can be described by the angles θ and φ. Correspondingly, its wave function can be expressed as

$$\Psi(\theta, \varphi) = \langle \theta, \varphi | \Psi \rangle = \sum_{l=0}^{\infty} \sum_{m=-l}^{l} c_{lm} \langle \theta, \varphi | lm \rangle, \tag{F.18}$$

where $c_{lm} \in \mathbb{C}$ and $\langle \theta, \varphi | lm \rangle = Y_{lm}(\theta, \varphi)$ are the spherical harmonics. The set of all states $|lm\rangle$ span the infinite-dimensional Hilbert space of wave functions on the sphere S^2. However, already the states $|lm\rangle$ with fixed $l \in \mathbb{N}$ and with $m \in \{-l, -l+1, \ldots, l\}$ alone span a finite $(2l + 1)$-dimensional representation of the so(3) algebra.

A representation is called *irreducible* if the algebra cannot be realized on any subset of states that are involved. The infinite-dimensional representation through the angular momentum operators $\hat{\vec{L}}$ is *reducible* and decomposes into a sum of irreducible representations characterized by the quantum number l. It turns out that the eigenvalues of the independent Casimir operators uniquely characterize the irreducible representations of a semi-simple Lie algebra.

The trivial, irreducible representation of the so(3) algebra is 1-dimensional and corresponds to $l = 0$ (with $m = 0$) consisting of the single state $|00\rangle$. In this trivial subspace of the Hilbert space, the generators are just given by

$$T^1 = T^2 = T^3 = 0, \tag{F.19}$$

which obviously satisfy the commutation relations. A non-trivial 3-dimensional representation corresponds to $l = 1$ and is spanned by the three states $|11\rangle$, $|10\rangle$, and $|1-1\rangle$. In this case, the so(3) generators are represented by 3×3 matrices

$$T^1 = \frac{1}{\sqrt{2}} \begin{pmatrix} 0 & 1 & 0 \\ 1 & 0 & 1 \\ 0 & 1 & 0 \end{pmatrix}, \quad T^2 = \frac{1}{\sqrt{2}} \begin{pmatrix} 0 & -i & 0 \\ i & 0 & -i \\ 0 & i & 0 \end{pmatrix}, \quad T^3 = \begin{pmatrix} 1 & 0 & 0 \\ 0 & 0 & 0 \\ 0 & 0 & -1 \end{pmatrix}. \tag{F.20}$$

It is straightforward to convince oneself that these matrices indeed obey the commutation relations of so(3). The dimension of this representation is equal to the number of generators, $n_G = 3$. Indeed, this representation is equivalent to the adjoint representation $T^a_{bc} = -i\epsilon_{abc}$. Although the representation matrices of eq. (F.20) do not obey this relation, they span the same algebra, because they are related to this form by a unitary transformation.

As we know from quantum mechanics, not only orbital angular momentum but also spin obeys the angular momentum commutation relations. In particular, in contrast to orbital angular momentum, spin can also be quantized in half-integer units. For example, the Pauli matrices $\vec{\sigma}$ give rise to the representation

$$T^1 = \frac{\sigma^1}{2} = \frac{1}{2} \begin{pmatrix} 0 & 1 \\ 1 & 0 \end{pmatrix}, \quad T^2 = \frac{\sigma^2}{2} = \frac{1}{2} \begin{pmatrix} 0 & -i \\ i & 0 \end{pmatrix}, \quad T^3 = \frac{\sigma^3}{2} = \frac{1}{2} \begin{pmatrix} 1 & 0 \\ 0 & -1 \end{pmatrix}. \tag{F.21}$$

This representation is usually associated with the algebra su(2), which, however, is identical with the one of so(3) (and also with the one of sp(1)). As we will see, while the algebras so(3) and su(2) are identical, the corresponding groups SO(3) and SU(2) are still different. In the following, we will associate general representations with integer or half-integer angular momentum with the algebra su(2). On the other hand, when we want to explicitly restrict ourselves to integer representations only, we will associate them with the algebra so(3).

In general, an irreducible $(2j + 1)$-dimensional representation of the su(2) algebra is characterized by its angular momentum $j \in \{0, 1/2, 1, 3/2, \ldots\}$. The spin $j = 1/2$ representation is the smallest non-trivial representation, also known as the *fundamental representation*. It is interesting to convince oneself that this representation is pseudo-real.

In fact, all su(2) representations are either real or pseudo-real. In particular, su(2) has no complex representations. For illustrative purposes, we also write down the 4×4 matrices of the spin $j = 3/2$ representation

$$T^1 = \frac{1}{2} \begin{pmatrix} 0 & \sqrt{3} & 0 & 0 \\ \sqrt{3} & 0 & 2 & 0 \\ 0 & 2 & 0 & \sqrt{3} \\ 0 & 0 & \sqrt{3} & 0 \end{pmatrix}, \quad T^2 = \frac{1}{2} \begin{pmatrix} 0 & -i\sqrt{3} & 0 & 0 \\ i\sqrt{3} & 0 & -2i & 0 \\ 0 & 2i & 0 & -i\sqrt{3} \\ 0 & 0 & i\sqrt{3} & 0 \end{pmatrix},$$

$$T^3 = \frac{1}{2} \begin{pmatrix} 3 & 0 & 0 & 0 \\ 0 & 1 & 0 & 0 \\ 0 & 0 & -1 & 0 \\ 0 & 0 & 0 & -3 \end{pmatrix}, \tag{F.22}$$

for which one can easily show that $T^a T^a = 3/2(3/2 + 1)\mathbb{1}$.

When we couple the orbital angular momentum $\hat{\vec{L}}$ and the spin $\hat{\vec{S}}$ of a particle to its total angular momentum $\hat{\vec{J}} = \hat{\vec{L}} + \hat{\vec{S}}$, we are working with the direct Kronecker product of two Lie algebras so(3) $\times$ su(2). While so(3) and su(2) (which, in fact, are the same algebra) are both simple, their product is semi-simple. Indeed, typical examples of semi-simple Lie algebras are direct products of simple ones.

F.5 Unitary Group SU(2) versus Orthogonal Group SO(3)

Let us now consider the SU(2) and SO(3) groups associated with the su(2) = so(3) algebra. The group U(2) consists of all unitary 2×2 matrices U, i.e. matrices that fulfill

$$UU^\dagger = U^\dagger U = \mathbb{1}. \tag{F.23}$$

The group SU(2) is a subgroup of U(2) consisting of those unitary 2×2 matrices that have determinant 1. It is straightforward to show that these matrices indeed form a group under matrix multiplication. The elements of SU(2) can be associated with elements of the su(2) algebra by exponentiation in the $j = 1/2$ representation, i.e.

$$U = \exp(\mathrm{i}\omega^a T^a) = \exp\left(\frac{\mathrm{i}}{2}\vec{\omega} \cdot \vec{\sigma}\right) = \cos\left(\frac{\omega}{2}\right)\mathbb{1} + \mathrm{i}\sin\left(\frac{\omega}{2}\right)\frac{\vec{\omega}}{\omega} \cdot \vec{\sigma}, \tag{F.24}$$

where $\omega = |\vec{\omega}|$ and $\omega^a \in \mathbb{R}$. A general SU(2) element can also be expressed as

$$U = \begin{pmatrix} A & B \\ -B^* & A^* \end{pmatrix}, \tag{F.25}$$

with complex elements $A = A_r + \mathrm{i}A_i$, $B = B_r + \mathrm{i}B_i \in \mathbb{C}$, and $\det U = |A|^2 + |B|^2 = 1$. With this condition, the two complex numbers A and B correspond to four real numbers (the real and imaginary parts of A and B) with the constraint $A_r^2 + A_i^2 + B_r^2 + B_i^2 = 1$. This implies that the group elements of SU(2) can be viewed as points on a 3-dimensional sphere S^3 (embedded in a 4-dimensional Euclidean space). Thus, the SU(2) group manifold is the sphere S^3.

Every group has a subgroup known as the *center* which consists of those group elements that commute with all other group elements. The center of the group SU(2) consists of the unit-matrix $\mathbb{1}$ and the matrix $-\mathbb{1}$, which is unitary, has determinant 1, and obviously commutes with all group elements. These two elements of the center form the Abelian subgroup $\mathbb{Z}(2)$.

Similarly, the group SO(3) consists of the real orthogonal 3×3 matrices with determinant 1, i.e.

$$OO^\mathsf{T} = O^\mathsf{T}O = \mathbb{1}, \quad \det O = 1. \tag{F.26}$$

Again, it is easy to convince oneself that these matrices form a group under matrix multiplication. In this case, the group elements can be obtained by exponentiating the algebra elements in the 3-dimensional adjoint representation $T^a_{bc} = -\mathrm{i}\epsilon_{abc}$, such that

$$O = \exp(i\omega^a T^a) = \exp \begin{pmatrix} 0 & \omega^3 & -\omega^2 \\ -\omega^3 & 0 & \omega^1 \\ \omega^2 & -\omega^1 & 0 \end{pmatrix}. \tag{F.27}$$

In contrast to SU(2), the center of SO(3) is trivial and consists only of the 3×3 unit-matrix $\mathbb{1}$. In particular, the matrix $-\mathbb{1}$ no longer has determinant 1 and thus does not belong to the group. Indeed, the non-trivial center of SU(2) is related to the fact that there are two types of su(2) representations: those with integer and those with half-integer spin, while the so(3) algebra only contains the integer spin representations.

Since the commutation relations of the su(2) and so(3) algebra are identical, it is not surprising that the SU(2) and SO(3) groups are also closely related. Indeed, it is instructive to convince oneself that the adjoint representation of SO(3) is related to the fundamental representation of SU(2) by

$$O_{ab} = \frac{1}{2} \mathrm{Tr} \left[U\sigma^a U^\dagger \sigma^b \right]. \tag{F.28}$$

In this way, the unit-element $U = \mathbb{1}$ is mapped to

$$O_{ab} = \frac{1}{2} \mathrm{Tr} \left[\sigma^a \sigma^b \right] = \delta_{ab}, \tag{F.29}$$

which corresponds to the unit-element $O = \mathbb{1}$ of SO(3). The inverse $U^\dagger$ of U is mapped to

$$\frac{1}{2} \mathrm{Tr} \left[U^\dagger \sigma^a U \sigma^b \right] = \frac{1}{2} \mathrm{Tr} \left[U\sigma^b U^\dagger \sigma^a \right] = O_{ba} = O_{ab}^\top, \tag{F.30}$$

which corresponds to the inverse of O. It is instructive to convince oneself that the group structure $U_1 U_2 = U$ is maintained by the mapping, such that $O_1 O_2 = O$.

Since there is a map from the group SU(2) to SO(3), one might expect that the two groups are identical. This is, however, not the case, because the mapping is not one to one, but rather two to one. In fact, both U and $-U$ are mapped to the same group element O of SO(3). In other words, SO(3) is insensitive to the center $\mathbb{Z}(2)$ of SU(2). This is consistent with the fact that SU(2) has the center $\mathbb{Z}(2)$ whereas SO(3) only has a trivial center (consisting just of the 3×3 unit-matrix $\mathbb{1}$). Since U and $-U$ are anti-podal points on the SU(2) group manifold S^3, the SO(3) manifold is the coset space $S^3/\mathbb{Z}(2)$ in which anti-podal points are identified with each other. SU(2) is known as the *universal covering group* of SO(3).

F.6 Unitary Group SU(n) and its Algebra su(n)

The unitary $n \times n$ matrices with determinant 1 form a group under matrix multiplication – the special unitary group SU(n). This follows immediately from

$$UU^\dagger = U^\dagger U = \mathbb{1}, \quad \det U = 1, \quad VV^\dagger = V^\dagger V = \mathbb{1}, \quad \det V = 1 \quad \Rightarrow$$
$$UV(UV)^\dagger = UVV^\dagger U^\dagger = \mathbb{1}, \quad \det(UV) = \det U \det V = 1. \tag{F.31}$$

In addition, associativity (*i.e.* $(UV)W = U(VW)$) holds for all matrices, a unit-element exists (the unit-matrix $\mathbb{1}$), and the inverse is $U^{-1} = U^{\dagger}$.

The group SU(n) is non-Abelian because in general $UV \neq VU$. Each element $U \in$ SU(n) can be represented as

$$U = \exp(\mathrm{i}H), \tag{F.32}$$

where H is Hermitian and traceless. The matrices H form the su(n) algebra. It has $n^2 - 1$ free parameters, and hence $n^2 - 1$ generators T^a, among which $n - 1$ commute with each other (they can be diagonalized simultaneously). Thus, the rank of su(n) is

$$r = n - 1. \tag{F.33}$$

Correspondingly, su(n) has $n - 1$ independent Casimir operators. The simplest non-trivial representation of su(n) is the fundamental n-dimensional representation in terms of generalized Pauli or Gell-Mann $n \times n$ matrices. For the fundamental representation, we adopt the standard ortho-normalization condition

$$\mathrm{Tr}\left[T^a T^b\right] = \frac{1}{2}\delta_{ab}. \tag{F.34}$$

The group manifold of SU(n) is, at least locally, a *product of spheres*

$$\mathrm{SU}(n) = S^3 \times S^5 \times \cdots \times S^{2n-1}. \tag{F.35}$$

In order to show this, we decompose a general SU(n) matrix U into an element W of the subgroup SU($n - 1$) and an element V of the coset SU(n)/SU($n - 1$)

$$U = VW, \quad W = \begin{pmatrix} 1 & 0 & 0 & \cdots & 0 \\ 0 & \tilde{U}_{11} & \tilde{U}_{12} & \cdots & \tilde{U}_{1\,n-1} \\ 0 & \tilde{U}_{21} & \tilde{U}_{22} & \cdots & \tilde{U}_{2\,n-1} \\ \cdot & \cdot & \cdot & & \cdot \\ \cdot & \cdot & \cdot & & \cdot \\ \cdot & \cdot & \cdot & & \cdot \\ 0 & \tilde{U}_{n-1\,1} & \tilde{U}_{n-1\,2} & \cdots & \tilde{U}_{n-1\,n-1} \end{pmatrix}, \tag{F.36}$$

where the embedded matrix $\tilde{U}$ is in SU($n - 1$). It is indirectly defined by

$$V = \begin{pmatrix} U_{11} & -U_{21}^* & -\dfrac{U_{31}^*(1+U_{11})}{1+U_{11}^*} & \cdots & -\dfrac{U_{n1}^*(1+U_{11})}{1+U_{11}^*} \\ U_{21} & \dfrac{1+U_{11}^*-|U_{21}|^2}{1+U_{11}} & -\dfrac{U_{31}^* U_{21}}{1+U_{11}^*} & \cdots & -\dfrac{U_{n1}^* U_{21}}{1+U_{11}^*} \\ U_{31} & -\dfrac{U_{21}^* U_{31}}{1+U_{11}} & \dfrac{1+U_{11}^*-|U_{31}|^2}{1+U_{11}^*} & \cdots & -\dfrac{U_{n1}^* U_{31}}{1+U_{11}^*} \\ \cdot & \cdot & \cdot & & \cdot \\ \cdot & \cdot & \cdot & & \cdot \\ U_{n1} & -\dfrac{U_{21}^* U_{n1}}{1+U_{11}} & -\dfrac{U_{31}^* U_{n1}}{1+U_{11}^*} & \cdots & \dfrac{1+U_{11}^*-|U_{n1}|^2}{1+U_{11}^*} \end{pmatrix} \in \mathrm{SU}(n). \tag{F.37}$$

One should convince oneself that V is indeed an SU(n) matrix, and that the resulting matrix $\tilde{U}$ is indeed in SU($n - 1$). The matrix V is constructed entirely from the elements

$U_{11}, U_{21}, \ldots, U_{n1}$ of the first column of the matrix U, which is normalized to $|U_{11}|^2 + |U_{21}|^2 + \cdots + |U_{n1}|^2 = 1$. This implies that the matrix V takes values on the sphere S^{2n-1} and hence, the coset manifold is

$$\mathrm{SU}(n)/\mathrm{SU}(n-1) = S^{2n-1}. \tag{F.38}$$

By successively factoring out odd-dimensional spheres, one reduces $\mathrm{SU}(n)$ all the way down to $\mathrm{SU}(2)$, whose group manifold is S^3, which thus proves eq. (F.35). This construction is also used in Section 19.2.

The center of $\mathrm{SU}(n)$ is the cyclic group

$$\mathbb{Z}(n) = \{\exp(2\pi\, \mathrm{i} m/n)\mathbb{1},\ m = 1, \ldots, n\}, \tag{F.39}$$

consisting of the unit-matrix $\mathbb{1}$ multiplied by a complex n-th root $\exp(2\pi\, \mathrm{i} m/n)$ of 1. These matrices obviously commute with all other group elements. In addition, they are unitary and have determinant 1; thus, they do belong to $\mathrm{SU}(n)$.

F.7 Group SU(3) and its Algebra su(3)

Let us consider the group SU(3) and its algebra su(3) in some detail. First of all, su(3) has $3^2 - 1 = 8$ generators. They can be chosen as $T^a = \lambda^a/2$, where λ^a are the *Gell-Mann matrices*

$$\lambda^1 = \begin{pmatrix} 0 & 1 & 0 \\ 1 & 0 & 0 \\ 0 & 0 & 0 \end{pmatrix}, \quad \lambda^2 = \begin{pmatrix} 0 & -\mathrm{i} & 0 \\ \mathrm{i} & 0 & 0 \\ 0 & 0 & 0 \end{pmatrix}, \quad \lambda^3 = \begin{pmatrix} 1 & 0 & 0 \\ 0 & -1 & 0 \\ 0 & 0 & 0 \end{pmatrix},$$

$$\lambda^4 = \begin{pmatrix} 0 & 0 & 1 \\ 0 & 0 & 0 \\ 1 & 0 & 0 \end{pmatrix}, \quad \lambda^5 = \begin{pmatrix} 0 & 0 & -\mathrm{i} \\ 0 & 0 & 0 \\ \mathrm{i} & 0 & 0 \end{pmatrix},$$

$$\lambda^6 = \begin{pmatrix} 0 & 0 & 0 \\ 0 & 0 & 1 \\ 0 & 1 & 0 \end{pmatrix}, \quad \lambda^7 = \begin{pmatrix} 0 & 0 & 0 \\ 0 & 0 & -\mathrm{i} \\ 0 & \mathrm{i} & 0 \end{pmatrix}, \quad \lambda^8 = \frac{1}{\sqrt{3}}\begin{pmatrix} 1 & 0 & 0 \\ 0 & 1 & 0 \\ 0 & 0 & -2 \end{pmatrix}. \tag{F.40}$$

The corresponding structure constants of the su(3) algebra are listed in Table F.1. The three generators T^1, T^2, and T^3 form an su(2) subalgebra. Since λ^3 and λ^8 commute, and since there is no other independent linear combination of generators that commutes with these two, the rank of su(3) is $r = 2$. According to Racah's theorem, this implies that su(3) has two independent Casimir operators. One of them is quadratic in the generators and simply given by $C_1 = T^a T^a$. The other Casimir operator is $C_2 = d_{abc} T^a T^b T^c$, where the coefficients d_{abc} are listed in Table F.2. The structure constants f_{abc} and the coefficients d_{abc} can be expressed as

$$f_{abc} = \frac{1}{4\mathrm{i}}\mathrm{Tr}\left[[\lambda^a, \lambda^b]\lambda^c\right], \quad d_{abc} = \frac{1}{4}\mathrm{Tr}\left[\{\lambda^a, \lambda^b\}\lambda^c\right]. \tag{F.41}$$

Table F.1 Non-zero structure constants f_{abc} of the Lie algebra su(3). As for any Lie algebra, the structure constants are completely anti-symmetric against permutation of the indices.

abc	123	147	156	246	257	345	367	458	678
f_{abc}	1	1/2	−1/2	1/2	1/2	1/2	−1/2	$\sqrt{3}/2$	$\sqrt{3}/2$

Table F.2 Non-zero coefficients d_{abc} of the Lie algebra su(3). These coefficients are invariant under permutation of the indices a, b, and c.

abc	118	146	157	228	247	256	338	344
d_{abc}	$1/\sqrt{3}$	1/2	1/2	$1/\sqrt{3}$	−1/2	1/2	$1/\sqrt{3}$	1/2

abc	355	366	377	448	558	668	778	888
d_{abc}	1/2	−1/2	−1/2	$-1/(2\sqrt{3})$	$-1/(2\sqrt{3})$	$-1/(2\sqrt{3})$	$-1/(2\sqrt{3})$	$-1/\sqrt{3}$

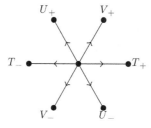

Fig. F.1 Effect of the shift operators $T_\pm$, $U_\pm$, $V_\pm$ on the triangular grid of SU(3) weight diagrams. The x-axis corresponds to the eigenvalues of T^3 and the y-axis to those of $T^8 = \sqrt{3}Y/2$.

The anti-commutator of two Gell-Mann matrices can be expressed as

$$\{\lambda^a, \lambda^b\} = \frac{4}{3}\delta_{ab}\mathbb{1} + 2d_{abc}\lambda^c. \tag{F.42}$$

In analogy to the raising and lowering operators of spin in the su(2) algebra, as illustrated in Figure F.1, we introduce the following shift operators

$$T_\pm = T^1 \pm \mathrm{i}T^2, \quad V_\pm = T^4 \pm \mathrm{i}T^5, \quad U_\pm = T^6 \pm \mathrm{i}T^7. \tag{F.43}$$

Besides these, there are the diagonal generators

$$T^3 = \frac{\lambda^3}{2}, \quad Y = \frac{\lambda^8}{\sqrt{3}} = \frac{2}{\sqrt{3}}T^8. \tag{F.44}$$

One then obtains the following commutation relations

$$\begin{aligned}
&\left[T^3, T_\pm\right] = \pm T_\pm, \quad \left[T_+, T_-\right] = 2T^3, \\
&\left[T^3, V_\pm\right] = \pm\frac{1}{2}V_\pm, \quad \left[V_+, V_-\right] = \frac{3}{2}Y + T^3, \\
&\left[T^3, U_\pm\right] = \mp\frac{1}{2}U_\pm, \quad \left[U_+, U_-\right] = \frac{3}{2}Y - T^3, \\
&\left[Y, T^3\right] = \left[Y, T_\pm\right] = 0, \quad \left[Y, V_\pm\right] = \pm V_\pm, \quad \left[Y, U_\pm\right] = \pm U_\pm, \\
&\left[T_+, V_+\right] = \left[T_+, U_-\right] = \left[U_+, V_+\right] = 0, \\
&\left[T_+, V_-\right] = -U_-, \quad \left[T_+, U_+\right] = V_+, \quad \left[U_+, V_-\right] = T_-.
\end{aligned} \tag{F.45}$$

Since the generators T^3 and Y commute, we can diagonalize them simultaneously and characterize the states of an su(3) multiplet by the corresponding eigenvalues. The generators T^1, T^2, and T^3 generate an su(2) subalgebra of su(3). Since these three generators all commute with Y, we can also simultaneously diagonalize the su(2) Casimir operator $C = T^1T^1 + T^2T^2 + T^3T^3 = T(T+1)\mathbb{1}$. Here, $\mathbb{1}$ denotes the $(2T+1) \times (2T+1)$ unit-matrix corresponding to an embedded $(2T+1)$-dimensional su(2) representation with "spin" T. The states of an su(3) multiplet can thus be further distinguished by the value of T. Hence, we characterize states of an irreducible su(3) representation $\{\Gamma\}$ by Y, T, and T^3, and we denote them as $|\{\Gamma\}YTT^3\rangle$. For example (using the flavor notation of up, down, and strange quarks), the states of the 3-dimensional fundamental representation $\{3\}$ are given by

$$u = \left|\{3\}\frac{1}{3}\frac{1}{2}\frac{1}{2}\right\rangle, \quad d = \left|\{3\}\frac{1}{3}\frac{1}{2}-\frac{1}{2}\right\rangle, \quad s = \left|\{3\}-\frac{2}{3}00\right\rangle. \tag{F.46}$$

Similarly, the states of the conjugate representation $\{\bar{3}\}$ (representing anti-quarks) are given by

$$\bar{u} = \left|\{\bar{3}\}-\frac{1}{3}\frac{1}{2}-\frac{1}{2}\right\rangle, \quad \bar{d} = \left|\{\bar{3}\}-\frac{1}{3}\frac{1}{2}\frac{1}{2}\right\rangle, \quad \bar{s} = \left|\{\bar{3}\}\frac{2}{3}00\right\rangle. \tag{F.47}$$

The raising and lowering operators $T_\pm$ shift the value of T^3 by ± 1, *i.e.*

$$T_\pm \left|\{\Gamma\}YTT^3\right\rangle = \sqrt{T(T+1) - T^3(T^3 \pm 1)}\left|\{\Gamma\}YTT^3 \pm 1\right\rangle. \tag{F.48}$$

The operators $V_\pm$ and $U_\pm$ also act as raising and lowering operators. In particular, we obtain

$$YV_\pm \left|\{\Gamma\}YTT^3\right\rangle = ([Y, V_\pm] + V_\pm Y)\left|\{\Gamma\}YTT^3\right\rangle$$
$$= (\pm V_\pm + V_\pm Y)\left|\{\Gamma\}YTT^3\right\rangle = (Y \pm 1) V_\pm \left|\{\Gamma\}YTT^3\right\rangle,$$
$$T^3 V_\pm \left|\{\Gamma\}YTT^3\right\rangle = ([T^3, V_\pm] + V_\pm T^3)\left|\{\Gamma\}YTT^3\right\rangle$$
$$= \left(\pm\frac{1}{2}V_\pm + V_\pm T^3\right)\left|\{\Gamma\}YTT^3\right\rangle = \left(T^3 \pm \frac{1}{2}\right) V_\pm \left|\{\Gamma\}YTT^3\right\rangle,$$
$$YU_\pm \left|\{\Gamma\}YTT^3\right\rangle = ([Y, U_\pm] + U_\pm Y)\left|\{\Gamma\}YTT^3\right\rangle$$
$$= (\pm U_\pm + U_\pm Y)\left|\{\Gamma\}YTT^3\right\rangle = (Y \pm 1) U_\pm \left|\{\Gamma\}YTT^3\right\rangle,$$
$$T^3 U_\pm \left|\{\Gamma\}YTT^3\right\rangle = ([T^3, U_\pm] + U_\pm T^3)\left|\{\Gamma\}YTT^3\right\rangle$$
$$= \left(\mp\frac{1}{2}U_\pm + U_\pm T^3\right)\left|\{\Gamma\}YTT^3\right\rangle = \left(T^3 \mp \frac{1}{2}\right) U_\pm \left|\{\Gamma\}YTT^3\right\rangle. \tag{F.49}$$

This implies that

$$V_\pm \left|\{\Gamma\}YTT^3\right\rangle = \sum_{T'} C_{T'YTT^3}\left|\{\Gamma\}Y \pm 1, T', T^3 \pm \frac{1}{2}\right\rangle,$$
$$U_\pm \left|\{\Gamma\}YTT^3\right\rangle = \sum_{T'} C'_{T'YTT^3}\left|\{\Gamma\}Y \pm 1, T', T^3 \mp \frac{1}{2}\right\rangle, \tag{F.50}$$

where $C_{T'YTT^3}$ and $C'_{T'YTT^3}$ are coefficients that can be determined by a straightforward calculation, and T' differs from T by a half-integer.

The weight diagrams of some su(3) representations are shown in Figure F.2. The x-axis corresponds to the eigenvalues of T^3, while the y-axis corresponds to those of $T^8 = \sqrt{3}Y/2$.

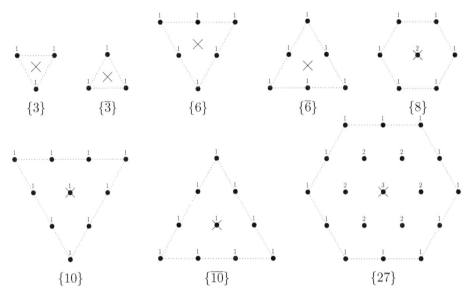

Weight diagrams of su(3) representations. The multiplicities of the states are indicated by small labels. The representations $\{8\}$, $\{10\}$, $\{\overline{10}\}$, and $\{27\}$ are trivial with respect to the center $\mathbb{Z}(3)$. The representations $\{3\}$ and $\{\overline{6}\}$ as well as $\{\overline{3}\}$ and $\{6\}$ belong to two distinct classes that transform non-trivially under the center subgroup.

F.8 Permutation Group S_N

Let us consider the permutation symmetry of N objects – for example, the fundamental representations of su(n). Their permutations form the group S_N. The *permutation group* has $N!$ elements – all permutations of N objects. The group S_2 has two elements: the identity and the pair permutation. The representations of S_2 are represented by *Young tableaux*, named after the British mathematician Alfred Young (1873–1940),

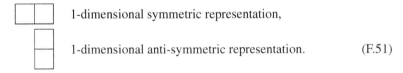

1-dimensional symmetric representation,

1-dimensional anti-symmetric representation.   (F.51)

The permutation properties of three objects are described by the group S_3. It has $3! = 6$ elements: the identity, 3 pair permutations, and 2 cyclic permutations. The group S_3 has three irreducible representations

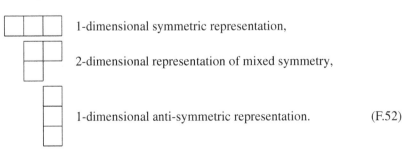

1-dimensional symmetric representation,

2-dimensional representation of mixed symmetry,

1-dimensional anti-symmetric representation.   (F.52)

The representations of the group S_N are given by the Young tableaux with N boxes. The boxes are arranged in left-bound rows, such that no row is longer than the one above it. For example, for the representations of S_4 one obtains

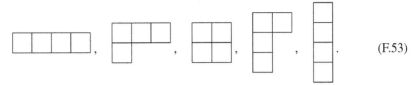

$$(F.53)$$

The *dimension* of a representation can be determined as follows. The boxes of the corresponding Young tableau are enumerated from 1 to N such that the numbers grow as one reads each row from left to right, and each column from top to bottom. The number of possible enumerations determines the dimension of the representation. For example, for S_3 one obtains

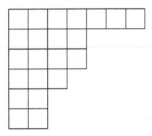

$$(F.54)$$

The squares of the dimensions of all representations add up to the order of the group, *i.e.*

$$\sum_{\Gamma} d_{\Gamma}^2 = N!.\tag{F.55}$$

In particular, for S_2 we have $1^2 + 1^2 = 2 = 2!$ and for S_3 one obtains $1^2 + 2^2 + 1^2 = 6 = 3!$.

A general Young tableau can be characterized by the number of boxes m_i in its i-th row. For example, the Young tableau

has $m_1 = 7$, $m_2 = 4$, $m_3 = 4$, $m_4 = 3$, $m_5 = 2$, and $m_6 = 2$. The dimension of the corresponding representation is given by

$$d_{m_1,m_2,\ldots,m_n} = N!\frac{\prod_{i<k}(l_i - l_k)}{l_1!l_2!\ldots l_n!}, \quad l_i = m_i + n - i,\tag{F.56}$$

where n is the number of rows in the Young tableau. Applying this formula to the following tableau of S_5

with $m_1 = 3$, $m_2 = 1$, $m_3 = 1$, and $n = 3$ yields $l_1 = 3 + 3 - 1 = 5$, $l_2 = 1 + 3 - 2 = 2$, $l_3 = 1 + 3 - 3 = 1$, and hence

$$d_{3,1,1} = 5!\frac{(l_1 - l_2)(l_1 - l_3)(l_2 - l_3)}{l_1!l_2!l_3!} = 5!\frac{3 \cdot 4 \cdot 1}{5!2!1!} = 6. \tag{F.57}$$

The permuted objects can be the fundamental representations of su(n). For su(2), we identify

$$\boxed{} = \{2\}. \tag{F.58}$$

Young tableaux with more than two rows have no realization in su(2) since among just two distinguishable objects no more than two can be combined anti-symmetrically. To each Young tableau with at most two rows, one can associate an su(2) representation. Such a Young tableau is characterized by the integers m_1 and m_2, e.g.,

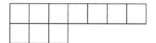

has $m_1 = 7$ and $m_2 = 3$. The corresponding su(2) representation has *spin*

$$S = \frac{1}{2}(m_1 - m_2), \tag{F.59}$$

which is also denoted by $\{m_1 - m_2 + 1\}$. The above Young tableau hence represents $S = 2$ – a spin quintet $\{5\}$.

F.9   su(n) Representations and Young Tableaux

The simplest non-trivial representation of su(n) is the *fundamental representation*. It is n-dimensional and can be identified with the Young tableau $\square$. Every irreducible representation of su(n) can be obtained as a tensor product by coupling N fundamental representations. In this way, each su(n) representation is associated with a Young tableau with N boxes, which characterizes the permutation symmetry of the fundamental representations in the tensor product. Since the fundamental representation is n-dimensional, there are n different fundamental properties (*e.g.*, u and d in the SU(2)$_I$ isospin or u, d, and s in the SU(3)$_f$ flavor symmetry of quarks). Hence, we can maximally anti-symmetrize n objects, and the Young tableaux for su(n) representations are therefore restricted to no more than n rows.

The dimension of an su(n) representation can be obtained from the corresponding Young tableau by filling it with factors as follows:

n	$n+1$	$n+2$	$n+3$	$n+4$	$n+5$	$n+6$
$n-1$	n	$n+1$	$n+2$			
$n-2$	$n-1$	n	$n+1$			
$n-3$	$n-2$	$n-1$				
$n-4$	$n-3$					
$n-5$	$n-4$					

The dimension of the su(n) representation is given as the product of all factors divided by $N!$ and multiplied with the S_N dimension $d_{m_1,m_2,\ldots,m_n}$ of the Young tableau

$$
\begin{aligned}
D^n_{m_1,m_2,\ldots,m_n} &= \frac{(n+m_1-1)!}{(n-1)!}\frac{(n+m_2-2)!}{(n-2)!}\cdots\frac{m_n!}{0!}\frac{1}{N!}N!\frac{\prod_{i<k}(l_i-l_k)}{l_1!l_2!\ldots l_n!} \\
&= \frac{\prod_{i<k}(m_i-m_k-i+k)}{(n-1)!(n-2)!\ldots 0!}.
\end{aligned}
\tag{F.60}
$$

We see that the dimension of a representation only depends on the differences $q_i = m_i - m_{i+1}$. In particular, for su(2) we find

$$
D^2_{m_1,m_2} = \frac{m_1-m_2-1+2}{1!0!} = m_1-m_2+1 = q_1+1 = 2S+1,
\tag{F.61}
$$

in agreement with the previous result eq. (F.59). For a rectangular Young tableau with n rows, e.g., in su(2) for

all $q_i = 0$, and we obtain

$$
D^n_{m,m,\ldots,m} = \frac{\prod_{i<k}(m_i-m_k-i+k)}{(n-1)!(n-2)!\ldots 0!} = \frac{(n-1)!(n-2)!\ldots 0!}{(n-1)!(n-2)!\ldots 0!} = 1,
\tag{F.62}
$$

and therefore a singlet. For $D^3_{1,1,1}$ this shows that in su(3) ⊟ corresponds to a singlet. It also explains why the dimension of an su(n) representation only depends on the differences q_i. Without changing the dimension, we can couple a representation with a singlet, and hence, we can always add a rectangular Young tableau with n rows to any su(n) representation. For example in su(3),

$$
\tag{F.63}
$$

We obtain the *conjugate representation* of a given representation by replacing m_i and q_i with

$$
\bar m_i = m_1 - m_{n-i+1}, \quad \bar q_i = \bar m_i - \bar m_{i+1} = m_{n-i} - m_{n-i+1} = q_{n-i}.
\tag{F.64}
$$

Geometrically, the Young tableau of a representation and its conjugate representation (after rotation) fit together to form a rectangular Young tableau with n rows. For example, in su(3)

and

are conjugate representations. In su(2), each representation is unitarily equivalent to its conjugate representation (*i.e.* the representations are real or pseudo-real). For example,

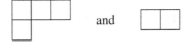

and

are conjugate representations, and

$$
\begin{array}{c}\boxed{}\end{array} \cong \begin{array}{c}\boxed{}\end{array} . \tag{F.65}
$$

This is not the case for higher n. The dimension of a representation and its conjugate representation are identical

$$
D^n_{\bar{m}_1,\bar{m}_2,\ldots,\bar{m}_n} = D^n_{m_1,m_2,\ldots,m_n}. \tag{F.66}
$$

For general n, the adjoint representation is given by $q_1 = q_{n-1} = 1$, $q_i = 0$ otherwise, and it is identical with its conjugate representation. The dimension of the adjoint representation is

$$
D^n_{2,1,1,\ldots,1,0} = n^2 - 1. \tag{F.67}
$$

F.10 Tensor Product of su(n) Representations

Next we want to discuss a method to *couple* su(n) representations by operating on their Young tableaux. Two Young tableaux with N and M boxes, respectively, are coupled by forming an *external product*. In this way, we generate Young tableaux with $N + M$ boxes that can then be translated back into su(n) representations. The external product is built as follows.

- The boxes of the first row of the second Young tableau are labeled with "a," the boxes of the second row with "b," etc.
- The boxes labeled with "a" are added to the first Young tableau in all possible ways that lead to new allowed Young tableaux, such that two "a" never end up in the same column. Then, the "b" boxes are added to the resulting Young tableaux in the same way.
- Now each of the resulting tableaux is read row-wise from top-right to bottom-left. Whenever a "b" or "c" appears before the first "a," or a "c" occurs before the first "b," etc., the corresponding Young tableau is deleted. The remaining tableaux form the reduction of the external product.

We now want to couple N fundamental representations of su(n). In Young tableau language, this reads

$$
\{n\} \times \{n\} \times \cdots \times \{n\} = \boxed{} \times \boxed{} \times \cdots \times \boxed{}. \tag{F.68}
$$

In this way, we generate *all* irreducible representations of S_N, *i.e.* all Young tableaux with N boxes. Each Young tableau is associated with an su(n) multiplet. It occurs in the product as often as the dimension of the corresponding S_N representation indicates, *i.e.* $d_{m_1,m_2,\ldots,m_n}$ times. Hence, we can write

$$
\{n\} \times \{n\} \times \cdots \times \{n\} = \sum_{\Gamma} d_{m_1,m_2,\ldots,m_n} \{D^n_{m_1,m_2,\ldots,m_n}\}. \tag{F.69}
$$

The sum runs over all Young tableaux with N boxes. For example,

$$\square \times \square \times \square = \boxed{} + 2\;\begin{array}{c}\square\square\\\square\end{array} + \begin{array}{c}\square\\\square\\\square\end{array}. \tag{F.70}$$

The translation into su(n) language reads

$$\{n\} \times \{n\} \times \{n\} = \left\{\frac{n(n+1)(n+2)}{6}\right\} + 2\left\{\frac{(n-1)n(n+1)}{3}\right\} + \left\{\frac{(n-2)(n-1)n}{6}\right\}. \tag{F.71}$$

The dimensions test

$$\frac{n(n+1)(n+2)}{6} + 2\frac{(n-1)n(n+1)}{3} + \frac{(n-2)(n-1)n}{6} = n^3 \tag{F.72}$$

confirms this result. In su(2), this corresponds to

$$\{2\} \times \{2\} \times \{2\} = \{4\} + 2\{2\} + \{0\}, \tag{F.73}$$

and in su(3) to

$$\{3\} \times \{3\} \times \{3\} = \{10\} + 2\{8\} + \{1\}. \tag{F.74}$$

F.11   Tensor Product of $\{3\}$ and $\{\overline{3}\}$ in su(3)

In order to further illustrate the coupling of su(n) representations, we now couple the fundamental $\{3\}$ representation of su(3) with its conjugate representation $\{\overline{3}\}$. From the considerations of the previous section, we obtain

$$\{3\} \times \{\overline{3}\} = \{1\} + \{8\}. \tag{F.75}$$

We now want to explicitly construct the singlet state and all states of the octet, by calculating the corresponding Clebsch–Gordan coefficients. First of all, we relate the states in the triplet (cf. eq. (F.46)) by shift operators

$$T_-u = d, \quad T_+d = u, \quad U_-d = s, \quad U_+s = d, \quad V_-u = s, \quad V_+s = u. \tag{F.76}$$

The representation $\{\overline{3}\}$ is obtained from $\{3\}$ by conjugation, which is realized by $\overline{T}^a = -(T^a)^{\mathsf{T}}$, such that, using eq. (F.47),

$$\overline{T}_-\bar{d} = -\bar{u}, \quad \overline{T}_+\bar{u} = -\bar{d}, \quad \overline{U}_-\bar{s} = -\bar{d}, \quad \overline{U}_+\bar{s} = -\bar{u},$$
$$\overline{V}_-\bar{s} = -\bar{u}, \quad \overline{V}_+\bar{s} = -\bar{d}. \tag{F.77}$$

We now construct the state in the octet that has the largest eigenvalue of T^3. This is the state

$$|\{8\}011\rangle = \left|\{3\}\frac{1}{3}\frac{1}{2}\frac{1}{2}\right\rangle \left|\{\overline{3}\}-\frac{1}{3}\frac{1}{2}\frac{1}{2}\right\rangle = u\bar{d}. \tag{F.78}$$

Starting from this particular state, we can now reach the other ones in the multiplet by applying the appropriate shift operations. In this manner, we obtain

$$-\sqrt{2}\,|\{8\}010\rangle = \left(T_- + \overline{T}_-\right)u\bar{d} = -u\bar{u} + d\bar{d} \quad \Rightarrow \quad |\{8\}010\rangle = \frac{1}{\sqrt{2}}\left(u\bar{u} - d\bar{d}\right), \tag{F.79}$$

as well as

$$\sqrt{2}\,|\{8\}01-1\rangle = \left(T_- + \overline{T}_-\right)\frac{1}{\sqrt{2}}\left(u\bar{u} - d\bar{d}\right) = \sqrt{2}d\bar{u} \quad \Rightarrow \quad |\{8\}010\rangle = d\bar{u}. \quad \text{(F.80)}$$

Applying the shift operator U_+ leads to

$$-\left|\{8\}1\frac{1}{2}\frac{1}{2}\right\rangle = \left(U_+ + \overline{U}_+\right)u\bar{d} = -u\bar{s} \quad \Rightarrow \quad \left|\{8\}1\frac{1}{2}\frac{1}{2}\right\rangle = u\bar{s}, \quad \text{(F.81)}$$

and in complete analogy

$$\left|\{8\}1\frac{1}{2}-\frac{1}{2}\right\rangle = d\bar{s}, \quad \left|\{8\}-1\frac{1}{2}\frac{1}{2}\right\rangle = s\bar{d}, \quad \left|\{8\}-1\frac{1}{2}-\frac{1}{2}\right\rangle = s\bar{u}. \quad \text{(F.82)}$$

We are still lacking the state $|\{8\}000\rangle$, which we reach by applying

$$\left(V_- + \overline{V}_-\right)\left|\{8\}1\frac{1}{2}\frac{1}{2}\right\rangle = \left(V_- + \overline{V}_-\right)u\bar{s} = s\bar{s} - u\bar{u} = \alpha\,|\{8\}000\rangle + \beta\,|\{8\}010\rangle. \quad \text{(F.83)}$$

Also demanding normalization and orthogonality, *i.e.* $\langle\{8\}000|\{8\}010\rangle = 0$, one fixes α and β and then arrives at

$$|\{8\}000\rangle = \frac{1}{\sqrt{6}}\left(u\bar{u} + d\bar{d} - 2s\bar{s}\right). \quad \text{(F.84)}$$

The last remaining state represents the singlet $|\{1\}000\rangle$. Again demanding normalization and orthogonality, one finally obtains

$$|\{1\}000\rangle = \frac{1}{\sqrt{3}}\left(u\bar{u} + d\bar{d} + s\bar{s}\right). \quad \text{(F.85)}$$

F.12 Orthogonal Group SO(n) and its Algebra so(n)

The real-valued $n \times n$ orthogonal matrices O with determinant 1 obey $OO^\mathsf{T} = O^\mathsf{T}O = \mathbb{1}$ and form the group SO(n) under matrix multiplication. The corresponding so(n) algebra consists of the purely imaginary, traceless, Hermitian $n \times n$ matrices. The number of such matrices is $n_G = n(n-1)/2$. The algebra so(4) = su(2) $\times$ su(2) is the direct product of two su(2) algebras, and thus, it is semi-simple (but not simple).

As we discussed before, the group SO(3) only has a trivial center, whereas its universal covering group SU(2) has the non-trivial center $\mathbb{Z}(2)$. The universal covering group of SO(n) is called Spin(n), such that Spin(3) = SU(2). Similarly, the universal covering group of SO(4) is Spin(4) = SU(2) $\times$ SU(2), which has the center $\mathbb{Z}(2) \times \mathbb{Z}(2)$. The center of SO(4) itself, on the other hand, is just $\mathbb{Z}(2)$ and consists of the 4×4 unit matrices $\mathbb{1}$ and $-\mathbb{1}$. Since for $n = 5$ the matrix $-\mathbb{1}$ does not have determinant 1, the group SO(5) has a trivial center, while its universal covering group Spin(5) has the center $\mathbb{Z}(2)$.

The so(6) algebra has $n_G = 6 \cdot 5/2 = 15$ generators. This is the same number as for su(4) which has $4^2 - 1$ generators. Moreover, one can show that the algebras of so(6) and su(4) are identical. As it was also the case for so(3) and su(2), the corresponding groups SO(6) and SU(4) are still different. In particular, the center of SO(6) is $\mathbb{Z}(2)$ while the center of SU(4) is $\mathbb{Z}(4)$. Therefore, the universal covering group of SO(6) is Spin(6) = SU(4), and SU(4)/$\mathbb{Z}(2)$ = SO(6). The higher so(n) algebras, with $n \geq 7$, are not equivalent to any su(n) algebra. The center of Spin(n) is $\mathbb{Z}(2)$ for odd n, $\mathbb{Z}(2) \times \mathbb{Z}(2)$ for $n = 4, 8, 12, \ldots$, and $\mathbb{Z}(4)$ for $n = 6, 10, 14, \ldots$.

Locally, the group manifold of Spin(n) is the product of spheres

$$\text{Spin}(n) = S^1 \times S^2 \times \cdots \times S^{n-1}. \tag{F.86}$$

Based on the so-called Hopf fibration, one can show that

$$S^3/S^1 = S^2, \quad S^7/S^3 = S^4, \quad S^{15}/S^7 = S^8, \tag{F.87}$$

such that at least locally

$$S^3 = S^1 \times S^2, \quad S^7 = S^3 \times S^4, \quad S^{15} = S^7 \times S^8. \tag{F.88}$$

Consequently, one obtains

$$\text{Spin}(3) = S^1 \times S^2 = S^3 = \text{SU}(2),$$
$$\text{Spin}(6) = S^1 \times S^2 \times S^3 \times S^4 \times S^5 = S^3 \times S^5 \times S^7 = \text{SU}(4). \tag{F.89}$$

The $n(n-1)/2$-dimensional adjoint representation of so(n) transforms as an anti-symmetric tensor under rotations in n dimensions. Similarly, there is an $[n(n+1)/2 - 1]$-dimensional representation that corresponds to a symmetric traceless tensor. In addition, so(n) has an n-dimensional *vector representation*. Since in three dimensions the vector cross product again generates a vector instead of an anti-symmetric tensor, for so(3) the vector representation is equivalent to the adjoint.

The so(n) algebras also have *spinor representations*. While so(3) = su(2) only has a single 2-dimensional spinor representation {2}, which corresponds to an ordinary spin-1/2, so(4) = su(2) $\times$ su(2) has two 2-dimensional spinor representations {2} $\times$ {1} and {1} $\times$ {2}, which are both pseudo-real. The algebra so(5) has a single 4-dimensional fundamental spinor representation, while so(6) = su(4) has two inequivalent 4-dimensional spinor representations, which correspond to the fundamental representation {4} of su(4) and its conjugate {$\bar{4}$}. In fact, the so(n) algebras with $n = 6, 10, 14, \ldots$ are the only ones that have complex representations.

Continuing this scheme, the algebra so(7) has a single 8-dimensional spinor representation, while so(8) has two inequivalent 8-dimensional pseudo-real spinor representations (in addition to its 8-dimensional vector representation). Similarly, so(9) has one 16-dimensional spinor representation, while so(10) has two inequivalent 16-dimensional, complex spinor representations, which are conjugate to each other. One of these representations plays an important role in Grand Unified Theories of the electroweak and strong interaction. It contains one generation of chiral Standard Model fermions (including quarks and leptons) as well as a sterile neutrino (cf. Section 26.11).

F.13 Symplectic Group Sp(n) and its Algebra sp(n)

The group Sp(n) is a subgroup of SU($2n$) which leaves the skew-symmetric matrix

$$J = \begin{pmatrix} 0 & \mathbb{1} \\ -\mathbb{1} & 0 \end{pmatrix} = i\sigma^2 \otimes \mathbb{1}, \tag{F.90}$$

invariant. Here, $\mathbb{1}$ is the $n \times n$ unit-matrix and σ^2 is the imaginary Pauli matrix. The elements $U \in \mathrm{SU}(2n)$ that belong to the subgroup $\mathrm{Sp}(n)$ satisfy the constraint

$$U^* = JUJ^\dagger. \tag{F.91}$$

Consequently, U and U^* are related by the unitary transformation J. Hence, the $2n$-dimensional fundamental representation of sp(n) is pseudo-real. The matrix J itself also belongs to $\mathrm{Sp}(n)$. It is easy to show that the matrices that obey the constraint eq. (F.91) form a group under matrix multiplication. The constraint implies the following form of a generic $\mathrm{Sp}(n)$ matrix

$$U = \begin{pmatrix} W & X \\ -X^* & W^* \end{pmatrix}, \tag{F.92}$$

where W and X are complex $n \times n$ matrices. Since U must still be an element of $\mathrm{SU}(2n)$, these matrices must satisfy $WW^\dagger + XX^\dagger = \mathbb{1}$ and $WX^\mathsf{T} = XW^\mathsf{T}$. Note that the eigenvalues of U come in complex conjugate pairs. Since center elements are multiples of the unit-matrix, in this case eq. (F.92) immediately implies $W = W^*$. Hence, the center of $\mathrm{Sp}(n)$ is $\mathbb{Z}(2)$ for all n. In fact, $\mathrm{Sp}(n)$ is its own universal covering group and thus acquires no further central extension.

Writing $U = \exp(\mathrm{i}H)$, where H is a Hermitian traceless matrix, eq. (F.91) implies that the generators H of the algebra sp(n) satisfy the constraint

$$H^* = -JHJ^\dagger = JHJ. \tag{F.93}$$

This relation leads to the following generic form,

$$H = \begin{pmatrix} A & B \\ B^* & -A^* \end{pmatrix}, \tag{F.94}$$

where A and B are $n \times n$ matrices. The Hermiticity condition $H = H^\dagger$ implies $A = A^\dagger$ and $B = B^\mathsf{T}$. Note that, since A is Hermitian, H is automatically traceless. The Hermitian $n \times n$ matrix A has n^2 real degrees of freedom and the complex symmetric $n \times n$ matrix B has $(n+1)n$ degrees of freedom. Hence, the dimension of the sp(n) algebra is

$$n_G = n^2 + (n+1)n = (2n+1)n. \tag{F.95}$$

There are n independent diagonal generators of the maximal Abelian Cartan subalgebra. Hence, the rank of $\mathrm{Sp}(n)$ is $r = n$.

The $n = 1$ case is equivalent to so(3), while the $n = 2$ case is equivalent to so(5). The group $\mathrm{Sp}(n)$ has the center $\mathbb{Z}(2)$, while $\mathrm{SO}(3)$ and $\mathrm{SO}(5)$ only have a trivial center. Hence, the group $\mathrm{Sp}(1)$ corresponds to the universal covering group $\mathrm{Spin}(3) = \mathrm{SU}(2)$ of $\mathrm{SO}(3)$, and the group $\mathrm{Sp}(2)$ is the universal covering group $\mathrm{Spin}(5)$ of $\mathrm{SO}(5)$. Although both sp(n) and so($2n+1$) have the same number of $(2n+1)n$ generators, the two algebras are inequivalent for $n \geq 3$.

Since sp(2) has rank $r = 2$, the weight diagrams of its representations can be drawn in a 2-d plane. The weight diagrams of some sp(2) representations, including the fundamental spinor representation {4}, the vector representation {5}, and the adjoint representation {10}, are shown in Figure F.3.

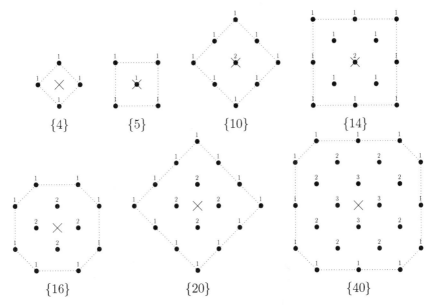

Fig. F.3 Weight diagrams of sp(2) representations. The multiplicities of the states are indicated by small labels. The representations {5}, {10}, and {14} are trivial with respect to the center $\mathbb{Z}(2)$, whereas {4}, {16}, {20}, and {40} are not.

The group manifold of Sp(n) is the product of spheres

$$Sp(n) = S^3 \times S^7 \times \cdots \times S^{4n-1}, \tag{F.96}$$

which implies

$$Sp(1) = S^3 = SU(2), \quad Sp(2) = S^3 \times S^7 = S^1 \times S^2 \times S^3 \times S^4 = Spin(5). \tag{F.97}$$

On the other hand, from $S^5 \times S^6 \neq S^{11}$ we infer

$$Sp(3) = S^3 \times S^7 \times S^{11} = S^1 \times S^2 \times S^3 \times S^4 \times S^{11}$$
$$\neq S^1 \times S^2 \times S^3 \times S^4 \times S^5 \times S^6 = Spin(7). \tag{F.98}$$

An overview of compact Lie group manifolds is given in Table G.2.

F.14 Exceptional Group G(2) and its Algebra g(2)

In this section, we discuss some basic properties of the Lie group G(2) – the simplest among the exceptional groups G(2), F(4), E(6), E(7), and E(8) – which do not fit into the main sequences SU(n), Spin(n), and Sp(n). While there is only one non-Abelian semi-simple Lie algebra of rank $r = 1$ – namely, the one of su(2) = so(3) = sp(1) – there are four of rank $r = 2$. These rank 2 algebras generate the groups SU(3), SU(2) $\times$ SU(2) = Spin(4), Spin(5) = Sp(2), and G(2), which have 8, 6, 10, and 14 generators, respectively.

The group G(2) is of particular interest because it has a trivial center but still is its own universal covering group.

It is natural to construct G(2) as a subgroup of SO(7) which has rank 3 and 21 generators.[2] The 7×7 real orthogonal matrices O of the group SO(7) have determinant 1 and obey the constraint

$$O_{ab}O_{ac} = \delta_{bc}. \tag{F.99}$$

The G(2) subgroup contains those matrices that, in addition, satisfy the cubic constraint

$$T_{abc} = T_{def}O_{da}O_{eb}O_{fc}. \tag{F.100}$$

Here, T is a totally anti-symmetric tensor whose non-zero elements follow by anti-symmetrization from

$$T_{127} = T_{154} = T_{163} = T_{235} = T_{264} = T_{374} = T_{576} = 1. \tag{F.101}$$

The tensor T also defines the multiplication rules for octonions. Equation (F.101) implies that eq. (F.100) represents 7 non-trivial constraints which reduce the 21 degrees of freedom of SO(7) to the 14 parameters of G(2). It should be noted that G(2) inherits the reality properties of SO(7): All its representations are real.

We make the following choice for the first 8 generators of g(2) in the 7-dimensional fundamental representation

$$\Lambda^a = \frac{1}{\sqrt{2}} \begin{pmatrix} \lambda^a & 0 & 0 \\ 0 & -\lambda^{a*} & 0 \\ 0 & 0 & 0 \end{pmatrix}. \tag{F.102}$$

Here, λ^a (with $a \in \{1, 2, \ldots, 8\}$) are the usual 3×3 Gell-Mann generators of su(3) which indeed is a subalgebra of g(2). Here, we use the normalization $\text{Tr}\left[\lambda^a \lambda^b\right] = \text{Tr}\left[\Lambda^a \Lambda^b\right] = 2\delta_{ab}$. This representation contains the complex representations $\{3\}$ and $\{\bar{3}\}$ of su(3). However, it is unitarily equivalent to a real representation. In the chosen basis of the generators, it is manifest that under SU(3) subgroup transformations the 7-dimensional representation decomposes into

$$\{7\} = \{3\} + \{\bar{3}\} + \{1\}. \tag{F.103}$$

Since g(2) has rank $r = 2$, only two generators can be diagonalized simultaneously. In our choice of basis, these are the su(3) subalgebra generators Λ^3 and Λ^8. Consequently, just as for su(3), the weight diagrams of g(2) representations can be drawn in a 2-dimensional plane. The weight diagrams of some g(2) representations, including the fundamental representation $\{7\}$ and the adjoint representation $\{14\}$, are shown in Figure F.4. One notes that the weight diagram of the representation $\{7\}$ is indeed a superposition of the weight diagrams of a $\{3\}$, $\{\bar{3}\}$, and $\{1\}$ of su(3). Since all g(2) representations are real, the $\{7\}$ representation is equivalent to its complex conjugate. The $\{3\} + \{\bar{3}\}$ contained in the $\{7\}$ representation of g(2) corresponds to a real, reducible, 6-dimensional representation of su(3).

---

[2] Generally, the groups O(n) and SO(n) have $n(n-1)/2$ generators, and their rank is $r = n/2$ if n is even, or $r = (n-1)/2$ if n is odd.

As usual for su(3),

$$T^+ = \frac{1}{\sqrt{2}} \left(\Lambda^1 + i\Lambda^2\right) = |1\rangle\langle 2| - |5\rangle\langle 4|, \quad T^- = \frac{1}{\sqrt{2}} \left(\Lambda^1 - i\Lambda^2\right) = |2\rangle\langle 1| - |4\rangle\langle 5|,$$

$$V^+ = \frac{1}{\sqrt{2}} \left(\Lambda^4 + i\Lambda^5\right) = |2\rangle\langle 3| - |6\rangle\langle 5|, \quad V^- = \frac{1}{\sqrt{2}} \left(\Lambda^4 - i\Lambda^5\right) = |3\rangle\langle 2| - |5\rangle\langle 6|,$$

$$U^+ = \frac{1}{\sqrt{2}} \left(\Lambda^6 + i\Lambda^7\right) = |1\rangle\langle 3| - |6\rangle\langle 4|, \quad U^- = \frac{1}{\sqrt{2}} \left(\Lambda^6 - i\Lambda^7\right) = |3\rangle\langle 1| - |4\rangle\langle 6|,$$

$$(F.104)$$

define shift operations between different states $|1\rangle, |2\rangle, \ldots, |7\rangle$ in the fundamental representation. The remaining 6 generators of G(2) also define shifts

$$X^+ = \frac{1}{\sqrt{2}} \left(\Lambda^9 + i\Lambda^{10}\right) = |2\rangle\langle 4| - |1\rangle\langle 5| - \sqrt{2}|7\rangle\langle 3| - \sqrt{2}|6\rangle\langle 7|,$$

$$X^- = \frac{1}{\sqrt{2}} \left(\Lambda^9 - i\Lambda^{10}\right) = |4\rangle\langle 2| - |5\rangle\langle 1| - \sqrt{2}|3\rangle\langle 7| - \sqrt{2}|7\rangle\langle 6|,$$

$$Y^+ = \frac{1}{\sqrt{2}} \left(\Lambda^{11} + i\Lambda^{12}\right) = |6\rangle\langle 1| - |4\rangle\langle 3| - \sqrt{2}|2\rangle\langle 7| - \sqrt{2}|7\rangle\langle 5|,$$

$$Y^- = \frac{1}{\sqrt{2}} \left(\Lambda^{11} - i\Lambda^{12}\right) = |1\rangle\langle 6| - |3\rangle\langle 4| - \sqrt{2}|7\rangle\langle 2| - \sqrt{2}|5\rangle\langle 7|,$$

$$Z^+ = \frac{1}{\sqrt{2}} \left(\Lambda^{13} + i\Lambda^{14}\right) = |3\rangle\langle 5| - |2\rangle\langle 6| - \sqrt{2}|7\rangle\langle 1| - \sqrt{2}|4\rangle\langle 7|,$$

$$Z^- = \frac{1}{\sqrt{2}} \left(\Lambda^{13} - i\Lambda^{14}\right) = |5\rangle\langle 3| - |6\rangle\langle 2| - \sqrt{2}|1\rangle\langle 7| - \sqrt{2}|7\rangle\langle 4|. \quad (F.105)$$

The generators themselves transform under the 14-dimensional adjoint representation of g(2).

From the weight diagram of the representation {14} (cf. Figure F.4), one sees that under the SU(3) subgroup transformation the adjoint representation of g(2) decomposes into

$$\{14\} = \{8\} + \{3\} + \{\bar{3}\}. \quad (F.106)$$

Let us also discuss the *center* of G(2). It is interesting to note that the maximal Abelian (Cartan) subgroup of both G(2) and SU(3) is $U(1)^2$, which must contain the center in both cases. Since G(2) contains SU(3) as a subgroup, its center cannot be larger than $\mathbb{Z}(3)$ (the center of SU(3)) because the potential center elements of G(2) must commute with all G(2) matrices (not just with the elements of the SU(3) subgroup). In the fundamental representation of G(2), the center elements of the SU(3) subgroup are given by

$$Z = \begin{pmatrix} z\mathbb{1} & 0 & 0 \\ 0 & z^*\mathbb{1} & 0 \\ 0 & 0 & 1 \end{pmatrix}, \quad (F.107)$$

where $\mathbb{1}$ is the 3×3 unit-matrix and $z \in \{1, \exp(\pm 2\pi i/3)\}$ is an element of $\mathbb{Z}(3)$. By construction, the three 7×7 matrices Z commute with the 8 generators of the SU(3) subgroup of G(2). However, an explicit calculation shows that this is not the case for the remaining 6 generators. Consequently, the center of G(2) is trivial and contains only the identity. The above argument applies to any representation of G(2). In other words, the universal covering group of G(2) is G(2) itself and still it has a trivial center.

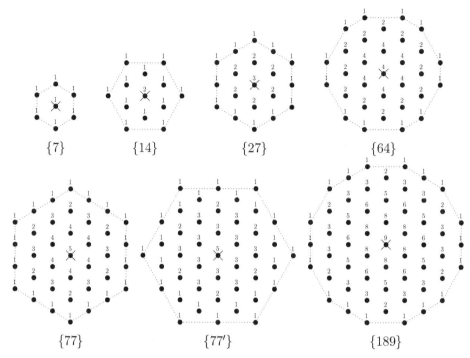

Fig. F.4 Some weight diagrams of irreducible g(2) representations. The multiplicities of the states are indicated by small labels. There are two inequivalent representations, {77} and {77'}, of the same dimension.

In SU(3), the non-trivial center $\mathbb{Z}(3)$ gives rise to the concept of *triality*: Each su(3) representation falls in one of three categories. For example, the trivial representation {1} and the adjoint representation {8} of su(3) have trivial triality, while the fundamental {3} and anti-fundamental {$\bar{3}$} have non-trivial opposite trialities. Since its center is trivial, the concept of triality does not extend to G(2). In particular, as one can see from eqs. (F.103) and (F.106), the g(2) representations decompose into mixtures of SU(3) representations with different trialities. This has interesting consequences for the results of g(2) tensor product decompositions. For example, in contrast to the SU(3) case, the product of two fundamental representations

$$\{7\} \times \{7\} = \{1\} + \{7\} + \{14\} + \{27\}, \tag{F.108}$$

contains again the fundamental representation. Some further tensor product decompositions are given by

$$\{7\} \times \{14\} = \{7\} + \{27\} + \{64\},$$
$$\{14\} \times \{14\} = \{1\} + \{14\} + \{27\} + \{77\} + \{77'\}. \tag{F.109}$$

Here, {77} and {77'} are the two inequivalent 77-dimensional representations of g(2), which are illustrated in Figure F.4.

The group manifold of G(2) is the product of the group manifold of SU(3) with a 6-dimensional sphere S^6, *i.e.*

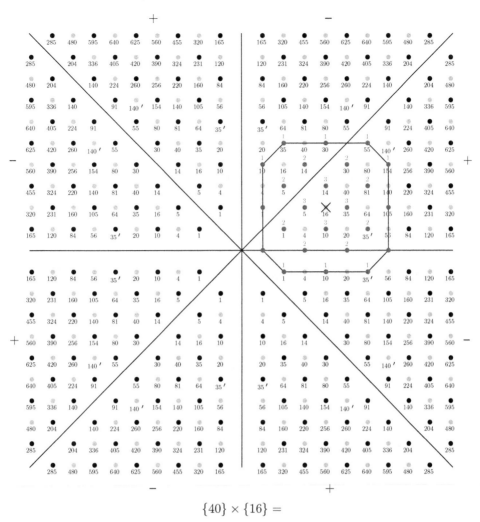

$$\{40\} \times \{16\} =$$

$$\{154\} + \{105\} + 2\{81\} + \{55\} + 2\{35\} + \{35'\} + \{30\} + \{14\} + \{10\} + \{5\}$$

Fig. F.5 The grid of sp(2) representations that is used in the graphical Antoine–Speiser scheme (Vlasii *et al.*, 2016). The weight diagram of the representation $\{40\}$ is centered at the position of the representation $\{16\}$ in order to reduce the products of these irreducible representations.

$$G(2) = SU(3) \times S^6 = S^3 \times S^5 \times S^6. \tag{F.110}$$

From this, one obtains

$$SO(7) = S^1 \times S^2 \times S^3 \times S^4 \times S^5 \times S^6 = G(2) \times S^1 \times S^2 \times S^4 = G(2) \times S^3 \times S^4$$
$$= G(2) \times S^7. \tag{F.111}$$

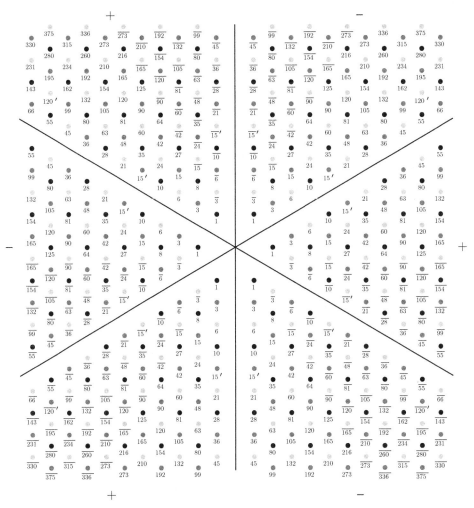

Fig. F.6 The grid of su(3) representations to be used in the graphical Antoine–Speiser scheme to reduce products of irreducible representations (Vlasii *et al.*, 2016).

F.15 Graphical Method for Tensor Product Reduction

Jean-Pierre Antoine and David Speiser have developed a simple graphical scheme that allows us to couple representations of rank 2 Lie algebras (Antoine and Speiser, 1964a,b). In this scheme, the different representations are associated with the points of a grid that is divided into sectors associated with alternating + and − signs, and separated by empty lines that carry no representations. When the origin of a weight diagram of a representation is placed on top of another representation in a + sector of the grid, one can read off the reduction of the product of the two representations into a sum of irreducible representations. This is done by listing all representations on the grid that are covered by the corresponding

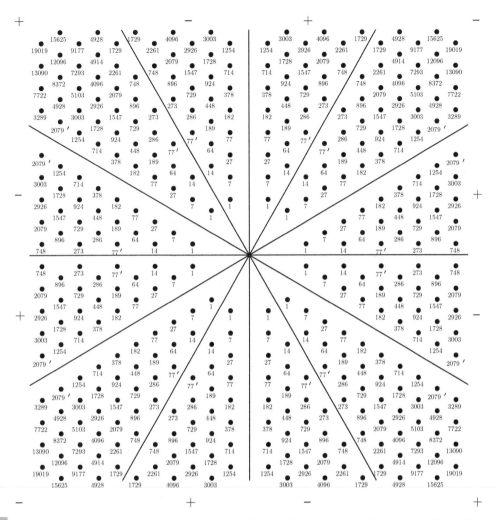

Fig. F.7 The grid of g(2) representations to be used in the graphical Antoine–Speiser scheme to reduce products of irreducible representations (Vlasii *et al.*, 2016).

states in the weight diagram, taking into account both the multiplicity of the state and the sign of the sector of the grid. Empty lines separate the sectors. Points of the weight diagram that fall on top of those empty lines do not contribute to the reduction.

This simple and elegant scheme generalizes to all compact semi-simple Lie algebras. For algebras with rank r, both the corresponding grid and the corresponding weight diagrams are r-dimensional. Hence, the scheme is most practical for the algebras of rank $r = 2$, which are su(3), so(4) = su(2) × su(2), so(5) = sp(2), and g(2).

The Antoine–Speiser scheme is illustrated in Figure F.5 for sp(2) (Vlasii *et al.*, 2016). The representation {40} is centered at the position of {16} on the grid, covering grid points in one + and two − sectors. The contributions from the − sectors (enclosed in square brackets) cancel some representations from the + sector, *i.e.*

$$\{40\} \times \{16\} = \{55\} + 2\{30\} + \{154\} + 3\{14\} + 2\{81\} + 3\{5\}$$
$$+ 3\{35\} + \{105\} + 2\{1\} + 3\{10\} + 2\{35'\}$$
$$- \Big[\{35\}+\{30\}+\{10\}+2\{14\}+2\{5\}+\{1\}\Big] - \Big[\{1\}+\{10\}+\{35'\}\Big]$$
$$- \{154\}+\{105\}+2\{81\}+\{55\}+2\{35\}+\{35'\}+\{30\}+\{14\}+\{10\}+\{5\}.$$

$$\text{(F.112)}$$

The grids for su(3) and g(2) are provided in Figures F.6 and F.7, respectively. The reader may want to copy the weight diagrams of Figures F.2, F.3, and F.4 on a transparency and place them on the corresponding grid in order to reduce the product of irreducible representations.

Appendix G Homotopy Groups and Topology

Homotopy groups play an important role in quantum field theory, and in this book, in particular, in Chapters 19, 20, 24, 25, and 26. Chapter 24 addresses this subject in the framework of Nambu–Goldstone boson fields, and Chapter 19 by referring to gauge fields, which is the basis of the discussion of the $U(1)_A$-problem in Chapter 20 and of the strong CP-problem in Chapter 25.

As we discuss in Chapter 19, gauge transformations can be viewed as maps from space or space–time into the corresponding gauge group. Such maps are classified topologically: Maps that can be continuously deformed into each other are topologically equivalent; thus, they fall into the same *homotopy class*. The homotopy classes are characterized by a *topological winding number,* which takes values in the *homotopy group*. The non-trivial topology of certain fields has an important impact both in particle and in condensed matter physics.

G.1 Maps from S^d to S^n

Let us first consider maps from a circle S^1. This circle[1] can be viewed as a compactified 1-d space (or interval) $x \in [0, L]$ with periodic boundary conditions. We can also interpret S^1 as the group manifold of $U(1)$. Indeed, $U(1)$ gauge transformations $U(x) = \exp(i\alpha(x)) \in U(1)$ in a 1-d compactified space represent maps $x \to U(x)$ from S^1 to S^1.

Such maps fall in topologically distinct equivalence classes. Two maps are equivalent if they can be continuously deformed into each other. The equivalence classes are known as *homotopy classes*, which are characterized by the *topological winding number*

$$n[U] = \frac{1}{2\pi i} \int_{S^1} dx \, U(x)^* \, \partial_x U(x) = \frac{1}{2\pi} \int_0^L dx \, \partial_x \alpha(x) = \frac{1}{2\pi} [\alpha(L) - \alpha(0)]$$
$$\in \Pi_1[U(1)] = \mathbb{Z}. \tag{G.1}$$

Periodic boundary conditions mean that $U(L) = U(0)$, which implies $\alpha(L) - \alpha(0) = 2\pi n[U]$, where $n[U] \in \mathbb{Z}$ is an integer winding number that characterizes the homotopy class of the gauge transformation. The winding numbers $n[U]$ form the homotopy group $\Pi_1[U(1)] = \mathbb{Z}$.

Winding numbers are additive under multiplication of the corresponding gauge transformations; *i.e.* for $U(x) = U_1(x)U_2(x)$, we obtain

$$n[U] = n[U_1] + n[U_2]. \tag{G.2}$$

---

[1] The symbol S^1 is sometimes used for a unit circle, but in this context we do not specify its radius. In the same way we handle its generalization to a d-dimensional sphere S^d.

- The constant gauge transformation $U(x) = 1$ is topologically trivial, $n[U] = 0$. This gauge transformation maps all points x in the compactified space onto the same point. All continuous deformations of this map have $n[U] = 0$ as well; thus, they are topologically trivial too.
- The identity map $U(x) = \exp(2\pi i x/L)$ covers each point in the image S^1 exactly once, with a positive orientation; thus, it has $n[U] = 1$.
- As a generalization, we can specify $U(x) = \exp(2\pi i n x/L)$, $n \in \mathbb{Z}$, as a representative in each homotopy class. This map covers each point in the image S^1 exactly $|n|$ times, with the orientation $\text{sign}(n)$, such that $n[U] = n$.

The sphere S^2 is *not* a group manifold. Hence, we cannot interpret maps from S^2 to S^2 as gauge transformations. Still, such maps occur, for example, in the 2-d O(3) model, which is a low-energy effective field theory for spatially 1-dimensional anti-ferromagnetic chains of quantum spins $1/2$. The direction of the staggered magnetization vector – the order parameter of an anti-ferromagnet – at a point x is described by a 3-component unit-vector $\vec{e}(x) \in S^2$. When we compactify space–time to a 2-d sphere, $x \in S^2$, the staggered magnetization defines a map $x \to \vec{e}(x)$ from S^2 to S^2. Such maps again fall into homotopy classes, characterized by a winding number, or topological charge,

$$n[\vec{e}] = \frac{1}{8\pi} \int d^2x\, \epsilon_{\mu\nu}\, \vec{e} \cdot \left(\partial_\mu \vec{e} \times \partial_\nu \vec{e}\right) \in \Pi_2[S^2] = \mathbb{Z}. \qquad (G.3)$$

The winding number $n[\vec{e}]$ counts topological tunneling events, known as *instantons* at the semi-classical level. For anti-ferromagnetic chains of quantum spins with half-odd-integer spin value, configurations with an odd value of $n[\vec{e}]$ contribute negatively to the functional integral. As first conjectured by Duncan Haldane (1983), this explains the vanishing massgap of these systems (Affleck and Haldane, 1987).

The Euclidean action of the 2-d O(3) model reads

$$S[\vec{e}] = \frac{1}{2g^2} \int d^2x\, \partial_\mu \vec{e} \cdot \partial_\mu \vec{e}. \qquad (G.4)$$

Let us evaluate the non-negative integral

$$0 \le \int d^2x \left(\partial_\mu \vec{e} \pm \epsilon_{\mu\nu}\partial_\nu \vec{e} \times \vec{e}\right)^2$$

$$= \int d^2x \left[2\partial_\mu \vec{e} \cdot \partial_\mu \vec{e} \pm 2\epsilon_{\mu\nu}\, \vec{e} \cdot \left(\partial_\mu \vec{e} \times \partial_\nu \vec{e}\right)\right] = 4g^2 S[\vec{e}] \pm 16\pi\, |n[\vec{e}]|, \quad (G.5)$$

to obtain a *Schwarz inequality*, from which we conclude that

$$S[\vec{e}] \ge \frac{4\pi}{g^2}\, |n[\vec{e}]|. \qquad (G.6)$$

The inequality is saturated only when the integrand in eq. (G.5) vanishes, *i.e.* when

$$\partial_\mu \vec{e}(x) + \sigma \epsilon_{\mu\nu}\partial_\nu \vec{e}(x) \times \vec{e}(x) = 0. \qquad (G.7)$$

Here, $\sigma = \pm 1$ determines the sign of the winding number $n[\vec{e}]$. Field configurations that satisfy this (anti-)self-duality equation are known as *instantons* for $\sigma = 1$ or *anti-instantons* for $\sigma = -1$. A concrete example of such configurations with winding number $n[\vec{e}] = \sigma n \in \mathbb{Z}$ with $0 < n \in \mathbb{N}$ (such that $n = |n[\vec{e}]|$) is given by

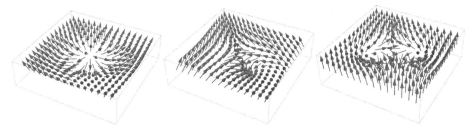

Fig. G.1 Examples of instanton and anti-instanton configurations of eq. (G.8) in the 2-d O(3) model. These configurations have winding number $n[\vec{e}] = 1, -1, -2$, respectively. The same configurations describe the spatial dependence of skyrmions and anti-skyrmions in the $(2 + 1)$-d O(3) model.

$$\vec{e}_n(r, \chi) = \left(\frac{2r^n \rho^n}{r^{2n} + \rho^{2n}} \cos(n\chi), \frac{2r^n \rho^n}{r^{2n} + \rho^{2n}} \sigma \sin(n\chi), \frac{r^{2n} - \rho^{2n}}{r^{2n} + \rho^{2n}} \right), \qquad (G.8)$$

where we have used polar coordinates for the space–time point $x = r(\cos \chi, \sin \chi)$. The scale parameter $\rho > 0$ determines the *size* of the (anti-)instanton. There are many other multi-(anti-)instanton configurations. Some examples of instanton and anti-instanton configurations are illustrated in Figure G.1.

When we consider the staggered magnetization field $\vec{e}(x)$ for a spatially 2-dimensional anti-ferromagnet, it maps the 3-dimensional Euclidean space–time (which can be compactified to S^3) onto S^2. Such maps fall into topological classes characterized by the *Hopf number* (Hopf, 1931)

$$H[\vec{e}] \in \Pi_3[S^2] = \mathbb{Z}. \qquad (G.9)$$

This has fascinating consequences in condensed matter physics, because it implies that in a 3-dimensional space–time, particles need not necessarily be either bosons or fermions, *i.e.* exchange symmetric or anti-symmetric. In fact, in two spatial dimensions, quasi-particles can be *anyons*, with *any* fractional spin[2] and *any* statistics (Leinaas and Myrheim, 1977; Wilczek, 1982). The latter manifests itself by a term $i\Theta H[\vec{e}]$, which enters the Euclidean action, with $\Theta \in \,]-\pi, \pi]$. Thus, an anyon exchange entails a phase factor $\exp(i\Theta)$, where $\Theta = 0$ and π correspond to bosons and fermions, respectively.

For example, the quasi-particle excitations in a *fractional quantum Hall sample* are Abelian anyons with fractional spin, statistics, and electric charge, which can be viewed as fractionalized electrons, *e.g.*, with electric charge $-e/3$. However, we should not think of an individual electron being divided into three anyons. Fractionalization is a collective phenomenon that can happen only if a macroscopic number of electrons behaves coherently at very low temperatures. Other spatially 2-dimensional condensed matter systems – known as quantum Hall ferromagnets – support *skyrmions*, which are topological particle-like excitations whose winding number is again given by $n[\vec{e}] \in \Pi_2[S^2] = \mathbb{Z}$ (Bär *et al.*, 2004). Again, a Hopf term endows the skyrmions with fractional statistics and turns them into Abelian anyons.

---

[2] This can be understood from the simple fact that in 2-d space there is only one rotational axis. Hence, the angular momentum algebra is reduced to a single operator, and there is no commutation relation that enforces discrete spin eigenvalues.

The presence of non-Abelian anyons, where permutation causes not just an Abelian phase factor but a non-Abelian unitary matrix factor, is also conceivable (Moore and Read, 1991; Wen, 1991). In 2022, such objects were not yet firmly established by experiment, but there are hints related to fractional quantum Hall states at filling fraction 5/2 (Stern, 2010).

Next, let us consider maps from S^3 to S^3, which is the group manifold of SU(2). For example, gauge transformations $U(x) \in$ SU(2) are maps from S^3 to S^3, if $x \in S^3$ is a point in a compactified 3-d space. In this case, the corresponding winding number is given by

$$ n[U] = \frac{1}{24\pi^2} \int_{S^3} d^3x\, \epsilon_{ijk} \mathrm{Tr}\left[\left(U^\dagger \partial_i U \right) \left(U^\dagger \partial_j U \right) \left(U^\dagger \partial_k U \right) \right] \in \Pi_3[\mathrm{SU}(2)] = \mathbb{Z}. \quad (\mathrm{G.10}) $$

It is instructive to verify that the expressions in eqs. (G.3) and (G.10) always assume integer values.

The pions in QCD are described by an effective field theory; see Chapter 23, whose field $U(x) \in$ SU(2) again maps 3-d space (compactified to S^3) into the SU(2) group manifold S^3. In this case, $n[U]$ counts topological *skyrmion excitations* in the pion field, which represent baryons, cf. Section 24.1. Hence, $n[U]$ corresponds to the baryon number, and it is amazing that baryons do appear in this effective theory, which is formulated entirely in terms of a meson field.

When we consider a compactified Euclidean space–time S^4, the pion field defines a map from S^4 to S^3. The important property $\Pi_4[S^3] = \mathbb{Z}(2)$ implies that in a 4-d space–time particles must be either *bosons or fermions*; anyons are not possible in $d = 4$.

The homotopy group $\Pi_4[S^3] = \mathbb{Z}(2)$ is also relevant in the electroweak sector of the Standard Model, where it characterizes the topology of SU(2)$_L$ gauge transformations governing the weak interaction. Since there are topologically non-trivial SU(2)$_L$ gauge transformations in $d = 4$, the Standard Model is affected by *Witten's global anomaly*, which is discussed in Section 15.5.

It is remarkable that $\Pi_4[\mathrm{SU}(2)] = \Pi_4[S^3] = \mathbb{Z}(2)$ (see Tables G.1 and G.2), whereas $\Pi_4[\mathrm{SU}(n)] = \{0\}$ for $n \geq 3$. The latter implies that in $d = 4$ the corresponding SU(n) gauge transformations can all be continuously deformed into each other. Hence, there is no global anomaly for SU(n) with $n \geq 3$.

On the other hand, chiral perturbation theory as the low-energy effective theory of 3-flavor QCD describes the light meson octet (pions, kaons, η-meson) as pseudo-Nambu–Goldstone bosons, which are described in terms of matrix-valued fields $U(x) \in$ SU(3), cf. Section 23.1. For a general number of quark flavors $N_\mathrm{f} \geq 2$, since $\Pi_3[\mathrm{SU}(N_\mathrm{f})] = \mathbb{Z}$, again there are skyrmions, which represent baryons.

Due to $\Pi_4[\mathrm{SU}(N_\mathrm{f} \geq 3)] = \{0\}$, for more than two flavors the effective theory does not have a global anomaly. Nevertheless, anomalies are present and are expressed by the *Wess–Zumino–Novikov–Witten (WZNW) term* (Wess and Zumino, 1971; Novikov, 1981, 1982; Witten, 1983a), which is discussed in Chapter 24,

$$ S_\mathrm{WZNW}[U] = \frac{1}{240\pi^2\mathrm{i}} \int_{H^5} d^5x\, \epsilon_{\mu\nu\rho\sigma\lambda} \mathrm{Tr}\left[\left(U^\dagger \partial_\mu U \right) \left(U^\dagger \partial_\nu U \right) \left(U^\dagger \partial_\rho U \right) \left(U^\dagger \partial_\sigma U \right) \left(U^\dagger \partial_\lambda U \right) \right]. $$
$$ (\mathrm{G.11}) $$

Here, H^5 is a 5-d hemisphere whose boundary $\partial H^5 = S^4$ is the compactified 4-d Euclidean space–time. As we describe in Section 24.6, the WZNW term can be constructed only

Table G.1 The homotopy groups $\Pi_d[S^n]$ for the dimensions $d, n = 1, 2, \ldots, 6$.

	S^1	S^2	S^3	S^4	S^5	S^6
Π_1	$\mathbb{Z}$	$\{0\}$	$\{0\}$	$\{0\}$	$\{0\}$	$\{0\}$
Π_2	$\{0\}$	$\mathbb{Z}$	$\{0\}$	$\{0\}$	$\{0\}$	$\{0\}$
Π_3	$\{0\}$	$\mathbb{Z}$	$\mathbb{Z}$	$\{0\}$	$\{0\}$	$\{0\}$
Π_4	$\{0\}$	$\mathbb{Z}(2)$	$\mathbb{Z}(2)$	$\mathbb{Z}$	$\{0\}$	$\{0\}$
Π_5	$\{0\}$	$\mathbb{Z}(2)$	$\mathbb{Z}(2)$	$\mathbb{Z}(2)$	$\mathbb{Z}$	$\{0\}$
Π_6	$\{0\}$	$\mathbb{Z}(12)$	$\mathbb{Z}(12)$	$\mathbb{Z}(2)$	$\mathbb{Z}(2)$	$\mathbb{Z}$

thanks to the property $\Pi_4[\mathrm{SU}(N_\mathrm{f})] = \{0\}$. It enters the Euclidean functional integral in the form $\exp(\mathrm{i}\, n\, S_{\mathrm{WZNW}}[U])$. The coefficient n is a low-energy constant, to be specified below.

The contribution of this term should not depend on the way how the field $U(x, x_5)$ is extended into the fifth dimension. To see that it is free of such an ambiguity, we consider the corresponding winding number

$$
w[U] = \frac{1}{480\pi^3 \mathrm{i}} \int_{S^5} d^5 x \, \epsilon_{\mu\nu\rho\sigma\lambda} \, \mathrm{Tr}\left[\left(U^\dagger \partial_\mu U\right)\left(U^\dagger \partial_\nu U\right)\left(U^\dagger \partial_\rho U\right)\left(U^\dagger \partial_\sigma U\right)\left(U^\dagger \partial_\lambda U\right)\right]
$$
$$
\in \Pi_5[\mathrm{SU}(N_\mathrm{f} \geq 3)] = \mathbb{Z}. \tag{G.12}
$$

Hence for different extensions of U in the x_5-direction, the WZNW term can only differ by $2\pi \mathrm{i}\, n\, w[U]$; see Section 24.6. This ambiguity cancels in the Euclidean functional integral, which is ambiguous by a contribution $\exp(2\pi \mathrm{i}\, n\, w[U])$, if and only if the coefficient n is quantized in integer units, $n \in \mathbb{Z}$.

Like other low-energy constants in chiral perturbation theory, n encodes some property of QCD as the underlying theory. Hence, it should correspond to an integer parameter of QCD, which does not leave many options. In fact, it turns out to be the number of colors, $n = N_\mathrm{c}$, which sneaks into the effective theory in this manner, although quarks and gluons do not occur explicitly (Witten, 1983a,b).

Let us finish this section with an overview of the homotopy groups $\Pi_d[S^n]$, up to 6 dimensions, which is summarized in Table G.1 and complementary to Table 19.1.

- For $d < n$, the map can always be contracted to a point, so there is only the trivial homotopy class $\Pi_d[S^n] = \{0\}$.
- For $d = n$, the sphere S^n can be mapped onto itself with an arbitrary integer winding number, which corresponds to the homotopy group $\Pi_d[S^d] = \mathbb{Z}$.
- The tricky scenario is $d > n$. First, we note that $\Pi_d[S^1] = \{0\}$ for all $d > 1$. We have already mentioned the peculiar case of the Hopf number $\Pi_3[S^2] = \mathbb{Z}$, as well as the case $\Pi_4[S^3] = \mathbb{Z}(2)$. The latter homotopy group coincides, for instance, with $\Pi_4[S^2] = \Pi_5[S^2] = \Pi_5[S^3] = \Pi_5[S^4] = \mathbb{Z}(2)$. In higher dimensions, $d \geq 6$, this scenario can also lead to extended cyclic groups, like $\Pi_6[S^2] = \Pi_6[S^3] = \mathbb{Z}(12)$. These homotopy groups are all finite, unless n is even and $d = 2n - 1$ (Serre, 1951); the Hopf number is the first exception of this kind, followed by $\Pi_7[S^4] = \mathbb{Z} \times \mathbb{Z}(12)$.

G.2 Topological Charge in 2-d Abelian Gauge Theory

Let us consider an Abelian gauge field $A_\mu(x)$ in 2-d Euclidean space–time. Its gauge action takes the form

$$S[A] = \frac{1}{2e^2} \int d^2x \, F_{\mu\nu} F_{\mu\nu}, \tag{G.13}$$

where $F_{\mu\nu}(x) = \partial_\mu A_\nu(x) - \partial_\nu A_\mu(x)$ is the field strength and e is the electric charge, which has the dimension of mass in $d = 2$. The tensor $F_{\mu\nu}$ has only one non-trivial component, $F_{12}(x) = -F_{21}(x) = E(x)$, which represents the electric field. The Euclidean equation of motion $\partial_\mu F_{\mu\nu}(x) = 0$ implies $\partial_1 E(x) = \partial_2 E(x) = 0$, such that $E(x)$ is a constant at the level of classical field theory.

The dual quantity,

$$q(x) = \frac{1}{4\pi} \epsilon_{\mu\nu} F_{\mu\nu}(x) = \frac{1}{2\pi} E(x), \tag{G.14}$$

is also given by the electric field, and it plays the role of the *topological charge density*. Hence, the topological charge is obtained as

$$Q = \int d^2x \, q = \frac{1}{4\pi} \int d^2x \, \epsilon_{\mu\nu} F_{\mu\nu}. \tag{G.15}$$

As in eq. (G.1), the topological charge density can be written as a *total divergence*

$$q(x) = \frac{1}{4\pi} \epsilon_{\mu\nu} F_{\mu\nu}(x) = \frac{1}{2\pi} \epsilon_{\mu\nu} \partial_\mu A_\nu(x) = \partial_\mu \Omega_\mu^{(0)}(x), \tag{G.16}$$

where

$$\Omega_\mu^{(0)}(x) = \frac{1}{2\pi} \epsilon_{\mu\nu} A_\nu(x). \tag{G.17}$$

(In the mathematical terminology, $\Omega_\mu^{(0)}$ is denoted as a "0-cochain".) Using Gauss' theorem, Q can be expressed as

$$Q = \int d^2x \, \partial_\mu \Omega_\mu^{(0)} = \int_{S^1} d\sigma_\mu \, \Omega_\mu^{(0)} = \frac{1}{2\pi} \int_{S^1} d\sigma_\mu \, \epsilon_{\mu\nu} A_\nu, \tag{G.18}$$

where $S^1 = \partial \mathbb{R}^2$ is a large circle at the "boundary" of the Euclidean space–time $\mathbb{R}^2$.

Let us consider field configurations of finite action, which implies that $F_{\mu\nu}(x)$ vanishes at space–time infinity, *i.e.* on S^1. The field strength vanishes when the vector potential is gauge equivalent to 0,

$$0 = A'_\mu(x) = A_\mu(x) - \partial_\mu \alpha(x) \quad \Rightarrow \quad A_\mu(x) = \partial_\mu \alpha(x), \tag{G.19}$$

i.e. $A_\mu(x)$ is a *pure gauge configuration*, at least at $|x| \to \infty$. Hence, the topological charge takes the form

$$Q = \frac{1}{2\pi} \int_{S^1} d\sigma_\mu \, \epsilon_{\mu\nu} A_\nu = \frac{1}{2\pi} \int_{S^1} d\sigma_\mu \, \epsilon_{\mu\nu} \partial_\nu \alpha \in \Pi_1[S^1] = \mathbb{Z}. \tag{G.20}$$

Here, we have identified Q as the topological winding number $n[U]$ of the gauge transformation $U(x) = \exp(i\alpha(x))$ at space–time infinity that we started from in Section G.1.

Table G.2 Compact Lie groups with group manifolds that include the sphere S^3.

SU(n)	=	$S^3 \times S^5 \times \cdots \times S^{2n-1}$	($n \geq 2$)
Spin(n)	=	$S^1 \times S^2 \times \cdots \times S^{n-1}$	($n \geq 3$)
Sp(n)	=	$S^3 \times S^7 \times \cdots \times S^{4n-1}$	($n \geq 1$)
G(2)	=	SU(3) $\times S^6 = S^3 \times S^5 \times S^6$	

G.3 Homotopy Groups of Lie Group Manifolds

As we discussed in Appendix F, the group manifolds of compact Lie groups are products of spheres, at least locally. In particular, the examples that we display in Table G.2 all involve the sphere S^3.

The presence of a factor S^3 implies

$$\Pi_3[\text{SU}(n)] = \Pi_3[\text{Spin}(n)] = \Pi_3[\text{Sp}(n)] = \Pi_3[\text{G}(2)] = \mathbb{Z}. \tag{G.21}$$

By referring again to Gauss' theorem, in analogy to eqs. (G.17) and (G.18), these terms represent the flux through the 3-d "boundary" of the 4-d Euclidean space–time. As a consequence, all these 4-d non-Abelian gauge theories have homotopy classes labeled by a topological charge $Q \in \mathbb{Z}$, and therefore instantons (Belavin *et al.*, 1975) (a discussion for the gauge group SU(2) is given in Section 19.4). Hence, a θ-term $i\theta Q[U]$ naturally enters the Euclidean action, with a vacuum-angle $\theta \in]-\pi, \pi]$; see Chapters 19 and 25.

Appendix H Monte Carlo Method

A powerful numerical technique for solving problems in classical statistical mechanics and lattice quantum field theory is the *Monte Carlo method*. The idea is to compute expectation values by stochastically generating field configurations. Partition functions (in classical statistical mechanics) or functional integrals (in quantum field theory) are defined as extremely high-dimensional integrals, with the number of integrations (or sums for discrete variables) increasing in proportion to the number of lattice points. As a consequence, performing the integration with numerical brute force is completely hopeless. In the Monte Carlo method, field configurations are generated according to their importance. The method of *importance sampling* uses the Boltzmann factor $\exp(-\beta\mathcal{H}[s])/Z$ or $\exp(-S[\Phi])/Z$ as a probability to generate a spin configuration $[s]$ or a Euclidean quantum field configuration $[\Phi]$. In this context, it is important that the Boltzmann factor is real and positive. Monte Carlo calculations underlie all numerical simulations of lattice QCD. For simplicity, we discuss the Monte Carlo method in the context of $O(N)$ models. This covers $O(N)$ spin models in classical statistical mechanics as well as $O(N)$-symmetric quantum field theories on the lattice, including the $O(4)$-symmetric Higgs sector of the Standard Model.

In this appendix, we focus on the theoretical background underlying the Monte Carlo method, rather than on specific algorithms. In particular, we discuss the concept of *Markov chains*, as well as the importance of *detailed balance* and *ergodicity*, which allow us to prove that a Markov chain converges to the desired Boltzmann distribution as the unique stationary distribution.

A simple Monte Carlo algorithm, which is almost universally applicable, is the *Metropolis algorithm*. However, it is often not very efficient. In particular, it suffers from *critical slowing down* when the correlation length increases near a second-order phase transition in a system of classical statistical mechanics or when taking the continuum limit of a lattice-regularized quantum field theory. This problem is not easy to overcome by just relying on the sheer numerical power of present-day computers. Hence, the development of more efficient Monte Carlo algorithms is very important in order to obtain accurate numerical results. Usually, deep insights into the physics are required in order to learn how to simulate a model efficiently. Here, we will concentrate on the theoretical basis of the Monte Carlo method, without going much beyond the simple Metropolis algorithm.

As Alan Sokal pointed out in some of his lecture notes (Sokal, 1997): "Monte Carlo is an extremely bad method; it should be used only when all alternative methods are worse". This is because the unavoidable statistical error of Monte Carlo calculations decreases only with the inverse square root of the invested computational resources. However, for generic, strongly coupled systems in classical statistical mechanics or quantum field theory, "though Monte Carlo is bad, all other alternative methods are worse". This makes Monte Carlo the method of choice for non-perturbative studies of strongly coupled quantum field theories.

From a non-perturbative perspective, the development of efficient Monte Carlo algorithms is as important as the development of efficient methods for evaluating Feynman diagrams in the context of perturbation theory. Monte Carlo algorithm development is an important part of theoretical physics that should not be underrated. Here, we provide an introduction to the theoretical basis of this exciting field of research.

H.1 Concept of a Markov Chain

To be specific, let us consider an $O(N)$ model with N-component scalar field variables $\Phi_x \in \mathbb{R}^N$ associated with the sites x of a hypercubic space–time lattice with periodic boundary conditions. We may restrict ourselves to a fixed-length field with $|\Phi_x|^2 = v^2$, such that $\Phi_x \in S^{N-1}$. The functional integral then takes the form

$$Z = \int \mathcal{D}\Phi \exp\left(-S[\Phi]\right) = \prod_x \int_{S^{N-1}} d\Phi_x \exp\left(-S[\Phi]\right). \tag{H.1}$$

A standard nearest-neighbor form of the Euclidean lattice action is given by

$$S[\Phi] = a^d \sum_{x,\mu} \frac{1}{2a^2} \left|\Phi_{x+\hat{\mu}} - \Phi_x\right|^2 = \sum_{x,\mu} a^{d-2}\left(v^2 - \Phi_x \cdot \Phi_{x+\hat{\mu}}\right). \tag{H.2}$$

In a Monte Carlo simulation, one generates a sequence of field configurations

$$[\Phi^{(0)}] \to [\Phi^{(1)}] \to [\Phi^{(2)}] \to \cdots \to [\Phi^{(M)}], \tag{H.3}$$

which form a *Markov chain*, by applying an algorithm that probabilistically turns the configuration $[\Phi^{(i)}]$ into the next configuration $[\Phi^{(i+1)}]$, without referring to previous configurations. The initial configuration $[\Phi^{(0)}]$ may be picked at random or selected otherwise. Ultimately, the results do not depend on this choice. If one picks each initial field variable $\Phi_x^{(0)}$ at random from a uniform distribution over S^{N-1}, the probability distribution of the initial configurations is simply given by

$$p^{(0)}[\Phi] = \frac{1}{Z^{(0)}}, \quad Z^{(0)} = \int \mathcal{D}\Phi\, 1 = \prod_x \int_{S^{N-1}} d\Phi\, 1, \tag{H.4}$$

such that the distribution is properly normalized to

$$\int \mathcal{D}\Phi\, p^{(0)}[\Phi] = 1. \tag{H.5}$$

The Monte Carlo algorithm that generates the Markov chain is completely characterized by the transition probability density $w[\Phi^{(i)}, \Phi^{(i+1)}]$ to turn a configuration $\Phi^{(i)}$ into the next configuration $\Phi^{(i+1)}$. The Metropolis algorithm corresponds to a specific choice of the transition probability that will be discussed later. After the first Monte Carlo step, the initial distribution of field configurations $p^{(0)}[\Phi]$ is turned into

$$p^{(1)}[\Phi'] = \int \mathcal{D}\Phi\, p^{(0)}[\Phi] w[\Phi, \Phi']. \tag{H.6}$$

This can be viewed as a multiplication of the "vector" $p^{(0)}$ by the transition "matrix" w. Since the Monte Carlo algorithm definitely generates a new configuration (although it may coincide with $[\Phi]$), the proper normalization of the transition probability density is

$$\int \mathcal{D}\Phi'\, w[\Phi, \Phi'] = 1, \tag{H.7}$$

such that the new probability distribution of field configurations is again properly normalized

$$\int \mathcal{D}\Phi'\, p^{(1)}[\Phi'] = \int \mathcal{D}\Phi\, p^{(0)}[\Phi] \int \mathcal{D}\Phi'\, w[\Phi, \Phi'] = \int \mathcal{D}\Phi\, p^{(0)}[\Phi] = 1. \tag{H.8}$$

Another application of the Monte Carlo algorithm then leads to

$$p^{(2)}[\Phi''] = \int \mathcal{D}\Phi'\, p^{(1)}[\Phi']w[\Phi', \Phi''] = \int \mathcal{D}\Phi' \int \mathcal{D}\Phi\, p^{(0)}[\Phi]w[\Phi, \Phi']w[\Phi', \Phi'']$$

$$= \int \mathcal{D}\Phi\, p^{(0)}[\Phi] \int \mathcal{D}\Phi'\, w[\Phi, \Phi']w[\Phi', \Phi''] = \int \mathcal{D}\Phi\, p^{(0)}[\Phi]w^2[\Phi, \Phi'']. \tag{H.9}$$

Here, we have introduced the squared "matrix"

$$w^2[\Phi, \Phi''] = \int \mathcal{D}\Phi'\, w[\Phi, \Phi']w[\Phi', \Phi''], \tag{H.10}$$

which describes the transition probability density that corresponds to two steps of the algorithm. By iterating the procedure, one obtains a sequence of properly normalized probability distributions that are related by

$$p^{(i)}[\Phi'] = \int \mathcal{D}\Phi\, p^{(0)}[\Phi]w^i[\Phi, \Phi']. \tag{H.11}$$

As we will see, the Markov chain shall be constructed in such a way that the resulting probability distribution $p^{(i)}[\Phi]$ ultimately (*i.e.* for $i \to \infty$) converges to the desired Boltzmann distribution

$$p[\Phi] = \frac{1}{Z} \exp(-S[\Phi]). \tag{H.12}$$

After a possibly large number M_0 of iterations (*i.e.* of applications of the Monte Carlo algorithm), the Boltzmann distribution is reached with sufficient accuracy, and the system has practically "forgotten" about the initial configuration. Only the configurations generated after the *equilibration (or thermalization)* are used in the actual calculation.

To estimate the expectation value of some observable $\mathcal{O}[\Phi]$, one averages its values over the configurations of the Monte Carlo sample

$$\langle \mathcal{O} \rangle = \lim_{M \to \infty} \frac{1}{M - M_0 + 1} \sum_{i=M_0}^{M} \mathcal{O}[\Phi^{(i)}]. \tag{H.13}$$

In the limit $M \to \infty$, the calculation becomes exact. At a finite number $M - M_0 + 1$ of measurements, one makes a calculable statistical error that decreases in proportion to $1/\sqrt{M - M_0}$. This is related to the *central limit theorem*, which assures that – under quite general conditions – the mean value of a set of independent random variables has a Gaussian probability distribution (even if the individual variables are distributed in a different way). Hence, in order to increase the numerical accuracy by a factor of 2, one must run the Monte Carlo algorithm 4 times longer. The Boltzmann factor $\exp(-S[\Phi])$ is not explicitly included in the above average. It is implicitly included, because the Markov chain will be constructed in such a way that the configurations ultimately occur with the probability density of eq. (H.12).

H.2   Detailed Balance

In order to demonstrate that a particular Monte Carlo algorithm equilibrates to the correct Boltzmann distribution, it is sufficient to show that it is ergodic and obeys detailed balance. *Detailed balance* means that

$$\exp(-S[\Phi])\, w[\Phi, \Phi'] = \exp(-S[\Phi'])\, w[\Phi', \Phi]. \tag{H.14}$$

Here, $w[\Phi, \Phi']$ is again the transition probability for the algorithm to turn a configuration $[\Phi]$ into $[\Phi']$. When the probability distribution $p^{(i)}[\Phi]$ converges to an equilibrium distribution $v_1[\Phi]$ of field configurations, that distribution is an eigenvector of the transition "matrix" $w[\Phi, \Phi']$ with eigenvalue 1, *i.e.*

$$\int \mathcal{D}\Phi \, v_1[\Phi] w[\Phi, \Phi'] = v_1[\Phi']. \tag{H.15}$$

It is easy to see that the desired Boltzmann distribution, $p[\Phi] = \exp(-S[\Phi])/Z = v_1[\Phi]$, is indeed an eigenvector of $w[\Phi, \Phi']$ with eigenvalue 1, provided that the algorithm obeys detailed balance

$$
\begin{aligned}
\int \mathcal{D}\Phi \, p[\Phi] w[\Phi, \Phi'] &= \int \mathcal{D}\Phi \, \frac{1}{Z} \exp\left(-S[\Phi]\right) w[\Phi, \Phi'] \\
&= \int \mathcal{D}\Phi \, \frac{1}{Z} \exp\left(-S[\Phi']\right) w[\Phi', \Phi] \\
&= \frac{1}{Z} \exp\left(-S[\Phi']\right) \int \mathcal{D}\Phi \, w[\Phi', \Phi] \\
&= \frac{1}{Z} \exp\left(-S[\Phi']\right) = p[\Phi'].
\end{aligned}
\tag{H.16}
$$

When combined with ergodicity, detailed balance is a sufficient (but not necessary) condition for reaching the desired equilibrium Boltzmann distribution.

H.3   Ergodicity and its Implications

This and the next section are inspired by unpublished lecture notes by Ferenc Niedermayer. The mathematical results that underlie this section are due to Oskar Perron in 1907 and Ferdinand Frobenius in 1912.

Ergodicity means that, starting from an arbitrary initial configuration, the algorithm can, at least with some non-zero probability, reach any other field configuration in a finite number of steps. This condition is necessary, because the correct result for an expectation value can be obtained only if all field configurations are indeed accessible.

Assuming ergodicity, we will now show that only one eigenvector of the "matrix" w with eigenvalue 1 exists, and that the equilibrium distribution is therefore unique. Using detailed balance, it then follows that the Markov chain indeed converges to the desired Boltzmann distribution.

Mathematically, ergodicity implies that there exists a finite number of steps n, such that $w^n[\Phi, \Phi'] > 0$ for any pair of configurations $[\Phi]$ and $[\Phi']$. For simplicity, we will assume

$n = 1$; the generalization of the following considerations to $n > 1$ is straightforward. Let us consider the eigenvectors $v_\lambda[\Phi]$ of the transition "matrix" w,

$$\int \mathcal{D}\Phi \, v_\lambda[\Phi]w[\Phi, \Phi'] = \lambda \, v_\lambda[\Phi']. \tag{H.17}$$

First of all, since w is in general not symmetric, *i.e.* $w[\Phi, \Phi'] \neq w[\Phi', \Phi]$, its eigenvalues are not necessarily real and its eigenvectors need not be orthogonal to each other. However, detailed balance allows us to construct the symmetric "matrix"

$$\widetilde{w}[\Phi, \Phi'] = \sqrt{\frac{p[\Phi]}{p[\Phi']}}w[\Phi, \Phi'] = \sqrt{\frac{p[\Phi']}{p[\Phi]}}w[\Phi', \Phi] = \widetilde{w}[\Phi', \Phi], \tag{H.18}$$

which has the same eigenvalues λ as w and corresponding eigenvectors

$$\widetilde{v}_\lambda[\Phi] = \frac{v_\lambda[\Phi]}{\sqrt{p[\Phi]}}, \tag{H.19}$$

because

$$\int \mathcal{D}\Phi \, \widetilde{v}_\lambda[\Phi]\widetilde{w}[\Phi, \Phi'] = \int \mathcal{D}\Phi \, \frac{v_\lambda[\Phi]}{\sqrt{p[\Phi']}}w[\Phi, \Phi'] = \lambda \, \widetilde{v}_\lambda[\Phi']. \tag{H.20}$$

Since $\widetilde{w}$ is symmetric, the eigenvalues λ, which are shared by $\widetilde{w}$ and w, are indeed real. Furthermore, the eigenvectors $\widetilde{v}_\lambda$ are orthogonal and can be normalized, such that

$$\int \mathcal{D}\Phi \, \widetilde{v}_\lambda[\Phi]\widetilde{v}_{\lambda'}[\Phi] = \int \mathcal{D}\Phi \, \frac{1}{p[\Phi]}v_\lambda[\Phi]v_{\lambda'}[\Phi] = \delta_{\lambda,\lambda'}. \tag{H.21}$$

This is consistent with the normalization of $v_1[\Phi] = p[\Phi]$ because

$$\int \mathcal{D}\Phi \, \frac{1}{p[\Phi]}v_1[\Phi]^2 = \int \mathcal{D}\Phi \, p[\Phi] = 1. \tag{H.22}$$

From the eigenvalue equation (H.17), one obtains

$$\lambda \int \mathcal{D}\Phi' \, v_\lambda[\Phi'] = \int \mathcal{D}\Phi \, v_\lambda[\Phi] \int \mathcal{D}\Phi' w[\Phi, \Phi'] = \int \mathcal{D}\Phi \, v_\lambda[\Phi] \quad \Rightarrow$$

$$(\lambda - 1) \int \mathcal{D}\Phi \, v_\lambda[\Phi] = 0. \tag{H.23}$$

This implies that the eigenvectors with eigenvalue $\lambda \neq 1$ obey $\int \mathcal{D}\Phi \, v_\lambda[\Phi] = 0$, which means that they cannot be interpreted as probability distributions. Moreover, the eigenvalues of all eigenvectors with $\lambda \neq 1$ obey $|\lambda| < 1$ because

$$|\lambda| \int \mathcal{D}\Phi' \, \big|v_\lambda[\Phi']\big| = \int \mathcal{D}\Phi' \, \big|\lambda v_\lambda[\Phi']\big| = \int \mathcal{D}\Phi' \, \left|\int \mathcal{D}\Phi \, v_\lambda[\Phi]w[\Phi, \Phi']\right|$$

$$< \int \mathcal{D}\Phi' \int \mathcal{D}\Phi \, \big|v_\lambda[\Phi]\big|w[\Phi, \Phi']$$

$$= \int \mathcal{D}\Phi \, \big|v_\lambda[\Phi]\big| \int \mathcal{D}\Phi' \, w[\Phi, \Phi'] = \int \mathcal{D}\Phi \, \big|v_\lambda[\Phi]\big|. \tag{H.24}$$

When one divides both sides by $\int \mathcal{D}\Phi \, |v_\lambda[\Phi]| \neq 0$, one immediately obtains $|\lambda| < 1$. To derive the above triangle inequality, we have used ergodicity, *i.e.* $w[\Phi, \Phi'] > 0$, which leads to

$$\left|\int \mathcal{D}\Phi \, v_\lambda[\Phi]w[\Phi, \Phi']\right| < \int \mathcal{D}\Phi \, \big|v_\lambda[\Phi]\big|w[\Phi, \Phi']. \tag{H.25}$$

Eigenvectors $v_1[\Phi]$ with eigenvalue $\lambda = 1$ are *stationary* distributions. For them, the inequality (H.24) turns into an equality, which implies

$$\left| \int \mathcal{D}\Phi\, v_1[\Phi] w[\Phi, \Phi'] \right| = \int \mathcal{D}\Phi\, |v_1[\Phi]| w[\Phi, \Phi']. \tag{H.26}$$

This is possible if $v_1[\Phi] \in \mathbb{R}_{\geq 0}$ for all field configurations $[\Phi]$ (possibly excluding exceptional configurations that do not contribute to the functional integral because they are of zero measure). It would also be possible if $v_1[\Phi] \in \mathbb{R}_{\leq 0}$ for all $[\Phi]$. In that case, one can simply change the overall sign of the eigenvector $v_1[\Phi]$ and return to $v_1[\Phi] \in \mathbb{R}_{\geq 0}$. Consequently, in contrast to eigenvectors $v_\lambda[\Phi]$ for which $|\lambda| < 1$, the eigenvectors $v_1[\Phi]$ with eigenvalue 1 can be normalized to $\int \mathcal{D}\Phi\, v_1[\Phi] = 1$ and can thus be interpreted as proper probability distributions. Indeed, assuming detailed balance, we have already seen that the properly normalized Boltzmann distribution, $p[\Phi] = \exp(-S[\Phi])/Z$, is a stationary point of the Markov chain and thus an eigenvector of the transition "matrix" with eigenvalue 1, *i.e.* $p[\Phi] = v_1[\Phi]$.

We will now prove that there can only be one unique eigenvector with eigenvalue 1. We will show this by contradiction. For this purpose, let us assume that there were two distinct eigenvectors $v_1[\Phi] \neq v_1'[\Phi]$ with eigenvalue 1, such that

$$\int \mathcal{D}\Phi\, \left(v_1[\Phi] - v_1'[\Phi] \right) w[\Phi, \Phi'] = v_1[\Phi'] - v_1'[\Phi']. \tag{H.27}$$

As we have just seen, we can normalize both distributions such that

$$\int \mathcal{D}\Phi\, v_1[\Phi] = \int \mathcal{D}\Phi\, v_1'[\Phi] = 1 \quad \Rightarrow \quad \int \mathcal{D}\Phi\, \left(v_1[\Phi] - v_1'[\Phi] \right) = 0. \tag{H.28}$$

However, as we have shown before, any eigenvector $v_\lambda[\Phi]$ that obeys $\int \mathcal{D}\Phi\, v_\lambda[\Phi] = 0$ has an eigenvalue $|\lambda| < 1$. This would hence also apply to $(v_1[\Phi] - v_1'[\Phi])$, in contradiction to the assumption that both $v_1[\Phi]$ and $v_1'[\Phi]$ (and thus their difference) have eigenvalue 1. We thus conclude that there is only one unique eigenvector $v_1[\Phi]$ with eigenvalue 1 and that, based on detailed balance, it coincides with the desired Boltzmann distribution

$$v_1[\Phi] = p[\Phi] = \frac{1}{Z} \exp\left(-S[\Phi] \right). \tag{H.29}$$

H.4   Convergence to the Stationary Distribution

We have seen that an ergodic Monte Carlo algorithm has a unique stationary distribution. When the algorithm obeys detailed balance, this distribution is the desired Boltzmann distribution. Now we will show that the Markov chain indeed *converges* to the unique stationary distribution, irrespective of the arbitrarily chosen initial distribution $p^{(0)}[\Phi]$. Assuming that the eigenvectors form a complete set, we express the initial distribution as the superposition

$$p^{(0)}[\Phi] = \sum_\lambda c_\lambda v_\lambda[\Phi] = c_1 v_1[\Phi] + \sum_{\lambda \neq 1} c_\lambda v_\lambda[\Phi] = p[\Phi] + \sum_{\lambda \neq 1} c_\lambda v_\lambda[\Phi]. \tag{H.30}$$

We have used the fact that the coefficient of the stationary distribution $v_1[\Phi] = p[\Phi]$ is $c_1 = 1$. This follows from the normalization

$$1 = \int \mathcal{D}\Phi \, p^{(0)}[\Phi] = c_1 \int \mathcal{D}\Phi \, v_1[\Phi] + \sum_{\lambda \neq 1} c_\lambda \int \mathcal{D}\Phi \, v_\lambda[\Phi] = c_1, \qquad (H.31)$$

because $\int \mathcal{D}\Phi \, v_\lambda[\Phi] = 0$ for $\lambda \neq 1$.

It may not be obvious that the eigenvalues λ form a discrete spectrum. If the spectrum would be continuous, the sum over λ in eq. (H.30) would have to be replaced by an integral. However, since the field configuration space is a direct product of spheres, $\Phi_x \in S^{N-1}$ (associated with the individual lattice points x), it forms a compact manifold. This implies that the spectrum is indeed discrete.

Let us now perform i Monte Carlo steps and consider the resulting distribution

$$p^{(i)}[\Phi] = \int \mathcal{D}\Phi' \, p^{(0)}[\Phi'] w^i[\Phi', \Phi] = p[\Phi] + \sum_{\lambda \neq 1} \lambda^i c_\lambda v_\lambda[\Phi] \underset{i \to \infty}{\to} p[\Phi]. \qquad (H.32)$$

Since $\lambda \neq 1$ implies $|\lambda| < 1$, in the limit $i \to \infty$ one obtains $\lambda^i \to 0$, such that the Markov chain indeed converges to the desired Boltzmann distribution as the unique stationary distribution.

Ordering all the eigenvalues by decreasing magnitude,

$$1 > |\lambda_1| \geq |\lambda_2| \geq |\lambda_3| \geq \ldots, \qquad (H.33)$$

the rate of convergence to the stationary distribution is determined by the eigenvalue λ_1 (for which $|\lambda_1|$ is closest to 1). Only once $|\lambda_1|^i \ll 1$, for some $i \geq M_0$, the Markov chain has equilibrated sufficiently, and one can begin to measure and average observables $\mathcal{O}[\Phi]$, as described in eq. (H.13).

H.5 Metropolis Algorithm

A simple example of an algorithm that is ergodic and obeys detailed balance is the *Metropolis algorithm* (Metropolis *et al.*, 1953). In this algorithm, a new configuration $[\Phi']$ is chosen at random in the vicinity of the old configuration $[\Phi]$. For example, one visits the lattice sites, either at random or one after the other (lexicographically), and proposes to change the old field value Φ_x by a new value $\Phi'_x \in C_\delta(\Phi_x)$. As illustrated in Figure H.1 a), $C_\delta(\Phi_x) \subset S^{N-1}$ is a cone of opening angle δ around the old field value Φ_x. The newly proposed value Φ'_x is then drawn from the uniform probability distribution within $C_\delta(\Phi_x)$. The probability density q to pick Φ'_x is thus determined by the ratio of the volume $V(S^{N-1})$ of the entire sphere S^{N-1} and the volume $V(C_\delta(\Phi_x))$ of the cone, $q = V(S^{N-1})/V(C_\delta(\Phi_x))$, such that

$$\int_{C_\delta(\Phi_x)} d\Phi'_x \, q = 1. \qquad (H.34)$$

If the Boltzmann weight of the new configuration is at least as large as that of the previous configuration, the new configuration is accepted with probability $p_{\text{acc}} = 1$, *i.e.*

$$S[\Phi'] \leq S[\Phi] \quad \Rightarrow \quad w[\Phi, \Phi'] = q \, p_{\text{acc}} = q. \qquad (H.35)$$

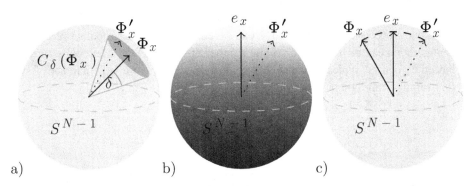

Fig. H.1 Different Monte Carlo algorithms for an $O(N)$-symmetric lattice scalar field theory. a) In a Metropolis step, a new field value $\Phi'_x \in C_\delta(\Phi_x) \subset S^{N-1}$ is proposed inside a cone $C_\delta(\Phi_x)$ of opening angle δ around the old field value Φ_x. b) In a heat-bath step, Φ'_x is drawn (independently of the old variable Φ_x) directly from the appropriate Boltzmann distribution centered at the normalized average e_x of the field values at the neighboring lattice sites. The gray scale symbolizes the Boltzmann distribution, with e_x being the most likely and $-e_x$ being the most unlikely value for Φ'_x. c) In a micro-canonical over-relaxation step, Φ_x is changed to Φ'_x by reflecting it at e_x.

Here, q is just the probability density to propose Φ'_x given Φ_x. On the other hand, if the Boltzmann weight of the new configuration is smaller, the configuration is accepted only with the probability $p_{\text{acc}} = \exp\left(-S[\Phi'] + S[\Phi]\right) \in [0, 1[$, *i.e.*

$$S[\Phi'] > S[\Phi] \quad \Rightarrow \quad w[\Phi, \Phi'] = q\, p_{\text{acc}} = q \exp\left(-S[\Phi'] + S[\Phi]\right). \tag{H.36}$$

Otherwise, the old configuration is kept. In this case, q is the probability density to propose Φ_x given Φ'_x (which coincides with the probability density to propose Φ'_x given Φ_x). The resulting change of the action is calculated by investigating the field values at the neighboring lattice sites $x \pm \hat{\mu}$. Then, following the Metropolis algorithm, it is decided whether or not Φ_x is indeed replaced by Φ'_x. In the latter case, the old value Φ_x is kept unchanged and the old configuration contributes once more to the average of observables. The general expression for the Metropolis acceptance rate is

$$p_{\text{acc}} = \min\{1, \exp\left(-S[\Phi'] + S[\Phi]\right)\} \in [0, 1]. \tag{H.37}$$

This algorithm obeys detailed balance. To see this, let us consider two configurations $[\Phi]$ and $[\Phi']$. Without loss of generality, we may assume that $S[\Phi'] \leq S[\Phi]$ such that $w[\Phi, \Phi'] = q$. Then, $S[\Phi] \geq S[\Phi']$ implies $w[\Phi', \Phi] = q \exp(-S[\Phi] + S[\Phi'])$, and hence

$$\exp\left(-S[\Phi]\right) w[\Phi, \Phi'] = \exp\left(-S[\Phi]\right) q$$
$$= \exp\left(-S[\Phi']\right) q \exp\left(-S[\Phi] + S[\Phi']\right)$$
$$= \exp\left(-S[\Phi']\right) w[\Phi', \Phi]. \tag{H.38}$$

In practice, the opening angle δ of the cone surrounding Φ_x is adjusted, in order to optimize the acceptance rate of the Metropolis step. When δ is very small, the action changes only very little and the acceptance rate is high. However, the new configuration is almost the same as the old one and the Markov chain is not moving rapidly through configuration space. When δ is too large, on the other hand, the action typically increases a lot and the Metropolis proposal is often rejected. As a result, one does again not move rapidly through configuration space either. An optimal choice of δ aims at an acceptance rate of around 65 to 80 percent.

When all field degrees of freedom Φ_x on the lattice have been updated in this way (provided one proceeds lexicographically), one has completed one Metropolis *sweep*. Any field configuration can be reached in a finite number of steps; *i.e.* the Metropolis algorithm is indeed ergodic. A typical Monte Carlo simulation consists of a large number of sweeps, for example, 1 million or more.

H.6 Error Analysis

Since any Markov chain generated in a Monte Carlo simulation has a finite length, the results are not exact but are affected by statistical errors. Consequently, an important part of a reliable Monte Carlo calculation is a careful error analysis. An ideal Monte Carlo algorithm (which usually does not exist in practice) would generate a Markov chain of statistically independent configurations. If the Monte Carlo data for an observable $\mathcal{O}$ are Gaussian distributed, the standard deviation from their average, divided by the square root of the number of measurements $(M - M_0 + 1)$ minus 1, *i.e.* their *standard error*, is given by

$$\Delta\mathcal{O} = \frac{1}{\sqrt{M - M_0}}\sqrt{\langle(\mathcal{O} - \langle\mathcal{O}\rangle)^2\rangle} = \frac{1}{\sqrt{M - M_0}}\sqrt{\langle\mathcal{O}^2\rangle - \langle\mathcal{O}\rangle^2}. \qquad (\text{H.39})$$

In order to reduce the statistical error by a factor of 10, the number of independent equilibrated configurations $(M - M_0 + 1)$ must hence be increased by a factor of 100.

Practical Monte Carlo algorithms (like the Metropolis algorithm) are not ideal; *i.e.* they do not generate statistically independent configurations. In particular, the Metropolis algorithm is rather simple, but often not very efficient. Since the new configuration is generated from the previous configuration in the Markov chain, subsequent configurations are correlated. This implies that the actual statistical error is larger than the above naive estimate of the standard error would suggest. In order to detect the *autocorrelations* of the Monte Carlo data, it is useful to bin the data. For this purpose, one divides the set of $(M - M_0 + 1)$ data into $(M - M_0 + 1)/M_b = m \in \mathbb{N}$ bins of M_b subsequent measurements, and one averages the observable over the different bins

$$\overline{\mathcal{O}}_j = \frac{1}{M_b}\sum_{i=M_0+(j-1)M_b}^{M_0+jM_b-1}\mathcal{O}[\Phi^{(i)}], \quad j \in \{1, 2, \ldots, m\}. \qquad (\text{H.40})$$

The error is then estimated by treating the bin averages as individual independent data

$$\Delta\overline{\mathcal{O}} = \frac{1}{\sqrt{m-1}}\sqrt{\frac{1}{m}\sum_{j=1}^{m}\left(\overline{\mathcal{O}}_j - \langle\mathcal{O}\rangle\right)^2}. \qquad (\text{H.41})$$

Obviously, the average over all bins coincides with the total average over all the data. If the bin size M_b is too small, the averages are still correlated and the resulting error underestimates the true statistical error. When one increases the bin size M_b, the resulting error estimate increases until subsequent bin averages are indeed statistically independent. Once the error reaches a plateau (by increasing M_b), one obtains a reasonable estimate of the true statistical error. A more sophisticated error analysis is described by Ulli Wolff (2004).

H.7 Critical Slowing Down

In order to estimate the number τ of Monte Carlo sweeps that separate statistically independent field configurations, it is also useful to determine the *autocorrelation function* of some observable $\mathcal{O}$

$$\langle \mathcal{O}^{(i)} \mathcal{O}^{(i+t)} \rangle = \lim_{M \to \infty} \frac{1}{M - M_0 + 1 - t} \sum_{i=M_0}^{M-t} \mathcal{O}[\Phi^{(i)}] \mathcal{O}[\Phi^{(i+t)}] \sim c_0 + c_1 \exp(-t/\tau), \quad \text{(H.42)}$$

where τ is the *autocorrelation time* associated with the observable $\mathcal{O}$. In the rare case that $\tau \leq 1/2$, the data can be considered statistically independent, such that the standard error $\Delta \mathcal{O}$ (H.39) is reliable. Otherwise, a simple estimate of the actual error is $\Delta \mathcal{O} \sqrt{2\tau}$ (Madras and Sokal, 1988).

The autocorrelation time τ of the Metropolis algorithm increases drastically when one approaches a second-order phase transition where the correlation length ξ diverges. One encounters critical slowing down if

$$\tau \propto \xi^z, \quad \text{(H.43)}$$

where $z > 0$ is a *dynamical critical exponent* characterizing the efficiency of a Monte Carlo algorithm. For the Metropolis algorithm, one typically finds $z \approx 2$, which leads to a severe critical slowing down problem.

An improvement of the Metropolis algorithm is the so-called *heat-bath algorithm*. The new local variable Φ'_x is then drawn from all possible values, independent of the old variable Φ_x, respecting the Boltzmann distribution. For the standard lattice action (H.2), the probability distribution

$$p(\Phi'_x) \propto \exp\left(a^{d-2} \Phi'_x \cdot \overline{\Phi}_x\right), \quad \overline{\Phi}_x = \sum_{\mu=1}^{d} \left(\Phi_{x+\hat{\mu}} + \Phi_{x-\hat{\mu}}\right), \quad e_x = \frac{\overline{\Phi}_x}{|\overline{\Phi}_x|}, \quad \text{(H.44)}$$

is centered at the normalized average e_x of the field values at the neighboring lattice sites $x \pm \hat{\mu}$, as illustrated in Figure H.1 b). Like the Metropolis algorithm, the heat-bath algorithm suffers from severe critical slowing down because it also has $z \approx 2$.

When they are combined with the *micro-canonical over-relaxation algorithm*, which by itself is not ergodic, both the Metropolis and the heat-bath algorithm can be tuned to achieve a significantly reduced critical slowing down with $z \approx 1$ (Wolff, 1992). The micro-canonical over-relaxation algorithm, which is illustrated in Figure H.1 c), is deterministic and updates a field variable Φ_x to Φ'_x by reflecting it at the normalized average of its nearest neighbors, *i.e.*

$$\Phi'_x = 2 \left(\Phi_x \cdot e_x\right) e_x - \Phi_x. \quad \text{(H.45)}$$

Since $\Phi'_x \cdot e_x = \Phi_x \cdot e_x$, the action remains unchanged in an over-relaxation step. Hence, the micro-canonical over-relaxation algorithm obeys detailed balance. Moreover, since

$$\Phi''_x = 2 \left(\Phi'_x \cdot e_x\right) e_x - \Phi'_x = 2 \left(\Phi_x \cdot e_x\right) e_x - 2 \left(\Phi_x \cdot e_x\right) e_x + \Phi_x = \Phi_x, \quad \text{(H.46)}$$

a repeated over-relaxation step would lead back to the original value Φ_x. In combination with a Metropolis or heat-bath algorithm, it becomes ergodic and is thus guaranteed to converge to the desired Boltzmann distribution.

Thanks to Robert Swendsen, Jian-Sheng Wang, as well as Ulli Wolff, for $O(N)$ models (including the Ising model that corresponds to $N = 1$), *cluster algorithms* exist (Swendsen and Wang, 1987; Wolff, 1989), which are much more efficient near criticality. A cluster algorithm combines correlated field variables to connected clusters which are updated collectively. The probabilistic rules for forming a cluster are designed in such a way that the cluster update is never rejected. These algorithms essentially eliminate critical slowing down, because they have $z \approx 0$. The development of cluster algorithms, which are based on deep insights into the physical nature of field correlations, implied a significant breakthrough in the accurate understanding of the universal features of $O(N)$ models.

In lattice Yang–Mills theory, the heat-bath algorithm combined with over-relaxation is a standard simulation method. For lattice QCD, numerous algorithmic improvements have facilitated much more efficient simulations than what would be possible with a simple Metropolis algorithm. In particular, the Hybrid Monte Carlo algorithm is the standard workhorse in lattice QCD simulations (Duane *et al.*, 1987). However, critical slowing down still persists and makes it challenging to approach both the continuum and the chiral limit. Still, in contrast to past decades, simulations directly at the physical value of the pion mass are feasible nowadays.

H.8 Supercritical Slowing Down and Sign Problems

Critical slowing down near a second-order phase transition is not the only challenge for Monte Carlo calculations. Even after approaching the continuum limit, a quantum field theory may undergo a first-order phase transition as a function of a physical parameter such as the temperature. At a first-order phase transition, two drastically different phases, such as liquid water and ice, coexist (see Appendix I). Even cluster algorithms may have a hard time to drive a Monte Carlo simulation from one phase to the other. The severity of this problem increases exponentially with the area of an interface that separates the two coexisting phases, and is thus even worse than critical slowing down, whose severity is described by the power-law (H.43).

In some cases, algorithms may even be plagued by exponentially severe slowing down, which we denote as *supercritical*. This is often related to topological excitations. The interfaces that separate coexisting bulk phases at a first-order phase transition fall in this category. In Monte Carlo calculations of lattice QCD, topologically non-trivial gluon field configurations may again cause supercritical slowing down, also known as *topological freezing*.

Even more severe problems arise when the Boltzmann factor is not positive definite, because in that case $\exp(-S[\Phi])$ cannot be used as the probability density to generate the configurations in a Markov chain, and the method of importance sampling is not applicable. Such so-called *sign problems* are typical for quantum field theories as well as for quantum many-body systems, but not for classical statistical mechanics, because $\exp(-\beta \mathcal{H}[s])$ is always positive.

Now let us go beyond the initial assumption of this appendix and consider a functional integral in which the Boltzmann weight $\mathrm{Sign}[\Phi] \exp(-S[\Phi])$ consists of a positive contribution $\exp(-S[\Phi]) \geq 0$ and a sign $\mathrm{Sign}[\Phi] = \pm 1$, such that

$$Z = \int \mathcal{D}\Phi \, \text{Sign}[\Phi] \exp\left(-S[\Phi]\right). \tag{H.47}$$

In this case, importance sampling is not applicable since $\text{Sign}[\Phi] \exp(-S[\Phi])$ is in general not positive. A naive attempt to circumvent the sign problem is to apply importance sampling only to the positive weight $\exp(-S[\Phi])$ and to include the sign factor $\text{Sign}[\Phi]$ in the measured observables,

$$\langle \mathcal{O} \rangle = \frac{1}{Z} \int \mathcal{D}\Phi \, \mathcal{O}[\Phi] \, \text{Sign}[\Phi] \exp\left(-S[\Phi]\right) = \frac{\langle \mathcal{O} \, \text{Sign} \rangle'}{\langle \text{Sign} \rangle'},$$

$$\langle \text{Sign} \rangle' = \frac{1}{Z'} \int \mathcal{D}\Phi \, \text{Sign}[\Phi] \exp\left(-S[\Phi]\right) = \frac{Z}{Z'}, \quad Z' = \int \mathcal{D}\Phi \, \exp\left(-S[\Phi]\right),$$

$$\langle \mathcal{O} \, \text{Sign} \rangle' = \frac{1}{Z'} \int \mathcal{D}\Phi \, \mathcal{O}[\Phi] \, \text{Sign}[\Phi] \exp\left(-S[\Phi]\right). \tag{H.48}$$

Here, quantities with a prime refer to a modified system with the positive Boltzmann weight $\exp(-S[\Phi])$, which can indeed be simulated using importance sampling.

However, this naive approach is doomed to fail for any severe sign problem, because the signals to be measured are completely overwhelmed by their statistical errors. In particular, the expectation value of the sign,

$$\langle \text{Sign} \rangle' = \frac{Z}{Z'} = \exp\left(-\left(f - f'\right)\beta V\right), \tag{H.49}$$

is exponentially suppressed with the Euclidean space–time volume βV. Here, $f - f' > 0$ is the difference of the free energy densities of the original system that suffers from the sign problem and the modified system for which importance sampling is possible. Let us estimate the statistical error of $\langle \text{Sign} \rangle'$, even assuming an optimal Monte Carlo algorithm without any autocorrelation problems, such that

$$\Delta \text{Sign} = \frac{1}{\sqrt{M - M_0}} \sqrt{\langle \text{Sign}^2 \rangle' - \langle \text{Sign} \rangle'^2} = \frac{1}{\sqrt{M - M_0}} \sqrt{1 - \langle \text{Sign} \rangle'^2}$$

$$\approx \frac{1}{\sqrt{M - M_0}}. \tag{H.50}$$

Here, we have used the fact that $\text{Sign}[\Phi]^2 = 1$ and we have neglected the exponentially small expectation value $\langle \text{Sign} \rangle'$ inside the square root. This implies that the signal-to-error-ratio is given by

$$\frac{\langle \text{Sign} \rangle'}{\Delta \text{Sign}} \approx \sqrt{M - M_0} \, \exp\left(-\left(f - f'\right)\beta V\right). \tag{H.51}$$

Hence, in order to retrieve the signal from the statistical noise, the Markov chain must generate at least

$$M - M_0 + 1 \propto \exp\left(2\left(f - f'\right)\beta V\right) \tag{H.52}$$

independent configurations. Since this number increases exponentially with the space–time volume, in practice it is completely hopeless to solve a severe sign problem in this naive manner. Sign problems indeed confront us with the most dramatic instances of supercritical slowing down.

Severe sign problems prevent for instance the numerical simulation of the fermionic repulsive Hubbard model away from half-filling in condensed matter physics and thus hinder a deeper understanding of high-temperature superconductivity. Similarly, a very severe complex action problem, with $\exp(-S) \in \mathbb{C}$, prevents Monte Carlo simulations of lattice QCD at low temperature T and even moderate values of the chemical potential μ_B for the baryon number. As a result, we can currently not address the QCD phase diagram in the μ_B-T plane from first principles either. This affects, for example, our understanding of the core of neutron stars, where exotic forms of dense quark matter are suspected to exist. Yet another class of sign problems is related to the complex Boltzmann factor $\exp(i\theta Q)$ related to θ-vacuum states (where Q is the topological charge).

H.9 Complexity Classes and the Severity of Sign Problems

The severity of computational problems is classified by *complexity classes*. The class P contains those problems that can be solved on a Turing machine with an effort that increases at most polynomially with the system size. A Turing machine is a mathematical idealization of a classical computer, as we use on a daily basis. The *complexity class NP* contains problems which could be solved in polynomial time on a hypothetical so-called non-deterministic computer (not to be confused with a quantum computer). A "non-deterministic" computer, which can be viewed as a mathematical generalization of a Turing machine, can simultaneously follow both branches of an if-statement, but cannot merge those branches any more. This is equivalent to having an arbitrarily increasable number of "processors", but without any communication between them. While it is impossible to build such a machine in the real world, it serves as a useful mathematical concept in complexity theory. Problems in the class NP are also characterized by the fact that their solution can be checked (but not necessarily obtained) on an ordinary Turing machine in polynomial time.

A problem is denoted as "NP-hard", if any problem in NP can be mapped to it with polynomial effort. Some of the hardest problems in the class NP (such as the traveling salesman problem) are known as "NP-complete problems". Solving any of these problems in polynomial time on a Turing machine would imply NP = P. Deciding whether or not this is the case is one of the millennium problems for whose solution the Clay Mathematics Institute offers a prize of 10^6 \$. While this problem remains unsolved, a vast majority of complexity theorists believes that NP $\supset$ P, because there have been countless unsuccessful attempts to solve NP-hard problems with a Turing machine in polynomial time.

Some sign problems indeed fall in the computational complexity class of NP-hard problems (Troyer and Wiese, 2005). This does not imply that algorithm development for solving sign problems is doomed to fail. Unless one supposes that NP = P, one should just not try to find a generically applicable method that solves all sign problems at once. Instead, one should aim at solving specific sign or complex action problems, including the ones that arise in the Hubbard model away from half-filling or in lattice QCD at $\mu_B \neq 0$. While this seems extremely difficult, it may not be hopeless.

Indeed, some severe sign problems have been solved using so-called meron-cluster algorithms (Bietenholz *et al.*, 1995; Chandrasekharan and Wiese, 1999). These algorithms decompose a field configuration into clusters that can be updated independently. Summing over the cluster orientations analytically, which is sometimes possible, increases the statistics by an amount that grows exponentially with the space–time volume. When the sign factorizes into a product over the individual clusters, and when the cluster signs average to non-negative results, one can indeed apply importance sampling and solve the sign problem completely. In view of the NP-hardness of some sign problems, it should not be surprising that the meron-cluster algorithm is, however, not universally applicable.

H.10 Quantum Computation and Simulation of Real-Time Evolution

As if all this would not be difficult enough, the numerical computation of the *real-time evolution* of any large, strongly coupled quantum system poses even bigger challenges, because the weight, $\exp(iS_M) \in U(1)$, in a real-time path integral (based on the action S_M in Minkowski space–time) is a rapidly oscillating complex phase. The parts of an isolated, strongly coupled quantum system with many degrees of freedom become more and more entangled during their real-time evolution. This may lead to highly non-classical states, similar to the ones of a Schrödinger cat, whose health is entangled with the potentially deadly environment it is confined to. Although the situation may typically be less dramatic, it seems rather hopeless that classical computers will be able to solve problems involving the real-time evolution of large, isolated quantum systems.

While classical computers have a hard time keeping track of the rapidly oscillating phases in the real-time path integral, quantum computers, which are under intense development, but still at an early stage, rely on quantum hardware and have entanglement naturally built in. A universal quantum computer is a quantum analogue of a Turing machine, which is defined by mathematical rules as pioneered by Paul Benioff (1980) and David Deutsch (1985).

A quantum computer consists of qubits which are two-state systems, equivalent to quantum spins $1/2$. In contrast to a classical bit, a qubit can be in a quantum mechanical superposition of the states 0 and 1. The circuit model of quantum computation applies a sequence of standardized unitary quantum gate operations either to single qubits or to pairs of them. The quantum gate set is constructed in such a way that any unitary time evolution of the quantum computer can be realized, at least in principle. The stroboscopic sequential application of unitary quantum gates is known as digital quantum computation. After an initial state preparation, during the computation, a non-deterministic quantum computer takes advantage of entanglement and quantum mechanical superposition, before the result of the computation is finally read out by a measurement. Like a typical quantum experiment, a quantum computation should usually be repeated many times and will in general produce a statistical distribution of outcomes. Building a sizable quantum computer is extremely challenging, because isolating it from its environment in order to avoid decoherence, while still being able to control it, is very difficult.

Peter Shor has designed a quantum algorithm that outperforms a classical computer in the problem of factoring large integers (Shor, 1994), with potential applications to breaking classical encryption schemes and thus compromising data security, once sufficiently powerful quantum computers would become available. A quantum computer running Shor's algorithm is not guaranteed to always provide the right answer, but it does so with a reasonable probability. This is sufficient because it is easy to check a proposed prime number factorization, while it is extremely hard to find the correct one.

Another successful quantum algorithm by Lov Grover solves the problem of searching an unstructured database of size N with an effort proportional to $\sqrt{N}$ (Grover, 1996).

As Richard Feynman pointed out, a quantum computer would be extremely useful for simulating quantum systems that are difficult to understand otherwise (Feynman, 1982). For this purpose, even quantum simulators, which are special-purpose quantum computers, are very useful. In contrast to the envisioned universal quantum computers, the existing quantum simulators do not have error correction and are thus of limited reliability, but still most welcome in the absence of powerful alternatives.

Quantum simulators and prototype quantum computers are delicate devices whose operation requires detailed knowledge of experimental physics. Should a theoretical physicist not be reluctant to rely on such experiments? Certainly, the status of classical computers, which are operated by theoretical physicists on a daily basis, may encourage us to be open-minded about the new developments. A theoretical physicist who is running a simulation of a lattice field theory on a classical computer is probably not thinking of herself as performing an experiment, despite the fact that she is operating a highly sophisticated piece of equipment. It is more likely that she just relies on the device as embodying the mathematical concept of a Turing machine. Before quantum computers reach a corresponding status, a theoretical physicist should be careful and critical when using such devices, but may still use them, just like classical computers, as a most welcome extension of our mental capabilities. After all, our brain, which is possibly the most advanced piece of "hardware" in the observable Universe, is not always operating completely reliably either, and we do not hesitate to use it as best as we can.

A *digital* quantum simulator can, for example, be realized with trapped ions which are driven through a designed time evolution by applying external laser pulses in a stroboscopic manner (Blatt and Roos, 2012). An *analog* quantum simulator, on the other hand, is a well-controlled quantum system that implements a specifically designed Hamilton operator and realizes its evolution in continuous real time. A breakthrough experiment implemented the bosonic Hubbard model in an analog quantum simulation using ultra-cold bosonic atoms in an optical lattice formed by counter-propagating laser beams (Greiner *et al.*, 2002). Extending such experiments to fermionic atoms at sufficiently low temperatures, and thus quantum simulating the fermionic Hubbard model, holds the promise to deepen our understanding of high-temperature superconductivity.

It is interesting to ask whether particle physics could also benefit from quantum computations or quantum simulations. In heavy-ion collisions, strongly interacting quarks and gluons undergo an intricate real-time evolution that is far beyond our capabilities for classical computations from QCD first principles. So, one usually resorts to a hydrodynamical description. Similarly, the QCD phase diagram at non-zero baryon density is currently as inaccessible as the fermionic Hubbard model away from half-filling. Quantum simulators and quantum computers, which do not suffer from sign or complex action problems, are very promising in this context. Indeed, quantum simulator constructions have already been proposed or even realized for toy model gauge theories (for reviews, see Wiese (2013); Zohar *et al.* (2016); Bañuls *et al.* (2020)). Extending these to QCD applications is a promising longer term challenge for the hopefully not too distant future (Wiese, 2014).

Appendix I   Phase Transitions and Critical Phenomena

This appendix summarizes some basic concepts which are used to describe phase transitions and critical phenomena. In particular, we sketch general aspects of second-order phase transitions, critical exponents, and universality. The framework of the discussion is classical spin models – in particular the Ising model – as introduced in Chapter 1. Based on the scaling hypothesis, we derive a set of scaling laws which relate the critical exponents.

I.1   Phase Transitions and Critical Points

Phase diagrams of thermodynamic systems specify the parameter values at which phase transitions take place. The most generic type is abrupt transitions (first order), while smoother transitions (second order) occur at specific *critical points* (they can also form critical lines, or in general hyper-surfaces). The term *critical phenomena* refers to the behavior in the vicinity of a critical point. Figure I.1 illustrates two prominent examples for phase diagrams.

- The figure on the left sketches the phase diagram of *water*. The phase transition that separates liquid and steam is of first order, characterized by an abrupt change of the density of water molecules. When we follow the coexistence curve of liquid and steam for rising temperature, we reach an endpoint at a critical temperature and pressure of $T_c = 647$ K, $p_c = 220.6$ b. At this point, the distinction between liquid and gas ceases, because both exist with the same density. The line of first-order phase transitions ends at a critical point that corresponds to a second-order phase transition. In the vicinity of the critical endpoint, water displays universal behavior. In particular, density fluctuations appear on all length scales, thus leading to the universal phenomenon of *critical opalescence*.
- On the right, we show the phase diagram of the 3-d Ising model which was introduced in Chapter 1. When we set the external magnetic field to $B = 0$, the system spontaneously magnetizes below a critical temperature T_c, while at $T \geq T_c$ the magnetization M vanishes. If we switch on a magnetic field at temperatures below T_c, the magnetization M changes abruptly when B changes sign. This is analogous to the abrupt density increase from steam to liquid when the pressure is increased below T_c in the phase diagram of water. This also represents a first-order phase transition. Again, this phase transition ends at a critical point. In fact, the critical endpoints in the phase diagrams of water and of the 3-d Ising model belong to the same *universality class*. They are characterized by the same *critical exponents*, as we will discuss below. In particular, some aspects of critical opalescence can be described quantitatively by the very simple Ising model, despite the fact that the microscopic dynamics of water molecules

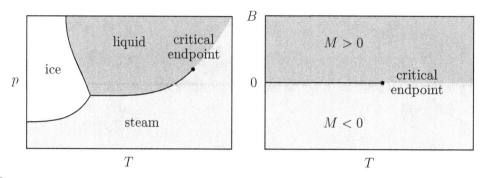

Fig. I.1 Examples for first-order transition lines, and for critical points (marked by bullets) in the phase diagrams of water as a function of temperature T and pressure p (on the left), and of the 3-d Ising model in an external magnetic field B (on the right).

is much more complicated than that of Ising spins. The long wavelength fluctuations that arise near second-order phase transitions are insensitive to microscopic details.

These examples illustrate that second-order phase transitions tend to occur at specific points, which can only be reached by a careful fine-tuning of the relevant parameters, such as the temperature.

The lines leading to a critical point have a fundamental feature in common: The system has the choice between different coexisting states (water of high or low density ρ, or the sign of the magnetization M). In both cases, this freedom of choice ends at the critical point. The situation below and above T_c can be distinguished by an *order parameter*, in this case

$$\rho - \rho_c, \; M = \begin{cases} \neq 0 & T < T_c \\ = 0 & T \geq T_c \end{cases}. \tag{I.1}$$

Here, ρ_c is the density of water at the critical point.[1] Generally, an order parameter drops to zero as one reaches criticality within the ordered phase. One characterizes the way *how* it vanishes at this point by critical exponents.

I.2 Critical Exponents

For the further discussion, we refer to the Ising model. At temperatures below T_c (and in the absence of an external magnetic field B), it exhibits spontaneous magnetization, as we already mentioned.

In order to quantify the way how the magnetization vanishes at T_c, we now invoke the fundamental quantities of statistical mechanics. A system is described by its *free energy F*

$$F = -T \log Z \tag{I.2}$$

where Z is the *partition function* that we know from Chapter 1. We recall that the unit of temperature is chosen such that the Boltzmann constant k_B is set to 1.

---

[1] Deviations from ρ_c at $T > T_c$ are small compared to the wide gap between ρ_{steam} and ρ_{liquid} below T_c.

The dependence of the free energy on the parameters T and B near the critical point characterizes the critical behavior. We first consider the magnetization as T approaches T_c from below, at $B = 0$,

$$T \lesssim T_c : \quad M = -\frac{\partial F}{\partial B}\bigg|_{T,B=0} \propto (T_c - T)^\beta. \tag{I.3}$$

The last expression is an ansatz, which introduces β as our first example for a critical exponent. Taking a further derivative yields the (magnetic) susceptibility χ along with another critical exponent γ,

$$\chi = \frac{\partial M}{\partial B}\bigg|_{T,B=0} = -\frac{\partial^2 F}{\partial B^2}\bigg|_{T,B=0} \propto |T - T_c|^{-\gamma}. \tag{I.4}$$

So far, we chose some temperature T near (but different from) T_c. Alternatively, we can consider the isothermal behavior at $T = T_c$, but $B \neq 0$. The way how M vanishes as B approaches zero determines the next critical exponent δ,

$$B \gtrsim 0 : \quad M|_{T=T_c} \propto B^{1/\delta}. \tag{I.5}$$

Here, we consider how M approaches zero in a direction in the phase diagram that differs from the one that eq. (I.3) refers to. Schematically, the behavior defining these critical exponents is illustrated in Figure I.2.

We saw in eq. (I.4) that a second derivative of the free energy diverges at a critical point. This property, which characterizes a rather smooth phase transition, inspired the term "second order". First-order phase transitions are more abrupt (as we anticipated in the introduction to this appendix), because there typically even a first derivative of F is discontinuous. We are referring to the (traditional) *classification by Paul Ehrenfest*, where the order of a phase transition corresponds to the number of derivatives of F that are needed to encounter a discontinuity.

In this book, we deal with second-order phase transitions whenever we refer to the continuum limit of field theory on the lattice. A physical first-order phase transitions is addressed, *e.g.*, in Chapter 14, when we discuss the deconfinement phase transition of gluons in SU(3) Yang–Mills theory.

As another characteristic of second-order phase transitions, we now start from the *free energy density f* and define the critical exponent α through the specific heat C,

$$C = -T\frac{\partial^2 f}{\partial T^2} \propto |T_c - T|^{-\alpha}. \tag{I.6}$$

In Section 1.4, we have introduced the correlation length ξ. The way how it diverges when the temperature approaches criticality defines another critical exponent ν,

$$\xi \propto |T - T_c|^{-\nu}. \tag{I.7}$$

In eqs. (I.4), (I.6), and (I.7), we could in principle distinguish exponents γ, α, ν for the case $T \gtrsim T_c$, and γ', α', ν' for $T \lesssim T_c$. However, γ and γ', α and α', as well as ν and ν' turn out to be the same – both in experiments and in theoretical investigations – so we ignore this distinction. The existence of γ' at $T > T_c$ is not in contradiction with the fact that M vanishes at $T > T_c$, because we take its derivative with respect to B.

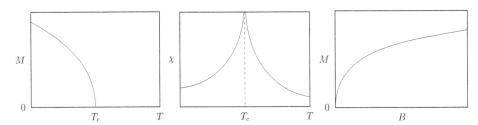

Fig. I.2 The qualitative behavior of the magnetization M (on the left) and the (magnetic) susceptibility χ (in the center) in a temperature interval that includes the critical temperature T_c. On the right, we illustrate the typical dependence of M on an external magnetic field B at T_c.

The corresponding exponents can also be introduced for the critical point which marks the end of a liquid–gas coexistence line shown in Figure I.1. In this correspondence, combinations of the pressure difference $|p - p_c|$ and the density difference $|\rho - \rho_c|$ adopt roles similar to the modulus of the external magnetic field and the magnetization, B and M. Remarkably, the critical exponents are the same for the critical points of the 3-d Ising model and of water.

I.3 Universal Critical Behavior

Besides water, there are many different materials that show critical behavior. For example, ferromagnets like iron, nickel, or cobalt magnetize spontaneously below a critical temperature T_c. We can distinguish different materials with different critical temperatures T_c, different crystal structures, etc. For instance, different ferromagnets have crystalline unit-cells of different sizes, so they could at best be related by a scale transformation. Therefore, dimensioned quantities – such as T_c – cannot be expected to be similar. On the other hand, a common property may concern the critical exponents, which necessarily are dimension-less, as all exponents. Indeed, over broad classes of critical points one observes the same critical exponents.

Ferromagnetic materials can be described by a number of approaches. We describe a sequence of increasingly simple types of descriptions:

- Atomic nuclei and electrons interacting by the Coulomb force: In this framework, ferromagnetism emerges through spin–orbit couplings between electrons in different bands. This formalism provides detailed insights into the microscopic origin of ferromagnetism, but it tends to be complicated.
- A simplification considers electrons on the sites of a crystal lattice with some effective potential. The latter captures the essential aspects of the interaction with a few parameters, which can be derived in condensed matter physics.
- It is even simpler to consider classical spins on lattice sites. Their effective interaction is a rough approximation to the coupling between magnetized domains. Here, critical phenomena are described by long-range collective effects of classical spins. In this respect, the short-ranged details are not important. Even this drastic simplification captures some collective phenomena perfectly well.

To be more precise on the last point, we anticipate that we will deal with short-range couplings between the spins, which may, however, induce long-range effects by the collective spin behavior. A variety of short-range couplings leads to the same long-range effects; it is in this sense that the short-ranged details are not of importance.

So, we stay with the last and simplest type of model: We introduce a lattice – for simplicity of a cubic structure – and label its sites by indices x. To each site, we attach a *classical spin*, which is just a unit-vector $\vec{s}_x = (s_x^1, s_x^2, \ldots, s_x^N) \in \mathbb{R}^N, |\vec{s}_x| = 1$.

As one considers the exact values of the critical exponents, it turns out that they agree precisely within certain classes of critical phenomena. This *universality* property relates apparently different systems and models, and it motivates the search for a universal description of the critical phenomena involved.

One expectation is that classical spin models capture many fundamental features of universality. This expectation was made explicit by Robert Griffiths (1967) with the formulation of the *universality hypothesis*, which we simplify as follows:

Box I.1   **Universality hypothesis**

As long as the interactions are local, the critical exponents in classical spin models only depend on the spatial dimension d and on the symmetries of the relevant order parameters.

The *locality* property requires that the couplings between spins should typically decay at least exponentially with the distance separating them. In practice, one usually considers the case of *ultra-locality* where these coupling strengths drop to zero for all distances beyond a few lattice spacings or even just beyond *one* lattice spacing. Based on the universality hypothesis, the models then fall into distinct *universality classes*. In general, the universality class – and therefore the value of the critical exponents – does not depend on the details of short-ranged couplings. As discussed in Chapter 5, the renormalization group provides the basis for understanding universal phenomena.

I.4 Scaling Hypothesis

The scaling hypothesis supposes that *one single scale* matters for a set of critical phenomena. This scale can be set by the correlation length ξ – it is therefore used as *the* measure for all divergences at a critical point.

This hypothesis attaches to all quantities a *scaling dimension*, which is the power of its dependence on ξ. In general, the scaling dimension differs from the (standard) engineering dimension. This unique scale links the critical exponents, and the relations among them are denoted as *scaling laws*. An immediate consequence of this hypothesis is that the relation $\nu = \nu'$ also entails $\alpha = \alpha'$ and $\gamma = \gamma'$.

Let us sketch a further example: The free energy F itself remains continuous at the critical point. Hence, it has the scaling dimension 0, $F \propto \xi^0$. In units of the correlation length, the free energy density near the critical point is therefore given by $f = F/\xi^d \propto \xi^{-d}$; *i.e.*, f has the scaling dimension $-d$. Now we obtain the temperature dependence of the specific heat,

$$C = -T\frac{\partial^2 f}{\partial T^2} \propto \frac{\xi^{-d}}{|T - T_c|^2} \propto |T - T_c|^{\nu\, d - 2}. \tag{I.8}$$

Confronting this relation with the definition of the critical exponent α, eq. (I.6), we read off *Josephson's scaling law* (Josephson, 1967)

$$\alpha = 2 - \nu\, d. \tag{I.9}$$

Let us introduce the (dimensionless) *reduced temperature*

$$\tau = 1 - T/T_c. \tag{I.10}$$

We now take a closer look at the scaling transformation of the free energy F in the vicinity of a critical point, located at $\tau = B = 0$, where we return to a uniform magnetic field B. Based on the scaling hypothesis, the free energy can be expressed as

$$F(\tau, B) = \frac{1}{\lambda} F(\lambda^u \tau, \lambda^v B), \tag{I.11}$$

with an arbitrary rescaling factor $\lambda > 0$. We also leave the values of the exponents u and v open. The assumption that we make is that τ and B asymptotically exhibit a *monomial* dependence on λ (they scale with a single power of λ). This assumption is by no means trivial, it is an application of the scaling hypothesis.

Once we have assumed this property, there are special choices for λ for which F turns into a function of only *one* variable. We show two possibilities to achieve this, and we denote the corresponding functions of a single variable as ϕ_1 and ϕ_2,

$$\lambda = \tau^{-1/u} \quad : \quad F(\tau, B) = \tau^{1/u} F(1, B/\tau^{v/u}) = \tau^{1/u}\phi_1(B/\tau^{v/u}), \tag{I.12}$$

and alternatively

$$\lambda = B^{-1/v} \quad : \quad F(\tau, B) = B^{1/v} F(\tau/B^{u/v}, 1) = B^{1/v}\phi_2(\tau/B^{u/v}). \tag{I.13}$$

We start from eq. (I.12) and relate u and v to the critical exponents β, γ, and α,

$$M|_{B=0} = -\left.\frac{\partial F}{\partial B}\right|_{B=0} = -\tau^{(1-v)/u}\phi_1'(0) \propto \tau^{\beta} \quad \Rightarrow \quad \beta = \frac{1 - v}{u}$$

$$\chi|_{B=0} = -\left.\frac{\partial^2 F}{\partial B^2}\right|_{B=0} = -\tau^{(1-2v)/u}\phi_1''(0) \propto \tau^{-\gamma} \quad \Rightarrow \quad \gamma = \frac{2v - 1}{u}$$

$$C|_{B=0} = -T\left.\frac{\partial^2 f}{\partial T^2}\right|_{B=0} \propto -\frac{T}{T_c^2}\frac{1}{u}\left(\frac{1}{u} - 1\right)\tau^{1/u-2}\phi_1(0) \propto \tau^{-\alpha} \quad \Rightarrow \quad \alpha = 2 - \frac{1}{u}. \tag{I.14}$$

The exponents u and v can be eliminated from the three eqs. (I.14) by means of the linear combination

$$\alpha + 2\beta + \gamma = 2. \tag{I.15}$$

This is *Rushbrooke's scaling law* (Rushbrooke, 1963).

To include also the critical exponent δ, we have to work at $\tau = 0$. Now eq. (I.13) proves useful,

$$M = -\left.\frac{\partial F}{\partial B}\right|_{\tau=0} = -\frac{1}{v}B^{1/v-1}\phi_2(0) \propto B^{1/\delta} \quad \Rightarrow \quad \delta = \frac{v}{1 - v}. \tag{I.16}$$

Along with eqs. (I.14), this additional relation provides another two possibilities to eliminate u and v,

$$\alpha + \beta(\delta + 1) = 2, \qquad \gamma - \beta(\delta - 1) = 0. \tag{I.17}$$

These are the *scaling laws by Griffiths* and *Widom*, respectively (Griffiths, 1967; Widom, 1964).

The renormalization group provides theoretical justification for the scaling hypothesis. In particular, it relates critical exponents to the anomalous dimensions of various quantities. An anomalous dimension measures the deviation of a scaling dimension from the corresponding engineering dimension. Via the renormalization group, critical exponents become accessible to perturbative investigations using the ε-expansion, with $d = 4 - \varepsilon$, around 4 dimensions.

I.5 Critical Exponents and Scaling Laws: An Overview

For the sake of a quick overview, we summarize the definitions, values, and scaling laws of the critical exponents involved. We refer to the following definitions:

- Magnetization: $\quad M = -\dfrac{\partial F}{\partial B}\Big|_{T,B=0} \propto (T_c - T)^{\beta}$

- Susceptibility: $\quad \chi = \dfrac{\partial M}{\partial B}\Big|_{T,B=0} \propto |T_c - T|^{-\gamma}$

- Isothermal magnetization: $\quad M|_{T=T_c} \propto B^{1/\delta}$

- Specific heat: $\quad C = -T \dfrac{\partial^2 f}{\partial T^2}\Big|_{B=0} \propto |T - T_c|^{-\alpha}$

- Correlation length: $\quad \xi \propto |T - T_c|^{-\nu}$

In Table I.1, we first list experimental values for the ferromagnets iron (Kaul, 1985) and nickel (Seeger *et al.*, 1995), and for a typical liquid–gas endpoint. We add the exponents for the Ising model in various dimensions, and in the so-called mean field approximation (MFA), which becomes exact in the limit of infinitely many dimensions. Triviality of $\lambda\phi^4$ theory implies that the critical exponents of the 4-d Ising model are those of free field theory. For the 3-d Ising model, the most accurate results are obtained with the conformal bootstrap (El-Showk *et al.*, 2014), while for the 3-d O(2) (Campostrini *et al.*, 2006) and the 3-d O(3) model (Campostrini *et al.*, 2002), Monte Carlo simulations performed with the very efficient Wolff cluster algorithm are most accurate (Wolff, 1989). Experimental investigations of the 3-d O(2) universality class have been performed with superfluid ^{4}He on the International Space Station in order to reduce disturbances due to gravity (Lipa *et al.*, 2003).

The scaling hypothesis assumes the critical phenomena to depend on a single scale, which relates the critical exponents. We could *a priori* distinguish exponents for divergences at $T \lesssim T_c$ and at $T \gtrsim T_c$, and consider $\nu = \nu'$, $\alpha = \alpha'$, $\gamma = \gamma'$ as a first set of scaling laws. In Table I.2, we list five more scaling laws which appear in the literature; the first four of them have been derived in Section I.4. The reader is encouraged to verify

Table I.1 Values for the critical exponents discussed in this appendix, obtained experimentally, analytically, or numerically. We give examples from a variety of universality classes.

		β	γ	α	δ	ν
Fe ($T_c - 1044$ K)		0.385(5)	1.333	−0.103(11)	4.350(5)	
Ni ($T_c = 632$ K)		0.395(10)	1.345(10)	−0.091(2)	4.58(5)	
Liquid–gas		0.35	1.37(20)	≈ 0	4.4(4)	0.64
Ising model	$d = 2$	1/8	7/4	0	15	1
	$d = 3$	0.326419(3)	1.237075(10)	0.11008(1)	4.78984(1)	0.629971(4)
	$d = 4$	1/2	1	0	3	1/2
	MFA	1/2	1	0	3	1/2
O(2) model	$d = 3$	0.3486(1)	1.3178(2)	−0.0151(3)	4.780(1)	0.6717(1)
O(3) model	$d = 3$	0.3689(3)	1.3960(9)	−0.1336(15)	4.783(3)	0.7112(5)

Table I.2 Some scaling laws, *i.e.*, relations among the critical exponents.

$\nu d + \alpha$	=	2	Josephson
$\alpha + 2\beta + \gamma$	=	2	Rushbrooke
$\alpha + \beta(1 + \delta)$	=	2	Griffiths I
γ	=	$\beta(\delta - 1)$	Widom
$\gamma(\delta + 1)$	=	$(2 - \alpha)(\delta - 1)$	Griffiths II

that only three of them are independent. Josephson's law is peculiar because it involves the space-dimension d; for this reason, it is denoted as a "hyperscaling law". These laws hold for the models that can be solved exactly. They also hold within error bars for the experimental values and for the models that were studied numerically or analytically by means of suitable expansions.

For completeness, we add that the connected correlation function on the isothermal curve at T_c vanishes with a *power-law* for increasing separation $|x - y|$. Its power defines yet another critical exponent η, according to

$$\langle \vec{s}_x \cdot \vec{s}_y \rangle_c |_{T_c} \propto |x - y|^{2+\eta-d}. \tag{I.18}$$

This is an example for an anomalous dimension. The exponent η is linked to the above critical exponents by further scaling laws, which can be found in the textbook literature, *e.g.*, Ma (1976); Pfeuty and Toulouse (1977); Amit (1978); and Herbut (2007).

References

Aaij, R., *et al.* 2013. First observation of CP violation in the decays of B_0^s mesons. *Phys. Rev. Lett.*, **110**, 221601.

Aaij, R., *et al.* 2019. Observation of CP violation in charm decays. *Phys. Rev. Lett.*, **122**, 21180.

Abbas, A. 1990. Anomalies and charge quantization in the standard model with arbitrary number of colours. *Phys. Lett. B*, **238**, 344–347.

Abbas, A. 2000. *On the number of colours in Quantum Chromodynamics*. Preprint arXiv:hep-ph/0009242.

Abbott, L. F., and Farhi, E. 1981. Are the weak interactions strong? *Phys. Lett. B*, **101**, 69–72.

Abbott, L. F., and Sikivie, P. 1983. A cosmological bound on the invisible axion. *Phys. Lett. B*, **120**, 133–136.

Abe, K. *et al.* (Belle Collaboration). 2001. Observation of large CP violation in the neutral B meson system. *Phys. Rev. Lett.*, **87**, 091802.

Abe, K. *et al.* (Super-Kamiokande Collaboration). 2017. Search for proton decay via $p \rightarrow e^+\pi^0$ and $p \rightarrow \mu^+\pi^0$ in 0.31 megaton·years exposure of the Super-Kamiokande water Cherenkov detector. *Phys. Rev. D*, **95**, 012004.

Abe, Y. *et al.* (Double Chooz Collaboration). 2012. Reactor $\bar{\nu}_e$ disappearance in the Double Chooz experiment. *Phys. Rev. D*, **86**, 052008.

Abel, S. *et al.* 2020. Measurement of the permanent electric dipole moment of the neutron. *Phys. Rev. Lett.*, **124**, 081803.

Adkins, G. S., Nappi, C. R., and Witten, E. 1983. Static properties of nucleons in the Skyrme model. *Nucl. Phys. B*, **228**, 552–566.

Adler, S. 1969. Axial-vector vertex in spinor electrodynamics. *Phys. Rev.*, **177**, 2426–2438.

Adler, S. L., and Bardeen, W. A. 1969. Absence of higher-order corrections in the anomalous axial-vector divergence equation. *Phys. Rev.*, **182**, 1517–1536.

Affleck, I. 1985. The quantum Hall effects, σ-models at $\theta = \pi$ and quantum spin chains. *Nucl. Phys. B*, **257**, 397–406.

Affleck, I. 1988. Field theory methods and quantum critical phenomena. Pages 563–640 of: Brézin, E., and Zinn-Justin, J. (eds), *Fields, Strings and Critical Phenomena*. Proceedings of the Les Houches Summer School, Session XLIX.

Affleck, I., and Haldane, F. D. M. 1987. Critical theory of quantum spin chains. *Phys. Rev. B*, **36**, 5291–5300.

Aguilar-Arévalo, A. and Bietenholz, W., Neutrinos: Mysterious Particles with Fascinating Features, which led to the Physics Nobel Prize 2015, Rev. Cubana Fis. 32 (2015) 127–136.

Aharonov, Y., and Bohm, B. 1959. Significance of electromagnetic potentials in the quantum theory. *Phys. Rev.*, **115**, 485–491.

Aharonov, Y., and Bohm, B. 1961. Further considerations on electromagnetic potentials in the quantum theory. *Phys. Rev.*, **123**, 1511–1524.

Ahmad, Q. R. *et al.* (SNO Collaboration). 2001. Measurements of the rate of $\nu_e + d \to p + p + e^-$ interactions produced by ^{8}B solar neutrinos at the Sudbury Neutrino Observatory. *Phys. Rev. Lett.*, **87**, 071301.

Ahn, J. K. *et al.* (RENO Collaboration). 2012. Observation of reactor electron antineutrinos disappearance in the RENO experiment. *Phys. Rev. Lett.*, **108**, 191802.

Aizenman, M. 1981. Proof of the triviality of ϕ_d^4 field theory and some mean field features of Ising models for $d > 4$. *Phys. Rev. Lett.*, **47**, 1–4.

Aizenman, M., and Duminil-Copin, H. 2021. Marginal triviality of the scaling limits of critical 4D Ising and ϕ^4 models. *Ann. Math.*, **194**, 163–235.

Aker, M. *et al.* (KATRIN Collaboration). 2019. Improved upper limit on the neutrino mass from a direct kinematic method by KATRIN. *Phys. Rev. Lett.*, **123**, 221802.

Akiba, T., Kikuchi, H., and Yanagida, T. 1988. Static minimum-energy path from a vacuum to a sphaleron in the Weinberg–Salam model. *Phys. Rev. D*, **38**, 1937–1941.

Albrecht, A., and Steinhardt, P. J. 1982. Cosmology for Grand Unified Theories with radiatively induced symmetry breaking. *Phys. Rev. Lett.*, **48**, 1220–1223.

Alekseev, A. Y., Cheianov, V. V., and Fröhlich, J. 1998. Universality of transport properties in equilibrium, the Goldstone theorem, and chiral anomaly. *Phys. Rev. Lett.*, **81**, 3503–3506.

Altarelli, G., and Parisi, G. 1977. Asymptotic freedom in parton language. *Nucl. Phys. B*, **126**, 298–318.

Alvarez-Gaumé, L., and Witten, E. 1984. Gravitational anomalies. *Nucl. Phys. B*, **234**, 269–330.

Amit, D. J. 1978. *Field Theory, the Renormalization Group and Critical Phenomena.* McGraw-Hill.

An, F. P. *et al.* (Daya Bay Collaboration). 2012. Observation of electron-antineutrino disappearance at Daya Bay. *Phys. Rev. Lett.*, **108**, 171803.

Anderson, C. D. 1933. The positive electron. *Phys. Rev.*, **43**, 491–498.

Anderson, P. W. 1963. Plasmons, gauge invariance, and mass. *Phys. Rev.*, **130**, 439–442.

Antoine, J.-P., and Speiser, D. 1964a. Characters of irreducible representations of the simple groups. I. General theory. *J. Math. Phys.*, **5**, 1226–1234.

Antoine, J.-P., and Speiser, D. 1964b. Characters of irreducible representations of the simple groups. II. Application to classical groups. *J. Math. Phys.*, **5**, 1560–1572.

Aoki, S. *et al.* (Flavour Lattice Averaging Group, FLAG). 2020. FLAG Review 2019. *Eur. Phys. J. C*, **80**, 113.

Aoki, Y. *et al.* (Flavor Lattice Averaging Group, FLAG), 2022, FLAG Review 2021, *Eur. Phys. J. C.*, **82**, 869.

Arnold, P. B., and McLerran, L. D. 1987. Sphalerons, small fluctuations and baryon number violation in electroweak theory. *Phys. Rev. D*, **36**, 581–595.

Arnold, V. I. 1978. *Methods of Classical Mechanics.* Springer.

Asaka, T., Buchmüller, W., and Covi, L. 2001. Gauge unification in six dimensions. *Phys. Lett. B*, **523**, 199–204.

Ashman, J. *et al.* (European Muon Collaboration). 1988. A measurement of the spin asymmetry and determination of the structure function g_1 in deep inelastic muon-proton scattering. *Phys. Lett. B*, **206**, 364–370.

Atiyah, M. F., and Singer, I. M. 1963. The index of elliptic operators on compact manifolds. *Bull. Amer. Math. Soc.*, **69**, 442–433.

Aubert, B., *et al.* 2001. Observation of CP violation in the B^0 meson system. *Phys. Rev. Lett.*, **87**, 091801.

Bañuls, M. C., *et al.* 2020. Simulating lattice gauge theories within quantum technologies. *Eur. Phys. J. D*, **74**, 165.

Bailey, S., Cridge, T., Harland-Lang, L. A., Martin, A. D., and Thorne, R. S. 2021. Parton distributions from LHC, HERA, Tevatron and fixed target data: MSHT20 PDFs. *Eur. Phys. J. C*, **81**, 341.

Baker, C. A. *et al.* 2006. Improved experimental limit on the electric dipole moment of the neutron. *Phys. Rev. Lett.*, **97**, 131801.

Banks, T., and Zaks, A. 1982. On the phase structure of vector-like gauge theories with massless fermions. *Nucl. Phys. B*, **196**, 189–204.

Bär, O. 2003. On Witten's global anomaly for higher SU(2) representations. *Nucl. Phys. B*, **650**, 522–542.

Bär, O., and Campos, I. 2000. Global anomalies in chiral gauge theories on the lattice. *Nucl. Phys. B*, **581**, 499–519.

Bär, O., and Wiese, U.-J. 2001. Can one see the number of colors? *Nucl. Phys. B*, **609**, 225–246.

Bär, O., Imboden, M., and Wiese, U.-J. 2004. Pions versus magnons: From QCD to anti-ferromagnets and quantum Hall ferromagnets. *Nucl. Phys. B*, **686**, 347–376.

Bardeen, J., Cooper, L. N., and Schrieffer, J. R. 1957. Theory of superconductivity. *Phys. Rev.*, **108**, 1175–1204.

Bazavov, A. *et al.* (HotQCD Collaboration). 2019. Chiral crossover in QCD at zero and non-zero chemical potentials. *Phys. Lett. B*, **795**, 15–21.

Becchi, C., Rouet, A., and Stora, R. 1974. The Abelian Higgs-Kibble model. Unitarity of the S operator. *Phys. Lett. B*, **52**, 344–346.

Becchi, C., Rouet, A., and Stora, R. 1975. Renormalization of the Abelian Higgs-Kibble model. *Commun. Math. Phys.*, **42**, 127–162.

Becchi, C., Rouet, A., and Stora, R. 1976. Renormalization of gauge theories. *Ann. Phys.*, **98**, 287–321.

Becher, P., Böhm, M., and Joos, H. 1983. *Eichtheorien der starken und elektroschwachen Wechselwirkung*. Teubner.

Becher, T., and Leutwyler, H. 1999. Baryon chiral perturbation theory in manifestly Lorentz invariant form. *Eur. Phys. J. C*, **9**, 643–671.

Belavin, A. A., Polyakov, A. M., Schwartz, A. S., and Tyupkin, Yu. S. 1975. Pseudoparticle solutions of the Yang–Mills equations. *Phys. Lett. B*, **59**, 85–87.

Bell, J. S. 1955. Time reversal in field theory. *Proc. R. Soc. Lond.*, **231**, 479–495.

Bell, J. S., and Jackiw, R. 1969. A PCAC puzzle: $\pi^0 \to \gamma\gamma$ in the σ-model. *Nuovo Cimento A*, **60**, 47–61.

Bell, T. L., and Wilson, K. G. 1975. Finite-lattice approximations to renormalization groups. *Phys. Rev. B*, **11**, 3431–3445.

Benioff, P. 1980. The computer as a physical system: A microscopic quantum mechanical Hamiltonian model of computers as represented by Turing machines. *J. Stat. Phys.*, **22**, 563–591.

Berezin, F. A. 1966. *The Method of Second Quantization*. Academic Press.

Berezinskiĭ, V. L. 1970. Destruction of long-range order in one-dimensional and two-dimensional systems having a continuous symmetry group I. Classical systems. *Zh. Eksp. Teor. Fiz.*, **59**, 907–920 [Sov. Phys. JETP 32 (1971) 493–500].

Berezinskiĭ, V. L. 1971. Destruction of long-range order in one-dimensional and two-dimensional systems possessing a continuous symmetry group. II. Quantum systems. *Zh. Eksp. Teor. Fiz.*, **61**, 1144–1156 [Sov. Phys. JETP 34 (1972) 610–616].

Bernard, V., and Meißner, U.-G. 2007. Chiral perturbation theory. *Annu. Rev. Nucl. Part. Sci.*, **57**, 33–60.

Bernard, V., Kaiser, N., Kambor, J., and Meißner, U.-G. 1992. Chiral structure of the nucleon. *Nucl. Phys. B*, **388**, 315–345.

Berry, M. V. 1980. Exact Aharonov–Bohm wavefunction obtained by applying Dirac's magnetic phase factor. *Eur. J. Phys.*, **1**, 244–248.

Bethe, H. A. 1931. Zur Theorie der Metalle. *Z. Phys.*, **71**, 205–226.

Bhattacharya, T. *et al.* (HotQCD Collaboration). 2014. QCD phase transition with chiral quarks and physical quark masses. *Phys. Rev. Lett.*, **113**, 082001.

Bietenholz, W. 1999. Solutions of the Ginsparg–Wilson relation and improved domain wall fermions. *Eur. Phys. J. C*, **6**, 537–547.

Bietenholz, W. 2002. Convergence rate and locality of improved overlap fermions. *Nucl. Phys. B*, **644**, 223–247.

Bietenholz, W. 2011. Cosmic rays and the search for a Lorentz invariance violation. *Phys. Rep.*, **505**, 145–185.

Bietenholz, W. 2021. From Ramanujan to renormalization: The art of doing away with divergences and arriving at physical results. *Rev. Mex. Fís. E*, **18**, 020203.

Bietenholz, W., and Hip, I. 2000. The Scaling of exact and approximate Ginsparg–Wilson fermions. *Nucl. Phys. B*, **570**, 423–451.

Bietenholz, W., and Prado, L. 2014. Revolutionary physics in reactionary Argentina. *Physics Today*, **67**(2), 38–42.

Bietenholz, W., and Wiese, U.-J. 1996a. A perturbative construction of lattice chiral fermions. *Phys. Lett. B*, **378**, 222–226.

Bietenholz, W., and Wiese, U.-J. 1996b. Perfect lattice actions for quarks and gluons. *Nucl. Phys. B*, **464**, 319–350.

Bietenholz, W., Pochinsky, A., and Wiese, U.-J. 1995. Meron-cluster simulation of the θ vacuum in the 2D O(3) model. *Phys. Rev. Lett.*, **75**, 4524–4527.

Bijnens, J. 1993. Chiral perturbation theory and anomalous processes. *Int. J. Mod. Phys. A*, **8**, 3045–3105.

Bjorken, J. D. 1969. Asymptotic sum rules at infinite momentum. *Phys. Rev.*, **179**, 1547–1553.

Blatt, R., and Roos, C. F. 2012. Quantum simulations with trapped ions. *Nature Phys.*, **8**, 277–284.

Bloch, F., and Nordsieck, A. 1937. Note on the radiation field of the electron. *Rev. Phys.*, **52**, 54–59.

Bloom, E. D. *et al.* 1969. High-Energy inelastic $e - p$ scattering at 6° and 10°. *Phys. Rev. Lett.*, **23**, 930–934.

Blundell, S. J., and Blundell, K. M. 2008. *Concepts in Thermal Physics*. Oxford University Press.

Bödeker, D., and Buchmüller, W. 2021. Baryogenesis from the weak scale to the grand unification scale. *Rev. Mod. Phys.*, **93**, 035004.

Bogoliubov, N. N., and Parasiuk, O. S. 1957. Über die Multiplikation der Kausalfunktionen in der Quantentheorie der Felder. *Acta Math.*, **97**, 227–266.

Bogoliubov, N. N., Struminski, B. V., and Tavkhelidze, A. N. 1964. *To the composite models (in Russian)*. Preprint JINR D-1968.

Bogoliubov, N. N., Tolmachov, V. V., and Širkov, D. V. 1958. A new method in the theory of superconductivity. *Fortschr. Phys.*, **6**, 605–682.

Bollini, C. G., and Giambiagi, J. J. 1972a. Dimensional renormalization: The number of dimensions as a regularizing parameter. *Nuovo Cimento B*, **12**, 20–26.

Bollini, C. G., and Giambiagi, J. J. 1972b. Lowest order "divergent" graphs in ν-dimensional space. *Phys. Lett. B*, **40**, 566–568.

Borasoy, B. 2004. The number of colors in the decays π^0, η, $\eta' \to \gamma\gamma$. *Eur. Phys. J. C*, **34**, 317–326.

Borasoy, B., and Lipartia, E. 2005. Can one see the number of colors in η, $\eta' \to \pi^+\pi^-\gamma$? *Phys. Rev. D*, **71**, 014027.

Borgs, C., and Nill, F. 1987. The phase diagram of the Abelian lattice Higgs model. A review of rigorous results. *J. Stat. Phys.*, **47**, 877–904.

Borsanyi, S. *et al.* (Wuppertal-Budapest Collaboration). 2010. Is there still any T_c mystery in lattice QCD? Results with physical masses in the continuum limit III. *JHEP*, **1009**, 073.

Brandt, F., Dragon, N., and Kreuzer, M. 1989. All consistent Yang–Mills anomalies. *Phys. Lett. B*, **231**, 263–270.

Brandt, F., Dragon, N., and Kreuzer, M. 1990a. Completeness and nontriviality of the solutions of the consistency conditions. *Nucl. Phys. B*, **332**, 224–249.

Brandt, F., Dragon, N., and Kreuzer, M. 1990b. Lie algebra cohomology. *Nucl. Phys. B*, **332**, 250–260.

Brandt, F., Barnich, G., and Henneaux, M. 2000. Local BRST cohomology in gauge theories. *Phys. Rep.*, **338**, 439–569.

Breidenbach, M. *et al.* 1969. Observed behavior of highly inelastic electron-proton scattering. *Phys. Rev. Lett.*, **23**, 935–939.

Bressi, G., Carugno, G., Onofrio, R., and Ruoso, G. 2002. Measurement of the Casimir force between parallel metallic surfaces. *Phys. Rev. Lett.*, **88**, 041804.

Brézin, E., and Zinn-Justin, J. 1976. Spontaneous breakdown of continuous symmetries near two dimensions. *Phys. Rev. B*, **14**, 3110–3120.

Brock, R. *et al.* 1995. Handbook of perturbative QCD: Version 1.0. *Rev. Mod. Phys.*, **67**, 157–248.

Brown, F. R., Christ, N. H., Deng, Y., Gao, M., and Woch, T. J. 1988. Nature of the deconfining phase transition in SU(3) lattice gauge theory. *Phys. Rev. Lett.*, **61**, 2058–2061.

Bruno, M. *et al.* (ALPHA collaboration). 2016. The strong coupling from a nonperturbative determination of the Λ parameter in three-flavor QCD. *Phys. Rev. Lett.*, **119**, 102001.

Buchbinder, I. L., and Shapiro, I. L. 2021. *Introduction to Quantum Field Theory with Applications to Quantum Gravity*. Oxford University Press.

Buchbinder, I. L., Odintsov, S. D., and Shapiro, I. L. 1992. *Effective Action in Quantum Gravity*. Taylor & Francis Group.

Buchholz, D. 1982. The physical state space of Quantum Electrodynamics. *Commun. Math. Phys.*, **85**, 49–71.

Buchholz, D. 2013. Massless particles and arrow of time in relativistic quantum field theory. In: *Raum und Materie*. Ev. Studienwerk Villigst.

Buchholz, D., and Roberts, J. E. 2014. New light on infrared problems: Sectors, statistics, symmetries and spectrum. *Commun. Math. Phys.*, **330**, 935–972.

Buchmüller, W., and Wyler, D. 1986. Effective Lagrangian analysis of new interactions and flavour conservation. *Nucl. Phys. B*, **268**, 621–653.

Buchmüller, W., Greub, C., and Minkowski, P. 1991. Neutrino masses, neutral vector bosons and the scale of $B - L$ breaking. *Phys. Lett. B*, **267**, 395–399.

Buchmüller, W., Peccei, R. D., and Yanagida, T. 2005a. Leptogenesis as the origin of matter. *Ann. Rev. Nucl. Part. Sci.*, **55**, 311–355.

Buchmüller, W., Di Bari, P., and Plumacher, M. 2005b. Leptogenesis for pedestrians. *Ann. Phys. (N. Y.)*, **315**, 305–351.

Buras, A. J., Ellis, J., Gaillard, M. K., and Nanopoulos, D. V. 1978. Aspects of the grand unification of strong, weak and electromagnetic interactions. *Nucl. Phys. B*, **135**, 66–92.

Cabibbo, N. 1963. Unitary symmetry and leptonic decays. *Phys. Rev. Lett.*, **10**, 531–533.

Callan, C. G. 1970. Broken scale invariance in scalar field theory. *Phys. Rev. D*, **2**, 1541–1547.

Callan, C. G. 1983. Monopole catalysis of baryon decay. *Nucl. Phys. B*, **212**, 391–400.

Callan, C. G., and Gross, D. J. 1969. High-energy electroproduction and the constitution of the electric current. *Phys. Rev. Lett.*, **22**, 156–159.

Callan, C. G., and Harvey, J. A. 1985. Anomalies and fermion zero modes on strings and domain walls. *Nucl. Phys. B*, **250**, 427–436.

Callan, C. G., and Witten, E. 1984. Monopole catalysis of skyrmion decay. *Nucl. Phys. B*, **239**, 161–176.

Callan, C. G., Coleman, S., Wess, J., and Zumino, B. 1969. Structure of phenomenological Lagrangians. II. *Phys. Rev.*, **177**, 2247–2250.

Callan, C. G., Dashen, R. F., and Gross, D. J. 1976. The structure of the gauge theory vacuum. *Phys. Lett. B*, **63**, 334–340.

Callaway, D. J. E. 1988. Triviality pursuit: Can elementary scalar particles exist? *Phys. Rep.*, **167**, 241–320.

Campbell, J., Huston, J., and Krauss, F. 2018. *The Black Book of Quantum Chromodynamics: A Primer for the LHC Era*. Oxford University Press.

Campostrini, M., Hasenbusch, M., Pelissetto, A., Rossi, P., and Vicari, E. 2002. Critical exponents and equation of state of the three-dimensional Heisenberg universality class. *Phys. Rev. B*, **65**, 144520.

Campostrini, M., Hasenbusch, M., Pelissetto, A., and Vicari, E. 2006. Theoretical estimates of the critical exponents of the superfluid transition in ^{4}He by lattice methods. *Phys. Rev. B*, **74**, 144506.

Capitani, S. 2003. Lattice perturbation theory. *Phys. Rep.*, **382**, 113–302.

Carroll, S. M. 2001. The cosmological constant. *Living Rev. Rel.*, **4:1**, 1–56.

Casimir, H. B. G. 1948. On the attraction between two perfectly conducting plates. *Proc. Kon. Ned. Akad. Wetensch. B*, **51**, 793–795.

Casimir, H. B. G., and Polder, D. 1948. The influence of retardation on the London–van der Waals Forces. *Phys. Rev.*, **73**, 360–372.

Caswell, W. E. 1974. Asymptotic behavior of non-Abelian gauge theories to two loop order. *Phys. Rev. Lett.*, **33**, 244–246.

Catterall, S., Kaplan, D. B., and Ünsal, M. 2009. Exact lattice supersymmetry. *Phys. Rep.*, **484**, 71–130.

Cè, M., Consonni, C., Engel, G. P., and Giusti, L. 2015. Non-Gaussianities in the topological charge distribution of the SU(3) Yang–Mills theory. *Phys. Rev. D*, **92**, 074502.

Cè, M., García Vera, M., Giusti, L., and Schaefer, S. 2016. The topological susceptibility in the large-N limit of SU(N) Yang–Mills theory. *Phys. Lett. B*, **762**, 232–236.

Chakravarty, S., Nelson, D. R., and Halperin, B. I. 1988. Low temperature behavior of two-dimensional quantum antiferromagnets. *Phys. Rev. Lett.*, **60**, 1057–1060.

Chakravarty, S., Nelson, D. R., and Halperin, B. I. 1989. Two-dimensional quantum Heisenberg antiferromagnet at low temperatures. *Phys. Rev. B*, **39**, 2344–2371.

Chandrasekharan, S., and Wiese, U.-J. 1999. Meron-cluster solution of fermion sign problems. *Phys. Rev. Lett.*, **83**, 3116–3119.

Chin, C., Grimm, R., Julienne, P., and Tiesinga, E. 2010. Feshbach resonances in ultracold gases. *Rev. Mod. Phys.*, **82**, 1225–1286.

Clerk Maxwell, J. 1865. A dynamical theory of the electromagnetic field. *Philosophical Transactions of the Royal Society of London*, **155**, 459–512.

Clowe, D. *et al.* 2006. A direct empirical proof of the existence of dark matter. *Astrophys. J.*, **648**, L109–L113.

Coleman, S. 1985. *Aspects of Symmetry*. Cambridge University Press.

Coleman, S., Wess, J., and Zumino, B. 1969. Structure of phenomenological Lagrangians. I. *Phys. Rev.*, **177**, 2239–2247.

Coleman, S. R. 1973. There are no Goldstone bosons in two dimensions. *Commun. Math. Phys.*, **31**, 259–264.

Collins, J. C., Soper, D. E., and Sterman, G. F. 1985. Factorization for short distance hadron-hadron scattering. *Nucl. Phys. B*, **261**, 104–142.

Collins, J. C., Soper, D. E., and Sterman, G. F. 1989. Factorization of hard processes in QCD. *Adv. Ser. Direct. High Energy Phys.*, **5**, 1–91.

Creutz, M. 1977. Gauge fixing, the transfer matrix, and confinement on a lattice. *Phys. Rev. D*, **15**, 1128–1136.

Creutz, M. 1983. *Quarks, Gluons and Lattices*. Cambridge University Press.

Creutz, M. 2011. Minimal doubling and point splitting. *PoS LATTICE2010*, 078.

Crewther, R. J. 1977. Chirality selection rules and the U(1) problem. *Phys. Lett. B*, **70**, 349–354.

Crewther, R. J., Di Vecchia, P., Veneziano, G., and Witten, E. 1979. Chiral estimate of the electric dipole moment of the neutron in Quantum Chromodynamics. *Phys. Lett. B*, **88**, 123–127.

Cunqueiro, L., and Sickles, A. M. 2022. Studying the QGP with jets at the LHC and RHIC. *Prog. Part. Nucl. Phys.*, **124**, 103940.

Currie, D. G., Jordan, T. F., and Sudarshan, E.C.G. 1963. Relativistic invariance and Hamiltonian theories of interacting particles. *Rev. Mod. Phys.*, **35**, 350–375.

D'Ambrosio, G., Giudice, G. F., Isidori, G., and Strumia, A. 2002. Minimal flavour violation: an effective field theory approach. *Nucl. Phys. B*, **645**, 155–187.

Dashen, R. 1969. Chiral SU(3)⊗SU(3) as a symmetry of the strong interactions. *Phys. Rev.*, **183**, 1245–1260.

Dashen, R. 1971. Some features of chiral symmetry breaking. *Phys. Rev. D*, **3**, 1879–1889.

Davis, R., Harmer, D. S., and Hoffman, K. C. 1968. Search for neutrinos from the sun. *Phys. Rev. Lett.*, **20**, 1205–1209.

de Gennes, P-G. 1999. *Superconductivity of Metals and Alloys*. Perseus Books.

Debye, P. 1912. Zur Theorie der spezifischen Wärme. *Annalen der Physik*, **39**, 789–839.

DeGrand, T., and DeTar, C. 2006. *Lattice Methods for Quantum Chromodynamics*. World Scientific.

Del Debbio, L., Giusti, L., and Pica, C. 2005. Topological susceptibility in the SU(3) gauge theory. *Phys. Rev. Lett.*, **94**, 032003.

Deutsch, D. 1985. Quantum theory, the Church-Turing principle and the universal quantum computer. *Proc. R. Soc. Lond. A*, **400**, 97–117.

D'Hoker, E., and Farhi, E. 1984a. The decay of the skyrmion. *Phys. Lett. B*, **134**, 86–90.

D'Hoker, E., and Farhi, E. 1984b. Skyrmions and/in the weak interactions. *Nucl. Phys. B*, **241**, 109–128.

Di Francesco, P., Mathieu, P., and Sénéchal, D. 1997. *Conformal Field Theory*. Springer.

Dimopoulos, S., Raby, S. A., and Wilczek, F. 1981. Supersymmetry and the scale of unification. *Phys. Rev. D*, **24**, 1681–1683.

Dine, M., Fischler, W., and Srednicki, M. 1981. A simple solution to the strong CP problem with a harmless axion. *Phys. Lett. B*, **104**, 199–202.

Dirac, P. A. M. 1928. The quantum theory of the electron. Pages 610–624 of: *Proceedings of the Royal Society A: Mathematical, Physical and Engineering Sciences*, vol. 117.

Dirac, P. A. M. 1930. A theory of electrons and protons. *Proc. R. Soc. A*, **126**, 360–365.

Dirac, P. A. M. 1931. Quantised singularities in the electromagnetic field. *Proc. Roy. Soc. (London) A*, **133**, 60–72.

Dirac, P. A. M. 1933. The Lagrangian in quantum mechanics. *Phys. Zeitschrift der Sowjetunion*, **3**, 64–72.

Dirac, P. A. M. 1964. *Lectures on Quantum Mechanics*. Belfer Graduate School of Science, Yeshiva University.

Dokshitzer, Y. L. 1977. Calculation of the structure functions for deep inelastic scattering and $e^+ - e^-$ annihilation by perturbation theory in Quantum Chromodynamics. *Zh. Eksp. Teor. Fiz.*, **73**, 1216–1240 [Sov. Phys. JETP 46 (1977) 641–653].

Donoghue, J. F. 2012. The effective field theory treatment of quantum gravity. *AIP Conf. Proc.*, **1483**, 73–94.

Donoghue, J. F., and Holstein, B. R. 2015. Low energy theorems of quantum gravity from effective field theory. *J. Phys. G*, **42**, 103102.

Drell, S. D., Weinstein, M., and Yankielowicz, S. 1976. Strong coupling field theories. 2. Fermions and gauge fields on a lattice. *Phys. Rev. D*, **14**, 1627–1647.

Duane, S., Kennedy, A. D., Pendleton, B. J., and Roweth, D. 1987. Hybrid Monte Carlo. *Phys. Lett. B*, **195**, 216–222.

Dürr, S., Fodor, Z., Hoelbling, C., and Kurth, T. 2007. Precision study of the SU(3) topological susceptibility in the continuum. *JHEP*, **04**, 055.

Dürr, S. *et al.* 2008. Ab initio determination of light hadron masses. *Science*, **322**, 1224–1227.

Dyson, F. 1949a. The radiation theories of Tomonaga, Schwinger, and Feynman. *Phys. Rev.*, **75**, 486–502.

Dyson, F. 1949b. The S matrix in Quantum Electrodynamics. *Phys. Rev.*, **75**, 1736–1755.

Ehrenfest, P. 1933. Phasenumwandlungen im üblichen und erweiterten Sinn, classifiziert nach den entsprechenden Singularitäten des thermodynamischen Potentiales. *Verhandelingen der Koninklijke Akademie van Wetenschappen*, **36**, 153–157.

Einstein, A. 1905. Zur Elektrodynamik bewegter Körper. *Annalen der Physik*, **17**, 891–921.

Einstein, A. 1915. Die Feldgleichungen der Gravitation. *Sitzungsberichte der Preußischen Akademie der Wissenschaften zu Berlin*, 844–847.

Einstein, A. 1916. Die Grundlage der allgemeinen Relativitätstheorie. *Annalen der Physik*, **354**, 769–822.

El-Showk, S., Paulos, M. F., Poland, D., Rychkov, S., Simmons-Duffin, D., and Vichi, A. 2012. Solving the 3D Ising model with the conformal bootstrap. *Phys. Rev. D*, **86**, 025022.

El-Showk, S., Paulos, M. F., Poland, D., Rychkov, S., Simmons-Duffin, D., and Vichi, A. 2014. Solving the 3d Ising model with the conformal bootstrap II. *c*-Minimization and precise critical exponents. *J. Stat. Phys.*, **157**, 869–914.

Elitzur, S. 1975. Impossibility of spontaneously breaking local symmetries. *Phys. Rev. D*, **12**, 3978–3982.

Engels, J., Karsch, F., Satz, H., and Montvay, I. 1982. Gauge field thermodynamics for the SU(2) Yang–Mills system. *Nucl. Phys. B*, **205**, 545–577.

Englert, F., and Brout, R. 1964. Broken symmetry and the mass of gauge vector mesons. *Phys. Rev. Lett.*, **13**, 321–323.

Enz, C. P., and Meyenn, K. (editors). 1994. *Wolfgang Pauli*. Springer.

Faddeev, L. D., and Popov, V. 1967. Feynman diagrams for the Yang–Mills field. *Phys. Lett. B*, **25**, 29–30.

Fermi, E. 1934. Versuch einer Theorie der β-Strahlen. I. *Z. Phys.*, **88**, 161–177.

Fernández, R., Fröhlich, J., and Sokal, A. D. 1992. *Random Walks, Critical Phenomena, and Triviality in Quantum Field Theory*. Springer.

Feshbach, H. 1958. Unified theory of nuclear reactions. *Ann. Phys. (NY)*, **5**, 357–390.

Feynman, R. P. 1948. Space–time approach to non-relativistic quantum mechanics. *Rev. Mod. Phys.*, **20**, 367–387.

Feynman, R. P. 1949a. The theory of positrons. *Phys. Rev.*, **76**, 749–759.

Feynman, R. P. 1949b. Space–time approach to quantum electrodynamics. *Phys. Rev.*, **76**, 769–789.

Feynman, R. P. 1963. Quantum theory of gravitation. *Acta Phys. Polon.*, **24**, 697–722.

Feynman, R. P. 1969. Very high-energy collisions of hadrons. *Phys. Rev. Lett.*, **23**, 1415–1417.

Feynman, R. P. 1982. Simulating physics with computers. *Int. J. Theor. Phys.*, **21**, 467–488.

Feynman, R. P., and Hibbs, A. 1965. *Quantum Physics and Path Integrals*. Mc. Graw Hill.

Fierz, M. 1939. Über die relativistische Theorie kräftefreier Teilchen mit beliebigem Spin. *Helv. Phys. Acta*, **12**, 3–37.

Fisher, M. E., and Barber, M. N. 1972. Scaling theory for finite-size effects in the critical region. *Phys. Rev. Lett.*, **28**, 1516–1519.

Fock, V. 1926. Über die invariante Form der Wellen- und Bewegungsgleichungen für einen geladenen Massenpunkt. *Z. Phys.*, **39**, 226–232.

Fodor, Z., Holland, K., Kuti, J., Nogradi, D., and Schroeder, C. 2008. New Higgs physics from the lattice. *PoS LATTICE2007*, 056.

Fradkin, E., and Shenker, S. H. 1979. Phase diagrams of lattice gauge theories with Higgs fields. *Phys. Rev. D*, **19**, 3682–3697.

Fredenhagen, K., and Marcu, M. 1986. Confinement criterion for QCD with dynamical quarks. *Phys. Rev. Lett.*, **56**, 223–224.

Fritzsch, H., and Minkowski, P. 1975. Unified interactions of leptons and hadrons. *Ann. Phys. (N. Y.)*, **93**, 193–266.

Fritzsch, H., Gell-Mann, M., and Leutwyler, H. 1973. Advantages of the color octet gluon picture. *Phys. Lett. B*, **47**, 365–368.

Fröhlich, J. 1973. On the infrared problem in a model of scalar electrons and massless, scalar bosons. *Ann. Inst. Henri Poincaré, Section Physique Théorique*, **19**, 1–103.

Fröhlich, J. 1982. On the triviality of $\lambda \phi_d^4$ theories and the approach to the critical point in $d_{(-)} > 4$ dimensions. *Nucl. Phys. B*, **200**, 281–296.

Fröhlich, J., and Pedrini, B. 2000. New applications of the chiral anomaly. Pages 9–47 of: Fokas, A., Grigoryan, A., Kibble, T., and Zegarlinski, B. (eds), *Mathematical Physics 2000*. Imperial College Press.

Fröhlich, J., and Studer, U. M. 1993. Gauge invariance and current algebra in nonrelativistic many body theory. *Rev. Mod. Phys.*, **65**, 733–802.

Fröhlich, J., Morchio, G., and Strocchi, F. 1979. Charged sectors and scattering states in Quantum Electrodynamics. *Ann. Phys. (NY)*, **119**, 241–284.

Frolov, S. A., and Slavnov, A. A. 1994. Removing fermion doublers in chiral gauge theories on the lattice. *Nucl. Phys. B*, **411**, 647–664.

Fujikawa, K., and Suzuki, H. 2004. *Path Integrals and Quantum Anomalies*. Clarendon Press.

Fukuda, H., and Miyamoto, Y. 1949. On the γ-decay of neutral meson. *Prog. Theor. Phys.*, **4**, 347–357.

Fukuda, Y. *et al.* (Super-Kamiokande Collaboration). 1998. Measurements of the solar neutrino flux from Super-Kamiokande's first 300 days [Erratum: Phys. Rev. Lett. 81 (1998) 4279]. *Phys. Rev. Lett.*, **81**, 1158.

Fukugita, M., Okawa, M., and Ukawa, A. 1989. Order of the deconfining phase transition in SU(3) lattice gauge theory. *Phys. Rev. Lett.*, **63**, 1768–1771.

Furry, W. H. 1939. On transition probabilities in double beta-disintegration. *Phys. Rev.*, **56**, 1184–1193.

Gao, H., and Vanderhaeghen, M. 2022. The proton charge radius. *Rev. Mod. Phys.*, **94**, 015002.

Gasser, J., and Leutwyler, H. 1984. Chiral perturbation theory to one loop. *Ann. Phys. (NY)*, **158**, 142–210.

Gasser, J., and Leutwyler, H. 1985. Chiral perturbation theory: Expansions in the mass of the strange quark. *Nucl. Phys. B*, **250**, 465–516.

Gasser, J., Sainio, M. E., and Švarc, A. 1988. Nucleons with chiral loops. *Nucl. Phys. B*, **307**, 779–853.

Gasser, J., Leutwyler, H., and Rusetsky, A. 2021. On the mass difference between proton and neutron. *Phys. Lett. B*, **814**, 136087.

Gattringer, G., and Lang, C. B. 2009. *Quantum Chromodynamics on the Lattice*. Springer.

Gell-Mann, M. 1961. *The Eightfold Way: A theory of strong interaction symmetry*. Synchrotron Laboratory Report CTSL-20.

Gell-Mann, M. 1964. A schematic model of baryons and mesons. *Phys. Lett. B*, **8**, 214–215.

Gell-Mann, M., Oakes, R. J., and Renner, B. 1968. Behavior of current divergences under $SU_3 \times SU_3$. *Phys. Rev.*, **175**, 2195–2199.

Gell-Mann, M., Ramond, P., and Slansky, R. 1979. Complex spinors and unified theories. Pages 315–321 of: van Nieuwenhuizen, P., and Freedman, D. Z. (eds), *Supergravity*. North Holland Publishing Company.

Georgi, H. 1975. The state of the art – Gauge theories. *AIP Conf. Proc.*, **23**, 575–582.

Georgi, H., and Glashow, S. L. 1974. Unity of all elementary particle forces. *Phys. Rev. Lett.*, **32**, 438–441.

Georgi, H., Quinn, H.R., and Weinberg, S. 1974. Hierarchy of interactions in unified gauge theories. *Phys. Rev. Lett.*, **33**, 451–454.

Gerard, J. M., and Lahna, T. 1995. The asymptotic behavior of the $\pi^0\gamma^*\gamma^*$ vertex. *Phys. Lett. B*, **356**, 381–385.

Gervais, J. L., and Sakita, B. 1971. Field theory interpretation of supergauges in dual models. *Nucl. Phys. B*, **34**, 632–639.

Ginsparg, P. H., and Wilson, K. G. 1982. A remnant of chiral symmetry on the lattice. *Phys. Rev. D*, **25**, 2649–2657.

Giuliani, A., and Poves, A. 2012. Neutrinoless double-beta decay. *Adv. in High Energy Phys.*, **2012**, 1–38.

Giusti, L., Rossi, G. C., Testa, M., and Veneziano, G. 2002. The $U_A(1)$ problem on the lattice with Ginsparg–Wilson fermions. *Nucl. Phys. B*, **628**, 234–252.

Giusti, L., Rossi, G. C., and Testa, M. 2004. Topological susceptibility in full QCD with Ginsparg–Wilson fermions. *Phys. Lett. B*, **587**, 157–166.

Glashow, S. L. 1980. The future of elementary particle physics. *NATO Sci. Ser. B*, **61**, 687–713.

Glashow, S.L. 1961. Partial-symmetries of weak interactions. *Nucl. Phys.*, **22**, 579–588.

Glimm, J., and Jaffe, A. 1987. *Quantum Physics: A Functional Integral Point of View*. Springer.

Göckeler, M., Laursen, M. L., Schierholz, G., and Wiese, U.-J. 1986. Topological charge of (lattice) gauge fields. *Commun. Math. Phys.*, **107**, 467–481.

Goldberger, M. L., and Treiman, S. B. 1958. Form factors in β decay and μ capture. *Phys. Rev.*, **111**, 354–361.

Goldstone, J. 1961. Field theories with "superconductor" solutions. *Nuovo Cimento*, **19**, 154–164.

Goldstone, J., and Wilczek, F. 1981. Fractional quantum numbers on solitons. *Phys. Rev. Lett*, **47**, 986–989.

Golfand, Yu. A., and Likhtman, E. P. 1971. Extension of the algebra of Poincaré group generators and violation of P invariance. *JETP Lett.*, **13**, 323–326.

Greiner, M., Mandel, O., Esslinger, T., Hänsch, T. W., and Bloch, I. 2002. Quantum phase transition from a superfluid to a Mott insulator in a gas of ultracold atoms. *Nature*, **415**, 39–44.

Gribov, V. N. 1978. Quantization of non-Abelian gauge theories. *Nucl. Phys. B*, **139**, 1–19.

Gribov, V. N., and Lipatov, L. N. 1972. Deep inelastic $e - p$ scattering in perturbation theory. *Sov. J. Nucl. Phys.*, **15**, 438–450.

Griffiths, R. B. 1967. Thermodynamic functions of fluids and ferromagnets near the critical point. *Phys. Rev.*, **158**, 176–187.

Gross, D. J., and Wilczek, F. 1973. Ultraviolet behavior of non-Abelian gauge theories. *Phys. Rev. Lett.*, **30**, 1323–1346.

Grover, L. K. 1996. A fast quantum mechanical algorithm for database search. Pages 212–219 of: *STOC '96: Proceedings of the 28th Annual ACM Symposium on the Theory of Computing.* Association for Computing Machinery.

Grzadkowski, B., Iskrzyński, M., Misiak, M, and Rosiek, J. 2010. Dimension-six terms in the Standard Model Lagrangian. *JHEP*, **85**, 1–85.

Guralnik, G. S., Hagen, C R., and Kibble, T. W. B. 1964. Global conservation laws and massless particles. *Phys. Rev. Lett.*, **13**, 585–587.

Guth, A. H. 1981. Inflationary universe: A possible solution to the horizon and flatness problems. *Phys. Rev. D*, **23**, 347–356.

Haag, R. 1958. Quantum field theories with composite particles and asymptotic conditions. *Phys. Rev.*, **112**, 669–673.

Haar, A. 1933. Der Massbegriff in der Theorie der kontinuierlichen Gruppen. *Ann. Math.*, **2**, 147–169.

Haldane, F. D. M. 1983. Nonlinear field theory of large-spin Heisenberg antiferromagnets: Semiclassically quantized solitons of the one-dimensional easy-axis Néel state. *Phys. Rev. Lett.*, **50**, 1153–1156.

Han, M., and Nambu, Y. 1965. Three-triplet model with double SU(3) symmetry. *Phys. Rev. B*, **139**, 1006–1010.

Hasenfratz, A., and Hasenfratz, P. 1980. The connection between the Λ parameters of lattice and continuum QCD. *Phys. Lett. B*, **93**, 165–169.

Hasenfratz, P. 1998. Lattice QCD without tuning, mixing and current renormalization. *Nucl. Phys. B*, **525**, 401–409.

Hasenfratz, P., and Niedermayer, F. 1993. Finite size and temperature effects in the AF Heisenberg model. *Z. Phys. B*, **92**, 91–112.

Hasenfratz, P., and Niedermayer, F. 1994. Perfect lattice action for asymptotically free theories. *Nucl. Phys. B*, **414**, 785–814.

Hasenfratz, P., Laliena, V., and Niedermayer, F. 1998. The index theorem in QCD with a finite cut-off. *Nucl. Lett. B*, **427**, 125–131.

Heisenberg, W. 1932. Über den Bau der Atomkerne. *Z. Phys.*, **77**, 1–11.

Hepp, K. 1966. Proof of the Bogoliubov-Parasiuk theorem on renormalization. *Commun. Math. Phys.*, **2**, 301–326.

Herbut, I. 2007. *A Modern Approach to Critical Phenomena.* Cambridge University Press.

Hernández, P., Jansen, K., and Lüscher, M. 1999. Locality properties of Neuberger's lattice Dirac operator. *Nucl. Phys. B*, **552**, 363–378.

Higgs, P. W. 1964a. Broken symmetries, massless particles and gauge fields. *Phys. Lett.*, **12**, 132–133.

Higgs, P. W. 1964b. Broken symmetries and the masses of gauge bosons. *Phys. Rev. Lett.*, **13**, 508–509.

Hilbert, D. 1915. Die Grundlagen der Physik. *Nachrichten von der Königlichen Gesellschaft der Wissenschaften zu Göttingen, Mathematisch-Physikalische Klasse*, 395–407.

Hill, C. T., and Simmons, E. H. 2003. Strong dynamics and electroweak symmetry breaking. *Phys. Rep.*, **381**, 235–402.

Hohenberg, P. C. 1967. Existence of long-range order in one and two dimensions. *Phys. Rev.*, **158**, 383–386.

Holdom, B. 1981. Raising the sideways scale. *Phys. Rev. D*, **24**, 1441–1444.

Holland, K., and Wiese, U.-J. 2001. The center symmetry and its spontaneous breakdown at high temperatures. Pages 1909–1944 of: Shifman, M. (ed), *Festschrift in honor of B.L. Ioffe*. At the Frontier of Particle Physics. World Scientific.

Holland, K., Minkowski, P., Pepe, M., and Wiese, U.-J. 2003. Exceptional confinement in G(2) gauge theory. *Nucl. Phys. B*, **668**, 207–236.

Hopf, H. 1931. Über die Abbildungen der dreidimensionalen Sphäre auf die Kugelfläche. *Mathematische Annalen*, **104**, 637–665.

Hoyle, F. 1954. On nuclear reactions occurring in very hot stars. I. The synthesis of elements from carbon to nickel. *Astrophys. J. Suppl. Series*, **1**, 121–146.

Hund, F. 1974. *The History of Quantum Theory*. Barnes & Noble Books.

Ioffe, B. L., Okun, L. B., and Rudik, A. P. 1957. The problem of parity non-conservation in weak interactions. *JETP*, **5**, 328–330.

Isgur, N., and Karl, G. 1977. Hyperfine interactions in negative parity baryons. *Phys. Lett. B*, **72**, 109–113.

Isgur, N., and Karl, G. 1979. Ground-state baryons in a quark model with hyperfine interactions. *Phys. Rev. D*, **20**, 1191–1194.

Ising, E. 1925. Beitrag zur Theorie des Ferromagnetismus. *Z. Phys.*, **31**, 253–258.

Jackiw, R., and Rebbi, C. 1976. Vacuum periodicity in a Yang–Mills quantum theory. *Phys. Rev. Lett.*, **37**, 172–175.

Jaffe, R. 1995. Where does the proton really get its spin? *Physics Today*, **48**, 24–30.

Jegerlehner, F. 2008. *The Anomalous Magnetic Moment of the Muon*. Springer.

Jenkins, E., and Manohar, A. V. 1991. Baryon chiral perturbation theory using a heavy fermion lagrangian. *Phys. Lett. B*, **255**, 558–562.

Jenkins, E. E., Manohar, A. V., and Stoffer, P. 2018. Low-energy effective field theory below the electroweak scale: operators and matching. *JHEP*, **1803**, 16.

Jersák, J., Lang, C. B., Neuhaus, T., and Vones, G. 1985. Properties of phase transitions of the lattice SU(2) Higgs model. *Phys. Rev. D*, **32**, 2761–2768.

Ji, X., Yuan, F., and Zhao, Y. 2021. What we know and what we don't know about the proton spin after 30 years. *Nat. Rev. Phys.*, **3**, 27–38.

Jones, D. R. T. 1974. Two loop diagrams in Yang–Mills theory. *Nucl. Phys. B*, **75**, 531–538.

Josephson, B. D. 1967. Inequality for the specific heat: II. Application to critical phenomena. *Proc. Phys. Soc.*, **92**, 276–284.

Jost, R. 1957. Eine Bemerkung zum CTP-Theorem. *Helv. Phys. Acta*, **30**, 409–416.

Joyce, M., and Shaposhnikov, M. 1997. Primordial magnetic fields, right electrons, and the Abelian anomaly. *Phys. Rev. Lett.*, **79**, 1193–1196.

Juge, K. J., Kuti, J., and Morningstar, C. 2003. Fine structure of the QCD string spectrum. *Phys. Rev. Lett.*, **90**, 161601.

Julia, B., and Zee, A. 1975. Poles with both magnetic and electric charges. *Phys. Rev. D*, **11**, 2227–2232.

Kadanoff, L. P. 1966. Scaling laws for Ising models near T_c. *Physics*, **2**, 263–272.

Kaiser, R., and Leutwyler, H. 2000. Large N_c in chiral perturbation theory. *Eur. Phys. J. C*, **17**, 623–649.

Kapec, D., Pate, M., and Strominger, A. 2017a. New symmetries of QED. *Adv. Theor. Math. Phys.*, **21**, 1769–1785.

Kapec, D., Perry, M., Raclariu, A.-M., and Strominger, A. 2017b. Infrared divergences in QED, Revisited. *Phys. Rev. D*, **96**, 085002.

Kaplan, D. B. 1992. A method for simulating chiral fermions on the lattice. *Phys. Lett. B*, **288**, 342–347.

Kaplan, D. B., Katz, E., and Ünsal, M. 2003. Supersymmetry on a spatial lattice. *JHEP*, **05**, 037.

Karsten, L. H., and Smit, J. 1978. Axial symmetry in lattice theories. *Nucl. Phys. B*, **144**, 536–546.

Karsten, L. H., and Smit, J. 1979. The vacuum polarization with SLAC lattice fermions. *Phys. Lett. B*, **85**, 100–102.

Kaul, S. N. 1985. Static critical phenomena in ferromagnets with quenched disorder. *J. Magn. Magn. Mater.*, **53**, 5–53.

Kawamura, Y. 2000. Gauge symmetry reduction from the extra space S^1/Z_2. *Prog. Theor. Phys.*, **103**, 613–619.

Kawamura, Y. 2001. Split multiplets, coupling unification and an extra dimension. *Prog. Theor. Phys.*, **105**, 691–696.

Khriplovich, I. B. 1969. Green's functions in theories with non-Abelian gauge group. *Sov. J. Nucl. Phys.*, **10**, 235–242.

Kibble, T. W. B. 1976. Topology of cosmic domains and strings. *J. Phys. A: Math. Gen.*, **9**, 1387–1398.

Kim, J. E. 1979. Weak-interaction singlet and strong CP invariance. *Phys. Rev. Lett.*, **43**, 103–107.

Kleinert, H. 1990. *Path Integrals in Quantum Physics, Statistics and Polymer Physics*. World Scientific.

Kleinert, H., and Schulte-Frohlinde, V. 2001. *Critical Properties of ϕ^4 Theories*. World Scientific.

Knechtli, F., Günther, M., and Peardon, M. 2017. *Lattice Quantum Chromodynamics: Practical Essentials*. Springer.

Kobayashi, M., and Maskawa, T. 1973. CP-violation in the renormalizable theory of weak interaction. *Prog. Theor. Phys.*, **49**, 652–657.

Kogut, J. B., and Susskind, L. 1975. Hamiltonian formulation of Wilson's lattice gauge theories. *Phys. Rev. D*, **11**, 395–408.

Kogut, J. *et al.* 1983. Deconfinement and chiral symmetry restoration at finite temperatures in SU(2) and SU(3) gauge theories. *Phys. Rev. Lett.*, **50**, 393–396.

Kolb, E., and Turner, M. 1990. *The Early Universe*. Vol. 69. Front. Phys.

Kosmann-Schwarzbach, Y. 2011. *The Noether Theorems: Invariance and Conservation Laws in the Twentieth Century*. Sources and Studies in the History of Mathematics and Physical Sciences. Springer.

Kosterlitz, J. M. 1974. The critical properties of the two-dimensional xy model. *J. Phys. C*, **7**, 1046–1060.

Kosterlitz, J. M., and Thouless, D. J. 1972. Long range order and metastability in two dimensional solids and superfluids. *J. Phys. C*, **5**, L124–L126.

Kosterlitz, J. M., and Thouless, D. J. 1973. Ordering, metastability and phase transitions in two-dimensional systems. *J. Phys. C*, **6**, 1181–1203.

Kronfeld, A. S., and Wiese, U.-J. 1991. SU(N) gauge theories with C-periodic boundary conditions (I). Topological structure. *Nucl. Phys. B*, **357**, 521–533.

Kuti, J., Polónyi, J., and Szlachányi, K. 1981. Monte Carlo study of SU(2) gauge theory at finite temperature. *Phys. Lett. B*, **98**, 199–204.

Kuzmin, V. A., Rubakov, V. A., and Shaposhnikov, M. E. 1985. On the anomalous electroweak baryon number non-conservation in the early Universe. *Phys. Lett. B*, **155**, 36–42.

Lamoreaux, S. K. 1997. Demonstration of the Casimir force in the 0.6 to 6 μm range. *Phys. Rev. Lett.*, **78**, 5–8.

Landau, D. P., and Binder, K. 2014. *A Guide to Monte Carlo Simulations in Statistical Physics*. Cambridge University Press.

Landau, L. 1957. On the conservation laws for weak interactions. *Nucl. Phys.*, **3**, 127–131.

Lang, C. B., Rebbi, C., and Virasoro, M. 1981. The phase structure of a non-Abelian gauge Higgs field system. *Phys. Lett. B*, **104**, 294–300.

Langacker, P. 2012. Grand unification. *Scholarpedia*, **7**, 11419.

Lee, B. W., and Zinn-Justin, J. 1972a. Spontaneously broken gauge symmetries I. *Phys. Rev. D*, **5**, 3121–3137.

Lee, B. W., and Zinn-Justin, J. 1972b. Spontaneously broken gauge symmetries II. *Phys. Rev. D*, **5**, 3137–3155.

Lee, B. W., and Zinn-Justin, J. 1972c. Spontaneously broken gauge symmetries III. *Phys. Rev. D*, **5**, 3155–3160.

Lee, B. W., and Zinn-Justin, J. 1973. Spontaneously broken gauge symmetries IV. *Phys. Rev. D*, **7**, 1049–1056.

Lee, T. D., and Yang, C. N. 1956a. Charge conjugation, a new quantum number G, and selection rules concerning a nucleon-antinucleon system. *Nuovo Cimento*, **3**, 749–753.

Lee, T. D., and Yang, C. N. 1956b. Question of parity conservation in weak interactions. *Phys. Rev.*, **104**, 254–258.

Lehmann, H., Symanzik, K., and Zimmermann, W. 1955. Zur Formulierung quantisierter Feldtheorien. *Nuovo Cimento*, **1**, 205–225.

Leinaas, J. M., and Myrheim, J. 1977. On the theory of identical particles. *Nuovo Cimento B*, **37**, 1–23.

Leutwyler, H. 1965. A no-interaction theorem in classical relativistic Hamiltonian particle mechanics. *Nuovo Cimento*, **37**, 556–567.

Leutwyler, H. 1990. How about $m_u = 0$? *Nucl. Phys. B*, **337**, 108–118.

Leutwyler, H. 2012. Chiral perturbation theory. *Scholarpedia*, **7**, 8708.

Leutwyler, H. 2014. On the history of the strong interaction. *Mod. Phys. Lett. A*, **29**, 1430023.

Linde, A. D. 1986. Eternally existing self-reproducing chaotic inflationary Universe. *Phys. Lett. B*, **175**, 395–400.

Lipa, J. A., Nissen, J. A., Stricker, D. A., Swanson, D. R., and Chui, T. C. P. 2003. Specific heat of liquid helium in zero gravity very near the lambda point. *Phys. Rev. B*, **68**, 174518.

Livio, M. 2013. *Brilliant Blunders: From Darwin to Einstein – Colossal Mistakes by Great Scientists That Changed Our Understanding of Life and the Universe.* Simon & Schuster.

London, F., and London, H. 1935. The electromagnetic equations of the supraconductor. *Proc. of the Royal Society A: Mathematical, Physical and Engineering Sciences*, **149**, 71–88.

Lüders, G. 1954. On the equivalence of invariance under time reversal and under particle–antiparticle conjugation for relativistic field theories. *Kong. Dan. Vid. Sel. Mat. Fys. Med.*, **28**, 1–17.

Lüders, G. 1957. Proof of the TCP theorem. *Ann. Phys. (N. Y.)*, **2**, 1–15.

Lüscher, M. 1977. Construction of a selfadjoint, strictly positive transfer matrix for Euclidean lattice gauge theories. *Commun. Math. Phys.*, **54**, 283–292.

Lüscher, M. 1981. Symmetry-breaking aspects of the roughening transition in gauge theories. *Nucl. Phys. B*, **180**, 317–329.

Lüscher, M. 1982. Topology of lattice gauge fields. *Commun. Math. Phys.*, **85**, 39–48.

Lüscher, M. 1986a. Volume dependence of the energy spectrum in massive quantum field theories I. Stable particle states. *Commun. Math. Phys.*, **104**, 177–206.

Lüscher, M. 1986b. Volume dependence of the energy spectrum in massive quantum field theories II. Scattering states. *Commun. Math. Phys.*, **105**, 153–188.

Lüscher, M. 1998. Exact chiral symmetry on the lattice and the Ginsparg–Wilson relation. *Phys. Lett. B*, **428**, 342–345.

Lüscher, M. 1999. Abelian chiral gauge theories on the lattice with exact gauge invariance. *Nucl. Phys. B*, **549**, 295–334.

Lüscher, M. 2000. Lattice regularization of chiral gauge theories to all orders of perturbation theory. *JHEP*, **0006**, 028.

Lüscher, M. 2002. Chiral gauge theories revisited. *Subnucl. Ser.*, **38**, 41–89.

Lüscher, M. 2004. Topological effects in QCD and the problem of short distance singularities. *Phys. Lett. B*, **593**, 296–301.

Lüscher, M. 2010. Properties and uses of the Wilson flow in lattice QCD. *JHEP*, **08**, 071.

Lüscher, M., and Weisz, P. 1985. On-shell improved lattice gauge theories. *Commun. Math. Phys.*, **97**, 59–77.

Lüscher, M., and Weisz, P. 1987. Scaling laws and triviality bounds in the lattice ϕ^4 theory: (I). One component model in the symmetric phase. *Nucl. Phys. B*, **290**, 25–60.

Lüscher, M., and Weisz, P. 1988. Scaling laws and triviality bounds in the lattice ϕ^4 theory: (II). One component model in the phase with spontaneous symmetry breaking. *Nucl. Phys. B*, **295**, 65–92.

Lüscher, M., and Weisz, P. 2004. String excitation energies in SU(N) gauge theories beyond the free-string approximation. *JHEP*, **07**, 014.

Lüscher, M., and Weisz, P. 2021. Renormalization of the topological charge density in QCD with dimensional regularization. *Eur. Phys. J. C*, **81**, 519.

Ma, S.-K. 1976. *Modern Theory of Critical Phenomena.* Addison-Wesley Publishing.

Madras, N., and Sokal, A. D. 1988. The pivot algorithm: A highly efficient Monte Carlo method for the self-avoiding walk. *J. Stat. Phys.*, **50**, 109–186.

Majorana, E. 1937. Teoria simmetrica dell'elettrone e del positrone. *Nuovo Cimento*, **14**, 171–184.

Maki, Z., Nakagawa, M., and Sakata, S. 1962. Remarks on the unified model of elementary particles. *Prog. Theor. Phys.*, **28**, 870–880.

Manohar, A., and Georgi, H. 1984. Chiral quarks and the non-relativistic quark model. *Nucl. Phys. B*, **234**, 189–212.

Manohar, A., and Moore, G. W. 1984. Anomalous inequivalence of phenomenological theories. *Nucl. Phys. B*, **243**, 55–64.

Mermin, D., and Wagner, H. 1966. Absence of ferromagnetism or antiferromagnetism in one- or two-dimensional isotropic Heisenberg models. *Phys. Rev. Lett.*, **17**, 1133–1136.

Metropolis, N., Rosenbluth, A. W., Rosenbluth, M. N., Teller, A. H., and Teller, E. 1953. Equation of state calculations by fast computing machines. *J. Chem. Phys.*, **21**, 1087–1092.

Mikheyev, S. P., and Smirnov, A. Y. 1985. Resonance enhancement of oscillations in matter and solar neutrino spectroscopy. *Sov. J. Nucl. Phys.*, **42**, 913–917.

Millikan, R. A. 1913. On the elementary electrical charge and the Avogadro constant. *Phys. Rev.*, **2**, 109–143.

Minkowski, H. 1908. Die Grundgleichungen für die elektromagnetischen Vorgänge in bewegten Körpern. *Nachrichten von der Gesellschaft der Wissenschaften zu Göttingen, Mathematisch-Physikalische Klasse*, 53–111.

Minkowski, P. 1977. $\mu \to e\gamma$ at a rate of one out of 10^9 muon decays? *Phys. Lett. B*, **67**, 421–428.

Minwalla, M., Van Raamsdonk, M., and Seiberg, N. 2000. Noncommutative perturbative dynamics. *JHEP*, **0002**, 020.

Misner, C. W., Thorne, K. S., and Wheeler, J. A. 1973. *Gravitation*. W. H. Freeman and Company.

Miyamoto, Y. 1965. Three kinds of triplet model. *Prog. Theor. Phys. Suppl.*, **E65**, 187–192.

Mohapatra, R.N., and Senjanovic, G. 1980. Neutrino mass and spontaneous parity nonconservation. *Phys. Rev. Lett.*, **44**, 912–915.

Mohideen, U., and Roy, A. 1998. Precision measurement of the Casimir force from 0.1 to 0.9 μm. *Phys. Rev. Lett.*, **81**, 4549–4552.

Montvay, I. 1985. Correlations in the SU(2) fundamental Higgs model. *Phys. Lett. B*, **150**, 442–446.

Montvay, I., and Münster, G. 1994. *Quantum Fields on a Lattice*. Cambridge University Press.

Moore, G., and Read, N. 1991. Nonabelions in the fractional quantum Hall effect. *Nucl. Phys. B*, **360**, 362–396.

Morningstar, C. J., and Peardon, M. J. 1999. The glueball spectrum from an anisotropic lattice study. *Phys. Rev. D*, **60**, 034509.

Murayama, H., and Shu, J. 2010. Topological dark matter. *Phys. Lett. B*, **686**, 162–165.

Nambu, Y., and Jona-Lasinio, G. 1961a. Dynamical model of elementary particles based on an analogy with superconductivity. I. *Phys. Rev.*, **122**, 345–358.

Nambu, Y., and Jona-Lasinio, G. 1961b. Dynamical model of elementary particles based on an analogy with superconductivity. II. *Phys. Rev.*, **124**, 246–254.

Narayanan, R., and Neuberger, H. 1993a. Infinitely many regulator fields for chiral fermions. *Phys. Lett. B*, **302**, 62–69.

Narayanan, R., and Neuberger, H. 1993b. Chiral fermions on the lattice. *Phys. Rev. Lett.*, **71**, 3251.

Narayanan, R., and Neuberger, H. 1994. Chiral determinant as an overlap of two vacua. *Nucl. Phys. B*, **412**, 574–606.

Narayanan, R., and Neuberger, H. 1995. A construction of lattice chiral gauge theories. *Nucl. Phys. B*, **443**, 305–385.

Necco, S., and Sommer, R. 2002. The $N_f = 0$ heavy quark potential from short to intermediate distances. *Nucl. Phys. B*, **622**, 328–346.

Neuberger, H. 1998. More about exactly massless quarks on the lattice. *Phys. Lett. B*, **427**, 353–355.

Nielsen, H. B., and Chadha, S. 1976. On how to count Goldstone bosons. *Nucl. Phys. B*, **105**, 445–453.

Nielsen, H. B., and Ninomiya, M. 1981a. Absence of neutrinos on a lattice. 1. Proof by homotopy theory [Erratum: Nucl. Phys. B195 (1982) 541–542]. *Nucl. Phys. B*, **185**, 20–40.

Nielsen, H. B., and Ninomiya, M. 1981b. Absence of neutrinos on a lattice. 2. Intuitive topological proof. *Nucl. Phys. B*, **193**, 173–194.

Noether, E. 1918a. Invarianten beliebiger Differentialausdrücke. *Nachrichten von der Gesellschaft der Wissenschaften zu Göttingen*, 37–44.

Noether, E. 1918b. Invariante Variationsprobleme. *Nachrichten von der Gesellschaft der Wissenschaften zu Göttingen*, 235–257.

Novikov, S. P. 1981. Multivalued functions and functionals. An analogue of the Morse theory. *Sov. Math. Dokl.*, **24**, 222–226.

Novikov, S. P. 1982. The Hamiltonian formalism and a many-valued analogue of Morse theory. *Usp. Math. Nauk.*, **37**, 3–49.

Okubo, S. 1962a. Note on unitary symmetry in strong interactions. *Prog. Theor. Phys.*, **27**, 949–966.

Okubo, S. 1962b. Note on unitary symmetry in strong interactions. II: Excited states of baryons. *Prog. Theor. Phys.*, **28**, 24–32.

Okun, L. B. 1989. The concept of mass. *Physics Today*, **42**, 31–36.

Onsager, L. 1944. Crystal statistics I. *Phys. Rev.*, **65**, 117–149.

O'Raifeartaigh, C., and Mitton, S. 2018. Interrogating the legend of Einstein's "Biggest Blunder". *Physics in Perspective*, **20**, 318.

Osterwalder, K., and Schrader, R. 1973. Axioms for Euclidean Green's functions. *Commun. Math. Phys.*, **31**, 83–112.

Osterwalder, K., and Schrader, R. 1975. Axioms for Euclidean Green's functions II. *Commun. Math. Phys.*, **42**, 281–305.

Padmanabhan, T. 2003. Cosmological constant: The weight of the vacuum. *Phys. Rep.*, **380**, 235–320.

Patella, A. 2017. QED corrections to hadronic observables. *PoS LATTICE2016*, 020.

Pati, J. C., and Salam, A. 1974. Lepton number as the fourth color. *Phys. Rev. D*, **10**, 275–289.

Pauli, W. 1940. The connection between spin and statistics. *Phys. Rev.*, **58**, 716–722.

Pauli, W. 1957. On the conservation of the lepton charge. *Nuovo Cimento*, **6**, 204–215.

Pauli, W., and Villars, F. 1949. On the invariant regularization in relativistic quantum theory. *Rev. Mod. Phys.*, **21**, 434–444.

Peccei, R. D., and Quinn, H. R. 1977a. CP conservation in the presence of pseudoparticles. *Phys. Rev. Lett.*, **38**, 1440–1443.

Peccei, R. D., and Quinn, H. R. 1977b. Constraints imposed by CP conservation in the presence of pseudoparticles. *Phys. Rev. D*, **16**, 1791–1797.

Peebles, P. J. E., and Ratra, B. 2003. The cosmological constant and dark energy. *Rev. Mod. Phys.*, **75**, 559–606.

Peierls, R. 1936. On Ising's model of ferromagnetism. *Proc. Cambridge Phil. Soc.*, **32**, 477–481.

Pendlebury, J. M. *et al.* 2015. Revised experimental upper limit on the electric dipole moment of the neutron. *Phys. Rev. D*, **92**, 092003.

Pepe, M., and Wiese, U.-J. 2007. Exceptional deconfinement in G(2) gauge theory. *Nucl. Phys. B*, **768**, 21–37.

Perlmutter, S. *et al. (Supernova Cosmology Project)*. 1999. Measurements of Omega and Lambda from 42 high redshift supernovae. *Astrophys. J.*, **517**, 565–586.

Peshkin, M., and Tonomura, A. 1989. *The Aharonov–Bohm Effect*. Springer.

Peskin, M. E., and Schroeder, D. V. 1997. *An Introduction to Quantum Field Theory*. Addison Wesley.

Pfeuty, P., and Toulouse, G. 1977. *Introduction to the Renormalization Group and to Critical Phenomena*. John Wiley and Sons.

Philipps, A., and Stone, D. 1986. Lattice gauge fields, principal bundles and the calculation of topological charge. *Commun. Math. Phys.*, **103**, 599–636.

Pich, A. 1995. Chiral perturbation theory. *Rep. Prog. Phys.*, **58**, 563–609.

Politzer, H. D. 1973. Reliable perturbative results for strong interactions. *Phys. Rev. Lett.*, **30**, 1346–1349.

Polley, L., and Wiese, U.-J. 1991. Monopole condensate and monopole mass in U(1) lattice gauge theory. *Nucl. Phys. B*, **356**, 629–654.

Polyakov, A. M. 1974. Particle spectrum in the quantum field theory. *JETP Lett.*, **20**, 194–195.

Polyakov, A. M. 1975. Compact gauge fields and the infrared catastrophe. *Phys. Lett. B*, **59**, 82–84.

Polyakov, A. M. 1987. *Gauge Fields and Strings*. Harwood Academic Publishers.

Polyakov, A. M. 2005. Confinement and liberation. Pages 311–329 of: 't Hooft, G. (ed), *50 Years of Yang–Mills Theory*. Proc. Eighth Nobel Symposium. World Scientific.

Pontecorvo, B. 1957. Inverse beta processes and nonconservation of lepton charge. *Zh. Eksp. Teor. Fiz.*, **34**: 247 [Sov. Phys. JETP 7 (1958) 172–173].

Popov, V. 1983. *Functional Integrals in Quantum Field Theory and Statistical Physics*. Math. Phys. and Appl. Math., vol. 8. Reidel.

Preskill, J., Wise, M. B., and Wilczek, F. 1983. Cosmology of the invisible axion. *Phys. Lett. B*, **120**, 127–132.

Privman, V., and Fisher, M. E. 1984. Universal critical amplitudes in finite-size scaling. *Phys. Rev. B*, **30**, 322–327.

Raby, S. 2017. *Supersymmetric Grand Unified Theories: From Quarks to Strings via SUSY GUTs*. Springer, Lecture Notes in Physics, Volume 939.

Rajagopal, K., and Wilczek, F. 1993a. Static and dynamic critical phenomena at a second order QCD phase transition. *Nucl. Phys. B*, **399**, 395–425.

Rajagopal, K., and Wilczek, F. 1993b. Emergence of coherent long wavelength oscillations after a quench: Application to QCD. *Nucl. Phys. B*, **404**, 577–589.

Ratra, P., and Peebles, P. J. E. 1988. Cosmological consequences of a rolling homogeneous scalar field. *Phys. Rev. D*, **37**, 3406–3427.

Reisz, T. 1988a. A power counting theorem for Feynman integrals on the lattice. *Commun. Math. Phys.*, **116**, 81–126.

Reisz, T. 1988b. Renormalization of Feynman integrals on the lattice. *Commun. Math. Phys.*, **117**, 79–108.

Reisz, T. 1989. Lattice gauge theory: Renormalization to all orders in the loop expansion. *Nucl. Phys. B*, **318**, 417–463.

Riemann, B. 1859. Über die Anzahl der Primzahlen unter einer gegebenen Größe. *Monatsberichte der Berliner Akademie.*

Riess, A. G. *et al.* 1998. Observational evidence from supernovae for an accelerating universe and a cosmological constant. *Astron. J.*, **116**, 1009–1038.

Roepstorff, G. 1994. *Path Integral Approach to Quantum Physics*. Springer.

Ross, G. G. 1985. *Grand Unified Theories*. Westview Press.

Rothe, H. J. 1992. *Lattice Gauge Theories: An Introduction*. World Scientific.

Rubakov, V. A. 1981. Superheavy magnetic monopoles and proton decay. *JETP Lett.*, **33**, 644–646.

Rubakov, V. A. 1982. Adler–Bell–Jackiw anomaly and fermion-number breaking in the presence of a magnetic monopole. *Nucl. Phys. B*, **203**, 311–348.

Rubakov, V. A., and Shaposhnikov, M. E. 1983. Do we live inside a domain wall? *Phys. Lett. B*, **125**, 136–138.

Rubin, V. C., and Ford, W. K. 1970. Rotation of the Andromeda Nebula from a spectroscopic survey of emission regions. *Astrophys. J.*, **159**, 379–403.

Rudaz, S. 1990. Electric-charge quantization in the standard model. *Phys. Rev. D*, **41**, 2619–2621.

Ruelle, D. 1962. On the asymptotic condition in quantum field theory. *Helv. Phys. Acta*, **35**, 147–163.

Rushbrooke, G. S. 1963. On the thermodynamics of the critical region for the Ising problem. *J. Chem. Phys.*, **39**, 842–843.

Ryder, L. H. 1996. *Quantum Field Theory*. Second edn. Cambridge University Press.

Sakharov, A. D. 1967. Violation of CP invariance, C asymmetry, and baryon asymmetry of the universe. *ZhETF Pis'ma*, **5**, 32–35.

Salam, A. 1968. Weak and electromagnetic interactions. Pages 367–377 of: Svartholm, N. (ed), *Elementary Particle Physics: Relativistic Groups and Analyticity*. Proc. Eighth Nobel Symposium. Almquvist and Wiksell.

Sardanashvily, G. 2016. *Noether's Theorems. Applications in Mechanics and Field Theory*. Springer.

Schechter, J., and Valle, J. W. F. 1980. Neutrino masses in SU(2) ⊗ U(1) theories. *Phys. Rev.*, **22**, 2227–2235.

Scherer, S. 2003. Introduction to chiral perturbation theory. *Adv. Nucl. Phys.*, **27**, 277–538.

Schroer, B. 1963. Infrateilchen in der Quantenfeldtheorie. *Fortschr. Phys.*, **11**, 1–31.

Schulman, L. 1981. *Techniques and Applications of Path Integration*. Wiley.

Schwartz, M. D. 2014. *Quantum Field Theory and the Standard Model*. Cambridge University Press.

Schwinger, J. 1948. On Quantum Electrodynamics and the magnetic moment of the electron. *Phys. Rev.*, **73**, 416–417.

Schwinger, J. 1969. A magnetic model of matter: A speculation probes deep within the structure of nuclear particles and predicts a new form of matter. *Science*, **165**, 757–761.

Seeger, M., Kaul, S. N., Kronmüller, H., and Reisser, R. 1995. Asymptotic critical behavior of Ni. *Phys. Rev. B*, **51**, 12585.

Serre, J.-P. 1951. Homologie singulière des espaces fibrés. Applications. *Ann. Math., Second Series*, **54**, 425–505.

Shamir, J. 1993. Chiral fermions from lattice boundaries. *Nucl. Phys. B*, **406**, 90–106.

Shapere, A. D., and Wilczek, F. 1987. Self-propulsion at low Reynolds number. *Phys. Rev. Lett.*, **58**, 2051–2054.

Shapere, A. D., and Wilczek, F. 1989. Gauge kinematics of deformable bodies. *Am. J. Phys.*, **57**, 514–518.

Shifman, M. A., Vainshtein, A. I., and Zakharov, V. I. 1980. Can confinement ensure natural CP invariance of strong interactions? *Nucl. Phys. B*, **166**, 493–506.

Shor, P. W. 1994. Algorithms for quantum computation: discrete logarithms and factoring. Pages 124–134 of: *SFCS '94: Proceedings 35th Annual Symposium on Foundations of Computer Science*.

Skyrme, T. H. R. 1961. Particle states of a quantized meson field. *Proc. Roy. Soc. Lond. A*, **262**, 237–245.

Skyrme, T. H. R. 1962. A unified field theory of mesons and baryons. *Nucl. Phys. B*, **31**, 556–569.

Slavnov, A. A. 1972. Ward identities in gauge theories. *Theor. Math. Phys.*, **10**, 99–104.

Smit, J. 2002. *Introduction to Quantum Fields on a Lattice*. Cambridge Lecture Notes in Physics.

Smoot, G. F. *et al.* 1992. Structure in the COBE differential microwave radiometer first-year maps. *Astrophys. J., Part 2 – Letters*, **396**, L1–L5.

Snyder, H. S. 1947. Quantized space–time. *Phys. Rev.*, **71**, 38–41.

Sokal, A. D. 1997. Monte Carlo methods in statistical mechanics: Foundations and new algorithms. Pages 131–192 of: DeWitt-Morette, C., Cartier, P., and Folacci, A. (eds), *Functional Integration: Basics and Applications*. NATO ASI Series, vol. 361. Springer.

Sommer, R. 1994. A new way to set the energy scale in lattice gauge theories and its applications to the static force and α_s in SU(2) Yang–Mills theory. *Nucl. Phys. B*, **411**, 839–854.

Sommer, R. 2014. Scale setting in lattice QCD. PoS LATTICE2013, 015.

Sommerfeld, A. 1916. Zur Quantentheorie der Spektrallinien. *Annalen der Physik*, **51**, 1–94.

Srednicki, M. 2007. *Quantum Field Theory*. Cambridge University Press.

Steinberger, J. 1949. On the use of subtraction fields and the lifetimes of some types of meson decay. *Phys. Rev.*, **76**, 1180–1186.

Stern, A. 2010. Non-Abelian states of matter. *Nature*, **464**, 187–193.

Straumann, N. 2002. On Pauli's invention of non-Abelian Kaluza-Klein theory in 1953. Pages 1063–1066 of: Gurzadyan, V. G., Jantzen, R. T., and Ruffini, R. (eds), *The Ninth Marcel Grossmann Meeting*. World Scientific.

Stückelberg, E. C. G. 1938. Interaction energy in electrodynamics and in the field theory of nuclear forces. *Helv. Phys. Acta*, **11**, 225–244.

Susskind, L. 1979. Dynamics of spontaneous symmetry breaking in the Weinberg-Salam theory. *Phys. Rev. D*, **20**, 2619–2625.

Susskind, L. 2006. *The Cosmic Landscape: String Theory and the Illusion of Intelligent Design*. Little, Brown and Company.

Svetitsky, B., and Yaffe, L. 1982. Critical behavior at finite-temperature confinement transitions. *Nucl. Phys. B*, **210**, 423–447.

Swanson, M. 1992. *Path Integrals and Quantum Processes*. Academic Press.

Swendsen, R. H., and Wang, J.-S. 1987. Nonuniversal critical dynamics in Monte Carlo simulations. *Phys. Rev. Lett.*, **58**, 86–88.

Symanzik, K. 1970. Small distance behaviour in field theory and power counting. *Commun. Math. Phys.*, **18**, 227–246.

Symanzik, K. 1983a. Continuum limit and improved action in lattice theories. 1. Principles and ϕ^4 theory. *Nucl.Phys. B*, **226**, 187–204.

Symanzik, K. 1983b. Continuum limit and improved action in lattice theories. 2. O(N) Nonlinear sigma model in perturbation theory. *Nucl. Phys. B*, **226**, 205–227.

't Hooft, G. 1971a. Renormalization of massless Yang–Mills fields. *Nucl. Phys. B*, **33**, 173–199.

't Hooft, G. 1971b. Renormalizable Lagrangians for massive Yang–Mills fields. *Nucl. Phys. B*, **35**, 167–188.

't Hooft, G. 1974. Magnetic monopoles in unified gauge theories. *Nucl. Phys. B*, **79**, 276–284.

't Hooft, G. 1976a. Symmetry breaking through Bell-Jackiw anomalies. *Phys. Rev. Lett.*, **37**, 8–11.

't Hooft, G. 1976b. Computation of the quantum effects due to a four-dimensional pseudoparticle. *Phys. Rev. D*, **14**, 3432–3450.

't Hooft, G. 1979. A property of electric and magnetic flux in non-Abelian gauge theories. *Nucl. Phys. B*, **153**, 141–160.

't Hooft, G. 1980. Naturalness, chiral symmetry, and spontaneous chiral symmetry breaking. Pages 135–157 of: *Recent Developments in Gauge Theories*. Springer.

't Hooft, G. 1986. How instantons solve the U(1) problem. *Phys. Rep.*, **142**, 357–387.

't Hooft, G, and Veltman, M. 1972. Regularization and renormalization of gauge fields. *Nucl. Phys. B*, **44**, 189–213.

Takahashi, Y. 1957. On the generalized Ward identity. *Nuovo Cimento*, **6**, 371–375.

Taylor, J. C. 1971. Ward identities and charge renormalization of the Yang–Mills field. *Nucl. Phys. B*, **33**, 436–444.

Tomboulis, E., and Woo, G. 1976. Soliton quantization in gauge theories. *Nucl. Phys. B*, **107**, 221–237.

Tomonaga, S. 1946. On a relativistically invariant formulation of the quantum theory of wave fields. *Prog. Theor. Phys.*, **1**, 27–42.

Trotter, H. F. 1959. On the product of semi-groups of operators. *Proc. Amer. Math. Soc.*, **10**, 545–551.

Troyer, M., and Wiese, U.-J. 2005. Computational complexity and fundamental limitations to fermionic quantum Monte Carlo simulations. *Phys. Rev. Lett.*, **94**, 170201.

Turner, M. S, and Widrow, L. M. 1988. Inflation-produced, large-scale magnetic fields. *Phys. Rev. D*, **37**, 2743–2754.

Tyutin, I. V. 1975. *Gauge invariance in field theory and statistical physics in operator formalism*. Lebedev Physics Institute preprint 39, arXiv:0812.0580.

Vanyashin, V. S., and Terentev, M. V. 1965. The vacuum polarization of a charged vector field. *Zh. Eksp. Teor. Fiz.*, **48**: 565–573 [Sov. Phys. JETP 21 (1965) 375–380].

Veneziano, G. 1979. U(1) without instantons. *Nucl. Phys. B*, **159**, 213–224.

Vlasii, N. D., von Rütte, F., and Wiese, U.-J. 2016. Graphical tensor product reduction scheme for the Lie algebras so(5)=sp(2), su(3), and g(2). *Ann. Phys. (NY)*, **371**, 199–227.

Volkov, D. V., and Akulov, V. P. 1972. Possible universal neutrino interaction. *JETP Lett.*, **16**, 438–440.

Ward, J. C. 1950. An identity in Quantum Electrodynamics. *Phys. Rev.*, **78**, 182–182.

Wegner, F. J. 1971. Duality in generalized Ising models and phase transitions without local order parameters. *J. Math. Phys.*, **12**, 2259–2272.

Weinberg, S. 1967. A model of leptons. *Phys. Rev. Lett.*, **19**, 1264–1266.

Weinberg, S. 1976. Implications of dynamical symmetry breaking. *Phys. Rev. D*, **13**, 974–996.

Weinberg, S. 1978. A new light boson? *Phys. Rev. Lett.*, **40**, 223–226.

Weinberg, S. 1979a. Baryon- and lepton-nonconserving processes. *Phys. Rev. Lett.*, **43**, 1566–1570.

Weinberg, S. 1979b. Implications of dynamical symmetry breaking: An addendum. *Phys. Rev. D*, **19**, 1277–1280.

Weinberg, S. 1979c. Phenomenological Lagrangians. *Physica A*, **96**, 327–340.

Weinberg, S. 1987. Anthropic bound on the cosmological constant. *Phys. Rev. Lett.*, **59**, 2607–2610.

Wen, X.-G. 1991. Non-Abelian statistics in the fractional quantum Hall states. *Phys. Rev. Lett.*, **66**, 802–805.

Wess, J., and Bagger, J. 1992. *Supersymmetry and Supergravity*. Princeton University Press.

Wess, J., and Zumino, B. 1971. Consequences of anomalous Ward identities. *Phys. Lett. B*, **37**, 95–97.

Wess, J., and Zumino, B. 1974. Supergauge transformations in four-dimensions. *Nucl. Phys. B*, **70**, 39–50.

Wetterich, C. 1988. Cosmology and the fate of dilatation symmetry. *Nucl. Phys. B*, **302**, 668–696.

Wetterich, C. 2011. Spinors in Euclidean field theory, complex structures and discrete symmetries. *Nucl. Phys. B*, **852**, 174–234.

Weyl, H. 1918. Gravitation und Elektrizität. *Sitzungsberichte der Königlich Preußischen Akademie der Wissenschaften zu Berlin*, 465–480.

Weyl, H. 1929. Elektron und Gravitation. I. *Z. Phys.*, **56**, 330–352.

Wick, G. C. 1950. The evaluation of the collision matrix. *Phys. Rev.*, **80**, 268–272.

Wick, G. C., Wightman, A. S., and Wigner, E. P. 1952. The intrinsic parity of elementary particles. *Phys. Rev.*, **88**, 101–105.

Widom, B. 1964. Degree of the critical isotherm. *J. Chem. Phys.*, **41**, 1633–1634.

Widom, B. 1975. Interfacial tensions of three fluid phases in equilibrium. *J. Chem. Phys.*, **62**, 1332–1336.

Wiese, U.-J. 1993. Fixed point actions for Wilson fermions. *Phys. Lett. B*, **315**, 417–424.

Wiese, U.-J. 2012. The confining string: Lattice results versus effective field theory. *Prog. Part. Nucl. Phys.*, **67**, 117–121.

Wiese, U.-J. 2013. Ultracold quantum gases and lattice systems: Quantum simulation of lattice gauge theories. *Annalen der Physik*, **525**, 777–796.

Wiese, U.-J. 2014. Towards quantum simulating QCD. *Nucl. Phys. A*, **931**, 246–256.

Wigner, E. P. 1960. The unreasonable effectiveness of mathematics in the natural sciences. *Comm. Pure Appl. Math.*, **13**, 1–14.

Wilczek, F. 1978. Problem of strong P and T invariance in the presence of instantons. *Phys. Rev. Lett.*, **40**, 279–282.

Wilczek, F. 1982. Quantum mechanics of fractional-spin particles. *Phys. Rev. Lett.*, **49**, 957–959.

Wilczek, F. 1987. Two applications of axion electrodynamics. *Phys. Rev. Lett.*, **58**, 1799–1802.

Wilczek, F. 2004. Four big questions with pretty good answers. Pages 79–97 of: Buschhorn G.W., Wess J. (ed), *Fundamental Physics – Heisenberg and Beyond*. Springer.

Wilczek, F. 2012. Origins of mass. *Cent. Eur. J. Phys.*, **10**, 1021–1037.

Wilczek, F., and Zee, A. 1979. Operator analysis of nucleon decay. *Phys. Rev. Lett.*, **43**, 1571–1573.

Wilczek, F., and Zee, A. 1984. Appearance of gauge structure in simple dynamical systems. *Phys. Rev. Lett.*, **52**, 2111–2114.

Will, C. M. 2014. The confrontation between General Relativity and experiment. *Living Rev. Relativity*, **17**, 4.

Willenbrock, S. 2004. Symmetries of the standard model. *Lectures given at Conference: C04-06-06.1, Proceedings*, 3–38.

Wilson, K. G. 1974. Confinement of quarks. *Phys. Rev. D*, **10**, 2445–2459.

Wilson, K. G. 1975. The renormalization group: Critical phenomena and the Kondo problem. *Rev. Mod. Phys.*, **47**, 773–840.

Wilson, K. G. 1976. Relativistically invariant lattice theories. Pages 243–264 of: Perlmutter, A. (ed), *New Pathways in High Energy Physics II*. Plenum Press.

Wilson, K. G. 1977. Quarks and strings on a lattice. Pages 69–142 of: Zichichi, A. (ed), *Proceedings of the 1975 International School of Subnuclear Physics Erice, Sicily; 11 Jul 1975*. Plenum Press.

Wilson, K. G. 2005. The origins of lattice gauge theory. *Nucl. Phys. Proc. Suppl.*, **140**, 3–19.

Wilson, K. G., and Fisher, M. E. 1972. Critical exponents in 3.99 dimensions. *Phys. Rev. Lett.*, **28**, 240–243.

Wilson, K. G., and Kogut, J. 1974. The renormalization group and the epsilon-expansion. *Phys. Rep.*, **12**, 75–199.

Witten, E. 1979a. Baryons in the $1/N$ expansion. *Nucl. Phys. B*, **160**, 57–115.

Witten, E. 1979b. Current algebra theorems for the U(1) "Goldstone boson". *Nucl. Phys. B*, **156**, 269–283.

Witten, E. 1979c. Dyons of charge $e\theta/2\pi$. *Phys. Lett. B*, **86**, 283–287.

Witten, E. 1982. An SU(2) anomaly. *Phys. Lett. B*, **117**, 324–328.

Witten, E. 1983a. Global aspects of current algebra. *Nucl. Phys. B*, **223**, 422–432.

Witten, E. 1983b. Current algebra, baryons, and quark confinement. *Nucl. Phys. B*, **223**, 433–444.

Witten, E. 1984a. Non-Abelian bosonization in two dimensions. *Commun. Math. Phys.*, **92**, 455–472.

Witten, E. 1984b. Some properties of O(32) superstrings. *Phys. Lett. B*, **149**, 351–356.

Wolfenstein, L. 1978. Neutrino oscillations in matter. *Phys. Rev. D*, **17**, 2369–2374.

Wolff, U. 1989. Collective Monte Carlo updating for spin systems. *Phys. Rev. Lett.*, **62**, 361–364.

Wolff, U. 1992. Dynamics of hybrid overrelaxation in the Gaussian model. *Phys. Lett. B*, **288**, 166–170.

Wolff, U. 2004. Monte Carlo errors with less errors. *Comput. Phys. Comm.*, **156**, 143–153.

Workman, R. L., *et al.* 2022. The Review of Particle Physics (2022). *Prog. Theor. Exp. Phys.*, **2022**, 083C01.

Wu, C. S., Ambler, E., Hayward, R. W., Hoppes, D. D., and Hudson, R. P. 1957. Experimental test of parity conservation in beta decay. *Phys. Rev.*, **105**, 1413–1415.

Yanagida, T. 1980. Horizontal symmetry and masses of neutrinos. *Prog. Theor. Phys.*, **64**, 1103–1105.

Yang, C. N., and Mills, R. 1954. Conservation of isotopic spin and isotopic gauge invariance. *Phys. Rev.*, **96**, 191–195.

Yennie, D. R., Frautschi, S. C., and Suura, H. 1961. The infrared divergence phenomena and high-energy processes. *Ann. Phys. (NY)*, **13**, 379–452.

Yeomans, J. M. 1992. *Statistical Mechanics of Phase Transitions*. Clarendon Press.

Yukawa, H. 1935. On the interaction of elementary particles. I. *Proc. Phys. Math. Soc. Jpn.*, **17**, 48–57.

Zee, A. 2010. *Quantum Field Theory in a Nutshell*. Princeton University Press.

Zhitnitsky, A. R. 1980. On possible suppression of the axion hadron interactions. *Yad. Fiz*, **31**: 497–504 [Sov. J. Nucl. Phys. 31 (1980) 260].

Zimmermann, W. 1969. Convergence of Bogoliubov's method of renormalization in momentum space. *Commun. Math. Phys.*, **15**, 208–234.

Zinn-Justin, J. 1975. Renormalization of gauge theories. Pages 1–39 of: H. Rollnik, H., and Dietz, K. (eds), *Trends in Elementary Particle Physics*. Lecture Notes in Physics, Springer, 37.

Zinn-Justin, J. 2009. Zinn-Justin equation. *Scholarpedia*, **4**, 7120.

Zohar, E., Cirac, J. I., and Reznik, B. 2016. Quantum simulations of lattice gauge theories using ultracold atoms in optical lattices. *Rep. Prog. Phys.*, **79**, 014401.

Zwanziger, D. 1968a. Exactly soluble nonrelativistic model of particles with both electric and magnetic charges. *Phys. Rev.*, **176**, 1480–1488.

Zwanziger, D. 1968b. Quantum field theory of particles with both electric and magnetic charges. *Phys. Rev.*, **176**, 1489–1495.

Zweig, G. 1964. *An SU(3) model for strong interaction symmetry and its breaking*. CERN Report No. 8182/TH.401.

Zwicky, F. 1933. Die Rotverschiebung von extragalaktischen Nebeln. *Helv. Phys. Acta*, **6**, 110–127.

Zyla, P.A. *et al.* (Particle Data Group). 2020. Review of Particle Physics (2020) and 2021 update. *Prog. Theor. Exp. Phys.*, **8**, 030001.

Author Index

Subject Index

Abelian gauge theory, 2-d, 646
accelerator neutrinos, 392
adjoint representation of a Lie algebra, 615
Adler–Bell–Jackiw anomaly, 417, 442, 508
Aharonov–Bohm effect, 6, 210, 564
angular momentum operator, 611
  of the photon field, 115
  of a Weyl fermion field, 139
anomalous dimension, 101
anomaly, 235
  axial, 235
  baryon number, 350
  cancellation, 344, 367, 372
  global $SU(2)_L$, 340
  gravitational, 346, 353, 367, 372
  lepton number, 350
  matching, 347
  triangle, 338
anthropic principle, 231, 317, 433, 592
anti-commutation relations, 137, 148
anti-ferromagnet, 250, 251
Antoine–Speiser scheme, 638
anyon, 507, 643
area law, 303
asymptotic freedom, 105, 231, 308, 309, 403, 596
atmospheric neutrinos, 391
axial anomaly, 235
axial current, 417
axion, 230, 233, 265, 266, 530
  astrophysical effects, 538
  cosmological effects, 538
  dark matter, 538
  field, 535
  mixing with pion, 536
  from O(32) superstring theory, 543

$B - L$ symmetry, 352, 364, 365, 554, 557, 570
Baker–Campbell–Hausdorff formula, 16, 34, 218
baryon
  asymmetry, 264, 555, 557
  chiral perturbation theory, 496
  decuplet, 458
  mass formula, 459
  octet, 458
baryon number, 233, 342, 506
  anomaly, 350, 351
  current, 418
  violation, 364, 556
B^0-$\overline{B^0}$ mixing, 387
Berezinskiĭ–Kosterlitz–Thouless phase transition, 252

β-function, 101, 104
  of $\lambda\phi^4$ theory, 103
  of QCD, 405
  of Yang–Mills theory, 105, 309
Bianchi identity, 607
Big Bang, 263
Bjorken limit, 476
Bjorken scaling, 480
Bjorken variable x, 476, 480
black-body spectrum, 117
Bloch wave, 51, 430
blocking transformation, 88, 90, 196
Bohr radius, 61
boost operator, 610
  of a Weyl fermion field, 139
Bose–Einstein condensate, 241
Bose–Einstein statistics, 46
BPHZ theorem, 84, 227
Breit frame, 478
Brillouin zone, 71
BRST symmetry, 203, 227
bubble nucleation, 264

C-periodic boundary conditions, 134
Cabibbo angle θ_C, 384
Callan–Gross relation, 480
Callan–Rubakov effect, 561
Callan–Symanzik equation, 101
canonical quantization
  of Abelian lattice gauge theory, 213
  of a fermion field, 136
  of non-Abelian lattice gauge theory, 215
  of the photon field, 110
  of a scalar field, 37
capillary wave, 312
Casimir effect, 315
Casimir operator, 611, 616
center of a group, 618
center symmetry, 323
character expansion, 303, 306
charge conjugation C, 141, 142, 176, 232, 333
charge quantization, 366, 372
charged current, 338, 383
charm-quark mass, 386
chemical potential, 145
Chern–Pontryagin density, 417
Chern–Simons density, 418
chiral
  condensate, 410, 489
  gauge theory, 225, 331

Printed in the United States
by Baker & Taylor Publisher Services